Biology
Dimensions of Life

Joelle Presson
University of Maryland

Jan Jenner

Mc
Graw
Hill

Boston   Burr Ridge, IL   Dubuque, IA   New York   San Francisco   St. Louis
Bangkok   Bogotá   Caracas   Kuala Lumpur   Lisbon   London   Madrid   Mexico City
Milan   Montreal   New Delhi   Santiago   Seoul   Singapore   Sydney   Taipei   Toronto

Higher Education

BIOLOGY: DIMENSIONS OF LIFE

Published by McGraw-Hill, a business unit of The McGraw-Hill Companies, Inc., 1221 Avenue of the Americas, New York, NY 10020. Copyright © 2008 by The McGraw-Hill Companies, Inc. All rights reserved. No part of this publication may be reproduced or distributed in any form or by any means, or stored in a database or retrieval system, without the prior written consent of The McGraw-Hill Companies, Inc., including, but not limited to, in any network or other electronic storage or transmission, or broadcast for distance learning.

Some ancillaries, including electronic and print components, may not be available to customers outside the United States.

This book is printed on recycled, acid-free paper containing 10% postconsumer waste.

1 2 3 4 5 6 7 8 9 0 QPD/QPD 0 9 8 7

ISBN 978-0-07-295267-4
MHID 0-07-295267-9

Publisher: *Janice Roerig-Blong*
Sponsoring Editor: *Thomas C. Lyon*
Developmental Editor: *Rose M. Koos*
Marketing Manager: *Tamara Maury*
Lead Project Manager: *Mary E. Powers*
Senior Production Supervisor: *Sherry L. Kane*
Senior Media Project Manager: *Jodi K. Banowetz*
Senior Media Producer: *Eric A. Weber*
Designer: *Rick D. Noel*
Cover/Interior Designer: *Elise Lansdon*
(USE) Cover Image: *©Steve Bloom/stevebloom.com*
Senior Photo Research Coordinator: *Lori Hancock*
Photo Research: *Jerry Marshall*
Supplement Producer: *Melissa M. Leick*
Compositor: *Precision Graphics*
Typeface: *9.5/12 Slimbach*
Printer: *Quebecor World Dubuque, IA*

The credits section for this book begins on page C-1 and is considered an extension of the copyright page.

Library of Congress Cataloging-in-Publication Data

Presson, Joelle C.
  Biology : dimensions of life / Joelle C. Presson, Janann V. Jenner. – 1st ed.
    p. cm.
  Includes index.
  ISBN 978-0-07-295267-4–ISBN 0-07-295267-9
  1. Biology. I. Jenner, Janann V. II. Title.

QH308.2P745 2008
570–dc22

                                                          2006046973

www.mhhe.com

About the Authors

Joelle Presson and Jan Jenner have spent years collaborating on writing this textbook. In fact, the book is what first brought them together, but the paths that brought them to this rewarding project were quite different.

Joelle Presson has been teaching at the University of Maryland since 1988. She earned a B.S. in psychology from the University of South Florida and a Ph.D. in neuroscience from the University of Oregon, under the mentorship of Barbara Gordon. During that time, she studied the effects of early experience on the development of the visual system in cats. In post-doctoral positions Joelle showed how the retinas and brains of fishes add new neurons throughout life, and in 1985 she began to study postembryonic neurogenesis in the auditory system of fishes.

Currently, Joelle's focus is on teaching nonmajors biology, and this book is an outgrowth of that teaching experience. She also has taught major's introductory biology, upper level neurophysiology, and neuropharmacology. Joelle is also the Assistant Dean of Undergraduate Academic Programs in the College of Chemical and Life Sciences at the University of Maryland, and in that role, Joelle works to develop the curriculum within Biology and across other disciplines.

Outside of her professional life, Joelle's passion is Kung Fu. When she is not at her desk or immersed in Kung Fu, Joelle often can be found reading science fiction, running on local trails, enjoying old Star Trek episodes, or making gourmet meals with her husband, son, and daughter.

Jan Jenner earned her Ph.D. in biology from New York University, studying herpetology under the mentorship of Herndon Dowling. Her studies focused on a taxonomic revision of a large group of neotropical snakes. She spent her teaching career at NYU learning to better communicate the ideas of introductory biology to more than 300 students each semester, and she also taught undergraduate field biology and ecology courses. Jan is the recipient of two teaching awards from NYU students. Jan also taught at the American Museum of Natural History in New York City, St. David's School (New York), and Talladega College (Alabama). Consequently, Jan has taught biology to every age group, from toddlers to college students and has learned that each has its own challenges, delights, rewards, and frustrations.

Recently Jan has divided her time between teaching biology and writing. She has written several middle school biology textbooks, served as developmental reviewer on hundreds of educational publications, and has written two other college biology textbooks. Her novel, *Sandeagozu*, follows the travels of a group of animals led by a python as they escape from captivity to seek their hearts' desires.

Jan lives in a gorgeous woods in rural Alabama with a talented husband, a pack of dogs, and several reptiles.

To Our Students

We hope that you enjoy our textbook. We've written it with you in mind and have tried our best to make it understandable and accessible.

To Our Families

We thank you for your steadfast support and encouragement and for your generous love. We never could have accomplished this without you.

Brief Contents

This application chapter is located on the book-specific website at www.aris.mhhe.com.

Contents

PART II

Reproduction of Cells and Inheritance

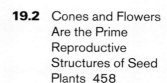

PART IV

Plant Biology

18 The Living Plant: Plant Structure and Function 428

19 The Thread of Life: Reproduction of Seed Plants 455

PART V

Human Biology

20 Senses, Nerves, Bones, Muscles 475

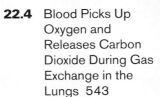

PART VI

Ecology

Applications

Application Chapters

An Added Dimension/ An Added Dimension Explained

Four Corners Disease

Cryopreservation

Mad Cow Disease

Superman and Spinal Cord Injury

Frankenstein, Energy, and the Origin of Life

Global Climate Change

Tuberculosis

Progerias

The Death of Times Beach

Eugenics and Fitter Families

What Promise Does Gene Therapy Hold?

Steve McQueen, Cancer, and Laetrile/Assessing Laetrile and Other Alternative Cancer Therapies

Dinosaurs and Birds

Are Human Races a Valid Idea or Not?

Poisons in the Sea

Salem—1692

Lindow and the Bog Bodies of Europe

The Green Fairy

The Intertwined Lives of Plants and Their Pollinators

Jennifer's Story/Understanding Jennifer's Addiction to Crack Cocaine

Pellagra

What Killed Len Bias?

Boy Meets Girl/What Determines Human Gender Identity?

Saving the Life of a Child/Jane Doe's Life Was Saved by Modern Reproductive Technology

Where Did HIV Originate?

Songbird Decline

The Death and Life of Lake Erie

What Does It Mean to Be Human?/Lessons in Humanity from Chimpanzees

Now You Can Understand

Risk Factors—What Do They Mean?

Antibiotic Resistance

Death by pH

Sick from Too Little Salt?

Nanotechnology

What Are Drugs?

Food Calories—What Are They, Really?

Antioxidants and Free Radicals

Why Cyanide Is a Deadly Poison

Antisense for Research and Medicine

Neural Tube Defects

It's a Boy. It's a Girl. Or Maybe Both

Side Effects of Accutane

Incest Taboos

The Genetics of Individual Differences

Does Chemotherapy Destroy the Human Immune System?

Can Vitamin Supplements Help Prevent Cancer?

Antibiotic Resistance

Biodiversity Hot Spots

Spores, Botulism, and Botox

How Antifungal Drugs Work

Forensic Molecular Botany

Poison Ivy

Pollen and Allergies

The Use of Brain Electrodes in Treatments of Disorders

Free Radicals, Reactive Oxygen Species, and Antioxidants

The Effect of Diuretics on Your Body

Preventing Pregnancy

Concerns About Unequal Sex Ratios

Anthrax Infections

Some Countries Are Close to Negative Population Growth

Ecologically, It Is Impossible to Do Just One Thing

Killer Fog and the Fight Against Air Pollution

What Do You Think?

Mental Telepathy

Just How Much Water Should You Drink?

Eat Chitin for Weight Loss?

Can Magnets Eliminate Pain?

Panspermia

Weight Loss and Fad Diets

Astrology

Pros and Cons of Embryonic Stem Cell Research

Are Pregnant Women Responsible for the Health of Their Unborn Children?

Does Heredity or Environment Exert More Influence on Human Intelligence?

Should There Be Mandatory Genetic Testing for Some Conditions?

Should New AIDS Drugs Be Fast Tracked?

The Future of Human Evolution

Are Wild and Farm-Raised Salmon Equivalent?

How Did the Bacterial Flagellum Evolve?

Big Foot and Little People

What Should Be Done About the High Cost of Paper?

Are Tree-Sitters Noble or Foolish?

Does Talking to Plants Help Them Grow?

"You Use Only 10% of Your Brain"

Should the Government Regulate Dietary Supplements?

Do Magnets Worn on the Skin Increase Blood Flow and So Relieve Pain and Heal Injuries?

Is Chemical Castration an Appropriate Way to Control Sex Offenders?

Human Cloning

Should HIV Testing Be Mandatory?

How Many Children Do You Plan to Have?

Wetlands Conservation

Human Impact on the Environment

Preface

Biology is a field of study with concepts that are highly dimensional. Each biological concept relates to the next and contributes to a clearer understanding of the entire field of biology and the world around us. Our vision for *Biology: Dimensions of Life* stems from our passion for experiencing and sharing these biological concepts. We want students to not only understand the framework concepts of biology, but also to appreciate the great impact of those concepts on their daily lives. For example, why is it so difficult for researchers to find a cure for cancer, and how does that research relate to the cell and its processes? Global warming and stem cell research are issues on the political docket, but how do students know who and what to believe about the issues?

Our goal of uniting concepts with applications has been guided by five principles as we developed *Biology: Dimensions of Life:*

- Writing Style That Converses with the Reader
- The Explanatory Power of Biology in the Context of Real-Life Topics
- The Unifying Theme of Evolutionary Biology
- The Guided Learning System
- Concepts Brought to Life Through Examples and Analogies

Each of these principles is explained here in further detail and constitutes what sets our book apart from other nonmajors textbooks.

What Sets This Book Apart?

Although a wide variety of introductory biology textbooks are available to adopters, our approach is unique as outlined below.

Writing Style That Converses with the Reader

Both of us are teachers, and in writing this book we brought our classroom voice to the text. Without allowing our personal style to distract from the content, we have written *Biology: Dimensions of Life* as a conversation with the students. We often use more casual language. We query students. We put students into the narrative by asking them to think about a topic or reread a passage. The writing is clear, engaging, and accessible. Our experience indicates that students will *read* this book.

The Explanatory Power of Biology in the Context of Real-Life Topics

Biology is more than just a collection of interesting facts and findings. From the beginning of human history, people have wanted to explain why things happen, and some of the most exciting and powerful explanations in biology are about complex and advanced topics—topics that are in the news and that students experience in their own lives. Our approach in *Biology: Dimensions of Life* is to provide students with a basic understanding of biological concepts, along with a deeper understanding of current topics that apply to their lives.

In *Biology: Dimensions of Life* the inclusion of relevant, applied material is handled primarily in two ways:

- *An Added Dimension* and *An Added Dimension Explained*
- Five *Application Chapters* incorporated throughout the textbook

An Added Dimension and An Added Dimension Explained

Every chapter opens and closes with a discussion of one relevant and interesting topic that can be better understood through the science of biology. The opening discussion, called *An Added Dimension,* explains the topic in a storylike format without the use of scientific jargon. Written in an interesting and friendly tone, the story welcomes students into the chapter. At the end of *An Added Dimension,* students are encouraged to consider questions about the topic and informed that the chapter will lead them to a better understanding of the questions raised.

Following the presentation of the chapter itself, the opening topic is discussed again in light of the concepts just presented. This discussion, called *An Added Dimension Explained,* is thorough, contains more scientific information than *An Added Dimension* at the beginning of the chapter, and includes references to key pages and figures.

The topics included in this feature are diverse. Some are current topics such as global climate change or gene therapy. Some ask the student to reach into history and understand how science has affected societal issues in the past, such as eugenics or pellagra. Others bring the student into exotic biological worlds, such as the strange relationship between fig trees and fig wasps. These scenarios can be readily incorporated into lectures, or you can use them as inspirations for your own topics.

Application Chapters

Of course, many of the topics of most interest to students, and important from a societal perspective, need more time and background knowledge than the *An Added Dimension* scenarios allow. We have developed *Biology: Dimensions of Life* around core biological concepts and have provided enough detail in all basic subject areas to give students a firm foundation in their study of biology. In addition, we have incorporated five *application chapters,* and a sixth chapter on the accompanying ARIS website, that allow students to tackle current topics at a deeper level. Students will benefit from the knowledge of how the biological concepts they just studied relate to topics in the news and in their lives. It is in these topics that the true explanatory power of biology can be experienced by students, as they use their knowledge to make decisions now and in the future.

Cancer, nanotechnology, bioremediation, antibiotic resistance, global climate change, emerging infectious diseases, cryonics, connections between brains and computers, and dementia are all topics that can be explained more thoroughly through science. Students who are not science majors can increase their understanding of such topics *if* the topics are chosen carefully, presented in a meaningful context that includes broader social and personal issues, and supported by sufficient coverage of the necessary basic material.

The inclusion of deeper, relevant material in a textbook requires careful organization. First, it is important that basic material be mastered before more integrative, applied material can be tackled. In our experience it is not possible to adequately teach basic concepts only within the context of applied topics. For example, it is difficult to teach the basics of cell biology in the middle of a discussion of cancer. An appreciation of deeper and more relevant topics must rest on a mastery of basic topics. Second, presentation of just the *biology* of applied or relevant topics is not enough to engage student interest or convey an understanding of why such topics are important. Students need to know something about the bigger issues surrounding biology. For example, students need to appreciate the degree to which cancer is a major public health problem, and the ways in which their own choices influence their chances of developing cancer as a context for understanding the biology and treatment of cancer. This approach to teaching may mean covering less and uncovering more. You—the instructor—will have to choose which topics to delve into and which topics to cover lightly or not at all. This text provides you with choices.

The most difficult part of writing the *application chapters* has been deciding which topics to include. There are so many topics that would be interesting and so few available pages to fill. We have included the more popular topics, but if you like this approach, please let us know which other cutting-edge topics you would like to see covered in new chapters.

The Unifying Theme of Evolutionary Biology

Instructors understand the importance of incorporating the theme of evolution throughout a biology text, and we have approached this by including explicit examples that are not usually found in nonmajors textbooks. In the evolution chapters, we emphasize that complex adaptations are elaborations on simpler evolutionary adaptations, and we carry this into the systems chapters. In the system chapters, we trace complex systems found in mammals to their evolutionary origins, which are often in prokaryotes. For example, we begin the discussion of sensory and neural systems with a brief mention of membrane protein channels in

prokaryotes. This may seem a bit detailed for nonmajors, but it can be introduced in just a paragraph or two, and if done consistently it emphasizes that complex biological mechanisms are built on ancestral forms.

The Guided Learning System

Your students will especially benefit from *Biology: Dimensions of Life*'s new **Guided Learning System.** This innovative set of features is designed to lead your students through the learning process as they work to understand the many new biological concepts and applications within each chapter. A few highlights of the **Guided Learning System** include:

- *An Added Dimension* and *An Added Dimension Explained*
- *Quick Checks*
- *Running Glossary*
- *Now You Can Understand*
- *What Do You Think?*
- *Quantitative Query*

A brief story from the news, *An Added Dimension,* starts the chapter in the context of something real; numbered sections help students organize the material; *Quick Check* questions assess student understanding of the concepts; marginal glossary terms aid students as they master new vocabulary; and a follow-up to the opening story, *An Added Dimension Explained,* brings student learning full circle. At the completion of each chapter, students are encouraged to apply what they have learned through the *Now You Can Understand* and *What Do You Think?* sections, and as they work through the chapter summary, and various types of assessment questions, including *Connecting the Concepts* and *Quantitative Queries.* You will find detailed descriptions of all pedagogical features in the Guided Tour on the next several pages.

Concepts Brought to Life Through Examples and Analogies

Teaching biology is more than presenting material. When teaching concepts you take time to explain, elaborate, give examples, provide analogies, do class exercises, and use other approaches to help students grasp difficult concepts. Throughout *Biology: Dimensions of Life* we strive to explain biology. We use precise narrative, analogies, examples, and other approaches to help students understand complex topics.

We hope you find *Biology: Dimensions of Life* interesting, informative, and most of all, a biology textbook that engages students. We are committed to producing the best materials available to help students understand and appreciate the study of biology—and to help you teach them. We welcome your suggestions for improving future editions of *Biology: Dimensions of Life.*

Warm regards,

Joelle C. Presson

Jan Jenner

Joelle C. Presson   Jan Jenner

Guided Tour

The Guided Learning System

Each chapter of *Biology: Dimensions of Life* includes a consistent pedagogical framework that guides students through the learning process. The pedagogical tools are designed to engage students, help them understand key concepts, and challenge them to analyze and apply the concepts.

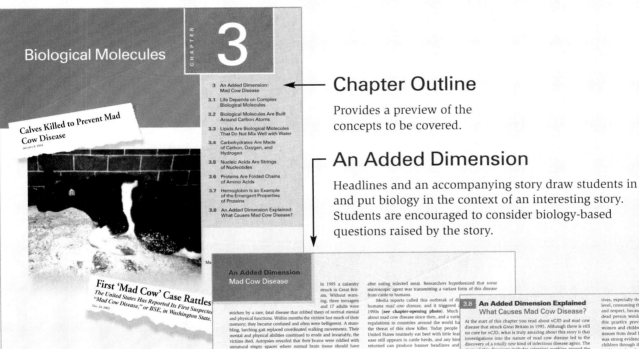

Chapter Outline

Provides a preview of the concepts to be covered.

An Added Dimension

Headlines and an accompanying story draw students in and put biology in the context of an interesting story. Students are encouraged to consider biology-based questions raised by the story.

An Added Dimension Explained

The final section of the chapter addresses the questions raised in the chapter-opening story, using and expanding upon what students have learned in the chapter. Students are encouraged to move beyond fact retention and apply the chapter concepts.

Instructive Section Heads

The main sections of each chapter begin with a numbered head that clearly summarizes the material to be covered.

Marginal Glossary Terms

Each key term is defined in the margin on the same page or spread as its discussion in the text. This is a quick reference for students as they work to master the terminology of biology.

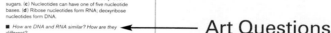

Art Questions

Questions appear with select figures and encourage students to study the art carefully, reinforcing the concepts explained in the text.

Quick Checks

Questions at the end of each section allow students to test their comprehension of key concepts. Answers to the questions are provided in the back of the book, so students can self-assess how well they understand the material before moving on to the next section.

Now You Can Understand

Students are prompted to consider a current topic or event in light of the biological concepts they just studied. This gives students an excellent opportunity to see how their understanding of biology helps them better understand something common.

What Do You Think?

Students are challenged to think about a current issue in the context of what they just learned and to investigate the issue further.

Chapter Review

CHAPTER SUMMARY

3.1 Life Depends on Complex Biological Molecules

Lipids, carbohydrates, nucleic acids, and proteins are the biological molecules shared by living organisms.

biological molecules 00

3.2 Biological Molecules Are Built Around Carbon Atoms

All biological molecules are built around carbon atoms. Hydroxyls, carboxyls, amino groups, and phosphates are functional groups that impart distinctive chemical characteristics to complex biological molecules.

amino group 00
carboxyl 00
functional group 00
hydrocarbon 00

hydroxyl 00
organic chemistry 00
phosphate 00

3.3 Lipids Are Biological Molecules That Do Not Mix Well with Water

Lipid molecules have large hydrocarbon portions, and some have water-soluble polar functional groups too. Lipids are components of cell membranes as well as energy storage molecules and hormones.

lipid 00

3.4 Carbohydrates Are Made of Carbon, Oxygen, and Hydrogen

Carbohydrates are molecules with at least three carbon atoms and have carbon, hydrogen, and oxygen atoms as well. Carbohydrates are energy storage molecules and structural molecules.

carbohydrate 00
disaccharide 00

monosaccharide 00
polysaccharide 00

3.5 Nucleic Acids Are Strings of Nucleotides

Nucleotide molecules have a sugar with a phosphate and nitrogenous base attached and form nucleic acids. Nucleotides form the nucleic acids, DNA and RNA. Both DNA and RNA contain instructions, but they differ in structure, nitrogenous bases, sugars, and in how the molecules function.

nitrogenous base 00
nucleic acid 00

nucleotide 00

3.6 Proteins Are Folded Chains of Amino Acids

Proteins are built from chains of amino acids, which are biological molecules with an amino group, carboxyl group, hydrogen atom, and a distinctive R-group attached to a central carbon atom. Amino acids are the building blocks of proteins, the molecules that do the work of

cells. All proteins have three levels of structure, and some have a fourth level.

amino acid 00
denaturation 00
peptide bond 00
primary structure 00

protein 00
quaternary structure 00
secondary structure 00
tertiary structure 00

3.7 Hemoglobin Is an Example of the Emergent Properties of Proteins

As sickle cell anemia shows, even a small change in the sequence of the amino acids of a protein can dramatically alter the way the protein acts.

3.8 An Added Dimension Explained: What Causes Mad Cow Disease?

Stanley Prusiner discovered that a previously unknown, infectious, deformed protein—a prion—causes vCJD and mad cow disease, as well as the related diseases kuru and scrapie.

prion 00

REVIEW QUESTIONS

TRUE or FALSE. If a statement is false, rewrite it to make it true.

1. Lipid molecules dissolve readily in water.
2. Proteins are the least complex of the biological molecules.
3. RNA is a nucleic acid; it is double stranded, incorporates deoxyribose sugar, and contains uracil.
4. In sickle cell anemia the hemoglobin molecule has one incorrect carbohydrate.
5. Prions are aberrant proteins that can cause mad cow disease.

MULTIPLE CHOICE. Choose the best answer of those provided.

6. The shape and charge distribution of a protein is critically important because
   a. the shape changes over time as the protein ages, giving it new properties.
   b. these properties govern the function of each protein.
   c. the charge and shape are identical for every kind of protein and provide an important instance of the unity of life.
   d. proteins are exceptions to the universality of the electrical force.
   e. proteins have no chemical bonds.

7. A protein that helps break up or combine molecules is called
   a. hemoglobin.
   b. collagen.
   c. an enzyme.
   d. a phosphate.
   e. a carbon ring.

Chapter Summary

Briefly restates the key points of the chapter, with page references back to key terms.

Review Questions

Multiple-choice, matching, and true or false questions allow students to assess their understanding of the concepts. Answers are provided in Appendix A.

Connecting Key Concepts

These questions go beyond memorization and require a deeper level of understanding. The questions encourage students to make connections between the concepts of the chapter. Answers are provided on the book-specific website at www.aris.mhhe.com.

8. A carbon atom can make a maximum of _____ covalent bonds.
   a. one
   b. two
   c. three
   d. four
   e. six

9. The chemical elements that are most common in living systems are
   a. C, S, Fe, H, N
   b. C, Cl, Mg, Fe, H
   c. C, H, O, N, P
   d. C, I, Na, K, O

10. Which of the following fats are considered "healthy" for you? Be sure you look at the type of fat and at the source; both must be correct
    a. saturated fats such as corn oil
    b. saturated fats such as lard
    c. trans fats found in shallow-water fish such as catfish
    d. polyunsaturated fats such as soybean oil
    e. trans fats in partially hydrogenated vegetable oils

11. In your diet you take in significant amounts of carbohydrates from all of the following except
    a. an apple slice.
    b. an ounce of lean beef.
    c. a small baked potato.
    d. a stalk of celery.
    e. a 1-ounce chocolate bar.

12. Lipids are used for
    a. storing energy in fat cells until it is needed.
    b. membranes that are in and around every cell in an organism.
    c. making some hormones, chemical messengers in the body.
    d. helping some nonpolar molecules to move into cells.
    e. All of the above are uses for lipids.

13–16. Match the small organic molecular subunit with the large organic biomolecule that is built from it (one choice will not match):

13. hydrocarbon chain
14. simple sugar
15. nucleotide
16. amino acid

a. protein
b. nucleic acid
c. carbohydrate
d. lipid
e. water

17–20. Match the molecule to one of its uses in living organisms (one choice will not match):

17. proteins
18. carbohydrates
19. lipids
20. nucleic acids

a. short-term energy storage
b. the basis of organic chemistry
c. accomplish most cellular processes
d. store energy and are important components of cell membranes
e. carry genetic instructions

CONNECTING KEY CONCEPTS

1. In what ways are the properties of complex carbohydrates an example of emergent properties?

2. What determines the three-dimensional shape of a protein?

QUANTITATIVE QUERY

Refer to the following graph. What is the youngest age group for which the percentage of women with high cholesterol is higher than the percentage of men with high cholesterol?

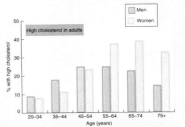

CRITICAL THINKING

1. In the 1800s Louis Pasteur discovered how to "pasteurize" liquids. Pasteur showed that if liquids are heated to a sufficient temperature, harmful bacteria are killed and the liquids are less likely to make people sick. Of course, you probably know that heating the bacteria "kills" them, but how? Using what you know about biological molecules, propose a mechanism that explains how heating kills the infectious bacteria in foods.

2. In this chapter you have read about the biological molecules that interact to produce life. Recall from Chapter 2, though, that in cells these interactions occur in a watery environment. One way to really understand this is to consider whether any other solvent could be substituted for water. You know that water is polar, and you also know about the charge distribution on the major biological molecules. Think about how each of these molecules would interact with or dissolve in water. Now think about the ethyl alcohol (ethanol) molecule, diagramed in Table 3.3. How would the interactions of the biological molecules differ if they were immersed in ethyl alcohol or if they were immersed in water?

For additional study tools, visit www.mhhe.com/presson1.

79

Quantitative Query

Math questions related to the biological concepts being studied provide a multi-disciplinary approach to learning. Students are asked to analyze graphs, and solve equations and word problems for a better understanding of biological concepts. Answers are provided on the book-specific website at www.aris.mhhe.com.

Critical Thinking

Brief scenarios of everyday occurrences or ideas challenge students to apply what they have learned to their life and their world.

Vivid Illustrations that Contribute to Learning

The illustration program for *Biology: Dimensions of Life* utilizes various types of art to communicate important concepts through visuals. Designed to harmonize with the text, and to be an educational tool on its own, the illustration program clearly presents the concepts through a combination of line art, explanatory labels, and photos.

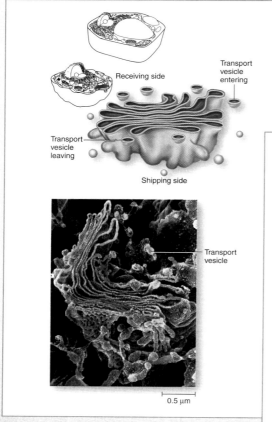

Combination Figures

Line art is combined with photos to give students the best of both perspectives: the realism of photos and the explanatory clarity of line drawings.

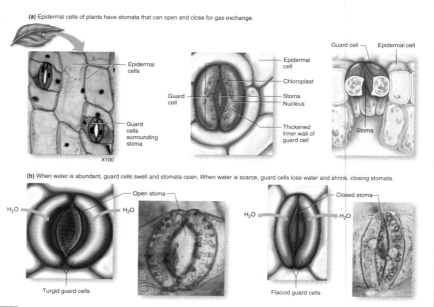

(a) Epidermal cells of plants have stomata that can open and close for gas exchange.

(b) When water is abundant, guard cells swell and stomata open. When water is scarce, guard cells lose water and shrink, closing stomata.

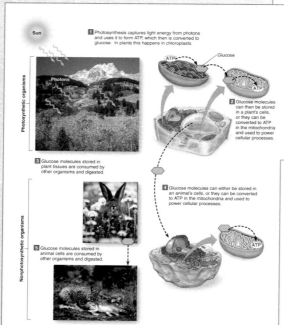

Process Figures

Processes are presented in steps and organized in an easy-to-follow format.

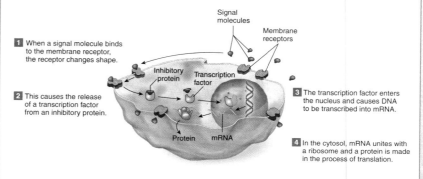

1 When a signal molecule binds to the membrane receptor, the receptor changes shape.

2 This causes the release of a transcription factor from an inhibitory protein.

3 The transcription factor enters the nucleus and causes DNA to be transcribed into mRNA.

4 In the cytosol, mRNA unites with a ribosome and a protein is made in the process of translation.

Macroscopic to Microscopic Figures

Macroscopic and microscopic views are combined to help students understand complex structures.

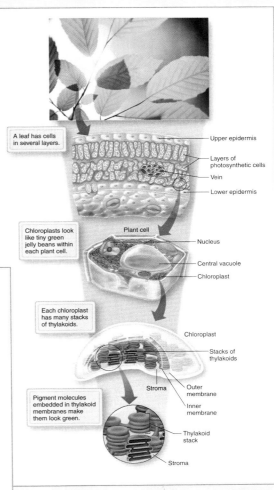

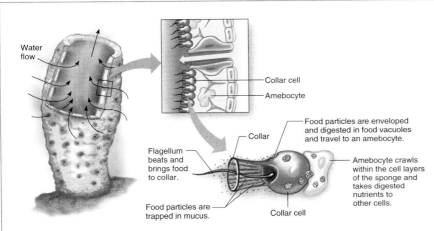

Instructional Integration of Text and Art

Line art and detailed labels are combined to appeal to various learning styles and to make the art highly effective.

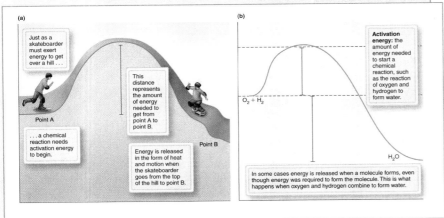

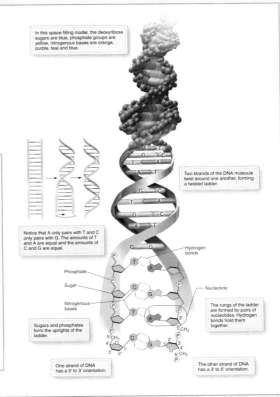

Teaching and Learning Supplements

For Instructors

McGraw-Hill Presentation Center

Build instructional materials where-ever, when-ever, and how-ever you want! ARIS Presentation Center is an online digital library containing assets such as photos, artwork, PowerPoints, animations, and other media types that can be used to create customized lectures, visually enhanced tests and quizzes, compelling course websites, or attractive printed support materials.

Nothing could be easier! Accessed from the instructor side of your textbook's ARIS website, Presentation Center's dynamic search engine allows you to explore by discipline, course, textbook chapter, asset type, or keyword. Simply browse, select, and download the files you need to build engaging course materials. All assets are copyright McGraw-Hill Higher Education but can be used by instructors for classroom purposes.

McGraw-Hill's ARIS: Assessment, Review, and Instruction System

McGraw-Hill's ARIS is a complete, online electronic homework and course management system, designed for greater ease of use than any other system available. Created specifically for *Biology: Dimensions of Life,* instructors can create and share course materials and assignments with colleagues with a few clicks of the mouse. For instructors, personal response system questions, all PowerPoint lectures, and assignable content are directly tied to text-specific materials in *Biology: Dimensions of Life.* Instructors also can edit questions, import their own content, and create announcements and due dates for assignments.

ARIS has automatic grading and reporting of easy to assign homework, quizzing, and testing. All student activity within McGraw-Hill's ARIS is automatically recorded and available to the instructor through a fully integrated grade book that can be downloaded to Excel.

For students, there are multiple-choice quizzes, animations with quizzing, videos with quizzing, and even more materials that may be used for self-study or in combination with assigned materials.

Go to www.aris.mhhe.com to learn more and register!

ScienCentral Videos

McGraw-Hill is pleased to announce an exciting partnership with ScienCentral, Inc., to provide brief biology news videos for use in lecture or for student study and assessment purposes. A complete set of ScienCentral videos is located within this text's ARIS course management system, and each video includes a learning objective and quiz questions. These active learning tools enhance a biology course by engaging students in real-life issues and applications such as developing new cancer treatments and understanding how methamphetamine damages the brain. ScienCentral, Inc., funded in part by grants from the National Science Foundation, produces science and technology content for television, video and the Web.

McGraw-Hill: *Biology Digitized Video Clips*

ISBN 13: 978-0-07-312155-0
ISBN 10: 0-07-312155-X
McGraw-Hill is pleased to offer adopting instructors a new presentation tool—digitized biology video clips on DVD! Licensed from some of the highest-quality science video producers in the World, these brief segments range from about five seconds to just under three minutes in length and cover all areas of general biology from cells to ecosystems. Engaging and informative, McGraw-Hill's digitized biology videos will help capture students' interest while illustrating key biological concepts and processes such as mitosis, how cilia and flagella work, and how some plants have evolved into carnivores.

Instructor's Testing and Resource CD-ROM

ISBN 13: 978-0-07-295270-4
ISBN 10: 0-07-295270-9
This cross-platform CD features a computerized test bank that uses testing software to quickly create customized exams. The user-friendly program allows instructors to search for questions by topic, format, or difficulty level; edit existing questions or add new ones; and scramble questions for multiple versions of the same test. An Instructor's Manual is also included on the CD-ROM.

Transparencies

ISBN 13: 978-0-07-295268-1
ISBN 10: 0-07-295268-7
This set of overhead transparencies includes key pieces of line art in the textbook plus tables. The images are printed with better visibility and contrast than ever before.

Photo Atlas for General Biology

ISBN 13: 978-0-07-284610-2
ISBN 10: 0-07-284610-0
This atlas was developed to support our numerous general biology titles. It can be used as a supplement for a general biology lecture or laboratory course.

PageOut

McGraw-Hill's exclusive tool for creating your own website for your introductory biology course. It requires no knowledge of coding and is hosted by McGraw-Hill.

Course Management Systems

ARIS content compatible with online course management systems like WebCT and Blackboard makes putting together your course web site easy. Contact your local McGraw-Hill sales representative for details.

For Students

ARIS (Assessment Review and Instruction System)

Explore this dynamic website for a variety of study tools.

- **Self-quizzes** test your understanding of key concepts.
- **Flash cards** ease learning of new vocabulary.
- **Animations** and **Videos** bring key genetic concepts to life and are followed by a quiz to test your understanding.
- **Essential Study Partner** contains hundreds of animations, learning activities, and quizzes designed to help students grasp complex concepts.

Go to aris.mhhe.com to learn more or go directly to this book's ARIS site at www.aris.mhhe.com.

Student Study Guide

ISBN 13: 978-0-07-328030-1
ISBN 10: 0-07-328030-5
This guide includes study aids developed specifically for *Biology: Dimensions of Life.* Written by Jolie Stepaniak of Wayne State University, each study guide chapter includes a chapter review, learning objectives and study questions for each section of the chapter, and a chapter test. Answers to all exercises are provided in the study guide to give students immediate feedback.

How to Study Science

ISBN 13: 978-0-07-234693-0
ISBN 10: 0-07-234693-0
This workbook offers students helpful suggestions for meeting the considerable challenges of a science course. It gives practical advice on such topics as how to take notes, how to get the most out of laboratories, and how to overcome science anxiety.

Acknowledgments

We have worked many years to bring this textbook to publication, and producing the finished work involved an extensive team effort. We acknowledge the commitment of our team to this project and thank them for their hard work.

Tom Lyon, Sponsoring Editor, has been with the project from the beginning. Tom believed in our vision of the book and worked hard to keep the project on track. Tom's enthusiasm has never seemed to flag and he always has been in our corner. We very much appreciate Tom's grasp of the interface between authors, publisher, sales force, and users. We feel exceptionally lucky to have had someone with such commitment, and we hope our collaboration continues on subsequent editions.

Tamara Maury, Marketing Manager, has been an enthusiastic supporter of our work and has provided invaluable insights into the requirements of the nonmajors market. Tamara combines market savvy with an intuitive sense of what users require of an introductory biology textbook. She has been most generous of her time and talents, and many of her wonderful ideas have been incorporated into the text.

Kathy Naylor was our first Developmental Editor. She gave us a solid start and helped us to initially develop our art and incorporate reviewers' comments into the first draft.

When the text was about a year from publication, Rose Koos assumed the responsibilities of Developmental Editor. Rose is a marvel of organization, a model of good judgment, and a joy to work with. Working as the editorial liaison with the McGraw-Hill Production team, Rose has been the eye of calm in the whirlwind of book production. We have relied on Rose to know when things had to be done and how to get them done.

Mary Powers, Lead Project Manager, guided the book into and through production, integrating the contributions of authors, artists, photo researchers, and compositors to ready the text for publication. Imagine threading a team of high-strung horses at high speed through a long series of narrow gates, and you have some idea of Mary's task. She did it beautifully.

Jane Peden, Senior Administrative Assistant, has been one of our "go to" people. Jane has been especially helpful in arranging travel, and in making sure our paperwork is in order and that we have the necessary resources.

Beatrice Sussman edited the manuscript, removed our clumsy language, and caught many of our errors. We thank her for making our text read more smoothly and concisely.

Lori Hancock and Jerry Marshall were responsible for photo research. They worked hard to fulfill our sometimes obscure photo requirements and have found many lovely photos to illustrate biological concepts.

Patrick Galliart, Catherine Tiene Gleason, William Wyatt Hoback, Kathleen Pelkki, and Jolie Stepaniak developed and wrote the various supplemental print and electronic materials that accompany our textbook. We appreciate their expertise and dedication.

Family, friends and colleagues have also provided a great deal of patient support throughout this project. On the home front, both of our families have had to adjust and accommodate to the time we needed to complete the book. All of our family members contributed in other important ways to the quality of our text. Jan's husband, Herndon Dowling, is one of the world's leading herpetologists and contributed insights into our chapters on evolution and animal diversity. In addition to reading many drafts of chapters, he also made many trips to town to mail manuscripts so that Jan could remain working at the computer for a little longer each day. Joelle's husband, Harold Hawkins, has expertise in broad areas of sensory science and provided a sounding board for many of the ways we presented that material. Joelle's children, Charles and Laura Hawkins, have been incredibly understanding of the time their mom has *not* been cooking dinner, doing laundry, or just hanging out with them. They also have contributed to the material in the book. Charles is deeply knowledgeable about physics, and so helped us to make sure we described physical concepts such as energy and electron orbitals correctly. Laura has expertise in English grammar and in archeology, and so was a resource for questions about writing style and some of the historical references in the book. And of course our families were there to sometimes take us away from the book and enjoy life.

Joelle's work on the book was especially supported by colleagues and supervisors at the University of Maryland. Joelle is a full time teacher/administrator in the College of Chemical and Life Sciences at UM. Both the Dean of the College, Norma Allewell, and the Associate Dean, Bob Infantino, have been supportive and flexible in allowing Joelle some time away from work to complete the project. Many colleagues at UM provided expert help on difficult chapters. Todd Cooke's expertise in plant biology was indispensable for our plant chapters. Albert Ades, who teaches our Biology of Cancer course, read the cancer chapter and ensured that our details in that arena were accurate. Others, including Jeff Jensen, William Higgins, and Charles Delwiche, also contributed expertise by answering random emails asking for help on some small but crucial points.

In addition to input from these colleagues, we have received invaluable feedback from numerous instructors who teach non-major students. Many of the comments and suggestions offered by these reviewers have been incorporated into *Biology: Dimensions of Life*, and as a result, have produced a stronger textbook. We are especially grateful for the time and expertise given to us by the Board of Advisors. These individuals reviewed materials at every stage of the process and helped us refine our work.

Reviewers

D. Daryl Adams, *Minnesota State University*
Sylvester Allred, *Northern Arizona University*
Kenneth D. Andrews, *East Central University*
Amir Assadi-Rad, *San Joaquin Delta College*
Jessica K. Baack, *Montgomery College*
Gail Baker, *LaGuardia Community College*
S.K. Ballal, *Tennessee Technology University*

Timothy A. Ballard, *University of North Carolina Wilmington*
Sarah F. Barlow, *Middle Tennessee State University*
Joseph Bettencourt, *Marist College*
Dan Bickerton, *Ogeechee Technical College*
Charles L. Biles, *East Central University*
Donna H. Bivans, *Pitt Community College*
Karen S. Borgstrom, *Moraine Valley Community College*
Richard A. Boutwell, *Missouri Western St. College*
Susan Bower, *Pasadena City College*
Clarence J. Branch Jr., *St. Augustines College*
Randy Brewton, *University of Tennessee*
Peggy Brickman, *University of Georgia*
Sharon Bringer, *John Wood Community College*
Carol A. Britson, *University of Mississippi*
James M. Britton, *Arkansas Stata University*
Steve Brumbaugh, *Green River Community College*
Neil J. Buckley, *The State University of New York*
Arthur L. Buikema, Jr., *Virginia Polytech Institute*
Nancy M. Butler, *Kutztown University*
Suzanne K. Butler, *Miami Dade College*
David Byres, *Florida Community College*
Kelly S. Cartwright, *College of Lake County*
Michelle Cawthorn, *Georgia Southern University*
Van Christman, *Brigham Young University*
Kimberly Cline-Brown, *University of Northern Iowa*
Jeffrey Scott Coker, *Elon University*
Jan R.P. Coles, *Kansas State University*
Gary Courts, *University of Dayton*
Donald W. Deters, *Bowling Green State University*
Diane Dixon, *Southeastern Oklahoma State University*
Lee C. Drickamer, *Northern Arizona University*
JodyLee Estrada Duek, *Pima Community College*
William E. Dunscombe, *Union County College*
Stephen D. Ebbs, *Southern Illinois University*
Patrick Enderle, *East Carolina University*
Steven E. Fields, *Winthrop University*
Edison Fowlkes, *Hampton University*
Carl F. Friese, *University of Dayton*
Andrew Goliszek, *North Carolina A & T University*
Sandra Grauer, *Limestone College*
Gretel M. Guest, *Alamance Community College*
Carla Guthridge, *Cameron University*
Peggy J. Guthrie, *University of Central Oklahoma*
Elgenaid Hamadain, *Jackson State University*
Charles T. Hammond, *Brescia University*
Jill Harp, *Winston Salem State University*
Lysa M. Hartley, *Methodist College*
Don Heck, *Iowa State University*
M.C. Hirrel, *University of Central Arkansas*
Victoria Hittinger, *Rhode Island College*
W. Wyatt Hoback, *University of Nebraska*

Eva Horne, *Kansas State University*
Jeremiah N. Jarrett, *Central Connecticut State University*
Carl Johansson, *Fresno City College*
Chong Jue, *Queensborough Community College*
Glenn H. Kageyama, *California State Polytechnic University*
Jeffrey S. Kaufmann, *Irvine Valley College*
Lee Kavaljian, *California State University*
Paul Kelly, *Salem State College*
Todd Kelson, *Brigham Young University*
Amine Kidane, *Columbus State Community College*
Alan Kolok, *Nicholls State University*
Stephen G. Lebsack, *Linn Benton Community College*
Brenda G. Leicht, *University of Iowa*
Kristen Lenertz, *Black Hawk College*
Kevin J. Lien, *Portland Community College*
Paul Mangum, *Midland College*
John E. Marshall, *Pulaski Technical College*
Craig E. Martin, *University of Kansas*
Gregory J. McCormac, *American River College*
Joseph R. Mendelson III, *Utah State University*
David R. Mercer, *University of Northern Iowa*
Jon Milhoun, *Azusa Pacific University*
Lynda R. Miller, *Southwest Tennessee Community College*
Beth A. Montelone, *Kansas State University*
Jerry Montvilo, *Rhode Island College*
Melanie O'Brien, *De Anza College*
Alexander E. Olvido, *Virginia State University*
Onesiumus Otieno, *Oakwood College*
Donald J. Padgett, *Bridgewater State College*
Joshua M. Parker, *Community College of Southern Nevada*
Brian K. Paulson, *California University of Pennsylvania*
Gary W. Pettibone, *Buffalo State College*
William J. Pietraface, *The State University of New York College*
Michael Plotkin, *American River College*
Robert Pope, *Miami Dade Community College*
Elena Pravosudova, *Sierra College*
James V. Price, *Utah Valley State College*
Louis P. Primavera, *Hawaii Pacific University*
Karen Raines, *Colorado State University*
David A. Rintoul, *Kansas State University*
Roger E. Robbins, *East Carolina University*

Michael L. Rutledge, *Middle Tennessee State University*
Mark W. Salata, *Independent Science Education Consultant*
Shamili A. Sandiford, *College of Dupage*
Mark Ajgaonkar Schoenbeck, *University of Nebraska*
Cara Shillington, *Eastern Michigan University*
Marilyn Shopper, *Johnson County Community College*
Jennifer L. Siemantel, *Cedar Valley College*
Willetta Simms, *Asuza Pacific University*
Mrs. Sheena Smith, *American River College*
Minou D. Spradley, *San Diego City College*
Gil Starks, *Central Michigan University*
Julie Voke Sutherland, *College of Dupage*
Mary Talbolt, *San Jose State University*
R. Brent Thomas, *University of South Carolina Upstate*
Janis G. Thompson, *Lorain County Community College*
Catherine Tiene Gleason, *Empire State College*
Willetta Toole-Simms, *Azusa Pacific University*
Lenny Vincent, *Fullerton College*
Jennifer M. Warner, *University of North Carolina*
Craig Weaver, *Bervard Community College*
Susan Weinstein, *Marshall University*
Lisa A. Werner, *Pima Community College*
Allison Wiedemeier, *University of Missouri*
Christine Wilcox, *University of Idaho*
Claudia M. Williams, *Campbell University*
Lura C. Williamson, *University of New Orleans*
Heather Wilson-Ashworth, *Utah Valley State College*
Calvin Young, *Fullerton College*
Karen Zagula, *Wake Tech Community College*
Debbie A. Zetts Dalrymple, *Thomas Nelson Community College*
Michelle Zurawski, *Moraine Valley Community College*

Board of Advisors

Jessica Baack, *Montgomery College*
David Byres, *Florida Community College —South Campus*
Michelle Cawthorn, *Georgia Southern University*
Eva Horne, *Kansas State University*
Dave Loring, *Johnson County Community College*
Jennifer M. Warner, *University of North Carolina*
Nicole Welch, *Middle Tennessee State University*

The Framework of Biology

Mount Vernon Man Dies from Hantavirus

November 9, 2003

Local Woman Dies of Hantavirus

Local News

May 10, 2003.

Panic and Concern: Four Corners Disease. In May 1993 a fatal epidemic erupted in the Four Corners area, triggering an intensive effort to identify the cause of the disease and find a cure.

On May 14, 1993, a young Navajo man was brought to the medical center in Gallup, New Mexico (**see chapter opening photo**). He was feverish and gasping for air and his condition rapidly worsened. As a medical team struggled to save him, his blood pressure dropped, his breathing became more strangled, and quite suddenly he died.

Because the young man died of unknown causes, an autopsy was ordered. The pathologist who performed it noted something odd. Just five days earlier he had seen another case just like this one: a young woman who had died from similar, flulike symptoms. Odder still, the pair turned out to be an engaged couple, and the young man had been brought to the emergency room while traveling to his fiancée's funeral. Because the young couple's deaths were mysterious, swift, and similar, the hospital notified the New Mexico Department of Health. They warned doctors throughout the state that a fatal epidemic might be developing in the Four Corners region—an area where Arizona, Colorado, New Mexico, and Utah join. Doctors were asked to be alert for reports of other, similar deaths. Within days what the media were calling Four Corners disease had claimed more victims. All were young and otherwise healthy; all had had flulike symptoms; all the deaths were swift and agonized.

News of Four Corners disease spread panic and concern among the general population, and authorities at the Centers for Disease Control and Prevention (CDC) in Atlanta, Georgia, responded swiftly. Within just *one month* workers identified the cause of the disease and tracked it to its source. By mid-June the outbreak was understood in a larger context, and people in the region were given practical advice on how to prevent, detect, and get early treatment for the disease. This success story is an outstanding example of how science solved a problem and provided fast answers that saved lives.

If you were a member of a team assigned to solve a new disease, how would you proceed? What assumptions about nature and life would guide your work? What rules of logic would help you to draw conclusions? How would you decide which of the many possible explanations to believe and act on? In this chapter you will be introduced to the scientific approach to solving problems that has been so successful in revealing life's secrets. ■

1.1 Biology Touches Every Aspect of Your Life

Science is an intellectual pursuit that has a profound influence on your life. If you compare how you live with life in the 1800s, changes wrought by technology are obvious. Sanitation, convenience foods, cars, computers, cell phones, and e-mail are just a few examples of everyday applications of scientific discoveries that make your life so different from your great-great-grandparents'. Many other far-reaching applications of scientific discoveries that affect how you live come from **biology,** the study of life. Medical advances are an example. When you go to the doctor for antibiotics to treat an infection, you probably don't think about the biologists who made your treatment possible. From scientists of the 1700s who first peered into microscopes and saw tiny life-forms to present-day researchers who reveal how those life-forms can cause diseases, the treatment of infectious diseases is based on biological science.

Discoveries in biology have made headlines in recent years. These include new antibiotics; contraceptives; drugs for high cholesterol and high blood pressure; treatments for clogged arteries, cancer, diabetes, and AIDS; genetic tests for a variety of diseases; and techniques that can ensure that your child does not inherit a serious genetic disease. All of these recent medical advances are rooted in biology. **Table 1.1** shows other less well-publicized areas where biology affects modern life. For example, microbes routinely are used to help clean up toxic waste sites and oil spills. Insights from studies of how living things depend on their environments can help us to predict what Earth's future will be like if humans continue to overpopulate, deplete, and pollute environments. Other applications of biology allow scientists to construct machines that can sense the environment, make decisions, mimic movements of people and other animals, and even surpass human mental abilities. In these and many other ways, biological science has daily impacts on your life.

Biology is a sprawling science. It encompasses a wide range of topics from electrons to ecosystems. This range is reflected in the work of people who "do" biology. For instance, many biologists focus their studies on cells. Such investigations include the chemicals that make up cells, how cells use energy, build complex structures, or reproduce. Studies of **organisms,** defined as entire living things, reveal how cells work together in plants, fungi, or animals. Other biologists concentrate on the ways different organisms interact with one another and with their environments, such as the interactions within families of elephants, the effects of pollutants on various species of algae, how birds migrate to their winter homes, or how new species evolve. All of these studies and many more are part of the science of biology.

biology the study of life

organism the entire body of a living thing

Table 1.1   Biology in the News

Headline	Description of the story
How Green Is Our Energy? *New York Times,* Sept. 26, 2004	The price of using fossil fuels is increased air pollution and increased respiratory disease such as asthma.
Experimental Test Finds Early Stages of Colon Cancer *Washington Post,* Feb. 1, 2002	"A new screening test appears able to find many colon cancer cases in their early, most curable stage by detecting extremely small traces of cancer genes in patients' stool."
2 Genes Make Cancer Risk Soar *Washington Post,* Oct. 23, 2004	The genes a woman carries are more important than family history in determining the risk of breast cancer.
Firm's Anti-Nerve-Gas Device Approved *Washington Post,* Jan. 30, 2002	"Meridian Medical Technologies Inc. of Columbia and the U.S. Army announced yesterday that they won approval from the Food and Drug Administration to sell a medical device that can be used to protect military personnel against chemical attack."
Time in a Bottle: Anti-Aging Boosters Claim Their Products Can Turn Back the Clock. *Independent Scientists Aren't Buying It* *Washington Post,* Jan. 29, 2002	Scientists dispute claims that products reverse or prevent aging. Scientists explain that the basic process of aging is inevitable, and no treatment can prevent or reverse it. Scientists distinguish between the effects of "disuse" and true aging.
The History of Chromosomes May Shape the Future of Diseases *New York Times,* Aug. 30, 2005	Macaque monkeys and humans shared a common ancestor 25 million years ago, but genetic relationships may provide information about human diseases.
A Glimpse at the Future of DNA: M.D.'s Inside the Body *New York Times,* Apr. 29, 2004	The world's smallest medical kit has been developed: a computer, made of DNA, that can diagnose disease and automatically dispense medicine to treat it.
DNA Leads to Suspect in 1988 N. Va. Killings Evidence Is Linked to Killer on California's Death Row *Washington Post,* Sept. 28, 2005	DNA taken from a convicted killer matches samples from the two crime scenes.
Persistent Red Tide Takes Toll on Florida Sea Life and Tourism *New York Times,* Oct. 8, 2005	For 9 straight months the algal blooms along Florida's southwest coast have created a severe instance of "red tide." Nearly 2,000 square miles of coastal waterway is now a dead zone, with no life other than bacteria.
Building Better Bodies *New York Times,* Aug. 25, 2005	New findings on genes that enhance muscle strength in mice and cattle raises concerns that athletes and others may try to use genetic manipulations to improve their own performance.
Explaining Differences in Twins *New York Times,* Jul. 5, 2005	Identical twins possess exactly the same set of genes. Yet as they grow older, they may begin to display subtle differences. New research shows that these differences can be explained by the processes of epigenetics.

Four Major Themes Run Through Your Study of Biology

In the chapters ahead you will encounter four major themes that run through biology. One theme is that complex things happen when simpler things interact, an idea known as *emergent properties.* A second theme can be expressed as *unity and diversity.* This means that all living things share certain fundamental structures and functions, yet each individual organism is unique and slightly different from all others. A related biological theme is *evolution by means of natural selection,* a process that explains the unity and diversity of life. This principle describes how some traits remain the same, while others change over generations. It also explains how these changes over the course of Earth's history have led to the evolution of totally new kinds of organisms. As a final theme, wherever feasible this book will explain the *scientific processes* that generate biological knowledge. Watch for these four themes throughout the chapters ahead. They will help you to make better sense of the new information you will learn.

In this introductory chapter you will explore some of the most important findings and conclusions in biology and consider how scientists use scientific approaches to reach their conclusions. This chapter is not a summary of all 29 chapters of the text. Rather, it provides the framework for your study of biology and gives you the basic vocabulary and concepts needed before you begin your journey into the study of life.

QUICK CHECK

1. What are the four major biological themes that you will encounter in this book?

1.2 Life is Defined by a Set of Features That All Living Things Share

Biology is the study of life, but what exactly is life and how are living things different from nonliving things? For instance, what is there about you, the bacteria in yogurt, or a potted plant that is different from a watch, a yogurt container, or a flowerpot? Although living things are remarkably diverse, all have common features that collectively make each of them alive. Non-living things might have some of these features, but only a living thing will have all of them. Here are some of the most obvious ones.

- Life happens only within cells. All organisms are made up of one or more cells; each cell is limited by a physical boundary.

- Living things are highly complex and organized.

- Living things strictly control their internal environments and keep internal conditions within certain limits.

- Living things respond to and interact with the external environment.

- Living things use energy to power their internal processes, build internal structures, and grow.

- Living things carry instructions for conducting their internal processes and use these instructions to reproduce.

(a)

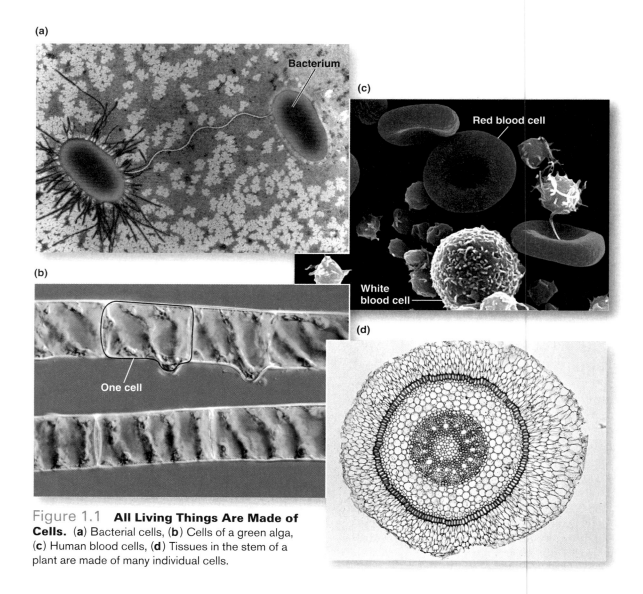

(c)

Red blood cell

White blood cell

(b)

One cell

(d)

Figure 1.1 **All Living Things Are Made of Cells.** (**a**) Bacterial cells, (**b**) Cells of a green alga, (**c**) Human blood cells, (**d**) Tissues in the stem of a plant are made of many individual cells.

Let's briefly consider each of these characteristics that collectively differentiate something that is alive from something that is not alive.

Life Happens Inside Cells That Are Complex, Highly Organized, and Homeostatic

One universal aspect of life on Earth is that it happens inside cells. A **cell** is the basic structural unit of life. Cells come in many sizes and shapes, and organisms consist of one or more cells. **Figure 1.1** shows how varied cells can be; their variety reflects the amazing diversity of life.

How do you know if something is a cell? One characteristic feature of a cell is the boundary that separates the life within the cell from the nonlife outside of the cell. You are familiar with the boundaries of living things such as your own skin, the bark of a tree, or the shell of a crab, but you may be less familiar with the boundary around each cell, called the **cell membrane** (**Figure 1.2**). It protects the delicate processes of life from the environment surrounding the cell and allows a cell to communicate with the environment.

It is the processes that go on inside of a cell that make it alive. These processes are complex and highly organized. What does this mean? *Complex* means that cells have many different structures and a multitude of chemical processes; *organized* means that these structures and processes are not in disarray. It might help to think of a room before and after you straighten it. Of course your room is complex. Not only does it contain your "stuff"—clothes, books, and furniture—but also it houses the complex activities that happen in your room, what biologists would call your "processes." For instance, you might study, listen to music, watch TV, and pet your cat—all in this same room. Before you tidy up, things are scattered everywhere. Once you have straightened the room, the things in it still are complex, but they are more organized because you have shelved the books, put away your clothes, and gathered up your dirty laundry. This may even inspire you to focus time just on your cat, then read awhile, and then listen to music. You can see that both situations—before and after straightening—are complex, but the room is organized only after you have cleaned it up.

In a similar way, life is both complex and organized. Every living cell has an ordered, complex, internal structure. Of course, you cannot see subcellular structures with unaided vision, but you can observe the complexity and organization characteristic of life at other levels. An oak tree provides a good example. Its roots, trunk, branches, twigs, buds, leaves, flowers, and acorns give the tree a complex structure (**Figure 1.3**), but there is order in this complexity. Roots are underground, the trunk supports the branches, and the branches repeatedly fork and diverge to form the canopy that reaches upward. Leaves have characteristic structures and are arranged in a pattern on the tree. Flowers are reproductive structures that appear only at a specific season of the year. Pollinated flowers may mature into seed-bearing acorns, and in organized ways seeds within acorns

cell the smallest structural unit of life

cell membrane the boundary that separates a cell from other cells and from the environment while it allows a cell to communicate with the environment

Red blood cells

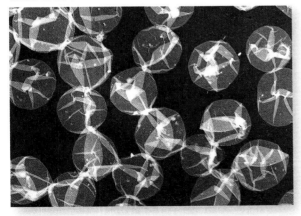

Membranes of red blood cells

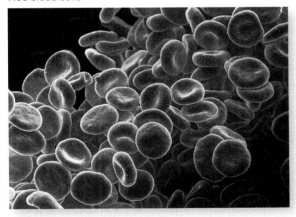

Figure 1.2   **A Membrane Surrounds a Cell.** When the contents of red blood cells are removed, their cell membranes become visible.

Figure 1.3   **An Oak Tree Has a Complex and Organized Structure.**

grow into new mature oak trees. If you take a few moments to look at the natural world, you will find other examples of the complexity and organization of life.

Life accomplishes this order and organization because the activities that go on inside an organism are strictly controlled. Again, think of your room. If you want to *keep* your room and your activities organized, you must control where you put each item and what you do at each moment. For instance, a book could go either on the desk or on the shelf and still be organized, but it can't go into the pile of laundry. Cells and organisms also have ways to control structures and activities to ensure that the ordered life of a cell does not fall into chaos.

To manage life's chemical processes, living things also must regulate the quantities and characteristics of the chemicals they contain. The inside of a cell is mostly water, but many chemicals are dissolved in that water—and these are critically involved in the structures, processes, and activities that collectively are called *life.* While external environmental conditions may fluctuate wildly, the internal environment of a cell cannot. Factors such as the amount of water, amount of salt and other chemicals, and acidity are carefully regulated. Some organisms also regulate more complex traits such as body temperature or food intake.

The delicate balance of internal aspects that all living organisms maintain is called **homeostasis.** If this internal chemical balance is upset—that is, if homeostasis is disrupted, an organism easily can die. Homeostasis is reflected in the way many systems of the human body work. Control of body temperature is a good example (**Figure 1.4**). If a mammal, such as a human, is suddenly put in a very cold environment, body temperature will not fall dramatically. Body systems respond to the cold and maintain body temperature close to normal. Only after a long time in the cold will body temperature begin to fall.

Living Things Respond to and Interact with Their Environments, Use Energy, and Reproduce

To survive, every organism must interact with its environment. Plants must locate and grow toward a source of light or water. Other organisms, must locate and take in food (**Figure 1.5**). An individual might move toward a potential mate, or its offspring, or need to know how wet, or hot, or salty the environment is. You might need to know if the person walking behind you is an innocent stranger,

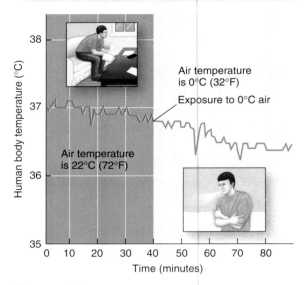

Even 40 minutes after exposure to 0°C air, human body temperature remains relatively constant and has fallen only about 0.5°C. Internal body mechanisms, including shivering, help maintain a steady body temperature.

Air temperature is 0°C (32°F)

Exposure to 0°C air

Air temperature is 22°C (72°F)

Human body temperature (°C)

Time (minutes)

Figure 1.4   **Maintenance of a Steady Body Temperature Is a Homeostatic Mechanism.**

■ *Why does body temperature remain close to normal when exposed to hot air or freezing temperatures?*

an old classmate, or someone to run from. All cells must sense and respond to their environments.

Another important point about living organisms is that they are active. Structures inside of cells move, cells move, and organisms that are made of many cells move. To carry out all of these activities living organisms must have a source of **energy.** For now energy will be defined as something objects have that gives them the ability to do work. In its need for a source of power, life is no different from a lawnmower, an electric toothbrush, or a cell phone. The Sun is the ultimate source of energy for nearly all organisms. While plants, some bacteria, and algae can harvest the Sun's energy directly, the rest of us obtain energy by eating plants or other living things (**Figure 1.6**). Living organisms spend much time getting energy and converting it to a form their cells readily can use.

When people work together to complete a complex set of tasks, it helps if they have instructions. In a similar way cells need instructions to carry out the processes of life. The instructions for the activities that go on in a cell are contained in large molecules called **DNA,** an acronym for deoxyribonucleic

homeostasis the stable internal conditions of an organism that support life's cellular and chemical processes

energy the ability to do work

DNA (deoxyribonucleic acid) the molecule that carries the instructions for life

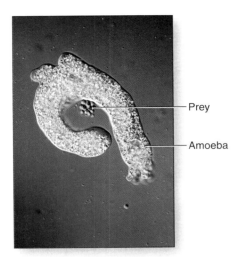

Figure 1.5 **Amoeba Surrounds and Engulfs Prey and Then Begins to Digest It.**

Figure 1.6 **Some Organisms Such as Birds Obtain Energy by Eating Other Living Things.**

Figure 1.7 **DNA Molecules Carry the Instructions for the Processes of Life for All Organisms.**

acid (**Figure 1.7**). Although the details of the DNA molecule differ, the instructions it carries are made of the same chemicals and work in the same way in all organisms. The structure of DNA and the consistent way it works to direct living processes are the basis for the unity of life. As you will read in later chapters, though, DNA molecules differ slightly between individuals, and these subtle variations are responsible for the incredible diversity of life.

In addition to providing instructions for a cell's activities, DNA gives organisms the special ability to reproduce. The DNA molecule can be copied exactly, and the copy can be given to new cells, which can outlive the parent cell. Without this special ability, life would not continue over time. When DNA is successfully copied and a new cell is made, the process ensures that the new cell has the same instructions and operates by the same rules as did its parent cell.

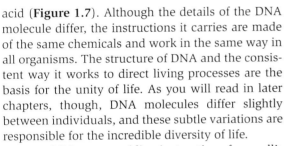

QUICK CHECK

1. What is the function of the cell membrane?
2. Describe homeostasis, using adjustments in body temperature as an example.

1.3 Levels of Organization Are Characteristic of Life

The complex interactions within and between living organisms can seem baffling, but you can begin to comprehend them by considering something that biologists call *levels of organization*. This idea is that living things can be grouped into increasingly complex units. An analogy might be helpful here. Imagine that you work for a large drug company trying to develop a drug to treat heart disease. Your job is to run chemical analyses on compounds that field agents find in the forests and jungles of the world. But of course you do not work alone. You are part of a research unit where every person's job is related to heart disease; each week all the people in your unit meet to trade ideas and discuss their work. The heart disease research unit is part of a larger group that includes biologists and medical doctors who will test the potential treatments in animals and in people. The entire heart disease workforce also would include people who might market and sell any treatments developed. And this large heart disease workforce is just one of many groups in the large drug company. Each level has its own complexity, which builds as you go from one level of organization to the next.

Life is organized in a similar way. The simplest level of organization involves the interactions of the chemicals that make up life. The groupings continue through different types of cells, to the cells that interact within larger organisms, to organisms that interact with one another. The most complex level of organization includes every organism and the nonliving environmental components that support them. There are many levels of organization in between these endpoints.

Individual Organisms Are Made of Chemicals, Cells, Tissues, and Organs

The first level of life is not actually alive but is the foundation on which life is built. Life is made of chemical building blocks called **chemical elements** (**Figure 1.8**). Each element is a basic kind

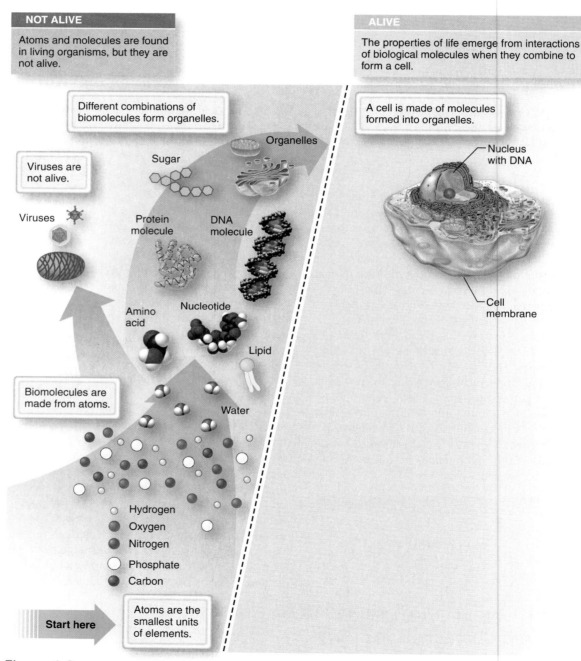

NOT ALIVE

Atoms and molecules are found in living organisms, but they are not alive.

ALIVE

The properties of life emerge from interactions of biological molecules when they combine to form a cell.

Different combinations of biomolecules form organelles.

A cell is made of molecules formed into organelles.

Viruses are not alive.

Organelles

Sugar

Viruses

Protein molecule

DNA molecule

Nucleus with DNA

Amino acid

Nucleotide

Lipid

Cell membrane

Biomolecules are made from atoms.

Water

○ Hydrogen
● Oxygen
● Nitrogen
○ Phosphate
● Carbon

Start here Atoms are the smallest units of elements.

chemical element a basic kind of matter

Figure 1.8   **From Atoms to a Cell, Life Has Levels of Organization.**

■ *Why isn't a DNA molecule alive?*

of matter. The smallest unit of an element that has all the properties of that element is an **atom,** and life is composed of just a handful of elements. Hydrogen (H), carbon (C), oxygen (O), and nitrogen (N) are a few examples of elements that are important chemical building blocks of life. Of course, life is more intricate than the atoms of hydrogen or oxygen. Atoms combine to form more complex structures such as *molecules.*

The next level of organization brings us closer to the realm of life (Figure 1.8). Molecules and atoms are not by themselves alive, but in the right combinations and under the right chemical and physical conditions they combine to form **organelles,** subcellular groups of molecules that have specific functions within cells. Cells are the basic units of life. Anything less than a cell is not alive. A **virus** is a small structure made of biological molecules, but it is not a cell. A virus cannot independently use chemical energy to build its own structures or carry out life's processes. So a virus is not alive. You will learn more about viruses in upcoming chapters. The point here is that as far as we know, only cells are alive.

Organisms that have just one cell are **unicellular organisms.** Despite their small size, unicellular organisms can have incredible complexity (**Figure 1.9a**). Organisms made of two or more cells are called **multicellular organisms.** Just as people in an organization specialize in certain jobs, different cells of multicellular organisms can specialize in certain jobs. Because of this, multicellular organisms can do interesting and complex things that unicellular organisms cannot do. Figure 1.9b shows *Trichoplax,* one of the simplest known multicellular organisms. *Trichoplax* has just five types of cells, but they aren't organized into tissues. A **tissue** is a group of similar cells gathered together into a unit that performs a specific function. For instance, blood is a tissue that transports nutrients and water. Muscle is a tissue that contracts and moves body parts.

In more complex multicellular organisms different kinds of tissues are grouped into **organs** to perform even more complex tasks (**Figure 1.10**). For example, individual bones are organs that are part of the skeletal system. Bones are composite structures that, like all organs, are formed of several kinds of tissues such as connective tissues, blood, blood vessels, and nerves. Connective tissue binds the bone to the rest of the body, while blood and blood vessels supply nutrients to bone cells and remove wastes. Bone cells secrete deposits of calcium around themselves, forming patterns of concentric rings. Just as atoms are not alive by

(a) A unicellular organism like *Euglena* has just a single cell with many complex organelles.

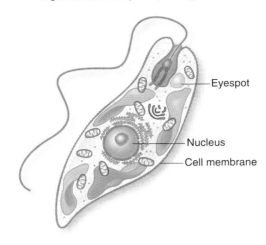

Eyespot

Nucleus

Cell membrane

(b) *Trichoplax* is a simple multicellular organism that has many cells arranged around a fluid-filled cavity. Each cell has complex organelles.

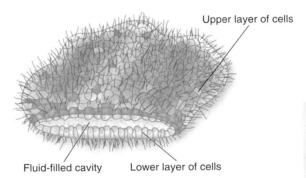

Upper layer of cells

Fluid-filled cavity   Lower layer of cells

(c) At its largest, *Trichoplax* is about ten times the size of *Euglena.*

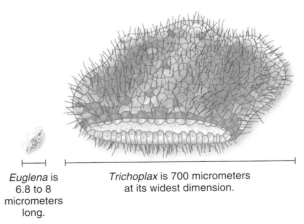

Euglena is 6.8 to 8 micrometers long.

Trichoplax is 700 micrometers at its widest dimension.

Figure 1.9   **A Complex Unicellular Organism and a Simple Multicellular Organism.** *Euglena spirogyra* has just one cell with complex organelles. *Trichoplax adhaerens* is one of the simplest multicellular organisms.

atom the smallest unit of an element that has all of the properties of that element

organelle subcellular groups of molecules that have specific functions within cells

virus a small, nonliving structure that incorporates biological molecules

unicellular organism an organism made of just one cell

multicellular organism an organism made of two or more cells

tissue a group of similar cells gathered together into a unit that performs a specific function

organ a grouping of different types of tissues that work together to perform a specific task

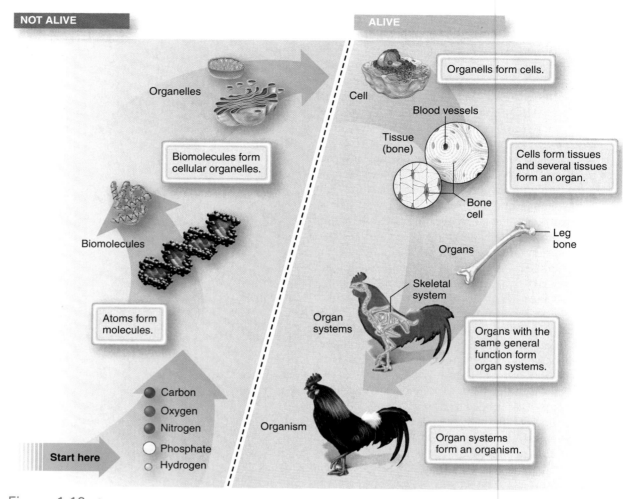

NOT ALIVE

ALIVE

Organelles

Organells form cells.

Cell

Blood vessels

Biomolecules form
cellular organelles.

Tissue
(bone)

Cells form tissues
and several tissues
form an organ.

Bone
cell

Biomolecules

Leg
bone

Organs

Skeletal
system

Organ
systems

Atoms form
molecules.

Organs with the
same general
function form
organ systems.

Carbon

Oxygen

Nitrogen

Organism

Start here

Phosphate

Hydrogen

Organ systems
form an organism.

Figure 1.10   **Levels of Organization: From Atoms to an Organism.**

organ system a group of organs that work together to perform a specific body function

population a group of individuals of one kind of organism living in one geographic locality

community interacting populations of different species that live in a local area

ecosystem communities and the nonliving aspects of their environments

biome a geographic region dominated by distinctive climate patterns and distinctive vegetation

biosphere all of Earth's ecosystems and biomes

themselves, tissues and organs are not by themselves whole organisms. Organs that perform related functions are grouped into an **organ system** that accomplishes a major body function. For instance, all of the bones in the body are all part of the skeletal system that has the job of supporting and protecting the body's soft tissues. **Table 1.2** lists the major organ systems of the human body. Finally, an organism consists of the whole body of a living thing, so you are an organism and a rooster, a *Euglena*, a *Trichoplax*, an oak tree, a bacterium, and a mosquito are organisms too.

Individual Organisms Interact as Part of One or Several Groups

If you took a careful census of the organisms that live in your backyard, you would probably discover thousands of different kinds, most of them small, and many of them insects. Not only are there lots of different organisms, but they can be viewed as components of larger interacting biological systems. Let's consider some of the interactions in the lives of one such insect, a leaf cutter ant that lives in rain forests in Central and South America (**Figure 1.11**). Ants interact with each other, with other kinds of organisms like the plants whose leaves they gather, and with the nonliving parts of their environment like soil, water, and air. A group of individuals of one kind of organism living in one geographic locality is called a **population.** All of the populations of a single kind of organism form a *species*. All individuals of a species have similar characteristics that help define them as one species.

One of the important lessons from modern biology is that individual organisms do not exist in isolation. Instead, every individual depends on others for its survival. Leaf cutter ants are part of a colony whose members accomplish different tasks and support each other. Worker ants use their sharp-edged jaws to snip off bits of plants and carry these back to the colony. There the raw plant matter is chewed and the mulch is added to the colony's underground garden of fungus that provides food for the colony. The tiniest workers tend the fungus garden. As a whole the colony enriches the soil it lives in, and so it has an impact on other organisms that live there. As the ants dig their colony into the soil, they bring up and distribute soil that has a higher mineral content than surface soil. The ants, in turn, are food for other animals like anteaters, armadillos, and some birds.

Interactions between individuals and their environment produce more complex organizations. Because of limits set by geography, an individual usually interacts only with a small subset of its own species. In many cases a population is the most important group when considering the interactions between individuals. For instance, while all the populations of *Atta cephalotes* leaf cutter ants form one species, populations of leaf cutters in widely separated geographic regions are not likely to interact. In contrast, populations of *different* species can interact to form a larger unit called a **community.** In the case of the ants, the community includes the plants they snip to grow their fungus gardens, the fungi they eat, and animals that prey on and parasitize them. Communities interact with other nearby communities, forming larger units that impact one another around the globe. An **ecosystem** is a larger unit that consists of communities plus their nonliving environments. In turn, ecosystems are grouped into larger regional ecosystems called **biomes** that are controlled by broad climatic patterns and characterized by typical forms of vegetation and by animals with distinctive adaptations to these environments. Leaf cutter ants are part of an ecosystem, the tropical rain forest biome, and the biosphere. Finally, all biomes are included in the **biosphere.**

Interactions at Each Level of Organization Produce the Complexities of the Next Higher Level

Living things do so many unique and amazing things that it may be hard to grasp the idea that life is controlled by the same rules as the rest of the universe. Nevertheless, the same universal laws that govern the interactions of chemicals in stars, power plants, and test tubes also govern chemical interactions within living organisms. How is it that life appears to be so different from chemical compounds found in jars and bottles on a chemist's

Table **1.2** Major Human Organ Systems and Their Functions	
Organ system and major components	**Function**
Integumentary Skin, hair, nails	Protection from water loss, mechanical injury, and infection
Muscular Skeletal muscles	Movements of body parts and of whole body
Skeletal Bones, ligaments, tendons	Support of body and protection of internal organs
Digestive Mouth, pharynx, esophagus, stomach, intestines, liver, pancreas, gallbladder, anus	Digestion and absorption of food, and elimination of food wastes
Circulatory Heart, blood vessels, blood	Movement of substances to and from individual cells
Respiratory Trachea, breathing tubes, lungs	Gas exchange: obtaining oxygen and releasing carbon dioxide
Lymphatic and immune Lymphatic vessels, lymph nodes, lymph, immune cells, bone marrow, thymus, spleen, white blood cells	Defense against infection and cleanup of dead cells
Excretory Kidneys, ureters, bladder, urethra	Elimination of metabolic wastes and regulation of chemistry of blood
Endocrine Pituitary, thyroid, pancreas, ovaries, testes, and other glands that secrete hormones	Coordination of body functions by hormones
Nervous Brain, spinal cord, nerves, sense organs	Sensing internal and external stimuli and coordination of entire body's functions
Reproductive Ovaries, testes, and related organs	Reproduction

Figure 1.11 **Leaf Cutter Ants.**

shelf? The answer lies in the principle of **emergent properties**: the concept that simple components can interact to produce something that is much more complex.

You experience emergent properties in your daily life. Think of the many parts of a car, as shown in **Figure 1.12**: engine, brakes, steering wheel, and so on. At a basic level a car is made of parts manufactured from steel, plastic, aluminum, rubber, and glass. By themselves these parts don't do much, but they are shaped and assembled into the car's component systems. For instance, your car has an engine, an electrical system, and a set of wheels associated with other parts that form a steering system. Each of the car's systems has interesting properties, but again, they are not too useful until they are combined and are working together. By itself, none of the parts could produce a vehicle that moves at high speeds, carries passengers, and can stop on demand. But together these individual component systems interact to make a car that does all of these things. The properties of a car emerge when all of its components interact. With this analogy in mind, let's return to the various levels of life's organization.

Just as unassembled auto parts do not have the properties of an intact car, by themselves the chemicals and molecules inside a cell do not have the properties of life. As you saw in Figure 1.10, life emerges from the interactions of atoms and molecules within a cell. Similarly, individual cells of a multicellular organism cannot do all of the things the whole organism can do. For instance, a single cell of your body cannot grasp a tool, watch an ant, or write a song. It takes the entire body to produce these complex behaviors. Repeating the same theme, new properties emerge when individuals interact within populations and when different populations interact within communities. While a single tree can produce shade, a grove of trees has a broad, leafy canopy that provides shelter and protection for the community that lives on, within, and beneath the canopy. Similarly, a single chimpanzee can defend itself from a predator, but a troop of chimpanzees can cooperatively defend an expansive territory. Individuals within the troop can specialize in more complex tasks such as mating or raising young. The ecosystem that the community of the leaf cutter ant colony belongs to has its own properties. The ecosystem influences the characteristics of the soil. It has characteristic patterns of sunlight, temperature, moisture, and rainfall, and provides nutrients, shelter, and other complex services for the organisms that live within the ecosystem.

emergent properties
the principle that simple components interact to produce more complex systems with more complex traits and behaviors

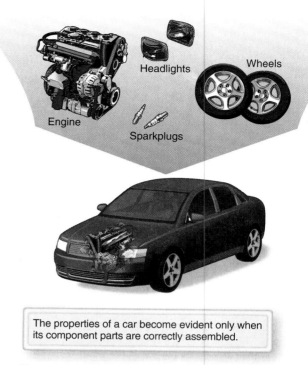

Engine · Headlights · Wheels · Sparkplugs

The properties of a car become evident only when its component parts are correctly assembled.

Figure 1.12 **The Concept of Emergent Properties.**

QUICK CHECK

1. List the levels of organization of life, starting with atoms and ending with the biosphere.
2. In what ways does a car show emergent properties?

1.4 Evolution and Natural Selection Have Produced Life's Diverse Forms

The study of modern biology has revealed many astounding things about the living world and one of them is the incredible diversity of life. Each day scientists discover new kinds of life that thrive in the most amazing places. Living things are small, large, soft, hard, colorful, drab, and have diverse adaptations that help them to survive. Even at the level of individuals there is diversity. Every human is unique, and this also is true of every other living thing. Diversity is fundamental to the nature of life. Yet, diverse organisms solve the problems of staying alive in a limited number of ways. This is a critical insight. Life is diverse—and at the same time life is unified by shared chemicals, processes, and structures.

These two themes—unity and diversity—characterize living organisms, but how can living things be so different and yet so similar? The idea of evolution provides the answer. The name most associated with evolution is Charles Darwin. He understood that **evolution** is a change in a species over generations. Because evolutionary changes lead to new species, evolution produces biological diversity.

One of the most convincing ways to see evolution is to trace the history of life on Earth (**Figure 1.13**).

Notice that different kinds of organisms emerged gradually over time, and that the transitions from one general group of organisms to another make sense. The earliest living things—probably formed about 3.8 billion years ago—were unicellular organisms with lots of biochemical diversity but not very much structural diversity. Later, about 3 billion years ago, unicellular organisms became more complex with intricate internal and external structures. The first multicellular organisms were not zebras or pine trees. Instead they were fairly

evolution a change in a species over generations

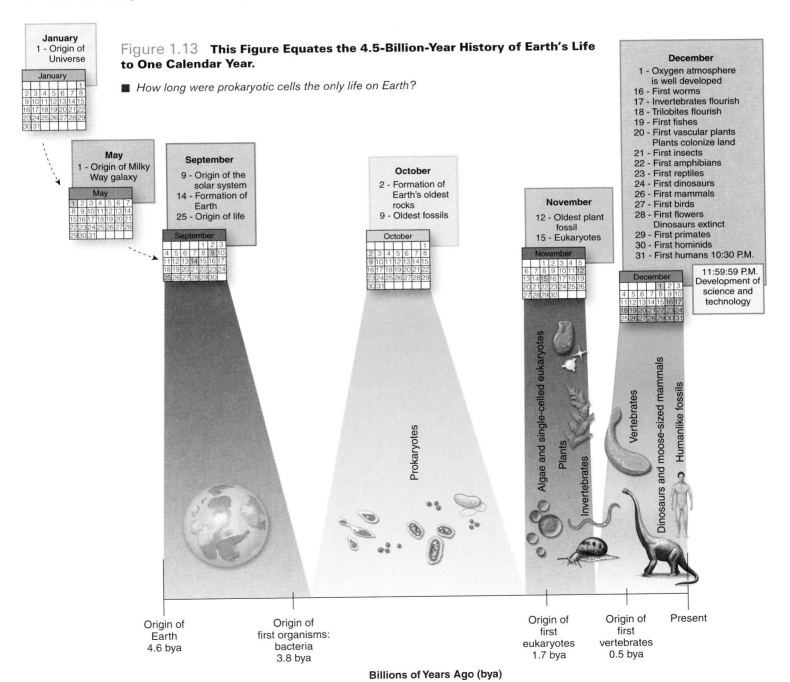

Figure 1.13 **This Figure Equates the 4.5-Billion-Year History of Earth's Life to One Calendar Year.**

■ *How long were prokaryotic cells the only life on Earth?*

January
1 - Origin of Universe

May
1 - Origin of Milky Way galaxy

September
9 - Origin of the solar system
14 - Formation of Earth
25 - Origin of life

October
2 - Formation of Earth's oldest rocks
9 - Oldest fossils

November
12 - Oldest plant fossil
15 - Eukaryotes

December
1 - Oxygen atmosphere is well developed
16 - First worms
17 - Invertebrates flourish
18 - Trilobites flourish
19 - First fishes
20 - First vascular plants Plants colonize land
21 - First insects
22 - First amphibians
23 - First reptiles
24 - First dinosaurs
26 - First mammals
27 - First birds
28 - First flowers Dinosaurs extinct
29 - First primates
30 - First hominids
31 - First humans 10:30 P.M.

11:59:59 P.M. Development of science and technology

Prokaryotes

Algae and single-celled eukaryotes

Plants

Invertebrates

Vertebrates

Dinosaurs and moose-sized mammals

Humanlike fossils

Origin of Earth
4.6 bya

Origin of first organisms: bacteria
3.8 bya

Origin of first eukaryotes
1.7 bya

Origin of first vertebrates
0.5 bya

Present

Billions of Years Ago (bya)

simple organisms not too different from the green algae that grow on the sides of your fish tank. The earliest animals had soft bodies, were small, and did not have limbs. Later more complex animals and plants evolved, and by half a billion years ago there were lots of complex life-forms. To this point all life was still restricted to the oceans, rivers, and lakes of the world, but there were invertebrate animals, vertebrate animals, and highly structured aquatic plants. Around this time organisms moved out of the water and colonized land, where new species evolved. Vertebrates with limbs, invertebrates with complex social organizations, plants with intricate life cycles, and many other diverse species evolved even more biological diversity.

How did this tremendous diversity of species come about? How does evolution happen? One answer is found in the concept of evolution by means of **natural selection.** The idea is simple, yet enormously powerful. In the process of evolution by means of natural selection, the traits of species change over time (or evolve) because in each generation only some individuals survive to reproduce the next generation. Therefore, those individuals that survive to produce offspring determine the traits of the next generation. In natural selection the qualities of nature—meaning various aspects of an individual's environment—are the factors that govern who survives to reproduce and pass on their DNA to the next generation (**Figure 1.14**). You will learn a great deal more about evolution by the mechanism of natural

natural selection the effects of environmental factors that allow some individuals in a population to survive and produce offspring, while others do not

selection and Charles Darwin's contributions to biology in Chapter 13.

One thing to note about this history of life is that life does not get *better* over time, but it does get more complex and more diverse. It would be a stretch to say that humans are in any way better organisms than bacteria. After all, bacteria can survive in harsh environments that quickly would kill a person. In addition, bacteria and other small, simple organisms will probably be around long after humans are extinct. But it is fair to say that humans are more complex than bacteria. Evolution has produced more complexity over time.

The modern understanding that evolution has produced Earth's incredible biological diversity is the cornerstone of all modern biology. Why? Because of this important insight: everything we know about biology says that *all organisms are genetically related.* Think about what that statement means. You know what it means genealogically to find a long-lost cousin: you are related to that cousin because you share a common ancestor. The same is true for all living things, alive or extinct. This conclusion that all living things are genetically related is supported by biological knowledge that has been accumulated over the past 200 years.

QUICK CHECK

1. What are the two biological themes that characterize living organisms?
2. How does natural selection produce evolution?

Figure 1.14 **Penguin Chicks in an Antarctic Storm.** Severe weather is one environmental factor that selects which individuals survive.

1.5 Biology Applies the Methods of Science to the Study of Life

Life can be understood by the same means used to reveal how the rest of the universe works—the methods of science. Simply put, science is a way to answer questions about the world around us. People have always asked questions, tried to understand their surroundings, and attempted to predict what will happen. Modern science is a more formal expression of these shared approaches to knowledge.

Art, music, philosophy, literature, ethics, and religion are also ways in which people ask questions and look for answers about life, but they are different from science. The central feature of a scientific statement is that it must be tied to an *observation of the real world.* Because the aim of science is to understand the world around us, science requires that ideas be tested against the world to find out if they are correct. More than just making observations, though, science reflects the human tendency to explain observations. In science explanations are given in terms of natural causes—other things that can be observed and described. These seemingly obvious comments about science have important implications, so let's consider them in greater detail.

Much scientific research is focused on getting a thorough description of the things and events in the world. To do this you must accept that there is a real world that can be studied. More importantly, you must understand that the nature of the world is independent of a person's beliefs. For example, no matter how firmly you believe that the Earth is flat or round, your belief would have no impact on whether the Earth is flat or round. Science is much more than just a description of the world, though. The real motivation behind doing science is to use knowledge to predict what will happen tomorrow. For instance, farmers want to predict the cycles of floods and droughts; governments want to predict storms and other possibly harmful natural events; parents want to predict the health of their children. In short, humans want some advantage in the face of an uncertain future—and the predictive power of science is the best way to get that advantage. A full description of events sometimes can be used to predict what will happen next, but the real predictive power of science comes from understanding the causes of events—in other words to understand how and why things happen. Where do scientists look for causes? An important assumption in science is that all observable events can be explained by other observable events. If an explanation for an event relies on something supernatural, or something that can never be observed, or can be observed only by people with certain beliefs, then it is not a scientific explanation.

Scientific Statements Reflect the Variation in the Universe

You may have experienced frustration when a newspaper article declares that science has discovered some fact, only to find three years later that this "fact" is not so reliable. What is going on here? Isn't a "scientific fact" something that is always true? The problem is that while there are reliable principles, causes, and events in the world, there also is variation. Because almost everything that is interesting to study has many dimensions and multiple causes, discovering the consistencies can be difficult. Consider, for example, the simple finding that smoking cigarettes causes lung cancer, one of the most reliable observations in modern medicine, yet this cause-and-effect relationship is not 100% certain. Not everyone who smokes gets lung can-

cer, and not everyone who does not smoke remains free of lung cancer. This is because lung cancer is a complex event affected by many different factors, one of which is cigarette smoke.

Probability is another way to describe the chances that something will happen. Probability is a statement of the likelihood of an event and ranges from zero to 1.0. The probability game in **Table 1.3** will help you to understand how science helps to predict the likelihood of something happening. An event with a 0.0 probability will never happen, while an event with 1.0 probability will definitely happen and has a 100% certainty. Some statements in science such as the speed of light or the structure of a hydrogen atom have close to a 1.0 probability. Other statements such as "the Earth is the center of the solar system" are known to be false and so have a probability of just about zero. Many of the statements in science, especially in biology, have a probability somewhere in

probability a number that represents the chances that something will happen; expressed as a decimal between 0.0 and 1.0 or as a fraction or a percentage

Table **1.3**   The Probability Game

This jar contains 15 tickets. If you agree to play the game, you will have a chance to draw a ticket. You must abide by the directive on the ticket. Will you play the game?

No? Well, it may help if you know what is on the tickets. Here are their messages:

- You win $1,000!
- A month's worth of food is donated to the family of your choice.
- You must pay $1.
- For one week you must do the laundry for everyone in your house.
- For one week you must stop talking to your best friend.

Now, will you play the game?

Is it likely that you will get a good message on one draw? What are the chances? Would it help if you know how the messages are distributed on the tickets?

- One ticket reads: You win $1,000!
- Two tickets read: A month's worth of food is donated to the family of your choice.
- Two tickets read: You must pay $1.
- Five tickets read: For one week you must do the laundry for everyone in your house.
- Five tickets read: For one week you must stop talking to your best friend.

Now will you play the game? Is there any more information to give you? No.

No one can predict the future, but you can calculate your chances of success or failure. As you see from this game, when you know more about what is in the jar, you know more about the odds of success. This is the kind of information that science provides. It tells you more about what is in the jar—or in the world around you.

between 0.0 and 1.0. This does *not* mean that science is a flawed way of looking at the world. Rather, it means that the world is a complex place, and nearly every event can have multiple causes. Repeated and verified scientific studies on a subject allow you to base your choices in life on the chances that certain events will follow those choices. The choice to smoke or not to smoke cigarettes is just one example.

Events or groups of events that can be expected to happen with near 1.0 probability and with mathematical precision can be expressed as **scientific laws.** A law can be formulated when the events in question can be given precise numbers, and these numbers can be used instead of words to describe what is happening. In some cases laws apply universally—as far as we know—such as the speed of light and its relationship to mass in Albert Einstein's famous equation $E = mc^2$. Sometimes laws apply only in some situations, but still are reliable. For instance, while Newton's laws of gravity do not apply precisely in all situations, they do describe the movement of objects with a great deal of accuracy.

Science Involves a Disciplined Way of Studying the Universe

Now it's time to consider how scientists ensure that scientific conclusions accurately reflect the world. There are two important aspects to consider: the processes of science and the culture of science. First, the processes of science usually involve thoughtful, careful observations or experiments. Second, the culture of science includes a set of expectations for behavior that increases the probability that scientific statements reflect the truth about the universe. Let's look more deeply at each of these aspects of science, starting with the sequence of steps involved in a scientific study. **Figure 1.15** uses the example of studying the relationship between tobacco smoking and cancer to show how the scientific method works.

- A *question* or an *observation* often sparks a scientific study.

- Additional *observations* lead to a preliminary answer to the question.

- The preliminary answer is formalized as an **hypothesis,** a statement that the researcher thinks explains the phenomenon or answers the question.

- The investigator uses the hypothesis to make a *prediction* that then will be tested by further observations and experiments.

- The investigator will test the prediction in a formal study or experiment, trying hard to disprove or *falsify* the hypothesis.

- The results of the test will be used to reject, modify, or support the hypothesis. This usually leads to a restatement of the hypothesis or the statement of a new hypothesis.

- The modified or new hypothesis is tested by a new round of observations.

Science often starts with observations and questions that lead to a hypothesis. You might develop your own questions after observing living things, or by thinking while you sit in your chair, or from a discussion with a friend. Your questions become science when you actually try to resolve them by testing them. But before you can even try to find an answer, you must clearly and precisely state your question and your plan for finding an answer, so that there is no confusion about what you mean. Initially your questions may be vague and uncertain, such as "Why did my aunt die of lung cancer?" As your thoughts get more organized, you can state your hypothesis more clearly and start to think of some answers. The question about your aunt might then be turned into this statement: "My aunt died from lung cancer because she smoked a pack of cigarettes a day for 30 years." When you state your proposed answer clearly in a way that can be tested, you have developed a hypothesis.

Once your question and hypothesis have been clearly stated, you must think of a way to test them. Sometimes you must devise specific situations and use special instruments to determine if your hypothesis is likely to be correct. Even more important, to be sure that you are not just fooling yourself, you must try hard to show that your hypothesis is *wrong*. This probably sounds strange. But if you try hard to show that your hypothesis is wrong—and it is wrong—you actually have learned something valuable. Alternatively, if you diligently try to show that your hypothesis is wrong and it seems to be right, you can have more confidence in this conclusion. In the case of your aunt, you would search for other possible reasons why she might have died from lung cancer. For instance, was she exposed to toxic chemicals or radiation? Did she have a history of associated lung diseases? These other reasons would be stated as hypotheses and tested along with the hypothesis that exposure to cigarette smoke did cause her cancer.

The scientific studies of a question rarely stop with the initial answers. Scientific conclusions go through stages. As more studies are done that support a hypothesis—and do not cause it to be

scientific law a reliable and precise mathematical description of an event or set of events

hypothesis a possible explanation of an event or an answer to a scientific question that can be tested by formal studies or experiments

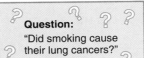

Observation:
"My cousin died of lung cancer and he smoked a pack of cigarettes every day for 30 years."

Another observation:
"I know of two other people who smoked and died of lung cancer."

Question:
"Did smoking cause their lung cancers?"

Figure 1.15 **The Sequence of Steps in a Scientific Study.**

■ *What makes a good hypothesis?*

Hypothesis:
"Smoking always causes lung cancer."

Prediction based on the hypothesis:
If I keep data on people who smoke until they die, every one of them will develop lung cancer.

Testing to falsify the hypothesis:
Observe and keep data on 7,500 adult smokers until they die.

If all smokers die of lung cancer, then the hypothesis is supported.

If all smokers do not die of lung cancer, then the hypothesis is not supported.

If the hypothesis is supported:
The investigator may ask a related question and continue the process of scientific inquiry.

Further questions could include:

"What is the risk of developing cancer from inhaling second-hand smoke?"

"Will a person develop lung cancer if they smoke for 10 years and then quit?"

If the hypothesis is not supported:
The investigator may develop other hypotheses.

Further hypotheses could include:

"Smoking increases the probability of developing lung cancer."

"Lung cancer is inherited."

rejected—the conclusions become more reliable and are more likely to be correct. As experiments continue, the hypothesis may become broader, and apply to more situations. Once a set of related hypotheses are strongly supported and apply broadly, eventually the explanation is called a **theory.** You probably do not think of the word *theory* in this way, but to science a theory is a well-tested explanation. The theory of continental drift in geology is an example, as is the theory of evolution by the process of natural selection in biology. So when you hear someone say, "That is just a theory," if they are talking about a scientific theory, then the idea is something you can count on as being supported by years of scientific research and the best explanation that is available.

Most science is done at the level of formulating and testing a hypothesis, so before leaving the topic of how science is done, let's use an example to make the development of a hypothesis less abstract. Let's assume that the trees in the northwest corner of your local park are covered with withered leaves in the middle of the summer (**Figure 1.16**). You are worried that these trees might be dying, so you team up with your local park service scientists to answer the question: "What is happening to the trees?"

Your investigation begins with careful observations. You note what kinds of trees are involved. You examine the area in their immediate vicinity and compare it with the rest of the park. You note how much sunlight the withered trees get and examine the soil beneath them. You try to notice any animals that are around them. To this point

your observations are not guided by any ideas; instead you are just gathering as much information as possible. As your information comes together, you realize that the only notable difference between the withered trees and those in the rest of the park seems to be a small wet area of ground in the midst of the withered trees where there are a number of dead young trees (Figure 1.16a).

Aha! Now you have an idea. Perhaps there is something leaking underground—maybe a sewage pipe? This idea is the *beginning* of a hypothesis, but you need more information to make your hypothesis more precise. From architectural drawings in the local library, you find that there *is* a sewage pipe from a nearby apartment complex (Figure 1.16b). Now you can make your hypothesis more precise. It becomes: "The trees are being damaged or killed by something that is seeping out of this pipe" (Figure 1.16c).

theory a scientific explanation that is strongly supported by a large body of scientific research and is a highly likely explanation of a broad set of related events

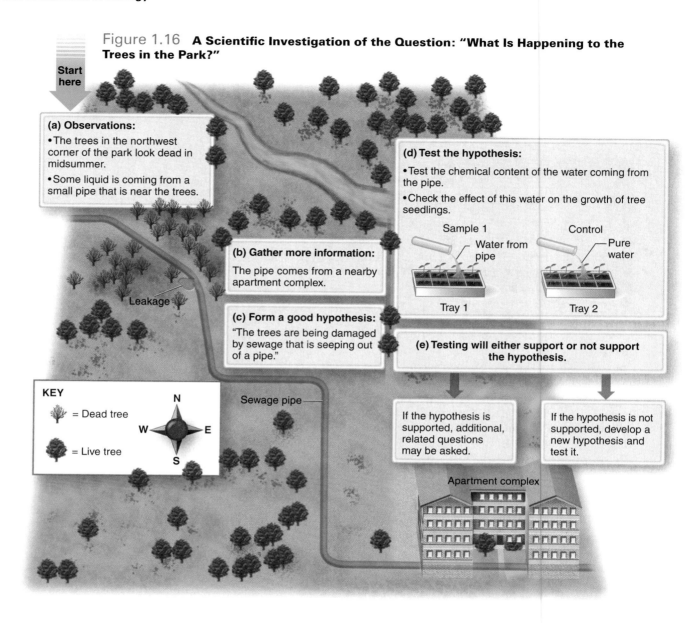

Figure 1.16 **A Scientific Investigation of the Question: "What Is Happening to the Trees in the Park?"**

Start here

(a) Observations:
• The trees in the northwest corner of the park look dead in midsummer.
• Some liquid is coming from a small pipe that is near the trees.

(b) Gather more information:
The pipe comes from a nearby apartment complex.

(c) Form a good hypothesis:
"The trees are being damaged by sewage that is seeping out of a pipe."

(d) Test the hypothesis:
• Test the chemical content of the water coming from the pipe.
• Check the effect of this water on the growth of tree seedlings.

Sample 1
Water from pipe
Tray 1

Control
Pure water
Tray 2

(e) Testing will either support or not support the hypothesis.

If the hypothesis is supported, additional, related questions may be asked.

If the hypothesis is not supported, develop a new hypothesis and test it.

KEY

= Dead tree

= Live tree

N W E S

Leakage

Sewage pipe

Apartment complex

The next step is to devise experiments that will test the hypothesis. For instance, your hypothesis predicts that the pipe is actually leaking a harmful substance, so you must test for that. You also might grow a patch of seedlings using water from the pipe to see if it causes them harm (Figure 1.16d). You might think of other ways to test your predictions and your hypothesis, but the critical point is that a hypothesis must be tested, and each test will help support or reject your hypothesis. (Figure 1.16e). For instance, if tests reveal that the water in the pipe contains no harmful chemicals, or if experiments show that the contaminated water does not harm the growth of tree seedlings, then you either will have to revise your hypothesis or reject it and form another one (Figure 1.16f). Once you are rea-

sonably confident of your results, you will share them with other people, perhaps through a town meeting or even publication in the local paper. In this way your work may have a positive impact on your local park and could probably help other scientists to understand stresses on trees in other urban environments.

Science Uses Different Kinds of Studies

Three kinds of studies commonly used by biologists include observational studies, correlational studies, and controlled experiments. Let's take a closer look at each of these. **Observational studies** are descriptions of organisms and the real problems they solve

observational study
a thoughtful, planned examination of organisms in their natural environment

in their natural environments. Scientific observations may involve sophisticated viewing and recording strategies and are more rigorous than casual or untrained observations. The field worker must be highly observant and careful not to interfere with the organisms under study. A scientific observer must take care to be objective and not influence his or her own observations. This is true whether the organisms are birds, mammals, plants, or even bacteria. To return to the example of the withered trees in the park, the initial careful observations of the trees and their surroundings are observational studies that provide a wealth of useful descriptive information that can be followed by other kinds of research.

The results of observational studies often are used to form a hypothesis and make a prediction. By zeroing in on specific information they help the investigator to begin to try to falsify the hypothesis. If you found sewage leaking from the pipe, you then could conclude that the sewage and the withered trees go together in space and time. By searching for other possible contaminants in the leakage, you might be able to conclude that excess metals or industrial pollutants were not present and therefore did not go together with the withered trees. This is good and necessary information. Based on this kind of study, however, you could not conclude that the leaking sewage *caused* the leaves to wither. The sort of finding that merely demonstrates that events occur together in space and time is called a **correlation.** Finding a correlation between two events is not enough to conclude that one event causes another.

To conclude that there is a causal relationship between two events, you must perform a **controlled experiment.** In this type of study you expose some experimental subjects—trees or mice or people, for example—to the factor you suspect is a cause and withhold exposure from another identical group of subjects. In the withered trees example you would grow a group of young trees with exposure to the fluid leaking from the pipe, while another group of young trees would have no exposure to it. The second group is treated exactly the same as the first, except it receives none of the leaking fluid. The second group serves as the **control group** and demonstrates what happens in the absence of the factor being investigated. Only a controlled experiment can allow you to conclude that the leaking fluid actually causes the withered leaves. Finally, just a single controlled experiment is not enough to support or refute a hypothesis. The controlled experiment must be repeated several times to make sure that the results are reliable.

The Culture of Science Maintains Scientific Integrity This discussion of the process of science gives you hints about the culture of science. Scientists follow some fairly strict rules and practices to ensure that science is reliable and trustworthy. First, they must strive to be objective and must try to prove themselves wrong. This may be difficult, especially when jobs and careers depend on successful research, but all well-trained scientists strive to be skeptical of their own or others' ideas. Any scientific finding is given intense review and criticism, and it is common for a finding or a study to be repeated before it can be published in a scientific journal. Nearly all findings are repeated by other scientists as they try to build on published knowledge. This is part of the reason that some "findings" reported in the newspapers are contradicted several months or years later.

Scientists must be honest about the results of their studies. While it is true that a false report is not likely to survive if other scientists try to repeat it and fail, the whole enterprise of science rests on the assumption that scientists honestly report the results of their studies. All scientists are trained in this code of honesty. The rare scientist who fakes data is seen as a threat to the entire enterprise. There have been cases of federal agents seizing lab notebooks and prosecuting scientists who faked results. Because of the way that new scientific research builds on previous experiments, frauds eventually are discovered and discredited. Fraudulent work leads to loss of respect, funding, and employment.

Science Versus Pseudoscience Sometimes it is hard to tell if something is science or just ideas masquerading as science. **Pseudoscience** is not easy to define but could be viewed as a statement that sounds scientific—but on closer examination is found to be based on faulty or incomplete scientific evidence. In addition, pseudoscience is more likely to push a particular idea rather than try to find out what the world is really like. Examples of pseudoscience are common, but it is often hard to see them without doing some background investigation. Many of the health claims for products sold on the Internet may be based on pseudoscience. One way to investigate the validity of these claims is to turn to the actual research literature and read the scientific studies that the claims are based on. Information also can be found on the websites of research universities, government organizations, and well-established private organizations. The American Cancer Society, for example, has information on alternative therapies for cancer that can

correlation when two events reliably happen together in space and time

controlled experiment a scientific study in which some subjects (the experimental subjects) are assigned to experience a specific experimental condition, while similar subjects (the controls) do not experience the experimental conditions

control group any group or condition that is included in a study to rule out other interpretations

pseudoscience a statement that sounds scientific but is based on faulty or incomplete evidence

be helpful. Often you will find that one or two studies make a suggestion that then is touted to sell a product. In some cases you will find some evidence to support the claims, but it may be incomplete. Science cannot keep up with the willingness of people to latch onto beliefs, so you will not always find a scientific answer to whether a particular claim is pseudoscience. Especially where health is concerned you certainly want to have as much information as possible before deciding on treatment options.

Biologists Use Special Techniques and Tools in Their Study of Life

Like other modern scientific fields, biology relies on special techniques and tools. It is not possible to describe them all, but here a few examples will give you a feel for the range of available tools.

Mathematical Tools Are an Important Foundation of All Sciences   Many of the observations of the universe involve counting or measuring, so the results must be expressed in numbers. Sometimes it is useful to examine, graph, or categorize the numbers obtained from your measurements, but often simply reporting the numerical results of a study can be confusing and difficult to interpret. If you make the same measurement on many different subjects or samples you will not get the exact same number each time. Human body weight is a simple example (**Figure 1.17**). If you measure the mass of 30 people to the nearest kilogram or pound, you will get a wide range of results. People have difficulty interpreting long lists of numbers, so they use mathematical tools to summarize them. The **mean,** or average of a group of numbers, is one example of a mathematical summary statistic. It is easier to grasp that one

group of people has an average mass of 72 kilograms (160 pounds) than it is to look at a list of 30 measurements and draw conclusions about the mass of the people in the group.

Mathematical tools also can help you to interpret the results of a study. Interpretation is necessary when you want to know if two groups of subjects are the same or different on some measurement. Suppose you want to know if average heights for men and women differ. Of course, you cannot measure the height of every human alive, so you might go into a local football stadium during a game and measure the heights of 300 men and 300 women. You find that your sample of men is on average taller than your sample of women. But is this the correct interpretation of your finding? Because every individual, human or otherwise, is different from every other individual, there is always some chance that differences between groups are just a reflection of random individual variation. By chance you may have sampled a lot of short women and a lot of tall men. The mathematical tools called **inferential statistics** can help you to judge the validity of your results. With inferential statistics you can calculate the probability that two averages or two other measurements of a group represent two different populations, rather than being two samples taken from the same population or group. You may not realize that ads often use results of studies analyzed with inferential statistics, but the next time you encounter a claim that is something like "eating a low-fat diet *significantly* reduces the risk of heart disease," you can know that inferential statistics were used. In this case inferential statistics help determine that a low-fat diet is highly likely to have a real and measurable effect on your probability of having heart disease.

Chemical Tools Are Used to Identify and Analyze Biological Chemicals   A variety of tools and techniques can tell you about the chemicals present in a biological sample, and some chemical tests are quite simple. For example, some chemicals will change color when they react with other chemicals. Starches—complex molecules that are plentiful in potatoes and pasta—will turn a dark color when mixed with iodine. So you can test a sample with iodine, and if it turns dark, there is starch in the sample. Other chemical techniques are more complex. For example, a sophisticated group of techniques and tools can be used to identify and analyze the DNA in an organism's cells. Collectively called **molecular biology techniques,** these procedures can detect even tiny differences in the DNA

---

mean   the average of a group of numbers; to calculate the mean you divide the sum by the total number of values

inferential statistics mathematical tools that can determine the probability that results of a study are real or just chance differences

molecular biology techniques tools and procedures used to isolate and identify the DNA in a sample of cells

compound microscope, or **light microscope** a scientific instrument that uses lenses to bend light rays in a way that makes objects appear closer to your eye and therefore larger

electron microscope (EM) a scientific instrument that uses a beam of electrons to give greater magnification than can be obtained with a light microscope

---

Weights of 30 People (in pounds)	
154	129
146	135
299	176
179	169
100	65
162	193
179	185
145	180
235	140
204	192
172	153
110	120
105	148
85	282
112	147

Mean = 160.03

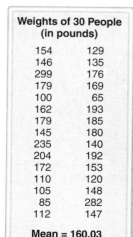

Figure 1.17   **Summary Statistics Like the Mean Make Long Lists of Data Easier to Understand.**

taken from two people and determine if they may be genetically closely related.

Microscopes Reveal Worlds That Are Unseen by the Naked Eye

Most living organisms are quite small, and the structures within cells are too tiny to be seen with the unaided eye. For example, a human blood cell is 7.5 micrometers in diameter. A micrometer is one-millionth of a meter. The period at the end of this sentence is about 500 micrometers, or half a millimeter in diameter—and about 67 red blood cells could line up across its middle. Microscopes of various sorts allow you to see tiny objects. Since the early 1600s, when it began to be developed, the microscope has allowed biologists to investigate forms of life and living structures too small to be seen by the unaided eye. Microscopes are some of the most influential and important tools of modern biology.

The **compound microscope,** or **light microscope,** is the oldest and most familiar kind of microscope (**Figure 1.18a**). The compound microscope comes in many forms, but in all of them the system of internal lenses accomplishes the same thing—the lenses bend light rays in a way that mimics bringing an object closer to your eye. This makes the object appear larger, and allows you to see more detail.

Compound microscopes can magnify objects up to 1,200 times. At higher magnifications the image gets larger, but it is not possible to see greater details. For instance, cells can be seen easily with a compound microscope, but many of their small internal structures are not visible. Other types of microscopes make subcellular details visible. One of the most useful is the **electron microscope (EM),** which uses a beam of electrons instead of light rays to produce an image. With a magnification of up to 80,000 to 100,000 times, electron microscopes produce black-and-white photographs that show the internal or external details of cells or other objects. Electron microscopes come in two kinds. A transmission electron microscope (TEM) magnifies thin slices of objects by passing an electron beam through the slice (Figure 1.18b). A scanning electron microscope (SEM) bounces a beam of electrons off the surface of a small object, which reveals remarkably fine surface details (Figure 1.18e).

(a) Compound microscope

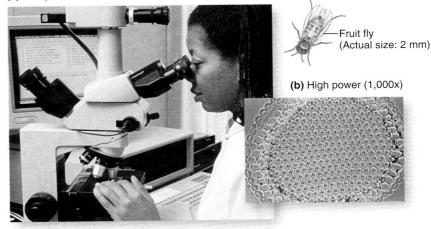

Fruit fly
(Actual size: 2 mm)

(b) High power (1,000x)

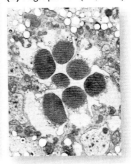

(c) Transmission electron microscope (TEM)

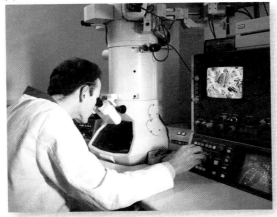

(d) High power (18,000x)

(e) Scanning electron microscope (SEM)

(f) High power (18,000x)

Figure 1.18 **Different Microscopes Reveal Different Images of the Eye of a Fruit Fly. (a)** The lenses of a compound microscope bend light and mimic bringing an object closer to your eye. This makes the object appear to be larger so you can see more detail. **(b)** The eye of a fruit fly viewed with high power of a compound microscope. **(c)** The beam of electrons generated by a transmission electron microscope passes through extremely thin sections of a sample, throwing areas of light and shadow onto a photographic plate. **(d)** Electron micrographs of the eye of a fruit fly. **(e)** A scanning electron microscope reveals extremely fine details of its surface. **(f)** High-power scanning electron micrograph of the eye of a fruit fly.

■ *What are the advantages and disadvantages of each kind of microscope?*

1.6 **An Added Dimension Explained**
What Causes Four Corners Disease?

In this chapter you have seen that science uses a set of organized steps to answer questions about the world. The organized steps are not a dogmatic list of requirements. They are a set of guidelines that must be adapted to fit each new situation. Let's return to the chapter-opening problem concerning Four Corners disease and see how scientific methods helped to solve it. While this disease scare happened many years ago, the detailed methods used to find the answer are applied to disease outbreaks whenever they occur. Similar approaches have been used to figure out recent diseases such as severe acute respiratory syndrome (SARS), bird flu, and Ebola hemorrhagic fever.

Scientific solutions to questions usually start with careful observations, and investigation of Four Corners disease followed this pattern. Hospital workers were stunned and puzzled when they saw the first case of this mysterious illness in May 1993. They had no immediate hypotheses but ran many tests on the sick young man, hoping to find evidence of a known disease that they could work to cure. They did not find the answer in these first rounds of tests, but the results did give the researchers enough evidence to develop some ideas about what was going on. When the young man died, the health professionals involved became more determined to find out what killed him.

One of their priorities was to find out if this was an isolated case or if there were other people who showed similar symptoms. His specific symptoms had been documented, and workers searched for others with the same symptoms. Not only did they find that his fiancée had died in a similar fashion, in a short time they also found others in the same region with a similar disease. Hospital researchers assembled a detailed description of the common symptoms. They kept records of how this mysterious disease progressed, noting symptoms as well as who recovered and who died. These detailed observations formed the base of the studies that isolated the cause of Four Corners disease.

Armed with this background information, some researchers were ready to state a specific hypothesis. Others in the region made wild guesses, but the health officials and scientists involved kept their hypotheses tied to known information. The first and most important hypothesis was that a virus caused the disease. A study of published records showed that a family of viruses called hantaviruses had caused similar disease symptoms in other locations around the world, but a hantavirus had never been isolated in the United States. So rather than suspect a new and unusual possibility, researchers began with the most likely hypothesis and predicted that a hantavirus would turn out to be the culprit. To test this hypothesis they examined tissue samples using chemical tests that would reveal whether the victims had been exposed to any hantavirus. But they also tested for exposure to many other kinds of viruses as well—and in so doing, they tried to prove themselves wrong. The results showed that the only viral exposure that all victims shared was to a hantavirus.

Another important test of the hantavirus hypothesis was the connection to rats, mice, and other rodents. In other known cases rodents carried this virus, so if a hantavirus was the culprit, rats or mice also should be involved. To determine if this prediction was correct, researchers looked for signs of rodent infestation in households with infected inhabitants. But this would not be enough evidence to support the hypothesis. They also had to search for signs of rodent infestation in control houses—neighboring households where there was no disease. For example, if all houses in the Four Corners area had mice or rats, then it would be less likely that rodents were the disease carriers. Signs of deer mouse infestation were found exclusively in infected households (**Figure 1.19**).

Modern health workers were not the only ones to make careful observations and hypotheses to explain Four Corners disease. The Navajo people who live in the Four Corners area had long recognized this disease, and scientists used Navajo oral histories to better understand why hantavirus disease appeared at some times but not at others. Navajo oral history documents outbreaks of the disease in 1918 and 1933, when wet weather had allowed local mouse populations to increase. Navajo traditions admonished people to keep deer mice out of their homes and to keep food supplies clean of mouse contamination.

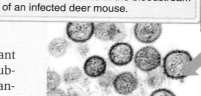

Deer mice carry the hantavirus that causes Four Corners disease or Hantavirus Pulmonary Syndrome (HPS).

TEM of hantaviruses in the bloodstream of an infected deer mouse.

Hantaviruses are found in urine and feces of an infected mouse. When a person inhales hantaviruses from mouse urine or mouse droppings, he or she may become infected with HPS.

Figure 1.19 **The Connection Between Hantaviruses, Deer Mice, Humans, and HPS.**

Scientists strive to provide explanations that cut across different disciplines and levels of explanation. This relates back to the point made earlier that all of the disciplines in science describe the same universe. The investigation of Four Corners disease is no different. For example, scientists wanted to know why this hantavirus disease erupted at this particular time. Why were there so many deer mice in the spring of 1993? Here the search for the cause of this outbreak of deadly hantavirus in the Four Corners region becomes related to broader environmental factors. Every 3 to 7 years a warm current of water develops in the Pacific Ocean and affects weather patterns in all of the Americas. Because the current often coincides with the Christmas holidays, it is called *El Niño*. Environmental biologists and meteorologists confirmed that during the winter of 1992 El Niño produced abundant snow and rain in the desert Southwest. That meant a lush growth of green plants in a place that usually is arid, and consequently it meant lots of food for young deer mice. The wet weather produced by El Niño (**Figure 1.20**) resulted in a rapid increase in the deer mouse population. Consequently, more deer mice invaded human houses, and human exposure to hantavirus in mouse droppings and dried urine was more widespread. **Table 1.4** summarizes the steps that helped confirm the hantavirus hypothesis and relates them to what you have learned about scientific procedures.

The study of the hantavirus and all of its ramifications did not stop in 1993 because wherever deer mice and certain other rodents are common, people continue to be infected. The number of people infected each year is low, but infected people still can get quite sick. Even with current treatments there are no medicines to kill the virus, and nearly half of infected people still die. This ongoing problem means that scientists will continue to study the hantavirus, trying to find a way to treat the disease it causes. If you search a medical database, you will find that scientists continue to learn more about this virus and how it causes disease, now called HPS (short for hantavirus pulmonary syndrome). Scientific studies almost never end when the immediate question is answered. The curiosity of individual scientists and the need for answers drive the search for deeper explanations.

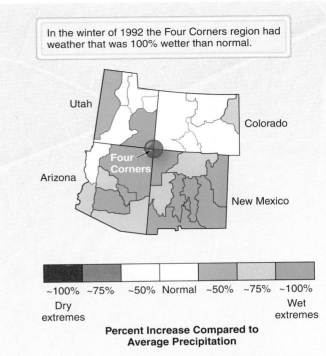

In the winter of 1992 the Four Corners region had weather that was 100% wetter than normal.

~100% ~75%   ~50% Normal ~50%   ~75% ~100%
Dry   Wet
extremes   extremes

**Percent Increase Compared to
Average Precipitation**

Figure 1.20   **During El Niño Years the Four Corners Region had Extremely Wet Weather.**

QUICK CHECK

1. What infectious agent was the cause of Four Corners disease?
2. Why did researchers suspect that Four Corners disease might be spread by rats and mice?

Now You Can **Understand**

Chapter 1 introduced you to three general foundations of the science of biology. First, you considered an overview of the breadth of biological science. Second, you learned a bit about what science is and how it differs from other intellectual disciplines. Third, you discovered some of the important tools that biologists use as they seek to understand the living world. You can use the insights gained here to better understand some of the events and debates taking place around you. While you certainly will have a more sophisticated appreciation of these topics by the end of your biology course, here are some examples of issues that you should have a greater understanding of having studied Chapter 1. You can find more information about these topics by exploring resources at your local library or on the Internet.

Risk factors—What Do They Mean?

Almost every newspaper or Web news summary has information about risk factors. Often these risk factors refer to common human diseases, such as heart disease, diabetes, and cancer. Risk factors are characteristics that give a higher or lower probability of succumbing to a disease or condition. To be useful risk factors must be evaluated realistically. For example, a recent study showed that eating a "Mediterranean diet" reduces your risk of dying from heart disease by over 70%, compared with people who eat a "regular" American diet. That is a dramatic effect. The risk of dying from heart disease or other complication is quite high if you eat a diet rich in animal fats and simple starches and if you are overweight and do not exercise. In comparison, the risk of dying from eating apples sprayed with pesticides is quite low. This does not mean that you should not worry about pesticides in your food or

Table 1.4	Steps That Solved the Mystery of Four Corners Disease and How They Relate to the Scientific Process	
Date in 1993	**Events**	**Effect in the scientific process**
May 14	Unexplained death and autopsy of first patients: a young Navajo man, his fiancée, and four others who died of respiratory illness	Careful observation revealed that this illness was not routine and was linked to other unexplained deaths. Routine tests ruled out common disorders. *First hypothesis: these deaths could be explained by tests for several well-known diseases.*
Late May	Indian Health Service (IHS) investigation into poisons as alternative cause of the deaths	*Second hypothesis: poison, not a pathogen, caused the deaths.* Visits to homes of victims showed no evidence of any exposure to poisons. *Poison hypothesis rejected.*
May	CDC is alerted to the presence of a mysterious, fatal disease in the Four Corners region and joins the investigation	*Third hypothesis: a known but rare virus is responsible for the deaths.* Chemical tests revealed whether a person had been exposed to a particular virus. Most tests were negative. Tests of hantavirus strains showed a weak positive result.
June	CDC critically evaluates hantavirus hypothesis	*Fourth hypothesis: hantavirus carried by local rodents causes Four Corners disease.* Hantavirus is always associated with rodents. This hypothesis is tested by investigating role of local rodents in disease.
Summer	Even before the hypothesis was confirmed, U.S. State Health Departments notify doctors of this threat to health and warn people to avoid rodents	Practical public health action protects public safety.
November	CDC Special Pathogens Unit continues to work on the case	Attempts to characterize virus: 1. Matches to known hantavirus with antibody tests give only weak matches. 2. Attempts to grow and characterize viruses isolated from Four Corners rodents in laboratory fail. 3. Molecular biology techniques copy viral DNA and identify virus as new form of hantavirus. *The rodent-hantavirus hypothesis is supported.*
Late in 1993	Investigators look further to support the hypothesis, investigating local rodent population patterns and tissues from victims of disease	Records from local Navajo tribes and from local university researchers confirm that the disease increases when local rodent populations increase and is associated with certain warm weather patterns. Further tests of tissues from victims of Four Corners disease confirm that they are infected with a local strain of hantavirus. *The rodent-hantavirus hypothesis is further supported.*
	Public health alerts slow the outbreak of disease	While no cure has been found for hantavirus, public health information allows people to prevent the disease by avoiding rodent habitats.

that eating pesticides might not make you sick. Rather, you should be sensitive to the actual level of risk that you face in various aspects of your life.

Antibiotic Resistance

You may have heard much about antibiotic resistance in the news. You will learn more about this topic in Chapter 26, but now you can understand antibiotic resistance as a variant of natural selection. Because human intervention is involved, it is a form of unintentional human-induced selection. When you take antibiotics to cure a bacterial infection, like strep throat or tuberculosis, the drug may not kill some of the germs. Bacteria that resist the drug's effects survive and multiply. They pass their trait of resistance to the antibiotic on to their offspring. If antibiotics are overused, large populations of resistant bacteria evolve, leaving people with no effective treatments when they do get sick. This growing prob-

lem reflects how some agent of selection—in this case the antibiotic—can produce a change in a population of bacteria—in this case resistance to the lethal effects of that antibiotic. This is an excellent example of how evolution works.

What Do **You** Think?

Mental Telepathy

Science provides a way to increase the chances that your beliefs about the world reflect what the world actually is like. Yet most people continue to hang on to some personal beliefs about the world, even in the face of strong scientific evidence to the contrary. Mental telepathy is an example. Many scientific studies have investigated whether mental telepathy can be demonstrated. There are some weak positive results in some special circumstances, but most studies find no evidence for mental telepathy. Yet many people firmly believe that mental telepathy is real, and many believe they have experienced it.

Here is an exercise to give you a perspective on your personal beliefs. First, think about and write down some of the strong beliefs you hold but which you suspect or know do not have scien-tific support. Second, over the next few weeks talk to your friends and colleagues about some of the things they believe in. When you find that your friends have beliefs that you do not hold, do you treat your own beliefs differently than those of your friends? Are your beliefs more or less credible if someone else shares them? What do you think?

Chapter Review

CHAPTER SUMMARY

1.1 Biology Touches Every Aspect of Your Life

Biology has profound impacts on your daily life. The biological themes of emergent properties, unity and diversity of life, evolution by means of natural selection, and the processes of science weave through all biological knowledge.

biology 2

organism 2

1.2 Life Is Defined by a Set of Features That All Living Things Share

Organisms are complex, organized, cellular, homeostatic entities that respond to and interact with their environments. They use energy to power their internal processes and use the instructions in DNA to con-duct these processes and to reproduce.

cell 5

cell membrane 5

DNA (deoxyribonucleic acid) 6

energy 6

homeostasis 6

1.3 Levels of Organization Are Characteristic of Life

Cells, tissues, organs, organ systems, and organisms form one set of biological levels. Above the level of individual organism are popula-tion, community, ecosystem, biome, and biosphere.

atom 9
biome 10
biosphere 10
chemical element 8
community 10
ecosystem 10
emergent properties 12
multicellular organism 9

organ 9
organelle 9
organ system 10
population 10
tissue 9
unicellular organism 9
virus 9

1.4 Evolution and Natural Selection Have Produced Life's Diverse Forms

All life is related. Earth's diverse forms of life have resulted from the process of evolution by means of natural selection.

evolution 13

natural selection 14

1.5 Biology Applies the Methods of Science to the Study of Life

Scientists follow a disciplined path toward gathering knowledge. Their methods involve thoughtful observations, reference to the scientific lit-erature, and careful stating and testing of hypotheses using controlled experiments. A scientific theory has been tested time and again and has been substantiated. Biologists use mathematical and physical tools to gather and process information. Different microscope tech-nologies reveal aspects of life that are invisible to the unaided eye.

compound (or light)
  microscope 20
control group 19
controlled experiment 19
correlation 19
electron microscope (EM) 20
hypothesis 16
inferential statistics 20

mean 20
molecular biology
  techniques 20
observational study 18
probability 15
pseudoscience 19
scientific law 16
theory 17

1.6 An Added Dimension Explained: What Causes Four Corners Disease?

In a relatively short period of time, investigators were able to use the methods of science to trace Four Corners disease to a hantavirus trans-mitted to humans from deer mice and to tie the outbreak of hantavirus to wider environmental events.

REVIEW QUESTIONS

TRUE or FALSE. If a statement is false, rewrite it to make it true.

1. Because every living organism is unique, it is not possible to discover common principles in the study of life.

2. The levels of biological organization, from simplest to most complex, are atom, molecule, cell, tissue, organ, organ system, organism, population, community, ecosystem, and biosphere.

3. The major chemical elements needed by all living things include silicon, hydrogen, oxygen, and nitrogen.

4. All the individuals of a single kind of organism living in one geographic area form a population.

5. A compound microscope uses lenses to bend light and magnify tiny objects.

MULTIPLE CHOICE. Choose the best answer of those provided.

6. Which of the following is *not* a theme in the study of biology that you were introduced to in this chapter?
   a. heat, cold, acidity are the fundamental determinants of life
   b. atoms combine to form molecules, which combine to form cells
   c. emergent properties, unity and diversity, evolution through natural selection
   d. scientific method, forming hypotheses that are testable, careful measurements, statistical analysis
   e. populations form a community; communities form an ecosystem; ecosystems form a biome; biomes form the biosphere

7. You are observing a burning candle. As you watch the flame, you wonder how someone who didn't know better could demonstrate whether the flame is alive. Which of the following is a characteristic of life that is not a characteristic of fire?
   a. The flame responds to and interacts with the environment.
   b. The flame uses energy to power itself.
   c. The flame can grow.
   d. The flame is made up of one or more cells.
   e. The flame can reproduce.

8. What is the role of the cell membrane?
   a. energy production and cellular instructions
   b. protection from the environment and communication with the environment
   c. cellular reproduction and cellular longevity
   d. protection from the environment and energy production
   e. none of the above

9. One example of homeostasis is that you
   a. eat food because it tastes good.
   b. shiver because you are cold.
   c. stay well because you take antibiotics.
   d. get sick because you are invaded by pathogenic bacteria.
   e. eat more and you gain weight.

10. Which of the following carries instructions for the processes of life?
    a. energy
    b. the cell membrane
    c. proteins
    d. individual atoms
    e. DNA

11. What is an atom?
    a. the nucleus of a cell that has DNA
    b. a complex molecule that carries information
    c. an organelle that is made of molecules
    d. the smallest unit of an element that has all the properties of that element
    e. the largest unit of a molecule that has all the properties of that molecule

12. What is one important feature of a virus?
    a. It is the same as a cell.
    b. It is larger than a cell.
    c. It is a small nonliving structure made of biological molecules.
    d. It is a small living structure made of biological molecules.
    e. It is a small structure that has internal organelles and carries out all of the processes of life.

13. Which of the following is a species of ants?
    a. all of the individuals of a particular kind of ant
    b. all of the ants that live in a particular location
    c. all of the ants that live in a particular environment
    d. only the ants that meet one another in daily interactions
    e. all of the above

14. What is one distinguishing feature of ecosystems?
    a. An ecosystem includes just the living organisms in an individual's environment.
    b. An ecosystem contains all of the biomes in the biosphere.
    c. Biomes are smaller than ecosystems.
    d. Ecosystems are all living systems outside the biosphere.
    e. Ecosystems include living communities and their nonliving environmental components.

15. Which of the following is not an important concept related to the evolution of life?
    a. All life is genetically related.
    b. Living organisms are diverse.
    c. Life evolves by the process of natural selection.
    d. The diversity of life emerged gradually over the history of the Earth.
    e. Every individual within a species is identical.

16. Central features of science are
    a. belief systems.
    b. observations of the real world.
    c. intuitions about the real word.
    d. well-worded philosophical arguments.
    e. statements that rely on authority figures in science.

17. A testable explanation or prediction is called
    a. a hypothesis.
    b. a theory.
    c. a correlation.
    d. a fact.
    e. none of the above.

18. Which of these statements is true? Control groups usually
    a. receive exactly the same treatment as experimental groups.
    b. are quite small.
    c. are quite large.
    d. are unnecessary.
    e. demonstrate the effect of no experimental treatment.

19. What are inferential statistics?
    a. mathematics that allow people to infer their own conclusions from the results of experiments
    b. mathematics that summarize a large set of numbers
    c. the misuse of statistics to promote erroneous conclusions
    d. mathematical tools to determine the probability that given results happened by chance
    e. mathematical tools to determine if certain patterns of numbers happened in the past

20. Hantaviruses have always been associated with
    a. hot climates and deserts.
    b. cold winters.
    c. poor food sanitation.
    d. antibiotic overuse.
    e. mice and rats.

1. What are the properties of life that all cells share?

2. Outline a scientific process you might use to determine if taking vitamin C can prevent people from catching colds.

QUANTITATIVE QUERY

1. Calculate the mean of this set of numbers:

10	35
15	2
2	2
30	45
25	4
1	3
1	35

   Is the mean a good representation of this set of numbers? Why or why not? What would be a better summary of these numbers?

THINKING CRITICALLY

1. A few years ago newspapers reported that people feared living near high-tension power lines because they thought it increased their chances of getting cancer. Imagine you are a member of your local county commission. Many families that live near power lines have stories about illness in their families. Angry and fearful residents are demanding that high-tension power lines be restructured and moved away from their houses. As a council member you must decide whether the county should undertake this expensive project, which will decrease funding for schools and other community projects. How will you decide whether to support the power line restructuring project?

2. Study the experiment presented in **Table 1.5.** What hypothesis is the experiment testing? Explain the way in which this hypothesis relates to the reason the person gives for using the supplement. What is wrong with the design of the experiment? Given the way that the experiment was carried out, are the experimenter's conclusions verified? Develop a hypothesis that will lead to a more useful experiment. Design a study that will answer the experimenter's question about the effectiveness of the supplement.

3. What is meant by the term *emergent properties*? Define the term, and then give an example that is not in the text—for instance, baking a loaf of bread.

For additional study tools, visit www.aris.mhhe.com.

Table 1.5 What Is Wrong with This Experiment?

Your friend is using a nutritional supplement that he hopes will improve his performance at martial arts. As a curious scientist you ask him why he thinks this supplement will improve his agility, stamina, and speed—all essential goals of the serious martial arts student.

Your friend replies that he has found the following study on a website for a distributor of this product. It has convinced him that the product is worthwhile. In this study:

- Five volunteers offered to take the supplement for 6 weeks.

- Their ability to perform a weight-lifting routine was measured at the beginning and end of the study.

- At the end of the study every participant had improved in both the amount of weight they could lift and in the number of repetitions they could do with each weight.

- Your friend believes that this is dramatic proof of the supplement's effectiveness.

Is your friend getting a good bargain with this nutritional "short cut"? Or is he wasting his money?

Life Emerges from Chemistry

ATOMS AND MOLECULES

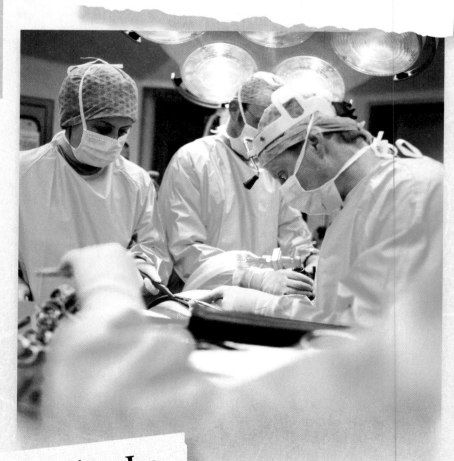

Cryonics: Freezing for the Future?
July 18, 2002

Cryonic freezing of humans gives some people
the hope of extending life.

Putting a Legend on Ice
July 10, 2002

Birth and death are unavoidable events in the rhythm of life. Each type of organism has a characteristic life span, and life spans vary from a day to thousands of years. Some adult mayflies live a day, many small frogs live for just a year, while mice live 2 or 3 years. At the long end of the life-span spectrum, some trees and fungi live for centuries, and some tortoises survive for over 200 years. The oldest known trees are bristlecone pines, which are over 4,000 years old. By comparison, the human life span is neither short nor exceptionally long. Although some long-lived people live to be well over 100 years, the average human life span in developed countries is about 70 to 80 years. For many, though, this near-century is not enough. The Internet and magazines are filled with products that claim to slow aging and increase life span, and many people buy them.

Each of us must someday face our own death, but what if you could be brought back to life after dying? Seriously. If you *had* the money, what would you pay for even a slim chance of being revived in 50 or 100 years? Some people hope and believe that *cryopreservation* will allow them to be frozen when they die and be brought back to life in the future. Cryopreservation is a process that freezes living tissue with the intention of thawing it out and restoring life at some time in the future (see chapter-opening photo). For a hefty fee you can arrange for a cryopreservation organization to freeze your body to an extremely cold tempera-ture shortly after your death. What happens if you sign up for this service?

The scenario would go something like this: upon your death, medics from the cryopreservation organization arrive and immediately begin to oxygenate your blood, using CPR. Then your body quickly is transported to their central facility, where it is infused with a protective liquid and gradually cooled to about −200°C (−328°F). You remain frozen in this ultracold state while basic research into *cryobiology*—the study of life at cold temperatures—makes slow but steady advances. Then, when all the necessary technology is in place, the cryogenics company will thaw you out and restore you to life—or so they say. The claim is that scientists will have learned how to bring you back to life and also will be able to cure the diseases that killed you. What can cryopreservation organizations tell clients about the prospects for reviving them in the future? Unfortunately, not much. At this point in scientific understanding, frozen human bodies or parts of bodies cannot be revived. Nevertheless, cryogenic organizations focus on what we do know about cryopreservation and tout the possibility that in the not too distant future biological research will develop methods to revive dead, frozen humans.

What are the scientific obstacles to cryopreservation? Why *can't* a frozen person be revived? You will return to these questions at the end of the chapter, but to understand that discussion, you must know about the chemistry of water and how water facilitates the chemical interactions of life. To move to this level of understanding, you first must learn a bit of basic chemistry. ■

2.1 New Properties Emerge When Substances Interact

You may be wondering why you must begin your study of biology by learning chemistry. *Chemistry* is the study of the properties of matter and how those properties relate to the structure of atoms. Much of the study of chemistry is focused on how and why chemicals interact with one another. What does chemistry have to do with life? Of course, living things are made of chemicals. More significant, though, is that new and surprising things can happen when simple chemicals interact. The new qualities often are called *emergent properties* (see Section 1.2). The complex structures and process of life introduced in Chapter 1,—such as the internal structures of cells, homeostasis, and energy use—are the result of chemical interactions. For example, the flashes of fireflies on a warm summer night produce an amazing display of biological fireworks (**Figure 2.1**) that result from simple chemical interactions.

To get an intuitive feeling for the idea of emergent properties, think back to your elementary school days, when you may have played with LEGO plastic building bricks. Although these simple components come in just a few different basic shapes, you can assemble them into complex and interesting structures. You provide a source of energy to join the pieces together and produce something different from the sum of the simple blocks. Your creation might have a new property. For instance, it might have a seat to carry toy people, or it might be able to move across the floor on wheels. If you added an energy source, it might even roll around, wave its arms, or spin propeller blades. Individual LEGO bricks or even buckets of them have none of these captivating new properties.

But if life is just a "bowl of chemicals," why doesn't life emerge from the mix of nonliving chemicals in kitchens or in laboratories? As far back as the mid-1800s scientific studies have

Figure 2.1 An Emergent Property. Firefly display emerges from a combination of chemicals within the light organs of fireflies.

When the right chemicals are combined within the light organ of a firefly, and when many fireflies gather on a warm summer night, a spectacular display emerges.

shown that in our day-to-day world, life comes only from other life—it never arises spontaneously. Scientists do, however, have good reasons to conclude that life *did* emerge from basic chemical interactions under the mix of conditions that existed on Earth about 4 billion years ago. The chemical reactions in those earliest cells were handed down and modified—generation after generation—and inside of cells life does, indeed, result from complex chemical interactions. Let's begin this journey into the chemical stuff of life by considering some basic questions, starting with: What is the universe made of?

QUICK CHECK

1. Why begin your study of biology with the study of chemistry?
2. Use a chemical example to summarize the concept of emergent properties.

2.2 | Matter Is Made of Atoms

The universe is made of matter and energy. This includes your body and everything that you can see, touch, hear, smell, or taste. So the chair you are sitting on, the sandwich and soda you had for lunch, your body, the air you breathe, and every other object in your environment is nothing other than matter and energy. You will read more about energy in Chapter 5. For now let's concentrate on matter. In commonsense terms matter is something tangible that is influenced by the pull of gravity and provides a resistance to being moved. In scientific terms **matter** is any material substance that takes up space

matter any material substance that takes up space and has mass

and has mass. And what is mass? *Mass* is a measurement of the amount of matter in an object. Mass is measured in metric units such as grams and English units such as ounces and pounds, although in scientific discussions the metric system almost always is used.

The universe is made up of matter, but how is one kind of matter different from another kind of matter? For instance, how is a piece of wood different from a piece of plastic? Are all forms of matter equally simple or complex? If you've ever taken something apart to figure out how it works, you know that objects sometimes can be understood by breaking them down into simpler parts. Chemists have done innumerable experiments to learn what simpler parts common substances are composed of, and you can do some of the simpler ones yourself. For example, let a glass of tap water, pond water, or seawater—or one of each—sit out on a counter until all of the liquid is gone. What remains? You should see a whitish film in the bottom of the glass. What can you conclude from this simple experiment? One conclusion is that the original "water" was actually made of a liquid component that has evaporated and a solid component that has remained in the glass.

Other experiments show that all of the thousands of substances you experience in life are made of just 92 basic forms of matter that occur naturally on Earth. Each basic kind of matter is an **element,** a substance that cannot readily be broken down into other substances by common processes such as evaporation, filtering, or burning. Gold, silver, copper, and iron are common chemical elements. If you heat any of these to a high temperature, you will end up with a puddle of melted gold, silver, copper, or iron. In contrast, burning, evaporation, heating, or filtering can break down other substances into the elements that compose them. For example, if you burn a tree and chemically analyze the ashes, you will find that they contain carbon, oxygen, hydrogen, and a few other elements. You also may be familiar with sulfur, phosphorus, helium, chlorine, sodium, aluminum, nickel, silicon, radium, uranium, and neon because these elements often appear in your daily life (**Figure 2.2**).

Elements Are Types of Atoms

What is the smallest amount of gold that can be measured? One speck of gold dust? A milligram? It turns out that an atom is the smallest unit of any element that retains the properties of that element. Every element—and thus all matter—is made of atoms.

The distinctive properties of each chemical element are determined by the internal structure of the atoms that make up that element. Atoms of various elements are different from one another because they have different numbers of smaller parts, called *subatomic particles*. Although physicists have identified and named a whole slew of subatomic particles, only three are important for this discussion: protons, neutrons, and electrons (**Figure 2.3**). Protons and neutrons are found in the atom's center or nucleus, while electrons are in continual motion around the nucleus.

Each kind of subatomic particle can be described in terms of its charge and mass. You have already read about mass, but what is charge? *Charge* describes how subatomic particles behave with respect to each other. Charge is not something by itself. Charge is a property of matter. A particle of matter can carry a positive charge, a negative charge, or no charge. Two particles with identical charges will be pushed away from each other, while two particles with opposite charges—positive and negative—will be pulled toward each other. A particle with no charge is neutral and will not pull or push another particle.

Now that you understand mass and charge, you can consider the characteristics of each

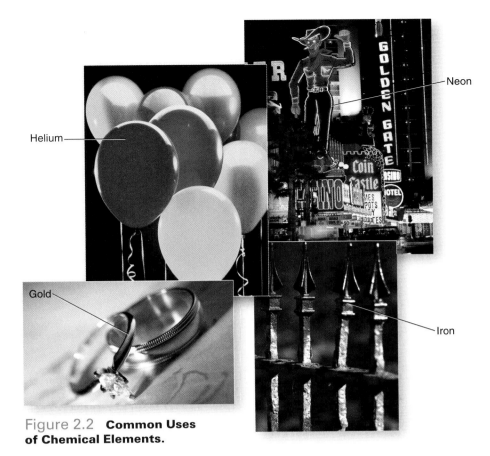

Figure 2.2 **Common Uses of Chemical Elements.**

subatomic particle. A **proton** is a positively charged subatomic particle with a mass so small that it is meaningless to most people (1.67×10^{-24} grams, which could also be written as 0.00000000000000000000000167 grams). A **neutron** is an uncharged subatomic particle that has about the same mass as a proton. Finally, an **electron** is a negatively charged subatomic particle that moves around the nucleus and has a mass that makes the proton look like a giant. It would take almost 2,000 electrons to equal the mass of one proton.

Let's return to the question of how atoms of various elements differ. Atoms of each element have a characteristic *number* of protons, neutrons, and electrons. For example, hydrogen is the smallest and simplest element. Hydrogen has just one proton, no neutrons, and one electron. Helium is a bit larger: it has two protons, two neutrons, and two electrons. An atom of carbon has six protons and six neutrons in a nucleus circled by six electrons. You can see a pattern here: with hydrogen as the exception, these atoms have the same number of protons, neutrons, and electrons. This pattern breaks down in larger atoms, where there are

element a substance that cannot readily be broken down into other substances by ordinary processes such as burning, evaporation, or filtration

proton a positively charged subatomic particle in an atom's nucleus

neutron an uncharged subatomic particle in an atom's nucleus

electron a negatively charged subatomic particle that moves around an atom's nucleus

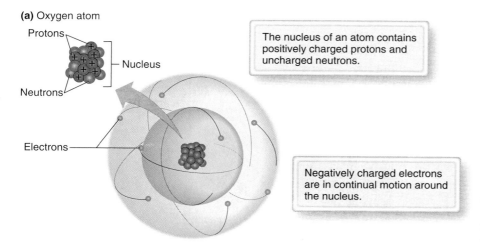

(a) Oxygen atom

Protons

Nucleus

Neutrons

Electrons

The nucleus of an atom contains positively charged protons and uncharged neutrons.

Negatively charged electrons are in continual motion around the nucleus.

(b) Diagrammatic representation of an oxygen atom

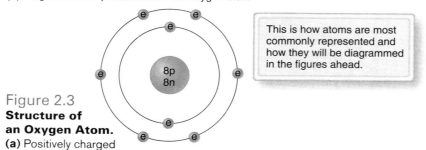

8p
8n

This is how atoms are most commonly represented and how they will be diagrammed in the figures ahead.

Figure 2.3
Structure of an Oxygen Atom.
(a) Positively charged protons and uncharged neutrons are within the atom's nucleus, while negatively charged electrons continually move around the nucleus in a cloud. **(b)** A diagrammatic representation of an oxygen atom.

■ *How does the path of electrons around the nucleus of an atom differ from the path of planets around the Sun?*

usually more neutrons than protons, but every atom is characterized by a specific number of protons, neutrons, and electrons.

You might be wondering which of the three subatomic particles actually determines an atom's element. Does an atom of carbon *always* have exactly six electrons, protons, and neutrons? It turns out that the answer is no. One way that an atom of a particular element can vary is in the number of electrons it contains. The resulting atom is still the same element, but because it has lost or gained an electron it now has a net positive or negative charge. An atom with a net charge is called an **ion.** A negatively charged ion has gained an electron and is called an **anion.** By comparison, a positively charged ion has lost an electron and is called a **cation** (**Figure 2.4**). Atoms of the same element also can have different numbers of neutrons. Atoms with the same number of protons but different numbers of neutrons are called **isotopes.** For example, an atom with six protons is a carbon atom, but it could have six or seven or even more neutrons. The number of protons is the one critical feature that defines an atom as being of one particular element.

The traits of the various elements are related to the numbers of subatomic particles that their atoms contain, and so it is useful to list the elements in a way that reflects these traits. The Periodic Table

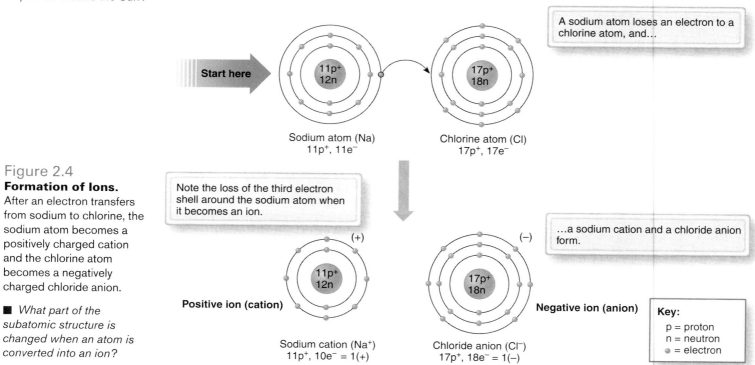

A sodium atom loses an electron to a chlorine atom, and...

Start here

11p+
12n

17p+
18n

Sodium atom (Na)
11p+, 11e−

Chlorine atom (Cl)
17p+, 17e−

Figure 2.4
Formation of Ions.
After an electron transfers from sodium to chlorine, the sodium atom becomes a positively charged cation and the chlorine atom becomes a negatively charged chloride anion.

■ *What part of the subatomic structure is changed when an atom is converted into an ion?*

Note the loss of the third electron shell around the sodium atom when it becomes an ion.

(+)

(−)

...a sodium cation and a chloride anion form.

11p+
12n

17p+
18n

Positive ion (cation)

Negative ion (anion)

Sodium cation (Na+)
11p+, 10e− = 1(+)

Chloride anion (Cl−)
17p+, 18e− = 1(−)

Key:

p = proton
n = neutron
● = electron

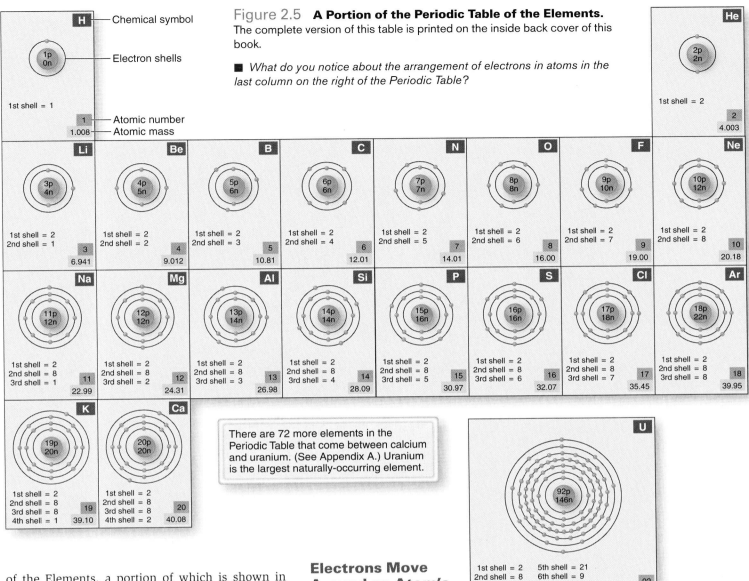

Figure 2.5 **A Portion of the Periodic Table of the Elements.**
The complete version of this table is printed on the inside back cover of this book.

■ *What do you notice about the arrangement of electrons in atoms in the last column on the right of the Periodic Table?*

There are 72 more elements in the Periodic Table that come between calcium and uranium. (See Appendix A.) Uranium is the largest naturally-occurring element.

of the Elements, a portion of which is shown in **Figure 2.5,** presents elements in the order of the number of protons they contain. Each box has the standard abbreviation, or chemical symbol, for that element. The *atomic number* tells how many protons an atom of each element contains; below the atomic number is the *atomic mass,* which is the total number of neutrons and protons. One odd thing you might notice is that most atoms have an atomic mass that is not a whole number. For instance, how can carbon have 12.011 neutrons plus protons? The answer is that atomic mass is the *average* mass of the various isotopes of an element. Carbon, for example, is represented with an atomic mass of 12.011 because it occurs as isotopes with six, seven, or even eight neutrons. The isotope with six neutrons is called C^{12}, or carbon-12, and this is the most abundant isotope of carbon. The other isotopes are C^{13} and C^{14}.

Electrons Move Around an Atom's Nucleus in Distinct Orbitals

All through this book you will read about chemical interactions that produce some of life's most interesting properties, and electrons are key players in these interactions. An understanding of how electrons behave in atoms will help you to visualize these interactions and will be the foundation for much of the information that is to come.

Electrons are always in motion as they move around an atom's nucleus. You may think that electrons "orbit" the nucleus like the planets orbit the Sun—but like many analogies this one is only loosely accurate. The orbits of the planets around the Sun are fairly regular, and you can predict with good accuracy where a particular planet will be at any given time. If this were not possible, NASA

ion an atom (or group of atoms) that carries a net electrical charge

anion a negatively charged ion

cation a positively charged ion

isotope a form of an element defined by the number of neutrons in the nucleus

could not have landed a rocket on Mars to unload the Mars Rover robots. In contrast, the electrons that move around an atom's nucleus do not have such predictable paths. Because you cannot pinpoint where a specific electron will be at a given moment, you could not launch a probe from the nucleus and program it to reliably hit a particular electron. However, you can say where an electron *is likely* to be at any given time. To attach a more scientific term to this concept, the space around the nucleus in which an electron moves—and so is likely to be found—is called an **orbital. Figure 2.6** shows the orbital for the electron of a hydrogen atom. Each dot represents where the electron of a hydrogen atom might be at any given time. Most of the dots cluster around the nucleus, which means that the electron is most likely to be found close to the nucleus. The dots thin out away from the nucleus, and at some distance from the nucleus there are essentially no dots. This whole space is the orbital for the single electron of a hydrogen atom.

This simple picture is fine for hydrogen, but in an atom with more than one electron, there can be more than one orbital. Each orbital has its own distinct shape and distance from the nucleus. How then does an electron get "placed" into a specific orbital? Here is an important point. Each orbital holds just two electrons, and electrons fill orbitals in an order related to the atom's size. The orbitals closest to the nucleus get filled first, and only larger atoms normally have electrons in the orbitals that are farther away from the nucleus. So, for example, the electron in hydrogen is in the first possible orbital around the nucleus. The *two* electrons in the helium atom are both in that first orbital. Lithium is the third element in the Periodic Table; its first orbital is filled with two electrons, and its third electron goes into the *second* possible orbital. This process of filling orbitals

applies to every element in the Periodic Table, giving every atom a defined configuration of electrons around its nucleus.

If you look back at Figure 2.5, though, the drawings of these atoms do not seem to reflect orbitals. Why is that? The answer is that orbitals are organized into groups called *shells.* Each shell contains a defined number of orbitals, and so each shell can hold a defined number of electrons. You see this reflected in Figure 2.5. The first shell has just one orbital, so it holds only two electrons. The second shell has four orbitals, so it holds eight electrons. Atoms with electrons in the second orbital include lithium through neon. The third shell is a bit more complex and contains subshells with nine orbitals and can hold up to 18 electrons. One thing to notice from Figure 2.5 is that *an orbital can be less than full.* For example, hydrogen has only one electron in its shell, even though it could hold two, and carbon has four electrons in the second shell, which could actually hold eight. Spend a minute now to reread the italicized phrase above and let this observation sink in. In sections ahead, when you read about how atoms interact to form more complex chemical structures, the idea that an orbital can be *less than full* will become critically important.

It is essential that you understand one final idea about electrons and their orbits, and this has to do with energy. Chapter 5 provides a deeper discussion of energy, but for now it's enough to say that any object in motion carries energy. Because electrons are in motion around the nucleus, electrons also carry energy. Even though all electrons have the same mass and the same charge, all electrons do not have equal energy. This is not such a strange idea. If your neighbor's four-year-old son throws a baseball at your car, it is not likely to carry enough energy to dent the metal. But if your 200-pound cousin who is a weight lifter throws the same ball at the same car, it easily could carry enough energy to do damage. So the same baseball can carry different amounts of energy. The same holds true for electrons. In this case the factor that predicts how much energy an electron will carry is its distance from the nucleus. Electrons that are farther from the nucleus carry more energy than electrons that are closer to the nucleus.

Internal Electrical Forces Hold Atoms Together

As previously mentioned, electrons are in constant motion around the atom's nucleus. Why don't the electrons just fly off into space? What holds an atom's subatomic particles together? A number of forces are involved, but the **electrical force**

orbital the space around the nucleus where a particular electron is likely to be found

electrical force the force that causes particles with opposite charges to be mutually attracted, and the force that causes particles with a similar charge to repel one another

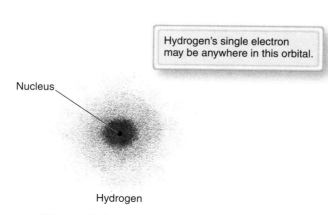

Hydrogen's single electron may be anywhere in this orbital.

Nucleus

Hydrogen

Figure 2.6 **Orbital of the Single Electron of Hydrogen.** The dots show the positions where the electron could be found at any given moment.

between positively charged protons and negatively charged electrons is one important cohesive factor. Electrical force causes particles carrying opposite electrical charges to be attracted to each other and causes particles carrying the same electrical charge to repel each other. In an atom the attractive force between the positive charge on the protons and the negative charge on the electrons keeps the electrons circling the nucleus, rather than flying off into space. In most atoms the numbers of protons and electrons are usually equal. Therefore, even though an atom is made up of smaller particles that have positive and negative charges, the charges balance one another. As a consequence, a typical atom—unless it is an ion—has no net electrical charge.

QUICK CHECK

1. What is an element?
2. What is an ion?

2.3 Bonds Hold Atoms Together in More Complex Structures

Now that you have a grasp of basic atomic structure, it's time to take the next big step and consider how atoms can come together to form more complex structures. When atoms of different elements join together, the resulting substance is called a **compound. Chemical bonds** are the "glue" that holds the atoms of a compound together. Bonds can form between two atoms of the same element or two atoms of different elements. What is a bond? There are different kinds of chemical bonds,

but in each case a *bond* is an interaction between the positively charged protons in the nucleus of one atom and the negatively charged electrons around the nucleus of another atom. When the protons of one atom attract the electrons of another atom, this electrical interaction holds the two atoms together.

The simplest bonds to understand are those that form between two oppositely charged ions. Sodium (Na) and chlorine (Cl) readily form ions. A sodium ion (Na^+) and a chloride ion (Cl^-) carry a positive charge and a negative charge, respectively. The opposite charges of the two ions are mutually attractive, and they hold the ions together. The bond that forms between two oppositely charged ions is called an **ionic bond.** The result of the formation of such an ionic bond is an ionic compound—in this case, sodium chloride, or common table salt (NaCl) (**Figure 2.7**). Large numbers of sodium and chloride ions can be held together by ionic bonds, forming a crystal or lattice structure.

Atoms that carry no electrical charge also can form bonds with each other, again based on electrical interactions. How can this be? When atoms get close together, the negatively charged electrons usually repel one another. But if two atoms get close enough, which usually requires the input of some energy, the protons in the nucleus of each atom will be attracted to the electrons of the other. When this happens, some of the electrons that are farthest away from the nucleus actually form a new orbital and surround *both* nuclei. This means that the two atoms *share* these outer electrons. The shared electrons form a bond between the two atoms and hold them together. A bond that involves shared electrons between two electrically

compound a substance made of the atoms of two or more elements

chemical bond the electrical interactions that hold two atoms together

ionic bond a bond that forms when a pair of oppositely charged ions are attracted to each other and held together by electrical forces that result from a transfer or gain or loss of an electron

A sodium atom loses an electron to a chlorine atom.

The sodium cation and chloride anion are attracted to each other and form an ionic bond.

The sodium and chloride ions held together by ionic bonds form a salt crystal.

Ionic bond

(+)   (−)

11p+
12n

17p+
18n

11p+
12n

17p+
18n

Sodium atom   Chlorine atom

Sodium cation (Na^+) + Chloride anion (Cl^-)

Na   +   Cl   →   NaCl

Na^+ Cl^-

Figure 2.7 **Structure of a Salt Crystal.** The electrical attraction between oppositely charged ions forms an ionic bond and joins the two ions in an ionic compound.

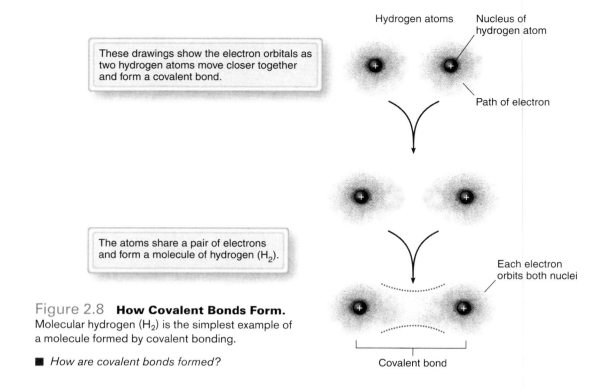

These drawings show the electron orbitals as two hydrogen atoms move closer together and form a covalent bond.

Hydrogen atoms

Nucleus of hydrogen atom

Path of electron

The atoms share a pair of electrons and form a molecule of hydrogen (H_2).

Each electron orbits both nuclei

Covalent bond

Figure 2.8 How Covalent Bonds Form.
Molecular hydrogen (H_2) is the simplest example of a molecule formed by covalent bonding.

■ *How are covalent bonds formed?*

Table 2.1	The Chemical Structures of Some Familiar Molecules	
Substance	**Space-filling model of the molecule**	**Structural formula showing chemical bonds**
Water (H_2O)		H—O—H Single covalent bond
Carbon dioxide (CO_2)		O=C=O Double covalent bond
Methane (CH_4)		
Sucrose ($C_{12}H_{22}O_{11}$)		

neutral atoms is called a **covalent bond. Figure 2.8** shows what a covalent bond between two hydrogen atoms might look like. Depending on the number of electrons in their outer shells, atoms can be involved in one or more covalent bonds.

When atoms combine through covalent chemical bonds, the resulting structure is called a **molecule.** Molecules can have any number of atoms, from just two to millions, and these atoms can be of the same or of different elements. **Table 2.1** diagrams some simple molecules. Most covalent bonds involve the sharing of two electrons, one from each atom, and this is called a *single bond.* Other bonds involve more than one pair of shared electrons. For instance, when two oxygen atoms bond with one carbon atom, each oxygen atom shares two electrons with the carbon atom, and the carbon atom contributes two electrons to the bond with each oxygen atom. Most covalent bonds involve one electron from each atom, but in carbon dioxide, as this molecule is called, each bond involves two electrons from each atom. This is called a *double bond.* In Table 2.1 the lines that connect the atoms represent bonds. A single line represents a single bond, while a double line represents a double bond. Three lines would represent a triple bond.

Some Atoms Form Molecules More Readily than Others Do

The next important idea that explains the formation of molecules is that some elements form bonds readily, while others do not. This limits the range of elements that molecules and compounds are made of—and has an impact on the range of elements that make up living things. For example, the elements helium, neon, argon, krypton, xenon, and radon are called *noble gases* because they don't form molecules or ionic compounds, and these elements are not found in the molecules of living things. Other atoms—including hydrogen, oxygen, and carbon—readily form bonds that produce molecules, which is one reason these are the major elements found in organisms.

One feature that helps to explain how readily atoms of a given element will form bonds is the number of electrons in the outer shell. Recall that each electron shell can hold some maximum number of electrons. Atoms are less likely to form chemical bonds with other atoms if their outermost shells are full. For example, both hydrogen and helium have a single shell that contains one orbital that can hold up to two electrons. In hydrogen this shell contains only one electron, while in helium the orbital contains its maximum number of electrons: two.

Thus helium atoms do not form bonds with other atoms. In contrast, hydrogen readily forms molecules, and most hydrogen atoms are bonded to other atoms. This same logic can be applied to other elements in the Periodic Table. For example, lithium has two electron shells, but its second shell contains only one electron of the eight that it could carry. Therefore, lithium readily forms bonds with other atoms, especially ionic bonds. Carbon is another good example because its second shell contains four electrons of the eight that it could carry. As you will read in Chapter 3, this allows a carbon atom to share as many as four electrons with other atoms.

New Properties Emerge When Atoms Combine to Form Molecules

Now you are ready to approach the idea of emergent properties of molecules and compounds. One important thing about molecules and compounds is that, like LEGO constructions, they often have properties dramatically different from the elements that compose them. Let's revisit the example of table salt (NaCl). Even though a salt crystal is made of just two elements, its properties are far different from those of either sodium or chlorine.

Elemental sodium is a highly chemically active metal that is as soft as cheddar cheese (**Figure 2.9**).

covalent bond a bond between two neutral atoms that involves the sharing of two or more electrons

molecule a chemical structure that results when two or more atoms combine through covalent chemical bonds

Elemental sodium is a soft metal.

Elemental chlorine is a yellowish-green gas.

Sodium atoms and chlorine atoms form sodium chloride or table salt.

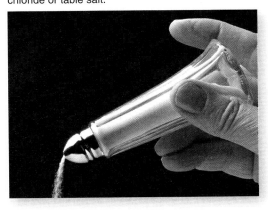

Figure 2.9 **Elemental Sodium Is Different from Elemental Chlorine.** None of the properties of these elements is seen in the ionic compound they make: table salt.

If you put a crumb of sodium into water, it will float for a moment and then may explode as the sodium reacts with the water and air. By contrast, elemental chlorine is an extremely toxic, greenish-yellow gas that is heavier than air. Chlorine gas became infamous in World War I, when it was used as a weapon to kill thousands of soldiers during trench warfare. You are probably familiar with chlorine because it is the active ingredient in household bleaches and in many swimming pool germicides. Even though table salt contains both sodium and chlorine, it has none of the properties of these elements. At room temperature table salt is a white, crystalline solid with a distinctive taste. As you can see, when sodium and chlorine combine, a substance with new properties emerges.

Table sugar or sucrose (its chemical name) is another good example of how different a molecule can be from the chemical elements that compose it—in this case, carbon, hydrogen, and oxygen. Pure carbon is a solid, nonmetallic element that can form coal or diamonds. Hydrogen and oxygen are both reactive gases. When carbon, hydrogen, and oxygen atoms come together to form a sucrose mol-

ecule, distinct properties emerge: sucrose is a solid crystal, it dissolves readily in water, and it tastes sweet. The complex properties of sucrose are related to the precise way that carbon, oxygen, and hydrogen atoms bond to form the sucrose molecule (**Figure 2.10**). A pile of carbon, oxygen, and hydrogen would not be sucrose, and it would have none of its properties.

Because there are 92 different kinds of naturally occurring atoms, it might seem impossible to understand how all of the molecules found in living organisms are formed. The situation is manageable, however, because atoms of only five elements—oxygen, carbon, hydrogen, nitrogen, and phosphorus—make up almost 99% of organisms, including humans, and only 19 other elements are commonly found as the remaining 1% or so (**Table 2.2**). Collagen, the most abundant protein complex in your body, is an example of a biological molecule made from just a few kinds of elements (**Figure 2.11**). (You will read more about proteins in Chapter 3.) Carbon, oxygen, hydrogen, nitrogen, and a little sulfur are the only kinds of atoms in collagen, but their precise spatial arrangement gives the collagen

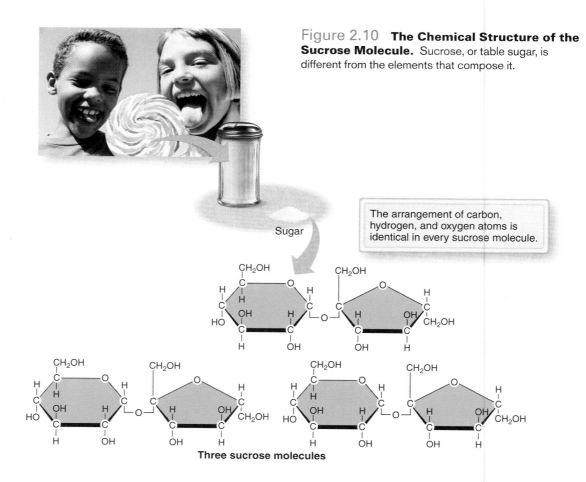

Figure 2.10   **The Chemical Structure of the Sucrose Molecule.** Sucrose, or table sugar, is different from the elements that compose it.

The arrangement of carbon, hydrogen, and oxygen atoms is identical in every sucrose molecule.

Sugar

Three sucrose molecules

Table 2.2 Major Chemical Elements in the Human Body: Amounts, Functions, and Sources

Element	% of body	Role in human body	Sources
Oxygen (O)	65.0	Component of molecules; necessary for cells to produce energy	Atmosphere; water in the process of photosynthesis
Carbon (C)	18.5	Major component of organic molecules	Foods
Hydrogen (H)	10.0	Component of molecules; makes important bonds between and within molecules	Foods, water
Nitrogen (N)	3.0	Component of proteins and nucleotides	Proteins in meats, seafoods, dairy products, whole and enriched grains
Calcium (Ca)	1.2	99% found in bones and teeth, remainder in body fluids; needed for muscle contraction, transmission of nerve impulses, and blood clotting	Dairy products such as milk, cheese, ice cream
Phosphorus (P)	1.1	Component of ATP, a compound that is a major energy source for cells, and of nucleic acids	Meat, eggs, dairy products
Sulfur (Su)	0.25	Component of muscle proteins	Meat, seafood, eggs, dairy products, plant proteins
Potassium (K)	0.35	Present in body fluids; regulates water balance; principal positively charged ion; necessary for muscle contraction and transmission of nerve impulses	Many fruits and vegetables
Sodium (Na)	0.15	Present in body fluids; influences the amount of water present in blood and so influences blood pressure; necessary for muscle contraction and transmission of nerve impulses	Table salt (sodium chloride) and in many processed foods
Chlorine (Cl)	0.15	Present in body fluids; functions in water balance; necessary for muscle contraction and transmission of nerve impulses	Table salt (sodium chloride)
Magnesium (Mg)	0.05	50% of body's magnesium is in bones, where it contributes to bone strength	Leafy green vegetables
Iodine (I)	0.004	Necessary for thyroid gland function	Seafood, iodized table salt
Iron (Fe)	0.0004	In red blood cells, enables hemoglobin to bind oxygen molecules	Beef, liver, whole-grain and enriched cereal products, dried fruits, egg yolk

Copper (Cu), manganese (Mn), molybdenum (Mo), selenium (Se), silicon (Si), and zinc (Zn) are also needed in tiny amounts.

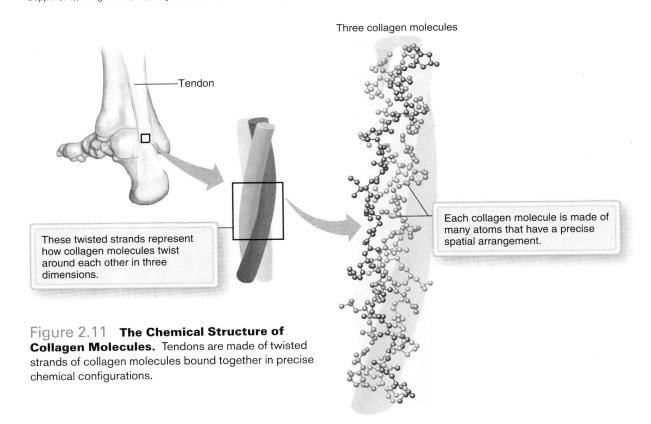

Three collagen molecules

Tendon

These twisted strands represent how collagen molecules twist around each other in three dimensions.

Each collagen molecule is made of many atoms that have a precise spatial arrangement.

Figure 2.11 **The Chemical Structure of Collagen Molecules.** Tendons are made of twisted strands of collagen molecules bound together in precise chemical configurations.

molecule many distinctive characteristics. In organisms, collagen molecules are arranged in long strands. Collagen's resilient fibers are key components of the connective tissues that bind tissues and organs together. For instance, collagen fibers in tendons connect muscles to bones, and collagen fibers in ligaments bind bones together. Other collagen fibers form the flexible parts of nose and ears and help joints to move smoothly. Collagen is just one example of how complex the molecules of living organisms can be.

Hydrogen Bonds Can Form Between Molecules

Covalent bonds are responsible for the formation of complex molecules, but the complexities of molecular structures do not end there. Other kinds of bonds allow different parts of large molecules or completely separate molecules to form even more complex structures. Like covalent and ionic bonds, these other types of bonds are based on electrical interactions. One important type of bond is a **hydrogen bond** that forms between chemical groups that carry a partial electrical charge, usually between hydrogen atoms on one chemical group and an oxygen atom on another chemical group. Hydrogen bonds are weaker than covalent bonds. Hydrogen bonds and other weak electrical interactions allow molecules to assemble and form the complex internal structures of cells.

For a better understanding of hydrogen bonds, let's look more closely at water, a common molecule important to all living organisms. A water molecule (H_2O) consists of one oxygen atom (O) covalently bonded to two hydrogen atoms (H_2) (**Figure 2.12**). The bonds form at particular angles, so that the shape of the entire water molecule resembles the head of a cartoon mouse. The large oxygen atom is the mouse's head and the small hydrogen atoms are its ears.

The most important feature of a water molecule is that certain regions have a partial electrical charge. While a water molecule has no *overall* charge, it does have a region with a slight negative charge and a region with a slight positive charge (see Figure 2.12). These charged regions occur because the large, positively charged nucleus of the

Figure 2.12 **The Chemical Structure of a Water Molecule.**

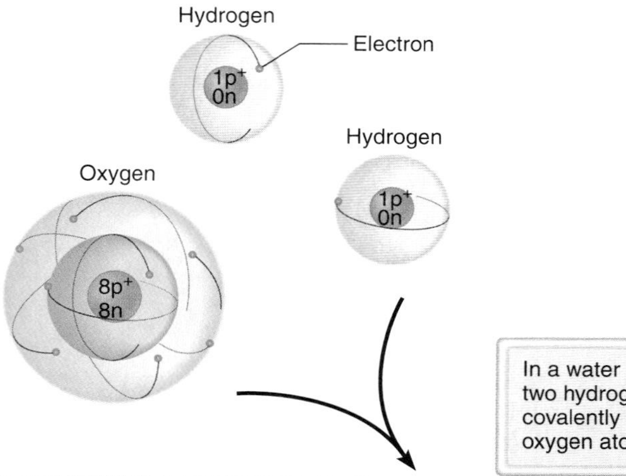

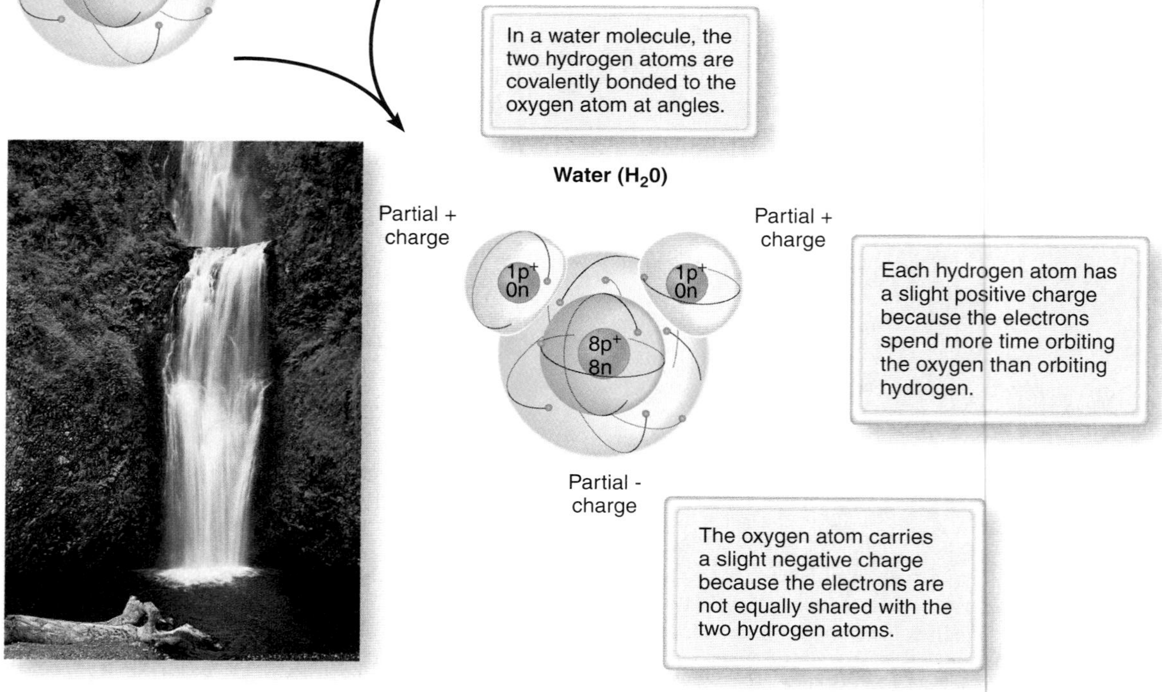

hydrogen bond a weak temporary chemical bond that forms between a partially positively charged hydrogen atom in one polar molecule or chemical group and a partially negatively charged atom in another nearby molecule or chemical group

oxygen atom pulls more strongly on the surrounding electrons than does the tiny nucleus of either hydrogen atom. So the orbiting electrons spend more time close to the oxygen nucleus, leaving the hydrogen nuclei shy of electrons. The bonds within water are still covalent bonds—but their electrons are drawn more toward the oxygen atom, so they are called **polar covalent bonds.** Because of the distribution of electrons, the region around the oxygen nucleus of a water molecule has a slight negative charge, while the region around the hydrogen atoms has a slight positive charge. If the image of the head of a cartoon mouse helps you to remember the structure of the water molecule, think of this: the chin of the mouse has a partial negative charge, while each ear has a partial positive charge. This uneven charge distribution means that water is a **polar molecule.**

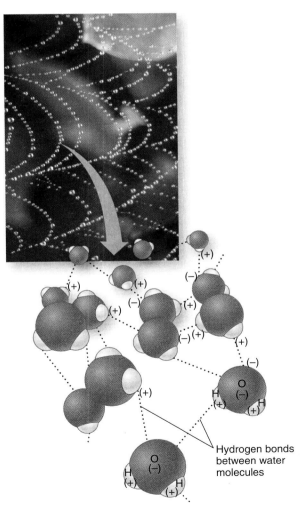

Figure 2.13   **Hydrogen Bonds Hold Water Molecules Together.**

Now you can understand the formation of hydrogen bonds between water molecules. Recall that negative charge attracts positive charge. Imagine that a group of water molecules are in a beaker on your desk (refer to **Figure 2.13** to help you visualize how they interact). The slight negative charge around the oxygen nucleus of one water molecule is attracted to the slight positive charge on a hydrogen atom of an adjacent water molecule. This attraction is a hydrogen bond, and it holds water molecules together. Hydrogen bonds usually are represented by dashed or dotted lines to indicate that hydrogen bonds in water are weak, temporary bonds. As you will see, the polarization of its electrical charge gives the water molecule emergent properties crucial to life.

Chemical Bonds Can Be Broken with an Input of Energy

The bonds that hold atoms and molecules together give any molecular structure its stability. At the same time these bonds can be broken; if this happens, molecules or more complex structures will fall apart. Energy is required to break these chemical bonds. For example, when you cook food, chemical bonds are broken and the food becomes softer and easier to chew. If you burn food, most of the chemical bonds will be broken and you'll be left with char or ashes. If the bonds are strong, then the structure will be hard to break apart; if the bonds are weak, the structure will be easier to break apart. Covalent and ionic bonds are relatively strong, whereas hydrogen bonds are relatively weak. This means, for example, that it takes less energy to break the hydrogen bonds that hold different water molecules together than it takes to break the covalent bonds of a single water molecule.

Atoms Are Rearranged in Chemical Reactions That Can Transform One Substance into Another

One of the important lessons from chemistry is that basic elements cannot, in most cases, be changed from one to another. On the other hand, atoms and molecules can combine or be rearranged to form different substances. These interactions are called **chemical reactions.** In any chemical reaction the substances that interact are called the **reactants**, and the substances that are produced are called the **products.** A discussion of water molecules will help you to understand chemical reactions. Each water molecule is composed of two hydrogen atoms and one oxygen atom. Hydrogen and oxygen themselves are found as molecules, each made of

polar covalent bond
a covalent bond in which the shared electrons are pulled more toward one atom involved in the bond than the other

polar molecule
a molecule with regions of negative and positive charge but no net charge across the whole molecule

chemical reaction
an interaction between atoms or molecules that transforms substances from one kind into another through making and breaking chemical bonds

reactants the substances that interact to produce a different substance during a chemical reaction

products the substances that are produced in a chemical reaction

two atoms: H_2 and O_2. The chemical reaction in which water is formed from hydrogen and oxygen can be written like this:

$$2H_2 + O_2 \rightarrow 2H_2O$$

This equation tells you that two molecules of hydrogen (H_2) and one molecule of oxygen (O_2) interact to produce two molecules of water ($2H_2O$). Hydrogen and oxygen molecules are the reactants and water is the product. During this chemical reaction the chemical bonds between the two atoms in H_2 and in O_2 are broken, and the chemical bonds within H_2O are formed.

The chemical interactions inside of cells involve thousands of different kinds of chemical reactions. Each involves reactants that are transformed into products. In some cases the reaction produces products that are simpler than the reactants. Larger molecules can be broken down into smaller ones in chemical reactions. In all cases chemical reactions transform substances and result in a product that is different from the reactants.

This discussion of atoms, bonds, and molecules should allow you to connect chemistry more closely to biology. Now you can appreciate that the processes of life are the results of chemical interactions.

QUICK CHECK

1. What is the difference between a covalent bond and a hydrogen bond?

2. What is a polar covalent bond?

2.4 The Chemicals of Life Interact in Water

Have you ever thought about the conditions that promote life? If you were searching for life on other planets, what single feature would be the best clue to its possible presence? One of the primary requirements for life is liquid water—without water there could be no life as we know it. Before you can appreciate why water is so important to life, you must understand some of the unusual properties of the water molecule.

Water Is an Unusual Molecule

Several of water's properties are listed in **Table 2.3**. Each of these properties plays an important role in supporting life's processes. First, let's describe each property and then consider why these properties are important for life.

Water is the only common substance on Earth that exists in all three physical forms at moderate temperatures—as a solid (ice), a liquid, and a gas (water vapor). While water is liquid over a wide range of temperatures (from 0°C to 100°C [32°F to 212°F]), it also readily forms a solid or a gas at temperatures commonly found on Earth. You may have experienced all of these at once on an early spring day when the air is misty and a mixture of rain, snow, and sleet is falling (**Figure 2.14**). But most important for this discussion is the fact that water is a liquid at the common temperatures found in many of Earth's environments. For instance, the oceans that cover most of Earth's

---

Table 2.3 Important Properties of Water		
Property of water	**Explanation**	**Importance to biological organisms**
Water is a liquid at moderate temperatures (0°–100°C [32°–212°F]).	Hydrogen bonds make, break and re-form readily in liquid water.	Water is a medium in which other chemicals can readily interact.
Water can hold a lot of heat relative to its volume.	Water's hydrogen bonds can hold a lot of heat.	Because water resists temperature changes, it moderates temperatures within organisms and within the biosphere.
Water molecules cohere.	Hydrogen bonds hold water molecules together, creating surface tension.	Lightweight organisms are supported on water's surface.
Water can dissolve many other substances.	Water is a polar molecule.	Dissolved substances can interact in liquid water; a wide array of dissolved substances are available to living organisms.
Water clings to many other surfaces, wetting them.	Water molecules form hydrogen bonds with other substances at moderate temperatures.	Water wets surfaces and will climb in small tubes such as those that conduct water within plants.

surface are not frozen, nor are they gaseous; they are liquid.

Water resists temperature changes and can hold a lot of heat relative to its volume without changing temperature. As an example, consider water in a metal cooking pot. In response to a given amount of energy, such as that provided by the heating element on your stove, the water will increase in temperature less than will the metal pot. This is not because the pot is closest to the heating element. Rather, it is because the metal translates energy into temperature change more readily than water does. This property of water is important for keeping temperatures on Earth stable. Without large bodies of water that capture and store heat energy, Earth would have greater fluctuations in temperature between night and day and between winter and summer.

When you turn on your kitchen faucet, the water flows because water molecules stick together. This *cohesion* of water molecules produces the third interesting property of water called *surface tension*. At the interface where water meets air, the water sticks to itself more strongly than it does to the air above it. This produces a kind of molecular "skin" at the air-water interface. You have observed surface tension if you've watched an insect scoot across the surface of a pond or a fallen leaf drift downstream (**Figure 2.15**). Water's surface tension prevents the insect or the leaf from falling into the water.

The last property of water to consider is its ability to dissolve a wide variety of other substances. Water can dissolve so many substances that it often is called the *universal solvent*. Think of the familiar things that dissolve in water: the sugar in your morning coffee, the boiled vegetables and meat that release their "nutrients" to make a broth, or a peppermint candy that dissolves in the watery saliva within your mouth. Of course, not everything dissolves in water. Oils and fats do not mix with water, and even this property is a reason why water is required for life.

The ability of water to dissolve so many different chemicals helps you to understand that water is an ideal environment for life. In watery environments specific ions and polar molecules meet, combine, and form new molecules. This kind of complex molecular dance of chemical reactions happens inside of cells and produces life. The molecular dance could not happen without water.

To fully comprehend how water allows chemical interactions, you should know how water comes to have its unique properties. The next few sections will give you a deeper understanding of

Figure 2.14 **Snow, Steam, and Water.** Sometimes three forms of water are present simultaneously.

Figure 2.15 **Skating on Water's Surface.** Surface tension due to the cohesiveness of water molecules prevents this insect from breaking the water's surface.

water at a molecular level. You will discover that these important features of water are emergent properties that arise from its molecular structure.

The Polar Structure of Water Produces Its Unusual Properties

The special properties of water are based on the polar nature of water molecules and their tendency to form hydrogen bonds. One of the most important things about a hydrogen bond is that it is weaker than a covalent bond or an ionic bond. This means that hydrogen bonds are made and broken easily, which has a big impact on the properties of water. At moderate temperatures (between 0° and 100°C [32° to 212°F]) hydrogen bonds will hold most of the water molecules together, causing them to form a pool of liquid water in a beaker. If the beaker is heated, the water molecules will move around more. This breaks hydrogen bonds

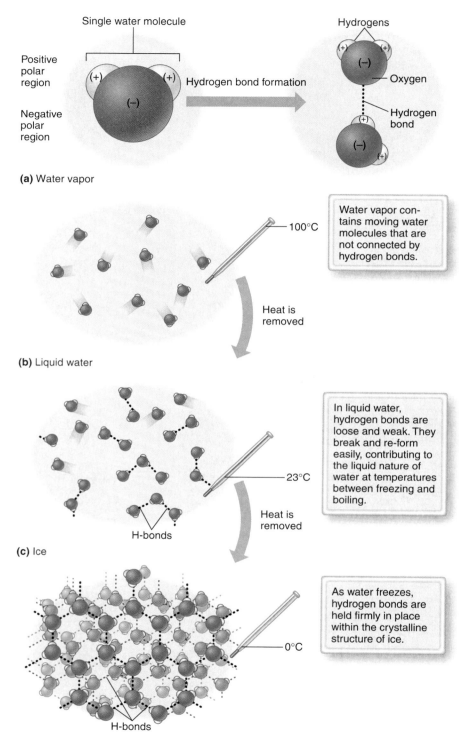

(a) Water vapor

Water vapor contains moving water molecules that are not connected by hydrogen bonds.

(b) Liquid water

In liquid water, hydrogen bonds are loose and weak. They break and re-form easily, contributing to the liquid nature of water at temperatures between freezing and boiling.

(c) Ice

As water freezes, hydrogen bonds are held firmly in place within the crystalline structure of ice.

Figure 2.16   **Hydrogen Bonds and the Arrangement of Water Molecules in the Three States of Water.**

■ *Predict what would happen to the hydrogen bonds as the water approaches the boiling point. The freezing point.*

and allows the water molecules to pull away from each other. If the water gets hot enough, it will boil and turn to water vapor, a gas in which the water molecules have separated and are dispersed in the air (**Figure 2.16a**). If you put the beaker into a freezer, heat is removed and as the water cools, the water molecules will move around less and less, allowing them to form more stable hydrogen bonds (Figure 2.16b). Eventually they move so little that they crystallize into ice (Figure 2.16c).

Water's unique properties largely can be accounted for by the polarity of water molecules and the resulting hydrogen bonds that form between them. For instance, water flows as a cohesive liquid because at normal room temperatures, hydrogen bonds between the polar water molecules easily form, break, and re-form. Surface tension and water's ability to absorb heat are also due to the nature of hydrogen bonds. The polarity of water molecules explains why so many substances can dissolve in it. In fact, any atom, molecule, or compound that is also polar (such as alcohol) or composed of ions (like salt) will dissolve in water. To understand water's important role as a universal solvent more fully, let's reexamine a relatively simple situation: the events that occur when you add salt to water.

You have seen that table salt is a crystal in which large numbers of sodium and chloride ions are linked together by ionic bonds. The repeating arrangement of the two ions gives the crystal its regular shape. When salt crystals are placed in water (**Figure 2.17**), the negatively charged oxygen atoms of water molecules are attracted to positively charged sodium ions, while the positively charged hydrogen atoms of water molecules are attracted to negatively charged chloride ions. As a result, the sodium and chloride ions are pulled away from each other and are dispersed throughout the water.

Of course, not all substances dissolve in water. For example, oils and pure hydrocarbons such as gasoline do not mix with water. These substances do not dissolve in water because their molecules are **nonpolar** substances (**Figure 2.18**). Unlike a water molecule, a nonpolar molecule has an even distribution of charge. There are no regions of positive and negative charges on a nonpolar molecule. A nonpolar or uncharged substance will not dissolve in water and will tend to remain separate from water even if the mixture is stirred, shaken, or heated. Scientists use the words *hydrophilic* and *hydrophobic*, respectively, to refer to substances that do and do not dissolve in water. Hydrophilic substances are "drawn to" water (hydrophilic means "water loving"), while hydrophobic substances "avoid" water (hydrophobic means "water hating"). But why is this ability to act as a solvent important for life?

Chemical Interactions Within Living Organisms Involve Molecules and Atoms Dissolved in Water

The internal environment of a living organism is an active place where molecules, ions, and atoms continually interact. For instance, within the cells of your body chemical interactions release energy that, in turn, is used in other chemical interactions. Your muscles move parts of your body when chemicals within the muscles interact, causing the muscles to shorten. Certain molecules in your liver interact with toxic substances that you have eaten, drunk, or breathed in and render them nontoxic. For now, the point is that all of these chemical interactions take place *in water*. None of these essential processes could take place if these biological molecules were not dissolved in the right amounts in a watery environment. A heap of table sugar that you've spilled on the kitchen counter is not alive and is not a part of life. But table sugar and other molecules can contribute to a living organism when they are dissolved—in the right amounts—in water. Water creates an environment where polar and charged atoms and molecules can move around and interact. These chemical interactions produce the basic processes of living organisms. So, if there were not a polar solvent, like water, life as we know it would not be possible.

Life's Chemical Interactions Cannot Occur in an Environment That Is Too Acidic or Too Basic

The chemical interactions that go on inside a cell are sensitive to the total chemical environment. If the internal chemistry changes in certain ways, life's delicately balanced chemical interactions will become unbalanced, and consequently the organism can get sick or die. One critical aspect of the internal environment of life is its level of acidity. You know acidic substances from your everyday experiences. For instance, vinegar and lemon juice are highly acidic. In everyday language the opposite of acidic is basic or alkaline. In a chemical sense the baking soda you put into cake batter is a basic substance, as are ammonia, bleach, and drain cleaner. Pure water is neither basic nor acidic, but is neutral. Living things generally have a watery internal environment just slightly to the

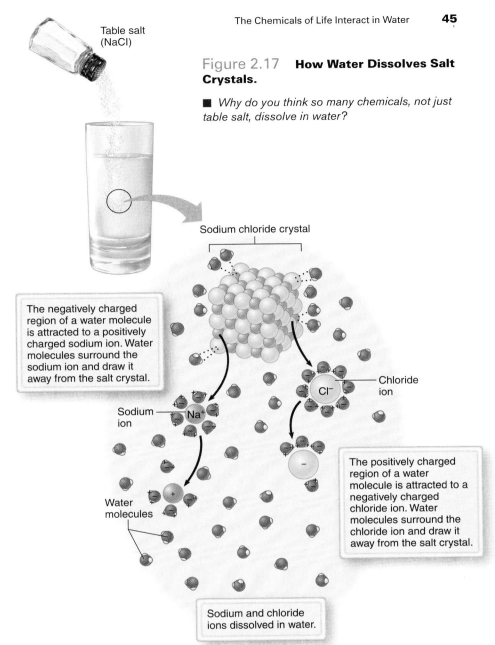

Table salt (NaCl)

Figure 2.17 **How Water Dissolves Salt Crystals.**

■ *Why do you think so many chemicals, not just table salt, dissolve in water?*

Sodium chloride crystal

The negatively charged region of a water molecule is attracted to a positively charged sodium ion. Water molecules surround the sodium ion and draw it away from the salt crystal.

Chloride ion

Sodium ion

Na⁺

Cl⁻

The positively charged region of a water molecule is attracted to a negatively charged chloride ion. Water molecules surround the chloride ion and draw it away from the salt crystal.

Water molecules

Sodium and chloride ions dissolved in water.

basic side of neutral. All of this is informative, but what makes something acidic or basic?

The first step in understanding acids and bases is to return to the nature of water. As you know, plain, pure water is made of H_2O molecules. The situation is more complex, though, because in any volume of water a few of the water molecules are *not* complete. This is because the strong electrical interactions in water cause a few water molecules to lose hydrogen ions (H^+). Think about what a hydrogen ion really is: a hydrogen atom without its electron is just a naked proton. So if a water molecule loses a H^+ ion, what is left of the original water molecule? The answer is a hydroxyl group

nonpolar the property of having an even charge distribution across the atoms of a molecule

Figure 2.18 **Spatial Arrangements of Atoms and Electrons in Polar (a) and Nonpolar (b and c) Molecules.**

(a) Water is a polar molecule. Polar molecules have regions of slight negative and slight positive charges because the distributions of their electrons are unbalanced.

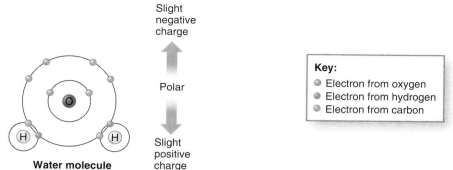

Slight negative charge

Polar

Slight positive charge

Water molecule

Key:
- Electron from oxygen
- Electron from hydrogen
- Electron from carbon

(b) Methane and carbon dioxide are nonpolar molecules. Nonpolar molecules are uncharged because the distributions of their electrons are balanced.

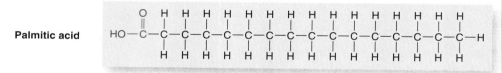

Methane molecule

Carbon dioxide molecule

(c) Palmitic acid is a nonpolar molecule. The balanced arrangement of atoms in this nonpolar molecule reflects the even distribution of its electrons.

Palmitic acid

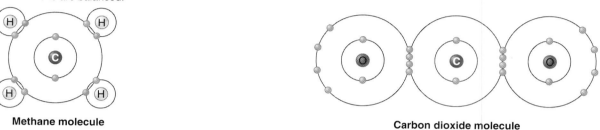

(OH) that has an extra electron. This extra electron makes it a hydroxyl ion (OH⁻).

In plain water the number of H^+ ions floating around is balanced by the OH^- ions, but when certain substances are dissolved in water, the concentrations of these ions can change. For example, hydrochloric acid (HCl) molecules will dissociate—come apart—in water to form H^+ ions and Cl^- ions; this dramatically increases the concentration of H^+ ions as compared with plain water (**Figure 2.19**). A solution with a higher concentration of H^+ ions than plain water is an **acid.** Other substances will effectively lower the H^+ ion concentration and increase the OH^- concentration of water. An example of a substance that does this is sodium hydroxide (NaOH), the active ingredient in oven or drain cleaners. In water NaOH dissociates to form Na^+ and OH^-. The extra OH^- ions not only increase the concentration of that ion, they can join with free H^+ ions to form water molecules, reducing the H^+ ion

concentration. A solution with a lower concentration of H^+ ions than plain water is a **base.**

The **pH** scale is a convenient way to describe how acidic or basic a solution is and ranges from 0 to 14, as shown in **Figure 2.20.** The most acidic solution measurable, which has the highest concentration of H^+ ions, is given a pH of 0. A neutral solution like pure water has the same concentration of H^+ and OH^- ions and has a pH of 7. The most basic solution measurable on the pH scale has the lowest concentration of H^+ ions, with a pH of 14. Each step toward the top end of the scale has a concentration of H^+ ions that is 10 times lower than the step below it.

The pH of a cell or its environment is an extremely important factor because the chemical reactions that life depends on occur only within a narrow pH range. pH is so important that all organisms, including you, have complex biochemical mechanisms that keep the pH of internal systems close to 7.4. If the pH of your internal body fluids

acid a solution with a higher concentration of H^+ ions than plain water

base a solution with a lower concentration of H^+ ions than plain water

pH how acidic or basic a substance is

Plain water has equal numbers of hydrogen and hydroxyl ions. Note that some H₂O molecules are incomplete.

When hydrogen chloride (HCl) is added to water, the HCl molecules split.

As the HCl molecules split, the number of hydrogen ions in the solution increases. These ions decrease the pH and make the water more acidic.

Key:
- Hydrogen ion (H⁺)
- Water molecule
- Hydroxyl ion (OH⁻)
- HCl (hydrogen chloride)
- Chloride ion (Cl⁻)
- Molecule splits

Figure 2.19   **How Water Becomes Acidified.**

rises to 7.8 or drops to 7.0, you will die. These sound like small differences in pH, but they actually are dramatic fluctuations in H^+ ion concentrations. Each step on the pH scale is actually 10 times more acidic than the step above it. For example, a solution with a pH of 1 has 10 times more H^+ ions than a solution with a pH of 2. Maintenance of the proper pH is critical to the survival of any organism.

QUICK CHECK

1. Describe four properties of water that are important for living things.
2. How are solutions with a pH of 2 and a pH of 10 different?

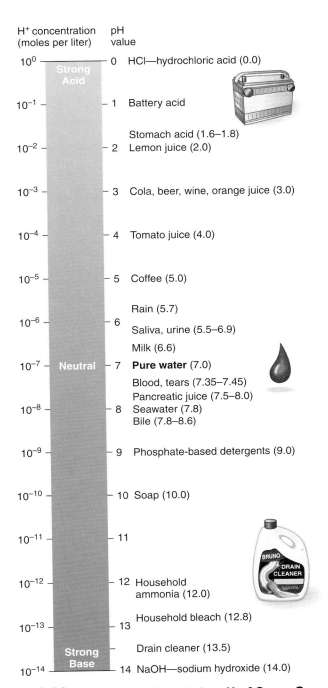

Figure 2.20   **The pH Scale and the pH of Some Common Substances.** The pH ranges from 0 to 14. The low numbers represent acids; pH 7 is the neutral point; and numbers above 7 represent bases. Notice that dangerous, corrosive substances are at both extreme ends of the pH scale.

2.5   An Added Dimension Explained
The Science of Cryopreservation

This chapter opened with questions about what happens to life at cold temperatures. The answers to these questions are related to the properties of water. Living organisms are made largely of water. So the success of cryopreservation depends on the ability to freeze and melt the water in living tissues while maintaining the chemical interactions that sustain life. You will have to learn a bit more about water to appreciate the problems associated with doing this.

Think back to Section 2.4 and recall that the state of water—whether it is a gas (water vapor), a liquid, or a solid (ice)—depends on how many hydrogen bonds, on average, each molecule of water has formed. In liquid water each water molecule is bonded, on average, to approximately three others, and the water molecules slide past one another as hydrogen bonds are made and broken. This allows water molecules to pack fairly tightly together. In other words, in liquid water the molecules are dense (**Figure 2.21a**). When water freezes, nearly all of the water molecules are locked into a rigid crystal structure in which each water molecule is bonded to four

Figure 2.21   Ice Damages Cells.
When water freezes, dissolved substances are displaced and may damage cellular structures.

(a) Pure liquid water

Pure water has no other substances dissolved in it. Hydrogen bonds are temporary. They form, break, and re-form.

Hydrogen bonds

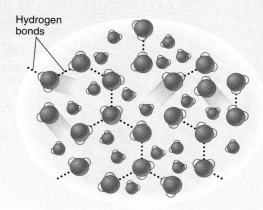

(b) Ice from pure water

Ice that forms from pure water has a crystalline structure, with each water molecule bonded to four other molecules. Note the empty spaces that make ice less dense than liquid water.

Hydrogen bonds

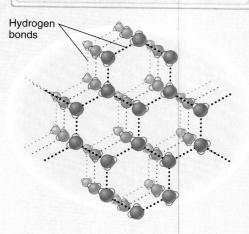

(c) Water within cells

Water within a cell has many other chemicals dissolved in it. Within and around the cell are delicate, membranous structures.

Dissolved chemicals

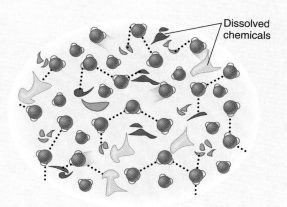

(d) Ice formation within cells

When ice crystals form within a cell, the dissolved substances are pushed out of solution by the formation of hydrogen bonds between water molecules. Because ice crystals are sharp, they also are likely to damage cellular structures.

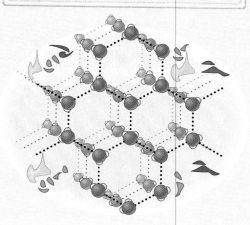

others. Look carefully at this crystal structure (Figure 2.21b). Notice that ice has empty spaces, and it is actually less dense than liquid water. On average, the water molecules in ice are farther apart from one another than are the molecules in liquid water. This explains why water expands when it freezes, and why ice—whether it is an iceberg or an ice cube—floats on top of denser liquid water.

These aspects of water and ice have important implications for what happens when organisms experience cold temperatures. First, recall that the interior of cells contains dissolved atoms and molecules held to water molecules by hydrogen bonds (Figure 2.21c). When a cell is cooled, the hydrogen bonds between the water molecules become more stable and solid. As a result, the dissolved molecules and atoms are pushed out of their interactions with water (Figure 2.21d). As you might expect, this dramatically disrupts the chemical interactions and even the internal chemical structures a living cell relies on. If a cell freezes, all of its water is in the crystalline form, and the normal interactions of all the cell's other atoms and molecules are disrupted. You might think that warming the cell would get it back to normal, but there is no guarantee this would restore the normal structures or interactions of the atoms and molecules typical of a living cell. Freezing can damage cells in other ways too. Recall that water expands when it freezes, which can exert pressure on a cell's delicate outer membrane and rupture it. In addition, ice crystals can have irregular shapes with sharp edges. These can permanently damage both internal cellular structures and the cell membrane. For all of these reasons, when life freezes, irreversible death usually follows.

Given all these problems, why are the cryogenic organizations so convinced that a frozen person can be revived? Two areas of scientific research provide this cautious optimism. First, scientists have discovered species that can survive temperatures at which most others would freeze and die. Two examples include a species of deep-sea fish that thrives in icy cold oceans, and some species of frogs that actually freeze solid in the winter and revive in the spring. If scientists can unlock the mechanisms that allow these species to survive being frozen, the knowledge might be used to revive frozen humans. The other area of related research comes from the ability of scientists to freeze and store individual cells. Samples of cells grown in laboratory culture dishes can be frozen at ultracold temperatures for long periods of time—and when they are later thawed, a small percentage will survive. These same techniques also are used to freeze samples of eggs or sperm or even very young embryos that are used in *in vitro* fertilization procedures for humans and other species. All of these approaches to cryogenics rely on flooding the cells or tissues with cryoprotectants, substances that prevent water from forming regular crystals while keeping the atoms and molecules dissolved in water in solution at cold temperatures. For example, the frogs that literally turn to ice during the winter use a form of sugar, called glucose, as a cryoprotectant. Laboratories that freeze cell cultures or small embryos often use molecules such as glycerol as a cryoprotectant.

Unfortunately, the use of cryoprotectants has not been perfected for mammals. The organizations that freeze bodies claim to use cryoprotectants in their procedures, but they admit the techniques only minimize, but do not prevent, the formation of ice crystals. While such processes may reduce cellular damage, the only thing such organizations really are offering is the hope that future techniques will be available to repair the damage produced by the freezing process. Is this likely? Only time will tell.

QUICK CHECK

1. In terms of interactions between water molecules, what is the difference between water as water vapor, liquid water, and ice?

2. What are the problems with human cryopreservation that have not yet been solved?

Now You Can **Understand**

Death by pH—Life is filled with hazards.

Daily living brings risks, and in modern civilization those risks are not always obvious. You now understand that life cannot exist if the acidity—or the pH—of your blood or cells becomes too low or too high. In fact, even a small change in pH will impair your body's chemical processes. Changes in pH can result from some common sources like the paint remover in your garage, the gases given off by spoiling grain in a farmer's silo, the solvents in glue, and chemical imbalances as a result of being anesthetized for surgery. Even being in an enclosed small space with no fresh air can change the body's pH. A change toward the more acid state, called acidosis, is the most common pH problem and the effects of many different toxins, poisons, and illnesses are related to it.

Sick from Too Little Salt?

The average American diet contains a lot of salt—sodium chloride. Both sodium and chlorine are required for normal cell functions. Many animals, including humans, have a strong "taste for" salt and seek it out in their food. As a result, many Americans eat too much salt. It also is possible though, to have too little salt. Salt is essential for nerve and muscle function, and any condition that causes you to lose salt could lead to muscle fatigue, nausea, and seizures. Could you ever be in this situation? It can happen during extreme long-term physical activity, especially in the heat. Marathon runners or people whose jobs put them in the heat all day with hard physical labor are at risk of losing too much salt. Salt is excreted with sweat—if a person drinks a lot of water under hot conditions, he or she will lose even more salt. Does this sound familiar? Popular sports drinks contain salts, rather than just pure water, to counteract this effect.

What Do **You** Think?

Just How Much Water Should You Drink?

Water is the elixir of life. You must have it. You will die without it. Water is the universal solvent that life depends on, but how much do you need to drink every day to stay healthy? Does eight glasses a day pop into your mind? This number has been touted for a long time. Some sources suggest that if you don't drink enough water, you will gain weight—and if you drink a lot, you will burn fat. But will it hurt you to drink too much water? What is the evidence for how much water you should drink? How do you decide what to believe about water consumption? Search the Web for information about how much water you need. What kinds of information do you find? How do you evaluate this information and make your own decision about how much water you drink?

Chapter Review

CHAPTER SUMMARY

2.1 New Properties Emerge When Substances Interact

The phenomenon of emergent properties explains how biology results from chemical interactions. Under the conditions on Earth today, life comes only from preexisting life.

2.2 Matter Is Made of Atoms

All matter is composed of chemical elements that are made up of atoms. Electrical forces hold the subatomic atom's particles together. The number of electrons in an atom's outermost orbital greatly influences how readily it forms bonds with other atoms. Anions and cations are negatively charged and positively charged ions, respectively.

anion 33	isotope 33
cation 33	matter 30
electrical force 34	neutron 31
electron 31	orbital 34
element 31	proton 31
ion 33	

2.3 Bonds Hold Atoms Together in More Complex Structures

Chemical bonds hold the atoms of molecules together. Ionic, nonpolar covalent, and polar covalent bonds involve transfer, equal sharing, and unequal sharing of electrons, respectively.

chemical bond 35	molecule 37
chemical reaction 41	polar covalent bond 40
compound 35	polar molecule 40
covalent bond 37	products 41
hydogen bond 39	reactants 41
ionic bond 35	

2.4 The Chemicals of Life Interact in Water

Water is an unusual polar molecule with many characteristic properties, including high surface tension, cohesiveness, adhesiveness, the ability to hold a great amount of heat, and the ability to dissolve a wide range of substances. Water is essential to life. pH is a measure of the hydrogen ion concentration of a substance.

acid 46	nonpolar 45
base 46	pH 46

2.5 An Added Dimension Explained: The Science of Cryopreservation

Current cryopreservation technology has not yet overcome the problems caused by the damage that results from freezing large multicellular organisms. These problems stem from the properties of water as it transforms into ice.

REVIEW QUESTIONS

TRUE or FALSE. If a statement is false, rewrite it to make it true.

1. An ionic bond is formed between two molecules as they share electrons.

2. Water is called the universal solvent because it is everywhere.

3. A covalent bond involves one atom donating two electrons to another atom.

4. An atom of water is the smallest amount of this chemical element.

5. Electrons of the element hydrogen are identical with those of the element lead.

MULTIPLE CHOICE. Choose the best answer of those provided.

6. What is the difference between an element and a molecule?
   a. A molecule has fewer atoms than an element.
   b. Elements are found in living organisms but molecules are not.
   c. Elements have more complex properties than molecules.
   d. An element is a basic kind of matter, while a molecule is made of more than one element.
   e. An element is made of atoms but a molecule is not.

7. Which of the following statements is *not* true about electrical charge?
   a. A particle can carry positive charge, a negative charge, or no charge.
   b. Charge is a property of matter.
   c. Negatively charged particles are attracted to positively charged particles.
   d. There are four basic kinds of charges.
   e. Like charges are pushed away from one another.

8. Ions always have
   a. a negative charge.
   b. a positive charge.
   c. no charge.
   d. a charge that can be positive or negative.
   e. You cannot tell in advance if an ion will have a charge or not.

9. The number of electrons in an atom's outermost shell determines
   a. how heavy it is.
   b. how reactive it is.
   c. how large it is.
   d. how many neutrons it contains.
   e. all of the above

10. A hydrogen bond can be formed between
    a. two atoms to form a molecule.
    b. two atoms to form an ion.
    c. two different polar molecules.
    d. two parts of a nonpolar molecule.
    e. three atoms.

11. Electrons circle the nucleus of an atom
    a. and are still.
    b. and are arranged at random.
    c. in defined orbitals.
    d. all of the above
    e. none of the above

12. In a(n) _____ bond two electrons orbit both nuclei of two atoms and are shared.
    a. ionic
    b. single covalent
    c. triple covalent
    d. hydrogen
    e. double covalent

13. In a(n) _____ bond an electron is transferred and electrical forces hold the two nuclei together.
    a. ionic
    b. single covalent
    c. triple covalent
    d. hydrogen
    e. double covalent

14. An atom has two orbitals and a total of nine electrons. Two electrons are in the first orbital, in the first shell. The second shell has seven electrons; its outermost orbital has one electron. Because of its distribution of electrons, the atom will be
    a. highly chemically reactive.
    b. chemically inert.
    c. basic.
    d. a noble gas.
    e. chemically stable.

15. The qualities of table salt are far different from those of the elements that compose it. This phenomenon is an example of
    a. covalent bonding.
    b. polarity.
    c. a complex biological molecule.
    d. emergent properties of NaCl.
    e. surface tension of NaCl.

16. What happens to water molecules when water is heated?
    a. The covalent bonds within each water molecule are broken.
    b. The covalent bonds that hold different water molecules together are broken.
    c. More hydrogen bonds are formed.
    d. More ionic bonds are formed.
    e. Hydrogen bonds that hold different water molecules together are broken.

17. What is one role that water plays in a living organism?
    a. Water holds an organism together.
    b. Water creates an environment that holds polar and charged molecules in place.
    c. Water is not important because any solvent could be the basis for life.
    d. Water creates an environment where polar and charged molecules can interact.
    e. Water creates an environment that keeps polar and charged molecules away from each other.

18. If you added some lemon juice to a liquid you would make it
    a. more basic, by increasing the hydrogen ions and raising the pH.
    b. more acidic, by increasing the hydrogen ions and lowering the pH.
    c. more ionic.
    d. more acidic, by decreasing the hydrogen ions and lowering the pH.
    e. more basic, by increasing the hydroxyl ions and raising the pH.

19. Which of the following is the best explanation of how a spoonful of salt dissolves in water?
    a. The water molecules are pulled apart by the salt molecules.
    b. The salt molecules slide between the water molecules.
    c. The salt molecules form ionic bonds with the water molecules.
    d. Attractions between the ions and the polar regions of the water molecule disperse the solid into the liquid.
    e. The salt molecules form carbon-to-carbon bonds with the water molecules.

20. What is one common feature of all chemical bonds?
    a. surface tension
    b. collagen
    c. the electrical force
    d. polarized molecules
    e. water

CONNECTING KEY CONCEPTS

1. What role do electrons play in the processes of life?
2. How does the structure of a water molecule account for its property as a nearly universal solvent?

QUANTITATIVE QUERY

1. The atomic mass of hydrogen is 1.008, and the mass of oxygen is 15.99. What is the mass of a single water molecule?

CRITICAL THINKING

1. Try this experiment at home. Fill one ice tray or other small container with pure water. Now mix three teaspoons of sugar with a quarter cup of water, and fill another small container or ice tray section with this mixture. Leave both in the freezer overnight. Describe the state of the cold mixtures that you find the next morning. Explain what you found, using what you know about the molecular structure of water.

2. Lead is common and inexpensive, while gold is rare and valuable. For hundreds of years people have tried to change lead into gold. They have used different ingredients, different chemical processes, incantations (speeches rather like "abracadabra"), and rituals. What did they really need to do to change lead into gold?

3. You and your friend, Steve, are on a camping trip with a small group of friends. For dinner you have grilled steaks and buttered corn. You and Steve have kitchen duty and must wash the greasy plates. He brings a bucket of lake water and starts to rinse the plates. Explain, in terms of what you have learned so far, what happens as he rinses and then washes the plates with water and dish soap.

4. In this chapter you read that life is the result of complex properties that emerge when simpler chemicals combine. The growing understanding of emergent properties is part of the reason many scientists believe life is a natural process—one that once came out of natural laws that humans can understand. Others, however, believe life cannot be explained by known chemical and physical principles. These people believe there is something more to life. In your own words, summarize what you understand the concept of emergent properties of life to mean. Are there any features of life that cannot be understood in terms of chemical emergent properties?

For additional study tools, visit www.aris.mhhe.com.

Biological Molecules

Calves Killed to Prevent Mad Cow Disease

January 8, 2004

Mad cow disease.

First 'Mad Cow' Case Rattles US

The United States Has Reported Its First Suspected Case of "Mad Cow Disease," or BSE, in Washington State.

Dec. 24, 2003

Mad Cow Disease

In 1995 a calamity struck in Great Britain. Without warning, three teenagers and 17 adults were stricken by a rare, fatal disease that robbed them of normal mental and physical functions. Within months the victims lost much of their memory; they became confused and often were belligerent. A stumbling, lurching gait replaced coordinated walking movements. Their mental and physical abilities continued to erode and invariably, the victims died. Autopsies revealed that their brains were riddled with unnatural empty spaces where normal brain tissue should have been. Doctors and health officials were concerned and somewhat baffled by the first few cases, but they soon realized this new disorder was similar to an inheritable, fatal form of dementia called *Creutzfeldt-Jakob disease (CJD)*. The pattern was not exactly like CJD, though, and so the new disease was named *variant CJD (vCJD* for short). Although they now had a name for it, the investigators still did not know what caused this frightening and mysterious tragedy.

By combining information from various sources, researchers concluded that vCJD is similar to an animal disease called *bovine spongioform encephalopathy* (BSE). Let's put that mouthful into more understandable terms. Bovine means the disease occurs in cattle; spongiform means "like a sponge"—full of holes; encephalopathy means that affected cattle had abnormal brains. By 1996 Britain was in the midst of a fatal BSE epidemic. The best explanation for vCJD in humans was that people can contract the disease after eating infected meat. Researchers hypothesized that some microscopic agent was transmitting a variant form of this disease from cattle to humans.

Media reports called this outbreak of disease in cows and humans *mad cow disease*, and it triggered a panic in the late 1990s (**see chapter-opening photo**). Much has been learned about mad cow disease since then, and a variety of governmental regulations in countries around the world have greatly reduced the threat of this slow killer. Today people in Britain and the United States routinely eat beef with little fear, but mad cow disease still appears in cattle herds, and any hint that mad cow has returned can produce banner headlines and cause widespread concern.

The case of mad cow disease and vCJD provides a good lesson in how basic science and public policy can work together to respond to a public health threat. The only way to truly understand and respond to the threat of vCJD is to know what causes the disease and how that agent produces the disease symptoms. Finding these answers required that scientists understand the chemistry of the infectious agent. Researchers had to consider many questions. What kinds of infectious agents would they look for? Were these agents alive? How could they be stopped, destroyed, or changed into something harmless? What chemicals were components of these agents? At the end of this chapter you will revisit these questions, but before you can understand BSE and vCJD, you need some basic information about the special molecules that make up living things. ■

| 3.1 | **Life Depends on Complex Biological Molecules** |

Life is built from a limited number of chemical elements. Carbon, hydrogen, oxygen, nitrogen, calcium, and phosphorus are the most abundant and prominent elements in living things. As many as 20 or more other elements—including sodium, potassium, iron, chlorine, and magnesium—are less prominent, but they still play critical roles. Chemical elements combine to produce the complex and interesting structures and behaviors of living creatures. If you could reduce a living organism to its chemical elements, the organism would be transformed to a pile of inert white dust. How do these elements combine to produce life? A first step toward answering this question is to understand that atoms combine to form molecules—and that molecules have new and distinct properties single atoms do not possess (see Section 2.3). In this chapter you will read how atoms combine to form the specific **biological molecules** that give living organisms their unique characteristics. Biological molecules are not found in nonliving things.

The most important biological molecules contain hundreds, thousands, or millions of atoms. These large, complex biological molecules are built by combinations of smaller biological molecules that serve as molecular building blocks. Combinations of basic carbon atoms and small functional groups of atoms, in turn, build these smaller biological molecules. You will learn the details of this modular approach to building biological molecules in this chapter.

Life Is Based on Four Kinds of Large Biological Molecules

You might think it will be a monumental task to learn about and understand the great variety of biological molecules found in living organisms—but prepare for a pleasant surprise. The great variety of large biological molecules can be categorized into just four major classes: lipids, carbohydrates, nucleic acids, and proteins (**Table 3.1**). Let's start with a brief introduction to these important

biological molecules
complex chemicals that are unique to living organisms but not commonly found in nonliving natural systems

Table 3.1   Dietary Sources and Roles Played by the Four Kinds of Biological Molecules

Biological molecule	Found in these foods	Roles within cells	Additional roles within the human body
Lipids	Oils, butter, fatty meats, mayonnaise, fried foods	Key components of cell membranes Long-term energy storage	Store energy Used to make hormones
Carbohydrates	Sugars, starches, rice, grains, pasta, flours, cereals, fruits, vegetables	Energy storage and quick release of energy Structural support	Most accessible stored energy source
Nucleic Acids	Found in small amounts in all foods Found in nuclei of all cells	Carry, transmit, and execute instructions for making new proteins; carry information for cell reproduction; direct formation of proteins and direct cell reproduction	
Proteins	Meats, fish, poultry, dairy products, beans, nuts, dark green vegetables, whole grains	Perform a variety of jobs within cells including: formation of cellular structures, transport, communication, and enzyme actions	Form muscles and perform most of the body's work

biological molecules and the pivotal roles each plays in cellular and body functions and in the human diet.

Lipids are the molecules that you know as fats and oils. You encounter them in substances like butter, fatty tissues in meats, and in oils like corn oil or olive oil. Lipids play two important roles in living systems. First, and most universally, lipids are important components of cell membranes. Second, lipids are used to store chemical energy so that cellular processes can continue during lean times. Lipids also play other roles in the human body. For example, many hormones that control body functions are lipids.

Lipids are a major part of the human diet, and many Americans consume more lipids than are optimal for good health. Dietary lipids are the fatty foods, such as red meats, salad oils, butter, margarine, and the fats in many prepared foods. Deep-fried foods like french fries, doughnuts, or fried shrimp absorb oil in cooking, and so they are higher in lipids than the same foods baked, broiled, or steamed. Mayonnaise and some baked goods are high in lipids too. In contrast, most fruits and vegetables are low in lipids, while other foods like nuts, avocados, and eggs have intermediate lipid content.

Carbohydrates are a mixed group of biological molecules that you know as the sugars and starches found in sweets, pastas, and flours. Fruits, vegetables, and grains all have high carbohydrate content. In living organisms carbohydrates play at least two important roles. First, many carbohydrates provide a source of quick energy: if you need a quick energy boost, you probably reach for something high in sugar like an apple, muffin, or candy bar. It might surprise you to learn that a second function of carbohydrates is to provide structural support for cells of some organisms. For instance, the stiff bark of a tree and the hard shell of a lobster are both made of carbohydrates.

Nucleic acids are the third general kind of biological molecule. You may know nucleic acids as DNA and RNA, short for deoxyribonucleic acid and ribonucleic acid, respectively. These large molecules can contain billions of atoms. DNA and RNA provide the information needed for a cell to carry out all of its various functions and nucleic acids allow cells to transfer that information to the next generation when cells divide. Nucleic acids usually are not considered important components of the diet. This is because you obtain supplies of them whenever you eat foods that include cells of other organisms, and because your body can make them from their basic subunits.

Proteins have the most diverse and complex chemical structures of all of the biological molecules and perform many different jobs within cells and organisms. For example, protein is the major component of muscle and functions to move bones from one position to another. If you lift weights and work out to increase your strength, what you are really trying to do is increase the protein content of your muscles. So it should not surprise you that one dietary source of protein is meat, the muscle tissue of other animals. Because of the diverse and important roles that proteins play in biological

organic chemistry the study of the interactions of carbon atoms and carbon-containing molecules

organisms, many foods other than meats have significant amounts of proteins. For example, whole grains, nuts, dark green leafy vegetables, beans, and dairy foods all contain proteins.

QUICK CHECK

1. List the five most common elements found in the major biological molecules.
2. Name the four major biological molecules and name one role that each plays in the body.

3.2 Biological Molecules Are Built Around Carbon Atoms

Only a handful of elements make up biological molecules, but how do these few kinds of atoms combine to make such large, complex molecules? The answer to this question starts with carbon. The four kinds of large biological molecules all contain hydrogen and oxygen, but carbon is the element that is crucial to their structures. Because the chemistry of carbon is so important to life's processes, an entire branch of chemistry, **organic chemistry,** is devoted to the study of molecules that contain carbon. So all biological molecules are organic molecules. Organic chemistry includes the study of many familiar, synthetic molecules such as plastics and nylon, both of which are made largely of carbon. But synthetic organic molecules are not nearly as complex and diverse as are the organic molecules of living organisms. To understand why carbon is so central to life, you need to take a closer look at the carbon atom.

One Carbon Atom Can Bind to Four Other Atoms at Once

What feature of the carbon atom makes it so important to living organisms? The atomic structure of carbon holds the key to this question (**Figure 3.1**). While carbon has six electrons, its outer shell contains only four of the eight electrons it could hold. A carbon atom has four "empty places" in its outermost electron shell (Section 2.3). Therefore, one carbon atom can bond to up to four other atoms at once, sharing one electron with each of four other atoms. This ability of carbon to bind to four other atoms results in the complexity of biological molecules.

It is possible for carbon to bind only to itself and form large carbon crystal structures like

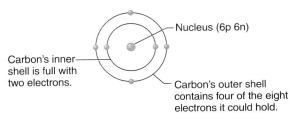

Figure 3.1   **The Structure of a Carbon Atom.** A carbon atom has six electrons distributed in two shells.

■ *How many bonds can a carbon atom form?*

diamonds or graphite. Most carbon molecules, however, include bonds to elements other than carbon. Hydrogen commonly is bonded to carbon, and a molecule with only carbon and hydrogen atoms is called a **hydrocarbon.** The simplest hydrocarbon is methane, in which one carbon atom binds to four hydrogen atoms. Methane is a gas that burns easily and is a major component of the natural gas used to heat homes. Certain bacteria produce methane as a waste product—and some of these live in the intestines of animals, like ants, cows, pigs, and even humans. **Figure 3.2** shows two different ways of drawing a methane molecule; each emphasizes a different aspect of methane's chemical bonds. In most drawings in this text you will see molecules drawn as they are shown in Figure 3.2b. There the letters represent the kinds of atoms and the lines represent the bonds.

Larger hydrocarbons can form when each carbon atom binds to two other carbons and two hydrogens, forming long chains of carbon and hydrogen atoms. The carbon-to-carbon bonds in a long-chain hydrocarbon contain a great deal of energy, which can be released when the molecule is

(a) The space-filling model shows the volume occupied by the molecule.

(b) The simplified bond-line model emphasizes the kinds of atoms involved in the molecule and uses simple lines for the bonds.

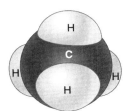

Figure 3.2   **Different Views of a Methane Molecule:** **(a)** the space-filling model and **(b)** the bond-line model.

burned. For instance, the gasoline you pump into your car is made of long chains of hydrocarbons, while the butane in lighters is made of shorter chains of hydrocarbons (**Table 3.2**). Many important carbon molecules have ring structures. In the ring form of a hydrocarbon, the carbons usually attach to one another in the ring and the hydrogens hang off of these carbons. Regardless of the number

hydrocarbon a molecule that contains only carbon and hydrogen atoms

Table **3.2**	Hydrocarbons Are Molecules Built Around Carbon Atoms
Some hydrocarbons are linear molecules; others have ring structures.	
Example	**Molecular structure**
High-octane gasoline	Octane
Butane lighter	Butane
Lemons	Limonene
Rubber tree	Natural rubber

of carbon atoms, or whether the chain has a ring structure, hydrocarbons have one important feature in common. All hydrocarbons are nonpolar and, of course, that means they do not mix with water.

Notice that a substance whose molecules are made of carbon and hydrogen has different properties than a substance made of just carbon. Diamond and graphite are solids that are not particularly reactive with other compounds. In contrast, hydrocarbons are liquids (like benzene) or gases (like propane, ethane, or butane), and at moderate temperatures hydrocarbons burn easily. This is another example of emergent properties. Keep in mind that the properties of an organic molecule are determined by the kinds of atoms bonded to the carbon atoms and the arrangement of their bonds.

Functional Groups Determine the Properties of Organic Molecules

You can better understand the properties of organic molecules by learning about some chemical groups that often form bonds with carbon. Let's consider the methane molecule again and see what happens when one hydrogen atom is replaced by an oxygen atom, which, in turn, is bound to a hydrogen atom. You could talk about the oxygen and hydrogen separately, but in the molecule these atoms behave as a group—and they influence the molecule as a group. A **functional group** is a group of atoms that is bound to an organic molecule and influences the properties of that molecule. A functional group determines the functional properties of the atom or molecule to which it is bonded. You can better understand this with an analogy. Consider a power drill you might use to assemble a piece of furniture. The basic action of the drill is to spin, but you can change the function by changing the drill's attachments. For instance, if you insert a small bit, the drill will bore a small hole—while if you insert a large bit, it will bore a large hole. If you insert a screwdriver bit, the drill will insert or remove screws. Functional groups also affect the properties of a molecule they are attached to. Just a few functional groups account for many of the interesting properties of organic molecules. Let's look at how each of these alters the function of organic molecules.

A **hydroxyl** is a functional group composed of oxygen and hydrogen; in chemical notation it is written as –OH (**Table 3.3**). The oxygen atom of a hydroxyl group can bind to a carbon, making a more complex molecule. Like a water molecule, a hydroxyl group is polar, so any molecule to which a hydroxyl is attached has a polar region. And, as you would expect, the presence of a hydroxyl group causes a molecule to form hydrogen bonds with water and perhaps even to dissolve fully in water.

Table 3.3 shows an example of the conversion of a hydrocarbon to a molecule with a polar region. The simple hydrocarbon, ethane, is a nonpolar molecule with just two carbons. Ethane is a flammable gas that is a component of the natural gas you may use to heat your home. When a hydroxyl group replaces one of ethane's hydrogen atoms ethane becomes ethanol, also known as the ethyl alcohol in alcoholic beverages. Even though there is only one atom different between the two molecules, ethanol has different properties than ethane. Ethanol is a liquid at room temperature and because of its hydroxyl group ethanol mixes with water. In addition, ethanol also has many profound effects on biological systems. So the presence of just a single hydroxyl group dramatically changes the properties of the two-carbon molecule, ethane.

Let's consider the **carboxyl** group next. In Table 3.3 you see that a carboxyl is a carbon atom bound to an oxygen atom and a hydroxyl. Accordingly, a carboxyl group is written as –COOH. The bond to the single oxygen atom is drawn with two lines because it is a double bond. Many of the molecules that include a carboxyl group are called carboxylic acids. These are acids because the H^+ ion is readily given up in water, lowering the pH of the solution and making it more acidic. Common carboxylic acids are acetic acid in vinegar and citric acid in lemons, oranges, limes, and other citrus fruits. Carboxylic acids have several interesting properties: they are *highly* polar and can form strong hydrogen bonds. Because of these characteristics, carboxylic acids dissolve in water—and when they do, the water becomes more acidic.

A **phosphate** (PO_4) is a functional group with a central phosphorus atom bound to four oxygen atoms (Table 3.3). Like other functional groups that contain oxygen, a phosphate group is polar. Because any of its four oxygen atoms also can bind to a carbon atom, a phosphate can bond to a hydrocarbon or another organic molecule. Alternatively, an oxygen atom can bind to another phosphate group, making a string of phosphates as shown in adenosine triphosphate (ATP), a molecule with three phosphate groups. Phosphate groups are found attached to many different kinds of biological molecules, and the energy contained in their bonds provides the power for cellular processes.

Amino groups are derived from ammonia (NH_3), a harsh gas that is highly soluble in water (Table 3.3). Liquid ammonia is used in household

---

functional group a group of atoms that is bound to an organic molecule and influences the properties of that molecule

hydroxyl a functional group with a hydrogen atom and an oxygen atom

carboxyl a chemical group with one carbon, two oxygens, and sometimes a hydrogen atom

phosphate a functional group with a central phosphorus atom bound to four oxygen atoms

Table 3.3   Some of the Functional Groups of Biological Molecules and Their Properties

Functional group and its properties	Representative molecule(s) and everyday examples
Hydroxyls (–OH) or (HO–) are polar and water soluble; they contribute water solubility to a molecule. $-O-H-$	 Ethane → Ethanol 
Carboxyls (COOH) or (CO_2H) are highly polar, acidic, and give up H^+ ions in water.	Acetic acid 
Phosphates (PO_4) are polar and contribute water solubility. The phosphate bond holds useable energy. $O-P=O$	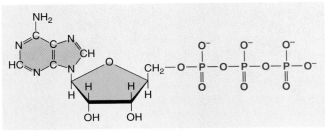 Adenosine triphosphate 
Amino groups (NH_2) or NH_3) are polar and basic.	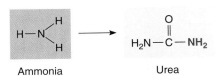 Ammonia → Urea 

cleaners and other industrial processes, but it is toxic. When a molecule of ammonia binds to carbon, one hydrogen atom is lost; the NH_2 group that remains is an **amino group.** An amino group attached to an organic molecule is not necessarily toxic—and amino groups are some of life's most important and common functional groups.

Now that you understand the basic structures of organic molecules and some of the important functional groups, you are ready to tackle the more com-plex structures of biological molecules including lipids, carbohydrates, nucleic acids, and proteins.

QUICK CHECK

1. What feature of a carbon atom allows it to form the large, complex biological molecules found in cells?
2. Name four functional groups found in organic compounds.

amino group a functional group derived from ammonia that has a nitrogen and two hydrogen atoms

3.3 Lipids Are Biological Molecules That Do Not Mix Well with Water

Chapter 2 presented water as the medium necessary for the chemical interactions of life. Recall that because water molecules are polar, other molecules that are charged and/or polar dissolve in water, where they may interact with one another. So it may surprise you that the first group of biological molecules you will consider contains molecules that are mostly nonpolar. As you have read, lipids are substances commonly known as fats and oils. A **lipid** is not very soluble in water. As you will see, this simple definition hides some of the most interesting properties of lipids.

One of the important jobs of lipid molecules is to form cell membranes. Membranes are actually made of both lipids and proteins, but lipids have the particular role of providing a barrier around the cell. Lipids are like the protective walls of a house that keep weather and intruders out. By preventing polar and charged substances from moving freely in and out of the cell, lipids protect the cell from random and uncontrolled factors in its environment. Proteins embedded in a lipid membrane are like the doors of a house that allow entry and exit and allow a cell to communicate with its environment in a controlled way.

In addition to the cell membrane, certain kinds of cells have internal membranes made largely of lipids. These internal membranes are like interior walls that partition a house into rooms. Internal cell membranes separate the chemicals involved in different biological processes. You will read more about these functions of lipids in Chapter 4. Finally, lipids can act as hormones. In Chapter 24 you will read that hormones are chemicals produced in one part of the body that act in another part. Lipid-based hormones include the general class of steroids such as cortisol, testosterone, and estrogen. Because the lipid hormones are similar in structure to the lipids of cell membranes, they readily cross to the inside of cells, where they influence internal cellular processes.

The key to understanding the function of lipids is to understand their structures. The lipid molecules of living organisms have a large nonpolar hydrocarbon region and a smaller region that is polar. One of the simplest lipids is a *fatty acid* (**Figure 3.3a**), which is a long hydrocarbon chain with a polar region at one end. A *triglyceride* is formed from three fatty acid molecules and a glycerol molecule (Figure 3.3b). Fatty acids are found in all sorts of foods such as butter, oils, and nuts—but in animal tissues fatty acids are combined into triglycerides, the main storage molecule for lipids in fat cells of animals. One of the most important lipids in cells is a *phospholipid*, a lipid molecule with a large phosphate group attached to two long strands of hydrocarbons (Figure 3.3c). Phospholipids commonly are found in cell membranes. Cholesterol is a lipid made of ring-shaped hydrocarbons. Except for a single hydroxyl group, cholesterol is nonpolar (Figure 3.3d). Cholesterol plays many roles in animal bodies. It is a lipid found in cell membranes. Cholesterol also can be chemically transformed into the steroid hormones recently mentioned.

Lipids Are Important for Many Aspects of Health

Lipids are important for many aspects of overall health, but the lipids you consume in your diet have a big impact on two major health concerns: obesity and heart disease. Because of their long hydrocarbon chains, lipids provide a lot of energy. So a diet heavy in lipids—which are listed as fats on nutritional labels of packaged foods—can supply more energy than the body needs. Because excess dietary fat is stored in fat cells, a diet high in fats often leads to obesity.

The relationship between dietary lipids and heart disease is a bit more complex. You can read more about this relationship in Chapter 22, so only a summary is presented here. Deposits of cholesterol in the walls of blood vessels that supply blood to the heart have been linked to heart disease. You might think that dietary cholesterol would be the major factor that influences how much cholesterol is in your bloodstream, but that is not the case. Because some of the fats you eat are converted to cholesterol, the amount of fat in your diet has a greater effect on the amount of cholesterol in your bloodstream than does dietary cholesterol. The whole story is not that simple, though. It turns out that some kinds of dietary fats promote heart disease, while others actually can help blood vessels and the heart to stay healthy. There are two basic kinds of fats: *saturated* and *unsaturated*. As you can see in **Table 3.4** the difference between them has to do with the presence of double bonds. In saturated fatty acids each carbon atom is bonded to two or three hydrogen atoms; in unsaturated fatty acids some carbons are involved in double bonds. Unsaturated fatty acids with double bonds have fewer hydrogen atoms than they would if all the bonds were single bonds. Saturated fats are solids at room temperature—you know them as butter, lard, or the fat deposits of meats

lipid a molecule that is not highly soluble in water

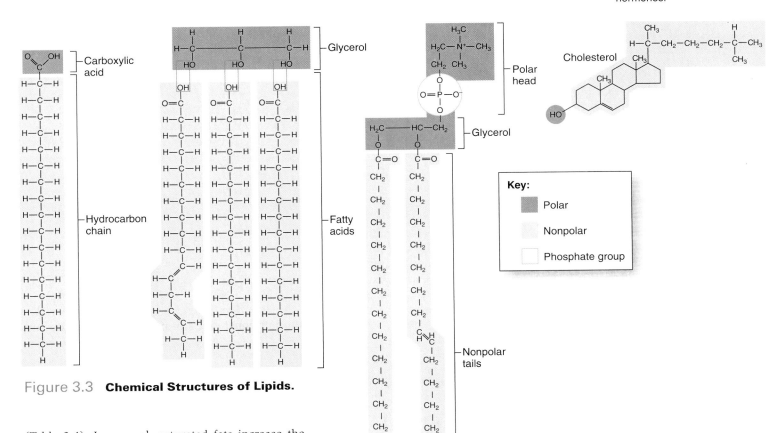

(a) A fatty acid molecule has a polar carboxyl acid group on one end of a hydrocarbon chain.

(b) A triglyceride is made of three fatty acids. The polar end of the triglyceride is a three-carbon molecule.

(c) A phospholipid has two fatty acid tails linked to a large polar head by a glycerol molecule. The polar head contains a phosphate group.

(d) A cholesterol molecule has four hydrocarbon rings. Cholesterol is found in arterial plaques and is the basis for all steroid hormones.

Key:
- Polar
- Nonpolar
- Phosphate group

Figure 3.3   **Chemical Structures of Lipids.**

(Table 3.4). In general, saturated fats increase the risk of heart disease; unsaturated fats can actually lower the risk of heart disease.

There are several kinds of unsaturated fats, all named for the number of double bonds on their carbon chains. *Monounsaturated fats* have just one double bond in the hydrocarbon chain, and they are liquid at room temperature (Table 3.4). Olive oil, canola oil, and peanut oil contain high amounts of monounsaturated fatty acids. *Polyunsaturated fats* have two or more double bonds in the hydrocarbon chain (Table 3.4). Polyunsaturated fats are liquid at room temperature and corn oil, soybean oil, seeds, whole grains, and fatty fishes like salmon and tuna contain high amounts of polyunsaturated fats. *Trans fats,* also called partially hydrogenated vegetable oils, recently have been added to nutritional labeling because they are linked to heart disease. Their name refers to the way that hydrogen atoms are bonded on both sides of the carbon chain, (Table 3.4). Chemical processing of a polyunsaturated vegetable oil produces a solid trans fat that can be used in baking instead of butter. Because this partially hydrogenated veg-

etable oil will not turn rancid, it gives cookies, chips, and baked goods made with it a longer shelf life. Unfortunately, trans fats are similar to saturated fats in their health risks, and there is much current pressure to decrease the amount of trans fats in processed foods.

Your understanding of lipids and their roles in body processes can help you to understand some common medical tests and dietary recommendations. For instance, when you have your cholesterol checked, the test also reveals levels of triglycerides in your blood. Both high cholesterol and high triglyceride levels are correlated with high

Table 3.4 Kinds of Fats

Type of fat	Chemical structure	Examples
Saturated fats are solid at room temperature.	Palmitic acid The fatty acid chains of **saturated fats** have the maximum number of hydrogen atoms.	Coconut oil, palm kernel oil, butter, cream cheese, beef fat
Unsaturated fats are liquid at room temperature. **Monounsaturated fats**	Oleic acid (omega-9) The fatty acid chains of **monounsaturated fats** have one double bond.	Olive oil, canola oil, peanut oil
Polyunsaturated fats	Alpha-linolenic acid (omega-3) The fatty acid chains of **polyunsaturated fats** have two or three double bonds.	Safflower oil, sunflower oil, soybean oil, corn oil
Trans fats	The fatty acid chains of **trans fat** have one "up" and one "down" hydrocarbon on either side of a double bond.	Fats in many processed foods

risk of heart disease. Nutritionists, including those representing the U.S. government, suggest that dietary fats should make up 10 to 30% of your diet, rather than the 40 to 50% typical of the diets of some Americans. Many studies show that ultra-low-fat diets result in a dramatic lowering of the risk of heart disease. Some popular diets aimed just at weight loss suggest that you safely may eat all of the fat you like as long as you restrict your intake of carbohydrates. Some nutritionists, however, are suspicious of this approach. Although sci-ence may have a difficult time keeping up with the rapid parade of fad diets, most scientists predict that if Americans continue to eat a high percentage of animal fats, the cost in disease and lives will increase.

QUICK CHECK

1. What is a lipid?
2. What is the structure of a phospholipid? What major role do phospholipids play in cells?

3.4 Carbohydrates Are Made of Carbon, Oxygen, and Hydrogen

Carbohydrates often are in the news, especially when the news is about diets. For years Americans have rushed from one diet to another. Some diet plans say that you should eat a lot of carbohydrates, while others say that you should avoid them. In Chapter 4 you will learn a bit about the role carbohydrates play in providing food energy, but in this section you will learn what carbohydrates really are at a chemical level.

In your common experience carbohydrates are sugars and starches. The table sugar you add to tea or coffee; the fructose, corn syrup, or glucose found in many products; as well as the starches in breads, cereals, rice, or pasta are all carbohydrates. And you read earlier that the stiffness of trees and the shells of insects and their relatives are due to carbohydrates. These substances are distinctly different from methane, gasoline, or even oils because carbohydrates are more complex than hydrocarbons. While a hydrocarbon contains only carbon and hydrogen, the most basic kind of a **carbohydrate** is a molecule that has at least three carbons and also contains hydrogen and oxygen. Other organic molecules also might contain carbon, hydrogen, and oxygen, but in a carbohydrate there is usually one oxygen atom for every carbon atom. The most basic carbohydrates have carbon:hydrogen:oxygen in a ratio of 1:2:1. Notice this means that there is one oxygen atom for every carbon atom. Carbohydrates have more oxygen relative to carbon than other organic molecules. **Figure 3.4** shows the chemical structures of several basic carbohydrates important in living organisms. Take a minute to compare the chemical structures of these three sugars. Glyceraldehyde is a carbohydrate with three carbons; ribose has five carbons; glucose has six carbons. Notice that molecules of larger carbohydrates like ribose and glucose can be in either a linear form or a ring form. In the ring form typical of living organisms, one of the oxygen atoms is in the ring, bound to two carbon atoms (Figure 3.4b, c). Because ring forms are the normal structure found in cells, carbohydrates are illustrated in this manner throughout the rest of this book.

An important feature of carbohydrates is their ability to link rings together to form larger molecules. Large, complex biological molecules often are formed by small subunits that bond together and form long chains. The building blocks of larger carbohydrates are called **monosaccharides.** The word

(a) Glyceraldehyde, a three-carbon carbohydrate

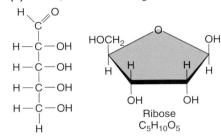

(b) Ribose, a five-carbon sugar

(c) Glucose, a six-carbon sugar

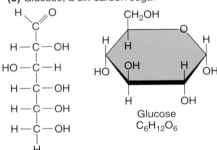

Figure 3.4 **Small Carbohydrate Molecules.** (**a**) Glyceraldehyde is a linear carbohydrate with three carbons. (**b**) Ribose and (**c**) glucose can be in linear or in ring forms.

saccharide means "sugar" and *mono* means "one," so these are sugars with just one subunit or building block. When two or more subunits combine the carbohydrate is called a **disaccharide** ("two sugars") or a **polysaccharide** ("many sugars"). **Figure 3.5** shows some of the more common and important monosaccharides and disaccharides found in living organisms. If you read labels on packaged foods (Figure 3.5c), you will find the names of these sugars. Common table sugar, or sucrose, is a disaccharide made of one molecule of fructose and one molecule of glucose. The more interesting properties of carbohydrates, though, emerge in polysaccharides.

Polysaccharides are especially important in cells. **Table 3.5** shows short stretches of four polysaccharides: glycogen, amylopectin, cellulose, and chitin. Glycogen and amylopectin are similar energy storage molecules of animals and plants, respectively. They differ in that the glycogen molecule branches more than the amylopectin molecule

carbohydrate a sugar or starch molecule that contains carbon, hydrogen, and oxygen—often in a ratio of 1:2:1

monosaccharide a carbohydrate with just one subunit or building block

disaccharide a carbohydrate with two subunits

polysaccharide a carbohydrate with many subunits

Figure 3.5 **Common Monosaccharides and Disaccharides.**
(**a**) Three examples of monosaccharides.
(**b**) Disaccharides form when two monosaccharides bond. (**c**) Mono- and disaccharides are ingredients of many packaged foods.

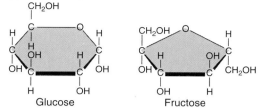

(a) Three examples of monosaccharides.

Glucose   Fructose   Galactose

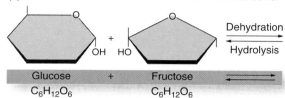

(b) Disaccharides form when two monosaccharides bond.

Dehydration

Hydrolysis

Glucose + Fructose ⇌ Sucrose + Water
$C_6H_{12}O_6$ $C_6H_{12}O_6$ $C_{12}H_{22}O_{11}$ H_2O

(c) Many foods contain mono- and disaccharides.

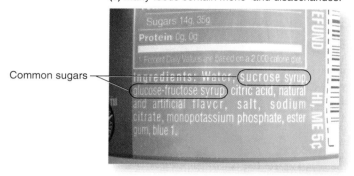

Common sugars

does. Cellulose is quite similar, but if you look carefully you see that in cellulose the glucose molecules have alternate orientations. Chitin is similar to cellulose, except there is a complex nitrogen group bound to each glucose molecule. The lesson to be learned here is that small differences in chemical structure can have big implications for the properties of a molecule.

Dietary Carbohydrates Come in Different Forms

Americans typically eat a lot of carbohydrates. Foods like bread, pasta, rice, french fries, and doughnuts make up a large percentage of many people's diet. Recent fad diets have encouraged people to eat little—or no—carbohydrates, while other diets encourage people to eat lots of carbohydrates. To shed some light on all this, it is worth considering the role of different kinds of carbohydrates in your diet. First, you probably know that you cannot digest every form of carbohydrate. Grass, for example, is largely cellulose and you can't digest cellulose. Corn plants are a tall, domes-

ticated species of grass whose large, edible seeds form along a cob within "ears." Most of the tissues of a corn plant have tough cellulose fibers that humans cannot digest. That is why you shuck ears of corn, discard the husks, and eat the corn kernels, which are high in sugars and digestible starches. Other animals, like cows and sheep, can digest grass and eat fodder made from chopped-up corn stalks and leaves because they carry bacteria in their digestive tracts that digest cellulose for them. You can't digest chitin either, so you crack lobster claws, discard the shells, and eat the meat. Nevertheless, some forms of indigestible carbohydrates—called fiber, or *roughage*—are important parts of your diet and help to keep the digestive tract active and healthy. Fresh fruit, raw vegetables, and whole grains are good sources of fiber.

Most dietary carbohydrates provide a plentiful source of energy. Many plant polysaccharides such as amylose are easy to digest, as are the mono- and disaccharides of many carbohydrates. However, not all carbohydrates in the food you eat are nutritionally equivalent. The primary point is to consider what the food source contains besides digestible carbohydrates. Take, for example, two pieces of bread—one made from white flour and the other made from whole-grain flour. Both pieces of bread contain the carbohydrate amylose, but the white bread contains little else. In the flour used to make white bread, the original grains of wheat were stripped of almost everything except amylopectin. Dietary carbohydrates stripped of other nutrients are often called *refined* carbohydrates. White rice and white flour are examples of refined carbohydrates.

In contrast, whole-grain flour or rice contains some indigestible fiber such as cellulose as well as

Table 3.5 Polysaccharides Store Energy and Are Used to Form Structures

Polysaccharide storage molecule	Chemical structure	
Glycogen is the storage polysaccharide used by animal cells. Glycogen is stored as granules in liver and muscle cells.	This is where the polymer branches.	
Amylopectin is a major storage polysaccharide of plants. It is stored in roots and in tubers like potatoes.		
Cellulose is the major structural molecule in plants. Cellulose strengthens plant cell walls. It is found in woody tissues of shrubs and trees.		
Chitin forms the hard outer shells of insects, spiders, and crustaceans.		

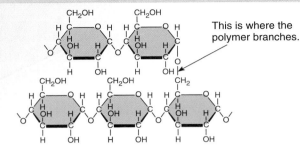

proteins, vitamins, and other nutrients that were in the original grain that was ground to make the flour. So whole-grain bread is a more varied source of nutrition and contains relatively less easily digestible carbohydrate than does the white bread. All this adds up to important differences for your nutrition. White bread has more excess carbohydrates and less other nutrients than whole-grain bread. Because white bread supplies more energy than you need immediately, the extra energy is stored in your body as fat. The bottom line is that while most nutritionists are leery of low-carb diets, they do agree that the carbs you do eat should come from natural, complex foods such as fruits, vegetables, and whole grains. So skip the doughnuts and bagels for breakfast, and choose blueberries on whole-grain cereal instead.

QUICK CHECK

1. What is a carbohydrate?
2. What is the difference between the refined carbohydrates and whole grain carbohydrates?

3.5 Nucleic Acids Are Strings of Nucleotides

Nucleic acids provide the instructions for all cellular processes. Despite the complexity of this job, there are only two major classes of nucleic acids in living organisms: DNA and RNA. DNA stands for *deoxyribonucleic acid* and RNA stands for *ribonucleic acid* (see Section 1.2). A simple way to describe their roles is to say that DNA carries instructions for making the proteins in a cell, and RNA executes those instructions. How can just two molecules carry and execute the instructions for all organisms? The overall structure of each type of nucleic acid is the same in every cell, regardless of what kind of organism it is, and the way that the instructions are carried and executed is the same in nearly all cells. The differences between organisms are determined by subtle chemical details in the individual DNA and RNA molecules found in any given cell. The result of these similarities is that all living organisms are similar in many fundamental ways, even though they may look different. You can think about this as a bit like the operating systems on computers. Different kinds of computers can use different kinds of systems, such as MacOS, or Windows, or Linux—but all of these operating systems carry out similar functions and all are fundamentally the same, even though their details are different. DNA and RNA are nucleic acids involved in the instructions for cellular processes. In Chapter 7 you will learn more about the forms of RNA and details of DNA structure and function.

Nucleotides Are Made from a Sugar, a Phosphate, and a Base

Nucleic acids are made of repeating units of nucleotides, smaller building block molecules. Nucleotides are much more complicated molecules than are hydrocarbons or carbohydrates. A **nucleotide** is a three-part molecule that contains a phosphate group and a nitrogenous base attached to a sugar ring (**Figure 3.6a**). Let's look at each of these features in turn.

At the center of a nucleotide is a five-carbon monosaccharide. This central five-carbon sugar can have one of two forms, and in Figure 3.6b you can see that the ribose sugar in RNA has one more oxygen atom than the deoxyribose sugar in DNA. A phosphate group is attached on one side of the sugar, and a **nitrogenous base** is attached on the other side. This nitrogenous base is a critical feature of a nucleotide. The distinctive chemical structure of a nitrogenous base includes a nitrogen atom in a carbon ring that also contains hydrogen and oxygen.

Only five nitrogenous bases are important in living organisms: adenine, guanine, cytosine, thymine, and uracil (Figure 3.6c). Notice that nitrogenous bases have two different shapes: they can be either double rings or single rings. The double ring structures are called *purines,* and the single ring structures are called *pyrimidines.* Adenine and guanine are purines; cytosine, thymine, and uracil are pyrimidines. These nitrogenous bases combine with sugars and phosphates to make nucleotides. While there are five different nitrogenous bases, in cells there are four different ribose-based nucleotides and four different deoxyribose-based nucleotides. As you soon will see, these five nucleotides combine to form the nucleic acids of a cell.

Reexamine the five nitrogenous bases in Figure 3.6c. Because of their crucial roles in inheritance and cell functions, you should pause to learn the names of these five nitrogenous bases and take special notice of their chemical structures. Try to learn which have double rings and which have single rings. Read on once you have become familiar with the names and structures of the five nitrogenous bases and with the structure of a nucleotide.

nucleotide a three-part molecule that contains a phosphate group and a nitrogenous base attached to a sugar ring

nitrogenous base an organic ring structure containing nitrogen, carbon, hydrogen, and oxygen

nucleic acid a chain of nucleotides

(a) Basic plan of a nucleotide.

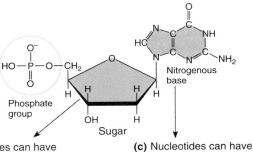

Phosphate group

Sugar

Nitrogenous base

(b) Nucleotides can have different sugars.

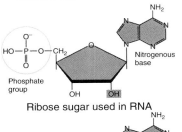

Phosphate group

Nitrogenous base

Ribose sugar used in RNA

Deoxyribose used in DNA

(c) Nucleotides can have different nitrogenous bases.

DNA or RNA

Guanine   **Adenine**
Nitrogenous bases with double rings (purines)

DNA or RNA   DNA only   RNA only

Cytosine   **Thymine**   **Uracil**
Nitrogenous bases with single rings (pyrimidines)

Figure 3.6 Nucleotides Combine to Form Nucleic Acids. (**a**) The general plan of a nucleotide. (**b**) Nucleotides can have either ribose or deoxyribose sugars. (**c**) Nucleotides can have one of five nucleotide bases. (**d**) Ribose nucleotides form RNA; deoxyribose nucleotides form DNA.

■ *How are DNA and RNA similar? How are they different?*

(d) There are two different nucleic acids.

RNA can be in the form of a ribbon, or a clover-leaf shape, but it is always a single strand of nucleotides.

DNA is a double-stranded molecule that is coiled into a helix.

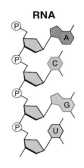

RNA

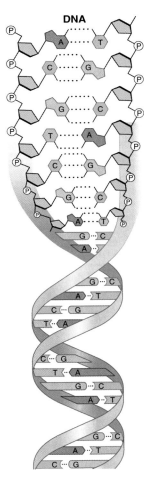

DNA

A String of Nucleotides Is a Nucleic Acid

Nucleic acids are long strings of nucleotides, and DNA and RNA molecules form when strings of nucleotides bond together. A tiny piece of each kind of nucleic acid is shown in Figure 3.6d. The chemical structures of these molecules are directly responsible for their functions, so it is worth spending some time to understand their structure. Figure 3.6 highlights some of the similarities and differences between RNA and DNA, so refer to this figure while studying the next paragraphs.

Consider the RNA molecule first. RNA is a long string made only of ribose nucleotides, which gives the nucleic acid its name: ribonucleic acid. Only four of the five nitrogenous bases are found in RNA: adenine, cytosine, guanine, and uracil. Thymine is never part of an RNA molecule. A strand of RNA can take on a variety of shapes, from a long ribbon to a complex, folded structure.

Now look at the DNA molecule. The nucleotides in DNA are built around a deoxyribose sugar—a five-carbon sugar with one less oxygen than ribose has. DNA also contains only four of the five possible nitrogenous bases, in this case ade-nine, cytosine, guanine, and thymine. DNA does not contain uracil. There are other differences between DNA and RNA. While RNA is usually single-stranded, in cells DNA is a double-stranded molecule. DNA is made of two strands of nucleotides that wind around each other in a double helix shaped like a twisted ladder. The two sides of the ladder are held together by hydrogen bonds between the pyrimidine and purine nitrogenous bases. One difference between RNA and DNA that you can't see in Figure 3.6 is that RNA molecules are usually much shorter than DNA molecules. RNA molecules range from several hundred to several thousand nucleotides long, while DNA molecules are millions of nucleotides long. In both RNA and DNA the exact identity and order of the nucleotides can vary, and so there are many different RNA and DNA molecules in any given cell.

QUICK CHECK

1. Name and describe the five nitrogenous bases in nucleotides.
2. List and briefly explain at least four differences between DNA and RNA.

3.6   Proteins Are Folded Chains of Amino Acids

How do cells carry out the various functions of life such as growth, development, and responsiveness? The answer is proteins. This last group of important biological molecules does the work of cells. The reason proteins can do so many different jobs is because they have the most varied and complex structures of all the biological molecules. The number of different kinds of proteins is hard to estimate, but there are certainly over 100,000 and there may be millions. How can proteins be such complex and diverse molecules? The answer lies in their basic subunits: amino acids.

Figure 3.7 shows the basic structure of an **amino acid.** An amino acid always has a central carbon bound to four atoms or groups of atoms. The molecule gets its name from the *amino* group, NH₂, which always is bonded on the central carbon at a position that is opposite to a carboxyl group at the other end of the molecule. A hydrogen atom always is found at the third position around the carbon atom. These three aspects of an amino acid are identical in every amino acid molecule. The functional group bonded at the fourth position around the central carbon of an amino acid accounts for the diversity of proteins. It is called an *R-group.* It is appropriately named because R stands for reactivity; an amino acid's R-group gives the molecule its unique reactivity with other atoms and molecules. The R-group could contain any number of functional groups, but in most organisms there are only 20 options for what is found in this position. This gives cells just 20 different amino acids to use.

The chemical structures of three of the 20 amino acids characteristic of organisms are shown in **Figure 3.8.** (The complete list of amino acids is illustrated in Appendix 2.) Look closely at these structures. First, notice that in this figure the amino and carboxyl groups are shown as charged—this is the configuration they have in living cells. Also you can see that some amino acid R-groups are charged, some are polar, and some are neither charged nor polar. Whether an amino acid's R-group is charged, polar, or neutral affects how that amino acid interacts with other molecules.

You may have heard that in humans, as in other organisms,

amino acid a molecule with a central carbon to which an amino group, a hydrogen atom, a carboxyl group, and an R-group are attached

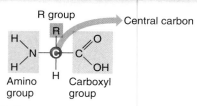

Amino acid molecules have a central carbon atom. Attached to it are an amino group, a hydrogen atom, and a carboxyl group. Each amino acid has an R-group that is unique to each of the 20 amino acids found in living organisms.

Figure 3.7   The Basic Plan of an Amino Acid.

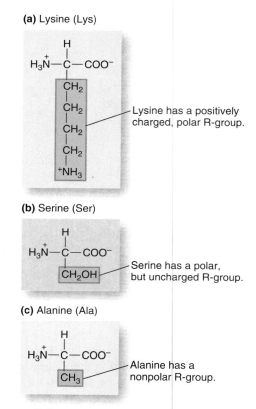

(a) Lysine (Lys)

Lysine has a positively charged, polar R-group.

(b) Serine (Ser)

Serine has a polar, but uncharged R-group.

(c) Alanine (Ala)

Alanine has a nonpolar R-group.

Figure 3.8   Distinctive R-Groups Give Each Amino Acid Unique Chemical Properties.

there are essential amino acids that must be in the diet (**Table 3.6**). While your body can make many of the amino acids from other molecules, nine amino acids cannot be made, and you must obtain these from your diet. If you eat any meat, you will have a sufficient supply of the essential amino acids. If you are a vegetarian, but eat dairy products like milk and eggs, your diet will supply all the essential amino acids. If you are a vegan vegetarian, however, you will need to do some research to ensure that you are getting all the essential amino acids.

Table 3.6	The 20 amino acids found in humans and their abbreviations
Essential amino acids for adults	**Nonessential amino acids**
Histidine (His)	Alanine (Ala)
Isoleucine (Ile)	Arginine (Arg)
Leucine (Leu)	Asparagine (Asn)
Lysine (Lys)	Aspartic acid (Asp)
Methionine (Met)	Cysteine (Cys)
Phenylalanine (Phe)	Tyrosine (Tyr)
Threonine (Thr)	Glutamic acid (Glu)
Tryptophan (Trp)	Glutamine (Gln)
Valine (Val)	Glycine (Gly)
Serine (Ser)	Proline (Pro)

Amino Acids Are the Building Blocks of Proteins

You probably know about proteins from a nutritional perspective. Protein is concentrated in cheeses, meats, poultry, fish, peanuts and other legumes, and eggs; lesser amounts of protein are in grains and vegetables. Protein is the primary component of muscle tissues. When you crack open an egg, the gooey egg white is a mass of a single kind of protein called albumin. Hair is made of a protein called keratin that can take on different forms that make hair straight or curly, coarse or fine. Keratin also forms fingernails and the claws, hoofs, horns, scales, and feathers of other animals.

A **protein** is a large biological molecule made of one or more chains of amino acids. Proteins accomplish nearly all of the cellular jobs in a living organism. Each kind of protein has a unique function. Take a moment to think about that last sentence for a minute. How can *each* protein have a different function? Each kind of protein is different because each has its own individual three-dimensional shape with its distribution of charges in the bulges and pockets, twists, pleats, and corkscrews that form the molecule. The shape and charge of a protein determine what it will do in a cell. This is one of the best examples of the now-familiar principle that complex properties emerge when simple chemicals combine. Each protein gets its unique charge and shape from the basic chemicals—the amino acids—in its structure.

The chains of amino acids that form proteins can be distinctive in several ways. One variable feature is the length of the amino acid chain, which can be just a few amino acids long or several hundred amino acids long. Another way that proteins vary is the specific amino acids they contain. Although it is possible that a protein could form from repeating chains of amino acids—for example, a string of all lysines, or alternating lysines and valines—such simple sequences are not often found in living cells. Most proteins usually contain all or nearly all of the 20 amino acids common to life. **Figure 3.9** illustrates this concept, using myoglobin as an example. In a chain of 153 amino acids, 19 out of the available 20 are used. Finally, the order of amino acids is important. Even two proteins made of exactly the same number and kinds of amino acids can be different if the amino acids are arranged in a different sequence. Therefore, it is easy to see that proteins are so varied because each kind has a distinct number and sequence of amino acids. Let's look at the implications of the amino acid sequence of a protein in greater depth.

Proteins Can Have Four Levels of Structure Just as letters of the alphabet form words, words form sentences, and sentences form paragraphs, proteins have different levels of structure. Most proteins have three levels of structure, and some proteins have a fourth. **Figure 3.10** shows these levels of structure, and you will refer to this figure throughout this section.

The **primary structure** is the specific sequence of amino acids. Figure 3.10a shows a short chain of amino acids that form a portion of the primary structure of a large protein. The covalent bond that holds the amino acids together occurs between the carboxyl carbon atom of one amino acid and the

protein a large biological molecule made of one or more chains of amino acids

primary structure the amino acid sequence of a protein

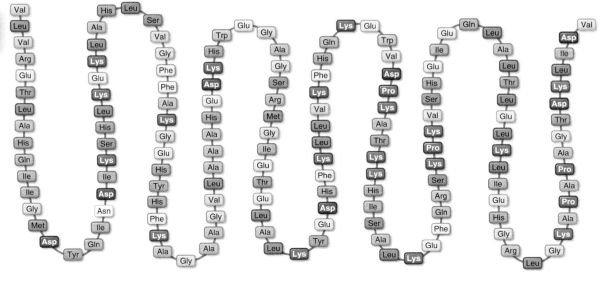

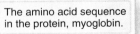
The amino acid sequence in the protein, myoglobin.

Figure 3.9 **The Amino Acid Sequence in Myoglobin.**

■ *If another protein molecule had these same amino acids, but arranged in a different order, would you expect it to have the same function as myoglobin? Why or why not?*

Figure 3.10 **Levels of Protein Structure.**

(a) The **primary structure** of a protein is the sequence of its amino acids.

H_2N—Ala—Thr—Cys—Tyr—Glu—Gly—COOH

(b) The **secondary structure** of a protein is formed by hydrogen bonds between nonadjacent carboxyl and amino groups.

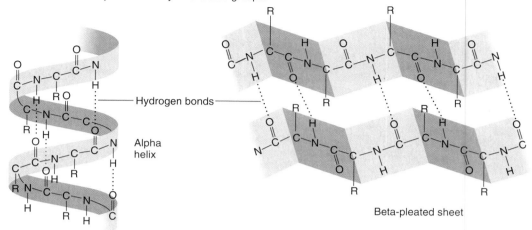

Alpha helix

Beta-pleated sheet

(c) The **tertiary structure** of a protein results from interactions between R-groups.

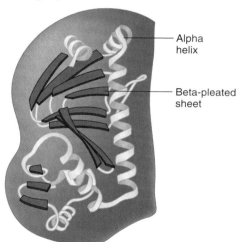

Alpha helix

Beta-pleated sheet

(d) The **quaternary structure** of a protein results from hydrogen and ionic bonds between separate tertiary structures.

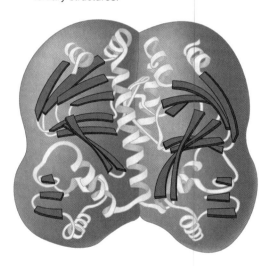

---

peptide bond the bond in a protein between the carboxyl carbon atom of one amino acid and the amino nitrogen of another amino acid

secondary structure the folding of an amino acid chain into a helix or pleated sheet that results from hydrogen bonds between the carboxyl group of one amino acid and the amino group of another

amino nitrogen atom of the next amino acid. Scientists call this a **peptide bond** (**Figure 3.11**). During the chemical reaction that forms a peptide bond, two hydrogen atoms are lost from one amino group and one oxygen atom is lost from a carboxyl group. In other words, one water molecule is lost during the formation of each peptide bond.

The shape of a protein is produced when the chain of amino acids folds. Figure 3.10b shows that there are two levels of folding. A first round of folding transforms parts of the chain into one of two shapes: either a helix that is like a springy phone cord or a pleated sheet like the "squeeze box" of an accordion. This shape is the protein's **secondary**

structure. If you want to better understand the difference between these two shapes, hold the ends of a long strip of paper. You can form a helix by twisting the ends, or if you fold the strip like an accordion, you'll form a pleated sheet.

Secondary structures come about through the formation of hydrogen bonds between carboxyl and amino groups of nearby amino acids along the existing primary structure. Recall that both of these functional groups are polar. In a way that is similar to the bonds that form between water molecules, hydrogen bonds form between the oxygen on the carboxyl group of one amino acid and the hydrogen on the amino group of another amino acid a few positions down the chain. These hydrogen bonds form at regular intervals all along the length of the primary structure. The regularity of these bonds produces the regular coils of the helix or the regular folds of the pleated sheet (see Figure 3.10b). Collectively, these bonds pull the chain of amino acids into the secondary structure. The exact sequence of amino acids in the string determines whether the secondary structure is a helix or pleated sheet and whether a length of amino acid chain has a secondary structure. In some proteins nearly the entire chain has a secondary structure. Other proteins may have lengths that are randomly coiled, without a secondary structure. Finally, a protein may contain stretches that form pleated sheets, other stretches that form helices, and still other areas that have no secondary structure at all (**Figure 3.12**).

The **tertiary structure** of a protein is the result of another folding process. Although there are generally only two configurations possible for secondary structures, helix or pleated sheet, each kind of protein has a unique tertiary structure (see Figure 3.10c). This makes sense when you recall that each kind of protein has a unique sequence of amino acids and thus a unique sequence of R-groups. A protein's tertiary structure emerges when the pleated sheet or helix folds due to interactions between the charges on R-groups along the amino acid chain. As a result of all the interactions between R-groups, plus other complex factors, the amino acid chain is pulled together and pushed apart. The protein crumples and snaps into its unique three-dimensional shape.

Some functional proteins are composed of a single amino acid sequence that has folded into its final tertiary structure. For other proteins the final functional molecule is composed of two or more amino acid chains bound together. When two or more amino acid chains bind, forming a huge, composite protein molecule, the resulting complex is called the **quaternary structure** of the final functional protein. The final structure is called a protein, even though it is more complex than a single amino acid chain. Hemoglobin is one good example of a protein with quaternary structure (**Figure 3.13**). Hemoglobin transports oxygen and carbon dioxide in the blood, and it is made of four individual amino acid chains.

You probably think that you've learned more than you'll ever need to know about proteins—and you're just about right. But now you get the payoff for working so hard to understand how most proteins are formed. Now that you have read about how proteins get their three-dimensional shapes, you can appreciate the most important lesson about proteins: *the function of most proteins is determined by their final structure.* Because of complex tertiary folding, each protein has characteristic pockets, bulges, and other internal and external topographical features. In addition, each protein has a unique distribution of electrical charges. This is because many of the charged R-groups are not bound to other R-groups, but instead they are free. So parts of the protein have unpaired charges. These charged parts of the protein molecule can react with other chemicals in the environment. Because each of the approximately 100,000 kinds of proteins in the human body has a distinct shape and charge distribution,

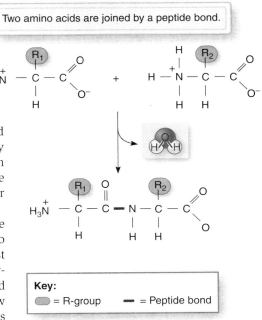

Two amino acids are joined by a peptide bond.

Key:
◯ = R-group   — = Peptide bond

Figure 3.11   **Peptide Bonds Join Amino Acids into Larger Molecules.**

Pleated sheets are shown in yellow. Helices are shown in red. Areas of proteins with no secondary structure are shown in white.

Figure 3.12   **A Protein May Contain More Than One Kind of Secondary Structure.**

tertiary structure the unique three-dimensional shape of a protein determined by the folding of the protein's secondary structure

quaternary structure a composite molecule made of two or more protein molecules

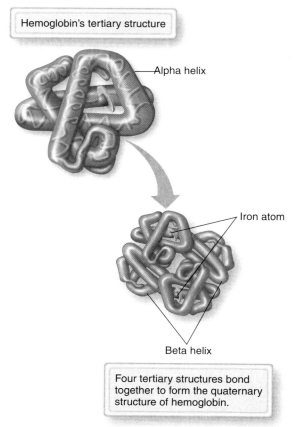

Hemoglobin's tertiary structure

Alpha helix

Iron atom

Beta helix

Four tertiary structures bond together to form the quaternary structure of hemoglobin.

Figure 3.13 **The Structure of a Hemoglobin Molecule.** The hemoglobin molecule contains four iron atoms and can hold eight oxygen atoms.

each engages in a set of chemical interactions that no other protein can duplicate. In this way proteins have properties that emerge from interactions between the chemicals from which they are made.

One thing to note is that the three-dimensional structure of a protein depends on environmental conditions. If these conditions are not exactly right, a protein can also lose its three-dimensional shape. Not surprisingly, the "right" conditions are normal body conditions. In a human, for example, those would include a temperature of about 36°C (98°F) and a pH of about 7.4. If a protein experiences an extreme environmental change, the hydrogen bonds that hold the protein together will be broken, and its three-dimensional shape will be lost. This process is called **denaturation**. This is what happens when you cook meat or boil an egg and the protein's texture and color change. Highly acidic or extremely basic conditions also can denature proteins. Some exotic recipes, for example, use acidic lime juice or salt to denature proteins in meat and give you delicacies such as salmon prepared as gravlax or scallops prepared as seviche.

denaturation when the hydrogen bonds that hold a protein together are broken, and its three-dimensional shape is lost

The chemicals used to permanently curl or straighten hair denature keratin by breaking its hydrogen bonds. After the chemicals have been neutralized, the keratin forms new hydrogen bonds in its new curly or straightened shape. The sensitivity of proteins to their environmental conditions is one reason why the cells in the body must maintain the right temperature and pH. Extremes of either may cause proteins to denature, which probably would result in death.

Proteins Perform at Least Five Sorts of Work

Cells do an amazing variety of things to stay alive and prosper. Cells take in energy and transform it into forms of energy they can use. They build complex molecular structures and carry out complex chemical processes. Each cell in a human body has its own specialized jobs to carry out, and the variety of cellular structures and processes reflect these jobs. For example, muscle cells contract—and in so doing, they move the body through space. Likewise, cells of salivary glands manufacture and secrete saliva, and cells of retinas respond to different wavelengths of light by firing nerve impulses that eventually result in vision. Proteins do most of these cellular jobs.

Structural proteins provide most of the physical structure of a cell. They determine cell shape and in large part determine the structure of a cell's organelles. Structural proteins also can be found outside of cells. The connective tissues that hold bones and muscles together are a prime example of structural proteins located outside of cells. *Enzymes* are another important class of proteins and are involved in making and breaking chemical bonds. Enzymes build and break down cellular molecules and structures and they are involved in many other chemical processes too. *Contractile proteins* are a special class of structural proteins that allow movement, such as the contractile proteins in muscle cells. Some proteins function in *cell communication*. These proteins actually are embedded in the lipids of cell membranes and allow a cell to communicate with its environment. Finally, there are *transport proteins* that bind to atoms or molecules and move them from one place to another. Hemoglobin is an example of a transport protein.

3.7 Hemoglobin Is an Example of the Emergent Properties of Proteins

Hemoglobin, found in red blood cells, is a protein that carries oxygen and carbon dioxide around the body. It gives blood its red color and is a good example of the emergent properties of proteins. Hemoglobin has a quaternary structure of four separate protein subunits—four amino acid chains—that are loosely bound together. Each amino acid chain is folded to make a single, central pocket that contains an atom of iron. The four iron atoms in one complete hemoglobin molecule allow it to bond with four molecules of oxygen (O_2) which then can be delivered to cells in various tissues of the body. Once in the tissues, hemoglobin releases oxygen and picks up carbon dioxide, which in turn is carried to the lungs and released. New oxygen is then picked up by hemoglobin and the process continues. Hemoglobin is a critical molecule for maintaining a supply of oxygen to cells and tissues.

The ability of hemoglobin to bind and release oxygen and carbon dioxide is determined by the three-dimensional shape of the amino acid sequence of the *entire* hemoglobin molecule. The individual amino acids cannot bind iron, and they cannot bind or release oxygen or carbon dioxide. The secondary helix structure of the protein subunits of hemoglobin doesn't contain iron or bind and release oxygen or carbon dioxide either. This ability is made possible only after each of the four helices has folded into its tertiary structure. Transport abilities are further enhanced when the four subunits come together to form the quaternary structure of a complete molecule of hemoglobin. Because the tertiary structure of hemoglobin is determined directly by the primary amino acid sequence, the ability of hemoglobin to carry and release oxygen and carbon dioxide is an emergent property.

The importance of hemoglobin's primary structure and its sequence of amino acids become evident when certain amino acids in the sequence are not correct. This is just what happens in the disease called *sickle cell anemia.* This is an inherited disease in which the DNA instructions for making hemoglobin are incorrect. Specifically, the flawed hemoglobin of sickle cell anemia contains a valine molecule in one position where a glutamic acid molecule should be (**Figure 3.14a**). The result of this single amino acid error is a hemoglobin molecule that forms rigid fibers instead of globular structures (Figure 3.14b), especially when the fluid around the red blood cells contains little dissolved oxygen. These elongated hemoglobin fibers deform the nor-mal dented disc shapes of red blood cells into angular sickle shapes (Figure 3.14c). Like sticks caught in a drain, the pointy red blood cells get stuck in the smallest blood vessels, causing various tissues of the body to become deprived of oxygen. Muscle weakness, brain damage, kidney damage, pains in the limbs, and heart failure can follow. What might seem to be a tiny chemical error in a protein's amino acid sequence can have major repercussions for the protein's function and for the individual.

QUICK CHECK

1. Describe the hemoglobin molecule.
2. How does the single amino acid change in a hemoglobin molecule affect individuals with the inherited disease sickle cell anemia?

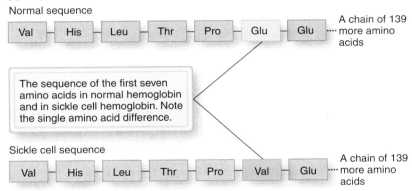

(a) Sequence of amino acids

Normal sequence

Val — His — Leu — Thr — Pro — Glu — Glu ···· A chain of 139 more amino acids

The sequence of the first seven amino acids in normal hemoglobin and in sickle cell hemoglobin. Note the single amino acid difference.

Sickle cell sequence

Val — His — Leu — Thr — Pro — Val — Glu ···· A chain of 139 more amino acids

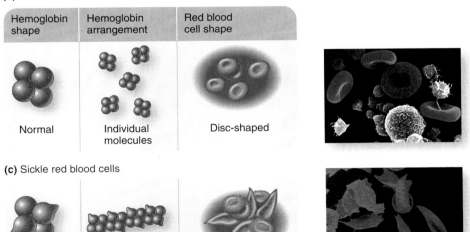

(b) Normal red blood cells

Hemoglobin shape	Hemoglobin arrangement	Red blood cell shape
Normal	Individual molecules	Disc-shaped

(c) Sickle red blood cells

Altered	Long chains of molecules	Sickle-shaped

Figure 3.14 **Sickle Cell Anemia.** (**a**) Just a single amino acid difference causes all of the symptoms of sickle cell anemia. (**b**) Normal human red blood cells have the shape of dimpled discs. (**c**) Red blood cells of a person with sickle cell anemia collapse into pointy shapes when they travel to tissues where oxygen is in short supply.

3.8   An Added Dimension Explained
What Causes Mad Cow Disease?

At the start of this chapter you read about vCJD and mad cow disease that struck Great Britain in 1995. Although there is still no cure for vCJD, what is truly amazing about this story is that investigations into the nature of mad cow disease led to the discovery of a *totally new* kind of infectious disease agent. The story of this discovery includes scientists working around the world on seemingly unrelated problems—and is a good example of how researchers cast a broad net to solve a specific problem.

One place to begin the story is with sheep in Great Britain, where farmers had coped with a disease called *scrapie* since the 1700s. Some symptoms of scrapie are similar to those of BSE. Sheep with scrapie lose motor coordination and become irritable and disoriented. The disease gets its name because the sheep scratch their bodies along fences, rub off their wool, and continue to scrape at their skins. By the 1960s researchers had good evidence that scrapie was an infectious disease that somehow passed between animals, although humans never seemed to be affected by it.

The next big clue on the road to understanding BSE and vCJD came from studies of the Fore people of New Guinea, half a world away from Britain (**Figure 3.15**). Starting early in the 1900s the Fore—who numbered only about 8,000 individuals—began to suffer from a strange form of dementia called *kuru*. The symptoms of kuru and the autopsied brains of kuru victims were similar to the pattern seen in vCJD. How did the Fore contract kuru?

The Fore, especially the women, practiced a form of ritual cannibalism. As strange as it may sound to Western society, the Fore women and their children ate the bodies of their dead rela-

Figure 3.15   **The Fore People of New Guinea.**

tives, especially their brains. The reasons are complex. At one level, consuming the bodies of relatives was a form of reverence and respect, because the Fore believed that the strength of the dead person would be passed to the living. At another level, this practice provided a needed source of protein for Fore women and children. Studies demonstrated that infected brain tissues from dead Fore people could infect other primates. This was strong evidence that kuru was being passed to women and children through the practice of ritual cannibalism. Eventually, as the practice of cannibalism has faded from the Fore culture, kuru also has faded.

Back in Great Britain, the rapid spread of BSE in cattle herds alarmed farmers and governments, and the conclusion that it had somehow spread to cattle from sheep was hard to dismiss. The feed that modern farmers gave to cows was fortified with animal protein that often came from sheep. And sometimes the sheep used to make protein additives for cattle feed had been sick with scrapie. Some cattle feed was even fortified with protein from sick cows. It seemed likely that British cows were getting BSE from eating infected cattle feed.

As the pattern of vCJD became clear, it was hard to avoid the conclusion that people contract vCJD by eating infected beef. The evidence was enough to convince health officials of a connection between BSE and vCJD—but a strong scientific conclusion needs more evidence. Specifically, scientists would not be convinced until the infectious agent that passed vCJD to humans from BSE cows was isolated, characterized, and identified.

Nearly all scientists were convinced that some common infectious agent would be responsible, such as bacteria or viruses. Unfortunately, laboratory studies did not support this hypothesis. Viruses, bacteria, and other known infectious agents can be inactivated—made noninfective—if they are heated or otherwise exposed to extreme treatments that denature their proteins and DNA. In contrast, the infected material from BSE cows could not be inactivated, even by extreme treatments like boiling and sterilizing. Furthermore, no trace of such an agent could be found, although many researchers intensively searched for it. Also the vCJD patients did not display the kinds of immune system reactions that a typical infectious agent would produce. These mysteries left many scientists scratching their heads, but at this point a remarkable scientist entered the arena.

After he had lost a patient to CJD early in his career as a physician, Stanley Prusiner devoted his scientific life to pursuing the agent that causes CJD. Prusiner's persistent, thorough work won over most of his colleagues, and in 1997 he was awarded the Nobel Prize in Physiology or Medicine. Here is part of his story.

First, Prusiner developed a procedure for isolating and concentrating a sample of the infectious agent from the brains of diseased individuals by using a strain of mice as research subjects

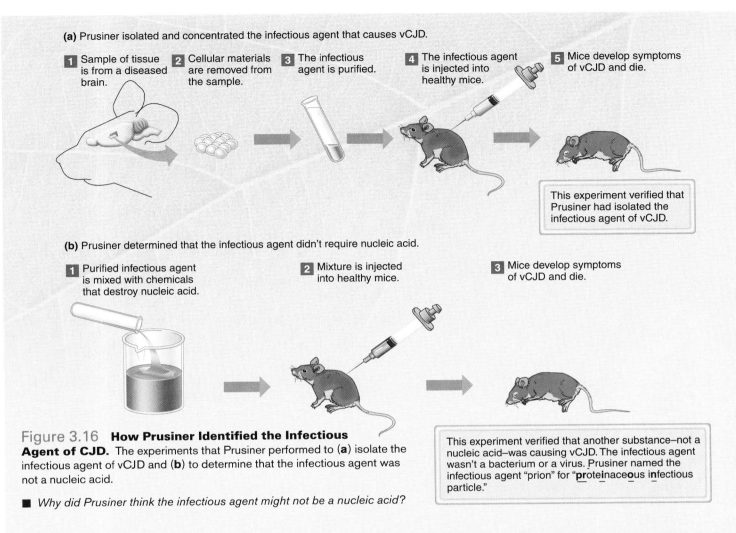

(a) Prusiner isolated and concentrated the infectious agent that causes vCJD.

1 Sample of tissue is from a diseased brain.

2 Cellular materials are removed from the sample.

3 The infectious agent is purified.

4 The infectious agent is injected into healthy mice.

5 Mice develop symptoms of vCJD and die.

This experiment verified that Prusiner had isolated the infectious agent of vCJD.

(b) Prusiner determined that the infectious agent didn't require nucleic acid.

1 Purified infectious agent is mixed with chemicals that destroy nucleic acid.

2 Mixture is injected into healthy mice.

3 Mice develop symptoms of vCJD and die.

Figure 3.16 **How Prusiner Identified the Infectious Agent of CJD.** The experiments that Prusiner performed to **(a)** isolate the infectious agent of vCJD and **(b)** to determine that the infectious agent was not a nucleic acid.

This experiment verified that another substance—not a nucleic acid—was causing vCJD. The infectious agent wasn't a bacterium or a virus. Prusiner named the infectious agent "prion" for "**pr**otein**a**ce**o**us **in**fectious particle."

■ *Why did Prusiner think the infectious agent might not be a nucleic acid?*

(**Figure 3.16**). Prusiner knew his procedure worked because he could cause healthy mice to develop the disease by injecting them with his sample. When he subjected some of his samples to chemical procedures that destroy nucleic acids, those procedures did not change the ability of the samples to produce the disease in mice. From these studies, Prusiner became convinced that there was no nucleic acid in the infectious agent that caused these fatal dementias. Additional experiments pointed clearly to proteins. To describe the proteins that cause these fatal dementias, Prusiner coined the term **prion,** for **p**roteinaceous **i**nfectious particle. And here is the surprising story about what we know, so far, about prions.

One of the human body's thousands of different kinds of proteins sits on the surface of the cell membrane of each brain cell. Most of this normal protein, whose function is still unknown, is folded into a helix. For reasons that are also unknown, the amino acids in the protein sometimes fold into a pleated sheet instead of a helix. The amino acid sequence in the two versions of the protein is exactly the same, and there are no other chemical modifications of the protein. When the protein folds into a pleated sheet, the brain begins to degenerate in the region of the abnormal proteins, producing abnormal brain tissue in which large holes form (**Figure 3.17**).

But a key feature of these diseases is that they are passed from one person to another or from one mammal to another in infected meat or infected brain tissue. How does infection occur? First, if a prion becomes detached from a cell's surface, it is then free to travel within the original animal or in the body of another animal that eats meat infected with prions. Even more amazing is that a prion somehow can direct a normal protein to fold in the abnormal way. When infection occurs, the invading prions begin altering normal proteins in the infected individual, transforming them into prions. As they become detached from cell membranes,

prion an infectious protein that can transform other proteins into defective, nonfunctional molecules

—Continued next page

Continued—

(a) Normal mouse brain tissue

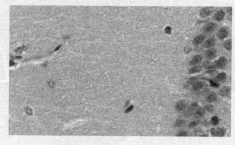

(b) Mouse brain tissue infected with prions

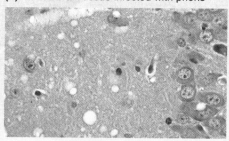

Figure 3.17 **Brain Tissue Samples.** (**a**) Normal mouse brain tissue and (**b**) brain tissue from a mouse infected with prions.

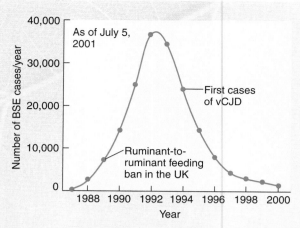

Figure 3.18 **The Epidemic of BSE in Great Britain.** The course of the epidemic of bovine spongiform encephalopathy (BSE) in Great Britain in 1987–2000.

the prions migrate through the brain, altering additional normal proteins and leaving destruction in their wake. The method by which prions multiply explains why it can take months or years after the initial infection for symptoms to show: the prions move gradually through the brain, slowly converting normal proteins to new prions as they go.

Prions can be present in the brains of many types of mammals and can be transmitted from one organism to another, even from one species to another. In the case of mad cow disease, it is likely that cattle in Great Britain became infected after eating feed contaminated with remains of infected sheep. When some people acquired the prions by eating infected beef, they developed symptoms similar to Creutzfeldt-Jakob disease and died of vCJD.

The scientific knowledge gained throughout this investigation was translated into public policy. In the late 1980s, even before the link to humans was known, people realized that cows probably contracted BSE after eating commercially prepared feed that contained ground-up tissue from other infected animals. In 1989, to eliminate this source of infection, Great Britain banned these cattle feeds. Once officials recognized the connection between BSE and vCJD, this ban was more strictly enforced, and between 1988 and 2000 the number of BSE-infected cattle dropped dramatically (**Figure 3.18**). In the United States and other countries similar bans were estab-

lished, and the import of British beef was prohibited. In Britain all cattle over 30 months old were killed, and their carcasses were burned. It is estimated that over 4 million animals were destroyed.

The controls on cattle feed seem to have dramatically reduced the number of BSE cattle in Great Britain, and the ban on imports of British beef seems to have prevented large outbreaks of BSE and vCJD in other countries. As of January 2004 more than 150 cases of vCJD had been diagnosed, but most of those were in Great Britain or involved people who had been in Britain in the 1980s. Still, it is not certain how many more cases may yet appear. No one knows if BSE is the only source of vCJD in humans. Governments and health officials still are concerned about the threat of BSE and vCJD.

Despite the loose ends and unanswered questions, the prion/mad cow disease story is important because it provides an excellent example of how the complex characteristics of living things—in this case, the behavior and thought processes of cows, sheep, and humans—can be altered by the malfunction of only *one* protein among a hundred thousand. While it is tempting to assume that human nature is too complex and subtle to be explained by simple chemistry, this interaction of deformed protein and brain cells shows how deeply chemistry is coiled into the fabric of life—including human behavior. Clearly, there can be no life without chemistry.

QUICK CHECK

1. Describe what a prion is.

2. How does the vCJD disease travel from an infected organism to another organism?

3. How did cows, which are grazing animals, eat diseased meat and get BSE?

Now You Can **Understand**

Nanotechnology

The study of biology has had a dramatic impact on many aspects of human life. Treatments for infectious diseases, treatments for cancer, modern water treatment plants, and household detergents are just a few examples. For the most part these applications are considered positive and helpful, but increasingly people are beginning to worry that the application of knowledge about life might prove more harmful than helpful. The next major wave of biological applications is already generating heated debates. The new technologies are collectively called *nanotechnology,* a process that makes complex structures on a molecular scale.

While the term *nano* refers to the tiny size of the objects produced, most nanotechnology applications come from knowledge of organic molecules—in particular, basic carbon structures, DNA, and proteins. Carbon nanotubes, for example, are long, incredibly thin tubes of pure carbon. Single nanotube filaments are invisible to the naked eye but are stronger than steel, yet flexible. The knowledge that led to the development of nanotubes is a direct result of studies of carbon compounds.

Nanotechnology applications driven by knowledge of DNA and proteins are even more astounding. In research laboratories scientists have made DNA computers, DNA tweezers that can grasp other molecules, and molecular ink that can be etched away by enzymes to produce printed patterns on a molecular scale. Tiny molecular motors have been created that can move objects. Many scientists believe that nanotechnology devices are going to revolutionize medicine. Medical diagnosis will be based on tiny chips that read your genes and proteins, and one day treatments for a variety of diseases will involve an injection of nanobots rather than an injection of a drug.

While many scientists are optimistic that nanotechnology will improve our quality of life, other people are certain that, either through accidental or deliberate terror, nanotechnology could cause catastrophic disaster. Which view is correct? Or are both likely to be true? One of the characteristic features of biological molecules is that they interact with each other and with other molecules. It seems likely then that biologically based nanotechnology will have some unpredictable effects. Now—based on your understanding of biological molecules—you should be able to follow the public discussions of what those potential effects might be. This is important, because it won't be long before you will take part in social and political discussions and decisions about whether nanotechnology should be banned, encouraged, or strictly regulated.

What Do **You** Think?

Eat chitin for weight loss?

The Internet has become a marketplace for new ideas, some of them good and some of them of little value. Because it is a marketplace, this means that many people are trying to make money from their ideas. Therefore, as a consumer, it is important to be as knowledgeable as possible about Internet sources and be able to determine which ones to trust and which to treat with caution. Many Web-based businesses offer weight-loss products—a common one is chitin. You already have read that chitin is a carbohydrate found in the external skeleton of many animals such as crustaceans and insects. Chitin is not digestible, but like cellulose, chitin can provide dietary fiber. How would you go about evaluating the claims by sellers of chitin products? One step would be to look in the scientific literature to find formal, objective studies of chitin in weight loss. You can do this at a government website called PubMed—the National Library of Medicine's search service—which provides summaries of all of the medically related scientific literature. A search in PubMed finds many studies on chitin, but the few studies of chitin and weight loss disagree with one another on chitin's effectiveness. So, for now at least, direct scientific studies cannot help you. Another approach is to do a standard search in a search engine such as Yahoo or Google. Here you have to be more careful because many websites are commercial sites or personal sites, and these are not as likely to provide objective information. Some of these sites indicate "decades" of research supporting the health value of chitin. How can this be true if there are few studies in PubMed on chitin and weight loss? One possible explanation is that the research is published in foreign journals, many of which are not included in PubMed. Or the commercial websites could be stretching the truth a bit. Either way, you are left with your own knowledge of biology to assess whether chitin is worth your money. Based on the fact that chitin is an indigestible fiber, your first guess would be that claims for weight loss are probably unfounded, but given the lack of actual evidence you may want to withhold judgment for the time being. But the decision is yours. What do *you* think?

Chapter Review

CHAPTER SUMMARY

3.1   Life Depends on Complex Biological Molecules

Lipids, carbohydrates, nucleic acids, and proteins are the biological molecules shared by living organisms.

biological molecules   54

3.2   Biological Molecules Are Built Around Carbon Atoms

All biological molecules are built around carbon atoms. Hydroxyls, carboxyls, amino groups, and phosphates are functional groups that impart distinctive chemical characteristics to complex biological molecules.

amino group   59
carboxyl   58
functional group   58
hydrocarbon   57

hydroxyl   58
organic chemistry   56
phosphate   58

3.3   Lipids Are Biological Molecules That Do Not Mix Well with Water

Lipid molecules have large hydrocarbon portions, and some have water-soluble polar functional groups too. Lipids are components of cell membranes as well as energy storage molecules and hormones.

lipid   60

3.4   Carbohydrates Are Made of Carbon, Oxygen, and Hydrogen

Carbohydrates are molecules with at least three carbon atoms and have carbon, hydrogen, and oxygen atoms as well. Carbohydrates are energy storage molecules and structural molecules.

carbohydrate   63
disaccharide   63

monosaccharide   63
polysaccharide   63

3.5   Nucleic Acids Are Strings of Nucleotides

Nucleotide molecules have a sugar with a phosphate and nitrogenous base attached and form nucleic acids. Nucleotides form the nucleic acids, DNA and RNA. Both DNA and RNA contain instructions, but they differ in structure, nitrogenous bases, sugars, and in how the molecules function.

nitrogenous base   66
nucleic acid   66

nucleotide   66

3.6   Proteins Are Folded Chains of Amino Acids

Proteins are built from chains of amino acids, which are biological molecules with an amino group, carboxyl group, hydrogen atom, and a distinctive R-group attached to a central carbon atom. Amino acids are the building blocks of proteins, the molecules that do the work of

cells. All proteins have three levels of structure, and some have a fourth level.

amino acid   68
denaturation   72
peptide bond   70
primary structure   69

protein   69
quaternary structure   71
secondary structure   70
tertiary structure   71

3.7   Hemoglobin Is an Example of the Emergent Properties of Proteins

As sickle cell anemia shows, even a small change in the sequence of the amino acids of a protein can dramatically alter the way the protein acts.

3.8   An Added Dimension Explained: What Causes Mad Cow Disease?

Stanley Prusiner discovered that a previously unknown, infectious, deformed protein—a prion—causes vCJD and mad cow disease, as well as the related diseases kuru and scrapie.

prion   75

REVIEW QUESTIONS

TRUE or FALSE. If a statement is false, rewrite it to make it true.

1. Lipid molecules dissolve readily in water.
2. Proteins are the least complex of the biological molecules.
3. RNA is a nucleic acid; it is double-stranded, incorporates deoxyribose sugar, and contains uracil.
4. In sickle cell anemia the hemoglobin molecule has one incorrect carbohydrate.
5. Prions are aberrant proteins that can cause mad cow disease.

MULTIPLE CHOICE. Choose the best answer of those provided.

6. The shape and charge distribution of a protein is critically important because
   a. the shape changes over time as the protein ages, giving it new properties.
   b. these properties govern the function of each protein.
   c. the charge and shape are identical for every kind of protein and provide an important instance of the unity of life.
   d. proteins are exceptions to the universality of the electrical force.
   e. proteins have no chemical bonds.

7. A protein that helps break up or combine molecules is called
   a. hemoglobin.
   b. collagen.
   c. an enzyme.
   d. a phosphate.
   e. a carbon ring.

8. A carbon atom can make a maximum of _____ covalent bonds.

   a. one
   b. two
   c. three
   d. four
   e. six

9. The chemical elements that are most common in living systems are

   a. C, S, Fe, H, N
   b. C, Cl, Mg, Fe, H
   c. C, H, O, N, P
   d. C, I, Na, K, O

10. Which of the following fats are considered "healthy" for you? Be sure you look at the type of fat and at the source; both must be correct.

    a. saturated fats such as corn oil
    b. saturated fats such as lard
    c. trans fats found in shallow-water fish such as catfish
    d. polyunsaturated fats such as soybean oil
    e. trans fats in partially hydrogenated vegetable oils

11. In your diet you take in significant amounts of carbohydrates from all of the following except

    a. an apple slice.
    b. an ounce of lean beef.
    c. a small baked potato.
    d. a stalk of celery.
    e. a 1-ounce chocolate bar.

12. Lipids are used for

    a. storing energy in fat cells until it is needed.
    b. membranes that are in and around every cell in an organism.
    c. making some hormones, chemical messengers in the body.
    d. helping some nonpolar molecules to move into cells.
    e. All of the above are uses for lipids.

MATCHING

13–16. Match the small organic molecular subunit with the large organic biomolecule that is built from it (one choice will not match):

13. hydrocarbon chain   a. protein

14. simple sugar   b. nucleic acid

15. nucleotide   c. carbohydrate

16. amino acid   d. lipid

                             e. water

17–20. Match the molecule to one of its uses in living organisms (one choice will not match):

17. proteins   a. short-term energy storage

18. carbohydrates   b. the basis of organic chemistry

19. lipids   c. accomplish most cellular processes

20. nucleic acids   d. store energy and are important components of cell membranes

                             e. carry genetic instructions

1. In what ways are the properties of complex carbohydrates an example of emergent properties?

2. What determines the three-dimensional shape of a protein?

Refer to the following graph. What is the youngest age group for which the percentage of women with high cholesterol is higher than the percentage of men with high cholesterol?

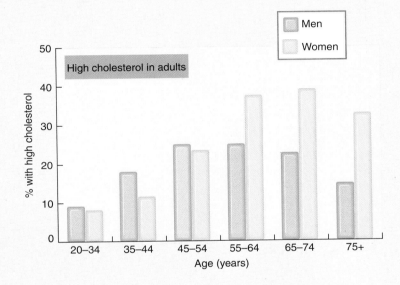

1. In the 1800s Louis Pasteur discovered how to "pasteurize" liquids. Pasteur showed that if liquids are heated to a sufficient temperature, harmful bacteria are killed and the liquids are less likely to make people sick. Of course, you probably know that heating the bacteria "kills" them, but how? Using what you know about biological molecules, propose a mechanism that explains how heating kills the infectious bacteria in foods.

2. In this chapter you have read about the biological molecules that interact to produce life. Recall from Chapter 2, though, that in cells these interactions occur in a watery environment. One way to really understand this is to consider whether any other solvent could be substituted for water. You know that water is polar, and you also know about the charge distribution on the major biological molecules. Think about how each of these molecules would interact with or dissolve in water. Now think about the ethyl alcohol (ethanol) molecule, diagramed in Table 3.3. How would the interactions of the biological molecules differ if they were immersed in ethyl alcohol or if they were immersed in water?

For additional study tools, visit www.aris.mhhe.com.

Riding Accident Paralyzes Actor Christopher Reeve

June 1, 1995

Spinal Cord Injuries: Actor Christopher Reeve's Tragedy Highlights the Devastating Effects

June 6, 1995

Actor Christopher Reeve suffered a paralyzing accident.

Superman is an enduring hero of American pop culture, and anyone who has read a *Superman* comic book can recite the litany of his superpowers. Thanks to his origin on the Planet Krypton, the Man of Steel makes even NFL tackles seem puny. He is faster than a speeding bullet and more powerful than a locomotive; he bends steel in his bare hands and leaps tall buildings at a single bound. Superman has X-ray vision, hyperacute hearing—and, best of all, with his red cape streaming in the wind, Superman can fly. He is a bona fide good guy who fights for truth, justice, and the American way. In 1978 Christopher Reeve was Hollywood's Superman incarnate, and he fit that exceptional image both on screen and off. Young, handsome, and strong, Reeve flew into movie history (**see chapter-opening photo**). He battled his archenemy Lex Luthor and saved heartthrob Lois Lane, cub reporter Jimmy Olsen, and the inhabitants of Metropolis from Luthor's evil plots. Reeve's first *Superman* movie was followed by three sequels, as well as 13 other feature films and many TV movies that established him as a Hollywood celebrity.

Reeve did have a life beyond the silver screen. He was a political activist who supported issues related to the environment, funding for the arts, education, and gun control. Reeve was also a talented athlete. An expert skier, sailor, and scuba diver, Reeve did his own movie stunts and twice flew a small plane solo across the Atlantic. He also loved *eventing*, a sport that combines a highly controlled form of horsemanship with bone-shaking cross-country jumping and precise show jumping. In 1995 Reeve's comfortable, successful life plunged into tragedy. His horse stopped short of a jump, and Reeve tumbled forward out of the saddle. Because his hands were tangled in the horse's bridle, Reeve landed headfirst and fractured his spine. He was paralyzed from the neck down and was unable to breathe. An emergency medical team saved his life, and surgery patched his neck bones back together. Sadly, they could do little for his damaged spinal cord. After the accident he could not move or even feel his body.

If you had a similar accident, the news would not be good. The doctors would tell you that any recovery would happen within a few months. It would be unlikely that you would fully recover. Christopher Reeve did not accept this future. He fought to improve his own condition and to change the future for others with spinal cord injury. Years after the best medical knowledge said his paralysis was permanent, although Reeve still was in a wheelchair he had recovered sensation in over 70% of his body. He was able to move all of his joints when immersed in a swimming pool. Out of water he could make small finger movements. With the aid of a new medical device, he was able to breathe without the aid of a respirator. Reeve even returned to acting. In an interesting bit of casting he appeared as the mysterious, brilliant, and beneficent Dr. Virgil Swann in *Smallville*, a TV sci-fi drama about the life of teenaged Clark Kent, a.k.a. Superman. Reeve dramatically bettered his own quality of life and was an inspiration for many others struggling with incapacitating illnesses and injuries. Sadly, in 2004 Christopher Reeve died of heart failure. He was 52 years old.

Why is it nearly impossible for medical treatments to repair or correct spinal cord injuries? Will people with spinal cord injuries ever be able to regain normal movement? When will there be effective treatments? These are urgent questions that you will return to at the end this chapter. By that time you will have learned much about the structures of cells and their interactions within cells. This new knowledge will enable you to more fully comprehend the problems involved in repairing a spinal cord injury. ■

4.1 Living Things Are Made of Cells

Imagine a realm that has been mapped in great detail but visited mostly at a distance. It is a mysterious destination, yet its architecture is known. Detailed photographs document the layout of its rooms, the placement of its furniture, and even the contents of its closets. This is a cell, the realm of life. Although you are made of trillions of these tiny places, you can visit them only by using the high-power technology of microscopy.

For the last 300 years biologists have been probing cells in various ways, trying to figure out what is inside of them, how they work, and what roles they play in living things. The modern understanding of cells has helped to banish many infectious diseases and has led to new screening tests for diseases as well as new medicines, therapies, and cures. Knowledge of the detailed structure of cells has even influenced the quality of the foods you eat and has led to sanitary practices that improve your health and enhance your living conditions. In sum, the scientific understanding of the structure and function of cells has dramatically changed your life.

An Appreciation of the Sizes of Cells Helps You to Visualize Them

Your body contains trillions of cells, yet because cells are so small, you can't see them or feel them or sense their presence. Except for a few exceptionally large cells like frog and toad eggs, and some unusually large single-celled organisms, most cells are invisible to the unaided human eye. Magnification of about 40 to 100 times or more is needed to see cells, and even higher magnifications are required to see the structures within cells. One way

Figure 4.1 **Sizes of Organisms, Cells, and Cellular Components.**

■ *If the period at the end of this sentence is 1 mm in diameter, how many human photoreceptors could fit across it?*

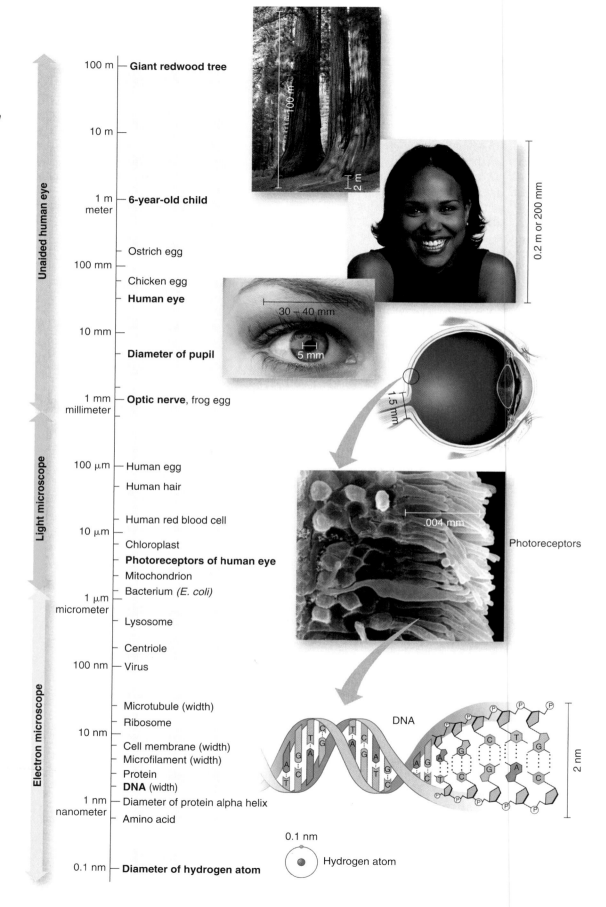

Unaided human eye

100 m — **Giant redwood tree**

10 m

1 m
meter — **6-year-old child**

100 mm — Ostrich egg

— Chicken egg
— **Human eye**

10 mm — **Diameter of pupil**

1 mm
millimeter — **Optic nerve**, frog egg

Light microscope

100 μm — Human egg

— Human hair

— Human red blood cell

10 μm — Chloroplast
— **Photoreceptors of human eye**
— Mitochondrion
— Bacterium *(E. coli)*

1 μm
micrometer — Lysosome

— Centriole

100 nm — Virus

Electron microscope

— Microtubule (width)
— Ribosome

10 nm — Cell membrane (width)
— Microfilament (width)
— Protein
— **DNA** (width)

1 nm
nanometer — Diameter of protein alpha helix
— Amino acid

0.1 nm — **Diameter of hydrogen atom**

100 m

0.2 m or 200 mm

30 – 40 mm

5 mm

1.5 mm

.004 mm

Photoreceptors

DNA

2 nm

0.1 nm

Hydrogen atom

to get an understanding of the sizes of cells is to start with large familiar organisms and move from there down to cells. Because scientists almost always use the metric system, let's consider these sizes in the scale of meters (**Figure 4.1**). A meter is about 39 inches, or 3.2 feet. One thousand meters is called a kilometer. Other commonly used metric measurements are centimeter (1/100th of a meter), millimeter (1/1,000th of a meter), micrometer (1/1,000,000th of a meter), and nanometer (1/1,000,000,000th of a meter).

Let's relate these measurements to the sizes of organisms and cells. A redwood tree can grow to over 100 meters in height ($\sim$ 300 feet), while an adult human might be 1.8 meters tall (6 feet), and a 6-year-old child might be 1 meter tall ($\sim$ 3 feet). The distance between your chin and the top of your head is about 0.2 meters (8 inches), which is 200 millimeters. Your eye is about 30 to 40 millimeters across ($\sim$ 1.25 to 1.5 inches), while your pupil might be about 5 millimeters in diameter. The nerve that takes visual information to your brain is about 1.5 millimeters in diameter. The cells that respond to light inside the eye are about 0.004 millimeters in diameter, which is expressed more easily as about 4 micrometers. Inside of the light-receptive cells the big biological molecules like proteins and DNA are in the nanometer range. Of course, knowledge about cells and their internal structures became possible only with the invention and development of microscopes, so it is worthwhile to consider some important milestones in the history of microscopes and in the study of cells.

Leeuwenhoek and Hooke Discovered the World of the Cell

Exploration of the small world of cells began almost 400 years ago. In 1637 Anton van Leeuwenhoek invented and crafted small magnifiers that had single, round lenses and focused on objects at the tip of a moveable needle (**Figure 4.2**). With these early magnifiers Leeuwenhoek could see green algae in lake water and small animals and single-celled organisms in a variety of substances. Leeuwenhoek was the first to describe bacteria, other single-celled organisms, live sperm, and small multicellular algae and animals. Robert Hooke made the next important advances. Hooke is best remembered for his description of *cellulae* in a thin slice of cork (Figure 4.2). He named these empty cubical structures after the tiny compartments in monasteries where monks slept; today you know them as "cells." Although the observations of Leeuwenhoek and Hooke opened the door

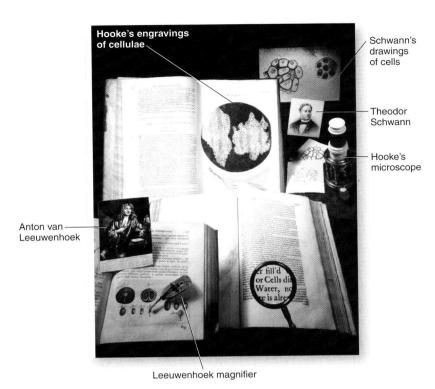

Figure 4.2   **Milestones in Microscopy.** Leeuwenhoek observed tiny "animalcules," Hooke described "cellulae," Schwann developed the cell theory.

to the microscopic realm, it took an additional 165 years before scientists understood that cells are the unit of life.

Why did it take so long to reach this conclusion? First, sophisticated microscopes were required to demonstrate that each cell is a separate and distinct structure. It was not until the 1830s that improved compound light microscopes revealed the structures inside of cells. The electron microscope, developed in the 1930s, revealed even smaller internal cellular structures than the light microscope. The electron microscope resolves structures about a nanometer (1 nm) in diameter, which is 1,000 times smaller than a micrometer. Combined with biochemical experimentation, the various kinds of light and electron microscopes have produced the explosion of knowledge about cells.

Second, the full use of powerful light and electron microscopes required associated technology. High magnifications require that specimens be preserved and sliced into extremely thin sections that are mounted on slides for observation. The *microtome* is a machine that works like a deli slicer, but instead of cutting salami, it slices prepared biological specimens into extremely thin sections. The microtome was not invented until 1870. For easier viewing, sample specimens usually are stained with dyes, and these began to be developed in the 1850s. When various cellular structures were dyed with different stains, researchers could more readily see them under the microscope. The final barrier to the exploration of cells was philosophical.

While the scientists of Hooke's time concentrated on making detailed descriptions of the natural world, they did not develop hypotheses that could be tested. At that time scientists focused just on making observations. This practice began to change after 1620, when the philosophical ideas of Francis Bacon began to influence scientists. Bacon suggested that scientific studies should begin with accurate observations. Then, based on these observations, a scientist should suggest a hypothesis that could be tested with experiments. Does this sound familiar? It should, because it is the heart of the scientific method you read about in Section 1.5.

It was many years before Bacon's method flowered into research on cells and cell structures. In 1838 Matthias Schleiden communicated his ideas about the importance of the nucleus to the life of a cell to his friend, Theodor Schwann (Figure 4.2). A year later Schwann published many examples of cells and produced his hypothesis that all tissues of plants and animals are made of cells. Unlike Hooke's cellulae, which were defined by their exterior walls, Schwann defined cells by the presence of a nucleus *within* the cell. Today Schleiden and Schwann are both given credit for the **cell theory,** the idea that all organisms are made of cells. The cell theory is one of the cornerstones of the science of biology. Like the theory of evolution (see Chapter 13) and the theory of inheritance (see Chapter 11), the cell theory has been tested and supported countless times. It also has been modified by other workers. For example, in 1858 Rudolf Virchow contributed a significant additional idea to the cell theory. After investigating the question of where cells came from, Virchow concluded that all cells develop only from other, preexisting cells.

From this slow beginning, scientists have produced a detailed, intricate picture of the internal structure of cells. Studies of cells from all of the major groups of organisms have reinforced the biological themes of unity and diversity that you were introduced to in Section 1.1. Although all cells have the same basic features, cells are incredibly varied. These variations reflect the diversity of life's millions of species as well as the rich variety of cellular specializations within a single, multicellular organism. In upcoming sections you will consider the similarities among cells and then explore how cellular differences contribute to the variations between organisms.

cell theory the cell is the universal, basic structural unit of organisms; cells arise only from preexisting cells

QUICK CHECK

1. What did Leeuwenhoek and Hooke contribute to our knowledge of cells?
2. State the cell theory.

4.2 The Cell Membrane Protects Internal Processes and Allows the Cell to Interact with the Environment

The cell membrane defines the boundary of a cell. Let's begin with the cell membrane and then move to the structures within cells. The chemical processes that define life can take place only in a watery environment, like that found within cells (see Section 2.4). Surrounding a cell is another watery environment that also is filled with chemicals. The pace and organization of chemical interactions on either side of the cell membrane are quite different. To get a better grasp of this difference between chemical interactions within a cell and in the fluid that surrounds it, let's travel to the basketball court of a sports arena where a major league game is under way.

As they scatter into formation, running up and down the court, the members of one team dribble the ball. They pass it back and forth and try to score a basket while the other team leaps to block shots and steal the ball. As you watch this game in your imagination, notice two interesting things that have parallels in how cells function. First, the competing teams engage in a complex series of interactions controlled by the rules of the game. Second, the game happens within the boundaries of the basketball court. Players can come in and out of the game across the boundary, but only when the rules allow it. The game is disrupted if anyone who is not a team member crashes over the boundary. If frenzied fans or teammates leap onto the court, the game is stopped. The coaches may give signals from outside the court, but they cannot cross into the game.

How does this imaginary basketball game relate to a cell membrane? Within the basketball court each player has a purpose, and each team works in a coordinated way. This is similar to the organized chemical structures and processes found inside cells. The situation is different outside of the court. There coaches yell and signal to their teams while fans cheer and boo, scream at the referees, and leap up and roar when their team scores. When they are bored or unhappy with the course of the game, the fans toss things

Phospholipid
— Polar head
— Nonpolar tail

In a bowl containing oil and water, phospholipids can orient themselves in three ways.

Figure 4.3 **Arrangements of Phospholipids in Water.** (**a**) Phospholipids can arrange themselves as a surface film, (**b**) in single-layered spheres, or (**c**) in double-layered spheres.

■ *What characteristic of water causes phospholipids to orient with their polar heads in contact with water and their nonpolar tails away from it?*

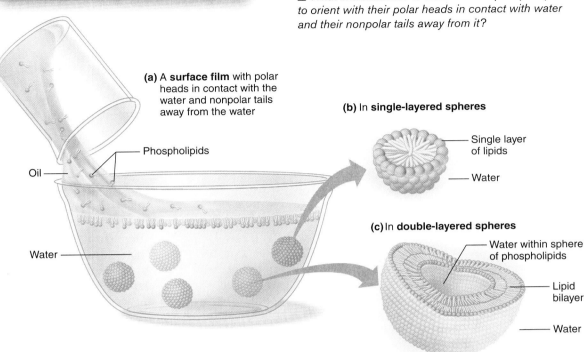

(**a**) A **surface film** with polar heads in contact with the water and nonpolar tails away from the water

Phospholipids

Oil

Water

(**b**) In **single-layered spheres**

Single layer of lipids

Water

(**c**) In **double-layered spheres**

Water within sphere of phospholipids

Lipid bilayer

Water

around the stadium, blow noisemakers, watch the cheerleaders and the team mascots, and eat and drink. This is more like the chemical environment outside of a cell.

Just as the basketball game must be insulated from outside interruptions, life needs some protection from the random, less predictable events of the outer environment. The cell membrane provides this protection by separating the processes of life from the outside world. Operations inside a cell, however, cannot occur by themselves. Just as the players and referees inside the basketball court must interact with coaches, timekeepers, and team members outside of the court, a cell must interact with the environment. Not only does the cell membrane protect the processes inside a cell, but also it allows a cell to interact with its environment. The chemical structure of the cell membrane accomplishes these two tasks.

The Lipid Bilayer Prevents Polar and Charged Ions and Molecules from Moving Freely Across the Cell Membrane

The basic structure of a cell membrane is a *lipid bilayer*. As the name implies, a lipid bilayer is a thin sheet made of two layers of lipids. This structure emerges spontaneously when certain kinds of lipids mix with water. Although a cell's lipid bilayer is made of different kinds of lipids, such as cholesterol, saturated fatty acids, and unsaturated fatty acids, phospholipids are its most prominent lipids. As a reminder, the structure of a phospholipid is shown in **Figure 4.3.** You can see that the nonpolar region of a **phospholipid** is a pair of long hydrocarbon chains that look like tails. The polar region of a phospholipid molecule is more globular in shape and so is called the *polar head* of the molecule.

Imagine how phospholipids might orient themselves in a cup of water. Recall from Section 3.3 that lipids are oily molecules that do not readily dissolve in water. The polar heads of phospholipid molecules are hydrophilic and are attracted to water molecules, while the nonpolar tails of the phospholipid molecules are hydrophobic and are not attracted to water molecules. Can you visualize the configuration that will result? There are actually three possible arrangements of molecules.

First, phospholipids can form a filmy layer on the water's surface (Figure 4.3a). The polar heads are in contact with the water, while the nonpolar tails are pointing away from the water. Second, phospholipids surrounded by water can form a

phospholipid a kind of lipid that incorporates a phosphate group and includes a polar head region and a nonpolar tail region

tight sphere (Figure 4.3b). Again, the polar heads of the phospholipids are in contact with water, but the nonpolar tails are aligned in the center of the sphere, away from the water. Finally, if some water is trapped inside such a sphere, phospholipid molecules can form a double layer (Figure 4.3c). Their polar heads are immersed in the water within the sphere and in the water that surrounds the sphere. Where are the nonpolar tails of the phospholipid molecules? Being hydrophobic, they avoid the polar water molecules by being sandwiched between two layers of polar heads. Take a second look at this last structure. Find the polar and nonpolar components of both layers of the bilayer and read on only when you are certain that you understand why this configuration forms.

This double-layered structure is a **lipid bilayer,** and it is the basic structure common to all cellular membranes. While the tightly packed polar heads of phospholipids of a lipid bilayer enable the membrane to interact with the watery environment both inside the cell and outside the cell, the intertwined nonpolar tails provide an effective barrier to any polar or charged molecule that might try to cross the membrane. Because of the chemical composition of the lipid bilayer, important biological molecules cannot get out of the cell, and many undesirable chemicals cannot get into the cell.

There is one major exception to this otherwise effective barrier: although *polar, or charged,* molecules are normally prevented from passing through the lipid bilayer, some *nonpolar* substances can pass through the membrane. Multicellular organisms take advantage of this to use lipid substances for cellular communication. For example, steroid hormones are lipids that can cross cell membranes and have an effect on the activities inside the cell. In addition, harmful lipid-soluble substances can use the same route into cells. For example, some insecticides are lipid soluble. They can pass through skin, cross membranes of deeper cells, and disrupt cellular processes. This is one good reason to wear gloves—an extra protective barrier—when handling toxic compounds like insecticides.

Membrane Proteins Allow Polar and Charged Substances to Cross the Cell Membrane in a Controlled Way

You might think the lipid membrane provides a boundary that seldom can be penetrated—but this is not the case. In controlled ways cells take in chemicals like water, nutrients, and ions from the environment—and they release chemicals like water, waste products, and chemical signals back into the environment. Cells also gain information from the environment about such things as toxins, food sources, and the presence and activities of other cells. They accomplish this controlled interchange with the environment in many ways, but nearly all of them involve **membrane proteins** embedded in the lipid bilayer of the cell membrane (**Figure 4.4**). Scientists estimate that as much as 50% of the cell membrane is protein. Many membrane proteins span the lipid bilayer, with some portions that stick out of the cell membrane into the environment and other portions that extend into the interior of the cell. Just as the doors in the walls of a house provide access to its interior spaces, such membrane-spanning proteins link a cell to the surrounding environment.

But wait, how can a protein extend into the polar, watery environment inside and outside the cell *and* span the nonpolar region of the lipid bilayer? Recall that proteins are complex molecules with regions that have different charges. Proteins that span a cell membrane have both polar and nonpolar regions. Their nonpolar portions are embedded within the membrane, while their polar portions extend into the watery environment on either side of the membrane. Because membrane proteins span the cell membrane, they can interact with the cell's external environment and can serve as paths that link the inside of the cell to its surrounding environment.

Water and Small Ions Move Across the Cell Membrane Through Protein Channels

Every cell is surrounded by small ions dissolved in water, and with the help of membrane proteins these ions flow in and out of the cell in a controlled way. A **membrane channel** is a protein that spans the cell membrane and allows water and ions to move into or out of the cell through the tiny hole in its center (**Figure 4.5**). It may help you to visualize membrane channels as holes in tubular beads. The passage of ions across membrane channels is strictly controlled. Each kind of membrane channel is specific for a particular ion or set of ions. One important thing about ion channels is that each one can be either open or closed. When a channel is closed, the ions that normally might move through that channel cannot cross the membrane; when the channel is open, the ions for that specific channel are free to move (Figure 4.5). As you will see, protein channels control the concentration of ions inside and outside of a cell.

Water is a special case. For decades scientists have known that water flows freely across most cell

lipid bilayer a double layer of lipid molecules oriented so that the molecules' polar heads are surrounded by water molecules on both the outside and the inside of the bilayer, while their nonpolar tails are intertwined and away from water molecules

membrane protein a protein located among the lipids in cell membranes

membrane channel a membrane protein that has a hole down the center through which ions and water will flow

(a)

Outside of cell

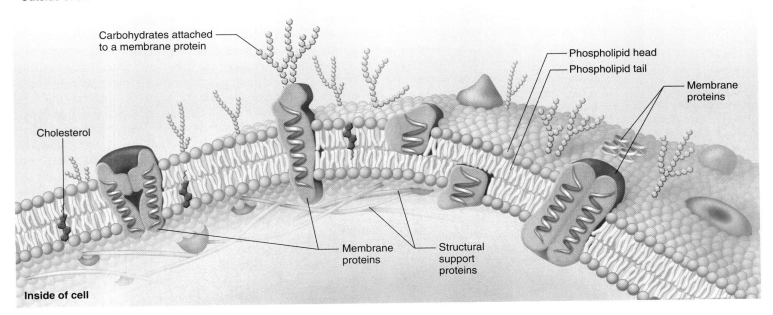

Carbohydrates attached to a membrane protein

Cholesterol

Phospholipid head

Phospholipid tail

Membrane proteins

Membrane proteins

Structural support proteins

Inside of cell

(b)

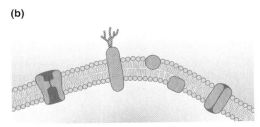

Figure 4.4 **Features of the Cell Membrane.** (**a**) The cell membrane is composed of a phospholipid bilayer (shown in yellow) studded with different kinds of proteins (shown in purple) and cholesterol molecules (shown in orange). The tertiary structures of the protein molecules are shown; (**b**) These simplified drawings will be used to show membrane proteins in upcoming figures.

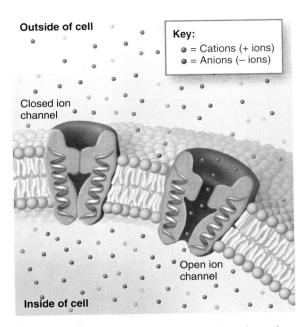

Outside of cell

Key:
- = Cations (+ ions)
- = Anions (− ions)

Closed ion channel

Open ion channel

Inside of cell

Figure 4.5 **Ion Channels.** When ion channels are open, ions can move into or out of a cell. This particular ion channel only allows positively charged ions to enter the cell.

membranes, as if there were no lipid membrane at all. How can this be? In the 1980s researchers discovered ion channels for water embedded in cell membranes. These channels allow only water molecules to cross the membrane, and in most cases they are always open. The discovery of water channels was so important that Peter Agre and Roderick MacKinnon, the two main researchers involved, were awarded the 2003 Nobel Prize in Chemistry for their work.

Diffusion and Osmosis Determine the Direction of Movement of Chemicals Through Membrane Proteins

Membrane proteins that allow water and ions to flow across the membrane are an important way that cells can control their internal environment. But how does this actually work? If a channel that allows water or ions to flow across the membrane is open, do the water molecules or ions move into the cell or out of it? The answer is dictated by the process of diffusion, a basic physical process that applies to both living and nonliving systems.

diffusion the movement of a substance from a place where it is more concentrated to a place where it is less concentrated

facilitated diffusion the diffusion of ions across the cell membrane through a protein channel

active transport the process in which a protein uses energy to move ions into or out of a cell toward the side of higher concentration

Because diffusion is such an important and basic process, let's consider it a bit more.

Imagine squirting an eyedropper of black ink into a cup of water (**Figure 4.6**). In a short while the ink spot is no longer visible. It seems to have disappeared into the water. This happens by the process of **diffusion,** defined as the movement of atoms or molecules from an area where they are more concentrated to an area where they are less concentrated. Diffusion is actually driven by simple particle motion. Atoms and molecules are constantly moving—they move faster if the surrounding environment is warmer and slower if it is colder. If a group of atoms or molecules is concentrated in one small region, in time their motion will cause them to become evenly distributed across the available space. This is what happens when the ink fades into the water. Although you may not be familiar with the word *diffusion*, you encounter the process every day. For instance, fragrances and odors diffuse into the air, a lump of sugar diffuses into your coffee, milk diffuses into your tea, and smoke diffuses into your hair and clothing as you sit by a campfire. Diffusion is an important process in living systems, and once you understand it, you will better understand how many of your body's systems work. For instance, as you will read in Chapter 23, within the smallest passages of your lungs, oxygen diffuses into your bloodstream, while carbon dioxide diffuses out of it.

Facilitated diffusion (**Figure 4.7**) is the diffusion of ions across a cell membrane through a protein channel. If the concentration of the substance is higher on the outside of the cell membrane than on the inside, the substance will move into the cell. If the concentration is higher on the inside, the substance will move out of the cell. If the concentration of the ion is the same on both sides, ions will still move through the protein channels, but the concentration of the substance on either side of the membrane will not change. In other words, the ions will be in a state of equilibrium across the membrane. Notice that because it depends only on a difference in concentration, no energy is required for facilitated diffusion. As you will see, energy sometimes is required to set up the concentration difference in the first place.

Now consider this: sometimes ions need to move across the cell membrane *toward* an area of

Figure 4.6 **Diffusion.** When an eyedropper of ink is squirted into a beaker of water, the ink diffuses into the water and quickly becomes evenly dispersed by random molecular motion.

■ *What could increase the rate of diffusion?*

Figure 4.7
Facilitated Diffusion.
(**a**) The cell membrane prevents many substances from diffusing across it and entering a cell. (**b**) When a membrane protein channel is closed, diffusion of this substance does not occur. (**c**) When a membrane protein channel is open, diffusion can proceed, and this substance moves from the area where it is more concentrated (outside of the cell) to the area where it is less concentrated (within the cell).

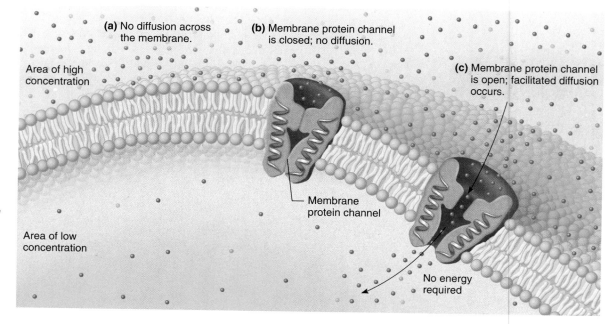

(a) No diffusion across the membrane.

(b) Membrane protein channel is closed; no diffusion.

(c) Membrane protein channel is open; facilitated diffusion occurs.

Area of high concentration

Membrane protein channel

Area of low concentration

No energy required

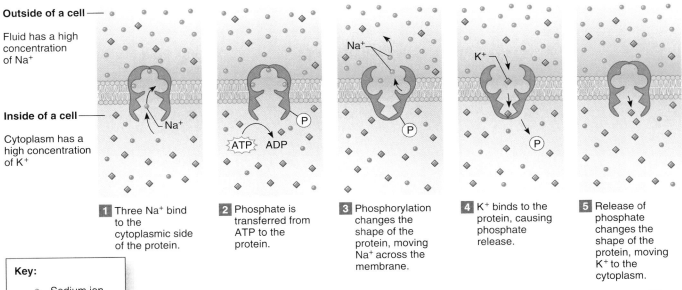

Outside of a cell —

Fluid has a high concentration of Na⁺

Inside of a cell —

Cytoplasm has a high concentration of K⁺

1 Three Na⁺ bind to the cytoplasmic side of the protein.

2 Phosphate is transferred from ATP to the protein.

3 Phosphorylation changes the shape of the protein, moving Na⁺ across the membrane.

4 K⁺ binds to the protein, causing phosphate release.

5 Release of phosphate changes the shape of the protein, moving K⁺ to the cytoplasm.

Key:
- ● Sodium ion
- ◆ Potassium ion

Figure 4.8 **Active Transport.** In active transport, a membrane protein uses the energy of ATP to move a substance from an area of low concentration to an area of high concentration. The pump can move three sodium ions out of the cell for every two potassium ions that it moves into the cell.

high concentration. Why? The normal functions of many cells depend on an *imbalance* in ion concentration across the membrane, and so cells must be able to maintain that imbalance. In animals, for example, most cells maintain a low concentration of sodium inside the cell compared with the outside of the cell. How do cells do this? The cell membrane of most cells contains a protein that acts as a pump to move sodium and potassium ions against the flow of diffusion. The membrane pump uses the energy in ATP molecules to push sodium out of the cell and pull potassium into the cell (**Figure 4.8**). Because the sodium-potassium pump works against the concentrations of these two ions, it is an example of **active transport,** the process that uses energy to move ions across the cell membrane toward the side of higher concentration.

Water moves freely across the cell membrane. Because there are many protein channels for water, and because most of them are always open, it might seem as though the cell membrane posed no barrier to water at all. The movement of water across the membrane is influenced, though, by ions dissolved in water inside and outside of the cell. Just like any other substance, water moves *from an area where it is more concentrated to an area where it is less concentrated.* To understand this process, it might be helpful to consider a half-full beaker of water that is completely separated into two compartments by a membrane (**Figure 4.9**). Although water can flow across this membrane, the membrane does not allow sodium or chloride ions to cross. At the start of this experiment, pure distilled water is on both sides of the membrane

(a) A semipermeable membrane permits water to cross but is impermeable to ions such as sodium and chloride. With pure water on either side of the membrane, water molecules move at the same rate in both directions.

Water molecules cross the membrane in either direction

(b) Three tablespoons of salt (NaCl) are added to one side of the beaker. The salt dissolves as the water molecules bind to the sodium and chloride ions. These water molecules surrounded by ions cannot cross the membrane.

Lower concentration of water

Higher concentration of water

Water bound to sodium and water bound to chloride

(c) The higher concentration of ions on one side of the beaker produces a lower concentration of water on that side. Water moves from the area where it is most concentrated to the area where it is less concentrated. As a result the water on the side of the beaker with dissolved ions increases in volume.

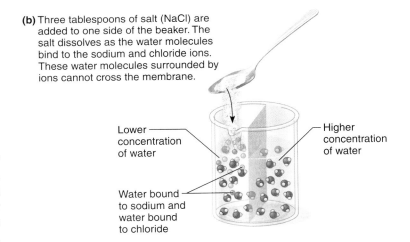
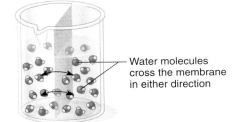

Figure 4.9 **Osmosis.** In osmosis, water molecules diffuse through a semipermeable membrane. Water molecules move from the area where they are most concentrated to the area where they are least concentrated.

(Figure 4.9a). Then you add three tablespoons of table salt (sodium chloride, NaCl) to one compartment of the beaker (Figure 4.9b). In effect, adding salt means that the concentration of water is *lower* in the salty compartment than it is in the pure water compartment. Now think carefully about how water will diffuse and predict what will happen. Will water move from the salty side into the pure side? Or will water move from the pure side to the salty side? If you apply the principle of diffusion, you will correctly conclude that water will flow from the area where water molecules are most concentrated (the pure water side) to the area where water molecules are less concentrated (the salty water side). If the movement is great enough relative to the diameter of the beaker, you would see the level of liquid on the salty side of the beaker rise because of the flow of water across the membrane (Figure 4.9c).

The diffusion of water across a cell membrane is so important that it is given its own name: osmosis. **Osmosis** is the diffusion of water across a membrane that does not allow the free movement of other substances. This kind of membrane is called *semipermeable,* and so a more formal definition of osmosis is diffusion of water across a semipermeable membrane. Of course, a cell membrane is a semipermeable membrane. Osmosis has important implications in the life of a cell, and mastering this basic principle is essential to learning how cells work.

If too much water flows into an animal cell, it will swell up and the chemical reactions inside the cell will become abnormal. If water continues to flow into the cell, it will burst. Consider what might happen if the cell were suddenly immersed in distilled water, a pure substance that has no ions or dissolved substances. If the appropriate membrane channels are open, ions like sodium, potassium, or chloride would move out of the cell, diffusing from areas of higher concentration (within the cell) to areas of lower concentration (the surrounding distilled water). But in their normal resting state most cells do not allow ions to cross freely in and out of the cell. As a result, in this environment water would move into the cell from its area of highest concentration (in the distilled water) to its area of lower concentration (within the cell) and the cell would swell and perhaps burst. Scientists take advantage of this process to produce "ghost cell membranes" for study (see Figure 1.2b). Alternatively, if the same cell is immersed in salty water, water will move out of the cell and it will shrivel. In either case, delicate chemical reactions inside the cell will be adversely affected.

Concerns about osmosis and diffusion are not just abstract concepts; these processes have practical consequences. For instance, when a hospital patient must have fluid given intravenously (IV), that fluid must have the precise levels of dissolved ions that normally are contained in the fluid surrounding human cells. If the fluid is not identical with tissue fluids, diffusion and osmosis could result in water and ions flowing improperly across the patient's cell membranes. The next time you visit a hospital patient who has an IV drip, look on the plastic bag of fluid for the word *isotonic.* Isotonic means that the fluid in the IV drip solution has the same concentration of water and ions as body fluids and, therefore, will not harm the body's cells.

Now that you understand diffusion and osmosis, the next section will discuss how the protein channels that allow passage of ions and water open or close at the right times.

QUICK CHECK

1. Make and label a drawing that shows how phospholipid molecules and proteins are arranged in a cell membrane.
2. List and briefly state the duties of membrane proteins.
3. How are diffusion and osmosis similar? How are they different?

4.3   Cells Interact with the Environment via Membrane Proteins

By now you have the idea that a cell is enclosed in an active, protective membrane composed of lipid and protein molecules. One important role of membrane proteins is to provide conduits for water and ions to cross the membrane. Another important role is to provide communication with the environment. How do membrane proteins accomplish something as complex as communication? Information about the environment typically is brought to a cell by **signal molecules.** These are molecules from the environment that change the chemical interactions within a cell. Many kinds of chemicals can be signal molecules including modified amino acids, nitrogenous bases, and small proteins. Signals can come from other cells or from some environmental condition. For example, insulin is a signal molecule that tells a cell to take up glucose molecules and link them together to form glycogen.

Cells in tissues all over the body make and release signal molecules that tell other cells what to

osmosis the diffusion of water across a semipermeable membrane

signal molecule a small molecule in a cell's environment that carries information about some aspect of the environment

do—as a result, cells are drenched in signal molecules. For instance, nerve cells make and release signal molecules that tell muscles to contract; cells in the thyroid gland release signals that tell other cells to increase or decrease energy use; cells in the adrenal glands release signals that tell the heart to beat faster and the blood vessels to constrict. Each cell in the body responds to a selective set of these signals. The net result is that each cell in the body knows exactly what to do and when to do it. How does a cell respond to a signal? This is a major job of the membrane proteins. Information is transmitted to a cell when a signal molecule attaches to a particular membrane protein that protrudes from the outer surface of the cell membrane. A membrane protein that can attach to a specific cellular signal molecule is called a **receptor.** Let's explore the interaction between signals and membrane receptors in more detail.

One feature of signal molecules is that many of them do not cross the membrane and thus do not enter a cell. How can a signal transfer information to a cell without actually entering the cell? Recall that a protein's function is determined by its charge and shape. A signal molecule is much smaller than a receptor protein, but like a protein, each signal molecule has its own shape and distribution of charges. Because of its shape and charges, a signal molecule "fits" into a specific spot on a receptor protein and can bind to it. The signal molecule and

protein form weak bonds, such as hydrogen bonds, that hold them together for a brief period of time. When a signal molecule binds to a receptor protein, the shape and charge of the receptor protein change. It is the *changed* receptor protein that actually has an effect on the processes inside the cell. Proteins can transfer information from the outside of the cell to the inside in one of two ways: facilitated diffusion or signal transduction. In facilitated diffusion the membrane receptor is an ion channel. When a signal molecule binds to an ion channel receptor, the channel opens and ions flow into or out of the cell. When the signal molecule is not bound, the channel is closed and ions do not flow. Therefore, in facilitated diffusion, the binding of a signal molecule to a membrane ion channel gives the ion the "green light" and it diffuses across the membrane (**Figure 4.10**). Each type of signal molecule binds to a particular type of membrane ion channel receptor. For example, some membrane receptors allow sodium and potassium ions to flow when a signal molecule is bound to the receptor. Others allow just chloride ions or calcium ions to flow. In either case, when the signal binds to the receptor the concentration of a particular ion increases or decreases inside the cell. This change determines how the cell functions.

Signal transduction is a different process in which the binding of the signal molecule to a membrane receptor produces changes in the internal

receptor a membrane protein specialized to interact with a signal molecule and transmit the signal's message to the inside of the cell

signal transduction binding of a signal molecule to a membrane receptor produces changes in the internal chemical reactions of a cell

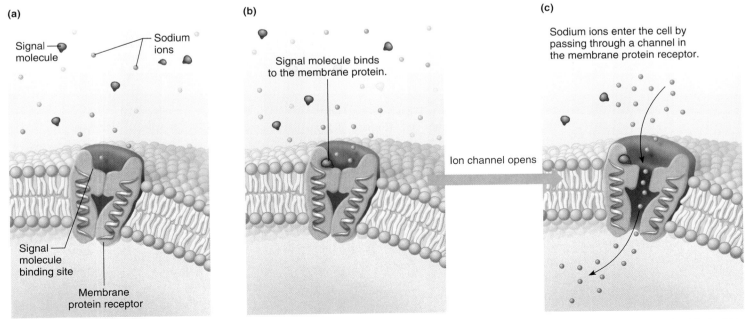

(a)

Signal molecule

Sodium ions

Signal molecule binding site

Membrane protein receptor

(b)

Signal molecule binds to the membrane protein.

Ion channel opens

(c)

Sodium ions enter the cell by passing through a channel in the membrane protein receptor.

Figure 4.10 **How a Membrane Channel Opens in Facilitated Diffusion.** (**a**) The membrane protein receptor is embedded within the lipid bilayer of a cell membrane. Notice that no signal molecule is bound to the receptor. (**b**) Signal molecule (shown in red) binds to the protein receptor. (**c**) The membrane protein receptor changes shape, the channel in the membrane opens, and ions flow into the cell.

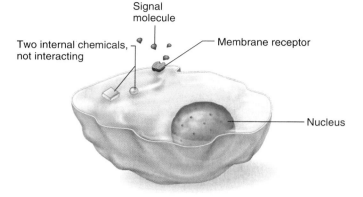

In signal transduction binding of a signal molecule causes an internal chemical change.

(a) Receptor and internal chemical not bound together

Signal molecule

Two internal chemicals, not interacting

Membrane receptor

Nucleus

(b) Binding of signal molecule and membrane receptor causes the two separate chemicals to bind and interact. This new internal chemical causes further internal chemical changes in the cell that then cause a change in cell shape.

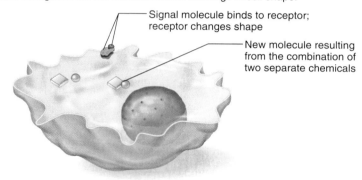

Signal molecule binds to receptor; receptor changes shape

New molecule resulting from the combination of two separate chemicals

Figure 4.11 **Signal Transduction.** In signal transduction the binding of a signal molecule results in chemical changes within the cell, but nothing actually enters the cell.

■ *How does this process differ from facilitated diffusion?*

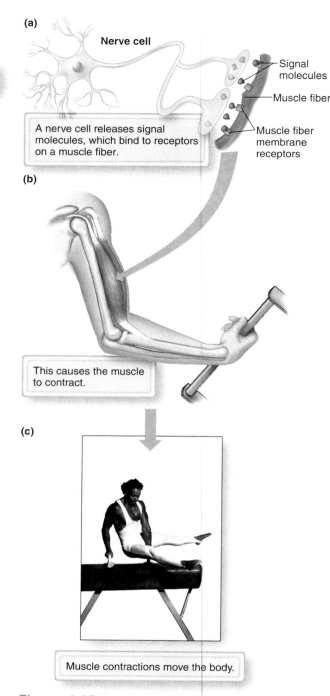

(a)

Nerve cell

Signal molecules

Muscle fiber

Muscle fiber membrane receptors

A nerve cell releases signal molecules, which bind to receptors on a muscle fiber.

(b)

This causes the muscle to contract.

(c)

Muscle contractions move the body.

Figure 4.12 **Membrane Receptors Influence Muscle Contractions.** Controlled muscle contractions can move the body into spectacular gymnastic movements.

chemical reactions within a cell, even though nothing flows in or out of the cell (**Figure 4.11**). Although the membrane receptor in signal transduction is not itself an ion channel, the change in the membrane receptor's shape alters the interactions of certain molecules inside the cell. Regardless of whether a signal binds to an ion channel receptor or to a signal transduction receptor, cell membrane proteins allow the environment to dictate what happens inside a cell.

Many body functions are controlled by a signal molecule that binds to a membrane receptor. Consider a gymnast who is bounding off the floor onto a pommel horse (**Figure 4.12**). His movements are the result of muscle contractions that occur when signal molecules are released by nerve cells and bind to receptors on muscle cells. In this case the receptors are also ion channels. When the signal molecule binds to the muscle cell membrane receptor, the channel opens and, because there is normally more sodium on the outside of the cell than on the inside, sodium flows into the cell according to the laws of diffusion. The extra sodium ions set off a chain reaction that ends with muscle contraction. Notice that the signal molecule does *not* enter the cell, and yet it causes the dramatic events associated with muscle contraction.

All of the complex thoughts and feelings generated by the nerve cells in your brain are also the result of signal molecules that are released from one cell and bind to a receptor protein on another cell. The processes of development and growth that produced your full-grown body, your sense of smell, the stress response that automatically turns on when you are in a threatening situation, your perception of pain, and many other body functions all involve signal molecules that bind to receptor proteins on cell membranes.

QUICK CHECK

1. Define a signal molecule. What is a membrane receptor?
2. Describe the process of facilitated diffusion.
3. Describe the process of signal transduction.

4.4 The Inside of a Cell Is Highly Organized

The inside of a cell is full of chemicals that interact with one another. In addition to biological molecules (see Sections 3.3 to 3.5), cells contain water and small ions such as sodium, potassium, calcium, and chloride, as well as trace amounts of many other elements such as iron, sulfur, magnesium, and zinc. Cells also contain other small organic molecules necessary for life's processes. You are familiar with some of these small organic molecules and know them as vitamins such as vitamin C and niacin.

At this point, you may think that the inside of a cell must be a pretty chaotic place, with so many chemicals interacting and all of life's processes going on at once. How does a cell keep track of everything and ensure that things happen at the right times? It turns out that the inside of a cell is a remarkably organized place. A cell's chemical processes are organized by concentrating the molecules and atoms necessary for a particular process in a particular place. But is every cell organized in the same way? As you will see in the next section, the answer is no, but despite their differences, cells have many fundamental features in common.

Cells Are Either Prokaryotes or Eukaryotes

If you were to examine cells of all living organisms, you would make the surprising discovery that the cells of organisms are organized in one of two major patterns, called *prokaryotic* and *eukaryotic*. These two basic types of cells differ in their internal layout and complexity (**Table 4.1**). To understand these

Table **4.1** A comparison of prokaryotic and eukaryotic cells		
Jobs in a cell	**Prokaryote structures that do this job**	**Eukaryote structures that do this job**
Capturing energy in chemical bonds	Enzymes in cytoplasm	Chloroplasts
Extracting energy from chemical bonds	Enzymes in cytoplasm	Mitochondria
Constructing proteins	DNA in nucleoid	DNA in nucleus
	mRNA	mRNA
	Ribosomes	Ribosomes in cytoplasm
Completing, modifying, and packaging proteins	Ribosomes	Ribosomes
	Enzymes in cytoplasm	RER
	Enzymes in cytoplasm	Golgi complex
Breaking down used and damaged proteins	Enzymes in cytoplasm	Lysosomes
Instructions for cellular processes	DNA in nucleoid	DNA in nucleus
Preparation and export of lipids	Enzymes in cytoplasm	Golgi complex

two cell types, consider that the molecules within any cell are designated to do certain jobs. Some of the important jobs in a cell are:

- capturing energy in chemical bonds and/or extracting energy from chemical bonds;
- building new proteins, using cellular instructions;
- packaging proteins and other small molecules for distribution inside or outside the cell;
- breaking down used and damaged molecules into simpler components; and
- providing instructions for cell processes and for making new cells.

In all cells these basic jobs are carried out by similar chemicals that work in similar ways. In prokaryotes, though, the spatial organization of these jobs is relatively simple when compared with eukaryotes. Let's explore these similarities and differences more deeply.

Although eukaryotes are relatively more complex in structure than prokaryotes, it is easier to define the two types of cells by first briefly describing eukaryotic cells. These complex cells have highly organized internal structures that can be studied using a light or electron microscope. Each kind of subcellular structure of a eukaryotic cell performs a particular function or job. For instance, cells of eukaryotes have specific structures that carry out energy reactions and other structures that package

organelle a molecular assemblage within a cell that carries out a specific function

nucleus the area surrounded by the nuclear membrane that contains the genetic material of a eukaryotic cell

nuclear membrane the barrier that separates DNA from the rest of a eukaryotic cell

and transport proteins and other biological molecules. Because these structures resemble tiny organs, they are called **organelles.** Many organelles are separated from the rest of the cell by membranes, and the lipid/protein membranes around organelles are similar to the cell membrane that separates the entire cell from its environment. The most prominent organelle is the **nucleus,** the region of a cell that contains the genetic material, DNA. The nucleus is surrounded by a **nuclear membrane,** and because it is the one organelle common to all eukaryotic cells, it sometimes is used as the defining feature of eukaryotes. The most common definition of a **eukaryotic cell,** though, is one that contains organelles surrounded by membranes.

In many ways a prokaryotic cell is defined by what it does not have. A **prokaryotic cell** does *not* have highly structured organelles, such as a nucleus surrounded by a membrane. **Figure 4.13** shows an example of a prokaryotic cell. This description should *not* lead you to conclude that prokaryotes are simple. They share many features with eukaryotes and carry out many of the same chemical processes. For example, both cell types have membranes made of lipids and proteins, and both have similar sorts of membrane receptors. Like eukaryotes, prokaryotes have special structures called ribosomes that carry out the manufacture of proteins. And, while prokaryotic DNA is not

enclosed in a nucleus, it is found in a special region of the cell called a **nucleoid.**

Prokaryotes are complex and highly successful organisms, and there are far more prokaryotic cells on Earth than there are eukaryotic cells. Prokaryotes can be divided into two major groups, called bacteria and archaea. Although both groups contain single-celled organisms with relatively simple cellular structures, the two groups show important differences in biochemical makeup and distribution. Bacteria are common in your environment, while archaea often live in more extreme environments that you rarely, if ever, encounter such as extremely high temperatures or high salt concentrations. Despite their differences, common usage still calls all prokaryotes "bacteria."

You probably think of bacteria as organisms that can cause disease. While this can be true, most prokaryotes are harmless—and many are beneficial and essential for the survival of other kinds of life. For example, your intestines are home to millions of beneficial bacterial cells that contribute to your overall health. By comparison, only a few kinds of bacteria cause diseases. Because the vast majority of cells are prokaryotes, in this sense Earth actually belongs to them. Prokaryotic cells were the first cells to evolve. Many experts think that over billions of generations some prokaryotic cells merged to become more complex cells, eventually giving rise to the many kinds of modern eukaryotes. The adaptable prokaryotes, however, survived and even flourished during periods when Earth changed dramatically and many eukaryotic organisms went extinct.

While prokaryotes are certainly more numerous in terms of individuals, eukaryotes are more familiar to us because they are larger and easier to see. Eukaryotes include plants, animals, and fungi. Eukaryotes display some of the most complex and engaging characteristics of life. While some eukaryotes are single-celled, most of the familiar eukaryotes are multicellular. The complex behaviors of multicellular organisms rely on the ability of eukaryotic cells to develop specialized functions. For these reasons, the remainder of this chapter will focus on characteristics of eukaryotic cells. Prokaryotes will be examined in greater detail in Chapter 15.

Eukaryotic Cells Share a Common Set of Organelles

It may help you to understand the organelles of a typical cell if you consider an analogy: the tasks of a cell might be likened to the tasks that must be carried out in a futuristic, self-contained human colony on the surface of Mars. First, it is essential that the

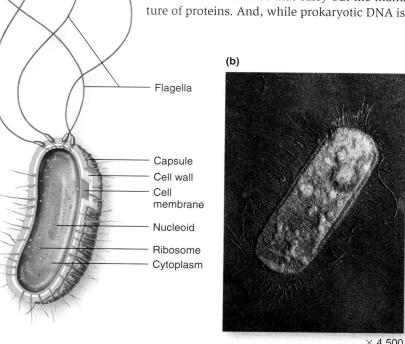

(a)

Flagella

Capsule
Cell wall
Cell membrane
Nucleoid
Ribosome
Cytoplasm

(b)

× 4,500

Figure 4.13   **Features of a Prokaryotic Cell. (a)** A prokaryotic cell is surrounded by an outer capsule, a cell wall, and a cell membrane. It lacks membranes around or within organelles. Prokaryotic DNA is located in one region of the cell (the nucleoid), while ribosomes are widespread within the cytoplasm. Some bacteria have flagella, structures for movement. **(b)** An electron microscope view of a bacterium. Colors have been artificially added to the photo.

colony have a boundary that encloses and protects the inhabitants. The outer boundary must allow certain necessary things to enter, such as sunlight and radio signals, while it must exclude harmful things, such as poisonous gases and violent storms. The protective boundary also needs portals or doors for the colonists to go in and out. In addition, something must keep the protective boundary upright and maintain its shape, such as internal beams or a thick external wall. Space is limited within the colony, so the facilities for similar functions should be close together. Accordingly, the colonists sleep and live in a residential area, while other areas are dedicated to essential processes such as energy capture, growing food, water purification, and machine repair. Finally, as with any group of people working toward a common goal, the Mars settlers need a control center to monitor activities, adjust the internal environment, and retain instructions for all these life-sustaining processes. As you will see, cells must solve all of these problems, too, and their organelles allow them to do so. Let's use an animal cell and a plant cell to demonstrate typical organelles of eukaryotic cells (**Figures 4.14** and **4.15**).

The Cell Membrane   As the physical boundary of a cell, the cell membrane defines the cell and protects it from the environment, acting much like the protective outer wall of a Mars settlement. Recall that lipids in a cell membrane act as a barrier to polar and charged substances. In contrast, proteins span the cell membrane and provide avenues for the cell to interact with polar and charged substances. These proteins act like the doors in the Mars colony's outer boundary and allow substances to move in and out of the protected cell.

Cytosol   Inside the cell membrane the organelles of a cell are suspended in the **cytosol,** a jellylike substance composed of water and many dissolved proteins, small molecules, and ions. Chemical processes that do not occur within organelles take place in the cytosol. Together the organelles and cytosol are called the **cytoplasm.**

Cell Wall and Cytoskeleton   Just as the wall of the Mars colony must have a rigid supporting framework that maintains its shape, a cell depends on rigid molecules to maintain its shape. Plant cells have an external, rigid, supportive **cell wall** made of carbohydrates that lies outside of the cell membrane (Figure 4.15). Cell walls are not unique to plant cells. Cells of algae, fungi, and prokaryotes also have cell walls, but these are formed of molecules subtly different from those in plants. Cell walls serve many important functions in plants. They are like protec-

tive packaging that prevents delicate cellular structures from damage, and they give structural strength to the whole organism. The woodiness of some plants is due to the strength of cell walls. Plant cells stick together much more strongly than animal cells do because of specialized attachments of cell walls of neighboring cells. These also add strength to the whole plant and offer protection from viruses, bacteria, and other invasive agents. As you will read in Chapter 15, the cell walls of bacteria are an important key in medicine. Many antibiotics kill bacteria by disrupting their cell walls.

Animal cells do not have cell walls, but instead have an internal scaffold made of long, thin, stiff protein "rods" that extend throughout the cell (**Figure 4.16**). This internal protein scaffold is called a **cytoskeleton.** Several different types of proteins contribute to the cytoskeleton, each with its own structure and function. Three important kinds of cytoskeleton proteins are called *microtubules, intermediate filaments,* and *microfilaments.* Using special stains and a microscope, you can see the complex, weblike structures made by the various kinds of cytoskeleton proteins. Microtubules, as their name implies, are tiny, rigid tubes made up of protein subunits. Intermediate filaments are stringy molecules that contribute to a variety of cell functions. The specialized proteins that allow muscles to contract, for example, involve at least two types of microfilament proteins. The cilia and flagella on many types of cells often are involved in cell motion, and they contain another type of microfilament protein.

The cytoskeleton is a versatile organelle and because of its versatility, animal cells can have many different shapes (**Figure 4.17**). For instance, liver cells are roughly cubical; intestinal cells are long, narrow columns; red blood cells are shaped like dimpled discs; and some nerve cells have long extensions more elaborately branched and forked than the bare branches of most trees. Another important feature of the cytoskeleton is that it can be exceptionally dynamic, allowing a cell to change shape. You may be familiar with the single-celled eukaryotes called *amoebas,* which can move by extending portions of the cell in one direction or even in several directions at the same time. This is accomplished by simultaneously building up the cytoskeleton in one region of the cell while breaking down the cytoskeleton in other regions.

Another important organelle related to the cytoskeleton is the **centriole.** Each centriole is made of two small pieces of microtubule. Although centrioles are typical of animal cells, they are not present in cells of plants or fungi. Centrioles play an important role in cell division of animal cells, as you will see in Chapter 8.

eukaryotic cell a cell that has organelles surrounded by membranes, including a nucleus

prokaryotic cell a cell that lacks cellular organelles surrounded by membranes, including a nucleus

nucleoid the region of a prokaryotic cell where DNA is found

cytosol a jellylike substance within a cell that contains various molecules and ions

cytoplasm the organelles and cytosol taken together

cell wall a rigid supporting structure that is outside of the cell membrane

cytoskeleton a network of protein fibers that lends support and shape to a cell

centriole an organelle that is made of two small pieces of microtubule that function in division of animal cells

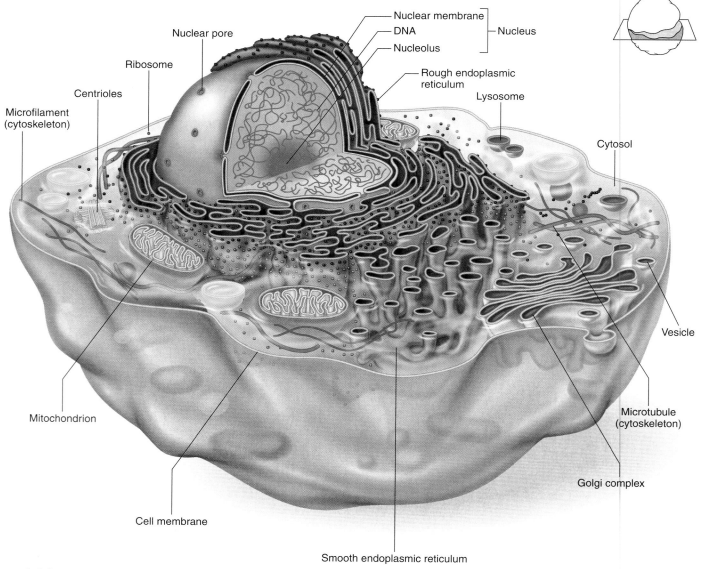

Figure 4.14   **Organelles of a Generalized Animal Cell.**   Animals are eukaryotes, and their cells contain many membranous organelles including the cytoskeleton, cell membranes, nucleus, nucleolus, rough endoplasmic reticulum, smooth endoplasmic reticulum, ribosomes, and Golgi complex.

nucleolus a dense area within the nucleus that contains DNA responsible for making ribosomal RNA

ribosome an organelle that bonds amino acids together to form a protein

Nucleus   The entire life of a eukaryotic cell and all its processes are directed by the nucleus (see Figures 4.14 and 4.15). This large, spherical organelle contains DNA, the molecule that carries all of the information needed for cellular processes and for cells to reproduce. Because DNA is transferred from one generation to the next, it controls the inheritance of genetic traits. You will learn much more about DNA structure and function in Chapter 7.

Nuclei of different kinds of cells can look different. One way they vary is in the presence or absence of a structure called a **nucleolus.** When some types of cells are stained with a blue or purple dye that stains DNA, the nucleolus stands out as a small,

dense, dark structure within the nucleus. This structure contains the DNA that provides the instructions for making special types of RNA that are used to make organelles called ribosomes.

Ribosomes   A eukaryotic cell has a complex system for manufacturing and distributing the products it makes. While the Martian colony might be manufacturing and exporting crystals or other useful products, cells make and distribute proteins. Sometimes proteins combine with other molecules to make larger cellular structures, including the organelles themselves. Proteins are made by tiny organelles called **ribosomes** (**Figure 4.18**). Amino acids are

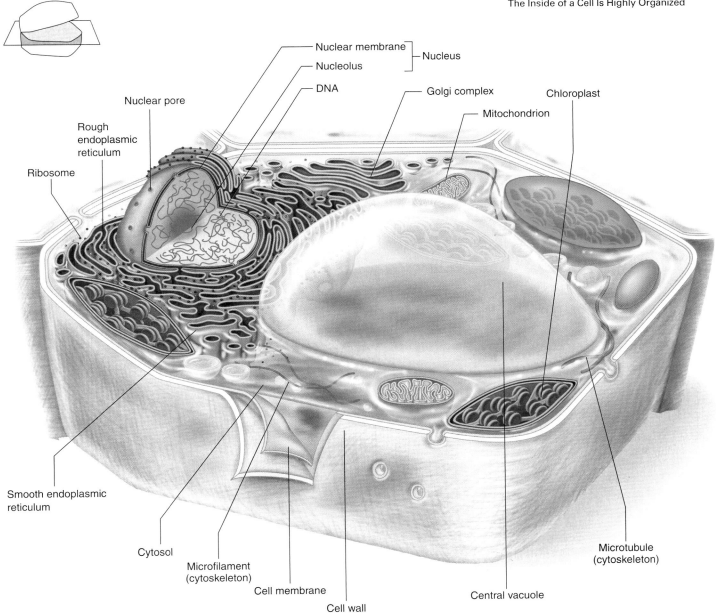

Figure 4.15 **Organelles of a Generalized Plant Cell.** In addition to all of the organelles of animal cells, plant cells also have the cell wall, chloroplasts, and central vacuole that animal cells lack.

■ *What organelles are found in plant cells, but not in animal cells?*

bonded together at ribosomes to form the long chains that are the primary structure of a protein (see Figure 3.15). Ribosomes are the one complex structure that prokaryotes and eukaryotes share, and ribosomes function in a similar way in both types of cells.

Endoplasmic Reticulum and Golgi Complex

Two related structures are involved in packaging and distributing products to various parts of the cell and outside the cell: the endoplasmic reticulum and the Golgi complex. The **endoplasmic reticulum (ER)** is a maze of membranous tunnels

that extends throughout the cell. It comes in two forms: smooth endoplasmic reticulum and rough endoplasmic reticulum (**Figure 4.19**). Both types are packed with enzymes embedded in the organelle's membranes and are within the internal space of the membranous tunnels. These enzymes are involved in manufacturing and modifying lipids and proteins.

   Smooth endoplasmic reticulum (SER) has the primary job of manufacturing and modifying lipids. For example, the lipids that form cell and internal membranes are made in the SER, which

endoplasmic reticulum (ER) a network of membranous tunnels in the cytosol that comes in two varieties, smooth endoplasmic reticulum and rough endoplasmic reticulum

smooth endoplasmic reticulum (SER) a network of membranous tubules that manufacture and modify lipids

Figure 4.16
Cytoskeleton Proteins.
Microtubules, intermediate
filaments, and microfilaments
are three important kinds of
cytoskeleton proteins. In this
photo a fluorescent dye
reveals a cell's webbing of
cytoskeleton.

The versatility of cytoskeleton
proteins allows cells to take on a
variety of complex shapes.

Tubulin
molecule

10 μm

Protein
molecules

Actin
molecule

23 nm
Microtubules

10 nm
Intermediate
filaments

7 nm
Microfilaments

**rough endoplasmic
reticulum (RER)**
a network of membranous
tubules that are closely
associated with ribosomes
and that modify proteins
by adding important
carbohydrate groups

Golgi complex a mass of
membranous spaces
where proteins are altered
before being exported from
the cell

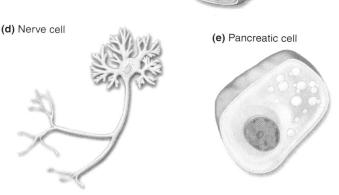

(a) Kidney cell

(b) Intestinal lining cells

(c) Red blood cells

(d) Nerve cell

(e) Pancreatic cell

Figure 4.17 **Diversity of Human Cells.** Each kind of cell's unique cytoskeleton
determines the cell's overall shape. (**a**) Kidney cells are shaped like cubes. (**b**) Cells of the
lining of the intestine are elongated and have protrusions at one end. (**c**) Red blood cells
are shaped like dimpled, jelly-filled donuts. (**d**) Nerve cells have complex shapes with many
extensive branches. (**e**) Cells that line the pancreas are shaped like bricks.

also is responsible for making many of the lipid-
based hormones. For example, steroid hormones
such as testosterone and estrogen are made from
cholesterol in the SER of specialized cells in the
testes and ovaries. Lipoproteins that carry cholesterol in the bloodstream are combinations of lipids
and proteins that also are made in the SER. Finally,
the SER plays a role in breaking down lipids that
are toxic to the body. The SER of liver cells is especially active in this role.

Endoplasmic reticulum has a special role to
play in the production of proteins that will end up
in the cell membrane or in membranes of
organelles. Many of the newly made proteins that
will be inserted into the cell membrane or any
other membrane must be modified before they can
carry out their functions. For example, proteins of
the cell membrane often have carbohydrate molecules attached to them, making a hybrid molecule.
The endoplasmic reticulum and ribosomes combine their functions to produce these modified proteins. To accomplish this, ribosomes attach to the
endoplasmic reticulum to form a different structure
called **rough endoplasmic reticulum,** or **RER.** As
the newly made proteins are produced, they are
inserted into the internal space of the RER, where
enzymes attach the appropriate carbohydrates. Figure 4.19 shows the variations in structure between
the smooth and rough endoplasmic reticulum.

The next question is what happens to the
lipids and modified proteins that come from the
SER and RER? This is where another organelle, the
Golgi complex, enters the picture. The **Golgi complex** is another set of membranous sacs that help
get lipids and proteins ready to take on their duties
in the cell (**Figure 4.20**). To visualize the Golgi
complex, imagine a set of disks stacked on top of
each other, sort of like a stack of pita breads. Each
disk is a membrane that surrounds an internal
space. The endoplasmic reticulum and Golgi membranes are connected by small membranous
spheres, called *vesicles,* shown in Figure 4.20. Vesicles pinch off from the membranous wall of the
endoplasmic reticulum and travel to the Golgi complex. Each vesicle contains proteins that were modified by the RER. The vesicle fuses with the Golgi
membrane, transferring its contents to the internal
spaces of the Golgi complex. The Golgi makes further modifications, usually of the carbohydrates
associated with proteins. At the end of this process,
vesicles pinch off of the membranous walls of the
Golgi complex, carrying completed proteins within
them. The vesicles then ferry the completed proteins to the cell membrane or to other internal locations in the cell.

Chloroplasts and Mitochondria   Two organelles are involved in how cells use the energy in chemical bonds to drive the processes that go on inside a cell: chloroplasts and mitochondria. Most of the energy used by living organisms comes from the Sun, but only certain kinds of cells can capture the energy in sunlight; among eukaryotes all such cells have special organelles called chloroplasts. A **chloroplast** is an organelle that captures the energy in sunlight and uses it to make ATP and other energy-carrying molecules (**Figure 4.21a**). Then these molecules provide the energy to make glucose. Unlike chloroplasts, **mitochondria** are found in nearly all eukaryotic cells (Figure 4.21b). Mitochondria have the crucial job of converting the energy stored in the bonds of glucose and various other molecules into ATP. Chloroplasts and mitochondria have internal mazes of membranes that provide internal spaces and surfaces where groups of enzymes carry out the transformation of energy from one form or molecule to another. You will learn more about the linked processes of energy capture and energy use in Chapter 5.

Lysosomes and Vacuoles   Two other organelles also are surrounded by membranes and packed with enzymes. **Lysosomes** are found in most types of eukaryotic cells, and they are filled with enzymes that break down biological molecules (**Figure 4.22**). This may seem odd, but cells contain many molecules that need replacement. Molecules may be old and damaged or new and malformed. These defective molecules are shipped to the lysosomes, and these organelles break them down into basic components that cannot damage delicate cellular processes.

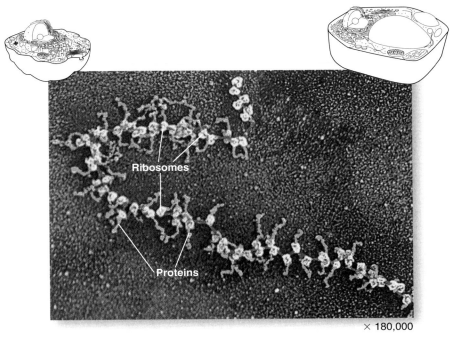

× 180,000

Figure 4.18   **Proteins Being Made on Ribosomes.** In the uppermost part of this photo, ribosomes (shown in pale blue) are arranged on a strand of mRNA (shown in red). In the lower part of the photo, chains of proteins (shown in green) extend from ribosomes.

Vacuoles are storage organelles that function differently in plant and animal cells. In animal cells vacuoles are used for temporary transport or storage. Some single-celled organisms enclose food within the membranes of vacuoles, where it is digested. Plant cells typically have a large central vacuole that contains liquid and serves as a water reservoir for the plant. When a plant wilts, water has been drawn out of the central vacuoles of its

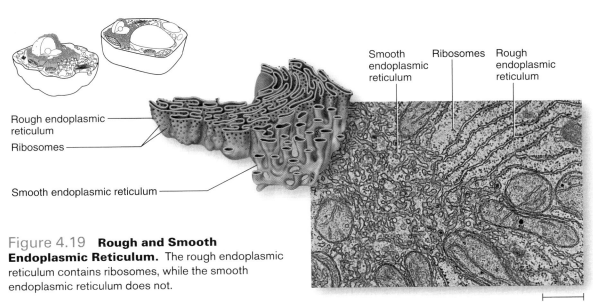

Figure 4.19   **Rough and Smooth Endoplasmic Reticulum.** The rough endoplasmic reticulum contains ribosomes, while the smooth endoplasmic reticulum does not.

Rough endoplasmic reticulum

Ribosomes

Smooth endoplasmic reticulum

Smooth endoplasmic reticulum   Ribosomes   Rough endoplasmic reticulum

0.2 μm

chloroplast an organelle within cells of green plants and certain other organisms where the energy in sunlight is captured and stored in molecules that provide the energy to produce glucose

mitochondria organelles where energy stored in glucose and various other molecules is used to produce ATP molecules

lysosome an organelle that contains enzymes that degrade wastes

vacuole a storage organelle

Figure 4.20 **The Golgi Complex.** This organelle is closely associated with the endoplasmic reticulum (ER). Vesicles from the ER carry proteins to the Golgi. Finished proteins leave the Golgi complex in other vesicles.

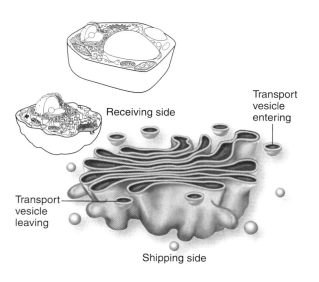

Receiving side

Transport vesicle entering

Transport vesicle leaving

Shipping side

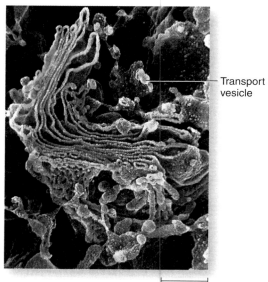

Transport vesicle

0.5 µm

Figure 4.21
Chloroplast (a) and Mitochondrion (b) Both organelles have internal membranous spaces where groups of enzymes transform energy from one form to another.

■ *How are the functions of chloroplasts and mitochondria different?*

(a) Chloroplast

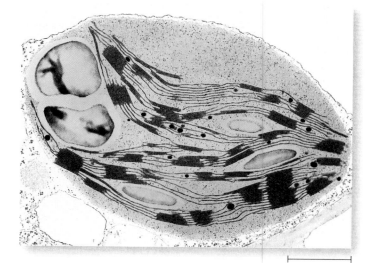

1 µm

(b) Mitochondrion

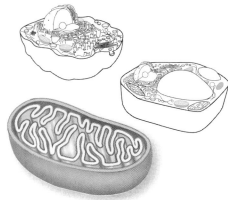

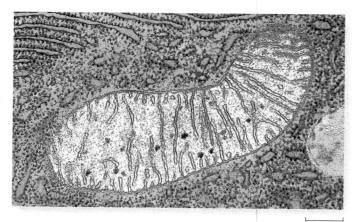

0.5 µm

cells. When you immerse limp lettuce, carrots, or celery in water, the plant tissues become crisp because the central vacuoles of their cells are refilled with water. In plants vacuoles also contain many kinds of enzymes and other chemicals as well. For example, opium poppies store organic compounds such as morphine and codeine in their central vacuoles. The green seed pods of the poppy plant, which form shortly after flowering, are especially high in these compounds. When a seed pod is cut, the sap that oozes out is collected and processed into opium. Other compounds stored in plant vacuoles include the pungent juice you squeeze out of garlic and the thick, gummy sap of tropical rubber trees.

Clearly, organelles are critical to the life of any cell, but their importance can be highlighted by considering just a few of the various human disorders that can arise when organelles do not function normally. **Table 4.2** lists a few of these.

Anything Less Than a Cell is Not Alive

In this chapter you have explored the characteristics of cells, the basic units of life. One question that you have not grappled with is: What are the limits of life? How do you know if something is alive or not? The list of life's characteristics that you read in Chapter 1 can provide a guide to defining life, but it takes a bit more thought to draw a line between living organisms and things that are not alive. Cells are the basic units of life, but is anything less than a cell alive?

Consider viruses. These tiny structures have many, but not all, of the features of cells, but are they alive? Most scientists would say that viruses are not alive. Why? **Figure 4.23** shows a typical virus. As you can see, a virus has a lipid membrane with embedded proteins. The virus also has internal proteins and nucleic acids that serve as its genetic material. So far this seems a lot like a cell. What is missing? One big thing that is missing is metabolism. A virus cannot take in, transform, and use its own energy. Because of this, a virus cannot make its own biological molecules and structures, and a virus cannot reproduce on its own. Every virus requires the internal energy metabolism and ribosomes of a cell to make new viruses. Look back at the definition of life. You can see that viruses do not have all of life's characteristics. Therefore, viruses are not alive.

Now what about multicellular organisms? Aren't humans, or mosquitoes, or pine trees alive? Yes, but a bit of thought tells you that the processes

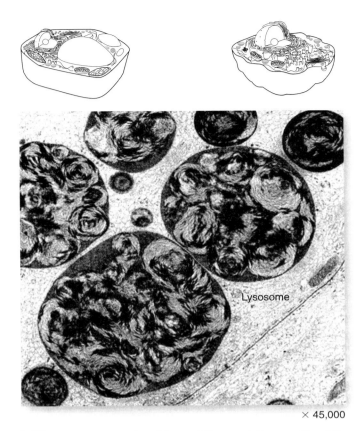

× 45,000

Figure 4.22 **Lysosomes Are Waste Disposal Organelles Where Molecules Are Recycled.**

Table 4.2	Inherited disorders that result from problems with organelles	
Organelle involved	**Name of disorder**	**Symptoms**
Endoplasmic reticulum	Glycosylation disorders	Delayed development, abnormal eye alignment, epilepsy
Golgi complex	Menke's disease and Wilson's disease	Delayed development, nervous system degeneration
Mitochondria	Carnitine deficiency	Abnormal muscle function; coma
	Luft's disease	
	Pearson's syndrome	Excess energy metabolism, sweating, weakness, fever
		Bone marrow and pancreas dysfunction
Cytoskeleton of nerve cells	Charcot-Marie-Tooth's disease	Degeneration of nerve cells, leading to muscle weakness
	Muscular dystrophy	Breakdown of muscle structure, leading to muscle weakness
Cytoskeleton of skin cells	Palmoplantar keratoderma	Excess cytoskeleton keratin in skin cells on soles of feet and palms of hands, resulting in painful cracks in skin

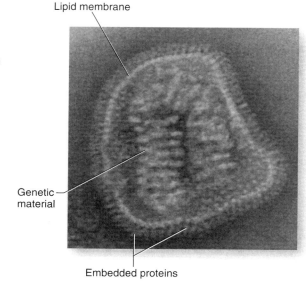

Figure 4.23 **The Structure of a Virus.** Viruses are membranes with embedded proteins surrounding genetic material.

Lipid membrane

Genetic material

Embedded proteins

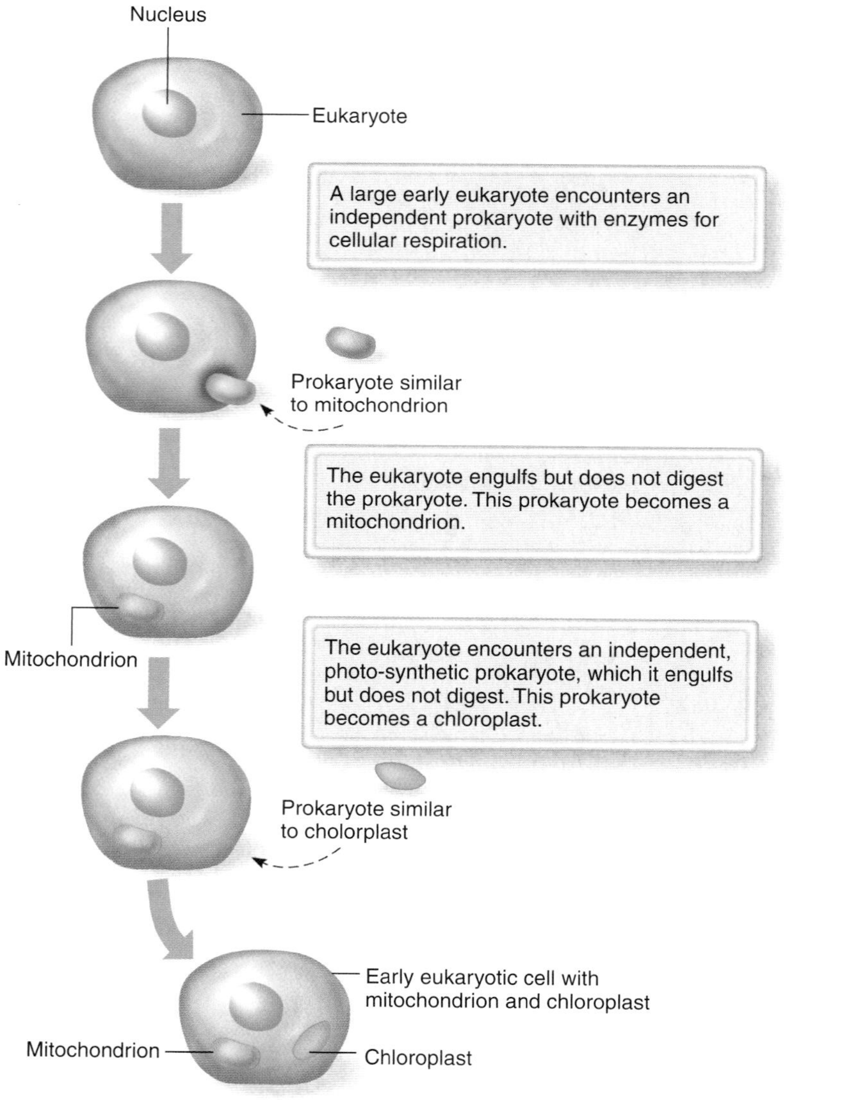

Nucleus

Eukaryote

A large early eukaryote encounters an independent prokaryote with enzymes for cellular respiration.

Prokaryote similar to mitochondrion

The eukaryote engulfs but does not digest the prokaryote. This prokaryote becomes a mitochondrion.

Mitochondrion

The eukaryote encounters an independent, photo-synthetic prokaryote, which it engulfs but does not digest. This prokaryote becomes a chloroplast.

Prokaryote similar to cholorplast

Early eukaryotic cell with mitochondrion and chloroplast

Mitochondrion

Chloroplast

Figure 4.24 **Origin of Eukaryotes.** Eukaryotic cells may have originated from a combination of prokaryotic cells.

■ *What evidence suggests that mitochondria and chloroplasts once were independent cells?*

of life occur within the cells of multicellular organisms. Even though they are within a living multicellular organism, the fluids and materials that surround cells are not alive. Therefore, it is only because you are made up of cells that you are alive.

Of course, this does not mean that the complexity of life stops with cells. In a multicellular organism, for example, cells develop distinct structures and functions, and they interact with one another in coordinated ways that unicellular organisms lack. These added features give multicellular organisms a range of abilities and characteristics that the individual cells they are made of do not possess. A single cell cannot walk or run, cannot write or sing, and cannot contemplate its own existence. At another level, individual organisms interact to produce more complex biological communities, as you read in Section 1.3. You may have heard people talk about "living ecosystems" or of Earth as the "living planet." Again, in each of these systems, the actual processes of life occur within cells. Without cells none of these systems would be alive. Nevertheless, interactions among cells and between organisms and components of their environments produce complexities that single cells do not possess.

Organelles Carry Evidence of the Origins of Eukaryotic Cells

The earliest forms of life, that date to 3.8 billion years ago, were clearly prokaryotes. It is hard to say when the first eukaryotic cells emerged, but it was probably about 2 billion years ago—nearly 2 billion years *after* the first prokaryotes. How did the more complex eukaryotes evolve? Where did the eukaryotic organelles originate? One strong possibility is that some of the organelles came from the fusion of two different kinds of cells. The best evidence for this scenario is the case of mitochondria and chloroplasts. **Figure 4.24** illustrates the idea that sometime during the first 2 billion years of life, a large prokaryotic or early eukaryotic cell engulfed or surrounded a smaller prokaryotic cell. The larger cell did not have the metabolic abilities of modern mitochondria, but the smaller one did. These combination cells were highly successful and eventually evolved into modern eukaryotes. This sounds like a strange idea, but there is evidence to support it.

One intriguing bit of evidence is that mitochondria of eukaryotes have their own DNA. Mitochondria go through cell division to produce more mitochondria, and they use their DNA to make mitochondrial proteins. Because of this, mitochondria seem to be separate cells living within the larger host cell. More evidence comes from the discovery that mitochondrial DNA is more similar to the DNA of certain kinds of bacteria than it is to the DNA in the nucleus of eukaryotic cells. All of this supports the idea that modern eukaryotic organelles arose from the fusion of two kinds of cells. Similar evidence is found for the evolution of chloroplasts. Other organelles such as the cytoskeleton, the nucleus, and flagella and cilia also may have originated in this way.

4.5 An Added Dimension Explained
Christopher Reeve and the Search for Treatments for Spinal Cord Injuries

In the opening section of this chapter, you read about Christopher Reeve and his spinal cord injury. Now that you have some background in how cells work, you can better understand what happens in spinal cord injury and can evaluate the prospects for successful treatment.

Spinal cord injury is damage to the spinal cord that causes loss of sensation and/or movement in parts of the body. In the United States around 250,000 people are living with spinal cord injuries, and several thousand new injuries happen every year. People in their twenties are the most likely to have a spinal cord injury, and the vast majority of them—over 80%—are males. To understand what happens in spinal cord injury, you have to learn a bit about the structure and growth of the spinal cord. As you see in **Figure 4.25a**, the *spinal cord* is actually an extension of the nervous tissue that makes up the brain. The spinal cord is made up of millions of nerve cells called *neurons.* The bones of the *spinal column* that surround and protect the spinal cord are called *vertebrae.*

The structure of a typical neuron is shown in Figure 4.25b. Notice that each neuron has a region called the *cell body* that contains the cell's nucleus and other organelles, and that two sets of fibers extend from the cell body. The bushy *dendrites* are located at one end of the cell body, and the long, thin fiber at the opposite end of the neuron is called an *axon*. Dendrites receive information from other neurons, and axons usually carry information to other neurons and other types of cells. *Sensory neurons* have long axons that carry information from the skin, joints, muscles, and other parts of the body to the spinal cord. These nerve fibers bring information about touch, heat, cold, pain, joint position, and other sensations. *Motor neurons* have long axons that send signals that tell muscle groups to contract. Inside the spinal cord long axons carry sensory information to the brain. Many of the commands about what muscles to contract come from areas in the brain, especially in a region at the top of the brain called the *motor cortex.* These neurons have axons that extend all the way down the spinal cord. These important long axonal fibers are the key players in spinal cord injuries.

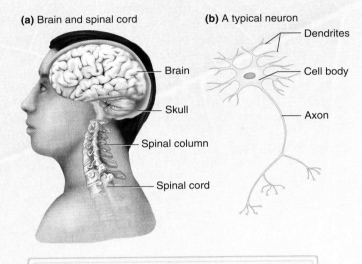

(a) Brain and spinal cord — Brain, Skull, Spinal column, Spinal cord

(b) A typical neuron — Dendrites, Cell body, Axon

The brain and spinal cord are collections of neurons. The neurons in the spinal cord connect the brain and body.

Figure 4.25 **Anatomy of the Spinal Cord, Spinal Column, Vertebrae, and Neuron.** (**a**) The brain and spinal cord are made of neurons that carry information in the form of nerve impulses. (**b**) Dendrites bring nerve impulses to cell bodies, and axons carry nerve impulses to other nerves, muscles, and organs.

In the days and weeks after a spinal cord injury the symptoms can be severe because temporary swelling damages local tissue. This type of damage can recover as the swelling goes down, but a traumatic injury also can damage or destroy the axons that carry sensory and motor signals through the spinal cord to the brain or to the peripheral muscles. Long-term permanent loss of function results when this happens. A person may be paralyzed and/or unable to feel. The area of the body affected by such an injury depends on where along the spinal cord the injury occurs (**Figure 4.26**). The region of the body below the injury usually is affected. If the injury is high, for example in the neck region, then nearly the entire body, except the head, is affected. Like Christopher Reeve, people with this kind of injury are *quadriplegic* and lose the use of all four

—Continued next page

Continued—

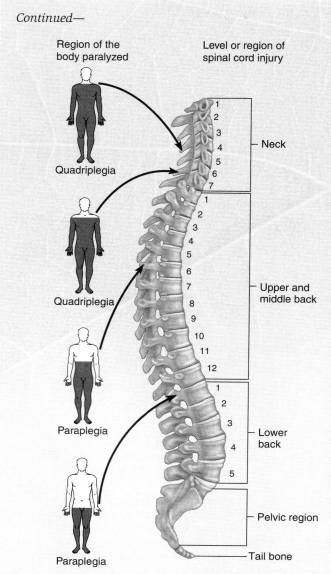

Region of the body paralyzed

Level or region of spinal cord injury

Quadriplegia

Quadriplegia

Paraplegia

Paraplegia

Neck

Upper and middle back

Lower back

Pelvic region

Tail bone

Figure 4.26   **Sites of Spinal Cord Injury.** The place where a spinal injury occurs influences the degree of the injuries that result from it.

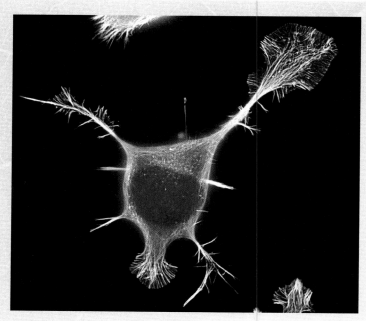

Figure 4.27   **Growth Cone.** A growth cone, made of cytoskeleton elements such as microfilaments and microtubules, is at the tip of a regenerating axon.

limbs. They often lose bladder and bowel control, sexual function, and sensation below the level of the injury. If the injury is at a lower level, then a smaller region of the body, below the injury is involved. Many aspects of an injury contribute to the degree of disability, but the big question is: Why don't the axons regrow?

In many situations axons do grow. For instance, axons grow during early development. And axons located outside of the spinal cord can regrow. So why don't axons within the spinal cord regrow? One of the earliest clues to the answer was the discovery that the axons within the spinal cord will regrow if they are given a different environment. In animal studies, for

example, the axons will regrow if they are isolated within a little tube that takes them out of the spinal cord environment. Isn't that odd? What is going on here?

The answer, it turns out, involves signal molecules and the cytoskeleton of an axon. Whether axons are growing for the first time during embryonic development or are regrowing after an injury, a structure called a *growth cone* forms at the end of the axon. A growth cone is a fanlike structure that crawls like an amoeba as the axon grows (**Figure 4.27**). The direction of growth cone movement determines the direction of axon growth. Axons grow by building and breaking down the cytoskeleton. Inside the growth cone are two kinds of cytoskeletal proteins which you already know about. Long filaments made of a protein called actin are found in the very tips of the growth cone, while long tubes of protein called microtubules are found in the center of the growth cone. Both of these protein elements can either be lengthened or dismantled in different parts of the growth cone. For example, if the cytoskeleton in the right side of the growth cone is lengthened, then the whole axon grows in that direction. This process of axon regrowth is often called *regeneration*.

One big piece of the spinal cord injury puzzle was solved when it was discovered that growth cones—whether embryonic or regenerating—collapse when they encounter certain cells normally found in the brain and spinal cord. These cells are called *astrocytes*, and when the spinal cord is injured, astrocytes form a scar around the injury. When new growth cones come up against these astrocyte scars, the growth cones collapse and the axons do not grow. Astrocytes are not found outside the brain and spinal cord. So in the rest of the body growth cones flourish and injured axons can regrow.

How does the axon "know" whether to grow in an environment? The answer is signal molecules. The actin and microtubule molecules that make up the cytoskeleton are controlled by signal molecules. Signal molecules in the environment of growth cones bind to membrane receptors, which, in turn, determine whether the cytoskeleton grows or is dismantled. Many different kinds of signal molecules have been identified. Outside the brain and spinal cord these signal molecules usually direct the cytoskeleton within the growth cone to grow and enlarge. Inside the brain and spinal cord the signal molecules produced by astrocytes work in the opposite way, and they cause the cytoskeleton to be dismantled. As a result, the growth cone collapses and the axon does not regrow. Some people hope that stem cells, discussed in Chapter 9, could provide an environment in which axons could regrow, but the research on stem cell treatment for spinal cord injuries is only just beginning.

The one mystery in all this is to explain how Christopher Reeve was able to recover as much function as he did, so long after his injury. The key was probably that Reeve pushed hard to exercise and stimulate his body as much as possible. Many studies of normal animals and humans have shown that intensive stimulation of the nervous system—through exercise, mental activities, music, and other ways—can cause the growth of dendrites and axons. While it is not known for sure that this can happen in an injured spinal cord, it is likely that all of this stimulation produces other signal molecules that may be able to overcome the negative signals from astrocytes. Inspired by Christopher Reeve's successes, future research will investigate this possibility.

QUICK CHECK

1. Why do so few people recover from spinal cord injuries?

2. What are the three major types of neurons in the body?

Now You Can **Understand**

What Are Drugs?

To coordinate their functions, cells of a multicellular organism must communicate with each other. In this chapter you have learned that cells do this by using chemical signals. Each signal interacts with a specific receptor protein, and this interaction depends on the charges and shapes of the signal and the receptor. It also is possible that other molecules could interact with this receptor. These molecules would have a similar charge and shape as the natural signal. This is a drug—a molecule from outside the body that interacts with a normal body protein. Drugs can be molecules from other kinds of organisms, or they can be "designed" by drug companies or researchers. Drugs can even be molecules that your body normally makes and uses but that are given to you in a pill, injection, or other external form. In any case, a drug can change the normal functioning of your cells because it interacts with your normal body proteins. You will learn more about specific drugs throughout this text.

What Do **You** Think?

Can Magnets Eliminate Pain?

You now have learned a great deal about cell biology. You know about the chemicals that make up cells, the chemical energy that drives cellular processes, and the complex cellular structures that organize these processes. At this point you can begin to think critically about how external stimuli might act on cellular processes. Of course, you have senses designed to detect light, sound, heat, and pressure, but what about more unusual stimuli such as magnetic fields? A brief Internet search will show there are scores of sites selling magnets that are supposed to relieve pain and improve recovery from injuries. Most sellers of magnets claim that a magnet placed over the skin increases blood flow in that region. Does this make sense? If you pull a muscle or have an arthritic joint, would wearing a magnet over the area eliminate the pain? How can you decide if the impressive claims are believable? Well, you could buy the magnets and see if they work for you. But that means gambling your money on the chance that the magnets may work. Another approach is to think the situation through logically, based on what you know about biology. Magnets interact with metals or with strong electric fields. Have you learned anything about cells that suggests they have high concentrations of metals with which a magnet could interact? This is what scientists call an empirical question. This means you must look at controlled scientific studies to see if it really works.

Chapter Summary

SUMMARY AND KEY TERMS

4.1 Living Things Are Made of Cells

Cellular organization is characteristic of life, but most cells are too small to be seen without magnification. Leeuwenhoek and Hooke made some of the earliest discoveries of the subvisible world of cells. Schleiden and Schwann suggested the cell theory, that all organisms are made of cells, and Virchow added that all cells originate from preexisting cells.

cell theory 84

4.2 The Cell Membrane Protects Internal Processes and Allows the Cell to Interact with the Environment

The phospholipid bilayer of the cell membrane protects the cell from unwanted chemical interactions with the environment. Membrane proteins embedded within the bilayer form portals into and out of the cell. Many membrane proteins span the membrane, while others are restricted to the exterior or interior surface of the lipid bilayer. Substances move into and out of cells by the processes of diffusion, active transport, osmosis, and facilitated diffusion.

active transport 89
diffusion 88
facilitated diffusion 88
lipid bilayer 86

membrane channel 86
membrane protein 86
osmosis 90
phospholipid 85

4.3 Cells Interact with the Environment via Membrane Proteins

The outer portions of proteins can interact with chemical signals in the environment, while the inner portions can interact with chemicals of the cytosol. Signal molecules alter the internal chemistry of the cell after they bind to receptor proteins and undergo shape changes. Signal molecules affect cellular processes through the phenomenon of signal transduction. Water moves across cell membranes in protein water channels. There also are proteins that allow movement of ions and molecules.

receptor 91
signal molecules 90

signal transduction 91

4.4 The Inside of a Cell is Highly Organized

Cells are organized as either prokaryotic cells or eukaryotic cells. Prokaryotic cells are simpler than eukaryotic cells and, in general, prokaryotic cells lack membrane-bounded organelles. The nucleus, cell membrane, cytosol, cytoplasm, cell wall, cytoskeleton, centrioles, nucleolus, ribosomes, smooth and rough endoplasmic reticulum, Golgi complex, chloroplasts, mitochondria, lysosomes, and vacuoles are eukaryotic organelles. The nucleus holds DNA with instructions for the amino acid sequence of proteins; ribosomes, ER, and Golgi complex synthesize, modify, and package proteins and lipids; the cytoskeleton maintains cell shape; and chloroplasts and mitochondria are involved in energy capture and energy release. Viruses are not alive because they cannot metabolize or reproduce independently. Although multicellular organisms are alive, nevertheless life's processes take place within the cells of multicellular organisms. Eukaryotic cells may represent the union of several kinds of prokaryotic cells.

cell wall 96
centriole 98
chloroplast 100
cytoplasm 96
cytoskeleton 97
cytosol 96
endoplasmic reticulum (ER) 98
eukaryotic cell 94
Golgi complex 99
lysosome 100
mitochondria 100

nuclear membrane 94
nucleoid 94
nucleolus 98
nucleus 94
organelle 94
prokaryotic cell 94
ribosome 98
rough endoplasmic reticulum (RER) 99
smooth endoplasmic reticulum (SER) 98
vacuole 101

4.5 An Added Dimension Explained: Christopher Reeve and the Search for Treatments for Spinal Cord Injuries

Within the spinal cord the regrowth of axons stops when cytoskeleton proteins encounter astrocytes.

REVIEW QUESTIONS

TRUE or FALSE. If a question is false, rewrite it to make it true.

1. Leeuwenhoek and Hooke were two pioneering scientists who studied the difference between eukaryotes and prokaryotes.

2. Membrane proteins prevent ions and polar molecules from crossing the cell membrane.

3. Active transport uses energy in transporting substances across the cell membrane.

4. Smooth ER has the primary job of manufacturing and modifying proteins.

5. Photosynthesis happens within the nucleus of a eukaryotic cell.

MULTIPLE CHOICE. Choose the best answer of those provided.

6. All living organisms are
   a. multicellular.
   b. unicellular.
   c. made of one or more cells.
   d. made of cells at least as large as this period.
   e. composed of prokaryotic cells.

7. Which of the following statements is *false?*
   a. A cell membrane is composed of a double layer of phospholipids and proteins.
   b. The heads of phospholipid molecules are nonpolar, while their tails are polar.
   c. Cell membranes are critically important because they shelter life's chemistry from the environment.
   d. Cell membranes are critically important because they allow a cell to communicate with its environment.
   e. Many of the organelles of eukaryotes are bounded by cell membranes.

8. How is osmosis different from diffusion?
   a. Osmosis refers to the movement of ions across the membrane, while diffusion is a special case of osmosis that refers to the movement of water across the cell membrane.
   b. Osmosis refers to free movement across a membrane, while diffusion refers to restricted movement across a membrane.
   c. Osmosis is governed by membrane proteins, while diffusion across the membrane is not governed by proteins.
   d. Osmosis refers to the movement of water across the cell membrane and is a special case of diffusion.
   e. Osmosis requires an input of energy, while diffusion does not.

9. Most membrane proteins convey information to the inside of cells by
   a. changing their shape.
   b. allowing steroids to enter the cell.
   c. directly influencing the magnetic forces inside the cell.
   d. changing the phospholipid layers.
   e. none of the above.

10. The energy most commonly used by cells is
    a. DNA and RNA.
    b. chloroplasts and mitochondria.
    c. ribosomes and other organelles.
    d. ATP and glucose.
    e. gravitational.

11. What is the major difference between prokaryotes and eukaryotes?
    a. Prokaryotes use photosynthesis, and eukaryotes do not.
    b. Prokaryotes have a nucleus, and eukaryotes do not.
    c. Prokaryotes do not have DNA, and eukaryotes do have DNA.
    d. Prokaryotes do not have a nucleus, and eukaryotes do have a nucleus.
    e. Prokaryotes use photosynthesis, while eukaryotes use respiration.

12. Substances diffuse
    a. from an area of lower concentration to an area of higher concentration.
    b. from an area of higher concentration to an area of lower concentration.
    c. from an area of lower concentration to any other area of low concentration.
    d. from an area of higher concentration to any other area of high concentration.
    e. randomly without regard for concentration.

13. What is the difference between cytoplasm and cytosol?
    a. Cytosol is the substance inside a cell that contains molecules and ions, while cytoplasm is the cytosol and organelles together.
    b. Cytoplasm is the substance inside a cell that contains molecules and ions, while cytosol is the cytoplasm and organelles together.
    c. Cytosol includes the cell membrane, while cytoplasm does not.
    d. Cytosol includes DNA, while cytoplasm includes RNA.
    e. Cytoplasm includes the extracellular environment, while cytosol does not.

14. Viruses are not considered to be alive because they don't
    a. reproduce on their own.
    b. take in nutrients.
    c. grow.
    d. use energy to maintain their own structures.
    e. All of the above are correct.

15. Researchers have come to the conclusion that cellular organelles, at least mitochondria and chloroplasts, were originally independent prokaryotic organisms. This is supported by the fact that mitochondria and chloroplasts
    a. can live alone, outside a cell.
    b. carry their own DNA.
    c. are both found in all cells.
    d. are found only in certain types of cells, such as muscles.
    e. are much larger than other organelles.

16–20. Match the type of cell with its appropriate fact or function:

16. cell wall
17. Golgi complex
18. ribosome
19. cytoskeleton
20. centriole

a. a network of protein fibers that support cell shape in some kinds of cells
b. an organelle that bonds amino acids together to form proteins
c. a rigid supporting structure outside the cell membrane of some kinds of cells
d. a stack of membranes where proteins are altered before being exported from the cell
e. an organelle made of small pieces of microtubule that function in the division of animal cells

CONNECTING KEY CONCEPTS

1. How can the cell membrane act as a barrier to the outside environment and yet allow a cell to communicate with the environment?

2. What is meant by the statement that the cellular processes in eukaryotes are more spatially organized than the cellular processes in prokaryotes?

QUANTITATIVE QUERY

One meter is approximately the distance from your fingertips on your outstretched arm to the tip of your nose. This distance can be expressed as 1,000,000 micrometers. A typical skin cell is about 60 micrometers in width. Approximately how many skin cells would fit from your fingertip to your nose?

THINKING CRITICALLY

1. What would happen to the flow of potassium across cell membranes if your cells suddenly were surrounded by a high potassium concentration?

For additional study tools, visit www.aris.mhhe.com.

5

Life Uses Chemical Energy

ENERGY AND LIFE

Dr. Frankenstein was obsessed with creating life.

Myth of the Mad Scientist
October 28, 1997

How Did Life Begin?
November 11, 2003

Chemist Adds Missing Pieces To Theory on Life's Origins
July 4, 1995

"None but those who have experienced them can conceive of the enticements of science. In other studies you go as far as others have gone before you, and there is nothing more to know; but in a scientific pursuit there is continual food for discovery and wonder. . . . Whence, I often asked myself, did the principle of life proceed? It was a bold question, and one which has ever been considered as a mystery. . . ."

With these words the author Mary Shelley has her protagonist, Victor Frankenstein, explain his fascination with the origin of life. Driven by a passion to understand life, Frankenstein discovers how to infuse life into a nonliving creature that he has stitched together from parts of corpses. The key, of course, is energy, that mysterious stuff of the universe that makes everything happen. Even in the 1800s Shelley understood that energy is required for life. Dr. Frankenstein needed only the "spark of life" to turn dead tissues into a living human being (chapter-opening photo). Just as important, the Frankenstein story tapped into the age-old belief that life could spring full-blown from things that are not alive. People saw maggots in rotting meat and small flies around the fruit in the kitchen; mice leapt out of piles of old rags and mayflies emerged from streams. It *seemed* obvious that life can come from nonlife. But does it?

There have always been critical thinkers who question what seem to be obvious truths. Not long after Shelley's classic novel was published, people began to explore other ideas about the source of life. For example, Louis Pasteur set out to scientifically test the idea that life can spring from nonlife. Pasteur's scientific approach has withstood many tests over the last century—in today's understanding of the world, it is now clear beyond a doubt that life can only come from other living things. A hunk of leftover steak, a pot of warm oatmeal, even a human body that recently was alive cannot generate new life unless it already contains life. Pasteur's edict stands in all biology textbooks: only life can produce life.

And yet, it all had to start somewhere. To say that life comes only from life just pushes the question back further in time. Where did life start? Did Mary Shelley have an inkling of the truth? Can life ever arise from nonlife? You will return to this question at the end of this chapter. For now, consider one important requirement for life: energy. ■

5.1 Life Requires Energy

Biological molecules can combine to form complex cellular structures such as the various organelles in a eukaryotic cell. But is this enough to qualify as being alive? If the chemical building blocks of life were put into a container and water added, would you have life? Of course, the answer is no, in part because one important ingredient would be missing. Life requires energy. In this chapter you will learn more about energy and how organisms use it. Then Chapter 6 will explain how living cells capture, store, and use energy.

Energy Makes Things Happen

Energy is around you, in you, used by you, and released by you. You must have energy to stay alive. But what is it? You can begin to understand energy if you think about how you experience it. For instance, transportation requires energy. When you drive a car, ride a bus or train, or fly in a plane, energy moves the vehicle that carries you. Other examples of energy also involve motion: an explosion of gunpowder sends a cannonball flying; a tennis player smashes a ball with a racquet and slams it over the net.

How far does this idea of energy as motion extend? It turns out that motion is an important aspect of energy. **Figure 5.1** shows some examples of the connection between energy and motion. Electricity, the energy that drives many home appliances, is the movement of electrons. When a friend asks what your plans are for the weekend, her voice produces sound energy. Sound is the motion of air molecules, and specialized cells within your ears detect that motion. Even light energy is the motion of photons, tiny particles of light energy that are detected by specialized light-sensitive cells in your eyes. Energy involves motion or the potential to produce motion. Biological molecules carry energy in the motion of their electrons.

To understand how electrons carry energy, let's consider **electromagnetic force,** one of four fundamental physical forces that govern the interactions between particles of matter. Electromagnetic force includes both electricity and magnetism. One example of electromagnetic force is the resistance that you feel when you try to push the north ends of two magnets together. The electrical part of the electromagnetic force is important in living systems. Objects that carry an electrical charge generate electrical forces. There are only two kinds of electrical charges: positive and negative; some objects are electrically neutral and carry no charge (see Section 2.2). If two particles with opposite charges are close to each other, they will be attracted and move toward each other as long as no barrier prevents this. In contrast, two particles of the same charge will repel one another. In your study of life's processes, you frequently will encounter chemical interactions that are driven by electrical forces.

electromagnetic force
the force that causes particles with opposite charges to be attracted to one another and particles with the same charge to repel one another

(a) When a copper wire is connected to a battery, electrons flow in the wire from one copper atom to the next, all around the circuit.

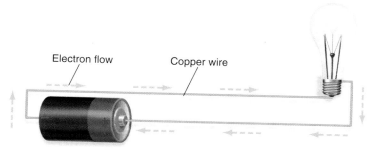

Electron flow   Copper wire

(b) A sound wave is created by the motion of air molecules. The motion of the air molecules is detected by the inner ear.

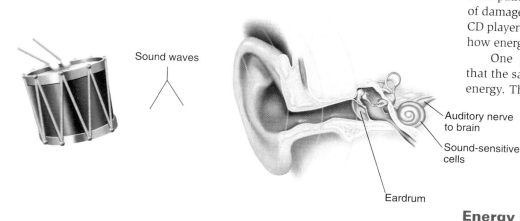

Sound waves

Auditory nerve to brain

Sound-sensitive cells

Eardrum

(c) Light energy is the motion of photons. That energy is detected by light-sensitive cells in the eye.

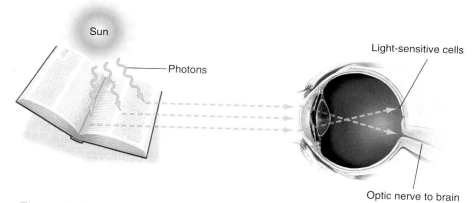

Sun

Photons

Light-sensitive cells

Optic nerve to brain

Figure 5.1 **Examples of Energy as Motion.** (**a**) The motion of electrons can be used to power electrical appliances. (**b**) Movement of air molecules is detected by the inner ear. (**c**) Photons can be detected by light-sensitive cells in the eye.

■ *How are sound waves and light waves alike? How are they different? Where is the energy in a sound wave? In a light wave?*

kinetic energy the energy carried by a moving object

What do forces have to do with energy? When an object moves in response to a force, energy is transferred to the object and the object carries the energy. This is true whether the object is a baseball or an electron. An object that carries energy can change other objects. As an example, consider a book. By itself that book hardly seems dangerous, but things change when the same book is carried up in a small airplane. Against the force of gravity, the energy generated by the engines carries the plane and the book aloft. As the plane gains altitude, you can say that energy is transferred to the book. Suppose the book falls out of the plane and plummets down toward the CD player that you left out on the back patio. Now the book has the power to do a lot of damage, as you will discover when you find your CD player smashed to bits. This is a good example of how energy is the power to do things.

One point demonstrated by this example is that the same object can carry different amounts of energy. The book carried little energy when it was on the ground, but it carried a lot of energy when it dropped from an airplane at cruising altitude. Electrons also can carry different amounts of energy at different times. Let's look into this idea in greater detail.

Energy Can Be Kinetic or Potential

From this discussion of energy you now can appreciate the formal definition of energy found in most textbooks. *Energy* is the potential to do work, which is another way of saying that energy is the potential to make things move (see Section 1.2). The energy carried by an object in motion is called **kinetic energy.** For instance, a car cruising at 75 miles per hour has kinetic energy; your body has kinetic energy as you move, whether that is sprinting or just winking. Electrons are always in motion, so they also carry kinetic energy. Even though the idea of kinetic energy can be abstract and hard to grasp, the amount of kinetic energy an object carries can be calculated precisely. The kinetic energy carried by an object is equal to one-half of its mass multiplied by the square of its velocity. Mathematically, this is summarized by the formula $KE = \frac{1}{2}(mv^2)$. While this formula may look intimidating, it actually is an excellent illustration of how scientific concepts can be expressed mathematically. Even if you don't like math, you will be surprised to see how easy it is to calculate the answer (**Figure 5.2**). In this equation kinetic energy is expressed in units called joules. Mass always is expressed in kilograms. Let's assume that a baseball thrown at your bedroom window has a mass of about 145 grams, or 0.145 kilograms. If your neighbor's child throws the ball at a velocity of 4 meters per second, the kinetic energy of the ball is $^1/_2$ (0.145×4^2), which equals 1.16 joules (Figure 5.2a). In contrast, if your weight-lifting friend throws the same baseball at 20 meters per second, the kinetic energy of the ball is 29 joules (Figure 5.2b). That is quite a difference. Similar calculations can be used to figure out how much energy a moving electron brings to a chemical reaction.

Even if an object is not actually in motion, it can carry energy as the potential to move, and this form of energy is called **potential energy.** For example, as you wait to hit a baseball, you store potential energy in the tension of your muscles and the position of your body as you prepare to swing. Objects store energy in many ways. For instance, a rock poised at the top of a cliff has potential energy. So does an arrow pulled back in a bow or a frog's tensed muscles as it gets ready to leap (**Figure 5.3**). When the rock is dislodged and crashes downhill, when the arrow is released from the bow, or when the frog jumps away, potential energy is translated into the motion of kinetic energy. Although it may seem odd, a book lying on your desk also has potential energy. The book's potential energy becomes evident when you nudge it over the edge of the desk and it crashes to the floor.

The Laws of Thermodynamics Describe How Energy Behaves

The properties of energy that you just have read about are universal. As far as we know, they apply to every time and place in the universe and can be stated as two of the laws of thermodynamics. The **first law of thermodynamics** says that energy can't be created or destroyed, only changed in form. The observation that one form of energy can be converted to another is a reflection of the first law of thermodynamics (**Figure 5.4**). The first law of thermodynamics also means that the total amount of energy in the universe stays constant. In practical terms, if you need energy for some process, you cannot make it. Instead, you must get it from something else. This is why you must eat to stay alive and why plants die when they are deprived of light.

The second law of thermodynamics is more abstract but is equally important for biological systems. To understand the second law, it helps if you understand that a highly ordered system of objects carries more energy than a disordered system. For instance, a pile of LEGO blocks carries less energy than the same blocks configured into a robot. Similarly, the ordered, complex molecules in your body carry more energy than the same kinds and numbers of atoms not formed into molecules. The sec-

Kinetic energy = $^1/_2$ the mass × the square of the velocity or
KE = $^1/_2$(mv²)

Mass of a baseball = 145 g or 0.145 kg

Kinetic energy of a baseball thrown at a velocity of 4 meters/second is:
$$KE = {}^1/_2(0.145 \times 4^2)$$
$$= {}^1/_2(0.145 \times 16)$$
$$= {}^1/_2(2.32)$$
$$= \textbf{1.16 joules}$$

Kinetic energy of a baseball thrown at a velocity of 20 meters/second is:
$$KE = {}^1/_2(0.145 \times 20^2)$$
$$= {}^1/_2(0.145 \times 400)$$
$$= {}^1/_2(58)$$
$$= \textbf{29 joules}$$

Figure 5.2 **Calculating Kinetic Energy.** Kinetic energy is expressed in joules, and mass is expressed in kilograms. A baseball thrown by a child (**a**) carries less energy than a baseball thrown by a larger person (**b**).

Potential   Kinetic energy

Figure 5.3 **Potential Energy and Kinetic Energy.** Potential energy stored in the frog's leg muscles becomes kinetic energy that propels its leap.

ond law of thermodynamics says that these ordered states of matter are *not* stable. Any organized thing will tend to lose order and become more disordered over time (**Figure 5.5**). The only way to stop disorder is to have an input of energy—which, according to the first law of thermodynamics, you must obtain from some other source. When energy is taken from one system to build order in a second system, the order in the first system will decrease. You probably know this all too well, especially when you think of how much effort it takes to keep your room or house clean and orderly. You provide the energy to maintain that order—in turn, you must

potential energy energy carried by a stationary object that has the potential to cause motion

first law of thermodynamics energy cannot be created or destroyed, only changed in form

second law of thermodynamics without an input of energy, the order in any system declines over time

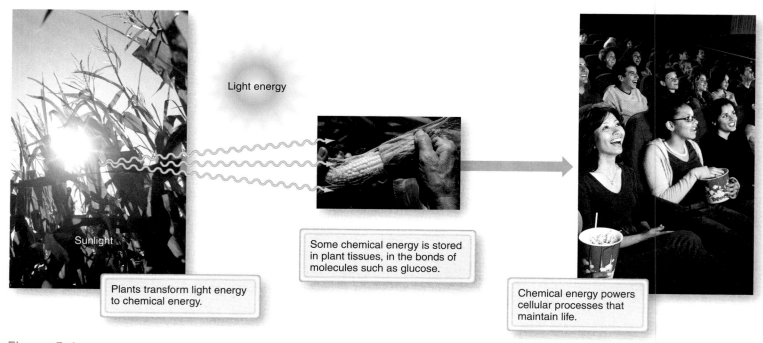

Light energy

Sunlight

Plants transform light energy to chemical energy.

Some chemical energy is stored in plant tissues, in the bonds of molecules such as glucose.

Chemical energy powers cellular processes that maintain life.

Figure 5.4   **First Law of Thermodynamics.** Energy can be converted from one form to another.

Figure 5.5   **Second Law of Thermodynamics.** Without continual inputs of energy, systems lose order and degrade.

have a source of energy to maintain your internal structures and processes. So, after you've cleaned your room you may need a snack.

On a grand scale the second law of thermodynamics predicts that the universe will eventually become a completely random collection of atoms that have no order at all and uniform energy throughout. This does not mean that it is impossible to produce an ordered system, such as a living cell. It means that order can develop only if there is a source of energy. The laws of thermodynamics have another implication: every time one form of

energy is transformed into another form of energy, some energy is lost as heat. For example, when the chemical energy in fossil fuels is converted to electricity, some energy is lost as heat. Heat energy usually dissipates to the environment and is wasted. This aspect of the laws of thermodynamics also can be seen in biological systems. For instance, when a rabbit eats a bunch of carrots it uses only a small portion of the energy available in the carrots. Much of the energy is lost as the rabbit's body heat that radiates out to the environment. When a bobcat eats the rabbit, heat energy also is lost in a similar way. You will see the implications of these successive levels of heat loss when you consider food chains and pyramids of energy in Chapter 27.

The first and second laws of thermodynamics set clear limits for how living organisms use energy. They dictate that during its lifetime an organism must have a source of energy to build and repair its complex structures. If an organism does not have a source of energy, it will die. When an organism dies, it no longer has the ability to obtain and use energy. As a result, its complex structures decay and its large molecules fall apart.

QUICK CHECK

1. What is energy?
2. What is the difference between potential and kinetic energy?
3. Restate the first and second laws of thermodynamics.

5.2 Cells Use Chemical Energy

Energy is differentiated by the kind of matter carrying the energy. For example, *electricity* is the movement or the potential movement of electrons. Electrons can move in a metal wire, as they do in your house, or they can move in another medium like water. *Light energy* is carried by photons produced by any light source such as a burning log, a glowing lightbulb, but more importantly for biological systems, by the Sun. Two other forms of energy important in living systems are heat energy and chemical energy. *Heat energy* is the random motion of any atom or molecule. If the atoms or molecules within a space are forced to move faster, they carry more energy, and more heat, and they feel warmer to the touch—even hot. This is what happens when water boils. Energy from a stove or fire causes the water molecules to move faster, and so the water boils. When the input of energy stops, the heat escapes to the surrounding environment and the water cools. Even though living organisms produce heat, it is not usable energy in biological systems. The heat our bodies produce keeps us warm and keeps our internal molecules moving around so that they can interact—but biological structures and processes in living organisms cannot be powered by heat energy.

Biological processes get energy from *chemical energy* stored in the chemical bonds of molecules. Making a chemical bond requires energy, and breaking a chemical bond can release energy. When a living cell builds an internal molecule or organelle, or when it carries out a metabolic process, the energy released from one set of chemical bonds is used to make or break another set of chemical bonds.

This might seem like a "robbing Peter to pay Paul" sort of system. If energy is just shuffled between molecules, how can living organisms grow and multiply? The answer relates to the fact that energy can be converted from one form to another. Most often photons from the Sun provide the initial input of energy in biological systems. Some organisms, such as green plants, can convert light energy to the energy stored in chemical bonds. In other words, they can use light energy to build molecules. These organisms then use the chemical energy derived from solar energy to drive their own biological processes. Even though only some kinds of organisms can transform light energy to chemical energy, nearly all living organisms rely on this conversion process as the source of the chemical energy necessary for life. Organisms that cannot capture and transform light energy consume organisms that can. For example, you gain from the work of a plant when you eat an apple and use the apple's chemical energy for your own needs.

Chemical Bonds Store and Release Energy

Where is the energy in chemical bonds? Recall that the chemical bonds that hold atoms together in molecules involve electrons (see Section 2.3). When two atoms form a molecule, an external force pushes two or more of their electrons into a new orbital. In their new orbital the electrons encircle the nuclei of both atoms. The chemical energy used to make the bond is located in the motion of the electrons around the two nuclei, and the electrons *carry the input of energy that was used to move them to the new orbital* (**Figure 5.6a**). This explains how a chemical bond holds energy. For as long as the molecule exists, the energy will stay in that bond. When this bond is broken, the electrons move back to their original orbitals and no longer carry the extra energy (Figure 5.6b).

This idea of storing and releasing energy from chemical bonds may be easier to understand if you relate it to a log burning in a fireplace (**Figure 5.7**). The log contains millions of large organic molecules such as cellulose, other carbohydrates, and proteins. All the bonds of these molecules contain energy. As long as the log is intact, the energy stays in those bonds, but when the log burns, the bonds holding those molecules together are broken. In a fire these bonds are broken rapidly, and their energy is released as heat and light. This energy feels warm and cozy, but it is not useful in biological processes.

Early in the history of life, however, cells evolved the ability to use the energy released from chemical bonds to do other things. At the most basic chemical level, cells are masterful chemical factories that use chemical energy to make and break other chemical bonds. The processes of life involve a continual rearrangement of atoms. New molecules are formed and existing molecules break apart. In one sense, the effort to understand life's processes is all about unraveling the molecular rearrangements that result in life's activities such as growth, cell division, movement, and response to environmental stimuli. Even complex and sophisticated activities like breathing, walking, thinking, and feeling are the result of molecular reactions and interactions driven by a source of chemical energy.

The way that chemical bonds form is a bit like maneuvering a skateboard up a steep hill. If you want to ride the skateboard from your house to

(a) Making a chemical bond

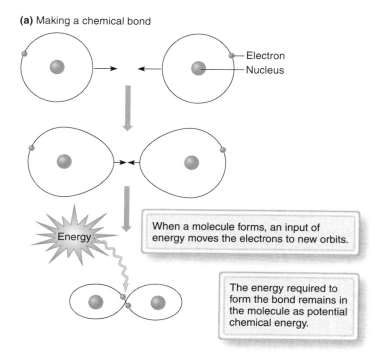

Electron
Nucleus

When a molecule forms, an input of energy moves the electrons to new orbits.

Energy

The energy required to form the bond remains in the molecule as potential chemical energy.

(b) Breaking a chemical bond

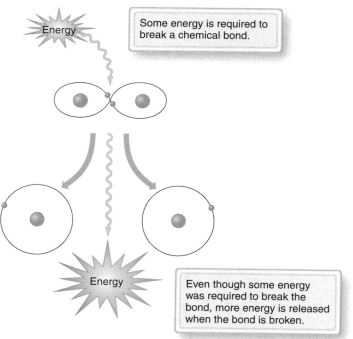

Some energy is required to break a chemical bond.

Energy

Energy

Even though some energy was required to break the bond, more energy is released when the bond is broken.

Figure 5.6 **Making and Breaking Chemical Bonds Involves Energy.** (a) A molecule is held together by chemical bonds that carry energy. (b) An input of energy is required to break a chemical bond with a subsequent release of energy.

■ *Think about the implications for life in the ability of molecules to store energy in their bonds. What sorts of things might an organism use this ability of molecules to do?*

A log contains millions of molecules that have innumerable chemical bonds. All of them contain potential chemical energy.

As a log burns, the chemical bonds are rapidly broken. Their energy is released as heat and light.

Figure 5.7 **Burning a Log Releases Energy Stored in Carbon-to-Carbon Bonds of Wood.**

■ *Using your knowledge of the molecular structure of various organic molecules, sketch the most probable molecular structure of the molecules found in firewood. Where in the plant did these molecules originate?*

your friend's house, and there is a hill in between, you must provide enough energy to push the skateboard up the hill to overcome the force of gravity, even if your friend's house is at a lower level than yours (**Figure 5.8a**). Once you have reached the top of the hill, the skateboard rolls downhill spontaneously. This is similar to the way that atoms form bonds. Just as you had to exert effort to push the skateboard up the hill, energy *always* is required to start a chemical reaction. This is true, even though the new molecule may have less or more energy than the original atoms or molecules.

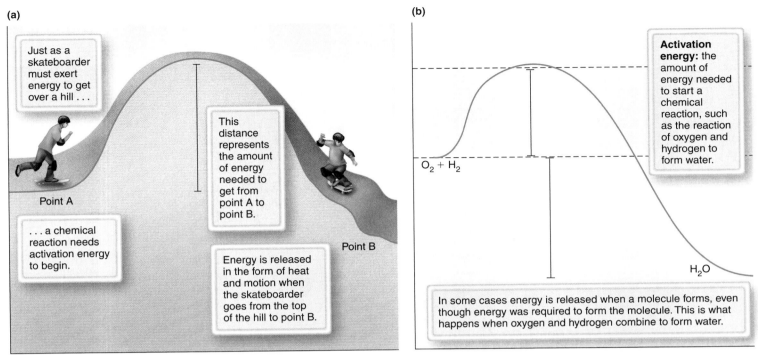

(a)

Just as a skateboarder must exert energy to get over a hill . . .

Point A

. . . a chemical reaction needs activation energy to begin.

This distance represents the amount of energy needed to get from point A to point B.

Point B

Energy is released in the form of heat and motion when the skateboarder goes from the top of the hill to point B.

(b)

Activation energy: the amount of energy needed to start a chemical reaction, such as the reaction of oxygen and hydrogen to form water.

$O_2 + H_2$

H_2O

In some cases energy is released when a molecule forms, even though energy was required to form the molecule. This is what happens when oxygen and hydrogen combine to form water.

Figure 5.8 **Skateboard Analogy and Activation Energy.** (**a**) Energy is needed to get from point A to point B, even though point B is downhill from point A. Once the skateboarder provides the energy to get up the hill, he can coast down to point B. The height of the hill is a measure of how much energy is needed to get from point A to point B. (**b**) When a water molecule forms from the reaction between oxygen gas (O_2) and hydrogen gas (H_2), energy is released because the total amount of energy in the water molecule is less than the sum of the energy in the O_2 molecule and the H_2 molecule. Still, energy is needed for the reaction to proceed. The amount of energy needed to start the reaction is the activation energy. The amount of energy that is released when one molecule of water forms is the total energy in $O_2 + H_2$ minus the energy in H_2O.

The same also is true when chemical bonds are broken. An initial input of energy is needed to break chemical bonds, even though energy is released when those bonds are broken.

The amount of energy needed to start the chemical reactions of making or breaking bonds is called **activation energy** and the amount required varies for different combinations of atoms. Some combinations of atoms take relatively little energy, and therefore they bond more frequently in nature. For example, a mixture of hydrogen and oxygen atoms readily forms bonds to make water, and just a small spark will set off this reaction (Figure 5.8b). A large amount of energy actually is released when water forms—in fact, so much energy is released that a mixture of oxygen and hydrogen can be used as rocket fuel.

The earlier example of the burning log also demonstrates that activation energy is required even when bonds are broken. The match used to light a fire provides the activation energy needed to start the wood burning. As the chemical bonds in the wood are broken, much more energy is released from the log than was provided by that initial match.

Living Organisms Use the Chemical Energy in the High-Energy Bonds of ATP

Because there are so many forms of life, and so many different chemical reactions involved in life, you might think there is a long list of chemical bonds used to store and release energy in living systems. Get ready for a pleasant surprise because living systems do not use just *any* chemical bond as a source of energy. Nearly all organisms use *just one type* of chemical bond to drive the processes of life—specifically, two particular bonds within a molecule of **adenosine triphosphate** (**ATP** for short). When a cell makes or breaks a chemical bond in a biological molecule, the energy to do this comes from the bond between the second and third phosphate groups of ATP. Nearly all cellular processes are powered by the energy derived from breaking the final phosphate bond of ATP. Much of the energy from breaking this bond is used to make or break another chemical bond.

ATP is the immediate source of chemical energy for a cell, but from the first law of thermodynamics you know that the energy to make ATP must come from somewhere. Where does it come from? As you

activation energy the amount of energy needed to start the chemical reactions of making or breaking bonds

ATP (adenosine triphosphate) a nucleotide molecule containing three phosphate groups that is a common energy source for living organisms

have read, the Sun is the primary source of energy for life. Of course, not all cells can make use of solar energy. For instance, you cannot lie out in the Sun and gather useful energy to run your cellular processes. But some cells can use the energy in sunlight to form ATP. These cells provide a reliable source of energy for themselves and for most other cells on Earth, including the cells of your body.

You can feel the energy of sunlight as the heat that it creates in your skin, but what exactly is the energy in sunlight made of? Chemical reactions inside the Sun produce huge quantities of heat and light energy that travel through space as photons and reach Earth. When photons strike an object, they cause its atoms and subatomic particles to become more energetic. More specifically, electrons in the sunlit object absorb the energy of the photons and are actually moved to a different, higher-energy, orbital (**Figure 5.9**). In this way the energy in a photon is transferred to an object that it hits. This is why skin begins to feel warm when it is exposed to the Sun. It also is why your skin can become burned, blistered, and damaged by too much exposure to solar radiation. In the next section you will consider an overview of how some cells can capture and use

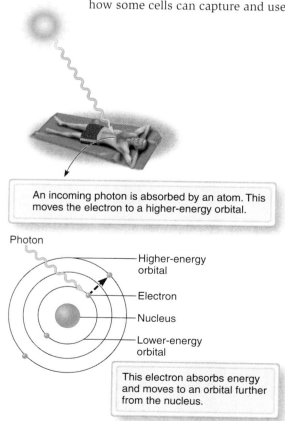

An incoming photon is absorbed by an atom. This moves the electron to a higher-energy orbital.

Photon

Higher-energy orbital

Electron

Nucleus

Lower-energy orbital

This electron absorbs energy and moves to an orbital further from the nucleus.

photosynthesis the process of using light energy to produce ATP molecules that provide the energy required to make glucose from carbon dioxide and water

Figure 5.9   **Objects in Sunlight Absorb the Energy of Photons.**

the energy in photons. Chapter 6 delves into this topic in greater detail.

QUICK CHECK

1. List and briefly describe four forms of energy.
2. What form of energy do biological organisms use to power cellular processes?
3. What is activation energy?

5.3 Photosynthesis Produces ATP and Glucose

Organisms that make use of the Sun's energy carry out the process of **photosynthesis** (**Figure 5.10**). During the two-step process of photosynthesis, the energy in sunlight is used to form the last phosphate bond in ATP. The ATP molecules produced by photosynthesis are not directly used to drive the biological processes in the photosynthesizing cell, however. This ATP contributes its energy to the second step of photosynthesis that converts carbon dioxide and water into glucose. More specifically, ATP energy helps form the carbon-to-carbon bonds of the glucose molecules.

You may be wondering why cells have *two* energy molecules. Why do cells, including photosynthetic cells, need both ATP and glucose? First, photosynthesizers can make ATP using the energy in photons only when the Sun is shining. They can't make ATP at night, and they can't make ATP as efficiently on cloudy days. In addition, single cells and multicellular organisms need to have energy in locations far from the place where ATP is produced. For example, in plants the energy produced by photosynthetic tissues must be moved to cells in structures like roots, stems, flowers, trunks, and branches that are not photosynthetic. ATP is not a good molecule for either of these purposes. ATP is a rather bulky molecule and because only *one* of its bonds is used for energy release, it isn't exceptionally efficient. A cell or organism would run out of space if it had to stockpile and transport ATP. Cells need an efficient source of chemical energy that they can use, move about according to demand, store for future use, and change into ATP as needed. Photosynthetic cells have evolved a solution to these problems by packing the energy from *many* ATP molecules into a *single* molecule: glucose. Glucose is a six-carbon sugar—although it is smaller than ATP, it can hold the energy of many ATP molecules. This makes glucose easier to transport to other parts of the cell or plant.

A photosynthetic organism can transfer the energy of ATP to glucose, but how is that energy harvested from glucose and put to use? All organisms, including photosynthetic organisms, have biochemical processes that use the energy in glucose to regenerate ATP. A single molecule of glucose yields up to a maximum of 38 ATP molecules. So cells in the roots, or stems, or even different parts of the same photosynthetic cells can break the chemical bonds in glucose and use that energy to produce ATP. This is an efficient system. The energy in sunlight powers the production of ATP. The energy in ATP is used to produce glucose. In turn, the energy in glucose is transferred to the ATP used to drive cellular processes.

Of course, not all organisms can use the energy in sunlight to make ATP, but nearly all organisms can use the energy in the bonds of glucose to make ATP. So nonphotosynthetic organisms can obtain the chemical energy in glucose by eating photosynthetic organisms. You do this every day when you eat cereal, bread, fruits, and vegetables. In this way photosynthetic organisms provide the chemical energy for themselves and for the rest of the Earth's life. **Figure 5.11** shows a simplified summary of how the energy captured from photons is transferred between molecules and organisms.

The final requirement for living systems is for long-term energy storage. Many plants have efficient strategies to store the energy of glucose for long periods of time. Just as a computer compresses files for long-term storage, or just as you cram out-of-season clothes into storage containers, cells have processes that store glucose molecules as long, complex carbohydrate molecules. Many plants use the starch amylopectin as a storage molecule (see Section 3.4). Animals can eat plants and use the sugars and starches stored in plant tissues. For instance, the tubers and roots that people like to eat—potatoes and carrots—are filled with plant starches and sugars. Animal cells can chain glucose molecules together to form glycogen molecules that then are stored in liver or muscle tissues until they are needed. In addition, some of the energy of glucose molecules can be kept in the longer-term storage molecules of lipids.

This is an efficient system that allows cells to move energy from one molecule to another, depending on whether the immediate need is to use energy or to store energy. Scientists use the word *metabolism* to refer to these chemical processes that build up and break down chemicals—therefore, the buildup and breakdown of glucose is called **glucose metabolism.**

(a) Photosynthesis occurs in the photosynthetic membranes of cyanobacteria and in the chloroplasts of eukaryotic photosynthesizers.

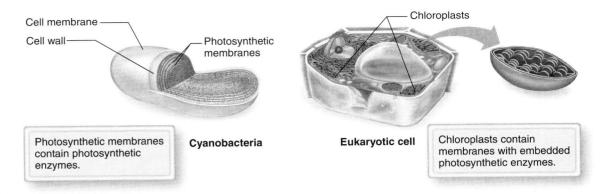

Cell membrane
Cell wall
Photosynthetic membranes
Chloroplasts

Cyanobacteria **Eukaryotic cell**

Photosynthetic membranes contain photosynthetic enzymes.

Chloroplasts contain membranes with embedded photosynthetic enzymes.

(b) Photosynthesis is a two-step process.

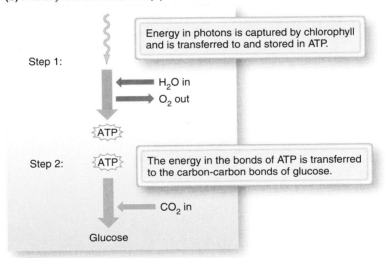

Step 1:

Energy in photons is captured by chlorophyll and is transferred to and stored in ATP.

H_2O in
O_2 out

ATP

Step 2:   ATP

The energy in the bonds of ATP is transferred to the carbon-carbon bonds of glucose.

CO_2 in

Glucose

Figure 5.10
Photosynthesis Overview.
(**a**) Photosynthesis is carried out by enzymes embedded in the internal membranes of cyanobacteria and chloroplasts.
(**b**) Photosynthesis is a two-step process where energy in ATP is transferred to and stored in the carbon-to-carbon bonds of glucose.

QUICK CHECK

1. What are the two major steps of photosynthesis?
2. What is the advantage if an organism makes both ATP and glucose molecules?
3. Make a diagram that shows how photosynthesis and glucose metabolism are related.

glucose metabolism the process of building up and breaking down glucose

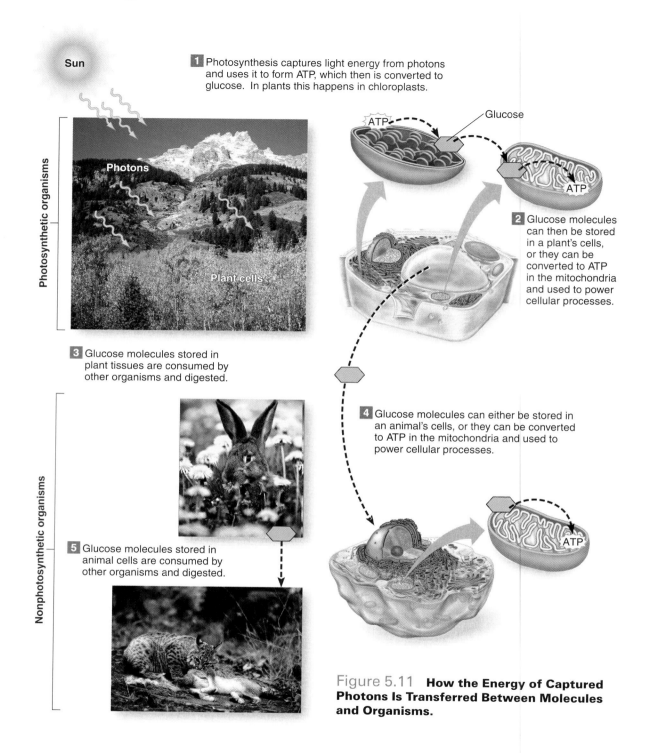

1 Photosynthesis captures light energy from photons and uses it to form ATP, which then is converted to glucose. In plants this happens in chloroplasts.

Sun

Photons

Plant cells

Photosynthetic organisms

Glucose

ATP

ATP

2 Glucose molecules can then be stored in a plant's cells, or they can be converted to ATP in the mitochondria and used to power cellular processes.

3 Glucose molecules stored in plant tissues are consumed by other organisms and digested.

4 Glucose molecules can either be stored in an animal's cells, or they can be converted to ATP in the mitochondria and used to power cellular processes.

Nonphotosynthetic organisms

5 Glucose molecules stored in animal cells are consumed by other organisms and digested.

ATP

Figure 5.11 How the Energy of Captured Photons Is Transferred Between Molecules and Organisms.

5.4 Enzymes Control the Pace of Life's Chemical Processes

Life's processes revolve around chemical bonds that hold biological molecules together, and chemical bonds can be broken to release usable energy.

This sounds simple enough, but there is still the tricky problem regarding the *rate* at which bonds are made and broken. Organisms must make some bonds at a much faster rate than bonds spontaneously form in nature. In addition, organisms also must break other bonds more slowly than is usual in many other natural processes. Let's examine each of these requirements.

Why is it necessary to speed up chemical reactions in living organisms? Molecules require energy to form, but even with a source of energy, the complex molecules that are so common in organisms are not likely to form spontaneously. The large numbers of lipids, carbohydrates, nucleic acids, and proteins built by a living organism *in a day* would not form in a bowl full of building block molecules even if you waited for *100 years* or longer. It also is important to control the sequence in which chemical reactions occur in a cell. The trick is to speed up and thus favor the occurrence of the *specific* chemical reactions that produce the molecules needed at any given time. Molecules produced at a fast rate will accumulate in a cell and dictate the cell's properties and activities. This gives cells a way to accumulate specific proteins, carbohydrates, lipids, and nucleic acids. Different kinds of cells within an organism can fine-tune this process to produce different sets of biological molecules. For example, muscle cells speed up the production of certain kinds of proteins and, as a result, are able to contract.

The job of speeding up and controlling chemical reactions—whether they are reactions in biological systems or any other reactions—can be accomplished by chemical catalysts. A **chemical catalyst** is a substance that speeds up a chemical reaction by decreasing the amount of energy needed to form or break a particular chemical bond. In other words, a catalyst lowers the activation energy needed for a particular chemical reaction to occur. Of course, energy is still needed, but with the aid of the catalyst much less energy is required. Returning to the skateboard analogy, a chemical catalyst is like putting a tunnel through the hill that you must scale on your skateboard (**Figure 5.12a**). It requires less energy to go through the tunnel than it does to go over the hill (Figure 5.12b). A catalyst does not actually take part in the chemical reaction, and it does not provide any energy for the reaction. This means a catalyst is not changed in the process and can be used again and again to speed up many repetitions of the same reaction. A catalyst simply increases the rate at which a particular chemical reaction will take place.

Lots of different kinds of atoms and molecules can serve as catalysts. For example, metals such as nickel are used as catalysts in many manufacturing processes. Cells have a special kind of catalyst, protein molecules called **enzymes.** Each kind of enzyme catalyzes only one or just a few chemical reactions. For example, in plant cells one kind of enzyme catalyzes the production of starch molecules from glucose. This enzyme dramatically reduces the energy needed for two glucose molecules to bind together. As with any protein, the shape and charge(s) of an enzyme molecule are responsible for its chemical activity. Many of the approximately 100,000 different proteins in human cells are enzymes; each catalyzes or speeds up a specific set of chemical reactions.

Enzymes produce phenomenal increases in reaction rates. An enzyme-catalyzed reaction proceeds approximately a *million to several billion*

chemical catalyst a chemical that speeds up a chemical reaction but does not take part in that chemical process

enzyme a protein or nucleic acid that acts as a catalyst to speed the chemical reactions inside a living organism by lowering the activation energy of a specific chemical reaction

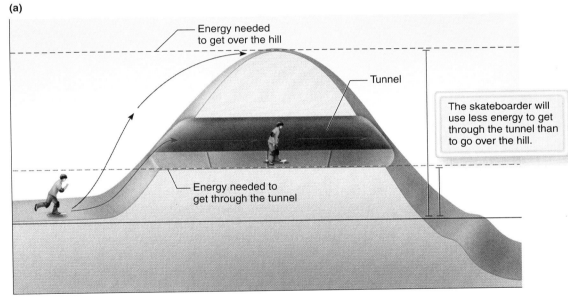

(a)

Energy needed to get over the hill

Tunnel

The skateboarder will use less energy to get through the tunnel than to go over the hill.

Energy needed to get through the tunnel

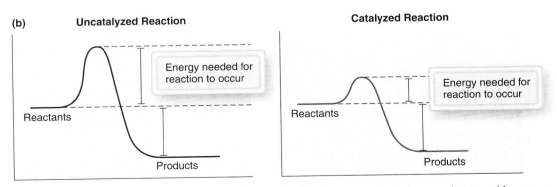

(b)

Uncatalyzed Reaction

Energy needed for reaction to occur

Reactants

Products

Catalyzed Reaction

Energy needed for reaction to occur

Reactants

Products

Figure 5.12   **Effect of a Catalyst.** (**a**) Just as it takes less energy to slide through a tunnel from one side of a hill to the other, (**b**) catalyzed reactions lower the amount of energy needed for a reaction to occur.

(a)

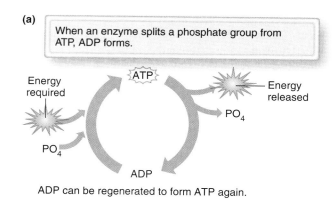

When an enzyme splits a phosphate group from ATP, ADP forms.

ADP can be regenerated to form ATP again.

(b)

A mousetrap baited with cheese is similar to an ATP molecule. The cheese is like the third phosphate group of ATP. A sprung mousetrap is like ADP.

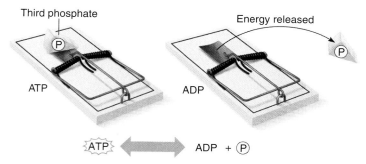

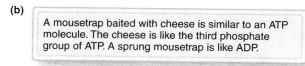

Figure 5.13 **Mousetrap Analogy for the ADP—ATP System.**

times faster than the same reaction without a catalyst. For example, if you placed lots of amino acids in a watery solution containing an appropriate energy source, the reaction to join only the first two amino acids in a protein might occur at a rate of one bond every *two years.* In contrast, in a living organism a group of enzymes catalyzes this reaction at a rate of about 20 bonds per *second.* This means that it takes only about 20 seconds to synthesize a protein chain containing 400 amino acids! These numbers illustrate how important enzymes are for life's processes. They also give a clue to how life may have evolved. Because life never could have developed on a large scale without them, enzymes must have been among the first biological molecules to evolve.

It might help to understand this concept of making and breaking chemical bonds in a controlled way if you reconsider the log that is still burning in our fireplace and the tree that it came from. A tree formed each cellulose molecule of the

adenosine diphosphate (ADP) a nucleotide molecule containing two phosphate groups that is generated when ATP is used by an enzyme to power a chemical reaction

wood one carbon-to-carbon bond at a time. The chemical process of making those chemical bonds in cellulose was speeded and controlled by enzymes. It may seem that trees grow slowly, but they would probably never grow at all without the help of enzymes. On the other hand, the energy in the bonds of glucose cannot be released too quickly. The energy in the bonds of glucose is released in a slow and regulated way, again under the control of enzymes. If all six of the carbon-to-carbon bonds in a single glucose molecule were simultaneously broken, the sudden release of energy could damage delicate biological molecules and cellular processes. This is what happens when you set a match to the log in the fireplace. The match supplies the activation energy needed to begin the process that rapidly breaks the chemical bonds of cellulose and other molecules in the wood. Burning releases the energy within chemical bonds as light and heat, which you see as flames and feel as warmth. So much energy is released that the entire log eventually is reduced to ashes.

Enzymes Use the Chemical Energy in ATP to Make and Break Chemical Bonds

Enzymes speed up and control the chemical reactions in a cell, but these enzyme-catalyzed reactions still must have a source of energy. Nearly all enzymes in all living things use ATP as a source of chemical energy. Enzymes can break the bond between the last two phosphates of ATP; the energy released from each bond can be used to power a chemical reaction. When this final phosphate bond is broken, a molecule of **ADP** (short for **adenosine diphosphate**) and one phosphate molecule remain (**Figure 5.13**). ATP can be regenerated from ADP by adding back the third phosphate, using a source of energy such as the energy in glucose or sunlight (Figure 5.13a). In this way ATP can be used and regenerated endlessly within a cell. You might think about the ADP—ATP system as similar to an old-fashioned mousetrap (Figure 5.13b). When the trap is set and loaded with cheese it is like ATP. The cheese in the trap is like ATP's third phosphate group. When the trap is sprung, the cheese and energy are released as the trap snaps back to the unloaded position (ADP). Cheese is added and the trap (ADP) is reset (ATP). As long as a mouse does not eat the cheese, you could keep loading the same piece of cheese again and again. The input of energy for the mousetrap system comes from you, when you push the trap to the loaded position.

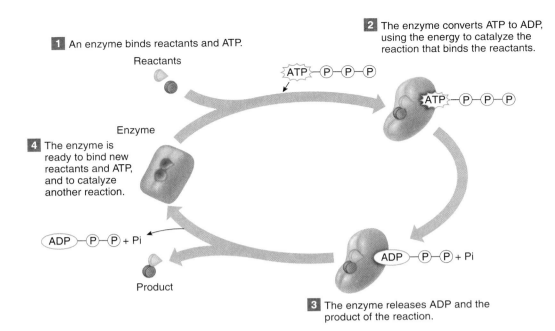

1 An enzyme binds reactants and ATP.

Reactants

Enzyme

4 The enzyme is ready to bind new reactants and ATP, and to catalyze another reaction.

ADP—P—P + Pi

Product

2 The enzyme converts ATP to ADP, using the energy to catalyze the reaction that binds the reactants.

ATP—P—P—P

ATP—P—P—P

ADP—P—P + Pi

3 The enzyme releases ADP and the product of the reaction.

Figure 5.14 **Enzymes, Reactants, and Product.**

■ *Why are enzymes so important to the life of cells?*

ATP and Reactants Bind to Enzymes at Specific Binding Sites

The function of a protein is related to its three-dimensional charge and shape (see Section 3.6). Enzymes are proteins, and so this principle also applies to enzyme function. **Figure 5.14** shows how the charge and shape of an enzyme allow it to use ATP to catalyze the production of a new molecule from two simpler molecules. The enzyme speeds up the conversion of the two original molecules, called *reactants*, to the final molecule or molecules, called the *product(s)*. Notice that the two reactants bind to special places on the enzyme called *binding sites*. The binding sites are local regions of the enzyme that have a charge and shape complementary to the reactants' charge and shape. The reactants fit in the binding site much like a key fits into a lock. The enzyme also has an ATP binding site. The binding site for the reactants is different for each kind of enzyme. On the other hand, the charge and shape of the ATP binding site is nearly the same on every kind of enzyme.

How does binding of reactants and ATP to an enzyme produce the products? When ATP binds to the enzyme, it puts a strain on the bond between the second and third phosphate groups and that bond is broken. The energy released from this bond is captured by the reactants, allowing a bond to form between them. At the same time the binding of the products brings them close enough together to encourage the formation of a new bond. This whole process produces the new molecule, the product.

Because the binding site(s) on an enzyme reflect the charge and shape of the reactants, each enzyme only can catalyze specific reactions. A chemical similar to the reactants in charge and shape can sometimes bind to the binding site and can prevent the reactants from binding. Such chemicals are not normally found inside cells, and so most of the time a cell's enzyme-catalyzed reactions proceed normally. Sometimes, though, a cell or organism can ingest, or in some other way take in, a chemical that will interfere with some specific type of enzyme. In some cases such chemicals are poisons. Plants produce such poisons to discourage animals from eating them—many plant poisons interfere with normal enzyme-catalyzed reactions in this way. In other cases chemicals that inhibit enzymes are used as drugs. An example is the drug disulfiram, also called Antabuse. This drug binds to one specific type of enzyme involved in the breakdown of ethanol to harmless chemicals. Ethanol is the form of alcohol in alcoholic drinks. Antabuse occupies the binding site of this ethanol-breakdown enzyme. As a result, a chemical called acetaldehyde accumulates, causing nausea and vomiting. When they take Antabuse, alcoholics will become quite ill if they drink alcohol.

QUICK CHECK

1. What is a catalyst?

2. What is the function of an enzyme?

3. What is the function of ATP and what organisms use it?

5.5   An Added Dimension Explained
How Did Life Originate?

In the deep history of the universe, over 4.5 billion years ago planet Earth formed. Chunks of rock coalesced to form a planet large enough to stay together and orbit the Sun. This planet-forming process was not a rare event. As our solar system developed, at least 10, and perhaps more, large bodies emerged from the debris orbiting the Sun. On Earth something unusual happened: life emerged. Life evolved, adapted, and flourished to cover the planet.

It is clear beyond doubt that on Earth today, life only comes from other living things and never originates from nonlife. Therefore, the Frankenstein story is as close to impossible as anything science can rule out. Yet this crucial event—life emerging from nonlife—did happen on a hostile and alien planet, the early Earth of 4 billion years ago. How did this happen? What were the first biological molecules and the first living cells? What source of energy was available for these early biological processes? What role did ATP have in these earliest cells?

One of the first important questions about the origin of life is: When did it occur? The answer to this question comes from the study of the earliest known fossils. A **fossil** is a preserved remnant of a dead organism, usually found embedded in rock. **Figure 5.15a** shows one of the oldest known fossils, dating to about 3.5 billion years ago. The oldest fossil forms clearly resemble bacteria, and the younger fossils resemble single-celled eukaryotes. From this kind of evidence scientists have concluded that life probably first arose shortly after the Earth formed and that the first life was about as simple as cells can be. One other conclusion from the fossil record is that at least some of the earliest life-forms were photosynthetic. They look nearly identical to the photosynthetic prokaryotes that are alive today.

Early Earth Was Different from Today's Earth

If you landed on that early Earth, you would not recognize it as the nurturing blue-green planet you know as home (**Figure 5.16**). The atmosphere was a mixture of carbon dioxide, methane, and ammonia gases—toxic to modern eukaryotes. Meteors rained down, smashing craters into Earth's surface. Parts of the planet were charged with enough energy to destroy anything alive today. Ultraviolet (UV) light from a young Sun, hot thermal undersea vents, and bolts of lightning provided intense amounts of energy, but in some places and times Earth may have been deadly cold. But—a bit of good news—hot gases from inside the Earth and meteorites from space produced large amounts of water vapor. In Earth's early atmosphere this condensed and rained back onto the planet. Washing over the land, water accumulated

fossil a preserved remnant of a dead organism, usually found embedded in rock

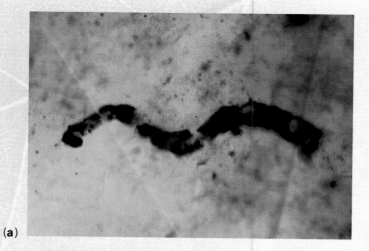
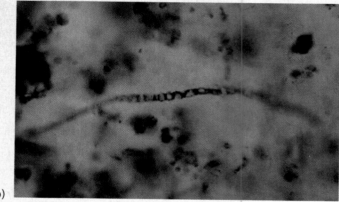

(a)

(b)

Figure 5.15   **Early Fossils.** (a) One of the oldest known fossils, dating to about 3.5 billion years ago. (b) A slightly younger fossil, dating to about 2.5 billion years ago.

Figure 5.16   **Artist's Concept of Early Earth.**

and produced Earth's rivers and oceans. Early Earth had a lot of water, and the first life appeared in this warm, wet, violent environment.

Many scientific studies have shown that this combination of conditions—water, carbon, other elements, and a source of intense energy—can produce biological molecules. The experiments of Miller and Urey in the 1950s were the first such studies. Miller and Urey used a methane/ammonia atmosphere over a simulated ocean. They sent electric sparks through this experimental atmosphere to simulate the abundance of lightning on the early Earth (**Figure 5.17**). After the apparatus had run for a week, the original clear water had become a murky solution that contained a variety of amino acids and other organic molecules. Other researchers have used different atmospheric combinations and also produced biological molecules. Still other workers have demonstrated the formation of nucleotides and short strings of amino acids when the watery solutions of molecules are dripped over hot rocks—a situation that could have occurred on early Earth. Finally, organic molecules have even been found in the "dust" of deep space, brought here by meteorites. Together these studies show that biological molecules can form under a variety of conditions, including those common on early Earth.

Once the array of biological molecules was present in Earth's oceans, groups of these molecules could have spontaneously formed cells. Many laboratory studies have demonstrated the spontaneous formation of something like cells. To date, getting a mixture of laboratory chemicals to replicate has been elusive, but it is clear that biological molecules can self-assemble into structures similar to cells.

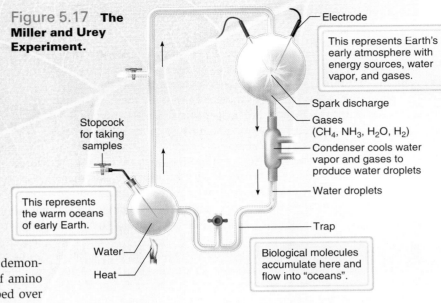

Figure 5.17 **The Miller and Urey Experiment.**

This represents Earth's early atmosphere with energy sources, water vapor, and gases.

Electrode

Spark discharge

Gases (CH_4, NH_3, H_2O, H_2)

Condenser cools water vapor and gases to produce water droplets

Water droplets

Stopcock for taking samples

Trap

This represents the warm oceans of early Earth.

Water

Heat

Biological molecules accumulate here and flow into "oceans".

The Use of ATP as an Energy Source Probably Evolved Early in the History of Life

The intense energy on early Earth allowed the production of biological molecules, but such intense energy also can destroy molecules. Cells would be damaged by UV light or by lightning. Once cells had formed, chemical bonds quickly became the dominant source of energy to support the various cellular processes. Some bacteria can directly use the chemical energy in the bonds of nonorganic molecules, such as the energy in the bonds of methane, hydrogen sulfide, and ammonia. It seems likely that some of the earliest cells used these nonorganic sources of chemical energy. This solution was not the most successful though, because today only a few kinds of bacteria use such a chemical source of energy.

Early in the evolution of life two related developments occurred that set the stage for how all later generations would acquire and handle energy. Sunlight is the one consistent source of energy on the early Earth, and on Earth today. A key evolutionary innovation in the history of life was coupling the energy in

Figure 5.18 **ATP Synthase.** This enzyme is found in membranes of all organisms and catalyzes the reaction of ATP from ADP.

All cells have ATP synthase, the enzyme that makes ATP from ADP and a phosphate group.

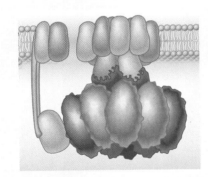

sunlight to the production of ATP. Early cells evolved a set of enzymes that could capture the energy in sunlight and transfer that energy to a large enzyme, called *ATP synthase* (**Figure 5.18**), which makes ATP from ADP and a phosphate group. Once this solution to the energy problem had evolved, it was passed down to new generations. This solution is so successful that nearly all life-forms depend on this link between sunlight and ATP production. Chapter 6 provides details of how this system works.

QUICK CHECK

1. What living organisms do the earliest fossils resemble?
2. Describe what the conditions on Earth were like 4 billion years ago.
3. What sources of energy may have driven the production of the earliest biological molecules on Earth?

Now You Can **Understand**

Food Calories—What Are They, Really?

All life requires energy, and to most people that probably translates to mean food. Your food consists of biological molecules that contain carbon-to-carbon bonds. Your body breaks the carbon-to-carbon bonds in the biological molecules in your food, releasing the energy they hold. While most organisms struggle to get enough food energy to survive, many humans have so much surplus food energy that it threatens their health.

Of course, there is more to nutrition than just understanding that food contains energy. It is useful to have a clear measurement of how much energy the chemical bonds in a food contain, and this is indicated by the familiar term, *calories*. Technically, 1 food calorie has enough energy to raise the temperature of 1 liter of water by 1°C. To put calories in less abstract terms, consider that a cup of broccoli has about 30 calories. Although there is a lot of variation, an average, active young woman needs around 2,200 calories daily. A cheeseburger has over 300 calories. It would take about a half hour of strenuous aerobic exercise for the body to use up the calories in that cheeseburger. So, if you eat a burger, fries, and a soda for lunch and then are fairly sedentary for the rest of the day, the calories from your lunch will provide all the energy that you will need for 10 to 12 hours. Being sedentary means activities like just going to class, studying, and watching TV. But, if you're like most people, before you have used up the calories from your lunch, you will have eaten dinner and perhaps have had a midnight snack. It is easy to see how most of us consume more calories than our bodies burn. The calories that are not used in metabolism are stored, often in fat deposits.

What Do **You** Think?

Panspermia

Life on Earth is the only life we know. For many scientists, the search for the origins of life has to be here on Earth—the only place we know for sure that life exists. One indisputable thing about human minds, though, is the ability for creative thought. From the time humans discovered their place in the solar system and universe, they have wondered whether life originated here or came to Earth from space—an idea known as *panspermia*. Science fiction stories with this scenario are common. Others take the suggestion quite seriously in the realm of nonfiction. Some groups believe that all humans are clones of some space-faring race. Even some scientists favor the idea of panspermia. Those who are interested in testing their ideas against reality postulate that bacterial life came to Earth on meteors and either seeded life here or significantly influenced the evolution of life here. What do you think about panspermia?

Chapter Review

CHAPTER SUMMARY

5.1 Life Requires Energy

Energy is the potential to do work. Kinetic energy is energy in motion, while potential energy has the capacity to do work. Energy can be imparted to objects or transferred from one object to another. The first law of thermodynamics states that one form of energy can be converted into another, but energy cannot be created or destroyed. The second law of thermodynamics states that the maintenance of complex structures and processes requires continual inputs of energy. Deprived of energy, an organism cannot build and repair its complex structures, and it dies.

electromagnetic force   109
first law of thermodynamics   111
kinetic energy   110

potential energy   111
second law of thermodynamics   111

5.2 Cells Use Chemical Energy

Chemical energy is stored in bonds that hold the atoms of molecules together. Photosynthetic organisms can store the Sun's energy in the bond that links the final phosphate group to an ADP molecule, forming an ATP molecule.

activation energy   115

ATP (adenosine triphosphate)   115

5.3 Photosynthesis Produces ATP and Glucose

Photosynthesizers can use the energy in the final phosphate bonds of ATP molecules to power the transformation of carbon dioxide and water to glucose. In a glucose molecule potential chemical energy is stored in the molecule's carbon-to-carbon bonds. When a cell needs energy, the energy in glucose is then converted back to the energy in ATP. The final phosphate bond of ATP and the carbon-to-carbon bonds of ATP provide moderate levels of energy. These bonds can be broken without requiring too much energy, and when they break they do not damage living systems.

glucose metabolism   117

photosynthesis   116

5.4 Enzymes Control the Pace of Life's Chemical Processes

Enzymes are biological catalysts that lower the activation energy needed for a reaction to proceed. In glucose metabolism, sets of enzymes within cells work in sequence to release energy in carbon-to-carbon bonds of glucose to form ATP. Cells also can use ATP as the energy source for powering enzyme-catalyzed reactions. Although the energy needed to drive life's processes comes from the Sun, no organism can directly use the Sun's energy.

ADP (adenosine diphosphate)   120
chemical catalyst   119

enzyme   119

An Added Dimension Explained: How Did Life Originate?

Many studies have focused on how life arose. Although none of the studies has led to a firm conclusion about how the first biological molecules arose on Earth, the evidence is clear that it is possible for biology to emerge from chemistry under conditions similar to those on early Earth.

fossil   122

REVIEW QUESTIONS

TRUE or FALSE. If a statement is false, rewrite it to make it true.

1. A spoon lying on a table has potential energy.

2. Life requires a continual supply of energy to build and repair complex structures and maintain life's chemical processes.

3. Organisms use the energy in glucose to directly drive the making and breaking of chemical bonds.

4. The energy in the bonds of glucose can be used to make ATP.

5. The energy in the final phosphate bond of ATP can be used to make glucose.

MULTIPLE CHOICE. Choose the best answer of those provided.

6. Which form of energy is used by organisms to directly power biological processes?

   a. light
   b. heat
   c. chemical
   d. electrical
   e. magnetic

7. The energy in the final phosphate bond of many ATP molecules can be stored in one molecule of

   a. glucose.
   b. DNA.
   c. an enzyme.
   d. a muscle.
   e. potential.

8. If something has the possibility to do work, it has

   a. enzymes.
   b. energy.
   c. ATP.
   d. glucose.
   e. photosynthesis.

9. A ball that has been hit by a bat displays

   a. light energy.
   b. chemical energy.
   c. electrical energy.
   d. kinetic energy.
   e. potential energy.

10. The energy in a fresh, unused AA battery is an example of

   a. electrical energy.
   b. magnetic energy.
   c. heat energy.
   d. kinetic energy.
   e. potential energy.

11. What is an enzyme?

   a. a biological catalyst
   b. an energy storage molecule
   c. an information storage molecule
   d. a structural protein
   e. a carbohydrate

12. An enzyme is what kind of biological molecule?

   a. lipid
   b. protein
   c. carbohydrate
   d. DNA
   e. nucleic acid

13. Biological catalysts

   a. can be used again and again, many times.
   b. are used up in the reactions.
   c. can be used three times before degenerating.
   d. work rapidly at first and then slow down as they are transformed into ADP.
   e. are the molecules that store energy for all living things.

14. The process of using energy from sunlight to make glucose from water and carbon dioxide is called

   a. metabolism.
   b. photosynthesis.
   c. replication.
   d. transcription.
   e. translation.

15. The process of photosynthesis generates _____.

   a. catalysts
   b. sunlight
   c. carbon
   d. ATP
   e. chlorophyll

16. Breaking the _____ bonds of ____ yields energy that can be used to make ATP.

   a. hydrogen, DNA
   b. hydrogen, glucose
   c. ionic, photosynthesis
   d. carbon-to-carbon, DNA
   e. carbon-to-carbon, glucose

17. What effect do enzymes have on activation energy?

   a. They increase it.
   b. They decrease it.
   c. They force it to stay the same.
   d. They divide it between two molecules.
   e. Enzymes have no effect on activation energy.

18. How do some poisons and drugs affect enzyme function? They

   a. unfold the tertiary structure of the enzyme.
   b. change the amino acid sequence of the enzyme.
   c. cause the enzyme to be secreted from the cell.
   d. digest the enzyme.
   e. bind to the reactant binding site of the enzyme.

19. Photosynthetic organisms include

   a. plants.
   b. algae.
   c. mosses and ferns.
   d. cyanobacteria.
   e. all of these.

20. The Miller-Urey experiments demonstrated that under conditions thought to mimic those of early Earth

   a. life can come from nonlife.
   b. spontaneous generation of life is inevitable.
   c. DNA forms.
   d. small organic molecules form.
   e. enzymes form.

1. What roles do the energy of sunlight and chemical energy play in living systems?

2. What role do enzymes play in living things? How might the evolution of life have been different without enzymes?

QUANTITATIVE QUERY

A rabbit munches on grass. Over its lifetime the grass that the rabbit eats contains 1,000 units of energy. The rabbit uses 500 units of this energy to build its biological molecules, and so 500 units of this energy are in the chemical bonds of the rabbit. The remaining 500 units of energy are lost as heat. A bobcat eats the rabbit. Of the 500 units of energy in the rabbit, the bobcat uses 100 to build biological molecules and loses 400 as heat. What percentage of the original 1,000 units of energy does the bobcat use to build biological molecules? Based solely on these energy considerations, what would be a more energy-efficient way of feeding a human population?

THINKING CRITICALLY

1. Have you ever heard of "spontaneous human combustion"? Primarily the subject of mystery novels, this unpleasant phenomenon is defined as the burning of a human body as the result of its own internal chemical reactions. A quick Internet search gives some fairly good explanations of why spontaneous human combustion is unlikely to the point of impossible. Most sources note that the human body is composed mostly of water—and is in fact quite difficult to burn. Based on reading this chapter, you should understand another reason why it is unlikely that a human body could spontaneously burst into flames as a result of its own internal chemistry. Think about this and write out your thoughts.

2. One of the central principles of modern biology is that all life is related. After reading this chapter, what aspect(s) of energy use by living organisms support(s) this principle? Take your answer a bit deeper by investigating the nature of the ATP binding site on enzymes.

3. The first law of thermodynamics states that energy can't be created or destroyed, but transformed into other types of energy. Start with the Sun's energy falling on the leaves of a tree and briefly list the energy transformations that end with your sitting in front of a campfire, cooking hamburgers and eating them.

For additional study tools, visit www.aris.mhhe.com.

6

The Engine of Life

PHOTOSYNTHESIS AND GLUCOSE METABOLISM

Global Warming: Is It Hot in Here?

July 13, 2004

Study Says Polar Bears Could Face Extinction

Warming Shrinks Sea Ice Mammals Depend On

November 9, 2004

Because polar bears depend on sea ice for hunting seals, global climate change threatens their ability to catch enough prey to survive arctic winters.

Drip, drip, drip. Drip. If you could tune your ears to Earth's ice, you would hear it melting. Until recently, Earth has been significantly covered with ice. The Arctic and Antarctic were capped by large ice sheets that grew and melted each year as cold and warm seasons passed. Glaciers sat atop the world's highest mountains, sucking up moisture in the cold months and releasing it when the warm seasons returned. Since prehistoric times ice has been a significant part of Earth's environment. Masses of ice influence ocean currents and so determine local climates around the world. And the temperature of ocean currents is influenced by how much water is frozen in polar ice. Arctic and Antarctic ice locks up ocean water and lowers global sea level, revealing low-lying coral islands in the South Pacific, where humans have lived for tens of thousands of years. And now, everywhere, the ice is melting.

But you may be a skeptic and may want to see the proof with your own eyes. You might start in Glacier National Park, Montana. In 1850 the park had 150 glaciers. Today it has fewer than 40, and scientists predict that in 50 years or less Glacier National Park will have *no* glaciers. Being a skeptic, you may think that this is just a fluke—nature fluctuates, and Glacier National Park is experiencing the climatic variation that normally happens over time. But take a look at Greenland, or Mount Kilimanjaro, the Andes Mountains, or the ice cover in the Arctic and Antarctic. Everywhere you look the trend is clear: the ice is diminishing in places that usually are covered with ice (**see chapter-opening photo**).

The ice is melting because Earth is getting warmer. This increase in temperature often is called global warming, but because the consequences go beyond temperature, most scientists call it *global climate change.* Rising temperatures are causing concern because if the glaciers are melting the water has to go somewhere. Levels of seas, rivers, and lakes are starting to rise. Some predictions are more extreme: Weather is likely to change and forecasts include more droughts, more floods, more mudslides, and stronger typhoons, tornados, and hurricanes. Global climate change could cause Northern Europe to become colder, while the polar regions might turn downright balmy.

What might be causing global climate change? You will return to this important question at the end of this chapter. To understand global warming, though, you may be surprised to learn that you first must understand some aspects of energy metabolism. ■

---

metabolic pathways chains of related chemical reactions

thylakoid a membranous sac within a chloroplast where light energy is captured and stored as chemical energy

6.1 Energy Flows in Metabolic Pathways That Link Cells and Organisms

Cells require a continuous input of energy to carry out life's processes, and chemical bonds provide this energy (Sections 2.3 & 5.2). The mechanisms that make and break chemical bonds form chains of related chemical reactions called **metabolic pathways.** Just as trails through a forest can serve as a guide from point A to point B, metabolic pathways guide the flow of energy inside of a cell. The impact of a metabolic pathway goes beyond a cell. Metabolic pathways in one cell may end with products that can be used by, or have an impact on, other cells or even other organisms. Cells and organisms are connected in metabolic relationships.

Figure 6.1 provides an overview of the major metabolic pathways. Recall that enzymes use the energy in the third phosphate bond of ATP to speed up the rate of specific chemical reactions. The laws of thermodynamics, though, dictate that the energy in ATP must come from elsewhere. Energy cannot be created or destroyed, only transferred from one source to another. ATP is made during the process of *photosynthesis;* the energy stored in ATP's chemical bonds comes from the photons in sunlight. The ATP made from sunlight is not available to all parts of the organism that made the ATP, so the energy in ATP is used to make glucose, a useful storage and transport molecule. Then glucose is used to regenerate ATP, making it available to be used by enzymes, in the process of *glucose metabolism.* Using the energy of the photons in sunlight, photosynthetic cells make their own glucose; other organisms get glucose

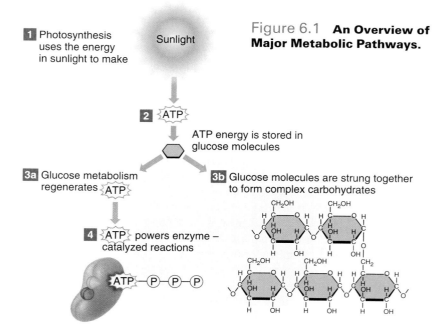

Figure 6.1 **An Overview of Major Metabolic Pathways.**

1 Photosynthesis uses the energy in sunlight to make

Sunlight

2 ATP

ATP energy is stored in glucose molecules

3a Glucose metabolism regenerates ATP

3b Glucose molecules are strung together to form complex carbohydrates

4 ATP powers enzyme–catalyzed reactions

ATP—P—P—P

secondhand from plants or other food sources. This chapter examines the components of these linked metabolic pathways in more detail.

6.2 Photosynthesis Harnesses the Energy in Sunlight

Life as we enjoy it would not be possible without photosynthesis. Certain bacteria, some single-celled eukaryotes, algae, and nearly all plants can capture the energy of photons. Because you are probably more familiar with plants than with other photosynthetic organisms, let's use plants as our example of how photosynthesis works.

The Biochemical Pathways of Photosynthesis Take Place Largely Inside of Chloroplasts

In plant cells photosynthesis occurs within chloroplasts (**Figure 6.2**). Compared with other plant parts, leaves usually contain the highest concentration of these small power transformers. If you peeled a thin film of tissue from a leaf and examined it under a compound microscope, chloroplasts would appear as bright green, jelly bean–shaped organelles. With the higher magnification of an electron microscope, you would see that each chloroplast is enclosed in a double membrane. Like other membranes these are formed of lipid bilayers with embedded proteins. Inside of the chloroplast are structures that look like stacks of coins. These "coins" are actually interconnected membranous sacs stacked on top of each other like pita breads. One sac is called a **thylakoid,** a name based on the Greek word for "pouch" or "sac." Thylakoid membranes contain molecules that capture the energy in sunlight as well as enzymes that use that energy to form chemical bonds. The fluid-filled space between the stacks of thylakoids and the chloroplast's inner membrane is called the **stroma.**

Pigment molecules embedded in thylakoid membranes are the starting point for photosynthesis. These pigment molecules absorb the energy in sunlight. To understand pigments, you need to know that photons of light act as waves. Photons oscillate, or vibrate, in a wavelike fashion.

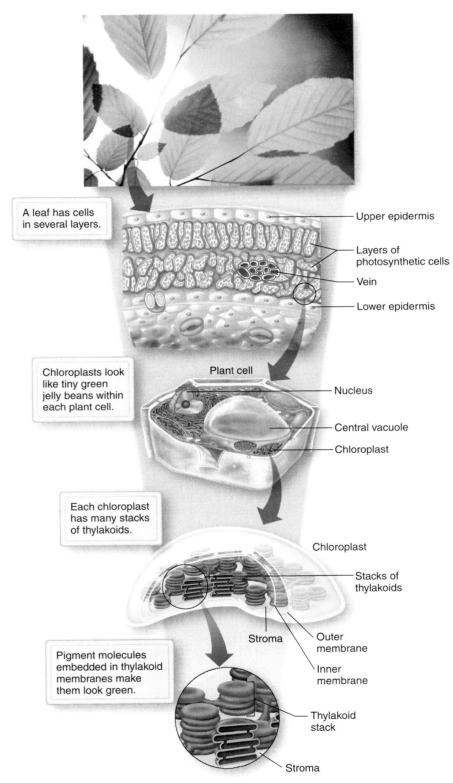

A leaf has cells in several layers.

Upper epidermis

Layers of photosynthetic cells

Vein

Lower epidermis

Chloroplasts look like tiny green jelly beans within each plant cell.

Plant cell

Nucleus

Central vacuole

Chloroplast

Each chloroplast has many stacks of thylakoids.

Chloroplast

Stacks of thylakoids

Stroma

Outer membrane

Inner membrane

Pigment molecules embedded in thylakoid membranes make them look green.

Thylakoid stack

Stroma

Figure 6.2 Chloroplasts Are Organelles Where Photosynthesis Occurs.

■ *What might be the reason that the thylakoids are in stacks?*

stroma the space between the stacks of thylakoid disks and the external membrane of the chloroplast

Different photons oscillate at different frequencies that are roughly correlated with the colors of light that you see (**Figure 6.3**). Photons that oscillate at a higher frequency, or shorter wavelength, are seen as blue or violet; photons that oscillate at a lower frequency, or longer wavelength, are seen as red; photons of medium wavelengths appear yellow to green. White light, such as that from the Sun or from a lightbulb, actually carries photons of all wavelengths. Pigments have color because each pigment absorbs light of some wavelengths and reflects light of other wavelengths. For example, when struck by white light, a green leaf reflects much of the green and yellow light and absorbs much of the blue and red light. The reflected wavelengths of light reach your eye, and you see the leaf as green.

Chlorophyll is the most prominent energy-capturing pigment in the thylakoid membrane (**Figure 6.4**). Because of its long hydrocarbon "tail," chlorophyll looks a bit like a lipid, but it has a large complex head region with a magnesium atom in its center. Other pigments, called *carotenes, anthocyanins,* and *xanthophylls,* also are present in the thylakoid membrane and other parts of the leaf; they also can be involved in capturing the energy of photons. During the fall months, as temperatures get colder and days get shorter, chlorophyll production slows down in deciduous trees, those that drop their leaves in winter. Existing chlorophyll molecules decay. At the same time production of other pigment molecules can increase before the trees eventually drop their leaves. When the green pigment breaks down and is withdrawn from leaves, other pigments that were hidden by chlorophyll become visible. As a result of all these dynamic changes in pigments, deciduous trees produce fall's spectacular yellow, orange, and red colors.

Photosynthesis Includes Light-Driven Reactions and Carbon Fixation Reactions

The overall process of making sugars in photosynthesis can be described by the following chemical equation:

$$6CO_2 + 6H_2O \rightarrow C_6H_{12}O_6 + 6O_2$$

This equation says that six molecules of carbon dioxide plus six molecules of water produce one molecule of glucose and six molecules of oxygen. If you count the carbons, oxygens, and hydrogens on either side of the equation's arrow, you will see that no atoms are lost in the process. Every atom of the carbon dioxide and water shows up in the products

chlorophyll the pigment that helps capture the energy in sunlight to provide the power for photosynthesis

(a) Visible light varies from violet to red. It is a small part of the electromagnetic energy from the Sun that strikes Earth.

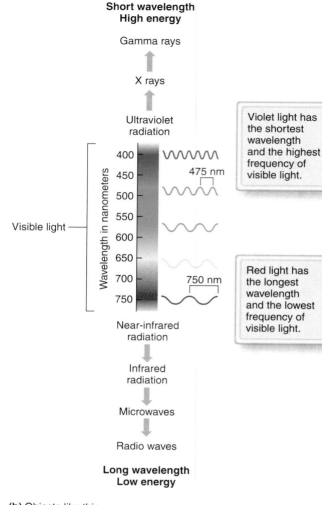

Short wavelength High energy

Gamma rays

X rays

Ultraviolet radiation

Visible light

Wavelength in nanometers
400
450 — 475 nm
500
550
600
650
700 — 750 nm
750

Near-infrared radiation

Infrared radiation

Microwaves

Radio waves

Long wavelength Low energy

Violet light has the shortest wavelength and the highest frequency of visible light.

Red light has the longest wavelength and the lowest frequency of visible light.

(b) Objects like this leaf get their colors from the visible light they reflect.

Sunlight

Reflected light

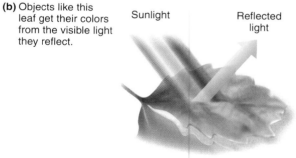

Figure 6.3 **Light Waves Are Composed of Different Wavelengths of Light.** (**a**) Visible light is composed of different wavelengths of electromagnetic energy. Wavelengths of each color of light have characteristic frequencies and amounts of energy. Shorter and longer wavelengths of electromagnetic energy are not visible. (**b**) This leaf looks green because it reflects green light and absorbs light of other colors.

■ *Describe what would happen to a typical plant if you tried to grow it in a dark room with no light. What would happen if you grew the plant under filtered green light only?*

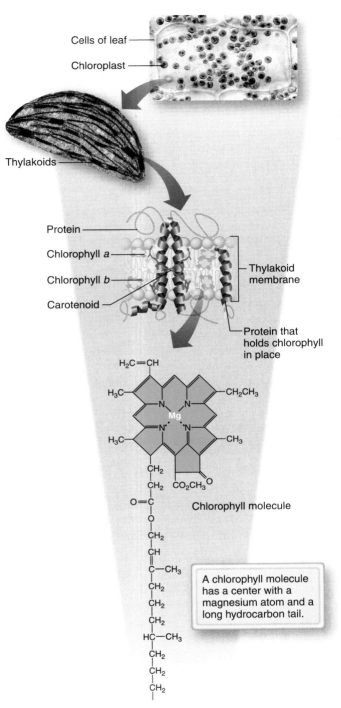

Figure 6.4 **Chlorophyll Molecules Are Embedded Within Membranes of Thylakoids.** Chlorophyll gives the green color to chloroplasts and to the green tissues of plants. Notice the magnesium (Mg) atom in the center of the chlorophyll molecule.

of the chemical reaction: glucose and oxygen. One factor not shown in this equation is the energy needed to drive the process. This energy comes from ATP molecules, and the energy to make ATP comes from photons of sunlight.

Like other metabolic processes, cells accomplish photosynthesis by using many small chemical reactions that are carried out by enzymes. The chemical reactions of photosynthesis can be grouped into two metabolic pathways. The *light-driven reactions* capture the energy in sunlight, while the *carbon fixation reactions* use the products of the light-driven reactions to power the formation of sugars. The carbon fixation reactions also are known as the *Calvin-Benson cycle,* named for the scientists who first discovered them. To help you understand the chemistry of photosynthesis, let's emphasize the general process of photosynthesis before filling in the details (**Figure 6.5**).

The light-driven reactions happen in the thylakoid membranes. They use sunlight to produce ATP and another important molecule, NADPH, which carries energy-rich electrons. Both ATP and NADPH contribute energy that the carbon fixation reactions use to produce carbohydrates. The carbon fixation reactions take place in the stroma of chloroplasts; glucose production begins in the stroma and ends in the cytosol of the plant cell. Carbon fixation reactions produce ADP and $NADP^+$, which become available to the light-driven reactions for the production of ATP and NADPH.

The Light-Driven Reactions of Photosynthesis Use Light Energy to Produce ATP and NADPH

Production of ATP and NADPH by the light reactions depends on two processes: transfer of energy to and from electrons and diffusion of hydrogen ions from a region of high concentration to a region of low concentration. First, let's revisit the relationship between electrons and energy, because this is where photosynthesis begins.

When energy from heat, light, or any other source strikes a molecule, the electrons in the molecule can absorb some of that energy. But an atom or molecule with an energetic electron is not stable. Like players passing a "hot potato," the electron must give up the extra energy it has gained. The electron can release energy as heat or light, it can transfer the extra energy to another electron, or the energetic electron itself can be passed to another molecule. In the light-driven reactions of photosynthesis, the passing of energy and electrons occurs along an **electron transport chain,** which is a series of proteins embedded in the thylakoid membrane. This series of proteins passes the energized electrons from one protein to the next, until the electrons reach a final electron acceptor. As the energy is passed from one protein

electron transport chain a series of proteins that handles energy-laden electrons by passing them from one protein to another and releasing some of their energy with each step

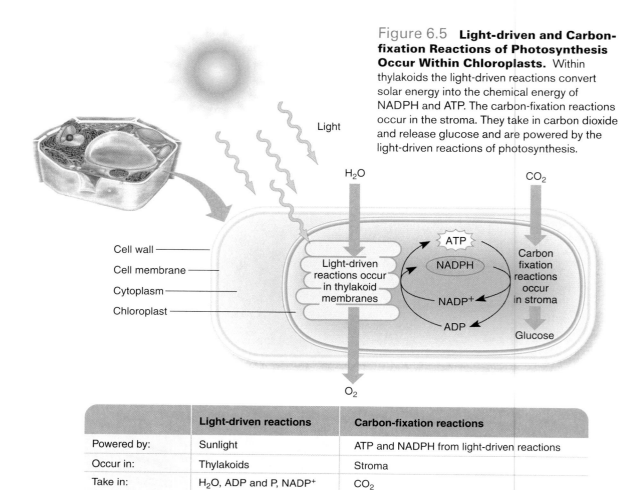

Figure 6.5 **Light-driven and Carbon-fixation Reactions of Photosynthesis Occur Within Chloroplasts.** Within thylakoids the light-driven reactions convert solar energy into the chemical energy of NADPH and ATP. The carbon-fixation reactions occur in the stroma. They take in carbon dioxide and release glucose and are powered by the light-driven reactions of photosynthesis.

	Light-driven reactions	**Carbon-fixation reactions**
Powered by:	Sunlight	ATP and NADPH from light-driven reactions
Occur in:	Thylakoids	Stroma
Take in:	H_2O, ADP and P, $NADP^+$	CO_2
Release:	O_2, ATP, NADPH	Glucose

to another, some of it is used to make ATP and NADPH.

The movement of electrons down this chain is like a tennis ball bouncing down a flight of steps. At the top of the stairs the ball has a lot of energy. As it bounces down the stairs, it loses a bit of energy at each step. Some is lost to the surrounding atmosphere and some to the stairs. By the time the ball reaches the bottom, it has lost all of its kinetic energy and rolls to a stop. The electron transport chain is like the stairs, with each step representing a protein in the chain. The electrons that pass along the electron transport chain are like the tennis ball. Therefore, you can think of the energy of electrons in the electron transport chain as flowing "downhill" until it reaches a stable, lower energy level. At each step some energy is lost as heat, some energy is used to prepare for ATP production, and some energy is transferred to the next protein. The next few paragraphs explore this process in more detail.

In plants the light-driven reactions, including the electron transport chain, are organized into two sets of chemical systems: photosystem II and photosystem I. The process starts with photosystem II and ends with photosystem I. This name reversal happened because even though photosystem I happens second in the overall process, it was discovered first. The two photosystems work in parallel and have a similar organization. Each has three functional components: an antenna system, a reaction center, and the electron transport chain (**Figure 6.6**). They can absorb photons simultaneously, but it is easier to understand photosynthesis if you start with the absorption of a photon by the antenna molecules of photosystem II and then follow the process through photosystem I.

The *antenna* is a large complex of hundreds of pigment molecules linked by proteins to form a structure that resembles a spider's web. Chlorophyll is the dominant pigment, but the antenna complex also contains other pigment molecules

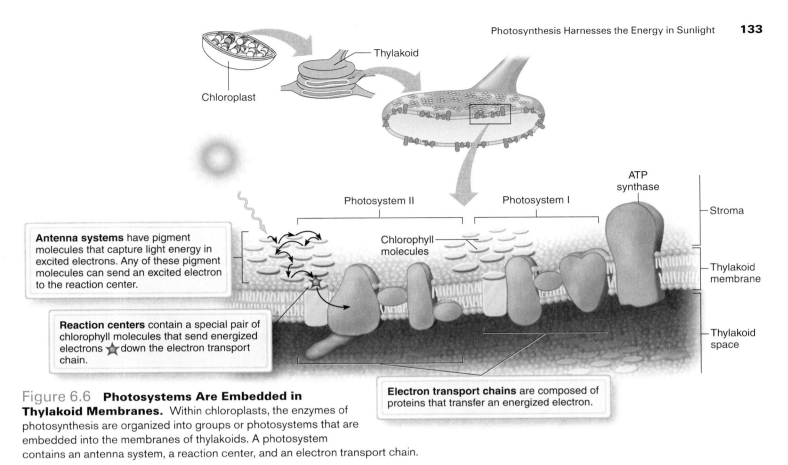

Figure 6.6 **Photosystems Are Embedded in Thylakoid Membranes.** Within chloroplasts, the enzymes of photosynthesis are organized into groups or photosystems that are embedded into the membranes of thylakoids. A photosystem contains an antenna system, a reaction center, and an electron transport chain.

■ *How are thylakoid membranes like cell membranes? How are they different?*

such as carotenes. The antenna does just what its name implies. Just as the radio antenna on your car captures radio waves, the antenna complex directly captures light energy. If light strikes any pigment molecule in the antenna, that light energy is absorbed and transferred to an electron. The result is an energized electron. The absorbed energy pushes the electron to a higher orbital, and it stays there until it gives up this newly acquired energy to another antenna molecule. All this energy is fed into the final energy acceptors in the antenna complex, a special pair of chlorophyll molecules in the *reaction center*.

The chlorophyll molecules in the reaction center actually give up electrons, rather than just energy, to the electron transport chain (**Figure 6.7**). Of course, these lost electrons must be replaced. Remember that the overall reaction for photosynthesis requires water and releases oxygen, and here water and oxygen enter the picture. Enzymes in the thylakoid membrane are able to split each water molecule, breaking it into an oxygen atom, two hydrogen ions, and two electrons (Figure 6.7). The electrons go to the chlorophyll molecules in the reaction center of photosystem II that have lost electrons. The oxygen atom combines with another oxy-gen atom—from another split water molecule—to form oxygen gas (O_2), which is released from the leaf into the atmosphere. What happens to the hydrogen ions?

One use of the energy released by the electron transport chain is to move hydrogen ions out of the stroma, the space around the thylakoids, into the thylakoid's interior space (Figure 6.7). These hydro-gen ions come, at least in part, from the water mole-cules that were split into hydrogen and oxygen. Of course, this builds up a high concentration of hydro-gen ions in the inner thylakoid space and these hydrogen ions are the key to ATP production. An important enzyme embedded in the thylakoid mem-brane, ATP synthase, allows the hydrogen ions to flow down their concentration gradients and move back into the stroma. ATP synthase has a pore, or channel, through which the hydrogen ions flow—this is an example of facilitated diffusion. Because energy was used to push hydrogen ions out of the stroma, energy is released when hydrogen ions flow through the ATP synthase molecule. ATP synthase uses this energy to transform ADP and a phosphate group into ATP. ATP synthase is found in nearly all living organisms and has the same job in all of them—to make ATP.

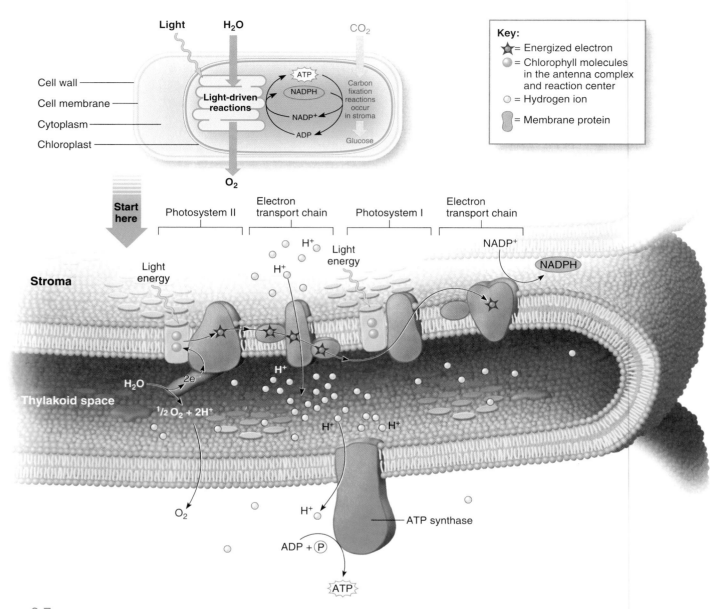

Figure 6.7 **Photosystems, Electron Transport, and ATP Production.** Photosystems II and I capture energy of photons, and transfer energized electrons to the electron transport chains. The two electron transport chains use activated electrons to move hydrogen ions into the thylakoid space. Hydrogen ions move through ATP synthase, providing the energy to produce ATP from ADP.

The transfer of energized electrons does not stop with photosystem II. The final electrons in photosystem II are transferred to a chlorophyll molecule in the reaction center of photosystem I. Photosystem I is quite similar to photosystem II. When photons strike the chlorophyll molecules of photosystem I, electrons move from the chlorophyll molecule in the reaction center of photosystem I to its electron transport chain. One difference is that photosystem I does not produce ATP, but it does produce an important energy-carrying molecule, NADPH (Figure 6.7).

NADPH carries the energized electrons to the carbon fixation reactions that are the next biochemical pathway of photosynthesis.

Once again let's return to the overall equation for photosynthesis:

$$6CO_2 \ + \ 6H_2O \rightarrow C_6H_{12}O_6 \ + \ 6O_2$$

Notice that so far we have accounted for the oxygen and water molecules, but we have not mentioned carbon atoms. The carbon fixation reactions use carbon dioxide to make glucose.

The Carbon Fixation Reactions Make Glucose

In the carbon fixation reactions a chemical cycle makes glucose. A molecule of carbon dioxide contains just *one* carbon. Because a molecule of glucose contains *six* carbons, six carbon dioxide molecules are needed to make one glucose molecule. Making glucose, though, is not just a simple matter of stringing the carbons from carbon dioxide molecules together. The cycle of reactions that fixes carbon takes in just one carbon dioxide molecule at a time. Note that although glucose is made on each turn of the cycle, it takes *six* turns of the cycle to account for all six carbons in a single glucose molecule. This cycle of chemical reactions begins inside the chloroplast, close to the thylakoid membranes where ATP and NADPH are generated. It ends in the cytosol that surrounds the chloroplast. ATP and NADPH are both used in this cycle.

The carbon fixation cycle *starts* when one molecule of CO_2 is added to a five-carbon sugar, called ribulose (RuBP) (**Figure 6.8**), to make a six-carbon molecule. The enzyme that catalyzes this addition of carbon dioxide to ribulose is known by the nickname *rubisco*. Rubisco is one of life's most important enzymes and about 40 million tons of rubisco could be extracted from the plants and other photosynthetic organisms alive today. The six-carbon structure that rubisco helps to generate, though, is *not* glucose. It is an unstable molecule that quickly breaks into a pair of three-carbon chains, one of which is shown in Figure 6.8. In a series of reactions the atoms in these two chains are rearranged and modified to produce two copies of a three-carbon product called glyceraldehyde 3-phosphate, G3P for short. The formation of *each* G3P requires the energy in one ATP molecule and also uses one NADPH molecule. Therefore, in the six turns of the

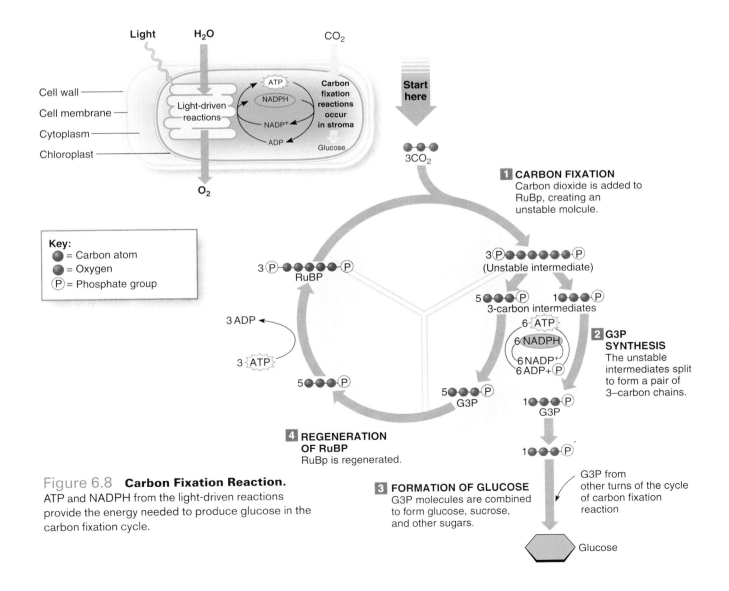

Figure 6.8   **Carbon Fixation Reaction.**
ATP and NADPH from the light-driven reactions provide the energy needed to produce glucose in the carbon fixation cycle.

cycle required to account for one glucose molecule, the cell would use 12 ATPs and 12 NADPHs.

Once they have formed, the two G3P molecules part company. One molecule combines with a two-carbon sugar to regenerate ribulose, which now is ready to accept another carbon dioxide molecule to keep the carbon fixation cycle going. The other G3P molecule goes *out* of the chloroplast to the cytosol of the plant cell. There it is combined with other carbon molecules to make glucose or other sugars. Some of these sugars are used by the cell as an energy source, but other sugars return to the chloroplast to combine with G3P and regenerate ribulose.

You have learned how the energy in sunlight is stored in the chemical bonds of ATP and NADPH and used to make glucose. As extraordinary as this process is, it is only half of the essential reactions that provide a cell with the chemical energy necessary for life. The next section turns to the other half of the chemical equation of glucose metabolism: the extraction of energy from glucose.

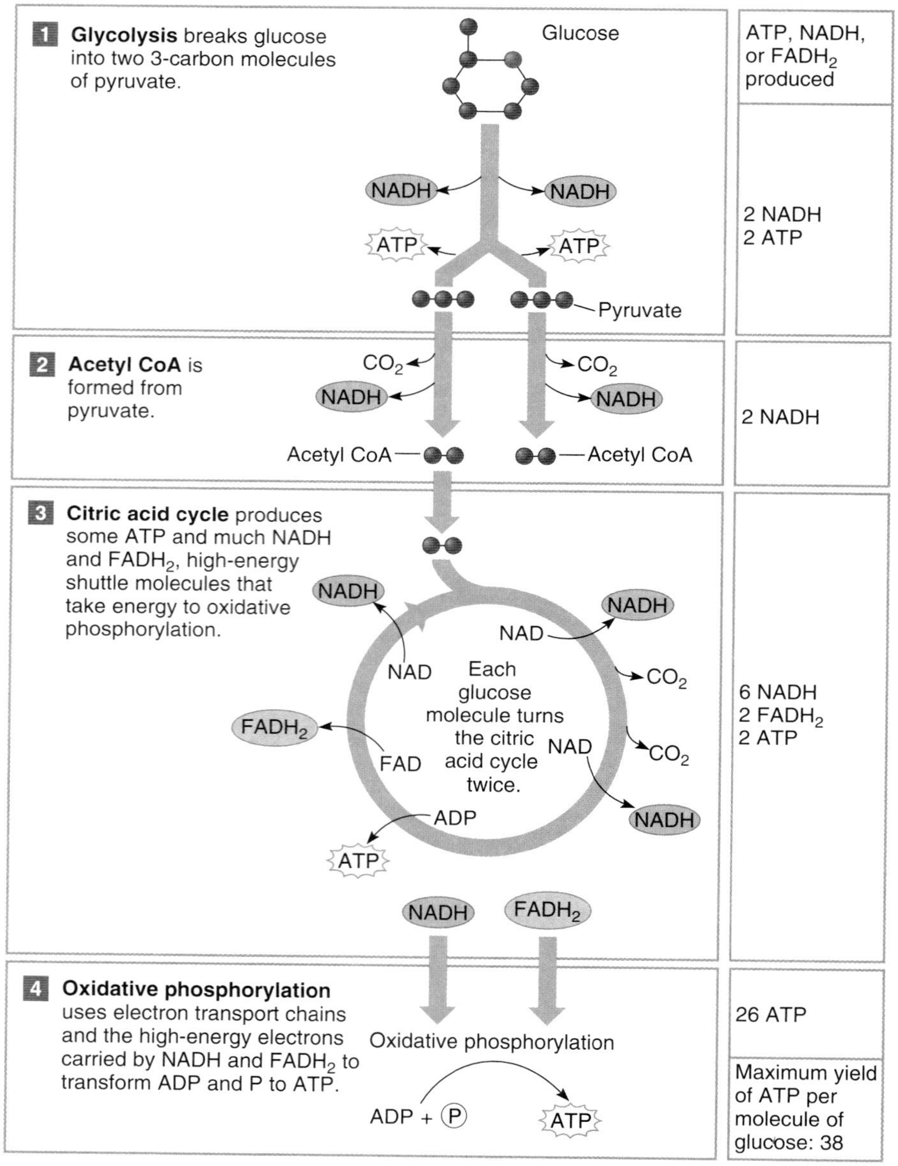

Figure 6.9 **Cells Produce ATP from Glucose.** Four linked molecular pathways break down glucose to produce energy and store it in molecules of ATP. The biochemical pathways are glycolysis, making acetyl CoA, the citric acid cycle, and oxidative phosphorylation.

QUICK CHECK

1. Sketch a chloroplast and label the major components. What does each major part of the chloroplast do?
2. Write down the chemical reaction that summarizes photosynthesis and describe what this reaction represents.
3. What is an electron transport chain?
4. What is meant by the term *carbon fixation?*

6.3 Cells Extract Energy from Glucose

Glucose metabolism provides energy in safe and usable amounts. Energy is a two-edged sword: it can be destructive if it is not controlled but extremely useful if it is controlled. For instance, the 100 million to 1 billion volts of electricity in a lightning strike would fry your home appliances, while the 110 volts of electricity that runs home appliances make life easier and more comfortable. The difference is control. In a lightning strike an enormous amount of electricity is explosively released, while household appliances use energy in small, controlled amounts. Glucose metabolism uses many small steps to transform the large amount of energy in a molecule of glucose to the smaller amounts of energy in molecules of ATP. With a few exceptions, all eukaryotic cells—including those of animals, protists, and fungi—carry out glucose metabolism. Even plants that make their own ATP from photosynthesis also convert the energy in glucose to ATP.

The general chemical reaction that harvests energy from glucose is the opposite of photosynthesis, which forms glucose:

$$C_6H_{12}O_6 + 6O_2 \rightarrow 6CO_2 + 6H_2O$$

Like photosynthesis, this chemical reaction is organized into smaller groups of chemical reactions (**Figure 6.9**). These include glycolysis, the generation of acetyl CoA (an intermediate carbon molecule), the citric acid cycle, and oxidative phosphorylation. Together these four biochemical pathways are known as glucose metabolism. In aerobic eukaryotic cells the net result of the metabolism of one molecule of glucose is up to 38 molecules of ATP. This shows how much energy one molecule of glucose contains.

Before considering the details of glucose metabolism, take a moment to study **Figure 6.10** and see where each phase of glucose metabolism takes place. Glycolysis occurs in the cytosol of a cell. Then the pyruvate molecules produced by glycolysis are transported into mitochondria. Acetyl

CoA is produced in the innermost spaces of the mitochondria, the *mitochondrial matrix*. It is used in the citric acid cycle, which also takes place in the mitochondrial matrix. Oxidative phosphorylation produces ATP on the inner mitochondrial membrane.

Glycolysis Breaks Glucose into a Pair of Three-Carbon Molecules

The first step in the release of energy from glucose is **glycolysis,** the chemical process that breaks the six-carbon glucose molecule into a pair of three-carbon pyruvate molecules. The simplified view of

glycolysis the chemical process that breaks glucose into two three-carbon molecules and releases two to four ATP molecules

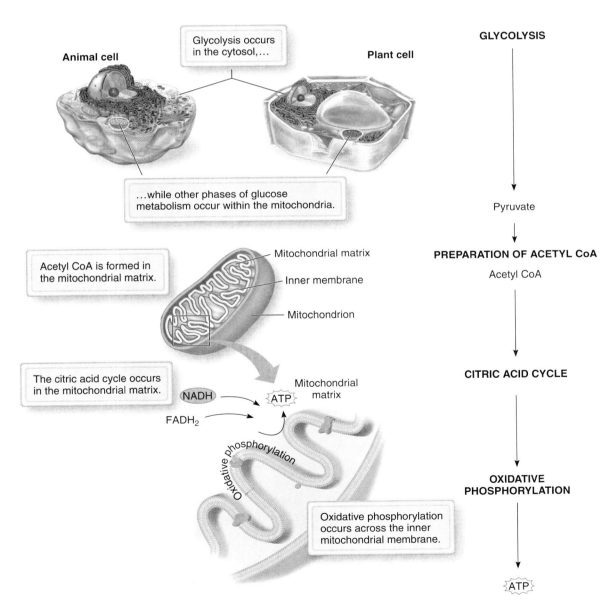

Animal cell

Glycolysis occurs in the cytosol,…

Plant cell

…while other phases of glucose metabolism occur within the mitochondria.

Acetyl CoA is formed in the mitochondrial matrix.

Mitochondrial matrix

Inner membrane

Mitochondrion

The citric acid cycle occurs in the mitochondrial matrix.

NADH

ATP

FADH$_2$

Mitochondrial matrix

Oxidative phosphorylation

Oxidative phosphorylation occurs across the inner mitochondrial membrane.

GLYCOLYSIS

Pyruvate

PREPARATION OF ACETYL CoA

Acetyl CoA

CITRIC ACID CYCLE

OXIDATIVE PHOSPHORYLATION

ATP

Figure 6.10 **Where the Phases of Glucose Metabolism Take Place.** Glycolysis happens in the cytosol. Acetyl CoA and the citric acid cycle take place in the mitochondrial matrix. Oxidative phosphorylation happens across the inner mitochondrial matrix.

citric acid cycle the chemical process that extracts energy from the two-carbon acetyl CoA molecule and provides energy-rich molecules for the process that makes ATP

glycolysis in **Figure 6.11** emphasizes carbon atoms and energy yields. Although only the major steps of this process are shown here, at least 10 chemical reactions are needed to transform glucose to pyruvate molecules, and a distinct enzyme governs

each reaction. Two ATPs are used in glycolysis, but four ATPs are produced. Therefore, glycolysis produces a net gain of just two ATP molecules. Note that glycolysis produces two molecules of NADH. These are used later in the process of oxidative phosphorylation.

The Citric Acid Cycle Releases the Energy of Acetyl CoA

In addition to releasing a molecule of carbon dioxide, preparation of acetyl CoA makes a molecule of NADH that will be used in oxidative phosphorylation. Acetyl CoA molecules enter the citric acid cycle, an intermediate process between glycolysis and oxidative phosphorylation. The citric acid cycle takes energy from the two-carbon chain of acetyl CoA (**Figure 6.12**) and makes molecules that can be used in oxidative phosphorylation to produce ATP. The citric acid cycle also is known as the *Krebs cycle*, named for Hans Krebs, the scientist who first discovered it.

The **citric acid cycle** starts when the two-carbon chain from acetyl CoA combines with a four-carbon sugar that came out of the previous turn of the cycle (**Figure 6.13**). The resulting six-carbon molecule produced in this first step is called *citrate*, and it gives the cycle its name. Through a series of chemical rearrangements, energy is harvested from the carbon-to-carbon bonds of citrate and stored in other molecules. A specific enzyme catalyzes each step in the cycle. During the process, the citrate molecule loses two carbon atoms as carbon dioxide. This regenerates another four-carbon chain that can combine with another molecule of acetyl CoA to start the citric acid cycle again. Because each molecule of glucose produces two acetyl CoA molecules, it takes two turns through the citric acid cycle to process the carbons from a single glucose molecule.

You may be surprised to see that the citric acid cycle itself produces only two molecules of ATP for every molecule of glucose. The focus of the citric acid cycle is to *gradually break* carbon-to-carbon bonds and *safely store* their energy in molecules of NADH and $FADH_2$. In this way, the citric acid cycle acts like an electrical transformer that reduces the voltage of the electricity carried by power lines (several thousand volts) to the voltage used in normal household current circuits (110 volts). NADH and $FADH_2$ from the citric acid cycle shuttle energy to the next step in the process that takes place across the inner membranes of mitochondria: ATP production.

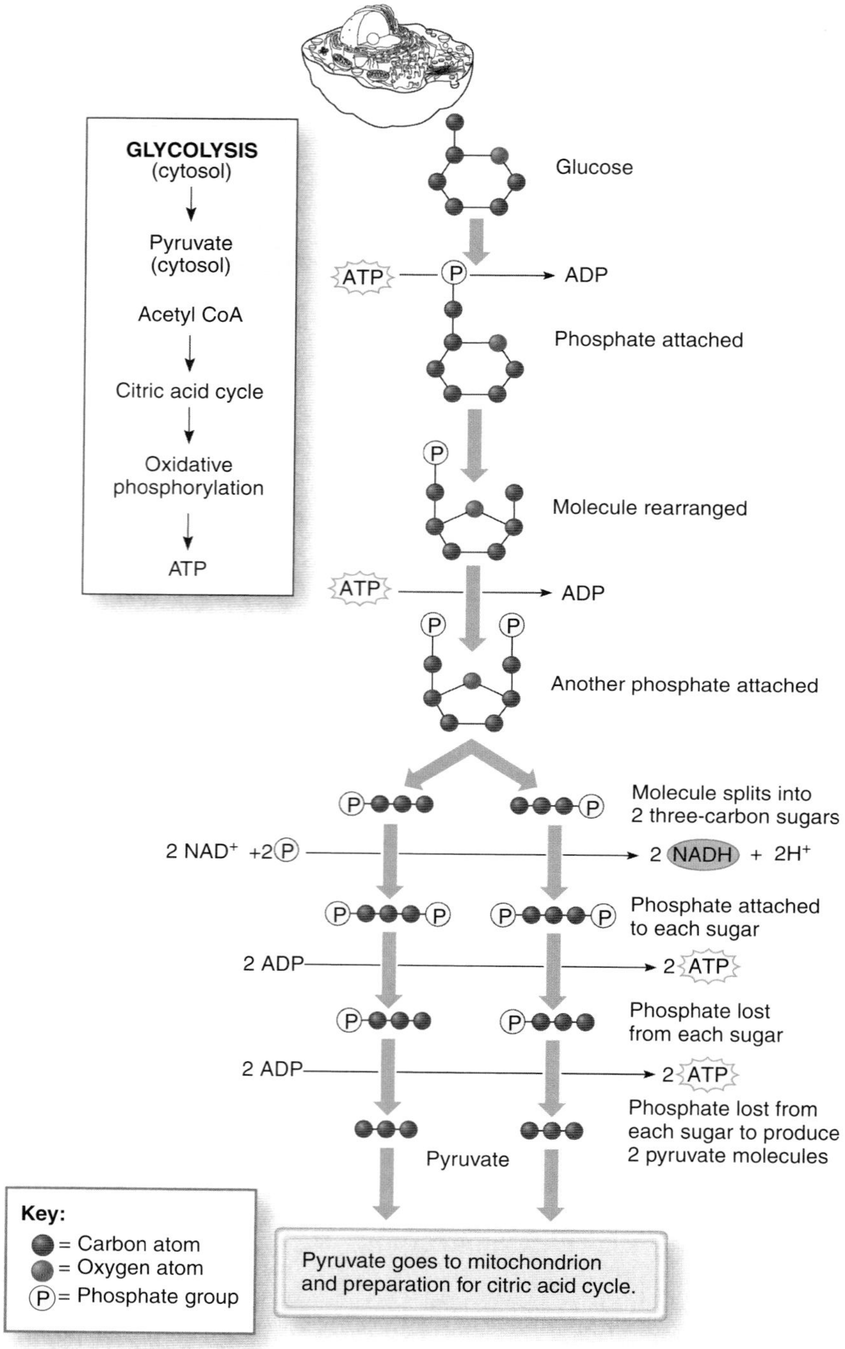

Figure 6.11   **A Simplified View of Glycolysis.** Glycolysis breaks glucose into two molecules of pyruvate.

ATP Is Made Across the Inner Membranes of Mitochondria

The final phase of glucose metabolism is **oxidative phosphorylation,** in which the energy-laden electrons from the molecules FADH and $NADH_2$ are used to transform ADP to ATP. Like photosynthesis, oxidative phosphorylation depends on an electron transport chain, and many copies of this chain are found in the convoluted membranes of the mitochondria (**Figure 6.14**). The energy intermediates, NADH and $FADH_2$, shuttle electrons carrying energy released from the carbon-to-carbon bonds of glucose. NADH and $FADH_2$ transfer electrons to an electron transport chain on the inner membrane of a mitochondrion. As the electrons move down the chain, they release energy in small steps. With each step, the enzymes involved use the released energy to move a hydrogen ion across the internal mitochondrial membrane. Again, this process is similar to the production of ATP during photosynthesis. In photosynthesis, however, the last electron acceptor is NADP and the process makes NADPH. In contrast, in oxidative phosphorylation the *last electron acceptor is oxygen.* This step is

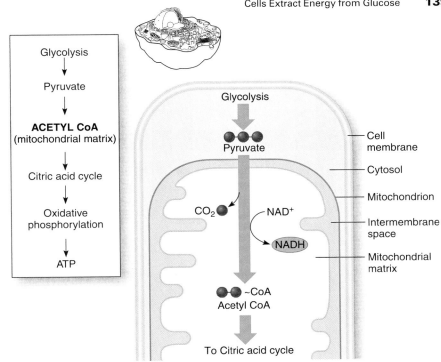

Figure 6.12   **Making Acetyl CoA.** Within a mitochondrion, pyruvate is converted to acetyl CoA. Carbon dioxide is released and NADH is produced.

Figure 6.13   **Details of the Citric Acid Cycle.** The citric acid cycle combines acetyl CoA with a 4-carbon molecule, forming citrate. In the citric acid cycle citrate is rearranged into intermediate molecules and two carbons are removed as molecules of CO_2. For each turn of the citric acid cycle, 3 NADHs, 1 $FADH_2$, and 1 ATP are generated. NADH and $FADH_2$ carry energy to oxidative phosphorylation.

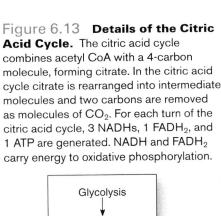

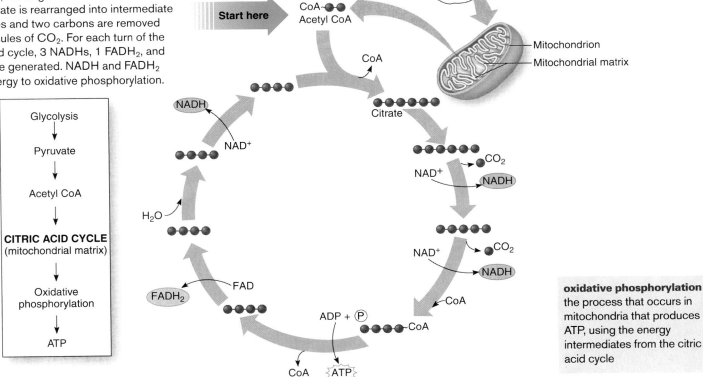

oxidative phosphorylation the process that occurs in mitochondria that produces ATP, using the energy intermediates from the citric acid cycle

Figure 6.14 **Oxidative Phosphorylation Occurs on Inner Membranes of Mitochondria.** High energy electrons from NADH and FADH$_2$ set up the high concentration of hydrogen ions in the intermembrane space necessary to produce ATP. Proteins of an electron transport chain and ATP synthase use the energy of NADH and FADH$_2$ to produce ATP. The electron transport chain requires oxygen and produces water.

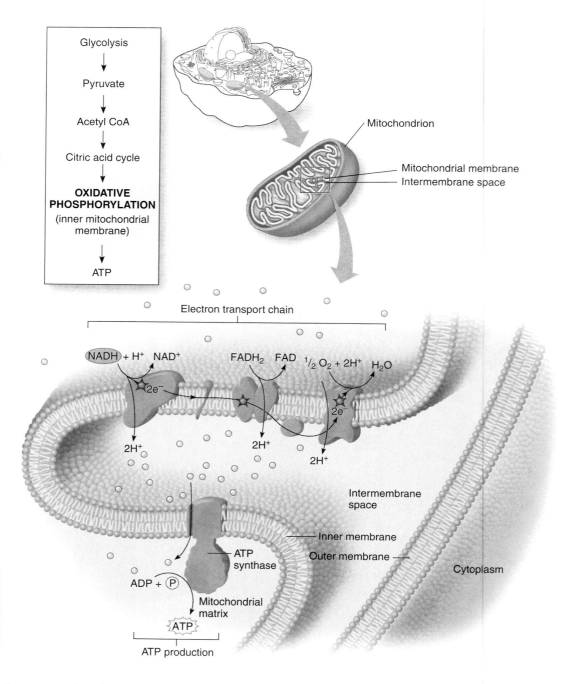

Key:
○ = Hydrogen ion
☆ = Energized electron
▮ = Membrane proteins
Ⓟ = Phosphate group

cellular respiration the combined processes of the citric acid cycle and oxidative phosphorylation

why you must breathe air that contains oxygen. Each oxygen atom combines with two transferred electrons and two hydrogen ions, to produce one water molecule.

So far, so good. To this point in oxidative phosphorylation, the energized electron has been relieved of its excess energy. How is ATP produced? The process is similar to ATP production in photosynthesis. As a result of electron transport, hydrogen ions now are concentrated in the *intermembrane space* between the two mitochondrial membranes. Look back to Figure 6.14. Essentially,

the concentration of hydrogen ions now possesses the energy of the energized electrons. At another location along the inner mitochondrial membrane, the pent-up hydrogen ions pass out of the intermembrane space by the process of facilitated diffusion. As in photosynthesis, they move through an ATP synthase enzyme that uses their energy to make ATP from ADP. ATP now can power the many cellular processes throughout the organism that require chemical energy.

Cellular respiration is the combined processes of the citric acid cycle and oxidative phosphorylation

(**Figure 6.15**). The linked but separate stages of cellular respiration release energy from carbon-to-carbon bonds, form ATP, use oxygen, and release carbon dioxide. Cellular respiration is the source of energy for life's biochemical processes. For instance, when you inhale, your bloodstream distributes oxygen to all of the cells in your body. They use as much oxygen as they need to produce ATP, and then they release the carbon dioxide produced in their metabolism of glucose. It is carried in the bloodstream and released when you exhale. Without oxygen, the whole process stops. Soon your cells would have no ATP, and your life processes would cease.

During your normal day-to-day life, carbohydrates are the major source of cellular energy for the production of ATP. Your diet, however, contains two other major sources of energy: proteins and lipids. Recall that lipids are made up of fatty acids. Complex metabolic pathways can convert the carbons in fatty acid chains to acetyl CoA molecules. These then are fed into the citric acid cycle for further metabolism. Proteins are broken down into amino acids, and these can be converted to pyruvate, or acetyl CoA, or other molecules that can enter the glucose metabolism pathways. As you can see, the body has many backup systems to ensure that its cells have a continuous supply of ATP.

QUICK CHECK

1. List the four steps to extract energy from glucose.
2. Summarize the process of oxidative phosphorylation. Tell where it happens, its major molecules and structures, and the outcome in terms of the ATP that it generates.
3. Describe what happens if you do not get enough carbohydrates in your diet to provide the raw materials for glucose metabolism.

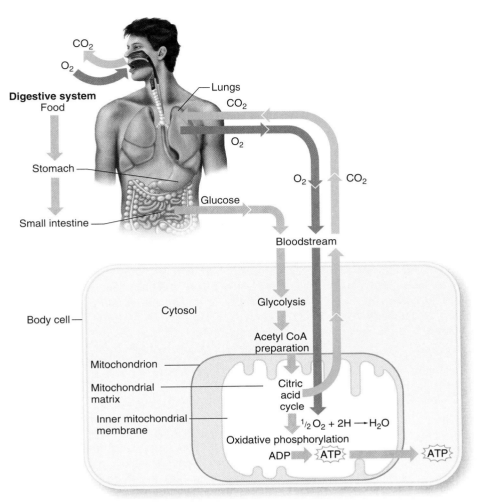

Figure 6.15 **A Summary of Cellular Respiration.** Food is digested down to nutrients that enter the bloodstream and are delivered to cells of the body. The four chemical phases of glucose metabolism chemically transform glucose to ATP. They require oxygen and release ATP, carbon dioxide, and water.

6.4 Metabolic Diversity Reflects the Evolution of Energy Pathways

Photosynthesis, glycolysis, and oxidative phosphorylation are the biochemical energy pathways of eukaryotes—but other organisms, especially prokaryotes, can use other energy pathways. **Figure 6.16** shows some of the important features that energy pathways have in common, as well as places where they can differ. Electron transport chains are the heart of all energy pathways, and all cells have at least one kind of electron transport chain. Although the electron transport chains may differ slightly in different species, their basic mechanisms are the same (Figure 6.16a). Recall that electron transport chains move hydrogen ions to produce a region of high hydrogen ion concentration and then allow the ions to diffuse back down their concentration gradient through an ATP synthase molecule. The ATP synthase molecule uses the energy of hydrogen ion diffusion to produce ATP (Figure 6.16b). The structure and function of this molecule also is nearly identical in all cells.

Next let's consider the other basic features of an electron transport chain. To operate, an electron transport chain must have a source of energy, a source of electrons, and a final electron acceptor. Cells have evolved a great deal of diversity in these features. For instance, in photosynthesis the energy source is energetic photons in sunlight, electrons come from water molecules, and the final electron acceptor is NADP$^+$ (Figure 6.16c). Compare this with the electron transport chain of cellular respiration. Here the energy source and the necessary electrons come from carbon compounds like glucose, and the final electron acceptor is oxygen. Metabolic pathways that use oxygen in this manner, and the organisms that use these pathways, are called **aerobic.** All eukaryotes and some prokaryotes are aerobic, meaning that they require oxygen to fully metabolize the energy in glucose or other carbon compounds. In some prokaryotes the energy source comes from carbon compounds, but the final electron acceptor is not oxygen, but some other molecule. These are **anaerobic** metabolic

pathways. Many prokaryotes are anaerobic—and to anaerobic organisms oxygen is toxic and deadly. Metabolic pathways also differ in the source of energy they use to make ATP. The chemical bonds of inorganic atoms also carry energy and electrons. Some prokaryotes can use molecules such as hydrogen gas (H_2), hydrogen sulfide (H_2S), or other inorganic molecules as the energy source for their electron transport chains.

Where and when did all this diversity of metabolic pathways arise? Organisms alive today provide hints of how metabolic diversity might have evolved (**Figure 6.17**). First, because electron transport chains and ATP synthase molecules are common to all living cells, scientists conclude that the first cells to emerge and reproduce successfully over generations used these two mechanisms to handle energy flow. The earliest cells probably used inorganic molecules as a source of energy because on the early Earth carbon molecules probably weren't plentiful. Because there was little oxygen in the early atmosphere, it is certain that the earliest cells were anaerobic. The earliest cells probably could make organic, carbon-based compounds such as glucose from carbon dioxide in the atmosphere. They may have been similar to simple anaerobic bacteria found all over the Earth today (Figure 6.17a).

Photosynthesis probably evolved early in the history of life. Sunlight is a more abundant energy source than inorganic or organic molecules. By using sunlight, cells could make plenty of organic molecules, reproduce rapidly, and build more complex cellular structures. The earliest photosynthetic cells probably did not release oxygen as a product of photosynthesis. They probably were similar to photosynthetic purple sulfur bacteria alive today (Figure 6.17b). It was probably not long, however, before oxygen-producing photosynthesis evolved on the early Earth. This type of photosynthesis is found in the prokaryotic cyanobacteria (Figure 6.17c) that resemble some of the earliest fossils

aerobic metabolic pathways and conditions that use oxygen

anaerobic metabolic pathways and conditions that do not use oxygen

Figure 6.16
Similarities and Differences in Metabolic Pathways.

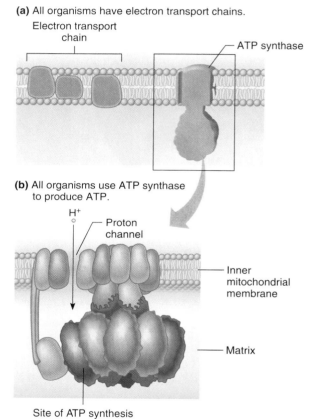

(a) All organisms have electron transport chains.

Electron transport chain

ATP synthase

(b) All organisms use ATP synthase to produce ATP.

H^+

Proton channel

Inner mitochondrial membrane

Matrix

Site of ATP synthesis

(c) Organisms use different sources of energy, sources of electrons, and final electron acceptors.

Pathways	Source of energy	Source of electrons	Final electron acceptor
Photosynthesis	Photons	Water molecules	NADP$^+$
Cellular respiration	Glucose and other carbon compounds	Glucose and other carbon compounds	Oxygen
Anaerobic pathways	H_2, H_2S, NH_3	H_2, H_2S, NH_3	NO_3, SO_4

dated to 3.5 billion years ago. After millions of years these early photosynthetic organisms caused oxygen to accumulate in the atmosphere. Once there was an oxygen atmosphere, aerobic metabolism of glucose could evolve.

Eukaryotes emerged after prokaryotes were well established on Earth. Eukaryotes probably evolved when some prokaryotes engulfed other prokaryotes, leading to the presence of nuclei, chloroplasts, mitochondria, and other organelles within eukaryotic cells (see Section 4.4). Now you can understand that eukaryotes *acquired* the processes of aerobic cellular respiration and photosynthesis. Modern eukaryotic cells have simply inherited the biochemical pathways that were used in energy metabolism by early eukaryote ancestors. You also can appreciate the big advantage gained by having cellular respiration based on sunlight, water, and oxygen. These metabolic pathways produce much more energy than do the other forms of metabolism found in prokaryotes. Coupled with the acquisition, or evolution, of internal structures such as cytoskeleton and endoplasmic reticulum, the availability of additional energy allowed eukaryotic cells to evolve into more complex structures than prokaryotes. Therefore, the evolution of complex multicellular life-forms was dependent on cellular respiration.

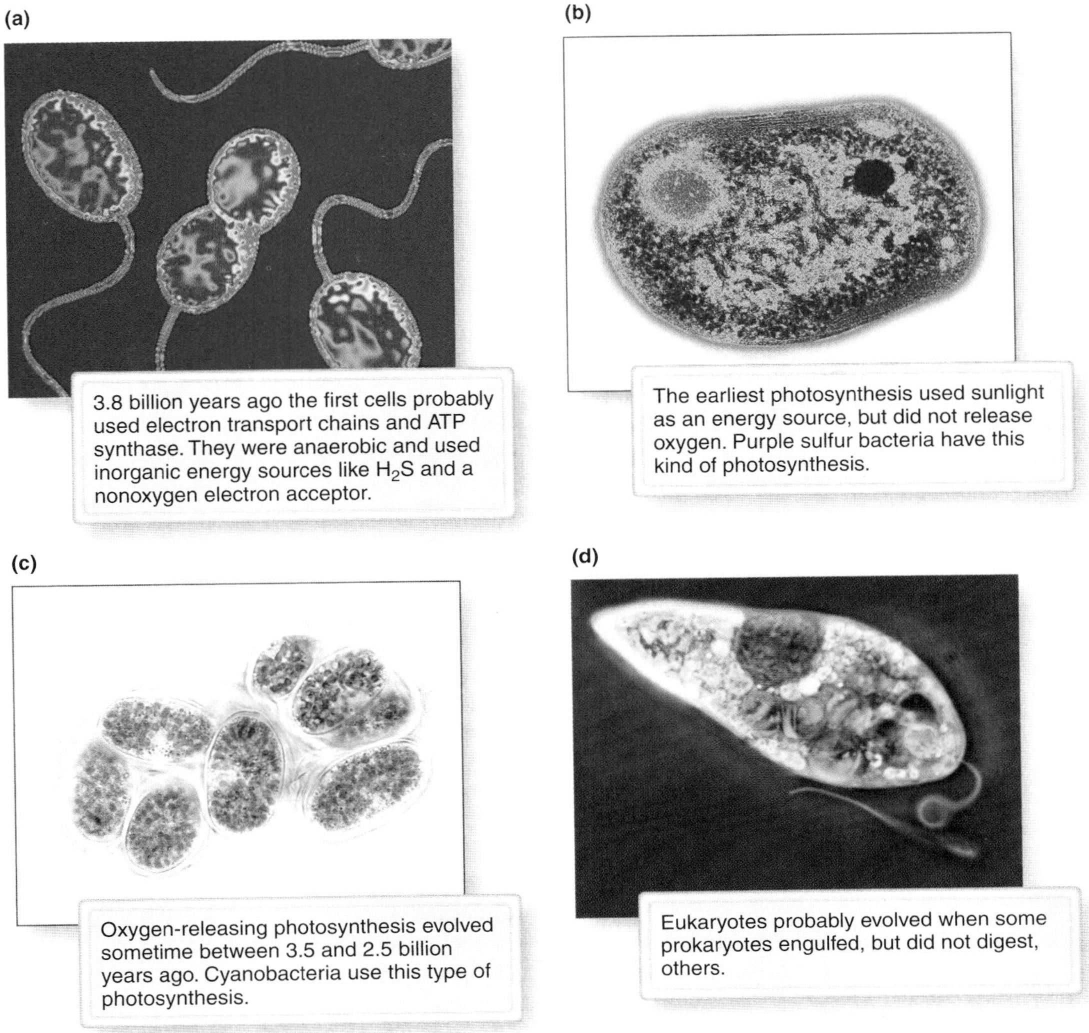

(a)

3.8 billion years ago the first cells probably used electron transport chains and ATP synthase. They were anaerobic and used inorganic energy sources like H_2S and a nonoxygen electron acceptor.

(b)

The earliest photosynthesis used sunlight as an energy source, but did not release oxygen. Purple sulfur bacteria have this kind of photosynthesis.

(c)

Oxygen-releasing photosynthesis evolved sometime between 3.5 and 2.5 billion years ago. Cyanobacteria use this type of photosynthesis.

(d)

Eukaryotes probably evolved when some prokaryotes engulfed, but did not digest, others.

Figure 6.17   **A Scenario for the History of Metabolic Diversity.** (**a**) The first cells probably were anaerobic. (**b**) The earliest photosynthetic cells did not release oxygen. (**c**) Photosynthetic cells that released oxygen evolved later. (**d**) Eukaryotes evolved much later.

■ *Why did the first cells use a non-oxygen electron acceptor?*

Metabolic Diversity in Eukaryotes Gives You Bread and Wine

Eukaryotes are so successful, in part, because of their aerobic metabolism, the citric acid cycle, and oxidative phosphorylation. Some eukaryotic cells, though, have the ability to survive on the energy provided by glycolysis alone. Metabolic pathways that do not use electron transport, such as glycolysis, are called **fermentation reactions.** Yeast cells have specialized fermentation reactions that produce alcohol and carbon dioxide gas as by-products. If oxygen is present, yeast cells will use oxygen to fully metabolize glucose to carbon dioxide and water. In the absence of oxygen, however, yeasts can still grow and survive by using alcoholic fermentation. This is why yeast is used in the chemical processes that make beer, wine, and bread. In bread dough, the yeast uses the carbohydrates in the dough and

produces carbon dioxide gas that bubbles through the dough and makes it rise. Carbon dioxide gas leaves large and small pockets in the baked bread. Yeast also produces alcohol, but it evaporates in the hot oven. When making wine, there must be plenty of sugar for the yeast cells to metabolize. In the liquid environment of the wine vat, any dissolved oxygen is quickly depleted and the yeast cells switch to fermentation that produces alcohol.

Muscle Cells Can Use Aerobic or Anaerobic Processes to Produce ATP

Muscle cells take advantage of fermentation. Consider the amount of energy they require when the body is in motion. If you dash down the street to

fermentation reaction
an anaerobic metabolic pathway, such as glycolysis, that does not use electron transport

catch the bus, compete in the 100-meter hurdle, or run a marathon, your muscles must have an ample supply of ATP energy to get the work done. An interesting aspect of muscle cells is that they can use either aerobic or anaerobic metabolic pathways to produce ATP. If you exercise at a regular but steady pace, your lungs can provide a steady supply of oxygen that allows aerobic exercise to continue for a long time. Early in a long, aerobic workout muscle cells use glucose to generate ATP through cellular respiration. When glucose runs out, muscle cells can use glycogen—a long carbohydrate chain—or lipid molecules as sources of energy in cellular respiration. Once glycogen and even local lipid supplies have been used up, the body will be out of ATP energy and must stop exercising and rest.

If you push your muscles extremely hard and fast, they can use an alternate metabolic pathway. When the oxygen provided by the lungs and circulatory system cannot provide sufficient oxygen to keep up with oxygen demands of muscles, the environment inside of muscle cells actually becomes *anaerobic*. Under these conditions animal muscle cells can produce ATP using fermentation reactions. Instead of producing alcohol as a by-product of fermentation, muscle cells produce lactic acid (**Figure 6.18**).

There are some interesting aspects to our understanding of anaerobic muscle metabolism. In the watery environment of a cell, lactic acid loses a hydrogen ion. As a result, the pH of the muscle cell changes, becoming more acidic. During and after a hard workout, lactic acid can build up in the muscles and bloodstream. Since the discovery of lactic acid buildup during exercise, scientists have suggested ideas about what role lactic acid might play in muscle fatigue and soreness. One early hypothesis was that lactic acid causes the muscle tiredness felt during exercise and the soreness experienced a day or two after exercise. This idea has been tested by measuring lactic acid levels in muscle cells. It turns out that during exercise lactic acid does not remain in muscle cells for a long time. Also, measurements of lactic acid in muscles do *not* show any relationship to fatigue. Scientists have other ideas about what causes muscle fatigue and soreness, such as buildup of phosphate and tiny tears in the muscle tissue, but lactic acid does not appear to be involved in muscle fatigue and soreness.

Photosynthesis and Cellular Respiration Maintain Earth's Atmosphere

Now that you understand the molecular aspects of photosynthesis and energy metabolism, take a step back and consider the consequences of these

(a) Under anaerobic conditions skeletal muscle switches to anaerobic metabolism.

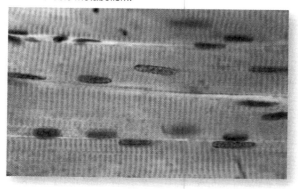

(b) Anaerobic metabolism produces lactic acid from pyruvate.

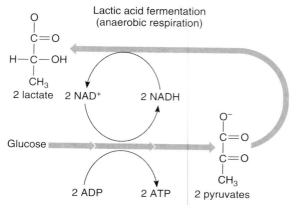

Figure 6.18   **Under Anaerobic Conditions Muscle Cells Can Produce Lactic Acid Instead of Pyruvate.**

linked processes for Planet Earth. Almost all living organisms today depend on the Sun for energy, but things were radically different 3.8 to 3.5 billion years ago. This was a time when the early oceans were charged with energy from intense solar radiation, lightning strikes, cosmic rays, and heat. Earth's early atmosphere was more like that of present-day Venus than present-day Earth. The best evidence is that the early atmosphere was composed mainly of nitrogen, hydrogen, and carbon dioxide, with little oxygen or water vapor. Today's atmosphere is quite different. **Figure 6.19** shows a timeline of the chemical changes in Earth's atmosphere over the Earth's history. About 2.5 billion years ago Earth's atmosphere began to undergo a gradual but radical change. The percentage of oxygen in the atmosphere began to rise to the current level of about 20%. How do we know? One way is by looking at rocks.

When oxygen is present in the atmosphere, it combines with iron and other minerals in rocks and other substances to form rusty red-colored oxides. Such "red bed" rocks begin to appear about 2 billion years ago in the geological record, which means that atmospheric oxygen must have begun to increase at about that time (**Figure 6.20**). Fossil evidence points to photosynthesis by cyanobacteria as the cause for this rise in atmospheric oxygen. Cyanobacteria certainly were among the most successful early organisms. Fossils show that huge mats of these prokaryotes probably covered much of Earth's surface. They likely resembled *stromatolites*, the mats or columns of cyanobacteria that today are found in a few shallow ocean environments (**Figure 6.21**). As oxygen accumulated, aerobic prokaryotes and eukaryotes that could use oxygen in the production of ATP evolved. These included many forms of algae and plants whose own photosynthesis brought oxygen levels up to their current concentrations.

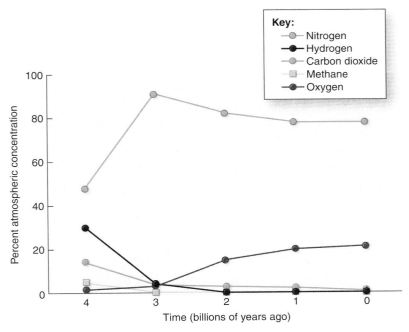

Figure 6.19 **Composition of Various Gases in Earth's Atmosphere from About 4.3 Billion Years Ago to the Present.**

QUICK CHECK

1. What is the difference between aerobic and anaerobic metabolism?
2. What is the result of fermentation metabolism in yeast?
3. How did photosynthetic organisms contribute to the composition of Earth's atmosphere?

Figure 6.20 **Red Bed Rocks.** The oxidized sediments in these rocks are evidence that oxygen was available in the biosphere.

Figure 6.21 **Stromatolites Are Large Marine Colonies of Prokaryotes.**

6.5 An Added Dimension Explained
What Is the Connection Between Metabolism and Global Climate Change?

The introduction to this chapter asked you to consider the concerns about **global climate change,** defined as the relatively rapid, significant and dramatic changes in the Earth's climate that are happening as a result of human activities. One major component of global climate change is the increase in Earth's surface temperature over the past several decades (**Figure 6.22**). Since 1940 Earth's average surface temperature has increased about half a degree Centigrade; for the past 10 years nearly every year has broken the previous year's record temperature. Some parts of the globe, such as the Northern Hemisphere, have warmed more than others. For example, the Arctic has increased nearly a full degree Centigrade over the same time. Of course, the Earth has gone through many temperature changes in the past, but this increase is more rapid than many others and probably has been caused by human activities.

To understand global warming, you first must understand some of the factors that contribute to Earth's temperature. The Sun is the major source of energy that warms the Earth. Only a small fraction of the photons that strike the Earth provide energy for photosynthesis. Most of the solar energy that reaches Earth's surface is absorbed by soil, rocks, and other objects, and it heats the Earth's surface. As long as the Sun is shining, this raises the temperature of the Earth. At night the region of Earth that is dark gets colder as the energy that was absorbed from sunlight is given up to the environment as heat. Where does that heat energy go? It could be lost to the cold of space, but the dark regions of the Earth do not get as cold as might be expected because some gets trapped by gases in the atmosphere. This phenomenon is called a *greenhouse effect* (**Figure 6.23**). Just as a glass greenhouse lets in sunlight, traps heat, and raises the temperature inside the greenhouse, Earth's atmosphere lets in sunlight and traps heat. If Earth had no atmosphere, the planet would be much colder.

Not all atmospheric gases are efficient at trapping heat. Nitrogen is not especially efficient, but water vapor, carbon diox-ide, and methane are good at trapping heat. For example, carbon dioxide dominates the atmosphere of Venus, a planet with a surface temperature of 462°C (864°F)—hot enough to melt lead. Atmospheric gases that trap heat and increase a planet's temperature are called **greenhouse gases.** Carbon dioxide, methane, nitrous oxide, and chlorofluorocarbons are the most effective greenhouse gases. Greenhouse gases such as carbon dioxide and water were a large percentage of the atmosphere of early Earth. Over time the percentage of nitrogen and oxygen have increased and the percentage of carbon dioxide has decreased. The result is an average Earth temperature that allows life to flourish and an atmosphere with enough oxygen to support complex aerobic organisms. If carbon dioxide had remained a large component of Earth's atmosphere, Earth would be much hotter and the evolution of life might have been limited, if not snuffed out altogether.

How is it that the atmosphere of early Earth lost so much of its carbon dioxide and gained oxygen? In part the answer lies in photosynthesis. Remember that there is a long history of abundant photosynthesizers on Earth. Photosynthetic organisms were among the first to evolve, and they were very successful. As these organisms grew and reproduced over millions of years, they took in carbon dioxide, released oxygen, and converted carbon dioxide to glucose. In addition to being used as a source of chemical energy, much of the glucose was used to build cellular structures—the cell walls, stems, leaves, trunks, and other structures of photosynthetic organisms. In this way early photosynthetic life changed the composition of Earth's atmosphere.

Eventually, organisms evolved that use the carbon molecules in plants and other photosynthetic organism as a source of energy. These organisms take oxygen out of the atmosphere and return carbon dioxide to the atmosphere. With time these opposing processes came into balance, and the atmosphere developed the low and relatively stable percentage of carbon dioxide it has had for hundreds of thousands of years.

What happens to the carbon trapped in a living organism when that organism dies? Today, when a photosynthetic organism dies, aerobic bacteria, fungi, and other kinds of organisms feed on its dead tissues and release nearly all of its carbon into the atmosphere as carbon dioxide. Whether it's an enormous oak tree or a blade of grass, the carbon structures that were built atom by

global climate change relatively rapid, significant, and dramatic changes in the Earth's climate that are happening as a result of human activities

greenhouse gas an atmospheric gas that causes an increase in a planet's surface temperature

fossil fuel a hydrocarbon fuel such as peat, coal, oil, and natural gas

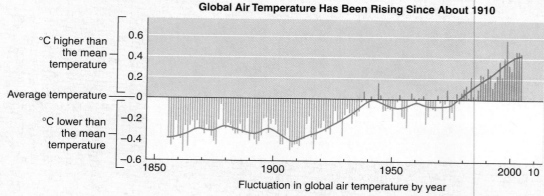

Figure 6.22 **Global Temperature Change, 1855 to 2003.**

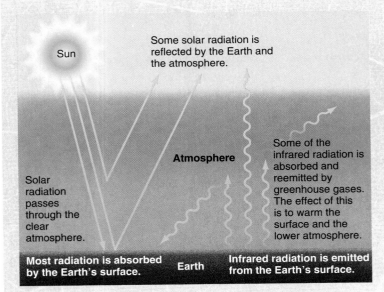

Figure 6.23 **Increases in Greenhouse Gases.** Like the glass in a greenhouse, carbon dioxide and other greenhouse gases in the atmosphere trap heat and reradiate it to Earth's surface. The greenhouse effect is intensified by burning fossil fuels, introducing more carbon dioxide and other greenhouse gases into the atmosphere.

atom through the plant's photosynthetic processes are broken down, atom by atom, and disappear. During the first 1 or 2 billion years of Earth's history, photosynthetic organisms dominated Earth. During this time much of the carbon contained in photosynthetic organisms was not returned to the atmosphere, but instead it was converted to other forms by physical and chemical processes. Some ended up in rocks on land and under the ocean. In addition, much of the carbon in these organisms was converted to peat, coal, oil, or natural gas.

Peat, coal, oil, natural gas, and the gasoline refined from oil are largely *hydrocarbons*. The chemical and physical processes that produce these substances removed most of the oxygen, nitrogen, and other atoms and left mostly hydrogen and carbon. In Section 3.2 you read about one important feature of hydrocarbons: they burn. We call peat, coal, oil, and natural gas **fossil fuels** because they come from long-dead organisms; when they are burned, they provide fuel that powers much of our modern way of life.

Now you are ready to consider what burning actually involves. As an example, the chemical reaction that occurs when a six-carbon hydrocarbon burns is:

$$2C_6H_{14} + 19O_2 \rightarrow 12CO_2 + 14H_2O$$

Notice that this chemical reaction is essentially the reverse of the process of photosynthesis. When a hydrocarbon burns, its chemical bonds are broken and its carbon and hydrogen atoms combine with oxygen to produce carbon dioxide and water. Notice also that this is essentially the same process as glucose metabolism—with one important difference: when a gram of glucose is burned in cells, metabolic pathways release energy slowly and in a controlled way. In contrast, when gaso-

line is burned in the engine of a car, the energy is released rapidly and in an uncontrolled way. In a second the carbon from the many long-dead animals goes up in a puff of exhaust that contains carbon dioxide.

Over the past 100 years humans have burned nearly half of Earth's large oil deposits. At this rate by the year 2050 nearly all large oil deposits will be exhausted. The carbon dioxide removed from the atmosphere over millions of years is being returned to the atmosphere over two hundred years or less. What does this have to do with the increase in the Earth's average temperature? This is where the principles of global warming come into play. The rapid burning of fossil fuels is causing a rapid increase in atmospheric carbon dioxide (**Figure 6.24**). Today's carbon dioxide levels are higher than carbon dioxide levels have been in the past several hundred thousand years. As long as humans continue to burn fossil fuels at such a rapid pace, carbon dioxide levels in the atmosphere will continue to rise.

Based on this complex set of observations, most climate scientists have come to three conclusions. First, the Earth is getting warmer. Second, higher levels of carbon dioxide and other greenhouse gases in the atmosphere have caused increased surface temperatures. Third, burning fossil fuels has caused most of the increase in greenhouse gases.

Predictions about how carbon dioxide and global climate change will play out over the next 100 years depend on many factors. It is likely that human activities will cause carbon dioxide levels to rise even further, but climate is a complex phenomenon. For example, increased temperatures may cause increased evaporation of water from the oceans and increased cloud cover. This, in turn, might cause less sunlight to hit the Earth's surface, and so might slow future rises in Earth's temperature. In addition, as it always has, life on Earth will play a big role in what happens to Earth's climates. For example, it is likely that increased temperatures will cause plants and other photosynthetic organisms to grow faster and so to take up more carbon dioxide. It might sound as though this would reverse the accumulation of carbon dioxide in the atmosphere, but carbon dioxide will be returned to the atmosphere when these organisms die.

Despite these complexities, carbon dioxide levels in the atmosphere probably will continue to rise, as will surface temperatures. Most scientists are convinced that as carbon dioxide levels rise, climate will become more variable. It is likely to get hotter, but also colder, drier, and wetter. Storms will become more severe. Discussions about what to do about these problems are taking place at all levels, from households to communities to nations. You will have opportunities to make your own decisions and to influence the decisions of your political leaders.

QUICK CHECK

1. What is a greenhouse gas?
2. How are greenhouse gasses and global climate change related?
3. What is a fossil fuel and how is burning fossil fuels affecting Earth's atmosphere?

—Continued next page

Continued—

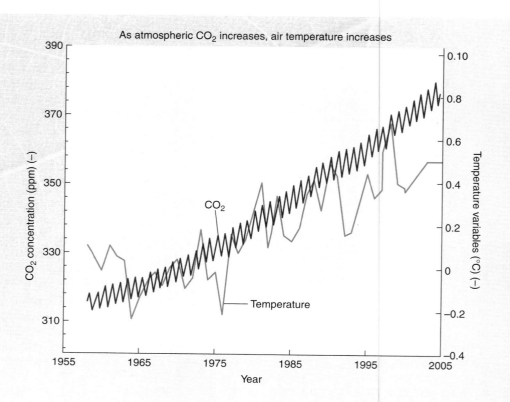

Figure 6.24 **Increases in Carbon Dioxide and Increases in Average Temperature from 1958 to 2005.** The red line shows the level of carbon dioxide in the atmosphere; the green line shows air temperatures. Both measurements fluctuate, but both measurements show marked increases. These measurements were taken at Mauna Loa Observatory in Hawaii, a location that is not affected by short-term variations characteristic of urban areas.

■ *What do you think can be done to reduce carbon dioxide emissions: (a) at the national or international level? (b) at the local or individual level?*

Now You Can **Understand**

Antioxidants and Free Radicals

You probably are aware of the general increase in interest in certain vitamins and other supplements over the past 10 years. Vitamin C, vitamin E, grape seed extract, echinacea, and many other supplements are touted as "antioxidants" that will help people to live long and healthy lives. But why would you need antioxidants? The quick answer is because you breathe oxygen.

Oxygen is a highly reactive molecule and it can do more in your body than accept electrons in mitochondria as a part of cellular respiration. Oxygen can react with other atoms and molecules, producing new kinds of molecules that, in turn, are especially reactive with DNA. These "free radicals," or "superoxides," can alter DNA in ways that disrupt normal cellular functions. Antioxidants are molecules that have been shown, usually in a test tube, to combine with free radicals and inhibit their reactivity. Do they work to keep you healthy? That question will be addressed in Chapter 12.

Why Cyanide Is a Deadly Poison

Cyanide is the stuff of murder mysteries and the deadly gas of gas chambers. How does it work? Cyanide is a small molecule. In its simplest form hydrogen cyanide is just HCN. This small molecule, however, can bind to one of the key enzymes involved in oxidative phosphorylation. When cyanide binds to an enzyme called cytochrome oxidase, electrons cannot find their way down the electron transport chain to oxygen where they normally would

form water. Without the electron transport chain, no ATP is made. This is important because the human body does not store ATP, and a person poisoned with cyanide will quickly die from lack of usable energy. This little-known bit of chemistry is a dramatic demonstration of how important cellular respiration is to survival.

What Do **You** Think?

Weight Loss and Diet Fads

Many Americans are obsessed with body weight. We eat too much and worry that we are too fat. It doesn't take any more knowledge of biology than you currently have to understand that it's all a matter of "energy in" versus "energy out." If you eat less and exercise more, you will maintain a healthier weight. In general, however, many people seem attracted to other approaches to weight loss. You may have heard of the grapefruit diet or the low-carb diet or the . . . water diet. Some people are confident that simply drinking lots and lots of water will "burn fat" and will lead to weight loss. Does this make any sense? How would you approach this from a scientific perspective?

One method would be to examine lipid metabolism in more detail and determine if water is needed in that process. It turns out the answer is yes. One molecule of water is used for every acetyl CoA molecule generated from a fatty acid chain. The next question to explore would be whether excess water might lead to increased fat metabolism. There are few, if any, scientific studies of this, but it is a general principle that glucose metabolism is regulated by how much energy cells need. As long as there is adequate water, it is unlikely that simply drinking more will increase fat metabolism. Nevertheless, there are many reasons to drink adequate amounts of water. There may be other ways that drinking water might reduce the amount of food you eat. Therefore, the diets that urge drinking lots of water probably are not harmful. Based on this information what do *you* think about the popular "water diets"?

Chapter Review

CHAPTER SUMMARY

6.1 Energy Flows in Metabolic Pathways That Link Cells and Organisms

Energy flows in metabolic pathways that are within cells, between cells, and between organisms.

metabolic pathways   128

6.2 Photosynthesis Harnesses the Energy in Sunlight

Photosynthesis captures the energy of sunlight and stores it in ATP and NADPH molecules that are used to form glucose molecules. The light reactions of photosynthesis occur within thylakoids of chloroplasts, where the sets of molecules of photosystems II and I capture energy, transfer energy from enzyme to enzyme, and generate oxygen molecules. A gradient of hydrogen ions builds up within the thylakoid and as hydrogen ions are moved out of the thylakoid membrane, ADP is transformed to ATP. Photosystem II ends by providing the energy for a hydrogen ion to move across the thylakoid membrane. Photosystem I transfers the energy of a moving electron to bond a hydrogen ion onto $NADP^+$ to form NADPH. The carbon fixation reactions of photosynthesis occur within chloroplasts, in the spaces around thylakoids. In these reactions, the NADPH and ATP from the light reactions are used to transform CO_2 into glucose. It takes six molecules of CO_2 to create one molecule of glucose.

chlorophyll   130
electron transport chain   131

stroma   130
thylakoid   129

6.3 Cells Extract Energy from Glucose

In cellular respiration the energy in the six carbon-to-carbon bonds of glucose is harvested and stored in ATP. Glycolysis transforms glucose into pyruvate, which is transformed into acetyl CoA, which enters the citric acid cycle. The products of the citric acid cycle, NADH and FADH, shuttle hydrogen ions to sets of enzymes embedded in the interior membranes of mitochondria. Similar to photosynthesis, these move the hydrogen ions across the inner membrane of the mitochondrion into its outer, membrane-bounded compartment. The energy from moving the hydrogen ions down their concentration gradient across the membrane is sufficient to attach a phosphate group onto ADP, forming ATP. This last process is called oxidative phosphorylation. The combination of the citric acid cycle and oxidative phosphorylation is cellular respiration. Proteins and lipids as well as glucose molecules can feed into the metabolic pathways of cellular respiration.

cellular respiration   140
citric acid cycle   138

glycolysis   137
oxidative phosphorylation   139

6.4 Metabolic Diversity Reflects the Evolution of Energy Pathways

The metabolic diversity of organisms is reflected in their electron transport chains, final electron acceptors, and the source of energy that they use. The earliest cells may have been anaerobic organisms that derived energy from inorganic chemicals. Oxygen-producing photosynthesis evolved later. Eukaryotes originated as prokaryotic ancestors engulfed other prokaryotes that had different metabolic strategies. The oxygen in Earth's atmosphere originated from photosynthesis over millions of years, first by cyanobacteria, then by algae and plants.

aerobic   142
anaerobic   142

fermentation reaction   143

6.5 An Added Dimension Explained: What Is the Connection Between Metabolism and Global Climate Change?

The linked processes of photosynthesis and cellular respiration have produced and maintain Earth's atmosphere. Global climate change seems to be a genuine phenomenon. It arises from greenhouse gases that trap heat in the atmosphere. Greenhouse gases originate from human activities, especially from burning fossil fuels. No one is yet certain of the outcome of global warming.

fossil fuel   147
global climate change   146

greenhouse gas   146

REVIEW QUESTIONS

TRUE or FALSE. If a statement is false, rewrite it to make it true.

1. Photosynthesis produces ATP.
2. The antenna system is composed of many chlorophyll molecules that absorb light energy and pass it on to the reaction center.
3. Oxidative phosphorylation occurs in chloroplasts.
4. Nitrogen is a major greenhouse gas.
5. Global warming is most likely caused by the increase in atmospheric carbon dioxide that is in turn caused by burning fossil fuels.

MULTIPLE CHOICE. Choose the best answer of those provided.

6. The final products of photosynthesis are
   a. CO_2 and H_2O.
   b. H_2O and ATP.
   c. NADPH and FADH.
   d. O_2, ATP, and glucose.
   e. acetyl CoA and rubisco.

7. Plants capture light in the _____ of _____.
   a. inner membrane, mitochondria
   b. cytoplasm, vacuole
   c. outer membrane, chloroplasts
   d. thylakoid membrane, chloroplasts
   e. stroma, chloroplasts

8. Chlorophyll captures the energy in _____ and that energy is used to make _____.
   a. glucose, ATP and NADH
   b. ATP, glucose and NADPH
   c. sunlight, ATP and NADPH
   d. sunlight, FADH and NADH
   e. water, oxygen and carbon dioxide

9. What source for carbon is used in carbon fixation reactions?
   a. glucose
   b. ATP
   c. AMP
   d. carbon dioxide
   e. lipids

10. In plants chlorophyll is found
    a. in mitochondria.
    b. in chloroplasts.
    c. embedded in thylakoid membranes.
    d. only a and b
    e. only b and c

11. The light-driven reactions, reactions
    a. generate a hydrogen ion concentration gradient and use the diffusion of hydrogen ions through ATP synthase to generate ATP.
    b. produce glucose.
    c. use the energy in ATP to make glucose molecules through the process of oxidative phosphorylation.
    d. use a chemical cycle.
    e. require CO_2.

12. The reaction center of the antenna complex
    a. is made of chlorophyll.
    b. is chlorophyll that loses an electron to the electron transport chain.
    c. is located in thylakoid membranes.
    d. passes an energized electron.
    e. all of the above.

13. The water molecule that is one of the starting ingredients of photosynthesis is split and the electrons
    a. are shot through the proton pump to synthesize molecules of ATP.
    b. move through the electron transport chain.
    c. replace those lost by photosystem II.
    d. reach the reaction center and energize it.
    e. provide the power for photosynthesis.

14. The process of oxidative phosphorylation
    a. is part of the light-driven reactions.
    b. makes ADP from ATP.
    c. is part of photosynthesis.
    d. generates a hydrogen ion concentration gradient and uses the diffusion of hydrogen ions through ATP synthase to generate ATP.
    e. occurs in chloroplasts.

15. Glycolysis produces _____ from a single glucose molecule.
    a. one pyruvate molecule
    b. two pyruvate molecules
    c. one ATP
    d. one oxygen molecule
    e. two oxygen molecules

16. The citric acid cycle produces
    a. 16 ATP molecules.
    b. NADH and FADH that are used in oxidative phosphorylation.
    c. oxygen.
    d. water.
    e. none of the above.

17. The early atmosphere of Earth contained large amounts of
    a. oxygen.
    b. carbon dioxide and methane.
    c. oxygen and methane.
    d. sulfur.
    e. ozone.

18. Electron transport chains
    a. gradually release the energy carried by electrons in a series of transport steps.
    b. transport ATP from the mitochondria to the cytoplasm.
    c. make ATP from hydrogen ion concentrations.
    d. gradually increase the energy carried by electrons in a series of transport steps.
    e. make glucose.

19. ATP synthase
    a. gradually releases the energy carried by electrons in a series of transport steps.
    b. transports ATP from the mitochondria to the cytoplasm.
    c. makes ATP from hydrogen ion concentrations.
    d. gradually increases the energy carried by electrons in a series of transport steps.
    e. makes glucose.

20. The light reactions of photosynthesis and oxidative phosphorylation are similar in that they both use
    a. electron transport chains.
    b. water.
    c. ATP to make glucose.
    d. glucose to make ATP.
    e. carbon dioxide to make oxygen.

CONNECTING KEY CONCEPTS

1. How are electron transport chains and facilitated diffusion linked in the overall process that produces ATP?

2. ATP is the molecule that drives nearly all of the enzyme-catalyzed reactions in a cell. Yet glucose is the source of chemical energy for most cells. What is the relationship between ATP and glucose and what role does each play in energy use by a cell?

Answer the following questions by referring to the following figure.

1. What is the approximate range of atmospheric carbon dioxide, in parts per million, between 450,000 and 50,000 years ago and today?

2. What is the approximate level of carbon dioxide in the atmosphere today?

3. Compared with the maximum level of carbon dioxide over the past 450,000 years, the current level of carbon dioxide is _____ times greater.

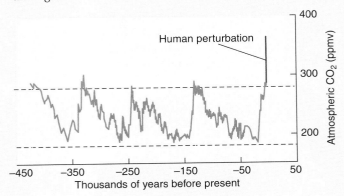

1. Go online to find out more about global warming—what people think about it and what some believe should be done. Read especially about the Kyoto Accord and the previous global warming treaties. Based on this exploration, how serious a challenge is global warming? Take the time to consider what should be done to mitigate the threat of global warming, and write a proposal that would achieve what should be accomplished over the next 10 or 20 years.

2. Houseplants are almost as popular with people as pets. Some people talk to their plants and are convinced it makes them grow. Assuming that it does happen, how do you think this might work?

3. Provide three possible strategies that human societies could use to reduce the rate of increase of carbon dioxide in the atmosphere. Explain how each might work and discuss the possible problems with your solution.

For additional study tools, visit www.aris.mhhe.com.

The Master Molecule of Life

DNA: STRUCTURE AND FUNCTION

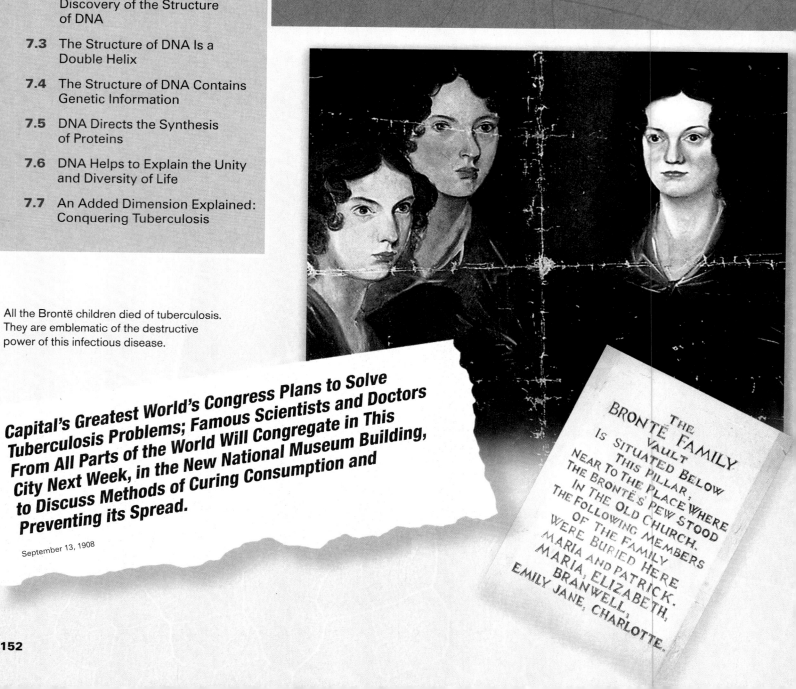

All the Brontë children died of tuberculosis.
They are emblematic of the destructive
power of this infectious disease.

Capital's Greatest World's Congress Plans to Solve
Tuberculosis Problems; Famous Scientists and Doctors
From All Parts of the World Will Congregate in This
City Next Week, in the New National Museum Building,
to Discuss Methods of Curing Consumption and
Preventing its Spread.

September 13, 1908

THE
BRONTË FAMILY
VAULT
IS SITUATED BELOW
THIS PILLAR;
NEAR TO THE PLACE WHERE
THE BRONTËS' PEW STOOD
IN THE OLD CHURCH.
THE FOLLOWING MEMBERS
OF THE FAMILY
WERE BURIED HERE
MARIA AND PATRICK,
MARIA, ELIZABETH,
BRANWELL,
EMILY JANE, CHARLOTTE.

An early death is tragic, but it is a catastrophe when more than one member of a family dies young. People often wonder, "What would those people have done with their lives?" The answer was clear for Ellis, Acton, and Currer Bell, siblings who lived in the early part of the 18th century. If they had lived longer they would have contributed more wonderful poetry and literature for the world to enjoy. Ellis died in 1848 at age 29; Acton died in 1849 at age 30; and Currer died in 1855 at age 39. Three other siblings died when they were 10, 11, and 31 years old. When Ellis and Acton died, the literary world did not know their true identities. It was not until 1850, when some critics charged that all of their literary works were written by just one of them, did Currer let the truth be known. The popular male authors were three sisters: Emily, Anne, and Charlotte Brontë.

All six children of one family dying from the same infectious disease is a tragedy that is hard to imagine today, but it was commonplace in Victorian times. Before the development of effective treatments, tuberculosis (TB) was a familiar, deadly, contagious disease. TB has been with humans a long time. Egyptian mummies, dated as early as 2500 BC show signs of TB in scars on their lungs and bones. Hippocrates, the famous Greek physician from the 5th century BC described the symptoms of TB. Until the 1800s, a diagnosis of TB was considered a death sentence. By some estimates in just the last 200 years TB has killed a billion people.

During the past 100 years the threat of TB seemed to be diminishing. Modern medicine tackled TB and had spectacular success in curing it. But today, the number of TB cases is rising again and is causing a new public health concern. Exactly what is TB? How was it nearly conquered in the mid-20th century, only to turn deadly again, and what can be done about the resurgence of TB? Finally, what does TB have to do with the topic of this chapter, the structure and function of DNA? You will return to these questions at the end of the chapter after you have delved more deeply into how DNA plays such a central role in the success and perpetuation of life. ■

7.1 DNA Is the Master Molecule of Life

Deoxyribonucleic acid, or DNA, is the genetic molecule and so is responsible for traits you have inherited from your family, including genetic diseases. DNA directs an organism's development from conception through death. Today, news stories commonly relate discoveries of genes for important diseases such as Alzheimer's disease, depression, or a gene that leads to extraordinary muscular strength. DNA is at the root of all of these discoveries.

One astounding lesson from the study of biology is that DNA plays its central role through just two basic functions. First, DNA holds the instructions for making proteins, the biological molecules that carry out nearly all cellular work. Second, DNA can be copied, so these instructions can be passed from one generation to the next. Here you will explore the chemical structure of DNA and discover how the two functions of DNA emerge from this structure. This chapter will help you to understand how the unity and diversity of life are tied to the structure of DNA.

Studies in the Early 1900s Led to the Discovery of the Structure of DNA

How did scientists discover the important roles played by DNA? A brief retelling of this story will give you a sense of the excitement surrounding the discovery of the structure of DNA. **Figure 7.1** shows the important events in the long history of this quest. In the early 1900s scientists understood the distinction between the nucleus and the cytoplasm, and by the 1930s techniques were good enough to demonstrate that the nucleus contains the hereditary material. Scientists also knew that the nucleus of a typical cell contains about 70% proteins and 30% nucleic acids. But which was the genetic molecule? The big debate was between those who favored proteins as the genetic molecule and those who favored nucleic acids.

The first major breakthrough came in 1928. British scientist, Fred Griffith, learned that genetic material could be transferred from one bacterial cell to another. Griffith also discovered that the transferred DNA could alter the traits of the recipient cells. Griffith was studying two strains of bacteria: one that caused pneumonia in mice and another that did not. When grown in a culture dish, the infectious bacteria always produced smooth-looking colonies, while the noninfectious bacteria colonies always had a rough appearance (**Figure 7.2**). As a part of his experiments Griffith killed the infectious bacteria within a test tube by boiling it. Then he injected a sample of the dead bacteria into mice. As Griffith expected, the dead bacteria did not cause disease. Next he injected a mouse with a mixture of live noninfectious bacteria and *dead* infectious bacteria. Griffith was surprised to discover that mice injected with this mixture developed pneumonia and died. He correctly concluded that something from the dead, infectious bacteria had transformed the live, harmless bacteria. In addition, when he grew colonies of bacteria from samples taken from these dead mice, the colonies looked smooth like the infectious bacteria, not rough like the harmless bacteria. This demonstrated that the new trait had been inherited and passed on to all the cells of the colony. If researchers could identify the molecule responsible for transforming harmless bacteria to infectious bacteria—or rough bacteria into smooth bacteria—they would have a good candidate for the genetic molecule.

Like the opening shot in a game of pool that scatters balls in all directions, it is hard to predict the impact of a scientific research project. Each researcher builds on the work of others, and not always in ways that are obvious. Although Griffith did not continue to investigate the genetic molecule, Oswald Avery, a U.S. scientist, saw the value of Griffith's approach for investigating the genetic molecule. Avery's research was able to point to DNA as the basis for inheritance. How did he do this? **Figure 7.3** shows the answer.

Avery realized that because the infectious and harmless bacteria looked different from one another (Figure 7.3a), it would be unnecessary to use mice to study Griffith's bacterial transforma-

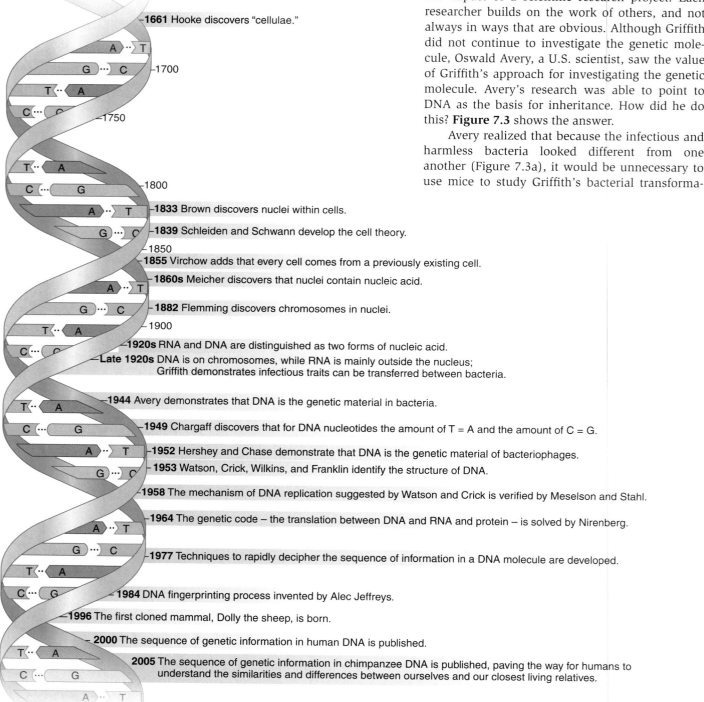

–1650

1661 Hooke discovers "cellulae."

–1700

–1750

–1800

1833 Brown discovers nuclei within cells.

1839 Schleiden and Schwann develop the cell theory.

–1850

1855 Virchow adds that every cell comes from a previously existing cell.

1860s Meicher discovers that nuclei contain nucleic acid.

1882 Flemming discovers chromosomes in nuclei.

–1900

1920s RNA and DNA are distinguished as two forms of nucleic acid.

Late 1920s DNA is on chromosomes, while RNA is mainly outside the nucleus; Griffith demonstrates infectious traits can be transferred between bacteria.

1944 Avery demonstrates that DNA is the genetic material in bacteria.

1949 Chargaff discovers that for DNA nucleotides the amount of T = A and the amount of C = G.

1952 Hershey and Chase demonstrate that DNA is the genetic material of bacteriophages.

1953 Watson, Crick, Wilkins, and Franklin identify the structure of DNA.

1958 The mechanism of DNA replication suggested by Watson and Crick is verified by Meselson and Stahl.

1964 The genetic code – the translation between DNA and RNA and protein – is solved by Nirenberg.

1977 Techniques to rapidly decipher the sequence of information in a DNA molecule are developed.

1984 DNA fingerprinting process invented by Alec Jeffreys.

1996 The first cloned mammal, Dolly the sheep, is born.

2000 The sequence of genetic information in human DNA is published.

2005 The sequence of genetic information in chimpanzee DNA is published, paving the way for humans to understand the similarities and differences between ourselves and our closest living relatives.

Figure 7.1   **Timeline of DNA Discoveries.**

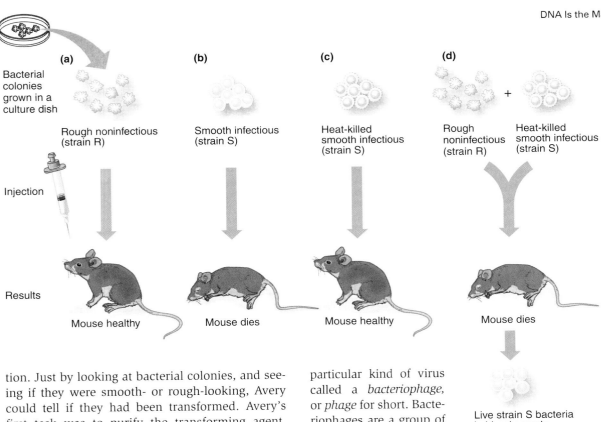

(a) Bacterial colonies grown in a culture dish

Rough noninfectious (strain R)

(b) Smooth infectious (strain S)

(c) Heat-killed smooth infectious (strain S)

(d) Rough noninfectious (strain R) + Heat-killed smooth infectious (strain S)

Injection

Results

Mouse healthy

Mouse dies

Mouse healthy

Mouse dies

Live strain S bacteria in blood sample from dead mouse

Conclusion: Something from the heat-killed infectious bacteria changed the live, harmless bacteria. It turned out to be DNA.

Figure 7.2 Griffith's Experiment. Griffith discovered that bacteria could be transformed from noninfectious to infectious.

tion. Just by looking at bacterial colonies, and seeing if they were smooth- or rough-looking, Avery could tell if they had been transformed. Avery's first task was to purify the transforming agent. Avery crushed a batch of infectious bacteria and removed all their solid parts—fragments of membrane and organelles. In a complex series of steps, Avery treated the remaining solution with ethanol and ended up with a small amount of a white, fibrous substance. This extraction process chemically removed all biological molecules except deoxyribonucleic acid, the substance that Avery suspected was the transforming agent. When a bit of this extract was added to the harmless, rough-looking bacteria, they were transformed into smooth, shimmery colonies (Figure 7.3b). Just to be certain that DNA was the transforming agent, Avery performed additional experiments. He treated the white substance with enzymes that destroy proteins, or with enzymes that destroy RNA. After either treatment the extract retained its ability to transform the rough bacteria into smooth bacteria (Figure 7.3c). This meant that neither protein nor RNA was the transforming agent. Lastly, Avery treated the extract with enzymes that destroy DNA. In this case the white extract lost both its fibrous texture and its ability to transform rough bacteria into smooth bacteria (Figure 7.3d). This set of experiments clearly indicated that DNA was the transforming agent.

While Avery was studying bacterial transformation, Alfred Hershey and Margaret Chase also found evidence that DNA is the genetic molecule. Their classic experiment focused on the reproduction of a particular kind of virus called a *bacteriophage,* or *phage* for short. Bacteriophages are a group of viruses that infect only bacteria (**Figure 7.4a,b**). In a typical phage a large headlike region sits on top of a "neck," and several spindly "legs" support the phage. Like other viruses, phages are not alive. They use the processes of a living cell to replicate themselves and make new viral particles. The legs of the phage attach to a bacterial cell, and the virus injects material from the head through the neck and into the cell (Figure 7.4b). Because this injected material allows the phage to replicate inside the host cell, whatever is injected must be the genetic material of the virus. Guided by the genetic material, many copies of the phage are made within the infected bacterial cell. Eventually, the bacterium may burst and release all of these phages. Each phage is ready to infect another bacterium.

Hershey and Chase wanted to know which molecule—protein or DNA—was the genetic material that phages inject into bacteria. To answer this question, they designed a clever experiment (Figure 7.4c). At the time of these studies researchers knew that proteins often contain a lot of sulfur and that nucleic acids contain a lot of phosphorus. The sulfur in proteins is involved in bonds that help to maintain their three-dimensional shapes; the phosphorus

(a) Avery conducted his studies using infectious smooth bacteria and noninfectious rough bacteria grown in culture tubes and culture dishes.

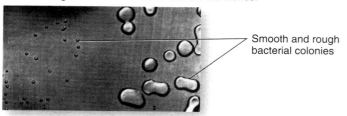

Smooth and rough bacterial colonies

(b) Avery demonstrated that rough bacteria can be converted to smooth bacteria in a culture tube.

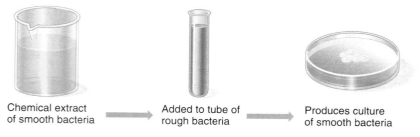

Chemical extract of smooth bacteria → Added to tube of rough bacteria → Produces culture of smooth bacteria

(c) Avery showed that neither RNA nor protein was the factor that caused rough bacteria to be converted to smooth bacteria.

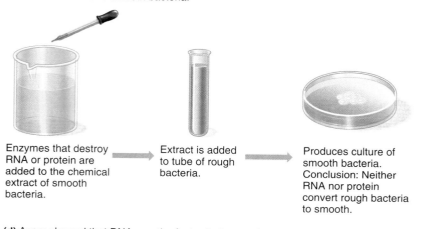

Enzymes that destroy RNA or protein are added to the chemical extract of smooth bacteria. → Extract is added to tube of rough bacteria. → Produces culture of smooth bacteria. Conclusion: Neither RNA nor protein convert rough bacteria to smooth.

(d) Avery showed that DNA was the factor that caused rough bacteria to be converted to smooth bacteria.

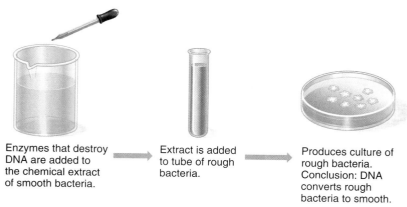

Enzymes that destroy DNA are added to the chemical extract of smooth bacteria → Extract is added to tube of rough bacteria. → Produces culture of rough bacteria. Conclusion: DNA converts rough bacteria to smooth.

Figure 7.3   **Avery's Experiments.** Avery's experiments demonstrated that DNA is the genetic material.

in nucleic acids is part of the phosphate group of each nucleotide. To track proteins and DNA through the life cycles of phages and bacteria, Hershey and Chase grew two batches of phages with bacteria in nutrient solutions that contained different radioactively labeled molecules. One batch of phages was grown with radioactive sulfur; the other batch was grown with radioactive phosphorus. The phages incorporated the radioactive atoms into their structures. Phages grown with radioactive sulfur would contain radioactive *proteins*, while phages grown with radioactive phosphorus would contain *radioactive nucleic acids*.

At this point each kind of radioactive phage was allowed to infect bacteria in a fresh culture that did not contain radioactive atoms. When bacteria from these cultures were isolated and examined with the electron microscope, some bacteria still had phages attached to their cell membranes. These phage "ghosts" were the empty coats of the viruses that remained after they had injected their genetic material into the bacteria. Hershey and Chase separated the ghosts from the bacterial cells. Then they measured the amount and type of radioactivity in the phage ghosts and in the bacterial cells.

Can you predict how Hershey and Chase planned to identify the genetic material? If protein was the genetic material, then the bacteria infected by phages labeled with radioactive sulfur would be radioactive at the end of the experiment. If DNA was the genetic material, then the bacteria infected by phages labeled with radioactive phosphorus would be radioactive. Hershey and Chase found this latter result. In addition, the ghost phages that had been labeled with radioactive phosphorus were not radioactive. This means they had injected radioactive nucleic acid into the bacteria. The ghost phages that had been labeled with radioactive sulfur were still radioactive, which meant they had *not* injected radioactive protein into the bacteria. This complex but clever experiment was strong support for the idea that DNA is the genetic molecule.

One final point should be made about the studies of Griffith, Avery, and Hershey and Chase. These experiments provided critical and fundamental knowledge of the molecules involved in inheritance. The conclusions drawn from these experiments have stood the test of nearly 75 years and have helped us to understand the basic traits of living organisms, how humans are similar to and different from other organisms, and how devastating human diseases such as cancer arise. Yet none of these studies involved humans. They did not even involve eukaryotes. The mechanisms of inheritance that these researchers discovered in bacteria and viruses apply to every living organism, including humans. These studies are just a few of the thousands of studies that demonstrate that all life is related.

7.2 Scientific Sleuthing: The Discovery of the Structure of DNA

The studies by Avery and by Hershey and Chase gave strong evidence that DNA is the genetic molecule and that proteins are not. At the time only a few groups of researchers realized that to understand how DNA contains genetic information, it is necessary to understand the structure of the DNA molecule. This insight led to the most important biological discovery of the twentieth century.

Three groups of researchers had important roles in this story. Working at the California Institute of Technology, Linus Pauling led the world's most well-known group of biochemists. Linus Pauling already had used accurate models of molecules to discover the helical secondary structure of proteins. At the University of London, Maurice Wilkins and Rosalind Franklin were well-respected DNA chemists who used the technique of X-ray crystallography (see Figure 7.6) to study molecular structures. At Cambridge University in England, Francis Crick and James Watson were two young scientists who worked obsessively on DNA. Because Watson and Crick respected Linus Pauling's work, they adopted the technique of building molecular models as an important part of their studies. Because they understood the significance of the information in X-ray images of crystals, they closely followed the work of Rosalind Franklin. All of these researchers made important contributions, but Watson and Crick are credited with the discovery of DNA's structure. In 1962 they shared the Nobel Prize in Physiology or Medicine for their work on DNA with Maurice Wilkins. This scientific breakthrough was the foundation for the astounding discoveries about DNA being made today and deserves to be examined in greater detail.

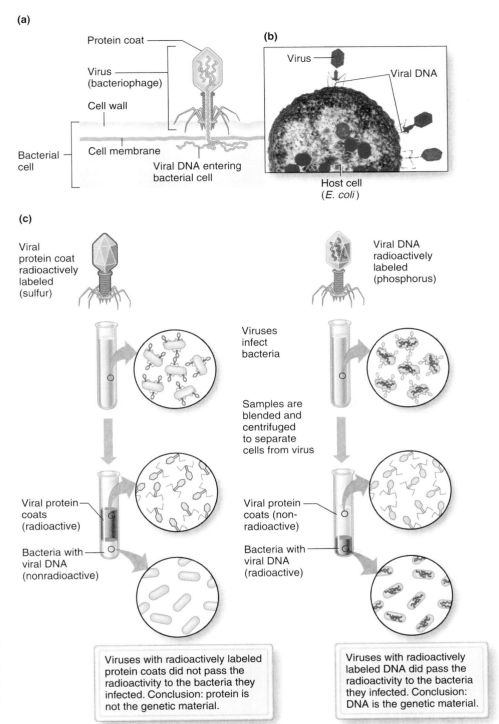

Figure 7.4 **Hershey-Chase Experiment.** This experiment with bacteriophages confirmed that DNA is the genetic material and showed that protein is not the genetic material. (**a**) Structure of a virus injecting DNA into a bacterium. (**b**) Electron micrograph of bacteriophages injecting DNA into a bacterial cell. (**c**) Using some viruses that had radioactively labeled protein coats and other viruses that had radioactively labeled DNA, Hershey and Chase showed that DNA is the genetic material.

In the fall of 1951 James Watson and Francis Crick (**Figure 7.5**) were convinced that the discovery of the structure of DNA would put their names into the scientific history books. Scientists knew what atoms nucleic acids were made of, but they were not sure how the various atoms were arranged. Watson and Crick used their knowledge of chemistry to figure out this arrangement. Their major approach was to assemble accurate metal replicas of the components of DNA into a large model of the entire molecule, to see how the pieces actually might fit together. They could have continued indefinitely, building models of different DNA configurations, but part of their genius was in realizing what else they needed to know. They knew that the missing information could be found in X-ray photographs of DNA crystals—this is where Rosalind Franklin's work enters the story.

In the process of X-ray crystallography, a crystal of a chemical is bombarded with X-rays. The electrons of X-rays bounce off of a molecule, in ways that are similar to how light rays bounce off of an object. Just like light rays, X-rays can form images on photographic film. One complication, however, is that an X-ray crystallography image is not an exact likeness of the molecule, but more like a complex shadow of the molecule. The shadow is meaningful only if you know how to interpret it.

Watson and Crick were experts at interpreting the images; Franklin was expert at obtaining X-ray images of DNA crystals (**Figure 7.6**). Her X-ray crystallography images gave Watson and Crick the critical information that allowed them to figure out the structure of the DNA molecule.

Putting all of this information together, Watson and Crick realized that DNA is a double-stranded helix with sugar-phosphate groups on the outside of the helix and nitrogenous bases on the inside of the helix (**Figure 7.7**). They also correctly concluded that adenine is always across from thymine and that cytosine is always across from guanine. On April 28, 1953, the prestigious journal *Nature* published Watson and Crick's description of their model along with separate papers by Wilkins and

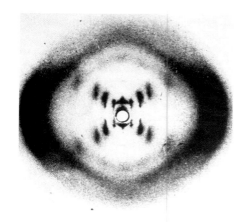

Figure 7.5 **Watson and Crick and Their DNA Model.** James Watson (left) and Francis Crick in 1953 with one of their original models of the DNA molecule.

Figure 7.6 **Rosalind Franklin and Her X-ray Crystallography Image of DNA.** When James Watson saw this X-ray photograph of a DNA crystal, he realized that the DNA molecule was in the shape of a helix, and that it was a double helix. The X of dark spots in the center of the image gave him this information.

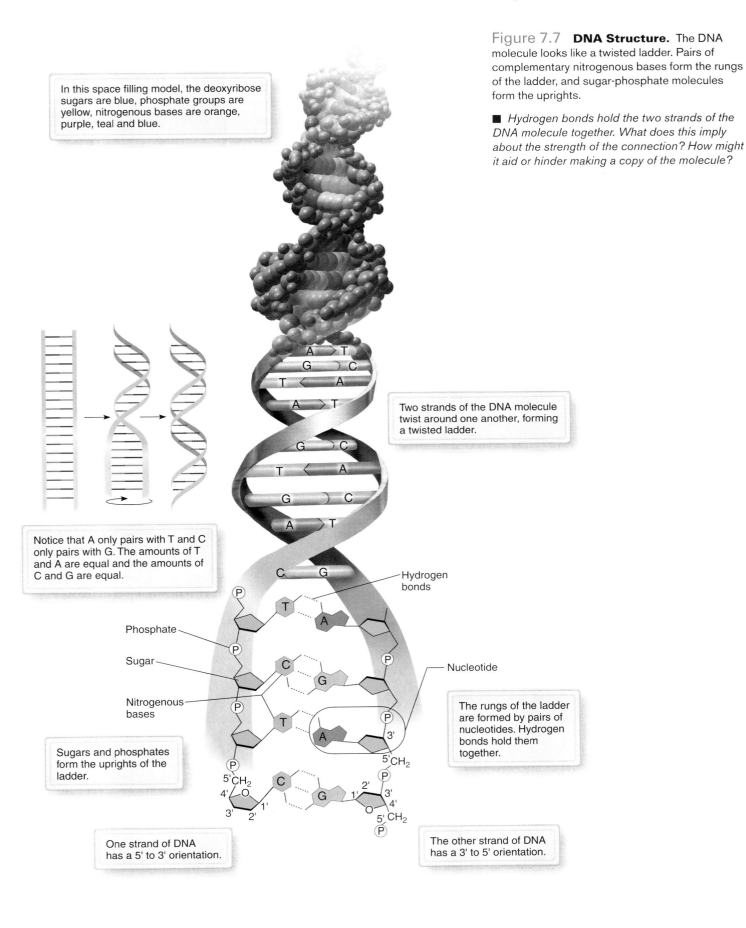

Figure 7.7 **DNA Structure.** The DNA molecule looks like a twisted ladder. Pairs of complementary nitrogenous bases form the rungs of the ladder, and sugar-phosphate molecules form the uprights.

■ *Hydrogen bonds hold the two strands of the DNA molecule together. What does this imply about the strength of the connection? How might it aid or hinder making a copy of the molecule?*

In this space filling model, the deoxyribose sugars are blue, phosphate groups are yellow, nitrogenous bases are orange, purple, teal and blue.

Two strands of the DNA molecule twist around one another, forming a twisted ladder.

Notice that A only pairs with T and C only pairs with G. The amounts of T and A are equal and the amounts of C and G are equal.

Hydrogen bonds

Phosphate

Sugar

Nitrogenous bases

Nucleotide

The rungs of the ladder are formed by pairs of nucleotides. Hydrogen bonds hold them together.

Sugars and phosphates form the uprights of the ladder.

One strand of DNA has a 5' to 3' orientation.

The other strand of DNA has a 3' to 5' orientation.

by Franklin. Watson and Crick included in their paper these understated but prophetic words:

We wish to suggest a structure for . . . deoxyribose nucleic acid (D.N.A.). This structure has novel features that are of considerable biological interest. . . . It has not escaped our notice that the specific pairing we have postulated immediately suggests a possible copying mechanism for the genetic material.

You will explore this important feature of DNA in a later section of this chapter.

You may be wondering how Watson and Crick gained access to Franklin's images, what contribution Wilkins made to the discovery, and why Franklin did not share the Nobel Prize. Wilkins and Franklin were both working on the imaging of DNA using X-ray crystallography and other techniques and often discussed their findings. On one fateful visit to King's College, Watson asked Wilkins about progress in the X-ray crystallography imaging studies of DNA, and Wilkins showed him some of Franklin's new work, but without Franklin's permission. Instantly, Watson and Crick were struck by the implications of one of Franklin's new images. This particular image allowed them to finally figure out the structure of DNA. So why was Wilkins awarded the Nobel Prize and not Franklin? It is hard to second-guess history but some things are clear. Apparently, Franklin was not happy at King's College, where she perceived a male-dominated culture that provided little support or appreciation for a talented female scientist. Shortly after the DNA papers were published, Franklin left King's College to take a position at another school and embarked on a successful career studying viruses. Sadly, her life was cut short and in 1958 she died of cancer. Wilkins continued his work at King's College, and while he did not produce the critical images that Franklin had, he did publish a number of other important papers on the structure of DNA. The Nobel Prize is always given to living scientists and, therefore, in 1962 it was awarded to Watson, Crick, and Wilkins. No one knows whether Franklin also would have been awarded the Nobel Prize, if she had lived. Nevertheless, many people believe that she deserved to win it.

Armed with the basic framework of the structure of DNA researchers were able to reveal how DNA directs the synthesis of proteins. Today, the entire sequence of an organism's nucleotides can be determined, and this information has revolutionized biological science and medicine. Because of all that has followed from it, the discovery of the structure of DNA is one of the most important scientific breakthroughs in the twentieth century.

1. Name the three research teams involved in the discovery of the structure of DNA and briefly explain the major contribution made by each group.

7.3 The Structure of DNA Is a Double Helix

Recall that DNA is a string of nucleotides, and a nucleotide is a three-part molecule made of a sugar, a phosphate group, and a nitrogenous base (see Section 3.5 and Figure 3.12). Typically, a DNA molecule has two strands of nucleotides (Figure 7.7). Just as Watson and Crick discovered, the strands of nucleotides wind around each other, forming a long double helix that resembles a twisted ladder. The sugar-phosphate components of the nucleotides are on the outside of the double helix. They form the uprights of the ladder and help stabilize the molecule. The rungs of the ladder are formed by pairs of nitrogenous bases that are held together by hydrogen bonds. Of course, hydrogen bonds are relatively weak and by breaking these hydrogen bonds, a DNA molecule can be unzipped fairly easily, resulting in two single strands of nucleotides.

Notice that in the complete double-stranded DNA molecule, the two single strands have opposite orientations. Look at the detailed drawings of nucleotides in the DNA molecule in Figure 7.7. The five carbon atoms in the sugar molecules are numbered 1′ (read as "1 prime") through 5′. In a nucleotide molecule the 1′ carbon is bonded to the nitrogenous base, and the 5′ carbon is bonded to the phosphate group. The 3′ carbon is bonded to the phosphate group of the *next* nucleotide in the strand. In Figure 7.7 the string of nucleotides on the left side of the DNA molecule are oriented from 5′ to 3′ (reading from the top of the figure down to the bottom), but the string of nucleotides on the right side of the DNA molecule runs from the 3′ to 5′ end (reading top to bottom). This is a good way to tell one side of the molecule from the other, and it gives scientists a way to talk about the two strands.

One of the important features of DNA is how the bases are paired across the ladder. Many experiments have demonstrated that Watson and Crick's proposed pattern for base pairing was correct: *adenine always binds to thymine and cytosine always binds to guanine.* This rule of base pairing governs the structure of DNA molecules in all living organisms. It gives a regular structure to all DNA molecules, and it is critical to the functions of DNA.

DNA Combines with Proteins to Form Chromosomes

The structure of the DNA molecule discovered by Watson and Crick is the nucleic acid molecule by itself, but in a cell DNA does not exist in this pure form. The pure DNA molecule is extremely long and thin, and if it were left in one continuous strand, the total DNA from just one human cell would be about 3.0 meters (9.84 feet) long. This extraordinary length of DNA must be stored compactly to fit within a cell. The first step in solving this problem is to divide the entire strand of DNA into several smaller pieces.

In a human cell DNA is found in 46 separate pieces, and each piece is a separate DNA molecule. Each kind of organism has its own characteristic number of DNA pieces. For example, fruit flies have 8 pieces, cattle have 60, snakes have 36, dogs have 78, corn plants have 20, and one species of adder's tongue fern has more than 1,200. Getting the DNA into pieces, though, does not solve the problem of fitting it into a cell, because each piece is still thousands of times longer than the diameter of the cell's nucleus. The individual DNA molecules must be compressed into even smaller packages; proteins do this.

How do proteins coil DNA? The process occurs in stages, and **Figure 7.8** diagrams what the DNA looks like at each level of coiling, while the photograph shows an actual picture of DNA taken by an electron microscope. While DNA can be seen with an electron microscope, the two helices cannot be distinguished with this technique. In the first stage of coiling DNA is wound around little "balls" of protein at regular intervals along the DNA molecule (Figure 7.8b). These balls of protein wrapped with DNA are called **nucleosomes** and look like beads along a string of naked DNA. But the nucleosome structure is still too unwieldy, and DNA undergoes still higher levels of coiling (Figure 7.8c). At each stage of coiling the DNA is compacted tighter and the proteins associated with DNA do this job. At the end of this process a piece of DNA that was about 3.0 meters (9.84 feet) long in its naked form has been wound up to form a chromosome that is a few micrometers or less in length and fits inside the nucleus (Figure 7.8d). This is like taking a strand of angel hair pasta that stretches across your plate and folding it into a tiny

piece of corkscrew pasta the size of a grain of rice. A **chromosome** is defined as a piece of DNA and associated protein.

Scientists use the words *condensed* and *dispersed* to describe how coiled a stretch of DNA is at any given time. A chromosome that is coiled as tightly as possible is fully condensed, while a stretch of DNA that is in the "beads on a string" configuration is fully dispersed. Depending on what the chromosome is doing, chromosomes are converted between condensed and dispersed states. For example, when a section of the DNA molecule is needed to instruct protein synthesis, that section can uncoil to the dispersed state. To save space, sections of DNA that are not immediately needed for directing protein synthesis may remain tightly coiled. Therefore, within a given chromosome, some parts of the DNA molecule may be condensed, while other parts are dispersed. If the whole chromosome has to be moved from one part of the cell to another, as happens in cell division, the chromosome is tightly condensed. This gives cell flexibility in handling its DNA.

QUICK CHECK

1. Sketch and describe the structure of DNA.
2. Briefly restate the process of chromosome coiling.

7.4 The Structure of DNA Contains Genetic Information

DNA has two important functions in a cell: directing the synthesis of proteins and passing on those instructions to a new cell during the process of cell reproduction. Because these two functions of DNA follow directly from its chemical structure, both functions are good examples of the principle that biological processes emerge from chemical structures and chemical interactions. This section considers how DNA passes life's instructions to the next generation. In the next section you will learn how DNA directs the synthesis of proteins.

The Double Helix Contains the Mechanism for Inheritance

The central mechanism for all types of inheritance is the ability of a single cell to make a copy of its own DNA. When it reproduces, a parent cell copies its DNA, and one copy is given to each of two new cells. As soon as they had completed their DNA

nucleosome a little ball of protein with DNA wound around it, a nucleosome is active at the first stage of coiling and condensing of DNA molecules

chromosome a piece of DNA and associated protein

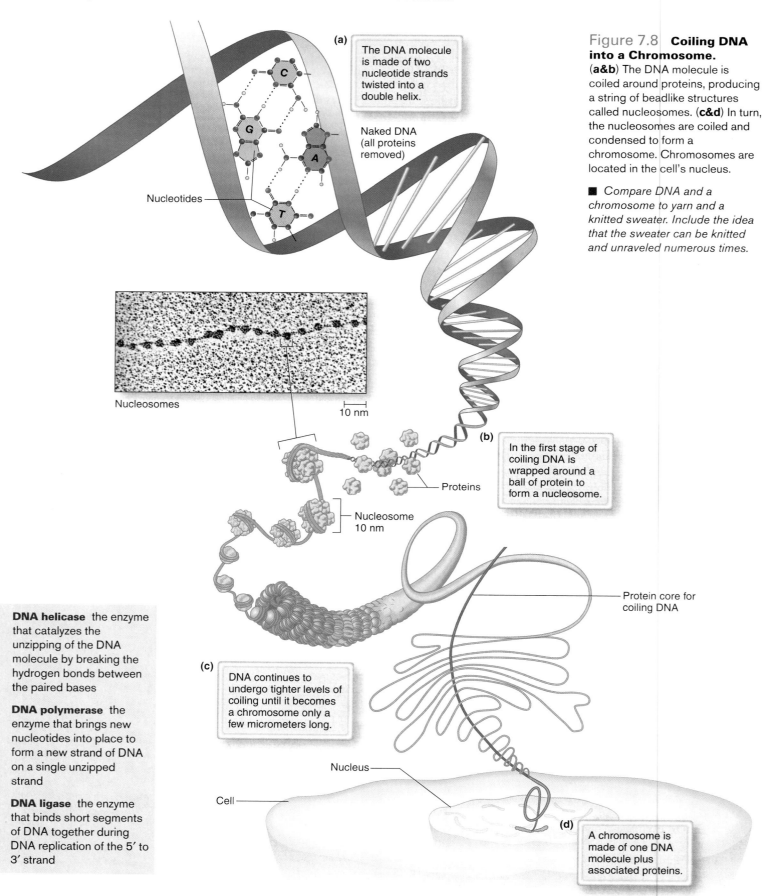

(a) The DNA molecule is made of two nucleotide strands twisted into a double helix.

C

G

A

T

Nucleotides

Naked DNA (all proteins removed)

Nucleosomes

10 nm

(b) In the first stage of coiling DNA is wrapped around a ball of protein to form a nucleosome.

Proteins

Nucleosome 10 nm

Protein core for coiling DNA

(c) DNA continues to undergo tighter levels of coiling until it becomes a chromosome only a few micrometers long.

Nucleus

Cell

(d) A chromosome is made of one DNA molecule plus associated proteins.

Figure 7.8 Coiling DNA into a Chromosome.
(a&b) The DNA molecule is coiled around proteins, producing a string of beadlike structures called nucleosomes. **(c&d)** In turn, the nucleosomes are coiled and condensed to form a chromosome. Chromosomes are located in the cell's nucleus.

■ *Compare DNA and a chromosome to yarn and a knitted sweater. Include the idea that the sweater can be knitted and unraveled numerous times.*

DNA helicase the enzyme that catalyzes the unzipping of the DNA molecule by breaking the hydrogen bonds between the paired bases

DNA polymerase the enzyme that brings new nucleotides into place to form a new strand of DNA on a single unzipped strand

DNA ligase the enzyme that binds short segments of DNA together during DNA replication of the 5′ to 3′ strand

model, Watson and Crick realized how DNA could be copied. Their hypothesis is presented in **Figure 7.9**, and it turned out to be correct. If a DNA double helix molecule is split down the middle, then all of the information needed to rebuild the entire molecule remains in each single strand. How? The copying mechanism results from the rule for pairing of nucleotides. Remember that A (adenine) always binds to T (thymine), and that C (cytosine) always binds to G (guanine). Therefore, if one strand in a DNA molecule has the sequence CCGTTCACC—when it is read from the 5′ end to the 3′ end—the other strand will have the sequence GGCAAGTGG—when it is read from its 3′ end to 5′ end. If the double helix is split down the middle to form two single nucleotide strands, then two new double-stranded DNA molecules can be made by placing complementary nucleotides across from each original nucleotide. This creates two replicas of the original double-stranded molecule (Figure 7.9).

The amazing thing is that a cell can copy all of its DNA in a matter of hours, something that might never happen spontaneously in a hundred years or more. As you might expect, enzymes and energy are responsible for hastening the process. The complex process of DNA replication involves seven or more enzymes and other proteins. Two of these enzymes are DNA helicase and DNA polymerase. **Figure 7.10** shows their roles. First, **DNA helicase** unzips the DNA molecule down the middle by breaking the hydrogen bonds between nucleotides (Figure 7.10a). Then a **DNA polymerase** molecule moves along each original DNA strand, attaching the complementary nucleotides to the original nucleotides (Figure 7.10b). Because two strands are being copied, two DNA polymerase molecules are active, one on each strand.

This simple description of DNA replication is a good representation for the 3′ to 5′ strand of the original DNA molecule. The DNA polymerase molecule reads in the 3′ to 5′ direction, and so puts the new nucleotide strand together in the 5′ to 3′ direction. For the 3′ to 5′ strand of the original DNA molecule, replication is a smooth and continuous process. Problems arise, however, for the 5′ to 3′ strand of the original DNA molecule. DNA polymerase cannot read this strand in one continuous sequence, but reads it in short segments. As a result, it makes short segments of new DNA from the 5′ to 3′ strand. **DNA ligase** enzyme binds these short segments together to make the complete DNA double helix.

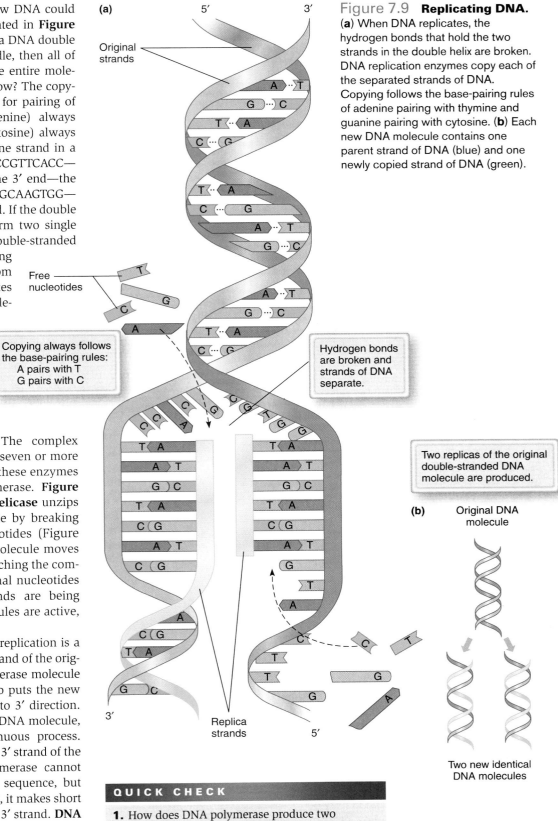

(a)

Original strands

A···T
G···C
T···A
C···G

T···A
C···G
A···T
G···C

Free nucleotides

A···T
G···C
T···A
C···G

Copying always follows the base-pairing rules:
A pairs with T
G pairs with C

Hydrogen bonds are broken and strands of DNA separate.

T A
A T
G C
T A
C G
A T
C G

T A
A T
G C
T A
C G
G
T
A

3′

Replica strands

5′

Figure 7.9 **Replicating DNA.** (**a**) When DNA replicates, the hydrogen bonds that hold the two strands in the double helix are broken. DNA replication enzymes copy each of the separated strands of DNA. Copying follows the base-pairing rules of adenine pairing with thymine and guanine pairing with cytosine. (**b**) Each new DNA molecule contains one parent strand of DNA (blue) and one newly copied strand of DNA (green).

Two replicas of the original double-stranded DNA molecule are produced.

(b)   Original DNA molecule

Two new identical DNA molecules

QUICK CHECK

1. How does DNA polymerase produce two complete copies of an original DNA molecule?

Enzymes active in DNA replication:

DNA helicase unwinds parental double helix.

DNA polymerase binds nucleotides and forms a new strand.

(a) DNA helicase binds to DNA and separates the strands.

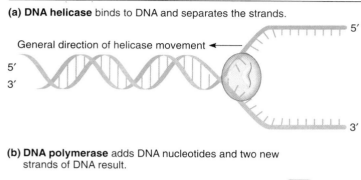

General direction of helicase movement

5′
3′

5′

3′

(b) DNA polymerase adds DNA nucleotides and two new strands of DNA result.

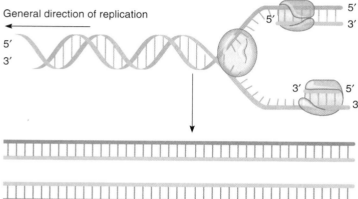

General direction of replication

5′
3′

5′
5′
3′

3′ 5′
3′

Figure 7.10   Enzymes in DNA Replication. (a) DNA helicase unzips DNA. **(b)** DNA polymerase adds DNA nucleotides to form two new strands of DNA.

7.5   DNA Directs the Synthesis of Proteins

DNA is truly the golden molecule. In the same structure, DNA directs cellular activity *and* passes these instructions on to the next generation. At one level the inheritance of DNA instructions is fairly simple, even if the details are complex. But how can DNA also direct the functions of a cell or an entire organism? Again the principle is simple, but the execution is more complicated. Remember that a cell's activities are largely determined by what proteins it produces. DNA directs the activities of a cell in only one way: by carrying the instructions to make proteins.

A Sequence of Nucleotides Codes for a Sequence of Amino Acids

Where in the DNA molecule are the instructions to make proteins? Because the sugars and phosphates are identical all along the molecule, the only place in DNA where the instructions for protein synthesis could be found is in the sequence of nitrogenous bases. Somehow the sequence of nucleotides must be used as information to build protein molecules. But is there enough information in the DNA sequences to provide the instructions for all of the thousands of proteins in a cell? The answer is yes. A given DNA molecule will have tens of thousands of bases running along each strand. Even though there are only four nitrogenous bases, the number of different sequences of nucleotides possible in a segment of any DNA molecule is incredibly large. You can see this if you look at all of the possible sequences in a strand of DNA that is just 10 nucleotides long. Mathematically, the question can be phrased: In how many ways can the four bases be arranged in a sequence that is 10 bases long? This can be calculated as $4 \times 4 \times 4 \times 4 \times 4 \times 4 \times 4 \times 4 \times 4 \times 4$, or 4^{10}, which equals a surprising 1,048,876 ways! In humans *billions* of nucleotides are lined up in sequence along DNA molecules, and in such long strands the number of different sequences that are possible is staggering. Enough information is in a cell's DNA nucleotide sequence to carry the instructions needed for even the most elaborate cellular processes.

But how is the information in DNA used to create proteins? The key to answering this question is to realize that, while DNA is a sequence of nucleotides, a protein is a sequence of amino acids. The process involves translation of one sequence of items (nucleotides in DNA) into another sequence of items (amino acids in a protein). This idea is similar to the codes that children might use to send secret messages. **Figure 7.11** shows an example of such a code in which every letter in the normal English alphabet is assigned a three-digit code number. Below the assigned numerical alphabet is a short sentence written in the code. Can you translate the sentence?

The process of going from DNA to protein is not very different from this secret code. There must be a code that translates from DNA language (nucleotides) into protein language (amino acids). What is the code? You can find a clue if you look at the number of amino acids commonly found in proteins versus the number of nitrogenous bases in DNA. In a protein each amino acid

A	B	C	D	E	F	G	H	I	J	K	L	M	N	O	P	Q	R	S	T	U	V	W	X	Y	Z
222	346	788	999	123	098	666	425	007	567	160	550	235	340	872	398	453	098	265	981	298	888	376	594	111	000

| 981 | 425 | 123 | 788 | 222 | 981 | 007 | 265 | 666 | 098 | 123 | 111 |

Figure 7.11 **An Example of a Three-Number Code.** Solving this code puzzle will give you an intuitive understanding of the triplet code that translates DNA and mRNA language into proteins. ANSWER: THE CAT IS GREY

position can be occupied by any one of 20 different amino acids. In DNA, on the other hand, there are only four possible nitrogenous bases for each position. This means the code can't be one nucleotide corresponding to one amino acid. If this were true, then there would be enough "code" for only four amino acids. Some other code must be at work. What if a sequence of two nitrogenous bases represented one amino acid? For example, AA might code for alanine, while CC might code for proline, and so on. This doesn't work either, however, and you can demonstrate that by making a list of all possible sequences of two nitrogenous bases. Your list would add up to only 16 possible sequences of two nitrogenous bases. Once again, this isn't enough to represent 20 different amino acids. If you try a sequence of three nitrogenous bases, however, you will have better luck. To make this list, start with AAA, AAT, AAC, AAG, and keep going until you have listed all the possibilities: 64 sequences of three nucleotides. This is many more than the number needed to represent the 20 amino acids. Experiments have shown that, indeed, this triplet code is the answer. So there it is—the secret to translating from DNA language to protein language: *a sequence of three nitrogenous bases codes for each single amino acid in a protein.* But why don't the numbers match up exactly, and what happens to the sequences of three nucleotides that are "left over"? The three nucleotide code is *redundant* because there is more than one three-nucleotide code that represents many of the 20 amino acids. For instance, UCU, UCC, UCA, and UCG are all translated as the amino acid, serine.

A segment of a DNA strand that carries the nucleotide code for one protein is called a **gene,** while the entire nucleotide sequence for a cell or organism is a **genome.** Each chromosome is a sequence of genes. Next you will read how the information in genes is translated into proteins.

RNA Plays Important Roles in the Process of Protein Synthesis

So far you have focused on the structure and roles of DNA, but the nucleic acid RNA also has an important role in protein synthesis. There are three important forms of RNA in a cell: messenger RNA, transfer RNA, and ribosomal RNA (**Table 7.1**). **Messenger RNA,** also called **mRNA,** is a single strand of nucleic acid that maintains a stringlike structure. The mRNA carries the information in DNA to the ribosome. **Transfer RNA,** usually called **tRNA,** carries specific amino acids found in the cytoplasm to the ribosome, where they are used to build new proteins. Because some of the nucleotides along the chain of a tRNA molecule form hydrogen bonds to one another, tRNA has a

gene a sequence of DNA that codes for a single protein

genome the entire nucleotide sequence of all the genetic material in an organism

messenger RNA (mRNA) ribonucleic acid that carries DNA's instructions for making proteins

transfer RNA (tRNA) an RNA molecule that transfers amino acids from the cytosol to a ribosome, where it ensures that the correct amino acid is used to make a specific protein

Table **7.1**	Major kinds of RNA	
Type of RNA	**Function**	
Messenger RNA (mRNA)	Carries DNA information to ribosome in cytosol	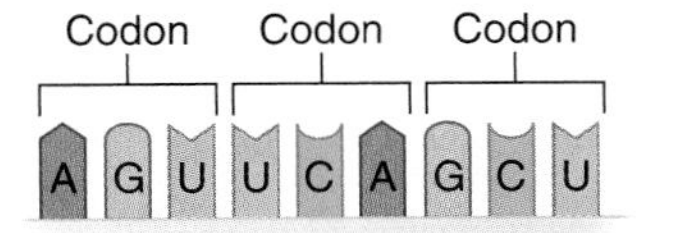 Codon Codon Codon A G U U C A G C U
Transfer RNA (tRNA)	Shuttles amino acids to ribosomes that are making proteins	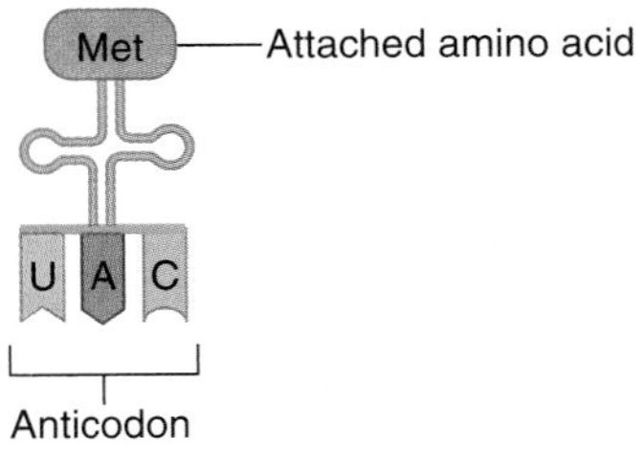 Met —Attached amino acid U A C Anticodon
Ribosomal RNA (rRNA)	Acts as a catalyst to bind amino acids into protein strands	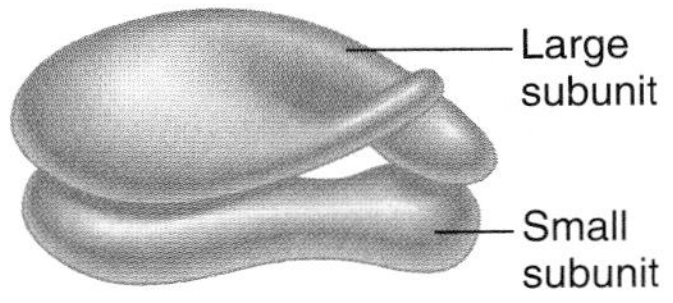 Large subunit Small subunit

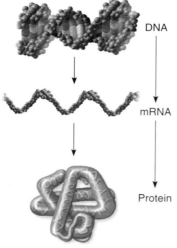

The flow of information in a cell goes from DNA to mRNA to protein.

DNA

mRNA

Protein

Figure 7.12 **Central Dogma.**

folded shape. Finally, ribosomes are made of protein and RNA, a version of RNA called **ribosomal RNA,** or **rRNA.** Scientists were surprised to discover that the rRNA molecules are actually catalysts and aptly named them **ribozymes.**

DNA Directs the Synthesis of Proteins by the Processes of Transcription and Translation

DNA directs protein synthesis in a two-step process. First, the instructions carried in DNA are copied into mRNA strands. Second, mRNA strands are used to produce proteins. The principle called the *central dogma* summarizes this flow of information (**Figure 7.12**). The central dogma states that DNA codes for RNA that, in turn, codes for protein. Note that the central dogma asserts that DNA has ultimate control over proteins, not *vice versa*. DNA can be used to make proteins, but the information in proteins does not feed back to produce or change DNA. As with

any general rule in biology, the central dogma has its limits. Some viruses, for example, use RNA as the genetic material and can copy RNA to DNA. In addition, some of the RNA copied from DNA is not used to make protein but instead is used for other cellular activities. Nevertheless, the formulation of the central dogma was a big step forward in understanding how DNA directs cellular processes, and it appears to apply to every living cell.

One important part of the central dogma is the fact that mRNA is the molecule that contains the information used by ribosomes to make protein. Therefore, it is the sequence of triplet nucleotides of mRNA that gets translated into a specific amino acid sequence by the ribosome. The three-nucleotide sequences that code for each amino acid are named **codons,** and there are 64 of these codons. The complete list of translations between each codon and its corresponding amino acid is called the **genetic code** (**Figure 7.13**). Because the genetic code is executed by mRNA, the code is stated using the four nucleotides of RNA: adenine,

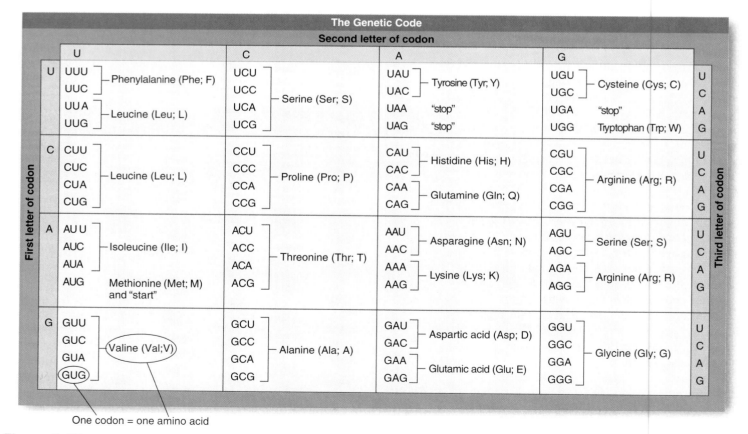

Figure 7.13 **The Dictionary of the Genetic Code.** A sequence of three RNA nucleotide bases codes for an amino acid and is called a codon. To see how this dictionary works, first notice how it is organized: the first base of each codon is along the left-hand margin, the second base is across the top, and the third base is down the right margin. Imagine any sequence of three nucleotide bases, for instance A U G, and see what amino acid AUG codes for. AUG codes for methionine, and it also is the signal for transcription to START. There are several STOP codes and none codes for an amino acid. Finally, count how many triplet codes translate into the amino acid arginine.

■ *Why is it necessary to have a STOP codon in an RNA? What would be the amino acid sequence of this sequence of codons: UUC UCA AAA CGA?*

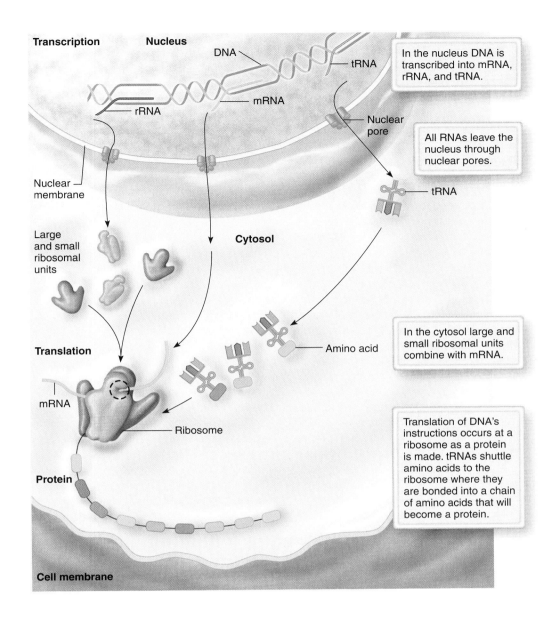

Transcription   **Nucleus**

In the nucleus DNA is transcribed into mRNA, rRNA, and tRNA.

All RNAs leave the nucleus through nuclear pores.

Nuclear pore

tRNA

Cytosol

In the cytosol large and small ribosomal units combine with mRNA.

Amino acid

Nuclear membrane

Large and small ribosomal units

Translation

mRNA

Ribosome

Translation of DNA's instructions occurs at a ribosome as a protein is made. tRNAs shuttle amino acids to the ribosome where they are bonded into a chain of amino acids that will become a protein.

Protein

Cell membrane

Figure 7.14
Transcription and Translation. In eukaryotes transcription occurs in the nucleus. mRNA travels out of the nucleus to the cytosol. There mRNA binds to a ribosome, and translation of DNA's code into amino acids that form a protein begins. rRNAs and tRNAs also are transcribed from DNA and move to their functions in the cytosol.

■ *Develop your own analogy for protein synthesis.*

ribosomal RNA (rRNA) RNA found in the structure of ribosomes

ribozymes RNA molecules that function as biological catalysts

codon a sequence of three nucleotides that codes for an individual amino acid

genetic code the complete list of translations between each triplet codon on mRNA and the amino acid each one represents

transcription transfer of information from DNA to messenger RNA

translation transformation of messenger RNA's instructions into a chain of amino acids

cytosine, guanine, and *uracil*—not adenine, cytosine, guanine, and *thymine*. For example, CUUAU-UCGC codes for the amino acids leucine, isoleucine, and arginine. As just mentioned, some of the codons are redundant. A few of the codons in Figure 7.13 have been highlighted to show that they have special meanings. For example, the code for the amino acid methionine indicates the beginning of the sequence for a particular protein. In other words, AUG, the codon for methionine (Met), is always the START codon. Other codons, for example, UAG, UAA, and UGA indicate the end of the instructions for one particular protein, and these are called STOP codons.

Now that you understand how amino acids are represented in the nucleotide sequences of DNA and mRNA, let's focus on the biochemical processes that accomplish protein synthesis. The two major processes of transcription and translation are involved when DNA directs the synthesis of proteins (**Figure 7.14**). **Transcription** is the process that transfers the information in DNA to mRNA, and it takes place in the cell's nucleus. The transcribed mRNA molecules then pass out of the nucleus through the nuclear pores. **Translation** uses the information in mRNA to direct the formation of a chain of amino acids by a ribosome, and happens outside the nucleus, in the cytosol. The assembled protein molecules are then either released into the cytosol, or transported by vesicles to the Golgi apparatus where they can be exported from the cell.

Figure 7.15 explains the transcription process in a typical eukaryotic cell in more detail. **RNA polymerase** is responsible for carrying out transcription from DNA to mRNA. RNA polymerase is a large enzyme complex and has several subunits that are similar to the set of proteins that function in DNA replication. RNA polymerase first separates the two strands of the DNA double helix so that the sequence of nucleotides is exposed. To do this, RNA polymerase breaks the hydrogen bonds that link the bases across the rung of the double helix, much like DNA helicase does during DNA replication. Once the two strands of DNA have been separated, RNA polymerase builds the complementary mRNA strand with free RNA nucleotides.

Recall that messenger RNA is a single-stranded nucleic acid. Therefore, just one side of the DNA double helix is used as a gene to transcribe mRNA. Once RNA polymerase starts to transcribe a particular gene, it stays on that one strand until the transcription of that gene is completed. Using the rules of base pairing, RNA polymerase builds the correct mRNA chain. For instance, when RNA polymerase encounters a C (cytosine) on the DNA chain, it puts a G (guanine) in the RNA chain. When a T (thymine) is encountered in the DNA chain, an A (adenine) is placed in the RNA chain, and so on. However, because RNA contains uracil (U) instead of T (thymine), the rule is slightly different than the base-pairing rule in DNA (A-T; G-C). Wherever there is an A (adenine) in the DNA chain, RNA polymerase places a U (uracil) in the RNA chain. The end result of transcription is a single strand of mRNA that contains a sequence complementary to the original gene, with U (uracil) substituted for T (thymine).

Now the newly transcribed mRNA is ready to guide translation, the second process in the central dogma (see Figure 7.14). During translation mRNA, which leaves the nucleus through a nuclear pore, interacts with a ribosome and tRNAs to produce an amino acid chain. In this process mRNA brings information about which amino acids to add to the developing protein chain. Different tRNA molecules bring the appropriate amino acids from the cytosol to the ribosome. A ribosome is an assemblage of biological molecules that functions as a molecular machine. The ribosome bonds the amino acids together to form the chain of amino acids. In eukaryotic cells the ribosome complex bonds two amino acids per second. If the ribosome is best described as a molecular machine in the protein building process, then tRNA could be considered as a transportation, or shuttle, molecule.

Transfer RNA (tRNA) gets its name because it transfers amino acids from the cytosol to a ribosome, where they are incorporated into the growing amino acid chain. To make a specific protein, tRNA molecules must bring the correct amino acids to the ribosome. This part of the process has several phases: first, a tRNA molecule connects with an amino acid in the cytosol and then moves it to a ribosome. tRNA interacts with both the ribosome and the mRNA that is temporarily attached to the ribosome. The structure of tRNA allows it to carry out both of these tasks. Just like mRNA, tRNA is a string of nucleotides, but because of its unique nucleotide sequence, tRNA folds into a molecule with a distinctive shape. **Figure 7.16** shows different views of a tRNA molecule. You can think of the simplest of these as rather like a three-leafed clover leaf (Figure 7.16a). Two of the more important parts of the tRNA are the "stem" of the molecule, which interacts with an amino acid, and the opposite end of the molecule, where the "middle leaf" interacts with mRNA in a ribosome.

RNA polymerase an enzyme that catalyzes the formation of mRNA from DNA

Figure 7.15 **RNA Polymerase Catalyzes the Formation of mRNA.** RNA polymerase separates the two strands of a DNA molecule and attaches nucleotides in the proper sequence. As RNA polymerase moves along the DNA molecule, the stretch of mRNA that it is making grows longer.

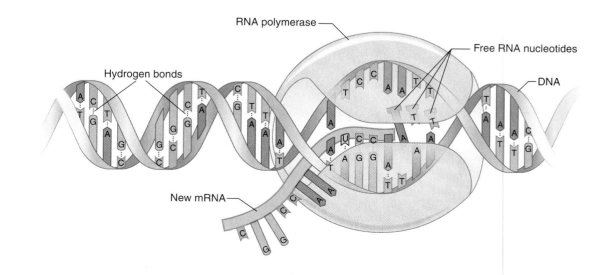

But how do tRNA molecules bond to amino acids? At the upper stem end of the tRNA molecule is an attachment site for an amino acid (Figure 7.16b). Interestingly, most tRNA molecules can bond to just one specific kind of amino acid. For instance, a tRNA might bond only to the amino acid valine, or only to leucine, or to tryptophan. This means that there are different versions of tRNA, and each is named for the amino acid that it bonds. For example, there is tRNA^{Gly} and tRNA^{Lys}, meaning "tRNA for glycine" and "tRNA for lysine," respectively. There is a specific tRNA molecule for each of the 20 kinds of amino acids commonly found in organisms.

Look again at Figure 7.16a. Note that a tRNA molecule carries a sequence of three nucleotides that has a special role in translation. Opposite from the amino acid attachment site is a three-nucleotide sequence that is the *complement* of the codon for the specific amino acid carried by that tRNA. This complementary nucleotide sequence is called an **anticodon.** For example, the amino acid methionine is represented in mRNA by the codon AUG; the tRNA molecules that carry methionine have the complement of this codon—UAC—as an anticodon located at the bottom of the tRNA molecule. Just as each tRNA is specific for just one amino acid, each tRNA has the anticodon for that particular amino acid.

Of course, the bonding of an amino acid to a tRNA molecule requires energy from ATP. This process is referred to as "charging the tRNA." Later this charging energy is used to bond this particular amino acid to a growing amino acid chain being made by the ribosome. The charged tRNA molecule is released from the enzyme. It is free to proceed to the next step of protein synthesis.

Now you are ready to examine what happens at the ribosome during translation (**Figure 7.17**). The small and large subunits of the ribosome come together only in the presence of mRNA (Figure 7.17a). Once the whole complex has assembled, three slots—or bays—form that will hold tRNA–amino acid molecules long enough for the ribosome to bind the amino acids together. The bays are called the A-site, the P-site, and the E-site (Figure 7.17b,c). Here is how the whole process works.

The first two codons on the mRNA strand are held in the P-site and the A-site, respectively (Figure 7.17d). For now, the third bay, the E-site, is empty, but it will be used later in the process. What determines which tRNA molecule rests in each site? This is where the codons and anticodons come into play. The codons and anticodons interact via hydrogen bonds. This interaction is governed by the familiar base-pairing rules. The first two incoming tRNA molecules with the appropriate anticodons bind weakly to the codons on the mRNA molecule sitting at the bottom of the P-site and A-site. For example, if the first two codons on the mRNA are AUG and CCC, then tRNA molecules having anticodons UAC and GGG would be the first two in the P- and A-sites of the ribosome. Take a minute to consult Figure 7.13 and find out which amino acids these two tRNA molecules would carry. Notice that the first codon, AUG, codes for methionine, the START codon, which always marks the beginning of translation.

The ribosome holds the mRNA and tRNAs in place long enough for the two amino acids to be linked by a covalent bond (Figure 7.17e). Within the ribosome enzymes and ribozymes catalyze the reaction that binds the two amino acids together, but this happens in a highly specific way. The amino acid in the P-site is removed from its tRNA and attached to the amino acid on the tRNA in the A-site (Figure 7.17f). The bond between the first tRNA and its amino acid is broken, and the released energy is captured and used to make a new bond between the two amino acids. As a result, the amino acids are strung from the tRNA molecule that is still in the A-site.

Once the first two amino acids are linked, the ribosome must continue to build the protein chain. The way this happens is rather amazing because the components of the ribosome complex move relative to one another. From the perspective of Figure

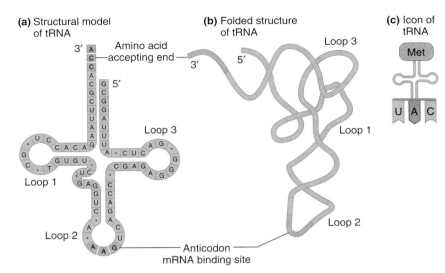

Figure 7.16 **Three Ways to Represent a tRNA Molecule.** (**a**) The cloverleaf model highlights a tRNA molecule. (**b**) The folded model shows the three-dimensional shape of a tRNA molecule. (**c**) This icon will represent tRNA in upcoming figures.

anticodon a three-base sequence that is the complement of the codon for the specific amino acid attached to a tRNA molecule

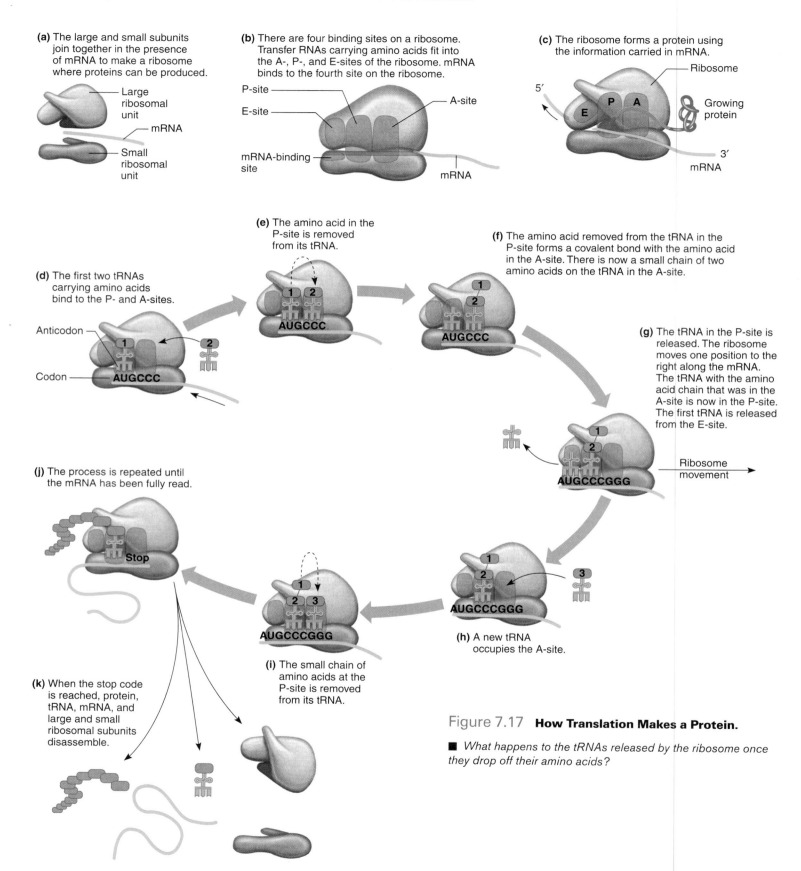

(a) The large and small subunits join together in the presence of mRNA to make a ribosome where proteins can be produced.

Large ribosomal unit

mRNA

Small ribosomal unit

(b) There are four binding sites on a ribosome. Transfer RNAs carrying amino acids fit into the A-, P-, and E-sites of the ribosome. mRNA binds to the fourth site on the ribosome.

P-site

E-site

A-site

mRNA-binding site

mRNA

(c) The ribosome forms a protein using the information carried in mRNA.

Ribosome

5′

E   P   A

Growing protein

3′

mRNA

(d) The first two tRNAs carrying amino acids bind to the P- and A-sites.

Anticodon

Codon

AUGCCC

(e) The amino acid in the P-site is removed from its tRNA.

AUGCCC

(f) The amino acid removed from the tRNA in the P-site forms a covalent bond with the amino acid in the A-site. There is now a small chain of two amino acids on the tRNA in the A-site.

AUGCCC

(g) The tRNA in the P-site is released. The ribosome moves one position to the right along the mRNA. The tRNA with the amino acid chain that was in the A-site is now in the P-site. The first tRNA is released from the E-site.

AUGCCCGGG

Ribosome movement

(j) The process is repeated until the mRNA has been fully read.

Stop

(i) The small chain of amino acids at the P-site is removed from its tRNA.

AUGCCCGGG

(h) A new tRNA occupies the A-site.

AUGCCCGGG

(k) When the stop code is reached, protein, tRNA, mRNA, and large and small ribosomal subunits disassemble.

Figure 7.17 **How Translation Makes a Protein.**

■ *What happens to the tRNAs released by the ribosome once they drop off their amino acids?*

7.17, the ribosome moves to the right, relative to the mRNA and the tRNAs, while the mRNA and the tRNAs move to the left. After the first two amino acids have been linked, the chain of two amino acids sits on the tRNA in the A-site. The tRNA in the P-site moves to the E-site and then is released. Simultaneously, the tRNA in the A-site, with the amino acid chain attached, is moved to the P-site. Now the A-site is empty, but sits over the third codon on the mRNA molecule (Figure 7.17g). A new tRNA now comes into the A-site, binds to mRNA via the codon-anticodon mechanism, and the process is repeated (Figure 7.17h). Once the ribosome has the mRNA's second and third codons in the P- and the A-sites, the third tRNA weakly binds the third codon. The two–amino acid chain on the tRNA, in the P-site, moves over to the new tRNA that now is in the A-site. Note that the tRNA in the P-site always gives its amino acid chain to the tRNA that is in the A-site (Figure 7.17i). The tRNA in the P-site then is released through the E-site, and the ribosome moves down a notch on the mRNA molecule—to your right as you look at the figure. Then the whole process repeats. When the ribosome moves, the tRNA carrying the growing amino acid chain is shifted from the A-site to the P-site. The A-site now is empty, ready to receive another tRNA carrying another amino acid.

This movement sometimes is hard to visualize, but you might imagine using a hole punch to put holes in a strip of paper. The hole punch represents the two units of the ribosome, and the paper is the mRNA. Each time you make a hole, the punch has to move down to the next position to make another hole. To make the analogy a bit more complete, imagine that the circles that come out of the hole punch stay connected, and you end up with a long string of paper circles. In a similar way the ribosome moves down the mRNA chain, adding one amino acid to the growing protein—rather than paper circles—at each location along the mRNA chain.

How does the whole process know when to stop? The last codon in the mRNA strand is always the special stop signal, a STOP codon. The codons UAA, UAG, and UGA mean "stop translation here." When any one of these codons is encountered, the whole mRNA, tRNA, ribosome complex comes apart and releases the complete amino acid chain (Figure 7.17j). The linear amino acid chain spontaneously folds into a three-dimensional protein, according to the principles of protein structure (see Section 3.6). With some final modifications made by other organelles, the newly made proteins are sent to their destinations and are ready to go to work.

This chapter presented the basic cellular process involved in transcription and translation.

The picture given here reflects the simplest aspects of these processes as it is seen in prokaryotes. The process in eukaryotes is essentially the same but has additional complications. For instance, in eukaryotes, mRNA almost always gets modified before it interacts with the ribosome. This means that in eukaryotes the mRNA that comes from the process of transcription is not exactly the same molecule that directs protein synthesis. Such posttranscriptional modifications to mRNA allow a cell to make an even greater variety of proteins from its DNA. Chapter 9 describes more about how transcription and translation are controlled and modified to produce this variety of proteins and cellular activities.

QUICK CHECK

1. What is the genetic code?
2. What are the three types of RNA and what role does each play in transcription and translation?
3. Briefly summarize the events of transcription.
4. Describe what happens at a ribosome during translation.

7.6 DNA Helps to Explain the Unity and Diversity of Life

Also hidden in the description of how DNA works are the reasons for both the unity of life and the diversity of life. The unity of life comes from two aspects of DNA. First, the genetic code is almost universal. Nearly every known kind of organism uses the same genetic code. Only a handful of species of prokaryotes use a slightly different code, but the principle of a triplet codon as the genetic code is true for every organism that has ever been studied. This means that all organisms are based on the same general principles. Second, all life is genetically related because over all the generations of life, DNA has been copied and handed down from cell to cell and from parent to offspring. Every living organism received its DNA from a previous generation. Going back to the origins of the first cells, every living organism is related to every other living organism.

One implication of this universal genetic code, and the mechanisms that read it, is that any cell can read the information in DNA from any other cell. For example, if a string of DNA from your best friend could be inserted into one of your cells—not an easy thing to do but not impossible—your cell would transcribe and translate the DNA from your friend. In other words, your cells *could* make proteins that normally are made by someone else.

Your immune system probably would attack the foreign proteins, but they could be produced. This unity extends to all cells, as far as we know. Furthermore, cells can transcribe and translate DNA from other species. Prokaryotic cells can even transcribe DNA of eukaryotic cells and vice versa.

The tremendous diversity of life is a reflection of the fact that the sequence of nucleotides in DNA can have nearly infinite variety. Even though the code is the same, the message that the code conveys can vary from cell to cell. Even small differences in nucleotide sequences can result in proteins with greatly different properties, leading to differences in the traits of the organism that carries them. This variation contributes to the millions of species on Earth today and explains why each individual is different in some way from every other individual. How do these differences in DNA sequence between individuals arise?

The most fundamental way that differences in DNA sequences arise is through mutations. A **gene mutation** is any change in the nucleotide sequence for a particular gene. For example, albinism can be caused by a variety of mutations, each of which changes just one nucleotide in the sequence of a particular gene that controls skin pigmentation (**Figure 7.18**). Mutations are caused by many different factors. For instance, random high-energy radiation from the Sun or other sources can cause a change in the DNA sequence. Errors in DNA replication also can cause nucleotide changes. If natural or synthetic chemicals gain access to DNA, they can change the nucleotide sequence or change the structure of existing nucleotides and cause mutations. The industrial chemicals DDT and PCBs are examples of synthetic chemicals that can change DNA; several kinds of natural oxygen molecules also may affect DNA. In multicellular organisms, mutations sometimes can occur during the processes that produce a new generation, and these

gene mutation a change in the nucleotide sequences in the DNA of a cell

mutations are inherited by the next generation. Other mutations happen only in the nonreproductive body cells. These mutations have an effect on the individual but do not affect its offspring.

A mutation can have one of three overall effects. It can cause little or no change in the overall function of an organism, it can be harmful, or it can be helpful. A mutation in a DNA sequence will cause a change in the proteins produced by the mutated cell. Life depends on an exceptionally delicate balance of chemical interactions, and a typical protein often will not function normally if its amino acid sequence is changed. For this reason, many mutations reduce the likelihood that the cell or organism will survive or live a normal life span. Many human diseases are the result of minor mutations in DNA (**Table 7.2**). Sometimes, however, mutations can lead to useful changes in protein function, or they can introduce an entirely new protein function. These positive mutations usually are small changes in one or a few nucleotides and can lead to new kinds of cells and even new kinds of organisms. For instance, the ability of mosquitoes to survive exposure to insecticides, the resistance of bacteria to antibiotics, and even complex behaviors like human speech have been associated with small mutations in DNA. In this way DNA mutations influence the diversity of life. The variation in organisms produced by mutations is acted on by natural selection to determine which mutations remain in the population and which are lost.

Scientists recently have discovered other ways that the genome of a cell or organism can change. One interesting way that DNA can change is by moving whole stretches of DNA, either within the genome of one organism or between the genomes of different organisms. Prokaryotes commonly exchange of pieces of DNA. This is how the harmless bacteria in Griffith's studies picked up DNA from the infectious strain and also became infectious. Within

Examples of Mutations in the Tyrosinase Enzyme Gene, Any One of Which Can Lead to Albinism in Humans and Other Animals.

Number of nucleotides along gene	Change in base at that location
140	G changes to A
230	G changes to A
240	G changes to A
336	C and A have been deleted
766	C changes to T
823	G changes to T
1,015	A changes to G
1,336	G changes to A

Figure 7.18 **Albinism or Lack of Pigmentation Are Caused by Mutations in the Gene for the Enzyme Tyrosinase.**
Several mutations that are changes in one nucleotide base can lead to albinism.

Table 7.2	Human diseases caused by errors in DNA			
Disease name	**Protein or enzyme involved**	**Protein function**	**Symptoms/prognosis**	**Particular groups affected**
Gaucher's disease	Glucocerebrosidase	Fat metabolism	Fat accumulates in certain tissues, causing pain, fatigue, anemia, death	Eastern European Jewish populations
Glucose galactose malabsorption	SGLT1	Absorption of small sugars from intestine	Diarrhea, dehydration, death	Prevalent in cases of familial intermarriage
Tay-Sachs disease	Beta-hexosaminidase	Breakdown of lipids in lysosomes	Damage to nervous system that includes blindness, paralysis, psychosis	Eastern European Jewish populations, French Canadians, southern U.S. Cajuns
Marfan syndrome	Fibrillin	Part of connective tissue	Affects many tissues throughout body	None known
Cystic fibrosis	Membrane chloride channel	Allows chloride ions into and out of a cell	Production of thick mucus in body tissues, which can eventually cause respiratory failure	People of European descent

an organism, genes can move from one location to another along the chromosomes, or stretches of nucleotides can switch places on chromosomes. Special enzymes cut these nucleotide sequences from one location and paste them into another. Sometimes this has no effect on transcription and translation, but other times these movements can disrupt how genes are transcribed and translated, or it can lead to genes working together in new ways. For example, viral proteins are adept at cutting into a sequence of DNA and inserting a viral DNA sequence into the host genome. Gene transfers are an important mechanism for generating genetic diversity. Humans also can be affected by gene transfers. Sometimes these effects are dramatic. For example, the transfer of a single gene can cause an XX individual to be a male. XX individuals normally are females.

Another interesting and important way that genomes can change over generations is through gene duplication. If you compare the genomes of different species, or even of different individuals, it is not uncommon to see copies of a particular gene. In some cases individuals or species can have the whole genome copied, resulting in two or more copies of every single gene. Most individuals with such gene copies are probably destined for an early death, but when gene duplication is not lethal, it can provide fertile ground for further mutations and for the evolution of new traits and species. Scientists estimate that as many as 1,500 of the approximately 25,000 human genes were duplicated at one point in human evolutionary history.

You now have explored the secrets of the master molecule of life, DNA that directs development and cellular functions in all known living organisms. The fertilized egg that started your life carried the DNA instructions that have guided your development in all the years since then. Every cell in your body carries those same instructions, and all of your abilities emerge from the proteins made following instructions from your own DNA. Scientists now are moving beyond the secret of the genetic code, striving to understand the genes that the code represents. Only when we know the function of each protein produced by the human genome can we begin to understand how the expression of proteins works together with the environment to produce a whole human being.

QUICK CHECK

1. In what way does DNA function provide evidence for the unity of life?

2. List and briefly describe three ways that DNA can change over generations.

7.7 An Added Dimension Explained
Conquering Tuberculosis

At the beginning of this chapter you were introduced to the Brontës—a family with extraordinary literary talent. Charlotte wrote the novel *Jane Eyre,* Emily wrote *Wuthering Heights,* and Anne wrote *The Tenant of Wildfell Hall.* Tragically, their lives were cut short by tuberculosis, and their brother and two other sisters also died of the disease. How does TB relate to the topics you have studied in this chapter? The link, as you will see, is the ribosome, but before we explore ribosomes, let's look at what causes TB and how it can be treated.

Tuberculosis is a bacterial infectious disease that primarily affects the respiratory system. It also can spread throughout the body to affect other organs. Early symptoms include a chronic cough, fever and chills, and weight loss. If the immune system is not successful in combating the infection, the tuberculosis bacteria can destroy, or consume, healthy tissue—which explains the name consumption. Death usually comes from widespread damage to organs, but the lungs are especially damaged.

Even in the Middle Ages physicians realized that TB is contagious. Doctors were warned not to visit sick patients. They noted that when one family member became ill, others in the family often succumbed too. Before the late 1800s everyone assumed that consumption was always fatal. In 1854 one particular TB patient was told by his doctor to move to a healthy area, eat well, rest, and get lots of fresh air. This patient followed this advice and was the first recorded patient to recover from TB. Later he built the first sanitarium for TB patients. Eventually, such retreats were established all over the world and for the next century a sanitarium was the only place where TB patients had much hope of recovery. Other treatments were invented, but none was especially successful.

The modern era of understanding TB came in the 1880s when Robert Koch identified the agent that causes TB as a tiny bacterium called *Mycobacterium tuberculosis* (**Figure 7.19**). Then in 1928 Alexander Fleming discovered the antibiotic effects of the common mold, *Penicillium.* The antibiotic penicillin was manufactured and became widely available in the early 1940s. These events ushered in the modern era of antibiotic treatment of bacterial diseases, but because penicillin is not particularly effective against *M. tuberculosis,* it did not immediately help treatment of patients with TB. Nevertheless, once it was understood that molds produce antibiotic substances, the race was on to find more of them. Albert Shatz and Jacob Waksman found that the bacterium *Streptomyces griseus* produces a substance that has an antibiotic effect on TB bacteria. You know that substance as streptomycin, the first antibiotic shown to be effective against TB. It was the first in a line of antibiotics that are collectively called the *aminoglycosides.* These include gentamycin, kanamycin, and many others.

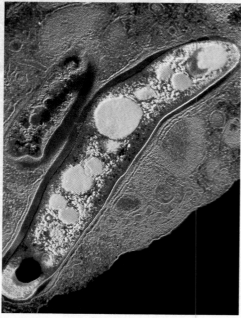

X10,000

Figure 7.19 **Mycobacterium tuberculosis.** This bacterium causes tuberculosis.

This might have been the happy ending if the saga of TB had ended with the discovery of streptomycin. It was soon apparent, however, that often within a few weeks of beginning treatment, *M. tuberculosis* become resistant to the antibiotic. Currently, TB is cured only when treatment involves *several* antibiotic drugs—up to four at once—and lasts for six months or longer. Even with this regimen, many strains of *M. tuberculosis* are resistant to the typical antibiotics used to treat TB. Deaths from TB are rising, and each year nearly 3 million people worldwide die from a disease we once thought was conquered. The modern success of antibiotics in treating bacterial infections is being undermined by the evolution of resistant strains, especially in the case of TB. Can we ever truly eliminate this disease? The only real answer is to go beyond the discovery of antibiotics to an understanding of how antibiotics work against TB and how bacteria have become resistant to them.

The modern path to understanding TB treatment starts with new studies that reveal the detailed structures of ribosomes and explain how they work. **Figure 7.20** shows a detailed molecular model of the small and large subunits of bacterial ribosomes. A ribosome is amazingly accurate at adding amino acids to the growing protein. Typically, a ribosome puts the wrong amino acid in a growing protein chain only once in every 3,000 amino acids. The first mechanism responsible for the ribosome's extreme accuracy in copying mRNA comes into play when the proper tRNA is bound to mRNA in the A-site. This arrangement changes the shape of the ribosome in a particular

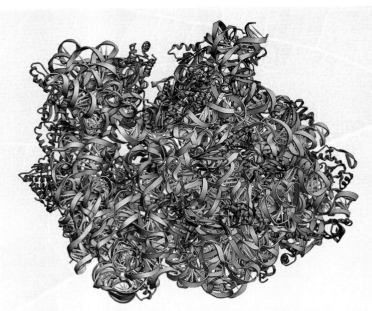

Figure 7.20 **Three-Dimensional Model of Ribosome.**

Inhibition of Protein Synthesis by Antibiotics

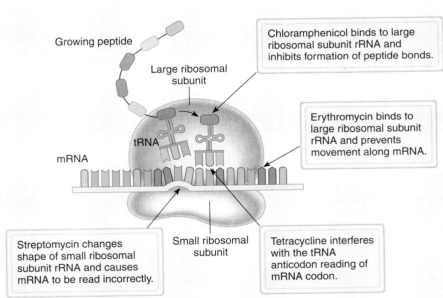

Growing peptide

Large ribosomal subunit

Chloramphenicol binds to large ribosomal subunit rRNA and inhibits formation of peptide bonds.

Erythromycin binds to large ribosomal subunit rRNA and prevents movement along mRNA.

tRNA

mRNA

Streptomycin changes shape of small ribosomal subunit rRNA and causes mRNA to be read incorrectly.

Small ribosomal subunit

Tetracycline interferes with the tRNA anticodon reading of mRNA codon.

Figure 7.21 **Antibiotics and the Ribosome.** Various kinds of antibiotics kill bacteria by preventing the correct translation of bacterial proteins at ribosomes. Human cells are not damaged by these same antibiotics because human cells have different ribosomes.

way. With the correct change of shape, translation proceeds, and a new covalent bond is formed between the existing chain and the next amino acid. If the wrong tRNA is in the A-site, however, the ribosome does not take on the right shape. The covalent bond is not formed and translation stops. Accuracy of translation also is affected by the overall shape of the ribosome. The large and small subunits can vary in shape and sometimes do not fit together perfectly. If the overall shape of the ribosome is not quite right, then the wrong tRNA can sit in one of the bays, and the wrong amino acid can be placed into the growing chain. But how does this knowledge help in the fight against TB?

Aminoglycoside antibiotics and others like tetracycline, erythromycin, and chloramphenicol kill TB bacteria by interfering with these functions of the ribosome (**Figure 7.21**). All of these drugs bind to specific regions on the rRNA and protein molecules that make up a ribosome. This changes the shape of these molecules and therefore affects the function of the ribosome. For example, streptomycin locks the overall shape of the A-site. In this configuration the bacterial ribosome makes many errors—as a result, the ribosome produces faulty proteins, and soon the bacterium dies. One interesting and important thing about these antibiotics is that they do not bind very strongly to eukaryotic ribosomes. This is why a person can take aminoglycosides and experience only minimal side effects. If these drugs were able to bind strongly eukaryotic ribosomes, they would kill human cells as efficiently as bacteria.

This information certainly will prove useful in designing new drugs that have the hope of overcoming bacterial resistance to the aminoglycoside antibiotics. Researchers will be able to learn how certain strains of bacteria are unaffected by the drugs and then will be able to design groups of drugs that will circumvent bacterial resistance.

QUICK CHECK

1. What is tuberculosis?
2. Why is the treatment of tuberculosis so difficult?

Now You Can **Understand**

Antisense for Research and Medicine

Over the next century the continued study of DNA structure and function will revolutionize medicine and yield advancements in treatments of disease. In upcoming chapters you will read about a variety of applications related to DNA, but one particularly interesting application is the development of antisense molecules. An antisense molecule is a nucleotide sequence of either RNA or DNA that has the complementary sequence to a normal gene of interest. Here is the fascinating part: remember that when a DNA sequence is used to instruct the synthesis of a protein, it is actually mRNA that takes the information to the ribosome and directs the protein synthesis process. Therefore, antisense molecules are usually DNA or RNA sequences that have the complementary nucleotide sequence to a particular mRNA strand. Some of the major uses of antisense molecules come from the fact that these will bond to their native mRNA counterparts. Why is that significant? If an antisense molecule binds to a specific mRNA molecule, *then the protein that the mRNA codes for will not be produced.* Therefore, antisense molecules are one way to shut off the expression of a particular gene. In other words, antisense molecules prevent the production of a particular protein—which could be an extremely useful tool with applications for both research and medicine. If the production of a specific protein is shut off in a cell, tissue, or organ, its function can be studied. In addition, many diseases—such as cancer and Alzheimer's—are caused by the overproduction or overactivity of certain proteins. Many companies are investigating the possibility of using antisense molecules to treat cancer and other life-threatening diseases.

What Do **You** Think?

Astrology

The science of DNA has the power to alter our understanding of our bodies and even our understanding of what it means to be human. You have learned that all of DNA's effects are indirect. DNA acts only to direct the synthesis of proteins; these interact within a cell, between cells, and with the environment to produce an entire organism. Now you know enough about DNA and cellular processes to consider other belief systems about how you become who you are. Astrology is one popular belief system. These ancient beliefs rest on the idea that the planets in Earth's solar system can influence how a human embryo develops. Thus, proponents of astrology believe that the position of the planets at the time of your birth helps to determine your traits.

There are two scientific aspects to assessing your own beliefs about astrology. The first is the simple question of whether it works. You may find, though, that this is a difficult question to answer. If you think back to Chapter 1, you cannot even start to answer the question until you define your terms and clarify your hypothesis. By *astrology* do you mean the horoscopes that run on daily Web pages? Or do you mean something more complex? Browse online websites and develop your own definition of astrology. Next, how might you now test whether astrology "works"? Would it be enough to track your own astrological charts and see if they predict your path in life? How specific are the predictions from these charts? Are they open to more than one interpretation? Assuming you can figure out how to interpret and test the predictions, would you need to see tests on lots of individuals? What level of evidence would satisfy you that astrology works?

Another approach is to see whether astrology makes sense in terms of what else is known about human development. One of the central questions for astrology is how the planets, moons, and Sun could influence either the structure of your DNA or the expression of proteins directed by your DNA. Some astrology proponents try to integrate the effects of planetary systems on DNA. In one way or another, most others propose that you are determined by factors other than DNA. At this point in your understanding of DNA and modern biology, what do *you* think about the usefulness of astrology in predicting the person you have become?

Chapter Review

CHAPTER SUMMARY

7.1 DNA Is the Master Molecule of Life

The discovery of the structure of DNA was a crucial stepping stone that has allowed tremendous advances in the understanding of life and its processes. As with other scientific breakthroughs, researchers learned from the work of previous experimenters. Experiments begun by Griffith and continued by Avery pinpointed DNA as the genetic material at a time when many thought that the genetic material had to be a protein.

7.2 Scientific Sleuthing: The Discovery of the Structure of DNA

Watson and Crick, Wilkins and Franklin, and Linus Pauling's research group all actively pursued the molecular structure of DNA, but Watson and Crick were the first to precisely identify it. Franklin's X-ray crystallography photograph of a DNA molecule was the critical bit of evidence that Watson and Crick used to build a correct model of the DNA molecule.

7.3 The Structure of DNA is a Double Helix

The DNA molecule has the shape of a double helix. It is formed of a pair of strands of DNA nucleotides and coils around like a twisted ladder. Within a cell, DNA is further coiled and condensed by proteins to form chromosomes.

nucleosome   161   chromosome   161

7.4 The Structure of DNA Contains Genetic Information

DNA transfers the information for life from one generation to the next. This is possible because the double-stranded nature of DNA allows the DNA molecule to be copied. To copy a molecule of DNA, a set of enzymes uncoils and unzips the two complementary strands and makes a new strand from each. The new strand forms, following the base-pairing rules. While the whole process requires seven or more enzymes, three of the most important enzymes are DNA helicase, DNA polymerase, and DNA ligase.

DNA helicase   163   DNA polymerase   163
DNA ligase   163

7.5 DNA Directs the Synthesis of Proteins

A stretch of nucleotides in DNA is transcribed into an mRNA molecule that leaves the nucleus. In the cytosol mRNA interacts with a ribosome; in the process of translation, the mRNA sequence of nucleotides is translated into amino acids. Three nucleotides on the mRNA strand code for each of the 20 amino acids commonly encountered in organisms. The tRNAs ferry amino acids to the growing chain of amino acids at a ribosome. When the STOP code is encountered, the ribosome spontaneously comes apart, and the complex of protein, ribosomal subunits, mRNA, and tRNAs all separate. Subsequently, the protein folds into its secondary structure and may enter the endoplasmic reticulum for further processing into a finished protein. The central dogma summarizes the way that information flows within a cell: DNA to mRNA to protein.

gene   165   codon   166
genome   165   genetic code   166
messenger RNA (mRNA)   165   transcription   167
transfer RNA (tRNA)   165   translation   167
ribosomal RNA (rRNA)   166   RNA polymerase   168
ribozymes   166   anticodon   169

7.6 DNA Helps to Explain the Unity and Diversity of Life

The genetic code is almost universal, shared by nearly all cells. The tremendous diversity of life is rooted in the infinite variety of DNA nucleotide sequences. Mutations are a source of genetic diversity, although some have no effects, and some have harmful effects. Transfer of gene sequences and gene duplication also change genomes.

gene mutation   172

7.7 An Added Dimension Explained: Conquering Tuberculosis

TB bacteria have become resistant to antibiotics that used to kill them. Detailed knowledge of how proteins are made on ribosomes will help researchers to develop new drugs to kill resistant bacteria.

REVIEW QUESTIONS

TRUE or FALSE. If a statement is false, rewrite it to make it true.

1. In the DNA molecule sugar molecules form the rungs of the ladder.
2. Rosalind Franklin provided the key information that Watson and Crick needed to solve the structure of DNA.
3. Watson and Crick did not realize the implications of the structure of DNA as a means for copying genes.
4. When DNA is replicated, DNA helicase unzips DNA, while DNA polymerase binds nucleotides into place to form a new DNA strand.
5. RNA polymerase is the enzyme that transcribes RNA into DNA.

MULTIPLE CHOICE. Choose the best answer from those provided.

6. Experiments by Avery showed that _____ were the hereditary molecules.
   a. lipids   d. nucleic acids
   b. carbohydrates   e. enzymes
   c. proteins

7. Watson and Crick's strategy was to solve the structure of DNA using
   a. careful chemical analysis.   d. X-ray crystallography.
   b. logic.   e. all of the above.
   c. molecular models.

8. In DNA the amounts of which nitrogenous bases are equal?
   a. adenine and guanine; cytosine and uracil
   b. guanine and thymine; cytosine and adenine
   c. adenine and cytosine; thymine and guanine
   d. adenine and thymine; uracil and guanine
   e. adenine and thymine; cytosine and guanine

9. If the structure of a DNA molecule is a twisted ladder, then the nitrogenous bases form the
   a. rungs.   d. top of the ladder.
   b. upright sides of the   e. bottom of the ladder.
      ladder.
   c. shelf for holding a bucket.

10. The shape of a DNA molecule is a
    a. single helix.   d. reverse spiral.
    b. double helix.   e. pleated sheet.
    c. triple helix.

11. If one strand of DNA reads ATTTCG, the complementary strand will read
    a. TAAAGC.   d. CAAAUG.
    b. ATTTCG.   e. TGGGGC.
    c. GCCCAT.

12. Which of the following molecules coil up DNA and condense it?
    a. nucleosomes that will become proteins
    b. proteins that will help form nucleosomes
    c. DNA polymerase
    d. DNA helicase
    e. ribosomes

13. In eukaryotes, transcription occurs in the
    a. cytosol.
    b. nucleus.
    c. ribosome.
    d. mitochondrion.
    e. endoplasmic reticulum.

14. Which of the following things happens in translation?
    a. Proteins reach their final form.
    b. DNA's instructions are carried out of the nucleus.
    c. mRNA's instructions are transformed into chains of amino acids.
    d. The DNA molecule is unzipped and copied.
    e. The DNA molecule prepares to divide into two new molecules.

15. The central dogma states that
    a. life comes from nonlife.
    b. chemicals interact to produce life.
    c. proteins make DNA, which makes mRNA.
    d. DNA makes proteins, which make mRNA.
    e. DNA makes mRNA, which makes proteins.

16. _____ made X-ray photographs of DNA crystals that gave Watson and Crick crucial insights into the structure of the DNA molecule and helped them win the Nobel Prize for its structure.
    a. Margaret Chase
    b. Maurice Wilkins
    c. Rosalind Franklin
    d. Linus Pauling
    e. Edwin Chargaff

17. Chromosomes are made of
    a. DNA and proteins.
    b. DNA and nucleotides.
    c. DNA and RNA.
    d. nucleotides and bases.
    e. DNA helicase and DNA polymerase.

18. The genetic code translates nucleotides into amino acids by reading the bases
    a. one at a time.
    b. two at a time.
    c. three at a time.
    d. four at a time.
    e. from the antisense end.

19. DNA ligase
    a. pulls apart strands of DNA.
    b. copies strands of DNA.
    c. repairs strands of DNA.
    d. joins two new DNA strands together.
    e. helps make mRNA.

20. _____takes the genetic code in DNA _____.
    a. mRNA, and makes identical copies for new cells
    b. mRNA, out to a ribosome in cytosol where proteins are made
    c. tRNA, and makes identical copies for new cells
    d. tRNA, out to a ribosome in cytosol where proteins are made
    e. rRNA, and makes identical copies for new cells

CONNECTING KEY CONCEPTS

1. How are the functions of DNA contained within its structure?
2. How does DNA account for both the unity and diversity of life?

QUANTITATIVE QUERY

Cytochrome *c* is an enzyme involved in the production of ATP during oxidative phosphorylation. In most species this protein is about 100 amino acids long. What would be the length in nucleotides of the mRNA sequence that codes for these 100 amino acids? If you look at the amino acid sequence in many different species of eukaryotes, about 30 of the amino acid positions in the protein are common in all species. What percentage of the total number of amino acids is this?

THINKING CRITICALLY

1. Many people have high hopes for the Human Genome Project and believe that many useful, important things will come out of this effort. Other people think that it will not be that useful, and that more money should be invested in understanding how whole organisms work rather than just what their DNA is like. Based on what you know so far about life and DNA, develop arguments in defense of both of these positions. Then decide what your own position is and defend it.

2. Discuss protein synthesis from the point of view of the recurrent theme of emergent properties.

3. Most illnesses that fall under the headings of "colds" or "flu" are caused by viruses. As you have learned, a viral particle consists primarily of a protein coat with a piece of DNA or RNA inside. If you are sick with the flu, should the doctor write you a prescription for an antibiotic such as streptomycin or tetracycline? Defend your answer.

For additional study tools, visit www.aris.mhhe.com.

Life Renews Itself

REPRODUCTION OF CELLS

Scientists Shed New Light on Aging Process

June 30, 2005

Gene mutation gradually causes devastating effects on cellular structure and function in Hutchinson-Gilford progeria syndrome

June 29, 2004

People with progerias have accelerated aging.

■ *Why do you think that the genetic disorder causing premature aging in individuals who are 20 to 30 years old is more common than the disorders causing aging in young children?*

When Brandon was born, his parents immediately suspected that something was wrong. Not only was their son awfully small, but his head was much too large for his body, and he already had two fully formed teeth. A few months later the baby hadn't grown as expected: his forehead was too wide, his nose was small and beak-shaped, and his eyes were set low in his face. He also had few eyelashes or eyebrows and was nearly bald; prominent veins pulsed beneath his scalp. Clearly, something was profoundly wrong. Soon Brandon was diagnosed with an extremely rare genetic disorder called *Wiedemann-Rautenstrauch syndrome (WRS)*. Brandon's parents were devastated—it appeared that their baby had been born old. Like a plant sprouting, blooming, and withering all in a few seconds of time-lapse photography, he would age extremely quickly and die early from diseases, such as heart disease or stroke, which usually kill aged people.

WRS is one of several genetic disorders that can be classified as **progerias,** disorders in which the body seems to age much too quickly. Some progerias strike early, aging even unborn fetuses in their mothers' wombs. Other progerias take a slower course, but all progerias guarantee a life that is shorter and less healthy than normal.

Hutchinson-Gilford progeria (HGP) is another example of these rare inherited disorders. Only one child in 6 million has HGP. Since it was first described in 1666 only about 100 cases have been identified worldwide. Children with HGP seem normal at birth, but by their second year they begin to age at an accelerated rate. Eventually, they become tiny, wizened old people (**see chapter-opening photo**). HGP children generally live only about 15 years before they too die from the afflictions of old age. People with Werner's syndrome, a third kind of progeria, seem normal for 20 or 30 years, but when they should be most active in their careers, education, and family life, the speeded-up aging begins for them. Their hair turns gray and begins to fall out, their skin becomes as stiff and dry as parchment, and they develop cancers, heart disease, osteoporosis, type II diabetes, and cataracts. Werner's syndrome is the most common kind of progeria. It strikes about 1 in 100,000 people, and its victims rarely live beyond 50 or 60 years of age.

There is no cure for any progeria, but if you live to be 70-plus years, the odds are that you, too, will experience a similar fate—only when it happens to you, the process will be called "normal aging." If given the option, most people would choose to avoid normal aging and stay young forever. Before scientists can hope to give humanity this option, they must develop testable hypotheses about what causes aging. You might think that much of this research would examine hypotheses related to cellular damage and death—and indeed research of this kind is ongoing. But you are learning about aging in a chapter on cell division and reproduction. As you will see, some evidence suggests that the premature aging syndromes just described are actually disorders of cell division. Before you can begin to understand this research, you first must learn how cells make new cells, which is the focus of this chapter. ∎

8.1 Life Continues Because Organisms Reproduce

For at least 3.8 billion years life on Earth has survived, flourished, and spread to nearly every place on the globe. Life's ability to reproduce is one reason for its success. Given favorable conditions and enough time, a single tree could make a vast forest, and a single-celled photosynthetic organism floating alone in the sea could fill Earth's oceans with its offspring. Over the course of about 2,000 years, modern humans have expanded from a population of about 2 to 3 million to more than 6 billion people.

Just as life happens within cells, the ability of organisms to reproduce depends on processes that happen inside of cells. Organisms can reproduce because the cells they are made of can reproduce. Cell division also allows a single fertilized egg to develop into a mature multicellular organism made of thousands, or millions, or billions of cells. In eukaryotic organisms—our focus in this chapter—there are just two types of cell division, known as *mitotic cell division* and *meiotic cell division*. These two types of cell division allow organisms to reproduce in one of two ways, *asexually* or *sexually*.

Most Organisms Reproduce Asexually or Sexually

All aspects of life show variation, and reproduction is no exception. Each species has its own special reproductive habits. The colorful flowers of many plants attract animals that carry pollen to other flowers and initiate reproduction. Pine trees produce cones that open to disperse pollen on the wind. Animals have intricate mating rituals that ensure that males and females get close enough together to mate. The males of Australian bower birds build and decorate twiggy dance halls to attract females. Fishes engage in elaborate displays that stimulate females to produce eggs and males to produce sperm. Even humans engage in elaborate, distinctive courtships to ensure reproduction. Other species have a different strategy. They dispense with individuals of separate sexes and sexual behaviors, and each individual independently reproduces offspring by dividing in two. Bacteria commonly reproduce in this way—but plants, fungi, many worms, sea stars, and even some snakes and lizards can reproduce from just a single parent. Despite all of this variation there are common themes in reproductive strategies. With some

progeria an umbrella term for a number of rare genetic disorders in which the symptoms of aging occur prematurely

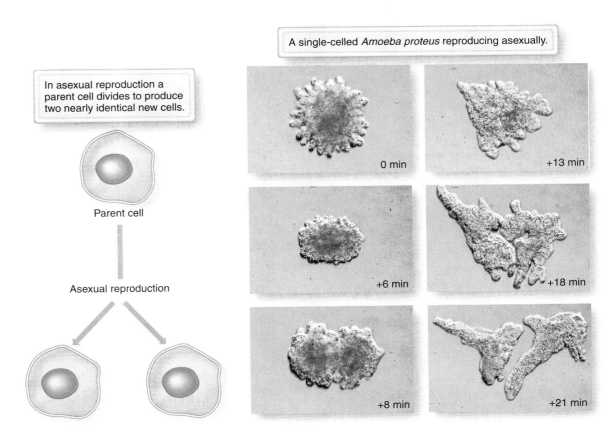

In asexual reproduction a parent cell divides to produce two nearly identical new cells.

Parent cell

Asexual reproduction

A single-celled *Amoeba proteus* reproducing asexually.

0 min

+13 min

+6 min

+18 min

+8 min

+21 min

Figure 8.1 **Asexual Reproduction.** Cells of actively growing tissues reproduce by dividing in two. So do bacteria and single-celled eukaryotes.

interesting exceptions, most organisms reproduce in just two ways: asexually or sexually.

Asexual reproduction occurs when a single parent cell or organism gives a copy of its own DNA to the offspring. An offspring produced by asexual reproduction is a genetic replica of the parent, except for any random mutations that might happen between generations. This is like copying files on your computer: you can make multiple copies of a file, and each will be identical to the original except for random copying errors. **Figure 8.1** is a simple diagram of asexual reproduction in a single-celled organism and shows that the offspring are nearly identical to the parent. Another way to describe asexual reproduction is to say that it is nature's way of making a **clone.** Because each offspring carries the same DNA as the parent, offspring also are nearly genetically identical to each other. During asexual reproduction only mutations and other copying errors can produce differences between offspring.

Prokaryotes reproduce using asexual reproduction (**Figure 8.2a**). The single-celled parent divides into two cells. Many eukaryotes also can reproduce asexually. Single-celled eukaryotes can divide into two nearly identical daughter cells. Even multicellular eukaryotes can reproduce asexually. When you take a small cutting from a rose bush or a

houseplant and grow a whole new plant, you are taking advantage of the plant's ability to reproduce asexually. Some kinds of animals, such as the aquatic hydra, reproduce in a similar way (Figure 8.2b). Even complex organisms such as flatworms or starfish can reproduce a whole organism from just a fragment of the parent's body. All of these are forms of asexual reproduction.

Sexual reproduction occurs when two parents each contribute half of the offspring's DNA. In other words, each parent contributes half of its own genome to the offspring (**Figure 8.3**). The result of sexual reproduction is an offspring with DNA that is distinctly different from that of either parent. Therefore, the offspring are *not* identical to their parents. The parents are usually male and female, but even in eukaryotes that do not have obvious male and female individuals, such as algae and fungi, sexual reproduction usually involves DNA from two different parents. There are some exceptions to this general rule. For example, because an individual plant may contain both male and female reproductive organs, some plants can mate with themselves. This is still sexual reproduction, because the offspring can have a combination of genes different from the parent's. Some animals such as worms, some snails, and even some reptiles also can do this. Sexual reproduction is only found in eukaryotes.

asexual reproduction a form of biological reproduction in which a single parent cell or parent organism produces an offspring whose genetic makeup is identical to the parent's

clone genetic replica of an individual organism or cell

sexual reproduction the method of reproduction used by multicellular organisms in which each of two parents contributes DNA to the offspring, resulting in a new, completely unique combination of DNA

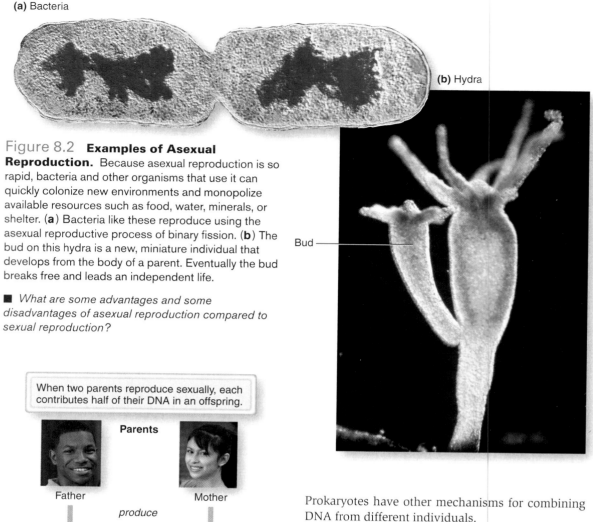

(a) Bacteria

(b) Hydra

Bud

Figure 8.2 **Examples of Asexual Reproduction.** Because asexual reproduction is so rapid, bacteria and other organisms that use it can quickly colonize new environments and monopolize available resources such as food, water, minerals, or shelter. (**a**) Bacteria like these reproduce using the asexual reproductive process of binary fission. (**b**) The bud on this hydra is a new, miniature individual that develops from the body of a parent. Eventually the bud breaks free and leads an independent life.

■ *What are some advantages and some disadvantages of asexual reproduction compared to sexual reproduction?*

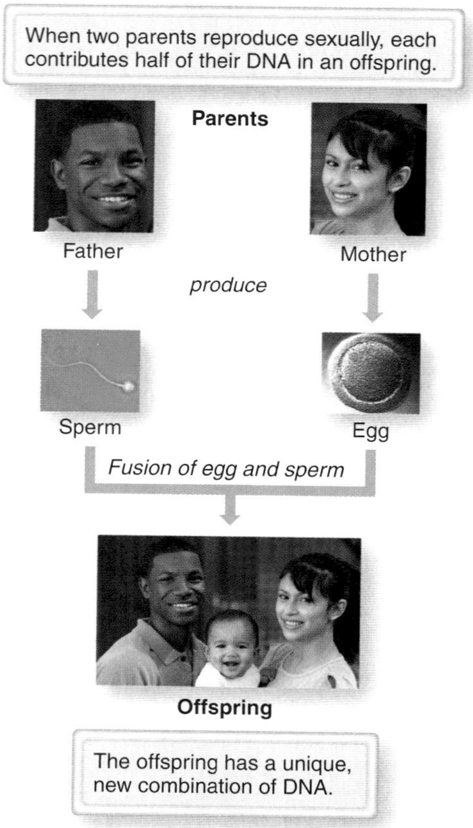

When two parents reproduce sexually, each contributes half of their DNA in an offspring.

Parents

Father

Mother

produce

Sperm

Egg

Fusion of egg and sperm

Offspring

The offspring has a unique, new combination of DNA.

Figure 8.3 **Overview of Sexual Reproduction.** Because sperm and egg have half the normal amount of DNA that is in body cells, the normal amount of DNA is restored when sperm and egg combine.

Prokaryotes have other mechanisms for combining DNA from different individuals.

Sexual reproduction produces greater genetic variation than asexual reproduction. You can see the contribution of sexual reproduction by just looking at your classmates. Each looks different. One glance at any human family will show that multicellular organisms usually do not produce offspring that are exact replicas of the parents (**Figure 8.4**), even though there are family resemblances. Variability is such a successful feature of life that all organisms have mechanisms that introduce genetic variation into their offspring. Even prokaryotes have special reproductive mechanisms that produce genetic variation, but these mechanisms do not produce as much genetic variation as sexual reproduction. The genetic variation produced by sexual reproduction leads to greater survival of a population over generations.

QUICK CHECK

1. What is the difference between asexual and sexual reproduction?

Figure 8.4 **Variation and Similarity in a Human Family.** Although the members of this family show many similar characteristics, the variation produced by sexual reproduction gives each child a blend of characteristics that is unique and different from that of either parent.

8.2 Eukaryotic Chromosomes Have Characteristics That Allow Sexual Reproduction

Some special problems are associated with providing offspring with DNA from two parents, and their solution comes from the nature of eukaryotic chromosomes. Chromosomes are complex structures composed of DNA and proteins (see Section 7.3). The total DNA of a eukarote is divided into several linear chromosomes and each eukaryotic species has a characteristic number of chromosomes (see Section 7.3). For example, humans have 46 chromosomes. The chromosomes in a given eukaryotic cell or species are not all distinctly different. Rather, they come in pairs. Let's explore this important idea further.

Alleles and Homologous Chromosomes Allow Eukaryotes to Carry out Sexual Reproduction

Each chromosome carries a sequence of genes, and each gene is a sequence of nucleotides that codes for the production of a specific protein (see Section 7.3). In turn, these proteins determine the traits of an individual. To make matters a bit more complicated, there are different versions of most traits.

Consider human skin pigmentation as an example. Human skin can range from palest pink to blackish-brown and all shades in between. The degree of brownness is determined by the amount of a skin pigment called *melanin*. A series of chemical reactions produces melanin, and each of the various enzymes involved in melanin production can exist in two or more slightly different forms. Because the different forms of a protein all play the same *role* in the chemical reactions that produce skin pigments, they all are called by the same name, for instance, tyrosinase. The different forms of tyrosinase, however, differ in their effectiveness in producing melanin. A person who carries only ineffective forms of tyrosinase will have light skin; a person who carries only effective forms of tyrosinase will have dark skin.

Of course, if proteins come in different forms, then the genes that direct protein synthesis come in different forms too. Different forms of a gene that code for different forms of the same protein are called **alleles** (**Figure 8.5**). Just as different recipes for chocolate chip cookies will produce slightly different versions of chocolate chip cookies, different alleles produce slightly different versions of the same protein. In the case of skin color, the two alleles produce protein that either produces melanin or does not. Each human has as many as four copies of the tyrosinase gene and the alleles for each one can either produce melanin or not. The end result is that

alleles different versions of genes that control the same trait but are not identical

Figure 8.5 **Alleles.**
This example shows the distribution of two different alleles for melanin genes on pairs of homologous chromosomes. One allele produces no melanin; the other allele produces melanin. Depending on the combination of alleles, a person may have dark skin or light skin.

■ *What other human traits can you think of that might show this pattern of multiple genes with two alleles in which the final expression of the trait is the combination of all gene alleles present?*

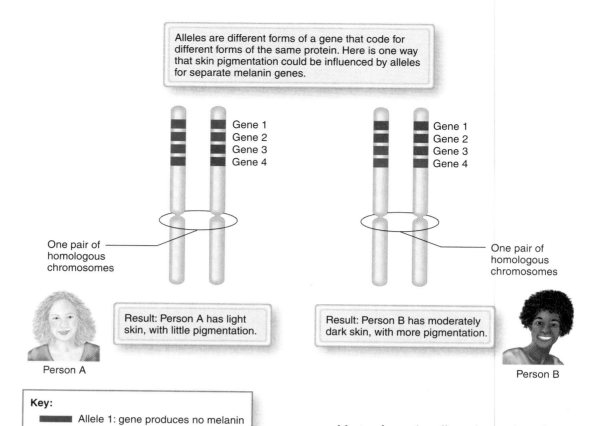

Alleles are different forms of a gene that code for different forms of the same protein. Here is one way that skin pigmentation could be influenced by alleles for separate melanin genes.

Gene 1
Gene 2
Gene 3
Gene 4

Gene 1
Gene 2
Gene 3
Gene 4

One pair of homologous chromosomes

One pair of homologous chromosomes

Result: Person A has light skin, with little pigmentation.

Result: Person B has moderately dark skin, with more pigmentation.

Person A

Person B

Key:

Allele 1: gene produces no melanin
Allele 2: gene produces melanin

homologous chromosome one of a pair of chromosomes; each carries one allele for a particular gene

diploid a cell or organism that carries two complete sets of each homologous chromosome

haploid a cell or organism that carries one complete set of each homologous chromosome

polyploid a cell or organism that carries more than two complete sets of each homologous chromosome

each individual can have very light skin, very dark skin, or something in between. This leads to an important finding: *in eukaryotes each individual has two alleles for each gene.* The alleles might be for identical forms of a protein, or they might be for different forms of the same protein. But each individual *has only two alleles for each gene.* Although there might be many possible alleles in a population of a species, in one individual the alleles for each gene come in pairs.

Where are these different alleles? In eukaryotes there are *two* versions of each chromosome and each has its own set of alleles (Figure 8.5). The two chromosome versions are called **homologous chromosomes.** For instance, the 46 human chromosomes are in 23 homologous pairs; the eight chromosomes of fruit flies are in four homologous pairs (**Figure 8.6**). This means that the 46 chromosomes in a human are *not* numbered 1 through 46. Instead, they are numbered 1 to 22 with a letter that designates each member of the homologous pair. In this text the chromosomes will be numbered 1^a, 1^b, 2^a, 2^b, 3^a, 3^b, and so on. The last pair of human chromosomes are the X and Y sex chromosomes, and you will learn much more about them in Chapter 23.

Most eukaryotic cells and organisms have two complete sets of homologous chromosomes—this state is called **diploid.** Some eukaryotic cells and even whole organisms, though, can live with just one set of homologous chromosomes. A cell or organism with one set of homologous chromosomes is called **haploid.** For most eukaryotic species the diploid state is the norm and most, if not all, multicellular eukaryotes have a diploid phase as part of their life cycle. Each species has its own characteristic diploid number, haploid number, and number of homologous pairs of chromosomes. For instance, the diploid number of cattle is 60 chromosomes in 30 homologous pairs; chickens have 78 chromosomes in 39 homologous pairs; onions have 16 chromosomes in 8 homologous pairs. Haploid numbers for these species are 30, 39, and 8, respectively. Cells or organisms with a haploid number of chromosomes play an important part in the life cycle of most, if not all, eukaryotes. For example, reproductive cells of eukaryotes, such as the egg and sperm cells, have the haploid number of chromosomes. Many plants have a complete, independent haploid organism as part of their life cycles. Chapter 19 presents more about the life histories of plants.

Yet another variation occurs when an individual has more than a diploid number of chromosomes. This is called **polyploid.** More than half of all flowering plant species probably originated as

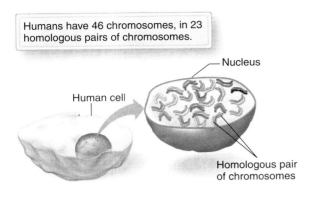

Humans have 46 chromosomes, in 23 homologous pairs of chromosomes.

Human cell

Nucleus

Homologous pair
of chromosomes

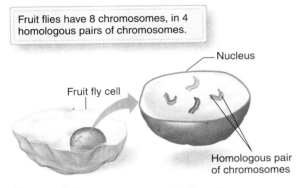

Fruit flies have 8 chromosomes, in 4 homologous pairs of chromosomes.

Fruit fly cell

Nucleus

Homologous pair
of chromosomes

Figure 8.6 **Homologous Pairs of Chromosomes.** Different organisms have different numbers of chromosomes.

■ *Why do you think the number of chromosomes is always an even number?*

polyploids, including most ferns, as well as wheat, tobacco, cotton, chrysanthemums, pansies, and bananas. Although polyploidy is rare in animals, it may have contributed to the evolution of the many species of fishes.

QUICK CHECK

1. Define the term homologous chromosomes.
2. What are alleles?
3. What is the difference between a haploid number of chromosomes and a diploid number of chromosomes?

8.3 Cell Division Is the Cellular Basis for Reproduction

When you think of reproduction, you might think of finding a partner and having a family of your own. You also might think of your pet cat having kittens, or dandelion seeds blown on the wind—or even the bacteria in the spoiled chicken salad that

you had for lunch, multiplying in your gut, making you sick. All reproduction is based on the ability of cells to divide. As the name implies, cell division involves one cell dividing into two cells. The two new cells that come from a single cell division are called **daughter cells.** If there is a need for more than two daughter cells, the division process is repeated until the number of cells needed is achieved.

In all kinds of organisms and in both asexual and sexual reproduction, cell division has three important features. First, the DNA in the original cell is copied, producing identical DNA molecules. So that the copies don't get mixed up inside the parent cell the two copies of the DNA are separated. Finally, the cell splits in a way that gives each new cell the right set of DNA. Cells can elaborate on these basic mechanisms, but all forms of cell division share these three features.

Different kinds of organisms use variations on this basic theme of cell division (**Figure 8.7**). Prokaryotes use the asexual reproductive process of **binary fission** to accomplish cell division. During binary fission the DNA in the circular chromosome of a bacterium is copied, and the copies are separated (Figure 8.7a). Each copy is attached to the inside of the cell membrane. The attachment points are located on opposite sides of the cell. Once the new chromosomes are well away from each other, the cell divides down the middle, producing two new daughter cells.

In eukaryotic cells **mitotic cell division** produces two daughter cells genetically identical to the parent cell (Figure 8.7b). Mitotic cell division is the cellular basis for asexual reproduction in eukaryotes. It also is the process that produces a multicellular organism from a single embryonic cell. For this reason, an understanding of mitotic cell division is crucial for understanding the normal development of a multicellular organism. Mitotic cell division involves copying the entire diploid genome of a multicellular parent cell, moving the copied chromosomes to opposite ends of the cell, and splitting the parent cell into two new cells. You will learn the details of this process later in this chapter.

To accomplish sexual reproduction, multicellular eukaryotic organisms make **gametes,** special reproductive cells that carry the haploid number of chromosomes, or half the DNA of normal body cells in a diploid organism. As far as scientists know, each individual uses only two kinds of gametes, not three or four or five. The two kinds of gametes can be called plus and minus, male and female, or sperm and ova. Male and female gametes are usually produced in separate individuals, making the individual male or female. In some

daughter cells the two cells produced when a parent cell divides

binary fission the process of a prokaryotic cell dividing into two; each daughter cell receives a copy of the parent's chromosome

mitotic cell division the process of cell division in eukaryotic cells that includes copying DNA, moving the copies away from each other, and dividing the parent cell into two

gamete a reproductive cell with a haploid number of chromosomes

(a) Prokaryotes divide using binary fission. **(b)** Most eukaryotic cells divide using mitosis. **(c)** Some eukaryotic cells divide using meiosis.

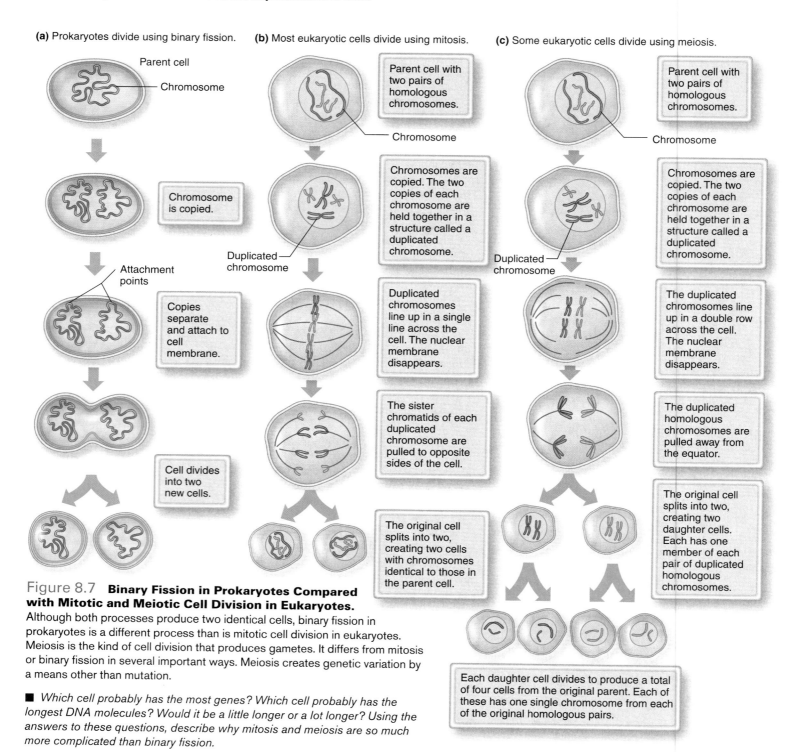

Figure 8.7   **Binary Fission in Prokaryotes Compared with Mitotic and Meiotic Cell Division in Eukaryotes.**
Although both processes produce two identical cells, binary fission in prokaryotes is a different process than is mitotic cell division in eukaryotes. Meiosis is the kind of cell division that produces gametes. It differs from mitosis or binary fission in several important ways. Meiosis creates genetic variation by a means other than mutation.

■ *Which cell probably has the most genes? Which cell probably has the longest DNA molecules? Would it be a little longer or a lot longer? Using the answers to these questions, describe why mitosis and meiosis are so much more complicated than binary fission.*

meiotic cell division the form of cell division that allocates half of the DNA in the parent cell to each daughter cell and forms gametes

cases though, male and female gametes can be produced in the same individual. An animal that can make both male and female gametes is called a *hermaphrodite.*

The production of gametes used in sexual reproduction requires that the daughter cells from a division receive half of the parental DNA. This precise portioning of DNA relies on a special form of cell division. **Meiotic cell division** is the process that allocates half of the DNA of the parent cell to each daughter cell. Meiosis produces haploid gametes (Figure 8.7c). In the process of fertilization the haploid gametes fuse to produce a new individual. In plants and animals the fused cell that results from fertilization is called a **zygote**. The zygote goes through mitotic cell division to produce an embryo, which continues to grow and develop into a mature individual (**Figure 8.8**). The single-celled zygote first divides into 2 cells. The next divisions produce 4, 16, 32, 64, etc, and more cells as the embryo that develops into a mature individual (Figure 8.8).

Table 8.1 shows the different outcomes of mitotic and meiotic cell division. The next sections describe how cells carry out mitotic versus meiotic cell division—this understanding will be useful when you consider the reproduction of organisms. Mitotic cell division will help you to understand your normal body functions and provide insights into how those functions sometimes can go awry.

QUICK CHECK

1. What is the difference in the outcomes from mitotic cell division and meiotic cell division?

8.4 The Cell Cycle Is the Orderly Progression of Cellular Activities of Eukaryotic Cells

Mitotic and meiotic cell division are crucial for the formation and development of new generations of multicellular organisms. But, of course, cells do more than just divide. Cells must find or make food, extract energy from food, sense the environment, move toward or away from things in the environment, and engage in many other activities. In multicellular organisms many cells have specialized jobs. For instance, muscle cells contract; nerve cells allow communication. While some cellular processes can occur simultaneously, cell division is strictly segregated from most other cellular functions. Cells alternate between general cellular functions and cell division. The term **cell cycle** refers to the orderly sequence of events that occur during the life of a typical eukaryotic cell (**Figure 8.9**). One cell cycle begins immediately after a daughter cell has been produced and ends when that cell splits into two new daughter cells.

The different activities of a cell during the cell cycle can be compared with the cycle of a washing machine, where one complete cycle is defined as the events that occur between the start of one wash

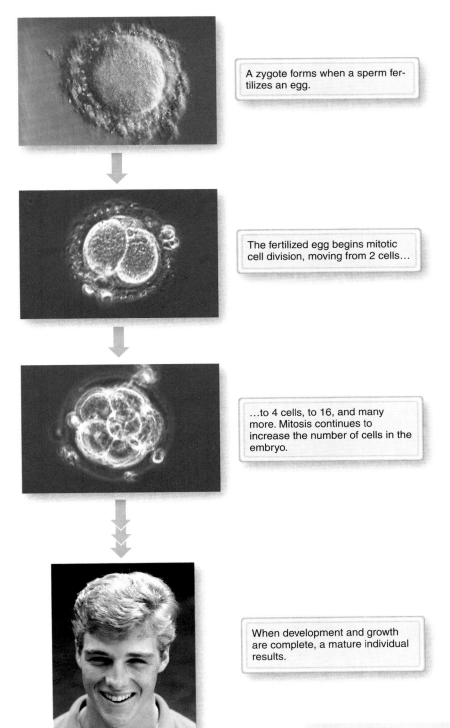

A zygote forms when a sperm fertilizes an egg.

The fertilized egg begins mitotic cell division, moving from 2 cells...

...to 4 cells, to 16, and many more. Mitosis continues to increase the number of cells in the embryo.

When development and growth are complete, a mature individual results.

Figure 8.8   Fertilization Produces a Zygote That Develops into a Mature Individual.

■ *Describe how these pictures could be your earliest baby pictures. At this stage how would you have been different from a sibling or another human?*

zygote the cell produced when two gametes fuse to form a new individual

cell cycle the orderly sequence of events in the life of a cell that includes the processes of cellular function and cell division

Table 8.1 A comparison of mitotic and meiotic cell division		
	Mitotic cell division	**Meiotic cell division**
What kinds of new cells are produced by this process?	Identical daughter cells	Gametes: sperm or eggs
How many new cells are produced from one parent cell?	2 daughter cells	4 daughter cells
How much DNA does each new cell have?	Same amount of DNA as parent cell	Half of DNA of parent cell; gametes are haploid
In humans how many chromosomes does each new cell have?	46 chromosomes	23 chromosomes, one of each homologous pair
Where does the process take place in humans?	In cells of body	Only in cells of ovaries or testes
Why is this process essential?	It allows a fertilized egg to grow and develop into a mature adult; it also allows damaged or lost tissues to be repaired or replaced. Some species can use this process to reproduce a new individual.	Meiotic cell division produces gametes, and sexual reproduction relies on the production of gametes.

G_1 (gap-1) the phase of the cell cycle between mitosis and DNA synthesis when a cell carries out diverse, mature cellular functions

S-phase (synthesis phase) the cell cycle phase immediately after G_1, when DNA is copied

G_2 (gap-2) the cell cycle phase immediately after S-phase, when the cell biochemically prepares itself for division

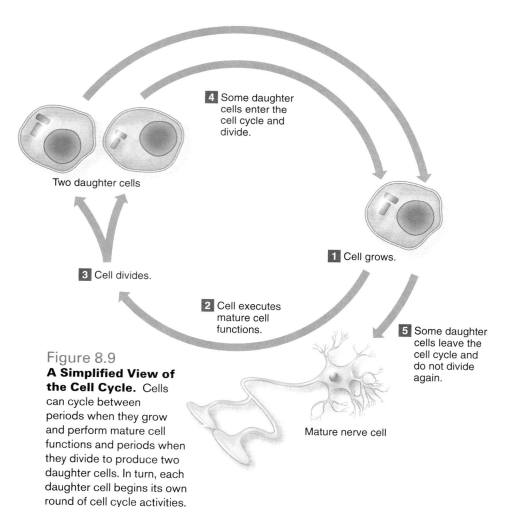

Figure 8.9
A Simplified View of the Cell Cycle. Cells can cycle between periods when they grow and perform mature cell functions and periods when they divide to produce two daughter cells. In turn, each daughter cell begins its own round of cell cycle activities.

Two daughter cells

4 Some daughter cells enter the cell cycle and divide.

3 Cell divides.

2 Cell executes mature cell functions.

1 Cell grows.

5 Some daughter cells leave the cell cycle and do not divide again.

Mature nerve cell

and the beginning of the next. Between these points, the washing machine goes through a controlled sequence of activities (**Figure 8.10**). After the water flows in, the clothes are sloshed around and washed by agitation. Then the water drains out and a new phase, the rinse, begins, during which the clothes are sloshed around in fresh water and soap is rinsed out. Then the machine enters the spin phase and much of the water is expelled from the clothes. Sometimes there is a second rinse and spin phase. Then the machine stops, you take out the clean, wet clothes, and the washer is ready to begin the cycle again. The cell cycle is an orderly progression of events, much like this.

Both mitotic and meiotic cell division occur as part of the cell cycle—it will be easier to understand both if we begin with mitotic cell division. A later section focuses on the cell cycle in relation to meiotic cell division.

The Cell Cycle Has Five Phases

Each cell cycle is divided into five phases (**Figure 8.11**). Each phase is identified by initials representing the activities that happen during each phase. The phases are called G_1 (gap-1), S (DNA synthesis), G_2 (gap-2), M (mitosis), and C (cytokinesis). As you can see from Figure 8.11, a cell moves from one phase to another in a cyclic fashion. There is not really a beginning or ending phase to the cell cycle, but let's consider G_1 as the beginning of one cycle, and start our introduction to the cell cycle there.

G_1 is short for "**gap-1,**" a term that refers to a "gap" in the cellular activities between the readily observed events of mitosis and DNA synthesis. Early scientists did not have the tools to discover what the cell did during this time, and so they called it a gap. We now know that G_1 is the only phase when a cell does things that are *not* related to cell division. During G_1 a cell has the time to grow larger, to develop a complex shape, and to carry out mature functions.

The remaining four phases of the cell cycle are part of cell division. Each phase is dominated by one major task that must be completed so that cell division may occur. The fact that four phases are needed for cell division shows you that this is not a simple process. The first thing a cell must do to get ready to divide is copy its DNA. The beginning of DNA synthesis marks the end of G_1 and the beginning of **synthesis,** or **S-phase.** During S-phase, each chromosome is copied, and two complete copies of each chromosome result. After S-phase comes **gap-2,** or G_2, another period that puzzled early scientists because the cell seemed to be inactive at this time. It is now known that during G_2 cells make important

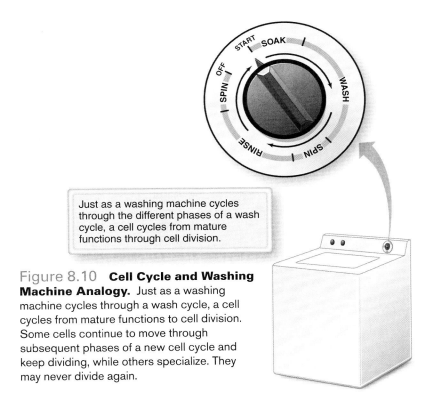

Just as a washing machine cycles through the different phases of a wash cycle, a cell cycles from mature functions through cell division.

Figure 8.10 Cell Cycle and Washing Machine Analogy. Just as a washing machine cycles through a wash cycle, a cell cycles from mature functions to cell division. Some cells continue to move through subsequent phases of a new cell cycle and keep dividing, while others specialize. They may never divide again.

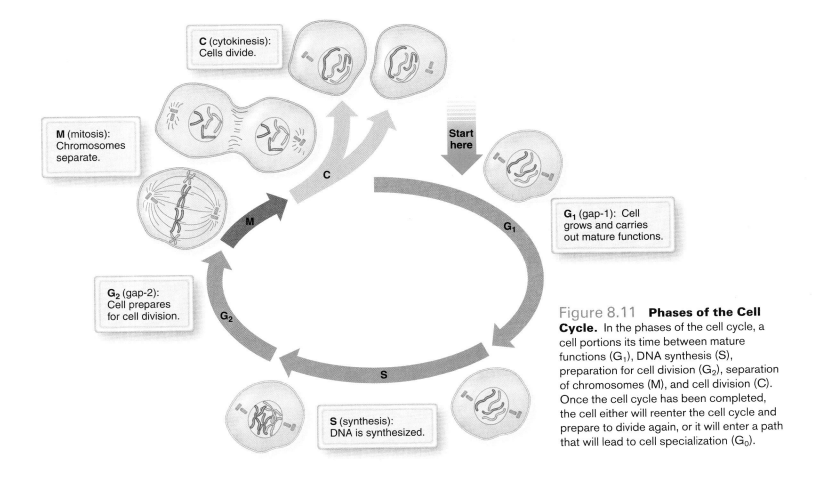

C (cytokinesis): Cells divide.

M (mitosis): Chromosomes separate.

Start here

G_1 (gap-1): Cell grows and carries out mature functions.

G_2 (gap-2): Cell prepares for cell division.

S (synthesis): DNA is synthesized.

Figure 8.11 Phases of the Cell Cycle. In the phases of the cell cycle, a cell portions its time between mature functions (G_1), DNA synthesis (S), preparation for cell division (G_2), separation of chromosomes (M), and cell division (C). Once the cell cycle has been completed, the cell either will reenter the cell cycle and prepare to divide again, or it will enter a path that will lead to cell specialization (G_0).

biochemical preparations for cell division; you will learn about these preparations shortly. In the context of the cell cycle, **mitosis,** or **M,** is the name for the fourth phase of the cell cycle. Mitosis refers only to the phase of the cell cycle when the copies of the chromosomes separate. During mitosis the two copies of each chromosome are moved to different parts of the cell. This process is so critical to successful division that the terms *cell division* and *mitosis* are sometimes used interchangeably. Finally, in the fifth and last phase, called **cytokinesis,** the cell divides into two new cells.

What happens after cytokinesis depends on the role that each daughter cell will play in the tissue or organism. If a daughter cell will replicate its DNA and divide again, it returns to G_1. In contrast, cells that remain for a long or indefinite time without dividing are said to be in G_0, which stands for **G-zero.** With this introduction to the phases of the cell cycle, you are ready to delve into each phase a bit more deeply.

G_1 and G_0 Require Cell-Specific Functional Proteins

G_1 and G_0 are periods when the cell performs important functions and, of course, proteins are required to accomplish these jobs (see Section 3.6). Some cell types rest in G_1 for a while, producing metabolic enzymes, organelles, and other molecules and structures needed for growth. Later they return to S-phase. Other cells remain in G_0 indefinitely. They must synthesize additional proteins needed for their particular job. For instance, skin cells produce a special type of cytoskeleton protein called *keratin,* nerve cells produce a cytoskeleton protein called *neurofilament,* and muscle cells produce the contractile proteins, actin and myosin. Liver cells produce a variety of special enzymes, and red blood cells produce the oxygen-carrying protein, hemoglobin.

S-Phase Is Devoted to the Duplication of DNA

At the end of G_1 a cell prepares for S phase by making special cell division proteins. All of the proteins needed to copy DNA—such as DNA helicase, DNA polymerase, and DNA ligase—must be synthesized. Also, the proteins that are part of the chromosome, such as those that form nucleosomes, must be made. S-phase cannot begin until a supply of these proteins is available. During S-phase a cell makes a full copy of each molecule of DNA in the nucleus, using nucleotides brought in from the cytosol. DNA molecules are uncoiled during synthesis and are in their dispersed state rather than

their condensed state. The process of copying all of a cell's DNA is slow. It can happen in a few hours, or it can take as much as 12 hours to complete.

When S-phase is over, a cell has *twice* the amount of DNA than it had before S-phase started. Put another way, after S-phase a cell has *two* complete sets of all nucleotide sequences that make up its genome. Consider a cell that, before S-phase, has four chromosomes that form two homologous pairs. After S-phase the cell has two copies of chromosome 1^a, two copies of chromosome 1^b, two copies of chromosome 2^a, and two copies of chromosome 2^b. You would think that after S-phase this cell would have eight chromosomes. In some ways it does, but for historical reasons scientists say that after S-phase the cell still has four chromosomes. How can this be?

The answer is related to another problem. To ensure that each daughter cell will receive a full set of the correct genetic instructions, the dividing cell must have a mechanism that keeps track of each chromosome and its copy. The two copies of each chromosome that exist after S-phase are not set free in the nucleus to float about separately. Instead, the identical copies of each chromosome are joined together at a point along the length of the two copies. That means that, in a sense, there are four chromosome structures before S-phase and four chromosome structures after S-phase. But after S-phase each chromosome has *twice* the DNA as its single—uncopied—counterpart had before S-phase. **Figure 8.12** diagrams what the single (Figure 8.12a) and joined chromosomes (Figure 8.12b) would look like if you could see them individually. In this example, the nucleus contains four chromosomes during G_1. After S-phase the cell actually contains eight chromosomes, but they are in the form of four joined sets.

To help to relieve some of the confusion, this text will call a structure made of joined chromosomes that has two copies of DNA a **duplicated chromosome** (Figure 8.12). As long as the two copies of DNA are joined in this duplicated chromosome, each DNA copy will be called a **sister chromatid.** A cell does not have sister chromatids during G_1, or during cytokinesis; they are present only after S-phase, during G_2, and through half of mitosis.

During G_2 the Cell Prepares for Mitosis

Although DNA replication is complete at the end of S-phase, the cell is not yet ready to segregate the chromosome copies and divide. A few more tasks must be accomplished before cell division, and these remaining tasks are completed in G_2. In some

mitosis (M) the cell cycle phase immediately after G_2, when the duplicated chromosomes are separated so that each daughter cell can receive an exact copy of the DNA from the parent cell

cytokinesis the cell cycle phase immediately after mitosis, when the cell with separated chromosomes is split into two daughter cells

G_0 (G-zero) the phase of the cell cycle immediately after cell division; some cells are permanently mature and never return to cell division

duplicated chromosome a joined set of chromosomes; each has two copies of a strand of DNA, or a pair of sister chromatids

sister chromatids the name given to the identical chromosome copies that are joined together after DNA replication in S-phase

cases a cell may need to continue growing to accumulate enough cellular material to divide between two daughter cells. As a part of this growth, organelles like mitochondria, ribosomes, endoplasmic reticulum, and Golgi bodies are produced during G_2. Of course, these organelles grow and are copied during G_1, but G_2 is another important growth phase. In addition, some complex organelles must be dismantled during G_2. For example, if a cell has a complex shape maintained by cytoskeleton proteins, these proteins must be taken apart. Usually, a cell does not maintain a complex shape during the actual process of division.

Another important activity during G_2 is to check that DNA replication has been done correctly. Cells in G_2 have biochemical mechanisms to ensure that their DNA has been copied only once. Cells also have mechanisms to check that the copies are accurate. If significant errors are found, the cell stops in its path toward division. If these errors can be fixed, then the cell can proceed to divide—but if the errors cannot be fixed, the cell either will remain stuck in G_2 or will die.

Mitosis Is the Process That Separates Sister Chromatids of Duplicated Chromosomes

If a cell passes through all of G_2, it is ready to physically divide into two cells. The next step on the path to cell division is to separate the sister chromatids that thus far have been joined together. **Figure 8.13** summarizes mitosis. Refer to this figure all through this section. You can see in this overview that during mitosis the duplicated chromosomes line up at the middle of the cell. Then the sister chromatids separate and are moved to opposite ends of the cell so that they can be distributed properly to daughter cells.

During the First Part of Mitosis the Copied Chromosomes Coil and Condense   The first thing that happens during mitosis is condensation of the duplicated chromosomes, which means that the chromosomes are coiled as tightly as possible (see Section 7.3). During condensation the two sister chromatids remain stuck together. The place on the chromosome where two sister chromatids are joined is called the **centromere.** A photograph showing all of a cell's chromosomes in this duplicated, condensed state is called a **karyotype** (**Figure 8.14**). Karyotypes often are used to determine if a developing embryo or fetus has the right number and configuration of chromosomes. Another important process in mitosis that happens at about the same time as chromosome condensation is the disintegration of

(a) Single, dispersed chromosomes before they are copied.

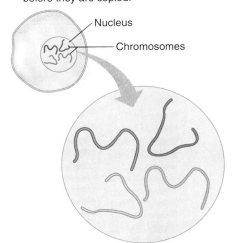

Nucleus

Chromosomes

(b) Duplicated, dispersed chromosomes after S-phase. Each copied chromosome is now made of two sister chromatids.

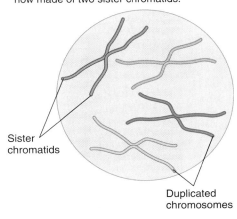

Sister chromatids

Duplicated chromosomes

Figure 8.12   Joined Chromatids. After S-phase duplicated chromosomes are made of sister chromatids that are joined at their midsections.

the nuclear membrane. This allows the chromosomes to be moved during the next stages of mitosis.

Later in Mitosis the Spindle Apparatus Pulls the Sister Chromatids Apart   Once the copied sister chromatids are coiled neatly, what does the cell do with them? Keep in mind that the point of this whole process is to get one complete set of DNA—one complete set of chromosomes—into each of two daughter cells. To do this, the two identical copies of each original chromosome must be pulled away from one another. This is made easier because the chromosomes are neatly coiled, but how does the separation actually happen?

The simple answer is that the duplicated chromosomes line up along the middle of the cell. Then the sister chromatids are pulled apart, leaving two individual chromosomes on opposite sides of the cell. The process that accomplishes this segregation of chromosomes starts with the centrioles. In animal cells (see Figure 4.14) the **centrioles** are a pair of cylindrical, ridged organelles that look a bit like cut pieces of celery stalks. During mitosis, the centrioles move to opposite sides of the cell, and out of each grows a temporary structure called the **spindle apparatus,** an array of protein fibers that spreads out from each centriole like a burst of fireworks (**Figure 8.15a**). The fibers converge on the equator of the cell where they interact with the sister chromatids. Each sister chromatid is contacted by spindle fibers from one side of the cell (Figure 8.15b). The spindle fibers pull the duplicated chromosomes into a line along the middle of the cell. They continue to pull on the sister chromatids, causing them to separate.

centromere the point on a duplicated chromosome where the two sister chromatids join

karyotype a photograph of chromosomes taken when each chromosome is duplicated; a karyotype shows sets of two completely condensed sister chromatids

centriole a pair of ridged, cylindrical organelles found in animal cells that give rise to the spindle apparatus during cell division

spindle apparatus an array of fibers that spread out from each pole of a cell and overlap at the equator of the cell, where they attach to sister chromatids

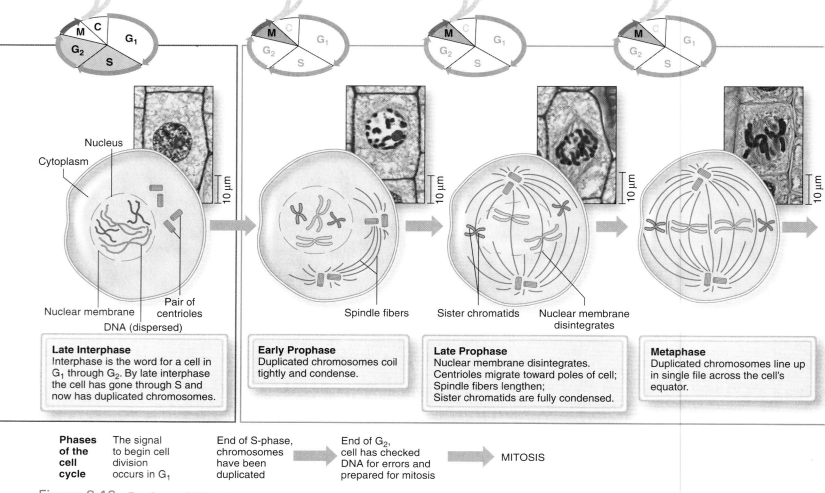

Nucleus

Cytoplasm

Nuclear membrane

DNA (dispersed)

Pair of centrioles

Spindle fibers

Sister chromatids

Nuclear membrane disintegrates

Late Interphase
Interphase is the word for a cell in G_1 through G_2. By late interphase the cell has gone through S and now has duplicated chromosomes.

Early Prophase
Duplicated chromosomes coil tightly and condense.

Late Prophase
Nuclear membrane disintegrates.
Centrioles migrate toward poles of cell;
Spindle fibers lengthen;
Sister chromatids are fully condensed.

Metaphase
Duplicated chromosomes line up in single file across the cell's equator.

Phases of the cell cycle

The signal to begin cell division occurs in G_1

End of S-phase, chromosomes have been duplicated

End of G_2, cell has checked DNA for errors and prepared for mitosis

MITOSIS

Figure 8.13 **Review of Mitosis.**

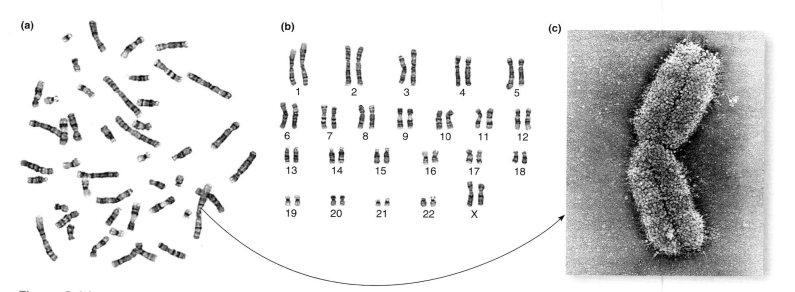

(a)

(b)

1 2 3 4 5

6 7 8 9 10 11 12

13 14 15 16 17 18

19 20 21 22 X

(c)

Figure 8.14 **A Human Karyotype.** (**a**) A stained preparation of human chromosomes. (**b**) To make a karyotype, the photograph of stained chromosomes is enlarged and individual chromosomes are cut out of the photo and arranged in pairs according to size. This is a karyotype of a female, so there are two X chromosomes.

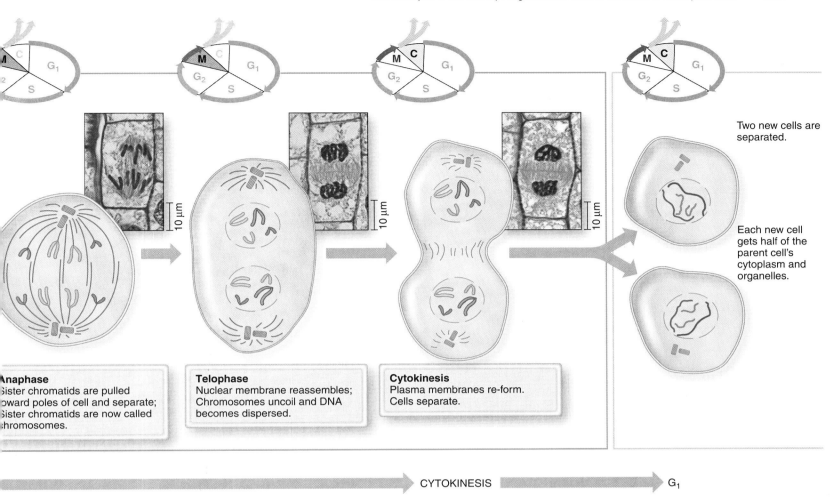

Anaphase
Sister chromatids are pulled toward poles of cell and separate; Sister chromatids are now called chromosomes.

Telophase
Nuclear membrane reassembles; Chromosomes uncoil and DNA becomes dispersed.

Cytokinesis
Plasma membranes re-form. Cells separate.

Two new cells are separated.

Each new cell gets half of the parent cell's cytoplasm and organelles.

CYTOKINESIS   G₁

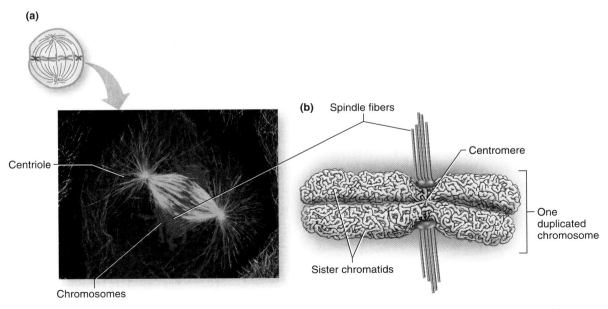

Figure 8.15 **Separating Chromosomes.** (**a**) Spindle fibers extend from centrioles to the equator of a dividing cell, where they align the duplicated chromosomes in a single line. (**b**) Enlarged view of the attachment of spindle fibers to chromatids.

The spindles keep pulling until the chromatids are segregated at opposite poles of the cell. Don't forget that each "chromatid" is actually a complete, single chromosome. At this point, the dividing human cell has 92 individual chromosomes—but now instead of being joined into 46 duplicated chromosomes, they are separated into two complete sets of 46 chromosomes that have moved to opposite ends of the dividing cell.

Now that you have read about the process of mitosis, return to Figure 8.13 and review the whole process. Notice that Figure 8.13 includes some special terms used to describe the stages before and during mitosis: interphase, prophase, metaphase, anaphase, and telophase (see Figure 8.13). *Interphase* is the name given to G_1, S-phase, and G_2. During G_1, the single chromosomes are dispersed, and at the end of S-phase, the duplicated chromosomes are dispersed. *Prophase* refers to the beginning of mitosis, when the duplicated chromosomes are becoming condensed. During *metaphase* the chromosomes are lined up along the equator, and you can distin-guish the sister chromatids that make up each dupli-cated chromosome. *Anaphase* and *telophase* repre-sent the process of pulling the sister chromatids apart to opposite sides of the cell. While these terms for the phases of mitosis can be useful, it is impor-tant to remember that the whole process is continu-ous. Cells do not pause at these particular phases of mitosis. The entire cell cycle is not complete until cells leave mitosis and go through cytokinesis.

Cytokinesis Splits the Parent Cell into Two Daughter Cells

After G_2 and mitosis a cell has two sets of key organelles and double the diploid amount of DNA. All that remains of the cell cycle is to split the parent cell into two daughter cells. This process is called cytokinesis. **Figure 8.16** shows cytokinesis in both plant and animal cells. As you can see, this is one process that proceeds differently in these two differ-ent kinds of cells. In animal cells special proteins form a band around the cell's equator. These proteins

(a) In dividing animal cells a contractile ring of microfilaments forms a cleavage furrow between cells in cytokinesis.

(b) In dividing plant cells a line of vesicles merges at the midline of the dividing cells. A cell plate forms between the two new cells and eventually two primary cell walls divide the new cells.

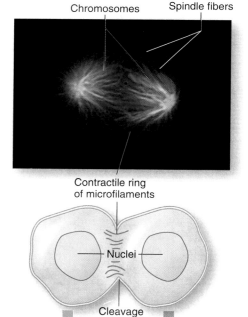

Chromosomes
Spindle fibers
Contractile ring of microfilaments
Nuclei
Cleavage furrow
Daughter cells

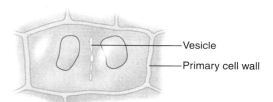

Vesicle
Primary cell wall

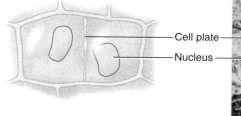

Cell plate
Nucleus

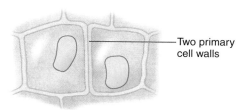

Two primary cell walls

10 μm

Figure 8.16 **Cytokinesis Differs in Animal and Plant Cells.** (**a**) In animal cells a contractile ring of microfilaments produces cytokinesis. (**b**) Cytokinesis in plant cells includes formation of new cell membranes and new cell walls.

■ *Describe the similarities and differences between plant and animal cell division.*

are similar to those used by muscle cells to cause muscle contraction. They contract and pinch around the center of the cell until the cell membrane meets. Where different regions of the membrane meet, they fuse and the cell splits into two daughter cells (Figure 8.16a). Plant cells take a different approach (Figure 8.16b). Not only must plant cells make a new cell membrane between the two new daughter cells, but also they must make new cell walls. This is accomplished by simply building these structures in the middle of a cell that has gone through mitosis. This job is accomplished by the plant cell's Golgi complexes that make little packets of material, called *vesicles,* that line up at the cell's equator. Soon they transform into the cell membranes and cell walls that divide the two new daughter cells. Then the nuclear membrane re-forms and the two new daughter cells proceed to G_1.

The Progress of the Cell Cycle Can Be Controlled in G_1 and G_2

How is a cell's progress through the cell cycle controlled? When and how does a cell know to move from one phase of the cell cycle to the next? In eukaryotic cells the signal to begin the cell division process comes in G_1. Once a cell leaves G_1, the rest of the cell cycle is carried out automatically unless errors halt the cell cycle. If serious mistakes in the DNA are found in either G_1 or G_2, cell division is stopped. The cell cycle can be stopped late in G_1 or in G_2, and for this reason these two phases of the cell cycle also are called *cell cycle checkpoints.*

The message to begin the division process and proceed from G_1 to S-phase actually comes from outside of the cell, in the form of signal molecules (see Section 4.3). Special kinds of signal molecules called **growth factors** start the cell division process. Growth factors are made and released by other cells. There are many different kinds of growth factors; each is released by a specific kind of cell and acts either on that same kind of cell or on other types of cells. Growth factors are important during embryonic development when cells are undergoing rapid and complex patterns of mitotic cell division. Growth factors also are important in some tissues and body systems of adults. For instance, cells in the lining of the gut and the immune system both have intense mitotic activity. In both of these systems cells in adult individuals continue to go through cell division.

Disorders of Mitotic Cell Division Can Lead to Serious Diseases

The process of mitotic cell division is critical for normal development. Mitotic cell division is strictly controlled throughout development so that cells divide at the right times and in the right places. Disorders of cell division can lead to serious diseases, such as cancer. Cancer is essentially a disease in which cell division occurs independent of any controls. Uncontrolled cell division produces masses of tissue growing in places where they do not belong. Hundreds of thousands of people die of cancer each year. Chapter 12 discusses cancer in detail.

Other disorders can result when there is not enough cell division. Dwarfism and other small-stature syndromes have complex causes, but limited cell division is certainly part of the problem. As you will read at the end of this chapter, even disease syndromes that seem to have little to do with cell division, such as the premature aging syndromes, are partly related to problems with enzymes involved in cell division.

QUICK CHECK

1. Define the term cell cycle. List the five stages of a eukaryotic cell's life cycle and briefly describe what happens in each.
2. How are the sister chromatids of the duplicated chromosomes separated during mitosis?
3. What role do growth factors play in cell division?

8.5 Sexual Reproduction Involves the Fusion of Haploid Gametes

For life to persist, it must reproduce itself and this means that cells as well as whole organisms must reproduce. Some multicellular organisms can reproduce asexually, but sexual reproduction is the norm among multicellular eukaryotic organisms. Sexual reproduction produces the genetic variation across generations that helps individuals and species to survive in a variety of environments. During sexual reproduction the parents produce haploid gametes that combine to form a new, unique individual. Gametes contain half of the DNA found in other body cells, and meiotic cell division accomplishes this precise reduction of the amount of DNA. Learning about meiotic cell division will be easier if you first understand the genetic requirements of sexual reproduction. What are the requirements for a gamete?

Gametes Contain Half the DNA of a Parent's Body Cells

It is essential that gametes carry exactly the right amount of DNA. A human baby must have 46 chromosomes, and a fruit fly maggot must have 8 chromosomes. Although small variations in nucleotide

growth factor a type of signal molecule that causes a cell to progress through the cell cycle and divide

sequences are acceptable, offspring need the exact number of genes and chromosomes required to form that particular kind of organism. Therefore, the mechanism of sexual reproduction must have a reliable way to give the full set of chromosomes to each offspring.

In other words, an offspring usually cannot develop normally *if it has too much or too little DNA.* **Figure 8.17** explores this idea, using a cell with six chromosomes as an example. Here the six chromosomes are numbered according to their homologous pairs—1^a, 1^b, 2^a, 2^b, 3^a, 3^b, and so on—reinforcing the concept that each parent carries two copies of each numbered chromosome. If gametes carry the same amount of DNA as all other body cells, each new generation produced by the union of egg and sperm will have *twice* the DNA of its parents (Figure 8.17a). Such a drastic change in the amount of genetic material occasionally can produce a whole new species, which has happened with many plant species. In most

other organisms and especially in most animals, having twice the amount of DNA is not likely to produce a viable offspring. Figure 8.17b shows another way that the chromosomes could be abnormal. Here each gamete has half the chromosomes of the parents, but the offspring do not get the right combination of chromosomes. In this example the offspring has two copies of chromosomes 1 and 2, one copy of chromosomes 5 and 6, but no copies of chromosomes 3 and 4. Clearly, this will not produce a normal, viable offspring. Therefore, each gamete must have exactly half the DNA of the parent cell. It also must have a portion of DNA that exactly complements the DNA in gametes from its mate. How do multicellular organisms accomplish this delicate balancing act? How does each gamete get not only the right amount of DNA, but also the right pieces of DNA to match the gametes from other individuals? The answer is that each gamete must carry one member of each homologous pair of chromosomes.

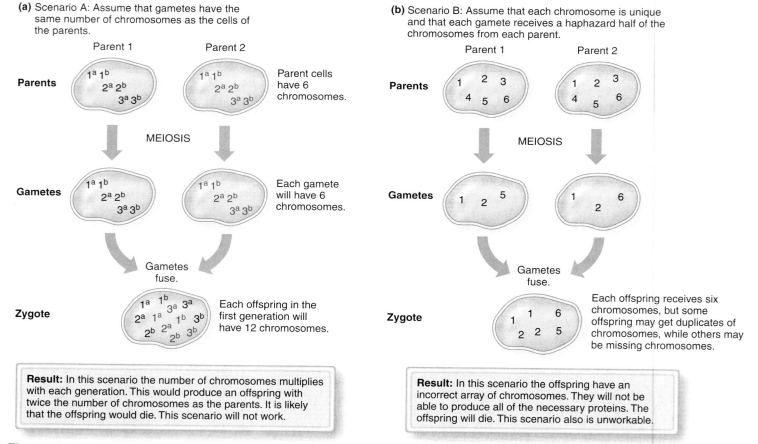

(a) Scenario A: Assume that gametes have the same number of chromosomes as the cells of the parents.

(b) Scenario B: Assume that each chromosome is unique and that each gamete receives a haphazard half of the chromosomes from each parent.

Result: In this scenario the number of chromosomes multiplies with each generation. This would produce an offspring with twice the number of chromosomes as the parents. It is likely that the offspring would die. This scenario will not work.

Result: In this scenario the offspring have an incorrect array of chromosomes. They will not be able to produce all of the necessary proteins. The offspring will die. This scenario also is unworkable.

Figure 8.17 **Two Unworkable Ways to Portion Chromosomes into Daughter Cells.** (**a**) Scenario A shows what would happen if gametes had the same number of chromosomes as the cells of their parents. As you can see, this scenario is unworkable. (**b**) In scenario B unique chromosomes are portioned haphazardly into gametes. This scenario is also unworkable.

You can now appreciate how homologous chromosomes assure that the offspring produced by sexual reproduction will have the correct set of genes. Because the 46 human chromosomes are really 23 homologous pairs, each gamete (each egg or sperm) receives just one member of each homologous pair—one "a" chromosome or one "b" chromosome for each chromosome number. **Figure 8.18** shows how homologous chromosomes are distributed when gametes are formed. Each sperm and egg will have its own sequence of 1^a, 2^b, and so on. Of course, the 1^a from the ovum will not necessarily carry the same alleles as the 1^b from the sperm; these labels are just convenient ways of keeping track of different homologous chromosomes in a given cell.

Next let's look at the process of meiosis itself. If you keep in mind that meiosis produces gametes that have one of each homologous chromosome, meiosis will be easier to understand.

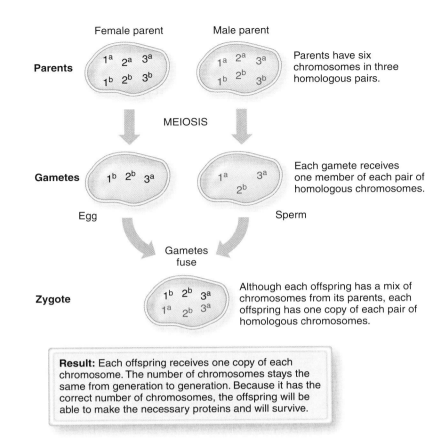

Figure 8.18   **Homologous Chromosomes Provide a Means of Portioning Chromosomes Correctly into New Cells.** With homologous chromosomes the chromosome number stays the same from generation to generation. The offspring receives the correct number of chromosomes, so it will be able to make all the proteins and cellular structures that are necessary for its life.

QUICK CHECK

1. What is a gamete?
2. What is the purpose of sexual reproduction?
3. Describe some of the problems that can occur with chromosome number in sexual reproduction.

8.6 Meiotic Cell Division Produces Gametes

Meiotic cell division is the process that accomplishes the parceling of homologous chromosomes into gametes. Just like mitotic cell division, meiotic cell division starts with S-phase and ends with cytokinesis. The difference is that instead of mitosis, meiotic cell division includes a phase called **meiosis.** Like mitosis, meiosis separates the duplicated, coiled homologous chromosomes, but it does so in a way that produces gametes instead of daughter cells identical to the parent cells. While mitosis can occur in nearly any body cell, meiosis happens only in specialized cells in an organism. In many animals the specialized cells that go through meiosis are in the gonads: the ovaries and testes. In plants the cells that produce gametes also are located in reproductive organs such as flowers or cones. Cells that will give rise to gametes have the same DNA as any other body cells, but the process of meiosis reduces that DNA by half.

If you think about the homologous pairs of chromosomes in a body cell, a simple way to produce gametes might occur to you. A cell could organize its chromosomes by homologous pairs, move one of each pair to opposite ends of the cell, and then divide the cell in two. This would be an "easy" way to make gametes, but it is *not* the way that it is done. Instead, just as in mitotic cell division, meiosis occurs only after a cell has copied all of its DNA during S-phase. This means that, just like mitosis, meiosis starts with duplicated chromosomes. This small but important point is crucial to understanding meiosis.

An important difference between mitosis and meiosis is that meiosis includes *two rounds* of cell division (**Figure 8.19**). The original cell ultimately gives rise to *four* daughter cells rather than the two daughter cells produced by a single round of mitotic cell division. The first meiotic division (meiosis I) separates duplicated homologous chromosomes, and the second meiotic division (meiosis II) separates sister chromatids. With this overview in mind,

meiosis the process of separating condensed coiled chromosomes that results in gamete formation

Figure 8.19
Overview of Meiosis.
Meiosis has two cell divisions. In the first division copied homologous chromosomes line up in a double row at the equator of the dividing cell, and one copy of each homologous chromosome is moved to each of two daughter cells. In the second division of meiosis sister chromatids separate, giving the gametes produced by meiosis half the number of chromosomes that were found in the parent cell.

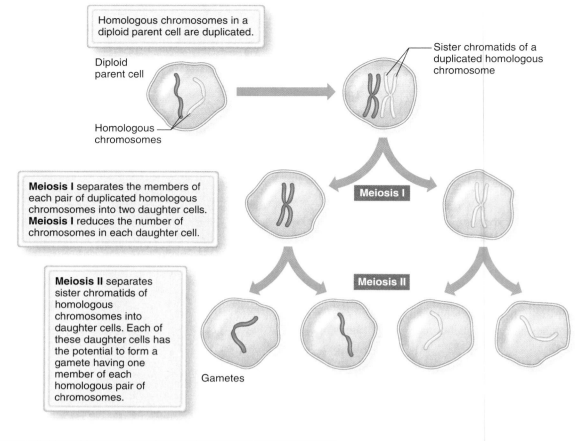

Homologous chromosomes in a diploid parent cell are duplicated.

Diploid parent cell

Homologous chromosomes

Sister chromatids of a duplicated homologous chromosome

Meiosis I separates the members of each pair of duplicated homologous chromosomes into two daughter cells.
Meiosis I reduces the number of chromosomes in each daughter cell.

Meiosis I

Meiosis II separates sister chromatids of homologous chromosomes into daughter cells. Each of these daughter cells has the potential to form a gamete having one member of each homologous pair of chromosomes.

Meiosis II

Gametes

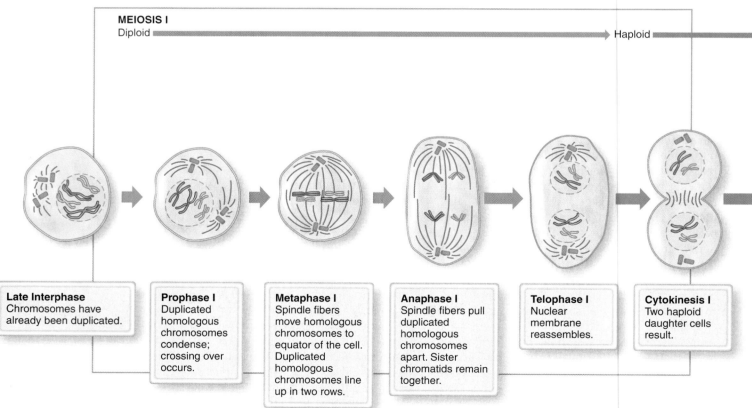

MEIOSIS I
Diploid ➤➤➤➤➤➤➤➤➤➤➤➤➤➤➤➤➤➤➤➤➤ Haploid ➤➤➤

Late Interphase
Chromosomes have already been duplicated.

Prophase I
Duplicated homologous chromosomes condense; crossing over occurs.

Metaphase I
Spindle fibers move homologous chromosomes to equator of the cell. Duplicated homologous chromosomes line up in two rows.

Anaphase I
Spindle fibers pull duplicated homologous chromosomes apart. Sister chromatids remain together.

Telophase I
Nuclear membrane reassembles.

Cytokinesis I
Two haploid daughter cells result.

Figure 8.20 **Details of the Process of Meiosis.** Although meiosis looks superficially similar to mitosis, meiosis only occurs in reproductive tissues, such as the ovaries or testes of animals and the ovaries or anthers of plants. Meiosis has two rounds of cell division and produces haploid gametes.

■ *Look back at Figure 8.4. Can you clarify how crossing over relates to the question of why children from the same parents look different?*

let's walk through the details of the process of meiosis using **Figure 8.20** as a guide.

During Meiosis Duplicated Chromosomes Line Up Differently Than Duplicated Chromosomes in Mitosis

After S-phase and G$_2$ the duplicated chromosomes in a cell undergoing meiosis condense to form the X-shaped structures in Figure 8.15. Just as in mitosis, the duplicated chromosomes in meiosis are formed of sister chromatids. Just as in mitosis, the nuclear membrane disintegrates, the microtubule spindle apparatus forms, and the duplicated chromosomes line up along the cell's equator. Up to this point the two processes are identical, but the next step in meiosis is different from what happens in mitosis.

One way to understand how meiosis parcels out homologous chromosomes is to note how the duplicated chromosomes are lined up between the fibers of the spindle apparatus. Recall that during mitosis, the duplicated chromosomes form one long line along the cell's equator and spindle fibers are attached to each sister chromatid. In contrast, during meiosis the duplicated chromosomes form *two lines* of duplicated homologous pairs. It's sort of like a dance line in which each duplicated chromosome is lined up across from its homologous partner. During this phase of meiosis duplicated homologous chromosomes are pulled away from each other and end up on opposite ends of the cell. Look carefully at the diagram of this in Figure 8.20 in which the cell is simplified and only has two chromosomes. At each pole of the cell is a complete duplicated chromosome, one from each homologous pair. Next, cytokinesis divides the original cell into two daughter cells. This is the first round of cell division during meiosis, sometimes called *meiosis I.* Now think about what chromosomes are in each daughter cell from this division: each cell from this first division has *two duplicated chromosomes,* one from each homologous pair. Therefore, each daughter cell still has the diploid number of chromosomes and the normal amount of DNA.

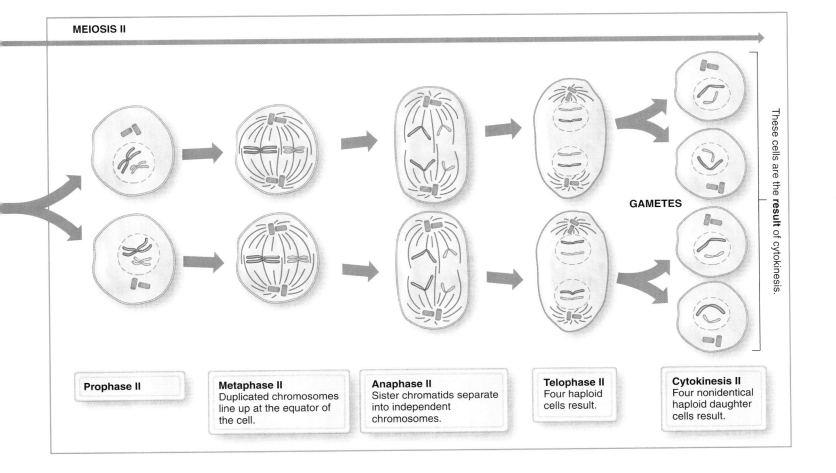

MEIOSIS II

GAMETES

These cells are the **result** of cytokinesis.

Prophase II

Metaphase II
Duplicated chromosomes line up at the equator of the cell.

Anaphase II
Sister chromatids separate into independent chromosomes.

Telophase II
Four haploid cells result.

Cytokinesis II
Four nonidentical haploid daughter cells result.

Meiosis Involves a Second Round of Cell Division That Gives One Allele to Each Daughter Cell

Are the two daughter cells from the first round of meiotic cell division now ready to be gametes? If each of those daughter cells fused with another cell just like it, would the fertilized egg have the normal amount of DNA for that species? Of course, the answer is no. The daughter cells from this first division of meiosis contain *duplicated* chromosomes, each made of two sister chromatids. If two of these cells fused, the fertilized egg would have duplicated chromosomes, not single, sister chromatids. Therefore, another step is required to produce gametes from these cells with duplicated chromosomes.

During the second division of meiosis (*meiosis II*), the sister chromatids are pulled apart, resulting in cells with single chromosomes. This process is much like mitosis. The duplicated chromosomes line up in single file along the equator. The spindle fibers that stretch across the cell attach to the centromere on each sister chromatid (Figure 8.20). As the spindle fibers shorten, the sister chromatids are pulled apart and are moved to opposite sides of the cell. Cytokinesis occurs, and the final result is four daughter cells, each with single chromosomes and the haploid amount of DNA. With some further chemical and structural refinements, some or all of these daughter cells can become gametes. If two such gametes (one from each parent) combine, the resulting offspring will have the normal number of chromosomes, two of each homologous pair.

Meiosis ensures that each offspring that results from sexual reproduction will have a unique set of genetic traits. **Figure 8.21** shows how each gamete can have a unique selection of the homologous chromosomes from the parent organism. Just four

possible gametes are shown, but there are many more possibilities. In organisms with even larger numbers of chromosomes, the number of possible different gametes is quite large. The traits of a particular offspring will depend on which alleles were contributed by the particular gametes that fused to form it.

Crossing Over Is Another Way That Meiosis Produces Genetic Variation

Meiosis produces individual gametes genetically different from one another because their homologous chromosomes carry different alleles. When gametes fuse, the individual offspring have different combinations of alleles. One of the surprises of sexual reproduction is that during meiosis alleles from the "a" and "b" homologous chromosomes can be traded between chromosomes. This amazing mechanism is called crossing over, and it happens early during the first round of meiotic division when the duplicated homologous chromosomes are just starting to line up. As the duplicated homologous chromosomes line up along the cell's equator—1ᵃ across from 1ᵇ, 2ᵃ across from 2ᵇ, and so on—each pair of alleles is directly lined up too. The two homologous chromosomes actually touch; at the places where they touch, alleles can break out and trade places (**Figure 8.22**). Enzymes cut the DNA and splice it back together. This remarkable phenomenon exchanges alleles between the homologous chromosomes that face each other at the cell's equator. Crossing over results in eggs or sperm that have chromosomes with different sequences of alleles than those in the rest of the parent's cells. **Crossing over,** then, is the process in which a series of alleles on homologous chromosomes are physically switched, producing a new combination of alleles on chromosomes.

crossing over the process in which the alleles on homologous chromosomes switch places, producing new combinations of alleles on each chromosome involved

Figure 8.21
Reassorting Chromosomes into Gametes. Here is just one example of the variation that results when gametes are formed. Note that none of the gametes has the same assortment of chromosomes.

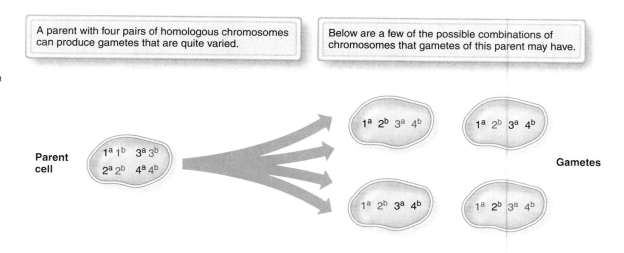

A parent with four pairs of homologous chromosomes can produce gametes that are quite varied.

Below are a few of the possible combinations of chromosomes that gametes of this parent may have.

Parent cell

1ᵃ 1ᵇ 3ᵃ 3ᵇ
2ᵃ 2ᵇ 4ᵃ 4ᵇ

1ᵃ 2ᵇ 3ᵃ 4ᵇ

1ᵃ 2ᵇ 3ᵃ 4ᵇ

1ᵃ 2ᵇ 3ᵃ 4ᵇ

1ᵃ 2ᵇ 3ᵃ 4ᵇ

Gametes

Errors Sometimes Occur During Meiosis

This description is the way that meiosis should happen, but errors can occur. One kind of error results when either homologous chromosomes or sister chromatids do not separate properly—in either meiosis I or meiosis II. The result of this kind of error is that a gamete—and so a zygote—can have too few or too many chromosomes. For example, a gamete—and so a zygote—could be missing one of the homologous versions of chromosome 2, or could have three copies of chromosome 2. Most zygotes with this kind of abnormality will die. In some cases, though, the individual survives, but has developmental abnormalities. **Table 8.2** gives examples of meiotic errors that produce a viable but abnormal child. Down syndrome is one of the most common examples, and occurs when an error in meiosis produces an individual with three copies of chromosome 21 (**Figure 8.23**), rather than the normal two. This is why Down's syndrome is also called *Trisomy 21*. This abnormality can be seen in a karyotype and is one reason why some women choose to have the karyotype of their unborn child examined. People with Down syndrome have an altered physical appearance. Notably, they have flattened faces, small noses, and upward slanted eyes. They lack muscle tone and their physical and mental development is not normal. Individuals with Down syndrome age at an accelerated rate and seldom live past age 50.

Either a sperm or an egg can carry an extra chromosome, but most cases of trisomy 21 result from eggs that did not form properly. The frequency of Down syndrome increases as the age of the mother increases. A 30-year-old mother has 1 chance in 1,000 of giving birth to a Down child, while a 45-year-old mother has 1 chance in 50. Unfortunately, it is still not known why older mothers have an increased risk of bearing a Down syndrome baby.

Mitosis and Meiosis Are Similar, But They Differ in Important Ways

Because mitosis and meiosis are similar in some ways, it can be easy to confuse them. Let's go over the important features of each process. Both are forms of cell division, but they serve different functions for a multicellular organism. Mitotic cell division results in two daughter cells identical to the parent cell. In a multicellular organism mitosis is used to turn the single fertilized egg into the thousands to trillions of cells in an adult organism. Mitotic cell division also is important for replacing cells that die or are damaged in an adult. In

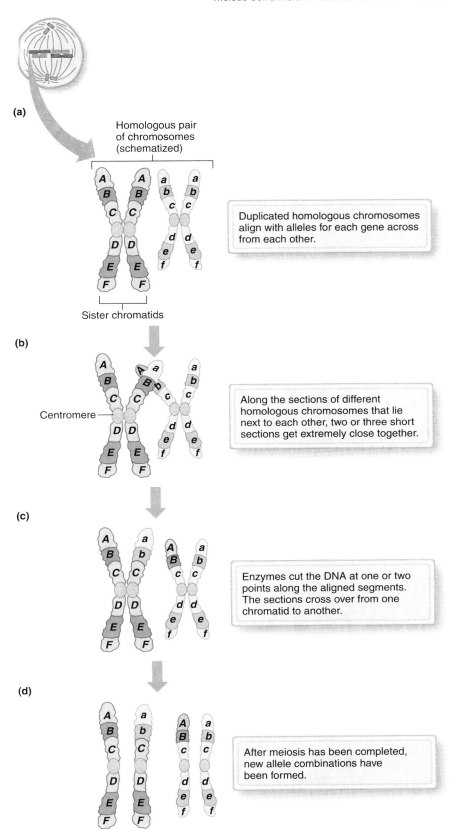

(a) Homologous pair of chromosomes (schematized)

Duplicated homologous chromosomes align with alleles for each gene across from each other.

Sister chromatids

(b) Centromere

Along the sections of different homologous chromosomes that lie next to each other, two or three short sections get extremely close together.

(c) Enzymes cut the DNA at one or two points along the aligned segments. The sections cross over from one chromatid to another.

(d) After meiosis has been completed, new allele combinations have been formed.

Figure 8.22 **Crossing Over.** When duplicated homologous chromosomes line up at meiosis I (**a**), sections can overlap (**b**), and be exchanged (**c**). Crossing over results in chromosomes with new combinations of alleles (**d**).

Table 8.2 Errors of meiosis

Name of disorder	Chromosome abnormality	Description
Down syndrome	Trisomy* 21	Mental retardation, heart malformations, dementia similar to Alzheimer's
Edward's syndrome	Trisomy 18	Mental retardation, abnormalities in head, feet, and kidneys
Patau's syndrome	Trisomy 13	Mental retardation, a variety of abnormalities, death usually by 3 months of age
Turner's syndrome	XO, lacking one sex chromosome	Lowered IQ, abnormal genitals, heart abnormalities

*Trisomy means having three homologous copies of that chromosome, rather than the normal two.

contrast, meiosis occurs only in cells that produce gametes. Meiosis results in gametes that have half of the DNA of the parent cell. Gametes from different parents can combine to form a new organism. Keeping these functional differences in mind will make it easier for you to understand how the events of mitosis and meiosis differ. The comparison of mitosis and meiosis in **Figure 8.24** will help distinguish between the two processes. Notice that both mitosis and meiosis start with a cell in G_1 that enters S-phase and duplicates its DNA.

■ The first major difference arises in the way that the chromosomes line up along the

equator after the chromosomes have condensed. In mitosis the duplicated chromosomes form a single line, but during the first division of meiosis, the duplicated chromosomes line up as homologous pairs and form a double line.

■ The process of pulling the chromosomes to either end of the cell is similar during mitosis and meiosis. The one difference is that after mitosis, a group of single chromosomes is segregated at each pole of the cell, while in meiosis the first chromosome separation results in a group of duplicated chromosomes at each pole.

■ In both mitosis and meiosis, cytokinesis divides the cell into two cells after the first chromosome separation.

■ After this first cell division, the two processes follow different paths. Cytokinesis marks the end of the division cycle for a cell going through mitosis; some daughter cells will enter G_1 until it is time for them to divide again, while others will enter G_0. In meiosis each daughter cell from the first cell division immediately goes into a second cell division. This time the duplicated chromosomes in each daughter cell line up single file along the middle of the cell and the duplicated

Figure 8.23 **Down Syndrome.** Down syndrome results when there are three copies of chromosome 21. This is a karyotype of a male. Notice the difference in the size of X and Y chromosomes.

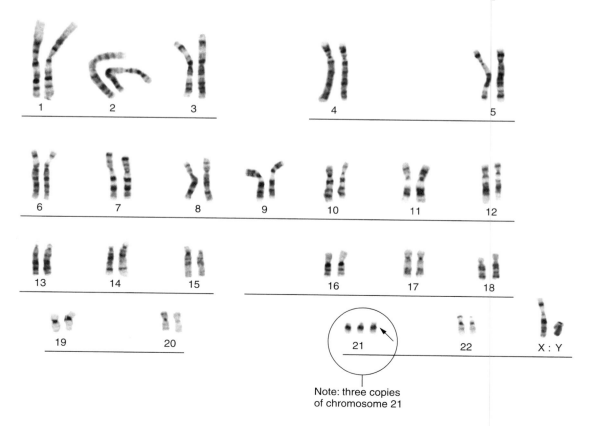

1 2 3 4 5

6 7 8 9 10 11 12

13 14 15 16 17 18

19 20 21 22 X : Y

Note: three copies of chromosome 21

	MITOSIS	MEIOSIS
Cell cycle phase at the start of the process:	G₁ cell enters S-phase and duplicates its DNA	
Arrangement of duplicated homologous pairs of chromosomes on the equator:	Single line	Double line
Spindle apparatus and separation of duplicated chromosomes:	Similar	
End of separation of duplicated chromosomes:	Single chromosomes are at each pole of the cell.	Duplicated chromosomes are at each pole of the cell.
Cytokinesis:	Similar	
Next event:	End of cell division cell. Cell returns to G₁ or enters G₀.	Meiosis II follows. Duplicated chromosomes line up in single file at equator; sister chromatids are pulled apart.
Daughter cells:	Two cells result and each has a diploid number of chromosomes.	Four cells result and each has a haploid number of chromosomes.
Function:	Mitosis allows development and growth of individuals.	Meiosis produces gametes with the correct number of chromosomes, and so ensures that offspring have the correct number of chromosomes.

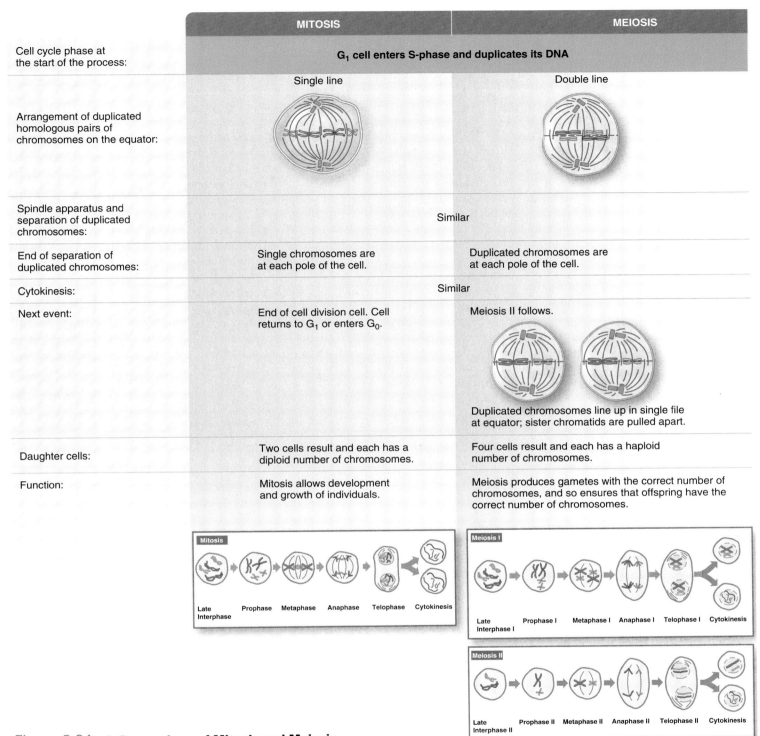

Figure 8.24 **A Comparison of Mitosis and Meiosis.**

chromosomes (sister chromatids) are pulled apart. Each of these daughter cells then divides in two. In mitosis two daughter cells are produced, and each has the diploid number of chromosomes, while in meiosis four daughter cells are produced, and each has the haploid number of chromosomes.

QUICK CHECK

1. How do the duplicated, coiled chromosomes get separated during meiosis so that gametes are formed?
2. What is crossing over?
3. What are the purposes of mitosis and meiosis?

8.7   An Added Dimension Explained
Are Cell Proliferation and Aging Related?

In the opening scenario of this chapter, you read about the accelerated aging and shortened life span of people with genetic progerias. But, of course, the material in this chapter dealt with cellular mechanisms that allow cells to reproduce and develop. The repeated rounds of mitosis, also called **cell proliferation,** increase the number of cells in a multicellular organism and result in the normal development of an embryo. What does cell proliferation have to do with aging?

The first question to ask is: What is aging? Aging is a complex process—there is no single answer, but researchers are putting together a picture of aging. As individual cells age, their functions deteriorate. One reason is that cells produce proteins that do not have quite the right amino acid sequences and therefore cannot carry out their functions efficiently or correctly. The accumulation of cellular problems eventually gets so great that a cell dies. In a multicellular organism the accumulation of cellular aging eventually compromises the functions of a tissue or an organ, and the whole organism deteriorates and dies. In simple terms, aging is the accumulation of cellular abnormalities that leads to a decline in functions and to the eventual death of the individual cell or organism. DNA directs protein synthesis and changes in DNA are the central cause of the deterioration of cell functions that contribute to the aging process.

In humans, the outward consequences of DNA and protein abnormalities related to aging are hair that turns gray, skin that becomes wrinkled as layers of fat beneath the skin grow thin, stiff joints, and weakened muscles and bones. DNA damage is associated with specific diseases, such as heart disease, osteoporosis (loss of bone strength), diabetes (loss of the ability to control blood-sugar levels), dementia (loss of mental functions), and cancers.

Of course, this combination of DNA damage, altered proteins, and cell death occurs throughout life. Even in young children, mature cells in the lining of the gut and the skin, for example, go through degenerative changes and die. But because young children retain many cells that can divide and replace those that are lost, children are resilient to these changes. You may have marveled at how quickly the skin of a youngster heals after a cut or scratch. Children have cells that divide rapidly, providing a supply of new cells to replace those that are lost or damaged. As a person grows older, the numbers and types of cells that can be replaced declines. The decline happens more rapidly in some tissues than others. For instance, nerve cells are rarely replaced; muscle cells can

be replaced under some circumstances, while skin and cells that line the gut are more readily replaced. Nevertheless, as a person ages, replacement of lost cells is slower.

Part of the evidence for a change in cell proliferation with age comes from studying cells in a **cell culture**—a population of cells growing in a laboratory glass dish or tube. The value of cell culture studies is that they allow a researcher to study the effects of complex factors one at a time, isolated from all the known and unknown factors that interact within the body. One kind of human cell commonly studied in culture is called a *fibroblast*. Fibroblasts secrete proteins such as collagen and elastic fibers in skin, muscles, and tendons. These give tissues their characteristic structures and resilience.

Fibroblasts divide frequently throughout a person's life, but their life span appears to have an inherent limit. If you take fibroblasts from a young child and grow them in a culture dish, they will divide only a certain number of times (**Figure 8.25**). Early in the life of the culture, each human fibroblast will divide and produce daughter cells. As the culture ages, the proportion of

cell proliferation repeated rounds of mitotic cell division that increase the number of cells in an organism, or in a part of an organism

cell culture a population of cells growing in a laboratory dish or tube

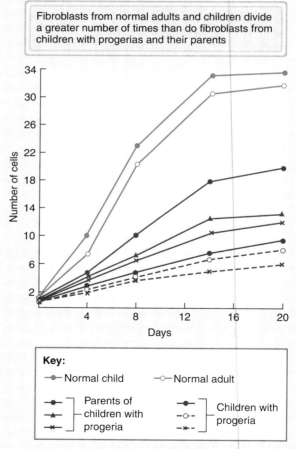

Fibroblasts from normal adults and children divide a greater number of times than do fibroblasts from children with progerias and their parents

Key:
- Normal child
- Normal adult
- Parents of children with progeria
- Children with progeria

Figure 8.25   **Growth Rates of Fibroblasts in Cell Cultures.** Fibroblasts of children with progerias don't divide normally or increase as fast as fibroblasts of normal children.

cells that divide declines, and each round of cell division takes longer. Eventually, after about 50 divisions, no cells divide and the whole culture becomes "old." The culture itself then dies, because without cell division the population is not renewed. Even more interesting, if you culture fibroblasts from an older person, their "life span"—that is, the number of times they divide—is shorter than the life span of fibroblasts from a younger person. These findings suggest that a slowing of cell proliferation may be one cause of the signs of aging.

Studies of the fibroblasts from progeria victims provide some of the most intriguing evidence that aging is related to declines in cell division. Fibroblasts from children with progeria divide for only a limited time when grown in cell culture, and the division of these fibroblasts resembles that of fibroblasts from a person who is 60-plus years old. One possible hypothesis to come from such studies is that the symptoms of progeria are partially caused by the limited ability of cells to divide.

Other lines of evidence support the hypothesis that both normal aging, and the abnormal aging seen in progerias, are related—at least in part—to limits on cell division. For example, some studies have shown that normal older people and young people with progeria had about 10% more abnormal genes than normal young people. The gene abnormalities found in these studies were *not* inherited, but instead were the result of damage that occurs to DNA as it is copied again and again, and as it is exposed to various agents such as chemicals and radiation. Among these abnormal genes, about a third carried instructions for manufacture of proteins involved in the control of cell division. The same cell division genes were affected in older people and young people with progeria. These findings add to the growing conclusion that problems with cell division are part of aging.

Another indication that flawed cell proliferation is one cause of aging comes from the isolation of the inherited genes that cause Werner's syndrome and Hutchinson-Gilford progeria (HGP). In Werner's syndrome at least one defective allele is for a DNA helicase. This enzyme unwinds DNA so that it can be copied (see Section 7.4), and checked for errors and repaired.

The presence of defective helicase probably impairs DNA duplication and thus impairs cell proliferation. It also allows errors in DNA to accumulate; these errors cause the cell to function abnormally. Because the signs of Werner's syndrome are so similar to those of normal aging, scientists are searching for evidence that as a person gets older, the gene for DNA helicase becomes defective in many cells. In HGP at least one important defective allele is for a cytoskeleton protein called *lamin*. This protein plays many roles in maintaining the structure of the nucleus, but it also is involved in DNA replication. Therefore, HGP also has a link to an abnormal form of a protein that is important for cell proliferation.

Could the link between cell proliferation and aging be used to help extend longevity? The answer seems to be "Yes," and the way to do it seems easy enough: find a way to stimulate cell proliferation or switch cells in G_0 back to G_1 and aging will be abolished! By now you should know that nothing biological is ever that simple, and that any simple manipulation will certainly produce unexpected results. In the case of aging and cell proliferation, there is one worrisome problem: cancer. When cells proliferate abnormally, cancer can result—so increasing cell proliferation could be a cause for concern, especially in older people who already run a greater risk of developing cancer. If old, damaged cells were stimulated to proliferate, cancer might be an even greater risk. At present it is not possible to say just how serious the risk of cancer would be if researchers developed treatments to increase cell proliferation. Most scientists, however, believe that there are tradeoffs between increased cell proliferation and cancer risk. You will have to stay tuned to the scientific developments in the field of cell proliferation, aging, and cancer, before you will be able to say whether premature or natural aging could be alleviated by treatments that increase cell proliferation.

QUICK CHECK

1. What is aging?
2. Discuss aging and function in human fibroblast cells.

Now You Can **Understand**

Neural Tube Defects

Modern medicine has reduced the impact of many diseases and disorders, but birth defects still affect a large number of babies. One of the most common birth defects involves a structure called the *neural tube,* an embryonic structure that develops into the spinal cord and brain. The neural tube begins development as a sheet of cells. These multiply, and the edges of the sheet curl up and fuse to form a tube. If the tube does not close properly at any point along its length, then part of the brain and/or spinal cord will not develop normally. This is called a *neural tube defect.* One well-known example is a defect in the spine called *spina bifida.* In some cases neural

tube defects are minor; in other cases they are extreme. Prior to the early 1970s, babies born with spina bifida, or other neural tube defects, were likely to die—since then surgical techniques have been developed that often can correct the problem. Nevertheless, neural tube defects still can be fatal.

In the United States about 1 in every 1,000 babies is born with a neural tube defect. In the past decade evidence has increased that if mothers take folic acid supplements during pregnancy, the risk of neural tube defects can be cut in half. Even though this is encouraging news, it is important to find out what causes these developmental disorders. Recent studies have suggested that abnormal patterns of cell division are involved. The evidence comes from studies of mice showing neural tube defects similar to those in humans. In a mouse with spina bifida, the embryo shows a lack of cell division in the tail end of the neural tube early in development. There also will be too much cell division in that same area later in development. Although folic acid is necessary for normal DNA replication, more studies need to be done to find out folic acid's role and how it relates to abnormal patterns of cell division.

It's a Boy. It's a Girl. Or Maybe Both.

You have read about how errors during meiosis can lead to trisomy 21, or Down syndrome. Most meiotic errors result in an embryo that is not viable and dies early. There are some meiotic errors, though, that do survive and develop. Recall that the sex chromosomes, X and Y, are not exactly homologous pairs. Nevertheless, during meiosis they line up as if they were. If meiosis proceeds normally, a person with XX is female, while a person with XY is male. If these chromosomes do not separate properly during the two cell division cycles of meiosis, then the embryo will have an abnormal number of sex chromosomes. There are several ways that this can happen, and the traits of a person born with such a syndrome depend on what chromosomes he or she has. A person can have three X chromosomes or have XXY, XYY, or YYY. In extreme cases the person's secondary sex characteristics never fully develop. In other cases, the physical differences are so subtle that the person does not guess that he or she has an extra sex chromosome. Frequently, the condition is discovered with a karyotype.

What Do **You** Think?

Pros and Cons of Embryonic Stem Cell Research

Embryonic stem cells are often in the news and are highly controversial. Researchers claim that studies of human embryonic stem cells could provide cures for the effects of many diseases like cancer, heart disease, stroke, and Alzheimer's disease. Some active political groups, though, are strongly against embryonic stem cell research. Are you familiar with the arguments on both sides? Do you know the current state of regulations that control stem cell research? Do you know what a stem cell is?

In this chapter you have read that cells in a young embryo divide rapidly and spend little time doing mature functions. As you get older, more of your cells are devoted to mature functions and fewer are able to divide. A practical side effect of this bit of biology is that in adults many damaged tissues cannot be replaced or repaired. Imagine that you had a supply of embryonic cells and could use them to generate new tissues and organs. Imagine if damaged heart tissue, or brain tissue, could be replaced, even in an elderly person. This is the hope of researchers in the field of embryonic stem cells because, under the right conditions, embryonic stem cells may be coaxed to produce any normal adult tissue.

The political hot potato, of course, is how researchers can obtain embryonic stem cells. How can researchers get a supply of human embryonic stem cells to experiment with and develop these potentially life-saving techniques? Human embryonic stem cells can come only from human embryos. Although some researchers are investigating other sources for stem cells, such as cells in the placenta that have many of the same qualities, currently the best source for stem cells remains the human embryo.

This may seem puzzling, but there is a vast source of human embryonic stem cells. Each year thousands of infertile couples undertake the procedure of *in vitro* fertilization to have a child of their own. Chapter 24 presents more about this procedure, but for now it is enough to know that this procedure generates thousands of human embryos that are never used to produce a child. These "leftover" embryos have between 8 and 16 cells, and they are kept frozen in case their "owners" ever want to produce a child from them.

Here you see the dilemma. Many researchers believe that if these embryos are available for study, major advances in medical treatments will result. Other individuals are strongly offended by the idea of using human embryonic cells in any way. What do *you* think about the issue of embryonic stem cell research?

Chapter Review

CHAPTER SUMMARY

8.1   Life Continues Because Organisms Reproduce

Mitotic and meiotic cell division are responsible for multicellular organisms and for reproduction of organisms. Asexual reproduction produces offspring that are exact copies of their parent, while sexual reproduction creates offspring that vary from their parents. In sexual reproduction, haploid gametes fuse and restore the diploid amount of DNA. Organisms produced by sexual reproduction are slightly different from either parent.

asexual reproduction   181
clone   181

progerias   180
sexual reproduction   181

8.2   Eukaryotic Chromosomes Have Characteristics That Allow Sexual Reproduction

Alleles code for different forms of a gene. While prokaryotes have one circular chromosome, eukaryotes have paired homologous chromosomes. Diploid human cells have 23 pairs of homologous chromosomes, while haploid human cells (egg and sperm) have 23 single chromosomes. Other species have haploid individuals or polyploid individuals.

alleles   183
diploid   184
haploid   184

homologous chromosome   184
polyploid   184

8.3 Cell Division Is the Cellular Basis for Reproduction

Mitosis, or binary fission, is the type of cell division used in asexual reproduction and in growth and development. It results in two daughter cells with the same number of chromosomes as the parent cell. Meiosis is more restricted and is the process that produces four haploid gametes. Gametes from male and female parents fuse to form the diploid zygote.

binary fission   185
daughter cells   185
gamete   185

meiotic cell division   186
mitotic cell division   185
zygote   187

8.4 The Cell Cycle Is the Orderly Progression of Cellular Activities of Eukaryotic Cells

The life of a cell includes the various phases of the cell cycle in which it alternates between mature cell function and reproduction. A karyotype of a dividing cell will show darkly stained chromosomes. In G_1 a cell carries out mature cell functions. In S-phase a cell's DNA is copied. In G_2 the cell biochemically prepares to divide. DNA is duplicated and the duplicated chromosomes are joined together at the centromere, forming sister chromatids. In mitosis the duplicated chromosomes are pulled apart by the spindle apparatus, and each daughter cell receives an exact copy of the parent cell's DNA. In cytokinesis the cells physically separate, and new cell walls (if present) and cell membranes appear around each daughter cell. Some kinds of cells continually cycle and reproduce, while other kinds of cells are in G_0 and do not reenter the cell cycle. In mitosis, after chromosomes are duplicated, they are pulled to opposite ends of the cell by the spindle apparatus.

cell cycle   187
centriole   191
centromere   191
cytokinesis   190
duplicated chromosome   190
G_1 (gap-1)   189
G_2 (gap-2)   189

G_0 (G-zero)   190
growth factor   195
karyotype   191
mitosis (M)   190
sister chromatids   190
S-phase (synthesis phase)   189
spindle apparatus   191

8.5 Sexual Reproduction Involves the Fusion of Haploid Gametes

Gametes have half the normal number of chromosomes and are produced by meiosis.

8.6 Meiotic Cell Division Produces Gametes

Meiosis involves a second round of cell division that gives each daughter cell one copy of each homologous chromosome. Crossing over introduces genetic diversity when the ends of duplicated chromosomes get swapped in meiosis. Down syndrome, or trisomy 21 is an error that happens during meiosis as chromosomes do not separate correctly. Mitosis and meiosis are similar, but they differ in important ways. In mitosis the condensed, duplicated chromosomes form a single line, while in meiosis they form a double line. Mitosis produces two daughter cells, and each has the normal number of chromosomes; meiosis produces four daughter cells, and each has half the normal number of chromosomes.

crossing over   200

meiosis   197

8.7 An Added Dimension Explained: Are Cell Proliferation and Aging Related?

Aging and progerias are related to deterioration of DNA, accumulation of errors that produce flawed proteins. Progerias are related to inherited errors that make nuclear membranes break down.

cell culture   204

cell proliferation   204

REVIEW QUESTIONS

TRUE or FALSE. If a statement is false, rewrite it to make it true.

1. Karyotype is a literal photograph taken of chromosomes in a single cell. To search for abnormalities the chromosomes are cut out of the picture and then arranged in a sequence of homologous pairs.

2. Crossing over is one type of genetically caused abnormality that occurs during meiosis; it increases variability in offspring.

3. Alleles are identical copies of a gene.

4. Haploid eggs and sperm are gametes.

5. Asexual reproduction involves the fusion of two haploid gametes into a single diploid zygote.

MULTIPLE CHOICE. Choose the best answer of those provided.

6. Mitosis
    a. produces two daughter cells with different DNA.
    b. produces three daughter cells with identical DNA.
    c. produces gametes.
    d. is the cellular basis for sexual reproduction.
    e. produces two daughter cells with identical DNA.

7. Meiosis occurs
    a. only in sperm and eggs.
    b. only in cells that are actively producing gametes.
    c. only in mature gametes.
    d. only in mature nerve cells and mature fibroblast cells.
    e. in many types of cells, including skin cells and in developing embryos.

8. Genetic variability
    a. is produced through reshuffling of genes on chromosomes in meiosis.
    b. is typical of organisms produced by sexual reproduction.
    c. helps populations of organisms to survive environmental changes.
    d. is increased by crossing over.
    e. all of the above

9. Which of the following statements is the best description of events in the cell cycle?
    a. In the cell cycle a cell alternates between mature functions and cell division.
    b. The cell cycle is the alternation of mitosis and meiosis.
    c. In the cell cycle a cell prepares to divide and eventually divides into somewhere between two and six new cells.
    d. Depending on the kind of cell, at the end of the cell cycle a cell may reenter G_1 or it may enter G_2, G_3, or go straight into G_4.
    e. The cell cycle is regular and dependable and takes about the same amount of time in all cells in an organism.

10. Asexual reproduction involves
    a. different versions of a gene creating different types of daughter cells.
    b. meiosis in ovaries or testes.
    c. segregation of chromosomes into different types of daughter cells.
    d. offspring that are nearly identical to the parent.
    e. all of the above.

11. Sexual reproduction involves
    a. different versions of a gene that create identical daughter cells.
    b. meiosis in ovaries or testes that create nonidentical daughter cells.
    c. segregation of chromosomes into two, four, or six identical daughter cells.
    d. fission of cells into two identical daughter cells.
    e. all of the above.

12. Spindle fibers
    a. assist with rigidity of the nuclear membrane.
    b. assist with cytokinesis.
    c. attach to chromosomes and pull sister chromatids apart during mitosis.
    d. attach to proteins and pull homologous chromosomes apart during mitosis.
    e. attach to ATP and assist during meiosis.

13. The number of pairs of homologous chromosomes in humans is
    a. 12 pairs, 24 total chromosomes.
    b. 23 pairs, 46 total chromosomes.
    c. 24 pairs, 48 total chromosomes.
    d. 46 pairs, 92 total chromosomes.
    e. a variable number, from 18 pairs to 27 pairs.

14. Alleles
    a. is another name for chromosomes.
    b. attach to chromosomes and assist during mitosis.
    c. attach to chromosomes and assist during meiosis.
    d. are different versions of a particular gene.
    e. contain groups of genes.

15. Chromosome(s)
    a. is another name for alleles.
    b. are made of spindle fibers and proteins and contain genetic information.
    c. are made of DNA and proteins and contain genetic information.
    d. are made of DNA and RNA.
    e. continue to divide in all cells throughout the life of an individual.

MATCHING

16–20. Match the phase of the cell cycle with the event that occurs during it.

16. G_1
17. mitosis
18. cytokinesis
19. S phase
20. G_0

    a. DNA is copied in preparation for cell division
    b. daughter cells split completely apart
    c. copied chromosomes are portioned into two new daughter cells
    d. a cell carries out its mature functions; it will divide at a later time
    e. a cell carries out its mature functions and never divides again

CONNECTING KEY CONCEPTS

1. How does mitotic cell division allow asexual reproduction in eukaryotes?

2. How does meiotic cell division allow sexual reproduction in eukaryotes?

QUANTITATIVE QUERY

Some types of bacterial cells can undergo asexual reproduction (fission) as frequently as every 20 minutes. If you began with a single bacterial cell how many bacteria would you have after 10 hours, assuming they all lived?

THINKING CRITICALLY

1. In the early twentieth century scientists were able to study cell division only when they could stain the condensed chromosomes during mitosis and examine them under a microscope. This method was not especially informative, though, because mitosis is brief. Even in a group of cells that are dividing rapidly, it is hard to catch a cell in mitosis. Researchers have discovered another way to determine which cells in a population are getting ready to divide. They have developed ways to visually and chemically identify cells that are in S-phase. Think about what you know about DNA duplication. If you could put a visible "tag" on any particular molecule, which tagged molecule would show you the cells that are in S-phase?

2. The news that a loss of cell proliferation might be related to aging will certainly lead to claims that certain substances can reverse this effect. Assume for the moment that a seller of "home remedies" claims that a concoction of herbs causes cells to increase their rate of cell division. The evidence cited for this claim is that fibroblasts grown in a dish containing this concoction continue to divide longer than cells grown in a standard nutrient solution. As a careful consumer, however, you want to know more.

    a. Based on what you know about how mitotic cell division works, generate a list of possible ways that this herb concoction might work.
    b. Consider the possible negative effects of taking this concoction.
    c. Describe a series of experiments that might tell you if the mixture is safe and effective for humans.

For additional study tools, visit www.aris.mhhe.com.

Constructing Life

THE CONTROL OF EUKARYOTIC GENE EXPRESSION

Many Tainted Towns
U.S. Answer for Times Beach Is to Buy But Dioxin Questions Cover Missouri
February 24, 1983

Because of dioxin contamination the town of Times Beach, MO, was bulldozed, cleaned up, and replaced by a park.

Dioxin Level High at a Missouri Site
Tests Report Contamination Is Worst Among 31 Locations Listed by U.S. Agency
June 21, 1983

In the early 1980s the Piatt family lived in the small town of Times Beach, Missouri, where they owned and operated a stable. Today, the Piatt family stables are gone. In fact, the entire town of Times Beach is gone. Only a park remains as a reminder that there was ever a human community here. How could an entire U.S. town have been wiped off the map? The story starts with horses.

Horses were part of the landscape in Times Beach. One constant problem for the area's several stables, and indeed for the roads in and around Times Beach, was the endless dust. In 1972 waste oil hauler Russell Bliss came up with a solution for the dust in his own stable—he sprayed it with waste oil, which kept the dust down for months. Encouraged by Bliss's success, the city of Times Beach contracted with Bliss to spray roads with used oil during the summer of 1972 and 1973. Other stable owners, including the Piatts, also hired Bliss to spray their properties. The spray certainly did the job, but within weeks the seemingly good solution rapidly turned into a nightmare as horses and other animals became ill and started dying. Eventually, the Piatt stables lost 62 horses, and other stables lost nearly as many. Large numbers of birds and other wildlife around the sprayed barns died too. Bliss denied that there was any problem with his oil spray, but the situation was so serious that the Centers for Disease Control and Prevention (CDC) eventually got involved in trying to solve the mystery. It took two years for the CDC to find an answer. After many inconclusive tests the CDC finally blamed the death of the horses and other animals on a toxic soil contaminant called *dioxin*.

Dioxin was known to be a potent cancer-causing agent, and people suspected it also might cause other health problems. The levels found in the barns and roads around Times Beach were 300 times the levels set as acceptable by the U.S. Environmental Protection Agency (EPA). Bliss claimed to know nothing about dioxin in his oil spray, so where had the dioxin originated?

Here the story leads to a company called the North Eastern Pharmaceutical and Chemical Company, or NEPACCO for short. This chemical manufacturing firm made one major product, hexachlorophene, which was used in a variety of other products. Dioxin is a by-product of hexachlorophene production, and in its chemical manufacturing processes NEPACCO generated massive amounts of dioxin. Because dioxin is a lipid, the waste was an oily sludge. NEPACCO was fully aware that dioxin was a toxic substance, and the company also knew that burning the material was the best and safest way to dispose of it. But they took a short cut with some of their dioxin waste and paid Russell Bliss to haul it away. The company claims that Bliss knew that the material was toxic and that he also knew that he was supposed to dispose of it properly. In contrast, Bliss claims he never was told that the waste was dioxin. Regardless of who is right—and it was never settled—Mr. Bliss's oil spray was the source of most of the dioxin that contaminated the Times Beach community.

Eventually, the EPA was called in to clean things up. Rather than battle with corporations for years, Times Beach was one of the first regions tackled by the EPA Superfund, a government supply of money set aside to clean up areas in the United States that are contaminated with toxic chemicals. Before cleanup could begin, the entire town had to be evacuated. The federal government bought the town, paid each landowner for his or her property, and helped them to relocate. The contaminated soil of Times Beach was dug up, processed, and burned in a specially built incinerator. The effort cost nearly 200 million dollars and was not completed until 1999, but today the town of Times Beach is gone, the surrounding area is clean, and now a park and nature preserve occupy the land once contaminated with toxic wastes.

What is dioxin? What effects does dioxin have on people and on animals? What risks do the people of Times Beach face in the future as a result of their exposure to dioxin? The answers to these questions are related to the topic of this chapter, the control of gene expression. Once you are more familiar with gene expression and how it is controlled, you can return to these questions to discover how dioxin interacts with, and affects, normal cellular processes. ∎

9.1 The Development of Multicellular Organisms Involves Cell Division and Cell Differentiation

Understanding the dramatic cellular transformations that allow a multicellular organism to develop is one of the most active and exciting research areas in biology. How can scientists study human development—a delicate process that takes place deep within the shelter of a mother's uterus? Because the basic principles of development are common to multicellular species, one approach is to study development in other organisms such as fruit flies, fishes, and various kinds of plants. Commonly studied species often are small, have short life cycles, and produce many embryos in a short period of time. Other species have unique characteristics, making them highly useful subjects for the study of development. For instance, young zebrafish have no pigmentation, and so the development of their cells, tissues, and organs is especially easy to observe.

A brief look at the development of zebrafish shows some of the features scientists work to explain (**Figure 9.1**). First is the increase in cell number. The fertilized egg is one cell, and gradually

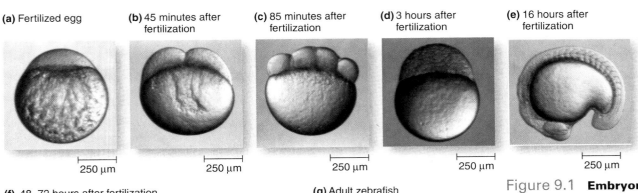

(a) Fertilized egg **(b)** 45 minutes after fertilization **(c)** 85 minutes after fertilization **(d)** 3 hours after fertilization **(e)** 16 hours after fertilization

250 μm   250 μm   250 μm   250 μm   250 μm

(f) 48–72 hours after fertilization

(g) Adult zebrafish

250 μm   250 μm

Figure 9.1 **Embryonic Development of Zebrafish.** (**a**) In the fertilized egg the embryo is at the "north pole" on top of the yolk. (**b**) At 45 minutes after fertilization the embryo has two cells. (**c**) At 85 minutes after fertilization the embryo has four cells. (**d**) After 3 hours the embryo has many cells. (**e**) After 16 hours the embryo is differentiated into many kinds of cells that make different structures and have different functions. Muscles and tail are clearly visible. (**f**) Hatching zebrafish—after 48 to 72 hours, much of the embryo's yolk supply has been used up. (**g**) Adult zebrafish.

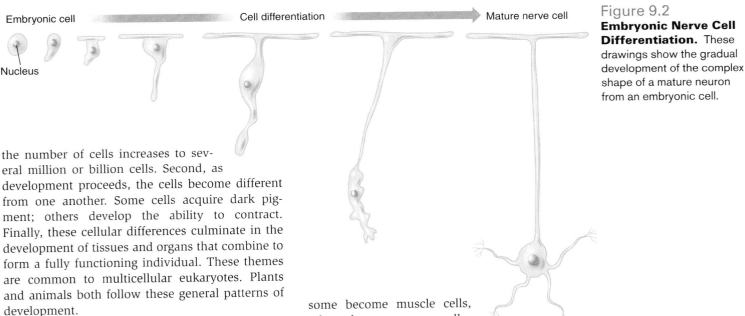

Embryonic cell   Cell differentiation   Mature nerve cell

Nucleus

Figure 9.2 **Embryonic Nerve Cell Differentiation.** These drawings show the gradual development of the complex shape of a mature neuron from an embryonic cell.

the number of cells increases to several million or billion cells. Second, as development proceeds, the cells become different from one another. Some cells acquire dark pigment; others develop the ability to contract. Finally, these cellular differences culminate in the development of tissues and organs that combine to form a fully functioning individual. These themes are common to multicellular eukaryotes. Plants and animals both follow these general patterns of development.

The two dominant mechanisms involved in animal development are cell division and cell differentiation (Figure 9.1). First, the process of mitotic cell division dramatically increases the number of cells over the course of development. In zebrafish, one day after fertilization a unicellular zygote is transformed into an embryo with several thousand cells. By a week after fertilization the young zebrafish has millions of cells. Notice, though, that if cell division were the only means of development, a mass of identical cells would result instead of a mature organism. As an embryo develops, cells not only increase in number, they also become *different* from one another. For instance,

some become muscle cells, others become nerve cells, some move deep inside the embryo to line the gut, and still others develop into blood cells. The development of specific cell traits is called **cell differentiation. Figure 9.2** shows the gradual differentiation of a nerve cell from an embryonic cell. The remainder of this chapter will explore how gene expression controls cell differentiation and so controls development.

QUICK CHECK

1. What is the difference between cell division and cell differentiation?

cell differentiation the development of specific cell traits that occurs during the growth of an embryo

9.2 During Development Cells with the Same DNA Gradually Come to Express Different Proteins

The process of cell differentiation does not happen all at once. Body parts and organs emerge gradually (**Figure 9.3**). Just as a complex task often is broken down into smaller steps, during development the emergence of body parts and organs occurs in small but organized stages. This stepwise process is easiest to understand where it first starts, at the beginning of development, when things still are relatively simple.

Mitotic cell division produces two daughter cells with identical DNA. After the first cell division, however, the single-celled zygote produces two daughter cells with different features. With each round of cell division, the process of differentiation gets progressively more specific. For example, some cells produce daughter cells that will differentiate into the head, or *anterior*, end of the animal, while the daughter cells produced in other regions of the embryo will differentiate into the tail, or *posterior*, end of the animal. Within the anterior region some cells become committed to being brain cells; others become committed to being cells of the heart, fingers, skull bones, or other structures. With each suc-

cessive round of cell division, the resulting daughter cells become more committed to differentiating into specific cells of particular structures.

You can see this sequential, step-by-step process in more detail in the development of a vertebrate limb, such as an arm or leg. The first step in vertebrate limb development is a series of small chemical changes in the cells on the embryo's trunk. These chemical changes cannot be seen by looking at the organism from the outside. You can see them if you apply special chemical stains that reveal particular proteins and other molecules inside cells. The first external sign of limb development is a small swelling called a *limb bud* that appears along the developing embryo's trunk (**Figure 9.4**). Only after the appearance of the limb bud do the limb's internal structures begin to form. And this happens in an orderly sequence. For instance, the first bones to develop are those closest to the body: the upper arm bone (or humerus) and the thigh bone (or femur). The limb bones located farther from the trunk are the next to appear, followed by the bones of the fingers and toes.

During Embryonic Development Cells That Carry the Same DNA Express Different Genes

All of the cells in the developing embryo have the same DNA, so how do cells become different from one another? The answer rests on the understanding that DNA contains the instructions for producing specific proteins (see Section 7.4). The proteins that a cell expresses determine the cell's traits. **Figure 9.5** illustrates this concept using the example of muscle cells. Muscle cells are different from other body cells because they produce special proteins that allow a muscle to contract. The genes that code for muscle proteins, such as actin and myosin, are transcribed and translated in the cells that will differentiate into muscle cells (Figure 9.5a&b). The presence of muscle proteins turns the undifferentiated embryonic cell into a muscle cell (Figure 9.5c). Making proteins is not really the beginning of the story of how cells differentiate, though. To produce the right body cells at the right

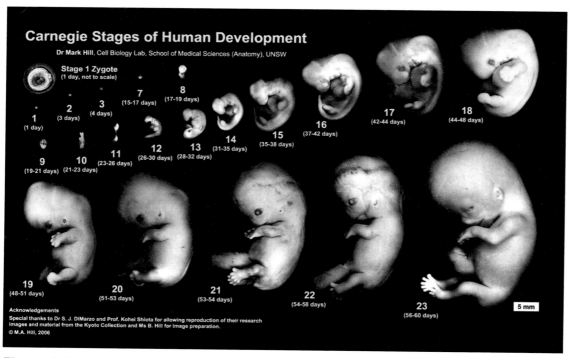

Figure 9.3 **Human Embryonic Development.** The body parts of an embryo develop gradually.

times and in the right places, transcription and translation must be controlled.

Early in development embryonic cells divide rapidly and do not make the proteins needed for mature cell functions. As development proceeds, each cell makes those proteins necessary for it to become a specific kind of mature cell. When a cell is mature, it uses only a *small portion* of its DNA to make only certain kinds of proteins. Scientists say that a cell *expresses* only a portion of its DNA. **Gene expression** is the process that produces specific protein(s) from DNA's full library of instructions. Section 7.2 defined *genome* as all of an organism's genes. A similar-sounding term, the

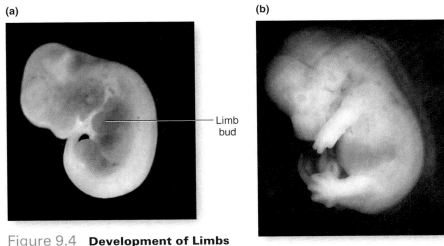

(a) **(b)**

Limb bud

Figure 9.4   **Development of Limbs in a Mouse.** (**a**) Limb buds develop first; (**b**) recognizable limbs develop later.

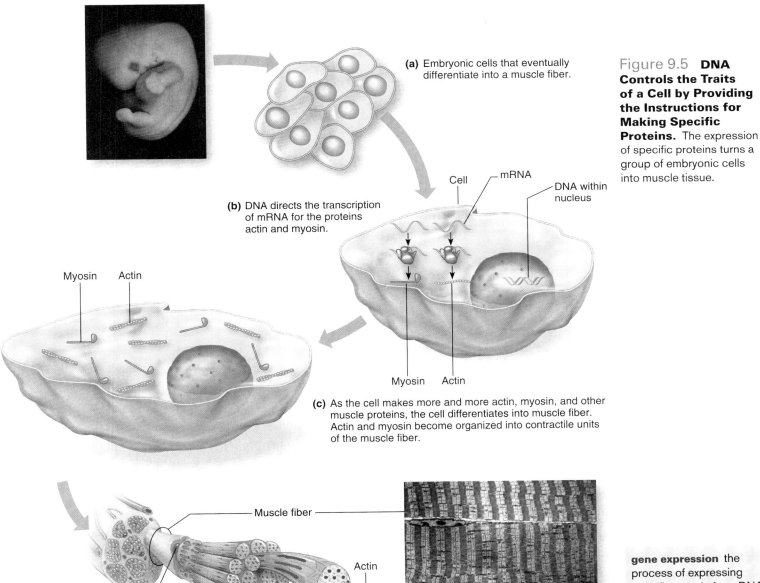

(a) Embryonic cells that eventually differentiate into a muscle fiber.

Figure 9.5   **DNA Controls the Traits of a Cell by Providing the Instructions for Making Specific Proteins.** The expression of specific proteins turns a group of embryonic cells into muscle tissue.

Cell   mRNA   DNA within nucleus

(b) DNA directs the transcription of mRNA for the proteins actin and myosin.

Myosin   Actin

Myosin   Actin

(c) As the cell makes more and more actin, myosin, and other muscle proteins, the cell differentiates into muscle fiber. Actin and myosin become organized into contractile units of the muscle fiber.

Muscle fiber

Actin

Nucleus

Myosin

gene expression the process of expressing specific protein from DNA

proteome, describes all of the proteins that a cell or organism expresses. For example, the human genome has about 25,000 genes that can produce about 100,000 or more proteins, the full human proteome. The mechanisms that control gene expression reduce the proteome of any individual cell to far fewer proteins than could be expressed by the whole individual. This is somewhat like the way you use the recipes in your cookbooks. While you may have recipes for making 267 kinds of cookies, you usually use just the recipes for the cookies needed for a given occasion. In the next two sections you will explore how gene expression is controlled.

QUICK CHECK

1. Briefly list the gradual steps during development that produce the limbs of a vertebrate.

2. What is the difference between the genome and the proteome?

9.3 The Structure of Eukaryotic DNA Allows Control of Gene Expression

A gene is a bit like a light switch, in that a gene can be turned on or off. When a gene is turned on, its protein is expressed; when a gene is turned off, its protein is not expressed. In the next few sections you will learn several ways that genes are controlled. First are two mechanisms that prevent transcription and so completely turn off whole sections of DNA: chromosome condensation and DNA methylation.

Chromosome Condensation Can Turn Off Whole Chromosomes or Pieces of Chromosomes

Several levels of coiling allow the complex of DNA and proteins to be coiled so that it fits into a cell (see Section 7.3). Fully condensed chromosomes are tightly coiled, while fully dispersed chromosomes are uncoiled. In fact, a given chromosome may at any time have both condensed regions and dispersed regions. Condensed chromosomes allow DNA to be readily segregated and delivered to each daughter cell during mitotic cell division, but this tightly coiled state serves another purpose too. Highly condensed regions of DNA typically are not transcribed. As cells differentiate during embryonic development, the portions of DNA that never will be used are coiled up and are tightly condensed. In

essence, these portions of the DNA library have been packed up and put into long-term storage. With an electron microscope you can see the overall level of chromosome condensation in a cell. As you can see in **Figure 9.6a** dispersed chromosomes make a stippled pattern in the nucleus. In rapidly dividing cells—such as embryonic cells—nearly all of the DNA is dispersed. The DNA in dispersed chromosomes is accessible to DNA polymerase and other S-phase enzymes. This allows the DNA to be copied in preparation for cell division. In contrast, usually only a portion of a mature cell's DNA is dispersed (Figure 9.6b). The dark patches are regions of highly condensed chromosomes, while the speckled areas are regions of dispersed chromosomes available for active transcription.

The nuclei of cells of female mammals show an interesting and extreme example of inactivating genes by chromosome condensation. While males have one X and one Y chromosome in each body cell, female mammals have two X chromosomes in

(a) In an immature cell, most of the DNA is dispersed. Only some DNA is condensed.

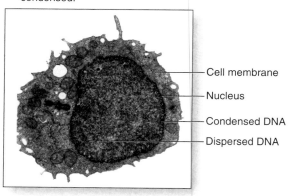

Cell membrane
Nucleus
Condensed DNA
Dispersed DNA

(b) In a mature cell, most of the DNA is condensed; less DNA is dispersed.

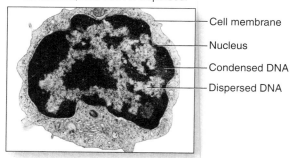

Cell membrane
Nucleus
Condensed DNA
Dispersed DNA

Figure 9.6 **More of the DNA is Dispersed in Immature Cells Than in Mature Cells.**

■ *Why is it important for an immature cell to have more dispersed DNA?*

proteome an umbrella term for all of the protein expressed by a cell

each body cell (**Figure 9.7a**). It turns out, though, that two active X chromosomes are too many, and if a female embryo is to survive, one X chromosome must be "turned off." Accordingly, one of the X chromosomes in each cell is condensed, making it inactive (Figure 9.7b). So if you are a female, in every cell of your body one X chromosome is shut down. The heavily condensed X chromosome can be seen with a light microscope (Figure 9.7c) and is called a Barr body, after the scientist who first noticed it in 1948. An even more interesting twist is that it is not the *same* X chromosome in each cell that is inactivated. Because the two homologous X chromosomes can carry different alleles for each gene, the process of turning off one of them can result in a mosaic pattern of certain traits. The irregular coloration of a calico cat's fur also is a result of the pattern of X chromosome inactivation (Figure 9.7d). A rare and strange condition in women, in which patches of the body carry no sweat glands, is also a result of the pattern of X chromosome inactivation. Barr bodies provided an early version of genetic testing, before modern DNA testing techniques were developed. To ensure that athletes who were competing as women were, indeed, women, this test once was standard practice in Olympic competitions.

DNA Methylation Is a Second Way to Turn Off Stretches of DNA

A second way to turn off segments of DNA is DNA methylation, a process that attaches methyl groups to certain cytosine molecules along DNA (**Figure 9.8**). RNA polymerase cannot bind to a length of DNA with methyl groups attached to it and so that length of DNA is not transcribed.

In an adult cell only a small percentage of DNA is expressed as protein. This means that a large percentage—as much as 80% or more—of the DNA in a mature, differentiated cell is condensed and/or methylated. There are ways to reverse methylation though. For instance, genes can change their methylation patterns during development. An example is hemoglobin, the protein that carries oxygen in red blood cells. Fetal hemoglobin is a slightly different molecule than adult hemoglobin and is produced by a different gene than adult hemoglobin. As development progresses, different hemoglobin genes are expressed at different times, and the methylation patterns of these hemoglobin genes change accordingly. Hemoglobin genes that are being expressed at a particular developmental stage are not methylated; those that are not expressed are methylated.

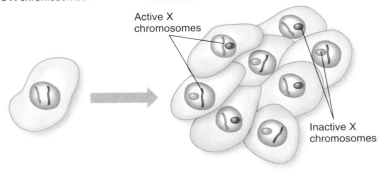

(a) The zygote of a female mammal has two X chromosomes.

(b) Early in development of the female embryo, one of the two X chromosomes becomes condensed and inactivated.

Active X chromosomes

Inactive X chromosomes

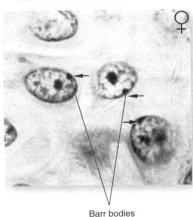

(c) The inactivated X chromosomes appear as dark blobs on the edge of the nucleus of each cell.

Barr bodies

Figure 9.7 **Barr Bodies Are Inactivated X Chromosomes.**

(d) In a calico cat, the tricolor coat results from cells expressing the allele for black coat color and the allele for yellow coat color. A different set of alleles governs white fur or patches of colored fur.

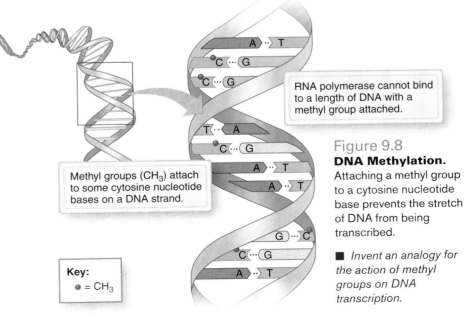

RNA polymerase cannot bind to a length of DNA with a methyl group attached.

Methyl groups (CH_3) attach to some cytosine nucleotide bases on a DNA strand.

Key:
● = CH_3

Figure 9.8
DNA Methylation.
Attaching a methyl group to a cytosine nucleotide base prevents the stretch of DNA from being transcribed.

■ *Invent an analogy for the action of methyl groups on DNA transcription.*

Some Non-coding DNA Has Important Roles in Gene Expression

To this point genes have been presented as if they are uninterrupted stretches of DNA that contain the nucleotide sequence for a single protein. This simple picture applies fairly well to prokaryotes—but not to eukaryotes. During the last half of the twentieth century, scientists discovered some intriguing mysteries about the structure of eukaryotic DNA that turn out to be related to the control of gene expression.

As the nucleotide sequences of eukaryotic genes were revealed, scientists were surprised that some nucleotide sequences did *not* code for functional proteins. Soon these long nonsensical stretches of nucleotide bases were termed *junk DNA*. This term, however, expressed more about the frustrations of scientists who were trying to understand DNA than about the DNA itself. Now it turns out that junk DNA is more interesting and important than it at first had seemed. To emphasize this point, in the following sections DNA sequences that sometimes are called junk will be called *noncoding DNA*. This is a shorthand way of saying that these sequences do not end up being translated into proteins.

One interesting class of DNA that does not code for protein does code for *functional RNA*. As the name implies, it is RNA that has a specific function in a cell, apart from providing the code for making protein. Some types of functional RNA would be the transfer RNA that brings amino acids to the ribosome, or ribosomal RNAs. Recently, other types of functional RNA have been discovered. So part of a eukaryotic genome directs the expression of functional RNA, rather than providing the code for protein.

Another type of noncoding DNA forms structures called telomeres. A **telomere** is a sequence of DNA located at either end of a chromosome. For instance, at the end of each human chromosome is a telomere of about 10,000 nucleotides that repeats the sequence GGGTTA. This repeated sequence can be marked with a dye and seen with a compound microscope (**Figure 9.9**). Why does a chromosome need a cap of noncoding DNA on either end? It turns out that DNA polymerase—the enzyme that copies DNA during S-phase—needs a stretch of DNA ahead of it. Telomeres at each end of a chromosome provide this stretch of DNA—without telomeres the last sequences on each chromosome would not get copied. Unless the telomeres are replaced, they get shorter with each cell division. Many cells express special enzymes that

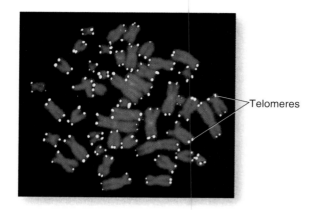

Figure 9.9 **Telomeres.** These caps of non-coding DNA allow chromosomes to be fully copied. They can be seen because of special stains.

reconstruct the telomere sequences. Telomeres and the enzymes that make them are important features of cancer cells.

Perhaps the most intriguing sort of noncoding DNA is found *within* genes (**Figure 9.10**). As this figure shows, the mRNA that carries the instructions for a protein is one continuous nucleotide sequence that interacts with a ribosome to produce one continuous string of amino acids. In eukaryotic cells, however, the stretch of DNA that codes for a protein is not continuous. Just as a TV program is broken up by commercials that have nothing to do with the plot of the show you are watching, a gene is broken up by sections of DNA that are not part of the code to produce a protein. In other words, the coding DNA sequence is interrupted by noncoding stretches of DNA. Surprisingly, this complex structure is the *typical* organization of eukaryotic genes. It is so common that special names are given to the coding and noncoding portions of a gene. The nucleotide sequences that are part of the actual gene are called **exons,** and the noncoding sequences are called **introns.** The process that cuts out introns is known as **mRNA splicing.** In this process an enzyme complex cuts up the full piece of mRNA. The introns are cut out, and the exon mRNA pieces are bonded—or spliced—together. The result is a functional mRNA molecule that can be used by ribosomes to make a functional protein.

mRNA splicing is an important way to alter the proteins that a cell produces. Depending on the conditions, different segments of the original mRNA molecule can be treated as introns. In other words, under some conditions one piece of the mRNA is cut out as an intron; under other conditions that same stretch of mRNA is used as an

telomere a stretch of noncoding DNA at either end of a chromosome

exon the portion of a gene that actually codes for a protein

intron a noncoding sequence of DNA located within the nucleotide sequence of a gene

mRNA splicing the process in which introns are cut out of a mRNA transcript and the cut pieces are spliced back together to make a functional mRNA molecule

In eukaryotes a gene is a series of introns and exons.

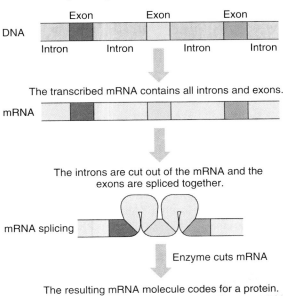

The transcribed mRNA contains all introns and exons.

The introns are cut out of the mRNA and the exons are spliced together.

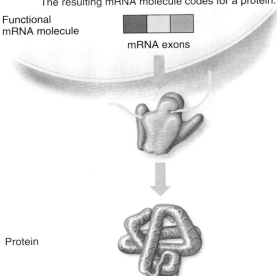

Enzyme cuts mRNA

The resulting mRNA molecule codes for a protein.

Figure 9.10 **Exons and Introns.** Exons are regions of genes that code for proteins. Exons are interrupted at intervals by introns.

exon, and other pieces are cut out as introns. What remains is always a continuous piece of mRNA that will be used to produce a functional protein. Determination of the gender of fruit flies is based on alternative mRNA splicing. In humans mRNA splicing plays a large role in producing the 100,000 or more different kinds of proteins.

So far you have read how the complexity of the genome produces complex patterns of gene expression, which, in turn, produce the proteome for a cell or organism. Eukaryotic cells have yet another way to change the protein that results from the transcription of a particular stretch of DNA—a process called *post translational modification of proteins.* Translation produces a string of amino acids that spontaneously folds to form a protein. This protein, though, can be altered by enzymes in the cytoplasm to produce a different protein or more than one protein.

From these mechanisms you can see how a eukaryotic organism can have fewer genes than it has proteins. As surprising as it might seem, mRNA splicing and post translational modifications are responsible for producing over half of the different proteins found in a human. The same sequence of DNA can code for a variety of proteins, so that the definition of a gene as a sequence of DNA that codes for one protein is not strictly accurate—although currently it is probably the best working definition.

> **QUICK CHECK**
>
> 1. How are methylation and chromosome condensation similar? How are they different?
> 2. Explain how only about 25,000 genes in the human genome can make about 100,000 human proteins.

9.4 Transcription Is Controlled by Transcription Factors

The control mechanisms discussed so far—chromosome condensation, DNA methylation, mRNA splicing, and posttranslational modifications of proteins—have critical roles to play in determining the proteome of a cell, but their influence is limited. None of these mechanisms can direct a cell to turn off or turn on transcription of a particular gene. Any cell will express different proteins at different times over its lifetime. This requires that a cell also turn transcription of genes on and off. So one of the most important controls on gene expression is initiation of transcription.

To understand how transcription of a gene starts, think back to the structure of the DNA sequences along a chromosome. Earlier you read that a eukaryotic gene contains both introns and exons. Beyond this complication, there are other noncoding regions that play a role in the control of transcription. These regions are located outside of the gene sequences. One important noncoding sequence called the **promoter** sits just ahead of a gene itself. The promoter is a DNA sequence to

promoter a noncoding DNA sequence located before the DNA sequence of a gene; to initiate the transcription of a gene, RNA polymerase first binds to the promoter and then transcribes the gene

which RNA polymerase can bind (**Figure 9.11**). RNA polymerase attaches to this promoter region and begins transcription of the gene. Every eukaryotic gene has a promoter region, and different promoter regions share certain DNA sequences. For example, most contain the sequence of nucleotides, TATATA, or something close to it.

By itself, though, RNA polymerase does not bind well to promoter sequences. So under most conditions any gene will have only a low level of transcription, at best. Something else is needed to get RNA polymerase securely bound to the promoter and get transcription going at a fast pace. How do cells solve this problem? The answer is shown in **Figure 9.12** and you should examine this figure carefully as you read the following explanation. Note that if a cell needs a particular protein in significant quantities, *other* proteins called **transcription factors** bind to special regions of DNA and increase the rate of transcription. Each gene is controlled by one or more transcription factors that bind to specific DNA sequences. When one or

> **transcription factor**
> a protein that binds to DNA and through interactions with RNA polymerase increases transcription of a particular gene

more specific transcription factors are bound to DNA, RNA polymerase binds more effectively to the targeted gene's promoter. Thus, when transcription factors are bound, lots of mRNA is transcribed from specific genes. Transcription factors are critically important in the control of gene expression—many of the 100,000 or so proteins of a human are transcription factors.

The binding of transcription factors is one important way that the amount of each protein in a cell is controlled. If you think for a minute, though, you will realize that this just pushes the question of control further back. How are transcription factors controlled? One common mechanism is shown in **Figure 9.13**. To understand transcription factor control, you need to appreciate that cells must respond quickly to new situations. The environment can change rapidly. A cell that does not respond quickly may soon be a dead cell. To rapidly respond to changing situations, cells keep many copies of important transcription factors on hand—but in chemically *inactive* forms. When a protein is needed, the cell does not have to make transcription factor molecules. Instead, it can quickly convert the inactive transcription factor to an active form. This is an efficient way to provide a cell with flexibility and speed of response, but what tells a cell to convert an inactive transcription factor to an active one? The answer is signal molecules (see Section 4.3). Recall that the cells in a multicellular organism are awash in signals. Signal molecules bind to membrane receptor proteins and stimulate the conversion of a specific transcription factor from an inactive to an active form.

1 RNA polymerase binds to DNA at a promoter region.

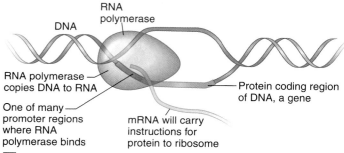

2 Once bound to the promoter, RNA polymerase begins to transcribe the gene into mRNA.

Figure 9.11   **The Role of a Promoter.** Transcription of DNA involves the binding of RNA polymerase to a promoter.

QUICK CHECK

1. Define the roles of the promoter, RNA polymerase, and a transcription factor in the initiation of transcription.

Figure 9.12

Transcription Factors Increase the Rate of Copying of a Gene into mRNA.

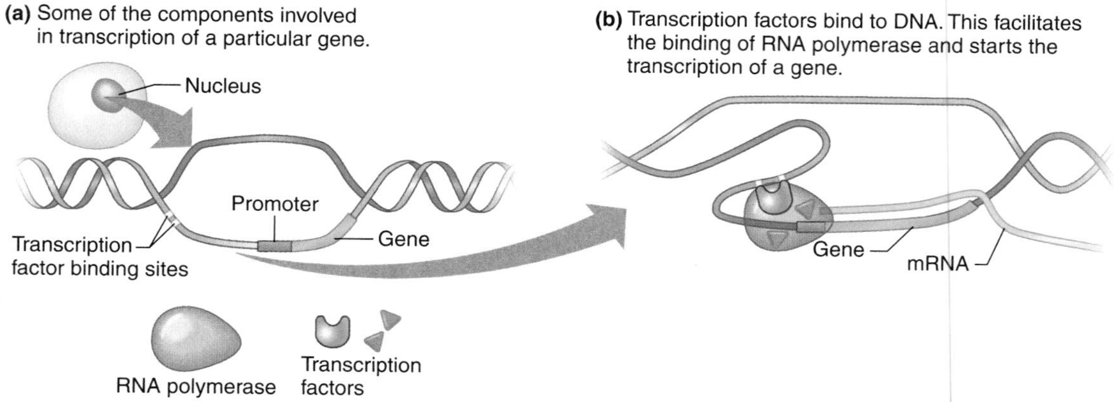

(a) Some of the components involved in transcription of a particular gene.

- Nucleus
- Promoter
- Gene
- Transcription factor binding sites
- RNA polymerase
- Transcription factors

(b) Transcription factors bind to DNA. This facilitates the binding of RNA polymerase and starts the transcription of a gene.

- Gene
- mRNA

1 When a signal molecule binds to the membrane receptor, the receptor changes shape.

2 This causes the release of a transcription factor from an inhibitory protein.

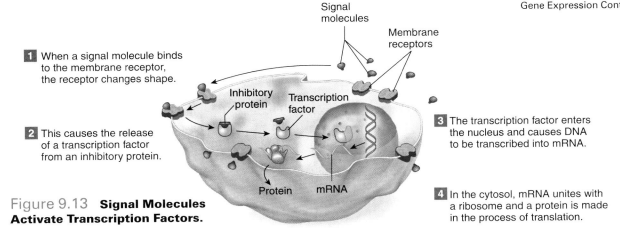

Signal molecules

Membrane receptors

Inhibitory protein

Transcription factor

3 The transcription factor enters the nucleus and causes DNA to be transcribed into mRNA.

Protein

mRNA

4 In the cytosol, mRNA unites with a ribosome and a protein is made in the process of translation.

Figure 9.13 **Signal Molecules Activate Transcription Factors.**

9.5 Gene Expression Controls Development

This chapter started with the puzzle of embryonic development in a multicellular organism. Look again at Figure 9.1, which illustrates the beauty of this process in the developing zebrafish embryo. Through the process of mitotic cell division, the fertilized egg gives every cell in the developing organism the same DNA. Nevertheless, cells develop unique identities, becoming muscle, bone, eyes, gills, teeth, and so on. These genetically identical cells become differentiated through the regulation of gene expression. A look at this process in a bit more detail will deepen your understanding and appreciation of the complexity this entails.

It is obvious that embryonic development happens in three dimensions, but let's focus on just a single dimension of embryonic development as an example of how the process works: the distinction between the head region of the embryo and the tail region of the embryo. You already know that after the first cell division the two cells of the embryo are different from one another. Where do these differences come from? Well, as strange as it sounds, the differences start in the egg itself. During egg formation—in the mother's ovaries—mRNA molecules needed early in embryonic development are produced in large quantities. This ensures that when fertilization does occur, the new zygote is ready to translate the genes necessary to get development started. Most of these mRNAs code for various transcription factors that turn on specific genes in the zygote's genome. At least in animals, however, the spatial distribution of these mRNAs is not uniform in the egg.

Figure 9.14 shows how this works for just one dimension of development in the fruit fly. When the fruit fly egg is formed, certain mRNA molecules accumulate at one end. One of these mRNAs codes for a protein with the odd name *bicoid*. Accordingly, the mRNA that is used to translate bicoid protein is

(a) The orange stain at one end of the fruit fly egg shows where bicoid mRNA is concentrated.

Highest concentration of bicoid in egg

(b) After the egg is fertilized, bicoid mRNA molecules are translated to proteins that also show a head-to-tail gradient.

Highest concentration of bicoid in zygote

(c) As the larva develops, the regions where bicoid proteins are the most concentrated become the head end of the fruit fly.

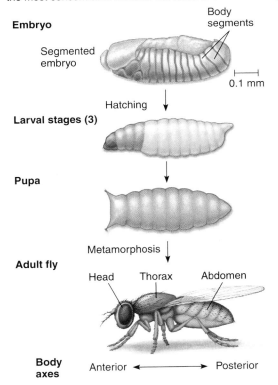

Embryo

Segmented embryo

Body segments

0.1 mm

Hatching

Larval stages (3)

Pupa

Metamorphosis

Adult fly

Head   Thorax   Abdomen

Body axes   Anterior ◄──────► Posterior

Figure 9.14 **Spatial Distribution of Bicoid Protein Establishes the Head End of a Fruit Fly Embryo.**

called *bicoid mRNA* and the gene that codes for bicoid protein is called the *bicoid gene*. Bicoid mRNA begins translating bicoid protein soon after fertilization. One tip of the egg carries a high concentration of bicoid mRNA (Figure 9.14a) and half of the egg contains a high concentration of bicoid protein (Figure 9.14b). The half of the egg with the high bicoid protein concentration will develop into the embryo's head end (Figure 9.14c). Other types of mRNA and protein also are contained in this half of the egg, but bicoid was one of the first discovered. The end of the egg that lacks bicoid becomes the tail end of the embryo. The tail end of the egg expresses other types of mRNA and protein.

How does this one protein, bicoid, play such an important role in development? The answer is that bicoid protein is a transcription factor. It acts to turn on the transcription of a family of other genes, many of which are themselves transcription factors. This means that by itself the *bicoid* gene does not directly code for the protein required to make the fly's head region. Rather, the bicoid transcription factor starts a cascade of gene expression that unfolds over the course of development. Similar processes involving other genes and other transcription factors begin in other regions of the egg. Each contributes to the development of a specific region of the adult fly. This picture of gene expression during development is understood most completely for fruit flies, but studies in other organisms, including mammals, show that these principles are universal.

Pax6 Is a Master Control Gene for Eye Development

A cascade of gene expression also can shed light on the development of such complex and diverse body structures as animal eyes. Animal eyes share two common features. First, all eyes have photoreceptor cells that respond to light. The simplest eyes have only one photoreceptor cell; more complex eyes have millions of photoreceptor cells organized into a tissue called a *retina*. Second, all eyes studied so far have pigment molecules that absorb scattered light. In addition, many, but not all, eyes have a lens that focuses light on the photoreceptor cells. In insects, molluscs, and vertebrates these features form complex eyes. A vertebrate eye would have all of the components shown in **Figure 9.15.** It may seem unlikely that they could have evolved from simpler sorts of eyes. Nevertheless, scientists have unraveled the complexities of how eyes evolved. **Figure 9.16** shows a simple eye that probably evolved into the complex eyes of vertebrates.

Because of the features that all eyes have in common, scientists have long suspected that the strikingly different kinds of eyes evolved from a common ancestral eye. But how can we know for sure? Some evidence comes from genetic abnormalities in eye development. Species ranging from fruit flies, to mice, to humans have a genetic abnormality in which the eyes do not develop normally, and all parts of the eye are smaller than normal. It turns out that these disorders all are the result of a mutation in a similar gene, called *Pax6*. When the *Pax6* gene is normal, cells in the outer layer of an embryo respond to a signal from underlying cells and develop into lens cells. At the same time, cells that express *Pax6* produce other signals that have an effect on the underlying cells. Eventually, this back-and-forth exchange of information leads to the development of an eye.

The next set of experiments on *Pax6* may astound you, but they are important because they emphasize the genetic relatedness of all living things (**Figure 9.17**). An enterprising group of scientists isolated the *Pax6* gene from mice and inserted it into the tail regions of developing fruit fly embryos. Of course, fruit flies are invertebrates and mice are vertebrates. Like other arthropods, fruit flies have *compound eyes* made of clusters of sometimes hundreds of individual components, each with a separate lens and its own photoreceptors, while mice have eyes that are similar to yours. Nevertheless, the mouse *Pax6* gene induced the development of a *fly eye* in the fly embryo's abdomen. Not only did *Pax6* direct the development of an eye in a wrong location, *Pax6* from a mouse directed the development of a *compound eye* in a dramatically different species: a fly! Like so many genes that are important in development, the protein coded for by the *Pax6* gene is a transcription factor that starts a cascade of gene transcription. It is only one of many genes involved in eye

atavism a feature of an individual that reflects an ancestral structure or form

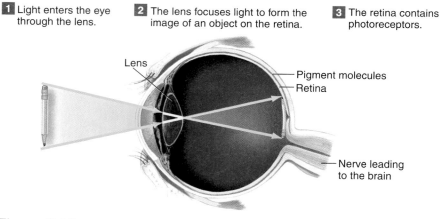

1 Light enters the eye through the lens.

2 The lens focuses light to form the image of an object on the retina.

3 The retina contains photoreceptors.

Lens

Pigment molecules
Retina

Nerve leading to the brain

Figure 9.15   **Components of a Vertebrate Eye.** A typical vetebrate eye has a lens, a retina, and a pigment layer with photoreceptors.

development, but it is an important gene found in all the animal species examined so far.

This story of *Pax6* demonstrates several important points. First, master control genes govern the development of complex structures in multicellular organisms, but they do not strictly dictate the final biological structure. Instead, master control genes set up a cascade of gene expression unique to each species. Master control genes are so important that they have been maintained during the process of evolution. They are the same in different species, even species that seem strikingly different. The existence of common master control genes shows that diverse organisms have a common genetic ancestry. Furthermore, these studies suggest that the animal eye probably evolved only once from a common ancestor with a *Pax6* gene. *Pax6* is just one of many master control genes that function in the same way across animal species.

Evolutionary Changes in Gene Expression May Explain Atavisms

One of the intriguing puzzles of evolution and development is the expression of **atavisms.** An atavism is a feature of an individual that reflects an ancestral structure or form. In more everyday terms, an atavism is a throwback. As you might imagine, atavisms include a wide range of traits that seem bizarre at first glance. For example, atavisms are sometimes seen in the limbs of horses. The legs of a horse contain the same bones as the limbs of any mammal, but through evolution they have been modified. Four of the horse's original five toes have been lost through evolution, and the remaining toe has become large and strong. The foot and ankle bones are elongated to form the lower part of the leg, and what looks like the horse's "knee" is actually its wrist joint. The two forearm bones found in mammals—the radius and the ulna—are fused in modern horses. Atavisms are seen in many of these features. For instance, the favorite horses of Julius Caesar and Alexander the Great are said to have had extra toes, rather than the single toe of most horses. So those horses were atavistic in this feature. This sort of genetic throwback is not uncommon in horses. A recent paper described a young horse with leg abnormalities that X rays revealed to be a distinct ulna and radius in one of its front legs. Whales are descendants of land animals that had four limbs. Occasionally, a modern whale has small but distinct hind leg bones.

Humans can show atavisms as well. One intriguing human atavism is called *hypertrichosis,*

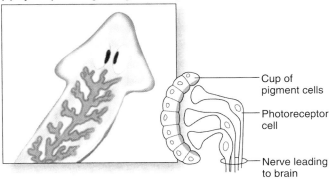

(a) **Eye cups** distinguish light intensity and direction.

— Cup of pigment cells
— Photoreceptor cell
— Nerve leading to brain

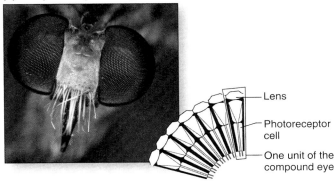

(b) **Compound eyes** distinguish shapes and are sensitive to motion.

— Lens
— Photoreceptor cell
— One unit of the compound eye

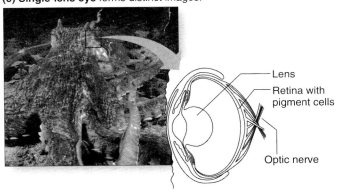

(c) **Single-lens eye** forms distinct images.

— Lens
— Retina with pigment cells
— Optic nerve

Figure 9.16 **Simple and Complex Eyes.** (**a**) Eye cups of planaria are simple eyes. (**b**) Compound eyes like those of a fly are made of multiple photoreceptive units. (**c**) An octopus has a single-lens eye similar to that of a vertebrate.

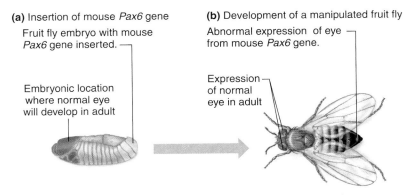

(a) Insertion of mouse *Pax6* gene

Fruit fly embryo with mouse *Pax6* gene inserted.

Embryonic location where normal eye will develop in adult

(b) Development of a manipulated fruit fly

Abnormal expression of eye from mouse *Pax6* gene.

Expression of normal eye in adult

Figure 9.17 ***Pax6* Gene Transplant Experiment.** When a mouse *Pax6* gene is inserted into a fly embryo, the alien gene causes a compound eye to develop in the fly.

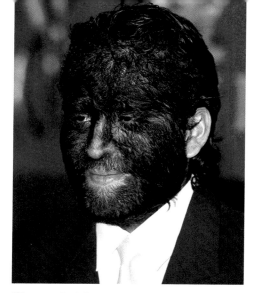

Figure 9.18 **An Example of an Atavism.** Hypertrichosis is a disorder characterized by excessive body hair.

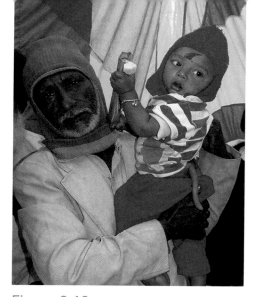
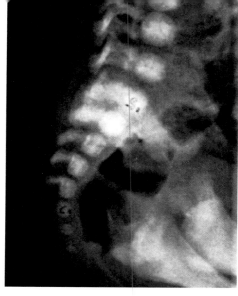

Figure 9.19 **Atavism: Human Tail.** (**a**) In this photo from United Press International, it appears that this child was born with a tail. (**b**) X ray of the spine of a 3-month-old boy who was born with a tail.

the medical term for lots of body hair (**Figure 9.18**). People with this atavism have hair all over their bodies, much like ancestral mammals. Individuals with hypertrichosis may be the basis for the tales about werewolves—humans who changed into wolves under a full moon. Another fascinating but rare atavism is the human tail (**Figure 9.19**).

Atavisms are abnormal structures—but they are similar to many *vestigial structures* found in the normal development of many species. Vestigial structures also can be evolutionary throwbacks—but unlike atavisms, vestigial structures appear in *most or all* individuals of a species. Some species have vestigial structures that persist all through adulthood. For instance, boas and pythons have tiny, internal hind limb bones that appear as thorny spurs near the tails of male snakes. These hind limb bones are thought to be remnants of the evolutionary process through which the lizardlike ancestors of modern snakes lost their legs. In other species vestigial structures are seen only during early development. For example, many vertebrate embryos have gill slits in their throats that close up as development proceeds. In humans one of the gill slits persists in the form of the eustachian tube, connecting the middle ear to the throat. To return to human tails, most humans do have a tail during one embryonic stage, but like gill slits it gradually recedes, and most people are born without a tail.

Understanding gene expression during development may provide answers to the mysteries of atavisms. The first step is to distinguish between *structural genes* and *regulatory genes*. Structural genes code for the proteins that do the work in cells. These proteins can be enzymes that make and break chemical bonds, or they can be proteins that make up the organelles and other cellular and tis-

sue structures. In contrast, regulatory genes code for proteins that control DNA—such as transcription factors. When genes were first discovered, most scientists thought that mutations in structural genes would be the most important mechanisms in evolution. That prediction probably is wrong: much evolutionary change probably occurs because of mutations in regulatory sequences. While this area of research is still in its early stages, it provides a good hypothesis to explain atavisms.

Let's consider those rare humans who are born with a tail. At this point there is no clear answer to how this happens, but a good working hypothesis can be presented based on studies of how tails grow—or fail to grow—in other vertebrates (**Figure 9.20**). Tail development begins with a special group of cells at the end of the spine called the *tail organizer*. These cells produce and release small proteins that act as signal molecules for surrounding cells. These signal molecules stimulate cell division, causing the tail to elongate and increase in girth. They also stimulate cell differentiation, causing the tail cells to become bone, or muscle, or connective tissue. The production of these small proteins is under the control of transcription factors. Human ancestors probably lost their tails not because of mutations in structural genes that produce tails—but from mutations in transcription factors that turn on the production of signals that orchestrate tail development. Humans probably still carry the genes for these tail transcription factors, but the genes do not produce functional proteins. Every once in a while a person develops a mutation that restores the original transcription factor gene. When the transcription factor is produced, it sets into motion a cascade of cellular events that still are present in humans—ready to

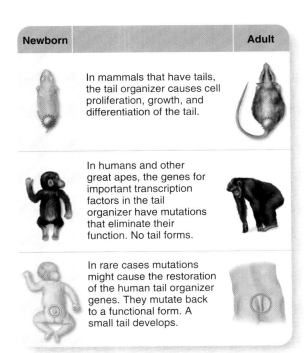

Newborn		Adult
	In mammals that have tails, the tail organizer causes cell proliferation, growth, and differentiation of the tail.	
	In humans and other great apes, the genes for important transcription factors in the tail organizer have mutations that eliminate their function. No tail forms.	
	In rare cases mutations might cause the restoration of the human tail organizer genes. They mutate back to a functional form. A small tail develops.	

Figure 9.20   **Possible Scenario for How a Human Tail Can Develop.**

produce a tail. Currently, this is just an intriguing hypothesis, but it illustrates how mutations in transcription factors might have profound effects on an organism's structure and function.

QUICK CHECK

1. What is the evidence that the *Pax6* gene plays a similar role in the development of eyes in diverse animals?

9.6   Normal Development Is the Result of the Progressive Differentiation of Embryonic Stem Cells

During development cells gradually differentiate into adult cells and tissues. Cells in an early embryo can produce any type of body cell. With each cell division daughter cells become progressively more committed to playing a particular role in the adult, and lose the ability to develop into any cell in the body. These committed cells eventually lose the ability to divide and are a particular cell type, such as nerve or muscle cells. Other cells retain the ability to divide and produce daughter cells. When these cells divide, one daughter cell

remains as a source of future daughter cells, while the other daughter cell begins to differentiate. Each time this daughter cell divides, its daughter cells are progressively more committed to a particular adult fate. The cells that stay behind as a source of more daughter cells are called *stem cells* (**Figure 9.21**). **Embryonic stem cells** are the most flexible in terms of the range of cells that can differentiate from them, and their daughter cells can produce a wide variety of cell types. **Adult stem cells** are more restricted, and their daughter cells can produce just a few cell types. Through the processes that control gene expression, the daughters of embryonic stem cells eventually become progressively more committed during development. In this way a normal embryo develops into a newborn individual and eventually into a mature adult.

Differentiation of stem cells is controlled by gene expression that, in turn, is controlled by signal molecules and transcription factors. As development proceeds, each cell becomes surrounded by a specific set of signal molecules and so produces a defined set of transcription factors that limit how its daughter cells will differentiate. Depending on the kinds of genes that already have been activated in a cell, its fate can sometimes be changed if it is exposed to different kinds of signal molecules. This can be done most easily if the stem cells are taken out of the body and put into a cell culture dish, where the environment can be strictly controlled by the experimenter. Researchers also have shown that stem cells can be induced to differentiate into different cell types if they are injected into different regions of an animal's body. For instance, stem cells injected into the brain sometimes can differentiate into neurons; similar stem cells injected into muscle may differentiate into muscle. This may make it sound like getting a stem cell to differentiate into a specific kind of cell is easy, but figuring out which signal molecules control what kinds of development takes a long time and requires many experiments.

Stem cells could have tremendous potential for the treatment of human diseases. Many diseases are caused by the death or malfunction of particular cell types. For example, juvenile diabetes is a debilitating disease that develops when cells in the pancreas, called *islet cells,* fail to release insulin. People who suffer from diabetes must take insulin injections, often several per day, and even then they can suffer a variety of disorders and an early death. Many people hold out the hope that stem cells could be injected into diabetic patients and replace the nonfunctional islet cells. Studies on mice and rats suggest that such a treatment might be possible. Heart disease can result from damaged heart tissue. A

embryonic stem cells embryonic cell that divides repeatedly to produce a long line of daughter cells, each of which has the ability to differentiate into many adult cell types

adult stem cells adult cell that divides repeatedly to produce a long line of daughter cells, each of which produces a restricted set of one or a few adult cell types

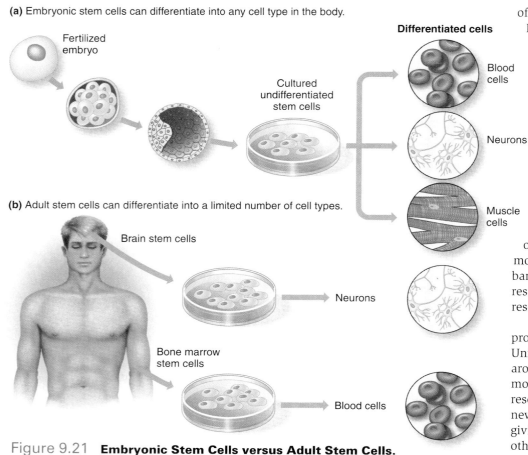

(a) Embryonic stem cells can differentiate into any cell type in the body.

Fertilized embryo

Cultured undifferentiated stem cells

Differentiated cells

Blood cells

Neurons

Muscle cells

(b) Adult stem cells can differentiate into a limited number of cell types.

Brain stem cells

Neurons

Bone marrow stem cells

Blood cells

Figure 9.21 **Embryonic Stem Cells versus Adult Stem Cells.**
(a) While embryonic stem cells have unlimited developmental potential, **(b)** adult stem cells have a limited range of developmental possibilities.

recent study in mice with damaged hearts showed that injected stem cells will differentiate into normal heart muscle cells and repair the damage.

This may sound as though exciting new treatments for many diseases are not far away, as scientists learn more about stem cells. Some important roadblocks, however, impede progress. One contentious issue is where these human stem cells should come from. Human embryos are the best source of human stem cells for research and the potential treatment of human disease. Thousands of frozen, unused human embryos are destroyed when they no longer are needed by couples using *in vitro* fertilization procedures to try to have a baby (see Chapter 24). It is nearly impossible, however, for research scientists to culture and study embryonic stem cells from these embryos, even if the couples who generated the embryos wanted to donate them to research instead of destroying them. Why? The answer has to do with the source of money that supports basic research such as stem cell research.

Basic research in the United States is largely funded by the U.S. government, through a variety of institutions such as the National Science Foundation and the National Institutes of Health. In 2001 the U.S. government declared that government funding for human stem cell research could only be used to study 64 existing human embryonic cell cultures, called *cell lines*. No government money could be used to generate or study any other human embryonic cell lines. Sixty-four cells lines may sound like a lot, but many of these cell lines are fraught with problems. Some are contaminated with mouse cells. Some do not function as embryonic stem cells any longer. Still others are extremely expensive. As a result most scientists predict that as long as this ban is in effect, progress on human stem cell research will be impossible, at least by U.S. researchers.

Human embryonic stem cell research is proceeding in other countries, and in the United States some people are trying to get around the federal government ban by using money from nonfederal sources. One U.S. researcher secured private money to develop new human embryonic stem cell lines and is giving away cells from these cultures to other researchers. In a dramatic development in 2004 the state of California voted to provide state money for human embryonic stem cell research.

Many who oppose the use of human embryonic stem cells for research and therapy advocate the use of adult stem cells for these purposes. The idea is that stem cells could be removed from a patient who has the disease. After being cultured in the right environment, these cells could be replaced into the patient to correct the disorder. Adult stem cells could solve many problems associated with the use of embryonic stem cells. First, because the therapeutic stem cells would come from the patient, not from an embryo, the ethical objections would not apply. Second, the risk of tissue rejection could be avoided. Tissue rejection often happens with organ transplants that come from an unrelated person, and a transplant recipient must take drugs to suppress the immune response that rejects the transplant. A treatment that involves embryonic stem cells would have a risk of rejection—but if the therapeutic stem cells come from the patient, there is little or no risk of rejection. Finally, in animal models embryonic stem cell treatments often increase the development of tumors. No one is sure why this happens, but it does not seem to happen with adult stem cells. These arguments point to adult stem cells as the answer, but it still is not clear just how useful adult stem cells might be. Much more basic research is needed before the therapeutic potential of either embryonic stem cells or adult stem cells will be realized.

QUICK CHECK

1. What is a stem cell?
2. What are the differences between embryonic and adult stem cells?

9.7 An Added Dimension Explained
Times Beach and the Effects of Dioxin

In the beginning of this chapter you read the sad story of Times Beach, Missouri. This small town was evacuated, incinerated, and rebuilt as a park, all because of a small lipid molecule called dioxin (**Figure 9.22**). What do we really know about dioxin? And what does dioxin have to do with the topics of this chapter, gene expression and development?

In the early 1980s scientists knew that dioxin causes cancer in animals. Laboratory studies showed that tumor-causing doses vary from species to species. In 1976 an industrial accident covered the town of Seveso, Italy, with dioxin contamination. Women who were 40 years old or younger at the time of the Seveso accident have had a much higher than normal rate of breast cancer. From these and other studies it appears that dioxin can cause cancer in humans, at least at very high doses.

Recently a greater concern has been voiced about dioxin's effects on developing embryos. In 1979 another industrial accident in Taiwan exposed a large harvest of rice to dioxin. Subsequently, concentrated dioxin was discovered in rice oil, commonly used in cooking in Asia. Some Taiwanese women ate a lot of the contaminated rice oil. One disturbing finding was that women who were pregnant at the time of exposure to dioxin-contaminated rice oil had children with a variety of disorders. The effects sometimes emerged years after exposure to dioxin. The male children of these women have a high rate of infertility when they reach adulthood, and their sperm are visibly abnormal. A similar conclusion about dioxin's effects on unborn children also was seen in Seveso, Italy—for decades after exposure to dioxin the men in the area fathered far fewer than normal male children.

(a) Space-filling model of dioxin molecule
(b) Structural model of dioxin molecule
Cl
O
Cl
Cl
O
Cl
Key:
= carbon atom
= oxygen atom
= hydrogen atom
= chlorine atom

Figure 9.22 **Structure of a Dioxin Molecule.**

Laboratory studies on rats and mice support the conclusion that dioxin affects embryonic development and also suggest that dioxin has more subtle effects. Male embryos exposed to low doses of dioxin when they are in the uterus show a variety of genital abnormalities. When they grow up, these rats will mate and can produce offspring, but they exhibit female as well as male behaviors. Overall, the dioxin-exposed male rats are feminized. Female rats also show subtle physical effects of dioxin exposure. Dioxin also affects the development of reproductive systems of a completely different class of animals: fishes. In many locations around the country researchers are finding fishes that have a mixed sexuality. The fishes have both testes and ovaries and do not reproduce normally. In other populations of fishes there are many more females than males. Scientists are fairly certain that these abnormal fishes are the result of dioxinlike compounds accumulating in the environment.

These findings have led scientists to investigate how dioxin produces these effects. Dioxin is one of a class of environmental pollutants called *estrogen disrupters*. These compounds either mimic or block estrogen effects in different tissues. Estrogen is a hormone that plays a major role in female reproduction. In males estrogen is produced in only small amounts but also plays an important role in normal development and function. Because estrogen is a lipid, it passes through the cell membrane and binds to a protein receptor located inside of a cell. When estrogen binds to its receptor, the resulting complex is a transcription factor (**Figure 9.23**). The estrogen-receptor complex binds to DNA and causes the transcription of a variety of proteins, depending on the cell. Two different estrogen receptors occur in tissues throughout the body and often have opposite effects on cell functions. The overall effect of estrogen on a tissue or organ depends on

—Continued next page

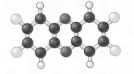

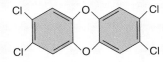

Continued—

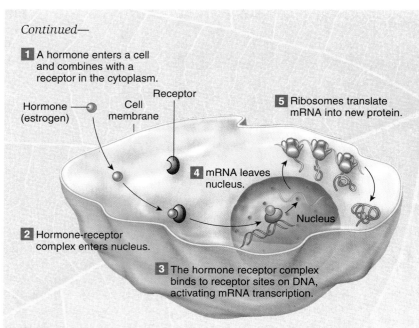

1 A hormone enters a cell and combines with a receptor in the cytoplasm.

Hormone (estrogen)

Cell membrane

Receptor

5 Ribosomes translate mRNA into new protein.

4 mRNA leaves nucleus.

Nucleus

2 Hormone-receptor complex enters nucleus.

3 The hormone receptor complex binds to receptor sites on DNA, activating mRNA transcription.

Figure 9.23 **Estrogen Stimulates Production of New Proteins.**

Table 9.1	Some examples of the effects of estrogen

Effect	Examples of tissues where effect is seen
Increased blood flow	Brain, uterus, breast
Increased cell proliferation	Breast, uterus, bone
Increased or decreased enzyme function	Brain, liver
Increased high-density lipoproteins	Liver

the balance of these two receptors. **Table 9.1** gives some examples of the effects of the estrogen-receptor complex on the activities of cells and whole organisms.

How does this relate to dioxin? The story is just beginning to emerge, but some things are clear (**Figure 9.24**). First, like estrogen, dioxin is a lipid that crosses the cell membrane. Like estrogen, dioxin binds to a protein complex in the cytosol. Once dioxin is bound, this complex becomes a transcription factor that, in turn, enters the nucleus and binds to DNA. The combination of dioxin and transcription factor complex binds to many sites along DNA and so leads to the transcription of a variety of proteins. The proteins that are transcribed help to explain some of dioxin's effects. Some of these proteins actually break down dioxin and get rid of it, a process that seems to be a more general mechanism to rid a cell of toxic compounds. During the breakdown of dioxin a variety of chemicals are released—some of these can cause cancer. So the cell gets rid of dioxin, but in doing so it may pay the price of developing cancer. Other proteins that are transcribed as a result of dioxin interact with the estrogen pathways in a cell. Some of these proteins enhance estrogen's effects; others block estrogen's effects. Scientists are

(a) Dioxin crosses the cell membrane, enters the cell, and binds to a transcription factor inside the cell. The dioxin-transcription factor complex travels to the nucleus.

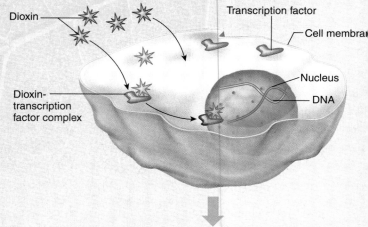

Dioxin

Transcription factor

Cell membrane

Nucleus

DNA

Dioxin-transcription factor complex

(b) The complex binds to DNA at more than one site and starts the transcription of several genes.

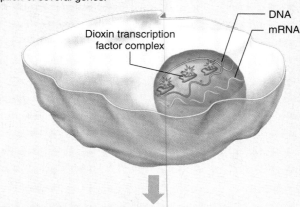

DNA

mRNA

Dioxin transcription factor complex

(c) Some of the genes code for proteins that break down dioxin and get rid of it. Other genes produce protein products that either mimic estrogen or interfere with estrogen function.

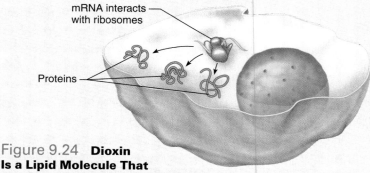

mRNA interacts with ribosomes

Proteins

Figure 9.24 **Dioxin Is a Lipid Molecule That Acts in Ways Similar to Estrogen.**

beginning to understand that dioxin activates a transcription factor complex—this transcription factor complex can explain the effects of dioxin on the whole organism.

QUICK CHECK

1. How does dioxin affect transcription?

Now You Can **Understand**

Side Effects of Accutane

A new treatment recently has become available for severe cases of acne. Isotretinoin (Accutane) can dramatically reduce the occurrence of severe acne, but as with many powerful medications it also can have side effects. In the case of Accutane young women must consider whether they might be pregnant when using the drug. How are these two topics—acne and pregnancy—connected? Now that you understand how the control of gene expression produces a new life, you can understand this link. Let's start with acne.

Researchers believe that Accutane acts by reducing the production of an oily substance in the skin that allows bacterial growth, which, in turn, produces a skin response that manifests itself as acne. If a woman is pregnant and using Accutane, these side effects can extend to her unborn child and cause it to be born with severe birth defects. Accutane is in a class of chemicals called *retinoids*. Vitamin A and retinoic acid are both retinoids. Retinoic acid is one of the important signal molecules that control gene expression during embryonic development. Retinoic acid is a signal involved in the development of the retina, the heart, the limbs, and many other structures. In ways that are not totally clear, Accutane either interferes with retinoic acid functions, or it mimics retinoic acid and so bathes the developing embryo in the signal. Either way, Accutane will disrupt the delicate balance of signals and gene expression in an embryo. The results are tragic: an aborted embryo, or a child who is born with many biological difficulties.

What Do **You** Think?

Are Pregnant Women Responsible for the Health of Their Unborn Children?

You have just learned a great deal about how genes control development. What use is this information? How does it impact you or the society around you? One intriguing implication of what you have just read is for understanding how to have a healthy baby. Certainly, the development of a healthy baby depends on the genes that baby carries—which the mother or father can do little about, at least if the pregnancy is achieved in the usual ways. (You can read about the unusual ways in Chapter 25.) An understanding of gene expression puts more responsibility on a pregnant woman for her baby's normal development. Certain behaviors by a mother can put her baby at risk. For instance, drinking alcohol risks fetal alcohol syndrome, smoking cigarettes risks low birth weight and other problems, consuming other drugs such as methamphetamine can cause a variety of birth abnormalities. How do these drugs produce their effects? These drugs act in many ways, but one common theme is that they all effect gene expression. If gene expression is abnormal, then the embryo cannot develop normally. Once gene expression is abnormal, its effects will be difficult to correct.

So what should be done with this knowledge? Abnormal babies come at a high cost—not only for the child and its family but also for the rest of society. Should society demand that pregnant women be tested for substance use? Should there be penalties for women who are warned of the effects of their behaviors on early gene expression? Should more money be spent on research to correct the abnormalities? How far should society go to increase the chances a baby will be born healthy? What do *you* think about this issue?

Chapter Review

CHAPTER SUMMARY

9.1 The Development of Multicellular Organisms Involves Cell Division and Cell Differentiation

While mitosis transforms a single-celled zygote into a multicellular organism, cell differentiation transforms a zygote into an adult with many different kinds of cells and different structures. Differentiation is a gradual process that involves the expression of different genes in different cells.

cell differentiation   211

9.2 During Development Cells with the Same DNA Gradually Come to Express Different Proteins

The end result of development is a mature organism that contains many different cell types that make up various tissues and organs. Another way to say this is that every cell in an individual expresses its proteome, which is only a fraction of all of the proteins its DNA can produce.

gene expression   213   proteome   214

9.3 The Structure of Eukaryotic DNA Allows Control of Gene Expression

A variety of mechanisms contribute to differential gene expression. Chromosome condensation and methylation structurally and chemically turn off the transcription of large segments of DNA. In addition, eukaryotic genes contain noncoding introns, which are sequences of nucleotides that are not translated into amino acid sequences. These are sometimes called junk DNA. Enzymes remove introns from transcribed mRNA, leaving only the coding sequence in a strand of mRNA that interacts with the ribosome and guides protein synthesis. This process means that eukaryotes can have more proteins than genes, because the coding sequences from a stretch of mRNA can be spliced together in different combinations. Humans have about 25,000 genes, but express about 100,000 proteins. Further chemical modifications of the amino acid chains that come off of the ribosome can lead to even more different kinds of final proteins.

exon   216   mRNA splicing   216
intron   216   telomere   216

9.4 Transcription Is Controlled by Transcription Factors

Which genes are transcribed is controlled by transcription factors. Transcription factors bind to DNA at a location called a control element and facilitate the binding of RNA polymerase. This means that transcription factors increase the rate of transcription of particular genes. Transcription factors are kept in an inactive form in the cell. They are activated when they are needed.

promoter   217   transcription factor   218

9.5 Gene Expression Controls Development

In a developing embryo or mature organism, transcription factors are activated by reception of signal molecules present in the cell's environment. The initial pattern of transcription factors in the early embryo is set up because the maternal egg already has an unequal distribution of different kinds of transcription factors, which are distributed to different daughter cells during the early rounds of embryonic cell division. The *Pax6* gene is found in all animal species and is a transcription factor that directs the formation of an eye. Atavisms and vestigial structures reflect ancestral structures. Structural genes code for proteins that do work in a cell; regulatory sequences of nucleotides code for proteins that control DNA. Atavisms result from mutations turning structural and regulatory sequences on and off.

atavism   221

9.6 Normal Development Is the Result of the Progressive Differentiation of Embryonic Stem Cells

Embryonic stem cells and adult stem cells have the potential to develop into a wide range of cells. They hold great promise for treatment of diseases. At this point U.S. government funds for stem cell research are restricted.

adult stem cell   223   embryonic stem cell   223

9.7 An Added Dimension Explained: Times Beach and the Effects of Dioxin

Dioxin is a toxic substance that activates transcription factors inside of cells. Dioxin is a lipid, so it crosses the cell membrane, binds to and activates certain transcription factors, and then stimulates or inhibits the transcription of certain genes. Many of the transcription factors that dioxin interacts with are normally either activated or inhibited by estrogen. As a result many of dioxin's effects are similar to the effects of estrogen.

REVIEW QUESTIONS

TRUE or FALSE. If a statement is false, rewrite it to make it true.

1. When chromatin is condensed, its proteins are transcribed more quickly.

2. Human females have one inactive X chromosome.

3. Methylation is one mechanism that inactivates the transcription of stretches of DNA.

4. All of the cells in a developing embryo carry DNA that has the same nucleotide sequence.

MULTIPLE CHOICE. Choose the best answer of those provided.

5. Human embryonic development can
   a. be understood by studying human embryos and nothing else.
   b. can never really be understood.
   c. be understood by studying the development of other mammals, but not nonmammalian animals.
   d. be understood by studying the development of a variety of eukaryotic species including plants, fruit flies, and fishes.
   e. be understood by using computer simulations alone.

6. A complex multicellular eukaryotic organism develops from a single fertilized egg. How does this happen?
   a. Mitosis and meiosis in the progeny of the fertilized egg.
   b. Because each cell in the embryo has different DNA.
   c. The processes of mitotic cell division and cell differentiation.
   d. The processes that control differential gene expression.
   e. c and d are correct.

7. Junk DNA is
   a. DNA that does nothing in a cell.
   b. DNA that is extruded from the cell during cell division.
   c. the difference between the DNA of humans and that of other species.
   d. noncoding DNA that has many important cellular functions.
   e. coding DNA that adds the unnecessary embellishments to a species.

8. The genes in a eukaryotic genome contain stretches of DNA that
   a. do not code for protein and are cut out of the transcribed mRNA.
   b. add important structural features to the proteins that are translated.
   c. identify the species that the gene comes from.
   d. determine the tissue that the protein functions in.
   e. all of the above

9. A Barr body is
   a. an inactivated, coiled X chromosome.
   b. an inactivated, coiled Y chromosome.
   c. an organelle found only in prokaryotes.
   d. another name for a particular kind of atavism.
   e. a clump of several inactivated, coiled chromosomes.

10. A single gene can produce more than one kind of protein by the process of
    a. transcription.   d. meiosis.
    b. translation.   e. alternative mRNA splicing.
    c. mitosis.

11. Where on the DNA molecule does RNA polymerase bind?
    a. the control element   d. two homologous sites
    b. the promoter   simultaneously
    c. the allele   e. the intron

12. Transcription factors become activated
    a. by signal molecules that bind to a membrane receptor and cause activation of the transcription factor.
    b. from the transcription of a specific DNA sequence that, in turn, activates the transcription factor.
    c. through factors released by mitochondria.
    d. by introns that bind to a membrane receptor and cause activation of the transcription factor.
    e. by exons.

13. How do different transcription factors end up in the cells of the head end and tail end of the fruit fly embryo?
   a. The cells in the head and tail end of the fruit fly embryo carry different DNA.
   b. The cells in the head end come from the sperm DNA, and the cells in the tail end come from the egg DNA.
   c. Transcription in the head end is slower than transcription in the tail end.
   d. Different concentrations of specific mRNAs are laid down in the head end and tail end of the egg before fertilization.
   e. Enzymes in the head end destroy some of the mRNA before it can be used to direct translation.

14. All animal eyes contain the same two basic cell types, which are
   a. prokaryote and eukaryote.
   b. mast cells and T cells.
   c. pigment cells and photoreceptor cells.
   d. photoreceptor cells and lens cells.
   e. photoreceptor cells and *Pax6* cells.

15. *Pax6* is a
   a. master control gene involved in the development of all animal eyes studied thus far.
   b. type of cell in all animal eyes.
   c. protein that inhibits eye formation.
   d. carbohydrate on the surface of retinal cells.
   e. gene that controls the development of human tails.

MATCHING

16–20. Match the columns.

16. telomeres
17. number of human genes
18. introns
19. number of human proteins
20. exons

a. DNA that codes for proteins
b. allow all functional DNA to be copied
c. about 100,000
d. junk DNA located within eukaryotic genes
e. 25,000

CONNECTING KEY CONCEPTS

1. How do the mechanisms of gene regulation allow the different kinds of cells in an organism to produce different kinds of proteins?
2. How does knowledge about the structure and development of eyes support the hypotheses that all animal species are genetically related to one another and that complex eyes evolved from simpler eyes?

QUANTITATIVE QUERY

The alternative mRNA splicing that you have read about in this chapter often produces different families of similar proteins. Each family or class is a group of different proteins that are similar but somewhat different. In other words, splicing might produce a family of membrane receptor proteins that are similar to one another but are different enough to have different functions. For example, the different proteins in a family might bind to the same signal molecule but produce different cellular effects as a result of that binding. Assume that in humans there are 25,000 genes and 100,000 different proteins organized into families. Assume also that each gene produces one family of proteins. On average, how many members would there be of each family of proteins?

THINKING CRITICALLY

1. Many human disorders are a result of problems with proteins. Abnormal proteins are produced, or too much protein is produced, or not enough protein is produced. How might knowledge of how transcription factors work lead to effective, practical treatments for such disorders?
2. Identical twins happen when a single developing embryo splits into two embryos early in development. Some researchers now have concluded that identical twins do not really have identical DNA. Based on what you have learned in this chapter, what could they mean by this?

For additional study tools, visit www.aris.mhhe.com.

Rules of Inheritance

CLASSICAL GENETICS

Eugenists Hail Their Progress as Indicating Era of Supermen

Scientists, in Congress Here, Advocate Birth Selection, Not Control, as Proper

1932

Winners of trophy for "Best Couple" 1926 Texas State Fair

Raymond Hudlow was awarded the Purple Heart and other medals for his war service.

An Added Dimension

Eugenics and Fitter Families

The stories behind the **chapter-opening photographs** are profound. All of these people were young at a time in U.S. history when "good breeding" was a major concern. Not good breeding of animals or crops, mind you, but good breeding of humans. The couple in the top picture was part of the American Dream, as it played out in the 1920s—this picture was taken at the Texas State Fair.

You might associate a county fair with Ferris wheels, roller coasters, cotton candy, corn dogs, and chances to win a huge stuffed bear by hitting a target. In addition, state and county fairs traditionally are places where Americans showcase their agricultural and domestic skills. Homemade canned goods, pies and cakes, crafts, and needlework are entered in competitions. Sheep, cows, pigs, rabbits, and chickens are judged for their genetic qualities. And the best entry in each competition wins the coveted blue ribbon.

What does this have to do with the couple in the photo? This couple won a blue ribbon and trophy at the 1926 Texas State Fair—but not for the best cherry pie, excellent crochet work, or pedigreed Hereford heifer. Instead, they won "Best Couple" based entirely on their own traits. In the early 1900s some believed that people should be judged for their traits, just like livestock. All across the country families were encouraged to enter the "Fitter Family" contests at their local county and state fairs. Contestants filled out long forms detailing their family traits. They were required to document that no "defective" traits, such as "feeblemindedness" or moral degeneracy, were present in their families. Then they had to undergo physical and psychological tests. These contests were popular, and families across the country submitted their pedigrees and family traits to the Fitter Family board of judges.

The Fitter Family contests were not just a lighthearted competition, though. They were part of a concerted effort in the United States and in many other countries to improve the genetic makeup of the human race, through a program known as eugenics. **Eugenics** is the philosophical belief that the human race can be improved if people with "good" traits are encouraged to breed and have children, while people with "bad" traits are discouraged from having children or even prevented from having them.

The man in the bottom chapter-opening photo experienced this dark side of the Eugenics Movement. He was 8 years old when he was surgically sterilized as a "mental defective." When the boy grew up, he fought in World War II, earning the Bronze Star, Purple Heart, and Prisoner of War medals for his acts of heroism. The irony of his story does not stop there, though. He was awarded these medals for fighting against Nazi Germany, where the ideas of eugenics were pursued most vigorously and brutally.

Today, most people recoil from the idea of a government-enforced eugenics effort—but if corn plants or cattle can be selectively bred, can't the same be done with people? What can science contribute to this debate? At the end of this chapter you will evaluate the claims of the early Eugenics Movement and their modern counterparts and form your own opinions about their validity. But before you think about these ideas further you must grasp the basics of genetics, as biologists understand them today. ■

10.1 The Science of Genetics Brings Together DNA, Cell Division, and Gene Expression

For longer than recorded history people have observed that an offspring resembles its parents. People also learned to save and plant seeds from the best food plants with the expectation that crops grown from these seeds would have the desirable traits of the parent plants. Humans were similarly clever about taming and breeding wild animals to produce useful and relatively docile domesticated species (**Figure 10.1**). Animals that provided food and clothing (pigs, cattle, rabbits, sheep, and various kinds of fowl), helped with chores (donkeys, cattle), were useful companions (dogs and cats), or provided transport (donkeys, horses, camels, and elephants) became prized possessions. The earliest successes at breeding useful plants and animals probably were integral parts of the beginning of human civilization.

Despite the obvious successes of agricultural breeding, it is hard to predict what traits the offspring of a mating will have based on informal observations. For thousands of years people relied on folk wisdom, passed from generation to generation, to predict the outcome of a selective mating. Sometimes this knowledge was useful but simplistic, like the old saying, "The apple doesn't fall far from the tree." Folk beliefs often were wrong—for example, the warning that "Pregnant women should never look at ugly or deformed people because they will give birth to deformed offspring." Eventually, people approached the problem of predicting the traits of offspring from a scientific perspective.

By the mid-1800s a more formal approach to understanding nature was improving human knowledge in many fields, including the selective breeding of plants and animals. People were developing clear hypotheses and systematically testing them, using the scientific method (see Section 1.5). First, researchers mated carefully chosen individuals in selective pairings called **crosses**

eugenics the philosophical and political belief system that asserts the human race can and should be improved by selective breeding

crosses the mating of selected individuals (usually of different kinds) to reveal the rules underlying inheritance

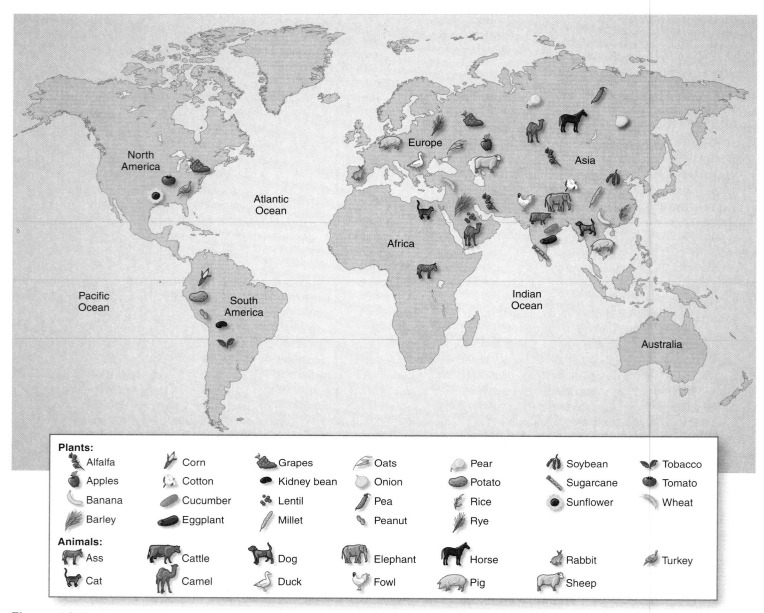

Figure 10.1 **Geographic Regions Where Animals and Plants Were Domesticated.**

(**Figure 10.2**). Analysis of the results (the offspring) of controlled crosses allowed people to form hypotheses about the rules that govern inheritance. The results of controlled crosses helped to evaluate, refine, and confirm hypotheses related to selective breeding. Gregor Mendel carried out the most systematic and insightful studies in this area of research.

Later generations of scientists have built on Mendel's work, providing comprehensive understanding of inheritance. It now is clear that the laws of inheritance Mendel discovered are based on the functions of DNA (Chapter 7) and on the mechanisms of cell division (Chapter 8). Today, you can understand the genetic basis of the inheritance of many human traits, including many diseases. Let's start by exploring Mendel's work in greater detail.

QUICK CHECK

1. Explain how a genetic cross is done.

(a) When parents with fine, soft wool are crossed…

…the cross can produce
a lamb with soft, fine wool.

(b) When parents with long-fibered wool are crossed…

…the cross can produce
a lamb with long-fibered wool.

Figure 10.2
**Crosses Are Matings
of Selected Parents.**

10.2 Gregor Mendel Was the First Systematic Researcher in the Field of Genetics

Gregor Mendel (1822–84) was one of the most able and thorough of the early scientists who studied inheritance, a field that today is known as **genetics.** He was an Augustinian friar at the Monastery of St. Thomas in Brunn (now called Brno), which then was in Austria, but now is in the Czech Republic (**Figure 10.3**). Although the monastery sent him to the University of Vienna for three years to become a certified teacher, Mendel never earned a degree. At the university he attended classes in physics, botany, and mathematics, all of which were useful in his later research into genetics. Although Mendel was an able student, he choked up on tests and was frustrated in his attempts to pass the required exams. Returning to his

Figure 10.3 **Gregor Mendel and the St. Thomas Monastery in Brno, Czech Republic.** Before Mendel became abbot of this monastery he experimented with the inheritance of traits in pea plants in its gardens.

genetics the scientific study of inheritance

monastery in 1853, the 31-year-old Mendel became an uncertified substitute teacher of physics and natural science at the monastery's school. Like other friars at his monastery, he also experimented in natural science. His bishop did not approve of mating experiments on animals, so from 1853 to 1862 Mendel worked in the monastery's garden and greenhouse, investigating the inheritance of traits in plants.

Mendel began his studies with careful observations. He grew, tended, and observed different varieties of garden peas. Mendel learned the patterns of how those traits appeared to pass from one generation to the next. To manipulate reproduction in garden pea plants, Mendel had to understand how the peas reproduce and how flowers function in reproduction (**Figure 10.4a**). Notice that the petals of pea flowers are not wide open like those of a daisy, but instead are folded over on one another (Figure 10.4b). The petals cover the reproductive structures of the flower, and each flower has both male and female structures (Figure 10.4c). The anthers contain pollen grains, the plant's immature male gametes; the female gametes are within the ovules that are inside the carpel. Male gametes travel to the female gametes and unite with ovules, fertilizing them. Once an ovule is fertilized, it will develop into a seed. In pea plants, the seeds are the peas that you eat cooked with butter and salt. The seeds are contained within a tough outer pod that protects them as they develop (Figure 10.4d). Eventually, pods split open and seeds are scattered (Figure 10.4e&f). A seed contains an embryo plant similar to the embryo that forms during animal development. That is to say, under the proper conditions, the embryo in the seed develops into an adult plant.

Mendel's knowledge of traits of pea plants allowed him to complete detailed studies of inheritance in pea plants. After two years of initial observations Mendel began systematic crosses. The basic principles that Mendel uncovered have stood the test of time—sometimes they are referred to as the Mendelian laws of inheritance.

Mendel Used Thorough and Diligent Methods to Uncover the Laws of Inheritance

How was Mendel able to discover any regularity in the patterns of inheritance? Many traits are inherited in complex ways, and the patterns of inheritance are not always easy to observe. For instance, imagine trying to understand the rules of inheritance by breeding cats and making note of the fur color of parents and littermates (**Figure 10.5**). If

(a) A pea plant in flower.

(b) Pea reproductive parts are completely enclosed within petals. Normally, pea flowers pollinate themselves.

(c) This pea flower has been cut open to show female and male reproductive structures. Notice the tiny row of ovules within the carpel and the pollen grains on the anthers.

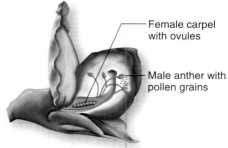

— Female carpel with ovules

— Male anther with pollen grains

(d) Once the ovules have been fertilized, the carpel and its ovules expand into a pod that contains peas.

(e) Each pea originates as a separately fertilized ovule. In this dried pod you see variations in peas. Some have smooth skins and others have wrinkled skins.

(f) Each pea can be planted and will grow into a new plant that will flower and repeat the life cycle.

Figure 10.4 **How Pea Plants Develop Flowers and Seeds.**

Figure 10.5 **Cat and Kittens.** The kittens in a litter can be many different colors.

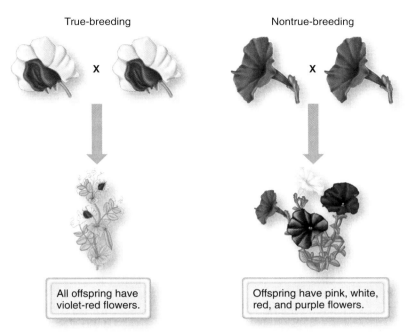

Figure 10.6 **True-Breeding Traits.** All offspring of parents with a true-breeding trait display that same trait.

you have ever seen a litter of kittens, you know that even if the parents look identical, in a single litter a mother cat can have kittens of many colors. There are rules that govern the inheritance of fur color in cats, but if you knew nothing about inheritance, the rules would be difficult to discover. When Mendel began his studies, he knew little about the rules of inheritance, yet he discovered a great deal about it. What were the secrets to Mendel's success?

One secret to Mendel's success was that he studied the inheritance of traits that he knew were **true-breeding** (**Figure 10.6**). True-breeding means that when two parents with an identical trait mate, the offspring have the parents' trait—and this continues in each succeeding generation.

Mendel spent two years confirming that different varieties of pea plants had true-breeding traits (**Figure 10.7**). For instance, Mendel kept one garden of white-flowered pea plants that always produced offspring with white flowers and another

true-breeding traits
identical parental characteristics that appear in offspring when both of their parents have that same trait

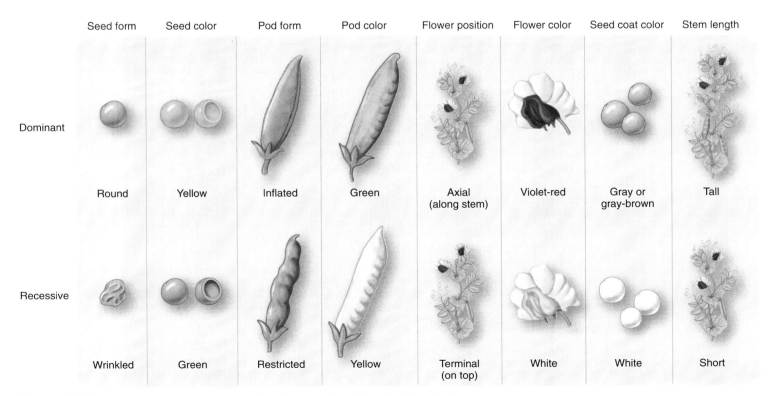

Figure 10.7 **True-Breeding Traits of Pea Plants That Mendel Studied.**

garden of violet-red-flowered pea plants that always produced offspring with violet-red flowers. He had other gardens of pea plants that were true-breeding for short stems (also called dwarf plants) or long stems (also called tall plants); or for peas with smooth seed coats or wrinkled seed coats. In all, Mendel conducted experiments on eight traits in 14 varieties of true-breeding pea plants. Imagine how careful Mendel must have been to separate and maintain his experimental pea gardens in this way.

Mendel's experimental methods had other qualities that contributed to his success. To eliminate the influence of random chance, experiments must be repeated many times. Mendel had to breed about 10,000 pea plants before the rules of inheritance became clear to him. In a remarkable break with biological methods of his day, Mendel mathematically analyzed five years of data. The traits of pea plants that Mendel chose to study, replications of his experiments, and use of mathematical analysis all contributed to Mendel's success in discovering the laws of inheritance.

Mendel also used meticulous methods in carrying out his crosses. Pea flowers normally fertilize themselves (**Figure 10.8a,b**). Sometimes, however, the pollen from one pea plant might reach a flower on a different pea plant, having been blown by the wind or carried by an insect. So when Mendel wanted a pea flower to self-pollinate, he would enclose the flower in a small cloth bag to exclude foreign pollen. To ensure a specific kind of cross-fertilization, Mendel pollinated each flower by hand. First, he used tweezers to pluck the pollen-producing anthers from a pea flower so it would not pollinate itself (Figure 10.8b). Then, using a paintbrush, Mendel dabbed pollen from a selected pea plant onto the flower's female reproductive organs. Finally, to prevent any subsequent contamination, he enclosed the pea flower in a bag (Figure 10.8b). The pollen that Mendel had applied would fertilize the pea flower, and its seeds would develop. When the seeds were mature, Mendel collected them, counted them, and noted any pertinent details of their appearance. He carefully stored each kind of seed from each experiment. The next season Mendel planted these seeds and grew a second generation of pea plants.

QUICK CHECK

1. If Mendel wanted to cross a dwarf pea plant with a tall pea plant, how would he have carried out the cross?

(a) A pea flower usually pollinates itself. In self-pollination, pollen from anthers falls onto the stigma and moves toward the ovules. Eventually, a sperm from a pollen grain reaches an ovule and fertilizes it.

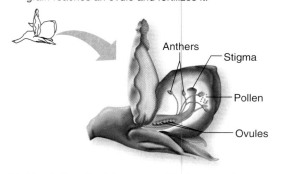

(b) Mendel's method for cross-pollinating pea plants.

1 Before their pollen is mature, the anthers are removed. This prevents self-fertilization. Pollen from another flower is dabbed onto the flower's stigma. Fertilized ovules from the cross-pollinated flower develop into seeds.

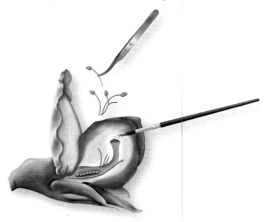

2 The cross-pollinated flower is enclosed in a small bag. This prevents pollen from another flower from getting onto the flower.

3 The next generation of pea plants develops. Each seed is a new individual.

Figure 10.8 **Mendel's Methods.** Mendel made his crosses of pea plants by removing anthers, applying pollen from a selected parent, and enclosing hand-pollinated flowers in bags.

Mendel Discovered Three Rules of Inheritance

Three important rules of inheritance came out of Mendel's work. These rules have been confirmed by over 100 years of subsequent research. Often these three rules are described as laws because they hold true in so many situations and with many organisms other than pea plants. These laws of inheritance include dominance, segregation, and independent assortment. In the next few sections you will examine each of these.

Mendel Discovered Dominant and Recessive Traits

Long before Mendel began investigating the genetics of pea plants, people knew about true-breeding traits. For instance, if a farmer wanted cows that produced abundant milk, he would breed the best milk producers until he had a herd of cows that produced lots of milk. Or if people wanted sweeter apples, a grower would plant seeds from a tree that had sweet apples until eventually his orchard had many trees of this kind. The predictions were not as clear, however, when it came to crossing organisms with different traits. Think about what the offspring might be if a true-breeding pea plant that produced violet-red flowers was crossed with a true-breeding plant that produced white flowers.

Figure 10.9 shows some of the possible outcomes that might have been predicted before Mendel's work. Although few early genetics researchers expected the outcome to be random, many probably would have predicted that blending was the primary rule of inheritance. Blending inheritance is the biological equivalent of mixing white and black paint to produce gray. For example, many researchers would have predicted that if you crossed true-breeding pea plants that had white flowers with true-breeding pea plants that had violet-red flowers, the offspring would produce light purple flowers. Mendel learned that blending was *not* the primary rule of inheritance for the traits he was studying.

In one experiment Mendel crossed true-breeding plants that would produce violet-red flowers with true-breeding plants that would produce white flowers. To keep the generations straight in his notes, Mendel developed some specific terms for each generation, and these terms are still in use today. The first generation of crossed individuals is the **parental,** or **P,** generation (**Figure 10.10**). The offspring of these parents are the **F₁,** for first filial generation. When the F_1 generation is allowed to cross, the offspring are called **F₂,** short for second filial generation. Filial may be a new word to you; it means son (*filius*) or daughter (*filia*) in Latin. So you and any siblings are the F_1 generation produced by your parents.

To test the hypothesis that blending of traits is the primary rule of inheritance, Mendel repeated

parental generation (P) the first parents crossed in a given genetic experiment

first filial generation (F₁) the offspring that result from the mating of the parental generation

second filial generation (F₂) the offspring that result when members of the F_1 generation are crossed

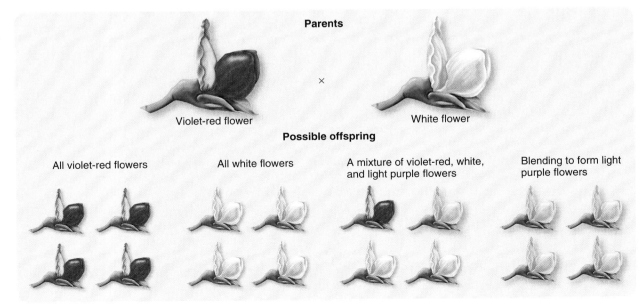

Parents

Violet-red flower   ×   White flower

Possible offspring

All violet-red flowers   All white flowers   A mixture of violet-red, white, and light purple flowers   Blending to form light purple flowers

Figure 10.9   **Potential Outcomes of Crossing a Pea Plant with Violet-Red Flowers with a Pea Plant with White Flowers.**

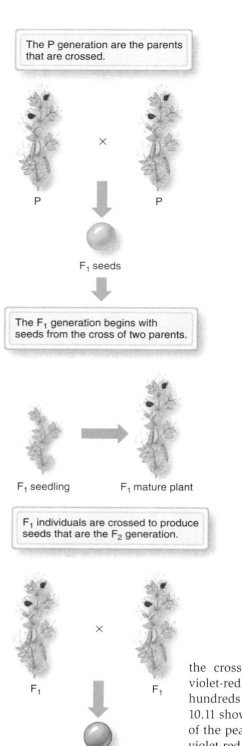

The P generation are the parents that are crossed.

P × P

F_1 seeds

The F_1 generation begins with seeds from the cross of two parents.

F_1 seedling F_1 mature plant

F_1 individuals are crossed to produce seeds that are the F_2 generation.

F_1 × F_1

F_2 seeds

F_2 seedling F_2 mature plants

Figure 10.10 **Names for Generations.** Parental, F_1, and F_2 are terms used to describe the generations in studies of inheritance.

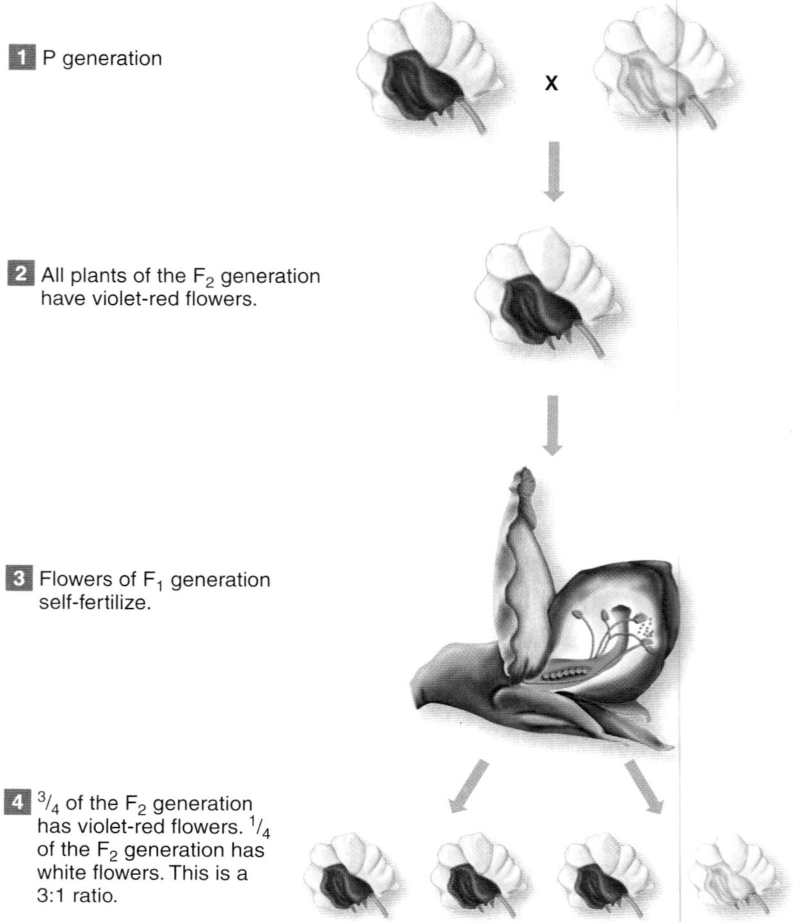

1 P generation

X

2 All plants of the F_2 generation have violet-red flowers.

3 Flowers of F_1 generation self-fertilize.

4 $^3/_4$ of the F_2 generation has violet-red flowers. $^1/_4$ of the F_2 generation has white flowers. This is a 3:1 ratio.

Figure 10.11 **A Trait That Disappears in the F_1 Generation Reappears in a 3:1 Ratio in the F_2 Generation.** When a white flowered pea plant is crossed with a violet-red flowered pea plant, all of the F_1 plants have violet-red flowers. After the F_1 plants self-fertilize, $^3/_4$ of them have violet-red flowers and $^1/_4$ have white flowers. This is a 3:1 ratio of violet-red to white flowers.

the cross of true-breeding parents with violet-red flowers and with white flowers hundreds of times (**Figure 10.11**). Figure 10.11 shows the results of these crosses: all of the pea plants in the F_1 generation had violet-red flowers. Based on these results Mendel could clearly reject some of the hypotheses presented in Figure 10.9. But Mendel's results raised many other questions.

One important question was: What happened to the trait for white flowers? Had it been lost forever, or did the violet-red color somehow mask the white color? It was not possible to answer these

questions from this initial study; therefore, Mendel went on to cross the F_1 generation with itself to produce an F_2 generation. The results were striking, and they are summarized in Figure 10.11. Mendel observed that about three-quarters of the F_2 pea plants had violet-red flowers and about one-quarter had white flowers. So the trait of white-colored flowers had returned! To put it another way, the F_1 generation that resulted from crossing violet-red and white flowered plants was *not* true-breeding.

This same pattern of results across the F_1 and F_2 generations showed up for all eight pairs of contrasting, true-breeding traits that Mendel investigated (**Figure 10.12**). In the F_2 generation each of these traits occurs at a ratio of 3:1. In other words, three-quarters of the F_2 offspring have one trait (like violet-red flowers), while one-quarter has the other (like white flowers).

How could you interpret these findings? Mendel had a simple explanation. He correctly suggested that two distinct genetic units control flower color and other traits in pea plants. Even though Mendel did not know about genes or alleles, he understood that the two genetic units that control each trait in an individual plant could be the same, or they could be different. For instance, a pea plant could have two alleles for violet-red flowers, two alleles for white flowers, or one allele for each flower color. **Figure 10.13** shows how alleles on homologous chromosomes correspond to Mendel's ideas about traits. Each chromosome carries a sequence of genes. The two chromosomes of one homologous pair carry alleles for each of those genes. Each pair of alleles codes for a protein that contributes to the same trait, but the alleles can code for different forms of the protein. The visible traits of the organism are determined by the combined action of the alleles on the two homologous chromosomes.

Mendel went one step further in his conclusions, though. Mendel's experiment suggested to him that when the two alleles are not in agreement—for instance, if one allele says "make proteins for white flowers" and the other says "make proteins for violet-red flowers," one allele completely controls the trait expressed by the individual. In the case of flower color, the violet-red allele wins and the flowers are violet-red, even though the plant also carries one allele for white flowers. If the trait associated with a particular allele is expressed regardless of the identity of the second allele, then the first allele is the **dominant allele.** In contrast, the trait that is associated with a **recessive allele** is expressed only if an individual carries *both* alleles for that trait. This is Mendel's *law of dominance.* Notice that any pea plant with white flowers must have two white alleles because the white allele is recessive. On the other hand, two different combinations of alleles will result in violet-red flowers: two violet-red alleles (both dominant), or one violet-red allele (dominant) and one white allele (recessive).

This underscores the fact that the appearance of a trait often is different from the genes that produce it. Biologists use two terms to distinguish between the traits of an organism and its genes. The alleles an organism carries are called its **genotype,** while the observable traits of an organism—which result from the interactions of alleles—are called its **phenotype. Figure 10.14** shows the genotype and phenotype for many of the traits of pea plants that Mendel investigated. As you can see, two plants can have the same phenotype with different genotypes.

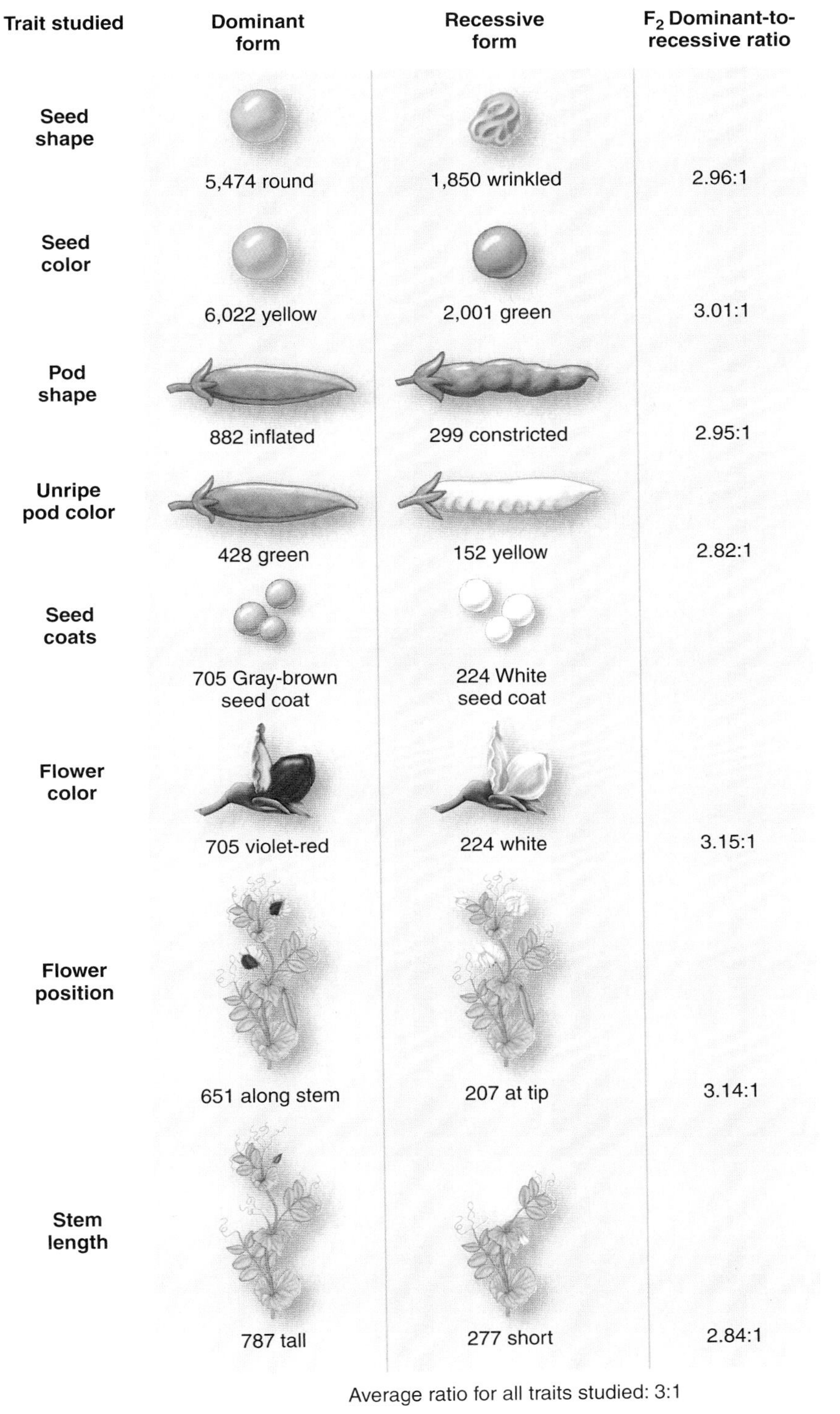

Trait studied	Dominant form	Recessive form	F$_2$ Dominant-to-recessive ratio
Seed shape	5,474 round	1,850 wrinkled	2.96:1
Seed color	6,022 yellow	2,001 green	3.01:1
Pod shape	882 inflated	299 constricted	2.95:1
Unripe pod color	428 green	152 yellow	2.82:1
Seed coats	705 Gray-brown seed coat	224 White seed coat	
Flower color	705 violet-red	224 white	3.15:1
Flower position	651 along stem	207 at tip	3.14:1
Stem length	787 tall	277 short	2.84:1

Average ratio for all traits studied: 3:1

Figure 10.12 Traits of Peas and Their F$_2$ Dominant to Recessive Ratios.

dominant allele an allele expressed regardless of the other allele being carried for that trait

recessive allele an allele expressed only when the other allele for that trait is identical

genotype the alleles that an individual carries

phenotype the traits that an individual expresses that result from expression and interaction of alleles

(a) Here is a pair of homologous chromosomes in a particular individual. One chromosome came from the male parent. The other chromosome came from the female parent.

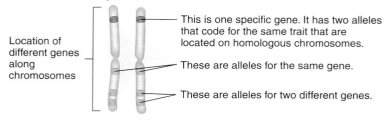

Location of different genes along chromosomes

This is one specific gene. It has two alleles that code for the same trait that are located on homologous chromosomes.

These are alleles for the same gene.

These are alleles for two different genes.

(b) Two alleles control the color of flowers on pea plants. One allele is on each homologous chromosome. There are three possible combinations of alleles for flower color, but only two possible visible traits: violet-red or white flowers.

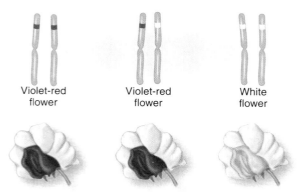

Violet-red flower

Violet-red flower

White flower

Figure 10.13 **Alleles and Traits.** There are two different alleles for each trait that Mendel studied.

Mendel Understood That Alleles Segregate During the Formation of Gametes

Even though Mendel did not know about alleles, genes, or chromosomes, he did understand that the basic genetic units segregate during the process of reproduction. You can understand this by returning to your knowledge of how alleles segregate during the process of meiosis, which produces gametes. In the original true-breeding parental generation, each pea plant that produced violet-red flowers had two violet-red alleles, one on each homologous chromosome. Similarly, each pea plant that produced white flowers had two white alleles, one on each homologous chromosome. When an organism carries two identical alleles for a trait, it is **homozygous** for this trait. The true-breeding pea plants with violet-red flowers are homozygous dominant, while the true-breeding pea plants with white flowers are homozygous recessive. During the formation of gametes in the parent plant, meiosis separates the homologous chromosomes, and therefore the two alleles associated with a trait, into different gametes (**Figure 10.15**). Even though Mendel did not specifically know about meiosis, he understood this concept and he expressed it as his *law of segregation*. The alleles

Characteristic	Homozygous dominant			Homozygous recessive			Heterozygous		
	Genotype		Phenotype	Genotype		Phenotype	Genotype		Phenotype
Seed shape	round	round	round	wrinkled	wrinkled	wrinkled	round	wrinkled	round
Seed color	yellow	yellow	yellow	green	green	green	yellow	green	yellow
Pod shape	inflated	inflated	inflated pod	constricted	constricted	constricted	inflated	constricted	inflated pod
Flower color	violet-red	violet-red	violet-red	white	white	white	violet-red	white	violet-red

Figure 10.14 **Genotypes and Phenotypes of Some of Mendel's Crosses.** Pea plants have two alleles for each of the traits that Mendel studied. These can be homozygous dominant, homozygous recessive, or heterozygous. Recessive traits only are expressed in the phenotype of homozygous recessive individuals.

carried by a parent are separated—or segregated—during the formation of gametes.

In Mendel's true-breeding pea plants each homozygous true-breeding parent carries two identical alleles, and so all of the gametes from each plant will carry the same allele for this trait. For instance, each gamete from the true-breeding violet-red flowered parents will carry a violet-red allele, and each gamete from the true-breeding white-flowered parents will carry a white allele. Therefore, when the white and violet-red true-breeding plants are crossed, the *only* combination of alleles that the F_1 offspring can inherit is one violet-red allele and one white allele. Every F_1 plant from this cross will have one of each allele. When an organism carries two different alleles for a trait, as it does in this specific cross, it is **heterozygous** for that trait. Because the allele for violet-red flowers is dominant over the allele for white flowers, all of the F_1 generation will have violet-red flowers. Although their flowers will look just like those from their violet-red flowered parent, their genotypes will be different.

How do these ideas apply to the F_2 generation? In Mendel's experiment with violet-red and white flowers, how did the white flowers come back? This, too, can be understood by considering the alleles and gametes in the experimental F_1 generation. All of the plants in the F_1 generation are heterozygous: all have one allele for white flowers and one allele for violet-red flowers (**Figure 10.16a**).

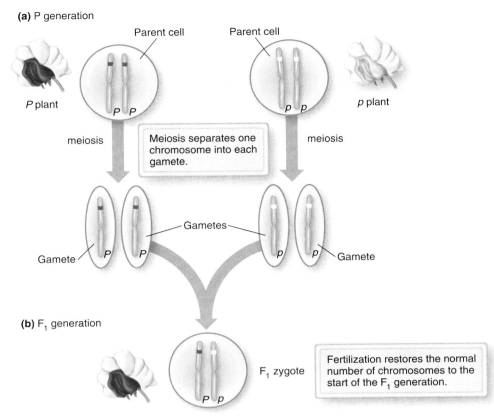

(a) P generation

Meiosis separates one chromosome into each gamete.

(b) F_1 generation

Fertilization restores the normal number of chromosomes to the start of the F_1 generation.

Figure 10.15   **Explaining Mendel's Results.** (**a**) Members of pairs of alleles become separated when gametes form. (**b**) The normal number of chromosomes is restored at fertilization. The zygote has one allele from each parent.

(a) F_1 generation: All plants are heterozygous for flower color.

(b) Gametes from the F_1 generation only can have an allele for violet-red or for white flowers.

gamete — gamete

(c) When the F_1 generation self-fertilizes, the F_2 offspring can have these combinations of alleles:

Figure 10.16   **Gametes and Alleles of F_1 and F_2 Generations of Violet-Red-Flowered and White-Flowered Pea Plants.**

(d) The F_2 generation is shown within the grid of this Punnett square:

Pp Female F_1 parent

P   *p* Female gametes

Male F_1 parent   *Pp*

Male gametes   *P*   *p*

	P	*p*
P	*PP*	*Pp*
p	*Pp*	*pp*

Phenotypic ratio 3:1
75% will have violet-red flowers.
25% will have white flowers.

Genotypic ratio 1:1:2
25% are homozygous and have two violet-red alleles (*PP*).
25% are homozygous and have two white alleles (*pp*).
50% are heterozygous and have one violet-red allele and one white allele (*Pp*).

■ Violet-red
□ White

homozygous when the pair of alleles that govern a particular trait are the same; one allele is located on each homologous chromosome

heterozygous when the pair of alleles that govern a particular trait are not the same; one allele is located on each homologous chromosome

A gamete from these plants can have either of these two alleles. Accordingly, half of the gametes in the F_1 generation will carry the allele for violet-red flowers (P), and half will carry the allele for white flowers (p) (Figure 10.16b). What are the possibilities when the F_1 generation is crossed? Although you can see the answer in Figure 10.16c, it is easiest to keep the alleles from each F_1 plant straight using a table called a *Punnett square* (Figure 10.16d). In this diagram the possible gametes from one parent are shown across the top of the table, and the possible gametes from the other parent are shown down the side. The genotypes of all potential offspring from this cross are written within the table's grid, and a Punnett square shows each possible combination of parental gametes. When the alleles are in a simple dominant/recessive relationship, using a Punnett square makes predictions about offspring relatively easy. In this example each F_1 parent will contribute one (P) allele or one (p) allele to each offspring. This means each F_2 offspring can have one of four possible sets of alleles:

- a white allele (p) from each parent,
- a violet-red allele (P) from each parent,
- a violet-red allele (P) from parent A and a white allele (p) from parent B, or
- a white allele (p) from parent A and a violet-red allele (P) from parent B.

If you imagine 100 such crosses, the Punnett square tells you that 25 (25%) of the offspring will be homozygous for white alleles, 25 (25%) will be homozygous for violet-red alleles, and 50 (50%) will be heterozygous. Given these genotypes, what will the offspring look like? Those with two white alleles (25%) will have white flowers, but all of the rest (75%) will have violet-red flowers. Look back a few paragraphs, and in Figure 10.12 you will see that these are exactly the percentages that Mendel found in his F_2 generation.

Mendel Found That the Alleles for Different Traits Usually Act Independently to Influence Traits

Mendel studied several traits in pea plants—therefore, it is not surprising that he looked at how the inheritance of one trait might affect the inheritance of another trait. Take a moment to think about this. For each trait that Mendel studied, one allele was dominant and one recessive. For example, round seed shape was dominant over wrinkled seed shape, and yellow seed color was dominant over green seed color. The alleles in the two populations of true-breeding parents are *RRYY* for round, yellow and *rryy* for wrinkled, green. Mendel wanted to know if yellow seeds always associate with round seeds, and green with wrinkled, or if the two traits are independent of one another. He came up with a clever way to find out. A cross of these two true-breeding parent populations will produce an F_1 generation with one dominant and one recessive allele for each trait, even though all of the F_1 generation will look like the round-yellow parents (**Figure 10.17a**). In other words, each individual plant of the F_1 generation will have *RrYy* alleles. The question of independence of the alleles can be answered by looking at what happens when the F_1 generation is crossed. In other words, what do plants of the F_2 generation look like? Are there any that are wrinkled and yellow or round and green?

Mendel did just this experiment, and the answer was unmistakably clear. Out of every 16 F_2 generation plants, 9 were round-yellow, 3 were round-green, 3 were wrinkled-yellow, and 1 was wrinkled-green (Figure 10.17b). From this experiment Mendel concluded that alleles for different traits act independently from one another. This is called Mendel's *law of independent assortment.*

Experimental crosses in which the researcher pays attention to two different traits, such as the cross of round-yellow and wrinkled-green peas, are called *dihybrid crosses.* Like the results from crosses involving just one trait, the results of dihybrid crosses can be understood by analyzing the cross with a Punnett square (Figure 10.17b). The possible alleles from the F_1 plants are shown around the outside of the Punnett square. With respect to seed color and seed shape, each F_1 plant can make four different gametes: *RY* (round-yellow), *Ry* (round-green), *rY* (wrinkled-yellow), and *ry* (wrinkled-green). Counting the offspring shown inside the Punnett square shows that the F_2 offspring will have these traits in a ratio of 9:3:3:1.

The simple rules of dominant and recessive traits discovered by Mendel are not the whole story in genetics. Nevertheless, these simple patterns do accurately describe the inheritance of many traits. **Figure 10.18** shows some of the human traits that are inherited in a dominant/recessive pattern. Many are diseases, but others are more benign traits, such as how your earlobes attach to your head.

QUICK CHECK

1. Use the new terms you have learned in this section to describe Mendel's crosses of yellow- and green-seeded pea plants and the results he obtained—that is, dominant/recessive, homozygous/heterozygous, genotype/phenotype.

When taken to the F₂ generation, a dihybrid cross of a pea plant shows a distinctive ratio of dominant and recessive traits.

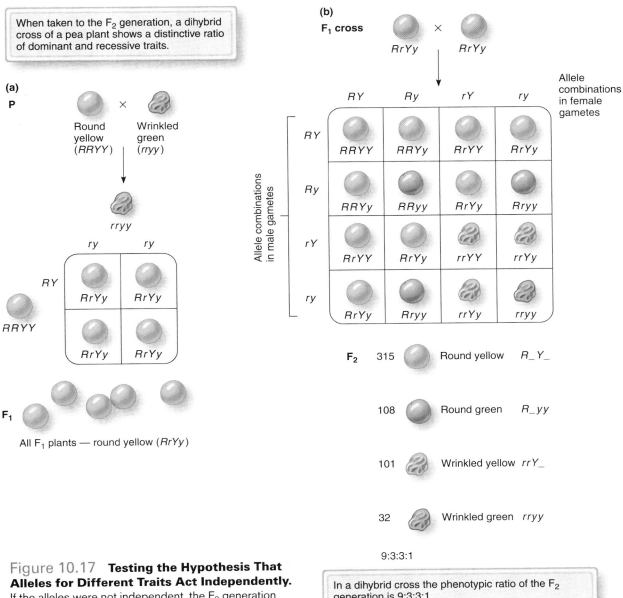

(b)
F₁ cross

RrYy × RrYy

Allele combinations in female gametes

Allele combinations in male gametes

(a)
P

Round yellow (*RRYY*) × Wrinkled green (*rryy*)

rryy

F₁

All F₁ plants — round yellow (*RrYy*)

F₂

315 — Round yellow — *R_Y_*

108 — Round green — *R_yy*

101 — Wrinkled yellow — *rrY_*

32 — Wrinkled green — *rryy*

9:3:3:1

Figure 10.17 Testing the Hypothesis That Alleles for Different Traits Act Independently. If the alleles were not independent, the F₂ generation would have been 3/4 yellow-round and 1/4 wrinkled.

In a dihybrid cross the phenotypic ratio of the F₂ generation is 9:3:3:1.

(a) Attached earlobes (r)

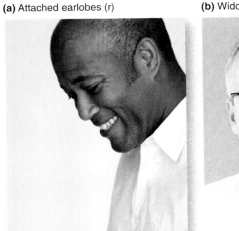

(b) Widow s peak (d)

(c) Curly/wavy hair (d)

Figure 10.18 Dominant (d) and Recessive (r) Human Traits.

243

10.4 More Complex Patterns of Inheritance Are an Extension of Mendel's Basic Rules

In a sense, Mendel was lucky because simple dominant and recessive alleles governed many traits of the pea plants he chose to study. When he crossed the parent generations from true-breeding strains of pea plants, he got just one trait in the F_1 generation. This simple dominant/recessive relationship between alleles describes the inheritance of many traits, but many other traits have more complex patterns of inheritance. These more complex patterns do not invalidate the rules that Mendel discovered but are extensions of Mendel's basic discoveries. Let's look at some of these other ways that traits can be inherited by considering some specific examples.

Some Alleles Show Incomplete Dominance or Codominance

If Mendel had studied the inheritance of color in chicken feathers instead of traits in pea plants, his task would have been more difficult. Chickens come in several colors. Like Mendel's varieties of true-breeding pea plants, a breeder can develop strains of chickens that breed true for feather color. For instance, a breeder can develop a true-breeding flock of white-feathered chickens or a true-breeding flock of black-feathered chickens. Black or white feather color appears to be governed by a single pair of alleles, so you might expect one feather color to be dominant and the other to be recessive. In other words, if you crossed a black hen with a white rooster, you might expect all of the F_1 generation to be either black or white. But life is full of surprises, and the study of genetics is no exception. When true-breeding black and true-breeding white chickens are crossed, all offspring have bluish-gray feathers. This feather color is intermediate between that of their parents (**Figure 10.19**). Geneticists call

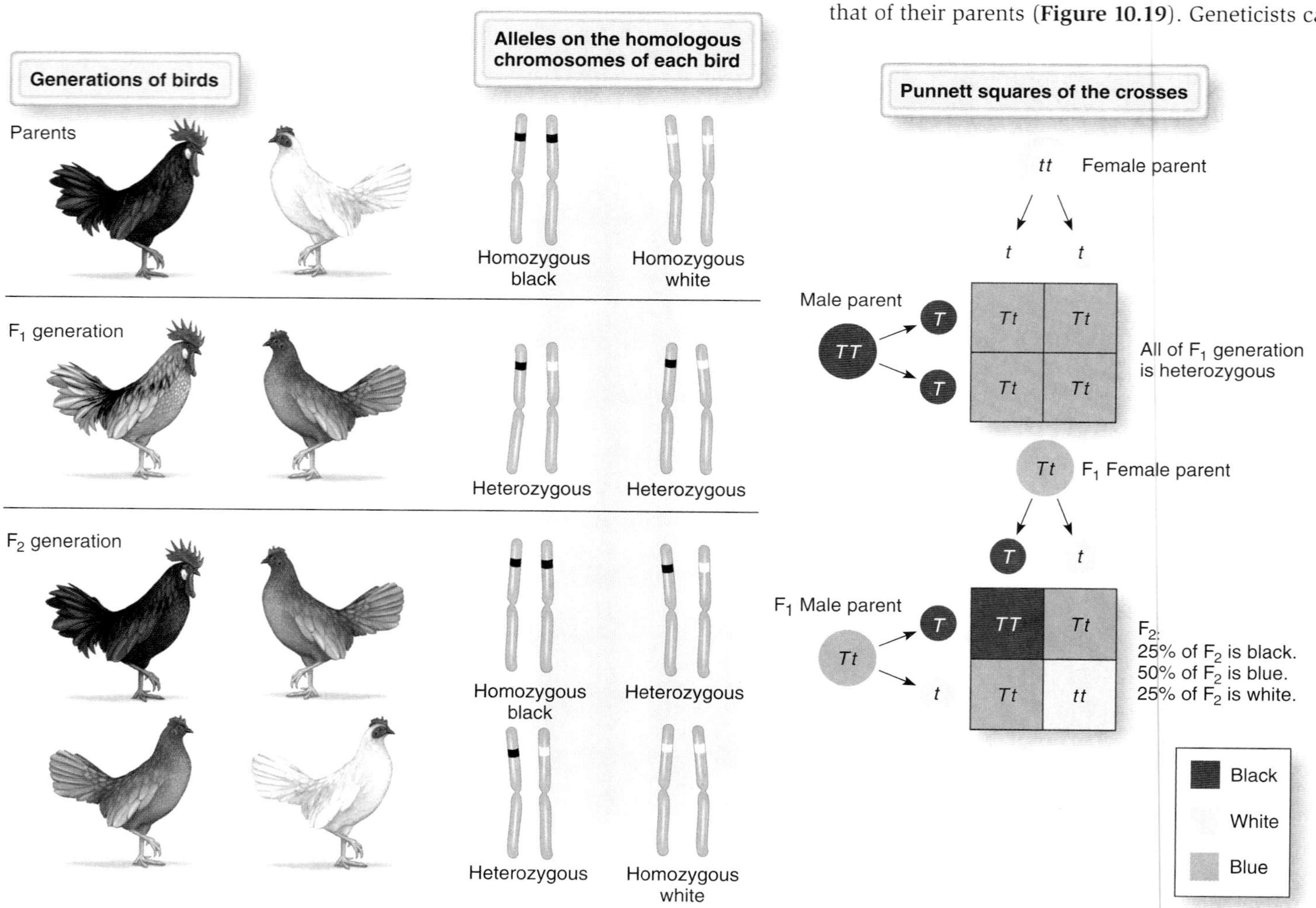

Figure 10.19 **Blue Chickens.** When a black chicken (TT) mates with a white chicken (tt) the offspring have blue feathers. Birds of the F_2 generation have black (TT) or white (tt) or blue (Tt) feathers. This is an example of incomplete dominance.

this effect **incomplete dominance.** There are usually three phenotypes with incomplete dominance: the two phenotypes expressed by the true-breeding parents and one intermediate phenotype.

Incomplete dominance sounds like support for the hypothesis that genes blend to produce the phenotype. Well, not quite. One prediction of the blending hypothesis is that the blue-gray birds would be true-breeding and crosses of blue-gray birds with themselves would produce only blue-gray birds. But this does not happen. In the F_2 generation the black and white traits reappear. While half of the F_2 chickens are blue-gray, one-quarter are black, and one-quarter are white. What does this mean?

These results can still be explained by alleles on chromosomes, summarized by the Punnett square and the diagrams in Figure 10.19. The two parents are homozygous for either black or white alleles for feather color. Each bird of the F_1 generation is heterozygous, and the feathers of F_1 birds are blue-gray; neither allele is dominant, and the result is an intermediate phenotype. In the F_2 generation half of the offspring are heterozygous and have the blue-gray feathers that demonstrate incomplete dominance, but one-quarter of the offspring are homozygous for black alleles, and one-quarter are homozygous for white alleles.

Yet another way for alleles to interact is for both alleles to be completely expressed. This kind of inheritance is called **codominance,** the condition in which the traits of both true-breeding parents are visible in the offspring. Codominance is *not* blending, and the traits of the offspring are *not* intermediate between those of their parents. Rather, each parental trait is clearly seen in the offspring. A good example of codominance is the expression of human blood types.

There are four different human blood groups: A, B, AB, and O. Unique carbohydrate molecules on the surfaces of red blood cells define each blood group. Blood type A has one kind of carbohydrate, type B has another, while type AB red blood cells have both kinds, and type O blood cells lack either of these carbohydrate molecules. The carbohydrate molecules are made with the help of enzymes, and these enzymes are determined by genes. People with type A blood can have one of two genotypes. They either have two alleles that code for type A enzymes, or they have one A allele and one O allele. People with type B blood are either homozygous for type B alleles or heterozygous for BO. People with type O blood are homozygous for O alleles. As you might expect, heterozygotes for A and B have alleles that code for both kinds of enzymes and they produce *both* A and B carbohydrates. This means that the A and B alleles are codominant and when both

are present, both are expressed. Notice the subtle difference between codominant traits and incompletely dominant traits: a pair of heterozygous codominant alleles expresses *both* phenotypes, while a pair of heterozygous incompletely dominant alleles produces an intermediate phenotype.

Using a modified Punnett square makes inheritance of blood type easier to understand (**Figure 10.20**). All possible alleles in the population are indicated above the columns and beside the rows, while the possible genotypes and phenotypes are shown in the grid. The actual proportions of A, B, AB, and O blood types in a population vary around the world (**Table 10.1**). In general, type O blood is

incomplete dominance the type of inheritance when two true-breeding parents have offspring with traits that are intermediate between those of their parents

codominance the type of inheritance in which the traits of both true-breeding parents are seen in each offspring

Although each person only has 2 alleles for blood type, there are three possible alleles for blood type in human populations: A, B, and O.

	A	B	O
A	AA **Type A**	AB **Type AB**	AO **Type A**
B	AB **Type AB**	BB **Type B**	BO **Type B**
O	OA **Type A**	OB **Type B**	OO **Type O**

Figure 10.20 **Genotype and Phenotype of Human Blood Groups.** The genotype is in the upper left corner of each box. The phenotype is in the middle of each box.

■ *Draw a Punnett square to show the possible blood groups in the children if the father is blood group O and the mother is blood group AB.*

Table **10.1**	The percentage of the ABO blood types in different world populations			
	BLOOD TYPE			
Population	**A**	**B**	**AB**	**O**
Overall human population	40%	10%	4%	46%
Amerindian	0–2%	0–2%	0–2%	90–100%
Ainu	30%	32%	18%	17%
European	40%	10–15%	5%	40–45%

most common (46%), while type AB is most rare (4%). In populations that have had little interbreeding, these percentages can be much different. In the rare populations of Native Americans that have not mixed with Europeans and Africans, type O blood is found in nearly 100% of any population. At the other end of the spectrum, the Ainu people of Japan—a small isolated population—have only 17% type O blood and 32% type B blood. Because human groups tend to migrate and interbreed, these extreme distributions are not common. Most populations show blood types similar to those of European ancestry. Among Europeans roughly 40 to 45% of the population have type O blood, 40% have type A, 10 to 15% have type B, and about 5% have type AB.

One Phenotype Can Be the Result of the Action of Multiple Genes, and One Gene Can Affect Multiple Phenotypes

The traits Mendel studied were easy to understand because each trait was controlled by just one gene. Many other traits in many species also follow this simple pattern, making it easy to figure out the rules of inheritance and relatively easy to identify the genes responsible for traits. For example, about 10,000 inherited human disorders are the result of the action of a single gene. **Table 10.2** lists just a few of these, and tells if the disease allele is dominant or recessive.

Most traits of multicellular organisms result from the actions of many different proteins. This means that many phenotypes result from the actions of more than one gene, a condition that is called **polygenic.** Now that you understand that DNA codes for proteins and not directly for traits, this should make sense to you. Proteins do things

polygenic the type of inheritance in which a trait is controlled by more than one gene; polygenic traits are expressed along a continuum

pleiotropy the name for when a single gene has effects on more than one phenotype

like catalyze cellular reactions and serve as receptors for cellular signals. Sometimes the traits of an individual result from the actions of just one protein, but this is the exception, not the rule.

For example, human skin color is determined by the amount of melanin produced in the skin, which in turn is controlled by three or four genes, all of which carry the instructions for an enzyme that catalyzes the synthesis of melanin, a brown pigment. Other genes, however, also are involved in the control of skin color. Genes for transcription factors, membrane receptors, and signal molecules involved in melanin production also affect skin color. The full story of the genetics of human skin color—and other polygenic traits—probably will be understood in the next 10 years or so but are not well understood now.

Not only can one phenotype be produced by the expression of more than one gene, one gene can influence more than one phenotype. This makes sense, given that each gene carries the instructions for one kind of protein. A given protein might be involved in only one cellular or body function, but it is more likely to be involved in many functions. When a single gene affects more than one phenotype, it is called **pleiotropy.** Albinism, a striking recessive mammalian trait, will make the concept of pleiotropy clearer.

An albino human usually has nearly pure white skin, white hair, blue irises, and pupils that look red rather than black. Besides this severe lack of pigmentation, albino humans and other animals often have significant vision problems. Their visual acuity is below normal, and their depth perception often is impaired. In most instances all of these effects are caused by a single allele. As in the case of normal variation in skin color, albinism is caused by a gene that codes for a protein involved in melanin production. Melanin not only produces skin color but also plays an important role in retinal development. The retina contains many darkly pigmented cells. If pigmented cells are not present, the retina's light-sensing and light-processing cells do not develop normally. In this way albinism is an example of pleiotropy: alleles for melanin affect several different and apparently unrelated functions.

Recessive Alleles Can Be Expressed Differently If They Are Carried on the X Chromosome

Mendel's studies revealed the patterns of traits if the alleles involved are in a dominant/recessive relationship. When two homozygous parents—one with two dominant alleles, the other with two

Table **10.2** Some inherited human disorders that are caused by the action of a single gene	
Disorder	**Dominant or recessive**
Adenosine deaminase (ADA) immunodeficiency disorder	Recessive
Albinism	Recessive
Cystic fibrosis	Recessive
Familial hypercholesterolemia	Dominant
Growth hormone insensitivity syndrome	Recessive
Huntington's disease	Dominant
Phenylketonuria	Recessive
Sickle cell anemia	Recessive

recessive alleles—mate, all of their offspring will be heterozygous and express the dominant phenotype. There is one situation, though, where this result is not seen in the F_1 generation, even with dominant and recessive alleles, and that is when the alleles are carried on the human X chromosome.

Recall that humans have 22 pairs of homologous chromosomes and a twenty-third pair of qualitatively different chromosomes, called the X and Y chromosomes. A female carries two X chromosomes, while a male carries one X chromosome and one Y chromosome. The Y chromosome is smaller than the X. It is so small that there are many alleles present on the X chromosome that are *not found* on the Y. This has important implications for the inheritance of traits carried on the X chromosome and a specific example will illustrate this concept.

Consider the trait of hemophilia, which is a disorder of blood clotting. Hemophiliacs bleed profusely when cut and suffer severe blood loss from even simple bruises. The allele responsible for hemophilia is recessive to the normal version of the allele that produces a blood-clotting protein called *factor VIII*. While either version of the allele is carried on the X chromosome, there is no such matching gene on the Y chromosome. Inheritance patterns for hemophilia are summarized in **Figure 10.21a&b** where you can see that if a female carries two normal alleles, or one normal and one hemophilia allele, she will have a normal phenotype. She will show illness only if she carries two hemophilia alleles. In contrast, if a male has one normal allele on his X chromosome, he has normal blood clotting, but if he has a single allele for hemophilia on his X chromosome, then he will show the disease. As you can see in Figure 10.21c, even though females carry the disease, males are more likely to have hemophilia. This pattern is called a **sex-linked** genetic trait.

Genes Interact with the Environment to Produce a Phenotype

This description may have given you the impression that genes rigidly determine the traits of an organism, but nothing could be further from the truth. Life is successful because it interacts with, and responds to, the environment. While genes carry instructions for making proteins, protein function is influenced by interactions with other proteins and by interactions with the environment. The environment controls the genes that are transcribed and translated (see Chapter 9). Even disorders that seem to be rigidly controlled by genes are influenced by the environment. For example, stud-

> Alleles for hemophilia are carried on the X chromosome. *HB* is the normal allele for Factor VIII; *hb* is the abnormal allele for Factor VIII.

(a) Although a female carrier has normal blood clotting, 50% of her ova have a dominant X^{HB} allele and 50% of her ova have a recessive X^{hb} allele.

HB   hb

(b) Here are the possible offspring if she mates with a man who is normal for Factor VIII.

$$X^{HB}\ X^{hb}$$

(c) The following offspring are possible if a female carrier has a child with a male who is normal for Factor VIII.

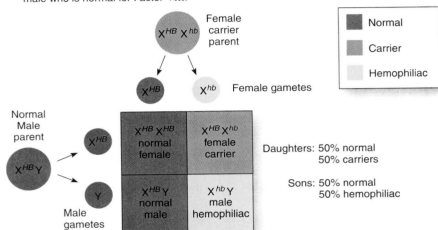

Daughters: 50% normal
50% carriers

Sons: 50% normal
50% hemophiliac

Figure 10.21   **Inheritance of Hemophilia.**

■ *Can males be carriers of hemophilia? Why or why not?*

ies in humans and animals support the hypothesis that the severity of mental impairment, in such disorders as Down syndrome or Huntington's disease, can be reduced if an affected person or animal is exposed to an enriching environment. The inheritance of fur color in certain rabbits and cats demonstrates some of these principles in a simple model system.

Like skin color in humans, fur color in animals is controlled partly by the amount of melanin produced. A completely white cat or rabbit has few or no alleles that result in melanin production, while a coal black cat or rabbit has many of them. One of the alleles that contributes to fur color produces a temperature-dependent form of an enzyme involved in melanin synthesis. This allele produces an enzyme that operates well at cool temperatures but

sex-linked trait a trait that is carried on a sex chromosome

Figure 10.22 **Color Change in Siamese Cats.** As Siamese kittens grow older their ears and tail tips get farther from their warmer bodies and their melanin alleles begin to be expressed. This shows up as dark tips on nose, ears, tail, and paws.

does not work at warmer temperatures. To see how this genotype determines the animal's phenotype—the color of its fur—let's follow the development of a cat or rabbit with two copies of such a temperature-dependent allele.

Cats and rabbits begin to grow fur while they are still in their mother's womb, and melanin production begins then as well. If the young animal has at least one normal allele for melanin, then it will grow colored fur. In contrast, if it has two alleles for enzymes that can never make melanin, it will grow white fur. On the other hand, if it has two temperature-sensitive alleles, something different happens. In the womb the animal is toasty warm, and the melanin enzymes do not work. Therefore, the newborn kitten or rabbit has light fur. As the animal grows, the tips of the ears and tail move farther from the center of its body, and therefore these regions become cooler. At cooler temperatures the temperature-dependent enzyme is functional, and melanin is produced. The result is an animal with a light-colored body and dark fur on the tips of nose, ears, tail, and toes. Sound familiar? This is a Siamese cat or a Himalayan rabbit (**Figure 10.22**).

How can scientists be certain that temperature controls the function of the melanin enzyme in animals like these? They make observations and do experiments. In one experiment (**Figure 10.23**) a group of Himalayan rabbits was reared at higher than normal temperatures. This experimental treatment raised the temperature of even the tips of the ears, tail, and paws above the temperature where the enzyme works—consequently, these animals were uniformly light-colored (Figure 10.23b). To push the protein in the other direction, another group of animals was raised in a normal temperature but with a "cool pack" placed somewhere on the body. These animals also had light-colored fur, except in the place where the cool pack rested. Dark-colored fur grew in the cooled places (Figure 10.23c). This simple example shows how genes, development, and environment can interact to produce a particular characteristic.

Both Nature and Nurture Influence Human Traits

If you pay attention to the popular presentations of science, you will find references to the debate about nature versus nurture. Nature refers to traits supposedly influenced by genes, while nurture refers to traits supposedly influenced by environment or experience. The reality is that all traits depend on the way genes act in a particular environment. A major aim of research in genetics is to determine how much, and in what ways, the envi-

(a) Rabbit raised at 25°C (77°F)
Result: Typical coloration

(b) Rabbit raised at temperatures warmer than 30°C (86°F)
Result: Uniformly light-colored

(c) Rabbit raised at 25°C, but with V-shaped cold pack
Result: Dark-colored fur grew in cooled areas.

Figure 10.23 **Effect of Temperature on Himalayan Rabbit Fur Color.**

ronment can modulate the way that genes are expressed.

The importance of environmental influences has been shown in many studies of early development. For example, if kittens are raised in an artificial environment where they see only vertical lines or are fitted with goggles that prevent them from using both eyes simultaneously, then their sensory abilities and brain structures are permanently altered. These changes are permanent and cannot be "unlearned." A similar thing happens in humans who have strabismus or "lazy eye," a condition caused by eye muscles pulling too strongly in one direction, causing the eye to rest in an off-center position

(**Figure 10.24**). If lazy eye is not corrected when a child is quite young, there can be permanent changes in vision. Studies indicate that similar rules apply to many important human traits that rely on brain functions. Intelligence, sensory perception, language, and many other brain-related traits do not develop in an impoverished environment. The flip side is also true: regardless of one's genotype an enriched environment can enhance brain development.

For many traits of interest to humans there are no animal equivalents to shed light on how the environment guides the expression of a person's genotype. One example of how scientists approach uniquely human phenotypes is the study of schizophrenia, a mental disorder in which a person appears to be out of touch with the reality that most other humans experience. The symptoms of schizophrenia can vary in different people, but they include hearing voices in your head, thought disturbances such as speaking in jumbled words and phrases, unfounded fears of persecution and danger, and uncontrollable hostility. Schizophrenia is rare and strikes only about 1% of the population in all cultures worldwide, but its impact is devastating. To make progress toward curing schizophrenia, it is critical to determine the extent to which genes and the environment each contribute to the disorder. One way to study this is to study schizophrenia in identical twins. They will have identical or nearly identical DNA—and, while most twins are reared in the same environment, some are separated at birth and reared in different environments by different parents. If schizophrenia is a completely inherited disease, then regardless of their environments, if one twin is schizophrenic, then the other twin should be schizophrenic too.

If one twin is diagnosed with schizophrenia, there is about a 50% chance—a number that comes from an average of many studies—that the other twin also will develop schizophrenia. This is a very high rate of coincidence. If you randomly chose any two unrelated people from the general population, there would be no way to predict whether one has schizophrenia based on the mental state of the other. Even within the same family, if one person has schizophrenia, the chances that someone else in the family has it usually are less than 15%. But if you know that two people are identical twins, and if you know that one twin has schizophrenia, there is about a 50% chance that the other twin also will have schizophrenia, regardless of the environment in which each was raised. This means that schizophrenia is partly determined by genetics. It also means, though, that something in the environment also plays a role. Researchers still do not know what environmental factor combines with the genetic predisposition to produce schizophrenia. It might be a

virus, a chemical, stress, lack of sunlight, or some combination of factors that produces schizophrenia. What is certain is that schizophrenia is inherited to some degree but not in a simple or absolute way.

Schizophrenia is just one example of a complex human trait determined by the interaction of genes and environment. The study of schizophrenia is somewhat straightforward because it is rare in the general population and has some distinct symptoms. Other traits such as intelligence or compassion are more difficult to define and study. Nevertheless, scientists have reason to believe that nearly all human traits develop as a result of a combination of genes and environment. In some cases, such as schizophrenia, the genetic contribution might be quite high, while in other cases genes may play a lesser role. But the interaction between genes and environment holds the key to understanding human nature.

Figure 10.24
Strabismus, or Lazy Eye.

> **QUICK CHECK**
>
> **1.** Describe how full dominance, incomplete dominance, and codominance are different.
> **2.** How do fur color in cats and schizophrenia in humans demonstrate the interaction between genes and the environment?

10.5 Knowledge About DNA Illuminates Genetics

The modern understanding of genetics is greatly influenced by the understanding of DNA and how it works. DNA's genetic instructions dictate which proteins a cell or organism will make (see Chapter 7). Therefore, an understanding of the relationship between genes and proteins is the only way to fully understand genetic inheritance.

Mutations in DNA Produce Different Alleles

The variation in the traits of individuals is a result of the different alleles for each gene. But where does this variety of alleles come from? The answer is DNA mutations. Recall that a mutation is an alteration in the nucleotide sequence for a stretch of DNA (see Section 7.6). A mutation can be just one nucleotide, or it can be several, but any change in the DNA sequence will produce a new allele. If the allele is successfully handed down from parents to offspring, it will persist in the population, and the trait it produces may be seen in individuals. The history of life on Earth is a history of the success and failure of various DNA mutations. Mutations happen regularly, from a variety of mechanisms, and they provide the alleles that generate genetic variation.

A possible scenario for how some humans came to have light-colored skin illustrates this concept. *Homo sapiens,* the human species, originated in Africa 150,000 to 200,000 years ago. Our ancestors were a migratory breed, and by 50,000 years ago *H. sapiens* inhabited far-ranging lands from Northern Europe to Asia and even Australia. For most of this history all *H. sapiens* were dark skinned. They carried alleles for the production of melanin, and there was only modest variation in skin color. Recent studies suggest that sometime between 20,000 and 50,000 years ago, a single human living in Europe acquired a mutation in one of the alleles that control melanin production and distribution. This person was born with lighter skin and passed this mutation on to his or her offspring. Having light skin was an advantage in the weaker sunlight of northern climates. Lighter skin may have allowed people to produce more vitamin D, which is synthesized in the skin in response to sunlight. Or there might have been some other advantage. Regardless, people with lighter skin survived in greater numbers and had more children compared with darker-skinned people. Eventually, nearly all humans in northern Europe had light skin, because they carried the advantageous mutated allele.

Of course, not all mutations lead to increased survival. Some mutations lead to neutral alleles that neither help nor harm the individual that carries them. Still other mutations are harmful. The 10,000 single hereditary disorders controlled by a gene are examples. In some cases these mutations are eliminated from the overall population because the individuals that carry them die before they can reproduce. For example, there are no individuals with mutations that shut down the metabolic pathways that produce ATP, because cells with such a mutation could not survive. Some alleles with harmful mutations continue in a population because the disease or disorder is not severe enough to prevent survival and reproduction. Human cultural practices and knowledge can protect individuals with disorders. This may allow them to breed and pass on the deleterious allele. In yet other instances an allele might be deleterious under some conditions but helpful under other conditions and so is maintained in the population. In addition, many of these mutations recur spontaneously and are never completely eliminated from the overall human population. Regardless of the outcome, mutations are the basic mechanisms that allow phenotypic variation among individuals of any species.

Understanding Sickle Cell Anemia

Sickle cell anemia is a well-understood inherited disease that shows how genetics works at different lev-els. **Anemia** is a general name for diseases in which red blood cells cannot supply oxygen to cells in the body. **Sickle cell disease** is an inherited form of anemia caused by a single allele that carries the instructions to make the blood protein hemoglobin (see Section 3.7). Abnormal hemoglobin causes red blood cells to assume abnormal shapes, especially when oxygen levels are low. Red blood cells of people with sickle cell disease collapse into pointy, flattened, half-moon or "sickle" shapes when oxygen levels are low (see Figure 3.19). A lower amount of oxygen is the normal condition in tissues with a high level of cellular respiration, such as muscle. Sickled red blood cells cause two problems: first, they tend to clog the small blood vessels called *capillaries,* which carry blood to tissues and cells. Like leaves blocking a gutter, sickled cells accumulate in tiny capillaries and plug the normal flow of blood. This deprives downstream cells and tissues of oxygen and nutrients and causes wastes to accumulate. Second, sickled red blood cells have a shorter life span than that of normal red blood cells. Sickled cells die quickly, leaving the patient with too few red blood cells. For both of these reasons, a person who has sickle cell disease lacks sufficient oxygen in various body tissues.

The symptoms of sickle cell disease are diverse and can affect any part of the body. Muscle weakness, heart failure, pain and fever, brain damage, and kidney failure are typical, but the lack of adequate blood supply can damage any or all of the body's organs and often leads to an early death. You can see from this description that the allele that causes sickle cell anemia has wide-ranging effects throughout the body.

At the level of proteins, sickle cell disease is caused by abnormal hemoglobin. Normal hemoglobin has a globular shape and is made of four components: two alpha subunits and two beta subunits (**Figure 10.25a**). Each subunit binds a molecule of oxygen (O_2) (Figure 10.25b). Hemoglobin picks up oxygen when the blood cells pass through capillaries of the lungs and releases oxygen when the red blood cells reach peripheral tissues. Sickle cell hemoglobin does not have the normal globular shape. It forms long, rigid strands of protein that cause the red blood cell to change shape (Figure 10.25c). Once a cell has become sickled, it does not return to a normal rounded shape, and it no longer can carry oxygen back and forth between the lungs and tissues. The sickled red blood cell soon dies an early death. As this happens to more red blood cells, the person with sickle cell anemia has too few red blood cells along with damaged capillary circulation. At the DNA level the sickle cell allele carries a mutation in one triplet codon that results in a change in one amino acid in each of the two

anemia any disease in which there are too few red blood cells to supply various tissues with sufficient oxygen

sickle-cell disease an inherited form of anemia associated with red blood cells that assume an abnormal sickle shape under conditions when the oxygen content of blood is low

(a) Each of the four subunits of a hemoglobin molecule contains a central iron atom in a heme group.

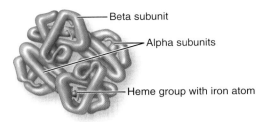

Beta subunit

Alpha subunits

Heme group with iron atom

(b) O₂ attaches to each heme group. Normal hemoglobin has a rounded shape.

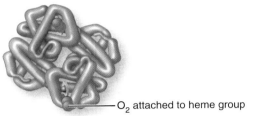

O₂ attached to heme group

(c) Hemoglobin molecules in a sickled red blood cell form long, inflexible strands when oxygen levels are low.

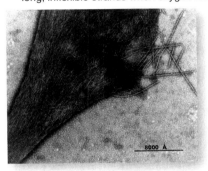

Figure 10.25 **Normal and Sickle Cell Hemoglobin.**

kinds of hemoglobin. The normal allele is transcribed and translated to produce normal hemoglobin; the sickle cell allele is transcribed and translated to produce sickle cell hemoglobin. The fact that both alleles in a heterozygous individual produce their specific type of hemoglobin is a clear example of codominance. This example emphasizes how the phenotype of a trait can be more fully understood if the genes and proteins involved have been studied.

Finally, you may wonder why the sickle cell allele is maintained in the human population. This allele is harmful under some conditions but helpful under others, and it shows how genetics and evolution are related. The sickle cell allele is found in people whose ancestors evolved in tropical Africa, especially in regions where a deadly variety of malaria is found (**Figure 10.26**). Malaria is caused by a single-celled eukaryotic parasite that spends part of its life cycle inside human red blood cells. In people who carry the sickle cell allele, infection with the parasite actually hastens the sickling process and changes the internal chemistry of the infected red blood cell. As a result, the body more rapidly detects and destroys infected red blood cells, thus killing the parasites within them. Therefore, having some sickled red blood cells offers partial protection from this lethal form of malaria.

Of course, someone with two sickle cell alleles will die anyway, so the protection from the malaria parasite afforded by their sickled red cells is hardly helpful to them. But some of the red blood cells of heterozygotes are collapsed into sickle shapes. Normal body defenses destroy the defective sickle-shaped cells and kill the parasites within them. These people walk in delicate balance: they have enough normal red blood cells to get oxygen to their tissues, and they have enough sickled red blood cells to kill some of the malaria parasites, and keep their levels tolerable. People who are heterozygous for sickle cell anemia are not as likely to die from this lethal form of malaria. Therefore, in certain tropical regions, people who are heterozygous for sickle cell anemia are more likely to survive and have children than are people who have no sickle cell alleles (who will probably die of malaria) or two sickle alleles (who will die of sickle cell anemia). In tropical Africa the advantage that results from having one sickle cell gene maintains the sickle cell allele in the human population.

beta subunits. This single amino acid replacement causes hemoglobin to form long, inflexible strands when oxygen levels are low.

At first glance sickle cell disease appears to be inherited in a simple Mendelian fashion. The disease is controlled by two alleles. The allele for normal hemoglobin is dominant; the allele for sickle cell is recessive. People who are homozygous for the sickle cell (recessive) allele have the disease, while heterozygotes appear not to have the disease. It turns out, however, that in extremely low-oxygen environments, such as at high altitudes, people who are heterozygous for hemoglobin can show some symptoms of the disease. The pattern of inheritance more closely resembles incomplete dominance because the heterozygous trait is intermediate between the homozygous dominant and homozygous recessive traits.

At the cellular and molecular levels the hemoglobin trait is an example of codominance. Depending on which alleles are present, a homozygous person has either all normal hemoglobin or all sickled hemoglobin, while a heterozygous person has both

QUICK CHECK

1. What is sickle cell anemia, how is it inherited, what causes it, and how does inheritance of the disease compare with inheritance of the cellular basis for the disease?

Figure 10.26
Geographic Distribution of Malaria.

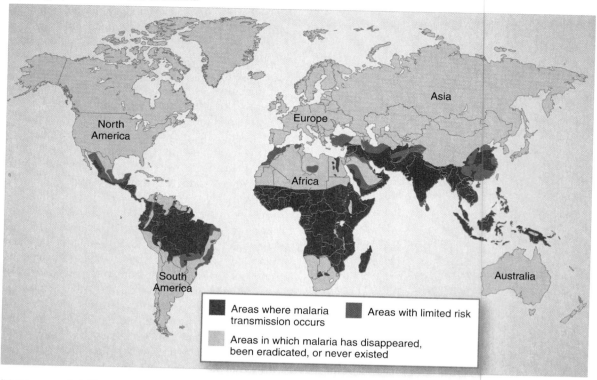

(a) Geographic distribution of malaria

North America

Europe

Asia

Africa

South America

Australia

Areas where malaria transmission occurs

Areas with limited risk

Areas in which malaria has disappeared, been eradicated, or never existed

(b) Genes and sickle cell anemia

Genotype	Hemoglobin phenotype	Relationship to malaria
HH	100% normal hemoglobin	No protection against malaria
HH′	50% normal, 50% sickle hemoglobin	Protection against malaria
H′H′	100% sickle hemoglobin	(Dies of sickle cell anemia)

10.6   An Added Dimension Explained
The Genetics of Eugenics

The introduction to this chapter left you pondering the efforts by the Eugenics Movement to produce a "fitter" human race. Would it be possible to change human traits by selective breeding? In terms of the laws of genetics, people are not different from peas or cows, and modern studies are locating genes that contribute to many complex human traits. You may hear of a "gene" for schizophrenia, or a "gene" for Alzheimer's disease. Now you can appreciate that carrying such a gene is not a 100% guarantee that either of those disorders will develop—because the relationship between a single gene and a complex trait such as intelligence or schizophrenia is not simple. But in the years after Mendel's work was made public, there was a widespread belief that individual genes completely control complex human traits. The Eugenics Movement of the early twentieth century

was based on the misconception that even traits such as laziness or "feeblemindedness" were inherited in a simple dominant/recessive fashion. Even though genetics was an emerging science, it was clear to Mendel and other geneticists that not all traits are inherited in a simple dominant/recessive manner. The proponents of eugenics, however, focused only on the science that supported their point of view.

The core belief of eugenics was that the human race would be improved if breeding were controlled like the breeding of prize horses or cats. Eugenics advocates used the emerging understanding of genetics to argue that genes determined *a person's social station or class*. Poverty, alcoholism, insanity, feeblemindedness, epilepsy, diabetes—indeed any trait that rendered a person "socially unfit"—was thought to be determined in a simple way by their genes. This conclusion was a huge leap beyond what the basic science of the day could support, but nevertheless social programs were built on this unsupported hypothesis.

To guide people in making breeding choices, lists of the traits of the "deformed class" were published (**Figure 10.27**).

The Eugenics Movement also had a profound effect on local and national laws. On both sides of the Atlantic, eugenics societies lobbied local, state, and federal governments urging them to enact policies and laws to promote the expansion of the upper classes and to limit the expansion of the lower classes. Both England and the United States passed laws that allowed sexual segregation and forced sterilization of the inmates of prisons, reform schools, and mental institutions. As a direct result of eugenics laws tens of thousands of people were involuntarily sterilized in the United States during the first half of the twentieth century. Immigrants were a particular target of legislation. Certainly, many factors have driven strict immigration laws in the United States, but eugenics played a prominent role in the first part of the twentieth century. In Nazi Germany the world saw the Eugenics Movement taken to a terrible extreme. The Nazi attempt to breed a "purified" Aryan race (defined as non-Jewish, Nordic) found justification for itself in the new science of eugenics. Eventually, the Holocaust would claim the lives of about 6 million Jews and about 5 million others judged to be undesirable or genetically flawed and racially "inferior" to people of pure Aryan stock.

But what of the biological basis for eugenics? To put the question another way, if we can breed prize milk cows, or superior wheat crops, why is it not reasonable to try to breed superior humans? The answers are complex, and at least three separate answers can be given. First, most complex human traits are not inherited in a simple dominant/recessive pattern. Second, the traits eugenicists were trying to breed in or breed out were not clearly articulated. And third, with adequate knowledge there could be success at breeding complex human traits, but it is probably not a good idea.

Given what is known about the inheritance of complex traits, it was naive to try to explain complex human traits using Mendelian dominant/recessive inheritance. Mendel's laws of inheritance are not wrong, but they do not describe the whole of inheritance. Although many traits follow a simple Mendelian pattern, many show incomplete dominance, codominance, pleiotropy, or polygenic inheritance. In addition, strong environmental factors influence more complex traits like intelligence, temperament, or mental illness.

The example of the inheritance of intelligence is revealing. The Eugenics Movement focused heavily on intelligence and pushed hard to have feebleminded individuals prevented from breeding. The inheritance of intelligence is complex, however. Intelligence often is measured by standard intelligence quotient (IQ) tests. One factor in determining a person's IQ is their parents' IQ, but scientific studies have shown that if two people who score low on various intelligence tests were to marry and have children, their children are likely to have more normal intelligence. The opposite also is true. If two people who score

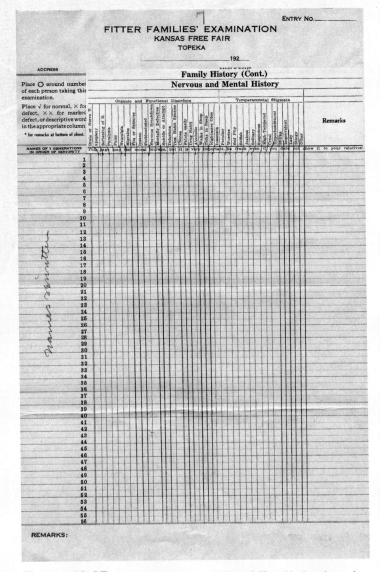

Figure 10.27 **Fitter Families' Checklist** Notice the traits, especially the "Temperamental Stigmata" used to evaluate less fit individuals. You may need a magnifying glass to read them, but it's worth the effort.

high on intelligence tests were to have children, their children are likely to have more normal intelligence.

The environment adds an additional layer of variation to already complex inheritance patterns for intelligence. For example, if children of people institutionalized for feeblemindedness are placed in adoptive homes their intelligence is strongly influenced by the quality of their new environment. Certainly, genes have an impact on the nature of an individ-

—Continued next page

Continued—

ual's intelligence, but given a normal environment—adequate nutrition, mental stimulation, and opportunities to learn—children of feebleminded parents often have perfectly normal intelligence. The effect of the environment also is seen in changes in IQ scores over time. Over the past century in every country studied, performance on IQ tests has improved. The standard IQ score is reset regularly so that the actual scores reported do not increase. The environment plays an important role in every complex human trait, but this is especially true for any measure of human intelligence.

A related problem with the Eugenics Movement was sloppiness in how human traits were defined. In Section 1.5 you read that a scientific hypothesis must be stated clearly and concisely. If a hypothesis is phrased loosely or vaguely, then testing that hypothesis will not produce reliable results. Eugenicists made this mistake in how they expressed their hypotheses. Complex traits such as "idiocy" or "laziness" cannot be defined in a simple way because these subjective labels mean different things in different situations. In the hypothesis that "laziness is inherited," the term laziness is vaguely defined—therefore, the hypothesis is difficult if not impossible to test.

A final point about eugenics: it *could be effective* if it were based on a deep knowledge of the inheritance of the desired traits. For example, carefully designed studies suggest that babies are born with different levels of reactivity to the world around them. Some babies respond quietly to novelty and stimulation; some seem to seek out novelty and stimulation; others respond with crying and agitation. These reactions can be precisely defined and hypotheses stated based on the babies' observable behaviors. In this context temperament can be defined in objective terms, and the genetic contribution can be studied. The contribution of temperament to other measures of a person's adjustment in life can be studied and the effects of different approaches to child rearing by the parents can be assessed. Based on such knowledge it might be possible to selectively breed humans for a quiet temperament or an excitable temperament. Of course, it takes a great deal of time and research to break down complex traits into their component parts. This approach will not yield quick answers to the questions of human inheritance, but it is more likely to yield reliable answers than the approach taken by eugenicists.

The important question is whether human society *should* pursue such selective breeding. Science has little to say about

Figure 10.28   Stephen Hawking. Even though Stephen Hawking has an incurable disease he has made monumental contributions to physics.

this question, and you may think it a far-fetched one. After all, it would be inconceivable for a modern democracy to impose such a selective breeding scheme. But society is facing this issue now. New techniques to treat infertility are giving parents choices about the genes carried by their children. For instance, they can choose to have a child free of a certain disease, or they can choose their child's sex. This topic is covered in detail in Chapter 24. Here it is enough to understand that if parents use modern reproductive technology, they can choose some genetic traits. Now *individuals*, rather than governments, are able to choose the genes their children will carry.

What if parents begin to choose male or female offspring or if society goes through genetic fads such as tall children or red-haired children? Few would argue that parents should choose a child with cystic fibrosis if technology can help them to avoid it, but what if that potential child carries other genetic traits that could make a large difference in the world? Consider the well-known case of Stephen Hawking, possibly one of the most brilliant physicists of the past 50 years (**Figure 10.28**). In his early twenties Hawking was afflicted with amylotrophic lateral sclerosis, or Lou Gehrig's disease, an incurable and progressive degeneration of nerves and muscles. Hawking now is confined to a wheelchair and has little independent movement. He speaks through a computer voice simulator. By any eugenic standard from the early 1900s, Hawking is physically "unfit," and contemporary parents might think twice if they had a choice about having a baby with the genes for this disease. Yet Stephen Hawking has made monumental contributions to modern theoretical physics in his studies of how gravity contributes to the formation of black holes. In addition, Hawking and his wife have raised three children. By his own account, Hawking leads a productive and happy life.

The converse of this idea is also worth considering. Is human knowledge and wisdom ready to take over the selective process that produces the diversity that makes us what we are? You will have a chance to return to these important questions in Chapter 11, where you will learn more about the potential applications of modern genetic knowledge.

QUICK CHECK

1. What scientific justification did the Eugenics Movement claim and in what way was this justification faulty?

Incest Taboos

Why do so many human cultures have taboos against sexual intimacy and marriage between close relatives? A hint of the answer comes from looking at what happens when closely related people mate and have children. Some of the most famous examples are the children of England's Queen Victoria and Prince Albert. Victoria assumed the throne in 1837, and she and Albert had nine children. In keeping with the customs of the time, Victoria and Albert arranged for their children and grandchildren to marry into the other royal families in Europe. This was not unusual. Europe's royal families had been intermarrying for many generations. This gave Victoria a great deal of political power, and intermarrying became one factor that caused European monarchies to become unstable. One of Victoria's sons, Leopold, had the "bleeding disease," now known as hemophilia. Although they did not know it, two of her daughters, Alice and Beatrice, also carried the gene and passed it on to some of their children. Alice had a hemophiliac son, Beatrice had two hemophiliac sons, and both had daughters who carried the gene. Two of these married into other European royal families: Alix married Czar Nicholas of Russia and Eugenie married King Alfonso of Spain. Both Russia and Spain had powerful monarchies, but the birth of hemophiliac sons in both families contributed to their downfall.

The potential negative effects of inbreeding are not restricted to royal families. For generations Amish people have married primarily within their own communities. One price they have paid is a higher rate of certain metabolic diseases and skeletal abnormalities. After reading this chapter on Mendelian genetics, you should have a better understanding of why inbreeding can lead to a higher rate of genetic problems. If a large family group carries a recessive gene for a disease, and if family members mate with one another, then prospective children are more likely to be homozygous for the allele and therefore may become sick or diseased. In contrast, mating with someone outside the family group assures that the allele is more likely to remain heterozygous in the offspring and not result in disease.

Does Heredity or Environment Exert More Influence on Human Intelligence?

The science of genetics and the politics of contemporary America can be an explosive mix. As you have seen in the Eugenics Movement, people often interpret current scientific findings in the context of their own political or social beliefs. The "genetic basis of intelligence" is a prime example. Beliefs about the inheritance of intelligence can influence views of social or governmental programs designed to improve intellectual abilities and performance. This is a case where the scientific evidence is not clear-cut on either side of the argument. Intelligence is not easy to define. But once you start to study a subject scientifically, you must define your terms and ideas carefully. Studies of intelligence have defined the term *intelligence* in specific and varied ways. Mice, for example, can be bred to be more "curious" and to explore their environment more than regular mice do. To some people this sort of exploratory curiosity is a part of intelligence. Other evidence comes from studies that track specific abilities across families. Musical ability or the lack of it, for example, is nearly identical in identical twins, even if they are reared in different families. Twin studies also suggest that general IQ is more similar in twins, even if they are reared apart, than in unrelated persons. Nevertheless, changes in a child's environment can have a profound impact on the child's level of achievement. This is especially true of children in poor social or economic conditions. Programs such as Head Start, which provide preschool enrichment for children from disadvantaged backgrounds, can dramatically increase measures of intellectual success such as graduating from high school. As you can see, the evidence can lead one to emphasize genes, or environment, or both as factors that determine human intelligence. What do *you* think about genes, intelligence, and societal responsibility?

Chapter Review

CHAPTER SUMMARY

10.1 The Science of Genetics Brings Together DNA, Cell Division, and Gene Expression

From practical experience it was common knowledge that offspring could resemble their parents in many ways. Domesticated species were developed by selective breeding in which chosen individuals were mated. Experimental crosses are the basis of genetics research.

crosses   231

10.2 Gregor Mendel Was the First Systematic Researcher in the Field of Genetics

Working with 14 varieties of garden pea plants, Gregor Mendel discovered the rules that govern inheritance and recognized dominant and recessive units of inheritance. Mendel's work was characterized by careful methods, many replications, and mathematical analyses of his results.

genetics   233   true-breeding traits   235

10.3 Mendel Discovered Three Rules of Inheritance

Mendel recognized that discrete units govern genetic traits and that these can be dominant or recessive. Individuals can be homozygous or heterozygous for a trait. Because dominant traits can hide recessive traits, genotype and phenotype can differ. Alleles are segregated into gametes, and alleles usually act independently of one another.

dominant allele   239
first filial generation (F₁)   237
genotype   239
heterozygous   241
homozygous   240

parental generation (P)   237
phenotype   239
recessive allele   239
second filial generation (F₂)   237

10.4   More Complex Patterns of Inheritance Are an Extension of Mendel's Basic Rules

While simple dominant/recessive inheritance controls the expression of many traits, other traits are expressed in an intermediate condition and are codominant or incompletely dominant. Polygenic traits are controlled by more than one gene and are expressed in a continuum of phenotypes. Single genes that affect more than one phenotype are pleiotropic. Other genes are incompletely dominant, and a heterozygote has an intermediate phenotype. In codominant traits all alleles are expressed, but an intermediate phenotype is not produced. Instead, both phenotypes are present in the heterozygote of incompletely dominant traits. Sex-linked characteristics are carried on the X and Y chromosomes. Because the Y chromosome is so much smaller than the X chromosome, some traits are found only on the X chromosome. These are invariably expressed and some, like hemophilia, can have negative consequences. Genes, development, and the environment produce many human characteristics.

codominance   245
incomplete dominance   245
pleiotropy   246

polygenic   246
sex-linked trait   247

10.5   Knowledge About DNA Illuminates Genetics

The relationship between genes in DNA and proteins dictated by those genes is fundamental to understanding genetic inheritance. A mutation in DNA will produce a different protein and thus a different trait. Mutations can be positive, neutral, or harmful and are the basic mechanism for phenotypic variation within a species. Sickle cell anemia is an example of a disease produced by the homozygous recessive condition of alleles for an abnormal nucleotide "recipe" for hemoglobin. Because heterozygotes have a survival advantage in environments that include a lethal form of malaria, the sickle cell allele is maintained in the human population.

anemia   250

sickle cell disease   250

10.6   An Added Dimension Explained: The Genetics of Eugenics

A simplistic application of Mendel's laws of inheritance led to ideas that human traits can and should be improved by selective breeding. Eugenics led to the forcible surgical sterilization of thousands of "feeble-minded" or "inferior" Americans; eugenics was behind the killing of approximately 11 million people by the Nazis during World War II. Many individuals still suffer from the effects of the Eugenics Movement that has been discredited as simplistic and unscientific. Nevertheless, it is an instructive misapplication of incompletely understood scientific data.

MULTIPLE CHOICE. Choose the best answer of those provided.

1. Sam has two groves of true-breeding apple trees that have been in his family for generations. One grove of trees produces green apples, and the other grove produces bright red apples. Sam crosses some trees from the two groups, and the offspring all produce red apples. What has Sam learned about the inheritance of apple color in these trees?
   a. Green is dominant over red.
   b. Red is dominant over green.
   c. Red is recessive.
   d. Green apple trees are not as hearty as red apple trees.
   e. The trait for green apples has been lost in his experimental trees.

2. Sally is heterozygous for the alleles associated with cystic fibrosis. What can you say about Sally's genotype?
   a. Sally carries two normal alleles.
   b. Sally carries two abnormal alleles.
   c. Her genotype will be directly passed to all of her children regardless of who the father is.
   d. Sally carries one normal allele and one abnormal allele.
   e. Sally carries one normal allele and two abnormal alleles.

3. What three laws of inheritance did Mendel discover?
   a. dominance, segregation, and independence
   b. dominance, codominance, and pleiotropy
   c. polygenic inheritance, pleiotropy, and incomplete dominance
   d. eugenics, the inheritance of anemia, and segregation
   e. none of the above

4. What cellular mechanism produces the segregation of alleles during gamete formation?
   a. mitosis
   b. transcription
   c. meiosis
   d. epigenetics
   e. translation

5. What are the genetic units that determine different versions of the same trait?
   a. proteins
   b. control elements
   c. chromosomes
   d. alleles
   e. telomeres

6. Sam returns to his crosses of his family apple trees. He has two populations of trees that breed true: one for apples that develop in bunches of three per stem and the other for apples that develop one per stem. When Sam crosses these he gets trees that have two apples per stem. This is an example of
   a. complete or full dominance.
   b. codominance.
   c. pleiotropy.
   d. polygenic inheritance.
   e. incomplete dominance.

7. Two hypothetical populations of cats have brown fur and orange fur. Researchers have discovered that this trait is inherited according to codominance. If this were all there was to the inheritance of coat color in these cats, what would the offspring look like if a brown cat were mated with an orange cat?
   a. all brown
   b. all orange
   c. sort of orangey brown
   d. orange and brown
   e. all white

8. What possible gametes can be produced in a pea plant that breeds true for wrinkled peas? Use the symbols R for round and r for wrinkled.

   a. R and R
   b. R and r
   c. r and r
   d. none of these
   e. all of these

9. What will be the genotype of individuals of the F_1 generation produced by a cross between parents that are both true-breeding wrinkled seeded and true-breeding white flowered? For the flower color trait assume p for white and P for violet-red. For the seed shape assume R for round and r for wrinkled.

   a. $rrpp$
   b. $rrPP$
   c. $RRpp$
   d. rp
   e. all of these

10. What will be the phenotype of this F_1 generation?

    a. round, white
    b. wrinkled, white
    c. wrinkled, violet-red
    d. none of these
    e. all of these

11. What is the genetic mechanism that produces alternative alleles?

    a. mutation
    b. crossing over
    c. methylation
    d. pleiotropy
    e. eugenics

12. What is one way that scientists can determine the extent to which a complex human trait has genetic and environmental components?

    a. identical twin studies
    b. raising human children in different environments
    c. selective human breeding studies
    d. cloning
    e. eugenic studies

13. The allele for sickle cell anemia has been maintained in human tropical populations because

    a. homozygous recessive individuals live longer and have more children.
    b. homozygous recessive individuals are more aggressive than other humans.
    c. heterozygous individuals are resistant to the malaria parasite.
    d. heterozygous individuals are protected against cancer.
    e. it has not been maintained, it was bred out of human populations 10,000 years ago.

TRUE or FALSE. If a statement is false, rewrite it to make it read true.

14. An individual who has the blood type AB is homozygous for this trait because the alleles that code for the A and B carbohydrates do not match.

15. What is called human intelligence is probably a polygenic characteristic that can be influenced by environment.

16. Alleles are varieties of genes that code for different proteins.

MATCHING

17–20. Match the terms and definitions. (One choice will not match.)

17. the belief that the human race can and should be improved by selective breeding
18. discovered the principles of inheritance
19. mating selected individuals to reveal the underlying rules of inheritance
20. offspring always show the traits of their parents

   a. dominant
   b. true-breeding traits
   c. Gregor Mendel
   d. eugenics
   e. crosses

CONNECTING KEY CONCEPTS

1. What were Mendel's basic discoveries, and what other modes of inheritance have since been recognized?

2. Using sickle cell anemia as an example, how does the modern understanding of how DNA works contribute to a more complete understanding of how traits are inherited, compared with just understanding the laws of inheritance discovered by Mendel?

3. The Eugenics Movement was not based on sound scientific principles—yet it probably is possible to selectively breed humans for complex traits. Explain this apparently contradictory statement.

QUANTITATIVE QUERY

Alcira is breeding rose bushes. She starts with a garden of true-breeding red plants and a garden of true-breeding yellow plants. She would like to have some orange-colored roses, so she crosses plants from her two gardens. To her delight all 10 offspring have orange flowers, showing that these flower color alleles are inherited by incomplete dominance. The next year all of her original true-breeding roses die out. So she crosses the orange flowers among themselves and ends up with 20 plants. How many of these plants will be red, yellow, and orange?

THINKING CRITICALLY

1. Consider this scenario. Your father had a long-term illness caused by a recessive gene. Your mother was genetically tested before they had children and therefore was certain that she did not carry this allele. Now you have fallen in love and have married, but as you begin to plan your family, your in-laws' genetic history makes it likely that your spouse's mother carried one allele for this same disease, while your spouse's father probably did not.

   a. What is your genotype with respect to the disease? Why do you not have the disease?
   b. What is the probability that your spouse carries an allele for the disease?
   c. If your spouse does carry the allele, what is the chance that your child will develop the disease?

2. In the past 50 years American farmers have been able to produce an increasing crop yield for each acre they plant. One of the many reasons for this increase in the nation's food production is the use of special hybrid strains of crops developed and marketed by big agricultural business. While hybrid crops do offer much greater productivity, they do have a downside: their seeds cannot be used for the next year's plantings. Instead, to continue to produce high yields, farmers must buy new seed. Give a genetic explanation for what is going on here. How can the big companies produce seeds that grow great plants, while the offspring of those same plants grow inferior plants?

For additional study tools, visit www.aris.mhhe.com.

Panel Urges Restricted Use of Gene Therapy
March 1, 2003

Dream Unmet 50 Years After DNA Milestone
Gene Therapy Debacle Casts Pall on Field
March, 2003

Jesse Gelsinger was a participant in experimental gene therapy.

Paul and Pattie Gelsinger lost their 18-year-old son, Jesse, on September 17, 1999. Any parent would be devastated by the loss of a child—and the Gelsingers clung to Jesse's hope that his death might help others to survive. In the difficult days and months after Jesse died, this slender hope began to fade as Paul and Pattie learned more about the medical procedures that had led to Jesse's death. Could his death have been avoided (**see chapter-opening photo**)? Had his death helped even one person?

It is difficult to answer these questions because not enough is known about ornithine transcarbamylase (OTC) deficiency, the inherited metabolic disease Jessie was born with. OTC is an enzyme involved in the formation of urea. When amino acids are converted to other biological molecules the nitrogen has to be removed and gotten rid of. OTC removes nitrogen from amino acids and incorporates it into urea, which is then excreted by the kidneys. If this enzyme is missing, another metabolic pathway forms ammonia, instead of urea. Ammonia is toxic to mammals, including humans. Because Jesse inherited an OTC deficiency, his life was threatened by accumulation of ammonia. Children born with OTC deficiency are not likely to live to adulthood. Victims' lives are marked by years of struggling to normalize their ammonia levels. Emergency trips to the hospital and near-death experiences are typical for people with OTC deficiency. Jesse was luckier than most. Even though he had to swallow nearly 50 pills each day, by the time he was in his teens his illness seemed to be under control with careful attention to his diet and drug regimen.

Despite the fact that his disease was under control, when Jesse learned of an experimental treatment for OTC he eagerly signed on as a participant. The treatment involved giving Jesse many copies of a gene that would produce normal OTC. Jesse signed up for a very early phase of the experiment, intended only to find out if the procedure was safe. It was not meant as a test of the curative power of the gene therapy. Even though Jesse knew that it was not likely that the procedure would improve his condition, he was enthusiastic about what would be learned from the process. Jesse understood there were risks, but he believed that if he died, the knowledge gained from his experimental treatment would improve the health of other children born with OTC deficiency.

Unfortunately, Jesse died just four days after the experimental treatment was administered, and his death highlights many questions about gene therapy. What exactly is gene therapy and how safe is it? What progress has been made in gene therapy since Jesse's death? What hope for treatments does it offer now, and what hope might it offer in the future? You will return to the prospects for gene therapy at the end of this chapter, after you have become acquainted with other forms of biotechnology. ■

11.1 Biotechnology Has Dramatically Changed Society and Human Lives

Scientific developments that are dramatic and far-reaching enough to be called scientific revolutions don't happen often. Newton's understanding of physics and Einstein's theory of general relativity have changed our lives and our understanding of the world. The practical use of electricity has had a major impact on human life. Similarly, the internal combustion engine has transformed society. Biological revolutions, such as the discovery that bacteria cause infectious diseases and that antibiotics kill these bacteria, also have had great impacts on people's lives. And although you may not realize it, you are living through what may prove to be the most profound scientific revolution that humans have ever experienced. This latest scientific revolution may make radical changes in our understanding of what it means to be human.

This new revolution began in 1953 with the discovery of the structure of DNA (see Section 7.2). Each year new information reveals more about DNA's ability to direct cellular functions and determine the traits of organisms. Together the use, study, and development of DNA-related applications and techniques are known as **biotechnology.** This term usually refers to the techniques used to isolate, characterize, manipulate, and control biological molecules such as DNA or proteins located inside or outside of a cell. Biotechnology has generated many applications used in industry and medicine. It also has generated a wealth of basic knowledge about how DNA controls cellular functions. Today, we are on the brink of so many breakthroughs in biotechnology that it is impossible to predict how biotechnology will change the world and our knowledge of it in another 50 years.

One revolutionary outcome of the application of biotechnology techniques is that humans now know what it means to be human at the level of DNA. The **Human Genome Project,** a nearly complete description of human DNA, was finished in the early years of this century. Researchers cataloged nearly all of the nucleotide sequences of human DNA and so composed the first human genetic "book." They determined the number of genes located on each of the 46 human chromosomes. Now scientists are investigating what these

biotechnology
techniques used to isolate, characterize, manipulate, and control biological molecules such as DNA or proteins

Human Genome Project
the research effort to identify all genes in the human genome and to determine the nucleotide sequences of the entire human genome

genes do. The Human Genome Project has brought many surprises already. For one thing, humans have far fewer genes that scientists had expected. Rather than having 100,000 genes to match the approximate number of human proteins, humans have only 25,000 to 30,000 genes. Genetic complexity in humans comes from the way that gene expression is controlled (see Chapter 9). As the genomes of other species are sequenced, we gain new insights into human genetics. For example, the genome of our closest living relative, the chimpanzee, has been sequenced. Studies now are focusing on which genes differ between humans and chimps and on what those genes do. It will not be long before you will be able to understand how evolution has produced humans and chimpanzees from our common ancestor.

QUICK CHECK

1. What is biotechnology?

11.2 Forensic DNA Techniques Match Individuals to DNA Samples

One of the first applications of biotechnology was to accurately identify individuals through their DNA—a use that is certain to expand in the future. Think for a moment of just some of the situations in which the correct identification of an individual may be important or useful. Of course, criminal cases rely on pinpointing whether a specific person committed a particular crime. Prisoners and military personnel are routinely required to undergo DNA testing. Paternity and maternity tests are crucial in child-support cases. In recent years some criminal investigators have even resorted to asking everyone in a community to provide DNA samples to match against crime scene samples. If you want to purchase a firearm or apply for a high-security job, some states require a background check that includes a DNA test. In today's political climate some feel the need to precisely identify who enters a country or boards an airplane or train. In the future there may be pressure to certify your identity by DNA analysis when you get a driver's license or even when you show up to take an exam in class! In any or all of these situations, DNA testing could be used to make sure that you are who you say you are. DNA testing can reveal if two samples of DNA are likely to have come from the same person.

forensics the application of scientific knowledge and techniques to legal issues

DNA fingerprint an analysis of a person's DNA that can be used to uniquely identify that person

DNA Testing Can Determine Guilt or Innocence in a Criminal Case

One of the earliest applications of DNA technology was in the field of **forensics,** the application of scientific knowledge and techniques to legal issues. Forensics helps to determine the guilt or innocence of a person accused—or already convicted—of a crime. The case of Eddie Joe Lloyd is just one example (**Figure 11.1**). In 1985 Mr. Lloyd was accused, tried, convicted, and sent to jail for a crime that he did not commit: the rape and murder of a young girl. The jury deliberated his case for just 30 minutes before delivering a guilty verdict. After the verdict the judge lamented that Michigan had no death penalty. He felt that death was the only punishment that fit the *facts* of this case. Mr. Lloyd was sentenced to prison for life.

If Eddie Joe Lloyd's trial had taken place a century ago, his fate would have been sealed because once a person had been convicted by a jury, it was difficult to prove that he or she was innocent. Knowledge of DNA has changed all of that. One of the earliest applications of biotechnology is known as *DNA fingerprinting*. In 2002 evidence from DNA fingerprinting freed Eddie Joe Lloyd from prison and cleared him of the murder charge.

A **DNA fingerprint** is an analysis of a person's DNA nucleotide sequences that distinguishes that person from all others. If the procedure is done correctly, it will generate a DNA fingerprint unique to

Figure 11.1 **Eddie Joe Lloyd as He Regained His Freedom.**

the person tested unless that person is an identical twin. Because of its power to distinguish between people, DNA testing has become a common tool in the courtroom. About 10,000 DNA tests are done each year on people accused of crimes. About 25% of these DNA tests demonstrate that the accused person is innocent. In other words, in about one out of four of these cases the DNA of the accused does not match the crime scene DNA.

You may wonder what advantage a DNA fingerprint has over a regular fingerprint. Both manual fingerprints and DNA fingerprints are unique for each individual. The key to using either form of identification is how the fingerprints are taken and analyzed. A careful and thorough analysis of either type of fingerprint can provide valid information. One reason that DNA fingerprinting has received so much attention is that in many cases DNA evidence is available when hand fingerprints are not (**Table 11.1**). A person can wear gloves or surgically obscure fingerprints, but the DNA in a dot of blood or semen, in a hair, or a flake of skin left at a crime scene cannot be altered. And of course, DNA analysis also can identify organisms other than humans. Often DNA found at a crime scene—such as that from pollen, bacteria, or even blades of grass—can be used to implicate or to free a suspect.

Figure 11.2 shows how a DNA fingerprint can be made. The first step is to obtain a pure DNA sample from the cells or tissues of interest. Most often the sample comes from live cells—but depending on the conditions, DNA can be recovered from dead tissue that is up to hundreds of thousands of years old. In the case of Mr. Lloyd a group called the Innocence Project helped to obtain a DNA sample from stored trial evidence. A DNA sample from Mr. Lloyd could have been obtained in a variety of ways, but the most common method is to swab the lining of a cheek to obtain a sample of surface cells.

Purifying DNA from a sample is a fairly straightforward procedure, especially if you have a large amount of tissue. Sample cells are placed in a solution inside a test tube. Then heat or chemicals are applied to rupture the cell and nuclear membranes and release the DNA, cytosol, and organelles. (Figure 11.2b). A problem arises at this stage because enzymes from the cytosol will destroy the DNA in the sample. To prevent this, before the membranes are ruptured, other enzymes are added that destroy the DNA-damaging enzymes. Now the sample contains DNA plus enzymes, as well as cellular materials such as fragmented organelles that are not needed for DNA fingerprinting. To get rid of the unwanted material, the tube is spun in a centrifuge, a smaller, more powerful version of a washing machine on spin cycle. The heavier components of the sample sink to the bottom of the tube, leaving the DNA in solution above them. It is removed for further processing. If there is a lot of DNA in the sample, it can be centrifuged again to produce a compact pellet of DNA that can be used in DNA fingerprinting.

One of the first methods for generating a DNA fingerprint used enzymes produced by bacteria called **restriction enzymes** (Figure 11.2c). These bacterial enzymes cut DNA at specific sequences, and bacteria produce them as a defense against the DNA of viruses. These DNA-snipping enzymes are called *restriction enzymes* because they limit—or

restriction enzyme
a bacterial enzyme that cuts DNA at a particular nucleotide sequence

Table 11.1 Comparison of a regular fingerprint and a DNA fingerprint

	Regular fingerprint	DNA fingerprint
Cost	Cheap	More expensive
Information about		
Body size	Yes	No
Gender	Yes	Yes
Occupation	Yes	No
Genetic diseases	No	Yes
Race	No	Yes
Likelihood of developing diseases	No	Yes
Can it tell identical twins apart?	Yes	Not at this time
Databases and cross-references	Yes	Stored in searchable databases that are shared across jurisdictions
Advantages to forensics	Fingerprints often are left at crime scenes on hard surfaces.	DNA samples often are inadvertently left at crime scenes. Evidence can be recovered from a broader range of environmental surfaces. Evidence may have DNA samples that connect a criminal with a crime. DNA evidence allows analysis of the probability that a conclusion is correct.

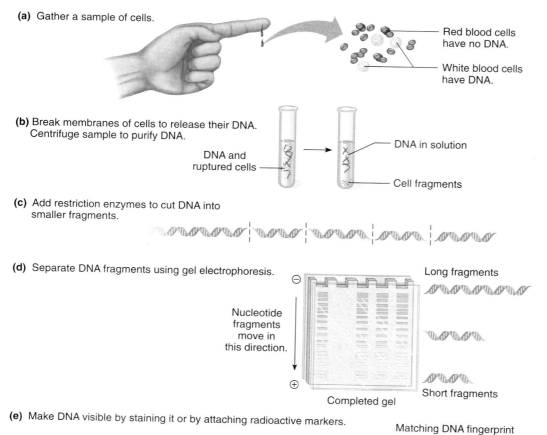

(a) Gather a sample of cells.

Red blood cells have no DNA.

White blood cells have DNA.

(b) Break membranes of cells to release their DNA. Centrifuge sample to purify DNA.

DNA and ruptured cells

DNA in solution

Cell fragments

(c) Add restriction enzymes to cut DNA into smaller fragments.

(d) Separate DNA fragments using gel electrophoresis.

Long fragments

Nucleotide fragments move in this direction.

Completed gel

Short fragments

(e) Make DNA visible by staining it or by attaching radioactive markers.

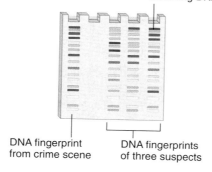

Matching DNA fingerprint

DNA fingerprint from crime scene

DNA fingerprints of three suspects

Figure 11.2 **One Way to Make a DNA Fingerprint.** DNA is obtained from a sample, and cut with restriction enzymes (**a–c**). A DNA fingerprint is obtained after a sample has been separated by electrophoresis and stained (**d, e**).

gel electrophoresis
a process used to separate different sized pieces of DNA

restriction fragment length polymorphism (RFLP) the pattern of different-sized DNA pieces produced by restriction enzymes that is unique to each individual

restrict—the ability of viral DNA to infect bacteria. These enzymes will cut *any* DNA, not just viral DNA. There are hundreds of different restriction enzymes, and each recognizes and cuts DNA at a specific nucleotide sequence. **Figure 11.3** shows examples of the actions of specific restriction enzymes. For instance, the enzyme *Eco*RI cuts between an A and a G, but it only makes this cut when AG occurs with the sequence CTTAAG, or its reverse complement, GAATTC.

Exposure of a sample of DNA to a mixture of restriction enzymes will transform a sample of large DNA molecules into a sample of short stretches of DNA of different sizes. These fragments can be sorted by size, and used to generate a DNA finger-

print. The DNA fragments are sorted using a lab technique called **gel electrophoresis.** A gel is a thin pad of a gelatin that has been poured onto a plastic plate. At one end of the gel samples of DNA from tissue and a reference sample with known DNA pieces are dropped onto different lanes (**Figure 11.4a**). Then an electric charge is applied across the gel. The negative pole of the battery is connected to the end of the gel where the DNA samples were applied. The positive pole of the battery is connected to the opposite end of the gel. Because DNA is a negatively charged molecule, the DNA fragments will move toward the positively charged end of the gel. Different fragments of DNA move through the molecules of the gel at different rates. Larger fragments of DNA will move more slowly and will be deposited at various locations near the top of the gel; smaller pieces of DNA will move more quickly and will be deposited at locations closer to the bottom of the gel (Figure 11.4b). Once the current has been applied for a standard length of time, the location of the deposited DNA is revealed using a special stain. Groups of different-sized pieces of DNA show up as bands along the length of the gel. **Figure 11.5** shows a stained gel with bands of DNA. This gel is one form of a DNA fingerprint. If DNA fragments of known size also are run in the gel alongside the DNA samples being tested, then the exact size of the fragments in each band can be determined. Gels can be constructed that result in different bands for segments of DNA that differ in length by just one nucleotide, so this technique is highly precise.

When restriction enzymes are used to produce a DNA fingerprint, the resulting banding patterns are called **restriction fragment length polymorphisms,** or **RFLPs** (pronounced "riff-lips"). The term *restriction* refers to the restriction enzymes; *fragment* refers to the fact that the DNA is cut into pieces; *length* refers to the length of the DNA segments produced by the restriction enzymes; *polymorphism* means "different form." If enough different restriction enzymes are used, the pattern of DNA bands will be similar for different tissues from the same person but will be different for tissues from other people. In Figure 11.5 you can see that the banding pattern obtained from tissue sample 1

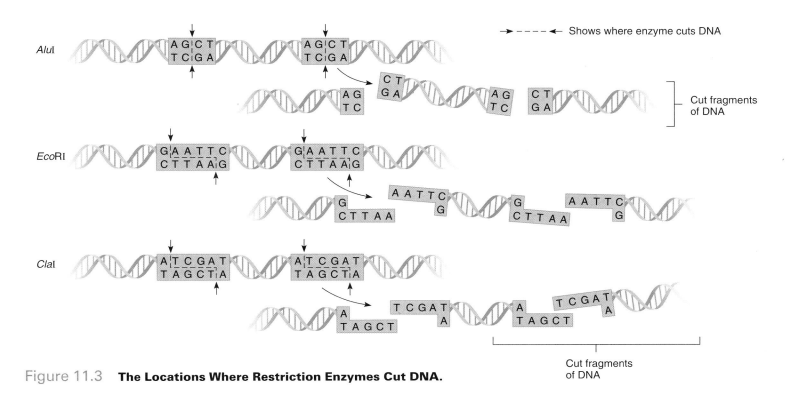

Figure 11.3 **The Locations Where Restriction Enzymes Cut DNA.**

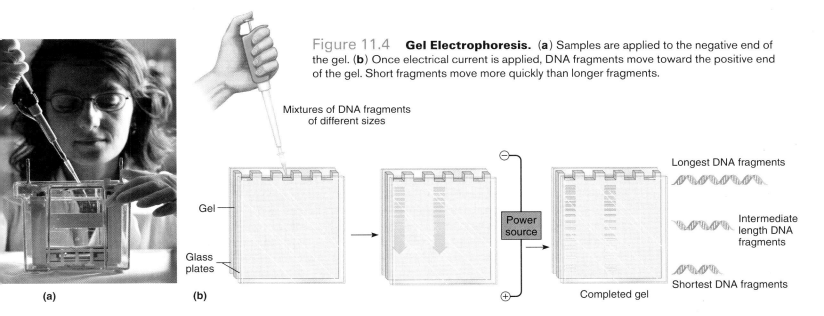

Figure 11.4 **Gel Electrophoresis. (a)** Samples are applied to the negative end of the gel. **(b)** Once electrical current is applied, DNA fragments move toward the positive end of the gel. Short fragments move more quickly than longer fragments.

is much different from the banding pattern obtained from tissue sample 2. Each particular DNA finger-print analysis comes with a certain probability that two different-looking prints might have come from the same individual. For instance, with a DNA fin-gerprint that has been well done, two DNA finger-prints with different banding patterns might have a 1 in 20 billion chance of coming from the same indi-vidual. This means that out of 20 billion such test results, only once would two prints with different banding patterns actually come from the same indi-vidual. With these kinds of odds you can be highly confident in the test results. If the analysis is not performed carefully, two DNA fingerprints with dif-ferent banding patterns might have a 1 in 20 chance of coming from the same person. These kinds of chances would make you question the testing pro-cedure and its results.

Figure 11.5 **Eight DNA Fingerprints.** The numbers refer to suspects. The bloodstain originated at a crime scene.

■ *Do you see a match? Do the DNA fingerprints of suspects match one another?*

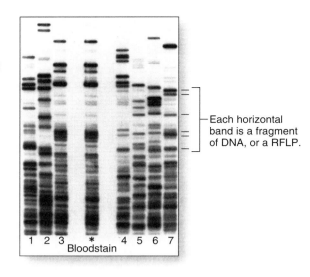

Each horizontal band is a fragment of DNA, or a RFLP.

1 2 3 * 4 5 6 7
Bloodstain

You might think that the differences between the DNA fingerprints of two people are due to different genes that code for functional proteins, but this is *not* usually the case. The DNA sequences that code for specific proteins are so important that they do not differ much between people. While there are techniques that can distinguish between the small differences in genes, most DNA fingerprinting techniques rely on the long repeated sequences of DNA that do *not* code for protein, (see Section 9.3). A particular noncoding sequence can be common among many, or even all, people—but the *number of repeats* of any given noncoding sequence can vary greatly between individuals. For instance, one person might have 20 repeats of a particular sequence, while another person might have 35 repeats. Restriction enzymes can be chosen that cut DNA at the beginning and end of several different repeated sequences—this turns out to be one reliable way to create a DNA fingerprint.

One drawback of the early DNA fingerprinting techniques was that they required lots of DNA. Thousands of cells were needed, but often a crime scene turns up just one eyelash, a tiny spot of blood, or a single cigarette butt that may have a skim of saliva containing cheek cells. A big technological advance came when one scientist discovered a way to make many copies of a strand of DNA in a test tube, using the technique called **polymerase chain reaction** (**PCR** for short). PCR makes it possible to create DNA fingerprints from just a tiny sample of evidence. The *polymerase* in PCR is a DNA polymerase molecule isolated from a prokaryote that lives in the hot springs of Yellowstone National Park. This DNA polymerase is similar to the DNA polymerase in other organisms, except that it is chemically active at very high temperatures that would denature most other DNA polymerase molecules. Using this particular DNA polymerase makes the PCR process more efficient and productive. Using PCR, the DNA in a cell that is attached to a single hair, the DNA in a dot of semen, the DNA in a few cheek cells on a cigarette butt, or the DNA in the white blood cells in a spot of blood can be amplified into enough DNA to make a DNA fingerprint.

DNA fingerprinting has had a big impact on criminal investigations. It demonstrated that Eddie Joe Lloyd was *not* guilty as charged. In 2002 Eddie Joe Lloyd was finally given his freedom after spending 17 years in jail for a crime he did not commit. In the United States Mr. Lloyd was the 110th person to be freed from prison based on evidence from DNA fingerprinting. Since then hundreds more prisoners have been freed and exonerated of crimes for which they had been unjustly convicted and sentenced to time in jail. And, of course, many criminals have been convicted based on a match between their own DNA fingerprint and that of a sample taken from a crime scene.

DNA Fingerprinting Is Not Foolproof DNA fingerprinting is a powerful technique, but in practice it is not 100% foolproof, often because of human errors. If a lab worker makes any mistake—even a small one—the results could be wrong. Mistakes could be things such as setting the temperature incorrectly, measuring out the wrong amount of important chemicals, or even mixing up samples. Contamination also can cause errors. This is especially important in PCR techniques because this procedure copies every piece of DNA in the sample several thousand times. So the results will be thrown off even if just a minute amount of DNA from another source gets into the sample. This could happen if a bit of the tissue sample from a crime scene is mistakenly mixed with the sample from the suspect. If this happened, the suspect's sample could match that of the victim, even if the suspect was innocent. These sources of error mean that if you are involved in any judgment or decision that involves DNA fingerprinting, you should look carefully at how meticulously the DNA evidence was gathered and stored and how well the DNA fingerprinting process was controlled.

polymerase chain reaction (PCR) a technique used to make thousands of copies of a small sample of DNA

QUICK CHECK

1. Briefly describe how restriction enzymes are used to generate a DNA fingerprint.
2. What is the polymerase chain reaction and how is it useful in the analysis of DNA?

11.3 DNA Studies Can Reveal Genetic Relatedness of Individuals, Populations, and Species

Another interesting application of biotechnology involves using DNA to determine whether different individuals are genetically related. Such tests can even determine the relationships between individuals who are no longer alive. **Table 11.2** lists several high-profile cases in which DNA tests have determined that a person was, or was not, related to someone else. Such tests do *not* use DNA fingerprinting techniques. Instead, DNA from mitochondria often is analyzed to provide convincing evidence about genetic relationships.

Recall that mitochondria are tiny organelles that complete the metabolism of glucose and other food molecules to produce ATP. Mitochondria once were free-living bacteria that billions of years ago came to reside within other, larger cells. Not only do mitochondria still carry their own DNA (**Figure 11.6**), but also mitochondria divide independently of the cell cycle. So new mitochondria inherit their DNA independently of nuclear DNA.

Two aspects of mitochondrial DNA make it useful for determining relationships. First, the

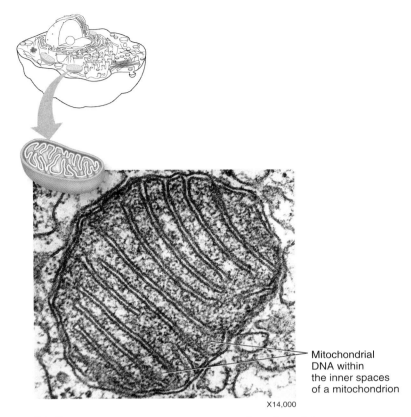

Mitochondrial DNA within the inner spaces of a mitochondrion

X14,000

Figure 11.6   **Location of DNA Within a Mitochondrion.**

Table 11.2   Cases of DNA testing to determine identity or relationships

Case	Samples tested	Conclusion
Who is in the Romanov grave?	DNA from bodies in the Romanov grave were compared with DNA from Prince Charles of England, who shares an ancestor with the Romanov family. **Gene test:** mitochondrial DNA	The Romanov grave contains the remains of Nicholas II, Romanov czar of Russia.
Was Anna Anderson really Anastasia, the daughter of Czar Nicholas of Russia?	Anna's DNA was compared with DNA from the Romanov grave and with families related to Romanovs. **Gene test:** mitochondrial DNA	Anna Anderson's DNA did not match Romanov DNA but did match that of relatives of Franziska Schanzkowska.
Who was the child sacrificed high in the mountains of Argentina 500 years ago?	DNA from an Incan girl sacrificed 500 years ago was compared with DNA from modern Peruvians. **Gene test:** mitochondrial DNA	A near-exact match was found between the sacrificed girl and a Peruvian now living in Washington, D.C.
Who is in Christopher Columbus' grave?	DNA from bones in Spain and the Dominican Republic will be tested against DNA from relatives of Columbus. **Gene test:** mitochondrial DNA	Tests are under way.
Did Thomas Jefferson father a child with his slave Sally Hemmings?	Testing of Hemmings' male descendants and the descendants of Thomas Jefferson's brother, Field. **Gene test:** Y chromosome test	The male descendants of one of Sally Hemmings' sons matches the DNA of the descendants of one of Field Jefferson's sons.

mitochondria in the cells of most multicellular animals originate only from the ovum of the female parent (**Figure 11.7**). This primarily is because the head of a sperm that fertilizes an ovum contains few, if any, mitochondria (Figure 11.7a). If mitochondria from a sperm do enter an egg at fertilization, they are eliminated during early development. In contrast, eggs are packed with mitochondria (Figure 11.7b). Second, because mitochondrial genes govern metabolic processes essential for survival, they do not change much from generation to generation. While mitochondrial DNA does mutate, cells that carry mitochondrial mutations usually die. As a result all cells in an individual have mitochondrial DNA that is nearly identical. Because of these two factors, the siblings of one generation have nearly identical mitochondrial DNA, and each has mitochondrial DNA highly similar to its mother's mitochondrial DNA. Mitochondrial DNA changes slowly over generations, and scientists have a good idea of the rate at which it does change. This means that people from the same family, even across many generations, will have highly similar mitochondrial DNA.

The techniques for testing mitochondrial DNA often involve determining the sequence of all, or some substantial part, of the entire **mitochondrial genome,** defined as the total genome in a single mitochondrion. Of course, it is possible to sequence the entire **nuclear genome** of a person, which is the total genome carried in the nucleus. But sequencing the nuclear genome is a costly, time-consuming process. Sequencing mitochondrial DNA is quicker and less expensive and can give valuable information about genetic relationships.

Genetic tests based on mitochondrial DNA are having significant impacts on people's lives. The military upheaval in Argentina in the 1970s and 1980s provides a dramatic example. During this time a military dictatorship disrupted thousands of Argentinean families. Many children were separated from their parents and often their biological mothers were killed. From the beginning of this social upheaval, groups of grandmothers worked to find the missing children and reconnect them with their biological relatives. Unfortunately the grandmothers often had no way to determine which children belonged with which families. Even as late as 1980 the only tests available were blood tests, and these are not completely definitive for determining human relatedness. Into this chaotic situation stepped Marie-Claire King, a researcher who had worked on DNA testing. She developed a mitochondrial DNA test that could reliably determine the grandchildren of each grandmother. The grandmothers established a DNA database so that when any child was found and tested, its family could be rapidly identified. The work of Marie-Claire King and the grandmothers of Argentina is an example of the profound social impact that the biotechnology revolution has had.

mitochondrial genome the total genome of a single mitochondrion

nuclear genome the total genome of an organism that is carried within the nucleus of each cell

Figure 11.7
Mitochondria in Sperm Compared to Mitochondria in Eggs.
Note the differences in the numbers of mitochondria in a sperm (**a**), compared with the numbers of mitochondria in an egg (**b**).

■ *If only mitochondrial DNA were used to make a genetic fingerprint, would two children who have the same parents have the same genetic fingerprint?*

(a) Sperm have a mitochondrion coiled around the tail.

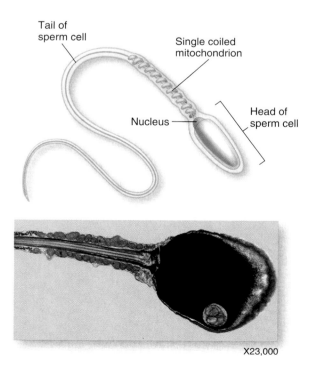

Tail of sperm cell
Single coiled mitochondrion
Nucleus
Head of sperm cell

X23,000

(b) Eggs are packed with mitochondria.

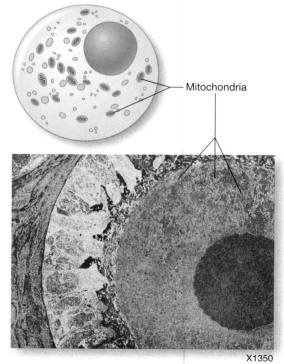

Mitochondria

X1350

One interesting case that was solved through mitochondrial DNA testing was the investigation into the Romanov family. Nicholas II and his wife Alexandra ruled Russia from 1894 to 1917 (**Figure 11.8**). In March 1917 Nicholas II abdicated the throne to his brother, but in July 1918 revolutionaries murdered both brothers and other family members too. Nicholas and Alexandra had five children, and the entire family was presumed dead. In 1991 after the fall of the Soviet Union, a possible Romanov gravesite was discovered, but the identity of the bodies was in dispute. In the mid-1990s mitochondrial as well as nuclear DNA was recovered from the bodies and was compared with that of a living Romanov relative, Prince Charles of England. The mitochondrial DNA strongly supported the hypothesis that Prince Charles and the remains of the body that had been identified as Nicholas II were indeed related through a maternal line. At some point in their family histories, Prince Charles and Nicholas II shared a common female ancestor.

This provided evidence that the bodies in the grave were Romanovs, but questions remained about two of the Romanov children. The biggest mystery was about Anastasia, whose remains could not be identified in the Romanov grave. The story of Anastasia (**Figure 11.9a**) is especially intriguing because a woman named Anna Anderson (Figure 11.19b) claimed that she was Anastasia. Anna Anderson did not speak Russian, but nevertheless she convinced many people that she was the lost Romanov daughter. Anna Anderson was cremated in 1984, but a tissue sample was found in a hospital, and testing of her mitochondrial DNA convinced researchers that she was not a Romanov. From the time that Anna Anderson first made her claims, some people suggested that she actually was a Polish factory worker named Franziska Schanzkowska. DNA tests of a grandnephew of Ms. Schanzkowska

—Anastasia

Figure 11.8 **Romanov Family.** The remains of the Romanov family were identified using tests of mitochondrial DNA.

suggested that Anna was, indeed, related to him and so supported the idea that Anna Anderson was actually Franziska.

Y chromosome DNA also is becoming a common source for genetic studies because it is inherited exclusively from the father and so offers another perspective on inheritance. Studies of Y chromosome DNA were the basis for the claim that it is likely that Thomas Jefferson fathered a child of his slave, Sally Hemmings (see Table 11.2).

One last example of the power of mitochondrial DNA testing is related to human evolution. Mitochondrial DNA tests can be used to calculate how long ago two individuals—whether they are humans or other eukaryotes—shared a common

(a)

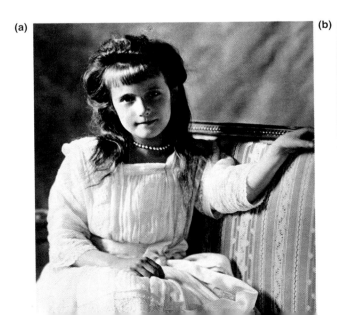

(b)

Figure 11.9
Anastasia and Anna Anderson. DNA testing has revealed that Anastasia (**a**) and Anna Anderson (**b**) are not the same person.

female ancestor. The logic of this analysis rests on the fact that the rate of mutations in mitochondrial DNA can be determined, and used as a kind of clock. For instance, if a segment of human mitochondrial DNA mutates at a rate of two nucleotides per generation, and if two individuals have mitochondrial DNA that differs by 10 nucleotides, they must have shared a common female ancestor five generations ago. The actual construction of a mitochondrial DNA clock that can be used to determine how long ago two humans had a common ancestor is much more complicated, but the techniques have produced some interesting conclusions. For example, when this type of analysis is done on Europeans, some researchers have determined that only a small number of women gave rise to the entire modern European human population. In addition, studies of human mitochondrial DNA agree with archeological studies in showing that all modern humans originally came from a small population of African ancestors who lived between 100,000 and 200,000 years ago.

DNA testing also can establish genetic relationships between different species and can provide information about their evolutionary history (see Chapter 13). DNA evidence supports the broad outlines of evolutionary history that have been gleaned from studies of fossils. Analysis of the important functional genes, such as those involved in metabolism or protein synthesis, also make it clear that all living organisms are genetically related and evolved from a small set of common ancestors. In addition, though, DNA studies have provided a deeper, more complete understanding of evolutionary relationships.

QUICK CHECK

1. What is the difference between mitochondrial DNA and nuclear DNA?
2. What kind of useful information can be gained by testing an individual's mitochondrial DNA?

11.4 Genetic Testing Offers Knowledge That Can Lead to Difficult Choices

DNA microarray a glass slide that carries thousands of dots of single-stranded DNA, used to detect complementary DNA or mRNA in a tissue sample

Genetic tests are being incorporated into routine medical practice. Many genetic tests are available that will tell whether an individual carries a gene for a particular disease. For instance, there are tests that identify genes involved in Alzheimer's disease, cystic fibrosis, breast cancer, Huntington's chorea, and sickle cell anemia. In each case the genetic test can distinguish between the normal and disease-related alleles. These tests can tell you if you carry none, one, or two alleles for a disease, and whether those alleles are dominant or recessive. Because there can be many different alleles for a gene that can lead to a particular disease, the development of these genetic tests has not been simple. For instance, researchers have discovered over 400 alleles of *BRCA1*, a gene related to the development of breast cancer and ovarian cancer. In another example cystic fibrosis can be caused by any one of scores of abnormal alleles for a particular membrane protein. So all genetic tests begin by identifying the nucleotide sequences of all of the known abnormal alleles that lead to the disease. Abnormal alleles are found by testing tissues from people who exhibit the disease in question. Once a library of abnormal allele sequences for a disease has been collected, this basic knowledge can be used in a variety of technical approaches.

One test developed in the 1990s is called a **DNA microarray. Figure 11.10** shows how this technique might be used to determine if a person carries any of the 400-plus abnormal alleles for the *BRCA1* gene. Copies of each of the 400-plus known abnormal gene sequences are manufactured and are robotically applied as small dots on a glass slide (Figure 11.10a). Each dot will have many copies of one abnormal DNA sequence. The genes in these dots are all single-stranded copies, and all of them have had the intron sequences cut out. So they are just the DNA sequences that code for the mRNA strands, which, in turn, code for a protein. This means that the single-stranded DNA in each dot is exactly complementary to the mRNA that would normally be made from that DNA. How can these dots of single-stranded DNA be used to find out if an individual carries a breast cancer gene? Think back to the base-pairing rules presented in Section 7.4. A sample of messenger RNA that matches any of the gene sequences in the dots will bind to the DNA in that dot because of this complementary relationship. A genetic test can be performed by taking a small tissue sample from a patient and extracting all of the mRNA (Figure 11.10b). Then the mRNA sample is poured over the microarray, and any matching sequences will bind to the DNA in each dot. The only other trick needed is to label the mRNA from the tissue sample with a dye that can be seen if it has bound to the DNA in a particular dot. Figure 11.10c shows a completed DNA microarray test for *BRCA1*. Computer analysis of this pattern determines the genes present, or expressed, in the tissue. Microarray technology is proving to be a powerful tool, both in research and in the development of other genetic tests.

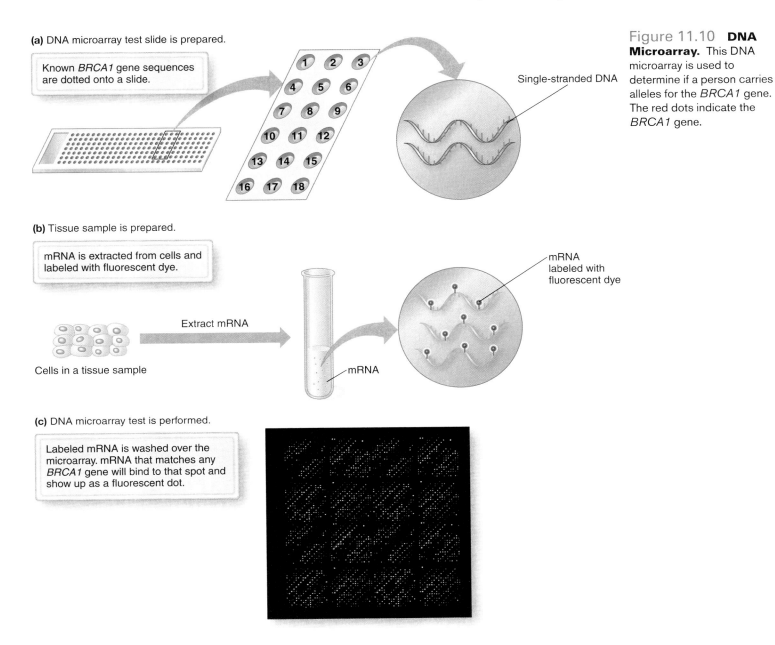

(a) DNA microarray test slide is prepared.

Known *BRCA1* gene sequences are dotted onto a slide.

Single-stranded DNA

(b) Tissue sample is prepared.

mRNA is extracted from cells and labeled with fluorescent dye.

Extract mRNA

Cells in a tissue sample

mRNA

mRNA labeled with fluorescent dye

(c) DNA microarray test is performed.

Labeled mRNA is washed over the microarray. mRNA that matches any *BRCA1* gene will bind to that spot and show up as a fluorescent dot.

Figure 11.10 **DNA Microarray.** This DNA microarray is used to determine if a person carries alleles for the *BRCA1* gene. The red dots indicate the *BRCA1* gene.

Would you want to know if you carry an allele related to a disease? Imagine that both you and your spouse are healthy, but that cystic fibrosis has occurred in your family. Cystic fibrosis (CF) is a genetic disease caused by recessive alleles. A genetic test would tell you what the chances are that you and your spouse would have a child with cystic fibrosis. Each of you could be either homozygous dominant—carrying no CF allele—or heterozygous—carrying one CF allele and one normal allele. Take a minute to complete Punnett squares that describe your potential offspring in each case. If, for example, the test results show that both of you are heterozygous, what are the chances that you would have a child with cystic fibrosis?

What if you are homozygous dominant and your spouse is heterozygous? **Figure 11.11** shows the answers. Of course, genetic tests like this cannot tell you whether to take the chance and try to have a child. That choice is yours, but as described in Chapter 25, there are ways to use these kinds of genetic tests to be relatively certain that your child will not carry a genetic disease, even if both parents are carriers.

As another example, you may want to know whether you carry genes for a particular disease and are at risk of developing it in the future. A positive test indicates an increased risk, but in most instances it is hard to pin down exactly how much risk is involved. Tests for inherited breast cancer

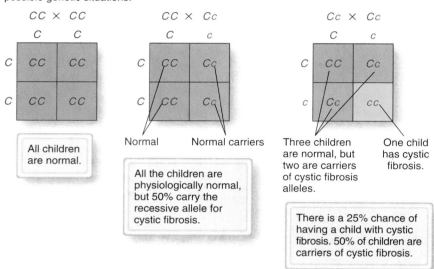

Key:

C = Normal allele

c = Cystic fibrosis allele

Figure 11.11 **Inheritance of Cystic Fibrosis.**

Parents who are healthy, but whose families show cystic fibrosis, could have three possible genetic situations.

All children are normal.

Normal   Normal carriers

All the children are physiologically normal, but 50% carry the recessive allele for cystic fibrosis.

Three children are normal, but two are carriers of cystic fibrosis alleles.

One child has cystic fibrosis.

There is a 25% chance of having a child with cystic fibrosis. 50% of children are carriers of cystic fibrosis.

are a good example. These tests detect certain cancer-related alleles, but they test only the ones we know about. They may not test for combinations of genes that could increase the risk of breast cancer. Other factors—such as age at first menstruation, exposure to cigarette smoke, unknown environmental chemicals, drug interactions, general health, and even stress levels—all may play roles in determining whether a person who carries alleles for inherited breast cancer will develop cancer.

QUICK CHECK

1. What is a DNA microarray and how are DNA microarrays useful in testing for genetic diseases?

11.5 Transgenic Organisms Are Important in Medicine and Agriculture

From the beginning of the biotechnology revolution, attention has been focused on the dream of altering the genome of cells to suit our purposes. An individual bioengineered to carry the genes of another species is called a **transgenic organism.** Why would anyone want to do that? Some of the

transgenic organism bioengineered organism that combines the genes of two or more organisms

plasmid a small loop of bacterial DNA that can be transferred to another bacterial cell

earliest and most useful applications of biotechnology have involved the use of proteins obtained from bioengineered organisms.

Transgenic Bacteria Are Now a Reliable Source of Human Insulin

Diabetes mellitus is a serious disorder that can affect nearly all body systems and lead to an early death. The early-onset form of the disease, called *Type 1 diabetes,* appears during childhood and is associated with fatigue, thirst, and weight loss accompanied by increased hunger and nausea. Type 1 diabetes is a lack of insulin, a signal molecule that binds to membrane receptors on body cells and stimulates the uptake of glucose into the cell. In the absence of insulin, glucose—also called *blood sugar*—builds up in the bloodstream and can damage many body tissues. What we call diabetes has been recognized for most of recorded human history, but until recently there has been no adequate treatment. In the early 1900s many folk remedies were tried such as oatmeal, milk, or rice diets. Opium was even used as a treatment for diabetes. Once it was discovered that insulin deficiency causes diabetes, insulin injections became standard treatment—however, there were limitations on its use. Insulin was expensive because it had to be extracted from the pancreas glands of slaughtered pigs and cows, and it required extensive purification before it could be injected into humans. In 1983 a new source of insulin became available that was less expensive and more abundant. This discovery has made a great difference in the lives of diabetics. The new source of insulin is transgenic bacteria.

One of the earliest transgenic approaches allowed biologists to insert new genes into bacterial cells. **Figure 11.12** outlines the procedure that accomplishes this transformation. Most of the DNA in a bacterium is in one long, double strand of DNA connected in a circle. In addition to this main chromosome, many bacterial cells have smaller circles of DNA called **plasmids.** They are separate from the main chromosome and are copied and expressed separately. Plasmids usually carry between 2 and 250 separate genes that help the bacterial cell survive in certain environments. The plasmid genes are not involved in crucial aspects of life, such as basic metabolism, but they can provide bacteria with special traits. For example, genes for antibiotic resistance usually are carried on a plasmid.

The intriguing thing about plasmids is that they can be transferred from one bacterial cell to another. This mechanism increases genetic variation in a bacterial population. The transferred piece of plasmid becomes a permanent part of the plasmid genome of the bacterial cell that receives it. Biologists can take advantage of this natural process and insert foreign DNA into a bacterial cell by splicing a bit of DNA into a plasmid. First, the plasmids from a sample of millions of bacterial cells are separated from the main bacterial chromosomes by gel electrophoresis. Once the plasmids are isolated, restriction enzymes are used to cut open the rings of DNA, and other enzymes are used to insert a desired gene into each. The genetically modified plasmids are mixed with other bacterial cells and, if conditions are just right, some cells will take them up.

Once a genetically engineered plasmid is inside a bacterium, it will act like native DNA. Even if the inserted genes came from a human, they are transcribed by the bacterium and new proteins are produced. A genetically engineered bacterium will pass the inserted gene to all of its progeny, and each of these will produce the foreign protein. Because bacteria reproduce rapidly, a single batch of genetically engineered bacterial cells will soon give rise to volumes of such cells. One of the first applications was the production of transgenic *Escherichia coli* that incorporated the human gene for insulin. Today insulin is produced in large vats full of transgenic *E. coli* (**Figure 11.13**). Other transgenic strains of bacteria are used to produce hepatitis B vaccine, human growth hormone, and an enzyme used in the manufacture of cheese.

Transgenic Plants and Animals Play Important Roles in Research and Agriculture

One lesson from transgenic organisms is that it does not seem to matter how closely related the donor and the recipient are. The DNA from one organism can be expressed in any other organism (**Figure 11.14**). This is a powerful bit of evidence that all life is based on the same basic molecular mechanisms, and that all life is genetically related.

One of the most common uses of transgenic technology is the production of genetically modified food crops, also known as *genetically modified organisms*, or *GMOs*. *Bt* corn, Flavr-Savr tomatoes, golden rice, and many other genetically modified crops have been developed, and some are in regular use. *Bt* corn is one of the best-known examples. It gets its name because it carries a gene from the bacterium *Bacillus thuringiensis*. This prokaryote produces toxins that kill the larvae of the European corn borer, a major pest for

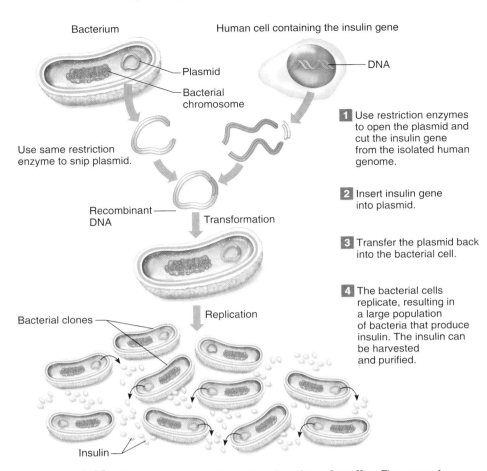

1 Use restriction enzymes to open the plasmid and cut the insulin gene from the isolated human genome.

2 Insert insulin gene into plasmid.

3 Transfer the plasmid back into the bacterial cell.

4 The bacterial cells replicate, resulting in a large population of bacteria that produce insulin. The insulin can be harvested and purified.

Figure 11.12 **Bioengineering *E. coli* to Produce Insulin.** The genes for human insulin can be inserted into *E. coli*. These bioengineered bacterial cells then produce human insulin.

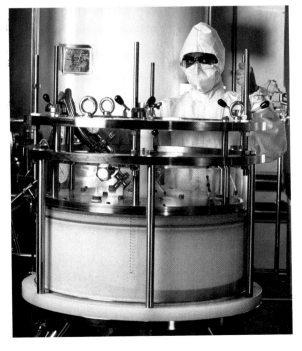

Figure 11.13
Commercial Production of Insulin. Insulin-producing transgenic *E. coli* are cultured in these vats.

Figure 11.14 **DNA from One Organism Can Be Expressed in Another.** This baby mouse carries firefly genes that make it glow in the dark.

growers of corn. To control the European corn borer, some farmers used to spray corn with live *B. thuringiensis.* But the effects of a single spraying did not last long. Before *Bt* corn was developed, most farmers used repeated applications of pesticides to control the corn borer. Now many farmers plant *Bt* corn. For farmers *Bt* corn and other genetically modified crops offer the promise of even greater productivity per acre of land than can be achieved with traditional crops. Other people anticipate further benefits of genetically modified foods. For example, a genetically modified form of rice called *golden rice* carries a gene to produce a substance that the body turns into vitamin A. Although the effects of golden rice on vitamin A level in humans remain to be studied, researchers are hopeful that this rice can help to prevent vitamin A deficiency, a common problem that causes permanent blindness in children in many parts of the world. Genetically modified crops can be designed to survive in adverse growing conditions, such as in soil with a high salt content, in cold or hot climates, or where local pests are out of control. Transgenic crops could greatly increase the world's food supply.

Many people, including some biologists, voice objections to genetically modified crops. Among their concerns are possible health risks for humans who eat genetically modified foods such as corn, possible negative environmental effects of genetically engineered crops, and the possibility that the foreign genes might invade wild plants in the environment. Are these fears realistic? At this time the scientfic evidence is not strong in either direction. There is little evidence for negative effects of genetically modified crops, but at the same time, many of the specific concerns of the critics cannot be answered with confidence.

While the world is waiting for scientists to sort out the safety of genetically modified foods, their legal status depends on where you live. Many countries have banned genetically modified organisms in human food, but in the United States several kinds of genetically modified foods have been approved. Most end up in the feed of farm animals such as pigs, cattle, and chickens. You *could* be eating genetically modified foods in your breakfast cereal, but it is difficult to tell because there is no labeling requirement for genetically modified foods. An example of the sensitivity of this subject happened in 2000 when transgenic corn intended for animal feed ended up in the corn chips sold by Taco Bell. Consumers were upset, chips had to be recalled, and safeguards were put in place.

Other transgenic organisms are being developed with varied agricultural as well as other applications. For example, one day you might wear spider silk clothing that originated as goat's milk, or police might use bulletproof vests made of silk processed from goat's milk. Spider silk is one of the strongest yet most flexible materials known. But spiders don't produce that much silk—so it has not found much practical use. Scientists at a Canadian research firm changed all that when they bioengineered dairy goats to produce spider silk in their milk. The possibility of processing silk from large quantities of milk from bioengineered goats holds great promise.

Transgenic Animals Allow Scientists to Study Human Diseases The last point to mention about transgenic organisms is how important they are in the study of human diseases. Before they are tested in human subjects, treatments for many human diseases are developed and tested in animal models, such as mice. In some cases there are natural strains of mice or rats that show traits similar to those of humans. For example, some strains of mice will become obese if allowed unlimited access to food; other strains of mice will drink large amounts of alcohol if given the choice between a weak alcohol solution and plain water. These strains can be used to study human obesity and alcoholism. In many cases, however, there are no naturally occurring rodent models, but in the last few decades scientists have developed transgenic mice that express a particular disease or an aspect of it. Of course, to produce a transgenic mouse, you first must have isolated a particular gene associated with a disease. Then the gene can be inserted into mouse zygotes, the zygotes can be implanted into a foster mother, and the transgenic animals develop. The amazing thing about transgenic mice is that if the gene insertion works, these mice will pass the gene on to their offspring, producing not only one transgenic mouse, but also a whole population of such mice. **Table 11.3** lists some of the transgenic mouse models of diseases now being studied.

You might be wondering whether transgenic humans could be produced. As you can read in Chapter 25, the techniques to fertilize human eggs in a dish are well developed, and there is no reason why the techniques used to produce transgenic mice could not be used on humans. The major technical problem is that even in animal models these techniques are successful only a small percentage of the time. Many transgenic fertilized eggs do not proceed through embryonic development—of those that do, many are abnormal and do not survive long after birth. For these reasons, the risk and uncertainty involved in producing a transgenic human embryo presently is too great. Even if this risk were overcome, some people would certainly raise ethical or moral objections to producing a human that carries the genes of some other species.

QUICK CHECK

1. What is a transgenic organism and how are transgenic organisms used?

11.6 Cloning Is an Application of DNA Technology

Each of the DNA technologies discussed so far has changed society and will have even more widespread effects in the future. In terms of public reaction, though, the possibility of cloning a whole organism has probably generated more headlines, interest, and concern than any other application of biotechnology. In this context a **clone** is a genetic replica of another individual. Identical twins, triplets, or other multiple births are essentially clones of one another that occur naturally. Today, it is possible to produce such clones using laboratory techniques.

The first cloned animals were albino frogs produced in the early 1970s. Dolly the sheep was cloned in 1996 (**Figure 11.15**). Since then cows, pigs, cats, and other animal species have been cloned. The big question is can humans be cloned? Despite the claims of religious cults or errant doctors, the cloning of a fully developed human is not

Table 11.3 Transgenic mouse strains being studied to understand human disease

Human disease under study	Specific gene inserted into mouse strain	Examples of scientific discovery from these studies
Alzheimer's disease	ApoE4	Abnormal *ApoE4* can produce neuronal abnormalities
Alzheimer's disease	APP	Treatment of *APP* transgenic mice with antibodies to certain proteins reduced the Alzheimer's-like symptoms
Atherosclerosis	ApoA-II	Reduced clogged arteries in mice fed a high-fat diet
Lymphoma	Bcl2	Caused increased risk of cancer of the lymph system
AIDS	Entire human immuno-deficiency virus protein-coding genome	Caused development of full-blown AIDS

Figure 11.15 **Dolly the Sheep.** Dolly made headlines because she was the first cloned mammal.

likely to happen in the near future. But human cloning could happen some day. Let's discuss how cloning is accomplished, and consider some of the difficulties and challenges of cloning.

The most common method to clone animals is called *nuclear transfer cloning.* **Figure 11.16** shows how a researcher in the early 1970s used nuclear transfer techniques to produce clones of an albino frog. In this process the nucleus from one cell is removed and replaced with the nucleus of another cell (Figure 11.16a). The receiving cell is an egg removed from a sexually mature female, and the donated nucleus comes from a mature adult cell. In this example the egg came from a pigmented female frog and the nucleus came from a gut cell of an

clone a genetic replica of another individual

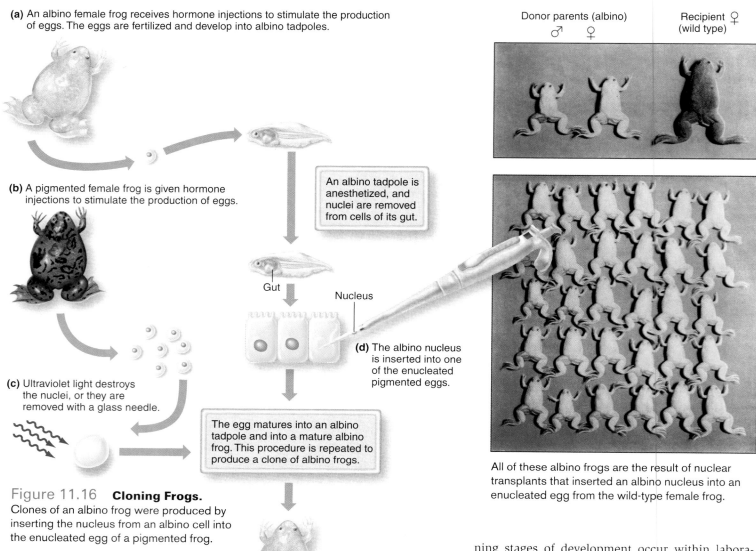

(a) An albino female frog receives hormone injections to stimulate the production of eggs. The eggs are fertilized and develop into albino tadpoles.

(b) A pigmented female frog is given hormone injections to stimulate the production of eggs.

(c) Ultraviolet light destroys the nuclei, or they are removed with a glass needle.

An albino tadpole is anesthetized, and nuclei are removed from cells of its gut.

Gut

Nucleus

(d) The albino nucleus is inserted into one of the enucleated pigmented eggs.

The egg matures into an albino tadpole and into a mature albino frog. This procedure is repeated to produce a clone of albino frogs.

Figure 11.16   **Cloning Frogs.**
Clones of an albino frog were produced by inserting the nucleus from an albino cell into the enucleated egg of a pigmented frog.

Donor parents (albino)   Recipient ♀
♂   ♀   (wild type)

All of these albino frogs are the result of nuclear transplants that inserted an albino nucleus into an enucleated egg from the wild-type female frog.

albino tadpole (Figure 11.16a). The DNA of the egg was destroyed using UV light, and the nucleus from the albino frog was gently injected into the egg (Figure 11.16b&c). Clones of the albino frog developed (Figure 11.16d). In frogs this is relatively simple because frog eggs are large and easy to handle and because frog eggs normally develop in pond water.

The apparently simple success of cloning frogs led people to attempt cloning in other species, especially mammals. It turned out, though, that cloning mammals is not easy at all. One problem is that in contrast to frogs, most mammals normally develop within a uterus. During cloning the begin-

ning stages of development occur within laboratory glassware. Many of the zygotes produced for mammalian cloning often do not even begin development. The nucleus can be damaged when it is transferred from one cell to another. There is some evidence that mammalian nuclei are more delicate than frog nuclei and are more easily damaged during the transfer process. The most important reason for the failure of mammalian clones may be the interaction between the recipient egg cytoplasm and the donor nucleus. Genes can be turned off by DNA methylation and other mechanisms (see Section 9.2). During the first few hours of development, the egg cytoplasm resets the methylation patterns on the DNA so that genes needed for development can be transcribed. Many researchers believe that this pattern of gene resetting does not happen normally during cloning, and so the embryos are abnormal.

The end result is that very few cloned mammalian embryos ever develop into a full organism.

(a) Creation of a cloned embryo

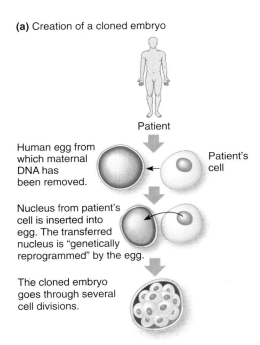

Patient

Human egg from which maternal DNA has been removed.

Patient's cell

Nucleus from patient's cell is inserted into egg. The transferred nucleus is "genetically reprogrammed" by the egg.

The cloned embryo goes through several cell divisions.

Figure 11.17 **How Stem Cells Could Be Obtained and Used in Medical Treatments.**

(b) Stem cells derived from the cloned embryo could form a variety of cells.

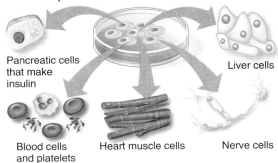

Pancreatic cells that make insulin

Liver cells

Blood cells and platelets

Heart muscle cells

Nerve cells

(c) Cells developed from embryonic stem cells could be used to treat many medical problems.

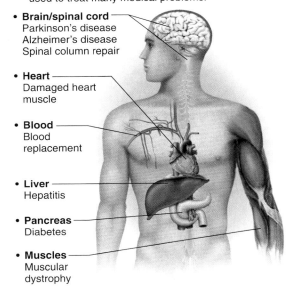

- **Brain/spinal cord**
  Parkinson's disease
  Alzheimer's disease
  Spinal column repair

- **Heart**
  Damaged heart muscle

- **Blood**
  Blood replacement

- **Liver**
  Hepatitis

- **Pancreas**
  Diabetes

- **Muscles**
  Muscular dystrophy

Many of those that do develop are deformed or die shortly after birth. Some experiments are more successful than others, but on average only about 3% of cloned mammalian embryos develop normally and are born with no complications. Even clones that are born and develop normally do not always lead normal lives. Many, including Dolly, have shortened life spans. The rates are even lower for primates. Only two monkeys have ever been successfully cloned despite many hundreds of attempts. Although researchers are beginning to sort out the problems associated with cloning mammals, especially primates, because of these low success rates it is unlikely that human cloning will be successful. At this point nearly all scientists believe it would be irresponsible to attempt to produce a human child by cloning.

This leads to the next question. Why would anyone want to clone a mammal, much less a person? The answer to this question depends on the distinction between reproductive cloning and therapeutic cloning. **Reproductive cloning** is the production of a fully developed, cloned individual, such as Dolly the sheep. Scientists are interested in reproductive cloning to have tailor-made organisms for study. For example, a transgenic clone of mice would have a controlled genome that would make it easier to study the effects of the transgenic gene on development and disease. On the other hand, some people are interested in reproductive

cloning to reproduce themselves, another person, or even a beloved pet. At least one company claims it will do so and has delivered cloned pets to devoted pet owners—at a cost of $50,000, at least in the year 2005. Many find this practice unethical. As you can imagine, the debates over cloning a human are much more intense.

Therapeutic cloning is much different. The aim of **therapeutic cloning** is not to produce an entire organism, but to develop cells or tissues that can be used to treat disease. The way that therapeutic cloning would work is shown in **Figure 11.17.** For instance, therapeutic cloning could help a person who has a damaged heart, and a significant portion of heart muscle cells that do not function properly. Short of a heart transplant, this person soon will die, but therapeutic cloning someday might provide a treatment. In therapeutic cloning this person would provide her own cellular nuclei for transplantation into an egg (Figure 11.17a). The egg can come from any human egg donor. The

reproductive cloning the production of a fully developed, cloned individual

therapeutic cloning developing cells or tissues that can be used to treat disease

cloned embryo is allowed to develop to around the 100-cell stage, only long enough to produce heart stem cells (see Section 9.6) that can be transplanted back into the patient's heart (Figure 11.17b). Once there they respond to local signal molecules and develop into fully functional heart cells (Figure 11.17c). The advantage to using her own nuclei as donors is that there will be no tissue rejection. The patient's heart is being healed with her own tissue. This may sound far-fetched, but a

similar study has already been done in rats, and it has been successful. Damaged rat hearts were significantly repaired with cells derived from therapeutic cloning.

> **gene therapy** a general term for the replacement of a defective gene with a normal one in the affected tissues

QUICK CHECK

1. How is nuclear transfer cloning accomplished?
2. What is the difference between reproductive and therapeutic cloning?

11.7 An Added Dimension Explained
What Promise Does Gene Therapy Hold?

From the early days of the DNA revolution, people have hoped that gene therapy would be the miracle treatment that would cure many diseases. **Gene therapy** is a general term for the replacement of a defective gene with a normal one in the affected tissues with the aim of correcting a disease or disorder. Despite tragedies such as the death of Jesse Gelsinger, gene therapy does hold promise. Although there are currently no FDA (Food and Drug Administration) approved gene therapy treatments, many animal experiments demonstrate that gene therapy can work, and many human clinical trials are in progress. *Clinical trials* are the experiments conducted on human volunteers that establish the safety and effectiveness of any new treatment for a disease. In the United States there have been over 700 clinical trials for gene therapy. Let's now consider why gene therapy could be better than conventional therapy, and see the obstacles to making gene therapy work.

The aim of most gene therapy is to provide a normal, functioning protein for people who have a protein that does not work normally. You might wonder why you could not just take such a protein in a pill, but providing normal proteins is one area where traditional drug treatment does not work. Proteins

depend on their three-dimensional shape to carry out their specific cellular job. This shape is held together by hydrogen bonds, and these are broken easily. If you take proteins in the form of pills, they are likely to be denatured by stomach acids. A denatured protein has lost its three-dimensional tertiary structure (see Section 3.6) and so cannot do the job it is intended to do. Even if the protein enters your bloodstream intact—say by an injection—the proteins won't be around for long because enzymes may attack and destroy them. Finally, even those proteins that survived in your blood would have a hard time getting into the cells where they do their job. For these reasons it is not easy to give proteins as "medicines." Gene therapy aims to avoid these problems by sending normal DNA sequences directly into the abnormal cells that cause the problem. The idea is that the normal DNA would be expressed, thus producing normal proteins, which would correct the problem. As appealing as this sounds, there are some serious obstacles to be overcome.

One obstacle is how to get the DNA into cells. Recall that like proteins, DNA is a large, charged molecule. It will not pass readily across cell membranes—and, like proteins, DNA would be damaged by digestion if taken as a pill. Several approaches to delivering therapeutic genes are under study (**Figure 11.18**). One way to get therapeutic genes into cells is to use viruses that normally insert their own DNA into cells when they infect them (Figure 11.18a). This method has been used to treat blood cell disorders in animals such as sickle cell anemia or abnormalities of cells of the immune system. Limited studies using viruses also have been done in humans with immune disorders. In this kind of gene therapy

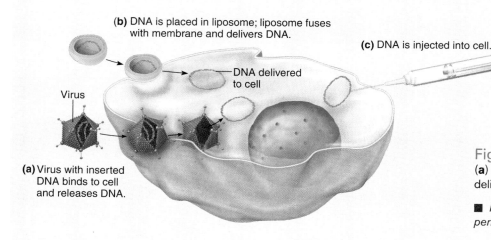

(b) DNA is placed in liposome; liposome fuses with membrane and delivers DNA.

(c) DNA is injected into cell.

DNA delivered to cell

Virus

(a) Virus with inserted DNA binds to cell and releases DNA.

Figure 11.18 Ways to Get DNA Into a Cell.
(a) A virus, **(b)** a lipsome, or **(c)** a fine glass pipette can deliver DNA to a cell.

■ *How might such a change to a person's DNA be permanent?*

abnormal cells that give rise to red blood cells are removed from the patients, and genetically modified viruses are used to deliver normal DNA to the abnormal cells. The genetically modified cells are then returned to the patient with the hope that they will develop into normal blood cells or normal immune system cells. Some genetically modified viruses also can be administered in a nasal spray or by other means, so the patient's own cells do not have to be removed. Again, this approach has been successful in animal studies and preliminary studies are under way in humans.

A different approach puts the therapeutic genes into liposomes, tiny lipid spheres (Figure 11.18b). Although liposomes have structures similar to cells, they do not contain any living processes. So liposomes can be "filled" with biological molecules or drug molecules and because of their lipid boundary, liposomes can deliver their contents across cell membranes. Some liposomes are even small enough to get through the nuclear pores, and so they can carry DNA into a cell's nucleus. In an even simpler approach using lipids, the normal DNA can be attached to lipid molecules. This increases the chances that the new DNA will cross the cell membrane. Finally, cells can sometimes take up naked DNA.

An example of how a gene therapy treatment would actually work comes from the attempts to develop a treatment for a disease called SCID (severe combined immunodeficiency). Patients with SCID have minimal if any immune responses to infectious agents, and so they are prone to severe illnesses. You may not appreciate how much your immune system does on a

daily basis, but people who lack a normal immune system can die from bacteria and viruses that never would make the rest of us even mildly ill. Before the early 1980s most children born with SCID died before they were 2 years old. A variety of genetic mutations can produce SCID, and the immune defect can be mild or severe. **Figure 11.19** shows the general outline of the gene therapy procedures used to treat SCID. Early efforts met with only modest success, but in the last few years a French team of researchers has reported remarkable success with gene therapy for SCID. Even more recently, however, at least two of these successful gene therapy patients have developed a form of leukemia. It still is not clear if their leukemia is related to the gene therapy treatment, but the earlier excitement over gene therapy for SCID has been dampened and further gene therapy has been halted until this problem is better understood.

Thus far, gene therapy for other diseases is still in the research stages. Clinical trials are under way for using gene therapy to treat cancer, Parkinson's disease, and other disorders. It remains difficult to get the necessary genes into the affected cells—currently researchers are unable to ensure that the gene will be expressed when it is needed. To be expressed, the DNA strands must be incorporated into the patient's native chromosomes, but there is no way to control where the new DNA will insert or what promoters it will become associated with. If the new DNA is not inserted into a chromosome, enzymes will likely destroy it. If the DNA is incorporated, there are concerns

—Continued next page

1 Isolate stem cells from bone marrow of child with severe combined immunodeficiency disease (SCID). These cells have a mutation in any one of several genes that can cause SCID. The mutation results in immune cells that do not function normally.

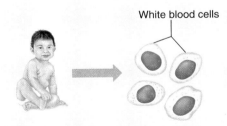

White blood cells

2 Insert a normal gene for the child's defective gene into a virus.

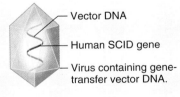

Vector DNA

Human SCID gene

Virus containing gene-transfer vector DNA.

3 Use the virus to insert the normal DNA into the defective stem cells from the SCID infant.

SCID gene

4 Grow genetically altered stem cells in culture, allowing them to divide and produce a large number of normal cells.

5 Infuse genetically altered stem cells back into infant. These altered stem cells produce normal white blood cells that give the infant a normal immune system. The treatment must be repeated periodically.

Figure 11.19   **Gene Therapy for SCID.**

Continued—

that it may not be expressed or that the newly introduced DNA could disrupt the expression of other normal genes.

Aside from the modest successes of gene therapy, other worries have arisen. These worries intensified when Jesse Gelsinger died. Jesse's death was caused by a massive and unexpected immune system reaction that he experienced after undergoing gene therapy. Many of the viruses used to carry the new gene into cells can themselves cause illness. In both humans and animals gene therapy can produce side effects such as flu and fever symptoms, decreased liver function, or even cancer. The number of animals or patients that develop such problems is small, so it is hard to know what has caused them. The case of Jesse Gelsinger and the SCID children who have developed leukemia show that the side effects of gene therapy can be severe and may be fatal. If the disease being treated is nearly always fatal, the risk may be worth taking. Knowledge about basic biological mechanisms and their applications for treating diseases is moving forward more rapidly than at any time in history. Many people in the scientific and medical community are still confident that gene therapy one day will be standard, successful treatment for many human disorders.

QUICK CHECK

1. What are three ways that genes can be gotten into cells in gene therapy?
2. What are the obstacles that must be overcome before the promise of gene therapy is realized?

Now You Can **Understand**

The Genetics of Individual Differences

This chapter has presented some amazing applications of knowledge about biological molecules. Most of these applications are based on techniques used to understand and manipulate DNA. In Chapters 7 and 9 you learned how DNA determines the traits of organisms, and in this chapter you found that DNA can be used to identify who you are and to whom you are related. Despite this progress, an aspect of the application of DNA knowledge is lacking. From the beginning of the DNA revolution, people have expected that knowledge about DNA could really *explain* human traits. You can probably think of many examples. Why is your sister so good at math, and you can't add a column of numbers without a calculator? Why do you have near-perfect pitch and a voice like a songbird, while your father sings like a crow? Why is your best friend so sensitive to medications, while another friend seems unaffected by them? Recall that complex traits such as these are often the result of the interactions of many genes, and the interaction of genes with the environment. These conclusions are clearly valid, but new techniques are finding ways to relate these individual traits with small differences in DNA sequences. In fact, now it seems likely that some individual differences are the result of variations in a single nucleotide along a particular gene.

A variation in a single nucleotide along a gene is called a *single nucleotide polymorphism,* or *SNP.* The study and application of SNPs is only just beginning—but by the time this book is in your hands, SNPs probably will explain a significant portion of individual differences. SNPs have been discovered that are related to the ancestry of a person. These include the degree of general resistance to drugs, responsiveness to certain drug treatments, likelihood of various immune system disorders, ability to metabolize alcohol, and many other genetically controlled traits. The discovery of SNPs and the techniques to find them is a major new development in the biotechnology revolution.

What Do **You** Think?

Should There Be Mandatory Genetic Testing for Some Conditions?

The biotechnology revolution brings with it difficult social issues. One is the question of mandatory testing. Many people immediately reject such an idea, without much thought as to why they are against it. Others embrace such an idea, perhaps without thinking of the consequences. Take the time now to think more deeply about mandatory tests. Here are some of the issues you might want to think about.

Mandatory testing is not a completely new idea. A child or teacher in public schools has to have a "TB test." This test shows whether you have been exposed to TB (tuberculosis) and so might be able to infect other people. Because TB is a highly contagious disease, schools have decided to screen carefully for its presence. A new test that has been recommended detects HIV. You may know that HIV is the virus that causes AIDS, a disease that destroys the immune system and is a major public health problem. A mother with AIDS can transmit the virus to her unborn child, but if it is known that she carries the virus, she can receive drugs that reduce the chances that her child will carry HIV. Some people argue that HIV testing should be mandatory for pregnant women, and it is mandatory in some states.

The development of tests for genetic diseases opens the door for other kinds of testing. For instance, it is possible to test adults, potential parents, and even early embryos, for a variety of inherited disorders. Such testing could provide preventive care and save money, time, and suffering. At the same time, genetic tests can be used to discriminate against individuals who test positive. For example, in the 1970s tests for sickle cell anemia led to documented discrimination by employers and insurance companies.

As you can see, the availability of genetic tests opens opportunities and presents problems. Think about these issues. Do some searching on the Internet to investigate other information and opinions. Is there any situation in which genetic testing should be mandatory? What do *you* think?

Chapter Review

11.1 Biotechnology Has Dramatically Changed Society and Human Lives

The biotechnology revolution is now under way. Techniques to manipulate, characterize, and control DNA have changed many aspects of life and may have even more radical practical applications in the future. The Human Genome Project was a most important milestone in mapping the human genome.

biotechnology 259   Human Genome Project 259

11.2 Forensic DNA Techniques Match Individuals to DNA Samples

DNA fingerprinting is one of the most commonly used biotechnological techniques. The advantage of DNA fingerprinting is that it allows unequivocal identification of an individual. Each individual has a unique DNA fingerprint derived from the sequence of nucleotides in their DNA. One technique makes DNA fingerprints using gel electrophoresis of DNA fragments obtained by treating a DNA sample with restriction enzymes derived from bacteria. RFLPs are unique to each individual and are derived from junk DNA, not from genes that code for proteins. Each individual has a unique number of repeated sequences of junk DNA. PCR amplifies small DNA samples. DNA fingerprinting is most commonly used to show that a person is innocent of a crime.

DNA fingerprint 260   restriction enzyme 261
forensics 260   restriction fragment length
gel electrophoresis 262   polymorphism (RFLP) 262
polymerase chain reaction (PCR) 264

11.3 DNA Studies Can Reveal Genetic Relatedness of Individuals, Populations, and Species

Analysis of mitochondrial DNA sequences shows genetic relatedness and is used to establish if two people are related. Y chromosome analysis is useful in paternity tests. Mitochondrial DNA analysis has traced all humans back to a small population of common female ancestors who lived in Africa 100,000 to 200,000 years ago. This correlates nicely with fossil evidence for the origin of humans.

mitochondrial genome 266   nuclear genome 266

11.4 Genetic Testing Offers Knowledge That Can Lead to Difficult Choices

Tests of abnormal alleles associated with genetic diseases allow identification of carriers. DNA microarrays are commonly used to identify abnormal alleles.

DNA microarray 268

11.5 Transgenic Organisms Are Important in Medicine and Agriculture

Genetically modified bacteria produce human insulin and other human proteins. Many other transgenic organisms have been developed that are used in medical and pharmaceutical research. Genetic modification of bacteria uses plasmids to deliver genes of interest to bacteria that subsequently produce the proteins of interest from these inserted genes. Transgenic plants contain genes that confer a variety of genetically controlled capabilities. *Bt* corn contains genes that make it toxic to insects. Other genetically modified plants have been developed with better flavor, longer shelf life, or frost resistance.

plasmid 270   transgenic organism 270

11.6 Cloning Is an Application of DNA Technology

Cloning different varieties of frogs has been successful, transferring nuclei from one variety to another. Cloning of mammals is much more difficult and still is in developmental stages. Discussion and research into therapeutic cloning or reproductive human cloning has been, and still is, a controversial topic. To date no one claims to have cloned a human.

clone 273   therapeutic cloning 275
reproductive cloning 275

11.7 An Added Dimension Explained: What Promise Does Gene Therapy Hold?

Because Jesse Gelsinger died soon after undergoing gene therapy, human clinical trials of gene therapy have been halted. Delivery of DNA into cells is a major obstacle that must be overcome before gene therapy can become widely used.

gene therapy 276

TRUE or FALSE. If a statement is false, rewrite it to make it true.

1. A DNA fingerprint is just as individualized as the pattern on the tips of your fingers and can be used to identify a single individual with great accuracy.

2. Restriction enzymes come from viruses and are used by viruses to help them infect bacteria.

3. Gel electrophoresis is used to sort out different-sized DNA fragments after restriction enzymes have cut up a large DNA molecule.

4. The biotechnology revolution began with the discovery of the structure of carbohydrates in 1975.

5. The Human Genome Project has demonstrated that there are at least 100,000 different human genes.

MULTIPLE CHOICE. Choose the best answer of those provided.

6. What is biotechnology?
   a. techniques used to isolate and purify membrane receptors
   b. the technology of biologically based machines used to harvest crops
   c. the term refers only to techniques of DNA fingerprinting
   d. the use of enzymes in manufacturing
   e. techniques to isolate, purify, manipulate and control biological molecules, especially DNA and proteins

7. What is the name for the use of scientific knowledge and techniques applied to legal issues?
   a. proteomics
   b. genomics
   c. forensics
   d. biotechnology
   e. the Human Genome Project

8. DNA fingerprints can be obtained from
   a. humans only.
   b. any eukaryote.
   c. bacteria only.
   d. humans and other mammals.
   e. mitochondrial DNA only.

9. What role do restriction enzymes play in generating a DNA fingerprint?
   a. They destroy the enzymes that would otherwise eat up DNA during the process of DNA purification.
   b. They cut up extraneous, contaminating DNA in the sample and so get rid of one source of error.
   c. They cut a long piece of DNA into smaller pieces.
   d. They cut DNA at identified points in the DNA sequence.
   e. c and d are correct.

10. What does PCR stand for?
    a. proteonomic chain reaction
    b. polymerase chain reaction
    c. partial chromosome restriction
    d. prokaryote chromosome restriction
    e. protein chain response

11. What does PCR do?
    a. makes many copies of a single strand of DNA
    b. binds together many small strands of DNA into one large strand
    c. matches the proteome of a cell with its genome
    d. uses prokaryotic restriction enzymes to identify eukaryotic chromosomes
    e. all of the above

12. What practical applications were mentioned in the chapter as uses for PCR?
    a. allowing DNA testing of a very small sample of DNA
    b. an alternative method for DNA fingerprinting
    c. a method to test genetic relationships between species
    d. a and b are correct.
    e. a, b, and c are correct.

13. On what basis does gel electrophoresis separate pieces of DNA?
    a. the number of nucleotides in the strand
    b. the number of adenines in the strand
    c. the number of guanines in the strand
    d. the shape of the strand
    e. the number of positive charges on the strand

14. Which of the following is true about a DNA fingerprint?
    a. It is a method of analyzing the DNA from a sample of tissue.
    b. It is a method to determine whether two samples come from the same person or two different people.
    c. It is a forensic technique that can be as reliable as regular fingerprints in determining whether a person committed a particular crime.
    d. It is a series of bands on a gel, each of which represents the amount of DNA fragments having a particular size.
    e. all of the above

15. Which of the following is one realistic reason that DNA fingerprinting can lead to the wrong conclusion about the identification of a particular person?
    a. The different tissues in the person being tested have different DNA.
    b. The DNA sample is contaminated.
    c. The two samples come from two unrelated people with exactly the same DNA sequences.
    d. The transcription factors used to cut the DNA into small pieces are not chosen properly.
    e. None of the above, because DNA fingerprinting is infallible.

16. Mitochondrial DNA studies can trace lineages
    a. through the mother's line only.
    b. through the father's line only.
    c. through the mother's or father's line.
    d. back three generations.
    e. through a single species but not between species.

17. What techniques have been most commonly used to reveal information about the genetic relationships between individual humans?
    a. PCR and microarrays
    b. full nuclear DNA sequencing and full mitochondrial DNA sequencing
    c. PCR and restriction enzyme analysis
    d. full mitochondrial DNA sequencing and Y chromosome sequencing
    e. all of the above

18. DNA microarrays are useful because they
    a. can analyze the nucleotide sequence of one person in great detail.
    b. have been used to determine the genetic relationship between species.
    c. can test for the presence of multiple genes in one sample.
    d. can be used to directly analyze the proteins in a sample.
    e. are commonly used to study mitochondrial DNA.

19. What role do plasmids play in generating transgenic organisms?
    a. They can be used to carry foreign DNA into an organism, particularly into bacteria.
    b. They can be used to cut DNA out of an organism.
    c. Plasmids are viral DNA that can carry foreign DNA into an organism during infection.
    d. Plasmids are early forms of DNA that can combine with newer forms of DNA to produce a new kind of organism.
    e. none of the above

20. Some of the methods used to carry out gene therapy include which of these risks?
    a. The genes might cause cancers such as leukemia.
    b. The genes might infect the germ layer and change the offspring.
    c. A strong immune response to the genes and the method used to transfer them can cause disease or death.
    d. The genes might cause a personality change.
    e. a and c are correct.

CONNECTING KEY CONCEPTS

1. The techniques of DNA fingerprinting and mitochondrial DNA analysis are two of the most commonly used applications of biotechnology. What are the details of these two techniques? What is each usually used for?

2. Biotechnology is having a great impact on American agriculture. What biotechnology technique is particularly relevant to farming? What procedure is used to accomplish this? Give a specific example of how this technique has been used thus far in the commercial production of food. List examples of farm products that could be seen in the future. What does this particular biotechnology application tell us about the genetic relationships between different species?

QUANTITATIVE QUERY

Consider this hypothetical example. A length of mitochondrial DNA that is 850 nucleotides long mutates at a rate of 4% mutations every million years. A population split from its ancestors to form a new species 2 million years ago. How many mutations would have occurred between the ancestor and the current species during that 2 million years?

THINKING CRITICALLY

1. In this chapter you read how DNA microarrays could be used to determine if a person carries a particular gene. The test can detect either DNA or mRNA in a tissue sample. How would the information obtained from testing an individual's DNA be different from the information obtained from testing a person's mRNA test?

2. It is possible that gene therapy given to an adult individual could reverse a genetic disease. If so, would that therapy also prevent any children the person had after the therapy from inheriting the disease? Explain why or why not, and what factors the answer would depend on.

For additional study tools, visit www.aris.mhhe.com.

Herbal Remedies Turned Deadly for Patients
Suits Fail to Bring Together Rules
September 5, 2004

Steve McQueen
used Laetrile and
died of lung cancer.

You Use That Stuff Too?
*You're Not the Only One Sold on Alternative
Approaches. Here Are Ten of America's Favorites—
and What the Science Shows*

June 29, 2004

In the 1960s and 1970s Steve McQueen was Hollywood's highest-paid, most popular movie star. McQueen's image was the epitome of Cool. Tough, determined, and self-assured, McQueen specialized in playing men of action who spoke little but were quick with their fists—dogged fighters who had quirky, winning personalities (**see chapter-opening photo**). His fans loved McQueen's rugged good looks, his sardonic humor, and those blue eyes, which could be innocent, calculating, or quizzical.

Although adored by millions of fans, Steve McQueen knew little love as a child. He came from a broken home and credited the California Boy's Republic, a private school for troubled boys, with helping him grow into a man. McQueen served four years as a Marine, and saved the lives of five other Marines during a training accident in the Arctic. At loose ends after his Marine Corps service, McQueen stumbled into acting classes. He went on to Broadway success and then moved to Hollywood where soon he had regular work on the television series, *Wanted: Dead or Alive*. McQueen's enduring films are *The Magnificent Seven, The Great Escape, Bullitt, The Thomas Crown Affair*, and *The Sand Pebbles*, for which he won an Academy Award nomination.

In 1979 McQueen's successful life was shattered. He was diagnosed with mesothelioma, a deadly form of lung cancer. His disease could have been caused by exposure to asbestos or from his many years of smoking cigarettes. At the time American medical doctors had no way to treat this form of cancer, and they gave Steve McQueen just months to live. So McQueen chose to head to Mexico for alternative cancer treatment. For three months he took animal extracts, high doses of vitamins, and the controversial treatment, laetrile—a drug derived from apricot pits. Unfortunately, nothing helped to cure his cancer—Steve McQueen died on November 7th, 1980.

Even though current medical treatments are adding years to patients' lives, many are tempted by the promises of alternative cancer treatments. While some people use alternative therapies in addition to conventional medicine, others are convinced that their cancers can be cured by alternative therapy alone. Jason Vale provides a more recent example of someone who not only took alternative treatment, but he also sold it to others.

Jason Vale was 18 years old when he learned that he had a rare form of bone cancer called Ewing's sarcoma. Because so few people develop Ewing's sarcoma, little is known about it, and there have been few advances in treatment. Nevertheless, by the time Jason was 35, it was hard to tell that he had ever had cancer at all. According to Jason and his supporters, his cancer had been diagnosed as "terminal," yet years later he had fully recovered. Jason started on the conventional treatments and, as expected, they made him weak and sick. Jason was so frustrated by the side effects of conventional medicines that he opted to stop taking them. Instead, he began eating apricot pits, the source of laetrile. Jason claims that the apricot pits cured his cancer and he began to sell apricot pits and laetrile over the Internet. To his supporters Jason Vale is dedicated to helping others who suffer from cancer and from traditional cancer treatments. But to the Federal Drug Administration (FDA), Jason is a criminal who used the Internet to promote and sell a fake cancer cure. In June 2004 he was sentenced to 63 months in prison. Since his incarceration Jason Vale's supporters have swamped the Internet with appeals for his release.

What is laetrile? Can it cure cancer? Are there other alternative treatments that are effective? These are important questions, but before you consider the answers you must understand what cancer is and the scientific basis for conventional cancer treatment. This chapter will allow you to use your knowledge of basic biological processes to understand this frightening, often fatal, disease. ■

12.1 Cancer Is a Major Killer

Second only to heart disease as a killer of Americans, cancer is one of the most threatening of all human illnesses. Each year in the United States over 500,000 people die from cancer, and more than a million new cases are diagnosed (**Table 12.1**). Although digestive system, prostate, breast, and lung cancers are most common, cancer can involve any part of the body. There are over 100 different kinds of cancer, and even rare forms kill thousands of people each year.

In response to the fear and suffering caused by cancer, the U.S. government and private organizations provide money for research to understand, prevent, and treat cancer. This "war on cancer" began in the early 1970s and continues even more intensively today. It has been a difficult campaign.

Progress has been slow and many potential treatments have not fulfilled their promise. There is finally some good news, however. Death rates for many kinds of cancer are finally declining, especially for people under the age of 65. Cancer death rates usually are reported as deaths per 100,000 people. In 1975 for every 100,000 people under the age of 65, about 85 people died of cancer. But in 2001 in this same age group, the mortality rate had dropped to about 68 people per 100,000 people. **Figure 12.1** shows these trends for breast cancer. Since 1975 the number of breast cancer cases per 100,000 has increased, but since about 1995 the number of breast cancer deaths per 100,000 has been dropping. Another way to see the success in treating cancer is the finding that the percentage of individuals who survive for at least five years after their diagnosis has increased for many types of

Table 12.1 Estimated incidence* and mortality of cancer in the United States (2001)

Type of Cancer	Estimated number of new cases per year	Estimated number of deaths per year
All Cancers	1,268,000	553,760
Prostate	198,100	30,719
Breast	192,200	41,394
Lung	169,500	156,380
Colon and rectum	135,400	56,887
Bladder	54,300	12,538
Skin (melanoma)	44,400	10,795
Leukemia	31,500	21,541
Pancreatic	29,200	29,802
Cervical	12,900	4,092
Hodgkin's lymphoma	7,400	1,323
Testes	7,200	335

*Incidence means the number of new cases per year.

Source: National Cancer Institute

Table 12.2 Five year survival rate*

Type of cancer	YEAR OF DIAGNOSIS	
	1975	2000
Breast	75%	87%
Lung	12%	15%
Colon and rectum	50%	64%
Pancreas	2.5%	4.5%
Skin (melanoma)	80%	90%
Prostate	67%	99%

Source: American Cancer Society

*Percent of cancer patients who survive five years after diagnosis of an invasive cancer that originated at these sites, averaged over all stages of cancer

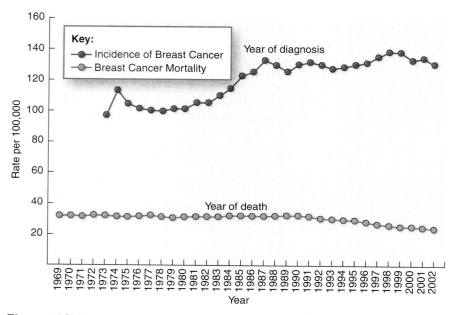

Figure 12.1 **Breast Cancer Death Rates.** As with many forms of cancer, annual breast cancer death rates for women in the United States have been declining.

cancer a disease involving a mass of cells produced by abnormally high levels of cell division; cancerous cells can detach from their neighbors and invade other tissues

tumor a mass of cells that results from abnormal cell proliferation

cancer (**Table 12.2**). Cancer probably never will be eliminated, but new knowledge could reduce the number of cancer deaths more dramatically in the not too distant future. This optimism is based on increased understanding of the cellular processes that lead to cancer.

QUICK CHECK

1. What is the trend in cancer death rates over the early years of the twenty-first century?

12.2 Different Types of Cancers Share Common Features

Unlike many other diseases, such as polio or tuberculosis, cancer is not a single disease. Cancer is not caused by a single infectious agent or by a sole environmental or genetic factor. Its symptoms and treatments depend on where in the body it originates, what organs it affects, and how long it has been present. These differences are reflected in the medical terms for different cancers. Names for all cancers end in the suffix -*oma,* which means "mass" or "tumor." Cancers are named for the kind of tissue they arise in. For example, epithelial tissues are found in many organs including breast, intestine, and lung. A cancer that originates in an epithelial tissue is called a *carcinoma.* A *melanoma* arises from cells that produce melanin; a *sarcoma* arises in bone tissue.

Despite these differences, various cancers do have common features. The definition of **cancer** is based on the two defining features of cancer cells. First, cancer cells have abnormally high rates of cell division. Second, cancer cells can detach from their neighbors and invade other tissues in the body. Abnormal cell division, also called *abnormal cell proliferation,* can produce an abnormal mass of cells. The tissue mass that results from abnormal proliferation is called a **tumor.** A tumor can develop from a single abnormal cell (**Figure 12.2**). This is why abnormal proliferation is at the heart of all life-threatening cancers. Multiplying unchecked, one cancer cell can turn into millions of cells, and a cancerous mass may overwhelm a tissue, an organ, or even a whole individual.

Some tumors are a more serious health risk than others. If a mass of cells is confined to the tissue where it originated, it is called a **benign tumor** and is not actually considered to be a cancer. The seriousness of a benign tumor depends on where it

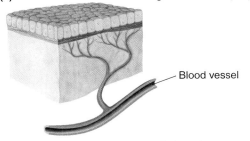

(a) One cell in a normal tissue begins to divide rapidly.

Blood vessel

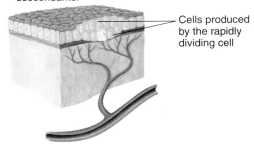

(b) The original rapidly-dividing cell gives rise to many descendants.

Cells produced by the rapidly dividing cell

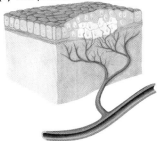

(c) This produces a tumor.

Figure 12.2 **Growth of a Tumor. (a–c)** A tumor can develop from a single rapidly-dividing cancer cell.

is located and how big it is. In many cases a benign tumor can be removed by surgery, and that ends the problem. In contrast, if a benign tumor is in a critical location such as the brain, it can be life threatening. Even small brain tumors can be difficult to remove and can press against brain regions that control important body functions. A tumor is cancerous if the cells in it have the ability to leave the original tumor and start tumors in other locations around the body. This is called a **malignant tumor.** The term that describes the process of establishing a new tumor from an old one is **metastasis,** and a tumor that has seeded new tumors in other parts of the body is said to have *metastasized.* Malignant, metastasized tumors are much harder to treat and so they are much more life threatening than are benign tumors. The remainder of this chapter concentrates on the characteristics, causes, and treatments of malignant tumors, also called cancers.

The fact that a malignant tumor can generate tumors in other parts of the body has led to a standard way to describe how far a cancer has progressed. An early-stage cancer has only spread to nearby tissues, while an advanced-stage cancer has

spread far and wide. **Figure 12.3** shows an example of one way of staging a cancer of the colon or large intestine. Colon cancer usually arises from cells in the inner layer of the large intestinal tract. These cells absorb water from food waste after food has been digested and absorbed by the small intestine. A tumor that will not move beyond this inner layer of the colon is benign. If abnormal cells are found beyond this inner layer, the tumor is assigned a stage of cancer, from I to IV. Stage I cancer has spread the least, and stage IV cancer has spread the farthest. The stages also indicate whether the cancer has spread to the lymph nodes (see Chapter 25) which are part of the immune system. They are connected to each other and to various organs and tissues by a system of lymphatic ducts. Cancer cells often travel in lymph ducts to metastasize in new locations.

> **QUICK CHECK**
>
> **1.** What is cancer?
> **2.** What is the difference between a benign tumor and a malignant tumor?
> **3.** What is a stage IV cancer?

12.3 Cancer Is a Genetic Disease

Scientists now understand much about the ways that cancer cells are abnormal and how these abnormalities explain cancer as a disease. This modern view of cancer will make little sense unless you understand that cancer is a genetic disease. This does not mean that cancer is always an inherited disease, although sometimes it can be inherited. All cancers are genetic because cancer is caused by mutations in DNA. A mutation is a change in the nucleotide sequence of DNA (see Section 7.5), and some mutations lead to altered proteins. This is the best description of what goes wrong in cancer cells: changes in DNA cause changes in proteins—which, in turn, change the traits of the affected cells. An inherited cancer will involve mutations in the cells that produce eggs and sperm. A noninherited cancer will be caused by mutations in other body cells.

It is important to understand that most DNA mutations do *not* lead to cancer. A mutation in a gene related to metabolism or brain function may have profound effects, but these are not related to cancer. Instead, cancer genes are those that are involved in a specific set of cellular functions. The most common abnormalities in cancer cells relate to cell proliferation, cell adhesion, apoptosis, and the ability to stimulate growth of blood vessels

benign tumor a mass of cells that is confined to a single area and can often be entirely removed by surgery

malignant tumor a tumor that contains cells that have the ability to leave their neighbors, travel through the body, and start a tumor in a new location

metastasis spread of cancerous cells from the site of the original tumor to other areas of the body

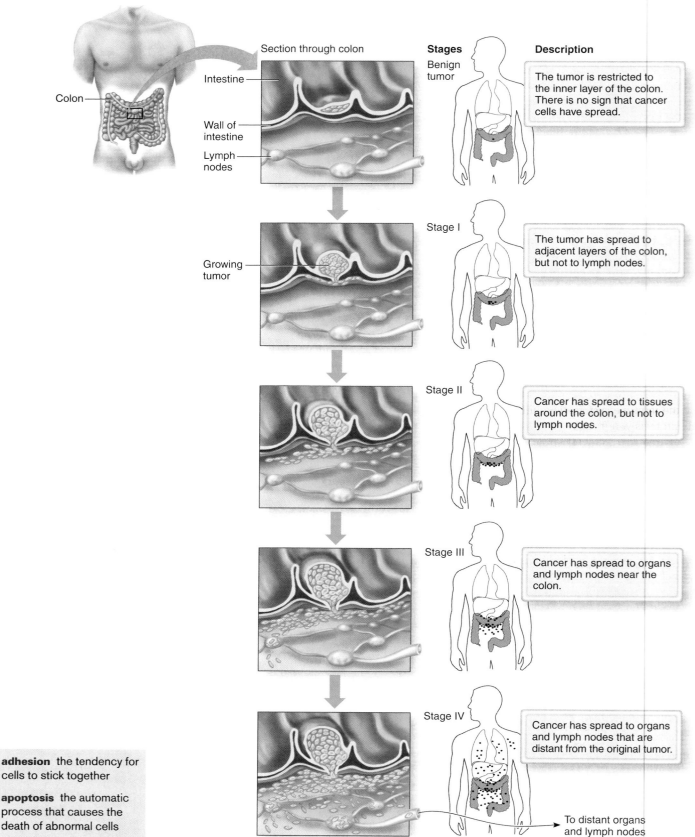

Section through colon

Stages

Description

Intestine

Colon

Wall of intestine

Lymph nodes

Benign tumor

The tumor is restricted to the inner layer of the colon. There is no sign that cancer cells have spread.

Growing tumor

Stage I

The tumor has spread to adjacent layers of the colon, but not to lymph nodes.

Stage II

Cancer has spread to tissues around the colon, but not to lymph nodes.

Stage III

Cancer has spread to organs and lymph nodes near the colon.

Stage IV

Cancer has spread to organs and lymph nodes that are distant from the original tumor.

To distant organs and lymph nodes

adhesion the tendency for cells to stick together

apoptosis the automatic process that causes the death of abnormal cells

angiogenesis the growth of blood vessels

Figure 12.3 **Stages of Cancer.** The commonly used stages of cancer reflect how far the cancer has spread from where it started.

(**Figure 12.4a–e**), although other abnormalities also can be present. Let's briefly look at each of these four central features of cancer cells.

Cancer cells divide more frequently than their normal counterparts, and this can lead to the development of a tumor. Cell **adhesion** is the term used to describe the tendency for cells in a tissue to stick together. Abnormal adhesion is the reason that cancer cells can leave their neighbors, travel to a different tissue, and start a tumor there. **Apoptosis** is another word for cellular suicide. Cells that are damaged or otherwise abnormal usually go through a systematic dying process. In contrast, many cancer cells lack this feature and so not only survive, but divide, and thereby pass their abnormal traits on to daughter cells. **Angiogenesis** is the growth of new blood vessels. Angiogenesis is triggered when cells in a tissue release growth factors that bind to cells in local blood vessels, causing them to divide and differentiate into new blood vessel structures. Cancer cells release large amounts of these growth factors and so stimulate blood vessels to grow into the tumor, providing it with the nutrients and other resources it needs to keep growing. One other important abnormality in cancer cells has to do with detection by the immune system. The immune system not only protects you from bacteria and viruses but also attacks and kills cancer cells (**Figure 12.5**). Especially in advanced stages of the disease, some cancer cells avoid detection by the immune system.

Because cancer involves cell proliferation and metastasis, a cancer cell must have mutations in genes that control cell division and cell adhesion. Many cancer cells have other important mutations as well, such as those involving apoptosis and stimulation of angiogenesis. Each of these general functions is the result of the actions of multiple genes, and so each is polygenic (see Section 10.4). Mutations in any of the genes involved in these functions can contribute to the development of cancer. But because each of these important cell functions is controlled by so many genes, a mutation in just one gene is unlikely to cause cancer. Other normal genes can keep a cell normal even with one or a few cancer mutations. The conclusion from these findings, which are supported by a great deal of evidence, is that cancer is a *multigenetic* disease. Mutations in one or two genes usually are not enough to cause cancer, and any given cancer will be the result of mutations in several genes.

QUICK CHECK

1. How can cancer be genetic but not always inherited?
2. List and briefly describe four cell functions that are abnormal in cancer cells.

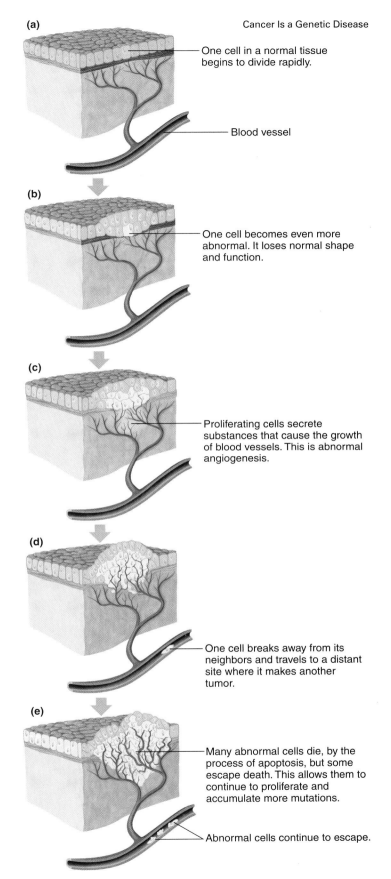

(a) One cell in a normal tissue begins to divide rapidly.

Blood vessel

(b) One cell becomes even more abnormal. It loses normal shape and function.

(c) Proliferating cells secrete substances that cause the growth of blood vessels. This is abnormal angiogenesis.

(d) One cell breaks away from its neighbors and travels to a distant site where it makes another tumor.

(e) Many abnormal cells die, by the process of apoptosis, but some escape death. This allows them to continue to proliferate and accumulate more mutations.

Abnormal cells continue to escape.

Figure 12.4 **Four Major Abnormalities Lead to the Development of Cancer.** (**a&b**) Abnormal cell proliferation, (**c**) abnormal angiogenesis, (**d**) metastasis, and (**e**) abnormal apoptosis are characteristic of cancer cells.

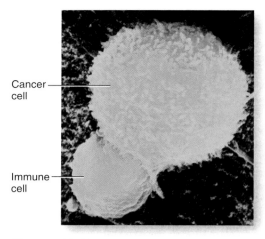

Figure 12.5 **An Immune Cell Killing a Cancer Cell.**

■ *How do you think immune cells distinguish a cancer cell from a normal cell?*

tumor suppressor protein
a protein that inhibits cell division by binding to and inhibiting transcription factors needed for cell division

12.4 Abnormal Proteins Are Responsible for the Abnormal Traits of Cancer Cells

While the root cause of cancer is genetic mutation, abnormal proteins are the direct cause of the cellular abnormalities of cancer cells. Genes carry the instructions for making proteins, and proteins carry out cellular work. Many different kinds of proteins can be abnormal in cancer cells. For example, an abnormal membrane receptor could cause a cancer cell to respond abnormally to a signal molecule. Or an abnormal enzyme could make too much or too little of a signal molecule that a cell normally releases. Still other abnormal proteins are defective versions of inhibitory factors that bind to membrane receptors, transcription factors, or signal molecules and inhibit their function.

Cell Proliferation Is the Central Abnormality in Cancer Cells

In a normal tissue cell division is strictly controlled. As a result, each tissue and organ has enough cells, but not too many. A tumor will develop if some cells in a tissue have mutations that lead them to divide too often. Two general classes of proteins are important in the control of cell proliferation: tumor suppressor proteins and proto-oncogene proteins. Mutations in the genes for these proteins often are found in cancer cells.

Tumor suppressor proteins inhibit a cell from moving from G_1 to S, and so inhibit cell division. They bind to and inhibit transcription factors that stimulate cell division. Recall from Section 9.4 that transcription factors control the production of proteins by binding to DNA and initiating transcription. When transcription factors are not needed, inhibitory proteins bind to them and prevent the transcription factors from binding to DNA. A transcription factor is activated when the inhibitory protein is removed from it. Tumor suppressor proteins are a class of inhibitory proteins that bind to the transcription factors involved in promoting cell division. *Retinoblastoma protein* is one example. Division of normal cells proceeds when these tumor suppressor proteins are removed from their transcription factors (**Figure 12.6a**). In many cancer cells the genes for tumor suppressor proteins are mutated and the resulting abnormal tumor suppressor proteins do not bind to their transcription factors (Figure 12.6b). The presence of mutated tumor suppressor genes contributes to abnormal cell divi-

(a) When retinoblastoma is normal it binds a transcription factor, inhibiting cell division.

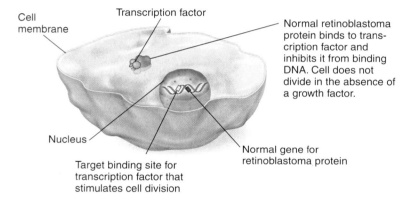

Cell membrane

Transcription factor

Normal retinoblastoma protein binds to transcription factor and inhibits it from binding DNA. Cell does not divide in the absence of a growth factor.

Nucleus

Target binding site for transcription factor that stimulates cell division

Normal gene for retinoblastoma protein

(b) When retinoblastoma is abnormal the transcription factor binds to DNA even when no signal molecule is present.

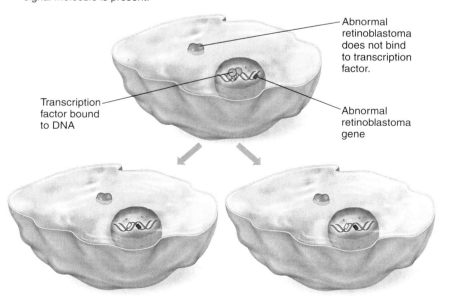

Abnormal retinoblastoma does not bind to transcription factor.

Transcription factor bound to DNA

Abnormal retinoblastoma gene

Figure 12.6 **Retinoblastoma Is a Tumor Suppressor Protein That Is Abnormal in Many Cancer Cells.** (**a**) Retinoblastoma normally binds to a transcription factor, preventing cell division. (**b**) Abnormal retinoblastoma does not bind to the transcription factor and the cell divides.

sion. Mutated tumor suppressor proteins are part of the cause of many types of cancer. Retinoblastoma protein originally was discovered in rare, inherited cancers of the retina. The normal version of this protein is found in nearly all cells, and an abnormal version is found in many types of cancers.

In contrast, **proto-oncogene proteins** normally promote cell division, and they are made following the instructions in proto-oncogenes. There are many proto-oncogene proteins, and each type has a particular role in moving a cell from G_1 to S-phase. For example, some proto-oncogene proteins inhibit or inactivate tumor suppressor proteins, preventing them from binding to transcription factors. Other proto-oncogene proteins are transcription factors that initiate the synthesis of cell cycle proteins. The prefix *-onco* is a Latin word that means "tumor." **Oncogenes** are mutated forms of proto-oncogenes that are found in cancer cells and are in part responsible for the excess cell proliferation in a tumor.

In normal cells both classes of proteins that control cell division—tumor suppressor proteins and proto-oncogene proteins—are, in turn, largely controlled by signal molecules from the external environment (see Section 9.4). During G_1 signal molecules interact with membrane receptors and move the cell into S-phase. A signal molecule that stimulates a cell to divide is called a **growth factor**. If a cell produces too many growth factor receptors, the cell will be overly responsive to growth factor and will divide more often than it should (**Figure 12.7**). For example, anywhere from 10 to 40% of breast cancer patients have cancer cells that express too many growth factor receptors. The growth factor receptor in these tumors is called *Her2*. Its discovery offers a new approach to breast cancer treatment considered later in this chapter.

Another protein that is especially important in determining whether a cell will move from G_1 to S-phase is p53, short for *protein 53*. p53 is activated by conditions that cause damage to DNA such as stress, radiation, or certain environmental chemicals. If DNA damage is moderate, p53 will activate a pathway that stops the cell from moving into S-phase. If DNA damage is more severe, p53 will stimulate the cell to go through apoptosis (**Figure 12.8a**). If the *p53* gene is mutated, however, a cell with abnormal DNA can continue to live and divide (Figure 12.8b). As a consequence, the abnormal DNA that it carries is passed on to all of its daughter cells. You can see why mutations in *p53* are so commonly found in cancer cells. As you will see later, this feature of cancer cells can make cancers especially difficult to kill.

(a) In a normal cell there are few *Her2* membrane receptors so cell division is controlled.

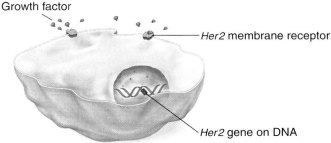

(b) Some cancer cells have excess *Her2* receptors. These cells divide at an increased rate when growth factors are present.

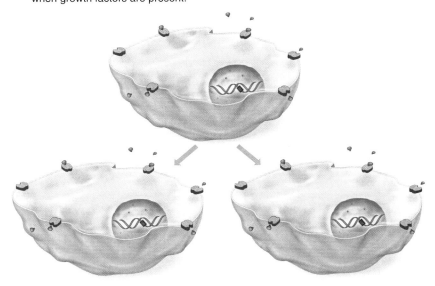

Figure 12.7   **The Role of *Her2* Receptors in Cell Division.** Overproduction of *Her2* growth factor receptor causes cells to proliferate. If a cell has excessive receptors for a growth factor it will multiply.

The central role of abnormal cell proliferation in the development of cancer helps to explain where in the body a cancer is likely to arise. In general, cells that are completely differentiated and do not ever go through cell division do not give rise to cancer. So mature neurons, muscle cells, bone cells, or skin cells usually do not become cancerous. How then can you have skin cancer or brain cancer or bone cancer? There are other types of cells in all of these tissues that do divide, and it is these cells that give rise to cancer. For example, the mature outer cells of human skin eventually die and are replaced by deeper cells in the skin that do divide. These deeper, dividing cells give rise to many types of skin cancer. If you revisit Table 12.1, you will see that the most common cancers are found in the breast, lungs, skin, and intestinal tract (colon). All of these tissues contain cells that normally divide.

proto-oncogene protein a protein that actively promotes cell division

oncogene a cancer gene that codes for an abnormal version of a protein that promotes cell division, or that overexpresses a protein that promotes cell division

growth factor a small molecule that is released by one cell and that stimulates both itself and surrounding cells to divide

(a) When a normal cell is damaged, p53 activates the apoptosis program and the cell dies.

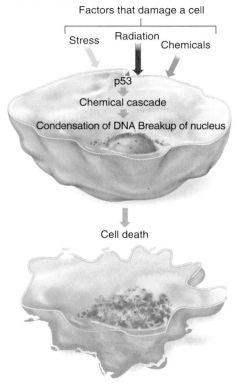

Factors that damage a cell

Stress   Radiation   Chemicals

p53

Chemical cascade

Condensation of DNA Breakup of nucleus

Cell death

(b) If a cell has abnormal p53, then damage may not lead to apoptosis, and the abnormal cell will proliferate.

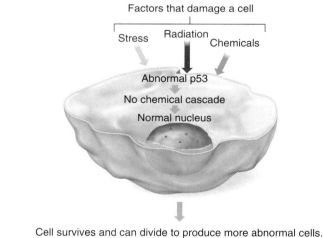

Factors that damage a cell

Stress   Radiation   Chemicals

Abnormal p53

No chemical cascade

Normal nucleus

Cell survives and can divide to produce more abnormal cells.

Figure 12.8   **p53 in Normal and in Cancer Cells.**   (a) While a damaged cell with normal p53 will enter apotosis and die, (b) a cell with abnormal p53 will not enter apoptosis and die.

Cancer Cells Often Lack Mature Structure and Function

A tumor is usually a useless mass of tissue because its cells do not have the ability to do their normal jobs. In other words, cancer cells do not show normal cell differentiation. For instance, the cancer cells in an intestinal tumor are not likely to carry out the normal functions of absorption and digestion of nutrients. Because of gene mutations they do not express the normal proteins required to do these jobs. This lack of function often is reflected in the abnormal shapes of cancer cells. For example, normal skin cells are flat-looking, normal gut cells look like elongated cubes, and normal brain cells have complex shapes with many branches that extend outward. The cells in a tumor in any of these tissues do not have these distinctive shapes. This characteristic of cancer cells is used in one common test for cancer, the Pap smear test, in which cells are removed from a woman's cervix by wiping it with a cotton-tipped swab. After the cells are smeared onto a glass slide, they are stained and examined under a microscope. Cancer is unlikely if the stained cells look like normal cervical cells, but if the stained cells have abnormal shapes, cancer is suspected and further tests are ordered. In **Figure 12.9** you can compare normal cells with the abnormally shaped cells that suggest the development of cervical cancer.

Cancer Cells Can Stimulate the Growth of Blood Vessels

Cells in any body tissue must have oxygen and glucose to power their cellular processes, and they must have cellular wastes removed. All of these substances are transported to and from cells in blood that travels to tissues in blood vessels. In normal tissues the growth of blood vessels, the process called *angiogenesis,* happens primarily during embryonic development and is limited during adulthood. Tumors are unusual in that they stimulate the growth of new blood vessels. Cancer cells secrete large amounts of growth factors that encourage the growth of blood vessels in and around the tumor. New research suggests that estrogen, the major female reproductive hormone, is an important signal that stimulates blood vessel growth, but many other growth factors are involved in angiogenesis as well. **Figure 12.10** shows a striking example of how extensive this growth of blood vessels into a tumor can be. As with other new knowledge about cancer cells, knowledge of the role of

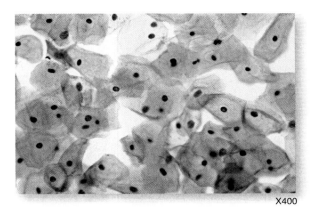

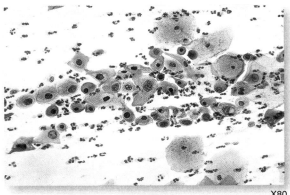

X400

X80

Figure 12.9 **Normal and Abnormal Cervical Cells.** The slide on the left shows normal cervical cells magnified through a microscope. Normal cells are uniform in size and shape. By comparison, the slide on the right shows irregular, disfigured cervical cells typical of cervical cancer.

angiogenesis in tumor growth is being used to develop new treatments.

Cancer Cells Escape From Adjacent Cells

Cells in any tumor can have all of the traits described in the preceding sections. To be called a cancer, though, a tumor must contain cells that are able to detach from neighboring cells, travel to another body location, and start a new tumor. Think about this statement for a moment and notice that it implies that normal body cells stick to their neighbors. Imagine if this were not so: if cells did not adhere to one another, tissues and organs would not form, and a multicellular organism would fall apart. Cells are held together by a complex set of proteins and carbohydrate molecules that stick out from a cell's surfaces. These proteins and carbohydrates interact with each other in specific ways to hold cells together. The exact configuration of these molecules differs for different kinds of cells—this is why the cells in each kind of tissue stick to each other and not to cells in other tissues. The tendency of cells to stick together is called *cellular adhesion.*

Mutations in the genes that control cell adhesion allow cancer cells to leave their neighbors and take up residence in other tissues (**Figure 12.11**). This is the major way that cancer spreads. The role of cell adhesion in cancer development helps to explain why some types of cancers are more invasive and spread more than other types do. For example, a form of skin cancer called *melanoma* can be extremely invasive. Melanoma begins as a dark bump on the skin that looks like an irregularly shaped mole. The cells that give rise to melanoma are *melanocytes.* These cells make melanin, the dark pigment that gives skin its color. Melanocytes divide frequently and are highly mobile. Although

(a)

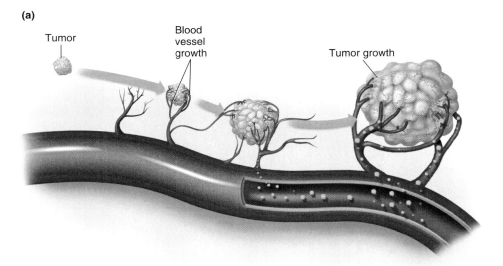

Tumor

Blood vessel growth

Tumor growth

(b)

Figure 12.10 **Angiogenesis.** (**a**) As a tumor grows it stimulates angiogenesis. (**b**) The blood vessels of a tumor become visible when they are injected with blue plastic and the tumor cells are removed. This tumor weighed half a pound.

Figure 12.11

Mutations in Cellular Adhesion Genes Lead to Metastasis. (**a**) Proteins that extend across cell membranes keep normal cells firmly attached. (**b**) Cancer cells lose their attachments to neighboring cells and escape.

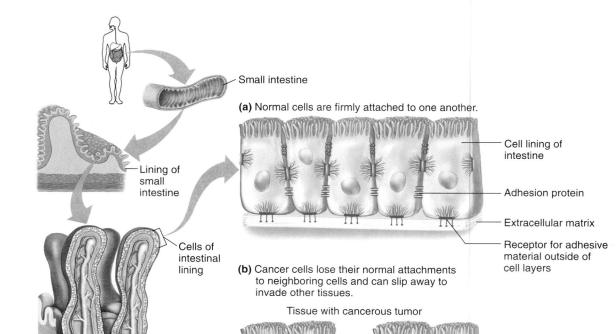

Small intestine

Lining of small intestine

Cells of intestinal lining

(a) Normal cells are firmly attached to one another.

Cell lining of intestine

Adhesion protein

Extracellular matrix

Receptor for adhesive material outside of cell layers

(b) Cancer cells lose their normal attachments to neighboring cells and can slip away to invade other tissues.

Tissue with cancerous tumor

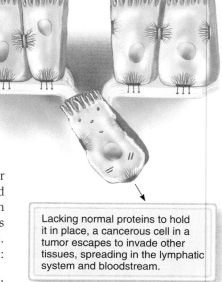

Lacking normal proteins to hold it in place, a cancerous cell in a tumor escapes to invade other tissues, spreading in the lymphatic system and bloodstream.

melanocytes move around in the deep layers of your skin, they do not normally move *out* of the skin. Because melanocytes divide frequently and move around a lot, it takes few mutations to cause melanocytes to become cancerous.

Many Cancer Cells Are Unaffected By the Mechanisms That Kill Other Abnormal Cells

One of the important ways that the body stays healthy and functions normally is that it continually finds and kills abnormal cells. Cancer cells and other abnormal cells are produced throughout life, but they are usually eliminated. In advanced cases of cancer, though, cancer cells escape the mechanisms that kill abnormal cells. Two important mechanisms kill abnormal cells: apoptosis and the immune system.

Apoptosis is the mechanism by which a cell kills itself. If at any point in the cell cycle a cell produces abnormal proteins, p53 and other proteins trigger processes that cause the cell to die. Many cancer cells are susceptible to this killing mechanism, but mutations in the *p53* gene produce cancer cells that cannot be killed by apoptosis. Once this happens, it increases the severity of the cancer.

The other important mechanism for killing cancer cells involves the immune system. You will read more about the immune system in Chapter 25, but you already may know that its job is to rid the body of foreign bacteria and viruses that cause infectious diseases. The immune system also kills cancer cells. The common feature of bacteria, viruses, and cancer cells is that they all display proteins that are not normal for the individual in which they are found. The immune system recognizes the presence of foreign proteins and attacks the cell or virus that carries them. In this way the immune system helps to rid the body of any cancer cells that do form. Many studies have shown that advanced cancer cells escape from immune system surveillance. Cancer cells that escape the defenses of the immune system are likely to produce even more cancer cells.

12.5 Risk Factors Affect Your Chances of Developing Cancer

One of the biggest concerns today is how to avoid getting cancer. Knowledge about cancer gives you information about your own cancer risk. Notice the term *cancer risk*. Like so many other aspects of living organisms, predictions about the development of cancer can be phrased only in terms of probabilities. As you now understand, cancer is a complex, disparate set of diseases involving many genes and proteins. So it is not surprising that cancer has many distinct and overlapping causes and risk factors. A risk factor is any aspect of you, or your experience, that increases your chances of developing cancer. You *can* understand the risk factors involved in cancer, and you *can* decrease or increase your chances of acquiring cancer based on the choices you make. What you *cannot* do is be 100% certain that you will or will not develop cancer.

Risk factors for cancer can be identified in a variety of ways. For example, studies at the cellular level can demonstrate that certain chemicals stimulate the development of cancerous cells in a culture dish. Such chemicals would then be considered as possible risk factors for human cancers. Other studies document the effects of chemicals, conditions, or inheritance on the development of cancers in animals. Again, any conditions that increase cancers in animals are likely to be risk factors for cancers in humans. Yet other studies look at the characteristics of people who get cancer and people who don't. Any characteristic found more often in people with cancer could be a risk factor. Regardless of the way that risk factors are identified, most risk factors must increase the chances of developing cancer by producing DNA mutations. In many cases the mechanisms by which these factors cause mutations are not known, but ongoing studies are actively looking for this link. Keep this connection in mind as you read this section on cancer risk factors.

Inherited Cancer Genes Account for Only a Small Percentage of Cases of Cancer

Inherited mutations would seem to be a major risk factor for cancer. Actually—averaged across different types of cancer—inheritance accounts for only 5 to 10% of cancers, although the risk from inheritance varies a lot between different cancers. **Table 12.3** shows some examples of inherited cancers, the percentage of cases of each of these inherited cancers, and one or more abnormal genes identified for each. Note that even if you carry one or even two abnormal alleles for an inherited cancer gene, it is not certain that you will develop cancer. To make this point, Table 12.3 also shows the percentage of people with each type of cancer allele who end up developing cancer. You can see that this last number is always less than 100%. Cancer is almost always caused by multiple genes and a mutation in one cancer gene—even if it is an inherited cancer gene—is usually not enough to produce disease.

Section 11.4 considered tests used to detect the presence of alleles associated with inherited diseases. Such tests are becoming increasingly available for a variety of cancers. For instance, at this

Table 12.3 Mutated genes in inherited cancers			
Type of cancer	**Mutated gene**	**Percent of people carrying this gene who will develop cancer**	**Percentage of cases that express this mutation**
Breast	*BRCA1*	80–85%	5%
Colon	*MLH1*	80%	10%
	MSH2		
Retinal	*Rb*	75–90%	30–40%

writing, tests are available for breast cancer, colon cancer, and melanoma; tests for other cancers will be available by the time this text is in print. Should *you* be tested for one of these inherited cancers? Only a small percentage of all cancers are due to inherited cancer genes, but if you carry a cancer gene, your chances of developing that particular cancer are higher than average. Cancer is easier to treat if it is detected at an early stage. If you know that you carry a gene for an inherited cancer, you may be more careful to have frequent cancer checkups, and so you are more likely to detect any cancer early in its development. Only you can decide whether or not you would want to know this information.

Some Environmental Agents Can Cause Cancer

Many people are concerned about getting cancer from environmental chemicals. An environmental chemical that causes cancer is called a **carcinogen.** Carcinogens are officially designated by government-sponsored research, but it is sometimes difficult to determine how that research applies to your risk from agents in your environment. For most carcinogens the designation comes from tests on non-human mammals, usually rodents such as mice or rats. If a compound produces cancer in rats, it is likely to cause cancer in people—but the dosages required to produce cancer could be much different, and it is difficult to know how much exposure is dangerous for humans. One source of information on environmental carcinogens is the International Agency for Research on Cancer, the IARC for short. Their website contains lists of environmental carcinogens that definitely cause cancer in humans as well as lists of chemical agents that probably cause cancer in humans.

Asbestos is an example of a carcinogen. It is a fibrous material that once was commonly used as insulation. **Figure 12.12a** shows highly magnified asbestos fibers as they look in isolation and within human tissues (Figure 12.12b). When these thin fibers are inhaled in large quantities, they can cause lung cancer and other kinds of cancers as well. Asbestos and other environmental carcinogens present serious cancer concerns. Much effort is expended to get rid of cancer-causing agents and to inform the public about their risks. To put this into perspective, though, overall the risk from environmental carcinogens is relatively small. For example, carcinogens from air and water pollution probably account for only 2% of all cancer cases. Food additives and industrial toxins account for another 2% of cancer cases. Counted separately

carcinogen a chemical that causes cancer

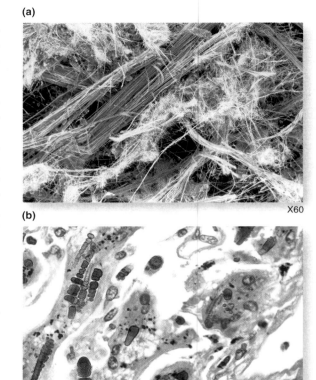

(a)

X60

(b)

X80

Figure 12.12 **Asbestos Fibers** Asbestos fibers (**a**) can lodge in tissues, (**b**) especially in the lungs, and can cause cancer.

from environmental carcinogens, work-related carcinogens account for about 4% of all cancer cases. Even though these numbers are small, cancers from environmental agents should be avoidable, and consequently much effort goes into testing for potential carcinogens. Chemicals suspected of causing cancer are strictly controlled, and any chemical with even a low risk as a carcinogen is banned as a food additive.

Any Factor That Increases Cell Proliferation Increases the Risk of Cancer

Any increase in the rate of cell proliferation produces an increased risk of cancer because cells with high cell proliferation rates can pass on mutations to many daughter cells. Exposure to any natural or artificial agent that increases cell proliferation will increase the chances of developing cancer. Estrogen is one such agent. Every month, in a cyclic pattern, a woman's estrogen levels rise and stimulate cell division in breasts and other reproductive organs. This is why a woman's risk of cancer increases the longer she is exposed to her own natural estrogens. Accordingly, women who reach puberty early, or reach menopause late, are more likely to develop breast cancer than are women with an overall shorter reproductive life.

The effects of hormone replacement therapy (HRT) on breast cancer risk can be understood in this context of lifetime estrogen exposure. HRT replaces the estrogen and progesterone lost after menopause. Many women take HRT to maintain the physiological state that has been typical of their adult lives. HRT also minimizes the side effects of menopause such as night sweats, hot flashes, and mood swings. Research shows that women who take HRT have a small but significant increase in the chance that they will develop breast cancer. On the other hand, some studies show that women on HRT live longer, or at least do not die sooner, than women who do not take HRT. As with many of the questions related to cancer, the full answer to the question of the risk of HRT will require much more research.

People With Suppressed Immune Systems Can Have an Increased Risk of Cancer

The immune system plays an important role in inhibiting the development of cancer. So it should not surprise you that people whose immune systems do not work effectively are susceptible to cancer. Inherited genetic factors, chronic infectious diseases, stress, HIV, exposure to intense UV sunlight, and many other factors can compromise the immune system, and this increases the risk of developing cancer. Very young people and very old people are more susceptible to cancer than others, in part because their immune systems are not as strong as those of people of other ages.

Exposure to Oxygen Increases Cancer Risk

Increased risk of cancer is one price that we pay for the use of oxygen in metabolism. The problem is that oxygen is a highly reactive molecule. It combines readily with other atoms and molecules to form a class of compounds called *reactive oxygen species* or *free radicals.* Some examples of reactive oxygen species are hydrogen peroxide (H_2O_2) and hydroxyl ions (OH^-), but there are many others as well. These reactive compounds do have important functions in cells, such as helping to destroy foreign bacteria and viruses, but they also cause damage to important biological molecules within cells. Among their effects, reactive oxygen species can cause mutations in DNA—and some of these mutations contribute to cancer.

Some Lifestyle Practices Influence the Probability of Developing Cancer

The biggest news in cancer research is that the vast majority of cancers appear to be related to lifestyle habits. Presently, biologists can only guess how lifestyle practices might affect DNA, but studies have revealed strong relationships between the presence or absence of particular habits and the chances of developing cancer. The three most prominent American lifestyle habits related to cancer are diet, exercise, and cigarette smoking. Let's briefly consider each of these.

Some Diets Are Associated With Higher Cancer Risks   Several aspects of diet appear to be related to cancer risk. The clearest studies that support this statement correlate what people eat with whether or not they get cancer. One dietary component that seems highly correlated with cancer is fat. High-fat diets, especially the fats found in meat and rich dairy products, have been related to several types of cancer. Cultures in which people typically eat low-fat diets tend to have lower cancer rates. While this relationship appears to be a strong one, it does not show up in every study. Because the causes of cancer are complex, sometimes different studies produce contradictory results. Nevertheless, taken together, the research suggests that diets low in fats, especially animal fats, reduce the chances of cancer.

Diets high in fruits and vegetables seem to have the opposite effect on cancer risk. People who eat lots of fruits and vegetables may have a lower cancer risk, although again not all studies show this pattern. Assuming the benefit is real, it might be explained by the variety of compounds in fruits and vegetables that seem to reduce the frequency of DNA mutations. Each kind of fruit or vegetable contains a different mix of such compounds, so consuming a variety of fruits and vegetables is an important feature of a diet that might protect against cancer. For example, tomatoes and tomato products contain a compound called *lycopene,* and some studies show lycopene can reduce cancer risk. How might these foods help prevent cancer? One hypothesis is that many fruits and vegetables are high in *antioxidants.* You will see this label on many food items in the supermarket. Antioxidants combine with reactive oxygen species and may prevent reactive oxygen species from damaging DNA and other biological molecules. Much more research is needed, however, to clarify the role of antioxidants in cancer. Some companies suggest that taking high doses of antioxidants in pill form

Figure 12.13   **Lance Armstrong.** Lance Armstrong won the Tour de France seven years in a row, after recovering from testicular cancer.

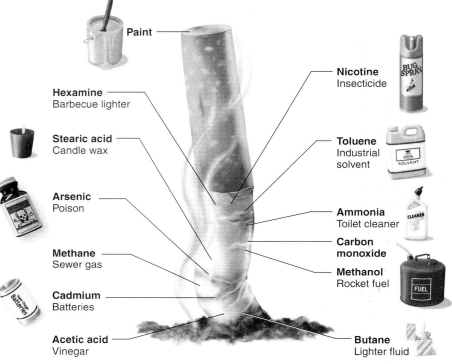

Chemical (known or probable carcinogen)	Amount per cigarette in inhaled smoke	Amount per cigarette in smoke emitted from a burning cigarette
Analine	346 ng	100,800 ng
Toluidine	161 ng	303 ng
Dioxins	1,500 ng	3,000 ng
Heavy metals	120 ng	960 ng
Benzofluorene	184 ng	751 ng
Ammonia	100 µg	17,000 µg

Key:

µg = Microgram or 0.000001 gram (a millionth of a gram)

ng = Nanogram or 0.000000001 gram (a thousand millionth of a gram)

Figure 12.14   **Ingredients in Cigarette Smoke.** The toxins and carcinogens in cigarette smoke make it dangerous for anyone exposed to it. Notice that second hand smoke has even more toxins than inhaled smoke.

can help reduce cancer risk, but so far research studies do not support this claim.

Exercise Reduces Cancer Risk   The image of long-term cancer survivor Lance Armstrong winning the grueling Tour de France bicycle race for the seventh time is inspiring for people with cancer, but his success provides an even deeper lesson (**Figure 12.13**). Many studies have found that exercise lowers the chances of developing cancer and improves the chances of recovering from cancer. Even after other cancer-related factors are accounted for (including age, gender, and smoking habits), people who are less active are more likely to develop cancer than are people who are more physically active. Most of the studies on exercise and cancer risk have focused on breast cancer, but a few studies on other cancers suggest that exercise is generally protective against cancer. Again, this does not mean that if you exercise you will not get cancer, nor does it mean that if you never exercise you are certain to get cancer. Exercise *reduces your chances* of developing cancer. Regular exercise can even delay the onset of cancer in people who have inherited cancer mutations such as *BRCA1* or *BRCA2*. Beyond the effects of exercise on cancer itself, many studies have documented that exercise improves the quality of life of cancer survivors and improves energy levels of those on cancer treatments.

Given these consistently positive effects of exercise, many researchers want to know how exercise has these effects. This is difficult to answer. Exercise is a complicated activity. It improves the functioning of muscles, lungs, and heart, but has other effects as well. Exercise changes the chemistry of many types of cells in the brain and has positive effects on many body organs. Some researchers hypothesize that exercise has positive effects on cancer risk because it enhances the functioning of the immune system, and the immune system kills many active cancer cells, especially in the early stages of cancer. These are all intriguing hypotheses, but there is not enough evidence yet to be confident of any of them.

Tobacco Smoke Causes Cancer   It seems that every day the news describes a new factor that supposedly increases or decreases the risk of cancer. Nevertheless, one factor dominates cancer rates in the United States and many other countries: the single biggest cause of cancer and cancer death in the United States is exposure to cigarette smoke. The smoke inhaled from a cigarette contains many individual toxic and carcinogenic chemicals (**Figure 12.14**). Even nonsmokers are at risk from cigarette smoke in the air. The smoke that wafts from the end of a lit cigarette contains an even higher concentration of toxic chemicals than does the inhaled smoke.

Smoking is primarily a lifestyle choice for most people, at least when they first begin to smoke. Not only does this choice increase the chances of getting cancer, it also increases the chances that family members and close associates who are exposed to cigarette smoke will get cancer too. **Figure 12.15** traces cancerous changes in lung tissue that are caused by smoking. Each year hundreds of thousands of cases of lung cancer are diagnosed worldwide, with over 160,000 new cases in the United States (see Table 12.1). The cost in human suffering and dollars is tremendous.

Exposure to Radiation Can Cause Cancer

One of the most serious environmental cancer risks comes from something most of us enjoy—the Sun. Even though the Sun provides the energy for life's chemical processes (see Section 5.1), some of the energy from the Sun's ultraviolet (UV) radiation is actually harmful. Humans cannot see UV radiation, but it is absorbed by the skin and can damage cells in deeper layers of skin that normally divide to replace cells that die and are shed from the skin surface. UV energy can cause mutations in the genes that control cell proliferation, adhesion, angiogenesis, and apoptosis. When this happens, a tumor can develop.

Skin cancer (**Figure 12.16**) comes in several forms, and some are much more serious than others. The least serious kinds of skin cancer, *squamous cell carcinoma* and *basal cell carcinoma,* result from abnormal proliferation of cells in the lower layers of the skin (Figure 12.16a,b). These tumors often look like small, reddish, rough spots on the skin. They are slow to develop and usually can be removed before they cause serious health problems. These carcinomas do not often spread because the normal skin cells from which they arise tend to stick together tightly. If basal cell or squamous cell carcinomas are left untreated, they may develop additional mutations and spread, but they are usually detected and treated before that happens.

Melanoma is a serious cancer (Figure 12.16c). Recall that melanoma gets its name because it arises from the melanocytes in skin. If melanocytes develop mutations, there is a good chance the cells will become cancerous and will spread widely. People with lots of moles are at higher risk for developing melanoma than are people who have no moles; people with fair skin are at greater risk of develop-

ing melanoma than are people with darker skin. Melanomas can be treated effectively if they are removed early, but if they are not discovered until late in the cancer process, the prognosis can be grave. The good news is that the risk of skin cancer can be reduced. The American Cancer Society estimates that many cases of skin cancers can be prevented if people use a strong sunscreen whenever they go out into the Sun.

Other kinds of radiation also can cause cancer. For example, while X rays give critical medical information, there is a concern that too many X rays could contribute to the development of cancer. This is one reason why a lead shield is used to protect other parts of your body when you get an X ray. The shield is especially important over the ovaries and testes, which produce eggs and sperm. Any mutation in an egg or sperm has a chance of being passed on to offspring, producing an inherited cancer.

There are many ways that radiation can cause cancer. One way involves the production of reactive oxygen species. Radiation is a form of energy, and adding an external source of energy simply increases the chance that the reactive oxygen molecule will form new molecules.

Many, But Not All, Cancers Are Preventable

You may be eager to learn about possible treatments and cures for cancer, a topic discussed later in this chapter. This section on lifestyle choices is important, though, because most experts agree that most cancers are caused by factors that people have choices about. **Figure 12.17** summarizes some of the lifestyle choices just described. The exact percentages vary depending on the cancer involved and what study you read. In general, though, studies agree that use of

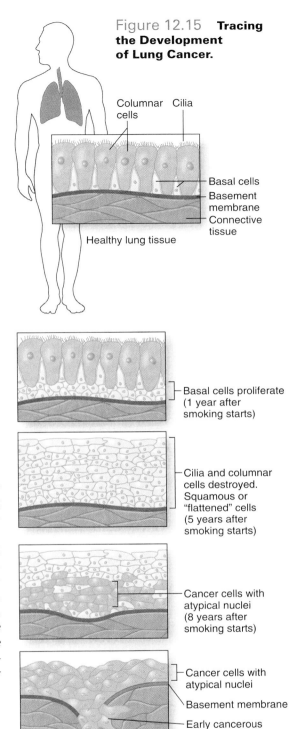

Figure 12.15 **Tracing the Development of Lung Cancer.**

Columnar cells
Cilia
Basal cells
Basement membrane
Connective tissue
Healthy lung tissue

Basal cells proliferate (1 year after smoking starts)

Cilia and columnar cells destroyed. Squamous or "flattened" cells (5 years after smoking starts)

Cancer cells with atypical nuclei (8 years after smoking starts)

Cancer cells with atypical nuclei
Basement membrane
Early cancerous invasion (20–22 years after smoking starts [first symptoms])

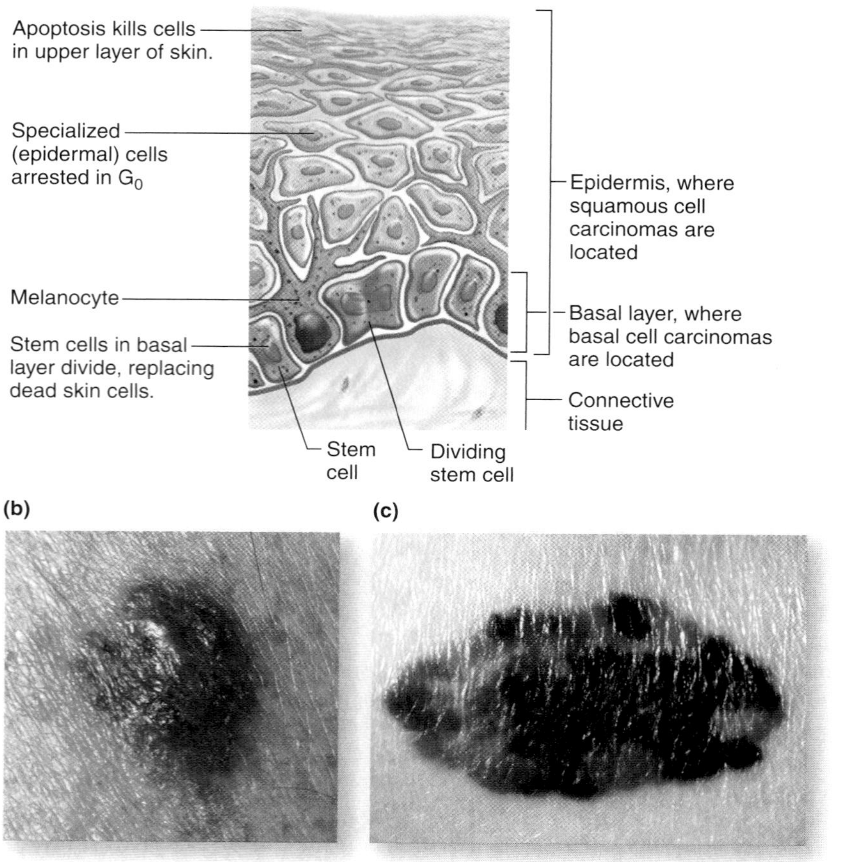

(a)

Apoptosis kills cells in upper layer of skin.

Specialized (epidermal) cells arrested in G_0

Melanocyte

Stem cells in basal layer divide, replacing dead skin cells.

Stem cell

Dividing stem cell

Epidermis, where squamous cell carcinomas are located

Basal layer, where basal cell carcinomas are located

Connective tissue

(b)   **(c)**

Figure 12.16   Skin Cell Layers and Cancers. (a) Cells, located deep in skin actively divide and produce new skin cells that move upward to replace skin cells lost from upper layers. (**b** and **c**) Skin cancers come in two basic forms: carcinomas and melanomas.

■ *What are some simple ways to avoid getting this form of cancer?*

tobacco, dietary factors, and lack of exercise account for the vast majority of cancers in the United States. There are no absolutes in science—if you quit smoking, change your eating habits, and exercise regularly, it is not guaranteed that you will evade cancer. But your best chance for not dying of cancer lies in making lifestyle changes that will prevent the disease.

Despite the optimism that cancer rates can be reduced by lifestyle choices, it is also a stark reality that cancer probably never will be eliminated. Even if every American ate the best diet, exercised, and avoided all known cancer risk factors, some people still would get cancer. It is probably impossible to fully protect DNA from accumulating mutations during an individual's lifetime. In fact, one major risk factor in cancer is age. Among 20-year-olds the rate of cancer per 100,000 people is fewer than 50. Among 80-year-olds the rate of cancer per 100,000 people is closer to 2,500. As medical techniques combat heart disease, diabetes, and other causes of death, it is likely that cancer will continue to be a risk, especially in older people.

QUICK CHECK

1. What are five risk factors that increase the chances a person will develop cancer?
2. What is a carcinogen?
3. What does it mean to say that many cancers are preventable?

Figure 12.17
Lifestyle Changes That Help You to Reduce Risk of Cancer.

■ *How many of these choices have you made? What about members of your family? Your friends?*

Reduce dietary animal fat.

Stop smoking, or better yet, never start.

Avoid UV light from sunlight and tanning beds.

To avoid or reduce the risk of cancer

Eat lots of fruits and vegetables.

Avoid obesity.

Get regular, vigorous exercise.

12.6 Early Detection and New Treatments Increase Prospects for Surviving Cancer

Most people who are confronted with a cancer diagnosis are primarily concerned about treatment options. Only 30 years ago treatments for cancer were limited, but today treatment options have expanded to the extent that many people can live years past a cancer diagnosis that would have meant rapid death just half a century ago. The future holds even greater promise. Because each kind of cancer is different in some ways from others, there never will be a single "magic bullet" to treat all cancers. But increasing knowledge about specific genetic mutations that lead to different cancers will lead to new treatments. Five-year survival rates for many types of cancers are increasing (see Table 12.2) and overall deaths per year from most types of cancer also are dropping (**Table 12.4**). These improvements are a result of early detection, traditional treatments, and new treatments developed as a direct result of modern knowledge of the biology of cancer.

Cancer Tests Have Increased the Rates of Early Detection and Cancer Survival

There is a much better chance of survival if cancer is detected early, when cancer cells are more likely to be restricted to a small area. Even if a tumor has developed the potential to spread to other areas of the body, it is less likely to do so from a tumor that is early in the cancer process. In addition, early-stage cancers are less likely to have completely lost the ability to be killed by the immune system or to go through apoptosis. This is why so much emphasis is placed on early detection. All of the available treatment strategies work better on cancer in its early stages. You can see this in **Table 12.5,** which compares five-year survival rates for patients with localized cancers with patients who have widespread cancers. For many types of cancer the chances of surviving for five years are near 100% after early treatment.

The traditional approach to early detection typically begins with a physical exam to detect tumors. During a physical exam, whether done by your doctor or yourself, the aim is to find lumps that might be tumors. In a procedure called a *biopsy,* tissue from a suspected tumor is removed and examined under a microscope. Microscopic examination allows the detection of abnormally shaped cells. If the abnormalities are modest, then the cancer is likely to be in an early state and more treatable. X rays and other kinds of sophisticated imaging systems can "see" lumps too small to feel.

A *mammogram* for breast cancer is one example of this type of screening technique, and it is also an example of the controversy that can surround cancer detection and treatment. Before the development of mammograms, women generally detected breast cancer by feeling a lump in the breast. By the time a tumor is large enough to be felt, however, it may be advanced enough to have metastasized. A mammogram is a form of X ray that can "see" a lump too small to be felt (**Figure 12.18**). Mammograms are not perfect, though. About 11% of mammograms will suggest that cancer exists when none is there. On the other hand, when a mammogram does detect a true mass, only 3% of these turn out

Table 12.4	U.S. death rates from cancer (1978–1998)	
	CANCER DEATHS PER 100,000 PEOPLE	
Type of cancer	**1975**	**2002**
All types	199.1	193.5
Stomach	8.5	4.2
Colon	28.0	19.6
*Lung, thoracic organs**	42.5	52.9
Breast	31.4	25.5
Bladder	5.5	4.4
Leukemia	8	8
Melanoma	2.0	2.6
Pancreas	10.6	10.5
Testicular	0.7	0.3

*A major factor in the increase in lung cancer deaths is the increase in the number of women who smoke cigarettes.

Source: National Cancer Institute

Table 12.5	Comparison of five-year survival rates for localized versus widespread cancers	
Type of cancer	**Localized**	**Widespread**
Breast	97%	25%
Lung	50%	2%
Colon	91%	10%
Pancreas	15%	2%
Skin (melanoma)	97%	16%
Prostate	100%	33%

Source: National Cancer Institute

(a)

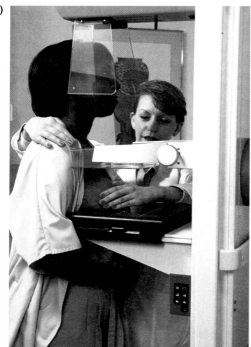

(b)

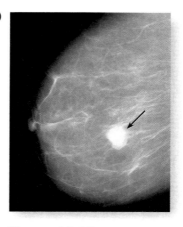

Figure 12.18
Mammography Can Detect Cancers Too Small to be Felt. (a) A mammogram is breast X ray. (b) A mammogram reveals a possible tumor.

to be malignant cancers. For these reasons, any positive mammogram result must be followed by further diagnostic tests. Despite these problems, because several studies have shown that mammograms do save lives, American medical doctors recommend that women over 40 or 50 years old have a mammogram every year.

Other new tests measure either DNA or proteins in cancer cells. One such test measures the blood concentration of a protein called PSA (short for Prostate Specific Antigen), which is released by normal prostate cells. Higher than normal PSA levels in the blood suggest the presence of a prostate tumor. Other new tests look directly at samples of DNA to determine whether mutations are present. A new DNA test that is used on cervical cells when an abnormal Pap smear is found can detect the presence of DNA from a sexually transmitted virus that causes many cases of cervical cancer. In years to come, DNA testing certainly will be an important part of cancer screening.

DNA tests offer more promise than just detection of cancer. Recent studies show that DNA tests can help to predict whether a person will respond to a given chemotherapy agent. Consider the drug Gleevec, a new medicine used to treat lung cancer. The early results from tests of this drug were not as beneficial as researchers had hoped. When researchers tried to figure out why, they discovered that some of their lung cancer patients had mutations in a gene that codes for a particular growth

factor receptor, called *epidermal growth factor receptor (EGFR).* Patients with a gene that produces too much EGFR are much more responsive to Gleevec than are lung cancer patients who do not show this mutation. Tests such as this one will help to make cancer therapies much more effective.

Traditional Treatments Include Surgery, Chemotherapy, and Radiation Therapy

Regardless of how cancer arises, the major concern for the patient is whether the disease can be treated and cured. Everyone wants to be told that no matter how difficult the treatment is, it will eliminate the cancer. Until just a few years ago only three kinds of traditional therapies were commonly available: surgery to remove as much of the tumor as possible followed by chemotherapy and/or radiation therapy to kill remaining cancer cells.

Traditional **chemotherapy** uses drugs (chemicals) that kill cancer by interfering with the cells' normal processes. One common feature of the diverse chemotherapy agents available is that they work by stimulating apoptosis in cells that are actively proliferating. Consider the many kinds of chemotherapy drugs that are similar in size and shape to nucleotide bases. The drugs are taken up by cells and are inserted into DNA in the place of normal nucleotide bases during the normal replication process (**Figure 12.19**). As a result, cells have abnormal DNA molecules that produce abnormal proteins. Any cell treated with such a chemotherapy agent that still has its normal mechanisms for detecting abnormal proteins or abnormal DNA will undergo apoptosis and die. So many chemotherapy drugs work indirectly by inducing cancer cells to cause their own deaths.

Radiation therapy uses high-energy waves, or particles, such as photons, electrons, or protons to kill cells in a tumor. Different methods of delivering radiation are used, depending on the kind of tumor(s), how widespread the tumors are, and where they are located. For example, radiation can be directed at a region of the body, or a radioactive substance can be placed in close proximity to a tumor. Like chemotherapy, radiation therapy also relies on apoptosis, and it preferentially kills proliferating cells. Radiation works by causing changes in DNA that lead to abnormal proteins and then to cell death (**Figure 12.20**). Cells that are actively copying their DNA are more susceptible to radiation damage than are cells in other phases of the cell cycle. In fact, the same radiation that can kill cancer cells at a high dose could cause cancer if it was given at a lower dose.

chemotherapy the use of a chemical, or drug, treatment to kill cancer cells

radiation therapy the use of high-energy rays to kill cancer cells

If you know anyone undergoing chemotherapy or radiation therapy, you are aware that these traditional treatments have serious side effects, and they are not always successful. Why do patients often lose their hair, experience nausea, suffer from infections, and seem so much sicker while undergoing these cancer treatments than they were before chemotherapy or radiation? One reason is that both chemotherapy and radiation therapy also damage normal cells. Although mature cells can be affected, both treatments have a greater effect on DNA that is in S-phase, when the DNA is dispersed, more exposed, and in the process of replication. For example, because normal cells in the digestive tract can be killed by chemotherapy or radiation therapy, a cancer patient having this kind of treatment may experience nausea. Because the normal, rapidly proliferating cells that maintain hair are susceptible to being killed by these same treatments, they also may die. The same is true for the cells in the deep layers of skin, and the cells in the immune system. Damage to the immune system is one of the most serious of the side effects of chemotherapy because it leaves a person open to many organisms that can cause infectious diseases. Chemotherapy also can damage the lungs and the heart. These kinds of cancer therapies can save lives, but sometimes their side effects are so damaging that a patient becomes too sick to complete the full recommended therapy. Chemotherapy and radiation therapy are often a race between the

Figure 12.19 **How Chemotherapy Works.** Chemotherapy drugs supply nucleotides that are almost, but not quite identical to normal nucleotides. When chemotherapy nucleotides are inserted into a growing strand of DNA, the cell detects the abnormal protein and apoptosis is triggered. As a result, the cell dies. While this has a positive effect on growing cancer cells, it has a negative effect on normal, healthy cells that often are damaged or killed by chemotherapy.

■ *What normal cells in your body will be affected by this treatment and in what way?*

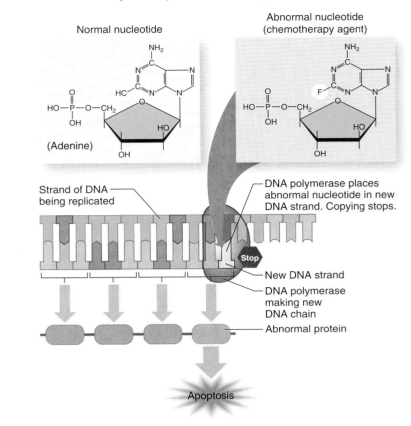

(a) This tumor contains a variety of cells in various stages of cancer progression.

(b) Radiation causes additional mutations, especially in cells in S-phase.

(c) Cells with intact apoptosis mechanism die, but cells with mutated apoptosis mechanism survive. The tumor is smaller, but is still aggressive.

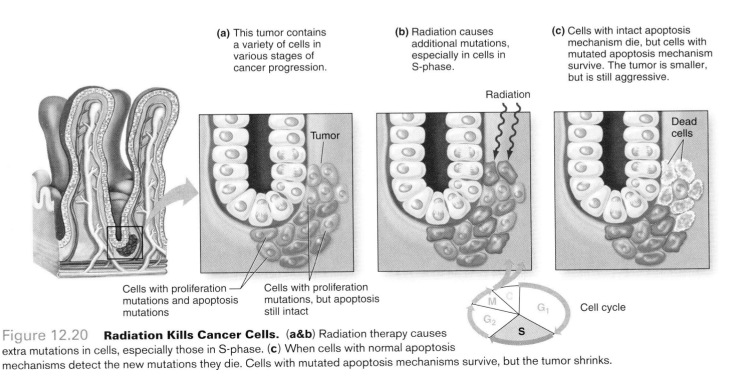

Figure 12.20 **Radiation Kills Cancer Cells.** (**a&b**) Radiation therapy causes extra mutations in cells, especially those in S-phase. (**c**) When cells with normal apoptosis mechanisms detect the new mutations they die. Cells with mutated apoptosis mechanisms survive, but the tumor shrinks.

(a) A cancer cell with excess *Her2* receptors divides more often than it should, contributing to the development of breast cancer.

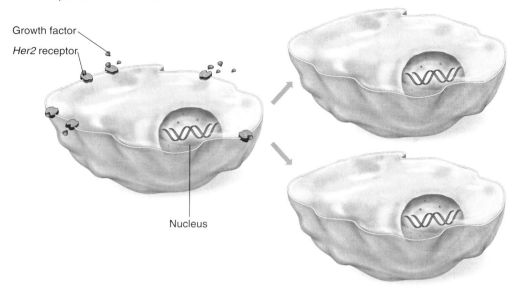

(b) Herceptin is an antibody that binds to, and so blocks, the *Her2* receptor so that growth factors cannot bind. This prevents cell division and so inhibits the development or progression of cancer.

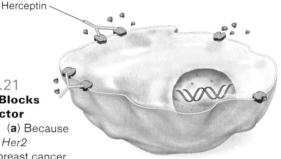

Figure 12.21
Herceptin Blocks Growth Factor Receptors. (**a**) Because it has excess *Her2* receptors, a breast cancer cell divides more frequently than normal. (**b**) Herceptin blocks *Her2* receptors and the cancer cell ceases to divide.

destruction of cancer cells and the destruction of normal cells.

Another limitation to traditional chemotherapy and radiation therapy is that some cancer cells have multiple mutations in the genes that control apoptosis and so do not cause their own deaths. Radiation and/or chemotherapy won't kill these cancer cells. For all these reasons, there are limits to the effectiveness of these traditional treatments. This situation should change in the future, however, as knowledge about the role of apoptosis in cancer development and cancer treatments increases.

Research Provides New Approaches to the Detection and Treatment of Cancer

One new treatment approach being used against breast cancer involves canceling the effect of growth factors by blocking their receptors. As previously noted, about one-third of breast cancers contain cells that produce excess amounts of a growth factor receptor called *Her2* (**Figure 12.21a**). The presence of excess receptors enables more growth factor molecules to be bound and this promotes cell proliferation. Herceptin, a drug for treatment of breast cancer, works by binding with *Her2* (Figure 12.21b). Herceptin effectively blocks the receptors from binding growth factor molecules. About one-quarter of patients with advanced breast cancer show significant improvement when *Her2* is blocked with Herceptin. The combination of new tests that detect *Her2* and treatment with Herceptin shows promise in treating some breast cancers.

Another new approach in cancer treatment involves directing treatments more specifically at cancer cells, so that other body cells are spared. For example, tamoxifen binds to and blocks the effects of the hormone, estrogen. Because estrogen stimulates cell proliferation in breast tissue, tamoxifen specifically inhibits cell proliferation in the breast, while most other tissues are unaffected. Another tissue-specific approach is the use of antibodies.

Antibodies are proteins produced by the immune system that bind to foreign proteins on invading bacteria and viruses. Once antibodies have attached to the proteins on an invader's surface, immune cells can kill the invader. Antibodies and immune cells also combine to fight cancer cells, so researchers have tried using laboratory-made antibodies to improve this process. Antibodies in a new drug called rituximab bind to a protein on the surface of cells in non-Hodgkin's lymphoma, a type of cancer that starts in the lymphatic system. This allows the immune system to kill some of the cancer cells and makes others more susceptible to standard chemotherapy and radiation therapy. In fact, Herceptin is actually an antibody against the *Her2* receptor. These kinds of highly specific cancer therapies appear to have a bright future.

Yet another promising approach to treating cancer is to inhibit the growth of blood vessels that provide the tumor with nutrients and energy. Recall that cancer cells release growth factors that stimulate the growth of blood vessels, also called angiogenesis. The new cancer drugs block the action of these growth factors. By themselves these drugs are not very effective, but when they are given in addition to more conventional chemotherapy, they can slow the growth of tumors and increase patient life span. These antiangiogenesis drugs will certainly get more attention in the future.

One of the most intriguing approaches to treating cancer uses messenger RNA to block the action of a defective gene instead of trying to stop an abnormal protein that is already having an effect in a cancer cell. Here the logic is that if the action of defective genes can be blocked, then the cells no longer will be cancerous. One example of this approach is directed at cells that have become invincible because they have an overactive *bcl-2* gene, a gene that produces a protein that inhibits apoptosis. Researchers have successfully blocked overactive *bcl-2* genes in animals by targeting the mRNA that is transcribed from the gene. Recall that mRNA is a strand of nucleotide bases. A complementary strand of nucleotide bases binds to the mRNA strand and prevents it from binding to ribosomes and thus producing protein. Researchers have produced such complementary *bcl-2* mRNA strands, called *antisense strands,* and have inserted them into cancer cells that have too much *bcl-2* mRNA. The excessive *bcl-2* is blocked and when cancer cells are deprived of *bcl-2,* they succumb to apoptosis. Because RNA is a large, polar molecule, the most difficult task in this treatment is to get the antisense RNA into the cells. Several ingenious approaches have been tried, and some have been successful enough to merit clinical trials of this treatment strategy in humans.

If you have cancer, or are close to someone who does, you may or may not feel that these new discoveries and treatments are helping you. It is likely, though, that these and other developments will greatly improve the treatment of many types of cancer in the coming years.

QUICK CHECK

1. How do chemotherapy and radiation therapy kill cancer cells?

2. What is *Her2* and what role does it have in breast cancer and breast cancer treatment?

12.7 An Added Dimension Explained
Assessing Laetrile and Other Alternative Cancer Therapies

Confronting cancer is one of the most difficult of all medical situations. The disease is nearly always deadly in the long run, and the treatments can seem nearly as bad as the disease. It is not surprising, then, that people look for some way out, and many alternatives to conventional medical treatments are for sale. An *alternative therapy* is one that is not usually used in Western medical practice and often is advertised as being superior to mainstream treatments. Common alternative approaches include unconventional diet regimens, body-cleansing therapies, herbal treatments, and special chemical formulas. Promoters of alternative therapies often claim to cure cancer, while most conventional Western scientists and physicians only claim to treat cancer. Alternative therapies typically lack established scientific support, while most conventional treatments have some scientific basis. Some alternative therapies, such as laetrile and others you will soon read about, have been studied and do not show any benefit for treating cancer. Others simply have not yet been tested. Some may prove promising with more research. Few if any are as effective as they are advertised to be, and some are dangerous.

Besides alternative therapies, there are approaches to cancer treatment called *complementary.* Complementary therapies do not in themselves treat cancer, but they can improve the quality of life for many cancer patients. Scientific and medical professionals are more likely to support and even advocate complementary therapies. These can include stress-reduction techniques such as meditation, spiritual and social support groups, exercise routines, and improvements in diet.

From a scientific perspective the important question is, Do alternative therapies work to treat cancer? Laetrile is one of the oldest alternative therapies, but it is a good example because the hypothesis for how it was supposed to work was clearly laid out, and it has been explicitly tested. In the early 1900s pharmacist Ernst T. Krebs began using laetrile extracted from apricot pits to treat cancer patients. (This is not the same Krebs who studied glucose metabolism.) Krebs had observed that the body metabolized laetrile to produce the poison, cyanide. Krebs proposed that the enzymes needed to produce cyanide from laetrile might be more abundant in cancer cells than in normal cells. He reasoned that the ingestion or injection of large quantities of laetrile would result in the release of cyanide in tumors, causing their destruction. The logic of his proposal was compelling, and without any governmental approval of the treatment, many clinics around the world began offering laetrile as part of an alternative cancer therapy. By the end of the 1970s laetrile had become so popular that an estimated 70,000 Americans had tried it. Proponents of laetrile therapy claim that it is 100% safe and nearly 100% effective in treating many forms of cancer. This seems astounding. If something works this well, why isn't everyone using it? Why wouldn't doctors prescribe it? Laetrile advocates would tell you that drug companies don't want people to use laetrile because more money can be made from conventional therapies. Your doctor would tell you that laetrile simply does not work to treat cancer, and it may be harmful. How can you evaluate these claims? You can look to scientific research.

Scientists in several dozen different research labs worldwide have designed and carried out studies to address three questions: (1) Is the biochemical basis for Krebs' proposal sound? (2) Is

—Continued next page

Continued—

laetrile safe? (3) Is laetrile effective in fighting cancer? The results of these studies were evaluated by other scientists before being published and are available in a variety of scientific magazines.

First, Ernst Krebs' hypothesis that tumor cells have high levels of the enzymes needed to release cyanide from laetrile is not likely to be correct. Several studies show that the greatest release of cyanide from laetrile occurs in the intestine and liver, not in cancer cells. The basis for a therapeutic effect of laetrile is not sound. Second, while proponents claim that laetrile is safe, several studies on both animals and humans report that laetrile users can suffer cyanide poisoning. Other chemotherapy agents cause serious side effects, though, and if laetrile effectively cured cancer, the risk of cyanide poisoning might be worth taking. Finally, is laetrile effective? Unfortunately, several studies in both animals and humans show no beneficial effect for laetrile. The proponents of laetrile therapy continue to argue that the studies were not done correctly. For the present, a scientific evaluation shows that it is unlikely that laetrile is an effective cancer treatment. So, based on the scientific studies available, it is not advisable to put your time and money into laetrile.

An Internet search will show scores, if not hundreds, of websites claiming that some product produces a safe and reliable cure for cancer. **Table 12.6** summarizes some alternative therapies. Because so many people try them, the government agencies that fund the research are starting to provide support to study them. Table 12.6 shows some interesting conclusions.

First, many alternative treatments have been evaluated and found to be not as effective as conventional treatments. Second, some treatments or regimens—such as a macrobiotic diet or the antineoplaston treatment promoted by *Buryzinski*—have not been evaluated. The antineoplaston treatment is interesting because *Buryzinski* publishes many papers and speaks in language that sounds scientific. His papers, however, are not published in well-respected scientific journals, and his scientific-sounding language does not quite make sense, if you read it carefully. Third, many of the treatments are based on models of cancer that do not agree with what is known about the disease—the kind of knowledge provided in this chapter. The treatments that claim to "boost the immune system" are a particularly subtle example of this. To date there is no known, reliable, documented way to boost your immune response. Even if you could, it is not likely that simply "improving" your immune system with herbs or formulas will cure a cancer that has already mutated beyond the reach of your existing immune response.

Finally, it is important to point out that some alternative treatments are promising areas of research. For example, treatment with shark cartilage is not effective, but shark cartilage is a source of a chemical that blocks the formation of blood vessels. The isolation of this chemical has led to a drug that is an important component of many conventional treatment programs. As previously noted, many alternative therapies claim to boost the immune system. While these specific claims are not likely to be valid, research into ways of boosting the immune system is an important avenue of research. The role of the

Table 12.6 Evaluation of some alternative therapies

Name of therapy	Rationale and treatment	Scientific evaluation
Livingston-Wheeler	All cancers are caused by bacterium called *Progenitor cryptocides*. Treatment consists of detoxification and enemas.	The bacterium has not been identified by any other researcher. Study matched 78 patients receiving this treatment with 78 patients receiving traditional therapy. All had similar types of advanced cancer. Average survival was 1 year for both groups. No advantage found for the alternative therapy.
Di Bella	Multiple approaches mixing conventional and alternative therapies. Therapies administered at special clinics.	Records of patients at clinics are incomplete and fragmentary. Several showed no advantage of Di Bella treatment compared with conventional alone.
Buryzinski	Cancer is caused by antineoplastons. Patients are treated with complex substance containing small proteins and amino acids.	Buryzinski has published many studies himself, but most researchers are critical of them. A large-scale study by the National Cancer Institute was halted because of difficulty with cooperation between Buryzinski and NCI researchers.
Chaparral	Native American general medicinal treatment.	One limited study found tumor reduction in 2 out of 44 patients.
Shark cartilage	Said to inhibit blood vessel formation in tumors.	Several studies show no effects compared with conventional therapy. However, shark cartilage and other vertebrate cartilage are sources of an effective drug that does inhibit blood vessel growth into tumors.
Macrobiotic	Cancer can be cured by particular diets. Diets emphasize organic grains such as rice, with some vegetables and minimal fruits and proteins.	The diet and lifestyle recommended can be healthy, but some versions are restrictive and do not provide balanced nutrition. Low-fat, low-meat diets have been associated with reduced cancer rates, but no studies are available on the efficacy of any diet in treating cancer.

immune system in cancer development is undisputed. Treatments that really do improve immune system function may yet prove helpful. For example, such a treatment might work to prevent cancer in high risk patients.

As a last point about alternative therapies, you can find provocative reports that specific individual people were cured of their cancer by a particular treatment. Claims that a single individual was cured are difficult to assess scientifically. Every individual is different, and there could be many reasons why a specific individual recovered from cancer, or did not recover.

Only by studying large numbers of people can scientists determine the probability that a particular treatment works, or does not work. Individual claims of recovery are not helpful to a scientific evaluation of a treatment, and they tell nothing about how you might respond to a treatment.

QUICK CHECK

1. Summarize the evidence that leads scientists to conclude that laetrile is neither effective at treating cancer nor safe.

Now You Can **Understand**

Does Chemotherapy Destroy the Human Immune System?

Based on what you have read so far, you now can give an educated answer. Chemotherapy does damage systems that rely on normal cell proliferation. During chemotherapy you may lose your hair, be nauseated, and have difficulty eating because normal, proliferating cells are killed. It is clear, though, that these functions return if chemotherapy is successful. Chemotherapy does not kill all cells in the hair follicles, and so hair grows back after chemotherapy. The same is true for the immune system. Chemotherapy *does impair* immune function, but the immune system can recover when chemotherapy stops.

It is unlikely that you will find a doctor who dismisses the role of the immune system in fighting cancer. In fact, new medicines are being developed and prescribed that help to boost the immune system during chemotherapy. But if you need chemotherapy, then your immune system has already fallen behind in the battle, and the chemotherapy could get rid of enough cancer cells to allow your recovered immune system to deal with the rest. Certainly you should always take care of your overall health and avoid things that might stress your immune system. This is especially true if you must go through chemotherapy. But there is no evidence that chemotherapy "destroys" your immune system.

Can Vitamin Supplements Help Prevent Cancer?

The evidence concerning supplements is conflicting. While some studies find support for the idea that taking regular high doses of vitamins A, E, and C can help reduce cancer rates, other studies show no such effect. The best evidence in support of supplements is that people who eat a healthful, balanced diet that features lots of fresh fruits and vegetables are somewhat less likely to get cancer. The idea behind taking supplements is that high doses of vitamins C and E may counter the effects of free radicals. Even though

simply taking pills is not likely to change your chances of developing cancer, it is unlikely that Americans will give up on easy, simple answers. Vitamins and other diet supplements are a big business, and Americans are responsive to advertising claims. Keep in mind that high doses of vitamins and other "natural" compounds do carry some risk. Just because a supplement is natural does not mean it's necessarily good for you. If you aren't under a doctor's care, you probably should avoid large doses of anything.

What Do **You** Think?

Should New AIDS Drugs Be Fast Tracked?

Most of us would agree that new drugs are needed for the treatment of cancer and for the treatment of other severe diseases such as AIDS. The governing organizations responsible for approving drugs, though, have a dilemma. The FDA is responsible for assuring Americans that the drug treatments they take are both *safe* and *effective.* No drug can be manufactured, sold, and prescribed in the United States without FDA approval. Science moves slowly because it takes intensive study to identify a drug treatment; it takes even more careful experimentation to show that a drug treatment is safe and effective. Many Americans have become critical of the government's slow review process and the burden of proof required by the FDA for the approval of a new drug. These frustrations have been heard by the FDA, which now has a "fast-track" process for the approval of drugs for treatments for "urgent unmet medical needs." Often this means that drugs intended for terminally ill patients can achieve approval faster than they would have in the past.

Many people applaud the new FDA rules, while others are more cautious. And when an approved drug turns out to have a hidden risk, some people get angry and upset with the drug companies. You can find more about the FDA fast-track process and the discussion surrounding it on the Internet. What do *you* think about the process of approving drug treatments?

Chapter Review

CHAPTER SUMMARY

12.1 Cancer Is a Major Killer

Cancer kills over half a million Americans each year in the United States and, although death rates are beginning to decline, cancer is the second leading cause of U.S. deaths.

12.2 Different Types of Cancers Share Certain Common Features

Cancer cells are characterized by abnormal cell division and the ability to metastasize. They lose mature cell functions, avoid cell death, and attract blood vessels. A tumor may be benign or malignant.

benign tumor   284   metastasis   285
cancer   284   tumor   284
malignant tumor   285

12.3 Cancer Is a Genetic Disease

Cancer is caused by mutations in DNA, especially in the genes that control cell division. While some factors that cause mutations in DNA are generated as by-products of metabolism, other mutations are inherited, or caused by environmental factors.

adhesion   287   apoptosis   287
angiogenesis   287

12.4 Abnormal Proteins Are Responsible for the Abnormal Traits of Cancer Cells

Abnormal tumor suppressor proteins and abnormal proto-oncogenes can lead to cancers. The abnormal characteristics of cancer cells include loss of mature cell shape, and function, abnormal growth of collateral blood vessels, loss of normal adhesion within a tissue, and abnormal apoptosis that allows them to survive when a normal cell would have killed itself.

growth factor   289   proto-oncogene protein   289
oncogene   289   tumor suppressor protein   288

12.5 Risk Factors Affect Your Chances of Developing Cancer

Most cancers come from exposure to carcinogens in cigarette smoke, a high-fat diet, or from lack of exercise. The key to cancer prevention is to make good lifestyle choices: stop smoking or never start smoking, get regular exercise, and eat a low-fat, high-fiber diet that includes a variety of fruits and vegetables.

carcinogen   294

12.6 Early Detection and New Treatments Increase Prospects for Surviving Cancer

Experts believe that most cancers are preventable. While traditional treatments for cancer have included surgery, radiation therapy, and chemotherapy, newer treatments are based on developing knowledge of cell biology. There probably never will be a single treatment for all cancers because cancers in different tissues are quite different. Because of new treatments people are living longer with cancer.

chemotherapy   300   radiation therapy   300

12.7 An Added Dimension Explained: Assessing Laetrile and Other Alternative Cancer Therapies

Alternative cancer treatments are being evaluated systematically. While some may prove helpful to a cancer patient, many, such as laetrile, have no objective effect on the progression of cancer.

REVIEW QUESTIONS

TRUE or FALSE. If a statement is false, rewrite it to make it true.

1. Death rates from breast cancer have decreased since 1975.
2. A benign tumor is one that has spread from its site of origin and so cannot be treated just with surgery.
3. Cancer is always an inherited disease.
4. Apoptosis is another word for cellular suicide.
5. Cells in normal tissues stay together by the process of adhesion.

MULTIPLE CHOICE. Choose the best answer of those provided.

6. The excess cell division that produces a tumor is also called cell
   a. proliferation.   d. transcription.
   b. mutation.   e. adhesion.
   c. suicide.

7. Metastasis is the process of
   a. cell division.
   b. cellular suicide as a result of abnormal cellular proteins.
   c. cell signaling that leads to abnormal cell communication.
   d. loss of cell adhesion that leads to spread of cancer.
   e. the mechanism that produces DNA mutations.

8. Which of the following cellular functions is *not* typically found to be abnormal in cancer cells?
   a. cell proliferation   d. apoptosis
   b. metabolism   e. stimulation of angiogenesis
   c. adhesion

9. Which two classes of proteins are important in controlling cell division in normal cells?
   a. cytochromes and electron transport proteins
   b. reactive oxygen proteins and antioxidant proteins
   c. tumor suppressor proteins and oncogene proteins
   d. proto-oncogene proteins and *Her2*
   e. tumor suppressor proteins and proto-oncogene proteins

10. A signal molecule that stimulates a cell to divide is called
    a. a protein.   d. a transcription factor.
    b. a growth factor.   e. p53.
    c. an oncogene.

11. p53 functions to:
    a. stimulate an abnormal cell to go through apoptosis.
    b. stimulate an abnormal cell to divide.
    c. stimulate a normal cell to go through apoptosis.
    d. stimulate a normal cell to divide.
    e. none of the above

12. Angiogenesis means:
    a. the death of blood vessels.
    b. the growth of blood vessels.
    c. the shrinkage of blood vessels.
    d. the dilation of blood vessels.
    e. none of the above

13. What role does the immune system have in cancer?
    a. It has no role.
    b. Immune system cells produce cancer cells.
    c. Immune system cells kill the bacteria that are the cause of all cancers.
    d. Immune system cells take apart cancerous DNA.
    e. none of the above

14. Reactive oxygen species within the body are primarily the result of
    a. cancer cell activity.
    b. apoptosis.
    c. the use of oxygen in glucose metabolism.
    d. the use of oxygen in cell proliferation.
    e. the use of oxygen in gamete production.

15. Cancerous cells
    a. cannot function like normal cells.
    b. look different from normal cells.
    c. secrete substances that cause blood vessels to grow.
    d. none of the above
    e. all of the above except d

16. Which of the following statements is *false?*
    a. Environmental carcinogens cause a small percentage of cancer cases.
    b. A small percentage of all cancers are inherited.
    c. Exposure to estrogen signals cells to divide.
    d. Cells in tissues that actively divide (lung, intestine, stomach) are more likely to develop cancer.
    e. Laetrile seems to be an effective cancer treatment.

17. What role does the use of oxygen in metabolism play in the development of cancer?
    a. The presence of oxygen increases risk of cancer.
    b. Oxygen forms reactive oxygen species which increase cancer risk.
    c. Oxygen forms reactive oxygen species which cause mutations in DNA.
    d. Oxygen plays no role.
    e. a, b, and c are correct.

18. Which of the following genes and proteins is *not* described in this text as being abnormal in cancer cells?
    a. *c-myc*
    b. *Her2*
    c. *Bcl-2*
    d. cytochrome oxidase
    e. p53

19. What kinds of body cells are most likely to give rise to a tumor?
    a. neurons
    b. kidney cells
    c. cells that are fully differentiated
    d. cells that normally divide
    e. cells that give rise to gametes

20. Which of the following is the best recommendation for avoiding cancer?
    a. Don't be exposed to cigarette smoke.
    b. Avoid fats from dairy products and red meat.
    c. Exercise regularly.
    d. Eat plenty of whole grains and fresh fruits and vegetables.
    e. all of the above

CONNECTING KEY CONCEPTS

1. Discuss the respective roles of DNA mutations and abnormal proteins in the development of cancer. Include reference to each of the four cellular functions presented in the text that are abnormal in cancer cells.

2. What does it mean to say that cancers discovered and treated early have a better outcome than cancers treated late. Use your knowledge of the cellular basis for cancer to describe why this is so.

QUANTITATIVE QUERY

1. For which of the cancers listed in Table 12.2 has the five-year cancer survival rate increased the most?

THINKING CRITICALLY

1. Government and private organizations have spent millions of dollars on campaigns to prevent people from becoming cigarette smokers. You can find some of these campaigns on The American Cancer Society website, among others. Based on this information, design a campaign that you think would be effective in preventing high school students from starting to smoke.

2. The American Cancer Society now lists many complementary and alternative therapies for cancer. Examine this list on their website. Choose one that you think is either promising or unlikely to be effective. Outline, in broad terms, the kinds of information you would need to evaluate the action of this treatment. Design an animal or human study to determine the effectiveness of the treatment.

For additional study tools, visit www.aris.mhhe.com.

CHAPTER 13

PART III
Evolution and Diversity of Life

Life Evolves

DARWIN AND THE SCIENCE OF EVOLUTION

Dinosaurs in Our Backyard

May 16, 1986

Could these birds at a backyard feeder actually be modern-day dinosaurs?

Dinosaur Egg Fossils Discovered in E. China

February 23, 2005

An Added Dimension
Dinosaurs and Birds

Were you a kid who was interested in dinosaurs? Did you pore over the dinosaur books and memorize the name of each species? For over 200 years dinosaurs have fascinated scientists too. Dinosaurs are the superstars of *paleontology,* the study of fossils.

People had found dinosaur bones as early as the 1600s, but they did not know what kind of animals they came from. In the light of what now is known about dinosaurs, one early idea was ahead of its time. In 1802 Edward Hitchcock found the three-toed tracks of several kinds of dinosaurs and concluded that they had been made by flocks of ostrich-sized birds.

Soon many more examples of large, reptilelike fossils were discovered. In 1841 the English naturalist Richard Owen coined the term *dinosauria,* Latin for "fearfully great lizard," for these newly discovered animals.

In Owen's time people thought dinosaurs were still alive somewhere on Earth, waiting for a naturalist to stumble onto them. From scientific studies we know that dinosaurs have been extinct for 65 million years. Early depictions of dinosaurs were also off the mark, showing them as sluggish creatures that lumbered about with their tails dragging. It was not until the mid-1900s that researchers put all the information together to realize that dinosaurs were agile creatures that could run with their tails held high. This switch in the basic concept of how dinosaurs lived started a debate about whether dinosaurs might have been warm-blooded. The debate raged until Robert T. Bakker found a fossilized dinosaur heart. It had four chambers like a crocodile, bird, or mammal heart, not three chambers like the heart of a turtle, snake, or lizard.

Recently, the debate about dinosaurs has heated up once again. Until the late twentieth century almost everyone agreed that catastrophic ecological and climatic changes following the collision of a meteor with Earth had caused the dinosaurs to go extinct 65 million years ago. Today, however, many paleontologists are convinced that the dinosaurs still are alive, although not in the way the earliest scholars thought. Even more surprising, they say that dinosaurs are so common that you can see them in your own backyard (**see chapter-opening photo**). They claim that birds are the evolutionary descendants of dinosaurs.

At the end of this chapter you will read the evidence that supports this conclusion. Before you can fully appreciate the idea that birds are the descendants of dinosaurs, you must become familiar with the basic principles of how life evolves. ■

13.1 Evolution Is the Central Idea of Biology

If you had to give just one reason why people are fascinated by living organisms, what would it be? Many scientists would point to the amazing **adaptations** that organisms exhibit. An adaptation is an inherited trait that allows an individual to survive and reproduce in its particular environment. Let's look at some of the unusual adaptations of just one species, the vampire bat. Their razor-sharp teeth slash a painless cut in the skin of a victim. Like a cat drinking milk, the bat laps up the blood that flows out of the cut (**Figure 13.1**). An anticoagulant in the bat's saliva keeps blood from clotting, while a second chemical dilates the veins and keeps the blood flowing. If a vampire bat is trapped on the ground, it can leap straight up or it can scurry away to safety. Where do amazing adaptations like these come from? This is the central question in the science of evolution and the topic of this chapter.

Biologists investigate the diversity of adaptations on many levels and from many perspectives including studies of molecules, anatomy and physiology of organ systems, ecological interactions, and behaviors. These different ways to study adaptations are united by one feature of living things: all organisms are genetically related. Just as your traits are best understood by considering your ancestry, adaptations of the vampire bat and of all organisms

Figure 13.1 **Vampire Bat.**

can best be understood in terms of genetic history. When viewed this way the seemingly unique traits of different kinds of organisms, including vampire bats, are not completely new. Mammals have teeth, saliva, and can walk. The special adaptations of vampire bats are similar to the traits of all mammals. Even a bat's ability to fly is not a completely new trait. A bat's arm bones and its exceptionally long finger bones form the supports of wings that are made by a web of skin.

adaptation an inherited trait that allows an individual with the trait to survive and reproduce in a certain environment

This example leads to an important conclusion. Every adaptation, no matter how bizarre or unusual, comes from an evolutionary modification of the traits of an ancestral population. For instance, the sophisticated brain mechanisms of perception and memory in animals depend on membrane protein channels (see Section 4.3). In animals these membrane proteins are part of complex information-processing circuits, but they are evolutionary modifications of the ways that the earliest cells responded to stimuli in their environments. These ancient mechanisms are still used by many kinds of living prokaryotes. Similarly, the specialized traits of vampire bats are modifications of traits of their mammalian ancestors. Their sharp teeth, special saliva, and wings are not totally new.

The understanding that the traits of descendants are different from, but related to, traits of their ancestors is expressed in the principles of *biological evolution.* Evolution is the change in the traits of a population of organisms over generations. Over a few generations such changes are likely to be small, but over many generations such changes can lead to the emergence of an entirely new kind of organism—a new species.

There is more to the modern understanding of evolution, though. Inherited traits are controlled by genes, and each gene comes in different forms called *alleles* (Section 10.3). Changes in the traits of a population over generations reflect changes in the distribution of alleles in the population. So the modern definition of **evolution** is a change in the traits of a population over generations, resulting from changes in allele frequencies within that population. For example, modern elephants have some traits that are similar to their extinct ancestors, woolly mammoths. These include large size, a long, flexible trunk, and unusual teeth. But modern elephants also have traits that are different. They are smaller than woolly mammoths, have less body hair, and have shorter tusks with different growth patterns. Populations of mammoths carried different alleles for body hair and tusks than do populations of modern elephants.

Charles Darwin was the first scientist to explain the relationship between ancestral and modern traits. Darwin did not express his ideas about evolution in the modern terms of allele frequencies, but his principles have provided the foundation for subsequent studies of how evolution works. It is important for you to understand what Darwin concluded, why he came to those conclusions, and how modern scientific evidence supports and expands his conclusions. Once you understand how the principles of evolution were realized, in Chapter 14 you can consider the mechanisms that produce evolution.

QUICK CHECK

1. Give an example of an adaptation and relate adaptations and evolution.

13.2 The Intellectual Climate of the 1800s Set the Stage for Charles Darwin's Ideas About Evolution

Charles Darwin (1809–1882) lived in England in the nineteenth century, and his scientific studies were carried out in the intellectual atmosphere of this era. The nineteenth century European worldview was a blend of Judeo-Christian and Western philosophy that had dominated Western culture for several centuries. According to this view, the universe and all life in it came from a single creative act of God. Each species was created as a distinct type that was inherently different from all other species. Furthermore, it was believed that no new species had been added since Creation. Finally, people assumed that species were static: their characteristics had not changed since Creation Day. This idea of a constant and unchanging universe also was held in other disciplines such as physics.

Nineteenth-century naturalists also believed that species could be ranked in a linear hierarchy, based on their closeness to God. Fishes, snakes, insects, and worms were thought to be simpler than mammals and birds, and so they were lower on the hierarchy. Plants were ranked lower than animals. In these schemes Man was considered to be the apex of life, with Woman coming in a bit below him. This organization of species is called the **scala naturae,** Latin for "scale of nature," or the Great Chain of Being. These early hypotheses of how organisms are related are contradicted by modern evidence, but they were the beginning of the scientific study of classification.

In spite of the dominance of this static view of the universe, the nineteenth century also was a time of exciting new ideas about the world. For example, the economist Thomas Malthus predicted that unchecked human population growth would lead to bitter competition for food and other resources. Scientists soon recognized that competition for resources plays a large role in the growth of populations in all species. The nineteenth century also was a time of exciting new ideas in geology.

evolution the change in the traits of a population of organisms over generations, resulting from changes in the frequencies of alleles found in that population

scala naturae a ranking of all organisms on a scale of closeness to God

Geologists were beginning to understand that Earth is much older than most people had thought. All of these ideas influenced Darwin's thinking.

Ideas about evolution were emerging at this time as well. Darwin's grandfather, Erasmus Darwin, had written on the subject, and a few earlier scholars had concluded that species are not fixed and unchanging for all time. Most of the early explanations of how species might evolve did not turn out to be correct. For example, Jean Baptiste de Lamarck proposed that when an individual changes as a result of experience—such as learning to play the piano—these changes are passed on to its offspring. Research has shown that acquired characteristics are not inherited by the next generation. In the end it was Darwin's ideas about evolution that turned out to be correct.

Because Darwin became a famous biologist, you might expect that he was a model student, but Darwin's academic career was not outstanding. At Cambridge University, Charles Darwin devoted much of his time to enjoying himself. Darwin's father was a medical doctor who hoped that Charles would follow in his footsteps, but Darwin chose seminary training instead. Darwin planned a future as a parson/country squire, a life that would give him plenty of time for riding, hunting, shooting, and collecting beetles. Although he did not excel at his studies at Cambridge, Darwin did forge friendships with some of the most inspiring scientists of his time. From these mentors Darwin learned that questions in science are worth pursuing in their own right. He developed a serious interest in geology, was an avid beetle collector, and deepened his knowledge of living things. Through his connections at Cambridge, and because he came from a privileged family, Darwin was invited to join a scientific expedition aboard H.M.S. *Beagle* as companion to the captain. Because of his interests he became the ship's unpaid naturalist. Initially, his father disapproved of the venture and refused to grant Charles the money for his expenses aboard ship. Luckily, an uncle convinced Dr. Darwin to change his mind and Charles set off on the *Beagle* voyage.

And the rest, as they say, is history. In 1831 Darwin began a nearly five-year-long voyage that eventually would change the course of scientific thinking. If Darwin had not joined the *Beagle* voyage, someone else probably would have developed a comprehensive explanation of evolution. In 1858 the British naturalist Alfred Russel Wallace sent a letter to Darwin that outlined a theory of evolution by means of natural selection. This was long after Darwin had returned from the *Beagle* voyage, after he had spent years reflecting on what he had seen

and developing hypotheses. In Wallace's letter Darwin was astounded to read what seemed to be a summary of his own ideas. At the urging of friends, Darwin and Wallace made a joint presentation of the theory of evolution by natural selection. Darwin receives most of the credit because he devoted his scientific life to studying and finding support for the idea of evolution by means of natural selection, while Wallace had a stroke of genius and then turned to other pursuits. In 1859 Darwin published *On the Origin of Species by Means of Natural Selection, or the Preservation of Favored Races in the Struggle for Life,* a book that presented his cohesive theory of how Earth's biological diversity evolved. You may know it by its abbreviated title: *The Origin of Species.*

> **QUICK CHECK**
>
> **1.** Describe at least two aspects of the nineteenth-century worldview that were widely accepted in Darwin's time.

13.3 Darwin Discovered the Fundamental Principles of Evolution

Darwin's (**Figure 13.2a**) understanding of the principles of evolution started with his experiences on H.M.S. *Beagle* (Figure 13.2b–d). The voyage was not easy or comfortable for Darwin. The *Beagle* was 90 feet long and 25 feet at her widest (27.4 meters long and 7.6 meters wide) and was crammed with 74 people and all the supplies and equipment necessary for a long sea voyage. The tiny cabin that Darwin shared with two other sailors was only 10 feet by 15 feet (Figure 13.2b). Darwin was six feet (1.8 meters) tall and had to stoop when he was belowdecks. He slept in a hammock above his worktable. Worse yet, for all five years of the voyage Darwin never got over being miserably seasick. Luckily, the ship docked often and his onshore collecting expeditions were extended and adventurous. During the voyage several different sorts of information influenced Darwin's thinking. These included his observations of living organisms, study of geological formations, and the large number of fossils he collected and studied.

The *Beagle* Voyage Showed Darwin the True Diversity of Life

The diversity of life that Darwin observed made a big impression on him, and he would spend the rest of his life struggling to understand it. Darwin

Figure 13.2 The Voyage of H.M.S. Beagle. (**a**) Charles Darwin at 22 years old. (**b**) The plan of H.M.S. Beagle, with Darwin's cabin highlighted. (**c**) H.M.S. Beagle. (**d**) The route of the Beagle voyage.

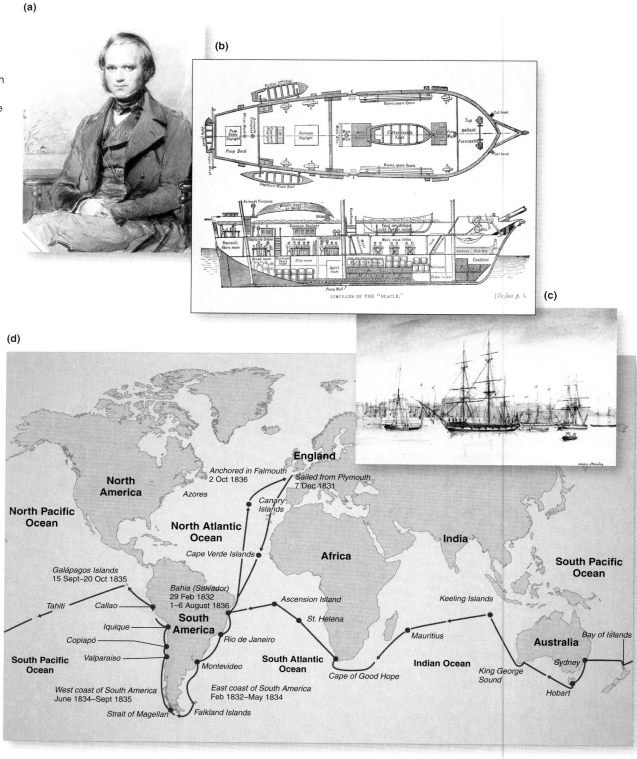

found new forms of life at every port, in every new environment, and even while the *Beagle* sailed. On the Galápagos Islands, a small volcanic archipelago in the Pacific Ocean off of the coast of Ecuador, Darwin observed some of the most intriguing examples of unique species (**Figure 13.3**). While Galápagos tortoises, lizards, hawks, finches, cormorants, beetles, and other forms of life were simi-

lar to those on the mainland, they also were subtly different. At this point in his travels Darwin could say with certainty that nearly every region he visited had its own distinct species.

Darwin also observed that each region had its own distinctive *style* of life-forms. Darwin noted that species were "clearly American" or "obviously African." One of the most striking examples of this is

Figure 13.3
Galápagos Animals Are Different from South American Animals. Although Galápagos marine iguanas (left) superficially resemble South American iguanas (right), they are distinct species with different habits and adaptations.

the distinction between the mammals of Australia and those of the rest of the world. Most of the native mammals of Australia are marsupials, whose young are born at an immature developmental stage and nurse within a fur-lined pouch on the mother's belly until they are mature enough to face the world. Outside of Australia placental mammals are predominant, and there are few, if any marsupial species.

Darwin saw other examples of regional similarities of animals. For instance, although South America had abundant herds of guanacos and llamas, as well as condors and large flightless birds called rheas, Africa had none of these. Instead, Africa had herds of various kinds of antelopes and the large, flightless birds were ostriches. From his experiences on farms back home in England, Darwin knew that organisms that look alike and live close together often are genetically related. He correctly concluded that species in nature that look alike and are geographically close to one another also must be genetically related. This conclusion, now supported by nearly 150 years of biological research, was the starting point for Darwin's understanding of how species evolve.

Darwin's Study of Fossils Expanded His Ideas About How Species Are Related

Darwin's thinking about how species might originate also was greatly influenced by the fossils that he found. A **fossil** is a preserved trace of a long-dead organism, often found in layers of rock or mineral. Fossils were well known in Darwin's time, but on the *Beagle* voyage Darwin began to look at fossils in a new way. In South America Darwin not only found familiar-looking animals that were fossilized, but also he found totally new fossil species. He was thrilled to discover fossils of a giant sloth the size of a hippo, an extinct type of elephant, and a fossil llama as big as a camel. Darwin realized that fossils in deeper layers of rock were older than fossils in more superficial rock layers. It seemed clear that some species disappeared and were replaced later in time by other species that, in their turn, also disappeared. In addition, even though some of the fossils that Darwin found were similar to modern species, they obviously were not the same as modern species (**Figure 13.4**). Putting all

fossil any trace of a living organism that has been preserved by natural processes

(a)

(b)

Figure 13.4 Fossils Found by Darwin and Their Living Counterparts. Fossilized remains of the glypton, (**a**) and a present-day South American armadillo (**b**).

of his insights together, it was difficult for Darwin to avoid the conclusion that extinct fossil species were the ancestors of their modern counterparts.

Studies of Geological Formations Showed Darwin That the Earth Changes Over Time

Another important influence on Darwin's thinking came from his study of geology. As the *Beagle* sailed from port to port, Darwin read the newly published *Principles of Geology* by Charles Lyell. In it Lyell proposed that the geological processes that shape Earth today—erosion, volcanoes, earthquakes, and floods—also operated in the past. Before Lyell it was not commonly understood that laws of nature operate in the same ways in different places and at different times. This led to the startling conclusion that the Earth is not static and unchanging.

Based on the scientific principles of geology, Lyell made an even more startling suggestion. Not only are deeper rocks and fossils older than more superficial ones, but the ages of rock layers can be determined. Lyell calculated that certain geological formations must be at least 100,000 years old. Other geologists at the time estimated that some rock formations were millions of years old. The idea that the Earth was much older than the accepted 6,000 years was startling. For Darwin it led to the conclusion that life also might be quite old.

Five Postulates Summarize Darwin's Ideas

Darwin's work fills volumes, and many people have attempted to summarize it. The well-respected biologist Ernst Mayr condensed Darwin's work into five postulates. These postulates, however, are not the final word on the understanding of evolution. Darwin's work provided a deep and comprehensive explanation for how species arise, but scientists have been refining and adding to Darwin's contributions ever since.

Darwin's postulates are as follows:

1. *Life evolves*—the species populating the Earth change over time.

2. *Gradualism*—the changes in species occur gradually over long periods of time.

3. *Multiplication of species*—the number of species that have ever lived increases over time.

4. *Common descent with modification*—all species have evolved from a single common ancestor through the modification of ancestral traits.

5. *Natural selection*—species change over generations through a natural process of selection based on normal genetic variation.

The first three of Darwin's ideas are related. Together they say that living forms change gradually over time, and that these changes result in the appearance of new species. One important point here is that "time" means many generations. Each generation has the possibility of being slightly different from its parents. Across many generations these small differences gradually can add up to the evolution of a new species. Once Darwin understood that new species evolve, it was a logical conclusion that the total number of species that have ever existed increases over time. Some species, and even whole groups of species, die out, while new species appear, flourish, and diversify, leading to other new species. In this way the mechanisms of evolution increase the diversity of life and underscore the idea that life is not static and unchanging.

Common descent with modification is one of the most dramatic of Darwin's ideas, especially because it has direct implications for the place of humans among living organisms. By common descent Darwin meant that all life on Earth is genetically related and has evolved from a single common ancestor. By modification Darwin meant that new traits result from modification of ancestral traits. Darwin visualized this as a tree of life and even made a sketch of this idea. His sketch is similar to a family tree that you might draw to show how you and other members of your family are all related as descendants of your great-grandfather. Darwin applied this idea to groups of organisms, and then to all organisms. For humans this path starts with the finding that all humans belong to the same species. It stretches back in time to relate to all eukaryotes and then even further back in time to an ancestor that links the eukaryotes and prokaryotes.

Natural Selection Is the Primary Mechanism That Drives the Process of Evolution

Darwin's explanation of the process of **natural selection** is perhaps his most important contribution because natural selection explains how new species emerge. Simply stated, natural selection occurs when individuals with particular characteristics survive, breed, and reproduce more than do individuals without those traits. This selective breeding influences the frequency of those traits in the next generation.

natural selection the process in which individuals that survive to breed in a given environment pass on their traits to the next generation

Natural Selection Changes Traits of Populations, Not Individuals It is critical to understand that *individuals* do not change as a result of natural selection. Of course, individuals do change during their lifetimes, but because the changes they acquire do not affect DNA in the egg and sperm, the changes do not affect the next generation. The process of natural selection relates to individuals in two ways: individuals can live to breed and pass on their traits, or individuals can die before they breed and so do not pass on their traits. The distribution of traits in the population will change as generations are born, breed, and die. Let's explore how natural selection changes traits in populations.

The process of natural selection involves several basic processes and phenomena. First, natural selection involves *breeding populations*, groups of individuals that are likely to mate with one another and have offspring. A species like humans that is distributed over most of Earth can have many breeding populations. For instance, even though in theory you could mate with any human of the opposite sex, you are more likely to have children with someone who lives in your state or country than with someone from another country. An individual cannot inherit a trait if the genes that contribute to it are not in the breeding population.

Second, natural selection can occur only when the individuals in a breeding population are different from one another (**Figure 13.5**), and when inheritance contributes to these differences. Of course, this is true in natural populations. There are few true-breeding populations in nature, and the traits expressed by organisms start with inherited DNA. A glance at how your classmates vary tells you that this also is true of humans. Except for identical twins or triplets, no two humans are alike, and in part their traits are determined by the alleles they inherit. Without genetic variation there would be no basis for evolutionary change.

Finally, natural selection occurs because of the overproduction of offspring. Darwin realized that in every generation more offspring are produced than can survive. You probably have noticed this, too. For instance, fishes produce hundreds of fertilized eggs in every reproductive cycle, but only a few survive to become adult fishes and produce the next generation. Although mammals do not produce nearly as many offspring as do fishes, nevertheless only a few of their young survive. Only those that do survive to breed will pass on their genetic traits to the next generation. The important question that Darwin answered is: what determines which individuals survive, and so breed?

Figure 13.5
Individual Variation.
Darwin observed that individuals of populations are not all alike. He concluded that some varieties are favored by natural selection, while others are selected against. These gray wolves vary in color, markings, and size.

Darwin put all this together and realized that environmental factors determine which individuals live and breed and which individuals die young and so do not breed. And, among those that do breed, factors in the environment allow individuals with certain traits to be more successful breeders than others. Many environmental factors can affect the survival and reproductive success of individual organisms, including food sources, weather and climate, geological changes, natural accidents, access to mates, and actions of predators and parasites. Individuals that survive environmental conditions to successfully breed and produce offspring will pass on their genes, and so their traits, to the next generation. Those individuals with more offspring will influence the traits of the next generation more than individuals with fewer offspring. So environmental factors control the traits found in a breeding population at any given time.

One important point is that natural selection does not result in "better" populations. Natural selection does not have goals, nor is it moving life toward a better state. Natural selection is a natural process that results in organisms suited to the environment in which they live; those organisms that do not fit the environment reproduce less successfully. The fit between the population and the environment is not perfect, and sometimes the match of organism and environment comes with a cost—especially if the environment changes. Natural selection simply favors the traits of the best breeders available in any given population. Now let's explore the evidence that shows how natural selection occurs and look at what is now known about how natural selection works.

QUICK CHECK

1. Summarize some of the experiences Darwin had while on the voyage of the *Beagle* that influenced his thoughts about how diversity arises.

2. How does natural selection lead to a change in the traits of a population?

13.4 Scientific Studies Document Natural Selection

Darwin did not observe natural selection in action but inferred it from his observations of diversity. Once his ideas were well known, however, scientists found evidence of natural selection and studied the process more deeply. This section considers examples of natural selection that produce changes in the traits of one species over generations. These kinds of examples often are called *microevolution* because they are evolution on a small scale. Chapter 14 examines how natural selection and other mechanisms combine to produce whole new species, a process often called *macroevolution*.

Natural Selection Is Seen in the Fitness of Species to Their Environment

In many cases the adaptation of a species to its environment seems perfect, but closer examination shows variation among individuals within the population. One example of what seems to be a near-perfect fit produced by natural selection is the golden-winged sunbird of East Africa. This delicate bird has a long, curved beak that seems an exact fit for the tubular mint flowers from which it gathers

nectar (**Figure 13.6a**). The match of its bill to the curve of the mint flowers is so close that a typical golden-winged sunbird is able to gather 90% of the nectar in a mint flower in just 1.3 seconds, a record unmatched by similar species whose bills are not so well adapted to the shape of mint flowers (Figure 13.6b). Still, beaks of golden-winged sunbirds vary. Some have bills that are slightly longer; some are slightly shorter, some are slightly more curved, and so on. This variation in bill shape means that some of the golden-winged sunbirds are more efficient at gathering nectar than are others.

Bill-shape variation provides the possibility for the population to change in response to environmental changes. Consider what would happen if part of the region where sunbirds live experienced a long period of colder and longer winters. A shorter growing season would mean that mint flowers would be smaller. Under these conditions birds with smaller bills would be more efficient at gathering nectar, would survive in greater numbers, and would have more offspring than birds with larger bills. Over several generations the typical bill for a golden-winged sunbird in this region would begin to change—and overall the population would have smaller bills than they did when the seasons were warmer. Of course, this can happen only if some of the birds in the original population had bills adapted to feed on smaller flowers. If individuals could not find flowers suited to their bill shape, the golden-winged sunbirds would die out and might become locally extinct.

Many Studies Trace Natural Selection in Action

In many cases it is hard to see natural selection happening, because the individuals of a species live such long life spans compared with human observers. For many species, though, life span is relatively short, and researchers can see changes in a population over generations. One of the most well-known examples involves moths in England.

Populations of peppered moths occur throughout Northern Europe and North America. These moths usually have light-colored wings with dark flecks, the feature that gives them their name. Peppered moths live in forests, often near inhabited areas—and there are written reports

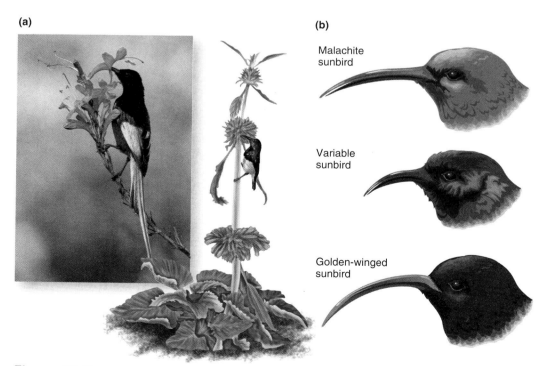

(a)

(b)

Malachite sunbird

Variable sunbird

Golden-winged sunbird

Figure 13.6 **A Close Match of Bill Shape and Flower Shape.** (**a**) The golden-winged sunbird can get nectar from mint flowers faster than two other sunbird species (**b**).

about populations of peppered moths from as early as the 1700s. Even though this species is identified as light colored, populations usually vary and include a few darker moths. These rare dark forms of peppered moth are found even in the earliest collections. Between the late 1800s and the mid-1900s, an interesting thing happened: across the Northern Hemisphere, peppered moth populations shifted from mostly light colored to mostly dark colored. By the mid-1950s 90% of the moths in some populations were dark. Then, in the latter part of the twentieth century, peppered moth populations switched back to being mostly light colored. Today peppered moths look like their light-colored eighteenth century ancestors. What has happened here?

The change in the color of the peppered moth populations closely parallels the level of industrial pollution during the twentieth century, and is called *industrial melanism.* This pattern is best documented in England, where the heavy use of coal produced soot that coated trees, rocks, buildings, and all other environmental surfaces. When the environment turned dark from soot, the populations of moths had more dark individuals. By

later in the twentieth century a shift from coal to oil and electricity and stricter smoke regulations cleared the air and environmental surfaces across Great Britain. The grime and soot gradually diminished, and the bark of trees where the moths rested during the day became lighter in color. The populations of moths became lighter too. This looks like a clear example of natural selection, but what was it about the darker environment of the early twentieth century that led to the shift in color of moth populations? An insightful series of studies led by British scientist Bernard Kettlewell supported the idea that predation by birds was a key factor (**Figure 13.7**). Before the environment was darkened, dark-colored moths were highly visible against lighter-colored tree bark—and accordingly, dark-colored moths got eaten by birds. When coal soot darkened the trees, the lighter-colored moths were more visible—and accordingly, light-colored moths got eaten. Then when pollution eased and the environment became lighter again, the dark moths were easy for birds to spot. Bird predation is probably not the full story of why the predominant color in

(a) 1956: Dark- and light-colored peppered moths

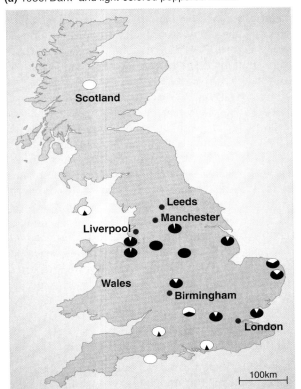

(b) 1996: Dark- and light-colored peppered moths

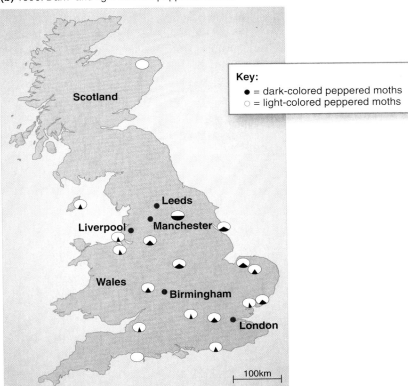

Figure 13.7 **Changes in Peppered Moth Populations.** The pie charts show the percentage of dark- and light-colored peppered moths at each location in 1956 (**a**) and in 1996 (**b**). Numbers of dark-colored peppered moths declined when industrial pollution lessened and the bark of trees returned to a natural, light color.

■ *How have peppered moth populations changed in the last 50-plus years? What has caused this change?*

moth populations changed from light, to dark, to light again. For example, other species of moths that are not preyed on by birds went through a similar color shift during this era in history. Regardless of what factors caused the shift, industrial melanism is an example of natural selection that has been well documented.

Many other examples document that populations change in response to the environment. House sparrows provide a good example. House sparrows are native to Eurasia, but in the 1850s several were released in Brooklyn, New York. By 1900 this hardy, adaptive, aggressive species had invaded suitable habitats around cities and towns all across North America and even had reached Vancouver, British Columbia, on the Pacific Coast of Canada. By 1914 house sparrows had reached Mexico City, and in 1974 they were found in Costa Rica. **Figure 13.8** shows house sparrows from different locations. Notice that the populations of house sparrows in different regions look subtly different, even though they are the same species. In general, northern populations are darker and larger than are southern populations. These are evidence of natural selection in action. As house sparrows spread out from New York City, they encountered different environments that favored the survival of house sparrows with slightly different traits.

A more thorough understanding of natural selection comes from studies of species of finches on the Galápagos Islands, collectively called Darwin's finches. One feature that distinguishes one species of Darwin's finches from another is the size and strength of the birds' beaks (**Figure 13.9**). The size and shape of a finch's beak is an inherited trait, and it is related to the type of food that each finch species eats. Species with larger, heavier beaks can break open and eat harder seeds, while finch species with smaller beaks are restricted to feeding on smaller, softer seeds. Of course, within any species, beak size and strength vary between individuals.

During most seasons there is enough rain to produce seeds of varying hardness on the Galápagos Islands, and so birds with a variety of beaks are able to survive. During droughts, however, seeds dry out and on average they become harder and more difficult to crack open. So birds with larger beaks are able to eat more seeds than are birds with smaller beaks. Of course, the birds that eat more are more likely to survive and to have offspring than birds with smaller beaks that get less to eat. The principles of natural selection lead to the hypothesis that after several seasons of drought the percentage of individuals in the population with slightly larger bills will increase.

Measurements of beak sizes from 1977 to 1984 provide direct support for this hypothesis. These were years of severe drought in the Galápagos Islands, and the dried-out seeds were hard and difficult to break open. **Figure 13.10** shows that at the end of two seasons of drought, the percentage of individuals in the population with larger beaks increased. And, since the size of the beak is determined by the alleles a bird carries, this change in beak size over generations reflects a change in the frequencies of alleles for larger or smaller bills within each population.

Eventually, the environment returned to its normal, moister state. Seeds became softer, and smaller seeds became more abundant. Accordingly, finches with smaller, weaker beaks could survive and reproduce in greater numbers. Think, though, about what would have happened if drought had become an environmental constant: seeds would have stayed hard, and each original finch species that survived would have evolved into a species with a slightly larger-sized beak.

Controlled Experiments Provide Support for Natural Selection

Laboratory experiments can be used to test hypotheses based on the principles of evolution. To do this kind of experiment, you need an organism with a rapid life cycle—one that breeds and matures quickly so you can observe many generations in a short time. The fruit fly *Drosophila melanogaster* is a good choice. Hundreds of fruit

Figure 13.8
Geographic Changes in House Sparrows.
The geographic variation in size and color of house sparrows in North America is evidence for natural selection. Large size is an adaptation to cold; pale color is an adaptation to match paler desert soil and rocks.

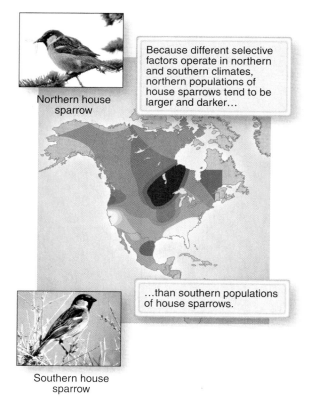

Northern house sparrow

Because different selective factors operate in northern and southern climates, northern populations of house sparrows tend to be larger and darker…

…than southern populations of house sparrows.

Southern house sparrow

flies can be housed in a single small bottle, and only two weeks after a female lays fertilized eggs, there will be a new generation of mature fruit flies.

One laboratory study of natural selection focused on the ability of *D. melanogaster* to metabolize alcohol. Alcohol is toxic to flies and people alike, and all animal species have enzymes that break down alcohol into less toxic substances. One such enzyme is *alcohol dehydrogenase*. Fruit fly populations have at least two forms of alleles for alcohol dehydrogenase, and one produces a more effective enzyme than the other. The ability of an individual fly to metabolize alcohol depends on which alleles it carries.

In one study a small amount of alcohol was added to the food of one population of flies, while another population received food that had no alcohol in it. As each generation hatched and matured, researchers randomly chose individuals from each group and allowed them to mate and produce the next generation. Every few generations the researchers tested some of the flies for the presence of the different forms of alcohol dehydrogenase. After 50 generations the population fed normal food showed no change in the frequency of the two forms of alcohol dehydrogenase (**Figure 13.11**). In contrast, the population that had alcohol in their food showed a steady increase in the presence of the enzyme that is more effective at detoxifying alcohol. What can you conclude from this experiment? When the environment changed to include food laced with alcohol, the individuals that had the more effective enzyme for detoxifying alcohol were more likely to survive. These survivors were available to be chosen when the

1. Medium ground finch
   Geospiza fortis

2. Large ground finch
   G. magnirostris

3. Sharp-beaked ground finch
   G. difficilis

4. Small ground finch
   G. fuliginosa

Large cactus finch
5. *G. omnirostris*

6. Cactus finch
   G. scandens

Figure 13.9   **A Comparison of Beaks of Some Species of Darwin's Finches.** Different species of Darwin's finches have beaks that vary in size and shape. Beaks are adaptations to feeding on specific kinds of foods. Larger beaks can crack harder seeds; smaller beaks can manipulate and open smaller, softer seeds.

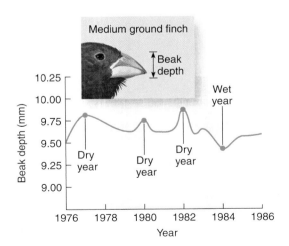

Figure 13.10   **During Dry Years Galápagos Medium Ground Finches Have Larger Beaks.** In dry years seeds are larger, harder, and fewer. In wet years seeds are smaller, softer, and more abundant. In dry years medium ground finches with slightly larger beaks can feed more abundantly and survive.

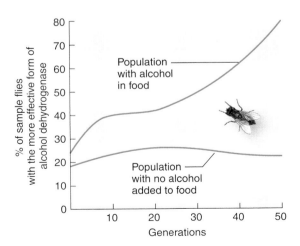

Figure 13.11   **Natural Selection in a Laboratory Experiment.** One laboratory population of fruit flies received food laced with alcohol. Another population received food with no alcohol in it. After 50 fruit fly generations, the percentage of individuals expressing the more effective form of alcohol dehydrogenase increased dramatically. In the control population, the percentage of flies with the more effective form of alcohol dehydrogenase was relatively unchanged after 50 generations.

researchers randomly gathered flies that would produce the next generation. This simple laboratory experiment is just one that demonstrates that the genetic makeup of a population changes in response to changes in the environment.

Figure 13.12 **Artificial Selection in Mustard Plants.** By selectively breeding plants with particular traits, all of these domestic vegetables have been bred from wild mustard plants.

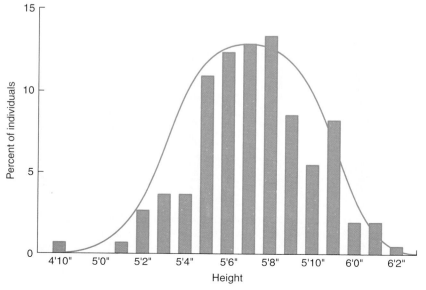

Figure 13.13 **Distribution of Height Among Men.** There were only 200 men in this sample. A larger sample might have more individuals at either end of the curve.

Artificial Selection Demonstrates the Effects of Selective Breeding When Darwin faced the problem of explaining how new species emerge, he had insights that were grounded in his experiences with English country life. Darwin knew that farmers could change the features of their domestic animals and crop plants by carefully controlling breeding. Controlled breeding that produces new strains of domestic organisms is called **artificial selection.** For example, dogs have been bred to produce canines as different as Chihuahuas and Neapolitan mastiffs; cats have been bred to produce the furless sphynx as well as fluffy Persians. As a result of selective breeding, every season brings new varieties of flowers to garden centers and different varieties of fruits and vegetables to supermarkets. Selective breeding was no less amazing in Darwin's time and in it he saw evidence that whether artificial or natural, selection can dramatically change the average traits of a population. **Figure 13.12** shows the results of one of the best modern examples of artificial selection. All of the vegetables illustrated here originated from the selective breeding of wild mustard plants.

Three Types of Natural Selection Have Been Identified

Natural selection can produce three different outcomes. First, natural selection could push the average value of a trait in one direction. Second, natural selection could keep the average value of a trait the same but eliminate extreme values of the trait. And finally, natural selection could produce two different groups within a population, each with its own average value of a trait. Before considering examples of each of these, it will help to consider what a graph of the values of a trait in a population would look like.

Figure 13.13 shows the distribution of height among men. The average value of height is about 5 feet 6 inches (1.6 meters), but some men are well under 5 feet tall and others are well over 6 feet tall. A graph shaped like this, in which most values cluster around the average and fewer values are at the extreme, is called a *normal distribution*. The effects of the different types of natural selection can be evaluated by changes in a normal distribution that take place over generations.

First, consider what happens if natural selection changes the average value of a trait. This type of natural selection is called **directional selection.** One striking example is happening right now in species of ocean fishes caught by fishing fleets. For many years humans have taken the larger fishes and have thrown smaller individuals back into the sea. As a result, the average size of the fishes that are caught

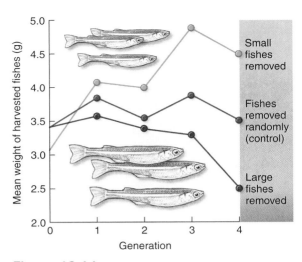

Figure 13.14 **An Experiment to Determine the Effect of Fishing on Size of Fishes.** Small fishes were removed from one population of captive fishes. Large fishes were removed from a second population. Fishes were randomly removed from the control population. After four generations of removing fishes, the average weight of fishes in the large and small populations had changed, while the control group was unaffected. This experiment shows that fishing practices can change the inherited traits of fish populations.

has been decreasing. Some studies suggest that these fishing practices are changing the genetic traits of populations of fishes (**Figure 13.14**). Researchers maintained three tanks of fishes over four generations. Each generation 90% of the fishes in each tank were removed but the fishes that were removed differed. In one tank a random 90% was removed; in the second tank the largest 90% were removed; in the third tank the smallest 90% were removed. After four generations of this practice, the fishes in the first tank grew to be about the same size as they had been in the first generation. The fishes in the second tank grew to be much smaller than they had been in the first generation. The fishes in the third tank grew to be much larger than they had been in the first generation. This demonstrates that by removing fishes with the alleles that allow growth to a large size, modern fishing techniques are unintentionally pushing the populations of ocean fishes toward a smaller size.

Stabilizing selection occurs when individuals with extreme values of a trait do not survive to breed and so the population clusters about the average values for these traits. One striking example is birth weight in humans (**Figure 13.15**). Babies that weigh less than 5 pounds or over 9 pounds are much more likely to die during birth than are babies that weigh between these values. This differential mortality leads to a narrow distribution of live birth weights. Since the late twentieth century medical advances have allowed higher rates of survival in

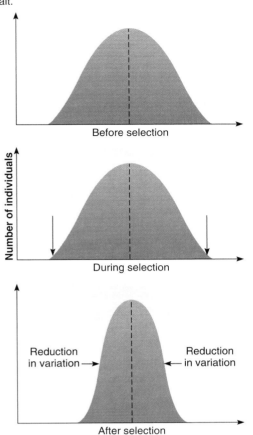

(a) Stabilizing selection reduces the amount of variation in a trait.

Before selection

During selection

Reduction in variation →   ← Reduction in variation

After selection

Value of a trait

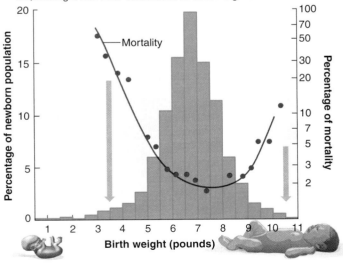

(b) Very small and very large babies are the most likely to die, leaving a narrower distribution of birth weights.

Mortality

Birth weight (pounds)

Figure 13.15 **Stabilizing Selection Is Selection for the Mean.**

both small and large babies. If this trend continues, in 25 or 50 years the birth weight distribution might look broader, which will mean that modern culture is changing human evolution.

Finally, **disruptive selection** pushes a population away from the average value of a trait, leading to a distribution that has two average values,

artificial selection the production of a new strain from an existing species by controlling which individuals mate and produce offspring

directional selection individuals at one extreme or another on a trait do not survive and breed, so the population average moves in one direction or another

stabilizing selection individuals with extreme values of a trait do not successfully breed, and so the population values tend to cluster around the average value

disruptive selection individuals with values of a trait around the mean do not survive and breed, so the population is pushed into two groups, each with extreme values of the trait or a *bimodal distribution*. **Figure 13.16** shows an intriguing example among black-bellied seedcracker birds in West Africa. The adults of this population have beaks that are either long or short but usually not of intermediate size. The reason seems to be that only two sizes of seeds are available—and birds tend to specialize, taking large seeds or small seeds. Birds with intermediate-sized beaks tend to die before they reach adulthood. If this pattern continues into future generations, it is possible that the small-beaked and large-beaked black-bellied seedcrackers will split into two different species.

Sexual Selection Is a Special Case of Natural Selection

Among animal species selection by the opposite sex is one aspect of the environment that determines which individuals mate and have offspring. Darwin was one of the first to suggest that mate choice could drive evolution. Females are especially choosy about which males they will mate with, and these choices can have a large impact on the traits of future generations, especially the males of the population. The peacock's feathers, the bright plumage of many male birds, and the bright colors of many male fishes are traits that may be driven by female mate choice (**Figure 13.17**). Males without these traits are less likely to find mates and do not breed. Why do females prefer showy males? The best hypothesis is that showy males are on average stronger and healthier than are drab or small males. Therefore, females who mate with showy males have stronger, healthier, more successful offspring.

Other Factors Can Change Allele Frequencies Across Generations

Natural selection is not the only process that can change allele frequencies, and so change traits across generations. Random factors also can change populations. *Genetic drift* is the general term for random changes in allele frequencies. To understand genetic drift, you must return to the idea of probabilities. Any population contains many alleles for each gene and many possible allele combinations. When the individuals of one generation mate and produce offspring, there is some probability that a particular allele will, or will not, appear in the next generation, just by chance. Because of the random chances involved in the formation of gametes and in fertilization, not every egg and sperm will end up contributing to a zygote and developing into a new individual. If an allele is lost in a population for random reasons, genetic drift has occurred. Or if an allele becomes more frequent in a population for random reasons, genetic drift has occurred. Genetic drift does not act to match a population to its environment. The changes in allele frequencies that occur as a result of genetic drift may be helpful to the survival of the popula-

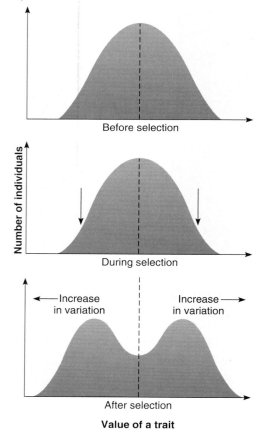

(a) Disruptive selection increases the amount of variation in a trait.

Before selection

During selection

Increase in variation ← ┆ → Increase in variation

After selection

Value of a trait

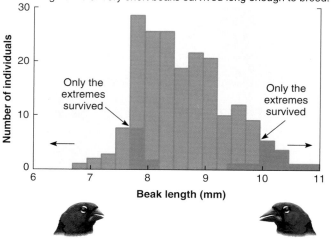

(b) For example, only juvenile black-bellied seedcrackers that had very long beaks or very short beaks survived long enough to breed.

Only the extremes survived

Only the extremes survived

Beak length (mm)

Figure 13.16 **Disruptive Selection.**

Figure 13.17 **Sexual Selection.** The gorgeous, but impractical tail feathers of a male peacock are the result of sexual selection.

tion, or they may not be helpful to the survival of a population, but they are not the result of selection.

One way that genetic drift can contribute to the evolution of a new population is the **founder effect,** which occurs if a small group of individuals from a larger population gets isolated and begins a new population. Any small group is not likely to carry the full range of alleles of the parent population, in the same proportions. If the small group remains isolated, over generations their allele frequencies will drift away from those of the parent group. Many examples of the founder effect have been described in humans and other species. The residents of a small South Pacific island called Pingelap know about the founder effect firsthand. Around 1775 nearly all of the several thousand residents of the island were killed by a typhoon and famine. Only about 20 people remained, and one particular man fathered many of the children who repopulated the island. This man had a rare form of inherited color blindness. In other human populations only about 0.7% of people have this disorder, while on Pingelap about 2.5% have the disorder. This is a striking example of a change in allele frequency due to the founder effect.

QUICK CHECK

1. Summarize one example of natural selection seen in nature and one example documented in laboratory studies.

13.5 Scientific Evidence Supports and Expands Darwin's Conclusions

The Origin of Species and Darwin's other writings provided a great deal of evidence to support Darwin's ideas about evolution. Ongoing research has provided even more support for Darwin's ideas and has expanded Darwin's original conclusions. Here are just a few examples.

Fossils Show That Life Has Changed Over Time

Fossils were important to Darwin's theory, and they remain important evidence in support of evolution today. One of the most important aspects of this evidence is the age of fossils. Fossils can be dated in a variety of ways, but radioactive dating is one method commonly used. All rocks on Earth, including fossils, contain radioactive atoms that give off radiation energy at a known rate. The amount of radioactive material in fossils can be used to determine the age of the fossils. Using a variety of dating methods, the oldest fossils are known to be 3.5 billion years old, and the oldest rocks on Earth formed about 4.0 billion years ago. The Earth originated about 4.5 billion years ago, so this means that life probably emerged on Earth within several hundred million years of its formation (see Figure 1.3).

Fossils have been studied for centuries, and new fossils are dug up every day. It would be ideal if paleontologists could find perfectly preserved fossils for every time period and place on Earth, but fossils form more readily under some conditions than others. Hard body parts such as wood, bones, shells, or teeth are more likely to be fossilized than are softer tissues like skin and muscles. Although the fossil record never will provide a complete history of life, the fossil record will continue to provide important information about how life has evolved. Each new fossil sheds more light on the history of life.

Fossils from different time periods in Earth's history show that the organisms alive today are different from the organisms that lived in previous times. On the other hand, many fossils look like relatives of modern forms—and so support the hypothesis that all life is genetically related. Even more astounding, the fossil record shows that some species remain nearly unchanged from when they first evolved. The fishes known as coelocanths, relatives of pine trees called ginkgos, and cockroaches are species that often are termed *living fossils*

founder effect the restricted gene pool of a small founding population influences the array of genes in future populations

Figure 13.18 **Living Fossils.** (**a**) Ginkgo. (**b**) Fossil ginkgo.

(a)

(b)

Figure 13.19 **Trends in the Evolution of Horses.**

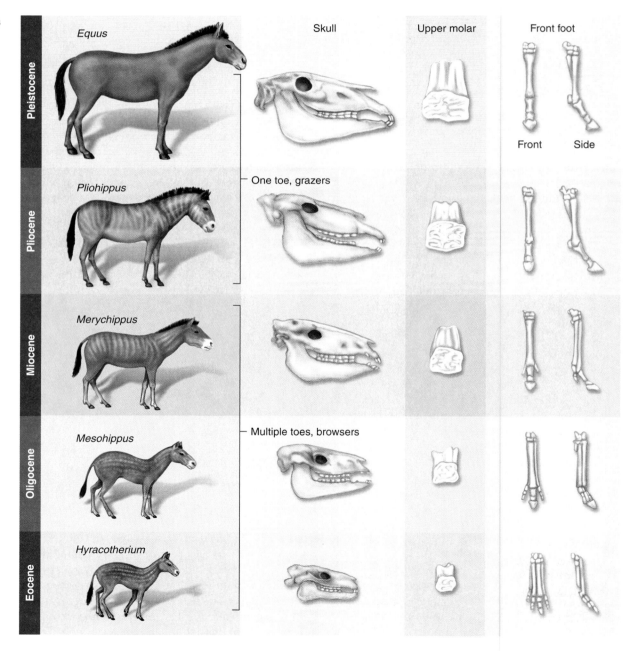

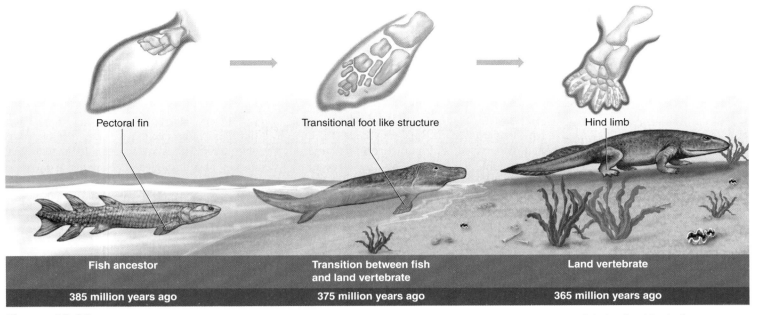

Pectoral fin

Transitional foot like structure

Hind limb

Fish ancestor

Transition between fish
and land vertebrate

Land vertebrate

385 million years ago

375 million years ago

365 million years ago

Figure 13.20   **The Evolution of Land Vertebrates from a Fish-Like Ancestor.** Fossils of a large fish that lived in shallow water show the evolutionary path from fishes to land vertebrates.

because they are nearly identical to their ancestors (**Figure 13.18**).

The fossil history of horses is an especially well-understood example of evolutionary change (**Figure 13.19**). The earliest recognizable ancestors of modern horses had feet with four splayed-out toes that made walking in damp forest environments easier. Although this dog-sized creature has only a few features of a modern horse, the fossil record shows many intermediate species that eventually led to the modern horse. The evolution of more than one feature of horse anatomy can be seen in the fossil record, but the evolution of the horse's hoof is particularly easy to follow. The basic mammalian pattern is to have five digits (fingers or toes) on each forelimb or hind limb. In contrast, modern horses have only a single large toe on each limb. Over many generations the bones of the central toes of the ancestors of modern horses became larger, while the bones of the other toes were reduced in size. Eventually, the center toe became the hoof of a modern horse.

As the toes and legs of horses changed, so did their teeth and skulls. These became larger and stronger as the various species of small early horse ancestors evolved into larger animals. Changes in toes, teeth, skulls, and size are correlated with a shift from life in a forest environment to life on an open plains environment. As the environment of horses changed, their diets changed too. With huge teeth, large jaws, and strong jaw muscles, modern horses are adapted to grazing on tough grasses. In contrast, the smaller teeth and more delicate jaws of earlier horse ancestors are best adapted to browsing on softer, leafy plants.

The Fossil Record Supports Darwin's Idea of Common Descent

When you look at the fossil record, the idea of common descent is difficult to escape. The existence of **intermediate forms** in the fossil record supports the conclusion that different kinds of organisms share a common ancestor. Intermediate forms are individuals that have traits of more than one species or group of organisms. Intermediate forms are exactly what you would expect to find if one species arose by modifications of an earlier species. Given the chancy nature of fossil formation, and the long odds against a fossil being unearthed for study, it is surprising that so many intermediate forms have been found. Scattered among the fossil record are numerous examples of species that can be traced through intermediate forms to a common ancestor. Let's look at one specific example—the evolution of modern mammals from early reptiles, a transition that took 150 million years and is particularly well documented in the fossil record.

Modern amphibians, reptiles, birds, and mammals are land vertebrates, and so they share as a common ancestor the first animals that moved from water to land. Land vertebrates are *tetrapods*, meaning they have four limbs, even though some species, such as snakes, have subsequently lost those limbs. Fossil evidence shows that a fish with limbs first moved from water to land about 375 to 400 million years ago. This early tetrapod ancestor was clearly a fish, with gills and scales, but had limbs that foreshadow the limbs of modern tetrapods (**Figure 13.20**). The fossils of this creature provide strong support that modern land vertebrates evolved from a fish ancestor.

intermediate form an individual, fossil or living, that has traits usually associated with more than one kind of organism

pelycosaur an extinct reptile from 300 million years ago that had the beginnings of mammalian skull and jaw characteristics

therapsid an extinct mammal-like reptile from about 300 to 275 million years ago that was probably ancestral to modern mammals

Although modern reptiles and mammals seem much different, they evolved from a common reptilelike ancestor that lived about 300 to 350 million years ago. **Figure 13.21** shows some of the evidence for this conclusion based on teeth and skulls. Teeth and skulls fossilize well, and many fossil teeth and skulls are available for study. The fossil record shows that before about 300 million years ago, there were no mammals—but there were many kinds of reptiles that are clearly ancestors of modern reptiles. Mammals do not appear fully formed in the fossil record, but rather mammalian traits appear gradually in fossil reptiles and accumulate until fossil animals that are clearly modern mammals appear around 70 million years ago.

The first sign of mammal-like skull features appears in a group of reptiles called **pelycosaurs** that lived about 300 million years ago (Figure 13.21). *Dimetrodon* and other pelycosaurs had a skull feature that links them to modern mammals: a tunnel on either side of the skull that lets longer, stronger jaw muscles pass through and attach to the top of the skull. Because no living reptiles have a similar tunnel, while mammals do, scientists place pelycosaurs as an early step on the evolutionary road that led from reptiles to mammals.

Most of the skeletal features of mammals evolved among **therapsids,** which often are called *mammal-like reptiles.* Therapsid skulls and jawbones are intermediate between those of modern reptiles and modern mammals. By 250 million years ago a group of mammal-like reptiles known

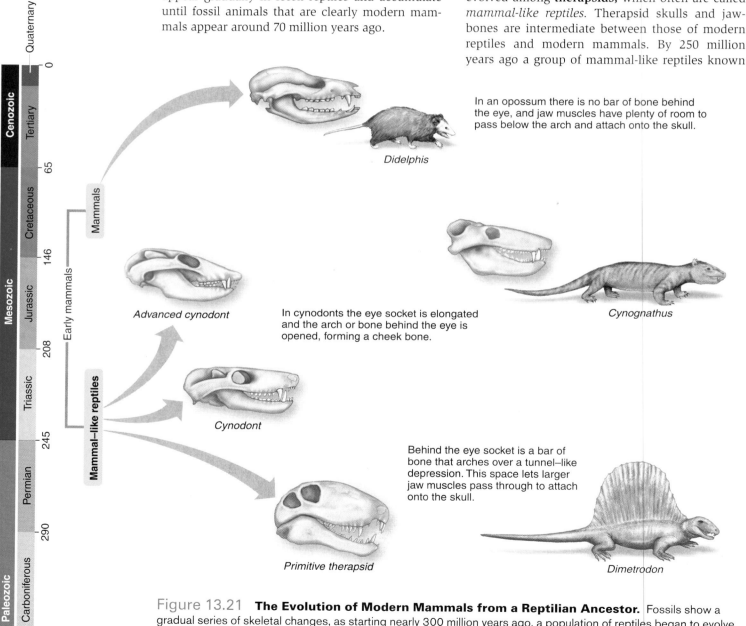

In an opossum there is no bar of bone behind the eye, and jaw muscles have plenty of room to pass below the arch and attach onto the skull.

Didelphis

In cynodonts the eye socket is elongated and the arch or bone behind the eye is opened, forming a cheek bone.

Cynognathus

Advanced cynodont

Cynodont

Behind the eye socket is a bar of bone that arches over a tunnel–like depression. This space lets larger jaw muscles pass through to attach onto the skull.

Primitive therapsid

Dimetrodon

Figure 13.21 **The Evolution of Modern Mammals from a Reptilian Ancestor.** Fossils show a gradual series of skeletal changes, as starting nearly 300 million years ago, a population of reptiles began to evolve into mammals.

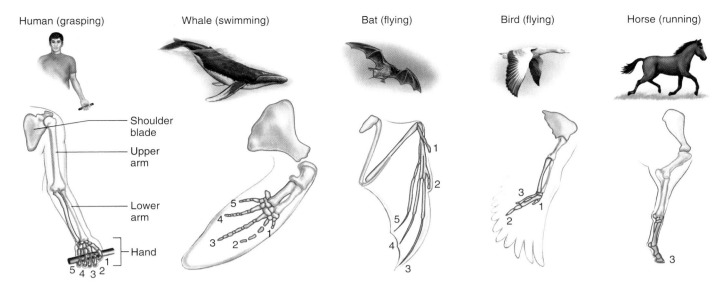

Figure 13.22 **Forelimbs of all Vertebrates Contain Similar Bones.**

as the **cynodonts** appears in the fossil record (Figure 13.21). The cynodonts shared even more characteristics with mammals such as teeth with different shapes, similar patterns of vertebrae and ribs, and even skeletal structures that suggest that their senses of hearing and smelling were like those of mammals. Although it is not shown here, there are fossils with increasingly mammalian features right up to the emergence of many of the modern groups of mammals about 70 million years ago. Through common descent, an animal that must have been something like a cynodont is the common ancestor of all modern mammals.

Anatomical Structures Support the Idea of Common Descent

The idea of common descent has one particularly important implication. If different species arise from a common ancestor, then new traits or new characteristics should be modifications of traits found in the ancestor. Examination of the bones of living and extinct vertebrate animals can test this hypothesis. If vertebrates are distinct, unrelated species, then the structures and functions of bones should be different in different vertebrate groups. In contrast, if vertebrates have evolved through common descent, then their bones should be modifications of a common basic theme. Darwin and later biologists understood this implication, so they carefully examined and compared the bones of many vertebrates. The results of these studies clearly support the idea of common descent. For example, the forelimbs of all vertebrates are made of the same set of bones (**Figure 13.22**). The forelegs of a horse, the flippers of a whale, the wings of a bat, and the arms and hands of a human all have the same kinds of bones, each modified to produce the final limb

structure. Totally new kinds of bones do not just spontaneously appear in an organism. Instead the vertebrate skeleton is modified to produce the bones of differently adapted vertebrate species.

One example is the thumb of the panda. Technically, only primates—animals like monkeys, chimpanzees, gorillas, and humans—have a true *opposable thumb,* one that can be moved to touch the pads of the other fingers. Pandas, however, have something that *looks* a bit like a thumb and *works* like a thumb too (**Figure 13.23**). Pandas use their "thumbs" to grab bamboo branches and maneuver them into their mouths. Pandas can manipulate objects just as primates do, but a panda's "thumb" is *not* formed from the same bones as the thumb of a primate. Instead the panda's "thumb" is a modified wrist bone. Pandas have five true digits on each paw, and each front paw has an extra, false digit, or "thumb."

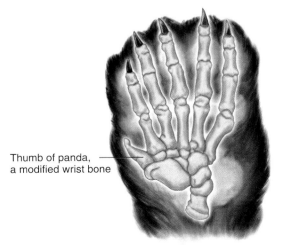

Thumb of panda, a modified wrist bone

Figure 13.23 **Bones in the Paw of a Panda.**
A panda's "thumb" is made of a modified wrist bone.

cynodont an extinct early mammal-like species from about 250 million years ago that had skeletal characteristics that closely resemble those of modern mammals

Common descent also is supported by studies of anatomical structures of other kinds of organisms. For instance, different groups of insects have vastly different mouthparts and antennae that are derived from the common mouthparts and antennae of their ancestors. The spines on cacti, the "flowers" of dogwood plants, and the jaws of insect-capturing Venus flytraps are all leaves that have been modified through evolution by natural selection. The evidence for common descent is seen again and again in both living and fossil species.

Modern DNA Studies Support Darwin's Principle That All Life Is Related

Everything that biologists have learned about DNA and proteins supports Darwin's idea that all organisms are related and that one species evolves from another. One important lesson from these studies is that gene mutations are an important source of the variation that natural selection works on. Gene mutations produce individuals that are different from their parents or their peers. Studies of gene mutations in DNA, and the proteins that come from these mutated genes, provide strong evidence for the process of evolution.

One of the first molecules to be studied in the context of evolution was cytochrome *c*, one of the proteins found in mitochondria. Cytochrome *c* is involved in the production of ATP during oxidative phosphorylation (see Section 6.3). Scientists have examined the amino acid sequence of cytochrome *c* in different species, including humans, and a striking pattern has emerged (**Figure 13.24**). Species that seem to be closely related to humans based on other measures—such as body structure—have nearly identical amino acid sequences in cytochrome *c*,

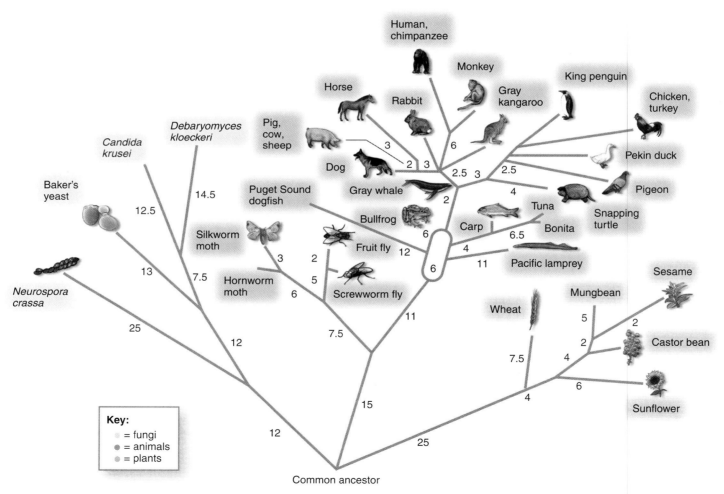

Figure 13.24 **An Evolutionary Tree Based on Similarities of Cytochrome *c* Proteins in Different Species.** When the sequences of amino acids in cytochrome *c* molecules of different species are analyzed and arranged in a graph, closely related species cluster together. The numbers indicate number of amino acid differences.

while species that are more distantly related to humans by other measures have less similar amino acid sequences of cytochrome *c*. At the same time the overall structure and function of the cytochrome *c* molecule is similar in all species. In fact, if the cytochrome *c* gene in yeast cells is destroyed and human cytochrome *c* is inserted, the yeast cells can use human cytochrome *c* to help produce ATP.

Studies of DNA and proteins show that new traits or characteristics are modifications of older ones. This is the same conclusion reached by studies of anatomical features like limb bones, leaves, or insect mouthparts. The evolution of body plans in animals is one of the best-known examples of this biological trend. Section 9.7 introduced the idea that the head-to-tail axis of an organism is laid down during development as a result of the pattern of gene expression. One group of genes expressed early in animal development is the set of homeobox genes, called *Hox* genes for short. During development *Hox* genes are turned on in complex sequences, and they determine how each section of the body develops. Incredibly, *Hox* genes are similar in all of the animal species studied. *Hox* genes are even lined up on chromosomes in the same order in all animals (**Figure 13.25**). *Hox* genes also are expressed in the same order during development in all animals and act in a similar way to determine the head-to-tail axis. For example, the *Hox* genes that determine the head end of a fly embryo are essentially the same genes that determine the head end of a human embryo.

One final point about how DNA mutations lead to evolution is important. Sometimes a tiny change in DNA can lead to large and important changes in the traits of an organism. When this happens, natural selection quickly can allow the organisms with the new trait to have more offspring and so contribute to new populations or even new species. One example is an allele involved in the ability of humans to use language. Humans with a mutation in this allele have problems in the areas of the brain that govern the production of speech. This allele is a version of a gene found in all mammals, but in humans the DNA sequence is slightly different from that found in other mammal species. For example, the protein made from this gene has only one amino acid difference between humans and chimps. All the evidence suggests that this small amino acid difference contributes to human use of language.

Evolutionary Change Can Happen at Different Speeds

Another insight into the mechanisms of evolution concerns the speed with which new species can originate. Darwin concluded that evolutionary change always occurs slowly over many thousands of generations, and many studies have supported this conclusion. On the other hand, it now is clear that evolutionary change can happen more rapidly than previously thought. Of course "rapidly" is a relative term. Because life has existed for

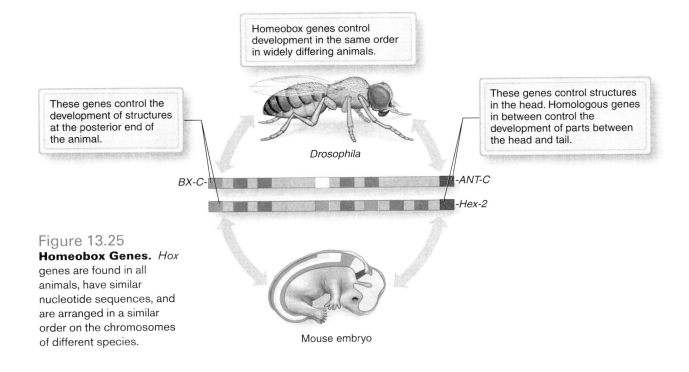

Figure 13.25
Homeobox Genes. *Hox* genes are found in all animals, have similar nucleotide sequences, and are arranged in a similar order on the chromosomes of different species.

Homeobox genes control development in the same order in widely differing animals.

These genes control the development of structures at the posterior end of the animal.

These genes control structures in the head. Homologous genes in between control the development of parts between the head and tail.

Drosophila

BX-C- -ANT-C
-Hex-2

Mouse embryo

nearly 4 billion years, rapidly can mean several hundred years or several million years.

The fossil record shows several examples where a group of organisms seemed to be stable for tens of millions of years, and then produced many species within just a few million years. Mammals are an example. The earliest mammal-like creatures emerged around 200 million years ago. They were small and *nocturnal,* meaning that they were active mostly at night. Over the past 60 million years many new species of mammals evolved with many new traits. But for the first 140 million years of their history, mammals remained small and nocturnal. During this time the dinosaurs diversified and were Earth's dominant animals. Then 65 million years ago some catastrophic event wiped out much of life. There is evidence that a meteor collided with Earth and caused planet-wide ecological disturbances and climate changes. The big dinosaurs went extinct fairly quickly. Many mammal species were wiped out, too, but in the absence of competing reptiles those that remained gave rise to many new and different groups of mammals. Twenty million years after the mass extinction of the dinosaurs, there were hoofed mammals as well as primates. There also were many unusual large mammals and many *diurnal* mammals that were active during the day. The age when mammals became Earth's dominant animals came about within a relatively short period of time—at least in geological time scales.

QUICK CHECK

1. Which of Darwin's ideas is illustrated by the studies of the similarities of cytochrome *c?*

13.6 Arguments Against Evolution Are Not Supported by Scientific Evidence

In the history of science few ideas have generated a more intense reaction than has evolution. Since the first publication of *The Origin of Species,* some people have refused to accept the principles of evolution. Most people accept the major advances in other realms of science—such as physics, cosmology, chemistry, medicine, cell biology, and even psychology—but some have difficulty accepting evolution and are vocal about it. In the United States some groups promote the idea that evolution is "just a theory"—and that there are other equally plausible explanations for how the diversity of life arose on Earth. Many of the proposed alternatives claim to be scientific, but a close look shows that they are not.

Recently, some groups have been arguing for a proposition called *Intelligent Design.* Proponents of Intelligent Design have tried to force high school teachers to present the idea as a scientific alternative to evolution. Scientists, on the other hand, argue that Intelligent Design is not a scientific hypothesis at all and cannot be tested by the methods of science.

The central idea of Intelligent Design is that many aspects of living organisms are too complex to be explained by natural processes, and therefore there must be some intelligent agent that designed these features. According to this argument the complex cellular machines that are the evolutionary foundation of biological diversity—such as a flagellum, or the ATP synthase molecule—are too complex to have evolved via the mechanisms of mutation and natural selection. And if science cannot explain these structures, then they must have been designed. The Intelligent Design argument often uses human-designed structures as analogies. The complex tools and machines that we use—cars, computers, power plants, for example—were obviously designed by an intelligent agent. Therefore life with all of its complexities must have been designed as well.

The scientific response to these assertions does *not* focus on whether there was a designer, or not. This is a matter of faith, not science. Rather, scientists point out that because you cannot test the hypothesis that there was a designer, it is not a scientific hypothesis. To say that the evolution of a cell is too complex for science to explain does not provide a test of the hypothesis that a designer was involved. Furthermore, as you have read in this book, the explosive pace of scientific discoveries about life is providing more and more solid evidence about how complex living structures did evolve. It is possible that before the end of the century scientists may be able to evolve complex cells in a laboratory. Even today, knowledge about how genes are expressed in different species during development is providing an explanation for how many specific complex biological structures probably evolved. None of this evidence says that there was *not* a designer, nor does it say there *was* a designer. The question of an Intelligent Designer cannot be answered by science.

QUICK CHECK

1. What are two primary assertions of Intelligent Design and why do scientists argue that Intelligent Design is not science?

13.7 An Added Dimension Explained
Are Birds Dinosaurs?

The idea that birds are a surviving lineage of dinosaurs is different from the long-held idea that birds and dinosaurs both split from some common reptilian ancestor millions of years ago. The idea that birds evolved from dinosaurs is not new. Evidence that birds actually are dinosaurs first was found back in the 1800s when the earliest discoverer of dinosaur tracks thought that they had been made by a flock of birds. In 1861 the link between dinosaurs and birds was strengthened by the discovery of a fossil called *Archaeopteryx*.

Archaeopteryx lived about 150 million years ago, when well-known dinosaurs such as *Tyrannosaurus* and *Brachiosaurus* were alive. Eight fossils of *Archaeopteryx* have been found, and some are nearly complete and highly detailed (**Figure 13.26**). *Archaeopteryx* shares many features with birds, but it is different from any modern birds (**Table 13.1**). *Archaeopteryx* had feathers, one of the prime characteristics used to define birds as a group, and *Archaeopteryx* also had the same general body form as modern birds. A closer look reveals many other features that set *Archaeopteryx* apart from birds. *Archaeopteryx* had a long tail, not a short, stubby tail like a bird; unlike most birds it had claws on its wings; and, unlike any modern bird, *Archaeopteryx* had a mouthful of teeth. Some early scientists thought that *Archaeopteryx* was the transition between birds and dinosaurs, but today scientists are fairly certain that *Archaeopteryx* was an early bird that retained many dinosaur features.

Despite many detailed studies of *Archaeopteryx*, the question of the relationship between birds and dinosaurs remained controversial until the late 1900s, when computer-based methods to determine genetic relationships became widely available. Using these modern methods, a clear link was demonstrated between birds and **theropod** dinosaurs, a group that includes *Tyrannosaurus rex*, the velociraptors, and other similar predatory species. Theropod dinosaurs were carnivores that walked or ran upright on their hind legs. You should not come away, however, thinking that *T. rex* was the ancestor of modern birds. *T. rex* was a theropod species that evolved late in the dinosaur lineage. Theropod dinosaurs have a long history and are closely related to the earliest dinosaurs that emerged 250 to 300 million years ago. Birds most likely evolved from an early theropod that was similar to *Velociraptor*. Let's explore the evidence that birds and dinosaurs are more similar to one another than either is to modern reptiles.

Perhaps the most obvious way that birds and theropods differ from modern reptiles involves the shape and use of their limbs. Most modern reptiles move on all fours, and their limbs are about the same length. Usually the limbs of reptiles sprawl out on either side of the body instead of being directly beneath the body. In contrast, theropods and birds are bipedal—they hop or walk on two legs. Their hind limbs are directly beneath the body's weight. Their front limbs are specialized to do other things. Many theropods grasped things with their forelimbs, while the forelimbs of birds are specialized for flight. Even the earliest dinosaurs had limbs that follow this pattern. The large pelvis and large lower vertebrae of theropods supported weight as theropods walked, hopped, or leapt upright on their hind legs. Modern reptiles rarely walk upright on their hind legs, but birds always do. Upright gait is one major feature that leads many biologists to propose that birds and theropod dinosaurs are more closely related than are birds and other kinds of modern reptiles.

theropod a predatory dinosaur that walked on two powerful hind legs, had small forelegs with sharp claws, and had a neck joint that allowed the head to swivel from side to side

Table 13.1 Features of *Archaeopteryx*	
Features similar to modern birds	**Features different from modern birds**
Feathers	Long bony tail
Overall skeletal structure	Full set of teeth
Presence of a breastbone	Breastbone is flat instead of having a deep keel
Short fingers	Claws on wings
Ring of bone supports the eye	

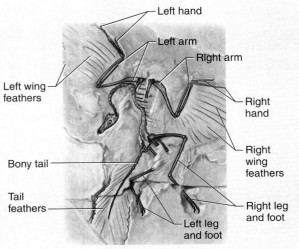 

Left hand
Left arm
Right arm
Left wing feathers
Right hand
Bony tail
Right wing feathers
Tail feathers
Left leg and foot
Right leg and foot

Figure 13.26 ***Archaeopteryx* Fossil.** This early bird had features characteristic of both birds and dinosaurs, such as teeth, claws on its forelimbs, and a long, bony tail.

—Continued next page

Continued—

Sharing just one trait probably isn't enough to convince you of a close relationship between dinosaurs and birds, so you should not be surprised that other traits also support the idea that birds evolved from theropod dinosaurs. Dinosaurs and birds have many similar skeletal traits, such as long, flexible necks and a flexible joint where the head meets the vertebral column. *T. rex* and his theropod ancestors could swivel their heads atop their long, flexible necks, allowing them to quickly survey the landscape. Their heads could swivel because of a special joint between the head and neck, which birds also have. Neither modern nor ancestral reptiles have this joint. Some of the later theropod dinosaurs also had a few teeth in a beaklike structure. If you examine other dinosaurs and other reptile groups, you will see some of these characteristics, but only the theropod dinosaurs have a close resemblance to modern birds.

Feathers on dinosaurs also support the idea that birds evolved from dinosaurs. Fossils of feathered dinosaurs have been found, and paleontologists think that even *T. rex* probably had a covering of downy feathers. Feathers may have evolved to conserve body heat, which is important for a warm-blooded animal. For many scientists the fact that many dinosaur species had feathers clinches the idea that birds evolved from dinosaurs.

Of course, flight is the feature that you most associate with birds. Not all birds can fly, but all birds have adaptations that make flight possible. These include modifications of the bones and muscles of the chest, shoulders, forelimbs, and hind limbs that make flight easier. Birds also have a number of adaptations that decrease body weight. These include lack of teeth; air sacs within their bodies; lightweight, hollow bones; and fewer internal organs. And, of course, birds have feathers that give them lift. Among all known reptiles, theropods share the greatest number of these features. But did dinosaurs fly? Several years ago most scientists would have said no, but today fossils suggest that some dinosaurs—or species that were intermediate between dinosaurs and birds—did indeed fly. One striking fossil is of an early species that has what seem to be two sets of wings (**Figure 13.27**).

Figure 13.27 **A Bird-Like Dinosaur.** A reconstruction of *Microraptor gui*, an early dinosaur with flight feathers on fore- and hind legs and a long, feathered tail. It may have glided, much like flying squirrels do today.

(a)

Figure 13.28 **Birds Evolved from Early Dinosaurs.** (**a**) Reconstruction of a dinosaur that was most similar to a modern flightless bird. (**b**) An evolutionary tree showing the relationships between modern birds and their dinosaur ancestors and cousins.

Recent fossil finds bring together all of these conclusions in a way that words cannot. **Figure 13.28a** shows a reconstruction of a dinosaur that was similar to modern birds. This creature lived about 130 million years ago and had many features of modern birds. It lived years after *Archaeopteryx*, though, and so this feathered dinosaur is not the ancestor of modern birds. Figure 13.28b summarizes the current understanding of the evolutionary relationships between dinosaurs, birds, and modern reptiles.

There is one last intriguing postscript to the dinosaur story. A few inspired scientists have wondered what *might* have happened to the dinosaurs if that meteor had not wiped them out 65 million years ago. Because birds were already well on their evolutionary path by this time, it is likely that birds would have continued to evolve into the modern species that we know

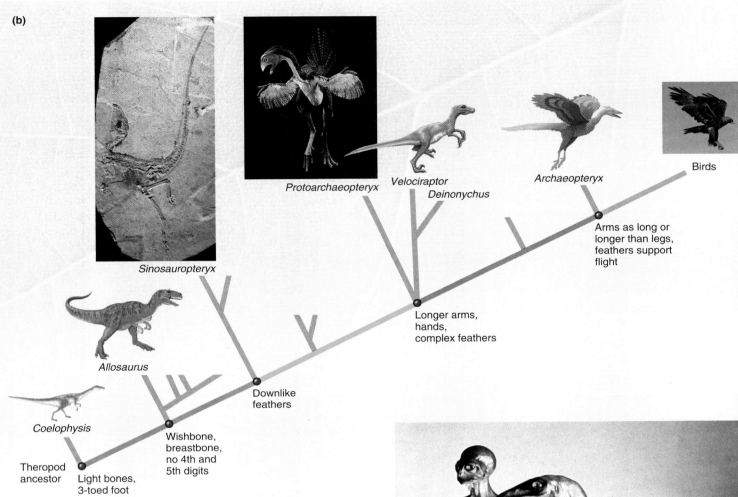

(b)

Sinosauropteryx

Protoarchaeopteryx

Velociraptor
Deinonychus

Archaeopteryx

Birds

Arms as long or
longer than legs,
feathers support
flight

Longer arms,
hands,
complex feathers

Allosaurus

Downlike
feathers

Coelophysis

Wishbone,
breastbone,
no 4th and
5th digits

Theropod
ancestor

Light bones,
3-toed foot

today. It is also likely that if the dinosaurs had not gone extinct, mammals would not have evolved into so many species. This means that humans might not have evolved. At least one researcher has suggested, though, that about the time of their demise certain theropods were evolving some striking traits: these theropods had larger brains, frontally placed eyes, and three-fingered hands that could manipulate objects. Sound familiar? If mammals had not replaced dinosaurs as Earth's dominant animal group, perhaps it would be the descendants of dinosaurs who were building civilizations, writing poetry, and learning about themselves and their world through science (**Figure 13.29**).

QUICK CHECK

1. To what kind of dinosaurs are birds thought to be most closely related? What is the basis of this relationship?

Figure 13.29 **Dinasauroid and *Stenonychosaurus* by Ron Seguin.**

Now You Can **Understand**

Antibiotic Resistance

Have you visited a doctor recently for a cold that was making you miserable? In spite of your misery, your doctor may have been unwilling to prescribe antibiotics. This is a common experience in medicine today, much more common than it was 20 years ago. You might find it frustrating, but there are good reasons why doctors now are more reluctant to prescribe antibiotics. First of all, when you get sick with a cold it is often caused by a virus, and an antibiotic drug will not kill viruses. Furthermore, in the long run the doctor could do more harm than good by prescribing antibiotics to treat an infection that you would have gotten over on your own. Why? The reason is **antibiotic resistance.** This refers to a situation in which an antibiotic that once was effective at killing a particular kind of bacteria no longer works. Antibiotic resistance is a result of evolution by natural selection. Can you see how?

If you are sick from a bacterial infection, it means that a particular type of bacteria is growing within your body. These bacteria form a population and, like any other population, the individual bacterial cells differ from one another. One way that they differ is in their ability to be killed by an antibiotic. Some are easily killed by a small amount of antibiotic, while it will take a moderate or a large amount to kill others. Still other bacteria may be essentially immune to the drug. Most doses of antibiotic will kill the majority of bacteria, but all doses will not kill bacteria that have resistance to the drug. This means that every time the antibiotic is used, many of the susceptible bacterial cells are killed off, leaving the more resistant cells free to reproduce even more rapidly. If this continues, the result is a strain of antibiotic-resistant bacteria that can spread rapidly and make people sick.

Widespread antibiotic resistance can result from overuse as well as underuse of antibiotics. An example of antibiotic overuse was seen in Eastern Europe during the 1970s and 1980s, when antibiotics were freely available and heavily used. Resistant strains of bacteria became a serious health problem, and governments responded by restricting antibiotic use. This shift led to a decline in drug-resistant strains in the population. An example of the effects of antibiotic underuse is seen with tuberculosis. Antibiotics must be taken daily for six to nine months to kill a significant percentage of the tuberculosis bacteria in an infected person. Because many people are unable to follow this recommended treatment, lethal strains of drug-resistant tuberculosis bacteria have developed. You also may have contributed to the development of antibiotic resistance. Have you ever not taken the full dose of a prescription of antibiotics? Perhaps you started to feel better and then simply forgot to take the medicine. Perhaps the antibiotics upset your stomach or otherwise made you feel sick. The dosages of antibiotics are set to kill nearly all of the infectious popula-

antibiotic resistance the phenomenon of bacteria becoming immune to an antibiotic that once would have killed them

tion, including those individual bacterial cells that have some resistance. If you stop taking an antibiotic before the prescribed dosage is complete, you could contribute to the evolution of antibiotic-resistant strains of bacteria.

What Do **You** Think?

The Future of Human Evolution

You probably have heard discussions or comments about whether humans are still evolving. Discussions of future human evolution can take many forms. You might find groups discussing whether extrasensory perception or greater human intelligence will "evolve" in the future. After reading this chapter, you now should have a greater understanding of what evolution means and how it happens. The basic mechanism of evolution is natural selection. Regardless of what DNA mutations might arise in individual people, or what abilities individuals might have, humans will not change unless there is differential survival and reproductive success. The only way for the current human population to evolve is for environmental conditions to favor the survival of individuals with certain traits over the survival of individuals with other traits. Do you see any ways that the current or recent human environment would favor some traits over others? If you do have some ideas about this, can you come up with ways to test your ideas scientifically?

Of course, culture has an important impact on survival. Any time a social group—human or otherwise—intervenes to save an individual who might have died, natural selection has been altered. Humans excel at protecting and nurturing those they care about, even if they are sick or injured. Modern medical care has expanded this practice, so many individuals who once might have died an early death now live and may reproduce. Some people argue that these cultural influences work against natural selection.

Yet another thought about modern human evolution is the idea that the crowded environments of cities introduce natural selection by infectious diseases. Bubonic plague, tuberculosis, and other infectious diseases spread rapidly in cities. In the past decade science has learned that some people have genes that give them immunity to certain infectious diseases—and so these individuals have survived to produce offspring. These findings have led some writers to propose that modern city living exerts a form of natural selection.

If you put all these thoughts together, can you formulate any predictions for the future of human evolution? Is it likely that humans of the future will, through evolution, become smarter, kinder, or reach a "higher plane of existence"? Given the influence of culture, are there other directions that human evolution might take? Are there any environmental influences that might currently affect the course of human evolution? Or have humans stopped evolving? What kinds of concrete information or evidence would you need to guide or evaluate your predictions? What do *you* think about the future course of human evolution?

Chapter Review

CHAPTER SUMMARY

13.1 Evolution Is the Central Idea of Biology

Adaptations are best understood in the context of the genetic history of an organism. Evolutionary changes result from different genes being represented in different frequencies in a population.

adaptation 309   evolution 310

13.2 The Intellectual Climate of the 1800s Set the Stage for Charles Darwin's Ideas About Evolution

Traditional Western ideas of the 1800s assumed that Earth and its organisms had been created, and that both were unchanging. After Darwin finished at Cambridge University, he was invited to travel on H.M.S. *Beagle* as gentleman companion to the ship's captain. Charles Darwin and Alfred Russel Wallace are given joint credit for the idea of natural selection as the mechanism for evolution.

scala naturae 310

13.3 Darwin Discovered the Fundamental Principles of Evolution

Darwin spent nearly five years on the *Beagle* voyage, exploring the diversity of life on different continents. His study of fossils, and geological evidence convinced Darwin that Earth was far older than generally assumed. All of Darwin's observations on the *Beagle* voyage fueled his knowledge of the natural world. After his voyage on H.M.S. *Beagle,* Darwin immersed himself in gathering evidence to support the idea that species evolve through the mechanism of natural selection. Darwin's ideas can be summarized as life gradually evolves, all species have evolved from a single common ancestor, the number of species that have ever lived increases over time, and natural selection is the driving force behind evolution.

fossil 313   natural selection 314

13.4 Scientific Studies Document Natural Selection

Studies of natural populations have demonstrated the effects of natural selection. The responses of colors of populations of peppered moths to environmental changes introduced by humans, and changes in populations of house sparrows, are examples of natural selection in the wild. Laboratory experiments also demonstrate natural selection. Populations of flies fed alcohol in their food shifted to have a more effective form of alcohol dehydrogenase than did control populations. Selection can be stabilizing, directional, or disruptive. Genetic drift also can by chance change gene frequencies within populations. Genetic drift is especially prevalent in small populations, a phenomenon called the founder effect. Natural selection is a common phenomenon that produces changes in the traits of populations over generations.

artificial selection 320   founder effect 323
directional selection 320   stabilizing selection 321
disruptive selection 321

13.5 Scientific Evidence Supports and Expands Darwin's Conclusions

Fossils support Darwin's theory. There are many examples of transitional forms in the lineage from reptiles to mammals. Artificial selection is like natural selection, except that in artificial selection, humans choose the individuals that breed. In natural selection the environment determines which individuals survive, thrive, and breed. Anatomical comparisons and DNA studies support Darwin's principle that all life is related. Studies of cytochrome *c* in various organisms show correlations to the fossil record. Modern developments have expanded on Darwin's ideas. Evolutionary change happens at different speeds.

cynodont 327   pelycosaur 326
intermediate form 325   therapsid 326

13.6 Arguments Against Evolution Are Not Supported by Scientific Evidence

The ideas of Intelligent Design suggest that a designer created complex features of organisms and complex life processes. These ideas are unscientific because they cannot be tested.

13.7 An Added Dimension Explained: Are Birds Dinosaurs?

Skeletal similarities between birds and theropod dinosaurs seem to indicate that birds may be modern-day theropod dinosaurs. Additional support for the idea comes from the presence of feathers, and warm-bloodedness in both groups.

antibiotic resistance 334   theropod 331

REVIEW QUESTIONS

TRUE or FALSE. If a statement is false, rewrite the underlined portion to make it true.

1. No known fossils show precisely what prehistoric animals or plants looked like.

2. *Archaeopteryx* has both reptilian and bird characteristics.

3. Charles Darwin originated the idea of evolution of life.

4. Alfred Russel Wallace came up with the same ideas that Darwin did, but some 20 years later.

5. The purpose of the voyage of H.M.S. *Beagle* was to collect fossil specimens.

6. In natural selection environmental factors act on individual variation to favor the survival and reproduction of some individuals, while choosing against the survival and reproduction of others.

7. Darwin's finches live in Australia.

8. Earth is about 4.5 billion years old.

9. The fossil record shows that life remains the same through time.

10. Two trends in the evolution of horses are <u>an increase in size and the loss of toes</u>.

11. Fossils show that <u>mammals evolved from reptiles</u>.

12. Mammalian traits evolved through modification of a <u>general primitive bird design</u>.

13. Chihuahuas, broccoli, seedless watermelons, and super-sweet ears of corn are just a few examples of the results of <u>natural selection</u>.

14. House sparrows are <u>uniform in color and size</u> all across the United States and down into Central America.

15. Populations of *Drosophila melanogaster* that are fed food with alcohol in it develop <u>lower than normal</u> frequencies of the enzyme that detoxifies alcohol.

16. The evidence thus far seems to indicate that <u>birds may be living dinosaurs</u>.

17. A change in the nucleotide sequence of DNA is defined as a <u>new species</u>.

18. Evolution is <u>invariably a slow, gradual process that takes hundreds of millions of years</u>.

19. Natural selection is <u>invariably a slow, gradual process that takes hundreds of millions of years</u>.

20. <u>New species</u> can emerge from a few migrants who invade an isolated territory.

CONNECTING KEY CONCEPTS

1. How does the modern scientific understanding of the history of the Earth and life on it differ from the view held by most Europeans in the 1800s?

2. What is natural selection? What is some of the evidence that natural selection changes the traits of populations over generations?

3. In what way does knowledge about DNA support and expand Darwin's principles of evolution?

THINKING CRITICALLY

1. Teosinte is a plant with broad leaves, seeds borne on a long stalk, and a sturdy tall stem. Its seeds are edible but small, and its leaves are fibrous and tough. Imagine yourself a human living before modern civilization began, perhaps in South America 10,000 years ago. You have the insight that if you carefully select the seeds from certain plants, you could eventually have a garden with a variety of plants that could make your life easier. What aspects of the plants would you select as you collect seeds, and what would be the goals of your artificial selection program?

2. Could you breed something that looks like a dinosaur from modern domestic chickens?

For additional study tools, visit www.aris.mhhe.com.

All Life is Related

UNDERSTANDING BIOLOGICAL DIVERSITY

Setting Notions of Race on Their Ear

February 6, 2006

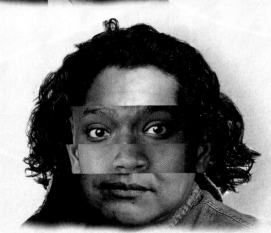

Is race a biological reality?

**Asian Populations Less Likely To Get Relief
From Chest Pain With Nitroglycerin**

January 27, 2006

Are Human Races a Valid Idea or Not?

Race is a pervasive concept in modern society. For instance, many application forms ask you to designate your race. The choices for race commonly include African American, Asian, Caucasian, Hispanic, Native American, and Other. Are you confident that you belong to one particular race? Or do you think that none of these categories quite describes you (**see chapter-opening photo**)?

Professor Samuel Richards presented these questions to his race and ethnic relations class at Penn State University. Most of the students were sure that they knew their race and most classified themselves as African American, Asian, Caucasian, Hispanic, or Native American. Only a few chose Other. Many students could identify their family background more precisely. For instance, some knew that their families had come from Ireland, Colombia, or India. This exercise began to get more interesting when Richards asked the students to take a simple DNA test that analyzes a person's genetic heritage. To the surprise of many, the tests indicated that nearly everyone in the class had far more mixed genetic heritage than they had known. For instance, some who had thought they were of northern European ancestry found that there was a significant Italian genetic component in

their heritage. The person from Colombia who thought she was of Spanish ancestry actually was a mix of European, Amerindian (the technical term for Native American), and African heritage. Finally, one African American who thought his ancestors all had come from West Africa learned that he had significant European ancestry.

DNA tests that claim to reveal genetic ancestry are becoming widely available. Now family disputes about Native American blood or mixed race heritage are being settled by DNA kits sold over the Internet. Newspaper articles frequently report new research that identifies a gene in a particular race that either increases or decreases a person's chances of developing a disease. From these developments you might conclude that the existence of genetically distinct human races is supported by scientific research. Actually, the hypothesis that race is a biological reality is highly controversial and many scientists have concluded that there is no biological reality to the commonly accepted concept of human races. How can scientists suggest that human races are not real when DNA studies seem to show that they are? You will return to these questions at the end of this chapter. First, though, it will help to understand how the basic unit of diversity—a species—is defined. It also will help if you have learned how species emerge and how the genetic relationships between species are studied. These are the topics of this chapter. ■

14.1 The Theory of Evolution Explains How Species Arise and How Species Are Related

In *The Origin of Species* Darwin proposed that all species are related, but it was a much bigger task to figure out the details of genetic relationships. Nearly 150 years later we are beginning to make progress in delineating the branching patterns of the tree of life. Darwin's hope of a more complete version of the tree of life now is emerging. In this chapter you will discover the progress that has been made in understanding the diversity of life.

An old Chinese saying declares that the first step toward understanding something is to name it. In biology the first step toward understanding the diversity of life is to name each kind of organism. The term *species* has long been used to define a basic kind of organism, and each species is given a scientific name that differentiates it from all other species. Every human culture has words to label different kinds of plants, birds, insects, spiders, and mammals. Although cultures differ in how observant they are of different organisms, the

species identified by different cultures are remarkably similar.

Although species are genuine biological categories, defining the boundary between one species and another can be difficult and tricky. When you try to identify the species of a particular individual, you often find that traits of closely related species overlap. Although other nineteenth-century naturalists glossed over these problems and tended to ignore individual differences, Darwin did not. He understood that some populations are so closely related that it is hard to define the boundary between one species and another. For instance, while an elephant is clearly different from a flea, it can be hard to decide how many different species of fleas there are or how many species of elephants there are (**Figure 14.1**). Because of this difficulty in deciding on the boundaries between species, scientists have used different ways to define a species.

The Greek philosopher Aristotle (384–322 BCE) was one of the first Western scholars to identify kinds of organisms. Using physical characteristics to distinguish one species from another, Aristotle noted details of external and internal anatomy of 520 species of Mediterranean animals and plants.

morphological species concept defines a species as a group of individuals that share distinctive anatomical characteristics

biological species concept defines a species as a group of individuals that naturally interbreed and produce fertile offspring

phylogenetic species concept a species is defined as the smallest group of organisms with similar features that comes from one common ancestor

He was one of the first to systematically employ the **morphological species concept.** *Morphology* is concerned with body shape and structure. The morphological species concept defines species based on common, distinctive physical characteristics such as details of size, shape, coloration, pattern, skeletal features, body organs, and body structures.

The morphological species concept was further developed by the eighteenth-century Swedish scholar Carl von Linné, also known by the Latinized version of his name, Carolus Linnaeus. His classification scheme also was based on morphology, but was more comprehensive than Aristotle's. For more than 200 years morphology continued to be the main basis for classifying organisms, but it did have limitations. Two organisms that look nearly identical can actually belong to different species (**Figure 14.2a**), while other organisms that look much different can actually belong to the same species (Figure 14.2b). Other ways to identify species have been developed, but the morphological species concept often is used as the primary method to identify an organism. For instance, you use the morphological species concept when you consult a field guide to identify the birds at your backyard feeder or use an identification key to identify trees by their leaves.

The morphological species concept was the only definition of a species in use until the early 1900s. By this time Mendel's work had been rediscovered and integrated with Darwin's understanding of evolution. In the 1940s Ernst Mayr reached the understanding that genes must be considered when defining a species. He formulated the most common definition of a species in use today. The **biological species concept** defines a species as a group of individuals that can naturally breed with each other, but do not naturally breed with individuals of other groups. This definition of a species focuses on the exchange of genes between different populations or groups. Think about this for a moment. If two populations such as white cats and calico cats can interbreed and have fertile offspring, then the offspring of these two populations share the genes of both populations. If two populations such as housecats and lions do not normally breed, then the genes of each population are isolated from one another.

In the 1970s a new approach was developed that considered the common ancestor of a group of individuals. In the **phylogenetic species concept** a species is defined as the smallest group of organisms

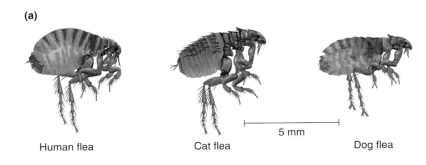

(a)

Human flea   Cat flea   Dog flea

5 mm

(b)

Figure 14.1 **How Many Fleas? How Many Elephants?** Although they look superficially similar, the differences between species of fleas (**a**) and species of elephants (**b**) become more obvious when you look more closely.

(a) Red-cheeked salamander and Imitator salamander are quite similar, but are different species.

(b) Male and female tiger swallowtails can look quite different, but are members of one species.

Figure 14.2 **Assigning Some Organisms to a Species Is Difficult.**
(**a**) Red-cheeked and Imitator salamanders look similar, but Red-cheeked is *Plethodon jordani* and Imitator is *Desmognathus ochrophaeus*. (**b**) These three butterflies are all *Papilio glaucus*. The middle butterfly is a male; the end butterflies are females.

with similar features that comes from one *common ancestor* (**Figure 14.3a**). This definition of a species relies on the evolutionary history of a group. The idea is that if many traits—morphological, genetic, behavioral, and many others—of many different individuals are analyzed and compared, then their genetic relationships will become apparent. This kind of analysis can determine whether two populations with similar traits have indeed evolved from a common ancestor. The phylogenetic species concept goes beyond defining a species and considers the evolutionary relationships of species. One dis-

tinctive tool used in applying the phylogenetic species concept is DNA analysis. Because DNA is the genetic material, the similarities and differences in DNA can provide information about phylogenetic relationships.

Figure 14.3b shows various groups of primates based on the phylogenetic species concept. You can see that humans are a single group—one species—that arose from a common ancestor. According to this species concept the different populations of humans are all the same species for two reasons: they share so many traits in common, *and* they share a common ancestor. Humans also share many traits with chimpanzees, but the two differ in enough traits to conclude that their family trees split about 7 million years ago. Since that time human and chimp ancestors have had different evolutionary histories. Modern humans emerged from a common ancestor about 200,000 years ago, while modern chimpanzees have their own common ancestor that lived several million years ago. Humans are more closely related to chimps than to gorillas, orangutans, and gibbons (Figure 14.3b). This kind of analysis can be carried out for groups of different sizes. It can be used to analyze relationships between a handful of populations or for samples that reflect all species.

These three different definitions of a species are used in different situations and under varying circumstances. They are not mutually exclusive. Rather they reflect different aspects of the processes that produce biological diversity. As more information becomes available about the traits of organisms—especially DNA and different alleles—the three definitions are converging. These three definitions reflect that in nature many processes lead one population of individuals to become genetically isolated and physically different from another population. To more fully understand what a species is, let's explore how a new species can form.

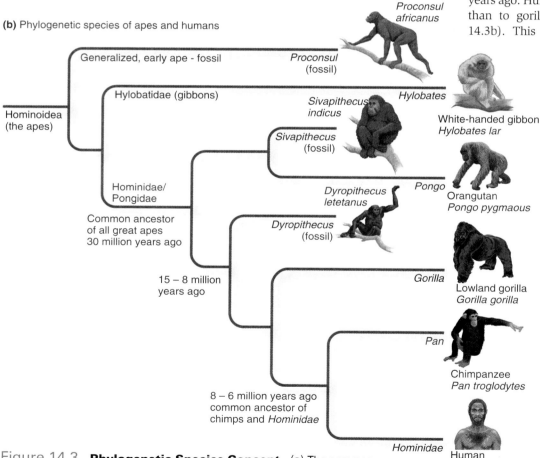

(a)

X is the common ancestor of Y & Z

X

All individuals of Species Y come from common ancestor Y

All individuals of Species Z come from common ancestor Z

Y — Species Y

Z — Species Z

(b) Phylogenetic species of apes and humans

Generalized, early ape - fossil

Proconsul (fossil)

Proconsul africanus

Hominoidea (the apes)

Hylobatidae (gibbons)

Hylobates

White-handed gibbon *Hylobates lar*

Sivapithecus indicus

Sivapithecus (fossil)

Hominidae/ Pongidae

Pongo

Orangutan *Pongo pygmaous*

Common ancestor of all great apes 30 million years ago

Dyropithecus letetanus

Dyropithecus (fossil)

15 – 8 million years ago

Gorilla

Lowland gorilla *Gorilla gorilla*

Pan

Chimpanzee *Pan troglodytes*

8 – 6 million years ago common ancestor of chimps and *Hominidae*

Hominidae

Human *Homo sapiens*

Figure 14.3 **Phylogenetic Species Concept.** (**a**) The common ancestor of a species, plus the species, are included in the phylogenetic species concept. (**b**) Phylogenetic species of apes and humans.

■ *Compare and contrast the definitions of species. Do you think one is better than the others? Why? Choose one definition and tell when it would be the best definition to use.*

A Variety of Mechanisms Separate One Species from Another

As researchers have worked to define species, they also have investigated the mechanisms that keep species separate. One central clusion is that two populations become two species when genes no longer flow from one population to the other. As two populations diverge and become two species, barriers evolve that prevent interbreeding. These barriers can be biochemical, cellular, physical, seasonal, behavioral, or geographic—biologists use the term **reproductive isolating mechanisms** for the processes or traits that prevent interbreeding. In some cases reproductive isolating mechanisms operate to prevent two kinds of individuals from mating. For example, two different species of fireflies do not interbreed because the males of each species make a particular pattern and color of flashes, and the females respond only to the flashes of their own species (**Figure 14.4**). This prevents the two species from mating. In other cases reproductive isolating mechanisms prevent the development of a viable, fertile offspring. This means that two species might indeed mate, but their offspring either do not develop or if they do develop, they are sterile. Although there are many ways that two species can be prevented from interbreeding, the result is always the same: reproductive isolating mechanisms produce genetic isolation. Genes do not flow between two reproductively isolated populations. Let's look at several factors that can lead to reproductive and genetic isolation.

Geographic isolation occurs when some physical barrier separates one population of a species from another. When a population becomes geographically isolated it is often a first step on the road to reproductive isolation. Floods, lava flows, earthquakes, landslides, windstorms, and intervention from other species can separate some individuals from the rest of the breeding population. A physical separation by itself, however, might not eliminate gene flow and so produce reproductive isolation. If the two groups *could* still interbreed, if they ever met, then they would not be separate species. But once two populations are unable to mate and breed, a new species can evolve. The mechanisms of genetic drift, mutation, and natural selection (see Section 13.4) can cause a geographically isolated population to become reproductively isolated.

A well-studied type of geographic isolation is the *founder effect* (see Section 13.4). In the founder effect a small group of individuals with limited genetic diversity is isolated from the larger population of the species (**Figure 14.5**). Because these few individuals carry only a small percentage of the alleles present in the parent species, they are likely to become the founders of a new species. The evolution of fruit flies in the Hawaiian Islands is a good example.

(a)

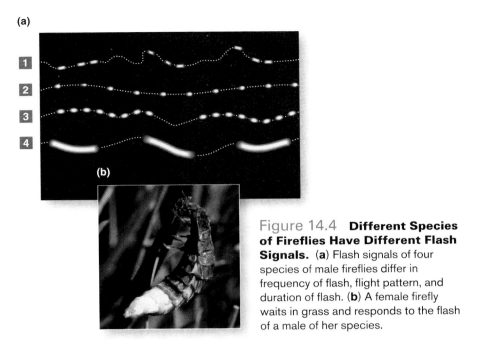

(b)

Figure 14.4 Different Species of Fireflies Have Different Flash Signals. (a) Flash signals of four species of male fireflies differ in frequency of flash, flight pattern, and duration of flash. **(b)** A female firefly waits in grass and responds to the flash of a male of her species.

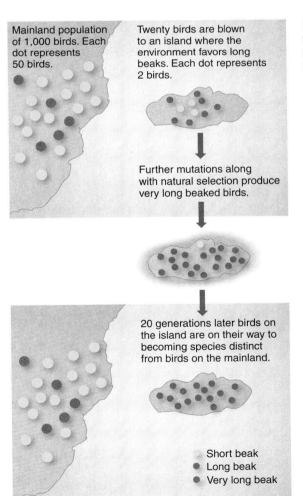

Figure 14.5 The Founder Effect in a Hypothetical Population of Birds.

Mainland population of 1,000 birds. Each dot represents 50 birds.

Twenty birds are blown to an island where the environment favors long beaks. Each dot represents 2 birds.

Further mutations along with natural selection produce very long beaked birds.

20 generations later birds on the island are on their way to becoming species distinct from birds on the mainland.

Short beak
Long beak
Very long beak

reproductive isolating mechanism a process or trait that results in barriers to reproduction and thus prevents interbreeding between species

geographic isolation an isolating mechanism that involves a geographic barrier that separates individuals of a single species into two or more different environments

The combined area of all the Hawaiian Islands is about equal to the area of New Jersey, yet the Hawaiian Islands are home to over 500 species of fruit flies (**Figure 14.6**). Hawaiian fruit flies range in size from giants about as large as houseflies to mini-flies a little larger than a particle of ground coffee. Hawaiian fruit flies have a wide variety of colors, shapes, and behavioral adaptations. Their mating behaviors involve complicated songs, postures, and wing whirring. Incredibly, the mating songs of some Hawaiian fruit flies are as loud as the buzzing of cicadas! And the lifestyles of different species are similarly diverse. For example, the larvae of some species mature in water; others thrive in decaying leaves or rotting fruits. Even more remarkable, some species of Hawaiian fruit flies have carnivorous larvae that attack and feed on other insects.

How could the spectacular diversity of Hawaiian fruit flies have originated? The best hypothesis is that a combination of successive geographic isolating mechanisms (**Figure 14.7**) produced this diversity. A few flies in each habitat found their way to new places on their home island or flew or were blown to other islands by strong winds. This isolated a few individuals in a new environment (Figure 14.7a). Once geographic isolation had occurred, natural selection molded the traits of future generations. This conclusion is supported by geological history. The Hawaiian Islands were formed relatively recently—between 1 and 5 million years ago—and some islands are younger than others. DNA analyses show that while all of the Hawaiian fruit fly species are closely related, the more recently formed islands are home to fruit flies that have evolved more recently. In addition, the DNA of fruit fly species that live in nearby geographic regions is more similar than the DNA of fruit fly species that live farther apart. The Hawaiian fruit fly story is a dramatic example of evolutionary mechanisms in action.

Geographic barriers are not the only ways that two populations can become genetically isolated. A change in the environment can lead one population to split into two breeding groups and eventually into two species. **Figure 14.8a** shows the well-documented example of hawthorn flies. This North American species feeds on hawthorn fruits that look like little apples. When the fruits first start to develop, the hawthorn fly lays its eggs on them. The larvae hatch and eat the fruit. Now the story gets more interesting. Remember Johnny Appleseed? Apple trees are similar to hawthorn trees and were introduced from England to New England about 300 years ago by Appleseed and other early settlers. It was not long before farmers learned that hawthorn flies lay their eggs on apple flowers just as they do on hawthorn flowers. Hawthorn flies became a major pest of apple trees. On the surface it looks like hawthorn flies simply have found a new food source, but the reality is more biologically complex:

Figure 14.6   **Hawaiian Fruit Flies Show Enormous Diversity.** The arrow points to a fruit fly that is about the same size as a common fruit fly.

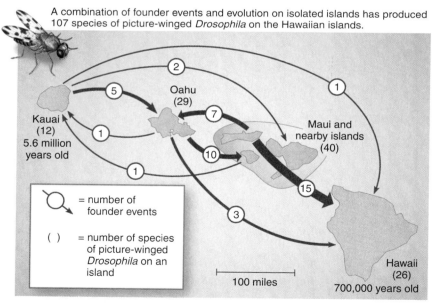

A combination of founder events and evolution on isolated islands has produced 107 species of picture-winged *Drosophila* on the Hawaiian islands.

Oahu (29)

Kauai (12) 5.6 million years old

Maui and nearby islands (40)

◯↘ = number of founder events

( ) = number of species of picture-winged *Drosophila* on an island

100 miles

Hawaii (26) 700,000 years old

Figure 14.7   **The Founder Effect Explains Many Species of Picture-winged *Drosophila* on the Hawaiian Islands.** Kauai, the oldest Hawaiian Island, has 12 species. Two arrived from other islands. Oahu has 29 species; 12 are traced to founder events, 17 evolved in place. Maui and nearby islands have 40 species, traced to 12 founder events. Twenty-eight species evolved on Maui and nearby islands. Hawaii, the youngest island, has had 19 founder events. Seven species of picture-winged *Drosophila* evolved on the island of Hawaii.

■ *Why do you think the main movement of Hawaiian fruit flies is from northwest to southeast?*

although they look nearly identical to hawthorn flies, apple flies (Figure 14.8b) are a new species. The United States now contains two similar species of flies: one mates and breeds on hawthorn fruits and another mates and breeds on apples. The two species mate at different times, do not mate with one another, and usually do not mate on one another's fruit. This is a clear example of environmental isolation leading to speciation.

Sexual selection is another way that allele frequencies in a population can change over generations (see Section 13.4), and sexual selection can lead to the formation of new species. In many animals mates are chosen by one partner or the other—most often females do the choosing. A change in male traits that can be selected by females, or a change in female preferences, can cause a population to become reproductively separated from the larger group. Eventually, they may become a separate species. An example is the evolution of the exotic stalk-eyed fly. This species of fly has eyes set on stalks that extend out from the head (**Figure 14.9**), and males have longer eye stalks than females. How did such a bizarre adaptation arise? Female stalk-eyed flies choose mates based on the length of the eye stalks of males. If females tend to choose long eye stalks, the population moves toward long eye stalks. If females choose short eye stalks, the population moves toward short eye stalks.

How might this have given rise to a new species? Phylogenetic studies show that the ancestors of the stalk-eyed flies did not have eyes that were stuck atop long stalks. Instead, their eyes looked like those of a housefly. But of course some flies had eyes that were set a bit farther apart than the eyes of other flies. At some point in the evolutionary history of these flies, some females began choosing mates that had eyes stuck out on stalks. Over time a population—and eventually a species—of stalk-eyed flies evolved.

Polyploidy is an intriguing reproductive isolating mechanism that occurs when an individual has some multiple of the typical diploid set of chromosomes. Polyploidy can arise from errors during meiosis. If the chromosomes do not separate properly during meiosis, then some gametes will have too many chromosomes, and any individual produced by such gametes will have too many chromosomes. A polyploid human, if such a person existed, would have 92 chromosomes, four times the haploid number. A polyploid human would not survive long after conception, but other kinds of organisms can survive and thrive as polyploids, especially plants. As many as 50% of flowering plant species are polyploid. Polyploidy is not as

(a) Hawthorn fly on hawthorn tree

(b) Apple fly

Figure 14.8 **Hawthorn (a) and Apple Flies (b).**

Figure 14.9 **Stalk-Eyed Fly.** This male's eyes are the red orbs at the tips of amazingly long stalks.

adaptive for animals, and most polyploid animals die before they reach maturity. There is evidence, though, that some species of fishes may have originated from polyploidy.

The Distinction Between Two Species Is Not Always Obvious

The emergence of a new species is not always obvious or clear cut. The divergence of one species into two often is accomplished in small steps. In many cases the isolating mechanisms are strong and species are strictly segregated, while in other cases isolating mechanisms may be weak and will not eliminate interbreeding completely. This can lead to blurry situations where two groups are not quite a single species, but not quite two. In these cases a small change in the environment or the genetics of a group might reverse the isolation. Or, two populations might remain indefinitely as species groups that are nearly separated, but have some interbreeding. When two species breed the offspring is a **hybrid**.

Lions and tigers are two distinct species that rarely meet in the wild but can interbreed and produce offspring in captivity. The two species of large

hybrid the result of a mating between two species

Figure 14.10 **A Liger Is the Result of the Mating of a Tiger and a Lion.** In the wild lions and tigers do not mate or hybridize, so they are separate species.

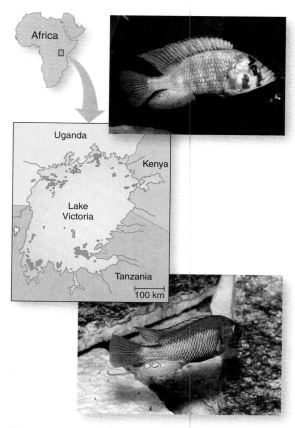

Figure 14.11 **Cichlid Fishes.** Two species of cichlids from Lake Victoria in East Africa.

cats are close relatives, and when they mate in captivity the offspring can thrive—and some are even fertile (**Figure 14.10**). This does not mean that lions and tigers are the same species. It means that they are closely related species, and that their reproductive isolation is not complete. In the wild where the natural barriers can work effectively, lions and tigers rarely, if ever, interbreed.

This fluidity of barriers to interbreeding also is seen in nature. An interesting example comes from the cichlid (pronounced "sick-lid") fishes in Lake Victoria, in eastern Africa (**Figure 14.11**). Historically, the combination of clear water and highly varied conditions in Lake Victoria allowed a large number of different species of cichlid fishes to evolve. The species often were kept apart by the coloration of the males, because the females of one species will mate only with males that have the bright colors characteristic of their own species. Recently, however, conditions have changed, and species of cichlids have been interbreeding. Contaminants introduced by human activity have caused the waters of Lake Victoria to be more turbid. In these cloudy waters female cichlids cannot easily distinguish males of their own species from males of other species. As a result, there has been an increase in interbreeding and a reduction in the distinctiveness of the colors of males. Under these human-influenced conditions, species that once were naturally distinct are beginning to merge.

14.2 Understanding Biological Diversity Involves Categorizing and Classifying Species

Classifying and categorizing come naturally to most people. Categorizing is a basic way to organize information or objects so that you can use them more effectively. For instance, when you take your laundry out of the drier, you probably do not just dump all of your clean clothes back into your closet. Instead, you make piles of similar items of clothing. All the underwear goes into one pile. Socks usually are matched and paired. And the folding and sorting process continues until you have piles of similar kinds of clothes. As you fold a T-shirt and put it into a pile with other Ts, you probably don't think, "Ah, my gray, too-small T-shirt with the barbecue stain and the ripped neck and the hole in the shoulder." Instead, in a wink you mentally match it up with other shirts that have short sleeves, no collars, no buttons, and round necks. A biologist would say that you have been classifying your laundry.

Biological **classification** is the process of sorting life-forms into groups. Identifying species is just the first level of sorting. It is followed by other sortings that group species into larger and larger categories. To return to the clothes analogy, skirts, shorts, and slacks could be lumped into one category of clothes that go on the bottom half of your body. Sweaters and heavy coats can be lumped into the category of winter clothes. Making larger categories of clothes is not straightforward, and the same can be said for making categories of organisms. Sorting species into larger categories can be a challenge. For one thing, there are millions of species to classify, and most are not even alive for direct examination. In addition, it is not always clear how to group species into larger categories. Whales obviously are in a different category from butterflies, but a glance at specimens in a museum tray suggests that many different kinds of butterflies have some overlapping and subtly different characteristics (**Figure 14.12**). Should these different kinds of butterflies be separated into different groups, or lumped together? Given this complexity, you should not be surprised that **taxonomists,** biologists who specialize in classification of organisms, have used a variety of schemes to try to classify species into larger groups.

Modern Classification Schemes Are Based on Phylogenetic Relationships

Biologists have been working at the classification of organisms for over two centuries—and their focus has changed as new information about species and their genetic relationships has become available. Early taxonomists were guided by the morphological species concept and focused on physical differences to sort species into larger groups. They used long, descriptive Latin phrases to name species. Linnaeus streamlined all of this (**Figure 14.13**). He developed a comprehensive system of classification based on similar morphological characteristics and gave a two-word name to each species. The Linnaean system is still used in common language and in scientific literature, so it is worth learning.

In the Linnaean system species with similar morphological traits are grouped into a genus (**Figure 14.14**). For example, all of the species that are

Figure 14.12 **Taxonomists Sort Species into Larger Groups.** These butterflies belong to two different genera. Numbers 1–4 are species of *Nymphalis* while the others are species of *Polygonia*.

similar to humans are grouped into the genus *Homo*. The scientific name given to each species is a two-word name made up of the names of its genus and its species. For example, modern humans are *Homo sapiens*. Other species of the genus *Homo* are *Homo erectus* and *Homo habilis*. So the name of the modern human species is *Homo sapiens*, not just *sapiens*. Once it is clear what the abbreviation stands for, the genus often is shortened to a first initial. So you may see *H. sapiens* in text that discusses the human species. As another example, the species name of the domestic dog is *Canis familiaris*, not just *familiaris*, and you may see that name written as *C. familiaris*. There are two last important technical points. Because scientific names are in Latin,

classification the process of sorting things based on their similarities and differences

taxonomist a biologist who specializes in classification of organisms

Watermelon

Before Linnaeus
Citrullus folio colocynthidis secto, semine negro

After Linnaeus
Cucurbita citrullus

Carnation

Before Linnaeus
Dianthus floribus solitariis, squamis calycinis subovatis brevissimus, corollis crenatis

After Linnaeus
Dianthus caryophyllus

Catnip

Before Linnaeus
Nepeta floribus interrupte spicatis pedunculatis

After Linnaeus
Nepeta cataria

Figure 14.13   **Linnaeus Assigned a Two-Word Name to Each Species.**

■ *Is there an advantage to using the scientific name of an organism over the common name?*

Species of *Homo*

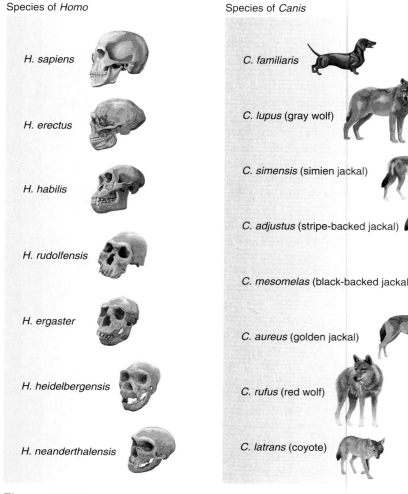

- *H. sapiens*
- *H. erectus*
- *H. habilis*
- *H. rudolfensis*
- *H. ergaster*
- *H. heidelbergensis*
- *H. neanderthalensis*

Species of *Canis*

- *C. familiaris*
- *C. lupus* (gray wolf)
- *C. simensis* (simien jackal)
- *C. adjustus* (stripe-backed jackal)
- *C. mesomelas* (black-backed jackal)
- *C. aureus* (golden jackal)
- *C. rufus* (red wolf)
- *C. latrans* (coyote)

Figure 14.14   **Species with Similar Morphological Traits Are Grouped into a Genus.**

phylogenetic classification a classification scheme that reflects genetic and evolutionary relationships

like all foreign language words that appear in English, scientific names are always italicized if they appear in type, or underlined if they are written by hand. Finally, note that the genus name is always capitalized and that the second word of a scientific name is never capitalized.

Different genera, the Latin plural for genus, are grouped into larger categories called families. In turn, families are grouped into orders, orders are grouped into classes, classes are grouped into phyla, and phyla are grouped into kingdoms (**Figure 14.15**). Linnaeus himself only identified two kingdoms, but the later revisions of his system identified five kingdoms: bacteria, protists, fungi, plants, and animals. As you will see shortly, the modern classification system does not focus on Linnaean kingdoms, but these names still are commonly used.

Until the mid-1900s this five-kingdom scheme—or some variation of it—was the only approach to classification available. Even so, researchers recog-

nized that it had major limitations. A big problem with the Linnaean system was that it depended on the judgments of researchers to sort species into higher categories. For instance, while one researcher might say that algae are protists, another might decide that algae are plants. The accumulation of knowledge about more traits has helped to resolve some of these problems, but the system still was somewhat arbitrary. It had no connection to how species are formed or their genetic relationships. As a result there was no commonly accepted way to settle disputes about classification. As knowledge of evolution and genetics became more widespread, classification efforts turned toward understanding the genetic and evolutionary relationships between species. This kind of classification is called **phylogenetic,** and it is related to the phylogenetic species concept. Certainly, the classification schemes developed in the 1800s and 1900s reflected phylogenetics. Organisms that are physically and behaviorally

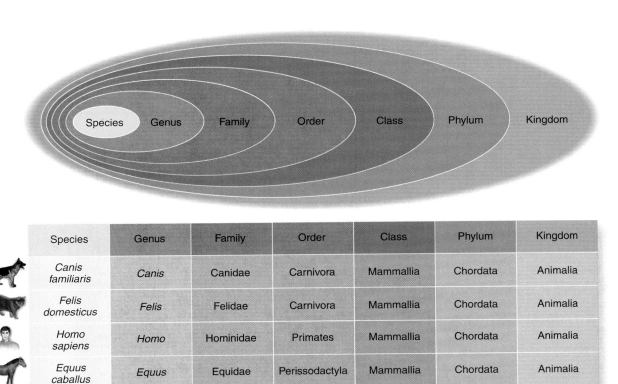

	Species	Genus	Family	Order	Class	Phylum	Kingdom
Domestic dog	*Canis familiaris*	*Canis*	Canidae	Carnivora	Mammallia	Chordata	Animalia
Domestic cat	*Felis domesticus*	*Felis*	Felidae	Carnivora	Mammallia	Chordata	Animalia
Human	*Homo sapiens*	*Homo*	Hominidae	Primates	Mammallia	Chordata	Animalia
Horse	*Equus caballus*	*Equus*	Equidae	Perissodactyla	Mammallia	Chordata	Animalia
Cow	*Bos taurus*	*Bos*	Bovidae	Artiodactyla	Mammallia	Chordata	Animalia
Chicken	*Gallus gallus*	*Gallus*	Phasanidae	Galliformes	Aves	Chordata	Animalia

Figure 14.15 **In the Linnaean System Larger Taxonomic Categories Contain Smaller Ones.**

similar often are genetically and evolutionarily related. This is why there is much validity to the old classification schemes. Newer schemes, however, combine physical characteristics with other traits such as physiology, development, biochemistry, and genetics to establish phylogenetic relationships between species.

One big advantage of phylogenetic classification is the simultaneous analysis of many features. The assumption is that species that share many traits must have a common ancestor, and so they should be grouped together. This approach can be applied to small groups of organisms as well as to larger groups. The result is a phylogenetic tree that links species into larger and larger groups that are related through common ancestors. **Figure 14.16** shows an example of a phylogenetic analysis of vertebrates. In this figure evolutionary time is represented on the *y* axis (vertical), and the degree of genetic relationship is represented on the *x* axis (horizontal). Figure 14.16 shows that mammals are most closely related to reptiles and birds. By the best estimates available, mammals and reptiles shared an ancestor 300 million years ago.

How could such a complicated phylogenetic tree be created? It turns out that computer programs now can incorporate all of the data and generate many possible phylogenetic trees—often 30,000 trees. One assumption that researchers make is that the simplest tree that accounts for all of the information is probably the correct tree. As new information becomes available, a new tree can be proposed. These computer programs can use information about DNA sequences as easily as any other trait, and this has produced some profound changes in our understanding of the relationships between species. One of the most important advances of this approach was to replace the five-kingdom system with one of three domains, an important idea considered next.

All Species Can Be Grouped into Three Domains

In the 1970s biologist Carl Woese (pronounced "Woes") realized the power of DNA sequences to shed light on evolutionary relationships. So Woese went looking for DNA sequences that would best serve this purpose. He chose the genes that code

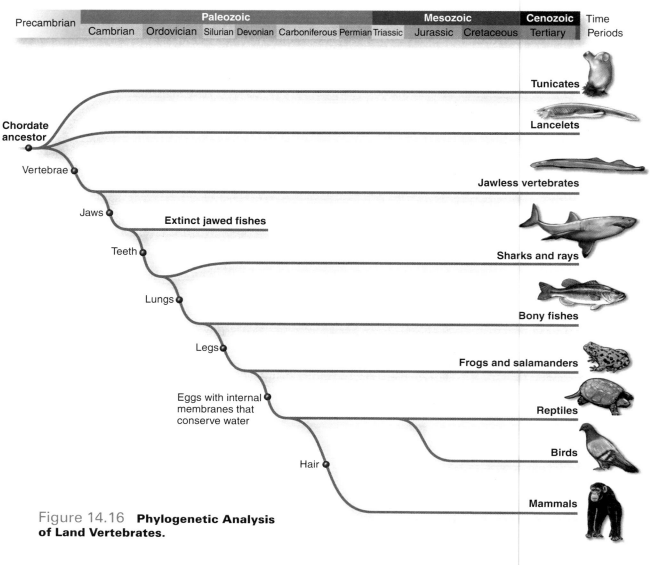

Figure 14.16 **Phylogenetic Analysis of Land Vertebrates.**

for the RNA found in the small subunit of the ribosome (see Section 7.4), also called ribosomal RNA, or rRNA. It might be strange to start with an RNA gene rather than a gene that codes for a protein, but Woese had some good reasons for this choice. First, ribosomes are found in every organism, living or extinct, from bacteria through people. So by using these RNA sequences, Woese could generate a tree for all of life. In addition, rRNA is so critical to the functions of a cell that it has changed exceptionally slowly over the approximately 4 billion years of the history of life. So the DNA that codes for rRNA is a good trait to use as the basis of a tree that includes all living things. In 1990 Woese published his *universal tree of life* (**Figure 14.17**). The big surprise was that instead of the five kingdoms, the tree clearly showed *three* major groups of organisms. Biologists call these major groups

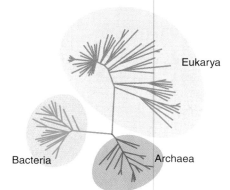

Figure 14.17 **Woese's Universal Tree of Life.** Each line represents a species that was sampled in this study of ribosomal DNA sequences. The distances between the lines represent the differences in their ribosomal DNA sequences.

domains, and the individual domains are named **Archaea, Bacteria,** and **Eukarya.** One of the surprises to come out of this analysis is the recognition of *two* domains of prokaryotes, Archaea and Bacteria. Until this time prokaryotes were thought to be one large, related group. Another important conclusion is that the Archaea are as closely related to the Eukarya as they are to the Bacteria. This three-domain system has been tested using many other molecules and traits and has been studied, debated, and challenged over the years. The basic conclusions of Woese's research have

withstood all of these tests, and the three domains of life are well supported by many scientific studies. Woese was the first researcher to describe the universal tree of life.

The genes for ribosomal RNA also can be used to understand the taxonomic relationships within each domain. **Figure 14.18** shows details of an analysis of ribosomal RNA genes from a large number of eukaryotes, and this figure suggests some especially interesting relationships. Notice that at least three of the traditional "five kingdoms" are revealed as true phylogenetic groups. The plants, the animals, and the fungi come out as three individual clusters. A second interesting suggestion is that the protists and algae, groups lumped together in the traditional scheme, are diverse sets of many phylogenetic groups. While subsequent studies have fine-tuned this picture, Woese's analysis of genes for ribosomal RNA gave a good first estimation of how organisms are related. Upcoming chapters will explore more details about these relationships. Before delving into these details, though, it is a good idea to have a general understanding of the major groups of organisms.

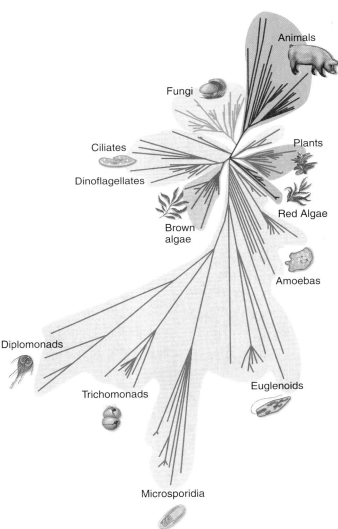

Figure 14.18   Details of the Universal Tree of Life Obtained by an Analysis of Ribosomal RNA Genes from Many Eukaryotes.

■ *Respond to this statement in light of the information in this figure. "Kingdom Protista should be eliminated from the classification system."*

> **QUICK CHECK**
>
> **1.** Describe the three-domain system and explain how it is different from previous classification schemes.

14.3 Traits of Major Phylogenetic Groups Reflect Life's Diversity

A major goal of biological studies is to understand the diverse adaptations of living organisms and unravel how they came about. In the latter part of this chapter, and in Chapters 15 through 17, the evolution of biological diversity will be considered in some detail. Before you begin that work, however, you should be familiar with the major groups of organisms. Let's focus on this topic in this section, starting with prokaryotes.

There are two domains of prokaryotes: Archaea and Bacteria. Both are single-celled organisms that can be shaped like rods, spheres, or corkscrews. Bacteria and Archaea are by far the most numerous organisms on Earth. If you were simply to weigh all of the organisms alive right now in the world, the Archaea and Bacteria would make up two-thirds of the mass of living things.

domain a phylogenetic group of organisms identified by DNA analysis

Archaea a domain of prokaryotes characterized by the nature of the DNA that codes for their ribosomal RNA and other traits

Bacteria a second domain of prokaryotes defined by the nature of the DNA that codes for their ribosomal RNA and other traits

Eukarya the domain of eukaryotic life defined by the nature of the DNA that codes for their ribosomal RNA and other traits

Figure 14.19
Everyday Objects Are Covered with Bacteria.
This tip of a pin has a coating of bacteria, artificially colored orange in this photo.

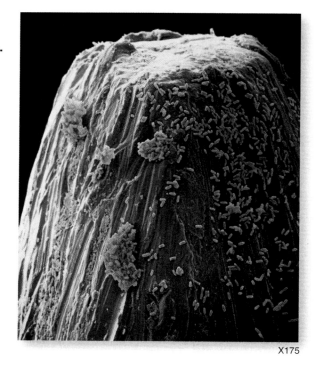

X175

Even common everyday objects are covered with bacteria (**Figure 14.19**). One dramatic physical similarity of Archaea and Bacteria is that they lack organelles surrounded by membranes, including a nucleus. Archaea and Bacteria have a wide variety of lifestyles (see Chapter 15).

All of the remaining organisms belong to the third domain of life, Eukarya. In addition to the distinctness of their ribosomal RNA, the Eukarya have

invertebrate an animal with no backbone or cartilage to protect the nerve cord

(a)

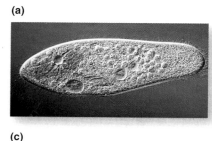

(b)

(c)

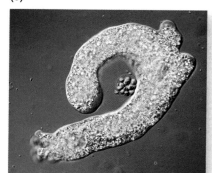

Figure 14.20   **Diversity of Single-Celled Eukaryotes.**
(**a**) *Paramecium*, (**b**) *Euglena*, (**c**) *Amoeba*.

organelles surrounded by membranes, including the nucleus. Eukaryotes range in size from microscopic, single-celled protozoans to 100-foot-long whales. Scientists dispute how the Eukarya should be further divided, but there are at least 20 groups that can be distinguished by DNA analysis, morphology, and habits. Some researchers call these groups kingdoms; others use other names for these large groupings. In the next few sections you will become familiar with the Eukarya, using common names for some of the major groups.

One distinguishing feature of eukaryotic species is whether they are unicellular or multicellular. Single-celled eukaryotes show complex, intricate, and fascinating features (**Figure 14.20**). These intriguing single-celled organisms are phylogenetically diverse groups that sometimes are called *protists*. Algae include unicellular species, but also include some of the simplest of the multicellular eukaryotes. You may know algae as pond scum or as seaweeds that wash up on the shore. By definition all algae are photosynthetic. DNA studies reveal that there are several different unrelated, groups of algae. The green algae are more closely related to *plants* than to any other group of eukaryotes, and green algae now are recognized as plants. Modern plants probably evolved from species similar to green algae. Plants are a group of complex, usually multicellular, photosynthetic eukaryotes (**Figure 14.21a**). Not only are most plants photosynthetic, but they share two types of photosynthetic pigments, chlorophyll *a* and chlorophyll *b*. Plants also have cell walls that are made of cellulose. Chapter 17 explores more about plants.

Fungi include the mushrooms, puffballs, and common yeasts (Figure 14.21b). *Animals* include simple organisms such as sponges, complex animals without backbones such as crustaceans and insects (Figure 14.21c), and the whole range of animals with backbones. Most fungi and all animals are multicellular, and both fungi and animals get energy from other organisms. One of the surprising findings from the new DNA studies is that fungi and animals are more closely related to one another than to the other kingdoms.

Animals are related to fungi, but in many ways they are much more diverse. **Table 14.1** shows the commonly recognized animal groups. Most animal phyla are **invertebrates,** animals without backbones, and invertebrates account for most of the spectacular diversity of animals. Among invertebrates the arthropods (including crustaceans, spiders, insects, and lots of other, smaller groups) have the most species and the greatest diversity. You may be surprised that only one phylum has animals

(a)

(b)

(c)

Figure 14.21 **Diversity of Multicellular Eukaryotes.**

Table 14.1 Commonly recognized animal phyla

Phylum	Examples and characteristics	
Placozoa	*Trichoplax adherens* Simplest animal, no distinct tissues, but has egg and sperm, amoebalike movement	
Porifera	Sponges Two layers of cells, but no tissues; attached to substrate; holes in body for water, nutrients, and gas exchange	
Cnidaria	Jellyfish, corals, sea anemones, *Hydra* Mostly marine organisms with tissues and organs; soft, cylindrical or umbrella-shaped bodies; have tentacles with stinging cells for catching prey	
Platyhelminthes	Flatworms, tapeworms, flukes Ribbon-shaped, soft body; have a mouth but no anus; often parasitic	
Mollusca	Squid, clams, snails, slugs, octopi, oysters Soft body, often with shell; complex tissues and organs; rasping mouthparts	
Nematoda	Roundworms Unsegmented worms, tube-within-a-tube body plan; gas exchange and circulation by diffusion	
Annelida	Earthworms, leeches Segmented bodies; complex internal organs; extensive muscular system; closed circulatory system	
Arthropoda	Insects, spiders, crustaceans, centipedes, millipedes Segmented bodies and appendages; external skeleton; metamorphosis	
Echinodermata	Sea stars, sea urchins, sea cucumbers 5-parted radial symmetry, water vascular system	
Chordata	Sea squirts, lampreys, fishes, amphibians, reptiles, birds, mammals Nerve cord along back; usually have enlarged brain; usually have skull and spinal column	

with backbones or *vertebrae* (singular: vertebra). Each vertebra has a central hole that allows the spinal cord to pass through it. So like the string that holds a bead necklace together, the spinal cord runs down the animal's back, traveling through the hole in each vertebra. Accordingly, animals that have backbones are called **vertebrates.** In phylum Chordata vertebrates are grouped with a few other species that have softer, cartilaginous structures around the spinal cord.

vertebrate an animal with a backbone around the spinal cord

QUICK CHECK

1. What are the characteristics of Bacteria, Archaea, and Eukarya?

14.4 The Diversity of Life Emerged Slowly

Earlier chapters presented pieces of the story of how life got started on Earth. Here you are going to return to this story and see how the magnificent diversity of life slowly emerged over the past 4 billion years.

Life Evolved in the Hostile Environment of Early Earth

Earth formed about 4.5 billion years ago. Early Earth was a hot planet of molten rock, bombarded by chunks of matter that whirled through the embryonic solar system. As Earth began to cool

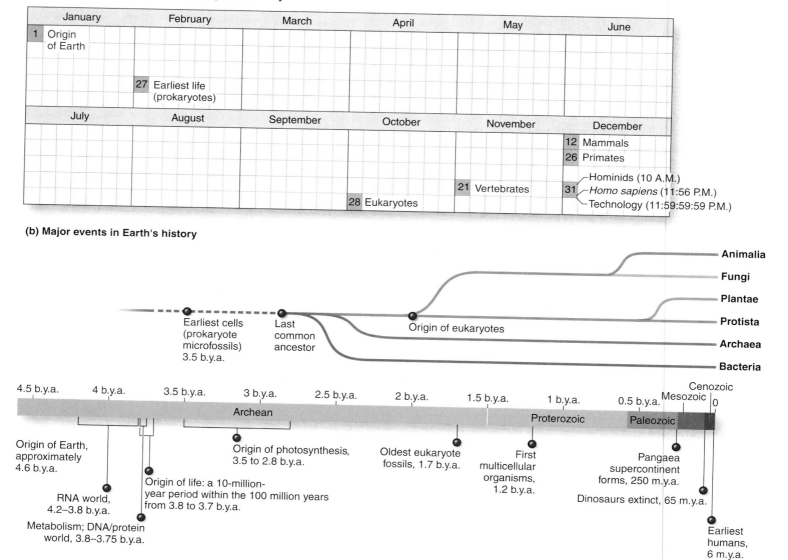

(a) Earth's history compared to a single calendar year

(b) Major events in Earth's history

Figure 14.22 **The Major Events in the Evolution of Life on Earth. (a)** Earth's history compressed into a single year. **(b)** Timeline of the evolution of life.

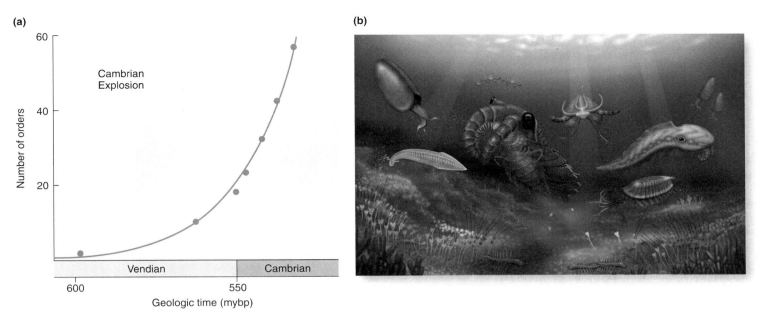

Figure 14.23 **Precambrian and Cambrian Life.** (a) An explosion of forms of life during the Cambrian Peroid. (b) Cambrian seas in what is now British Colombia were home to a diverse array of invertebrates. This painting reconstructs animals from the fossils of the Burgess Shale. Some of these species look like arthropods, but others are unique, with no known modern descendants.

and harden, something important happened: it rained. Water vapor came from the thick layer of volcanic vapors and perhaps from meteorite bombardments. In the atmosphere water vapor condensed and fell as rain that gathered to form lakes, rivers, and oceans. So early Earth was hot and wet.

Even with all that water, it may be hard to imagine life beginning in this tumultuous, inhospitable environment. Violent storms swept the oceans and the exposed landmasses, and there was nothing to block incoming cosmic rays and ultraviolet radiation. The atmosphere was far different from the blend of nitrogen, carbon dioxide, and oxygen that supports life today. But the basic elements needed for life (carbon, hydrogen, oxygen, nitrogen, sulfur) were present in Earth's early atmosphere. Other necessary elements were in Earth's crust, and still others may have originated in the meteors that crashed into the early Earth. Rainwater poured off the land, carrying dissolved minerals and salts into the accumulating oceans. And in this alien situation, life emerged.

Scientists do not know how the first cells formed, although some of their ideas were discussed in Sections 4.5 and 5.5. The earliest known fossils show fully formed cells and often filaments of many cells linked together. These look much like filaments of bacteria alive today. These earliest fossils demonstrate that life has been on Earth for at least 3.5 billion years—a span of time so immense that it is difficult to comprehend. The diversity of life emerged over these 3.5 billion years. The important milestones in the history of life are shown in **Figure 14.22**, in an analogy that compresses Earth's history into a single year.

When put into this time frame, the earliest life occurred at the end of February, but eukaryotes didn't appear until nearly Halloween. Vertebrates appeared near the end of November and mammals in mid-December. *H. sapiens* appeared 3.5 minutes before midnight of December 31, and all human history occurred in the last tenth of a second of the year. One important lesson in this story is that species have emerged in a sequence. While the fossil record is incomplete, it forces the conclusion that later species build on the traits of earlier species. Prokaryotes emerged before eukaryotes, single-celled organisms preceded diverse multicellular eukaryotes, and species with spinal columns appeared before vertebrate species with limbs. These patterns support the principle of common descent with modification.

Biological Diversity Exploded Once Eukaryotes Emerged

Prokaryotes emerged in the exotic environment of early Earth, and for the first 1 or 2 billion years prokaryotes had the planet to themselves. The appearance of photosynthetic **cyanobacteria** was a critical early development that occurred around 3 billion years ago. The oxygen produced as a by-product of photosynthesis eventually changed the composition of Earth's atmosphere. By 2 billion years ago, oxygen began to accumulate in the atmosphere, paving the way for the evolution of aerobic prokaryotes and eukaryotes.

About 1.8 billion years ago an extraordinary era began as eukaryotes appeared and with them came an explosion of life-forms (**Figure 14.23**).

cyanobacteria
photosynthetic bacteria able to use energy from sunlight to convert carbon dioxide and water to sugar

Figure 14.24 **Earliest Known Fossils of Multicellular Organisms.**

Eukaryotic diversity was built, however, on prokaryotic diversity. Early in eukaryotic evolution large cells engulfed smaller prokaryotic cells and made use of their special traits (Section 4.4). They were maintained as separate but valuable organisms inside their hosts. In time they were modified into organelles, such as mitochondria and chloroplasts. DNA evidence supports this idea. Mitochondria and chloroplasts have their *own* DNA that is more similar to the DNA of prokaryotes than it is to eukaryote DNA. Mitochondria provided the important ability to use oxygen in the metabolism of glucose. More oxygen meant more metabolic production of ATP. Because they had more energy available to them, eukaryotes could build more complex cellular structures.

The first eukaryotes were single-celled creatures, similar to today's protists. Multicellular algaelike organisms probably appeared soon after. The oldest multicellular animal fossils are about 600 million years old (**Figure 14.24**), although small animals probably emerged before this time. There is nothing on Earth today quite like them, but it is safe to say that they were soft-bodied, mobile organisms. At this point the pace of diversification into new species picked up dramatically—the period from 545 to 245 million years ago was an enormously productive time for life on Earth. This is the Paleozoic era (see Figure 1.13), which is divided into geological periods. During the Cambrian period, 590 to 505 million years ago, so many species crowd the fossil record that the phenomenon is called the *Cambrian explosion.* Every major phylum of modern animals appeared during this time. On the other hand, many

Cambrian species died out and are only known as fossils. The Cambrian animals were almost exclusively invertebrates, although the ancestors to vertebrates appeared during this era. Fossils of full-blown vertebrates, fishes with vertebrae and jaws, are first seen about 440 million years ago. Until 400 million years ago nearly all of the animals were aquatic, and many lived in Earth's oceans.

During the Paleozoic the oceans were the hotspot of biodiversity. Coral reefs appeared, and fishes diversified dramatically. Animals and plants made their first moves to land during this period. Fossils show insects, amphibians, primitive reptiles, and ferns flourishing 300 million years ago. In the latter part of the Paleozoic, life on land had its turn at an explosive period of evolution. Vertebrate diversity increased as amphibians, reptiles, dinosaurs, mammals, and birds appeared. Land plants developed roots, stems, leaves, flowers, and seeds within fruits. Aquatic evolution did not stop either. Species of fishes multiplied so much that fishes now are the most diverse vertebrates. Some mammals returned to oceans and rivers to produce modern whales, dolphins, and manatees. In the Mesozoic period, from 245 million to 65 million years ago, dinosaurs, birds, and flowering trees evolved and flourished. Mammals evolved but stayed small and nocturnal. The extinction of the dinosaurs marks the end of the Mesozoic and the beginning of the modern era, called the Cenozoic. The Cenozoic has been dominated by mammals, including humans. From this summary you can see how diversity has increased over the past 600 million years.

Several Mass Extinctions Have Occurred During the History of Life on Earth

This picture of increasing diversity is not complete, of course, because it has not included extinctions. Not only do extinctions remove some species, they also often change the environment in ways that lead to an explosion of new species. Occasionally there have been **mass extinctions,** which permanently wipe out a large percentage of species alive at that time. **Figure 14.25** shows the patterns of extinction of marine organisms in the last 600 million years. A huge extinction occurred about 250 million years ago in the Paleozoic, when nearly 90% of all marine species were lost. There seems to be a minimum basic rate of extinction that amounts to a loss of about 10 families per 25 million years. Extinction never seems to dip lower than this background rate. What do these rates of extinction mean for overall diversity? Can diversity grow fast enough to overcome them? Figure 14.25 shows that diversity begins to rise again about 10 million years after each mass extinction event. Diversity eventually surpasses its preextinction levels.

Another important point about mass extinctions is not illustrated by the graphs. After each mass extinction the loss of so many species, families, and phyla opens the way for the dominance of a new kind of organism. For example, after the mass extinction at the end of the Paleozoic era, when so many marine invertebrates were lost, the vertebrates emerged and flourished. Later the extinction of many ferns and their relatives made way for seed-bearing plants to diversify into new species and expand their geographic ranges. At the end of the Mesozoic era, when the dinosaurs were

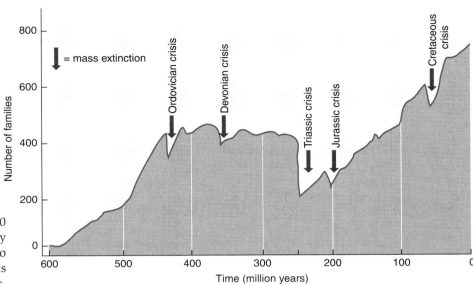

Figure 14.25 **Mass Extinctions in Life's History.** Several mass extinctions are recorded in the history of life. The last mass extinction, some 65 million years ago, caused the dinosaurs to go extinct.

lost, birds and mammals had a window of opportunity that resulted in the diversity of birds and mammals that you see today. To put this in another way, mass extinctions change the character of life on Earth. They alter the competitive landscape and allow a different set of organisms to flourish and blossom into many new species.

QUICK CHECK

1. Give examples from the history of life on Earth that show how changes in the conditions on Earth have allowed the evolution of different sorts of organisms.

mass extinction
a widespread extinction event that wipes out large numbers of species

---

14.5 An Added Dimension Explained
Are Human Races a Valid Idea or Not?

The idea that there were different kinds of humans was recognized in European writing in the early 1600s. At this time, though, the distinction between species and races was not well defined, and so by modern standards these writings are confusing. By Darwin's time, however, several distinct races were recognized such as Africans, Asians, and Native Americans. It is important to try to determine if the superficial features that distinguish the races as

defined today—such as skin pigmentation, hair texture, and facial characteristics—have a deeper biological significance. The answer is complex, and this is an active field of research. Nevertheless, some general conclusions about the validity of race as a biological concept can be reached, and you will consider them in this section. Here you will explore the hypothesis that the races commonly defined in the United States are genetically defined. These racial categories: Caucasian, African-American, Asian, Hispanic, Pacific Islanders, and Native Americans, roughly correspond to the content or region of origin for each group.

—*Continued next page*

Continued—

Many studies have searched for the degree of genetic variation between individuals from different races and compared that with the differences between individuals within each race. In these studies DNA from pairs of individuals is compared to determine how similar or different they are. Then a computer analysis determines the total genetic variation within each race and compares it with the total variation between races. In nearly every study the amount of variation in DNA is greater *within* each racial group than it is *between* races. For most nucleotide sequences—including those that code for proteins and those that do not—individual Caucasians are as different from *each other* as they are from Africans or Hispanics. This means that races are genetically more similar to one another than they are different. So these studies of overall genetic variation do *not* support the idea that the different races are genetically distinct. When the whole human genome is considered, you are as likely to find a close genetic match with someone of a different race as with someone of your own race.

Studies of genetic variation have revealed another important genetic relationship. For most nucleotide sequences that have been studied, nearly all of the genetic differences in Asian, European, and Amerindian populations are found *within African* populations. These and other genetic studies suggest that the ancestors of modern Africans are the ancestors of all modern humans. This intriguing conclusion is supported by the fossil evidence of human evolution (see Chapter 16).

In the last couple of paragraphs you have considered studies of the overall human genome that do *not* support the idea that the commonly accepted human races are genetically distinct. On the other hand, many recent studies, which focus on specific genetic sequences, seem to indicate that human races are genetically distinct. Alleles have been identified that increase the risk of certain diseases in specific racial populations. For instance, alleles that cause cystic fibrosis are found more frequently in people of European descent, while the sickle cell anemia allele is more frequently found in people of African descent. There are many other examples of these genetic correlations with race, such as human blood groups (see Section 10.4). These findings are one reason why tissue or organ transplants are more likely to be successful if the donor is from the same "race" as the recipient. On the other hand, many scientists would say that while these alleles correlate with a person's racial heritage, they do not define race. For example, some people of European ancestry can have a sickle cell allele, and some African Americans can have cystic fibrosis.

A more thorough search of the scientific literature shows an even more complex pattern of alleles and racial background. For one thing, differences in DNA between populations depend on which DNA sequences are studied. For example, using some DNA tests, African Americans are genetically different from Caribbean Africans, while Africans of Western African origins can be distinguished from those who originated in sub-Saharan Africa. East Asians can be distinguished from west Asians, and southern Europeans can be distinguished from northern Europeans. The list of identifiable human populations seems to grow with each new report. Depending on which alleles you study, you can conclude that there are five human races, or that there are 300. The conclusion of these studies is that there are indeed genetic differences between different human populations, but these groups cannot be rigidly defined—and they do not cleanly align with the commonly used racial designations.

One last line of evidence relates to the anecdotes that you read in the introduction to this chapter. Many studies have demonstrated that *few* people who belong to the standard races are racially "pure." For example, Hispanics are a complex mix of African, Amerindian, and European ancestry. African Americans who trace their ancestry back to the American slave trade carry anywhere from 3 to 30% European ancestry, and this varies by populations of African Americans. For instance, African American populations in certain northern regions have 25 to 30% European ancestry; African Americans living in the American South have 10 to 20% European ancestry. The isolated Gullah population on islands off the coast of South Carolina is an African-American community with very little European ancestry. Similarly, many European Americans have some African ancestry. Interestingly, there appears to be only a little Native American ancestry in white or black populations in the United States.

All of this may appear hopelessly complex. It seems as if there are indeed genetic differences between groups, but nevertheless, all humans have a great deal of genetic heritage in common. Differences depend on which DNA sequences are studied. How did such a hodge-podge of genetic diversity and similarity arise? It is very likely that the complex patterns of DNA sequences within and between populations can be explained by the history of *Homo sapiens.*

To understand modern human genetic diversity you must travel back nearly 200,000 years. A wealth of evidence from fossils, from archeological artifacts, and from DNA indicates that *H. sapiens* emerged in Africa about this time. The population that gave rise to modern *H. sapiens* may have been as small as just 700 individuals. One thing that our ancestors had in common was a tendency to travel. Probably around 100,000 to 50,000 years ago, *H. sapiens* began migrating out of Africa to Europe and Asia (**Figure 14.26**). The Asian populations may have migrated out across the Pacific in boats. They also may have crossed into the Americas on foot over the land bridge that once linked Siberia and Alaska. Asian groups also migrated to the Pacific Islands. You might recognize this process of migration and establishment of new populations as providing the beginnings of speciation. It is a good example of the founder effect. If these groups in Asia, Europe, and America had remained isolated, today they might be on the way to separating into new species. Even today, you can find relatively isolated populations that have not had contact with other human groups for a long time. But *H. sapiens* has continued to travel (Figure 14.26) and

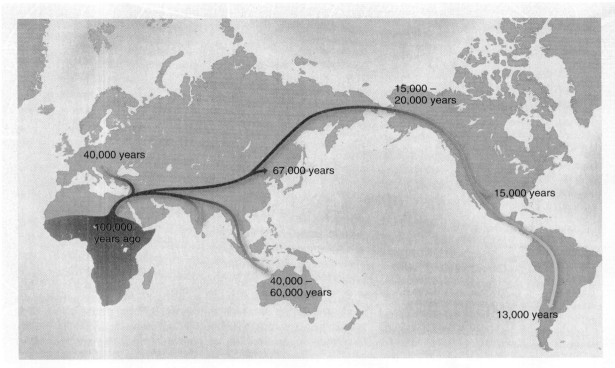

Figure 14.26
Migrations of Humans, Beginning 100,000 Years Ago.

has continued to interbreed. For example, the original migration of hunter-gatherer *H. sapiens* out of Africa was followed tens of thousand of years later by a major migration of farming *H. sapiens*. These often are referred to as the *Paleolithic* and *Neolithic* migrations. One important migratory event was the expansion of Europeans to all parts of the globe that started in the 1400s and persisted through the end of European colonialism in the 1900s. This migration was a bit different from previous ones. For one thing, colonial Europeans found native peoples already occupying every continent, and in some cases considerable interbreeding occurred between the Europeans and the native populations. This was particularly true of the Spanish colonial empire in Mexico, Middle America, and South America. A second difference from earlier migrations was the importation of slaves by the Europeans from Africa to the Americas. Not only did this force the migration of Africans, but also considerable interbreeding took place between Europeans and Africans in the New World.

Of course, twentieth-century technology is further promoting the movement of humans from one place to another. Easy travel fosters interbreeding of people from far-flung regions on the Earth. While it is certainly true that you are most likely to have children with people who are like you, and who live near you, the ease of modern travel means that in the future the human population is likely to be more homogeneous.

So what *are* human races? The evidence at this point shows that there are genetically defined human groups—but these groups are many, diverse, and overlapping. The typical race categories accepted by society today are *not* clearly defined genetic groups and are *not* a good summary of human genetic diversity.

QUICK CHECK

1. What does the scientific evidence say about the concept of race?

Now You Can **Understand**

Biodiversity Hot Spots

Earth's biodiversity is threatened. Although the threat comes from humans, many people are committed to limiting this threat and saving as much biodiversity as possible. But what can they do? How can biodiversity be saved? There are millions of species on Earth, so how do researchers and activists know which ones to focus on? One approach is to identify the biodiversity hot spots.

Basic research has made it clear that not all regions on Earth have the same degree of biodiversity. South America's Amazon rain forest, for example, carries a large percentage of Earth's biodiversity and probably has for as long as it has existed. In a loosely organized network, based in many different organizations and institutions, concerned people the world over are working to catalog the existing species in these biodiversity hot spots. The idea is that if we know what we might lose by destroying environments, then perhaps we will be less likely to do it. No one would try to rigidly define a biodiversity hot spot, but here are some of the areas that have a large degree of biodiversity:

1. The Tropical Andes (Venezuela, Colombia, Ecuador, Peru, and Bolivia)
2. Madagascar
3. Brazil's Atlantic Forest Region
4. The Philippines
5. Meso-American forests
6. Wallacea (eastern Indonesia)
7. Western Sunda (in Indonesia, Malaysia, and Brunei)
8. South Africa's Cape Floristic region
9. The Antilles
10. Brazil's Cerrado
11. The Darién and Chocó of Panama, Colombia, and Western Ecuador
12. Polynesia and Micronesian Island complex, including Hawaii
13. Southwestern Australia
14. The Eastern Mediterranean region
15. The Western Ghats of India and the island of Sri Lanka
16. The Guinean forests of West Africa
17. New Caledonia
18. Eastern Himalayas
19. Southeastern Australia and Tasmania

You can find more information about these areas online and might even find ways to contribute to the efforts.

What Do **You** Think?

Are Wild and Farm-Raised Salmon Equivalent?

The Endangered Species Act is one of several major efforts by the U.S. Congress to protect species on the verge of extinction. The fact that Congress enacts such laws shows that a majority of our elected representatives are convinced that biodiversity is worth protecting. The law requires that people change their habits to protect threatened species—not surprisingly, some people, businesses, and organizations are not happy to do so. It is not hard to find people who cannot understand why a spotted owl or tiny wildflower should prevent them from building a house, or putting up a shopping mall, or damming a river.

One species that the Endangered Species Act has been used to protect are the wild salmon of the Pacific Northwest. Wild salmon are so endangered that you probably will never see or even eat one. In supermarkets wild salmon have been replaced with farm-raised Atlantic and Pacific salmon. (There is debate about whether such farm-raised salmon are healthy, either for the environment or for the people who eat them, but the issue here is the Endangered Species Act.) In the Pacific Northwest commercial interests have been restricted to protect the environment of the wild salmon, which developers and others have complained about for years. Now the federal government has taken a new approach, declaring that when counts of wild salmon are done to determine how much protection the species needs, farm-raised salmon can be used in those counts. After all they are all salmon, aren't they? What's your opinion about how wild salmon should be counted? What do *you* think about the protection of wild Pacific Northwest salmon?

Chapter Review

CHAPTER SUMMARY

14.1 The Theory of Evolution Explains How Species Arise and How Species Are Related

The theory of evolution by means of natural selection provides the theoretical framework for all branches of biological science. It is a true scientific theory that has been tested many thousands of times. Establishing the genetic relationships of organisms and sorting them into categories based on relatedness is the work of classification. Aristotle defined species based upon morphological characteristics. Carl von Linné also used the morphological species concept. Ernst Mayr established the biological species concept that focuses on gene exchange. The phylogenetic species concept defines a species as the smallest group of organisms that comes from one common ancestor. Reproductive isolating mechanisms include modes of geographical isolation

such as the founder effect, sexual selection, and polyploidy. Hybrids reflect incomplete separation of two species. A species is defined as a naturally occurring, reproductively isolated group of organisms descended from a common ancestor.

biological species concept 339	phylogenetic species concept 339
geographic isolation 341	reproductive isolating mechanism
hybrid 343	341
morphological species concept 339	

14.2 Understanding Biological Diversity Involves Categorizing and Classifying Species

Categorizing and classifying is a common skill that is put into practice in biology by taxonomists. Linnaeus contributed a two-word scientific name and a comprehensive classification system of taxonomic categories. These include species, genus, family, order, class, phylum, and kingdom. Linnaeus identified two kingdoms. Later this was revised to five kingdoms. The contribution of phylogenetic classification reflects an analysis of genetic and evolutionary relationships in addition to morphological features. Computer analysis is an integral part of phylogenetic classification. Carl Woese used DNA sequences that code for ribosomal RNA as a basis for his universal tree of life. This analysis shows three domains of organisms and has caused the five-kingdom system to be reconsidered and revised. Archaea, Bacteria, and Eukarya are the three domains in Woese's analysis of DNA that codes for rRNA.

Archaea 349
Bacteria 349
classification 345
domain 349
Eukarya 349
phylogenetic classification 346
taxonomist 345

14.3 Traits of Major Phylogenetic Groups Reflect Life's Diversity

Archaea and Bacteria are prokaryotic domains. Archaea are more closely related to Eukarya than they are to Bacteria. Eukaryotes can be unicellular or multicellular. Green algae are grouped with plants. Fungi are more closely related to animals than to other eukaryotic groups.

invertebrate 350
vertebrate 352

14.4 The Diversity of Life Emerged Slowly

Life evolved on an early Earth that had far different environmental conditions than today's Earth. Human life is a relatively recent phenomenon in the history of life. Photosynthesis by cyanobacteria changed the atmosphere of early Earth by adding oxygen to it. Prokaryotes were the first forms of life, and eukaryotes evolved much later. Once organisms became multicellular, they diversified dramatically, forming many animal phyla. Mass extinctions are common in the history of life. Several million years after a mass extinction, biodiversity begins to rebound.

cyanobacteria 353
mass extinction 355

14.5 An Added Dimension Explained: Are Human Races a Valid Idea or Not?

Humans form a single species, *H. sapiens*. Because human races are not genetically distinct, human races are not biologically valid. Although some DNA sequences seem to be correlated with commonly accepted racial groupings, the genetic differences between racial groups are not cleanly defined and do not clearly align with the commonly accepted racial groupings. Most people have ancestry in several human groups. Migration and interbreeding have been characteristic of *Homo* species and modern *H. sapiens* continues this trend.

REVIEW QUESTIONS

TRUE or FALSE. If a statement is false, rewrite it to make it true.

1. The three domains are Archaea, Bacteria, and Prokaryota.

2. Extinctions have occurred repeatedly in the history of life.

3. The phylogenetic species concept includes the species and the ancestor that gave rise to it.

4. Invertebrates have exceptionally strong backbones.

5. Hawthorn flies and apple flies are a single species.

6. Polyploidy is a common kind of reproductive isolating mechanism in plants.

7. Only taxonomists use classification skills.

8. Life first evolved in an environment that was rich in oxygen.

9. Because human races are genetically distinct, human races are valid biological categories.

MULTIPLE CHOICE. Choose the best answer of those provided.

10. What is the best definition of a species?
    a. A species is a group of organisms with unique characteristics that they do not share with other organisms.
    b. A species lives together in one place.
    c. A species produces many offspring.
    d. A species often migrates from place to place.
    e. Members of a species can breed and produce viable, fertile young.

11. What has been produced by reproductive isolating mechanisms?
    a. About 500 species of fruit flies on the Hawaiian Islands are the result of reproductive isolating mechanisms.
    b. The flashy plumage of male birds is the result of reproductive isolating mechanisms.
    c. The increased breeding between cichlid species in Lake Victoria is the result of reproductive isolating mechanisms.
    d. Ligers result from reproductive isolating mechanisms.
    e. a and b are correct.

12. When female preferences change the physical appearance of males of the same species, which process has occurred?
    a. geographic isolation
    b. artificial selection
    c. sexual selection
    d. the founder effect
    e. classification

13. What is a phylogenetic classification based on?
    a. natural selection
    b. genetic and evolutionary relationships of species
    c. geographic isolating mechanisms
    d. viability of young
    e. artificial selection

14. Why is the DNA that codes for rRNA an excellent tool for phylogenetic investigations?
    a. The DNA that codes for rRNA changes slowly over evolutionary time.
    b. The DNA that codes for rRNA mutates rapidly.
    c. The DNA that codes for rRNA is found only in highly evolved species.
    d. All organisms have the DNA that codes for rRNA, including prokaryotes and eukaryotes.
    e. a and d are correct.

15. Bacteria and Archaea are both
    a. prokaryotic.
    b. eukaryotic.
    c. multicellular.
    d. in the same domain.
    e. closely related to Eukarya.

16. What does rRNA analysis of different kinds of algae show?
    a. Different kinds of algae are not closely related to each other, and green algae are related to plants.
    b. Different kinds of algae are closely related to each other.
    c. Different kinds of algae are closely related to animals.
    d. Different kinds of algae are closely related to fungi.
    e. Fungi should be reclassified as algae.

17. Who contributed the system of two-word scientific names for all known organisms and devised a classification system that had kingdom as the largest category and species as the smallest category?
    a. Aristotle   d. Charles Darwin
    b. Carl Woese   e. Ernst Mayr
    c. Carl von Linné

18. You use a field guide to identify species of flowers based upon their shapes, colors, number of petals, and position on stems. This is a classification system that is based on
    a. morphology.
    b. the biological species concept.
    c. the phylogenetic species concept.
    d. b and c are correct.
    e. presence of membranes around organelles.

19. An enormous lava flow and mudslide destroy an environment, cutting right through the geographic range of a species of small, wingless flies. Based on the example of what has happened in the evolution of Hawaiian fruit flies, predict the likely outcome of this event.
    a. The wingless flies will be unaffected by these geological changes.
    b. The geological changes have created two populations of wingless flies.
    c. Gene flow has been cut off between two new populations of wingless flies. Each may now evolve in isolation and there is the potential for new species to evolve.
    d. Female wingless flies in the two new populations will begin to prefer to mate with flies that can move faster and escape environmental catastrophes.
    e. b and c are correct.

20. What best explains the similarities between humans and chimpanzees?
    a. Chimps are descended from humans.
    b. Humans are descended from chimps.
    c. Both humans and chimps are descended from gorillas.
    d. Humans and chimps share a common ancestor.

CONNECTING KEY CONCEPTS

1. Consider this statement: the three different definitions of a species all reflect how species emerge, and so reflect different aspects of what a species is. Elaborate on this statement.

2. How do reproductive isolating mechanisms lead to the evolution of new species?

3. What has the analysis of the DNA that codes for ribosomal RNA told us about the phylogenetic relationships between organisms?

QUANTITATIVE QUERIES

1. Scientists have identified about 1.5 million species. They believe this represents about 15% of all living species. Based on these estimates, how many species probably inhabit the Earth today?

2. Scientists also conclude that as many as 1 billion—1,000,000,000—species have evolved since life first began on Earth, although of course most of them are now extinct. Using your answer from Quantitative Query 1, about what percentage of the total species that have ever lived are alive today?

THINKING CRITICALLY

1. How much do you know about the natural biological diversity in your own area? Use the Internet, nature centers, local museums, or older residents to find out what the natural landscape and biological diversity was like around your house or apartment before humans moved in. What species have been lost completely? Have any been pushed into less favorable environments? What issues relating to biodiversity have come before your local government in recent years? If you were on your city or county council, what steps might you take to rescue or restore some of the natural biodiversity in your area?

2. Many people hunt and eat wild animals to survive. This has always been the practice of humans, but in these modern times the hunting of bushmeat, as it is called, by millions of hungry people in Africa is a major reason for the extinction of many species. Imagine that you are working in a region where people hunt endangered species for food. Choose a region for your project, such as Africa or Asia or some areas of the United States. Devise a project that would help to provide alternative food sources for the people in your area. Before you delve into designing your project, consider what kinds of information you would need about the humans and animals that you will focus on. After you have outlined your own project, use the Web to find out what kinds of alternatives to bushmeat are being devised by people facing this kind of challenge in their own work.

3. Gather a handful of change—pennies, nickels, dimes, and quarters. Now, go outside and gather pebbles that are about the same sizes as the coins. Your task is to first examine the samples of coins and pennies. Decide on four categories for sorting coins and pebbles. Then sort the coins and pebbles. How is this exercise similar to and different from the work of a taxonomist?

For additional study tools, visit www.aris.mhhe.com.

CHAPTER 15

Varieties of Life

PROKARYOTES, SINGLE-CELLED EUKARYOTES, AND ALGAE

Red Tide Problem Persists on Cape

February 11, 2006

Red tides can carpet a beach with dead fishes. They also can turn fishes and shellfish poisonous.

Persistent Red Tide Takes Toll on Florida Sea Life and Tourism

October 8, 2005

Imagine that you are traveling along the coastal highway from Santa Barbara, California, up to Vancouver, Canada. The views from the coast highway are stunning, and walking along the shore brings you close to kinds of biological diversity that you may not have experienced. Slick, ropy, brown algae washed up along the beach tell you that out in the deep waters are the dense kelp forests where sea otters swim, and baby whales hide when their mothers go off to feed. If you're lucky, you'll see the otters playing in the water, looking like wet, whiskery dogs, or catch a glimpse of a whale farther out at sea. You probably will see scores of sea lions basking on the rocks. And when you get hungry there is great food. There are all kinds of fishes as well as oysters, clams, crabs, and mussels. You can gather your own or buy them at local seafood stands. If you take a month for an idyllic trip up the California coast into Washington State and north to the Canadian border, you will not be sorry.

Or maybe it won't be quite so perfect. Every once in a while—some say it happens more often now than in the past—the ocean serves up something that is not so nice (**see chapter-opening photo**). Sometimes the sea turns poisonous. Tourists are warned away from the shoreline. Beaches are closed, and local fishes and shellfish disappear from restaurant menus.

And fishes are not the only victims. In 1998 dead sea lions washed up on the beach in Monterey Bay. Autopsies revealed no signs of infection, and the deaths were mysterious. Then in 1990 in Puget Sound, a bay on the coast of Washington State, a disease wiped out the farm-raised salmon. All along the coast of Washington State, the clam-digging season draws as many as 30,000 people a day. During more than one year clamming season has been canceled because the clams were poisonous. These are just a few of the problems that western North America experiences with toxic substances in its ocean waters.

Problems like these are not restricted to the West Coast. Along the northeastern coast of the United States, from Virginia to New England, several large-scale deaths of marine mammals have occurred over the past 25 years. Across the globe in Malaysia, toxins often make shellfish inedible. Humans and other mammals are especially affected by these poisons. Every year in regions scattered around the globe a few people die after eating seafood—usually certain kinds of shellfish.

What causes these mysterious poisonings? What do people, shellfish, salmon, sea lions, and whales have in common that would cause die-offs and deadly poisonings? You will explore these questions at the end of this chapter after you have read about the some of the smallest varieties of life: prokaryotes and the single-celled eukaryotes and algae. ■

15.1 Scientists Continue to Identify New Species

When you think of Earth's spectacular biodiversity, what organisms come to mind? If you're like most people, you might think of tropical rain forests—but not undersea forests of seaweeds or the hordes of bacteria that live in your gut. You might think of herds of antelope on African plains—but not the microscopic life that abounds in a drop of pond water. The realm of prokaryotes, single-celled eukaryotes, and algae contains aspects of biodiversity that do not exist among the larger, more familiar eukaryotes. These organisms also play essential roles in the living world. Without them plants, fungi, and animals could not survive.

Scientists have studied and classified over 2 million species, but there are probably between 20 million and 50 million species on Earth. New estimates suggest there may be a million or more prokaryotic species alone. Their small size and habit of swapping genes make it difficult to assess the numbers of prokaryotic species. On the other hand, some scientists suggest that prokaryotes don't even *have* species in the same sense that

eukaryotes do. Studies of the biochemistry and DNA of prokaryotes can shed new light on this problem. Such studies show that what most people call a "species" of bacteria is usually quite a diverse population. For example, a sample of the colon bacterium *Escherichia coli*, commonly called *E. coli*, actually contains hundreds of different strains of bacteria. While most *E. coli* are harmless or even beneficial, sometimes a strain of *E. coli* is toxic. One dangerous strain of *E. coli*, called O157:H7, has been blamed for many deaths from food poisoning. The DNA and biochemistry of O157:H7 are distinctly different from that of the more common *E. coli*. While it is not yet clear if O157:H7 is a different species, it is obvious that gut bacteria that have been called "*E. coli*" actually are genetically diverse.

The identification of species of single-celled eukaryotes and algae pose similar problems: 30,000 species of single-celled eukaryotes and 24,000 species of red, brown, and golden algae have been identified, but certainly many more species exist. It often is hard to apply the biological species concept to these groups of algae because they do not mate and breed in the same ways that plants and animals do. The documentation of a population as a separate species requires a lot of work to demonstrate that it really *is* new. An interesting example is the discovery of two new amoeba species in 2003. *Amoeba* is a generic name for a group of single-celled eukaryotes that move by cytoplasmic extensions (see Figure 1.5). The researchers were looking for single-celled eukaryotes in soils that lack oxygen. Their efforts paid off when they found two kinds of single-celled eukaryotes that were different from any of the 30,000 species on record. The researchers used two of the species concepts to make this determination (Section 14.1). Initially, the morphological species concept was used to differentiate the two new species of amoebas. The newly discovered cells lack mitochondria and they have other unique morphological features. In an application of the phylogenetic species concept the researchers used DNA analysis to generate a phylogenetic tree of amoeba-like eukaryotes. The tree showed that these indeed were two new species.

QUICK CHECK

1. List three reasons that it is difficult to define prokaryotic species.

15.2 Diverse Adaptations Allow Organisms to Meet Common Challenges

Every organism must acquire and use energy, build biological structures, reproduce, get rid of wastes, maintain proper water balance, and carry out gas exchange. At first glance the specific solutions to these common problems may seem different from one species to the next. Closer inspection reveals that diverse solutions to the problems of survival are variations on common themes. These adaptations are never completely new because they are variations of the solutions that ancestral organisms used. Mutations produce small changes in proteins and sometimes these small changes give a new twist to an old solution. Subtle modifications of ancestral adaptations sometimes can produce large effects on the whole organism. And because all species are genetically related, you will see variations on the successful solutions used again and again, especially if you look directly at the proteins that are carrying out the solutions. Sometimes it seems that these solutions are not ideal. For exam-

ple, because the process of meiosis starts with DNA replication in S-phase (see Section 8.6), it seems unnecessarily complex. Meiosis starts this way because meiosis evolved from mitosis, which also begins with DNA replication in S-phase. Let's look at some of these challenges and solutions in a bit more detail.

Energy is absolutely necessary for life, and so every species must have successful biochemical metabolic pathways to handle energy (see Chapter 6). Most species obtain energy either from sunlight or from other organisms, but some species, especially bacteria and archaeans, have different solutions. The various adaptations of metabolic pathways to obtain and use energy are a good example of the principle that evolution does not invent entirely new solutions to problems. In all of these different kinds of organisms, the same basic kinds of proteins are used, and the same basic physical and chemical principles are exploited. For example, the molecules and processes of photosynthesis that produce ATP are similar to the molecules and processes of glucose metabolism that produce ATP.

Gas exchange also is another fundamental requirement. Most eukaryotic and many bacterial species use or produce carbon dioxide and oxygen in their metabolic processes, and so these gases must be able to get into, through, and out of an organism (see Section 4.2). All organisms take advantage of the basic laws of diffusion to accomplish gas exchange, but species have diverse structural adaptations for doing so. These include lungs, gills, tracheal tubes, and slender bodies that have tissues only a few cells thick.

Organisms must maintain the proper amount of water inside of each cell. If there is too much water inside of a cell, it can swell and burst; if there is too little water inside of a cell, it will shrivel. Either way, when a cell is out of water balance, its life processes can be impaired. As you will see, adaptations to control water balance are diverse. Contractile vacuoles in single-celled eukaryotes, stomata in plants, and kidneys in vertebrate animals all carry out this function. All of these adaptations exploit the process of osmosis (see Section 4.2). Furthermore, organisms control osmosis using proteins in their cell membranes, and these proteins are similar in all species.

In the process of natural selection the environment favors successful solutions over less successful ones (see Section 13.3). Every environment presents limits on the array of solutions that will be successful. For example, species that live in water have different adaptations than species that live in

Trout   **Maple tree**

Figure 15.1   **Life's Constraints and Challenges.** Life-forms as different as a trout and a maple tree face similar challenges and constraints. Both must have adaptations that allow them to survive where they live, adaptations that help them acquire energy, move gases to tissues, support their bodies, and maintain the correct amounts of water in their cells.

air (**Figure 15.1**). Fishes have gills to obtain oxygen from water, and plants have openings in their leaves that allow them to obtain carbon dioxide from air. Other species live in much more unusual settings such as within rock, in deep mud, within the cells or tissues of another organism, or in an unstable environment like a wave-swept shoreline. These environments drive different kinds of adaptations to the common problems of life.

15.3 Archaea and Bacteria Are Distinct Domains of Prokaryotes

Prokaryotes are what you commonly think of as bacteria, but biologists now recognize two major domains of prokaryotes: Archaea and Bacteria (**Figure 15.2**). It is impossible to distinguish between archaeans and bacteria just by looking, and this is one reason that all prokaryotes used to be lumped together as "bacteria." Under a light microscope members of the two groups look highly similar, and both are such small organisms that their internal structures are hard to see. Both domains contain unicellular organisms with cells that usually are less than 5 micrometers in diameter (see Figure 4.1), although some exceptionally large bacteria are up to 1 millimeter in diameter. Bacteria and archaeans have fairly simple shapes such as spheres, rods, or spirals (**Figure 15.3**).

The domains Archaea and Bacteria were first distinguished in studies of DNA sequences for the RNA in the small subunit of ribosomes, called rRNA. Even though their rRNA is different, ribosomes work in the same fundamental way in the two domains. Only about 50 nucleotide positions out of nearly 2,000 are different in the DNA that codes for rRNA in Archaea and Bacteria. The rRNA DNA sequences for species within each domain cluster together, showing that the Archaea and Bacteria form two distinct groups (see Section 14.3). DNA sequences that code for rRNA are about as different between Archaea and Bacteria as they are between Archaea and Eukarya. Clearly, the two groups of prokaryotes are different at the DNA level.

Other than DNA differences, how can you tell an archaean from a bacterium? A variety of biochemical studies have demonstrated other differ-

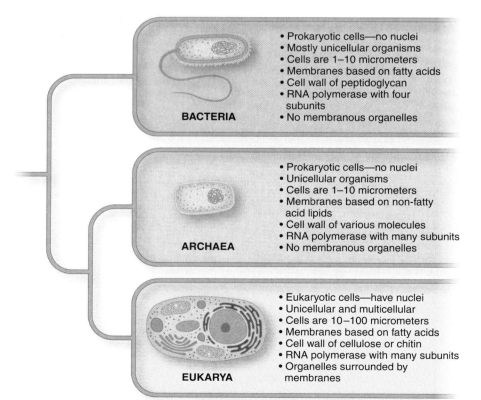

BACTERIA
- Prokaryotic cells—no nuclei
- Mostly unicellular organisms
- Cells are 1–10 micrometers
- Membranes based on fatty acids
- Cell wall of peptidoglycan
- RNA polymerase with four subunits
- No membranous organelles

ARCHAEA
- Prokaryotic cells—no nuclei
- Unicellular organisms
- Cells are 1–10 micrometers
- Membranes based on non-fatty acid lipids
- Cell wall of various molecules
- RNA polymerase with many subunits
- No membranous organelles

EUKARYA
- Eukaryotic cells—have nuclei
- Unicellular and multicellular
- Cells are 10–100 micrometers
- Membranes based on fatty acids
- Cell wall of cellulose or chitin
- RNA polymerase with many subunits
- Organelles surrounded by membranes

Figure 15.2   **A Comparison of Characteristics of the Three Domains of Life.**

(a) *Micrococcus*

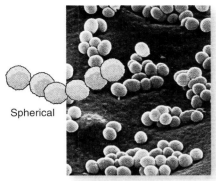

Spherical

(b) *Bacillus megaterium*

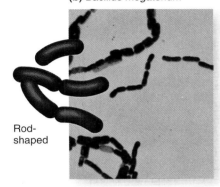

Rod-shaped

(c) *Rhodospirillum rubrum*

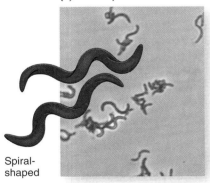

Spiral-shaped

Figure 15.3 **Shapes of Bacteria.** Bacteria can be **(a)** spherical like these *Micrococcus,* **(b)** rod-shaped like these *Bacillus megaterium,* or **(c)** spiral-shaped like these *Rhodospirillum rubrum.*

ences between archaeans and bacteria. These include different kinds of lipid molecules in their cell membranes, different molecules in their cell walls, differences in their RNA polymerase molecules, and different susceptibilities to various antibiotics. Consistent with the rRNA studies, other biochemical studies show that archaeans are similar to eukaryotes in many ways. Eukaryotes and archaeans have the same distinctive lipids in their cell membranes. Eukaryotes and archaeans

also do not make peptidoglycan, the molecule that bacteria use to form their cell walls. Finally, the RNA polymerase molecule of eukaryotes is similar to that of archaeans, and in some ways it is unlike the RNA polymerase molecule of bacteria.

Despite the biochemical and physical features that distinguish the three domains, it is important to remember that they are genetically related. All three evolved from a set of common ancestors, Earth's earliest forms of life, and they share many physical and biochemical features. For instance, species in the three domains have similar metabolic enzymes that extract and make use of energy. Many other biochemical processes in eukaryotes can be traced back to biochemical processes in prokaryotes. These include even complex processes in animals such as sensory perception, motor movements, and brain functions.

Bacteria and Archaeans Have Diverse Adaptations

Even though bacteria and archaeans have many biochemical differences, both are typically single-celled prokaryotes, and so they will be described together here. Prokaryotes are everywhere—in the air, in water, in ice, in soil, even deep within layers of rocks. Many types of bacteria live in or on other organisms, including you (**Figure 15.4**). For

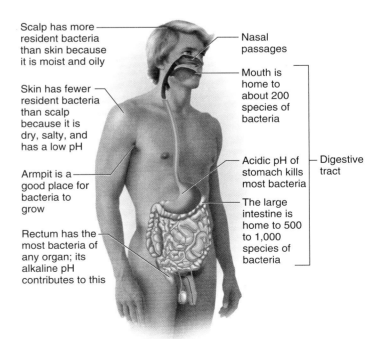

Scalp has more resident bacteria than skin because it is moist and oily

Nasal passages

Skin has fewer resident bacteria than scalp because it is dry, salty, and has a low pH

Mouth is home to about 200 species of bacteria

Armpit is a good place for bacteria to grow

Acidic pH of stomach kills most bacteria

Digestive tract

The large intestine is home to 500 to 1,000 species of bacteria

Rectum has the most bacteria of any organ; its alkaline pH contributes to this

Figure 15.4 **Populations of Resident Bacteria Are Found in Many Places in the Human Body.**

instance, they thrive on warm, moist, oily human skin and many are at home within portions of digestive, respiratory, and urogenital tracts of mammals, including humans. Even though some bacteria can make people ill, most bacteria that live on or within human bodies have a beneficial effect. Their presence is normal, and they contribute to human health. The bacteria in the human gut help digest food and often create an internal environment that is inhospitable to disease-causing bacteria, viruses, single-celled eukaryotes, and fungi. The normal bacteria in the human vagina keep its pH low, and this acidic environment inhibits the growth of harmful invaders.

Most prokaryotes are best adapted to thrive in wet or damp environments, but many species can survive in dry or cold environments. If the environment completely dries out or becomes cold, many prokaryotes survive by forming **bacterial spores** covered with tough protein coats that conserve the water inside of a cell and protect its biological molecules (**Figure 15.5**). Spores are a sort of microbial hibernation. They are not active and do not reproduce, but they maintain a minimal state of life over long periods of time. When the environment becomes more favorable for growth, the cell within the spore coat revives. Spores can survive for hundreds of years, and some studies claim to have revived spores that were millions of years old.

The properties of spores are important to remember when it comes to storing and cooking food. Food contains many kinds of bacteria, and probably archaeans as well, although the biggest concern is over bacterial contamination of food. Most bacteria are harmless, but some would make you sick if they had the opportunity to divide rapidly and produce quantities of bacterial cells and toxins. Because spores can survive extreme environmental conditions, they present problems during food preparation. While sterilization kills most spores, ordinary cooking destroys only some of the spores in food, leaving others to multiply. If you cook food that contains spores and then keep it at room temperature for a long time, the spores may become activated, and you could consume a dangerous dose of harmful bacteria and bacterial toxins. Freezing actually causes bacteria to form spores that stay dormant until the food is thawed. Spores revive when food is thawed, and bacteria begin to multiply after about 20 minutes at room temperature. Thawing at cool temperatures slows the rate of spore activation. Any live bacteria will be killed if the food is refrigerated while it is thawing, and then cooked thoroughly. If thawed food is refrozen, problems can arise. Because bacteria have begun to multiply within it, thawed food contains many more bacteria, bacterial spores, and higher levels of bacterial toxins. If you thaw this food and do not cook it thoroughly, you may have created conditions that will produce a large dose of harmful bacteria. This is why you should not refreeze previously frozen foods—and why you should thaw some foods like chicken, meat, and fish in the refrigerator (**Figure 15.6**).

Archaeans and bacteria can live in unusual environments, and many that live in the most extreme environments have earned the nickname *extremophiles*. For instance, halophiles, the "salt-loving" archaeans and bacteria, live in salty muds or other extremely salty environments. Like the internal fluids of all animals, human blood has a concentration of about 0.7 to 0.9% sodium chloride. In contrast, many halophiles are comfortable with salt levels that are about 9%; and some thrive in salt levels as high as 32%! Utah's Great Salt Lake is one place where such halophiles abound. You can tell that they are present because the water there sometimes turns an odd pink color (**Figure 15.7**). Halophiles and other extremophiles can survive in nonextreme environments, but there they are outcompeted and overgrown by other bacteria.

Some bacteria and archaeans also thrive at extreme temperatures that kill other forms of life. For instance, some archaeans are at home in cold ocean depths and Antarctic glaciers. Because archaeans

bacterial spore a resting stage of some bacteria that allows the organism to survive poor environmental conditions

Figure 15.5
A Bacerial Spore.
A spore forms when a bacterial cell encounters an adverse environment. (**a&b**) A spore is a bacterial cell enclosed by a protective outer layer. When environmental conditions improve the bacterial cell is released from the spore and resumes its normal life cycle.

(a)

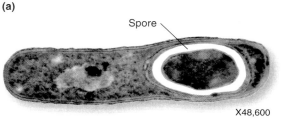

Spore

X48,600

(b)

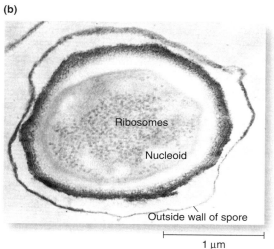

Ribosomes

Nucleoid

Outside wall of spore

1 μm

can live at such low temperatures, life could exist on planets like Mars that are much colder than Earth. In contrast, *thermophiles,* the "heat-loving" prokaryotes, thrive at temperatures ranging from 60°C (140°F) to over 100°C (212°F), the boiling temperature of water at sea level! Thermophiles live in the hot springs of Yellowstone National Park, in boiling hot waters of deep-sea thermal vents, and even in the hot coal refuse piles generated by some power plants. They can survive at these temperatures because their proteins are held together by stronger bonds than are found in typical proteins in other organisms. The proteins of extremophiles do not denature in high heat, salt, or acidity.

Knowledge about adaptations of extremophiles to particular environments has provided important insights about how life has evolved. In addition, the study of extremophiles has produced some useful technologies. The PCR technique that amplifies small samples of DNA (see Section 11.2) requires intervals of extremely high temperatures—accordingly, the DNA polymerase molecule that PCR uses must be stable at high temperatures. The DNA polymerase used in PCR was isolated from the bacterium *Thermus aquaticus,* a thermophile that lives in a hot springs in Yellowstone National Park, Wyoming.

Archaea and Bacteria Have Diverse Ways to Obtain and Use Energy

One reason that prokaryotes are so successful is the diverse ways that they obtain energy (**Figure 15.8**). Some prokaryotes are photosynthetic and capture

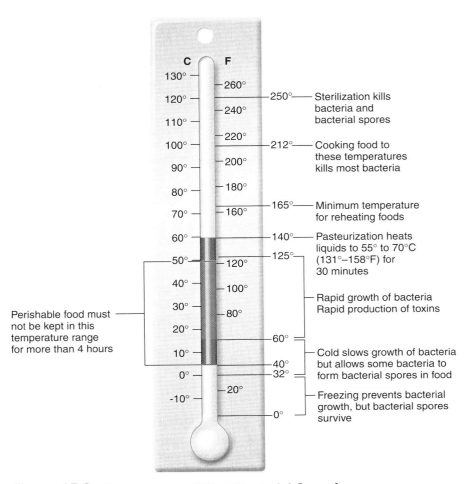

Figure 15.6 **Temperatures Affect Bacterial Growth.**

■ *What besides high temperatures can be used to kill bacteria?*

Figure 15.7 **Archaeans Turn the Great Salt Lake Pink.** The waters of Utah's Great Salt Lake sometimes turn pink or reddish because of plentiful archaeans.

Figure 15.8 **Cyanobacterium.** *Anabena* is a photosynthetic cyanobacterium commonly found in fresh water.

the energy in sunlight to make organic molecules that, in turn, are used as a source of chemical energy (Figure 15.8). Like green plants, some photosynthetic prokaryotes use water as a source of electrons for photosynthesis (see Section 6.2). The cyanobacteria have photosystems that are identical to those of plants; and just like plants, cyanobacteria produce oxygen as a result of photosynthesis. Like most eukaryotes, some bacteria require oxygen to produce ATP. In contrast, other prokaryotes use molecules like hydrogen sulfide (H_2S) as the source of electrons in the process of photosynthesis. As a result, sulfur bacteria do not release oxygen. Instead, they produce pure sulfur that they store in special packets inside the cell. Many other kinds of bacteria use the energy in the bonds of organic molecules to make ATP. Still other bacteria use the energy in the bonds of methane or other hydrocarbons to make ATP. Finally, some of the most unusual prokaryotes get their chemical energy from inorganic sources such as compounds of nitrogen, gold, or iron. These metabolic adaptations are never seen in the eukaryotic realm. It would be like you getting chemical energy by eating a bucket of iron ore!

Another aspect of the metabolism of some prokaryotes is their sensitivity to oxygen. Some prokaryotes can survive without oxygen and most anaerobic prokaryotes will die when exposed to it. Anaerobic environments include layers of mud beneath oceans, rivers, swamps, or lakes; the intestinal tracts of animals; ice caps and glaciers; deep ocean waters; or in the pockets between your gums and teeth.

flagellum (plural: flagella) a tail-like cytoplasmic extension used to propel a cell

bacterial conjugation the joining of two prokaryotic cells by a short cytoplasmic bridge through which DNA is exchanged

bacterial transformation process in which living prokaryotes take up DNA from the environment and incorporate it into their own genomes

Prokaryotes Have Diverse Cellular Structures

Prokaryotic cells are so tiny that it is nearly impossible to appreciate the diversity of their cellular structures without a powerful electron microscope. Compare the few features of prokaryotes that you can see with a light microscope (see Figure 15.3) with structural details that the higher magnification of an electron microscope reveals (**Figure 15.9**). Using an electron microscope you can see that prokaryotes lack the complex organelles of eukaryotes, and have fairly simple cellular structures. Most of the DNA of a prokaryote is in one large, circular chromosome localized in a central region of the cell. The prokaryotic chromosome is not as highly structured as a eukaryotic chromosome, and it is not tightly coiled. Additionally, there are smaller sets of genes called *plasmids* carried on circular pieces of prokaryotic DNA that are not integrated into the chromosome. Plasmids are widely used as tools of genetic engineering (see Section 11.5).

Prokaryotes have other structural adaptations, and you will see similar adaptations in other groups of organisms. They have one or many **flagella,** little whips that rotate, propelling the cell generally toward sources of food or light. The cell of a prokaryote has an outer barrier, the cell wall, that provides structural support and helps to control the flow of chemicals across the cell's boundary. Outside of the cell wall is a capsule that may be sticky or slimy. The capsule of disease-causing bacteria keeps them from being destroyed by immune cells and helps them to attach to the cells of the tissues that they attack. The plaque that accumulates on your teeth if you fail to brush them is an example of colonies of bacterial cells coated by slimy capsules (**Figure 15.10**).

As you might expect, prokaryotes with different metabolic adaptations have different internal structures. An electron microscope allows you to see some of the internal structural complexity of some prokaryotic cells (**Figure 15.11**). For instance, cyanobacteria are filled with internal membranes where photosynthesis takes place; prokaryotic cells that use the energy in inorganic molecules have a different internal structure.

Being Small Solves Many Problems

Because they are so small, prokaryotes do not need any special adaptations for gas exchange. Carbon dioxide and oxygen are nonpolar gases

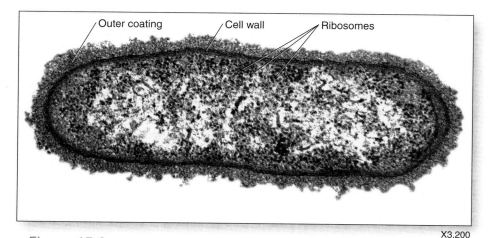

Outer coating · Cell wall · Ribosomes

X3,200

Figure 15.9 **The Detailed Structure of a Prokaryote.** Prokaryotes are complex cells, but they do not have nuclear membranes or other organelles enclosed in membranes.

that diffuse through the cell membrane and can move throughout the prokaryotic cell. Being small also simplifies the problem of water balance. If a prokaryotic cell takes up too much water, or loses too much water, membrane protein channels open to solve the problem. Water and ions flow through these channels until the proper water balance is restored.

Small size also contributes to many prokaryotic adaptations that you already have read about. For example, because prokaryotes are too small for gravity to have any significant impact on their structure, they can rely on the cell wall to maintain their shapes. Being so small limits the structural complexity of prokaryotes. They cannot develop the diverse shapes or complex functions that are typical of eukaryotic cells.

Prokaryotes Generate Genetic Diversity in Several Ways

Although prokaryotes do not have sexual reproduction in the same way that eukaryotes do, they have other ways to introduce genetic diversity into their populations. Prokaryotes primarily reproduce using binary fission, which is similar to, but not identical with, mitotic cell division (see Section 8.3). Fission produces genetically identical offspring. So how do prokaryotes become genetically diverse? One way is through mutations. Errors in DNA replication can lead to mutations. In addition, many environmental factors can cause changes in the sequence of nucleotides in DNA, and some mutations will allow an individual to survive in a new environment (see Section 13.5). Through binary fission successful mutations are passed from a prokaryote to its descendants.

Prokaryotes commonly swap genes, using the process of **bacterial conjugation,** which increases their genetic diversity. During conjugation two cells exchange pieces of DNA (**Figure 15.12a**). In this process, a short bridge of cytoplasm forms between two bacterial cells, and small pieces of DNA are exchanged through the bridge. Prokaryotes also can pick up DNA from the general environment. When prokaryotes die and release their DNA nearby, living prokaryotes can take it up and add it to their own genomes. This process is called **bacterial transformation** (Figure 15.12b), and Griffith's studies of transforming prokaryotes took advantage of this trait (see Section 7.1). These are the mechanisms for lateral gene transfer that you read about in Section 14.3.

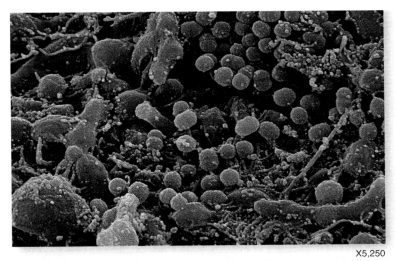

X5,250

Figure 15.10   **Dental Plaque.** Different kinds of bacteria multiply on unbrushed teeth, forming a film that coats them.

(a) Photosynthetic cyanobacteria have systems of internal membranes where photosynthetic enzymes are located.

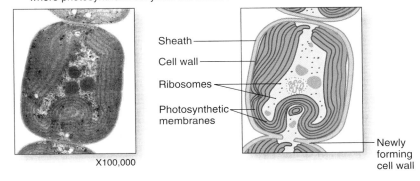

X100,000

Sheath
Cell wall
Ribosomes
Photosynthetic membranes
Newly forming cell wall

(b) Bacteria that metabolize nitrogen-containing compounds also have more complex internal membranes.

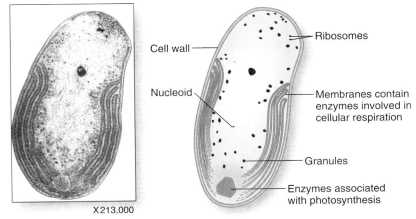

X213,000

Cell wall
Nucleoid
Ribosomes
Membranes contain enzymes involved in cellular respiration
Granules
Enzymes associated with photosynthesis

Figure 15.11   **Internal Structural Diversity of Bacteria.** (**a**) The internal structure of *Anabena,* a photosynthetic cyanobacterium. (**b**) The internal structure of *Nitrobacter,* a bacterium that metabolizes nitrogen.

■ *Give reasons for the differences in the internal structures of these cells.*

(a) Bacterial conjugation

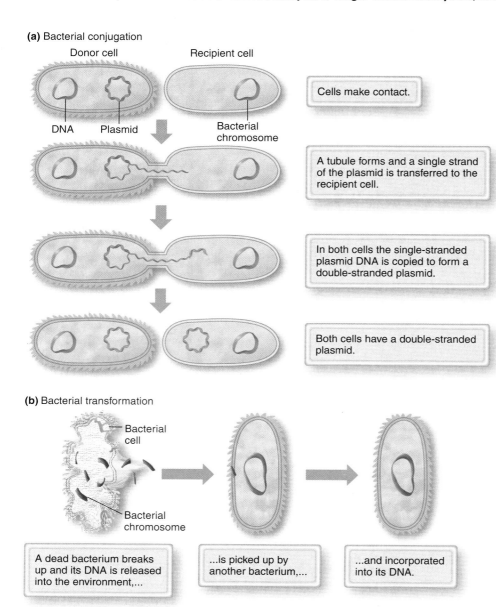

Donor cell

Recipient cell

DNA   Plasmid

Bacterial chromosome

Cells make contact.

A tubule forms and a single strand of the plasmid is transferred to the recipient cell.

In both cells the single-stranded plasmid DNA is copied to form a double-stranded plasmid.

Both cells have a double-stranded plasmid.

(b) Bacterial transformation

Bacterial cell

Bacterial chromosome

A dead bacterium breaks up and its DNA is released into the environment,...

...is picked up by another bacterium,...

...and incorporated into its DNA.

Figure 15.12   Bacterial Conjugation and Bacterial Transformation.
(a) A singled-stranded copy of the plasmid from one cell is transferred to another when bacteria conjugate. Later both cells replicate a double-stranded copy of the plasmid.
(b) In bacterial transformation a bacterium picks up a plasmid from the environment. Both processes increase the genetic diversity of bacteria.

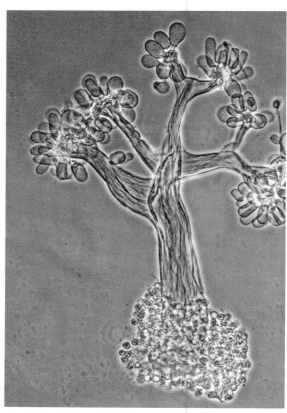

Figure 15.13   Myxobacteria. Most kinds of bacteria are unicellular, but some can form multicellular colonies at certain times in their life cycles. The swarming body of *Stigatella* is an example.

Some Prokaryotes Have Multicellular Life Cycle Stages

Throughout this text you have read about prokaryotes as single-celled organisms, but this is not entirely true. A few kinds of bacteria have brief multicellular phases in their life cycles. An example is the group known as *myxobacteria*. As single cells, these bacteria glide over a surface, rather than swimming with flagella. When energy sources become depleted, myxobacterial cells come together and form a multicellular structure that grows upward (**Figure 15.13**). Some of the cells differentiate into myxospores that are released into the environment. When a spore finds a favorable environment, it transforms into a gliding bacterial cell.

You might think that the multicellularity of myxobacteria is an interesting oddity, but it has a greater importance. Myxobacteria are yet another example of the observation that the complex characteristics of organisms evolve from simpler versions of them. Myxobacteria demonstrate that the potential for cell aggregation, cell differentiation, and multicellularity probably was present in the earliest cells. This means that these properties did not arise suddenly among the multicellular eukaryotes.

Gram Staining Helps to Identify Two Different Kinds of Bacteria

The Archaea and Bacteria can be subdivided into smaller groups of related prokaryotes. One of the oldest techniques used to classify bacteria is based on a staining procedure developed in 1884 by the

Danish physician Hans Christian Gram. The **Gram stain** separates bacteria into two major groups based on the structure of the cell wall. Bacteria with heavy cell walls turn purple when gram stain is applied. They are called gram-positive bacteria. The same technique turns bacteria with thinner cell walls light pink; these bacteria are called gram-negative bacteria (**Figure 15.14**).

Compared with modern molecular ways to identify and classify bacteria, the Gram stain may seem quite old-fashioned. It still, however, has an important place in diagnosing infections. One reason the Gram stain is still widely used is that it can be done quickly. Other tests require that a bacterial sample be grown in a culture dish for 24 hours or require genetic analyses that take several hours. The Gram stain can be done in 15 minutes. By itself the result of a Gram stain cannot identify a bacterium, or tell what antibiotic to use to treat a bacterial infection. But, together with other information about a patient's condition, a Gram stain can tell a lot. For example, if gram-negative bacteria are found in the spinal fluid of a patient with certain symptoms, the doctor can make a good guess the patient has spinal meningitis and can treat the patient accordingly. If the same patient's spinal fluid sample has few gram-negative bacteria, then spinal meningitis is unlikely.

Bacteria Have Many Crucial Roles in Biological Systems

Although the more exotic archaeans have little direct interaction with humans, most people are quite aware of bacteria because some strains or groups of bacteria are **pathogens**—that is, they cause diseases (**Table 15.1**). Pathogenic bacteria can have deadly or harmful effects, and you might wish that you could live in a germ-free environment. Nevertheless, it would be a poor idea to get rid of all bacteria because they perform crucial jobs in the overall scheme of life. For instance, without the bacteria that live in your colon, you would not stay healthy and fit because your *E. coli* supply the vitamin K that is necessary for normal blood clotting.

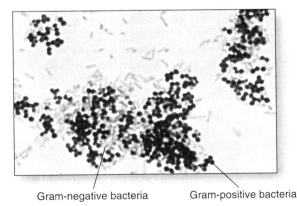

Figure 15.14 **Gram Staining.** Using the Gram staining technique, Gram-positive bacteria stain purple, while gram-negative bacteria stain pink.

Gram stain a staining technique that distinguishes between two kinds of cell walls of bacteria

pathogen an organism that causes a disease

Table 15.1 Fast facts about some pathogenic bacteria

Pathogenic bacterium and disease it causes	Fast fact
Bacillus anthracis Anthrax	Risk of infection is low, but illness is serious once infected.
Borrelia burgdorferi Lyme disease	About 10,000 cases each year; especially in areas where white-tailed deer are common; epidemic in islands off New York and Massachusetts.
Clostridium botulinum Botulism	Negligible risk because home canning of low-acid foods is now uncommon; genuine risk for infants who are fed honey.
Escherichia coli Strain O157:H7 Hemorrhagic colitis leading to kidney failure	People who eat undercooked beef and raw milk, especially children.
Mycobacterium leprae Leprosy or Hansen's disease	Around 12 million cases occur each year, worldwide; disabling, but not fatal.
Mycobacterium tuberculosis Tuberculosis	Overcrowded, stuffy conditions like jails and airplane cabins foster transmission of TB; incidence of TB is rising in the United States.
Neisseria gonorrhoeae Gonorrhea	90% of infections have none of the symptoms such as discharge or painful urination. About 3 million Americans are infected each year. Risk decreases with use of condoms.
Neisseria meningitidis Meningitis	A rapidly developing disease. Stiff neck, fever, brain hemorrhages, coma, and death can happen within a day. Survivors can have brain damage. Occurs in epidemics.
Rickettsia typhi Typhus	Characterized by a fever and rash; transmitted by rat fleas; typhus is fostered in crowded, dirty living conditions. Another reason to keep your house clean.
Salmonella species Food poisoning	This kind of food poisoning occurs 5–72 hours after eating an infected food. Common in raw meat, especially poultry and eggs.
Streptococcus species Sore throat, strep infection, scarlet fever, skin infection	Extremely common; the "flesh-eating" bacteria was a kind of strep.
Treponema pallidum Syphilis (the pox)	The occurrence of syphilis is rising in the United States, associated with drug abuse.
Vibrio cholerae Chlolera	Spreads whenever water or food is contaminated with human feces.
Yersinia pestis Bubonic plague (black plague, black death)	The route of transmission goes like this—infected rodents are bitten by fleas that then bite humans and pass on the bacteria. About 60% of untreated cases are fatal.

Table 15.2 lists a few of the many bacteria that have beneficial effects.

Bacteria have many roles that go beyond their importance to human health. One crucial role of bacteria is to provide other organisms with nitrogen in a form that they can use to make proteins and other biological molecules (see Section 3.1). Most organisms cannot make use of the abundant nitrogen that makes up about two-thirds of Earth's atmosphere. Atmospheric nitrogen is a molecule held together by a triple covalent bond. It takes a huge amount of energy to break that bond, but some species of bacteria can do it easily using enzymes. Without these nitrogen-processing bacteria, the rest of life could not make proteins, DNA, or other molecules that contain nitrogen.

Another important role of bacteria is to break down the tissues of dead organisms and recycle the nutrients and other chemicals that they contain. Without the recycling activities of bacteria, it would not be long before nutrients would be in short supply because they would remain locked within the tissues of dead organisms. Bacteria also are important photosynthesizers in many environments such as salt marshes and freshwater ponds and lakes. They form a nutrient-rich, green scum that is grazed by many other organisms, including single-celled eukaryotes, snails, tiny fishes, and larvae of crustaceans, crabs, lobsters, and shrimp. So bacteria have vital roles in keeping Earth's natural systems running smoothly. In one way or another all other organisms depend on the metabolic activities of bacteria.

QUICK CHECK

1. Why aren't archaeans grouped as bacteria? What do archaeans have in common with bacteria?

15.4 Genetic Relationships Are the Basis for Grouping Single-Celled Eukaryotes and Algae

Every eukaryotic cell has a membrane that encloses the nucleus as well as other complex organelles in the cytosol (see Figure 4.14). Single-celled eukaryotes and the algae are the simplest eukaryotes. For years biologists have debated about how to classify these organisms. Some placed both into a large group called protists; others reserved the name protist for single-celled eukaryotes and grouped the algae with the plants; still others placed algae into a separate group. DNA and biochemical analyses allow scientists to sort out these conflicting classification schemes, and it turns out that none is correct. Rather, each major group of multicellular eukaryotes—algae, fungi, plants, and animals—is genetically linked to one or a few groups of single-celled eukaryotes. The rest of the single-celled eukaryotes are a diverse mix whose genetic relationships are difficult to sort out. Let's concentrate here on the single-celled eukaryotes and trace their relationships with various kinds of algae.

Single-Celled Eukaryotes Live in Diverse Environments

Like prokaryotes, most single-celled eukaryotes inhabit watery environments. Many move about using flagella or *cilia,* which are smaller projections from the cell membrane that often cover the surface of a cell. Others have extremely plastic forms and move by oozing into extensions of the

Table 15.2	Fast facts about a few important bacteria and archaeans
Bacterium	**Fast fact**
Acetobacter	Turns wine into vinegar.
Bacilli	Produce many kinds of antibiotics.
Cellulomonas	Breaks down cellulose molecules, especially those of dead trees; only a few organisms can break down cellulose.
Cyanobacteria	Their photosynthesis adds oxygen to Earth's atmosphere; credited with producing Earth's current oxygen levels.
Escherichia coli	Has been extensively used in biological research and has enabled scientists to understand many of life's processes.
Lactobacillus	A prime fermenter; its activities produce alcohol, beer, wines, and cheeses; also makes citric acid, vitamin B_2, vitamin B_{12}, antibiotics, and MSG.
Methanogens	Have crucial roles in sewage-treatment plants; produce natural gas, methanol, formaldehyde, chloroform, and carbon tetrachloride.
Nitrobacter, Nitrospina, Nitrocystis, and Nitrosolobus	Function in the cycling of nitrogen; oxidize nitrite (NO_3) into nitrate (NO_2), making nitrogen available for plants, and indirectly, for other organisms.
Prochloron and Prochlorothrix	May be the descendants of prokaryotes that once gave rise to chloroplasts when they were engulfed by a cell that was on its way to becoming eukaryotic.
Rhizobium	Nitrogen-fixing bacteria that live within nodules on the roots of plants related to peas; these bacteria transform nitrites (NO_3) into nitrates (NO_2) that plants can take up and incorporate into their tissues.

cell membrane. Few, if any, single-celled eukaryotes live in the air, but like bacteria, they can form spores that allow them to survive drought or extreme temperatures.

Multicellular eukaryotes are not noted for their ability to live in extreme environments, but some species of single-celled eukaryotes do. Some species thrive in extremely salty, cold, warm, acidic, or alkaline environments. Certain species can even live in anaerobic environments. These extremophile eukaryotes are not necessarily genetically closely related to one another. Can you think of how adaptations to such diverse environments might have evolved in eukaryotes? Recall that chloroplasts and mitochondria originated when early cells engulfed prokaryotes. This same process can explain other physical features of eukaryotes, like flagella and cilia, as well as the ability of some eukaryotes to live in extreme environments. For example, anaerobic eukaryotes do not have mitochondria, but they do have a metabolic organelle called a *hydrogenosome* that resembles certain anaerobic prokaryotes. They may have obtained this organelle by engulfing, but not digesting, anaerobic prokaryotes.

Single-Celled Eukaryotes Have Diverse Structures

Single-celled eukaryotes have the basic organelles found in the multicellular eukaryotes, but many also show remarkable structural diversity. *Paramecium* is a common single-celled eukaryote that shows several complex cellular structures (**Figure 15.15**). *Paramecium* has one large **macronucleus** and at least one small **micronucleus.** The macronucleus takes care of day-to-day normal transcription and gene expression. The micronucleus is only used for "sexual" reproduction in which pairs of paramecia exchange micronuclei. After such an exchange, the macronucleus disintegrates, and the new micronucleus becomes integrated into the paramecium's DNA. Paramecia also have an **oral groove,** a depression that funnels food particles down to a place in the membrane where food enters the cell (Figure 15.15). As the cilia of *Paramecium* twirl it through the water, the cilia around the oral groove sweep other small cells into the oral groove. At the bottom of the oral groove, food particles are enclosed within food vacuoles. Lysosomes fuse with these and add digestive enzymes. The food vacuoles make a circuit through the cytosol of *Paramecium*, and digested food molecules diffuse out into the cytosol. Eventually the nutrients in the food vacuole have been exhausted and the undigested remains of the food are expelled at the **anal pore,** a specialized region of

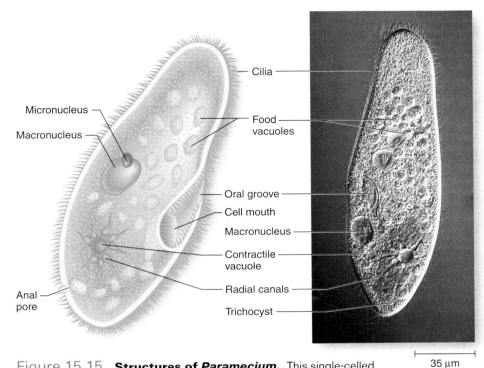

Figure 15.15 **Structures of *Paramecium*.** This single-celled eukaryote has a complex internal cellular structure with specialized organelles.

the cell membrane. So the combination of oral groove, food vacuoles, lysosomes, and anal pore acts as a digestive system within the single cell. Paramecia also have a single cell's version of an excretory system: a **contractile vacuole** surrounded by *radial canals* that help maintain the correct balance of water within the cell.

Single-Celled Eukaryotes Have a Variety of Feeding and Reproductive Strategies

Single-celled eukaryotes have ways to acquire energy that are as diverse as those of other eukaryotes. Many single-celled eukaryotes are photosynthetic, and the common *Euglena* is one example. Depending on environmental conditions, *Euglena* can be photosynthetic (**Figure 15.16a**) or can capture and ingest food. Other single-celled eukaryotes also must **ingest** their food, meaning they must bring it inside of the cell for digestion. Some collect food by **filter feeding,** a process that moves large amounts of water through the organism and sifts out the food particles. *Paramecium* is a filter feeder. *Amoeba* has a different strategy: it sends out cytoplasmic extensions that surround and engulf food and eventually enclose it within a food vacuole. Still other single-celled eukaryotes are predatory hunters

macronucleus contains the copy of the *Paramecium* genome used for day-to-day transcription

micronucleus contains the copy of the *Paramecium* genome used for sexual reproduction

oral groove a depression in the surface of *Paramecium* that funnels food particles down to a place in the membrane where food enters the cell

anal pore a membrane pore through which wastes are expelled from *Paramecium*

contractile vacuole an organelle in *Paramecium* that regulates cellular water balance

ingest the process of bringing food inside of an organism for digestion

filter feeding a process that moves large amounts of water through the organism and filters out food particles

(a) *Euglena*

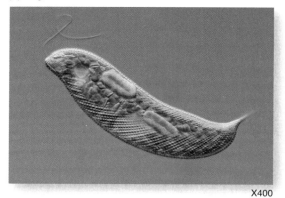

X400

(b) *Didinium*

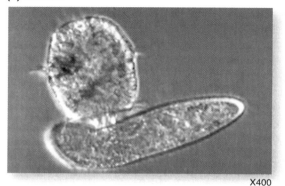

X400

(c) *Suctorian*

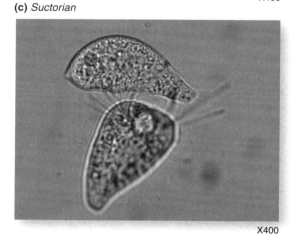

X400

Figure 15.16 Single-celled Eukaryotes Have Diverse Ways of Obtaining Energy. (**a**) *Euglena* is photosynthetic; (**b**) *Didinium* is a predator that punctures and engulfs other single-celled eukaryotes like this *Paramecium;* (**c**) A suctorian uses its tentacles to attach to another cell and suck in its cytoplasm.

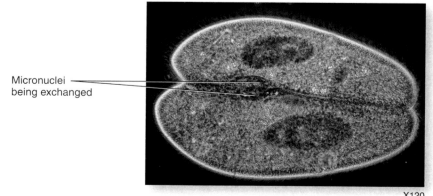

Micronuclei being exchanged

X120

Figure 15.17 Conjugation of *Paramecium*. When paramecia conjugate, their cell membranes partially disintegrate and they exchange micronuclei. After the cells separate, each undergoes fission and divides into two new cells.

(Figure 15.16b). Some send poisons into the environment that kill or immobilize their fast-moving prey. Others are like single-celled vampires: they adhere to their prey and suck out nutrients (Figure 15.16c). As in other aspects of life, single-celled eukaryotes have nutritional styles that are strikingly similar to the behaviors of multicellular eukaryotes.

Although most single-celled eukaryotes can reproduce asexually, using mitotic cell division, sexual reproduction also is common, and a single species may use both strategies. Sexual reproduction in single-celled eukaryotes is different from that of multicellular organisms. For instance, when two paramecia "mate," they join and exchange copies of the DNA carried in their micronuclei, a process that is similar to the process of conjugation in prokaryotes (**Figure 15.17**). As a result, each individual benefits from the genetic variations of its partner. Single-celled eukaryotes are not as sensitive to the total amount of DNA as are animals; many single-celled eukaryotes can carry multiple copies of their complete genome and still function normally. Meiosis does play a role in reproduction of single-celled eukaryotes. For example, photosynthetic single-celled eukaryotes called *diatoms* release structures similar to sperm or eggs that can fuse to produce a new complete diatom.

Single-Celled Eukaryotes May Have Remarkably Complex Support and Protective Structures

Even though single-celled eukaryotes are small, they have more complex structures and supports than do prokaryotes. Like all eukaryotes they have an internal cytoskeleton made of protein filaments and microtubules. In amoebas the cytoskeleton can be broken down and rebuilt in different parts of the cell. This is how amoebas move and how they surround their food. The distinctive shapes of many other single-celled eukaryotes also are dictated by the cytoskeleton.

Some single-celled eukaryotes have extremely complex and beautiful external structures. For instance, foraminiferans produce complex shells that are similar to shells of snails (**Figure 15.18a**). You can find the multichambered, highly variable shells of forams by examining grains of sand and bits of seaweed along the tide lines of many beaches. Forams capture and eat other organisms, and when they die their chalky shells settle to the ocean floor and may form thick layers. The famous white chalk cliffs of Dover, England, are made of compressed layers of shells of ancient forams. The glassy shells of radiolarians and diatoms are equally beautiful and amazingly complex (Figure 15.18b).

(a)

(b)

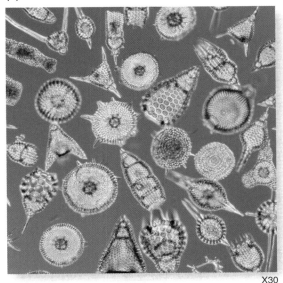

X30

Figure 15.18 **Foraminifera and Radiolarians.** (**a**) The chalky shells of forams have distinctive and unusual shapes. (**b**) The glassy shells of radiolarians have more precise geometry.

Figure 15.19 **Plankton.** Free-floating assemblages of single-celled eukaryotes, algae, and larvae of crustaceans are an important food source in oceans, lakes, and ponds.

Single-Celled Eukaryotes Play Important Biological Roles

Like prokaryotes, single-celled eukaryotes play crucial ecological roles that support other kinds of life. Many single-celled eukaryotes are photosynthetic and contribute to the oxygen in the Earth's oceans, lakes, and atmosphere. In oceans these single-celled eukaryotes multiply by the millions and are part of the free-floating community of tiny organisms commonly called **plankton** (**Figure 15.19**). Plankton is the food source for many larger organisms, including Earth's largest animals, the baleen whales. But, of course, single-celled eukaryotes also can cause disease. Malaria, African sleeping sickness, and paralytic shellfish poisoning are all caused by single-celled eukaryotes.

Algae Are Diverse Groups of Photosynthetic Organisms

Like the prokaryotes and single-celled eukaryotes, algae have been hard to classify. Studies of DNA, biochemistry, and genes of algae suggest that there are at least three major groups: green, brown, and red algae (**Figure 15.20**). Each is genetically linked to a particular group of single-celled eukaryotes. Because green algae are biochemically and genetically similar

plankton small organisms that float or drift suspended in oceans, lakes, rivers, and streams

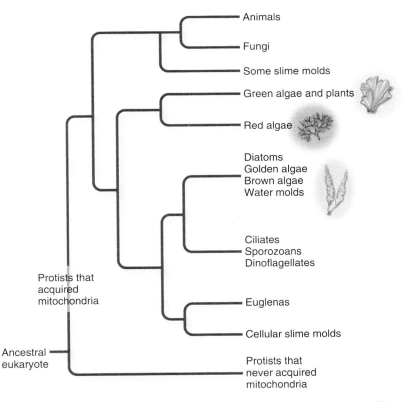

Figure 15.20 **Relationships of Algae and Other Eukaryotes.** This diagram shows that while green algae are grouped with plants, red algae and brown algae form separate, distinct groups.

to plants, the green algae will be discussed with plants in Chapter 16. Here you will read about the brown and red algae.

At a superficial level both brown and red algae resemble plants because they are multicellular, photosynthetic organisms. Biochemical analyses of brown and red algae reveal that they have only superficial similarities to plants and green algae. Brown and red algae have different chlorophylls and other pigments to capture the energy in sunlight. They also have different chemicals in their cell walls. While plants have internal tubules that carry water and nutrients, algae do not. Finally, a comparison of DNA of plants (including green algae), brown algae, and red algae shows that they are distantly related groups. Right now the best definition is that algae are photosynthetic eukaryotes that are *not* plants.

Because brown and red algae are photosynthetic organisms that live in water, usually in marine environments, they do have some features in common. Each group includes both single-celled and multicellular species. Although the multicellular species can grow to be huge, they are composed of only a few types of simple tissues. Algae do not have complex internal structures. Their bodies are thin and sheetlike or have indentations that give individual cells access to the environment. Accordingly,

gases can diffuse to all cells of an alga. All forms of algae have some sort of protective coating—a cell wall or a shell—that helps maintain structure and control water balance. Despite these similarities, brown and red algae are two distinct and distantly related groups of eukaryotes.

Brown algae and their relatives belong to a diverse group called *Chromista* that includes enormous kelps and single-celled diatoms. Kelps are the large brown algae with thick rubbery stalks and round rootlike bases that you may have seen washed up on beaches where the ocean water is cold. Viewed from underwater, kelp have quite a different appearance (**Figure 15.21**). Each kelp can be about 90 feet long (30 meters), and they form underwater kelp forests that are home to many other marine organisms including sea otters, marine turtles, sea urchins, crabs, shrimp, sea stars, and whales. Like so many other environments, many kelp forests have been dramatically reduced or have been damaged by human activities. Many scientists are convinced that global climate change, which is warming the oceans, is a major factor in the decline of kelp forests.

Red algae grow in cold to temperate waters around the world and are a major component of the coral reefs that are found in warm waters (**Figure 15.22**). Like brown algae, red algae can be unicellu-

Figure 15.21 **An Underwater Forest of Brown Algae.** Anchored on the bottom by round holdfasts, the blades of these giant kelps extend up as much as 100 meters (about 329 feet) toward the water's surface.

Figure 15.22 **Corraline Red Algae.** This kind of red alga looks like a miniature red tree and feels stiff and hard from deposits of calcium carbonate in its tissues.

lar or multicellular. They have some unique traits not found in any other eukaryotes—such as an unusual set of photosynthetic pigments that reside in an organelle called a *phycobilisome*. They also have cells with multiple nuclei. Despite these unique characteristics, the red algae probably are more closely related to green plants than are the brown algae.

Multicellular algae have more complex reproductive cycles than either prokaryotes or unicellular eukaryotes. Let's explore the reproduction of algae using the brown algae of the kelp forests as an example (**Figure 15.23**). The mature kelp is diploid and has two copies of each homologous chromosome. Specialized cells undergo meiosis and produce haploid cells that come in two kinds or sexes, but these haploid cells are *not* gametes that fuse to produce a zygote. Instead, these spores go through mitosis and each produces a multicellular *haploid* organism. The haploid individuals grow into filamentous structures that eventually produce eggs and sperm that can fuse to form zygotes. After many rounds of cell division, the zygote is transformed into a mature diploid brown alga.

This life cycle is an example of **alternation of generations** in which a haploid generation alternates with a diploid generation. To understand this more intuitively, imagine what human reproduction would be like if it involved alternation of generations. After the first round of meiosis the cells that would become egg or sperm in your body instead would go through a mitotic cell division and develop into a multicellular haploid organism. This organism would then produce eggs and sperm. These would unite to produce another mature diploid individual like yourself, which would be followed by another multicellular haploid organism that produced eggs and sperm, and so on. This is what many algae do. You will see alternation of generations again when you learn about the life cycles of plants.

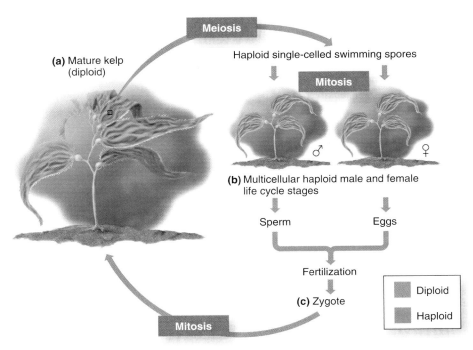

Figure 15.23 **Life Cycle of Kelp.** (a) A mature kelp is a diploid individual that undergoes meiosis to produce single-celled haploid spores. These develop into multicellular male and female haploid life cycle stages (**b**) that produce eggs and sperm. After fertilization the diploid zygote (**c**) divides by mitosis to form the adult kelp.

QUICK CHECK

1. How would your life change if all of Earth's algae were suddenly wiped out?
2. Make a sketch that shows the meaning of alternation of generations.

15.5 Prokaryotes, Single-Celled Eukaryotes, and Algae Were Among the Earliest Life-Forms

Studies of prokaryotes, single-celled eukaryotes, and algae raise intriguing questions about the early evolution of life on Earth. Bacteria and archaeans are both simpler than eukaryotes; the single-celled eukaryotes and algae are simpler than members of the rest of the domain Eukarya. How do bacteria and archaeans relate to the first life on Earth? When did multicellular eukaryotes arise? At least some fossils, dated to 3.5 billion years ago, look like short filaments of cells linked together in a string, and they seem nearly identical to forms of cyanobacteria alive today (see Figure 15.8). Larger early fossils called **stromatolites** are formed of millions of filaments of cyanobacteria and other prokaryotes compacted into stumpy columns (see Figure 6.22). While stromatolites are rare today, they thrive in clear, extremely salty, shallow waters off the coast of Australia. Fossils show that stromatolites may have been plentiful in the shallow oceans of early Earth. If so, it is not surprising that there were enough photosynthetic cyanobacteria to change the atmosphere of early Earth by introducing oxygen gas produced by eons of photosynthesis.

So bacteria clearly were among the earliest inhabitants of Earth. When did the Archaea first appear? This question is harder to answer because unlike filamentous cyanobacteria, archaeans have no identifying physical "signature" that allows them to be tracked in the fossil record. As a result, the fossil record is not much help in answering this question. Archaeans do have a characteristic *chemical*

alternation of generations a life cycle that alternates between a multicellular haploid generation and a multicellular diploid generation

stromatolite large columnar masses of cyanobacteria and other prokaryotes

signature, though. The lipids in the cell membranes of these prokaryotes are highly distinct, and some scientists believe they see traces of these lipids in rocks that are as much as 3.8 billion years old. So archaeans probably also were inhabitants of the early Earth. Right now it is not possible to tell which type of prokaryote emerged first, or if both emerged from an earlier form of life whose traces have yet to be found.

You might think that because eukaryotes have larger, more complex cells, evidence for the origin of the eukaryotes would be easier to come by—but the origin of Eukarya is still an open question. **Figure 15.24** summarizes how various groups of algae could have originated from early eukaryotes that engulfed, but did not digest, different kinds of free-living prokaryotes. From as early as 1.8 billion years ago, there are hints of fossils that look like Eukarya—and certainly eukaryotes were well established by 1 billion years ago. Some of these fossils are complex in ways that are similar to modern single-celled eukaryotes. Evidence from both the fossil record and from biochemical analyses indicates that many of the multicellular algae were present as early as 1.5 billion years ago, while other kinds of multicellular eukaryotes probably emerged around 700 million years ago.

QUICK CHECK

1. Was the first life bacteria, archaeans, or eukaryotes? Give reasons for your answer.

Figure 15.24 **One Hypothesis of How Different Groups of Eukaryotes May Have Evolved from an Early Eukaryote.** An early photosynthetic single-celled eukaryote that engulfed cyanobacteria would have obtained chlorophylls *a* and *b* and could have been ancestral to plants and euglenas. A photosynthetic single-celled eukaryote that engulfed cyanobacteria with chlorophylls *a* and *c* could have been ancestral to red algae. A photosynthetic alga with chlorophyll *c* and brown pigments that engulfed a red alga could have been ancestral to brown algae.

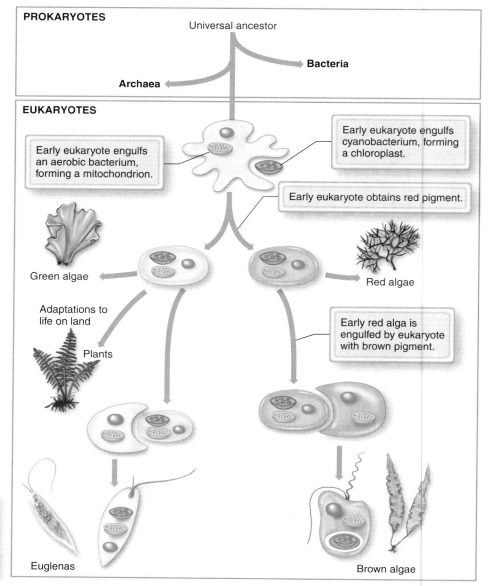

PROKARYOTES

Universal ancestor

Bacteria

Archaea

EUKARYOTES

Early eukaryote engulfs an aerobic bacterium, forming a mitochondrion.

Early eukaryote engulfs cyanobacterium, forming a chloroplast.

Early eukaryote obtains red pigment.

Green algae

Red algae

Adaptations to life on land

Early red alga is engulfed by eukaryote with brown pigment.

Plants

Euglenas

Brown algae

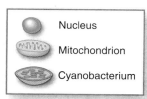

Nucleus

Mitochondrion

Cyanobacterium

15.6   An Added Dimension Explained
Poisons in the Sea

Records from the past 75 years show that there are few if any coastal regions of the United States that do not sometimes suffer from poisonous shellfish—and the same is true around the world. What is it about shellfish and ocean fishes that turns them poisonous? It turns out that single-celled algae are the answer.

From the earliest observations scientists suspected that the poisons in fishes and shellfish did not come directly from the animals themselves. Early accounts consistently reported that fishes and shellfish became poisonous when an algal bloom colored the water. An **algal bloom** is actually an explosion in the population of one or more species of photosynthetic, single-celled eukaryotes. Often called *red tides* because they sometimes turn the ocean water a reddish color, algal blooms can be found in almost any coastal region and can be caused by many different species of single-celled algae (**Figure 15.25**). If you have ever vacationed on the Florida coasts, you may have seen algal blooms. You may have smelled them too, because a typical algal bloom usually has a strong, repulsive odor. Sometimes fishes die off because the algae consume all the oxygen in the water. Not all algal blooms are associated with fishes that have become poisonous, but instances of poisonous fishes almost always are preceded by algal blooms.

It is not known why certain algae produce toxic compounds, but they are produced in higher concentrations under certain conditions. When the water is relatively warm, and there are lots of nutrients in the water, algae divide more rapidly than normal and large numbers of toxic cells result. When conditions are not favorable for growth, the algae produce multicellular resting stages called cysts similar in function to spores. Cysts have the highest concentrations of toxins. They usually sink to the bottom of the ocean, but storms or water currents can stir them up. Only about 2% of all algal species produce toxins, but they produce enough toxins to cause the extensive health and economic problems associated with *harmful algal blooms*.

The algal blooms, or stirred-up cysts, are excellent food sources for shellfish and small fishes. Many kinds of shellfish are filter feeders that pump large amounts of water through their bodies as they filter out and concentrate tiny, single-celled food particles. Tiny fishes nip the small algal cysts out of the water. If the algae are toxic, the shellfish and fishes ingest the toxin. In almost all species the toxins are concentrated in the gut—not in the muscle. This explains why some shellfish produce more problems than others. When people eat oysters, clams, or mussels, they eat the whole organism—and so they can get a hefty dose of toxin. Scallops also can be toxic, but people usually only eat the muscle of a scallop and so do not get much toxin.

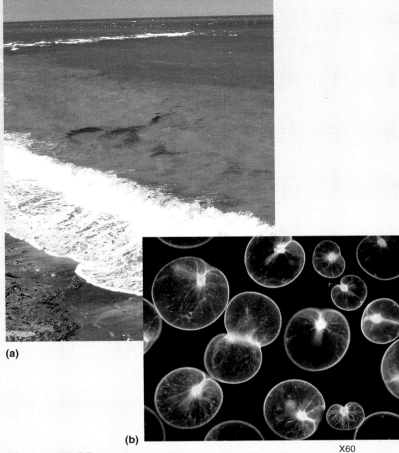

(a)

(b)

X60

Figure 15.25   **Red Tides and Species That Cause Them.** (**a**) A red tide; (**b**) *Noctiluca* is one of many species of single-celled eukaryotes that secrete toxins that cause red tides.

Two major kinds of toxins cause most of the trouble. The algae shown in **Figure 15.26a** produce a poison called *saxitoxin.* This chemical interacts with membrane protein receptors on nerve and muscle cells that are sodium channels (see Section 4.2). Saxitoxin blocks sodium channels and essentially stops nerve and muscle action. The first symptom of saxitoxin poisoning is a tingling and numbness in and around the mouth. If enough poison has been ingested, the numbness will spread down the neck to the fingers and toes and eventually will result in total paralysis. If poisoning progresses this far, the victim probably will die unless he or she can be kept on an artificial respiration machine until the toxin naturally clears from the body. Another toxin, produced by golden-brown algae (Figure 15.26b) is called *domoic acid.* This poison is responsible for

algal bloom an explosion in the population of one or more species of photosynthetic, single-celled eukaryotes

—Continued next page

Continued—

(a) A species that produces saxitoxin

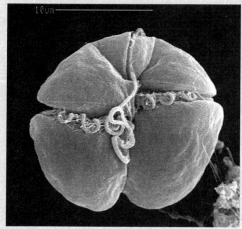

(b) A species that produces domoic acid

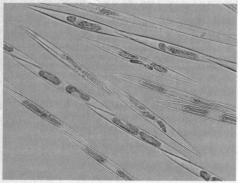

Figure 15.26 **Toxic Species of Algae.** (a) *Karenia brevis* produces saxitoxin, (b) *Pseudonitzschia* produces domoic acid.

the deaths of the sea lions in Monterey Bay in 1998. Domoic acid interacts with, and damages, neurons in the brain. The symptoms from this toxin also can be severe. Domoic acid first causes vomiting, diarrhea, and stomach cramps. Extreme headache, short-term memory loss, seizures, and coma follow.

The presence of toxins in algal blooms can be measured, and most coastal communities regularly monitor nearby waters. It is becoming increasingly common for coastal areas—like the coasts of Maine or Washington State—to place a complete ban on gathering shellfish for an entire season because of toxic blooms. Fish farms also are affected, and any fishes or shellfish that have fed on toxic algae must be destroyed. As you can see, harmful algal blooms have a big impact on animals and humans, and so the increase in the frequency of harmful algal blooms is troubling (**Figure 15.27**). The increased frequency of algal blooms probably is linked to increases in ocean temperatures and excess nutrients entering ocean waters. The warmer ocean temperatures are related to global climate change, while higher levels of nutrient runoff are related to greater development of coastal areas for golf courses, farms, and homes. This means that toxic shellfish are part of a much larger and more complex problem; future episodes of the problem of toxic fishes and shellfish are likely to get worse.

QUICK CHECK

1. What is the relationship between nutrient runoff and harmful algal blooms?

(a) Before 1972 red tides were known to occur, but were not common.

(b) Red tides became much more common and widespread after 1972 and their effects became more varied, affecting humans, and many species of fishes and marine animals.

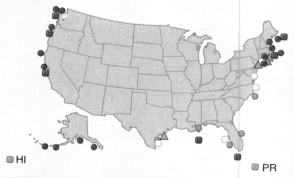

Figure 15.27 **Increase in Incidence of Red Tides.**
(a&b) The increase in the frequency of red tides and other harmful algal blooms seems to be caused by increased nutrients in runoff from land and increased temperatures that may be related to global climate change.

■ *Why do you think there were more problems after 1972 compared to before 1972?*

Key:
○ = Neurotoxic shellfish poisoning
● = Paralytic shellfish poisoning
▢ = Fish kills
▨ = Ciguatera shellfish poisoning
■ = Amnesiac shellfish poisoning
△ = Brown tide

Now You Can **Understand**

Spores, Botulism, and Botox

If you eat canned foods, you probably know that if a can is swollen or bulging, you should throw it out. You certainly should *not* eat any of its contents because they might contain a deadly toxin that has been produced by bacteria. This same toxin is the reason that children less than 1 year old should never be fed honey. Spores of the bacterium *Clostridium botulinum* cause both of these problems. Although these spores are destroyed by heat above 250°F, occasionally spores may not be killed when canned foods are processed. If the spores survive, the anaerobic conditions within a sealed can are perfect for their growth. Similarly, bacterial spores that are in honey, which typically is not processed at high temperatures, can germinate within a child's intestine—and the bacteria can survive and multiply in this low-oxygen environment. As they multiply, the bacteria release toxins. Although healthy adults are not affected by them, the toxins can kill a child. Partially cooked meat products like bacon, hot dogs, and sausages contain nitrites because these create a salty environment that discourages the germination of bacterial spores, including those of *Clostridium.*

Although cells of *C. botulinum* themselves are harmless, the toxin that they produce is one of the deadliest substances known, and an untreated victim of botulism usually dies. The toxin acts on nerves and results in vomiting, double vision, dizziness, abdominal pain, and difficult breathing. Minute amounts of this same toxin have a paralyzing effect when they are injected into facial muscles—and have become a common tool of cosmetic dermatologists. Botox (short for *botulinum* toxin) temporarily smoothes wrinkles, giving older faces a younger, fresher appearance.

What Do **You** Think?

How Did the Bacterial Flagellum Evolve?

Like many other organisms, bacteria move around in their environments, and as many as 50% of bacterial species have remarkable molecular machines for locomotion called flagella. Each flagellum has 100 or more different proteins that together rotate the flagellum up to 1,500 times a second. Intrigued by the complexity of bacterial flagella, scientists are studying them and are determined to explain how bacterial flagella work.

Bacterial flagella also are a focal point for arguments against the explanatory power of the theory of evolution. You already know that the theory of evolution is a well-founded scientific explanation for how biological diversity comes about. You also know that some people cannot accept the evidence that all species are related and that diversity comes about through the action of natural processes. Critics of evolution often look for phenomena that modern science cannot currently explain and use these as arguments against the theory of evolution. The bacterial flagellum has become a focus of this debate. The core of the argument is that the bacterial flagellum must have all of its parts to function and could not have evolved with the small steps that happen during natural selection. According to this argument, the bacterial flagellum must have appeared whole and functioning—all at once. So the flagellum must have been the work of some Intelligent Designer, not the result of the process of natural selection.

Scientists take a different approach to understanding the origin of the flagellum. Here are some of the current findings and ideas. First, many other complex biological structures are being explained by the principles of evolution. In one of these cases—the evolution of animal eyes—molecules and cells that had different functions in ancestral organisms have come together over the course of evolution to result in a new structure with a new function. Like other evolutionary developments, the changes that produce eyes are driven by the processes of genetic mutation, gene duplications, and natural selection. Scientists are working to test similar hypotheses about the evolution of the bacterial flagellum, but it is more difficult because the evolutionary ancestors of bacteria are not currently known. Nevertheless, similar principles seem to be at work. Versions of many of the 100 or so proteins in the flagellum are found in bacterial species that have no flagella, and in these species these proteins have slightly different jobs. For example, one key protein of the bacterial flagellum is highly similar to the ATP synthase molecule (see Section 6.2). Some scientists are testing the hypothesis that a dozen or so mutations in key proteins could have allowed the proteins of the flagellum to come together and function as a more complex unit.

What is your approach when you are faced with scientific uncertainty? Does the lack of a full account of the evolution of the flagellum lead you to conclude that an Intelligent Designer must be at work? Do you accept that the scientists are right and that the flagellum evolved through natural mechanisms? Or do you withhold your opinion until more is known? What do *you* think about the evolution of the bacterial flagellum?

Chapter Review

CHAPTER SUMMARY

15.1   Scientists Continue to Identify New Species

Bacteria, Archaea, single-celled eukaryotes, and algae have great biodiversity. Even though there are relatively few scientifically known species in each group, many unnamed species may exist. Comparative biochemical studies reveal the biodiversity in what were thought to be homogeneous populations of *E. coli*. Because prokaryotes are so small, reproduce asexually, freely swap genes, and take up DNA from the environment, the standard definition of a biological species is difficult to apply to prokaryotes.

15.2   Diverse Adaptations Allow Organisms to Meet Common Challenges

Organisms are constrained by environmental factors that include the medium and situation in which they live, the need to obtain and process energy, the need to maintain shape of cells and body in response to the force of gravity, the need to allow for the movement of carbon dioxide and oxygen, and the need to regulate the amount of water in cells. Adaptations of different groups reflect different solutions to these common environmental constraints.

15.3   Archaea and Bacteria Are Distinct Domains of Prokaryotes

Bacteria and Archaea are differentiated on the basis of their DNA as well as their biochemistry. Archaea and Eukarya are probably more closely related than are Archaea and Bacteria. Bacteria are crucial components of environments. They are important photosynthesizers and have essential roles in recycling of nutrients and cycling of some chemicals that are important to life. Bacteria are found nearly everywhere. Cells of bacteria and eukaryotes have many genes and enzyme systems that are identical. Bacteria are much more metabolically diverse than other domains. Although they reproduce asexually, they can increase their genetic diversity through mutation, bacterial conjugation, and bacterial transformation. Archaea are biochemically unique prokaryotes that were first identified from extreme environments.

bacterial conjugation 369	flagellum 368
bacterial spore 366	Gram stain 371
bacterial transformation 369	pathogen 371

15.4   Genetic Relationships Are the Basis for Grouping Single-Celled Eukaryotes and Algae

Single-celled eukaryotes are not necessarily closely related. They may have unusual organelles that are not found in multicellular eukaryotes. Although many single-celled eukaryotes are benign or beneficial, some cause virulent human diseases and have widespread ecological impacts. Red and brown algae are chemically distinct from plants, a group that includes green algae. Algae are distinguished by their photosynthetic pigments and the chemical composition of their cell walls. They have pivotal ecological roles as photosynthesizers in Earth's waters and make energy available for other, larger organisms that eat them.

alternation of generations 377	macronucleus 373
anal pore 373	micronucleus 373
contractile vacuole 373	oral groove 373
filter feeding 373	plankton 375
ingest 373	

15.5   Prokaryotes, Single-Celled Eukaryotes, and Algae Were Among the Earliest Life-Forms

Earth's first life was prokaryotic, but it is not certain whether it was bacterial or archaean. Prokaryotes were dominant on Earth for about 2 billion years before the first eukaryotic cells appeared. Algae appeared after single-celled eukaryotes.

stromatolite   377

15.6   An Added Dimension Explained: Poisons in the Sea

Toxins produced by blooms of some single-celled eukaryotes become concentrated within the intestines of shellfish. These toxins can kill fishes as well as humans and other mammals.

algal bloom   379

REVIEW QUESTIONS

TRUE or FALSE. If a statement is false, rewrite it to make it true.

1. Sleeping sickness and malaria are caused by bacteria.
2. Algae are plants that use chlorophyll *d* in photosynthesis.
3. The earliest fossils look a lot like bacteria.
4. As far as we know, bacteria and archaeans were Earth's earliest inhabitants and eukaryotes appeared later.
5. Red algae are unusual prokaryotes.
6. Archaeans live in extreme environments, but bacteria do not.
7. The division of prokaryotes into two domains is based on the kind of chlorophyll that they contain.
8. The smallest eukaryotes are at least twice as large as the average prokaryote.
9. Domains are larger, more inclusive groups than kingdoms.
10. Bacteria have macronuclei and micronuclei.
11. Bacteria, archaeans, various colors of algae, plants, and green algae have cell walls.
12. Gram stains allow us to identify different kinds of protists.
13. Most protists are decomposers.
14. Protists are prokaryotes.
15. Some protists make shells.

MULTIPLE CHOICE. Choose the best answer of those provided.

16. Why is it a bad idea to refreeze food that previously has been frozen and thawed?
    a. The food will have rotted.
    b. Domoic acid will have accumulated within the food.
    c. Bacteria will have formed spores in it.
    d. Bacteria will have hatched out of spores, multiplied, and put out their toxins into the food.

17. What happens in a life cycle that has alternation of generations? In alternation of generations
    a. a small organism alternates with a large organism.
    b. a single-celled organism alternates with a multicellular organism.
    c. a multicellular haploid organism alternates with a multicellular diploid organism.
    d. a prokaryotic organism alternates with a eukaryotic organism.

18. Macronucleus, micronucleus, oral groove, cilia, and contractile vacuoles are found in which prokaryote?
    a. *Paramecium*
    b. *Amoeba*
    c. *Bacillus*
    d. None of the above is correct.

19. How does *Amoeba* obtain food?
    a. *Amoeba* uses its oral groove and cilia to obtain food.
    b. *Amoeba* engulfs its food with its oozing body and digests it within its nucleus.
    c. *Amoeba* ingests food after trapping it within extensions of its plastic body.
    d. *Amoeba* is a photosynthetic single-celled red alga and makes its own food.

20. Which statement is *false* about algal blooms?
    a. Algal blooms can kill whales, fishes, shellfish, and humans.
    b. Algal blooms are linked to the presence of excess nutrients in offshore waters.
    c. Algal blooms are restricted to warm waters.
    d. Algal blooms are becoming more frequent.

CONNECTING KEY CONCEPTS

1. The Archaea and Bacteria are clearly different domains, based on analyses of their DNA. Yet it is still easy to discuss the two groups together. Explain why this is so.

2. In the minds of many researchers the Eukarya represent a major step in the evolution of complex organisms. What features of eukaryotes are so fundamentally more complex than those of Archaea or Bacteria?

QUANTITATIVE QUERY

A single bacterial cell lands in your soup. You suddenly realize that you are nearly late for an appointment, so you set the soup aside and return to it three hours later. Assuming that none of the daughter bacterial cells die, and that no more bacteria fall into the soup, how many bacteria will be present in your soup after three hours have passed? Use the following information to obtain the answer. The equation you need is

$$n = (k * 2)^{t/d}$$

where n is the number of bacteria you are trying to solve for; k is the number of bacteria at time zero; t is time elapsed in minutes, in this case 180 minutes; d is the time it takes one bacterial cell to divide into two, in this case assume 10 minutes; and 2 represents that each cell division produces two daughter cells.

THINKING CRITICALLY

1. In this chapter you have read a lot about phylogenetic relationships that were discovered by studying specific genes. Such studies compare the same gene across different species and measure the degree of nucleotide differences between the genes. Scientists know that they are the same gene because they code for a molecule that performs essentially the same function in every group of organisms studied. The 16S ribosomal RNA gene, for example, codes for an RNA molecule found in all ribosomes, in all organisms. What does the existence of such genes say about Darwin's principle that all organisms are related?

2. Scientists who are interested in the study of life on other planets are called astrobiologists. Many astrobiologists believe that there could be life on Mars, even today. Based on what you have read in this chapter, what kinds of Earth organisms would give researchers the idea that life might exist on Mars? What is it about these organisms that suggest life could exist in such an inhospitable place as Mars?

For additional study tools, visit www.aris.mhhe.com.

16

Varieties of Life
FUNGI AND ANIMALS

America's First 'Witch' Hanged in Connecticut

June 8, 1967

Salem Witch Trials. Is there a biological basis for the frenzy of witch-hunting that gripped New England in the late 1600s?

Science Historical Epidemiology

Puzzles of Past Blamed on Fungus in the Rye

November 20, 1989

The year was 1692; the date, January 20th. Two young girls in the village of Salem, Massachusetts were behaving strangely—and in this close-knit village strange behavior was feared and punished. Nine-year-old Elizabeth and 11-year-old Abigail swore blasphemous oaths, screamed, fell into convulsions, and had spells of trancelike unconsciousness. Soon more girls in the village were acting the same bizarre way. Nothing seemed to be physically wrong with these children, and the elders could think of only one explanation: the girls were under the influence of Satan. When the girls were pressed to explain this satanic influence, they answered, "Witches" (**see chapter-opening photo**). In Salem, Massachusetts, in 1692 people believed that witches were casting spells on innocent young girls. Hysteria over witchcraft spread through the community. Soon others claimed that they had been bewitched, had seen visions of witches, or been harmed by witches.

Under pressure from the village elders, Elizabeth and Abigail finally named Tituba, a Native American slave, as a witch. They also accused their neighbors Sarah Goode and Sarah Osbourne of being witches. Under interrogation Tituba confessed to seeing visions and claimed that a conspiracy of witches was operating in Salem. The two Sarahs insisted that they were innocent. Panic gripped the town as suspicion of witchcraft fell on more people. In the end more than 20 townspeople lost their lives. Some were hung for witchcraft, and others died in jail while waiting for judgment. Once the witches were hung, the townspeople were satisfied. Eventually, the erratic behavior and claims of witchcraft subsided, but the puzzle has remained for centuries. What happened in Salem, Massachusetts, in 1692?

Many factors influenced the witchcraft trials that occurred in the American colonies and in Europe in the 1600s and 1700s. People were afraid of anyone who was different. It was easy to conclude that devils or other supernatural entities were responsible for the unusual behaviors. Confessions were readily accepted, even if they came only after brutal and heavy-handed questioning. All of these factors contributed to the conclusion that witchcraft was the source of the strange behaviors. Nevertheless, the symptoms of the young girls in Salem suggest that something else was afoot. To twenty-first-century ears, these symptoms sound like a drug-induced state or poisoning, but what poison could have caused these behaviors? You will return to these intriguing questions at the end of this chapter—you may be surprised at the connection of the Salem witch trials to the topics of this chapter, the diverse kingdoms of fungi and animals. ■

16.1 Fungi and Animals Are Closely Related Kingdoms

In Chapters 14 and 15 you have begun to study how various forms of life are related. Deeper knowledge about phylogenetic relationships not only helps us to understand the world, but it also directly benefits humans. It has fostered many medical advances, including deeper understandings of genetic diseases and improved medical treatments. It is responsible for improvements in sanitation and living conditions in much of the world and has been the impetus for many improvements in agricultural crops. As you gain a better understanding of the evolution of life, you will acquire perspective on the place of humans in evolutionary history.

Recent molecular analyses of phylogenetic relationships have produced some surprises. One surprise is the relationship between fungi and animals. Analysis of the DNA that codes for ribosomal RNA makes two things about fungi and animals obvious (**Figure 16.1**). First, it is clear that fungi and animals are distinct groups; second, it also is clear that animals and fungi are more closely related to one another than either is to any other group. This chapter details how fungi and animals are related—and how each kingdom is unique.

There are important biochemical similarities between fungi and animals. For instance, the cell

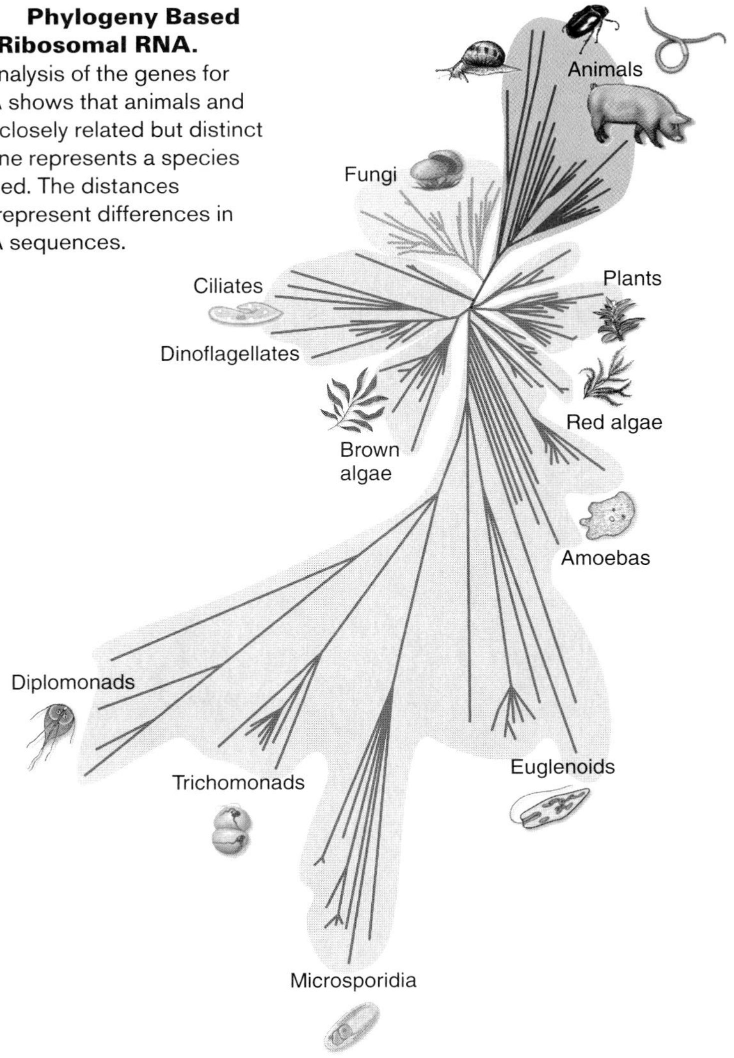

Figure 16.1 **Phylogeny Based on DNA for Ribosomal RNA.** Phylogenetic analysis of the genes for ribosomal RNA shows that animals and fungi form two closely related but distinct groups. Each line represents a species that was sampled. The distances between lines represent differences in ribosomal DNA sequences.

Table 16.1 Biochemical similarities between fungi and animals

	Fungi	Animals	Plants
Presence of chitin	Found in cell walls	Found in exoskeletons of arthropods	Absent
Cytoskeleton proteins	Similar	Similar	Different
Photosynthesis	Absent	Absent	Present
Method of nutrition	External digestion; absorption of nutrients	Ingest food, external digestion within gut; absorption of nutrients	Photosynthetic
Carbohydrate storage molecule	Glycogen	Glycogen	Starches, including amylopectin

walls of fungi and the exoskeletons of arthropods (see Section 3.4) both incorporate chitin, a tough, complex carbohydrate (**Table 16.1**). Chitin is an unusual molecule that is found only in fungi and arthropods. In addition, both animals and fungi use the molecule glycogen to store excess carbohydrates. In contrast, plants use a different carbohydrate molecule, amylopectin. Fungi and animals also have similar cytoskeleton proteins (see Section 4.4) which are different from those of other kingdoms. Fungi and animals share other important characteristics that have to do with nutrition. Neither kingdom is photosynthetic, and both feed on other organisms. Fungi and animals digest food *outside* of their bodies and absorb the nutrients. That may sound wrong to you when you think of the way that you—an animal—ingest food and digest it within your body, but in this chapter you will see how it is true. Based on this information, if you have sedentary habits you should really be called a couch mushroom—not a couch potato!

QUICK CHECK

1. What evidence supports the idea that fungi are more closely related to animals than to plants?

Figure 16.2
Relationships of Various Groups of Fungi. Four major groups of fungi are differentiated based upon their reproductive structures.

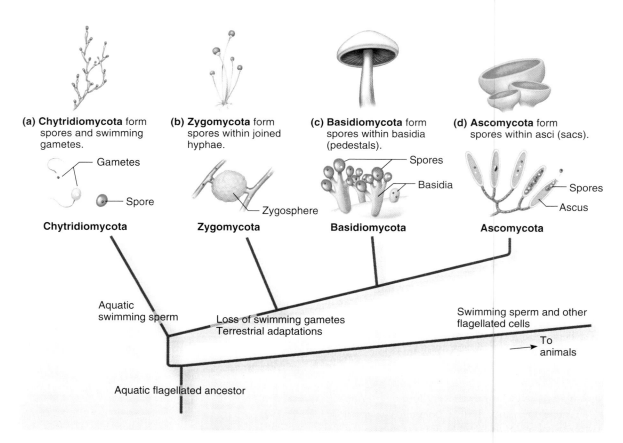

(a) Chytridiomycota form spores and swimming gametes.

Gametes

Spore

Chytridiomycota

(b) Zygomycota form spores within joined hyphae.

Zygosphere

Zygomycota

(c) Basidiomycota form spores within basidia (pedestals).

Spores

Basidia

Basidiomycota

(d) Ascomycota form spores within asci (sacs).

Spores

Ascus

Ascomycota

Aquatic swimming sperm

Loss of swimming gametes
Terrestrial adaptations

Swimming sperm and other flagellated cells

To animals

Aquatic flagellated ancestor

16.2 Fungi Have Unique Traits

You know fungi as mushrooms and toadstools, mold on bread, and yeasts that make bread rise, but there are many lesser known and diverse fungi. **Figure 16.2** shows the relationships of the major groups of fungi. Although this classification was based on the way that fungi reproduce, DNA analyses substantiate these four major phyla of fungi. Some fungi are unicellular, but most **fungi** are multicellular, non-photosynthetic eukaryotes that digest food outside of their bodies and absorb digested nutrients. **Table 16.2** gives fast facts about some fungi that you are likely to encounter.

Chytrids (pronounced "kit-rids") are among the simplest fungi and belong to the oldest phylum of fungi. Chytrids (Figure 16.2a) are unlike many other kinds of fungi, because they can live in water and have flagellated gametes that are like sperm cells. Most other fungi live on land in moist environments, and they do not have swimming gametes. *Zygomycota* (Figure 16.2b) includes the many kinds of molds that can grow in damp places within your home. The gray fuzzy mold that grows on bread and the black fuzzy mold that grows on forgotten refrigerated leftovers are the most familiar members of this group. *Basidiomycota* (Figure 16.2c) includes the kinds of mushrooms that you probably have eaten. The button mushrooms sold in supermarkets are in this group. So are toadstools, puffballs, and shelf fungi as well as the less familiar rusts and smuts that parasitize crop plants. *Ascomycota* (Figure 16.2d) includes the unicellular baker's yeasts and brewer's yeasts, as well as multicellular cup fungi, morels, and truffles. Some fungi that infect humans are also in this group, including the yeasts that cause diaper rash, vaginal yeast infections, and jock itch.

Fungi Have Simple But Unusual Body Structures

The bodies of multicellular fungi have a similar basic organization with just two major parts: the **mycelium** and the **fruiting body** (**Figure 16.3**). The mycelium supports and nourishes the fungus, while the fruiting

fungi mostly multicellular, nonphotosynthetic eukaryotes that digest food outside of their bodies and absorb digested nutrients

mycelium a mass of fine, threadlike fungal tissues that invade a food source and provide the fungus with nutrition

fruiting body a showy structure involved in the reproduction of a fungus

Table 16.2	Fast facts about fungi
Name of fungus	**Fast fact**
Chytrids	May be linked to the widespread and troublesome die-offs of frogs, resulting in extinction of some frog species
Many forms	Zygomycetes found in many mycorrhizae
Penicillium	Source of first antibiotic; the blue-green mold that grows on spoiled oranges
Dutch elm disease	Has killed elm trees in much of the United States
Many species of ascomycotes	Sources of statin drugs used to control blood cholesterol levels
Neurospora	Widely used in biological research
Candida albicans	Causes thrush, diaper rash, vaginal yeast infections
Chestnut blight	Has killed chestnut trees in much of the United States
Powdery and downy mildew	Parasitic infection of many horticultural plants
Many species of zygomycetes	Form lichens in partnership with blue-green algae
Many species of zygomycetes	Form mycorrhizae
Claviceps purpurea	Causes ergot in rye, wheat, and other grasses
Chanterelles and morels	Easy to identify edible wild mushroom
An ascomycote	May be Earth's largest organism
An ascomycote	Forms fungus gardens of leaf-cutter ants
Many species of basidiomycetes	Form mycorrhizae
Domesticated mushrooms, including button mushrooms, shitake, portabella, enoki	Delicious when sautéed in butter

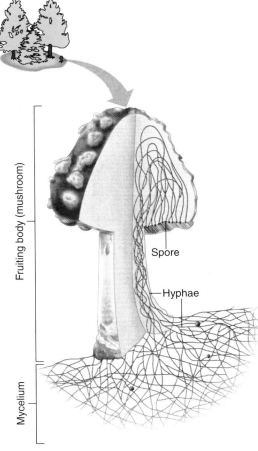

Figure 16.3 Body of a Fungus. Many fungi are multicellular organisms that have fairly simple body plans. The body of a mushroom has an underground mycelium and an aboveground fruiting body.

Fruiting body (mushroom)

Mycelium

Spore

Hyphae

Fruiting body and mycelium are the body parts of multicellular fungi. Both are made of hyphae.

■ *What body part of a mushroom are you* not *eating when you eat sauteed mushrooms?*

hyphae individual threads or tubes of a mycelium or fruiting body

body is involved in reproduction. The fruiting body is often the conspicuous mushroom or toadstool that you see above ground; the mycelium is hidden underground. The mycelium looks like a mass of fine white cottony threads, and it penetrates a source of food such as dead leaves, a rotting log, or the decomposing tissues of a dead animal. While a mycelium continually grows into its food source, fruiting bodies are produced more sporadically from a mycelium.

The internal cellular structure of fungi is unique. The entire body of a fungus is made of compressed tubes that look like fine strands of spaghetti (Figure 16.3). These strands are called **hyphae,** and both the rootlike mycelium and the fruiting body are made of them. Hyphae are unusual because some have few or incomplete cell walls. In most fungi perforated walls called *septa* incompletely divide hyphae into individual cell-like compartments. In other fungi there are no septa, and the hyphae are long tubes that contain hundreds to millions of nuclei but no individual cells. In both kinds of hyphae the cytoplasm and the organelles are free to roam around and interact.

It might seem that fungi violate the principle that all organisms are made of cells, but this is not the case. The multinucleate hyphae are modified cells. They make sense when you recall that the processes of mitosis and cytokinesis are separate events of mitotic cell division (see Section 8.4). As hyphae grow, cell division proceeds through mitosis, but cytokinesis is incomplete or absent. Similar noncellular structures with multiple nuclei also occur in animals. For example, skeletal muscles are made of large fibers that have multiple nuclei; cell membranes never form around these nuclei. Fungi have no specialized structures for gas exchange. Even though the overall structure of some fungi can be quite large, every hypha is in close contact with the surrounding air. So fungi can rely on the process of diffusion to move oxygen and carbon dioxide in and out of their bodies.

Fungi Digest Food Externally and Absorb Nutrients

Nutrition is another way that fungi are similar to animals. Both fungi and animals rely on other organisms for food. Both digest their food using enzymes to break down large food molecules like proteins and complex carbohydrates into smaller, more usable molecules like amino acids, monosaccharides, and disaccharides. Both fungi and animals have external digestion, meaning that they break down large food particles outside of the body. In animals the contents of the stomach and gut are technically outside of the body. In fungi hyphae grow into a food source and secrete digestive enzymes that break down the large food molecules (**Figure 16.4**). Resulting small food molecules are absorbed into hyphae.

Usually, the food source of a fungus is dead, but this is not always the case. Fungi can be parasitic and can get nutrition from the tissues of live organisms too. The common rust fungi that infest various plants are an example. Rusts have specialized hyphae that actually penetrate the walls of plant cells. Although the hyphae do not penetrate the cell membrane that lies within the wall of a plant cell, they do absorb nutrients from the plant cell across the cell's membrane. Even more amazing are the carnivorous soil fungi that capture and kill tiny roundworms that also live in soil (**Figure 16.5**). Hyphae of the fungus *Arthrobotrys* have loops that can catch and trap roundworms that crawl into them. Once a worm is trapped, hyphae rapidly grow out of the loops and *Arthrobotrys* secretes digestive enzymes into the worm to feed on its nitrogen-rich tissues.

(a) Mycelia grow into decaying tissues.

(b) Digestive enzymes are synthesized within a hypha and released into food source.

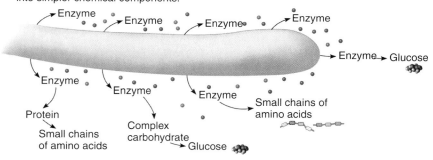

(c) Enzymes break down proteins and carbohydrates into simpler chemical components.

(d) Digested nutrients are absorbed into hypha.

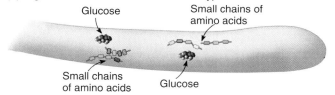

Figure 16.4 **Hyphae Release Digestive Enzymes.** (**a–c**) Hyphae excrete digestive enzymes that break down nutrient molecules. (**d**) Digested nutrient molecules are absorbed by hyphae.

Figure 16.5 **Worm Trapping Fungus.**
Hyphae of the fungus *Arthrobotrys* forms loops that have a scented lure. Nematode worms crawl into these loops, causing the hypha to tighten around the worm and eventually grow into it and kill it.

While the previous examples of parasitic or carnivorous fungi are remarkable, they are not the norm. Like many bacteria, most fungi are decomposers that play crucial roles in natural systems. If there were no decomposers, dead organisms would just pile up, and the nutrients in their tissues would not be recycled and made available to other organisms. Decomposers not only allow new generations space to live, they also spread a nutritional banquet for many other organisms. For instance, fungi do not absorb all of the nutrients that result from their enzymatic decomposition of a dead tree, and other organisms can feed on the surplus nutrients. Most plants take advantage of this by joining forces with fungi. Plant roots and fungi become intertwined in structures called **mycorrhizae.** Mycorrhizae give plants more nutrients than plants could take up from soil if they were growing without mycorrhizae (**Figure 16.6**). The importance of mycorrhizae is underscored when you consider that 98% of land plants have fungi associated with their roots. When plants are deprived of their mycorrhizae, they grow poorly because mycorrhizae help plants absorb phosphorus and nitrogen from soil.

Fungi Have Complex Reproductive Strategies

Reproduction of most kinds of fungi involves the production of spores. Fungal spores vary in size, but many are about the size of a human red blood cell (see Figure 4.1). Each spore is so light that it is lifted on the slightest air current, making spores an ideal way to disperse a fungus. Spores are produced in such great numbers that it is not an exaggeration to say that Earth is drenched in fungal spores. When a spore lands in a suitable environment, it germinates into a hypha that will grow and multiply, forming a new mycelium that eventually will produce a new, characteristic fruiting body.

With this in mind, let's briefly compare reproduction in the four major phyla of fungi (**Table 16.3**).

(a) Some mycorrhizae coat the roots of plants and help the roots to absorb nutrients more efficiently.

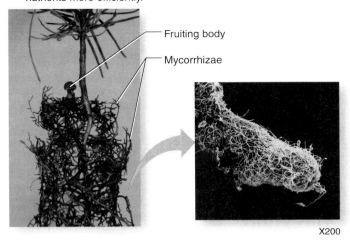

Fruiting body

Mycorrhizae

X200

(b) Mycorrhizae help plants grow better. These plants were grown without mycorrhizae (left) and with mycorrhizae (right).

Figure 16.6
Mycorrhizae Help Plants Absorb Nutrients More Efficiently. Mycorrhizae interact with the roots of plants and help them to absorb nutrients more efficiently.

Chytrids send out club-shaped structures that contain spores, and they also produce gametes similar to sperm cells that swim with lashing flagella. After two flagellated gametes have fused, the zygote becomes a resting stage that, like other sexually produced life cycle stages, has the advantage of increased genetic diversity to help it survive difficult environmental times.

Molds and their relatives in phylum Zygomycota release haploid asexual spores that germinate and develop new hyphae, which eventually grow new spore-forming structures. If two hyphae happen to touch as each penetrates into a substrate, an alternative reproductive structure forms that helps to increase the genetic diversity of molds and other zygomycotes. This hard-walled structure is called a *zygosphere,* and it contains nuclei from each partner. The nuclei pair, fuse, and then undergo meiosis to produce haploid spores that have greater genetic diversity than the parent did.

Morels, sac fungi, truffles, and their relatives in phylum Ascomycota produce their spores in a different sort of reproductive structure called an *ascus.* Within an ascus spores are lined up like

mycorrhizae symbiotic associations of plant roots and fungal hyphae that provide plants with an increased supply of nutrients

Table 16.3 Reproductive structures of fungi

Phylum and reproduction	Reproductive structure
Chytridiomycota *(100 genera)* Reproduction involves **flagellated gametes** as well as fruiting bodies that produce spores.	Flagellated gametes  Flagellated gamete
Zygomycota *(1,000 species)* Asexual reproduction uses spores released from sporangia; sexual reproduction features fused hyphae that form a **zygosphere** where meiosis occurs, forming haploid spores; these germinate into hyphae.	Zygosphere Zygosphere
Ascomycota *(32,000 species)* Spores produced in an **ascus** are the prime characteristic of sexual reproduction of this group.	Ascus 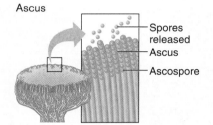 Spores released Ascus Ascospore
Basidiomycota *(22,000 species)* A **basidium** is the prime sexual reproductive characteristic of this group; basidia are housed on gills and each contains two haploid nuclei. These fuse and undergo meiosis to form four haploid spores that germinate into hyphae. Compatible hyphae combine and produce mycelia and the fruiting body.	Basidium Cap Gills 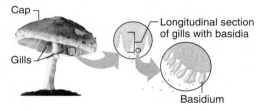 Longitudinal section of gills with basidia Basidium

unshelled peas in a pod. When the spores are ripe, they shoot out of all of the asci in a nearly simultaneous puff that can make a hissing noise and produces a dusty cloud of spores.

Mushrooms, puffballs, shelf fungi, and their relatives in phylum Basidiomycota develop spores in club-shaped structures located within the walls of gills on the undersides of mushrooms. Spores are released and germinate into hyphae. If hyphae of different mating types encounter one another, they can fuse and form structures where nuclei recombine.

QUICK CHECK

1. What is a mycelium? How does it function in the life of a fungus?

16.3 Humans Eat and Use Fungi

Many human cultures have traditions of foraging for wild mushrooms. Truffles are the most prized and fantastically expensive fungi. Truffles are small fruiting bodies of ascomycetes that grow underground on the roots of oak trees in France and Italy (**Figure 16.7**). Sows were traditionally used to sniff out and locate truffles, but these days they have largely been replaced by specially trained, keen-nosed truffling dogs. You may be wondering why a sow would be useful in locating these hidden fungi and apparently it all has to do with sex. Truffles contain a steroid compound also found in the breath of male pigs. The same compound also is a component of perspiration of human males and the urine of human females. Mammals of all kinds seem to be drawn to the scent of truffles and dig them out of the ground and eat them. Is there an advantage to the truffle in being eaten by a mammal? The spores of the fungus are within the truffle. They are not harmed by digestive enzymes and are deposited with the animal's wastes. So being eaten by a mammal is an advantage for truffles because it helps to increase the distribution of their spores.

Of course, truffles are just one species of fungus that humans and other animals like to eat. Box turtles, squirrels, and chipmunks like to eat certain mushrooms, and tiny fungus gnats specialize in consuming entire mushrooms. Female fungus gnats lay their eggs at the base of many mushrooms, and the mass of larvae eat their way upward, consuming the entire cap of the mushroom as they go. In

Figure 16.7 Black Truffles. These fruiting bodies of an ascomycote fungus grow underground on the roots of trees and are among the most prized gourmet foods.

■ *If truffles are underground fruiting bodies, where do you think the mycelia of these fungi are located?*

rain forests of Mexico, and Central and South America, one of the most ubiquitous and important organisms is the leaf-cutter ant that has "domesticated" a fungus. It is common to see parades of foraging leaf-cutter ants trailing along the forest floor. As she makes her way back to the colony, each worker carries a bit of leaf aloft in her jaws. Once the leaf fragments have been delivered underground, they are not eaten by the ants. Instead, the leaves are chewed into a paste that is added to an underground fungus garden tended by the ants. Spores are planted on the chewed-up leaves, and the ants eat the fungus that grows from them. In a somewhat similar fashion, humans plant fungal cultures into processed milk to produce delicious and rare cheeses. All of the blue-veined cheeses like Stilton, Gorgonzola, and blue have been partially digested by *Penicillium* mold.

Fermentation by Fungi Produces Alcohol and Carbon Dioxide

Not all of the ways that humans use fungi are nutritional. Judging from historical records, many human cultures have appreciated the effects of alcohol, a chemical produced by the metabolism of fungi. Alcohol is one result of the metabolic process of *fermentation* (see Section 6.4). In fermentation cells obtain energy from the chemical bonds in glucose, but they do this under anaerobic conditions. Fermentation is similar to glycolysis (see Section 6.3), but instead of yielding a pyruvate molecule, in some instances the end products of fermentation are ethyl alcohol and carbon dioxide. Yeast is a fast-dividing, microscopic-sized, single-celled fungus that carries out fermentation. Naturally-occurring yeasts are common. For instance, the whitish coating on grapes is actually a culture of yeasts that live on the grape skins (**Figure 16.8**).

Humans use yeasts to make beer, wine, and other alcoholic spirits. Beer and wine result from adding yeast to a mixture of fruit or grain that provides the necessary sugar for the yeast to ferment. Champagne is naturally bubbly because carbon dioxide is released by yeasts as they ferment the sugars in the juice of champagne grapes.

Bakers capitalize on fermentation of yeasts to produce light and airy breads. Yeasts give off carbon dioxide that makes bread dough rise. If you look closely at a slice of bread, you will see that it contains many holes (**Figure 16.9**). These are left by bubbles of carbon dioxide that were released by actively growing yeast cells as they fermented the sugar in bread dough. The alcohol produced by fermenting yeasts evaporates as the bread bakes, but alcohol is responsible for the aroma of baking bread.

Figure 16.8 **Yeast Growing on Grapes.** The pale, dusty-looking coating on these grapes is a natural yeast that normally grows on them.

Fungi Are Used for Medicines

Many human cultures traditionally have used wild mushrooms for medicines. In western societies the best known example is the discovery of the antibiotic drug, penicillin. The often-repeated story relates that in 1928 Alexander Fleming was investigating cultures of pathogenic bacteria. Some of his culture plates had become contaminated with mold, but before he threw them out, Fleming noticed something odd: the disease-causing bacteria did *not* grow in areas around the moldy spots (**Figure 16.10**). Fleming was sharp enough to recognize what he had discovered—something that had prevented the growth of bacteria. It took years of work in different laboratories, but in the early 1940s other researchers finally isolated the active antibiotic compound, and the era of antibiotics began. *Penicillium notatum* is the fungus that produces penicillin. Since the discovery and isolation of penicillin, the antibiotic properties of many more fungi have been discovered and used to make antibiotics to treat bacterial infections.

QUICK CHECK

1. How would a fungus such as *Penicillium* benefit from producing an antibiotic compound?

(a) The holes in bread are places where carbon dioxide gas collects in bread dough. CO_2 is released as a product of fermentation by yeasts.

(b) Bread making takes advantage of fermentation by yeasts.

Add yeast to warm water and sugar.

The yeasts ferments sugar and produces alcohol and CO_2. The CO_2 gas is released.

Add flour and salt to yeast, water, and sugar.

Knead dough to distribute ingredients.

Set in warm place to let bread dough rise.

Knead dough again and shape into loaves. Let loaves rise.

Bake.

The alcohol evaporates.

Yum! Fresh baked bread!

Figure 16.9   **Yeasts Make Bread Rise.** (**a**) The holes in a slice of bread are formed by bubbles of carbon dioxide gas given off by yeasts as they ferment the sugars in bread dough. (**b**) Fed by sugars in bread dough, yeasts release carbon dioxide that lightens bread dough as it rises. The aroma of baking bread is the scent of the alcohol formed by fermentation being evaporated.

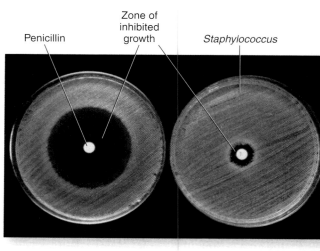

Penicillin

Zone of inhibited growth

Staphylococcus

Figure 16.10   **Penicillin Mold Inhibits the Growth of Bacteria.** Notice the clear ring near the penicillin where no bacteria are growing. The plate on the right shows the effect of penicillin on bacteria resistant to it.

16.4 Animals Are Adapted to Many Environments

Over a million species of animals have been scientifically identified. **Figure 16.11** shows nine of the currently recognized major animal phyla and some of their major features. Some of these characteristics are the presence or absence of true tissues, type of body symmetry, details of how animal embryos develop, internal architecture of animal bodies, including the presence of an internal body space lined and covered with mesoderm tissues. Each phylum has a set of characteristic physical features that define it. This section introduces you to some of these features.

Placozoans and sponges are the simplest animals. They lack true tissues and organs. Other animals have true tissues and are more complex. All of the animal phyla in Figure 16.11 except one are *invertebrates*, animals that have no bony vertebrae (see Section 14.4). Only in phylum Chordata are there *vertebrates,* animals with an internal bony skeleton that includes a backbone and skull to protect the nervous system. **Figure 16.12** shows some of the major groups of vertebrate animals and the important evolutionary adaptations that distinguish them. Early in their evolutionary history, some vertebrates acquired new traits such as upper and lower jaws, a skeleton and teeth that were strengthened by calcified tissues, and an embryo protected by the cushion of a fluid-filled amniotic membrane. The ancestral vertebrates with these traits passed them on to future generations—this is why all except a few vertebrate phyla have these traits today. Other important evolutionary innovations that define groups of vertebrates include the presence of lungs instead of gills; the evolution of limbs, mammary glands that secrete milk; and body coverings of scales, feathers, and hair or fur.

It is likely that animals arose from a small group of single-celled eukaryotes called *choanoflagellates.*

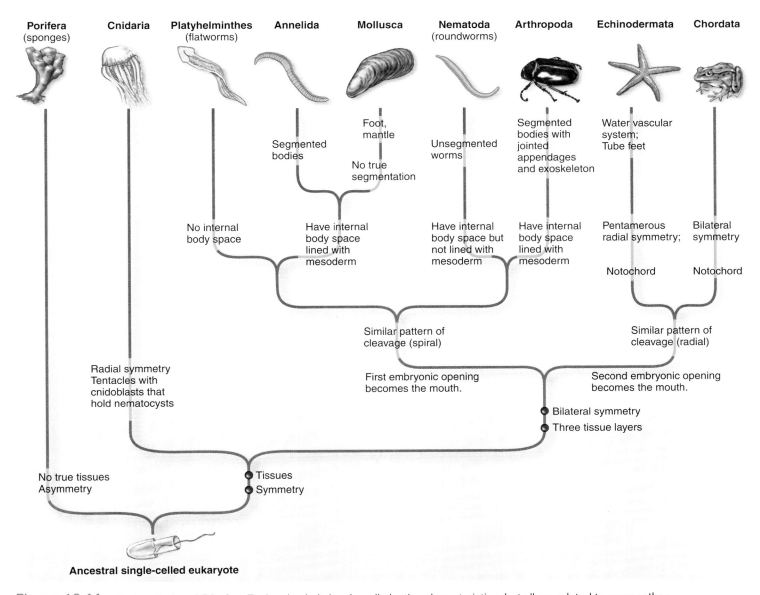

Figure 16.11   **Major Animal Phyla.** Each animal phylum has distinctive characteristics, but all are related to one another.

The simplest animals, such as sponges and *Hydra*, contain cells that look and function almost exactly like these specific single-celled eukaryotes (**Figure 16.13**). The diverse species in the other animal phyla evolved from ancestors that resembled these choanoflagellates.

What are some of the features that animals have in common? For one thing, all **animals** are multicellular, aerobic eukaryotes, and nearly all ingest other organisms to obtain energy. The exceptions are some photosynthetic animals, notably coral animals that make coral reefs. Their cells are filled with photosynthetic housemates that provide some of their nutrients. If they are deprived of these cells, coral animals are no more photosynthetic than

you are. Another feature of animals is that their bodies are made of layers of cells—you will read more about this feature shortly.

Animals live in diverse places that include all sorts of watery environments such as marine, brackish, or freshwater environments. Many are terrestrial, some burrow in the soil and sand, and a few spend most of their lives in the air. While animals are not as extreme in their environmental adaptations as are some prokaryotes, some animals do live in muggy tropics or hot deserts; others inhabit subzero landscapes of the poles.

Small animals can rely on diffusion to distribute gases and nutrients to all cells of their bodies. More complex animals must have mechanisms to

animal a multicellular, aerobic eukaryote that obtains energy by ingesting food

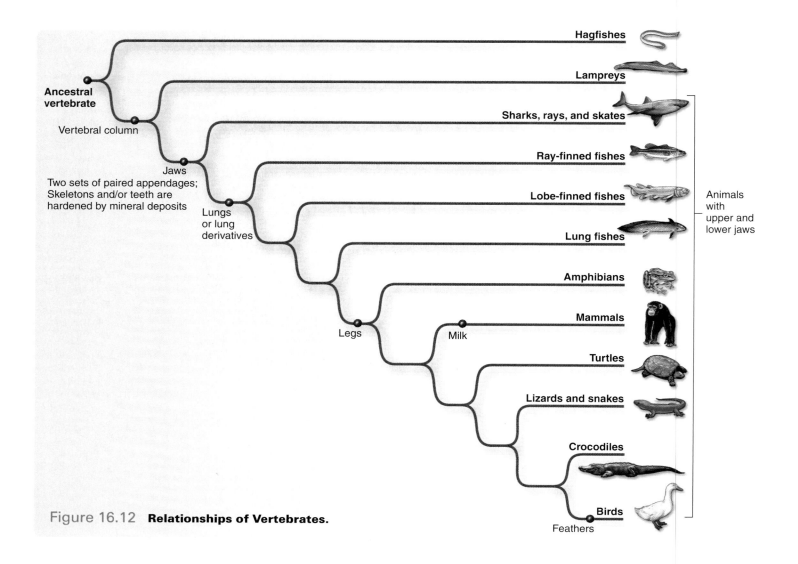

Figure 16.12 **Relationships of Vertebrates.**

Within the figure:

Hagfishes

Lampreys

Sharks, rays, and skates

Ray-finned fishes

Lobe-finned fishes

Lung fishes

Amphibians

Mammals

Turtles

Lizards and snakes

Crocodiles

Birds

Ancestral vertebrate

Vertebral column

Jaws

Two sets of paired appendages; Skeletons and/or teeth are hardened by mineral deposits

Lungs or lung derivatives

Legs

Milk

Feathers

Animals with upper and lower jaws

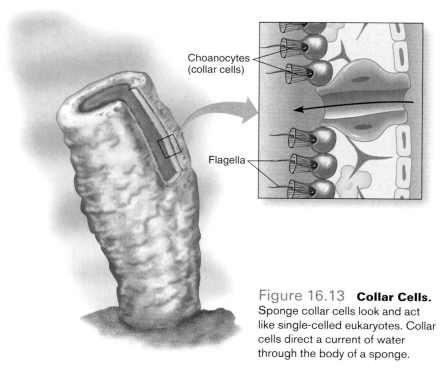

Choanocytes (collar cells)

Flagella

Figure 16.13 **Collar Cells.** Sponge collar cells look and act like single-celled eukaryotes. Collar cells direct a current of water through the body of a sponge.

move gases and nutrients. **Figure 16.14** shows a few examples. To allow gas exchange, many insects have a network of fine tubules that open to the environment in a pore on the body surface. The tubules also extend deep into body tissues. When an insect uses muscles to expand and contract its body, it physically pumps gases in and out. This bellows effect moves gases enough for diffusion to complete the job of gas exchange at the level of individual cells. Lungs do a similar job for many vertebrates. Muscle movements create and release pressure on lungs and allow air to rush into lungs or push air out of them. Other animals, including arthropods and aquatic vertebrates like fishes and salamander larvae use gills for gas exchange. Surprisingly, some vertebrates can exchange gases right through their skin and have neither gills nor lungs. An example is a group of lungless, land-dwelling salamanders. The ancestors of these species probably had lungs, but they were lost due to natural selection. Vertebrates also have an additional mechanism that gets oxygen and carbon dioxide to and from tissues. The oxygen

(a) Many animals use diffusion to move gases across the body surface, to and from body cells.

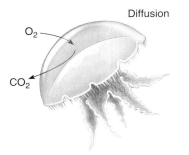

Diffusion

O_2

CO_2

(b) Insects have a network of fine tubules that open on the body surface and extend to individual cells. Oxygen and carbon dioxide diffuse in and out of these tubules.

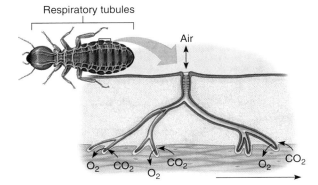

Respiratory tubules

Air

O_2 CO_2 CO_2 O_2 CO_2
O_2

(c) Some animals have external gills coupled with a circulatory system to deliver gases to cells.

(d) Some animals have internal gills.

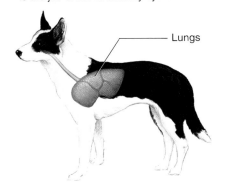

(e) Some animals have lungs for gas exchange with the environment. Gas exchange with body cells is the job of the circulatory system.

Lungs

Figure 16.14
Mechanisms for Gas Exchange. Animals have different mechanisms for gas exchange.

■ *Use your knowledge of diffusion to explain why a large animal like a dog, cannot rely on diffusion to obtain gases from the environment. Why does a fish or a dog need specialized respiratory tissues when a jellyfish does not?*

taken in by skin, gills, or lungs is taken up by the blood that is propelled around the organism by the beating of the heart.

Living Animals Exhibit Just a Few Basic Body Plans

All animals are multicellular, and the overall body plan is one way to distinguish major groups of animals. The simplest of all animal body plans are found in the placozoans and sponges (**Table 16.4**). They have just two layers of cells and only a few cell types. Given their limited cell types it is not surprising that placozoans and sponges do little more than feed and reproduce. The next most complex of the body plans are exemplified by the cnidarians (pronounced ni-DARE-ians), a group that includes the hydras, jelly-fishes, sea anemones, and corals. Cnidarians are aquatic and most are marine. Cnidarians have two body layers, but they also have a loosely organized third layer between these two layers, and they have many more cell types. Placozoans and sponges have no recognizable body symmetry, but cnidarians are **radially symmetrical.** This means that a line can be drawn through any diameter of the body and the two halves will be nearly identical.

A *Hydra* is a simple cnidarian (Table 16.4) whose body is a cylinder of cells organized into two layers. The cell layers surround a central, internal cavity. The cavity has only one opening through which food enters and solid wastes exit. So the opening is both mouth and anus. Hydras do have some specialized cells. For example, specialized cell types form highly mobile tentacles that have sting-ing cells called *nematocysts.*

In their life cycles many cnidarians alternate between a *polyp* body form that is fixed in place and a free-swimming jellyfish or *medusa* body form. In other cnidarians either the polyp or medusa body form dominates the life cycle. For instance, the reef-building coral animals are colonial polyps (**Figure 16.15**). Reefs are made of individual coral polyps that secrete external skeletons of calcium carbonate.

radial symmetry a kind of symmetry in which identical halves result from any cut that bisects the middle of an object or an organism

Figure 16.15 **Polyps of a Reef Building Coral.**

Table **16.4** Body plans and body symmetry of placozoans, sponges, and cnidarians

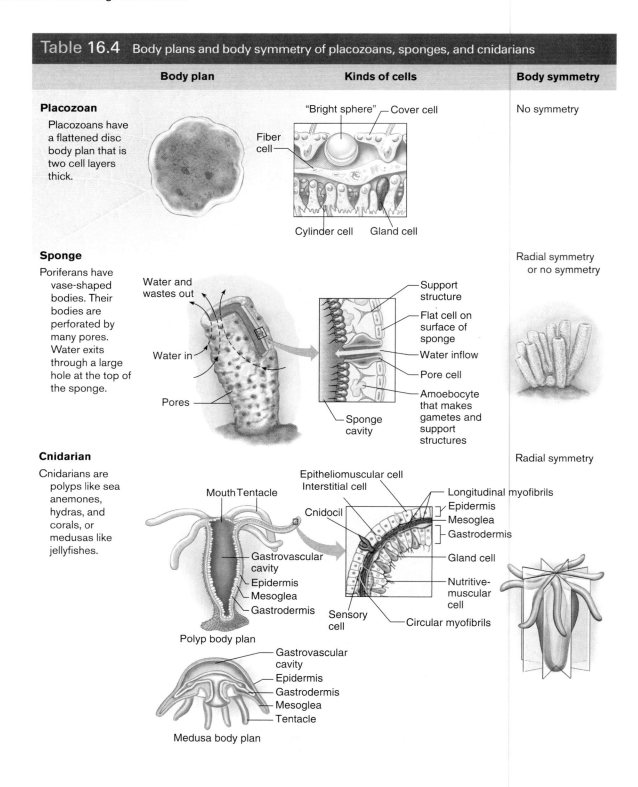

Body plan	Kinds of cells	Body symmetry
Placozoan — Placozoans have a flattened disc body plan that is two cell layers thick.	"Bright sphere", Cover cell, Fiber cell, Cylinder cell, Gland cell	No symmetry
Sponge — Poriferans have vase-shaped bodies. Their bodies are perforated by many pores. Water exits through a large hole at the top of the sponge. Water and wastes out, Water in, Pores	Support structure, Flat cell on surface of sponge, Water inflow, Pore cell, Amoebocyte that makes gametes and support structures, Sponge cavity	Radial symmetry or no symmetry
Cnidarian — Cnidarians are polyps like sea anemones, hydras, and corals, or medusas like jellyfishes. Mouth, Tentacle, Gastrovascular cavity, Epidermis, Mesoglea, Gastrodermis, Polyp body plan, Gastrovascular cavity, Epidermis, Gastrodermis, Mesoglea, Tentacle, Medusa body plan	Epitheliomuscular cell, Interstitial cell, Cnidocil, Longitudinal myofibrils, Epidermis, Mesoglea, Gastrodermis, Gland cell, Nutritive-muscular cell, Sensory cell, Circular myofibrils	Radial symmetry

Thin threads of cells connect one coral animal with its neighbors. Sea anemones also are polyps, but they tend to be larger, noncolonial, and do not form reefs. The Portuguese man-of-war is a large, colonial jellyfish that may include up to 1,000 individual cnidarians.

From the simplest flatworms to the most complex insects and vertebrates, the body plan of all other animals comes from three cell layers, not two. This arrangement of layers arises during embryonic development, and it allows for a much more complex body plan than the two layers of the cnidaria.

(a) Developmental changes that produce 3 embryonic cell layers

(b) Embryonic ectoderm, endoderm, and mesoderm each give rise to a distinct set of body tissues

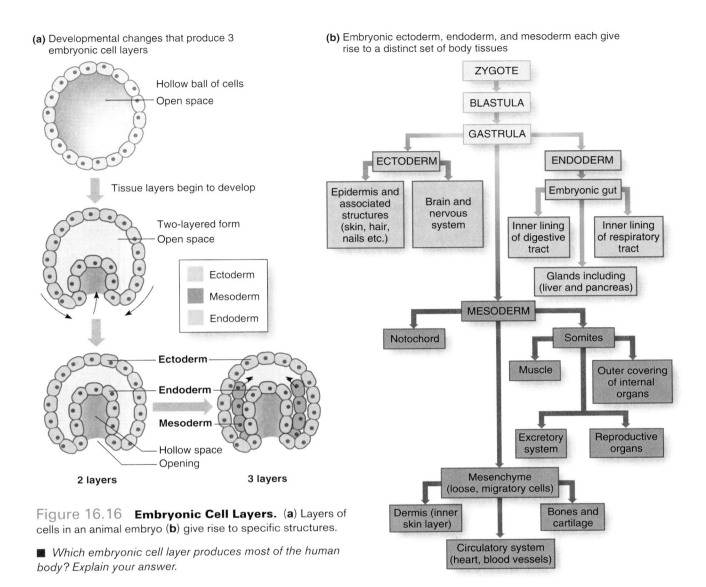

Figure 16.16 **Embryonic Cell Layers.** **(a)** Layers of cells in an animal embryo **(b)** give rise to specific structures.

■ *Which embryonic cell layer produces most of the human body? Explain your answer.*

It is hard to see the three layers in adult animals because the body structures have become so complex, but they are easier to see in embryos. The outer and inner embryonic layers are the **ectoderm** and **endoderm,** respectively (**Figure 16.16a**). **Mesoderm** is the third layer that is located between ectoderm and endoderm. This arrangement of outer and inner layers sets up a "tube-within-a-tube" body structure. The outer tube, made of ectoderm, forms the skin and associated organs (Figure 16.16b). The innermost layer, made of endoderm, forms the lining of the digestive tract. In between these two layers mesoderm contributes to the inner organs, muscles, and bones. Most organs have tissues that originated as cells from all three layers, but the existence of the layers sets up the basic body plan for most animals.

This tube-within-a-tube body plan has another important feature: it sets up a hollow core through the center of the organism (**Figure 16.17**). The inner tube, made of endoderm, surrounds a hollow space that has an opening at either end. In most animals the hole at one end is the mouth, and the hole at the other end is the anus. The hollow space is actually the digestive tract. Food goes in the mouth, moves through the digestive tract, and waste products exit through the anus. Another way to visualize this is to realize that the digestive tract is open to the outside world at both ends, and that the body surrounds this hollow core. The chemical process of digestion requires destructive enzymes and acids that break down large food molecules, and it is advantageous to secrete these to the *outside* of the body.

ectoderm outer layer of cells of an embryo

endoderm inner layer of cells of an embryo

mesoderm a layer of cells of an embryo that is sandwiched between ectoderm and endoderm

Figure 16.17 **Tube-Within-a-Tube Body Plans.**

(a) The tube-within-a-tube body plan features an external tube made of ectodermal cells and an internal tube made of endodermal cells. Mesodermal cells are in the spaces between the inner and outer tubes.

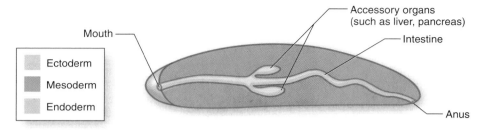

(b) Roundworms and earthworms are the simplest animals with the tube-within-a-tube body plan.

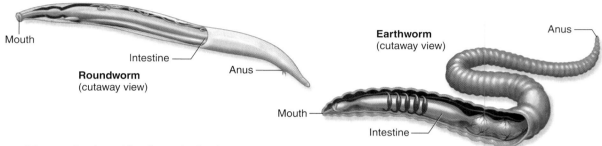

Most animals with three body layers have **bilateral symmetry,** in which only one plane will divide an object or organism into two nearly identical halves (**Figure 16.18**). So forks, guitars, and kayaks as well as worms, insects, earthworms, snails, and vertebrates all have bilateral symmetry. This arrangement means that different parts of the body can be identified and given unique names. For instance, you might call the mouth end the "head" and the anus end the "tail," but scientists call the head end the **anterior end** and the tail end the **posterior end.** Bilaterally symmetrical animals have other named body dimensions too. The belly or lower surface of a vertebrate is the **ventral surface,** and the back or upper surface is the **dorsal surface.** Finally, bilaterally symmetrical animals have a left side and a right side, defined from the animal's own perspective.

Two other trends in the body plans of bilateral animals are especially interesting. First, almost all bilateral animals that live in aquatic environments have no limbs, which are appendages like arms and legs that can bear the weight of an animal's body. In contrast, bilateral vertebrates that live on land often have limbs, and vertebrates usually have four limbs. The presence of four limbs is related to the evolutionary history of the vertebrates. Some of the early vertebrates that lived in water evolved four fleshy appendages that allowed them to drag themselves along the bottoms of lakes and rivers and along the shorelines. These fleshy fins actually contained bones that are similar to the bones in the limbs of land vertebrates. The evolution of four fleshy limbs, in turn, provided the genetic variation

for the evolution of terrestrial animals with four limbs. Later some groups of vertebrates like snakes, glass lizards, whales, and manatees lost limbs, but you can see the remnants of their ancestral limbs in some fossil and existing species.

A second important trend in the evolution of animals, especially in bilateral animals, is the evolution of the nervous system (**Figure 16.19**). Over the course of animal evolution sensory and nervous adaptations emerged that allow organisms to perceive and analyze information that comes from the environment. For radially symmetrical animals—that have no head or tail ends and that encounter the environment in 360°—it is an advantage to have a nervous system that is distributed throughout their body. In contrast, bilateral animals tend to move in one direction, and their anterior or head ends encounter the environment first. In these animals, over evolutionary history, the nervous and sensory tissues became increasingly elaborated and concentrated at the anterior end. The most elaborate neural structure of all animals is found in the brains of certain mammals, such as primates and marine mammals.

Animals Have Diverse Adaptations for Obtaining Food

Animals have many adaptations related to obtaining and ingesting food. Let's consider just a few of the major feeding adaptations of animals.

bilateral symmetry symmetry in which identical halves result only from a cut along the midline of an object or an organism

anterior end the head end of an animal

posterior end the hind end of an animal

ventral surface the belly or lower surface of an animal

dorsal surface the back or upper surface of an animal

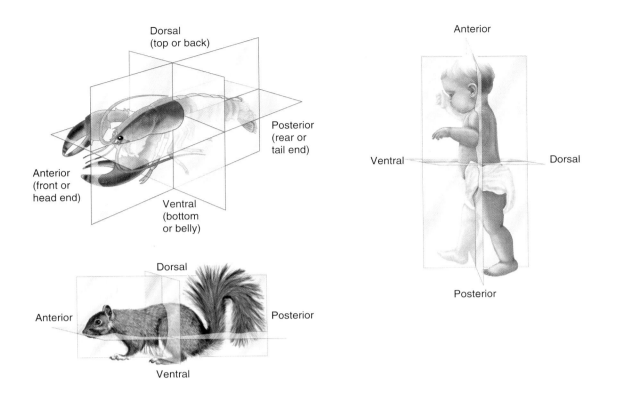

Dorsal
(top or back)

Posterior
(rear or
tail end)

Anterior
(front or
head end)

Ventral
(bottom
or belly)

Dorsal

Anterior

Posterior

Ventral

Anterior

Ventral

Dorsal

Posterior

Figure 16.18
Bilateral Symmetry.
An object or organism has bilateral symmetry when mirror images only can be produced by a cut that goes precisely down the middle of the object.

■ *You are crawling on all fours. Which body surface is nearest to the floor? Which encounters the environment first? What is the most posterior part of your body as you crawl?*

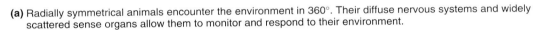

(a) Radially symmetrical animals encounter the environment in 360°. Their diffuse nervous systems and widely scattered sense organs allow them to monitor and respond to their environment.

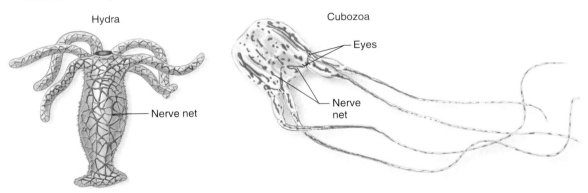

Hydra

Cubozoa

Eyes

Nerve net

Nerve net

Figure 16.19
Organization of the Nervous System Follows Two Basic Patterns. (**a**) It is advantageous for a radially symmetrical animal to have its nervous system arranged in 360°. (**b**) It is advantageous for a bilaterally symmetrical animal to have its nervous system concentrated in its anterior end.

(b) Bilaterally symmetrical animals usually encounter the environment head first. Their nervous systems are centralized in the anterior end of the body.

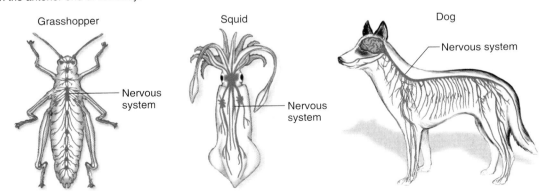

Grasshopper

Squid

Dog

Nervous system

Nervous system

Nervous system

filter feeder an animal that strains its food out of the environment, using a special feeding organ that water flows over

predator an animal that hunts other animals for its food

prey an animal that is hunted by a predator

Filter Feeders Strain Out Their Food For animals that live in water, one way to get food is to strain it out of the water in a process called **filter feeding.** The strategy of filter feeders is to have water flow over or through a special filtering structure that collects and concentrates food particles. Filter feeders have a variety of ways to ensure water flow. Some filter feeders such as barnacles attach themselves to a substrate in a place where there are strong currents. Others such as anemones use tentacles or cilia or other means to move, pull, or pump water over their filtering structure. Still other animals move through the water, filtering as they go. Filter feeders can be found in nearly all animal phyla. For instance, the sponges are all filter feeders (**Figure 16.20**). Clams, mussels, scallops, and most bivalve shellfish also are filter feeders. Flamingos are filter feeders, and you may be surprised to learn that some of the largest animals on Earth, such as the baleen whales, basking sharks, whale sharks, and manta rays also filter feed.

Grazers and Browsers Eat Plants or Algae
Grazers and browsers are animals that roam their habitats, continually munching on their preferred plants or algae. On land browsers tend to eat leaves, fruits, and flowers of shrubs and trees, while grazers eat low growing grasses and other plants. Grazers and browsers can be fishes, insects, crabs, snails and slugs, or mammals, but one common characteristic is that they must cover a wide territory as they forage to obtain enough food. Some plant-eaters are broad in their tastes and eat many kinds of plants; other plant-eaters have narrow tastes and restricted diets. For example, pigs, cows, and goats eat many kinds of grasses and other low plants and small trees, while giant pandas eat only bamboo. Not many animals can eat bamboo, so having an adaptation to eat bamboo gives giant pandas the advantage of sole access to their food source. Bamboo is not high in nutrients, though, and to make up for that, each day a giant panda must eat about 45 pounds (20 kilograms) of bamboo to survive.

Predators Actively Hunt and Kill Animals for Food Many animals feed on other animals. Those that do the hunting are called **predators,** while the animals that are eaten are called **prey.** *Hydra* uses nematocysts, specialized stinging cells to capture prey, and the hydra's tentacles stuff the prey into its mouth (**Figure 16.21**). Some predators such as fishes, sharks, lions, and cheetahs race after their prey. Their powerful jaws and enlarged, sharp teeth and claws grab and tear out hunks of flesh. Other predators sit and wait for their prey to come within reach and then rush at it and subdue it. Leopards and rattlesnakes use this predatory strategy. Spiders spin webs to snare their prey—then they inject venom to paralyze their catch and quickly wrap it up in silk. Spider venom has enzymes that predigest the prey, and spiders also use their fangs as soda straws to suck up the predigested soup that once was their prey's muscles and internal organs. Predators like lions hunt cooperatively and use the strategy of flushing game toward a hidden animal who rushes out to kill it. Others like wild dogs and wolves cooperatively chase game, waiting for it to tire or falter before they attack.

Animals Have Diverse Reproductive Strategies

As you might guess, all animals use sexual reproduction to produce offspring, but the story is much more diverse than this statement implies. Most of the animals with a two-layered body structure—such as placozoans, sponges, jellyfish, and comb jellies—also have asexual reproduction as part of their life cycle. Some flatworms such as planarians also can reproduce asexually. In some arthropods as well as in some vertebrates, eggs can develop into adult organisms without fertilization, a process called *parthenogenesis.* One method of asexual reproduction is *budding,* a process in which a small group of cells grows from the body of the parent and develops into a miniature, new organism.

Most animals undergo sexual reproduction at some point in their life

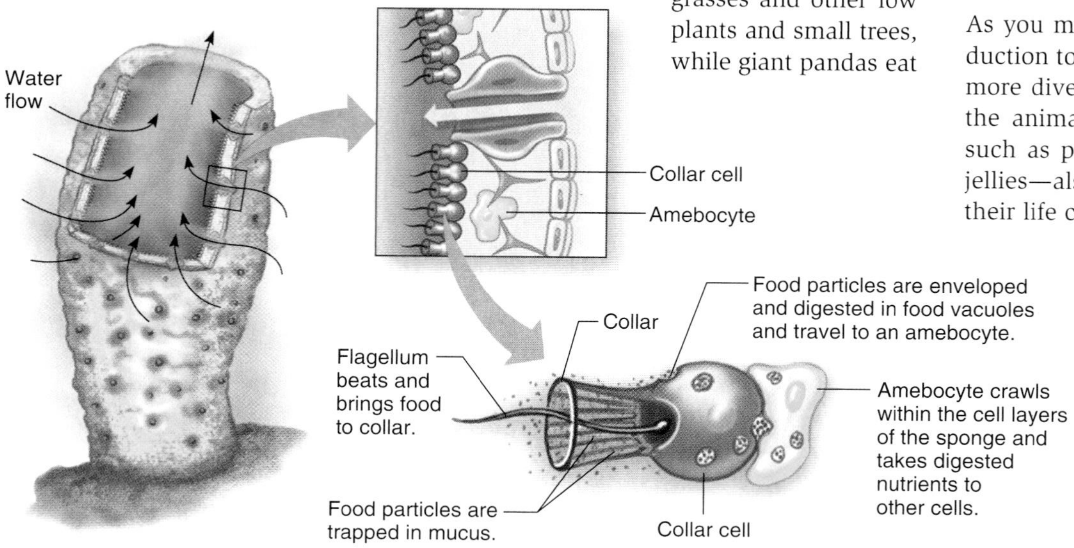

Water flow

Collar cell
Amebocyte

Collar

Food particles are enveloped and digested in food vacuoles and travel to an amebocyte.

Flagellum beats and brings food to collar.

Amebocyte crawls within the cell layers of the sponge and takes digested nutrients to other cells.

Food particles are trapped in mucus.

Collar cell

Figure 16.20 **How Sponges Filter Feed.**

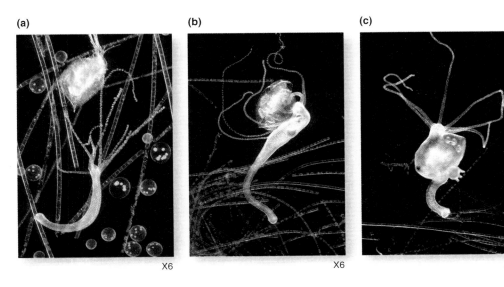

(a) X6   **(b)** X6   **(c)** X6

Figure 16.21
**A *Hydra* Attacks and
Consumes a Water Flea.**

hermaphrodite an
individual that can produce
both eggs and sperm

histories. Sexual reproduction involves the production of eggs—or ova—and sperm. Sperm are tiny cells that swim with flagella; they contain little cytoplasm and are mostly composed of DNA. Sperm are usually released to find eggs, and they must have an aquatic environment in which to swim to the eggs. This is why terrestrial mammals produce semen along with sperm. In contrast, eggs or ova are usually quite a bit larger than sperm because in addition to DNA, ova carry nutrients, proteins, and mRNA needed by the developing embryo.

In most animal species eggs are produced by females and sperm are produced by males. Some species of animals, though, combine male and female in a single individual—these are **hermaphrodites,** meaning that a given individual can produce both eggs and sperm. This does not necessarily mean that an individual mates with itself, but that any individual can play either the male or female role.

Some of the most complex reproductive life cycles of animals are found in cnidarians, and jellyfish provide a good example (**Figure 16.22**). Let's start this description with the mature *polyp*, a life cycle stage that is attached to some substrate. The polyp reproduces asexually by budding, but the bud does not produce another polyp. The same DNA that governs the colonial organisms now directs the development of a completely different-looking individual called a *medusa*. A medusa is a free-swimming life cycle stage that you will recognize as a jellyfish. This is an example of differential gene expression (see Section 9.2). The medusae grow by the process of mitotic cell division, and meiosis within the specialized tissues of testes and ovaries produces gametes. Eggs and sperm are released into the water where fertilization may occur if they find one another. The zygote settles

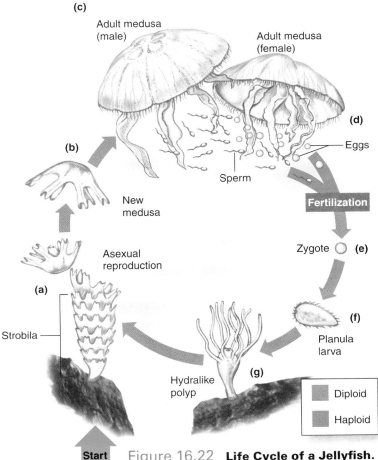

Figure 16.22 **Life Cycle of a Jellyfish.** (**a**) A polyp grows and asexually buds off small medusae or jellyfishes. (**b**) Young medusae grow and mature into adults that are male or female (**c**). Reproductive tissues of adults produce haploid eggs and sperm (**d**) that are released into the water where fertilization takes place and a diploid zygote results (**e**). The zygote divides mitotically and grows into a ciliated swimming larva (**f**). Eventually the larva settles onto a substrate and grows into a polyp (**g**) that eventually reproduces asexually.

■ *How are polyps different from medusae?*

on some substrate—such as an existing coral reef or a submerged piece of driftwood—and begins the development of a new polyp.

Having differently shaped bodies at different points in a life cycle is a common feature of animal development, even among vertebrates like most fishes and amphibians. Many animals enter a *larval phase* during which the individual grows and develops but does not reproduce. The larval stage or stages can have an entirely different morphology than the mature sexual stage. Many insects take this strategy to extremes in the process of *metamorphosis* (**Figure 16.23a**). Larval insects carry within them small groups of undifferentiated cells called *imaginal discs* (Figure 16.23b). After the larva has entered a resting stage within a cocoon or pupa,

the larval cells die and the imaginal disc cells divide and differentiate into a complete, new adult.

Regardless of how animals manage sexual reproduction, it invariably involves the two very different kinds of gametes, egg and sperm. The divergence of gametes into these two basic types appears to reflect two fundamentally different aspects of sexual reproduction. If you think of sexual reproduction from the perspective of survival of the offspring, it is obviously important for the zygote to have the resources it needs to start life, so from this perspective gametes should be large and packed with nutrients and macromolecules. Looked at from another perspective, however, sexual reproduction is successful if each individual casts its gametes widely and has the opportunity for many different matings. This increases the genetic diversity of the offspring and increases the likelihood that the species will survive environmental changes that inevitably occur over time. This constraint favors small, mobile gametes. In most animals evolution has resulted in the separation of these two functions. The nurturing function is concentrated in eggs; the dispersal function is left to sperm. In some animals this separation has been taken a step further, in that the individuals that carry out these two functions are separated into two genders.

One of the important aspects of reproductive strategy in animals is how many offspring are produced. Most animals produce large numbers of ova and sperm for fertilization, but only a few offspring actually survive. These species put little effort into parental care and rely on sheer numbers to ensure the survival of the species. Other animals, especially mammals, gestate their young for a long time and care for them over an extended period of time. Of all animals, humans probably invest the most in their young. Humans are not born until nine months after fertilization and in most cultures are cared for by their parents for another 15 to 20 years.

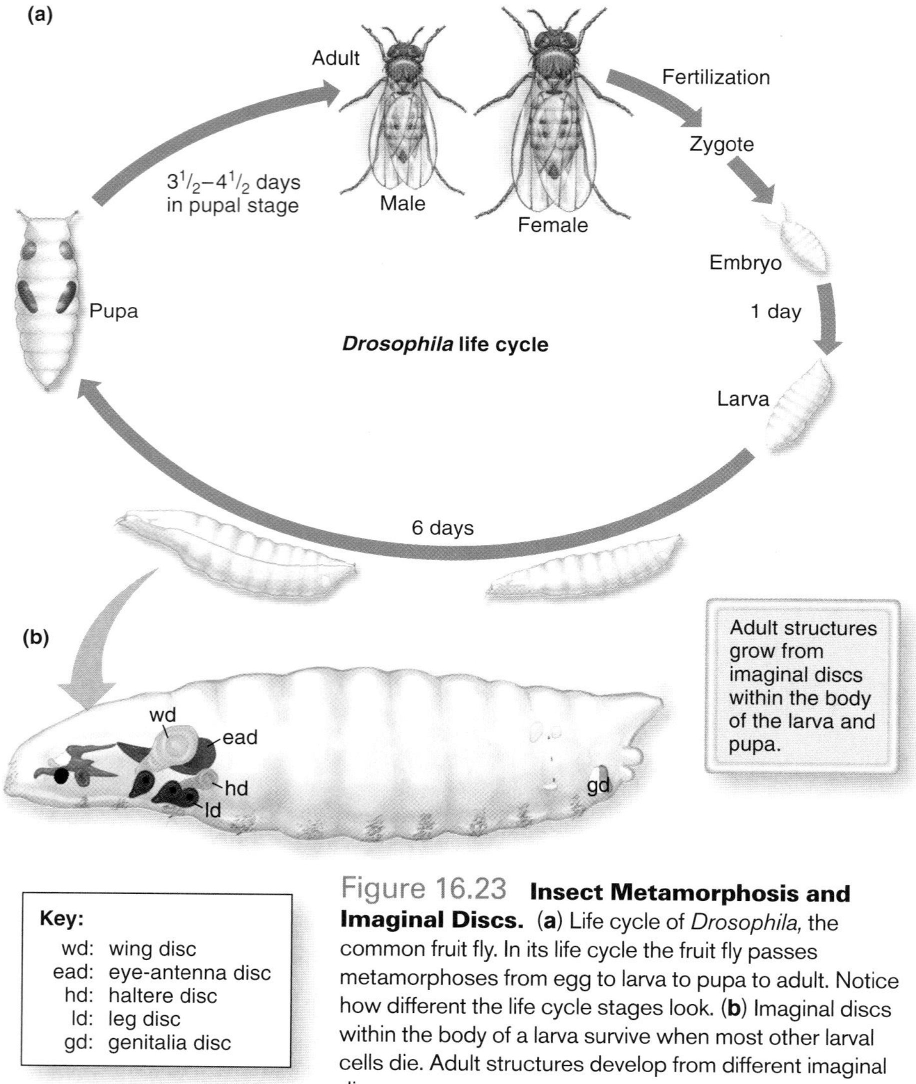

Drosophila life cycle

QUICK CHECK

1. What are some examples of animals that use asexual reproduction, and how does asexual reproduction work in one of these examples?

Key:
wd: wing disc
ead: eye-antenna disc
hd: haltere disc
ld: leg disc
gd: genitalia disc

Figure 16.23 **Insect Metamorphosis and Imaginal Discs.** (**a**) Life cycle of *Drosophila*, the common fruit fly. In its life cycle the fruit fly passes metamorphoses from egg to larva to pupa to adult. Notice how different the life cycle stages look. (**b**) Imaginal discs within the body of a larva survive when most other larval cells die. Adult structures develop from different imaginal discs.

16.5 Animal Parasites Can Make Humans and Other Animals Sick

All animals are susceptible to attacks by animal parasites and humans are no exception. Details of the lives of some common human parasites are presented in **Table 16.5.** Many of these are flatworms and roundworms that live within a host's intestines and live on the host's digested food. Some of these worms have a larval stage that can infect many body parts. They can be transmitted to humans who eat undercooked fish, beef, or pork that is infected with parasites. These parasites can cause serious illness. Diarrhea, weight loss, and lack of energy are three serious symptoms of a parasitic infection. If a person is healthy, such infections may clear up on their own by being passed with a person's solid waste, but for people who are not in excellent health, the consequences can be more serious.

Arthropods are another group of animals that can have health consequences for humans. They can be external parasites like head lice. Blood-sucking arthropods are of more serious concern when they are *disease vectors,* organisms that distribute infectious single-celled eukaryotes or bacteria with their bites. Mosquitoes, many kinds of flies, and ticks fall into this category. They can transmit a whole range of diseases including malaria, Lyme disease, West Nile virus, and bubonic plague. Almost all of the life cycles of these parasites involve another host besides humans. For bubonic plague the other host is rats. For Lyme disease, white-footed mice and white-tailed deer serve as the alternate hosts for deer ticks that also can bite humans and transmit the disease to them. Because the diseases transmitted by insect vectors have such complex life cycles, they often are difficult to treat.

QUICK CHECK

1. What are two kinds of animals that infect or infest humans?
2. What diseases can be transmitted by disease vectors such as ticks or mosquitoes?

16.6 Fungi and Animals Have Long Evolutionary Histories

Eukaryotes probably first appeared about 1.8 billion years ago. The single-celled eukaryotes became diverse in the years that followed their first appearance, and they probably had many different body plans, life histories, and adaptations. The multicellular kingdoms of the eukaryotes that we know today—the fungi, animals, and plants—evolved from some of these diverse ancient eukaryotes.

Information about the evolutionary history of fungi is far from complete. Data from studies of

Table 16.5 Some common human parasites		
Parasite	**Details of life cycle**	**Symptoms of infestation**
Pinworm	Live in human rectum; spread from hand to hand or by inhaling dust that contains pinworms; common in children.	Rectal itching
Hookworm	Juvenile worms live in soils; they burrow into human skin and feed on blood. They are spread by contact of skin with infected soil, especially by walking barefoot.	Weakness, low energy
Tapeworm	Infective cyst enters host in undercooked fish or meat. Worm hatches within intestine and attaches to intestinal wall. Adult worm absorbs nutrients from host.	Loss of weight, weakness, tapeworm segments in feces
Trichina worm	Infective cyst enters host in undercooked pork; adults lodge within muscles of host.	Muscle pain
Roundworm	Eggs enter host via the mouth; worms lodge in intestines.	Distended abdomen, worms in feces

DNA and proteins indicate that fungi existed as long as 1.4 billion years ago, but the first convincing fossils of fungi date to about 600 million years ago. These early fungi did not have complex fruiting bodies and were probably similar to modern chytrid fungi. Fossils showing sexual structures and fruiting bodies appear in rocks that are 400 to 200 million years old. What is striking is how similar these fossils look to modern forms (**Figure 16.24**). It seems that once the basic fungal form was established, it was successful and has not altered much.

Evidence from comparisons of DNA and proteins suggests that animals may have emerged a billion years ago. The oldest known animal fossils are not seen until about 600 million years ago, but many scientists predict that older animal fossils will be found. The fossils that are available have a very interesting tale to tell. Figure 14.24 reconstructs what these animals might have looked like. Most of these organisms were soft-bodied; they didn't have an external or internal skeleton as many more recent

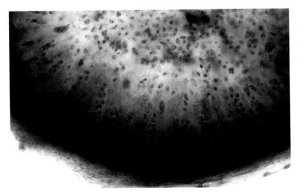

Figure 16.24 **Fossil Fungi.** These 400 million-year-old fungi have hyphae and fruiting bodies that are similar to modern fungi.

■ *Compare this photo to Table 16.3 and identify the phylum of this fossilized fungus. Explain the basis of your decision.*

groups of animals do. Given what you know about fossils, it should not surprise you that fossils of these soft-bodied animals are rare as compared with the abundant fossils of animals from later times that had hard parts that fossilize more readily. Nevertheless, the fossil record from this early period leads to some inescapable conclusions.

Some of these early animals were strange compared with life today. Yet, the start of the diversity of today's animals can be seen in these fossils. Some of them look remarkably familiar, even though they are not identical with modern forms. These findings point strongly to the conclusion that there was an explosion of diversity of animal body forms between 600 and 300 million years ago. Then, for reasons that are not completely clear, a mass extinction around 300 to 250 million years ago wiped out most of these diverse animal species. Only the ancestors of the species that are known today survived.

Early life was largely confined to the oceans and many complex invertebrates, fishes, and sharks evolved there. Eventually some species found their way to freshwater environments on land and had adaptations that allowed them to survive. From there species developed adaptations that allowed them to live completely on land. Arthropods made an early appearance on land 450 million years ago. Since that time insects have come to outnumber all other forms of life—they are found almost everywhere except in the oceans where their relatives the crustaceans abound. Reptiles appeared about 300 million years ago, and during the period from 200 to 65 million years ago reptiles including the dinosaurs greatly diversified and filled many different habitats. Some reptiles took to the skies on wings made of skin and bones. Other reptiles returned to the water and came to superficially resemble modern dolphins. On land some dinosaurs were tiny as chickens, while others were the largest land animals to ever live. The dinosaurs included the supersized carnivorous dinosaur *Giganotosaurus*, which was 44 to 46 feet long (13.5 to 14.3 meters) and stood about 12 feet tall (4 meters) at the hips. Because these once-dominant reptiles now are extinct, you might think of them as evolutionary failures, but dinosaurs ruled Earth for over 150 million years. What happened to the dinosaurs? Another mass extinction event, around 65 million years ago, wiped out most of them but, dinosaurs may be flitting around us as modern birds (Section 13.7).

Section 13.5 presented a detailed account of how mammals evolved from reptiles. The earliest animals that were clearly mammals lived about 200 million years ago. During the dominance of the dinosaurs mammals were small and nocturnal—they were active only at night. Based on fossil evidence, mammals did not come into their own until the dominant dinosaurs disappeared. No one is sure how or why mammals were not wiped out by whatever caused the extinction of the dinosaurs. Once the dinosaurs were out of the way, though, mammals evolved rapidly and took advantage of the available landscape. By 20 million years ago there were many orders of mammals, but today's varieties of modern mammals did not emerge all at once. Modern mammals fall into three groups. Many of the earliest mammals were egg-laying mammals, called **monotremes** (**Figure 16.25a**). Some fossil monotremes are nearly 150 million years old. Monotremes echo their reptilian ancestors and lay shelled eggs, but they also produce milk to nurse their young, and have fur. **Marsupial mammals** evolved later, with the earliest known fossils dating to 125 million years ago. Marsupials have short pregnancies, their young are born at an embryonic stage, and they are nursed within fur-lined pouches that are on females' bellies (Figure 16.25b). **Placental mammals** appeared sometime around the same time as marsupials, but in the end they were more successful. Today there are many more placental mammal species than marsupial species. Placental mammals have much longer pregnancies, their newborns are more fully developed than are newborn marsupials and like other mammals they suckle milk from their mothers' mammary glands (Figure 16.25c). One of the earliest placental mammals was a tiny shrewlike creature that lived about 125 million years ago.

Over the past 70 million years many species of mammals emerged, died out, and were replaced by other new mammalian species. Mammals that look more like their modern descendants appear about 40 million years ago. Gazelle-like creatures and the ancestors of horses evolved during this period. The first monkey species evolved about this time, while apes evolved from monkeys about 10 million years later. There were probably more ape species alive 20 million years ago than there are today. Out of all this diversity came human ancestors—it is that story that you will turn to next.

Humans Evolved in Africa

DNA evidence, supported by morphological evidence, shows that living humans are most closely related to living chimpanzees (**Figure 16.26**). Gorillas are the next closest group to humans and

monotreme a mammal that lays and incubates shelled eggs

marsupial mammal a mammal that nurtures its young within a pouch

placental mammal a mammal that has a placenta to nourish its young during pregnancy

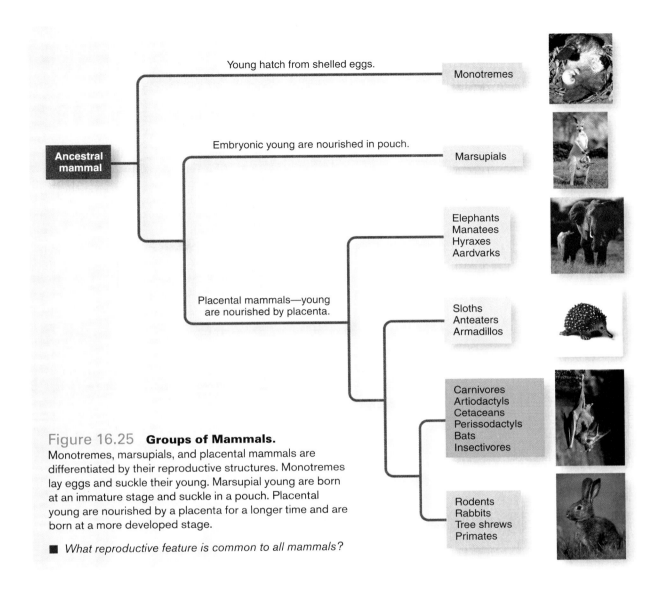

Figure 16.25   **Groups of Mammals.**
Monotremes, marsupials, and placental mammals are
differentiated by their reproductive structures. Monotremes
lay eggs and suckle their young. Marsupial young are born
at an immature stage and suckle in a pouch. Placental
young are nourished by a placenta for a longer time and are
born at a more developed stage.

■ *What reproductive feature is common to all mammals?*

chimps, followed by orangutans. Humans, chimps, gorillas, and orangutans form a closely related group that is more distantly related to gibbons and Old World monkeys. But don't make the error of thinking that humans evolved from chimps. Humans did not evolve from modern chimps, any more than you are a descendant of your modern-day cousins. The close relationship between chimps and humans means that if you are looking for human ancestors in the fossil record, some of those ancestors probably will share many traits with chimps. This is exactly the story recorded in fossils. Fossils from Africa support the conclusion that humans and chimps shared a common ancestor about 7 million years ago. About this time one line of descendants from the chimplike ancestor

evolved humanlike traits, including walking upright, while the other line evolved more chimp-like traits.

Figure 16.27 summarizes human evolutionary history that led to *Homo sapiens* after the lineages of chimps and humans diverged. By 3 million years ago a group on the human branch known as the *australopithecines* were widespread in Africa. Species of *Australopithecus* were between 3 and 4 feet tall, and they were definitely not chimpanzees. They had larger brain-cases than chimps, but their brains were much smaller than those of modern humans. Their teeth were more like human teeth than like chimpanzee teeth. Most importantly, from the shape of their pelvic bones, feet, and fossil footprints, it is clear that the australopithecines were upright walkers.

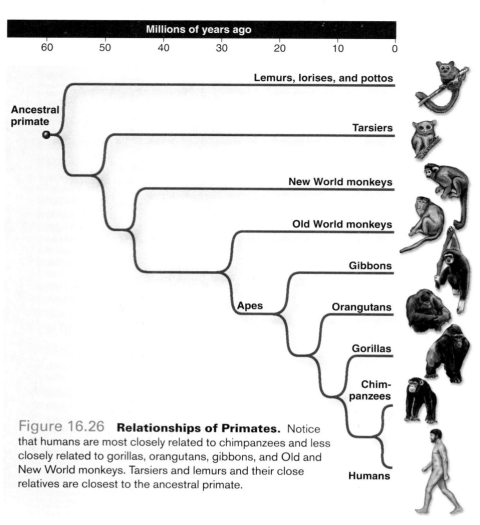

Figure 16.26 Relationships of Primates. Notice that humans are most closely related to chimpanzees and less closely related to gorillas, orangutans, gibbons, and Old and New World monkeys. Tarsiers and lemurs and their close relatives are closest to the ancestral primate.

Although chimpanzees sometimes do walk upright on their hind legs, they usually walk or run on all fours. So the stance and gait of australopithecines were much more human than chimpanzee.

Australopithecines of various kinds lived in Africa for a million years or more, and there were other early human species around too. Over time the fossils look more and more like modern humans and less like chimpanzees. Two million years ago at least three species of *Homo* and two species of the related genus *Paranthropus* lived together in Africa, but today only a single human species survives. Somewhere among the several *Homo* species are the direct ancestors of *H. sapiens.*

Homo erectus did many amazing things in their 2 million years on Earth. For one thing, they spread far and wide from Africa. Fossils from this group are found in Asia, in modern-day France, and even as far north as the Republic of Georgia. In Europe *H. erectus* relatives appeared as early as 500,000 years ago and diversified into *Homo neanderthalensis. H. erectus* and their descendants made sophisticated tools, probably had fire, and may have lived in small social groups. They may have protected their young and the infirm members of their clans. It is uncertain if they had language, though, and it seems they had little in the way of art.

What happened to all the other human species? Fossil evidence from Asia puts their disappearance as late as 50,000 years ago. In Europe they disappeared

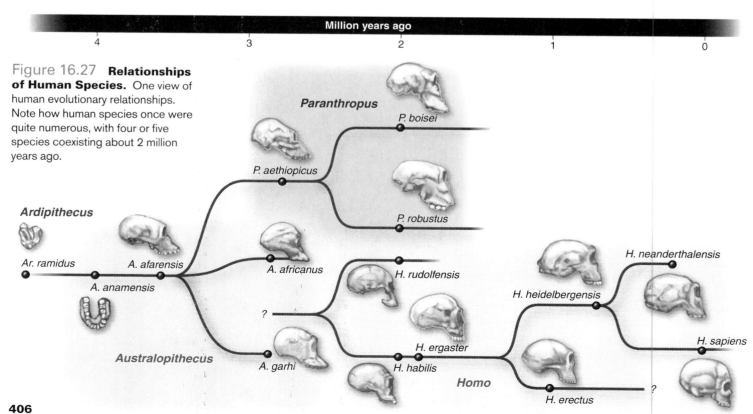

Figure 16.27 Relationships of Human Species. One view of human evolutionary relationships. Note how human species once were quite numerous, with four or five species coexisting about 2 million years ago.

perhaps as late as 30,000 years ago, and in some parts of southern Asia they may have existed as recently as 18,000 years ago. There are many hypotheses to explain their disappearance, but they all come back to *H. sapiens*. When *H. sapiens* moved in, the older groups died out.

Now you can turn your thoughts to the immediate ancestors of modern humans, the first *H. sapiens*. Here the molecular evidence comes into play, because biologists can use comparisons of modern human groups to estimate how long ago all living humans shared a common ancestor. These studies find that all living humans had an ancestor in eastern Africa sometime around 200,000 to 100,000 years ago. While the fossil record always is subject to new findings, it agrees with this conclusion. The oldest known *H. sapiens* fossils are from northern Africa and date to about 160,000 years ago. From there, this single population moved out across Asia and Europe and colonized much of the world, much like the previous *H. erectus* groups had.

Recently on the Flores Islands of Indonesia, researchers found the bones of small hominids that lived only 18,000 years ago (**Figure 16.28**). This is long after the many species similar to *H. erectus* were thought to have died out. Some scientists conclude that these tiny hominids are similar to *H. erectus*. They were given the scientific name *Homo floresiensis*, but many people affectionately call them Hobbits, after the Tolkien characters in the *Lord of the Ring* trilogy. Scientists have many ideas about how this species became so small. For example, it is not uncommon for species on island environments to become smaller over time. In fact, the island occupied by *H. floresiensis* does have fossils of other miniature mammals: pygmy elephants. Regardless of the answer to why these hominids became so small, their discovery demonstrates just how diverse life is—even human life.

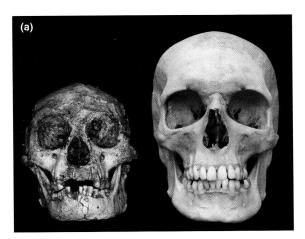

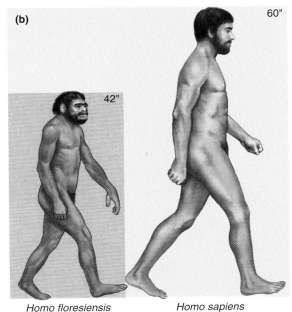

Homo floresiensis   Homo sapiens

Figure 16.28 *Homo floresiensis* and *Homo sapiens.* (**a**) This comparison of skulls is a dramatic demonstration of how small *H. floresiensis* was. (**b**) An artist's reconstruction of *H. floresiensis* and a comparison of the heights of *H. sapiens* and *H. floresiensis.*

QUICK CHECK

1. Why aren't humans descended from chimpanzees, gorillas, or orangutans?

—Continued next page

16.7 An Added Dimension Explained
What Happened in Salem in 1692?

The witch trials in Salem and across Europe have been the subject of many essays and discussions. Certainly, a great many social and religious factors were at play to cause the outbreak of witch trials. Modern studies suggest that there were not actually witches and that innocent people were probably unnecessarily persecuted and put to death. People who want to avoid such mistakes in the future are interested in understanding what really happened in these witch trials.

What do we know about the behavior of these so-called witches? One interesting finding is that prior to 1692 in all of Massachusetts only five people had been convicted of witchcraft. The sudden rash of witch trials—in Massachusetts, all across the northeastern colonies, and even in Europe—indicated that something out of the ordinary was happening in this era. The behaviors of the witches were remarkably consistent from location to location, even across the Atlantic Ocean. The bizarre behaviors and blasphemous speech might be attributable to a variety of causes, but other symptoms point to something biological. The accused witches had hallucinations, convulsions, and trances that are reminiscent of certain kinds of epilepsy. Many of these individuals had limb problems, which included jerky movements and infections in their arms and

Continued—

legs. Eventually, people would die from the condition itself, if they were not first killed for their behaviors.

Another relevant observation is the geographic distribution of the accused witches. Some areas were hotbeds of antiwitch fervor, while other areas were unaffected. Ireland, for example, showed little witch hunting, while Scotland seemed obsessed by it. What could account for all these diverse observations? One interesting hypothesis points to a fungus. This may sound unbelievable, but there is a fungus that produces many of the symptoms described in the witch trials. Ergot is a fungus—*Claviceps purpurea*—that grows exclusively on grains of rye and wheat (**Figure 16.29**). Most other types of grains are not infected by ergot. The fungus contains many biologically active compounds that interact with membrane protein receptors on the surfaces of cells, especially nerve cells, and alter their functions. Ergot causes blood vessels to constrict and can cause gangrene. As the limbs are degenerating, the muscles can twitch and cause erratic movements. Ergot causes the uterus to contract, and for centuries midwives have used small quantities of it to speed birth for women in labor. Ergot also stimulates parts of the nervous system in ways that can cause convulsions, similar to those seen in epilepsy. Finally, ergot is similar in structure to LSD, lysergic acid diethylamide, which has very strong effects on the nervous system, even in small quantities. LSD causes visual and auditory hallucinations similar to those described in the early witch trials. The potent actions of ergot on the nervous system have been the source of many modern drugs.

Ergot is more common in northerly, cold, moist climates. People who lived where witch trials were common relied heavily on rye in their diet. In Scotland, for example, people ate a lot of

Figure 16.29   **Ergot Growing on Grains of Wheat.**

rye, while people in Ireland ate little rye. Even small amounts of ergot, perhaps too small to notice or be concerned about, consumed over a period of time can cause symptoms. Not everyone with ergot poisoning was branded a witch. The pervasive disease called St. Anthony's fire was probably a result of ergot poisoning. Ergot poisoning was probably rampant in northern Europe from about 1000 to 1600 AD or longer, and up to 50,000 people may have died from various complications of ergot poisoning. By 1550 AD some physicians finally recognized that ergot was responsible, but it is not clear how far this knowledge had spread. The evidence suggests that the witch symptoms in Salem and elsewhere were caused by ergot poisoning.

You might think about the nature of the evidence that ergot caused the symptoms of the Salem witches. The evidence is strong and given the evidence, it is much more likely that the women were affected by some chemical than that they were actually inhabited by demons or were witches. Still, the evidence is correlational. Do you remember the difference between a correlation and an experiment from Chapter 1? It is not possible to do an experiment in which women are randomly given either ergot or a placebo and look for changes in behavior. These kinds of limitations are not uncommon in science, but it is still possible to come to reasonable conclusions based on the kinds of correlations seen in the Salem witch story.

QUICK CHECK

1. How does evidence from Ireland and Scotland lend support to the idea that ergot-infected rye is to blame for behaviors of people in Salem, Massachusetts, that were interpreted as evidence of witchcraft?

Now You Can **Understand**

How Antifungal Drugs Work

Pests plague humans. In each of these diversity chapters you have read about species that can infect humans or the food they eat. Controlling or eliminating these infections is a major goal of public health efforts, and knowledge about the similarities and differences between humans and other organisms is critical to these efforts. Many drugs or treatments are developed or understood based on the unique biology of each kind of pest. For instance, antifungal agents target the unique aspects of the fungal cell walls and cell membranes. Fungi have chitin in their cell walls and a compound called *ergosterol*, a lipid, in their cell membranes, so drugs that target the synthesis of these compounds can be fairly specific in combating fungal infections. Many such drugs would act to inhibit the enzymes used in making these biological molecules, more than acting on the molecules themselves. But humans are biochemically related to all living organisms, and so there are proteins, either enzymes or receptors, in humans that will be affected by these and all drugs. This is why drugs have side effects.

The negative effects of drug treatments lead scientists to search for more natural ways to combat infections. When the infection is in plants such as grasses of lawns and golf courses, there may be alternatives to drug or chemical treatments. One particular fungus, commonly called take-all fungus that infests turf grasses, can be controlled by introducing a less toxic fungus that competes with take-all. If the less toxic fungus flourishes, the impact of take-all fungus is reduced. In the future you will certainly find more specific and less harmful approaches to controlling pests that plague humans, all based on our increasing knowledge about Earth's biodiversity.

What Do **You** Think?

Big Foot and Little People

Humans are an exceptionally curious species—they look for the unusual and try to explain it. For some people this curiosity finds its outlet in scientific studies, but others are drawn to phenomena that seem to have no scientific support. People believe in and explore topics such as extrasensory perception, alien visitations, and ghosts. Persistent legends that fall into this category are tales of "unusual" creatures, and many of these are much like humans. At the large extreme there is a long history of legends known as Big Foot,

Sasquatch, and the Yeti. At the small extreme are the "little people" described in diverse cultures. The people of the Flores Islands in Indonesia, for example, claim that "little people" used to run naked in the forests and hunt "little elephants." If you search the Internet for such stories, you will find reams of supposed evidence to support them. Scientists do sometimes venture into the morass of contradictory evidence that people claim supports the existence of these creatures, and the evidence is always found wanting. This kind of evidence would never stand up under the typical scrutiny that scientific studies face and most of these tales are just that—tales.

Sometimes, though, folk tales do have a basis in something that is real—occasionally scientists find evidence that supports a tale and shakes the researchers out of their assumptions. For example, tales in the 1800s of "ape-men" in the jungles of Southeast Asia were originally rejected as just myths, but then the orangutan was discovered, and it is likely that this ape was the source of many of those myths. The *Homo floresiensis* fossils discovered in the early twenty-first century is another example. These findings seem to support the local myths of "little people hunting little elephants."

Some people will look at these scientific discoveries as support for *all* of the creature myths. If one myth has turned out to be true, then the rest are likely to be true—even in the absence of evidence. Based on what you have learned about science, is this conclusion scientific? Does the discovery of *H. floresiensis* provide any evidence about the existence of Big Foot? What do *you* think?

Chapter Review

CHAPTER SUMMARY

16.1 Fungi and Animals Are Closely Related Kingdoms

Biochemical similarities and similar ways of obtaining food link fungi and animals.

16.2 Fungi Have Unique Traits

Fungi feed using hyphae that invade a food source and absorb nutrients. Underground hyphae mass into a mycelium, while fruiting bodies made of hyphae grow out of the mycelium. Mycorrhizae are associations of fungi with plant roots. Most fungi are multicellular and haploid and have only brief diploid phases in their life cycles. Fungi reproduce asexually with spores. Each spore can germinate into a new mycelium. Fungi are grouped according to their methods of reproduction. Most fungi have brief phases of sexual reproduction that introduce genetic diversity.

fruiting body 387	mycelium 387
fungi 387	mycorrhizae 389
hyphae 388	

16.3 Humans Eat and Use Fungi

Fungi are extremely important as decomposers, but they may attack living tissues. Humans use fungi as foods and take advantage of fungal fermentation to produce alcoholic beverages and make breads and other baked goods light and fluffy. Many antibiotic drugs are derived from fungi.

16.4 Animals Are Adapted to Many Environments

Animals have a variety of body plans and differ in the complexity of the organization of their body layers, symmetry, placement of sense organs, and ways of moving respiratory gases. Placozoans and sponges have two cell layers, while all other animals have three. Symmetry can be radial, bilateral, or asymmetrical. Placement of sense organs is related to body symmetry. Bilateral animals have an anterior end with concentrations of nerve tissue. Radially symmetrical animals do not have heads. Filter feeding, grazing, browsing, and predation are some animal adaptations for obtaining food. Some animals are hermaphroditic, while others have separate sexes. Many animals have life cycles that feature metamorphosis. Animals vary in the amount of parental care given to their young.

animal 393	hermaphrodite 401
anterior end 398	mesoderm 397
bilateral symmetry 398	posterior end 398
dorsal surface 398	predator 400
ectoderm 397	prey 400
endoderm 397	radial symmetry 395
filter feeder 400	ventral surface 398

16.5 Animal Parasites Can Make Humans and Other Animals Sick

Animal parasites can infect and sicken other animals. Disease vectors are organisms that spread infectious single-celled eukaryotes or bacteria with their bites.

16.6 Fungi and Animals Have Long Evolutionary Histories

Fungi appeared between a billion and 600 million years ago and some seem to have retained their ancestral forms. The earliest definite animal fossils appeared about 600 million years ago. The earliest animals were soft-bodied and evolved in the oceans. Land animals appeared later, with arthropods being the first animals to colonize land. When many species of reptiles went extinct about 60 million years ago, mammals diversified into many orders and species. Mammals can be monotremes, marsupials, or placentals. Seven million years ago humans and chimps shared a common ancestor, and this is why the genes of humans and chimps are extremely similar. Humans have not descended from the modern apes, but they share a common genetic heritage with them. Species related to humans evolved in Africa and *Australopithecus* was an early African relative that was superseded by several species of *Homo*. All of these species disappeared once *H. sapiens* appeared, about 160,000 years ago.

marsupial mammal   404   placental mammal   404
monotreme   404

16.7 An Added Dimension Explained: What Happened in Salem in 1692?

The incidents in Salem in the 1600s that were ascribed to witches may have been caused by ergot poisoning.

REVIEW QUESTIONS

TRUE or FALSE. If a statement is false, rewrite it to make it true.

1. Sperm and ova are diploid.
2. A fungal hyphae may have haploid, or diploid nuclei.
3. Radially symmetrical animals tend to lack a head end.
4. All the fungi visible aboveground are haploid.
5. Like bacteria, fungi have enormous importance as decomposers.
6. Vertebrates have internal skeletons, while invertebrates do not.
7. Skin is made of ectoderm; gut lining is made of mesoderm.
8. Your belly button is located on your dorsal surface.
9. Humans have radial symmetry.
10. One of hydra's feeding adaptations is stinging nematocysts on tentacles.
11. Some of Earth's largest organisms are filter feeders.
12. Animals and fungi are both heterotrophs that ingest and then digest and absorb their food.
13. Australopithecines are a blend of human and chimpanzee characteristics.
14. Disease vectors move pathogens between hosts.

MULTIPLE CHOICE: Choose the best answer of those provided.

15. Fungi and animals are thought to be closely related because
    a. they share asexual reproduction.
    b. both use sexual reproduction.
    c. representatives of both kingdoms have chitin and similar proteins.
    d. representatives of each have similar DNA or rRNA
    e. c and d

MATCHING

16–20. Match these terms and descriptions. (One choice will not match.)

16. chytrids
17. yeasts, morels
18. mushrooms, toadstools, rusts, smuts
19. hyphae
20. mycorrhizae

a. Ascomycota
b. Zygomycota
c. Basidiomycota
d. have flagellated sperm; may link algae and fungi
e. secrete digestive enzymes and function in reproduction
f. association of fungi and plant roots

CONNECTING KEY CONCEPTS

1. What traits of fungi and animals provide the evidence for concluding that they are closely related? What traits cause them to be grouped separately?
2. How does the history of human evolution fit in with the overall evolution of animals on Earth?

THINKING CRITICALLY

1. Imagine you are working for your local park service and have been asked to consult on an outbreak of tree diseases in the region. What are the classes of agents that might be causing the diseases? How might you determine the causes?
2. World trade has brought much benefit to Americans and to other cultures as well. While many people are concerned about world trade from social and political perspectives, biologists raise some unique concerns about some of the risks of uncontrolled free trade. Have you seen any such concerns in newspaper articles or other news reports? From a biological perspective, what negative aspects of free world trade should our government and organizations be addressing? What can you personally do to limit these negative effects?

For additional study tools, visit www.aris.mhhe.com.

Varieties of Life

PLANTS

Search of Moors Stirs Specter of Old Murders

November 23, 1986

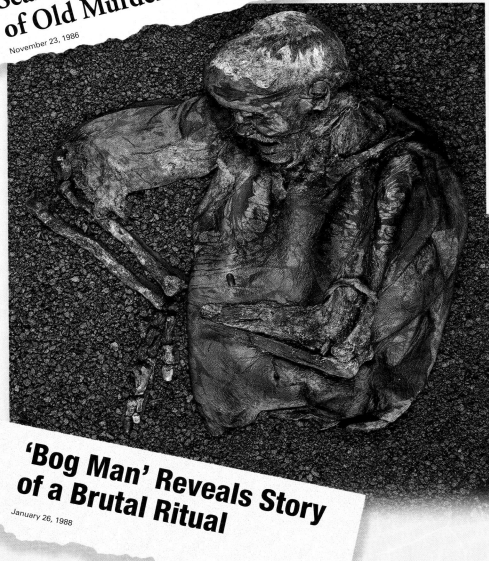

Lindow Man

'Bog Man' Reveals Story of a Brutal Ritual

January 26, 1988

Lindow and the Bog Bodies of Europe

Who was Lindow, and what events led to his brutal death nearly 2,000 years ago? To answer these questions, we look back to the time of the Druids, an ancient Celtic priesthood. The Druids recited prayers many times each day, organized rituals and festivals, and conducted traditional human sacrifices to honor the gods and goddesses—and in return the gods and goddesses protected the people. It was a bloody and grisly bargain, but it always had been this way—and Lindow knew that his sacrifice would serve a noble purpose. He was a king's son—a member of the privileged elite—and every royal knew that someday he or she might be sacrificed for the good of all. Lindow's brothers and sisters had made this ultimate sacrifice, and each had faced death with dignity and great calm. Court musicians celebrated their sacrifices and sang of their bravery, but Lindow knew that the songs they would sing about him would be different. They would be mocking songs—songs of ridicule because all would see that he did not want to die. Lindow feared death and most of all he dreaded the threefold sacrificial death reserved for royalty. Lindow could not escape his fear—there was no way out for him. Threefold was his violent death . . . and then the cold waters of the bog covered him (**see chapter-opening photo**).

Although local history is full of myths and legends relating to human sacrifices, until recently it was hard to know if tales like the fictional fragment that you just read could have been true. Then in 1983 Andy Mould, who worked for a peat-cutting operation in England, discovered the well-preserved remains of the central figure of our story, now known as Lindow Man.

This story is intriguing, but you may be wondering how Lindow Man connects to plants. As you soon will see, there is indeed a strong connection between plants and the story of Lindow Man. In the closing sections of this chapter, you will return to Andy Mould and Lindow Man, but first, read about the world of green algae and plants. ∎

17.1 Plants Sustain Life

Although most people don't pay too much attention to them, plants are pivotal components of Earth's environments.

Directly or indirectly plants provide food for most of Earth's organisms. Even if you're a dedicated meat lover, every day you eat plants and foods made from them—not just fruits, grains, vegetables, and nuts, but cookies, crackers, candies, cereals, breads, chewing gum, chocolate bars, colas, coffee, and tea, to list just a few. Other nonfood products are made from plants. For instance, your newspaper, the paper wrapper around a burger, a soda cup, this book, houses, cotton clothing, furniture, even pencils, are all made from plants. The aspirin that relieves your headache is based on a chemical in willow bark. Your cologne smells so good because it contains essential oils distilled from flowers and aromatic leaves. Medicine depends on opium-based drugs to kill pain, and opium comes from the sap of poppies. And this is just the start of a list of the nonfood items derived from plants.

Civilization itself came about partly because people learned how to grow and harvest plants and so provided the surplus food that enabled people to settle in one place (**Figure 17.1**). The Green Revolution of the 1970s has fed billions of people worldwide and is a result of improved cultivation of plant crops.

Plants also define the large ecological units on Earth called *biomes* (see Section 1.3). For instance, grasslands are biomes that depend on the presence of various kinds of grasses. Coniferous forests are biomes defined by stands of evergreens, and tropical rain forests are defined by the highest diversity of plants of any biome. In all of these biomes, characteristic plants provide oxygen, build and anchor soil, and hold water in soil. Without plants Earth's landscape would be desolate and either dry or flooded. Plants help to keep landscapes stable and help to cycle nutrients and essential gases.

Figure 17.1 **Agriculture Provided the Excess Food That Allowed the Growth of Civilizations.**

QUICK CHECK

1. Can you name an animal that does not directly or indirectly depend on plants for its existence? What would such an animal eat?

17.2 Green Algae and Land Plants Form One Phylogenetic Kingdom

From the earliest attempts to classify organisms, dating back to Carl von Linné in 1753, plants have been placed in their own kingdom, and modern DNA studies do not dispute this. A more difficult question has been: What are the limits of the plant kingdom? Some classification schemes have grouped fungi and all kinds of algae with plants, while other classification schemes have put them into separate

groups. DNA analyses can settle these questions with some certainty, and such studies make it clear that the green algae are genetically close to the land plants. Fungi and other kinds of algae fall into separate and distinct groups (**Figure 17.2**).

Other studies support the close phylogenetic relationship between green algae and land plants, and the conclusion from all of these studies is the same: plants and green algae should be grouped together.

From these studies, a **plant** can be defined as a photosynthetic eukaryotic organism that uses both chlorophyll a and chlorophyll b, has carotenoids as accessory pigments, has cell walls that contain cellulose, uses starch as a storage carbohydrate, and has flagellated cells with paired anterior flagella that propel with whiplash motions. **Figure 17.3** is a phylogenetic tree that shows the major groups of plants explored in this chapter.

plant a photosynthetic eukaryotic organism that uses both chlorophyll a and chlorophyll b, has carotenoids as accessory pigments, has cellulose in its cell walls, uses starch as a storage carbohydrate, and has paired anterior flagella on its flagellated cells

QUICK CHECK

1. Describe two lines of evidence that green algae and plants are closely related.

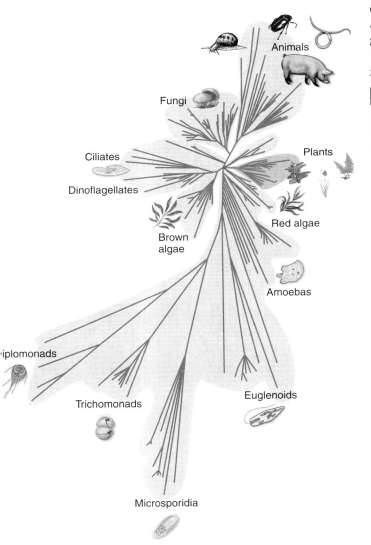

Figure 17.2 **Relationships of Plants to Other Eukaryotes.** Phylogenetic analysis of the genes for ribosomal RNA show that green algae and plants belong to the same group that is distinct from other kinds of algae. Each line represents a species that was sampled. Distances between lines represent differences in DNA sequences.

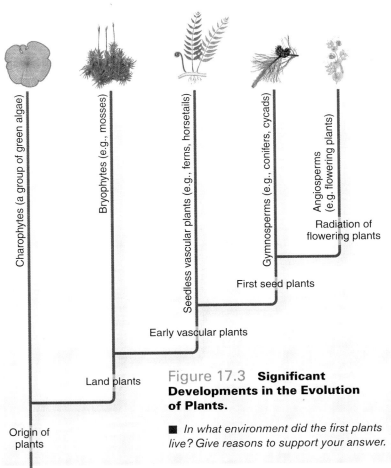

Figure 17.3 **Significant Developments in the Evolution of Plants.**

■ *In what environment did the first plants live? Give reasons to support your answer.*

cuticle a protective layer on the outer surface of plants that prevents water loss

stoma (plural: stomata) pores in the cuticle that are opened and closed by guard cells to admit carbon dioxide and release water vapor and oxygen

17.3 The Evolutionary History of Plants Is Dominated by the Transition from Water to Land

Fossils from 440 million years ago show that the first land plants evolved from green algae that lived in freshwater. These green algae may have been somewhat similar to *Coleochaete,* an alga that grows on underwater surfaces of water plants (**Figure 17.4**).

Figure 17.4 **The Earliest Green Plants May Have Been Similar to This Green Alga, *Coleochaete.***

X100

Pine

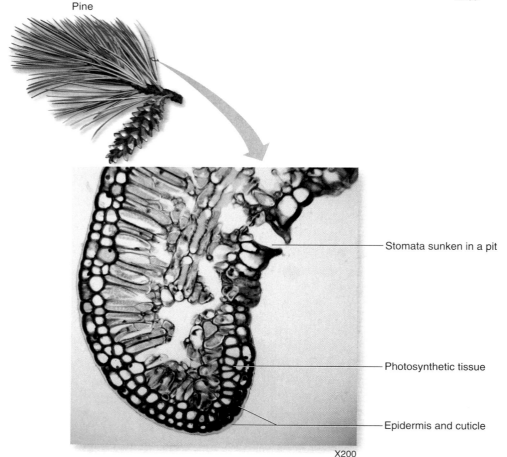

Stomata sunken in a pit

Photosynthetic tissue

Epidermis and cuticle

X200

Figure 17.5 **Cuticle.** Land plants have a thick cuticle layer that prevents water loss.

The ancestral green algae first may have been exposed to dry conditions along the edges of rivers and bodies of freshwater. When water retreated during low tides or droughts, green algae had the ability to survive periods of dessication and could survive and reproduce. Unexploited opportunities waited for the new land dwellers. While water was crowded with organisms, at this time the land was wide open and free of competitors and predators. Also, light is more abundant on land than it is in water. Because sunlight penetrates only the first few meters of water, and deeper waters are dim or completely dark, photosynthesizers are restricted to sunlit waters. In contrast, because land has no such limit on the availability of sunlight, photosynthesis can occur from the land surface up as high as a plant can grow. In water the supplies of oxygen and carbon dioxide can fluctuate on both daily and yearly cycles, while on land there are greater supplies of both of these gases. Finally, plant nutrients like nitrogen and phosphorus are more abundant on land than they are in water.

The transition from water to land was not easy, though, and it probably took millions of years before green algae found the combination of adaptations that would work. The biggest problem on land was the lack of water. A green alga that is removed from water quickly will dry out and die. Algae that eventually survived on land developed adaptations to protect them from water loss. These adaptations often were physical barriers that helped them to retain water. For instance, to keep a slice of bread from drying out, you surround it with a protective layer—plastic wrap, waxed paper, or a plastic bag—anything that will keep moisture from evaporating. As green algae colonized the land, selection would have favored those algae that had some form of a waterproof covering. Eventually land plants evolved a tough, waxy outer covering called a **cuticle** that prevents loss of water from plant cells to the air (**Figure 17.5**).

A new problem arose as the cuticles of green algae that had moved onto land became more efficient at retaining water. The barrier of thick cuticle that protected plants from water loss also hindered their exchange of gases with the environment. The evolutionary solution was the development of openings that allow the passage of carbon dioxide, oxygen, and water vapor. Except for some low-growing plants called liverworts, all of today's land plants have numerous small openings in the cuticle called **stomata** that are located on the undersides of leaves and on stems (**Figure 17.6a**). Each stoma is flanked by a pair of banana-shaped guard cells that go limp to block the opening or swell to open it (Figure 17.6b). Stomata close when the plant is

(a) Epidermal cells of plants have stomata that can open and close for gas exchange.

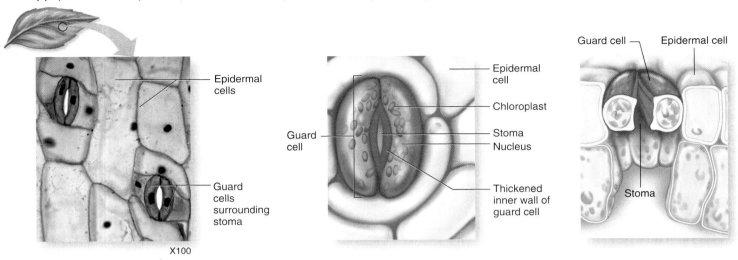

X100

(b) When water is abundant, guard cells swell and stomata open. When water is scarce, guard cells lose water and shrink, closing stomata.

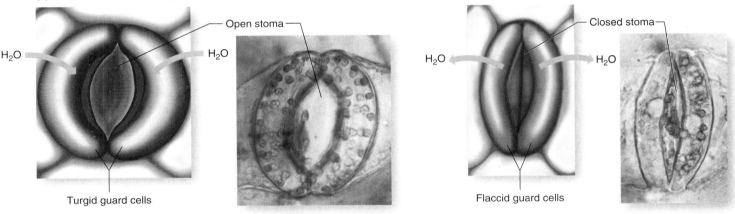

Figure 17.6   **Stomata.** **(a)** Stomata control gas exchange and loss of water vapor from leaf tissues. **(b)** Stomata open when water is plentiful and close when water is scarce.

■ *Make an analogy for a stoma that demonstrates that you understand the relationship between the structure of a stoma and guard cells and their functions.*

in danger of drying out and open when conditions are moist.

Stomata allow gases to move in and out of plants, but how do gases get to all of the cells in a plant? Plants have no pumping organs or bellows like the heart and lungs of animals that move fluids and gases. Rather, even the largest and most complex plants use the process of diffusion to exchange gases. If you examine the cut edge of a leaf with a compound microscope, you will see that there is a great deal of empty space between layers of cells of the upper and lower leaf surfaces (see Figure 17.5). Gases diffuse through these spaces and so have direct contact with plant cells. Oxygen, carbon dioxide, and water vapor diffuse into and around the spaces within leaves and diffuse into and out of cells according to their relative concentrations.

Once plants were ashore and equipped with a cuticle that had stomata, survival would have favored plants that could grow larger and taller and shade out their competitors. Having a large and compact body instead of a small, filmy, or thread-like body also would have helped plants to conserve water and minimize water loss. To keep from toppling over, tall plants require a firm anchor in the soil, a way to move water up to their highest tissues, and strong bodies that can resist the pull of gravity. **Vascular plants** have systems of internal tubules that are both internal waterworks and structural supports (**Figure 17.7**). One important adaptation was the evolution of **lignin** as part of the cell walls of vascular woody plants. Lignin is a complex molecule made from organic alcohol subunits. It is different from the carbohydrate cellulose in the

vascular plant plant that has internal systems of tubules that transport fluids and dissolved nutrients

lignin a complex organic alcohol molecule that is part of the cell wall of woody plants

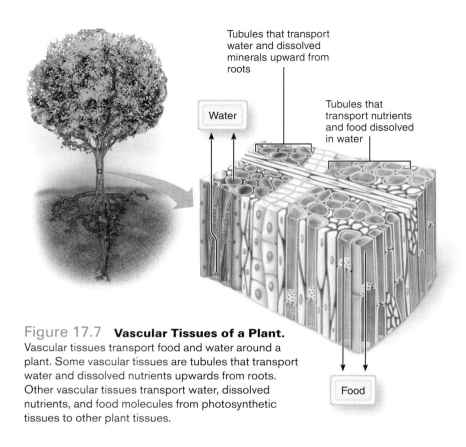

Figure 17.7 **Vascular Tissues of a Plant.** Vascular tissues transport food and water around a plant. Some vascular tissues are tubules that transport water and dissolved nutrients upwards from roots. Other vascular tissues transport water, dissolved nutrients, and food molecules from photosynthetic tissues to other plant tissues.

Labels in figure:
Tubules that transport water and dissolved minerals upward from roots

Water

Tubules that transport nutrients and food dissolved in water

Food

body plan of land plants holds part of the key to the ability of plants to grow tall. Let's consider plant body plans next.

QUICK CHECK

1. Explain why land plants need cuticles and stomata, while green algae do not.

17.4 Plants Have Diverse Body Structures

You may not think of plants as having bodies, but they do. The body structures of green algae and land plants are diverse, in part because plants range from tiny unicellular organisms to some of Earth's largest and most complex multicellular organisms. *Chlamydomonas* is an example of a simple, unicellular green alga that lives in freshwater (**Figure 17.8a**). *Chlamydomonas* has two flagella and a photosensitive eyespot, and both may assist it in moving toward lighted waters. The colonial and filamentous forms of green algae are a bit more complex. *Volvox* is a colonial green alga (Figure 17.8b). *Spirogyra* and *Ulothrix* (Figure 17.8c) are filamentous algae that form slimy masses in freshwater lakes and ponds, while *Cladophora* is a branched filamentous green alga that lives in both freshwater and saltwater. Both colonial and filamentous algae are intermediate between being unicellular and multicellular. Each individual is made of more than one cell, but other than the presence of reproductive cells, there is little or no cell differentiation. All of their cells are identical.

A **colony** is a group of cells that are held together by some external matrix but that retain the

colony a group of cells that are held together by some external matrix but that retain the identity of individual cells

cell wall and from the lipids and proteins in the cell membrane. Lignin is, in part, why an oak tree grows so much taller than a lily plant. In your diet lignin is part of the fiber that is listed on food labels.

So life on land offered direct access to more sunlight and thus also offered more energy for growth and development than did life in the water. As land plants evolved they acquired body plans that were adapted to the drier environment. The

(a) *Chlamydomonas*

(b) *Volvox*

(c) *Ulothrix*

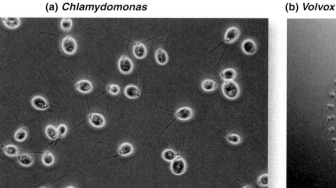

X600

X10

X50

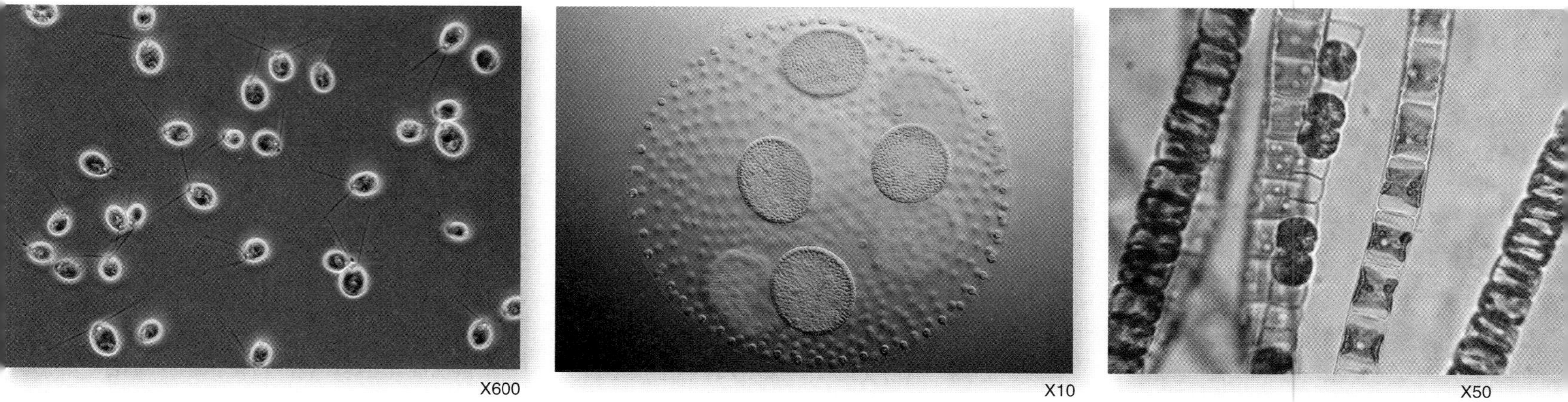

Figure 17.8 **Green Algae Are Diverse.** (**a**) *Chlamydomonas* are small, single-celled green algae with flagella. (**b**) *Volvox* is a colonial species of green algae. (**c**) *Ulothrix* is a filamentous green algae.

■ *What are the green, flat discs within* Volvox?

identity of individual cells. It is intriguing that individual cells of some colonial species of green algae, such as *Volvox*, look just like *Chlamydomonas*. *Volvox* is colonial but not multicellular. Cells of *Volvox* are attached by small cytoplasmic bridges that allow cellular communication. Some cells are specialized for reproduction and other flagellated cells are specialized for motility. To reproduce, cells divide and form tiny internal colonies. The parents eventually rupture and release the daughter cells.

Some green algae are multicellular in the true sense, meaning that they are composed of many cells with different functions. Despite their multicellularity, every cell in these green algae is exposed to the water, so the process of diffusion moves gases, nutrients, and wastes to and from every cell. Also, because the water keeps them buoyant, these multicellular green algae do not need special support structures. The familiar sea lettuce, *Ulva lactua*, is a multicellular, sheetlike green alga often seen washed up on the beach (**Figure 17.9**). It looks like a bright green, limp silk handkerchief. One Mexican species of green alga, *Codium magnum*, is especially impressive. It can grow to be more than 24 feet long (8 meters) and wider than the palm of your hand. *Codium* also lives in ocean water. Each

Figure 17.9 **Sea Lettuce.** This multicellular green alga often is washed up on beaches.

of its cells has many nuclei and lacks cell walls. So like the hyphae of fungi, all of *Codium*'s cells are interconnected.

Land plants have more differentiated body structures than green algae do. **Figure 17.10** introduces you to the basic organs of the most advanced land plants. Starting from belowground, **roots** have the primary functions of absorbing water and minerals from soil and anchoring the plant in the soil. **Stems** lift leaves up into the sun where they can

root a plant organ specialized to absorb fluids and minerals from the soil, anchor the plant, and sometimes serve as a storage depot for excess carbohydrates

stem a plant organ that supports leaves and contains vascular tissues that transport fluids from roots to aboveground portions of the plant

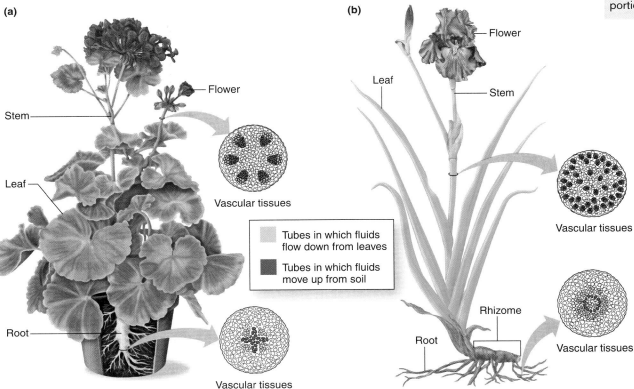

(a)

Stem

Flower

Leaf

Root

Vascular tissues

Vascular tissues

Tubes in which fluids flow down from leaves

Tubes in which fluids move up from soil

(b)

Flower

Leaf

Stem

Vascular tissues

Root

Rhizome

Vascular tissues

Figure 17.10 **Plant Organs.** Flower, leaf, stem, and root are organs of plants. A geranium (**a**) has a different arrangement of the tubules of vascular tissues in its root, stem, and leaves than does an iris (**b**).

■ *What are the major functions of a stem? of a root? of a leaf? of a flower?*

receive sufficient light for photosynthesis. Stems are living bridges that connect roots and leaves. *Rhizomes* are subterranean stems. Stems of advanced plants are called "true" stems because they are filled with tiny tubes of **vascular tissues** that function in transport of fluids within the plant. Similarly, true roots and true leaves also contain vascular tissues. Some vascular tissues move water and dissolved minerals upward from the roots. Other vascular tissues move sugars down from the leaves and distribute them to carbohydrate storage depots in stems or roots or to plant tissues that need nutrients. **Leaves** are flattened plant organs that are photosynthetic solar panels. Because leaves release water vapor, they also help the plant to move water upward from the roots. The water is used not only in photosynthesis, but it also helps to cool the plant. Plants have a variety of reproductive structures including spores, cones, flowers, pollen, fruits, and seeds. As you can see by comparing Figures 17.11 through 17.13, even though green algae and land plants are so closely related, you can usually tell that a plant is a land plant by examining its body structure and looking for typical plant organs.

vascular tissue a system of tubes within plant tissues that are used to transport fluids within the plant

leaf a flattened plant organ specialized for photosynthesis

> ### QUICK CHECK
>
> **1.** You are looking at a green mass that has washed up on a beach. What clues help you to decide whether it is green algae or a fragment of a land plant?

17.5 Plant Diversity Reflects the Variety of Plant Structures

About 500,000 species of land plants have been scientifically described, and every year new species are added to this list. How many species have yet to be discovered? The answer is uncertain, but estimates range up to an additional half million species. **Figure 17.11** shows a few of the major plant phyla and highlights their important physical characteristics.

Figure 17.11

A Phylogeny of Green Plants. Notice how lack of cuticle differentiates the green algae from other plants and how lack of stomata differentiates the liverworts. Similarly possession of vascular bundles differentiate ferns, gymnosperms, and flowering plants.

■ *What plant organs do green algae lack?*

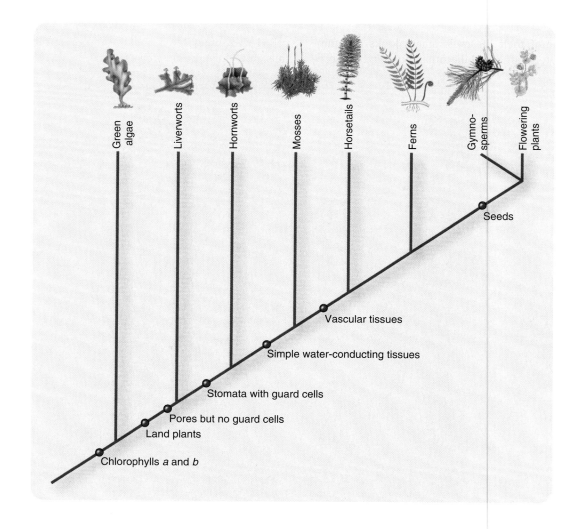

Liverworts, Hornworts, and Mosses Share Basic Features

Because they are small, low-growing plants, unless you are looking for them, liverworts, hornworts, or mosses may escape your notice. Liverworts, hornworts, and mosses collectively called *bryophytes,* have no vascular tissues, and so none of the bryophytes has what scientists call "true" leaves, stems, or roots. Instead bryophytes have thin threads called **rhizoids** that anchor them in the soil. Rhizoids do not absorb water, though, and bryophytes have two adaptations that help them obtain water. First, most bryophytes grow in habitats that are wet or moist, and they often grow in the shade of taller plants. Second, the bodies of bryophytes trap water that falls on them, and water moves by osmosis directly into their tissues.

Liverworts are less adapted for land than are other bryophytes. The word "wort" is Old English for "plant," and liverworts get their name because the flattened, lobed gametophyte of some species looks a bit like a liver (**Figure 17.12**); other species of liverworts look more like mosses. No matter what they look like, liverworts are the simplest land plants. Hornworts can live in drier environments, and one of their life cycle stages looks like miniature green antlers. Like other bryophytes, hornworts have no vascular tissues so they lack true leaves, stems, and roots, but they do have stomata. With nearly 10,000 species and a worldwide distribution, mosses are the most numerous and successful bryophytes. Some mosses even live in deserts, and many can spring back to life after long periods of drought. *Sphagnum* moss is one that you probably know. It is also called peat moss, and gardeners use it to improve the water-holding capacity of garden soils. *Sphagnum* lives in boggy areas, and scientists speculate that about 3% of Earth's surface is covered with this moss. *Sphagnum* and other mosses once were used for medical dressings and truly disposable baby diapers. *Sphagnum's* ability to hold water is amazing: it can sop up about 25 times its weight in water. In northern European countries, as well as in the British Isles, deep layers of *Sphagnum* accumulate in bogs. After bogs are drained, workers cut out the peat; dried peat is burned as fuel in fireplaces and in stoves.

Some Plants Have Simple Vessels

Club mosses and horsetails have better adaptations for life on land, but they also have features that restrict them to moist places. For instance, these plants have swimming sperm and like the bryophytes must have water for sperm to swim into the female reproductive structures. Club mosses and horsetails do have some specialized tissues that transport water and, unlike the soft tissues of liverworts and mosses, club mosses and horsetails have small, tough, evergreen leaves (**Figure 17.13**). Belowground these plants have horizontal stems called **rhizomes.** Roots of club mosses and horsetails grow down from rhizomes.

Although club mosses and horsetails are small plants now, about 400 million to 280 million years ago during the Carboniferous period they grew to be much larger. In those times extinct relatives of horsetails called calamites grew to be trees that were as much as 60 feet high (18 meters), with trunks a half a meter in diameter. Extinct relatives of club mosses were even larger, growing up to about 131 feet tall (40 meters). Forests of these trees blanketed much of eastern North America, and their remains survive to this day as carbonized

rhizoid nonvascular threadlike structure that anchors liverworts, mosses, and hornworts into soil

rhizome a horizontal belowground stem; shoots grow upward from it and roots grow downward from it

Figure 17.12 **Liverwort.** Female sex organs are located in the plant that has "drooping" umbrellas; male sex organs are in the "flat" umbrellas. Liverworts produce asexually with bundles of cells produced in little cups on the surface of the gametophyte.

Figure 17.13 **Club Moss.** Spores are produced in the erect, green structures.

Figure 17.14
Horsetail. The cone-like structures produce spores.

■ *How do horsetails, club mosses, and liver worts obtain food and energy?*

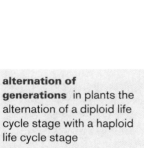

alternation of generations in plants the alternation of a diploid life cycle stage with a haploid life cycle stage

spore a haploid cell produced by meiosis that divides mitotically to produce the sporophyte generation of an alga or a plant

sporophyte a life cycle phase in a plant that has alternation of generations, a sporophyte is a diploid plant that produces spores by meiosis

gametophyte a life cycle phase in a plant that has alternation of generations, a gametophyte is a haploid plant that produces gametes by mitosis

fossils that form much of the coal mined in North America. Today, *Equisetum* is the only living genus of horsetails (**Figure 17.14**). It has silica in its cell walls, and pads of horsetails were used to scrub pots during Colonial times, giving horsetails the common name of scouring rushes.

Ferns Show More Extensive Adaptations for Life on Land

Ferns are the largest and most varied group of seedless vascular plants, and they reproduce using both spores and gametes. Ferns are worldwide in distribution, but most species of ferns live in the tropics. In tropical forests graceful tree ferns grow to be gigantic: up to 80 feet (24 meters) with leaves more than 16.5 feet long (5 meters). Ferns have true systems of vascular tissues that extend from their underground roots up to the tips of their fronds (**Figure 17.15**). These tubes supply aerial tissues

with water and are one of the reasons that ferns can grow to substantial sizes. Belowground a fern plant has a rhizome, with roots extending down from it.

QUICK CHECK

1. How can you tell a moss from a fern?

17.6 Plant Life Cycles Involve Alternation of Generations

Like brown and red algae, multicellular green algae have a life cycle that involves **alternation of generations.** A single species alternates between a multicellular diploid individual and a multicellular haploid individual. In other words, in multicellular green algae both the diploid and the haploid cells can mature into multicellular "adults." The life cycle of sea lettuce provides a good example (**Figure 17.16**). The diploid zygote forms by the fusion of free-swimming haploid gametes. In a pattern that is familiar to you, the zygote divides by mitosis to produce a diploid organism. Some cells within this diploid organism go through meiosis, but these cells are *not* gametes. Instead, they are haploid **spores,** and each divides mitotically to produce a mature haploid organism. Some cells in each haploid individual produce gametes. The gametes are shed into the surrounding waters where they fuse to produce a diploid zygote, and then the cycle repeats. The diploid plant in this cycle is called a **sporophyte** because it *produces* spores, while the haploid plant is called a **gametophyte** because it *produces* gametes.

Land plants inherited alternation of generations from green algae. In the evolutionary process that has produced today's land plants, the diploid sporophyte life cycle stages grew larger, while the haploid gametophyte life cycle stages grew smaller, eventually becoming tiny and embedded within the tissues of the sporophyte (**Figure 17.17**). The bryophytes show the beginning of this evolutionary trend. In liverworts and mosses the gametophyte is the largest and longest-lived life cycle stage, and the sporophyte is small and depends on the gametophyte for nutrition. In horsetails and ferns this trend begins to shift. The gametophytes of horsetails and ferns are small and short-lived, while their sporophytes are much larger. In seed plants the sporophyte generation is the dominant generation, and the gametophyte is tiny and lives within the tissues of the sporophyte's reproductive organs.

The life cycle of a fern clearly shows the transition to a drier environment. The fern sporophyte

Figure 17.15   **Fern.** Spores are produced on the undersides of these fern fronds.

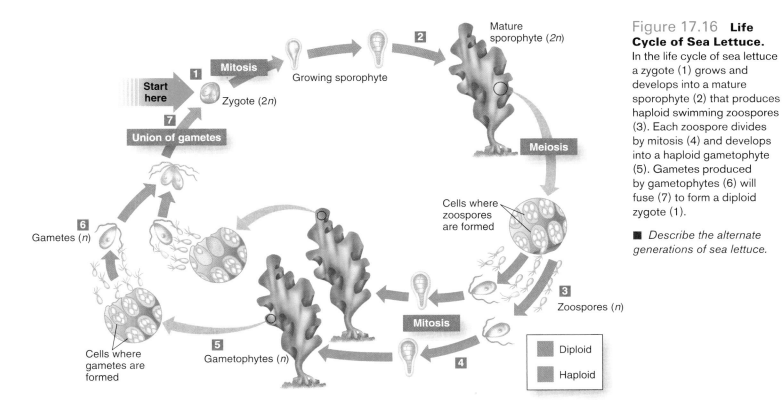

Figure 17.16 **Life Cycle of Sea Lettuce.** In the life cycle of sea lettuce a zygote (1) grows and develops into a mature sporophyte (2) that produces haploid swimming zoospores (3). Each zoospore divides by mitosis (4) and develops into a haploid gametophyte (5). Gametes produced by gametophytes (6) will fuse (7) to form a diploid zygote (1).

■ *Describe the alternate generations of sea lettuce.*

is the larger, more conspicuous life cycle stage (**Figure 17.18**). In contrast, the fern gametophyte is a delicate, heart-shaped structure that can be as large as the nail on your little finger and is just one cell thick. Eggs are produced near the "notch" of the heart, while sperm are produced near its tip. Fern gametophytes are generally found growing in moist conditions, and you can probably spot one if you look for damp places in a humid greenhouse where ferns are growing. The fern gametophyte's sperm swim into the gametophyte's female structures that produce and hold the eggs. There fertilization occurs, and the zygote develops into a sporophyte plant that grows out of the gametophyte. The embryo sporophyte develops and grows the fronds that are typical of a mature fern plant. On the reverse side of fronds are brownish raised dots called *sori*, where spore-producing structures cluster. Meiosis occurs within sori and produces mature spores that are released into the environment. The spores divide mitotically and grow into the haploid gametophytes of ferns.

The increase in the size of the sporophyte and the reduction in the size of the gametophyte are further adaptations to life on land. As the sporophyte phase of the fern life cycle became larger and acquired vascular tissue, it was able to grow taller than mosses, and so in prehistoric times ferns replaced mosses as the dominant land plants. While modern ferns do not grow to be the size of modern

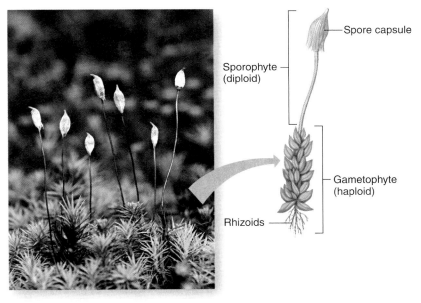

Figure 17.17 **Haircap Moss: Gametophyte and Sporophyte.** The green leafy portion of a moss plant is the haploid gametophyte phase of its life cycle. The diploid sporophyte grows out of the gametophyte and includes the stalk and the spore capsule. Meiosis in cells within the spore capsule produces haploid spores. These will grow into the next leafy gametophyte generation.

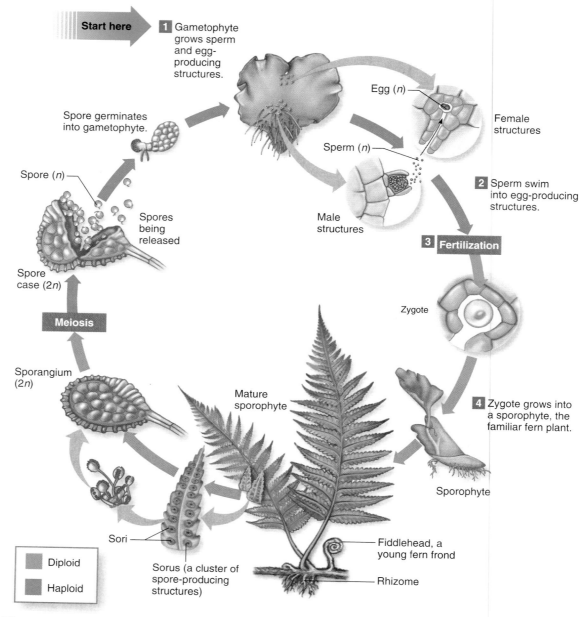

Start here → **1** Gametophyte grows sperm and egg-producing structures.

Spore germinates into gametophyte.

Egg (n)

Female structures

Sperm (n)

Spore (n)

Spores being released

Male structures

2 Sperm swim into egg-producing structures.

Spore case (2n)

3 **Fertilization**

Meiosis

Zygote

Sporangium (2n)

Mature sporophyte

4 Zygote grows into a sporophyte, the familiar fern plant.

Sporophyte

Sori

Fiddlehead, a young fern frond

Diploid

Haploid

Sorus (a cluster of spore-producing structures)

Rhizome

Figure 17.18   **Fern Life Cycle.**

evergreens and deciduous trees, 300 million years ago the ferns were a significant evolutionary development. Yet ferns lacked seeds and pollen—important adaptations to terrestrial life that natural selection would eventually produce.

Seeds and Pollen Are Important Evolutionary Adaptations of Plants to Life on Land

You plant them, you eat them, you spit them out of a mouthful of watermelon, you grind them into flour, and you sprinkle them onto cakes and ice cream sundaes. Seeds seem so humble and com-

monplace that it is hard to believe that they are an enormous evolutionary innovation. What exactly is a seed and why does it have such extraordinary evolutionary importance?

A **seed** is a time capsule that contains an embryo of an advanced plant (**Figure 17.19**). Protected by its seed coat and often surrounded by a supply of food, the dormant plant embryo waits until environmental conditions are right for its survival. It is protected from drying out and can survive cold, drought, and killing frosts. Within the seed the embryo carries on respiration but at an extremely low rate. Some seeds are short-lived, but others can germinate after extremely long dormant

seed a package that contains a plant embryo enclosed within protective tissues and often supplied with food

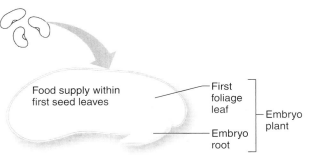

Food supply within first seed leaves

First foliage leaf

Embryo root

Embryo plant

Figure 17.19 **A Dissected Bean Seed.**
A seed contains the plant embryo and a supply of food.

periods. For instance, lotus seeds can germinate after 1,000 years in storage. Seeds have another advantage: they are a means of dispersal. Some seeds are equipped with parachutes, others float on water. Many seeds are embedded within fruits that animals consume. Later the seeds are eliminated with the animal's wastes, accompanied by a supply of fertilizer that will nourish the young plant.

Seed plants also have alternation of generations. The sporophyte generation of a seed plant is the conspicuous life cycle stage, while the gametophyte generation is tiny. Female gametophytes are embedded within tissues of the sporophyte and protected by it, while the tough coat of a pollen grain protects structures that will mature into the male gametophyte. This arrangement protects the tiny, delicate female gametophytes, gametes, and developing plant embryos from drying out. It also allows fertilization to occur in the absence of environmental water. In seed plants the eggs are housed within structures called **ovules** and sperm, and structures necessary for fertilization of the eggs are packaged within **pollen grains.** This system represents another remarkable evolutionary adaptation to life on land because pollen grains can withstand dry conditions. So pollen grains eliminate the need for a film of water for sperm to swim into female tissues. How does pollen move from male structures to female structures? Pollen usually is transported by animals, wind, or water.

Figure 17.20 shows a few plants that produce pollen and seeds. Gymnosperms include cycads, ginkgoes, and conifers that have no embryonic chambers around their seeds. Cycads aren't closely related to palm trees, but their leaves look like palms fronds surrounding large cones that look like pineapples. Ginkgoes are tall trees with fan-shaped leaves, and their seeds are embedded in strong-smelling fruits that look like pale green marbles. Conifers are evergreen trees that bear their seeds in cones. Spruces, firs, hemlocks, junipers, cypresses, yews, and pines are some familiar conifers. The oldest known tree, a bristlecone pine, is a conifer, as is the tallest tree, a sequoia. *Ephedra, Welwitschia,* and *Gnetum* are three genera of gymnosperms that seem to be related to the flowering plants, and the origins of the flowering

(a)

The gametophytes are within pollen cones and seed cones.

Male cones produce pollen.

Female cones produce ovules.

The sporophyte generation of a pine tree is the conspicuous life cycle stage.

(b)

(c)

Figure 17.20 **Sphorophyte and Gametophyte Generations.**
(**a**) The sporophyte of a pine is the large conspicuous plant; the pine gameotphytes are within male and female cones. (**b**) Cycad gametophytes are tiny and located within its showy cones; its sporophyte is the rest of the plant that looks like a palm tree. (**c**) The sporophyte of a sunflower is the green plant and most of the flower; the tiny sunflower gametophyte generation is held within male and female reproductive structures of the flower.

plants may lie within the ancestors of these genera. Angiosperms, or flowering plants, are the last group of seed plants, and they are different from gymnosperms in that their seeds are contained within specialized chambers called *ovaries.* Flowering plants are the most evolutionarily successful of all plants; 90% of all plant species are flowering plants.

QUICK CHECK

1. How can you tell the sporophyte generation of a magnolia tree from the gametophyte generation?
2. Explain the adaptations of a typical angiosperm like a dandelion to life on dry land.

ovule a structure of a seed plant that houses eggs

pollen grain a structure of a seed plant that matures to become the male gametophyte

17.7   An Added Dimension Explained
Lindow and the Bog Bodies of Europe

Before 1983 the claims of ritualistic human sacrifice and the three-fold death were mostly in the realm of myth, but in 1983 Andy Mould changed that. Andy worked for the Lindow Moss Peat Company in Cheshire County, England. He had the job of clearing the peat bog mines of large objects, such as logs or stones that might damage the machines that cut peat out of the bogs.

In 1983 Andy found a creepy thing in a bog where he was working. It looked like a soggy football, but turned out to be an 1,800-year-old human skull. The next year what seemed like a small log turned out to be a well-preserved foot. A local archaeologist discovered a chunk of skin projecting from the peat, and a scientific team was called in.

Painstaking work extracted more of the body of the person who belonged to the foot. The lower half of this body had been destroyed by the peat-cutting machine, but the upper torso remained intact. The body was naked, except for a fox-fur cap and a leather thong around the neck. The face still had the stubble of a beard that looked recently trimmed. The right arm was outstretched and the nails had been manicured. The man was clearly young, perhaps in his early twenties, with well developed upper body muscles. Further tests revealed that the young man had lived about 2,000 years ago in what is now Cheshire, England. He was probably a member of a ruling family. Several clues seem to show that he died a brutal and painful death. His face has a pained expression and, indeed he did suffer three deaths: a blow to the head, strangulation by a knotted leather rope, and a slit throat. Only then was his body submerged in the bog. The conclusion that the young man may have been a sacrificial victim is supported by the contents of his stomach: a simple last meal that included grain which had been heated to extreme temperatures. Celtic legends tell of blackened grains being used in a morbid sort of lottery in which the person who by chance draws the darkened loaf is the person marked for the threefold death.

The bog bodies of Europe, as remains like that of the Lindow Man are called, have been found in many northern locations. As many as 1,500 bog bodies have been discovered. The oldest dates to 8000 BC, but most date from around 1000 AD. At least one other, known as the Tollund Man, shows signs of having been a sacrificial victim (**Figure 17.21**). Tollund Man has many of the features of Lindow Man, but his expression is more relaxed, suggesting he did not struggle against his death.

The bog bodies are particularly interesting because of the way they have been naturally preserved (**Figure 17.22**). When any organism dies, bacteria, fungi, and other organisms begin immediately to feed on its tissues. This results in the decomposition and decay that ultimately reduces the organism to organic molecules. Under most conditions, hard parts of organ-

Figure 17.21   **Tollund Man.** Although the soft tissues of his body are so well preserved by the acidic waters of the bog where he was found, Tollund Man lived in Denmark between 405 and 100 BC.

isms such as shells, teeth, or bones can be fossilized, but soft parts decay. In the case of bog bodies, however, the reverse is true: usually the bones have dissolved or have become soft and pliable, while soft tissues such as skin and muscles have been preserved. What is going on here? The answer lies in the peat, which is mostly composed of *Sphagnum* moss.

While they are alive, mosses absorb minerals from the waters around them and excrete large amounts of hydrogen ions. Recall that hydrogen ions contribute to the acidity of a liquid. So as a result of the metabolic activity of the *Sphagnum* mosses, the waters where they live are low in minerals and high in acids. The mosses also alter the surrounding waters by excreting certain compounds called phenols that inhibit bacterial growth. These compounds include the tannins found in tea and red wine and the flavonoids found in many plants. Phenols

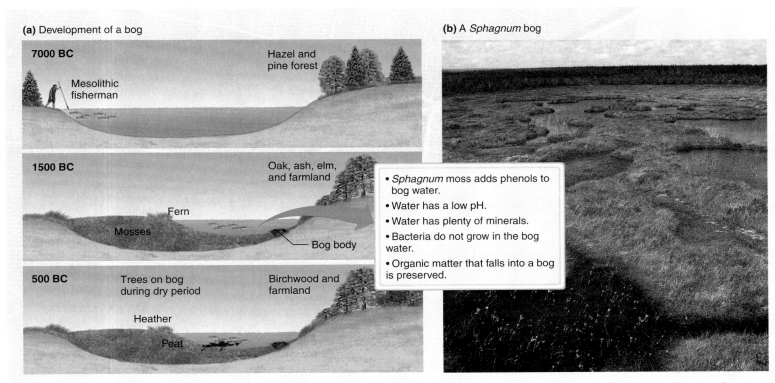

Figure 17.22 Bogs. (a) Over thousands of years *Sphagnum* moss creates conditions that foster the preservation of organic matter that falls into a bog. Dead plant material is compressed into layers of peat. (b). A *Sphagnum* bog.

■ *Based on your knowledge of pH, guess the pH of a* Sphagnum *bog.*

make plant tissues taste bad and protect them from animals that might eat them. Phenols also discourage the growth of pathogens that might cause infections in the living plant. The *Sphagnum* mosses add large amounts of these phenol compounds into the water around them. This combination of high acidity and high phenol content helps to form a bog.

Peat is the layered buildup of compressed but not decayed organic plant matter that is found in watery northern bogs. The acidic environment produced by the living mosses inhibits bacterial growth, and as a result the waters around the mosses are nearly sterile. So when mosses and other bog plants die and slip into the waters of the bog, they do not decay. Instead they build up, forming layer on layer of dead organic material. The dead plant tissues are compressed by the accumulating weight of layers above them. Eventually the carbon-rich material called *peat*

results. For centuries peat has been burned as a source of fuel in the British Isles and the northern European regions.

Peat bogs do not only inhibit the decay of plant tissues, but they also preserve any dead organisms that fall into a peat bog. The combination of the phenols, the high acidity, and the low mineral environment produced by the mosses preserves organic matter. This is why the Lindow Man, the Tollund Man, and many other bog bodies look as though they have died only recently. It also explains why the bog bodies are still here after 2,000 years to tell us the story of their lives.

QUICK CHECK

1. What combination of factors in the peat bog preserved Lindow Man?

Now You Can **Understand**

Forensic Molecular Botany

Forensics is everywhere today—especially on highly successful television series and in true crime movies. Forensics is the application of scientific findings and methods to the resolution of criminal or other important societal questions. Forensics is in play when scientists test weapons to determine whether a bullet found at the

scene of a crime was fired from a suspect's gun. The recent new information about biological systems has given a powerful new tool to forensics, and knowledge about the molecular genetics of plants has played an important role. A good example concerns a suspect in an Arizona murder case. A young woman had been murdered, and the prime suspect claimed to be innocent. An observant investigator noticed that in the bed of the defendant's truck were small seed pods from the kinds of trees found at the murder location. The investigator wanted to know if there was any way to determine if the seed pods in the truck matched those at the crime scene. Initially, the researchers on the case were skeptical, but in the end they matched the DNA of the seed pods not only to the species but also to the individual trees at the crime scene. This was enough to tip the balance in the trial, and the suspect was convicted. You can expect to see more forensic molecular botany in criminal trials in the future.

What Do **You** Think?

What Should Be Done About the High Cost of Paper?

Paper is everywhere. Even in the modern world of e-mails and other electronic transactions, we depend on paper. It piles up around us as printouts, newspapers, notebook paper, magazines, and books. Where does paper come from? What are the hidden costs of using so much paper?

You probably know that paper comes from trees. Paper actually is thin sheets of processed cellulose, which has been extracted from trees. Getting cellulose out of the trees, however, is not easy. The wood must be chopped fine as cornflakes and turned into a mushy pulp. Then comes the hard part because wood not only contains a lot of cellulose but also contains a lot of lignin, stuck tightly to the cellulose in the tree's cell walls, gluing them together. The only way

to remove the lignin from the cellulose is with a massive input of energy and chemicals. The removal of lignin from wood pulp is expensive and environmentally damaging. In many manufacturing facilities water laden with lignin is discharged into rivers and other waterways, and sulfur-containing compounds are sent into the atmosphere. Efforts are being made to develop cleaner paper manufacturing processes, and biologists may have part of the answer.

Modern science may have found a way to reduce pollution from paper manufacturing by using transgenic trees developed by using bacteria to insert genes that inhibit lignin production into tree embryos. This sounds like a great solution. The lignin content of these trees is 15% lower than normal, and so the cost and chemicals needed to produce paper would be 15% less.

Some people are not so sure that this is the ideal solution. There are some who argue that we do not know enough about transgenic organisms to start planting acre after acre of transgenic, low-lignin trees. In fact, there may never be enough scientific research to be certain of the potential effects of raising large numbers of these transgenic trees. The final decision about whether to proceed with low-lignin transgenic trees will not be made by scientists. Politicians, voters, advocacy groups, courts, lobbyists, and lawyers will all influence the decision, and you will be a part of this process. How will you decide what side to support? What do *you* think about using transgenic plants to reduce the environmental impact of making paper?

Chapter Review

CHAPTER SUMMARY

17.1   Plants Sustain Life

Because of biochemical similarities and structural similarities land plants and green algae are both grouped together as plants. Plants are of paramount ecological importance because so many other organisms directly or indirectly depend on them for food. Plants also define and help maintain biomes, and without them Earth would be dramatically different. Plants are extremely important economically, and a host of food and nonfood products are made from them.

17.2   Green Algae and Land Plants Form One Phylogenetic Kingdom

Green algae and land plants share structural and biochemical similarities.

plant   413

17.3   The Evolutionary History of Plants Is Dominated by the Transition from Water to Land

Modifications to prevent water loss allowed green algae to move from water to land and evolve into land plants. These modifications included large size, compact body form, and a cuticle with stomata.

Vascular tissues allowed land plants to achieve greater size. Today's ferns are the largest and most varied seedless vascular plants. They have life cycles that involve swimming sperm and spores.

cuticle   414   stoma (plural: stomata)   414
lignin   415   vascular plant   415

17.4   Plants Have Diverse Body Structures

Green algae include microscopic unicellular organisms, macroscopic multicellular organisms, and colonial organisms. Organs of plants include roots, stems, leaves, and reproductive structures and those of vascular plants have vascular tissues within them. Spores, cones, flowers, pollen, fruits, and seeds are some of the major reproductive structures of plants.

colony   416   stem   417
leaf   418   vascular tissue   418
root   417

17.5   Plant Diversity Reflects the Variety of Plant Structures

Liverworts, mosses, and hornworts are bryophytes, a group that lacks vascular tissues. Club mosses and horsetails have vascular tissues and swimming sperm.

rhizoid   419   rhizome   419

17.6 Plant Life Cycles Involve Alternation of Generations

Life cycles of plants show considerable variety, but one constant feature is the alternation of a diploid sporophyte generation with a haploid gametophyte generation. Trends in the evolution of land plants include features such as seeds and pollen that make plants increasingly independent of water.

alternation of generations 420
gametophyte 420
ovule 423
pollen grain 423
seed 422
spore 420
sporophyte 420

17.7 An Added Dimension Explained: Lindow and the Bog Bodies of Europe

Bogs preserve organic matter, and human bodies have been preserved within bogs, unusual aquatic environments created by the *Sphagnum* mosses of the bogs. The unusual preservation of bog bodies gives insights into prehistoric cultures in Britain and northern Europe.

REVIEW QUESTIONS

TRUE or FALSE. If a question is false, rewrite it to make it true.

1. If all of Earth's plants died off, animals would quickly adapt to the changed conditions.
2. A plant is a heterotrophic photosynthetic eukaryote.
3. There are no unicellular green algae.
4. There are no unicellular plants.
5. Gametes of green algae unite in water.
6. Liverworts, hornworts, and mosses have vascular tissue.
7. Liverworts, hornworts, mosses, and ferns have swimming sperm.
8. Eggs and sperm of liverworts, hornworts, mosses, and ferns unite in the water, and the fertilized egg swims into female tissues.
9. Seeds are a plant's means of dispersal that contain an embryo plant and stored food for the embryo plant.
10. Cycads and ginkgoes have no seeds.
11. Ninety percent of all plant species are conifers.

MULTIPLE CHOICE. Choose the best answer of those provided.

12. Plants provide most of the ___ in the biosphere.
    a. oxygen
    b. hydrogen gas
    c. carbon dioxide
    d. water
    e. nitrogen

13. Green algae and plants use this molecule to store excess food.
    a. chlorophyll *a*
    b. chlorophyll *b*
    c. Floridean starch
    d. starch
    e. laminarin

14. Plants and green algae are grouped as:
    a. protists.
    b. algae.
    c. plants.
    d. archaeans.
    e. fungi.

15. Using a light microscope you see a fairly large-sized, round, green-colored ball roll across your field of view. It has smaller green balls within it. You are looking at a sample of pond water. What organism did you just see?
    a. *Chlamydomonas*
    b. *Volvox*
    c. *Codium magnum*
    d. *Cladophora*
    e. *Sphagnum*

16. A rose plant including roots, leaves, and flower petals is the _____ generation, while the male and female reproductive structures of its flowers contain the ovules and pollen that form the _____ generation.
    a. sporophyte; gametophyte
    b. gametophyte; sporophyte

17. In plants alternation of generations features a _____ sporophyte generation alternating with a _____ gametophyte generation.
    a. haploid; diploid
    b. diploid; haploid

18. In life cycles of green algae the diploid zygote
    a. undergoes meiosis and produces four haploid algal cells.
    b. undergoes mitosis and produces a multicellular individual.

19. What is the advantage of stomata?
    a. They let in water.
    b. They take in nitrogen.
    c. They let in carbon dioxide.
    d. They prevent water loss.
    e. both c and d

20. A cuticle over the surface of a plant
    a. lets in water.
    b. lets in nitrogen.
    c. lets in carbon dioxide.
    d. prevents water loss.
    e. both c and d

CONNECTING KEY CONCEPTS

1. Some older classification schemes grouped all of the algae with land plants, while others grouped all of the algae together. Why is neither of these schemes accepted today? What is the currently accepted classification scheme and what evidence supports this scheme?

2. Summarize the diverse adaptations that allow land plants to survive outside of a watery environment.

THINKING CRITICALLY

1. Green algae—not red, brown, or golden algae—clearly are the direct ancestors of plants. Imagine how Earth might look different if a different kind of algae had become terrestrial and evolved into land plants. But why didn't the other forms of algae give rise to plants? What characteristics do green algae have that might have predisposed them to give rise to land plants rather than the other algae?

For additional study tools, visit www.aris.mhhe.com.

The Living Plant

PLANT STRUCTURE AND FUNCTION

Van Gogh's Madness
*Long After the Painter's Death,
the Diagnosis Debate Lives On*
November 22, 1998

Vincent van Gogh. This self portrait was painted after he cut off his own earlobe.

Van Gogh's Pen
October 29, 2005

Like many artists and writers in nineteenth-century Paris, Vincent van Gogh looked forward to evenings in cafés where he found company, interesting talk, women—and, of course, the *green fairy*. Most nights Vincent drank heavily, consuming glass after glass of the exceptionally alcoholic emerald-green liquid called absinthe, which turned opalescent chartreuse when mixed with water. Absinthe was a popular drink at the time, and van Gogh was one of many whose lives were disrupted—and sometimes destroyed—by an addiction to this unusual drink. Combined with all of Vincent's other problems, some believe that the green fairy may have shoved van Gogh on a swift downward spiral of alcoholism, addiction, and mental illness that culminated in his suicide.

Van Gogh made several attempts to escape from absinthe's hold on his life but never with any success. In February of 1888 Vincent fled Paris for Arles, a village in Provence in the South of France. There he was free of the dissolute bohemian café life and the distractions and expenses of Paris. At first Vincent was charmed by Arles and its inhabitants, and he painted with renewed enthusiasm. But money was scarce, and Vincent often went without food or ate only a little. Mostly he spent his days outdoors, working intensely, painting one masterpiece after another. Some evenings Vincent painted outside, with candles strapped to the brim of his hat and candles attached to his easel. But eventually, Vincent's evenings fell back to their old patterns. He returned to his café habit, where, of course, again and again he met the green fairy.

In late October of 1888 things became more promising when Paul Gauguin moved in to share Vincent's house, and there was hope that Vincent's loneliness and isolation would be lifted. But Vincent had periods of insanity, and their living arrangement ended when Vincent came at Gauguin with a razor. Gauguin "stared him down" and escaped injury, but that night Vincent's insane rage turned inward and he used the same razor to cut off his own earlobe (**see chapter-opening photo**). The next day Vincent was hospitalized, but the doctors could not diagnose his case, or cure him. Vincent was released from the hospital, and his troubles continued. He suffered from insomnia and hallucinations. His neighbors feared him. He was in and out of the hospital, and finally the police closed down his house. Eventually, Vincent voluntarily entered an asylum for mentally ill people, but his life there was no better. He tried to poison himself by swallowing paint. He also tried to drink turpentine. A sympathetic doctor offered to care for him, but no one was able to permanently reverse Vincent's downward slide. On July 27, 1890, he shot himself through the belly. It took him two days to die. He was just 37 years old.

The luminous art and sad life of Vincent van Gogh attract much speculation and interest. Was his insanity inborn, or a result of stress? Or was it absinthe? At the end of this chapter you will revisit the story of Vincent van Gogh and the deadly green fairy, and will see how an understanding of plants has contributed to solving this mystery. But before tackling those topics, let's expand your knowledge of the structure and function of plants. ■

18.1 Specialized Tissues Carry Out Plant Functions

Early land plants were not much different from multicellular green algae, and they lived only in moist environments. Over millions of years the processes of mutation, gene duplication, natural selection, and other evolutionary mechanisms produced the group of adaptations that now allow plants to survive on dry land. This section examines the structure of land plants and explores how different plant adaptations help plants to survive on land.

Nearly all land plants share the same body plan. Apart from reproductive structures, the body of a typical land plant has three conspicuous kinds of organs: roots, stems, and leaves (**Figure 18.1**). Stems and leaves are collectively called *shoots*. **Roots** typically grow deep into the soil and take up water and mineral nutrients. Roots also stabilize the plant so that it does not topple over. Figure 18.1 shows how the extent of the root structure compares to the plant parts that are aboveground. To grow tall, all terrestrial organisms must have some sort of supporting structure, and stiff tissues within **stems** allow many plant species to grow vertically. **Leaves** are the major source of chemical energy for a plant and are specialized to carry out photosynthesis. Leaves also allow gas exchange. They are the place where water vapor and oxygen move out of the plant and the carbon dioxide necessary for photosynthesis moves into the plant. Not all plants have broad, flattened leaves. In plants like cacti that lack leaves or have tiny leaves, stems take over the photosynthetic and gas exchange functions of leaves.

Epidermal Tissues Surround the Plant Body

To learn more about the internal organization of plants, let's pull up a plant, wash the dirt from its roots, cut extremely thin cross sections of the plant, stain the tissues with special dyes, and examine them with a light microscope. This process lets you see the cells and tissues that make up a typical plant. Covering the entire external surface of the plant is a layer of living **epidermal tissue** that protects the

root a plant organ that absorbs water and mineral nutrients and stabilizes the plant

stem a supportive plant organ that extends from roots to leaves

leaf a photosynthetic plant organ that also functions in gas exchange

epidermal tissue outer layer that covers a plant's surface and protects it from injury but is lost in woody plants

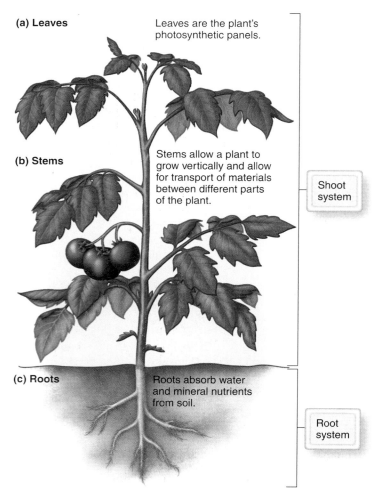

(a) Leaves — Leaves are the plant's photosynthetic panels.

(b) Stems — Stems allow a plant to grow vertically and allow for transport of materials between different parts of the plant.

Shoot system

(c) Roots — Roots absorb water and mineral nutrients from soil.

Root system

Figure 18.1 **Plant Organs.** The organs of a plant's body include leaves, stems, and roots. Plant organs are grouped into a shoot system and a root system.

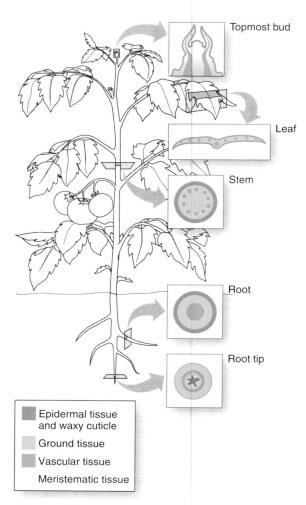

Topmost bud

Leaf

Stem

Root

Root tip

Epidermal tissue and waxy cuticle

Ground tissue

Vascular tissue

Meristematic tissue

Figure 18.2 **Plant Organs Are Composed of Four Kinds of Tissues.**

plant from injury and infections (**Figure 18.2**). It has a function similar to that of the outer layer of human skin. In plants, however, the epidermis is generally only one cell thick. Like human skin cells, epidermal cells are tightly joined to one another to prevent water loss. As they grow, many parts of woody plants lose their epidermal tissue as it is replaced by tissues that form bark. Note also that the sheath of epidermal tissue begins in the roots, encases the stem of the plant, and travels up and out to the leaves and buds. As you learn more about plant structure, this pattern of concentric sheaths of different tissues that begin in roots and extend through the rest of the plant will become a recurrent theme.

The *cuticle* is the waxy covering that helps keep the tissues of a plant from drying out. The waxy cuticle is excreted by the epidermal cells and covers nearly all of the plant surfaces that are

aboveground. Some plant species, particularly those that are drought adapted, have much thicker cuticles with many layers of waxes and carbohydrates. The cuticle is effective at preventing dehydration, but gases must diffuse in and out of a plant. The guard cells that open and close the stomata—openings in the epidermis—allow this to happen (see Section 17.3). The cuticle does not cover stomata. Guard cells are an important adaptation because without stomata and guard cells photosynthesis would cease for lack of carbon dioxide, and the plant would die from an inability to regulate its water content. The roots have a different external covering that prevents the entry of fungi and microorganisms but allows water and dissolved minerals to move into root cells by osmosis. Because they are located underground, roots are less prone to water loss than are aboveground stems and leaves.

Ground Tissue Contains Three Major Cell Types

The cells of **ground tissue** (Figure 18.2) carry out important metabolic functions for a plant, store nutrients, sometimes divide to produce more cells, and give a plant both flexibility and strength. Aside from the wood in some plants, ground tissue makes up most of the body of a plant. Ground tissue usually contains three cell types: parenchyma, collenchyma, and sclerenchyma (**Figure 18.3**). *Parenchyma cells* are the most common cell type in ground tissue. They have thin cell walls and often are separated by spaces that allow oxygen, carbon dioxide, and water vapor to flow around them. These versatile cells do many things for a plant, but one of the most important is their role in photosynthesis. Photosynthetic parenchyma cells are chock full of chloroplasts. In many plant species parenchyma cells provide food storage tissue. Depending on the species, the parenchyma of roots, stems, and even leaves may be modified in various ways as storage sites for surplus food. Humans raise plants to take advantage of this trait. For instance, carrots are modified roots packed with parenchyma cells that store starch. Asparagus are stems full of parenchyma cells used for storage, and onion bulbs are leaves with parenchyma cells specialized for storage. Finally, some parenchyma cells also have the ability to divide and produce all of the cell types of a plant. This is what happens when you take a "cutting" of a plant and use it to grow a whole new plant. Under the right conditions of tissue culture, some of the parenchyma cells revert to a less differentiated state, begin to divide, and can develop into a complete, mature plant.

Collenchyma is another cell type found in ground tissue (Figure 18.3). Collenchyma cells have thickened cell walls that contain lots of cellulose. Because they also are flexible cells, collenchyma impart strength and flexibility to actively growing tissues like young stems. Collenchyma cells often are found just beneath the epidermal layer of both stems and leaves. The thick strands in a stalk of celery are made of collenchyma cells. Some older studies have reported that the cell walls of collenchyma cells increase in thickness when a stem is repeatedly bent, as it might be in a strong wind. The interpretation of this finding is that the response of collenchyma cells helps to strengthen a plant in response to mechanical stress.

The third type of cell in ground tissue is *sclerenchyma* (Figure 18.3). These cells are stronger than collenchyma cells because they have a secondary cell wall that contains *lignin* (see Section 17.3). Lignin is a large, branched, organic molecule that gives a plant cell wall strength, beyond that imparted by cellulose. Lignin is especially concentrated in the woody parts of plants such as the trunk and major branches and roots of trees, which is why these parts of the tree are so hard and strong. Besides providing extra strength, lignin helps to protect a plant from pests such as insects and fungi.

Only a few organisms can digest cellulose and lignin, including some bacteria, some single-celled eukaryotes, and fungi. Nevertheless, many organisms derive all of their energy from eating plant materials. For example, most herbivores have colonies of bacteria and/or single-celled eukaryotes in their digestive tracts that break down cellulose and lignin. These organisms use some of the smaller organic molecules that result from this digestion, and the animals get the leftover molecules. Termites have a similar relationship with several single-celled eukaryotes that inhabit their intestines and allow them to derive nourishment from the wood that they eat. If their cellulose- and lignin-digesting roommates are removed, termites

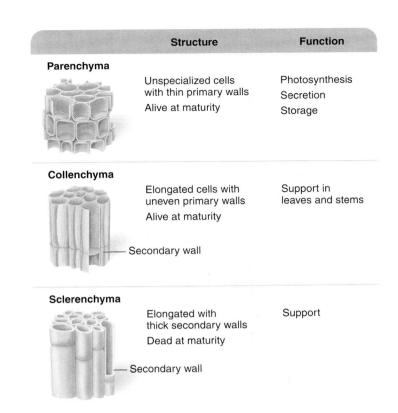

	Structure	Function
Parenchyma	Unspecialized cells with thin primary walls Alive at maturity	Photosynthesis Secretion Storage
Collenchyma	Elongated cells with uneven primary walls Alive at maturity — Secondary wall	Support in leaves and stems
Sclerenchyma	Elongated with thick secondary walls Dead at maturity — Secondary wall	Support

Figure 18.3 **Differences in Parenchyma, Collenchyma, and Sclerenchyma.** These three cell types make up ground tissue. Each has different characteristics and functions.

ground tissue tissue that carries out metabolic and other specific functions, and aside from wood, makes up most of the body of a plant: parenchyma, collenchyma, and sclerenchyma are the various kinds of ground tissues

still will gnaw wood but will be unable to derive nourishment from it and will die.

Sclerenchyma cells do most of their important work in a plant once they are dead. Plants are unusual organisms because many species have a significant percentage of dead tissues incorporated within their bodies. Although much of the outer epidermis of an animal is composed of dead cells, animals and fungi don't have functional dead tissues *within* their bodies. Sclerenchyma cells lose their cytoplasm and die when they reach maturity, but they do not decompose. The double cell walls of sclerenchyma cells that are reinforced with cellulose and lignin remain in place. Like the upright timbers in the frame of a house that support the walls and roof, the hollow cylinders of dead sclerenchyma cells give strength to plant organs.

Vascular Tissues Are Specialized to Move Fluids Within Plants

Vascular plants have internal vascular tissues to transport food and water (see Section 17.3). The cells that make up vascular tissue are organized into **vascular bundles** that run all through the body of a plant, from its deepest root, up through its stem, and to all parts of its highest leaf. There are two kinds of vascular tissues in vascular bundles: xylem and phloem. **Xylem** (pronounced "zeye-lem") transports water and dissolved nutrients taken up by the roots. **Phloem** (pronounced "flow-em") transports sugars and other important compounds dissolved in water. Together these xylem and phloem are the **vascular tissue.** Let's consider xylem first.

Although xylem is a tissue made of several different kinds of cells, let's focus on the xylem cells that transport water and minerals. Two kinds of conducting cells, called *tracheids* and *vessel elements,* form long channels running throughout the plant (**Figure 18.4a**). Like sclerenchyma cells, tracheid and vessel element cells have thick walls. They also are dead at maturity, so they become hollow pipelike cells that have secondary cell walls reinforced with lignin. Both tracheids and vessel elements have cell wall adaptations that allow water to flow from one cell to another. Tracheid cells have *pits,* regions where the secondary cell wall is absent and only the primary cell wall remains. Pits provide a path for water to flow from one tracheid cell in the tube to another. Vessel elements have *perforations,* regions where both the primary and secondary cell walls are absent, and these also provide a ready path for flow of water. The arrangements of the pits and the perforations allow water to move from the roots to the tips of the leaves. Xylem is obvious when you examine a lengthwise section of a stem. Parenchyma cells also are associated with xylem, and these cells allow tracheids and vessel elements to regenerate after an injury.

Fiber cells are the fourth kind of cell in xylem. They give strength to xylem tissue.

Phloem (**Figure 18.5a**), moves sugars and other biological molecules from one location to another in a plant. Two cell types are particularly important for sugar transport in phloem: *sieve tube elements* and *companion cells* (Figure 18.5b). Unlike the tracheids and vessel elements of xylem, both types of phloem cells retain their cytoplasm and are alive at maturity. The sieve tube elements form long tubes that carry sugar and other biological molecules from the leaves to other parts of the plant. Like vessel elements, sieve tube elements have perforated ends that allow substances to flow from one cell to another. So, in effect, the

vascular bundle the water-conducting and sugar-conducting tissues of plants

xylem a plant tissue that conducts water and minerals from the roots throughout the plant

phloem a vascular tissue that transports sugars and other biological molecules within a plant

vascular tissue strands of xylem and phloem that transport water, minerals, sugars, and other important molecules throughout the plant

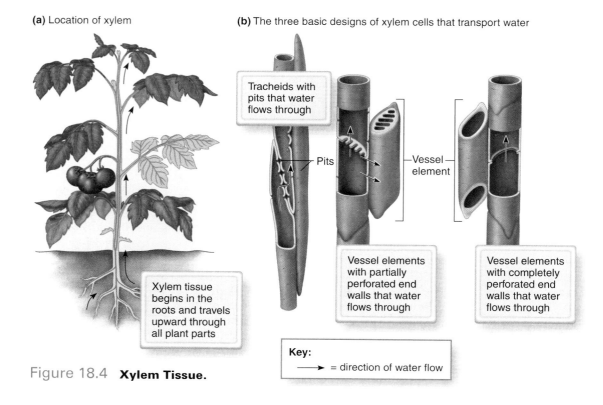

(a) Location of xylem

(b) The three basic designs of xylem cells that transport water

Tracheids with pits that water flows through

Pits

Vessel element

Vessel elements with partially perforated end walls that water flows through

Vessel elements with completely perforated end walls that water flows through

Xylem tissue begins in the roots and travels upward through all plant parts

Key:
⟶ = direction of water flow

Figure 18.4   **Xylem Tissue.**

cytoplasm of all sieve tube element cells in one phloem tube is united. Sieve tube element cells are unusual because they have no nuclei or organelles, but they do maintain a cytosol through which substances flow. Each sieve tube element has a close connection with an adjacent companion cell that has a nucleus and organelles. The companion cell provides the sieve tube element with the metabolic processes that it must have to remain alive. Companion cells also play a crucial role in the transport of sugars throughout the plant.

The vascular bundles, composed of both xylem and phloem, have different arrangements in different kinds of plants. Vascular bundles are scattered throughout the stem in some flowering plants (**Figure 18.6a**) such as corn, onions, palm trees, daffodils, and lilies, while they are located in a ring around a central core of ground tissue in other flowering plants (Figure 18.6b) such as beans, potatoes, oak trees, dandelions, and roses. Regardless of their arrangement, vascular bundles are continuous from the root, up the stem, and out into the leaves where they appear as veins. This will become especially relevant when you consider how water and sugar move throughout a plant.

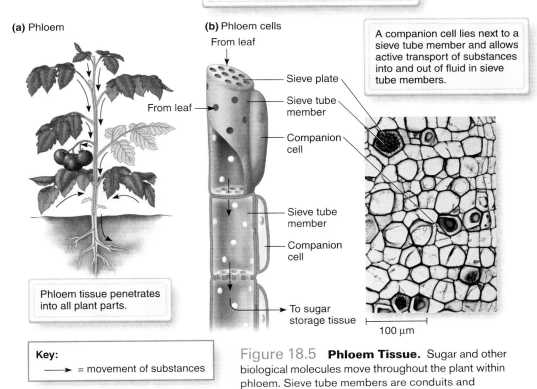

(a) Phloem

Phloem tissue penetrates into all plant parts.

Key:
⟶ = movement of substances

Sieve tube members have perforated end plates and their cytosol is contiguous with that of their companion cells.

(b) Phloem cells
From leaf

From leaf

Sieve plate
Sieve tube member
Companion cell
Sieve tube member
Companion cell
To sugar storage tissue

A companion cell lies next to a sieve tube member and allows active transport of substances into and out of fluid in sieve tube members.

100 μm

Figure 18.5 **Phloem Tissue.** Sugar and other biological molecules move throughout the plant within phloem. Sieve tube members are conduits and companion cells are their metabolic partners.

■ *What do companion cells that have sieve tube members lack?*

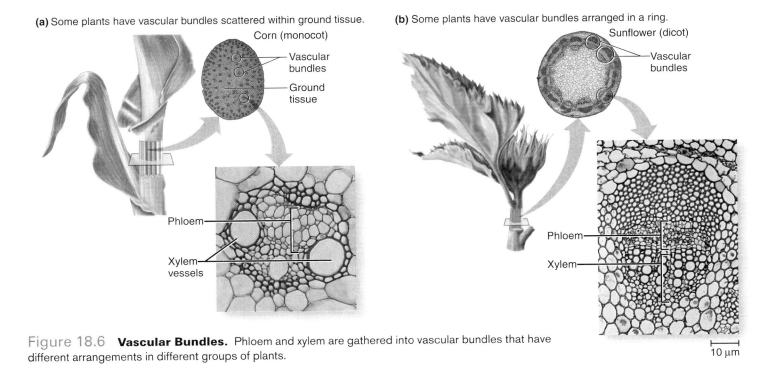

(a) Some plants have vascular bundles scattered within ground tissue.

Corn (monocot)
Vascular bundles
Ground tissue

Phloem
Xylem vessels

(b) Some plants have vascular bundles arranged in a ring.

Sunflower (dicot)
Vascular bundles

Phloem
Xylem

10 μm

Figure 18.6 **Vascular Bundles.** Phloem and xylem are gathered into vascular bundles that have different arrangements in different groups of plants.

433

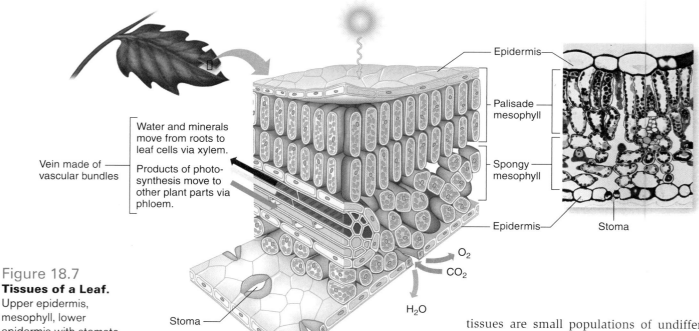

Vein made of vascular bundles

Water and minerals move from roots to leaf cells via xylem.

Products of photo-synthesis move to other plant parts via phloem.

Epidermis

Palisade mesophyll

Spongy mesophyll

Epidermis

Stoma

Stoma

O_2

CO_2

H_2O

O_2

H_2O

CO_2

Figure 18.7

Tissues of a Leaf. Upper epidermis, mesophyll, lower epidermis with stomata, and vascular bundles with xylem and phloem are the tissues of a leaf.

meristem plant tissue containing cells that divide and produce daughter cells that differentiate into mature plant cells, and so are responsible for plant growth; meristem tissues are similar to stem cells of animals

apical meristem meristem tissue at the tips of shoots or roots that allows for growth in length of shoots and roots

lateral meristem actively dividing layers of cells that produce the increase in the thickness of a tree trunk or branch

Epidermal, Ground, and Vascular Tissues Are Organized Within Leaves

The organization of tissues within a leaf allows you to see how the three tissues just described—epidermal, ground, and vascular—function together. The major job of a leaf is to carry out photosynthesis, and all three tissue types contribute to photosynthesis.

The epidermal tissue of a leaf is a thin sheet, often no more than one cell thick (**Figure 18.7**). Epidermal cells secrete the waxy cuticle that helps prevent water loss from the plant. The plant breathes through stomata in the epidermis (see Section 17.3). Beneath the epidermis is a ground tissue region called *mesophyll*. This photosynthetic layer is made up of two kinds of parenchyma: *palisade cells* and *spongy mesophyll cells*. The palisade cells are tightly packed to form a dense photosynthetic layer; the spongy mesophyll cells have spaces between them that allow gases and water vapor to penetrate deeply into the leaf. Vascular bundles made of xylem and phloem enter the leaf and form its veins. Much like your own blood vessels, large veins branch into smaller and smaller veins, allowing vascular bundles to reach all parts of the leaf.

Plants Grow in Meristem Tissues

The last important kind of plant tissue to consider allows a plant to grow throughout life, and collectively it is called the **meristem** tissue. Meristem tissues are small populations of undifferentiated cells that persist throughout the life of a plant. In many ways the cells within the meristem tissues are like stem cells in animal tissues. A subset of the meristem cells divide mitotically at a slow rate. The new cells that they produce divide more rapidly and eventually differentiate into one of many mature plant cell types. The meristem cells remain as a source that generates more plant cells. In this way the meristem tissues are responsible for plant growth.

Apical meristems are found at the tips of roots and shoots of many types of plants and allow plants to grow down into soil or up into the air or water. Growth from apical meristems is called *primary growth* (**Figure 18.8**). The apical meristem at the tips of shoots produces leaves, stem ground tissues, vascular tissues, and flowers. One apical meristem is found at the tip of each main branch of a bush or tree. As leaves sprout and the stem grows up from the apical meristem, a *lateral bud* is usually left behind. This is a dormant bit of meristem that can sprout side branches, under the right conditions. Just as in animals, the process of differentiation from a meristem cell to a mature plant cell is determined by patterns of gene expression (see Chapter 9). As you might imagine, many factors act on the cells of the apical meristem to determine the type of cell into which they differentiate. The availability of nutrients, environmental temperature, day length, and plant hormones all act on meristem cells to influence gene expression and so influence differentiation.

The apical meristems in roots usually have a different cellular organization and pattern of differentiation than do meristems in shoots (**Figure 18.9**). The root meristem is just inside the end of the root and is protected by a layer of mature cells called the *root cap.* At the tip of the root meristem is a population of cells that divide only rarely, and these are similar to stem cells in the skin or gut of an animal. Around the stem cells is a group of cells that divides rapidly, providing the new cells needed for root growth. As these dividing cells accumulate, they differentiate into sieve tube elements, vessel elements, and then other mature cell types.

The increase in the diameter of a trunk or branch is produced by another kind of meristem, called a **lateral meristem.** Growth that results from lateral meristem tissues is called *secondary growth.* Although not all plants have secondary growth, those plants that do have secondary growth can have one of two patterns of lateral meristems: *vascular cambium* and *cork cambium.* **Figure 18.10** shows that each of these meristem tissues is a thin cylinder of cells nested within other layers of the ground tissue. The cork cambium is just inside the outermost layer of the plant. Inside the cork cambium is phloem, and vascular cambium is inside the outermost layer of phloem. Let's look at vascular cambium and cork cambium in a bit more detail and see how they produce the thick trunk or hefty branches of a tree.

As its name implies, vascular cambium produces xylem and phloem in the growing trunk, branch, or root. The dividing cells located near the outside of the vascular cambium differentiate into phloem, so any tree has a ring of phloem not far inside the outer surface of the tree (**Figure 18.11**). Phloem is made of living vascular cells that transport sap, so if an animal tears off the outer layers on a tree, it often finds a moist, sweet layer underneath. The dividing cells located near the inside of the vascular cambium differentiate into xylem. As the xylem cells mature and die, they produce the layers of hard, dead xylem cells called *wood.* Layers of wood actually push the vascular cambium outward as the tree grows, so that this meristem tissue stays toward the outside of the trunk. As the vascular cambium is pushed outward, it crushes some rings of phloem, so as the layers of woody xylem grow thicker, the ring of phloem remains thin. *Bark* is the outermost layer on a tree. Bark is made up of the outer layer of hard, dead cells, an inner layer of cork tissue, and a cambium layer called cork cambium. Cork cambium produces cork cells, which have thick cell walls often fortified by suberin, a waxy substance that helps protect the tree against water loss, damage, and pests. These protective cork cells eventually die, producing the outer layer of dry, flaky tissue, which is the part of the bark that you see on the outside of the tree.

Apical Meristem of Shoot

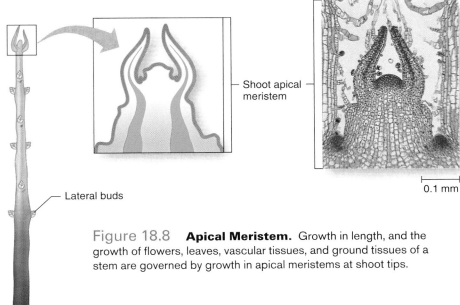

Figure 18.8 **Apical Meristem.** Growth in length, and the growth of flowers, leaves, vascular tissues, and ground tissues of a stem are governed by growth in apical meristems at shoot tips.

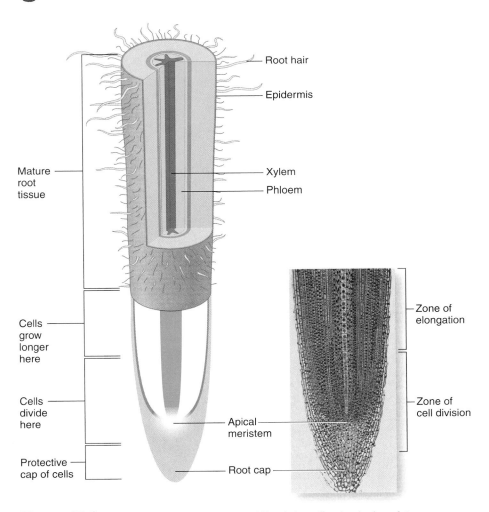

Figure 18.9 **Root Apical Meristem.** Mitosis in cells of apical meristems produces new cells. These become mature root cells.

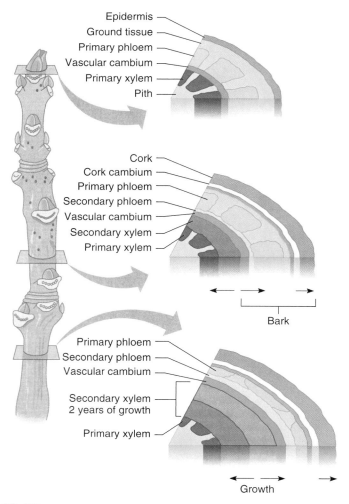

Epidermis
Ground tissue
Primary phloem
Vascular cambium
Primary xylem
Pith

Cork
Cork cambium
Primary phloem
Secondary phloem
Vascular cambium
Secondary xylem
Primary xylem

Bark

Primary phloem
Secondary phloem
Vascular cambium
Secondary xylem
2 years of growth
Primary xylem

Growth

Figure 18.10 **Apical and Lateral Meristems.** Primary growth in apical meristems allows a shoot to grow longer. Secondary growth in lateral meristems allows a shoot to grow wider.

Patterns of cell division in the lateral meristems across seasons and years produce the *tree rings* that you may have seen (Figure 18.11b). At a simple level, each tree ring represents the growth of a xylem layer during one year, and so you can tell the age of a tree by counting its annual growth rings. Tree rings are found in trees that grow seasonally in temperate and colder climates. You usually don't see annual rings in trees that grow in tropical regions. This is because tree rings reflect the different kinds of growth that happen at different seasons and tropical temperatures allow trees to grow year 'round. In spring when conditions are usually wet, the xylem cells produced by the vascular cambium are relatively large and produce a light-colored wood. In summer when conditions are drier, the xylem cells produced are smaller and more densely packed and the wood is darker. The result is alternating rings of dark and light wood in the trunk. Each light and nearby dark band in the cut tree represents one year of growth, although sometimes a microscope is needed to clearly see each ring. Annual tree rings have documented that some trees live to be thousands of years old.

Tree rings can provide more information than the age of the tree. The most recent annual xylem ring is the outermost ring that is nearest the vascular cambium. The oldest annual ring is nearest the center of the trunk. So annual rings provide a history of the tree's life. Years for which the lighter ring is wider than average usually mean that year was wetter than average. If both the light and dark rings are thin, it may mean the tree experienced an attack by insects during that year, that there was drought, or that the tree was shaded, perhaps by

(a) Vascular cambium produces xylem and phloem.

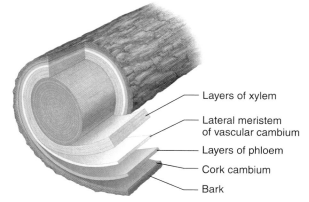

Layers of xylem
Lateral meristem of vascular cambium
Layers of phloem
Cork cambium
Bark

Figure 18.11 **Annual Rings.** (**a**) The lateral meristem of vascular cambium produces phloem toward the bark of a tree and xylem toward the heart of a tree. (**b**) An annual ring has two bands: lighter spring wood and darker summer wood.

(b) Each annual ring has a light part made of thin-walled xylem that the lateral meristem makes in the spring when it is actively growing. Each annual ring also has a dark part made in the summer when growth slows down.

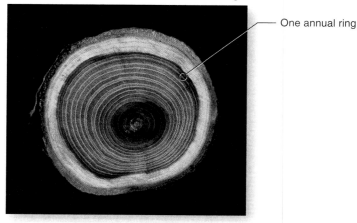

One annual ring

another tree. Scars or other abnormalities in the regularity of a ring can indicate a major fire during that season. Researchers often study annual rings on dead or logged trees, but it is also possible to take a deep, thin core sample from a living tree. These samples allow researchers to find further associations between the appearance of annual rings and the environment a tree experiences.

QUICK CHECK

1. What cell types make up ground tissue?
2. What structural features allow water to flow through the cells of xylem?
3. What does vascular cambium produce?

18.2 Plant Adaptations Move Water and Dissolved Substances

You probably know that many animals have an organ that actively pumps fluids through their bodies: the heart. As amazing as a heart is, it typically cannot pump blood higher than about 20 feet (about 6 meters) from the ground in a land animal—about the height of the top of a giraffe's head. This limits how tall an animal can be. Vascular plants are large and multicellular, but they do not rely on a heart to move water and nutrients around. Even so, the transport systems in plants can move fluid between the roots and the tips of a tree that is nearly 300 feet tall (over 90 meters). How do plants accomplish this amazing feat? A variety of plant adaptations are involved and next you will explore some of them.

Roots Have Adaptations to Increase Water Uptake from the Soil

Plants take up water when the soil where they are growing is wet. How does this happen? One answer is that roots have adaptations to increase their ability to absorb water and nutrients. On their outer surfaces many roots have **root hairs** (**Figure 18.12**). These look a bit like hair or fur, but are delicate extensions of the epidermis that increase the surface area of the root and allow it to absorb more water. Roots of most plants also have mycorrhizae, which are fungi living in close association with the plant. Mycorrhizae increase the efficiency of absorption, especially of certain plant nutrients (see Section 16.2). Roots and root hairs are not covered by the waxy cuticle that covers the rest of the plant. The cuticle prevents water loss and so also prevents water from coming into the plant. If they

Figure 18.12   **Root Hairs.** These furry extensions of a root's epidermis increase its surface area so it can absorb more water.

were covered by cuticle, roots could not carry out their function of absorbing water. As plants evolved, cuticle was expressed on the parts of the plant located aboveground but not on the parts of the plant located belowground.

Basic Chemical and Physical Processes Drive Transport in Plants

Before you look into how plants transport fluids in more detail, you need to have six basic processes firmly in mind. These are evaporation of water, cohesiveness of water, bulk flow of water, active transport of substances across a cell membrane, diffusion, and osmosis. Some of these processes have been treated earlier in this text, but let's review them here. Just as the transport of substances in the blood of animals depends on the properties and processes associated with water, the transport of substances in plants also depends on water. Not only does water have to get to all parts of an organism, other biological substances are dissolved in water, and so the properties of water are a big factor in how multicellular organisms solve the problem of transport. In plants evaporation, cohesiveness, osmosis, and bulk flow are all important mechanisms that influence the flow of water.

root hair a delicate extension of the epidermis of a root that increases its surface area and thus allows the root to absorb more water

Evaporation is the movement of the molecules in liquid water to the atmosphere, where they become single, isolated molecules of water vapor. Evaporation is more rapid when the concentration of water molecules in the atmosphere is low, or when the temperature of the liquid water is high. The *cohesiveness* of water molecules is the tendency of water molecules in liquid water to stick together. The hydrogen bonds that hold liquid water together (see Section 2.4) are weaker than covalent bonds, but they still are strong enough to pull on the water molecules in liquid water. *Osmosis* is the flow of water across a membrane that normally does not allow other substances to flow. So osmosis is how water flows across the cell membrane. Water flows freely through the many protein water channels in the cell membrane. The direction of water flow is governed by the concentration of dissolved substances outside and inside the cell. If the water outside the cell has a higher concentration of dissolved substances than the inside, water flows from the inside to the outside. If the outside of the cell has a lower concentration of dissolved substances, water flows to the inside of the cell.

Finally, *bulk flow* may be a new concept to you. Bulk flow is the movement of liquid water in response to some force or pressure. When you turn on a faucet or garden hose, there is a pressure that moves the liquid water out. The pressure that moves liquid water can come from many different sources and can either push water or pull water. The xylem and phloem in plants are long tubes not too unlike a garden hose. Water flows through the xylem and phloem by bulk flow. In xylem the dissolved minerals that plant cells need, such as potassium and phosphorus, are carried along with the water. In phloem the bulk flow of water carries sugars, hormones, and other biological molecules. With these concepts in mind you now can understand how water plus minerals move through xylem—and how water plus sugars and other biological molecules move through phloem. Leaves play a central role in both of these transport processes.

Leaves Pull Water and Dissolved Minerals Up Through the Xylem from the Soil

In plants water with dissolved minerals flows into the root hairs from the soil, and in a continuous stream moves up through the long xylem tubes to the stems and leaves. The key question is how does water move in the plant? In the absence of a heart, what forces are involved? The answer is that most

of the water that flows from soil through xylem is *pulled* up from leaves. Let's explore this further.

The flow of water from the soil through the xylem and to the leaves depends on the process of transpiration. In **transpiration** water evaporates from the exposed surfaces of cells within leaves and is lost to the environment as water vapor (**Figure 18.13**). Recall that inside a leaf there are spaces between the cells, and these spaces are filled with water or water vapor. When the air outside the plant is drier than the air inside the leaf water molecules evaporate from these spaces; that is, they diffuse out to the drier air. Evaporation is at a maximum on a warm, dry day when the sun is shining. This is not an insignificant movement of water. On a typical day more than 90% of the water that enters roots is lost to the air through transpiration. So the 48-foot-tall silver maple that absorbs 58 gallons (or 220 liters) of water per hour from the soil might transpire more than 52 gallons (or 198 liters) of water per hour into the air. An average tomato plant would lose about 27 gallons (or 104 liters) of water per day. How does the plant recover the lost water? The process of transpiration pulls water up through the xylem, all the way out to the soil that surrounds the plant. Let's explore how this works.

The cells in a leaf are covered with a thin film of water. These films actually are the ends of columns of water that extend down through the xylem all the way to the roots. Toward the outside these films of water may be exposed to air or water vapor, but toward the root side of the leaf they are in contact with other water molecules. Inside a typical angiosperm plant there are thousands to millions of these tiny water filled tubes—the tubes of xylem. As water evaporates through transpiration, the inside of the leaf becomes drier. This leads to increased surface tension along the films of water that coat cells inside the leaf. Surface tension is produced by the hydrogen bonds between water molecules (see Section 2.4) and is an aspect of the cohesiveness of water. Water molecules at the interface between air and water interact more strongly with one another than do water molecules not at the air/water boundary. This surface tension inside a leaf is so great that it pulls on water molecules all the way down the xylem to the roots of the plant and out into the soil. This pull is the force that causes water to flow up through the plant. It is an incredibly strong force that can pull water up the small xylem tubes a distance of several hundred feet.

You can see the movement of water upward in a plant in a simple demonstration. Place a stalk of celery in a glass of colored water. Over hours the colored water will move up the celery stalk and will

transpiration loss of water vapor from leaf tissues that pulls water up from the roots

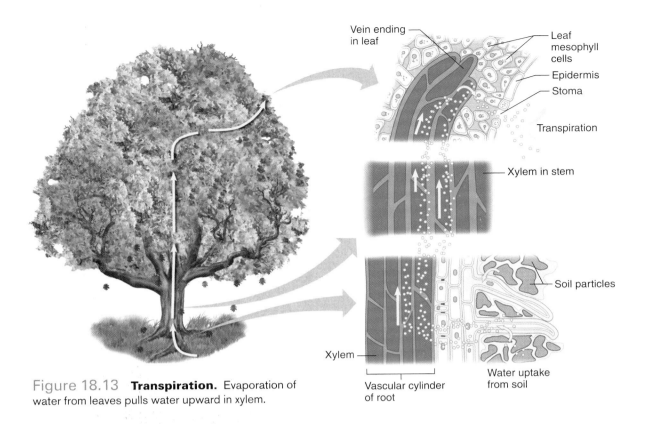

Figure 18.13 **Transpiration.** Evaporation of water from leaves pulls water upward in xylem.

color the strings of xylem as it goes. Water evaporates out of the leaves of the celery stalk, which is the process of transpiration. This pulls on the water molecules all the way into the water in the glass. Try this for yourself. If you want to go one step further, measure how far the water moves in an hour on a cool, humid day and compare that with how far the water moves on a hot, dry day. This will show you how environmental conditions determine the rate of transpiration through a plant.

Another way to think about the flow of water up a plant is the idea of bulk flow. The force created by transpiration in leaves results in a *negative pressure* on the liquid water in xylem. This is somewhat like what happens when you pull water up through a straw by sucking on one end of the straw. Sucking produces a negative pressure that pulls water up by bulk flow. Any ions or other substances dissolved in the water will get pulled up along with it. In this way transpiration not only provides the force that transports water in a plant, it also provides the force that transports minerals.

For the most part water and mineral flow in a plant is driven by passive mechanisms, but plants do exert some control over water flow. Control is provided by stomata (see Section 17.3). Each stoma is formed by two guard cells. The guard cells change shape under different conditions, and the shape of guard cells determines whether a stoma is open or closed. The stomata control two important processes in a plant: the loss of water through transpiration and the uptake of carbon dioxide for photosynthesis. To optimize photosynthesis, stomata are open during the day, and so transpiration is maximal during the day. At night stomata close up tight, and so transpiration does not occur after the sun goes down. In addition, stomata respond to the amount of water and ions in their environment. When the environment around an individual stoma is moist, the stoma will open. If the environment is drier, the stoma will close.

Water Also Can Move into a Plant by Osmosis

In the daytime transpiration is the dominant factor in moving water through a plant, but transpiration is turned off at night when stomata are closed. Under these conditions water can move into root hairs by the process of osmosis (see Section 4.2). When the soil is wet, the concentration of dissolved substances is greater in the water inside of a root hair cell than outside in the soil. As a result, water moves into the cells of the root hair by osmosis. The osmotic force is part of a phenomenon called *root pressure*, which pushes water up from the roots into the plant. Sometimes the flow of water from root pressure is so great that water oozes out of pores in the leaves; you can see this effect as drops of water at the edges of some leaves in the early morning. While root pressure is a strong force that pulls water into a plant, it cannot move water up more than a foot or so, and so osmosis and root pressure do not get water up to the leaves of a tall plant, even at night when there is no transpiration.

A Barrier Controls the Entry of Substances into Xylem

Like the cells in any multicellular organism, plant cells thrive only when the water outside them contains the right concentrations of various ions. Plant

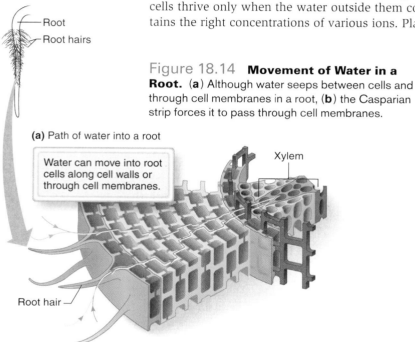

Figure 18.14 **Movement of Water in a Root.** (**a**) Although water seeps between cells and through cell membranes in a root, (**b**) the Casparian strip forces it to pass through cell membranes.

(a) Path of water into a root

Water can move into root cells along cell walls or through cell membranes.

Xylem

Root hair

(b) Casparian strip

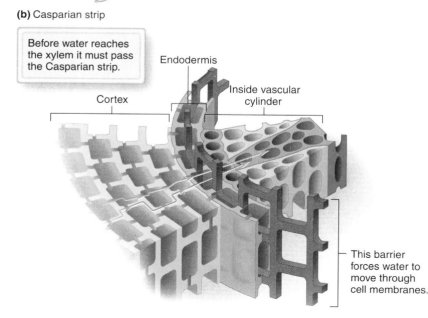

Before water reaches the xylem it must pass the Casparian strip.

Endodermis

Cortex

Inside vascular cylinder

This barrier forces water to move through cell membranes.

cells would not thrive if exposed to the typical variations in the mineral content of water in soil. Plants control the quality of their water at the point where water flows from the root hairs into xylem. They exert this control by manipulating the pathway that water takes as it enters xylem. Once inside the root hairs, water can move into the tissues of a root by two pathways: it can seep along cell walls and move between cells, or it can go across cell walls and cell membranes traveling into and out of cells (**Figure 18.14**). Before water and dissolved minerals reach the strands of xylem in the root, they reach a waxy barrier called the *Casparian strip*. This layer is secreted by a tightly packed layer of cells called the *endodermis*, and it fills the spaces between the endodermal cells. The Casparian strip forces water and dissolved minerals to move *through* membranes of endodermal cells, rather than between cells, before reaching the xylem. In effect it filters the water that has entered the plant. When water passes through cells, some dissolved substances will pass to the xylem more readily than others. For example, because of the nature of the proteins in plant cell membranes, potassium ions pass *through* cell membranes more readily than sodium ions do. The Casparian strip decreases the concentration of sodium ions of water in the xylem and increases the concentration of potassium ions. This is one way that a plant can control the quality of its internal water.

Sugars Are Pushed Through Phloem from Sugar Sources to Sugar Sinks

The process of photosynthesis produces sugar molecules that the cells of a plant use to generate ATP. It is the job of phloem to provide a conduit for transport of sugars and other biological molecules produced by leaves. The sugary fluid moved by plants often is called *sap*. You know that a plant has no internal pump to circulate fluids, so how does sap move within phloem vessels?

Figure 18.15 summarizes the process of sap movement in phloem. The first thing to notice is that sap moves from a region with a high sugar content to a region with a low sugar content. The regions of sugar concentration are called *sugar sources,* while the regions of low sugar concentration are called *sugar sinks.* You can think about a sugar source as the place sugar comes from, and a sugar sink as a place that sugar drains into. This means that sugar transport is not unidirectional. Phloem can transport sugars up, down, or laterally in a plant. The same structure can be a sugar source or sugar sink at different times. In the spring when new leaves are sprouting, a particular

Root
Root hairs

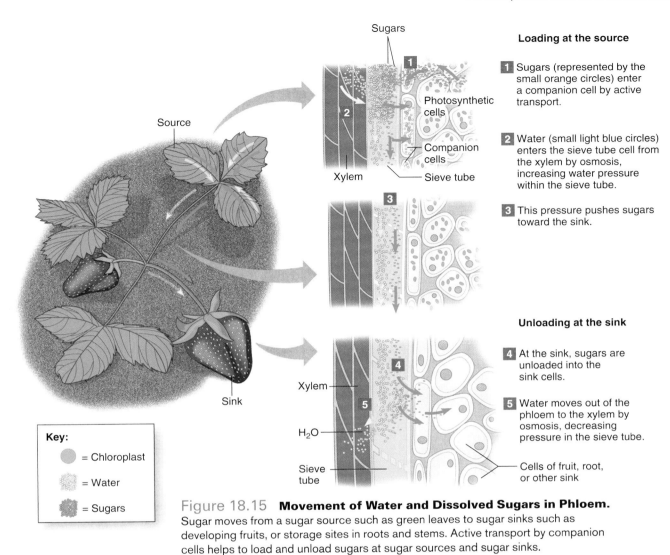

Loading at the source

1 Sugars (represented by the small orange circles) enter a companion cell by active transport.

2 Water (small light blue circles) enters the sieve tube cell from the xylem by osmosis, increasing water pressure within the sieve tube.

3 This pressure pushes sugars toward the sink.

Unloading at the sink

4 At the sink, sugars are unloaded into the sink cells.

5 Water moves out of the phloem to the xylem by osmosis, decreasing pressure in the sieve tube.

Sugars

Photosynthetic cells

Companion cells

Xylem

Sieve tube

Xylem

H_2O

Sieve tube

Cells of fruit, root, or other sink

Key:
- = Chloroplast
- = Water
- = Sugars

Figure 18.15   **Movement of Water and Dissolved Sugars in Phloem.**
Sugar moves from a sugar source such as green leaves to sugar sinks such as developing fruits, or storage sites in roots and stems. Active transport by companion cells helps to load and unload sugars at sugar sources and sugar sinks.

Source

Sink

leaf may be a sugar sink because it takes sugar from other regions of the plant to support its rapid growth. When the leaf is mature, it is a sugar source that provides sugar to other parts of the plant. The opposite pattern might be seen for a stem or root storage structure. In the spring when the plant is growing, a root or stem might serve as a sugar source, while in the height of summer that same root or stem might be a sugar sink. How does the plant carry out this complicated sugar distribution plan? Unlike the flow of water in xylem, the flow of sap involves energy from ATP. Let's go through this process, using a leaf as an example of a sugar source and a storage root such as a carrot as an example of a sugar sink.

The first important point is that sugar does not go directly from the photosynthetic cell to the phloem. Rather, sugar sources and sinks are actively produced by companion cells, which are associated with the sieve tube member cells of the phloem.

This requires moving sugar from a region of low concentration in the photosynthetic cell, to a region of high concentration in the companion cells. Recall from Section 4.2 that this kind of movement of molecules *against* a concentration gradient requires active transport and energy from ATP. As a result of active transport, sugar builds up in the companion cells. The high concentration of sugar causes water to flow into the companion cell from nearby xylem by osmosis. Here is where the special qualities of plant cells come into play. Plant cells have rigid cell walls, so when water flows in, the cell does *not* expand like an animal cell would. Rather, as water flows in, osmotic water pressure builds up within the companion cell. As the pressure builds, the only way out for the water is into the phloem, which has close connections to the companion cell. Of course, the sugar collected by the companion cell is dissolved in the water, so the sugar goes into the phloem by bulk flow. The pressure of water coming into the companion cell pushes water and sugar away, out into the phloem.

An analogy might help you to think about the flow of sugar and water. Imagine a rigid plastic milk jug connected to a hose. The jug is first loaded with sugar crystals, and then the water to the hose is turned on and flows in. As the jug fills, the sugar dissolves in the water, and pressure builds up in the jug. If there were no outlet, eventually water would stop flowing because pressure would push backward against the water flowing in the hose. Instead, the other end of our jug has a few small holes in it. As pressure builds, it will push water and dissolved sugar out the holes.

Again, active transport by companion cells is required to set up the sugar sink. Getting the carrot root cells fully packed with sugar requires sugars to be moved *into* a region of high concentration, and this involves active transport and the energy in ATP. Active transport moves sugar from the companion cells in this region into the root cells. This decreases the sugar content of the companion cell and allows sugar and water to flow from the sieve tube elements of the phloem vessel into the companion cell. So active transport and ATP keep sap moving into, through, and out of phloem, from sugar sources to sugar sinks.

QUICK CHECK

1. How are the structure and function of phloem different from the structure and function of xylem?

18.3 Plant Hormones Regulate and Integrate Plant Functions

Hormones? Plants have hormones? Now that you're accustomed to the idea that plants have specialized tissues and organs, it is time to introduce the hormones that integrate the functions of a plant. A **plant hormone** is a substance released by certain cells in the plant that causes a biochemical response in other cells in the plant. Nearly all of the meristems and growing tissues in a plant can and do produce hormones. Like hormones in animals, plant hormones control the timing and extent of major processes. For instance, hormones control when fruit develops and ripens, how and when roots grow and elongate, and how plants respond to gravity and light. Hormones also stimulate cell division and cause seeds to germinate. One kind of hormone acts in opposition to others and inhibits growth, closes stomata when water is in short supply, and causes

plants to become dormant. **Table 18.1** gives details of the actions of different plant hormones. Let's take a look at some of the most important ones.

You probably have observed that many plants respond to sunlight by growing toward it (**Figure 18.16**). Charles Darwin did some clever experiments to demonstrate that this effect was due to something in the very tip of a shoot. If he cut off the tip, or covered it, the plant gave no directional response to light. Later studies showed that this unidirectional growth is caused by the plant hormone **auxin.** This hormone is produced in a gradient. Auxin is most concentrated in cells at growing tips of plants; little auxin is produced in tissues that are far from growing tips. Auxin causes the cells that it contacts to lengthen. Sunlight causes auxin to move away from a plant part that is brightly lit and to flood cells on the *shaded* side of stems. These cells respond to auxin by growing in length, and their elongation causes the plant to bend toward the light. When light is directly above a plant, cells in all sides of its stems release the same amounts of auxin. In response the plant grows upward. Not all plants show this auxin-based response to sunlight. Many shade-loving plant species do not have this response and are marketed as houseplants.

Auxins also affect the branching of stems and roots. To have full, bushy plants instead of plants that are tall and leggy, gardeners "pinch back" the tips of shoots. This encourages side branches to develop. Removing these buds eliminates tissues on the tips of the shoots that produce auxin and allows tissues on other branches to respond to the auxin that they produce. In effect, pinching back has changed the distribution of the hormone auxin within the plant's tissues. Auxins also affect root growth, allowing rootlets to grow more luxuriantly. Gardeners take advantage of the effect of this hormone on root growth when they dip cuttings in rooting powders, which usually have auxin in their formulations.

Auxin has a variety of effects on plants beyond causing cells to elongate. Auxin affects almost all aspects of plant development. It stimulates the development of lateral roots by stimulating cell division. The dropping of leaves and flowers is inhibited by auxin, and auxin promotes fruit development. Farmers often apply auxin to plants to enhance the growth, development, and fruiting of crops, especially fruits and vegetables. Different kinds of auxins affect different kinds of plants. It is not clear why high doses have this effect, but it did not take long for people to think of using auxins as herbicides, chemicals that kill plants. Natural auxins turned out to be not too useful as herbicides because they quickly are broken down by any

---

plant hormone a substance produced by certain plant cells that causes other cells in the plant to respond biochemically

auxin a plant hormone that causes cells of stems to lengthen and affects other aspects of plant growth and development

plant. Artificial auxins have been developed that are similar to, but not identical with, natural auxins. These compounds persist in plants, and they are more effective against certain kinds of weeds than they are on grain plants such as wheat or corn. The use of herbicides is touted by some but criticized by others. Even though plants and animals are only distantly related, chemicals that kill plants can affect animals too. One of the more infamous auxin-related herbicides was Agent Orange. This compound was used in the Vietnam War to cause the leaves of plants to die and fall off, allowing U.S. military personnel to see Viet Cong fighters and camps more easily. Veterans of the Vietnam War are convinced that Agent Orange caused a variety of illness, but the controversy over whether this is true has never been completely resolved.

The part of a stem between leaves is called an *internode,* and **gibberellin** is a plant hormone that causes internodes to lengthen (**Figure 18.17**). Gibberellin also is used to produce fewer but larger flowers and fruits. For instance, those enormous grapes that you see in supermarkets probably were treated with gibberellins. Gibberellin is produced in internodes when light levels are low, and it causes the cells of internodes to elongate. In extreme conditions the combination of low light and abundant gibberellins can produce spindly, lanky, long-stemmed plants, a condition that you may recognize even if you don't know its technical name: *etiolation.*

Another plant hormone comes into play in the ripening of fruits. If you want to ripen fruit quickly,

Table **18.1** Actions of different plant hormones		
Hormone	**Where produced**	**Action**
Auxins	Developing leaves, seeds, and tips of shoots	Cells of seedlings, leaves, and embryos grow longer; inhibit growth of lateral buds.
Gibberellin	Developing seedlings and shoots	Cells grow longer and divide; seeds germinate; flowers open.
Ethylene	All plant parts, especially older or ripe parts	Fruits ripen; leaves and flowers die; leaves and fruits fall off.
Abscisic acid	Mature leaves	Shoots do not grow; seeds stay dormant.

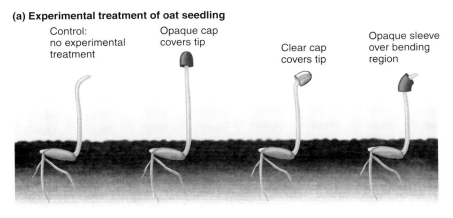

(a) Experimental treatment of oat seedling

Control: no experimental treatment

Opaque cap covers tip

Clear cap covers tip

Opaque sleeve over bending region

(b) Response of seedling to experimental treatment

Tip bends toward light.

Tip grows but does not bend toward light.

Tip bends toward light.

Tip bends toward light.

Figure 18.16 **Darwin's Experiments with Seedlings.** These experiments showed that something in the tips of seedlings caused them to bend toward light. It turned out to be auxins.

No gibberellins have been applied.

Gibberellins have been applied.

Fewer flowers and larger fruit

Delayed fruit harvest

Increased fruit size

Figure 18.17 **Effects of Gibberellins on Fruit and Flower Growth.** Gibberellins cause flowers to be fewer but larger than normal. Gibberellins also delay fruit harvest and increase the size of fruit.

■ *Look up the definition of animal hormones and compare their functions to plant hormones.*

gibberellin a plant hormone that causes internodes to lengthen

one trick that often works is to put it into a paper bag with an apple that has been chopped into pieces. The injured apple tissues release **ethylene,** a gaseous plant hormone that hurries the ripening of fruits. Unripe fruits that are picked and shipped often are treated with ethylene gas before being distributed to supermarkets. The taste of a naturally "vine-ripened" fruit or vegetable cannot be matched by gassing, which sometimes causes fruits to rot more quickly than usual. In fact, this effect of ethylene is the reason for the old saying "one rotten apple spoils the barrel." One overripe fruit that releases a lot of ethylene can hasten the ripening, and the rotting, of other nearby fruit.

A seasonal interplay between auxins and ethylene controls when leaves, fruits, and nuts fall from a plant. Leaves produce auxins in summer, and these keep leaves attached to stems. At the base of the slender stem that joins a leaf to a twig is a breakaway layer, and during summer the production of auxins keeps this layer intact. When temperatures drop, auxin production also drops, while production of ethylene increases. Both of these hormones affect the breakaway layer, causing it to weaken. Soon leaves and fruits fall from the tree. If you have ever been to an apple, peach, or cherry orchard in spring, you may have seen the gorgeous drifts of petals falling from the trees. Flower petals also have breakaway layers that simultaneously give way in response to seasonal fluctuations in auxins and ethylene gas. Falls of flower petals, fruits, nuts, pine needles, and pine cones are all orchestrated by fluctuations in these same hormones.

Abscisic acid is a plant hormone that acts in opposition to auxins and slows the growth of plant tissues. Abscisic acid has significant effects within seeds, where high levels of abscisic acid prevent them from germinating. Before the seeds of many plant species can sprout, they must be leached of abscisic acid. Other species rely on long periods of cold to break down the abscisic acid stored within seeds. This natural packaging of abscisic acid in seeds is one way to prevent them from germinating at inopportune times, like beneath a thick layer of snow when low temperatures quickly would kill the delicate seedling.

ethylene a plant hormone that causes fruits to ripen

abscisic acid a plant hormone that slows growth of a plant, causes seeds to become dormant, and closes stomata during dry conditions

QUICK CHECK

1. How does pinching back a plant change its distribution of auxin and affect its growth pattern?
2. What are the names of four plant hormones and how does each affect plant growth or plant function?

18.4 Plants Require a Simple Set of Basic Nutrients

Because people are animals it is easy for us to understand animal nutrition. Put simply, animals must eat other organisms to survive. Plants, on the other hand, make their own food and turn carbon dioxide, water, and light energy into sugars. This seems simple enough, but a deeper consideration of plant nutrition raises important questions. Organisms need more than just energy to survive. To synthesize necessary biological molecules, organisms must have certain chemical elements. For example, because amino acids, proteins, nucleotides, and nucleic acids all contain nitrogen, this element is an important nutritional requirement for all organisms. Phosphorus is part of ATP and also is found in many types of proteins. Potassium is essential to many biological processes and is especially important in communication between cells. Many other chemical elements are needed in small quantities because they are used by key enzymes or are small but critical components in key biological processes. The elements calcium, copper, zinc, molybdenum, magnesium, nickel, sulfur, and iron are all important in small quantities. Animals obtain all of the necessary chemical elements when they eat organisms, especially plants. But where do plants get the chemical elements that are necessary for life's chemical processes?

The simple answer is that soil is the source of the chemical elements needed by plants. Soil may sound like dirt to you, but actually it is a complex and highly variable substance. Many of the chemical elements in soil come from the breakdown of the rocks of Earth's crust. The earliest life used the Earth's crust as the source of these essential elements. As life continued, however, the compounds and elements that had accumulated in living organisms became available to future generations. Today soil contains both the basic chemical elements from the Earth's rocky crust as well as organic and inorganic molecules derived from the breakdown of dead tissues and waste products. Soil also contains uncounted billions of living microorganisms such as prokaryotes, single-celled eukaryotes, algae, and fungi that take in and chemically transform and excrete nutrients. In 1 gram of soil there are more organisms than there are human beings on Earth. Many small multicellular organisms like nematodes, termites, ants, segmented worms, slugs, and snails live within soil. They also act as decomposers and recycle nutrients (**Figure 18.18**). Plants take advantage of this rich source of nutrients to get everything they need to survive.

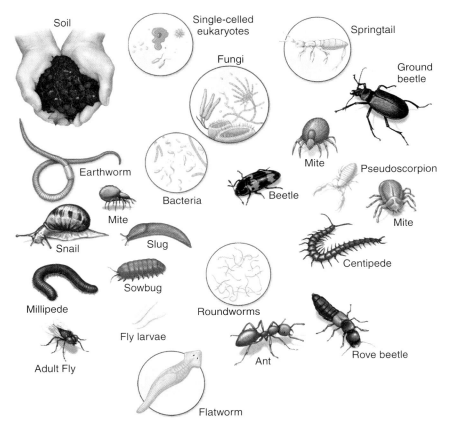

Soil

Single-celled eukaryotes

Springtail

Fungi

Ground beetle

Earthworm

Mite

Pseudoscorpion

Bacteria

Beetle

Mite

Mite

Snail

Slug

Centipede

Millipede

Sowbug

Roundworms

Fly larvae

Ant

Rove beetle

Adult Fly

Flatworm

Figure 18.18 **Soil Organisms.** A single handful of soil contains dense populations of bacteria, fungi, single-celled eukaryotes, and many kinds of soil invertebrates.

One piece of this story is not as simple, and that is nitrogen, a chemical element that every organism must have. You might think that because Earth's atmosphere is mostly composed of nitrogen, it would be easy for all organisms to obtain it, but this is not so. Although today's atmosphere contains about 80% molecular nitrogen (N_2), because each pair of nitrogen atoms is joined by an exceptionally strong triple covalent bond, atmospheric nitrogen is not readily available to organisms. Manufacturing processes have been developed to use atmospheric nitrogen to make nitrogen compounds that plants and animals can use, but it takes a huge amount of energy to split the triple covalently bonded molecules of molecular nitrogen. It turns out that billions of years ago, certain bacteria developed enzymes able to break the triple covalent bonds of molecules of nitrogen. They use the resulting nitrogen to make a variety of nitrogen compounds. The process requires 24 ATP molecules, and only some bacteria can do it. Most plants take up the nitrogen produced by bacteria from the soil. In some species of plants these bacteria take up residence in nodules on the roots of some plants and provide the plants with all the usable nitrogen they need (**Figure 18.19**). Called *nitrogen-fixing plants,* these include the legume plants such as peas, beans, clover, and soy. When animals eat the plants, nitrogen is available along with other nutrients in the plant.

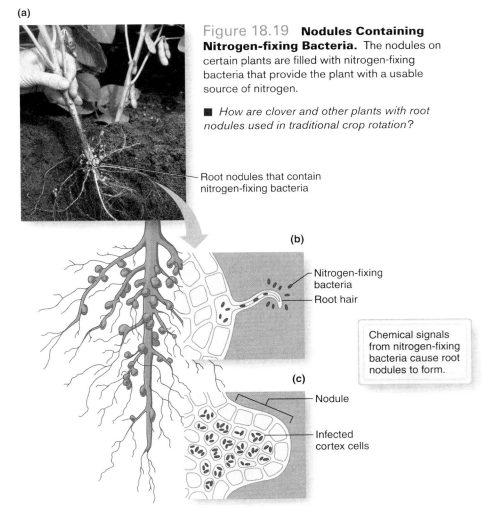

(a)

Figure 18.19 **Nodules Containing Nitrogen-fixing Bacteria.** The nodules on certain plants are filled with nitrogen-fixing bacteria that provide the plant with a usable source of nitrogen.

■ *How are clover and other plants with root nodules used in traditional crop rotation?*

Root nodules that contain nitrogen-fixing bacteria

(b)

Nitrogen-fixing bacteria

Root hair

Chemical signals from nitrogen-fixing bacteria cause root nodules to form.

(c)

Nodule

Infected cortex cells

Growth in nitrogen-poor soil   Growth in phosphate-poor soil   Growth in potassium-poor soil

Figure 18.20   **Consequences of Poor Nutrition in Plants.**

Not all soils provide the same levels of nutrients, and there can be serious consequences for the health of a plant if it grows in nutrient-poor soils. **Figure 18.20** shows some of the consequences for shortages of specific plant nutrients. Nutrient-poor soil is a factor that contributes to poverty in many regions of the world. Nutrient-poor soils can be corrected by the application of *fertilizers*, which are chemicals, mixtures, or compounds that contain nutrients. Humans have long realized that natural processes can add nutrients to the soil of farms and gardens. Early human cultures arose close to rivers in part because when a river floods over its banks it deposits rich nutrients on the surrounding land, enriching soil for growing crops. Humans also have learned that they can add nutrients to soil. In some cultures human fecal waste is saved and used as fertilizer.

The modern era of farming, which some people call the Green Revolution, started in the 1960s when industrialized societies learned to manufacture simple fertilizers. These simple fertilizers contain mostly nitrogen, phosphorus, and potassium. If you buy a container of fertilizer for your houseplants, you will see three numbers on the front of the package. The figures represent the percentages of nitrogen, phosphorus, and potassium that the fertilizer contains. For instance, an 8-8-8 fertilizer contains 8% of each of the major nutrients, while a 20-15-10 contains 20% nitrogen, 15% phosphorus, and 10% potassium. The percentage refers to the weight of the bag of fertilizer. So if you buy a 10-pound bag of 8-8-8, you will be getting just .80 pounds of each plant nutrient. The rest of the fertilizer will be made of inert ingredients. In contrast, if you buy a 10-pound bag of 20-15-10, you will be getting 20%, or 2 pounds, of nitrogen, 1.5 pounds of phosphorus; and 1 pound of potassium.

Large garden centers sell many different formulations of chemical fertilizers, so how do you know what formulation to buy? This depends on what kind of plants you are fertilizing and on the season. In general, though, nitrogen is needed for leaves and stems to grow well and to develop dark green color. Nitrogen is needed in the production of chlorophyll and helps plants to grow strong and healthy. Potassium increases the general health of a plant and promotes resistance to drought, disease, and heat. It also helps plants to take up water. Phosphorus is active in the early development of the plant. It also helps plants to grow to maturity and promotes the development of roots and blossoms.

QUICK CHECK

1. How do plants obtain nitrogen?

18.5   Plants Engage in Biological Warfare

For years biologists have known that plants are live green factories that produce sugars, starches, and oxygen gas. Only recently, biologists have learned how ingeniously plants use chemicals to manipulate other organisms and to protect themselves. Plants produce an enormous array of chemicals that can be classified into two general groups: primary and secondary metabolites. **Primary metabolites** are chemicals used in the basic life processes of photosynthesis, respiration, growth, and development. Chlorophyll, amino acids, nucleotides, lipids, and carbohydrates are examples of primary metabolites. In comparison, **secondary metabolites** are much more varied. These

primary metabolite
a basic metabolic compound produced by a plant that supports a plant's growth and development

secondary metabolite
a compound made by a plant that functions in plant defense and communication

help a plant to defend itself and communicate with other organisms. Although related plants produce the same kinds of secondary metabolites, there are no universal secondary metabolites that are produced by all plants.

Recent studies of how plants use secondary metabolites have changed how biologists think about the lives of plants. Plants are integrated into a web of chemical interactions with the organisms around them. For instance, plants release secondary metabolites to repel invaders, kill their enemies, or attract their friends and defenders. Plants send out secondary metabolites to recruit armies of aggressive defenders. And of course, the enemies fight back. These defensive and offensive interactions do seem a lot like a war. This section explores more about these chemicals that seem to raise garden-variety plants into the realm of science fiction. One thing is sure, when you've finished reading this section, you won't look at any plant—like that delicate African violet growing on the windowsill—as just a pretty photosynthesizer. Instead, you'll wonder just what chemical tricks that plant might be up to.

Secondary metabolites also are the foundation for much of the modern drug arsenal. Many of the secondary metabolites produced by plants have effects on mammals—including humans—and have been adapted to treat human and animal diseases and for other purposes. A few examples will make the point. The caffeine in your coffee or soft drink and the aspirin in your medicine cabinet are two of the most widely used plant compounds in human cultures. Morphine and other opiate drugs that control pain after surgery are derivatives of opium produced by poppies. Many cancer treatments are derived from plant compounds. Finally, many of the drugs that people abuse and whose use can ruin a promising life—such as marijuana, cocaine, and heroin—are derived from plant secondary metabolites.

Plants Have Physical Defenses

Of course, plants have physical defenses to prevent organisms from eating or otherwise injuring them, and thickened cuticle, thorns, spines, thick bark, and tough leaves are good examples (**Figure 18.21**). Physical defenses are barriers, similar to human skin, that prevent the outside world from getting into the plant. These kinds of barriers are not specific to a particular pest or condition but provide generalized protections. Cuticle is one example of a physical defense. While the waxes in cuticle play a role in water conservation, they func-

Figure 18.21 **Plant Primary Defenses.** Thorns are an example of primary defenses that a plant uses to protect against herbivores.

tion in defense against herbivorous insects. Waxy leaf surfaces are slippery and help deter leaf-eating insects or insects that suck plant juices. By making plant surfaces dry, waxes also may prevent the fungal spores that continually rain down on the plant from gaining access to its tissues. Of course, the fungi do not just passively accept this defense. A fungal spore that does germinate on a plant's cuticle exudes an enzyme that breaks down the molecules of cuticle, giving hyphae a way to grow into the plant and feed.

The physical barriers that a plant uses are not static. Wouldn't it be amazing if your skin would grow "tougher" in response to bacteria, viruses, and other pathogens? Plant cells can do this. For example, many plants increase their production of lignin in response to an immediate threat from a bacterial or viral pathogen.

Chemical Defenses Target the Biological Processes of Organisms in a Plant's Environment

Aside from physical defenses, plants produce many chemicals that are more vigorous in their actions. Some plant chemical defenses act like drugs against the plant's enemies. They can kill herbivores or pathogens, make them sick, or otherwise interfere with their normal biological functions. Plants also produce chemicals that in some situations will kill or retard the growth of other plant species. This is an amazing feat because all these defensive and offensive actions are accomplished by organisms that cannot move about freely and have no nervous systems. A detailed look at a few examples of plant chemical defenses will give you a better appreciation of this aspect of plant diversity.

Plant Toxins Can Kill Herbivores or Make Them Sick

Many of the herbivores that eat plants are small insects, and a healthy plant can produce enough chemical defenses to get rid of the pest entirely. Some plant chemical defenses are outright poisons that kill the attacking insects, and the active ingredients in many insecticides that people use are derived from these kinds of plant chemicals. For instance, look at the active ingredients listed on the label of an insect spray. If pyrethrin is listed, a flea fogger or cockroach spray incorporates a toxin derived from the leaves and flowers of chrysanthemum plants.

Many culinary herbs and citrus fruits have rich supplies of highly aromatic oils. While humans like to flavor foods and blend perfumes from essential oils, these compounds repel insects. Peppermint, spearmint, lemon, lavender, basil, oregano, sage, thyme, and orange are all examples of plants that use insect-repelling essential oils as plant defenses. Some essential oils also have antibacterial and antifungal effects, and some are toxic to vertebrates. A close look at the surface of a plant shows that plants often lodge essential oils in hairs that stick out from the epidermis. In effect, these oil-saturated hairs advertise that the plant's tissues contain toxic essential oils, and herbivorous insects avoid the plant. If a plant is attacked, let's say by aphids, it will become even more aromatic, and the extra supplies of essential oils attract flying predators and parasites like wasps and some beetles that arrive to attack the leaf-chewers.

Plants produce such a wide range of chemicals that can make herbivores sick that it is somewhat bewildering. Foxglove plants produce secondary metabolites that have toxic effects on heart muscle. Milkweed plants contain heart poisons that are toxic to most herbivores. Spurge plants and their relatives produce toxins that irritate mammals' skin and make them sick enough to die. If you think about how many different organisms plants must defend themselves from, it is not surprising that researchers are confident that there are thousands to tens of thousands of interesting plant chemicals to be studied and perhaps exploited by humans.

Plants Release Chemicals That Attract Predators of Herbivores

In the small-scale world of insects, there is a constant war between the plants, their herbivorous insect enemies, and the predatory insects that prey on these herbivores. In a saga that is filled with more intrigue than you probably ever imagined, alliances between plants and insects are made and sometimes broken. One aspect of these interactions is the way that plants enlist the aid of insects that prey on herbivores that damage them. Let's look at a couple of examples.

The mild-mannered cucumber plants in your garden are actually ready to defend against small arthropods called spider mites that attack them. To get rid of these pests, a cucumber plant releases a compound called beta-ocimene. It attracts a species of predatory mite that makes its way to the cucumber plant and eats the spider mites. Score for the cucumber! In another example, corn plants directly respond to herbivore damage. When a pest called a beet armyworm attacks a corn plant, it releases a compound called linalool. Within six hours of the attack the linalool attracts parasitic wasps that sting the armyworms and lay eggs in their bodies. The wasp larvae use the still-living body of the armyworm as a haven and food source. Eventually, the armyworm is killed when new wasps hatch out of its body.

Plants Release Chemicals That Kill Other Plants

Have you ever walked through an old-growth forest? The tall pines are majestic, and their trunks are as much as 3 feet in diameter. It is so quiet and dim beneath the canopy, and the ground often is free of undergrowth (**Figure 18.22**). Have you ever wondered why? Certainly the limited amount of light keeps some small plants from springing up, but there may be something else going on as well. It may be that the pines are active players in keeping the riffraff of competitive plants out of their neighborhood. It is difficult to clearly demonstrate this effect in a complex forest ecosystem, but laboratory studies have documented that plants produce chemicals that kill or stunt the seedlings of other plant species. In many natural systems the competitors develop a resistance to one another's herbicides and thrive even in high levels of the toxic compounds. But when a nonnative plant that produces a herbicide moves into a new neighborhood, the native plants can suffer. Some nonnative plants are masters at killing off the competition. Invasive plants are a big problem in U.S. parks and wild areas, because they often choke out native species.

Herbivores Often Overcome and Even Take Advantage of Plant Chemical Defenses

The herbivores and other pests that plants try to kill or chemically discourage do not take these plant defenses lightly. Evolution has produced

Decomposing needles release chemicals that kill other plants.

Figure 18.22   **An Old-Growth Pine Forest.**
The clear ground under the canopy of an old-growth forest may be due to growth inhibiting substances produced by the dominant trees.

Figure 18.23   **Pine Bark Beetles.**

many complex relationships in which organisms have evolved adaptations that allow them to overcome the chemical defenses of plants. As an example, let's consider the intertwined lives and the continual warfare of pine trees and pine bark beetles.

A walk in a pine forest brings that clean scent of pine resin, but to a bark beetle, the scent of pine resin means death (**Figure 18.23**). Pine resins contain secondary metabolic compounds that are toxic to pine beetles. When pines are being attacked by pine beetles, they produce additional supplies of these toxins that are concentrated in needles, twigs, and trunks of the trees. Of course, once a pine beetle has burrowed beneath the bark of a tree it can escape these chemical defenses. Secure beneath the bark, a pine bark beetle feeds on a fungus that quickly sprouts from the spores

that the beetle tracked into the tunnel it dug beneath the bark. The fungus eventually will kill the tree, but this will take years. Until the fungus overwhelms it, a pine tree has yet another defense against the invading beetle: it makes poisons and stores them in its wood. The growing fungal hyphae absorb these poisons, and pine bark beetles die when they eat the poisoned fungi. Other species of bark beetles have another trick: they can metabolically alter the tree's poison to produce a chemical that they release from the entrance to their tunnel in the pine tree's trunk. Incredibly, the converted poison acts as a sex attractant that lures potential mates to the safely hidden pine bark beetle. When a pine tree is infested with these kinds of bark beetles, it has still one other chemical card to play. It can manufacture more pine resins to make a gaseous signal that is released from the entrance of the pine bark beetle's tunnel. This works like a flare that marks the hidden entrance to the tunnel and alerts wasp predators to the bark beetle's hiding place. While flying around the pines, these predators home in on the scent, locate the hidden tunnel, and assassinate the pine bark beetle.

QUICK CHECK

1. What is the difference between a primary metabolite and a secondary metabolite?

2. Briefly describe three ways that secondary metabolites are used by plants.

An Added Dimension Explained
The Green Fairy

The introduction discussed the brilliant but tortured life of Vincent van Gogh. Many explanations for van Gogh's psychological and emotional turmoil have been suggested—but clearly toward the end of his life he was in the grip of the green fairy. Vincent van Gogh was not alone. Many of the artistic and literary figures of the late 1800s and early 1900s were affected by it as well, along with many others from all walks of life. Picasso, Manet, Degas, and Toulouse-Lautrec are just a few of the more famous artists whose lives were influenced, largely for the worse, by the green fairy. As one author put it: "The fairy with the green eyes has enslaved their brains, has stolen their souls." What was this

Figure 18.24   **Degas: Absinthe Drinker.**

green fairy? The answer can be seen in the common image found in the paintings of these artists (**Figure 18.24**). Among their distinctive works, each artist has painted one or more scenes that included a bottle or a glass of absinthe. Absinthe is the green fairy, and absinthe addiction is the most likely reason for much of van Gogh's bizarre behavior in the last few years of his life. Let's find out more about absinthe and discover how it relates to the topic of this chapter, the structure and function of plants.

First, what is absinthe? The word is applied to a variety of concoctions, but the absinthe of the late 1800s had a well-defined composition. Absinthe is a strong alcoholic drink that has been infused with a high concentration of the oil of the herb known as wormwood. In the late 1700s the Pernod family in France discovered a recipe for making absinthe, and the family built a hugely successful business by producing and selling the drink. You must remember that this was a time when many drug-laden drinks were concocted and sold for profit, in both Europe and the United States, and absinthe was just one of these.

During the late 1800s a whole absinthe culture had emerged among the social elite of France, especially in Paris. Absinthe drinking developed into a cocktail hour of sorts that formally ended the working day and began the leisurely evening. For most the "green hour," as it was called, involved drinking just one serving of absinthe before dinner. It was considered bad etiquette to drink more than one, but many people rapidly found themselves addicted, and it was not uncommon for a heavy drinker to have 10, 12, or even 20 absinthes a day. The addiction was partly due to the alcohol content, which typically could be as high as 60 to 70%, which translates to 120 to 140 proof, but the addictive nature of absinthe also was due to oil of wormwood, its other major ingredient.

The scientific name for wormwood is *Artemisia absinthium* (**Figure 18.25**). Like many other plants, *Artemisia* produces a complex set of oils that protect the plants from herbivores. Wormwood has been known for thousands of years, and its first mention in history is from ancient Egypt, in 1600 BC. The name wormwood comes from its use as a deworming agent, and in small doses it is effective in ridding humans and other animals of various kinds of intestinal worms. It also has pesticide effects and can be used to repel moths and other insect pests. Herbal medicine gives many uses for wormwood, although it is not possible to say if it is effective for all of these disorders. For instance, wormwood has been used to hasten labor in women and treat menstrual cramps and fevers. It also has been used to aid digestion and to treat diseases such as malaria. Interestingly, there are also references to the use of wormwood to counteract the effects of alcohol—that is, to cause drunken people to become sober.

Figure 18.25 **Wormwood.** Essential oils obtained from *Artemisia absinthium* are blended with strong alcohol in the manufacture of absinthe.

Terpenes are the primary chemical components of the essential oil of wormwood. Terpenes are organic compounds found in nearly all essential oils produced by plants, and there are many different varieties of terpenes. The pine oils produced by conifers and the menthols produced by many plants are typical terpenes. All terpenes have a variety of effects on other plants and on herbivores, some more severe than others. Although many terpenes are repulsive or toxic to other creatures, some are attractive, but all terpenes evolved as secondary metabolites that protect the plants that produce them.

So how does all this relate back to absinthe, the green fairy? Absinthe users find that the drink affects them much differently than alcohol does. Many people claim that absinthe produces clarity of thought, creativity, and a positive mood. Nineteenth-century artists and writers often found life depressing and unpleasant. While their creative efforts provided an outlet for feelings of despair, absinthe drinkers believed that absinthe was superior because it not only relieved them of negative feelings but also gave them a creative spark. Unfortunately, the objective evidence is that if absinthe provided any of these positive effects it did so with an enormous negative cost. Many horror stories of absinthe abuse can be found in the writings from this era. Not only was the drink highly addictive, but reports suggest that abusers suffered liver damage, convulsions, hallucinations, violence, and many committed suicide.

Is absinthe worse than other alcoholic drinks? Could absinthe have contributed to van Gogh's bizarre behavior? Researchers are only beginning to unravel this story. One of the important lessons from the studies of absinthe is how effective plant chemicals can be in affecting the brains of animals. The major active terpene in wormwood is a molecule called *thujone*. Early studies showed that thujone can kill brain cells in animals, but the exact mechanisms of its effects have been unclear until recently. Because thujone is similar in structure to the active ingredient in marijuana, many people assumed that the effects of absinthe were caused by actions on the brain that were similar to those of marijuana. Brain studies on the actions of thujone, however, showed that this was not likely. Instead, thujone interacts with a membrane receptor on certain nerve cells that control how active the brain is. These nerve cells tone down and inhibit brain activity, and thujone *blocks* this action. This accounts for the reports that absinthe users suffered convulsions and could be an explanation for the reports that absinthe causes hallucinations. This finding leads to the hypothesis that absinthe combines the relaxing and sedating effects of alcohol with the activating effects of thujone. Obviously, there are no controlled studies on the effects of such a combination, but future laboratory studies may shed further light on what van Gogh and others may have actually experienced from absinthe. What one can say is that drinking large amounts of absinthe would have different effects than drinking large amounts of other kinds of alcohol. Absinthe probably did not give van Gogh his artistic genius, but it probably did contribute to mental and physical traumas that led him to eventually take his own life.

QUICK CHECK

1. How does the active ingredient in absinthe, thujone, affect the brain? How is this different from the effects of alcohol on the brain?

Now You Can **Understand**

Poison Ivy

Have you ever gotten poison ivy or poison oak? Score one for a reaction produced by a secondary metabolite. **Figure 18.26** shows a typical poison ivy plant. Notice how perfect its glossy leaves look. Poison ivy and poison oak typically have leaves that have no ragged holes where insects have been gnawing. The reason is urushiol, an oil that poison ivy, poison oak, and poison sumac plants produce to defend themselves from leaf-chewing insects. When some people touch this oil, they can develop an inflammatory reaction produced by the immune system. Within a few days red, itchy blotches break out on the skin that has had exposure to urushiol. Skin that has had

Figure 18.26   Poison Ivy. Urushiol in poison ivy causes an itchy rash in many people. Urushiol is a secondary metabolite of poison ivy.

more exposure breaks out sooner and most intensely. The reaction does not spread through the liquid that oozes from the affected skin; only skin that comes in contact with the toxic oil will break out. So if you thoroughly wash your clothes and skin after you have touched poison ivy or poison oak, you can minimize the nasty symptoms.

What Do **You** Think?

Are Tree-Sitters Noble or Foolish?

Imagine that you are living in a tent that is 200 feet above the ground, supported on platforms that are attached to a redwood tree that is many hundreds of years old. Friends bring you food, water, maybe some books. They carry away your . . . wastes. You have no TV, no college classes, no paid job. But you live in the tree, month after month, for as long as a year. Why would you do such an odd thing? To save the life of the tree.

This is tree-sitting. Environmental activists object to the logging of extremely old trees, which sometimes are called *old-growth trees.* One of their strategies to prevent the trees from being cut down is to live in them. The idea is that logging companies will not cut down a tree if a human in the tree could be hurt when the tree falls. To many environmentalists this is the ultimate way to draw attention to the plight of old, majestic trees and the biodiversity that they nurture. For months at a time tree-sitters live without comforts or modern conveniences to draw attention to the fact that an old tree is destined to be cut down. To others the exercise is a futile waste of time because in the end the tree-sitters always lose. The trees belong to the lumber industry or to U.S. or state governments, and eventually the tree-sitters are carried down by force.

What is your view of the conflict between the tree-sitters and the logging companies? Is it worth a year of someone's life to send a message that emphasizes the value of old-growth timber, even if the tree is likely to fall in the end? What is your view of such extreme sacrifice in the name of some deeply held social commitment? What do *you* think of the tree-sitters?

Chapter Review

CHAPTER SUMMARY

18.1   Specialized Tissues Carry Out Plant Functions

Leaves, stems, and roots are the organs of a plant. Leaves and stems are grouped as shoots. These organs are composed of tissues that are arranged in sheaths. Different kinds of cells are specialized for strength, pliability, or the ability to transport water, plant nutrients, and food within the plant. Roots absorb water and plant nutrients; stems allow for vertical growth; leaves are specialized for photosynthesis and gas exchange. Epidermal tissue of a plant is generally only one cell thick and functions to prevent injury and water loss. Epidermal cells excrete the waxy cuticle that prevents water loss. Stomata open to allow gas exchange and close to prevent gas exchange.

Ground tissue makes up most of the body of a plant. Parenchyma is the most common kind of cell in ground tissue and is specialized for photosynthesis and surplus food storage. Collenchyma is a second kind of cell found in ground tissue. It has thickened, flexible cell walls. Sclerenchyma is the third kind of ground tissue and is stronger because of its double cell walls. At maturity sclerenchyma cells die but remain within the plant as woody tissue. Vascular tissue transports food and water. Xylem tissue transports water and minerals upward from roots. Tracheids and vessel elements are two kinds of xylem cells. Fiber cells give strength to xylem tissue. Parenchyma cells allow tracheids and vessel elements to regenerate. Leaves have epidermal, ground, and vascular tissues arranged in layers that are specialized for photosynthesis. Plant growth occurs in meristem tissues. Apical meristems allow for primary growth of roots and shoots. Lateral meristems allow for secondary growth. Vascular cambium and cork cambium are tissues of lateral meristems. Wood is made of xylem. An annual ring reflects one year's growth.

apical meristem   434
epidermal tissue   429
ground tissue   431
lateral meristem   435
leaf   429
meristem   434

phloem   432
root   429
stem   429
vascular bundle   432
vascular tissue   432
xylem   432

18.2 Plant Adaptations Move Water and Dissolved Substances

Pulled by the forces produced by transpiration, water enters a root across the membranes of the cells of root hairs. In the process of transpiration water vapor evaporates from leaf tissues through stomata. The Casparian strip forces water to move through endodermal cells. Pits and perforations allow water to flow into tracheids and vessel elements. Phloem moves sugars and other biological molecules from leaves to other plant parts. Sieve tube element cells and companion cells are found in phloem. Sugars accumulate in companion cells and pass to sieve tube elements by active transport from photosynthesizing cells. Dissolved sugars move throughout the plant by osmotic pressure and bulk flow.

root hair   437

transpiration   438

18.3 Plant Hormones Regulate and Integrate Plant Functions

Auxins cause stems to lengthen, causing plants to bend toward light. Auxins also affect branch and root growth and stimulate all aspects of plant development including promoting development of fruits. Gibberellin causes internodes to lengthen and also causes production of larger fruits and flowers. Ethylene causes fruits to ripen. Abscisic acid slows plant growth, makes seeds become dormant, and closes stomata when conditions are dry.

abscisic acid   444
auxin   442
ethylene   444

gibberellin   443
plant hormone   442

18.4 Plants Require a Simple Set of Basic Nutrients

For optimal health plants must have relatively large amounts of nitrogen, potassium, and phosphorus. Plants also must have other chemical elements in smaller amounts. Plants obtain nutrients from soils. Soil bacteria make nitrogen available to plant roots. Fertilizers are used to increase levels of nutrients in soils, and they mostly contain nitrogen, potassium, and phosphorus. Nitrogen is necessary for plants to form chlorophyll and promote the growth of leaves and stems. Potassium increases the general health of a plant and helps it to take up water and promotes the development of roots and blossoms. Phosphorus is important in the early development of a plant.

18.5 Plants Engage in Biological Warfare

Primary plant metabolites are chemicals used in basic life processes of plants; secondary plant metabolites are chemical weapons that plants use to defend their tissues against attacks and in communication with other plants. Plants use secondary metabolites to attract predators of herbivorous insects. Other secondary metabolites kill or retard the growth of other plants. Plants also have structural defenses. Many tra-

ditional and modern medicines and insecticides use secondary metabolites as their active ingredients.

primary metabolite   446

secondary metabolite   446

18.6 An Added Dimension Explained: The Green Fairy

Vincent van Gogh's life was severely affected by his addiction to absinthe. Oil of wormwood contains thujone, a secondary plant metabolite produced by wormwood plants (*Artemisia abstinthium*). Thujone enhances the action of nerve centers that control brain activity, leading to hallucinations. Absinthe probably contributed to van Gogh's inherited tendency toward depression, suicide, and epilepsy.

REVIEW QUESTIONS

TRUE or FALSE. If a statement is false, rewrite it to make it true.

1. Secondary metabolites often kill insects.
2. Transpiration and surface tension pull water upward in a plant.
3. Stomata open to let in carbon dioxide, and in doing so they release water that has traveled upward in phloem.
4. Spring wood is typically darker and denser than summer wood.
5. The companion cells of phloem are dead cells.
6. Growth in length of a plant is called secondary growth; growth in thickness of a plant is called primary growth.
7. Plants use secondary metabolites for growth.
8. Some plants recruit insects that act as defenders using secondary metabolites.

MULTIPLE CHOICE. Choose the best answer of those provided.

9. Which ground tissue cell is specialized for photosynthesis?
   a. sclerenchyma
   b. parenchyma
   c. collenchyma
   d. xylem
   e. phloem

10. Which tissue covers the outside of most of the body of a plant?
   a. epidermal tissue
   b. ground tissue
   c. xylem
   d. phloem
   e. meristems

11. Which tissue is responsible for the increase in the height of a plant?
   a. apical meristem
   b. lateral meristem
   c. xylem
   d. phloem
   e. ground tissue

12. Water flows upward in which vascular tissue?
   a. phloem
   b. xylem
   c. parenchyma
   d. sclerenchyma

13. Which does the woody trunk of a large tree contain *little* of?
   a. dead cells
   b. xylem
   c. vascular tissue
   d. phloem

14. When the stem of a houseplant bends toward light, auxin is causing cells on the
   a. shaded side of the stem to elongate.
   b. sunny side of the stem to elongate.
   c. shaded side of the stem to shorten.
   d. sunny side of the stem to shorten.

15–20. Matching the terms. (One answer will not match.)

15. roots, stems, leaves
16. have double cell walls
17. ground tissue with thin cell walls
18. phloem
19. meristems
20. cambium

a. parenchyma
b. transports sugars
c. plant tissues that can undergo mitosis
d. plant organs
e. sclerenchyma
f. conducts water down to roots
g. divides to produce new vascular tissue

CONNECTING KEY CONCEPTS

1. Compare the way that water, minerals, and nutrients move through xylem and phloem.
2. How does the lateral meristem produce wood?

QUANTITATIVE QUERY

It is often useful to be able to predict how much plants will grow. Plants usually do not grow at a constant rate but usually grow faster when they are younger. Plant growth also depends on many environmental factors such as amount of rain and the temperature. The best way to predict how much a plant will grow in the next period of time is to know how it grew in the last period of time. For example, over the next week a plant might grow 0.5 times the amount it grew last week.

1. Write a mathematical equation that describes how much the plant will grow in the next time period, based on what it grew in the last time period. Let x = the amount the plant grew in the last time period and y = the amount the plant is likely to grow in the next time period.
2. If a plant grew 2.2 centimeters last week, how much should you expect it to grow next week?
3. If a plant was 55 centimeters at the beginning of this measurement period—one week ago—how tall will it be one week from now?

THINKING CRITICALLY

1. A tropical rain forest is characterized by a very high level of humidity—and, of course, it rains a lot in this kind of forest. It seems reasonable to conclude that the trees grow because it rains, but in another sense it rains because the trees grow. If you cut down the rain forest, the region often becomes dry and the rains cease. What do you think accounts for this?

For additional study tools, visit www.aris.mhhe.com.

The Thread of Life

REPRODUCTION OF SEED PLANTS

Orchids Profit from False Advertising, Study Says

March 8, 2004

In a Humid Courtyard, the Orchid Beckons

February 3, 2006

False Advertising. A male long-horned bee is trying to mate with an orchid that imitates the shape, color, and scent of a female long-horned bee. As the male tries to mate with the flower, a pollinium is attached to his head. When it visits another bee orchid, the pollinium will be transferred to it.

Being firmly rooted in place presents a seed plant with a thorny problem: how can the plant find another individual of its species and mate? How can it move its pollen to the sexual organs of a compatible plant? The obvious solution is to get something to carry pollen, and many seed plants use wind or water to do this. There is no assurance, however, that pollen distributed by water or wind will reach its goal; most windblown or waterborne pollen is wasted as the plants dust the environment with it. A more efficient strategy is to deliver pollen only to the right plants. But how can a plant possibly do this? One answer is to get animals to do it for them. Seems impossible, doesn't it? Nevertheless, it happens every day—and, if the season is right, it may be happening outside your window right now.

The situation is like training a dog to roll over and play dead or fetch a ball. If you reward a dog for a behavior, it soon will learn to associate the reward with the signal for the behavior. And this is just what some plants do with their pollinators. Let's take a look at two examples.

Some tropical forest orchids offer fragrant-smelling chemicals. Males of a few species of bees collect and process these fragrances to make scents that female bees find irresistible. Male bees are drawn to the scent of these orchids and will fly through the forest nearly two miles to reach them. But the orchids make bees work to get the chemical. A bee must crawl upside down into a flower to scratch at small patches of orchid cells that excrete it. The bee licks up the fragrance, and later his body processes the fragrance into a slightly different scent that draws female bees to him.

The story might end there, but the orchid has not yet received any compensation for helping male bees to attract mates. The inner surfaces of the orchid blossom are smooth and glassy. Inevitably, the bee loses his grip on the blossom and falls into a water-filled pouch within the orchid. The only way he can escape is to crawl through a narrow passage. As he struggles free, the bee scrapes past a flower mechanism that claps a *pollinium*, a structure that contains pollen, onto his back. The bee is unable to remove the pollinium, and it can only be detached by a slightly different mechanism within a *second* orchid blossom of the same species. Will the bee fly to a second flower where the pollinium will be detached? Studies in the field have demonstrated that just as female bees find highly scented males irresistible, male bees cannot resist the scent of "their" orchid blossoms. The odds seem to be excellent that the orchid successfully has "trained" its pollinator to accurately deliver pollen to another individual of its same species.

Other species of plants called bee orchids offer their pollinators an even more amazing reward: the promise of sex itself. The shape, texture, and colors of bee orchids trick male bees into mistaking them for sexually receptive female bees (**see chapter-opening photo**). The flowers have some furry surfaces like bees do, as well as flower parts that look like antennae. To human eyes the bee orchids look like crude approximations of bees, but the bees seem to have a different view because male bees try to mate with the flowers. In the process a pollinium is glued onto his back or head. When the male bee notices that something has gone awry with this mating, he doesn't give up. Instead, he changes position. He buzzes his wings and sometimes even bites at the orchid petals. If he were a human, you might say that the male bee was expressing sexual frustration. Eventually, the bee flies away, but the pollinium is firmly stuck in place, detached only when he visits the next bee orchid. Again, another kind of orchid plant has manipulated the actions of a pollinator. The bee orchid gives no nectar or chemical reward to its pollinator, just the illusion that sex is available.

Even though the two groups of orchid bees display bizarre behaviors, evidence about their actions is mostly anecdotal and still is being investigated. At the end of this chapter you will read about one of the most intricate and best-studied sets of interactions between a flower and its pollinator and learn more precisely why scientists have concluded that pollinators and plants can be the driving force in one another's evolution. Before you read more about this topic, though, let's explore how seed plants reproduce. ■

19.1 Gymnosperms and Angiosperms Are Seed Plants

Seeds are some of the most important reproductive adaptations. Plants that produce seeds are called *seed plants*. A **seed** contains a plant embryo, protective structures, and sometimes a source of food to nourish the embryo's early growth (see Section 17.6). The two major groups of seed plants are gymnosperms and angiosperms. Gymnosperms include cycads, ginkgoes, and conifers, while angiosperms are plants that have flowers. You know angiosperms as trees and bushes, flowers in bouquets and wreaths, vegetables in supermarkets and gardens, and the many plants in manicured lawns as well as in untidy weed patches. And, of course, we also eat seeds or parts of seed plants and use them in other practical ways. Not only are foods derived from seed plants, but paper, cotton, furniture, and lumber come from seed plants. In addition, seed plants provide fragrances, cosmetics, tobacco, rubber, coffee, colas, wine, whiskies, beer, and medicines, as well as most of the foods of domestic animals. An understanding of how seed plants reproduce will expand your knowledge of life's diversity.

seed a dormant sporophyte embryo and its food supply stored within a protective coat

It is relatively easy to tell a conifer from a flowering plant (**Table 19.1**). For one thing, most conifers have narrow leaves aptly called needles, while most flowering plants have broad, flat leaves. Needles of conifers have a limited array of pigmentation: with few exceptions they are greens, gray greens, and blue greens. In contrast, angiosperm leaves and flowers can be as colorful as a carnival. Most conifers are *evergreen* and hold most of their leaves year 'round. In contrast, most angiosperms have *deciduous* leaves that seasonally drop from the plant. Microscopic examination of woody tissues reveals another fundamental difference: the xylem tissues of the two groups have different kinds of tubular cells that transport water. This difference means that as a group angiosperms are more efficient at transporting water than are conifers.

As seed plants, conifers and angiosperms share many reproductive features, and the most basic is their pattern of alternation of generations. Plants alternate diploid and haploid generations, and each generation is a multicellular individual (**Figure 19.1**) (see Section 17.6). It will help if you can remember this: all of the seed plants that are visible to the naked eye—from redwood trees to dandelions and orchids—are diploid sporophytes. In contrast, the haploid gametophyte generation of seed plants is small to microscopic in size and is protected within the reproductive structures of a sporophyte. Gametophytes of seed plants are totally dependent on the sporophytes that nourish and protect them. The gametophyte generation has two separate kinds of reproductive structures. One kind of gametophyte will produce male gametes, called *sperm*, while the other will produce female gametophytes, called *eggs*.

Once you have grasped the concept that sporophytes are diploid multicellular individuals that contain and shelter haploid multicellular gametophytes, you are ready to add important details to your knowledge of the life cycle of a seed plant (Figure 19.1). The sporophyte generation undergoes meiosis to produce haploid spores. These spores divide mitotically to form the gametophyte generation. Each gametophyte produces unicellular gametes by mitosis. A zygote results from the fusion of sperm and egg. Finally, the zygote divides by mitosis to produce an embryo that is enclosed within a seed. The embryo will develop and grow into the adult multicellular sporophyte generation. In land plants some gametophytes produce sperm, while other gametophytes produce eggs.

To make sure that you understand these important facets of life cycles of seed plants, reread this section, referring to Figure 19.1 as you review. Read on when you're certain that you understand the concept of alternation of generations as it applies to seed plants and that you understand the difference between the sporophyte generation and the gametophyte generation.

Seed Plant Reproduction Involves Two Kinds of Spores, as well as Pollen and Ovules

Gymnosperms and angiosperms exhibit many differences in the way that they reproduce. Because they are both seed plants, though, they also share

Table 19.1 Differences between conifers and angiosperms

	Conifers	Angiosperms
Leaf shape	Thin, flat needles Most leaves are evergreen Some species are deciduous	Most have broad, flat leaves Most leaves are deciduous Some species are evergreen
Colors	Mostly shades of green	Flowers and even some leaves are a wide array of colors
Woody tissue of xylem	Contains only tracheids	Contains tracheids and vessel elements
Reproductive structures	Cones with parts arranged in spirals	Flowers with parts arranged in rings
Protection of ovule and seeds	Seeds incompletely enclosed in protective tissues	Seeds completely enclosed in protective tissues
Pollinated by	Wind	Wind, water, animals

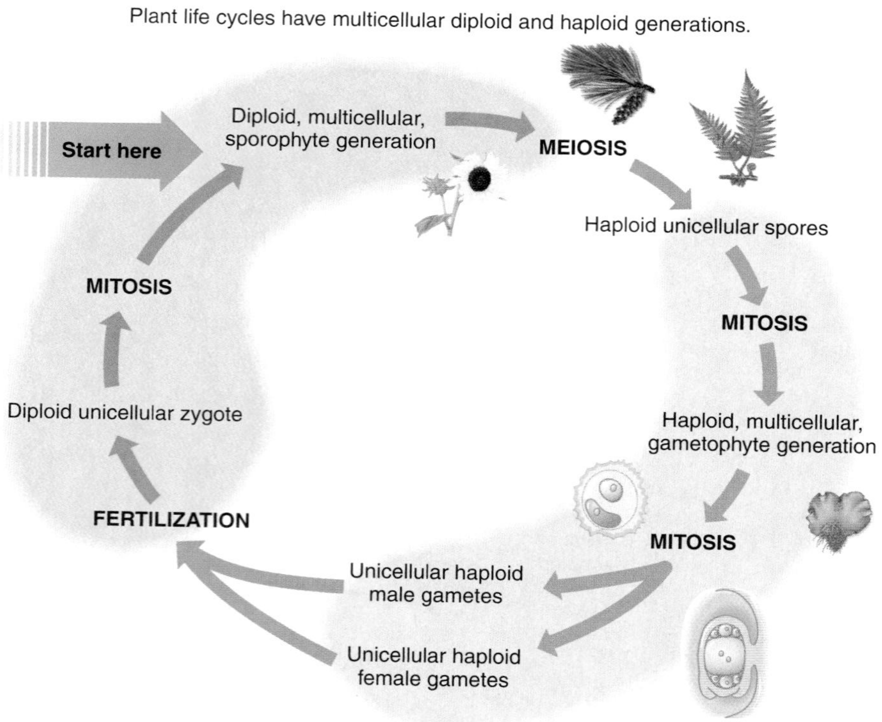

Plant life cycles have multicellular diploid and haploid generations.

Figure 19.1 **Generalized Life Cycle of a Plant.** Plants have alternating haploid (blue) and diploid (pink) generations.

microspore a spore in a reproductive structure that will develop into a pollen grain

pollen grain an immature male microgametophyte encased within a tough coating

ovule a reproductive structure that contains the megagametophyte generation, including the egg cell

megaspore a spore in an ovule that will develop into a megagametophyte that will give rise to an egg cell

fertilization union of sperm and egg

pollination transfer of pollen from the structures that produced it to structures where it will germinate

pollen tube a tube that grows out of a pollen grain; sperm travel through this tube to reach the egg

some reproductive features. Within their reproductive structures the sporophytes of gymnosperms and angiosperms have two kinds of cells that produce spores. **Microspores** are located within pollen-producing structures. They divide by mitosis to produce pollen grains, a plant adaptation that allows seed plants to reproduce without environmental water. A **pollen grain** has a tough coating and contains an immature male *microgametophyte* that at maturity will produce the male gamete(s). **Ovules** contain **megaspores** that divide by mitosis to produce the *megagametophytes*. Mitosis of a megagametophyte produces the female gamete or egg cell.

Like other multicellular organisms, reproduction in seed plants depends on **fertilization**. In the process of fertilization one sperm unites with the egg to form the zygote that will develop into the plant embryo. Seed plants use the process of **pollination** to begin the process that will accomplish fertilization. Pollination is the transfer of pollen from the structures that produce it to structures where it will germinate. When a pollen grain germinates, a **pollen tube** grows out of it and begins to extend toward an ovule. Sperm cells that were within the pollen grain travel within the pollen tube, eventually entering the ovule where sperm are released. Inside the ovule one sperm combines with one egg,

and the development of a new seed plant begins. With this introduction to the overall process of reproduction in seed plants, you are ready to compare reproduction in conifers with reproduction in angiosperms.

> **QUICK CHECK**
>
> 1. Make a diagram that demonstrates the concept of alternation of generations in the life cycle of a seed plant.
> 2. What are the reproductive structures that conifers and angiosperms share?

19.2 Cones and Flowers Are the Prime Reproductive Structures of Seed Plants

One of the ways to distinguish between conifers and angiosperms is by the structures in which the gametophyte generation develops. The reproductive structures of some gymnosperms are clustered within prominent **cones,** and so pines and their relatives are called *conifers.* Because conifers are a familiar form of gymnosperms, the remainder of this chapter will use conifers as an example of gymnosperm reproduction. In contrast, **flowers** are the specialized reproductive structures of angiosperms, which eventually will develop into fruits that enclose seeds. Here you will explore how cones and flowers function in plant reproduction.

Reproduction in Conifers Involves Seed Cones and Pollen Cones

If you've had the opportunity to watch a conifer over the course of a year, you may have noticed that cones of many species of conifers change size and orientation and eventually may fall from the branches. Although these changes vary with species, they reflect different milestones in the life cycle of a pine tree. Let's take a closer look at a pine branch and examine the cones.

Like many other organisms that reproduce sexually, pine trees have male and female gametes, and the cones that produce them are different. The cones that produce female gametes are called **seed cones** because seeds of pine trees develop within them, while the cones that produce male gametes in the form of pollen are called **pollen cones.** A seed cone is made of many flat, woody scales (**Figure 19.2**). If you remove a scale from an old cone and examine it, you will see the faint impression of a pair of winged seeds that once grew above

Figure 19.2 **A Look at Pinecones That Produce Seeds.**

■ *Why do you think a single pinecone has so many seeds?*

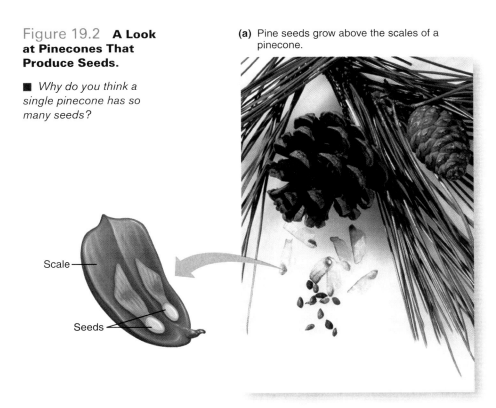

(a) Pine seeds grow above the scales of a pinecone.

Scale

Seeds

(b) A young cone sliced down the middle shows where seeds are located.

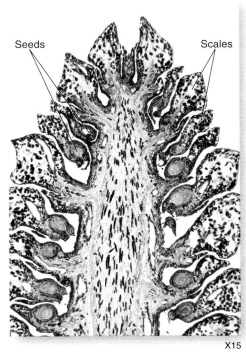

Seeds

Scales

X15

the scale (Figure 19.2a). If you look more closely, you even may see a few papery-winged seeds that are still lodged between the scales of the seed cone. When you slice a young seed cone down the middle, you can see how the growing seeds are arranged within it (Figure 19.2b).

Pollen cones are smaller than seed cones, and unlike seed cones they usually are located at the tips of branches. Pollen cones often look like small bunches of scaly worms that are covered with yellow dust (**Figure 19.3a**). Each speck of yellow dust is a pollen grain. If you shake a branch that has

cone a specialized reproductive structure of a conifer that produces pollen or seeds

flower a specialized reproductive structure of an angiosperm that produces pollen and/or ovules enclosed within a fruit

seed cone a reproductive structure of a conifer that produces ovules

pollen cone a reproductive structure of a conifer that produces pollen grains

(a) Pollen cones look like bunches of scaley worms.

(b) A pollen cone cut lengthwise shows where pollen is located.

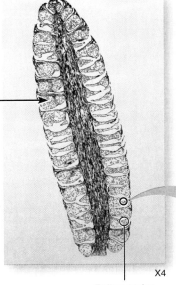

Pollen grains

X4

(c) A scanning electron micrograph of pine pollen.

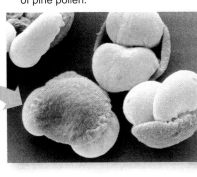

Figure 19.3 **Pollen Cones.** (**a**) Pollen cones occur in bunches, generally at or near the tips of branches. (**b**) A lengthwise section of an immature pollen cone shows that the chambers between its scales are packed with pollen grains. (**c**) A colored scanning electron micrograph of pollen reveals the wing-like bladders that help pollen grains to float on air. Magnification: x2010.

ripe pollen cones, you will release a cloud of dusty, yellow pollen (**Figure 19.4**). In the spring, wind blows clouds of pollen from pollen cones, and some of it will land on sticky surfaces of the ovules developing inside the seed cones. Within each ovule of a seed cone is a megagametophyte that contains two to five egg cells. After it has been fertilized, an ovule will develop into a seed like those in Figure 19.2b.

It may surprise you to learn that in seed plants pollination and fertilization are separate events. In many pines as much as a *year and a half* can elapse between pollination and fertilization. In pollination a pollen grain lands on an ovule, caught in a drop of sticky sap. The drop soon is withdrawn into the ovule, and it draws the pollen in with it. The pollen grain grows a pollen tube that conveys the sperm cells to the egg that is within an ovule

(**Figure 19.5**). In some species of pines the growth of the pollen tube is a slow process that can take up to 15 months. The pollen tube bursts when it reaches the ovule, releasing the sperm cells. Fertilization occurs when one sperm fuses with one egg. The zygote divides mitotically and develops into a multicellular embryo enclosed within a small, winged seed that has a thin, papery covering. A supply of food surrounds the embryo. The food supply and the seed coat develop from different parts of the ovule. This seed will be just one of many that are enclosed within the seed cone. A few months later, in the autumn of the second year after pollination, seeds are mature and the seed cones now also become mature. In many species the seed cones that were upright to better capture pollen now are pointing downward. When the scales of the seed cones open, it is easier for seeds to be shaken loose by the wind or carried away by animals.

Some pines differ from the generalized pine life cycle and have interesting adaptations that influence how they reproduce. Perhaps some of the most notable adaptations are those of the pitch pine (*Pinus rigida*). In northern populations of this species seed cones do not change orientation and do not open until they have passed through the intense heat of a forest fire. Although this might seem as though it would limit the reproductive capacity of pitch pines, actually it ensures a fresh stand of pitch pines even if the parent trees have perished in a forest fire.

Pine trees and other conifers represent an important transition in the evolution of plants, but from an evolutionary perspective the angiosperms

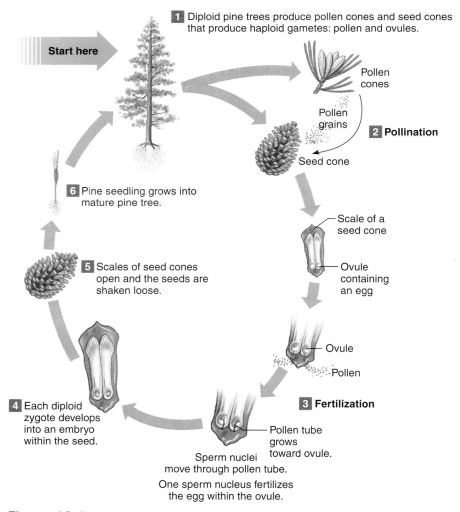

1 Diploid pine trees produce pollen cones and seed cones that produce haploid gametes: pollen and ovules.

Start here

Pollen cones

Pollen grains

2 **Pollination**

Seed cone

6 Pine seedling grows into mature pine tree.

Scale of a seed cone

Ovule containing an egg

5 Scales of seed cones open and the seeds are shaken loose.

Ovule

Pollen

3 **Fertilization**

4 Each diploid zygote develops into an embryo within the seed.

Pollen tube grows toward ovule.

Sperm nuclei move through pollen tube.

One sperm nucleus fertilizes the egg within the ovule.

Figure 19.4 **Pine Tree Life Cycle.** Like other land plants, pine tree life cycles include both diploid and haploid generations. Pollen and ovules are haploid. The fertilized ovule is the beginning of the diploid life cycle stage.

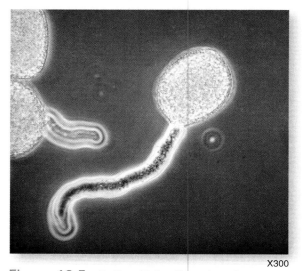

X300

Figure 19.5 **Pollen Tube Growing Out of a Pollen Grain.**

are the most successful of the land plants. From a human perspective flowers are arguably the loveliest of all plant adaptations, and it is easy to forget that flowers are angiosperm reproductive structures. Technically, a flower is a part of an angiosperm plant that produces pollen or ovules, or both. Flowers also are adapted to enhance plant reproduction. This discussion refines the concepts of sporophyte and gametophyte, but be prepared: flowering plants have complexities in their life cycles that will be new to you. At first it can be confusing, so allow plenty of time to work through the life cycle of an angiosperm.

The Life Cycle of an Angiosperm Features Flowers, Pollen, and Seeds

Although you may think of petals when you imagine a flower, a flower is a complex structure that serves many reproductive functions. Some flower parts produce gametophytes that, in turn, produce gametes. Other parts of a flower facilitate pollination, foster fertilization, or develop seeds. Flowers of some species have both male and female reproductive structures in a single flower, other species have separate male and female flowers, and still others have separate male and female plants. Flowers are made of modified leaves arranged in concentric rings. The outermost ring is made of several **sepals** that enclose and protect the flower's inner structures when they are developing (**Figure 19.6**). **Petals** are a second ring of modified leaves. Their primary function is to be a visual signal to attract pollinators. Petals often have fragrances that also advertise the flower's presence to

insects and other pollinators. The next ring of flower parts are **stamens,** the structures that produce pollen. Each stamen has a slender *filament* that holds *anthers* aloft. The innermost ring of reproductive structures within a flower is collectively called the **carpel,** and it consists of three parts: ovary, style, and stigma. In some flowers carpels are single structures, but in most flowers several carpels are fused together. A carpel actually is a modified leaf that produces ovules. Ovules contain the female gametophytes and are located within the ovary. The *style* extends up and out of the **ovary,** terminating in a specially shaped, flattened top called the *stigma*. Some botanists use the term **pistil** as an alternative to carpel, and it can refer to a single carpel or to a compound carpel.

Eggs are produced inside the carpel (**Figure 19.7a**). The first haploid cells that arise from meiotic cell division in the ovule of the carpel are called *megaspores*. Mitosis of megaspores produces the small multicellular megagametophye. The megagametophyte is considered a haploid plant, even though it has only seven cells and no plant organs. Unlike the life cycle stages of moss and fern, the gametophyte generations of seed plants never form independent plants. Instead, they are embedded within the tissues of the sporophyte generation—essentially one plant living within another, a little like nested Russian dolls. Within the ovary megaspores in the ovules undergo mitosis to produce more haploid cells. Some of these will become egg cells, while others are destined to become part of **endosperm,** a nutrient-rich tissue that provides food within a seed for the embryo plant.

sepal the outermost ring of modified leaves of a flower; sepals enclose and protect a flower bud

petal ring of modified leaves of a flower that gives a visual or olfactory signal to pollinators

stamen a pollen-producing male organ of a flower; it has a supporting filament and anther

carpel the innermost ring of modified leaves of a flower; carpels are the ovule-producing female structures of a flower, including the ovary and its contents, as well as style and stigma; can be a simple or a compound structure

ovary the portion of a carpel where ovules that contain eggs develop

pistil the portion of a flower that contains female reproductive parts; a simple pistil is made up of an ovary and its contents, and a style, and a stigma

endosperm a nutrient-rich tissue within an angiosperm seed that nourishes the plant embryo

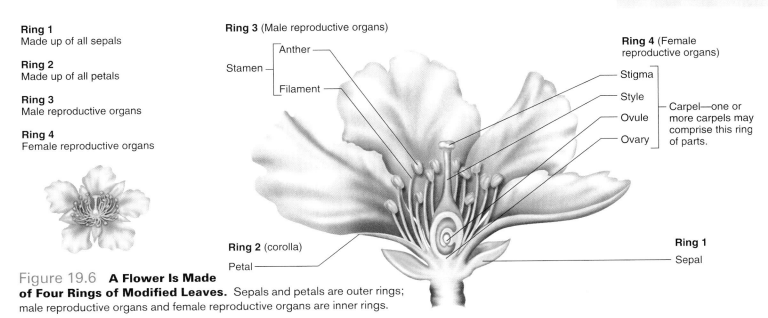

Ring 1
Made up of all sepals

Ring 2
Made up of all petals

Ring 3
Male reproductive organs

Ring 4
Female reproductive organs

Ring 3 (Male reproductive organs)

Anther

Stamen

Filament

Ring 4 (Female reproductive organs)

Stigma

Style

Ovule

Ovary

Carpel—one or more carpels may comprise this ring of parts.

Ring 2 (corolla)

Petal

Ring 1

Sepal

Figure 19.6 **A Flower Is Made of Four Rings of Modified Leaves.** Sepals and petals are outer rings; male reproductive organs and female reproductive organs are inner rings.

The flower is part of the mature sporophyte generation.

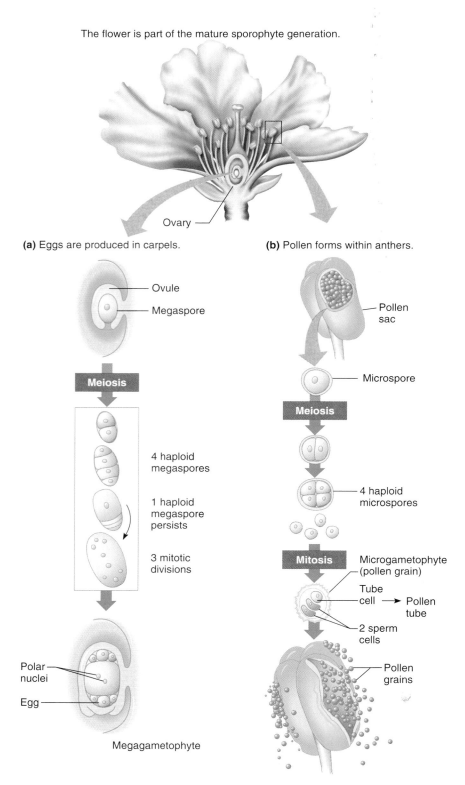

Ovary

(a) Eggs are produced in carpels.

Ovule
Megaspore

Meiosis

4 haploid megaspores

1 haploid megaspore persists

3 mitotic divisions

Polar nuclei

Egg

Megagametophyte

(b) Pollen forms within anthers.

Pollen sac

Microspore

Meiosis

4 haploid microspores

Mitosis

Microgametophyte (pollen grain)

Tube cell → Pollen tube

2 sperm cells

Pollen grains

sperm cells the pair of cells within a pollen grain; one will fertilize the egg to form the diploid zygote, while the other fertilizes the polar cells to form polyploid endosperm

Figure 19.7 **Eggs and Pollen Are Produced Within Carpels and Anthers.** (**a**) A megagametophyte containing an egg results from meiosis of a megaspore mother cell followed by several rounds of mitosis. (**b**) Microgametophytes or pollen grains result from meiosis of a microspore mother cell followed by mitosis.

Figure 19.7b illustrates how pollen forms within the anthers. Specialized parent cells within the anthers undergo meiosis to produce haploid male spores called microspores. The microspores undergo mitosis and become the microgametophytes of the plant. Meiosis in one parent cell produces four haploid daughter cells, the microspores. Each microspore divides once by mitosis to produce two daughter cells. One daughter cell divides again to produce two gametes, while the other daughter cell will produce a structure called a *pollen tube* that is involved in fertilization. These three cells are encased in a tough coating to produce a pollen grain. Angiosperms typically have two or three cells in a pollen grain. Even though it has only a few cells, the pollen grain is a haploid *microgametophyte* "plant" that exists within a flower. The entire pollen grain is not a gamete, but is an immature microgametophyte that carries the cell destined to divide, producing sperm cells. Next, let's consider what happens after pollination in greater depth.

In Angiosperms Double Fertilization Produces a Diploid Plant Embryo, as well as Tissues That Nourish the Young Plant

Once a pollen grain lands on the flower's stigma it germinates, and a pollen tube grows through the pollen grain's coat (**Figure 19.8**). Nourished by sugars on the stigma and in the style, the pollen tube grows longer and penetrates the tissue of the style. The pollen tube uses enzymes to digest its way down toward the ovary, which holds the ovules. Depending on the species of angiosperm, a pollen tube can grow amazingly quickly—as much as 0.04 to 0.12 inches (1 to 3 millimeters) an hour. So even if the flower's style is extremely long, such as the 3.9- to 7.8-inch-long (10- to 20-centimeter-long) style of corn silks, after only about 24 hours the pollen tube will have grown long enough to reach into the sac that holds the ovules. Once the pollen tube enters the sac, its tip bursts open to release the two sperm cells. The two **sperm cells** are the angiosperm's male gametes, and they travel to different places within the megagametophyte *embryo sac,* which holds the ovules. Fertilization ensues, but in angiosperms fertilization is different from that of most gymnosperms or other plants because it involves *two* separate fertilization events. It is called *double fertilization* because while one sperm nucleus fuses with the *egg cell* to form the diploid zygote, the other sperm nucleus travels to the middle of the megagametophyte to fuse with two *polar cells* located there. This forms an *endosperm nucleus* that most often is triploid but can have other chromosome numbers too. The endosperm nucleus divides mitotically to produce endosperm, a nutrient-rich tissue. Endosperm not only provides the food for the germinating seed, but also it provides much of the food that humans and other animals eat. So think about endosperm the next time you take a handful of popcorn, eat a portion of fried rice, or watch a horse munching oats, or a goldfinch eating seeds at a backyard bird feeder.

QUICK CHECK

1. Explain how pollination differs from fertilization. Illustrate your answer with events in the life of a pine tree.
2. List ways that pollination and fertilization of angiosperms and conifers differ.

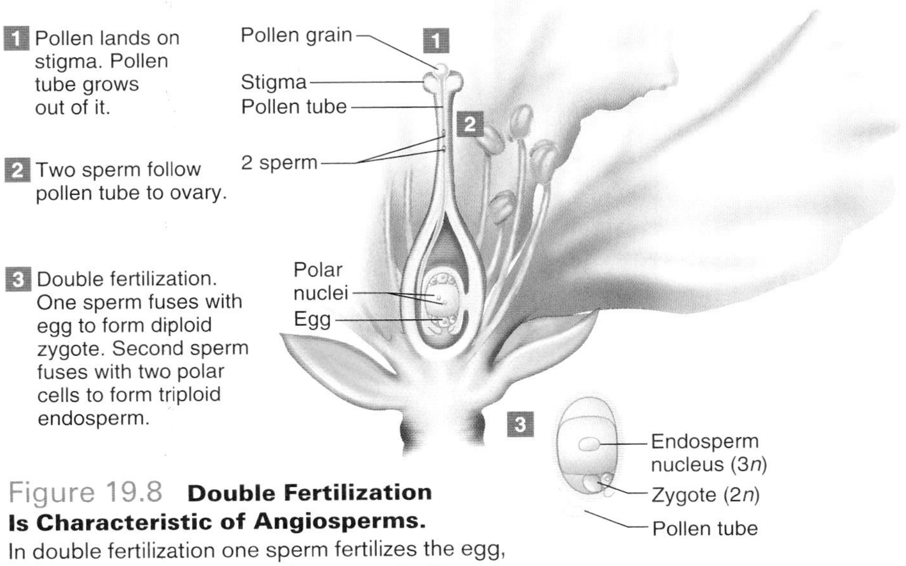

1 Pollen lands on stigma. Pollen tube grows out of it.

2 Two sperm follow pollen tube to ovary.

3 Double fertilization. One sperm fuses with egg to form diploid zygote. Second sperm fuses with two polar cells to form triploid endosperm.

Pollen grain
Stigma
Pollen tube
2 sperm

Polar nuclei
Egg

Endosperm nucleus (3*n*)
Zygote (2*n*)
Pollen tube

Figure 19.8 **Double Fertilization Is Characteristic of Angiosperms.** In double fertilization one sperm fertilizes the egg, forming a zygote, while another sperm fertilizes the polar nuclei, forming endosperm.

19.3 Plants Use Different Signals and Rewards to Lure Pollinators

Let's consider how flowers of different species have evolved to take advantage of the pollinating agents available in their environments.

Plants That Use Wind Pollination Have Tiny Flowers

Recognizing wind-pollinated angiosperms is easy because their flowers are quite small. **Figure 19.9** shows the reproductive structures of some wind-

pollinated plants. The pollen-producing structures of wind-pollinated flowers may be small or long and dangling like catkins or the tassels of corn plants. Ovule-producing structures of wind-pollinated plants often have sticky, expanded stigmas that may look like turrets, puffs, brushes, or feathers. Although you need a magnifying glass to see them well, the flowers of grasses show many interesting adaptations to wind pollination. Take a moment to examine the grass flowers in Figure 19.9. How many different adaptations to wind pollination

(a) (b) (c)

X12 X1.4

Figure 19.9 **Wind-pollinated Flowers.** (a) Female flowers of maple trees have large stigmas to catch pollen. (b) Anthers of delicate grass flowers scatter ripe pollen. (c) Each corn silk is part of a pistil that catches pollen.

can you see in these flowers? Wind pollination often occurs early in spring, before leaves of deciduous trees obstruct free movement of drifting pollen and before insects have become active for the year.

Flowers Lure Animal Pollinators

Plants pollinated by animals attract them with color, scent, and food rewards. Color and scent act like advertising to alert pollinators that a reward is available. Some flowers offer pollen as a reward, and some insects, particularly beetles, eat pollen. Other flowers offer sugary nectar as a reward, but the nectar often is located deep within the flower, in structures that are specially adapted to feeding habits of a narrow range of pollinators. Once you learn a few general patterns, you will be able to examine a flower and predict how it is pollinated. For instance, flowers with heavy fragrances attract and tend to be pollinated by beetles, insects that generally have a highly developed sense of smell and rely on smell more than on vision (**Figure 19.10**). Accordingly, flowers pollinated by beetles usually are white, tan, or dull colored, and their fragrances vary from fruity and spicy to slightly alcoholic—like fermenting fruits. Beetles eat pollen, petals, or nutritious clusters of cells within flowers. All of these are rewards that encourage beetles to visit another similar flower—ensuring cross-pollination. Lured by fragrance, the beetle enters a flower, finds and eats a tasty snack, and accidentally brushes against strategically positioned anthers that daub it with pollen. The hungry beetle then follows the scent to another, similar flower and transfers pollen as it explores the second flower.

In contrast, flowers pollinated by bees have bright colors, usually blue or yellow. Many bee-pollinated flowers have patterns in ultraviolet that bees can see but are invisible to humans. These flowers often have a landing platform and conspicuous stripes or dots that mark the way to a source of sweet nectar. Bees have long tongues, and the nectar of flowers they pollinate usually is within deep pockets, well out of the reach of smaller insects or insects with shorter tongues.

Moths usually are active at night when vision is not as useful as other senses, so flowers pollinated by moths generally have heavy fragrances that carry over long distances. Moth-pollinated flowers usually are white or pale colored and shaped like trumpets (Figure 19.10). They generally open late in the afternoon or only at night when their pollinators are active. Moths hover at flowers and sip nectar without bothering to land, so there seldom are landing pads on moth-pollinated flowers. Butterflies, on the other hand, are active during the day and can distinguish many colors. Accordingly, flowers pollinated

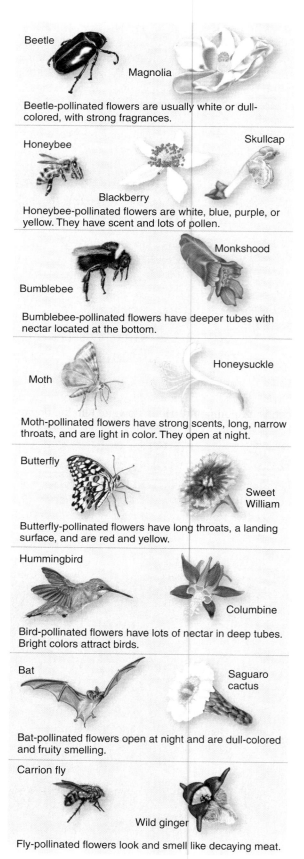

Beetle-pollinated flowers are usually white or dull-colored, with strong fragrances.

Honeybee-pollinated flowers are white, blue, purple, or yellow. They have scent and lots of pollen.

Bumblebee-pollinated flowers have deeper tubes with nectar located at the bottom.

Moth-pollinated flowers have strong scents, long, narrow throats, and are light in color. They open at night.

Butterfly-pollinated flowers have long throats, a landing surface, and are red and yellow.

Bird-pollinated flowers have lots of nectar in deep tubes. Bright colors attract birds.

Bat-pollinated flowers open at night and are dull-colored and fruity smelling.

Fly-pollinated flowers look and smell like decaying meat.

Figure 19.10 **Flowers Are Specialized to Attract Their Pollinators.**

by butterflies are highly colorful. Like moths, butterflies have extremely long tongues that they use to probe the deep, tubular nectaries of their flowers. Butterflies typically land before they begin to feed, and flowers they pollinate often have landing platforms.

Hummingbirds and sunbirds are the chief birds that pollinate flowers, and bird-pollinated flowers tend to be red or red and yellow (Figure 19.10). Because birds have a poorly developed sense of smell, flowers they pollinate do not need to expend the extra energy to produce fragrance. Bird-pollinated flowers usually have tubular shapes that the birds probe with their thin beaks; ample supplies of nectar are rewards for their pollinators.

Except for color, bat-pollinated flowers are similar to bird-pollinated flowers. As you might predict, however, bat-pollinated flowers open only at night, and are pale or dull colored (Figure 19.10). While bats are color-blind, they do have excellent senses of smell. Bat-pollinated flowers usually have strong, fruity odors and often give sweet rewards to their pollinators. Bats are not the only mammals to pollinate flowers. Lemurs, monkeys, honey-possums, mice, rats, and even raccoons all pollinate various flowers. Finally, flowers pollinated by flies often smell like putrid meat; some fly-pollinated flowers look like rotting meat too. Now that you have this information, examine the flowers in **Figure 19.11** and predict how each is pollinated.

QUICK CHECK

1. How does having a flower with a fragrance or sweet nectar increase the chances that a plant will be pollinated?

(a) **(b)**

Figure 19.11 **Predict How These Flowers Are Pollinated.** (**a**) Skunk cabbage has a strong, skunky scent, (**b**) Flowers of coral honeysuckle are brilliantly colored.

19.4 Pollen and Seeds Are Plant Adaptations to Reproduction Without Water

Pollen and seeds have allowed seed plants to become fully adapted to life on dry land. Let's take a closer look at pollen and seeds.

Pollen is produced within a flower's anthers, and anthers split or open and release the pollen. **Figure 19.12** shows high-magnification pictures of pollen grains; you can see that their small surfaces

(a) **(b)** **(c)**

Leg of a bee — Pollen basket of a bee — Pollen grains

Pollen grains

X500

Figure 19.12 **Pollen and Pollinator.**
(**a**) Notice the yellow pollen baskets on the hindlegs of this honeybee.
(**b**) In the higher magnification of a colored scanning electron micrograph, the pollen grains caught within the dense bristles of the bee's pollen baskets are visible. (**c**) A higher magnification scanning electron micrograph of the pollen from a pollen basket. The pollen of each species has a characteristic size, shape, and texture.

(a) Corn seedling

Seed with endosperm remains beneath soil.

(b) Bean seedling

First leaves

Seed leaves with endosperm unfurl aboveground.

Figure 19.13 **Endosperm Nourishes a Plant Embryo As it Grows Roots and Begins Photosynthesis.** The endosperm of a corn seedling is within the corn kernel (**a**), while the endosperm of a bean seedling (**b**) is in its first pair of thick seed leaves.

have intricate patterns. Here is something amazing: each *species* of seed plant has a uniquely sculptural pollen grain that only can pollinate plants of its own species. As the pollen grain matures, the microspore nucleus divides mitotically to produce a large vegetative cell that will grow a pollen tube and a smaller generative cell that will divide to produce two sperm cells, the male gametes. Pollen grains are adapted to be picked up and carried by the wind, water, or animals (Figure 19.12).

Seeds ensure that the plant embryo develops only when conditions are right. Sometimes it is too hot or dry for a seedling to survive. Sometimes it is too cold to support the growth of a seedling. A seed develops from a fertilized ovule and contains the plant embryo that develops from the fertilized egg. Some seeds also contain a source of food to nourish the embryo until it can grow roots and begin photosynthesis (**Figure 19.13**). The seed is enveloped in a protective coat that helps ensure the survival of the embryo until conditions are right for the seed to sprout. Many seed coats are sensitive to water and open when the seed lands in a suitably moist environment. Other seeds have such tough coats that they can pass through the digestive systems of animals without harm, and some seeds must have this treatment to soften their seed coats and allow germination. Seeds also have stores of abscisic acid, the plant hormone that fosters dormancy, and this delays germination of a seed until conditions are right.

Specializations of Seeds and Fruits Help Plants to Disperse into the Environment

Seeds come in a huge variety of forms that are specialized for dispersal by wind, water, or animals. Seed mortality is incredibly high, and most seeds land in uncongenial places, are eaten by animals, or fall prey to fungi. So seed dispersal is a chancy proposition. Why do plants disperse their seeds?

Think of what will happen to a parent plant that does *not* disperse its seeds away from itself: in

a short time they will grow into a thick patch of young plants that may choke out the parent as all vie for the same resources. In contrast, if some seeds are dispersed to different locales where they can grow and thrive, the parent and its progeny may survive. So it is to a parent plant's advantage to produce plenty of seeds and scatter them into the environment.

Depending on the species, the ovary wall can grow thick like the flesh of a peach, cantaloupe, or watermelon or thin and tough like a peapod. These are only a few examples of the many varieties of fruits (**Table 19.2**). Seeds remain within fruits until fruits are ripe and ready for dispersal. Dry fruits often have special mechanisms for dispersal. For instance, maple keys have papery wings that allow them to remain aloft and sail away from the parent plant, fruits of cockleburs have hooks that easily become tangled in fur of animals (**Figure 19.14**). When the animal grooms itself in a new location and removes the seeds, dispersal has been accomplished. The engineer that developed Velcro tape with its interlocking layers of hooks and loops was inspired by the design of a cocklebur. Fleshy fruits change from green to various colors, providing a visual advertisement of ripeness. Under the influence of plant hormones the taste, texture, and sugar content of fruits also become more attractive as they ripen and grow ready for dispersal. Animals eat the ripe fruit, and the seeds often pass through their digestive systems intact. In the time that the

Figure 19.14 **Cockleburs Get Tangled in Animal Fur and Are Dispersed.**

■ *Some plants spread their seeds by wind, others like cockleburs use animals. Which type of seed dispersal do you think would spread the seeds the greatest distance. Explain your answer.*

animal has been digesting the fruit, it may have traveled far from the parent plant, and the unharmed seed often is deposited with a nice bit of natural fertilizer—the animal's feces.

Seeds of plants that line waterways or beaches usually are contained in a fruit that has trapped pockets of air, and so these seeds float. The best-known water-dispersed seed is probably the coconut. A coconut fruit is about the size of a football, while the coconut seed that it contains is about the size of a softball. A coconut can travel on ocean currents for up to 3,000 miles, which explains why shores of tropical oceans are lined with graceful coconut palms (**Figure 19.15**).

Wind-dispersed fruits are tiny and come equipped with various kinds of parachutes and airfoils that help them remain aloft as they are blown away from the parent plant. While the smallest wind-borne seeds are the dustlike seeds of orchids, dandelion fruits are a familiar example of a wind-borne fruit with a parachute (**Figure 19.16**). On a dry day they can sail for miles, and if they happen to land in a favorable environment they may sprout. The odds do not favor survival of tiny wind-borne fruits, though, and plants that use wind for fruit and seed dispersal often compensate by producing huge numbers of tiny fruits that saturate the local environment. Tumbleweeds are plants with wind-dispersed seeds that take a slightly different strategy: the whole plant detaches from the ground and seeds are scattered as it rolls along.

Seeds of Conifers and Flowering Plants Are Much Different

The seeds of flowering plants and of conifers are much different in the extent of the coat of protective tissue that surrounds an ovule at the time of fertilization. In conifers and many other gymnosperms the tissues incompletely cover the seed, leading to the name gymnosperm, which means "naked seed." Although some gymnosperm seeds become enclosed in protective tissues as they mature, these seeds are never as enclosed as are seeds within fruits of angiosperms. The name angiosperm means "seed vessel," referring to the plant tissues that completely enclose the ovule at the time of fertilization.

There is another point of comparison between conifer and angiosperm seeds: the length of time that it takes for a seed to develop. Conifers have a much slower reproductive cycle: there can be up to a year and a half between pollination and fertilization, and seeds of some conifers can take as long as three years to mature. In contrast, some angiosperms can

Table 19.2 Types of fruits

Fleshy fruits	Characteristics	Examples	
Simple	One or more united carpels		
Drupe	Fleshy fruit; hard pit; one or more carpels with single seed each	Olive Cherry Peach Plum Coconut	
Berry	One or more carpels, each with many seeds; inner part fleshy	Grape Tomato Pepper Eggplant	
Pome	Tough core is derived from carpel walls; pulp derived from ovary wall and enlarged receptacle	Apple Pear	
Complex	Multiple separate carpels		
Aggregate	Derives from one flower with many separate carpels	Blackberry Strawberry Raspberry Magnolia	
Multiple	Develops from tight clusters of flowers whose ovaries fuse as the fruit develops	Pineapple	

Dry fruits	Characteristics	Examples	
Dehiscent	Mature fruit splits and releases seeds	Bean Pea Radish Milkweed	
Indehiscent	Mature fruit remains around seed(s)		
	Hard, thick pericarp	Hickory Acorn Chestnut	
	Thin pericarp	Parsley Carrot Maple Sunflower	

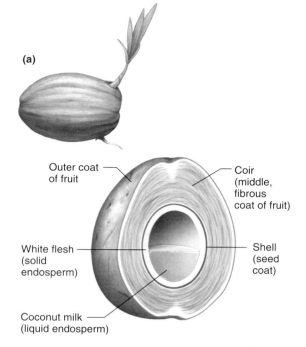

Figure 19.15
Floating Seed. (a) The thick fibrous coats of a coconut trap air and keep the seed afloat as it disperses from its parent. (b) High above this tropical beach, coconut seeds mature and grow ready for dispersal.

(a)

Outer coat of fruit

Coir (middle, fibrous coat of fruit)

White flesh (solid endosperm)

Shell (seed coat)

Coconut milk (liquid endosperm)

(b)

(a)

(b)

Figure 19.16   **Wind-Borne Fruits.**
(a) Dandelion fruits sail away from the parent plant.
(b) Maple keys twirl on the wind.

■ *What is an advantage of wind dispersal over animal dispersal? What is a disadvantage?*

go from germinating seed to mature flowering plant to the next generation of ripe seeds in just weeks. The more rapid generation time of angiosperms has contributed to their enormous success as Earth's dominant land plants.

vegetative propagation a form of asexual reproduction that allows the growth of a whole new plant from a portion of a parent plant

19.5 Angiosperms Also Use Asexual Reproduction

Although sexual reproduction is critically important for the long-term survival of plant species, angiosperms have other asexual ways to reproduce the diploid organism. The various asexual ways that flowering plants can reproduce are called **vegetative propagation.** In some species a whole new plant can grow from a small part of an adult plant. In most species this type of asexual production is the result of cell division and cell differentiation within meristem tissues. For instance, new potato plants are grown from the "eyes" that sprout when potatoes are left in storage for too long. Potatoes are modified stems, even though they grow underground. The cells in potatoes are specialized for storing starch, but like other stems they also have an apical meristem and many lateral bud meristems. These tissues are the eyes, and under certain conditions each eye can grow and develop into a complete mature potato plant.

You also may know that gardeners grow whole plants from cuttings of leaves, roots, or stems. Cuttings produce new plant organs when they are placed in soil, sand, or water (**Figure 19.17a**). Some plants like crabgrass and strawberries send out *runners,* horizontal stems that run aboveground and send up new plantlets. Some of the most invasive and aggressive plants like mints and wisteria use *rhizomes,* horizontal stems that are located belowground, to grow into thick patches that can quickly monopolize an area. Plants like kalanchoe have a different asexual strategy: they grow complete tiny plantlets along the rims of their leaves (Figure 19.17b). These drop off to root and grow into whole new plants. Roots of many plants such as cherries, apples, and blackberries send up *suckers,* sprouts that grow into new plants. A final form of vegetative asexual reproduction is *fragmentation,* where parts of an adult plant break off; if they end up at a favorable site, the fragments grow roots and establish a new plant.

Asexual reproduction quickly produces a clone of the parent plant, which is a tremendous advantage when that plant is growing in a favorable environment. Because the clone has the exact traits of the parent, the new plants grown from a runner, rhizome, sucker, or fragment also will be well adapted to this particular site and will have a good chance of surviving and reproducing. In contrast, plants produced by sexual reproduction vary genetically, and some may not be so well adapted to the current environment, even though they may be well adapted to any environmental change. In addition, most seeds and seedlings do not survive to become adult plants, so during sexual reproduction the plant expends a lot of energy to produce few offspring. Of course, the downside of asexual reproduction is that the colony of plants has limited genetic variability, and if the habitat changes, the colony may die out.

Some Plants Reproduce Sexually with Themselves, in a Process Called Selfing

Jewelweed is a plant that has a slightly different reproductive strategy: it hedges its bets with two kinds of flowers. Jewelweed (*Impatiens capensis*) has small, tubular orange flowers that tremble on thin stems like jeweled, dangling earrings (**Figure 19.18**). Jewelweed's showy flowers are shaped to attract specialized pollinators like hummingbirds and various large insects, whose long tongues can reach down to the nectar within each flower. But

(a)   **(b)**

New plantlets

Figure 19.17 **Vegetative Propagation.** (**a**) Cuttings of coleus quickly grow roots when submerged in water. (**b**) Kalanchoe grows tiny plantlets along the rims of leaves. Dropped from the parent, each may grow into a new plant.

■ *The plantlets of kalanchoe are all clones of the parent. What is a disadvantage of this type of asexual reproduction?*

Figure 19.18 **Jewelweed Flowers.** These showy flowers are pollinated by insects. In case the insects fail to arrive, the plant also has inconspicuous flowers that are self-pollinated.

these are not the only kinds of flowers that jewel-weeds have. Tiny flowers grow in the angles where leaves or lower branches join the stem. These inconspicuous flowers fertilize themselves and grow seeds within a capsule. Although there is genetic reshuffling within these flowers, seeds produced from them do not incorporate an influx of new genetic material that is typical of seeds produced by cross-pollinated flowers. The tiny self-fertilized flowers provide insurance that if animal pollinators do not visit the plants, at least some seeds will be produced to ensure the next generation.

> **QUICK CHECK**
>
> 1. Describe three examples of ways that plants reproduce asexually.
> 2. What is meant by the term selfing in plants?

19.6 An Added Dimension Explained
The Intertwined Lives of Plants and Their Pollinators

From an evolutionary perspective perfection is an elusive, unattainable state. Darwin's principles of evolution, and the evidence that supports them, lead only to the conclusion that some individuals in a population will be better adapted than others to survive in a given environment. Those individuals with better adaptations will outreproduce others in the population that lack these advantages. Over time the more favorable adaptations will become common within the population. Yet the casual observer, and even the seasoned naturalist, can be struck by what seems to be perfection in nature; many have been puzzled by how such perfection could have evolved. Some of the most enigmatic examples of apparent perfection are flowers and their pollinators. As you have read, many species of plants are pollinated only by certain species of pollinators, and the flowers and pollinators seem to be perfectly matched. The orchids and orchid bees described in the beginning of this chapter are just one example.

Early in the study of evolution researchers suspected that these were cases of **coevolution:** two species reciprocally influencing one another's evolution. Scientists see evidence for coevolution in cases where different populations of the same two species show local adaptations that are matched.

This idea of coevolution is highly appealing and provides an explanation for many relationships in nature that otherwise seem too perfect. But, of course, coevolution is proposed as a scientific hypothesis, and so it must be tested. Until recently, evidence for coevolution was circumstantial. Even with fossil evidence it was difficult to determine if two species had experienced a series of mutual changes over generations. More recently, the science of molecular genetics has provided a new way to look at coevolution, and one of the best-studied examples is a flower and its pollinator.

Natural selection works because individuals in a population vary. New variations are produced mostly by changes in DNA such as mutation or gene duplication. Sexual repro-duction can reshuffle the set of genes an individual inherits, but truly new proteins rely in large part on changes in nucleotide sequences of individual genes. So one way to assess the evolutionary "distance" between two species is to determine how different their DNA is. Combined with information about how often DNA mutations occur, differences in DNA also can tell you how long ago these differences arose. Using molecular information about DNA sequences, you can diagram an evolutionary tree that estimates how long ago the different species evolved. These kinds of analyses provide a way to test the hypothesis of coevolution. If two species have coevolved, there should be evidence for it in their phylogenetic trees. Let's see how this logic applies to species of fig trees and their pollinators, species of tiny wasps (**Figure 19.19**).

Have you ever eaten a fig? It is a fruit that matures from an inside-out flower and **Figure 19.20** helps you to see this. Figure 19.20 shows a slice through a fig. You can see that the internal structures are actually little flowers all pointed inward. Figs

coevolution the mutual evolutionary influences that pairs of species have on one another

Female wasp natural size

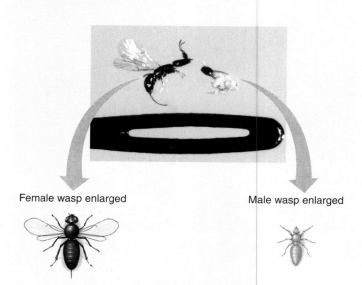
Female wasp enlarged   Male wasp enlarged

Figure 19.19   **Fig Wasps.** These fig wasps are enlarged so you can see them. The photo compares female and male fig wasps to the eye of a needle.

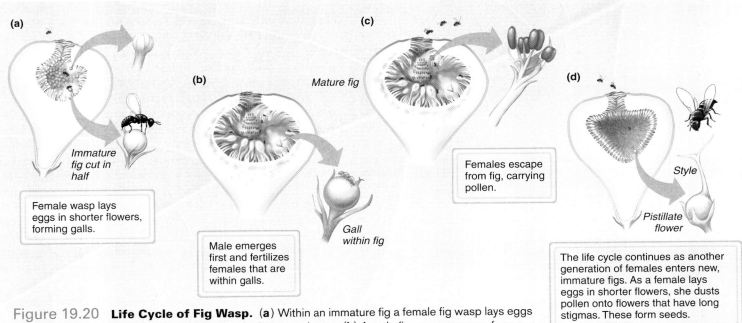

Figure 19.20 **Life Cycle of Fig Wasp.** (**a**) Within an immature fig a female fig wasp lays eggs in shorter flowers. Galls form around the developing wasp larvae. (**b**) A male fig wasp emerges from a gall and fertilizes females that have not yet hatched from galls. (**c**) A female escapes from the fig, carrying pollen from the longer flowers. (**d**) A female enters another immature fig. As she lays eggs she pollinates fig flowers.

grow in warm regions where they may flower all year. To produce seeds and make more fig trees, flowers of most species must be cross-pollinated. Rather than rely on random visits from diverse insects, fig trees rely solely on tiny wasps as pollinators. Each species of fig is pollinated by one or two species of wasps. In fact, the figs seem to go out of their way to provide a haven for these wasps, and the wasps, in turn, often seem to purposefully pick up fig pollen and carry it to another plant.

The whole process is rather astounding. To start, consider a female fig wasp that is carrying eggs. In looking for the proper place to lay her eggs, she heads for just one particular species of fig tree. The fig has a small but unusually shaped hole at one end, and the wasp's head and body fit into this hole. Other insects, even other species of fig wasps, have a harder time getting in. Our heroine slips in easily and begins the process of laying her eggs (Figure 19.20).

Within the fig are some flowers that produce pollen and other flowers that will produce seeds. The wasp ignores the flowers that produce only pollen and heads for the seed-producing flowers. These have a deep cavity that is just the right size and shape for her eggs. She begins laying eggs in these tiny flowers, but she does not lay eggs in all of them. Some of the flowers are shorter than others, and she lays eggs only in the shorter ones. Her job done, she usually dies shortly after laying her eggs. So far, this story is only about the advantage of the wasp-fig relationship to the wasp, and this apparent advantage continues for a while. The short flowers provide a safe place for young wasps to develop, and after about three

months new wasps begin to emerge from the tiny galls that have grown around them as they have passed through their larval and pupal stages. The males emerge first, and they are strange insects, indeed. They have no wings, small weak bodies, and large genital organs. Their primary job is to fertilize the female wasps that are still within galls. Females emerge from their galls already pregnant. Depending on the species, the males either die right after mating is completed, or before they die they perform one last task: digging a hole that will allow females to fly out of the fig.

Now the benefit for the fig begins to reveal itself. The female wasps move around a bit inside the fig. In some species females brush up against anthers and collect a dusting of pollen. In other species the female wasps actively collect pollen, loading up with it before escaping through the hole in the fig. Off each female goes, pregnant and searching for a new fig where she will lay her eggs. To get a better understanding of this coevolution story, follow the female into the fig again, but now you know that she carries pollen. As she moves around finding the short flowers to lay her eggs in, she pollinates the long flowers, which grow fig seeds that eventually produce new fig plants.

As you might expect in a biological system, though, the coevolution of figs and their pollinators is not perfect. There is a tension between the natural selection pressure on the fig plant to make more plants and on the wasps to make more wasps. So there is some variation in the system. Fig plants are sometimes

—Continued next page

Continued—

pollinated by other wasp species, and wasps do not always find fertile ground in their favorite fig species. Other insect species have evolved to take advantage of the system. There are non-pollinating wasps that sometimes can enter figs, lay eggs, and give nothing back to the fig. Can all of this be explained by coevolution?

The answer is not yet clear. Many studies of the evolution of the life cycles and physical traits of the figs and wasps suggested that there was coevolution. But this is a bit of a circular argument because the physical traits used to test the hypothesis of coevolution are often the ones that the hypothesis is trying to explain. Other studies have used DNA sequences to construct the evolutionary history of figs and their wasps and to compare the two groups. If wasps and fig trees were coevolving, scientists expected to see a one-to-one correspondence in phylogenetic trees of fig wasps and fig trees. In other words, the trees should have the same shape. There are overall similarities in the phylogenetic trees of fig trees and fig wasps, but also there are several instances where the evolutionary histories do not match well. One reason is that, contrary to one hypothesis, some species of fig wasps pollinate more than one kind of fig tree. To make things more complicated, some fig trees are pollinated by several species of fig wasps. More studies will have to be done on this topic, but at least some evidence does not support the neat and tidy story of coevolution of fig trees and pollinator wasps.

You might think that this is just a failure of an idea, but the case study of figs and wasps provides some important lessons in how science moves forward to find out what is really happening in nature. The coevolution hypothesis sounds like a plausible explanation for figs and their pollinating wasps, but in science a hypothesis must do more than sound good—it must be rigorously tested. Even after early studies of physical characteristics seemed to support the coevolution hypothesis, scientists continued to think critically and test the hypothesis further. This deeper digging now has shown problems for the coevolution hypothesis, but even this is unlikely to be the final answer. Scientists will continue to study this system until the true evolutionary history of figs and wasps is understood.

QUICK CHECK

1. What is the advantage to the fig of having its flowers located within the walls of its fruits where most pollinators cannot reach them?

Now You Can **Understand**

Pollen and Allergies

If you suffer from allergies, the very thought of pollen may make your nose start to run. The hay fever and rose fever that plague many people are side effects of pollen grains' small size and distinctive coats that irritate nasal passages and act as allergens. You will read more about allergies in Chapter 25; for now it is enough to know that windblown pollen grains are drawn into nasal passages by happenstance. For instance, in spring you inhale huge quantities of pine pollen, but it usually does not cause allergies. Pollen grains from other kinds of seed plants stimulate nasal passage cells to release histamine, which disrupts cell membranes, causing them to "leak." As a result, you have watering red eyes, and a drippy or a stuffy nose. So you get an allergic reaction to windblown pollens, not to pollens carried by insects.

What Do **You** Think?

Does Talking to Plants Help Them Grow?

Plants occupy an interesting place in our living environment. Sometimes you may not even think of them as alive because they seem so static and unresponsive. Then again, many people attribute humanlike traits to plants. Early researchers, even before Darwin, believed that plants had a nervous system, and you might hear people today talk about plants as if they had feeling or reactions. One popular reflection of this tendency was the fad of talking to plants or playing music to plants to promote their growth. Skeptics scoff at such notions, but advocates swear by them. Does it make sense to you that talking to plants or playing them music might help them grow? How would you demonstrate that such an effect really existed, and how might you explain it based on what you know about plants? What do *you* think about these more unusual aspects of human-plant interactions?

Chapter Review

CHAPTER SUMMARY

19.1 Gymnosperms and Angiosperms Are Seed Plants

As seed plants, gymnosperms and angiosperms have similarities and differences in their reproductive structures. These two groups differ in physical appearance, xylem cells, means of pollination, and amount of tissue that encloses their seeds. Both have alternation of generations that features the alternation of a diploid sporophyte generation with a haploid gametophyte generation. In seed plants sporophytes are larger than the gametophytes that they shelter, nourish, and protect. A seed contains a dormant sporophyte embryo surrounded by a food supply and a protective coat. Pollen-producing structures contain microspores that undergo meiosis to produce microgametophytes that produce pollen. Within each pollen grain is an immature male microgametophyte. Ovule-producing structures contain haploid megaspores. These divide by mitosis to produce the haploid megagametophyte(s). Mitosis of megagametophytes produces the egg cell. Pollination and fertilization are separate events, and a germinating pollen grain grows a pollen tube that allows sperm to travel to the eggs. Fertilization occurs when a sperm fuses with an egg. Pollen grains are a prime adaptation to reproduction without environmental water.

fertilization 458	pollen grain 458
megaspore 458	pollen tube 458
microspore 458	pollination 458
ovule 458	seed 456

19.2 Cones and Flowers Are the Prime Reproductive Structures of Seed Plants

Conifers develop pollen and ovules within pollen cones and seed cones, respectively; angiosperms develop pollen, ovules, and seeds within anthers and carpels. Conifers are wind-pollinated; angiosperms are pollinated by wind, water, insects and other animals. Pollination and fertilization can be lengthy processes in conifers. Flowers are made of concentric rings of modified leaves, including sepals, petals, stamens, and carpels. Meiosis of microspores produces immature microgametophytes within pollen grains. Meiosis of megaspores within ovules produces an egg cell and other cells that after fertilization will form the endosperm that nourishes the embryo plant within a seed. Endosperm is a polyploid nutritious tissue that results from fertilization of polar cells. The other sperm nucleus from a pollen grain fertilizes the egg cell to form the zygote. Most food that humans eat is actually endosperm of cereal grains.

carpel 461	pistil 461
cone 458	pollen cone 458
endosperm 461	seed cone 458
flower 458	sepal 461
ovary 461	sperm cells 462
petal 461	stamen 461

19.3 Plants Use Different Signals and Rewards to Lure Pollinators

Wind-pollinated plants have tiny flowers and may have dangling anthers and expanded stigmas. Animal-pollinated flowers offer food rewards and attract their pollinators with colored petals and fragrances. Flower shapes, petal color, and fragrances are adaptations to their specific pollinators.

19.4 Pollen and Seeds Are Plant Adaptations to Reproduction Without Water

Pollen and seeds are adaptations to reproduction on dry land. Pollen grains are adapted to be carried by pollinating agents. Seeds have adaptations that ensure that the plant embryo develops under the right conditions. Endosperm (if present), adaptations of the protective seed coat, and abscisic acid are all important in the survival of seeds. Plant hormones transform the ovary wall into fruits. Seeds of angiosperms are protected within fruits. Fruits and seeds have many adaptations for different modes of dispersal. Fruits can be distributed by animals, wind, and water. Angiosperm seeds typically have much more protective coverings than do conifer seeds. Angiosperm life cycles can be much quicker than conifer life cycles.

19.5 Angiosperms Also Use Asexual Reproduction

Seed plants use both asexual and sexual reproduction. Cuttings, runners, fragmentation, and development of whole new little plantlets are means of vegetative propagation. Asexual reproduction by runners, rhizomes, or fragments will quickly produce a clone of the parent. Asexual reproduction and sexual reproduction have advantages and disadvantages. Asexual reproduction is quicker but produces a clone of the parent, with restricted or no genetic diversity; sexual reproduction takes longer, it may require the cooperation of animals, and it has the advantage of introducing genetic diversity.

vegetative propagation 468

19.6 An Added Dimension Explained: The Intertwined Lives of Plants and Their Pollinators

In coevolution two species influence one another's evolutionary adaptations. Completely parallel patterns in the evolution of fig trees and the wasps that pollinate them have not been substantiated. At this point it is not clear that wasps and fig trees are coevolving.

coevolution 470

REVIEW QUESTIONS

TRUE or FALSE. If a statement is false, rewrite it to make it true.

1. Anthers form pollen grains that are sporophytes.
2. Pollen and sperm that swim in a film of water have allowed plants to become fully terrestrial.

3. Pollination and fertilization are simultaneous events.

4. Fruits allow seed plants to be spread into new places.

5. Most seeds survive and grow into new plants.

6. Seed cones produce pollen.

7. Flowers of insect-pollinated plants typically have red flowers that open only at night.

8. At fertilization the ovule of an angiosperm is not completely enclosed in tissue.

9. Carpels contain microgametophytes that will develop into seeds.

10. In seed plants the products of meiosis are microspores or megaspores.

MULTIPLE CHOICE. Choose the best answer of those provided.

11. You see a plant that has tiny, pale flowers. With a magnifying glass you notice that its anthers are long and dangling and tremble at the slightest movement of air. It is probably pollinated by
    a. bats. d. wind.
    b. monkeys. e. butterflies.
    c. bees.

12. You pull up a plant that has a long, horizontal underground stem. It sends up plantlets at regular intervals. This is a
    a. rhizome. d. spore.
    b. runner. e. sucker.
    c. gametophyte.

13. Vegetative propagation
    a. is typical of the gametophyte generation.
    b. creates a clone.
    c. is a form of sexual reproduction.
    d. is a slow process.
    e. is the same as selfing.

14. Which structure contains the immature microgametophyte of a seed plant?
    a. ovule d. stigma
    b. pollen grain e. style
    c. seed

15. Which part of a flower receives pollen grains?
    a. ovule d. style
    b. seed e. microspores
    c. stigma

MATCHING

16–20. Match the term to its definition (one choice will not match):

16. abscisic acid
17. stigma, style, ovary
18. red fruits
19. endosperm
20. tubular flowers with heavy perfumes

a. polyploid tissue within a seed
b. prevents seeds from germinating
c. attract attention of insects
d. carpel
e. attract attention of birds
f. specializations for pollination by moths

CONNECTING KEY CONCEPTS

1. How do the sporophyte and gametophyte generations compare in angiosperms? How is each generation formed?

2. What aspect of angiosperm reproduction do pollinators facilitate? What adaptations do angiosperms have that encourage pollinators to play this role?

THINKING CRITICALLY

1. Many plants are wind-pollinated, but insects and other animals pollinate many others. What are the advantages and disadvantages of each means of pollination?

2. In tropical rain forests animals including insects, birds, bats, monkeys, and other small mammals are the major pollinators. Why is wind pollination rare or absent?

For additional study tools, visit www.aris.mhhe.com.

Senses, Nerves, Bones, Muscles

A Pill to Treat Your Addiction? Don't Bet the Rent

February 14, 2006

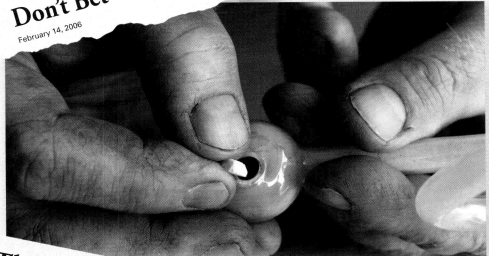

The Next Crack Cocaine?
As Meth Use Grows, Officials Fear Region Is Unprepared to Deal With It

March 19, 2006

A crack addict's nervous
system becomes dramatically
altered by the drug.

"You've got to throw her out," Jody cried. Her eyes darkened with fear as she pleaded with the counselor at the homeless shelter where she spent winter nights. Her friend could *not* stay at the shelter. She had broken the shelter's rules by doing crack cocaine.

The counselor felt sorry for Jody's friend. "It's so cold tonight. Let her stay. We can deal with this tomorrow."

But Jody insisted her friend must leave. Jody knew the pull of cocaine and the abyss that awaited her if she too gave in. Jody's voice was shrill with hysteria: "You've got to tell her to leave. If *she* can do crack here—in this shelter—how can *I* resist?"

And so the counselor complied and enforced the rules. Jody's friend gathered her belongings and disappeared into the frigid night.

Waiting for sleep, Jody felt no remorse. "I can stay clean now," she thinks.

Despite Jody's good intentions, staying clean was an elusive dream. After a few weeks she was back on the streets with her old friends, once again doing anything to get money for drugs. In a filthy, smelly room Jody began the ritual that had become so familiar (**see chapter-opening photo**). She carefully set up the cardboard box and arranged her utensils precisely. Careful not to drop a bit of rock, she loaded her pipe, held it at the perfect angle, and lit it.

"Hurry up, Jody," her companion urged, but Jody ignored him. She waited, just for a moment, anticipating what would come next. Then she inhaled, drawing the smoke deeply into her lungs. When she could take no more, she offered the pipe to her companion, but before the pipe left her hand, Jody's mind exploded in a shower of fireworks. Every fiber of her body seemed to open and then contract with pleasure. The feeling ebbed, waned, and faded, but it left warm embers behind. Jody became active, happy, and—unrealistically—confident that her future would be bright. Her bleak life faded in her mind as cocaine took over her perceptions. She would finish school. There was only a

year left. Then she'd get a good job and get a nice place for her son Mitchell and herself. She'd show her mother that she could take good care of Mitch. They'd be a real family again. Jody felt so happy that she could hardly sit still. Everything was going to be great. Just great.

Within hours those good feelings evaporated, but there was more crack. Jody had another hit, another amazing explosion of pleasure, and another brief burst of optimism. Each hit, though, was less powerful than the last. Finally, after hours of hitting, going high, coming down, and going high again, the cocaine turned nasty. The last hit produced anxiety, ugly fears, and anger. To cure the bad feelings, Jody turned to another old friend—bourbon—to wash everything away and bring sleep. After sharing a bottle with her companion, Jody unsteadily crawled away and collapsed. Her drug binge was over—at least for the moment—and even when Jody woke, she had no desire to hit again. "I'll never do that again," she promised herself. "It's no good. I'm going to get better. God help me. I don't want to live like this."

Even so, Jody's resolutions were not strong enough to fight her cravings. The longing for drugs would grow until she could no longer resist it. And then she'd once again return to some filthy room, seeking another intimate meeting with the pipe and the crack and the explosions in her brain.

Jody's story is fiction, but hundreds of thousands—perhaps millions—of people lead similar lives. Cocaine is one of the most destructive of addictive drugs, but other drugs carry similar risks. Use of cocaine has waned somewhat over the past few years, but use of methamphetamine—a very similar drug—is rising. People who abuse drugs come from all walks of life and are not much different from you or me or most people we know. Addicts know that it is so easy to be lured into the trap of drugs and that it is so hard to escape from it. Why are drug addicts caught in this dreadful cycle? Why can't they quit and stay drug free? Some answers can be found in the biology of the nervous system, and by the end of this chapter you will understand more fully how cocaine can so easily come to rule—and to ruin—a human life. Let's begin by exploring how the nervous system controls what you think, feel, and do. ∎

20.1 Sense Organs, Nervous System, Muscles, and Bones Coordinate Interactions with the Environment

Living things sense and respond to their environments. A bacterial cell detects glucose molecules and moves toward them; tree roots detect water and grow toward it; a rabbit catches the scent of fox and darts off through the grass. In vertebrates, behavioral reactions usually involve at least four specific body systems: the senses, the nervous system, muscles, and bones. Sense organs detect conditions in the environment or inside the body. They can detect objects by reflected light, emitted sounds, odors, tastes, or textures. The nervous system receives and analyzes sensory information. It decides what external or internal events need action and sends instructions to specific body systems to carry out the appropriate responses. Skeletal muscles contract or relax in response to signals from the nervous system. Skeletal muscles are attached to bones that provide a supportive framework for the body. Bones move when muscles relax or contract. Together, muscles and bones make the whole organism move through the environment.

To help you grasp what these systems accomplish, consider what happens if you oversleep, and yet manage to get to class on time. Your sense organs, nervous system, muscles, and bones interact in a coordinated reaction. From time to time in the sections ahead, you will return to this scenario and deepen your understanding of how these four systems work together.

QUICK CHECK

1. What are the four body systems that coordinate your response to the environment and what does each do?

20.2 Senses and Brains Have Prokaryotic Origins

In many ways the nervous and sensory systems of animals are the most complex and sophisticated organ systems. Yet, complex nervous and sensory systems had their start billions of years ago in membrane proteins of prokaryotes and single-celled eukaryotes. Let's look at two examples of proteins that give hints of how the complex nervous and visual systems of animals may have evolved.

Eukaryotic cells use membrane ion channels to communicate with one another and with their environments (Section 4.3). Ion channels are proteins, and when an ion channel is in an open state, a particular ion will flow across the cell membrane, in accord with the laws of diffusion. When protein channels are in a closed state, ions cannot flow across the membrane. In eukaryotes there are dozens of different kinds of protein ion channels, and each has its own subtle characteristics. The diversity of ion channels is one reason that animal nervous systems can analyze complex information and carry out so many diverse tasks.

Even though prokaryotes seem so much simpler than eukaryotes, they also communicate with their environments, and they use ion channels to do so. Prokaryotes have potassium channels that are probably the ancestor of many or all ion channels. One kind of potassium channel in prokaryotes opens when the cell membrane is mechanically stretched or deformed. Proteins that respond to mechanical stimulation are called *mechanoreceptors*. For instance, if the prokaryote bumps into an obstacle, the ion channel will open and potassium

(a) A prokaryote with membrane receptors that are protein channels for potassium closed.

(b) The prokaryote bumps into an object, such as a strand of algae. Its cell membrane is deformed. This stimulus causes protein channels for potassium to open.

(c) Potassium enters the cell ...

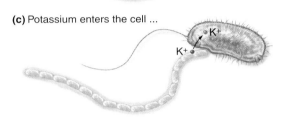

(d) ...and a flagellum rotates. This moves the prokaryote away from the algae.

Figure 20.1 **Membrane Receptors of Prokaryotes That Respond to Mechanical Stimulation.** When a membrane receptor is physically deformed (**a & b**), potassium channels open (**c**), and potassium enters the cell. This allows a flagellum to rotate (**d**) and the prokaryote moves away from the object.

ions will flow into the cell (**Figure 20.1**). The increase in internal potassium might cause flagella to beat, moving the cell away from the obstacle. Other kinds of membrane potassium channels in prokaryotes and single-celled eukaryotes respond to electrical signals, temperature changes, or other environmental variables. These simple membrane response systems were the foundation for more complex nervous systems.

Animals have evolved accessory structures like hairs, bristles, or filaments connected to the membrane mechanoreceptive potassium channels. These extend the reach of the mechanoreceptor.

Hairs and bristles extend the reach of a fly's mechanoreceptors.

X120

Figure 20.2 **Accessory Structures Extend the Reach of Mechanoreceptors.** Flies have an array of bristles on their feet that are connected to mechanoreceptors.

The fine hairs on your skin, the hairs covering the body of a fly (**Figure 20.2**), and the triggers of the stinging cells of a jellyfish are other examples of mechanoreceptors.

The eyes of animals also can be traced back to prokaryotic origins. Animal eyes use a membrane molecule called *rhodopsin,* which captures light energy. Rhodopsin evolved from *bacteriorhodopsin,* a molecule found in bacteria that captures light in the process of photosynthesis. Prokaryotes also use bacteriorhodopsin to detect light levels so they can move toward or away from a light source. Over many generations rhodopsin became associated with particular cell types in eukaryotes. For example, ciliated cells in the larvae of some jellyfish contain rhodopsin. These cells act as a rudder and change the direction that the larva swims in response to light. So in a sense the larva has an eye but no nervous system. Later in the history of life light-responsive cells connected to a nervous system.

Like sensory systems, nervous systems became increasingly large and complex as new species emerged. The radially symmetrical cnidarians have a simple, diffuse nervous system called a **nerve net** (see Section 16.4) that probably evolved early in the evolution of animal life. A nerve net is a large set of neurons connected together, but there is no region that has more nerve cells or more neural activity than any other. Later, bilateral animals such as flatworms evolved. Their nervous systems have many nerve cells clustered together in groups of neurons called **ganglia.** Within ganglia neurons can communicate with one another efficiently and can take on specialized functions. Most invertebrates have nervous systems composed of many ganglia (**Figure 20.3**), and even parts of the mammalian nervous system are organized as ganglia. In some animals a new organization evolved in which a large number of nerve cells are clustered in the head region, in a structure called a brain. Within the brain any cluster of neurons with a similar function is called a *brain nucleus.* Note that this is a different use of the word *nucleus* than the nucleus in an atom or the nucleus in a cell. Each brain nucleus contributes to a specific function. For example, the brain has auditory nuclei, motor nuclei, and nuclei that control the release of hormones. You will learn a great deal more about the

Figure 20.3 **An Invertebrate Nervous System.** Invertebrate nervous systems are made of several small groups of neurons, or ganglia. The nervous system of the sea hare, *Aplysia,* has ten ganglia (**a&b**) connected to nerve fibers that penetrate the body.

(a)

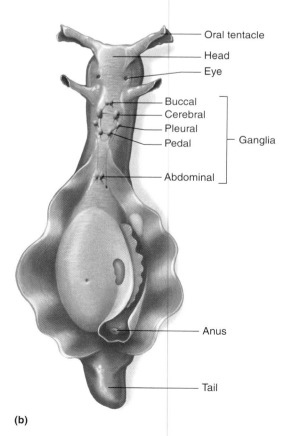

Oral tentacle
Head
Eye
Buccal
Cerebral
Pleural
Pedal
Ganglia
Abdominal
Anus
Tail

(b)

brain shortly. The main point here is that the complex nervous system of mammals can be traced back through evolutionary history to a simpler beginning.

The elegant muscle systems of animals also can be traced back to simpler systems in the earliest single-celled eukaryotes. *Actin* and *myosin* are the primary proteins involved in muscle contraction. These proteins interact to produce gliding movements in diatoms and other single-celled eukaryotes, and so muscle proteins evolved early in the history of life. The skeletal system also evolved from earlier, simpler, versions.

Now that you understand some of the evolutionary roots of sense organs, the nervous system, and the musculoskeletal system, let's look at sense organs in more detail.

QUICK CHECK

1. Describe two membrane proteins in prokaryotes that were the evolutionary ancestors of complex functions in animals.
2. How is the nervous system in a bilateral animal different from a nerve net?

20.3 Sensory Receptors Detect Environmental Conditions

Most vertebrates share the same basic set of sensory systems—usually vision, hearing, touch, temperature, balance, smell, and taste. In each sensory system cells called **sensory receptors** are uniquely adapted to respond to, and so detect, a specific aspect of the environment. Sensory receptors convert information about environmental conditions into signals that can be transmitted to nerve cells. *Sensory organs* such as eyes, ears, or skin contain sensory receptors and other accessory cell types that contribute to the detection of environmental conditions. A word of caution: do not confuse sensory receptor cells with the protein receptors lodged in cell membranes (**Figure 20.4**). Both kinds of receptors detect information, but they are not identical. Membrane protein receptors are protein molecules that are embedded in the cell membrane of a single cell. In contrast, sensory receptors are entire cells or groups of cells.

The Retina Detects and Analyzes Light

Visual information comes in the form of light that is reflected off objects in the environment. Light is

(a) Sensory receptors are entire cells or groups of cells.

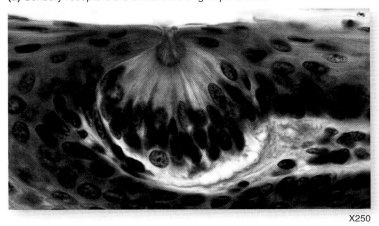

X250

(b) Membrane receptors are protein molecules in the cell membrane of one cell.

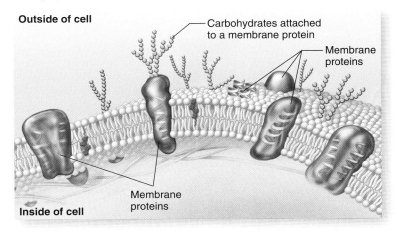

Figure 20.4 Sensory Receptors. (a) A taste bud is a group of many sensory receptors gathered together in a single structure. **(b)** Membrane receptors are proteins embedded in the phospholipid bilayer of a cell membrane.

detected and analyzed by the *visual system*, which includes the eye and certain nuclei and regions within the brain (**Figure 20.5a**). Light from an object in the environment enters the eye through the *pupil*, which is an opening in the muscles that form the *iris*. Pigments in the iris determine eye color. The iris controls the size of the pupil and therefore controls how much light enters the eye (Figure 20.5b). The *cornea* is a transparent covering on the outside of the eye; the lens is a flexible structure that sits just behind the iris. The retina is a sheet of tissue that contains photoreceptors and several types of neurons. The highly organized array of cells in the retina begins the process of analyzing visual information.

sensory receptor a cell or group of cells specialized to detect a particular kind of environmental information

(a) The visual system detects and analyzes light. It includes the eye and nuclei within the brain.

(b) Structures of the eye allow light to be focused on photoreceptors in the retina that respond and send information to the brain.

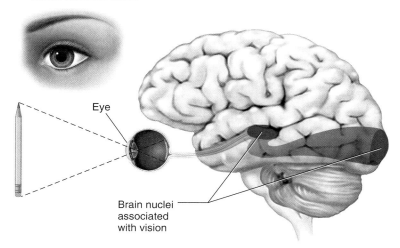

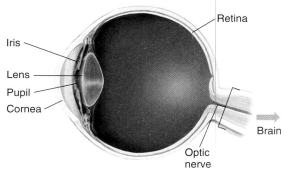

Figure 20.5 **Components of the Visual System.** (a) The eye, optic nerve, neurons, and brain nuclei make up the visual system. (b) Structures of the eye.

rod light-sensitive cell in the retina that detects different levels of light

cone light-sensitive cell in the retina that detects different colors of light

rhodopsin the pigment molecule found in retinal photoreceptors that detects light

cochlea the complex structure in the ear that detects and analyzes sounds

hair cell the specific sensory cell inside the cochlea that detects sounds

Rods and **cones** are the light-sensitive receptor cells of the retina (**Figure 20.6**). Rods respond differently to different levels of light, while three kinds of cones respond to different wavelengths of light and allow the visual system to distinguish different colors. Both rods and cones detect light using a chemical response. Light is actually small packages of energy called *photons.* Light-sensitive molecules in rods and cones can absorb photons in a way similar to the absorption of light by chlorophyll molecules of plants, algae, and bacteria. The light-absorbing molecule in rods is **rhodopsin,** while cones have other, similar molecules. The response of rhodopsin, and of the cone photoreceptor molecules to light are examples of signal transduction (see Section 4.3). When rhodopsin absorbs a photon, the molecule changes its overall charge and shape, and this produces a cascade of chemical changes inside the rod or cone. When you look at an object, such as the alarm clock by your bed, specific rods and cones across the retina are hit by photons and experience this internal chemical cascade. The responses of rods and cones form an image of the alarm clock across the retina.

Next information about the object is sent to the different types of neurons in the retina. Even though the retina is only the first stage in the analysis of visual information, these retinal neurons compare, contrast, and sum the responses of rods and cones (Figure 20.6b). Visual information is transferred from the photoreceptors to these neurons and then specialized neurons, the *retinal ganglion cells,* send information to the brain in a bundle called the *optic nerve.* The optic nerve sends visual information to specific nuclei and centers in the brain where other neurons do further analyses of the visual information. The end result of all this visual information processing is that you see an object in your environment, and in most cases you know enough about it to respond appropriately.

Each Sensory System Has Specialized Receptor Cells

The visual system is only one of many sensory systems that detects and analyzes environmental information. The function of each of these systems depends on specialized sensory receptor cells. **Figure 20.7** shows some of the specialized sensory receptor cells that detect other kinds of environmental information. Sound is the end result of the motion of air molecules. Nearly all kinds of animals have specialized cells that detect sounds, and those of humans and other mammals are found in a complex, coiled structure called the **cochlea,** located deep within the ear (Figure 20.7a). A sound wave that strikes the eardrum is magnified by three small bones within the ear and transmitted to the fluid within the cochlea. Vibrations in the fluid stimulate the **hair cells,** which are sensory receptor cells in the cochlea. Hair cells detect the energy in sound waves and send nerve impulses to the brain that are perceived as sounds.

Human skin is an unusual organ because of its unique role as an interface between the outside environment and the cells, tissues, and organs inside the body. In addition to protecting deeper tissues and organs from infections and other dangers, skin plays a major sensory role. All vertebrates have basically the same set of skin sensory receptors, whether the skin is scaly, bare, or heavily covered with feathers, fur, or hair. The major types of skin receptors are specialized to detect temperature, pain, touch, and pressure (Figure 20.7b). Taste is accomplished by groups of sensory receptor cells on the tongue called **taste buds** (Figure 20.7c). The cells of taste buds have specialized proteins that react to many of the complex molecules in your food. The receptor cells in the nose have similar membrane protein receptors, but these are specialized to detect airborne molecules (Figure 20.7d). These may be given off by potential food sources, potential mates, competitors, and predators. The perfume industry's success is based on the reliance of mammals, including humans, on chemical communication. Your cologne, scented aftershave, and shower gels send messages to those around you.

These basic senses have been modified and extended to give some species unusual sensory abilities. For example, honeybees can see wavelengths of ultraviolet light that are beyond the violet light that most animals can see. Many species of fish can detect electrical currents in water, and all known fishes have a sense organ called the *lateral line* that detects water currents. Snakes have heat receptors adapted to detect infrared, and so then can see warm objects in the dark in a way that is similar to night-vision goggles. Some birds and other organisms can even detect magnetic fields. As with so many aspects of life, there are common themes in the evolution of sensory and nervous systems as well as variations on those themes.

How does an animal determine what sensory information means, and how does it decide what to do about it? The answer involves some of the most complex and sophisticated processes in nature, and they are all carried out in the nervous system.

1. What is a sensory receptor cell, and how is it different from a membrane receptor?
2. What is the retina and what kinds of receptor cells does it contain?

(a) The retina is a layer of cells at the rear of the eyeball.

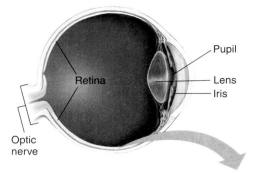

Pupil
Retina
Lens
Iris
Optic nerve

taste bud a multicellular sensory organ that detects molecules in food and produces the sensation of taste

(b) The retina is composed of layers of cells.

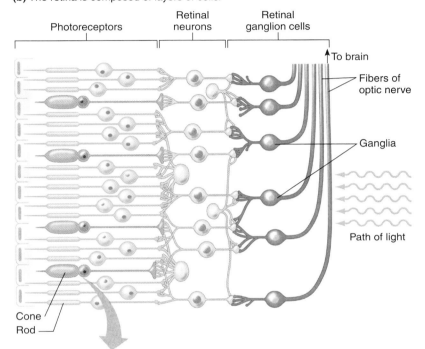

Photoreceptors
Retinal neurons
Retinal ganglion cells
To brain
Fibers of optic nerve
Ganglia
Path of light
Cone
Rod

(c) Rods and cones of the retina (SEM, false color).

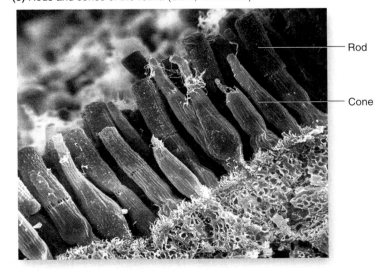

Rod
Cone

Figure 20.6 **Structure of Retina.** In the retina (**a**) photoreceptors, retinal neurons, retinal ganglion cells, and the fibers of the optic nerve are united in complex, organized neural pathways (**b**). These send information from the eye to the brain. (**c**) This scanning electron micrograph shows rods and cones.

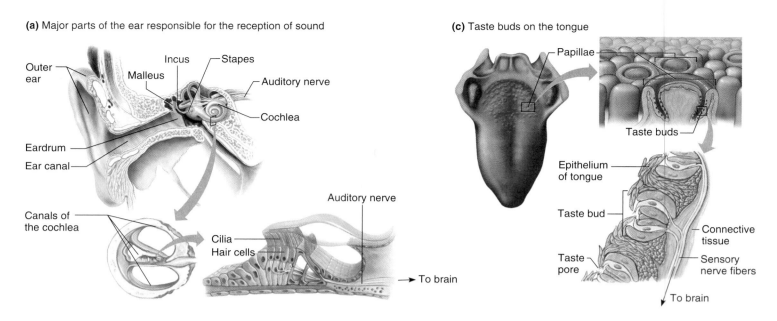

(a) Major parts of the ear responsible for the reception of sound

Outer ear
Incus
Stapes
Malleus
Auditory nerve
Cochlea
Eardrum
Ear canal
Canals of the cochlea
Auditory nerve
Cilia
Hair cells
To brain

(c) Taste buds on the tongue

Papillae
Taste buds
Epithelium of tongue
Taste bud
Connective tissue
Taste pore
Sensory nerve fibers
To brain

(b) Pain, touch, and pressure receptors in the skin

Section of skin
Epidermis
Dermis
Free nerve endings
Epithelial cells
Sensory nerve fiber
Epithelial cells
Touch receptor
Sensory nerve fiber
Pressure receptor
Sensory nerve fiber

(d) Receptors in the nose detect airborne molecules in the sense of smell

Olfactory tract
Olfactory bulb
To brain
Olfactory area of nasal cavity
Nasal cavity
Nerve fibers within the olfactory bulb
To brain
Cilia
Olfactory receptor cells

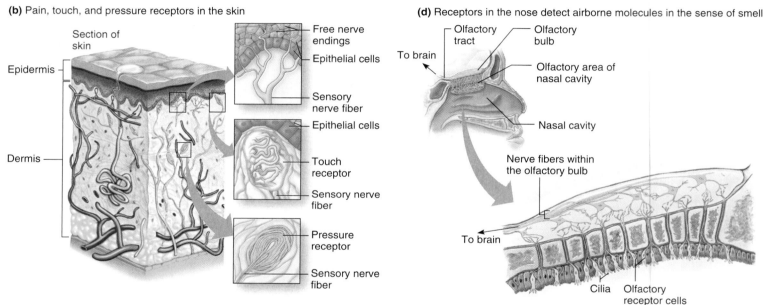

Figure 20.7 **Sense organs.** (a) Hair cells within the cochlea respond to vibrations of the eardrum that are magnified by the three bones of the middle ear and transmitted to the fluid-filled canals of the cochlea. Movements of the cilia of hair cells send nerve impulses to the brain by way of the auditory nerve. (b) Receptors in the skin respond to pain, touch, and pressure. (c) Cells within taste buds respond to chemical stimulation. (d) Olfactory receptor cells respond to molecules carried by air.

neuron the cellular unit of the nervous system

cell body the central portion of a neuron where mitochondria, ribosomes, endoplasmic reticulum, nucleus, and other organelles are located

dendrite an extension of the neuron cell body that usually receives input from other neurons

axon a thin extension that leads away from the cell body of a neuron and carries information to other neurons

20.4 Neurons Are Specialized Cells of the Nervous System

A **neuron** is the basic cellular unit of the nervous system. Neurons carry out the main function of the nervous system: to receive, analyze, and respond to information. Neurons have the same basic features found in all eukaryotic cells, but in neurons some of these features are developed in specialized ways. The central part of each neuron is a spherical region called the **cell body** (**Figure 20.8**). It con-tains the nucleus with DNA, ribosomes, mitochondria, and other familiar organelles that perform all of the usual functions of any cell.

To analyze information, neurons must commu-nicate with one another, and to do this, neurons have specialized structures beyond those found in many other types of cells. Typically, neurons communicate with other neurons, but they also receive informa-tion from sensory receptor cells or send information to muscle cells and other cell types. **Dendrites** are nerve cell structures that usually receive information from other cells, and **axons** usually send information

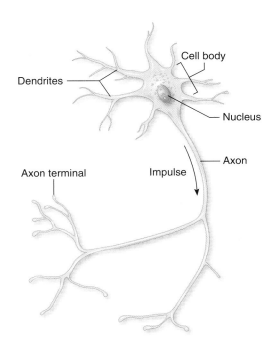

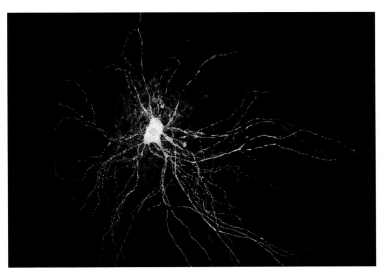

Figure 20.9   **Neurons Have Extraordinarily Complex Shapes.**

Figure 20.8   **Structure of a Neuron.**
Dendrites, cell body, and a single axon are the basic features of a neuron. Dendrites usually receive information from other neurons. The cell body carries out basic cellular functions and contains the nucleus. Axons send information to other neurons.

to other cells (Figure 20.8). The cell body, dendrites, and axon are parts of a single neuron. Let's look at dendrites and axons in more detail.

Dendrites are cellular branches that cluster at one end of a neuron. These branches fork and re-fork; as each extends away from the cell body, it decreases in size until its tips are too delicate to be seen even with a compound microscope. Many branches make the dendrites look rather like a tree. A single dendrite can receive thousands of inputs from other neurons at different places along its branches.

An axon is an extremely long, fine tube that extends from the cell body, often directly opposite from the dendrites (Figure 20.8). An axon can be as short as 100 micrometers or more than 3 meters (9.8 feet) in length. For example, one set of neurons that controls some muscle movements has cell bodies located at the top of the brain, just beneath the highest point on the skull. The tips of the axons of these neurons end halfway down the spinal cord. Like a telephone line that carries information from one phone to another, the function of an axon is to send information to other neurons that can be located a distance from the cell body of the sending cell. Near its target, the axon's tip branches and rebranches many times. These fine branches come

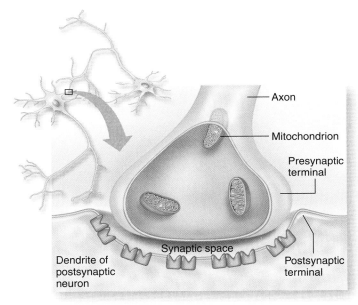

Figure 20.10   **Structure of a Synapse.** A synapse is the connection where one neuron sends information to another. Usually the information is sent from the end of an axon and received by the end of a dendrite. The sending structure is the presynaptic terminal, and the receiving structure is the postsynaptic terminal. A small synaptic space separates the pre- and postsynaptic terminals.

extremely close to the target cells, usually to the dendrites of other neurons, but do not actually touch them. Because of their dendrites and axons, neurons have some of the most complex shapes of any cell type in the body (**Figure 20.9**).

Information is passed from one neuron to another, or from a neuron to its target cell, at a structure called a **synapse** (**Figure 20.10**). A synapse

synapse the small gap between two neurons, usually between an axon terminal and a dendrite, where information is sent from one neuron to another

includes the tiny gap between the two neurons and the bit of cell membrane on either side of the gap. The small piece of membrane on the sending side of the synapse—usually the axon—is called the *presynaptic terminal,* while the small piece of membrane on the receiving side of the synapse—usually the dendrite—is called the *postsynaptic terminal.* Each synapse can carry just one item of information at any one time, but the nervous system contains tens of *billions* of synapses. Each neuron is involved in tens of thousands of synapses, and there are about 9 billion neurons in the nervous system.

Sensory Information Is Analyzed and Acted on by Groups of Cells in the Nervous System

The senses and the nervous system are so tightly entwined that in some cases it is not possible to distinguish the two. For instance, sensory receptors often are not distinct cells, but are the endings of neurons where cell bodies are in the nervous system. For purposes of definition, the *nervous system* includes all the tissue in between the sensory receptors and the muscles, organs, or glands that the nervous system controls, even if the sensory receptors are part of cells in the nervous system.

The major division in the naming of parts of the nervous system is the distinction between tissue located inside versus tissue located outside the skull and spinal column (**Figure 20.11**). The **central nervous system** includes all of the nervous tissue inside the skull and spinal column, and it has two parts. The **brain** is defined as the nervous tissue found inside the skull, while the **spinal cord** is the portion of the central nervous system outside the brain but contained within the vertebral column. The spinal cord contains many groups of neurons. Most of these neurons analyze sensory information from the skin or organize and send signals to the skeletal muscles. Many complex motor patterns are organized as *reflex pathways* in the spinal cord that do not require brain input for execution. For example, in some mammals walking motions can be carried out in response to stimulation of the feet, even without a connection between the spinal cord and the brain. Of course, the brain is required for these reflex pathways to be integrated into overall behavior. The brain exerts control over the neural systems in the spinal cord.

The portion of the nervous system that lies outside of the central nervous system is the **peripheral nervous system.** It is made up of the axons of *motor neurons* that connect directly to muscles and control their contractions. It also con-

central nervous system the parts of the nervous system that are enclosed in the spinal column and the skull

brain the large mass of nervous system tissue that sits atop the spinal cord, inside the skull

spinal cord the long rope of nervous system tissue that runs down the back of a vertebrate, inside the spinal column

peripheral nervous system the parts of the nervous system that are outside of the spinal column and skull

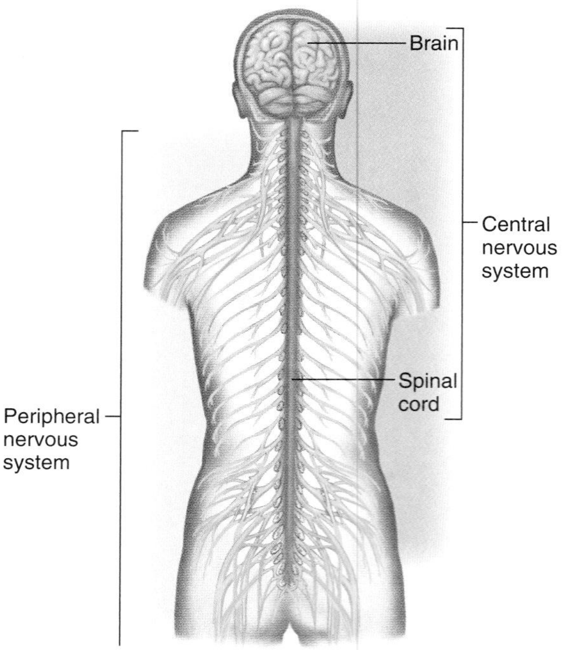

Figure 20.11 **Central and Peripheral Nervous Systems.** The central nervous system is within the skull and vertebral column, while the peripheral nervous system is outside of them.

■ *What is the advantage of having the nervous system divided into central and peripheral systems?*

tains axons of *sensory neurons* that carry sensory information from the skin to the spinal cord, as well as neurons involved in the control of peripheral glands such as the adrenal glands.

The distinction between the peripheral and central nervous systems is an example of the sometimes arbitrary naming scheme for nervous system structures. The "late for class" scenario introduced earlier can help to illustrate this point. When you open your eyes and look at the clock (**Figure 20.12**) retinal photoreceptors form a crude image across the retinas of the light and dark areas on the clock dial. Neurons in the retina do an initial analysis to determine where dark/light or colored borders appear in the overall image. Then axons in the *optic nerves* carry this information to the brain. The optic nerve is made up of all the axons of retinal ganglion cells, whose cell bodies are in the retina. Because the cell bodies of retinal ganglion cells are in the retina, they are part of the peripheral nervous system. The optic nerve is within the skull, so it is part of the brain and central nervous system. Later you will return to follow the path of information about the alarm clock as it flows through different parts of your brain, but for now let's resume a description of the some of the more important brain structures.

(a) When you look at the image on a clock…

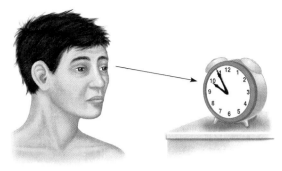

(b) … your photoreceptors form a crude image of light and dark areas on the clock.

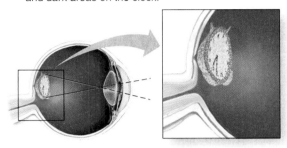

Figure 20.12 **Late for Class Scenario I.** The retina is part of the peripheral nervous system and the optic nerve is part of the central nervous system. The cell bodies of the neurons in the optic nerve are in the retina, and their axons are in the brain.

The Central Nervous System Is Made Up of Groups of Neurons That Are Dedicated to Particular Functions

The central nervous system is organized into groups of cell bodies, or nuclei, connected by long bundles of axons. In the spinal cord most nuclei are involved with direct control of the muscles or with sensory information from the body. Nuclei in the brain are concerned with motor and sensory information, but in addition they monitor other kinds of information such as analyzing blood glucose levels, regulating alertness and mood, and coordinating peripheral hormone levels.

The lowest region of the brain, just above the spinal cord, is called the **brain stem** (**Figure 20.13**). Nuclei in the brain stem control breathing and heart rate, help to determine whether you are asleep or awake, and help decide whether a particular situation is rewarding or punishing. If certain cells in the brain stem are damaged or killed—for instance, by an overdose of alcohol—the person will cease breathing and will die. Other neurons in the brain stem are part of the systems that analyze and respond to pain. The **hypothalamus,** a small neural complex at the base of the brain, regulates body functions such as eating, drinking, body temperature, and the glands that secrete hormones. Further toward the top of the brain, a region called

brain stem the structure at the base of the brain, just above the spinal cord, that governs basic body functions

hypothalamus a cluster of structures on the underside of the brain; helps to regulate functions such as body temperature, eating, drinking, and glandular secretions

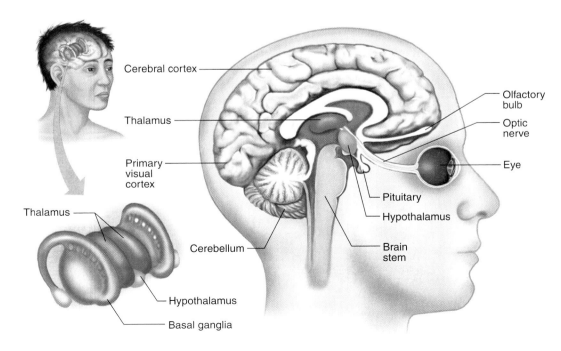

Figure 20.13   **Some of the Important Groups of Neurons, or Ganglia, in the Human Brain.**

the **basal ganglia** is involved in motor behavior and also in the perception of reward. Another important structure involved in the control of movement is the **cerebellum,** which sits over the brain stem like an overgrown walnut. An important structure in the organization of sensory systems is a region of concentrated cell bodies called the **thalamus.** Each group of cell bodies in the thalamus receives information, via axons, from a particular set of sensory receptors. Some regions of the thalamus receive auditory information, others receive tactile information, and still others receive visual information from the retinal ganglion cells. The neurons in the thalamus combine and analyze information and send their axons to another major region of the brain, the **cerebral cortex.** This complex structure carries out the most sophisticated of all information processing in the nervous system, and it contains so many neurons and synapses that the tissue is folded over on itself to fit in the skull. The cerebral cortex is just about all you see of the brain when the skull is removed. If large regions of the cerebral cortex are damaged, the person will lose consciousness and may fall into a coma. The cerebral cortex is more sensitive to drugs and other damage than are regions of the brain stem, which explains why a person can be in a coma but still be able to breathe and have simple motor reflexes.

To return to the "late for class" scenario, where among all these brain structures do you actually "see" the numbers on the clock? At least for the visual sense, scientists know a great deal about how the brain produces perception (**Figure 20.14**). The axons in the optic nerve send information about the overall distribution of light, dark, and color in the visual scene to a special area of the thalamus devoted to vision. Neurons in the thalamus do a similar analysis but over a larger region of visual space than the retinal cells. Axons from the thalamus send information to the *primary visual cortex,*

a region of cortex at the back of the brain. The primary visual cortex analyzes for larger visual features such as bars, long edges, and corners. It is not until the information goes to the visual neurons in the **temporal lobes,** the region of cortex that is under the temples, that more complex aspects of the scene are put together.

> **QUICK CHECK**
>
> 1. What are three basic parts of a neuron?
> 2. What is a synapse and what does it do?
> 3. List and describe four regions of the central nervous system.

20.5 Neurons Communicate Using Chemical and Electrical Signals

Let's look at how neurons communicate at synapses in greater detail. You will discover that neurons use both electrical and chemical signals to communicate. A good place to start is with the signals that travel through an axon.

Recall that an axon is a long-distance cable that allows neurons to send signals to other neurons that are some distance away. Axons accomplish this using an electrical signal called an **action potential.** Axons are found only in the neurons of animals, but action potentials are produced by other animal cells, cells of plants, algae, and single-celled eukaryotes. This is yet another example of the way that evolution adapts one function for a different purpose. For example, the carnivorous Venus's flytrap plant uses an action potential as a mechanism to cause the food-capturing leaves to close around an insect. As you will read later, action potentials cause muscles to contract. All action potentials are based on the flow of ions across cell membranes.

basal ganglia small group of neurons above the brain stem that is involved in motor behavior and perception of reward

cerebellum a large, convoluted structure sitting astride the brain stem that is involved in complex, learned motor behaviors

thalamus a structure roughly in the middle of the brain that is a processing station for sensory information

cerebral cortex a large convoluted structure that covers most of the rest of the brain in mammals; higher functions are governed by the cerebral cortex

temporal lobe a region of the cerebral cortex devoted to complex aspects of vision and to memory

action potential an electrical signal that travels along a neuron

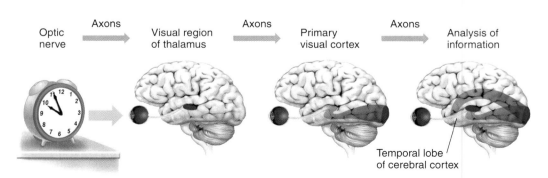

Figure 20.14   **Late for Class Scenario 2.** Visual information flows from the optic nerve to the thalamus to the primary visual cortex, and finally to the temporal lobe of the cerebral cortex. Each of these structures has a unique part to play in the perception of the numbers on the alarm clock.

In an axon an action potential is generated and carried by positively charged ions—sodium and potassium—so the action potential actually is an electrochemical process (**Figure 20.15a–e**). Sodium and potassium ions flow through membrane ion channels. These channels are normally closed, but they open when the voltage across the membrane changes. Normally, the inside of a cell is about 70 millivolts (mV) more negative inside than the outside of the cell membrane. This is similar to saying that the negative end of a battery is 9 volts more negative than the positive end. If the voltage difference between the inside and outside of the membrane becomes a bit less negative, usually due to a signal from another neuron, **voltage-dependent ion channels** in the axon open. Sodium ion

voltage-dependent ion channel a membrane ion channel that opens when the voltage across the membrane changes

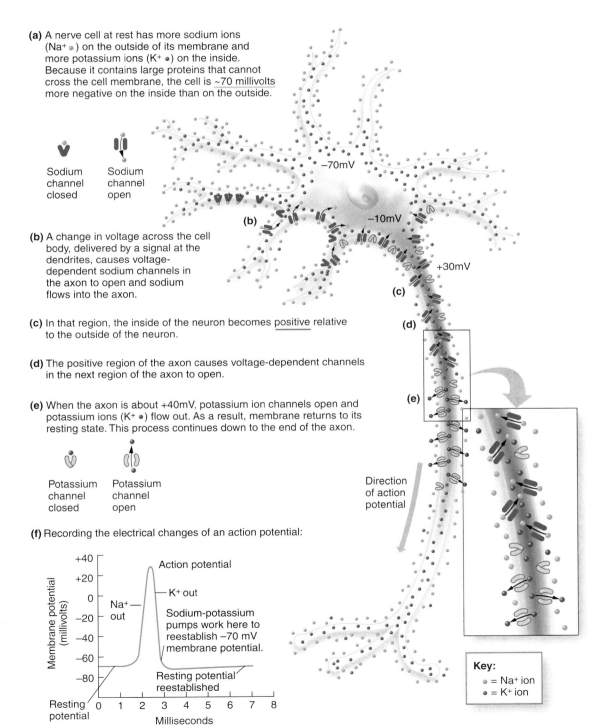

(a) A nerve cell at rest has more sodium ions (Na⁺ •) on the outside of its membrane and more potassium ions (K⁺ •) on the inside. Because it contains large proteins that cannot cross the cell membrane, the cell is ~70 millivolts more negative on the inside than on the outside.

Sodium channel closed   Sodium channel open

(b) A change in voltage across the cell body, delivered by a signal at the dendrites, causes voltage-dependent sodium channels in the axon to open and sodium flows into the axon.

(c) In that region, the inside of the neuron becomes positive relative to the outside of the neuron.

(d) The positive region of the axon causes voltage-dependent channels in the next region of the axon to open.

(e) When the axon is about +40mV, potassium ion channels open and potassium ions (K⁺ •) flow out. As a result, membrane returns to its resting state. This process continues down to the end of the axon.

Potassium channel closed   Potassium channel open

(f) Recording the electrical changes of an action potential:

Membrane potential (millivolts)

+40
+20
0
−20
−40
−60
−80

Action potential
K⁺ out
Na⁺ out
Sodium-potassium pumps work here to reestablish −70 mV membrane potential.
Resting potential reestablished

Resting potential

Milliseconds
0 1 2 3 4 5 6 7 8

−70mV
−10mV
+30mV

Direction of action potential

Key:
• = Na⁺ ion
• = K⁺ ion

Figure 20.15 **An Action Potential Moves Along a Neuron.** (**a**) A resting neuron. (**b–e**) Changes in voltage along the neuron membrane as an action potential moves along a neuron. (**f**) Events in an action potential can be seen as fluctuations of sodium and potassium ions that change the electrical charge of the interior of the neuron.

■ *By what process does sodium cross the membrane of a neuron at the beginning of an action potential?*

channels open first, and, driven by diffusion, sodium ions located on the outside of the membrane rush into the axon. Because sodium ions carry a positive charge, the inside of the cell membrane now becomes even more positively charged relative to the outside of the cell membrane. Once the inside is about +40 mV relative to the outside, the sodium ion channels close and now the potassium ion channels open. Because there is normally more potassium on the inside of the cell than the outside, potassium ions diffuse out of the cell. This returns the cell to being more negative on the inside than on the outside of the membrane. This rapid increase and decrease in the voltage across the membrane is an action potential.

An action potential can be measured by inserting a thin metal or glass electrode into a cell and hooking it up to a recording device such as a computer (Figure 20.15f). The whole action potential lasts only about 1.5 milliseconds (msec). A millisecond is one-thousandth of a second. Axons fire many action potentials in rapid succession. Usually, more action potentials indicate that the stimulus was more intense, while fewer action potentials mean that the stimulus was less intense.

This tells you how a single action potential is generated, but how does the signal travel from the cell body all the way to the tips of the axon terminals? The voltage change of the action potential itself causes the voltage-dependent ion channels in the next section of membrane to open, generating a new action potential there (Figure 20.15d). In this way each action potential at each location on the axon generates an action potential in the area of membrane just ahead, until the end of the neuron membrane has been reached (Figure 20.15e).

How does the membrane get the sodium back out of the axon and the potassium back in? Every cell has many **sodium-potassium pump** proteins in its membrane (see Section 4.2). The sodium-potassium pump uses the energy in ATP to pump sodium out of the cell and potassium into the cell (Figure 20.15f). In a neuron sodium-potassium pumps are constantly active, working to maintain the membrane potential at −70 mV. If the pumps get behind, the cell will not be able to generate another action potential, and this sets a limit on how frequently an axon can fire an action potential.

In most invertebrate animals the axons that carry action potentials are bare, like uncovered electrical wires. Because bare axons lose some of the sodium and potassium across the cell membrane, this slows the conduction of the action potential along the axon. Vertebrates have a useful adaptation that provides insulation for many axons and speeds up the conduction of an action potential. These axons are surrounded by many layers of cell membrane, called **myelin** (**Figure 20.16**). Myelin is like the plastic insulation on copper wires. It helps to prevent the loss of ions across the axon membrane. Myelin allows axons to conduct signals at a much faster rate. The myelin is not continuous all along the axon but is interrupted every millimeter or so by *nodes*, small stretches of axon that have no myelin. The action potential travels nearly instantaneously inside the axon between the nodes, and at each node it is regenerated by the flow of sodium and potassium ions. Many nervous system diseases are due to degeneration of myelin. Multiple sclerosis is one of these. When myelin coverings of even a small percentage of axons degenerate, the signals that control muscles are not coordinated, and the person loses

sodium-potassium pump a membrane protein that uses the energy in ATP to pump sodium out of the cell and potassium into the cell

myelin a wrapping of several layers of cell membrane around an axon; prevents loss of ions and increases the speed of the action potential traveling down the axon

Figure 20.16
A Myelin Sheath Insulates a Neuron.
Many of the axons in mammals are covered by myelin sheaths. Myelin sheaths are made of many concentric layers of cell membranes.

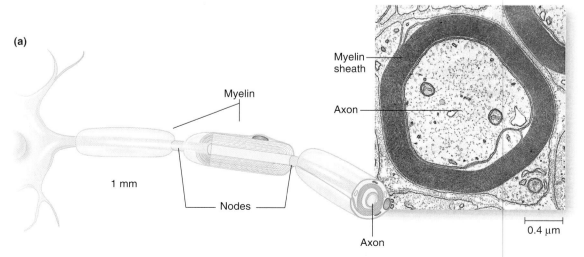

motor control. Recent studies suggest that many demyelinating diseases actually are caused by the patient's own immune system attacking and destroying the myelin.

At its end an axon branches into hundreds or more presynaptic terminals, and the action potential enters each of them. In this way every target cell gets the same message that was sent at the beginning of the axon. Now something rather unusual happens, and it is the key to the success of communication within the nervous system. Recall that the presynaptic terminal is separated from the postsynaptic terminal on the target cell by an extremely small space (**Figure 20.17a**). The action potential *cannot cross* that space. Instead, at each presynaptic terminal this electrical signal is changed into a *chemical signal* that diffuses across the space and affects the postsynaptic terminal. This is an important concept, so reread this last bit of information before moving on.

Figure 20.17 shows how information gets across the synapse. The chemical messenger molecules used by a presynaptic terminal are **neurotransmitters** stored within tiny membranous sacs, called **synaptic vesicles.** The presynaptic terminal retains these synaptic vesicles until it receives a signal to release them. What do you think that signal is? You are right if you thought that the signal is the action potential. When the electrical signal of an action potential enters the nerve terminal, it causes the synaptic vesicles to move toward the cell membrane. Some vesicles reach the cell membrane, fuse with it, and dump their contents—molecules of neurotransmitter—into the narrow gap between the two neurons (Figure 20.17b).

Once they are released into the synapse, the neurotransmitter molecules inevitably bump into the membrane of the postsynaptic terminal (Figure 20.17b). The postsynaptic membrane contains many embedded protein receptors called **neurotransmitter receptors** that can bind to neurotransmitters. So the message from the presynaptic terminal is relayed when the neurotransmitter molecule binds to a postsynaptic receptor and changes the shape of that receptor.

One of two things can happen when the neurotransmitter binds to its receptor on the postsynaptic membrane. In some cases an ion channel opens in the postsynaptic cell membrane (**Figure 20.18a**). Ions flow through it across the membrane and enter the synaptic region of the postsynaptic cell. In contrast, in other neurotransmitter/receptor pairs the binding of neurotransmitter and receptor changes the receptor's shape and starts a cascade

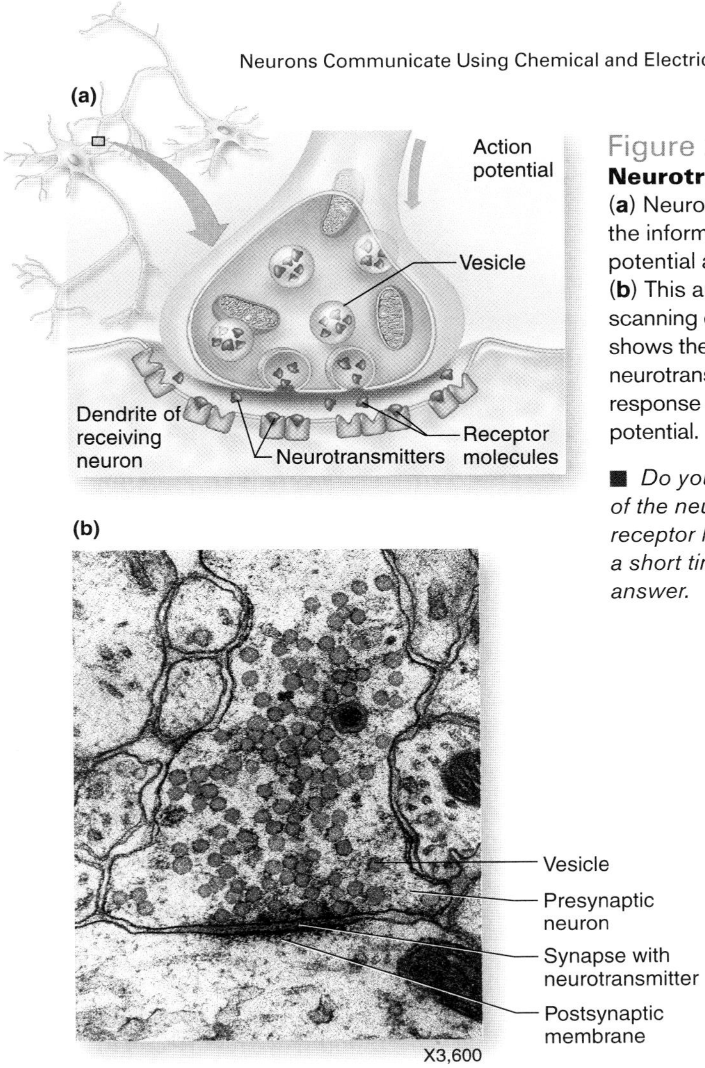

Figure 20.17
Neurotransmitters.
(**a**) Neurotransmitters carry the information in an action potential across a synapse. (**b**) This artificially colored scanning electron micrograph shows the movement of neurotransmitter vesicles in response to an action potential.

■ *Do you think the binding of the neurotransmitter to the receptor lasts a long time or a short time? Explain your answer.*

of biochemical changes inside the postsynaptic cell (Figure 20.18b). In either case the immediate result of neurotransmitter binding to a receptor is the generation of an electrical signal in the postsynaptic neurons. Depending on what kind of signal it is, this new electrical signal can do one of two things. In some cases the signal tells the postsynaptic cell that it should generate an action potential; in other cases it tells the postsynaptic cell not to generate an action potential. In other words, the neurotransmitter can excite or inhibit the postsynaptic cell.

How do these nerve signals correspond to what you see, remember, feel, and think? All of these mental functions are based on the electrical activities in neurons. For example, you "see" the numbers on your alarm clock when certain neurons in the visual cortex fire action potentials in response to excitatory signals from the thalamus; you recognize what time it is based on electrical activity patterns in the temporal lobes. But what about the panic and the call-to-action you feel

neurotransmitter a messenger molecule that will carry an action potential across a synapse

synaptic vesicle a tiny, membranous sac in a presynaptic terminal that stores and releases neurotransmitters

neurotransmitter receptor a specialized membrane protein receptor in a postsynaptic neuron that binds to a specific neurotransmitter

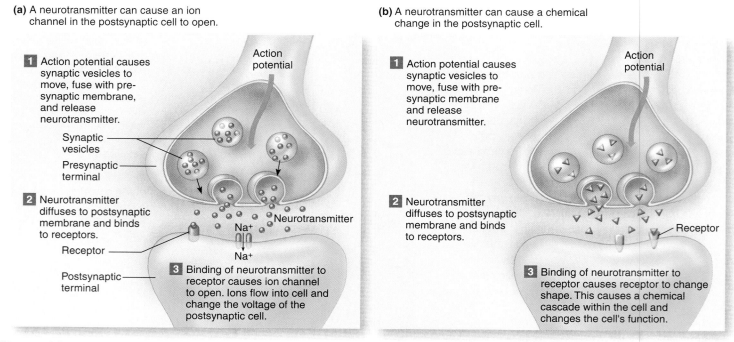

(a) A neurotransmitter can cause an ion channel in the postsynaptic cell to open.

1 Action potential causes synaptic vesicles to move, fuse with pre-synaptic membrane, and release neurotransmitter.

Synaptic vesicles

Presynaptic terminal

2 Neurotransmitter diffuses to postsynaptic membrane and binds to receptors.

Receptor

Postsynaptic terminal

Action potential

Neurotransmitter

Na^+

Na^+

3 Binding of neurotransmitter to receptor causes ion channel to open. Ions flow into cell and change the voltage of the postsynaptic cell.

(b) A neurotransmitter can cause a chemical change in the postsynaptic cell.

1 Action potential causes synaptic vesicles to move, fuse with pre-synaptic membrane and release neurotransmitter.

2 Neurotransmitter diffuses to postsynaptic membrane and binds to receptors.

Action potential

Receptor

3 Binding of neurotransmitter to receptor causes receptor to change shape. This causes a chemical cascade within the cell and changes the cell's function.

Figure 20.18   Possible Events at the Postsynaptic Membrane. The arrival of an action potential at a postsynaptic terminal causes the release of neurotransmitter. (**a**) The interaction of neurotransmitter with membrane receptors in the postsynaptic membrane can cause an ion channel to open. (**b**) Alternatively, the interaction of the neurotransmitter with the membrane receptors in the postsynaptic membrane can cause a chemical change in the postsynaptic cell.

when you see that it is nearly time for your class? Groups of neurons deep in the brain stem also get information from the visual system, and they feed information back to the entire brain (**Figure 20.19**). Once you are awake, neurons deep in the brain stem can activate all parts of your brain and throw you into immediate action. For instance, signals from the visual system are sent to neurons in the cortex that control the motor neurons in the spinal cord that directly control muscles. As a result, you are throwing on clothes, gathering up your books, and are out the door.

Many Drugs Produce Their Effects by Interacting with Parts of the Synapse

One important practical use of knowledge about the nervous system is an understanding of how certain drugs work. Many drugs that affect behavior, mood, or thought interact with various parts of the synapse. Chemicals that interact with the membrane proteins associated with the synapse can be grouped into two categories: agonists and antagonists (**Figure 20.20 b,c**). **Agonists** are drugs that bind to a neurotransmitter receptor and have the same or similar actions as the neurotransmitter itself. These drugs augment, enhance, and mimic normal neuronal activities. In

contrast, **antagonists** are drugs that bind to a neurotransmitter receptor but do not produce the normal actions of the neurotransmitter. Because antagonists occupy the receptor sites, the normal neurotransmitter cannot bind. So an antagonist blocks normal neuronal function. Other classes of drugs are similar to, but not quite the same as, agonists and antagonists. One important group is the **reuptake inhibitors** (**Figure 20.21**). The action of neurotransmitters ends when the neurotransmitter is taken back into the presynaptic terminal by a membrane protein called a *transporter*. If this transporter is blocked, then the neurotransmitter action is extended and enhanced. This is how reuptake inhibitors work.

Table 20.1 lists just a few of the neurotransmitter-related drugs commonly used in medicine. Some of these drugs, such as antipsychotics, restore imbalances. Others have profound and striking effects on an individual's normal perceptions, thoughts, and feelings. Actions of these drugs provide some of the strongest evidence that the nervous system produces the complex mental and behavioral phenomena that we define as human. Drugs that alter simple chemical interactions in the brain, such as the binding of a particular neurotransmitter, can dramatically change a person. These profound changes demonstrate that the nervous system plays a critical role in defining who we are.

agonist a drug that binds to a membrane receptor and has the same effect as the normal signal or neurotransmitter

antagonist a drug that binds to a receptor and has no effect, blocking the normal signal or neurotransmitter function

reuptake inhibitor a drug that binds to a reuptake transporter, blocking the normal reuptake and inactivation of a neurotransmitter

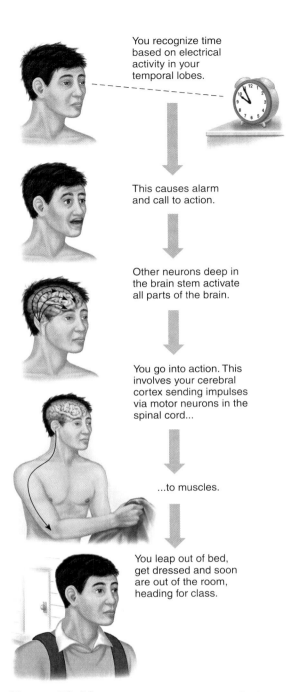

You recognize time based on electrical activity in your temporal lobes.

This causes alarm and call to action.

Other neurons deep in the brain stem activate all parts of the brain.

You go into action. This involves your cerebral cortex sending impulses via motor neurons in the spinal cord...

...to muscles.

You leap out of bed, get dressed and soon are out of the room, heading for class.

Figure 20.19   Late for Class Scenario 3.
The coordinated activity in many brain areas allows you to see the alarm clock, know you are late, and get you activated and out of bed.

QUICK CHECK

1. How does the flow of sodium and potassium ions produce an action potential?
2. What happens at a synapse?
3. How does a neuron "decide" whether to generate an action potential?

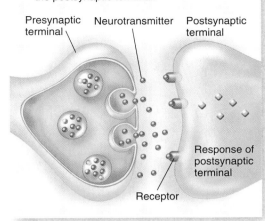

(a) Normal neurotransmitter binds to the receptor and produces a response in the postsynaptic terminal.

Presynaptic terminal   Neurotransmitter   Postsynaptic terminal

Response of postsynaptic terminal

Receptor

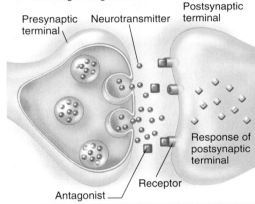

(b) An agonist is a drug that binds to the postsynaptic receptor and has the same effect as the neurotransmitter, but often the effect is of larger magnitude.

Presynaptic terminal   Neurotransmitter   Postsynaptic terminal

Response of postsynaptic terminal

Antagonist   Receptor

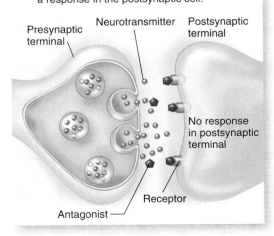

(c) An antagonist is a drug that binds to the postsynaptic receptor but does not produce a response in the postsynaptic cell.

Presynaptic terminal   Neurotransmitter   Postsynaptic terminal

No response in postsynaptic terminal

Antagonist   Receptor

Figure 20.20
Agonists and Antagonists Interact with Neurotransmitter Receptors.

■ *Epinephrine (or adrenaline) is a neurotransmitter that increases heart rate. What effect would an epinephrine antagonist have on heart rate?*

Table 20.1	Neurotransmitter-related drugs	
Drug	**Neurotransmitter action**	**Effect on brain function**
Cocaine	Blocks the transporter for the neurotransmitter dopamine	Euphoria, sexual rush, motor activation, increased concentration, suppressed appetite, psychosis at high doses, highly addictive especially at high doses
Phenothiazine	Antagonist at certain kinds of postsynaptic dopamine receptors	Antipsychotic used to relieve some of the symptoms of schizophrenia
Nicotine	Agonist for certain kinds of postsynaptic acetylcholine receptors	Mild euphoria, increased concentration, suppressed appetite, highly addictive
Heroin	Agonist for certain kinds of postsynaptic opiate receptors	Sedation, mild to intense euphoria, sexual rush, decreased concentration, reduces perception of pain, highly addictive
Sertraline	Blocks the transporter for the neurotransmitter seratonin	Used to treat depression and other psychiatric disorders

Figure 20.21
Reuptake Inhibitors.
These drugs block the protein that carries a neurotransmitter back into the presynaptic terminal.

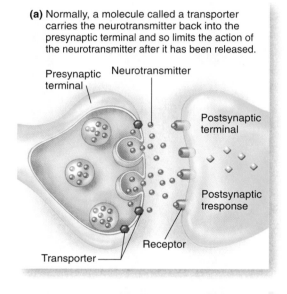

(a) Normally, a molecule called a transporter carries the neurotransmitter back into the presynaptic terminal and so limits the action of the neurotransmitter after it has been released.

Presynaptic terminal
Neurotransmitter
Postsynaptic terminal
Postsynaptic tresponse
Receptor
Transporter

(b) Some drugs block the transporter and so increase the action of the neurotransmitter on the postsynaptic receptor. This increases the postsynaptic response.

Presynaptic terminal
Neurotransmitter
Receptor
Postsynaptic terminal
Increased postsynaptic response

Drug that blocks transporter—reuptake inhibitor

20.6 Muscles Move Under Commands from the Nervous System

The information from your senses would do little good if it remained locked up within your brain. The nervous system not only analyzes information, but also it sends signals to muscles that direct them to move. Motor neurons that have their cell bodies and dendrites inside the spinal cord send out long axons that reach to the muscles. Before considering how motor neurons control muscle movement, let's learn a bit more about muscles.

Your body has over 600 individual skeletal muscles, from the tiny muscles that focus the lens of your eye to the massive hamstrings, three muscles on the backs of your thighs (**Figure 20.22**). But what exactly is a muscle? An individual muscle like your biceps is a tissue made up of many muscle fibers. Within each muscle fiber nuclei undergo mitosis, but there is no cytokinesis, and so each muscle fiber is multinucleate (**Figure 20.23**). A muscle *fiber* has a cell membrane, and so it is a large multinucleate cell. Muscle fibers do the work of moving your body by contracting and lengthening.

Muscle fibers have proteins that do all of the normal cellular jobs, and in addition muscle fibers have large amounts of two proteins: actin and myosin (**Figure 20.24**). In addition to giving a muscle fiber its shape, actin and myosin are the central components in the mechanism that allows a muscle fiber to relax and contract. In effect, actin and myosin form molecular machines. Within muscle fibers strands of actin and myosin are arranged in rows of overlapping, interdigitated bands. Actin bands are called *thin filaments* and myosin bands called *thick filaments.* Their arrangement gives muscle tissue a distinctive banded appearance when viewed with a microscope. The thin actin bands are anchored at regular intervals by vertically oriented structures called *Z lines* that are made of other proteins. This arrangement of protein filaments is possible only because muscle fibers aren't subdivided into individual cells. If each nucleus were enclosed by a cell membrane, the banded pattern of actin and myosin filaments would be disrupted. To understand how a muscle contracts, let's first look at muscle contraction at this level of thin and thick filaments; later we'll examine muscle contraction at a molecular level.

Figure 20.25 shows what happens to the thick and thin filaments and the Z lines when a muscle contracts. The thick and thin bands slide across each other. Because the thin actin fibers are attached to the Z lines, when the thin actin filaments slide over the thick myosin filaments, the Z lines come closer together. The thick bands of myosin do not actually move in this process, but as you will see, it is the action of the thick myosin

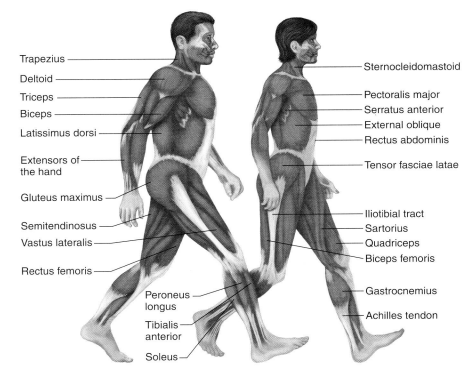

Figure 20.22 **Major Muscle Groups.** These are a few of the body's skeletal muscles. Each is made of many muscle fibers.

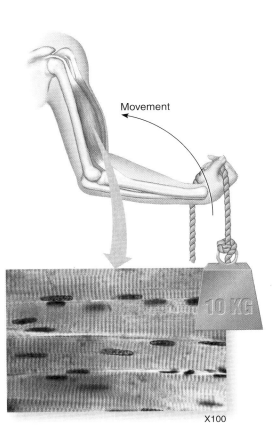

Figure 20.23 **Muscle Fibers Have Multiple Nuclei.**

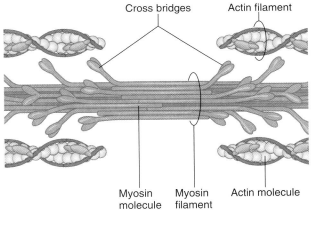

Figure 20.24
Skeletal Muscles Look Banded. The regular arrangement of myosin and actin filaments gives skeletal muscle a banded appearance when viewed with a microscope.

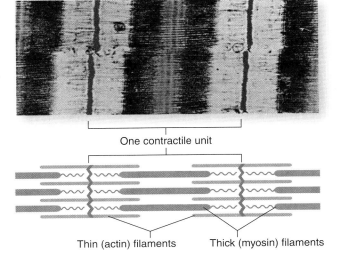

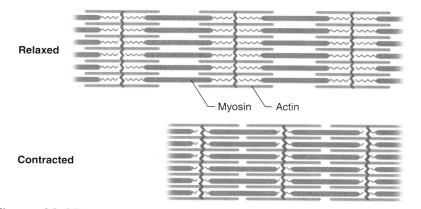

Relaxed

Contracted

Myosin   Actin

Figure 20.25   **Actin and Myosin Filaments Slide Across One Another When a Muscle Fiber Contracts.**

filaments that pulls the thin actin bands together. You can get a sense of this process by holding both hands out horizontally, with the left hand above the right. Move the left hand over the right, while keeping the right hand stationary. This is like half of the contracting unit shown in Figure 20.25. As your left hand moves, the distance between your wrists shortens, even though your right hand—the myosin filament—does not move. This shortening happens all along the muscle fibers that make up an individual muscle; as a result, the entire muscle shortens, or contracts. When this process is finished, the Z lines move apart, the muscle lengthens and relaxes, and the bands of thick and thin filaments return to their original positions.

Now let's take this description of muscle contraction one level deeper. How does the actin molecule actually move relative to the myosin molecule? The actin-myosin complex is just one example of a molecular motor—a complex of molecules that moves and does work. Like any other motor, a molecular motor requires an energy source. You should not be surprised that the source of energy for the actin-myosin motor is ATP (**Figure 20.26**). The structure to focus on in Figure 20.26 is the rounded head of the myosin molecule. In a resting muscle the golf-club-like head of the myosin molecule is sticking out nearly straight. When a muscle contracts, the head swings toward the actin molecule. As it moves, the head of the myosin molecule grazes the actin molecule. This change in position brings the head of

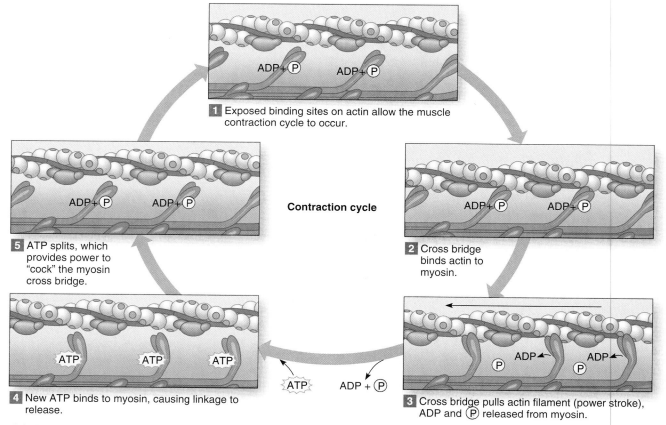

Contraction cycle

1 Exposed binding sites on actin allow the muscle contraction cycle to occur.

ADP+P   ADP+P

5 ATP splits, which provides power to "cock" the myosin cross bridge.

ADP+P   ADP+P

2 Cross bridge binds actin to myosin.

ADP+P   ADP+P

4 New ATP binds to myosin, causing linkage to release.

ATP   ATP   ATP

ATP   ADP + P

3 Cross bridge pulls actin filament (power stroke), ADP and P released from myosin.

ADP   ADP
P   P

Figure 20.26   **ATP Provides the Energy for Muscle Contraction.** ATP provides the energy for the myosin molecule to bind to actin and for the head of the myosin molecule to swivel. The binding of myosin molecules to actin molecules allows actin filaments to slide together.

the myosin molecule close enough so that a chemical bond forms between myosin and actin. As the head of the myosin molecule continues through its swing, it pulls the actin molecule with it. As a result, the actin molecule moves its position relative to the myosin molecule. This causes the Z lines to move closer together. If many myosin-actin pairs move in unison in a given muscle, the muscle will contract. Where does the energy come in? The head of the myosin molecule is an enzyme that breaks the final phosphate bond of ATP, releasing energy and causing the myosin head to move.

So far so good. Now let's bring in the control of muscle contraction. Muscles contract only when a signal is received from a nerve cell (**Figure 20.27**). The control of muscle contraction occurs at a synapse between nerve and muscle called the **neuromuscular junction.** When an action potential invades the terminal of the motor neuron, a neurotransmitter called **acetylcholine** is released from the axon. Acetylcholine binds to a membrane channel receptor and causes sodium and potassium ions to diffuse across the muscle fiber membrane.

When sodium flows into the muscle fiber, a complex series of events takes place. One feature of muscle fibers you have not yet encountered is the labyrinth of membranous sacs called the **sarcoplasmic reticulum.** The sarcoplasmic reticulum is similar to endoplasmic reticulum of other cells, but it stores and releases calcium ions (**Figure 20.28**). The influx of sodium that comes from the activation of acetylcholine receptors causes calcium to be released from the sarcoplasmic reticulum.

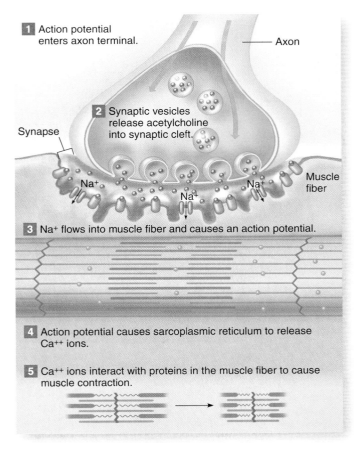

1 Action potential enters axon terminal.

Axon

2 Synaptic vesicles release acetylcholine into synaptic cleft.

Synapse

Na⁺   Na⁺

Na⁺

Muscle fiber

3 Na⁺ flows into muscle fiber and causes an action potential.

4 Action potential causes sarcoplasmic reticulum to release Ca⁺⁺ ions.

5 Ca⁺⁺ ions interact with proteins in the muscle fiber to cause muscle contraction.

Figure 20.27 **Arrival of An Action Potential at a Neuromuscular Junction.** (**1**) When an action potential arrives at a neuromuscular junction it causes the release of neurotransmitter (**2**). (**3**) When the action potential reaches the sarcoplasmic reticulum, it causes it to release calcium ions. (**4**) These allow actin and myosin filaments to slide across each other (**5**), and a muscle fiber contracts.

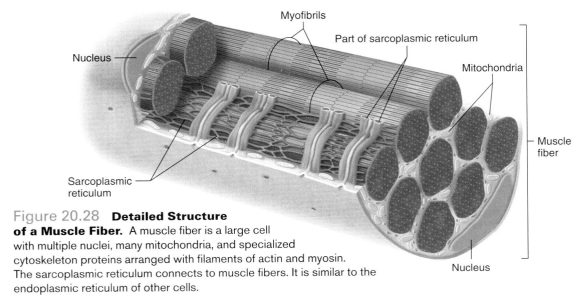

Myofibrils

Part of sarcoplasmic reticulum

Nucleus

Mitochondria

Muscle fiber

Sarcoplasmic reticulum

Figure 20.28 **Detailed Structure of a Muscle Fiber.** A muscle fiber is a large cell with multiple nuclei, many mitochondria, and specialized cytoskeleton proteins arranged with filaments of actin and myosin. The sarcoplasmic reticulum connects to muscle fibers. It is similar to the endoplasmic reticulum of other cells.

Nucleus

■ *What is the function of mitochondria in muscle contraction?*

neuromuscular junction
the synapse between a neuron and the muscle fiber that it controls

acetylcholine
a neurotransmitter used in motor synapses

sarcoplasmic reticulum
a labyrinth of membranous tubes that surrounds a muscle fiber and stores and releases calcium ions

What does all this internal calcium do in a muscle fiber? It allows the actin molecules to interact with the activated myosin heads. So in the absence of a nerve signal, no calcium released. As a result there is no interaction between actin and myosin and there is no muscle contraction.

At this point you have enough knowledge to take your understanding of the "late for class" scenario to the final stages. **Figure 20.29** shows how information from the temporal lobes—where you recognized the time on the clock and realized you were late—reaches the muscles to get you out of bed and moving fast. Information goes through a region called the motor cortex, which then sends axons all the way down to motor neurons in the spinal cord. These directly activate muscles. At the same time information from deep in the brain stem converges on the motor cortex and other motor regions to cause more rapid activation than otherwise would happen. These centers provide the urgency and speed to your movements.

QUICK CHECK

1. What are the two major proteins involved in muscle contraction and how does the process work? Refer to the events illustrated in Figures 20.25 and 20.26 to answer this question.

endoskeleton a hard structure inside an animal that provides body support and allows movement

bone a tissue made of living cells, extracellular collagen, and calcium deposits that forms the hard, supportive skeleton of vertebrate animals

collagen a large protein that forms long, elastic strands used in bone and many other body structures

joint a structure where two bones fit together

ligament a structure that connects one bone to another; made of collagen and other elastic proteins

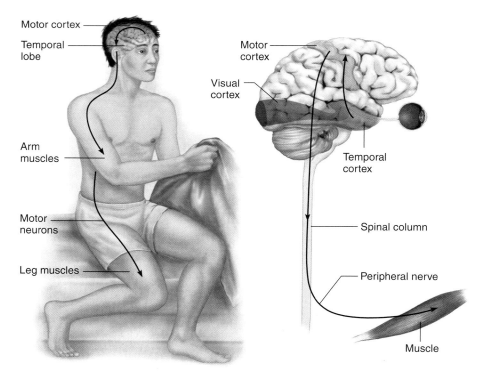

Figure 20.29 **Late for Class Scenario 4.** When you see the clock, the photoreceptors in your eyes send messages to brain centers. The temporal lobes realize that you are late. They send messages to the motor cortex that influence the spinal cord and muscles. As a result you hurry to get out of the house and off to class.

20.7 Bones Provide Internal Support and Allow Vertebrates to Move Despite the Pull of Gravity

Like the plants that grow up and out of soil, animals have internal structures that keep them upright. Your bones do most of this job for you. The human body has 206 separate bones that are organized into a skeleton. Because this skeleton is within your body, it is called an **endoskeleton,** in contrast to the *exoskeleton,* or external hard covering of insects, crustaceans, and other arthropods. The basic plan of the human skeleton is similar to that of all vertebrates. **Figure 20.30** shows skeletons of various vertebrates.

Bone is a successful support structure because it combines the resilience of collagen fibers with the hardness of calcium-based minerals. **Bone** is a hard tissue made of living cells, collagen, calcium, and other minerals. The properties of bone are unique because they are created from a matrix of tissue outside of cells. Bone cells secrete the substances that make bony tissue, and they are embedded within it (**Figure 20.31**). The matrix is made mostly of **collagen,** a large protein that forms long, elastic strands. Where does the hardness of bone come from? For years you have probably been told that the hardness of your bones comes from calcium in the foods you eat. You are encouraged to drink milk because "milk builds strong bodies" and, of course, milk has a high concentration of calcium. Mature bone cells send signals that stimulate the matrix to accumulate a compound called calcium phosphate. Calcium phosphate molecules form crystals within the matrix, and these are responsible for the strength of bone.

Bones are inflexible, but the joints in the skeleton allow movement in many different directions. The bones of the skeleton are held together at joints by tough connective tissues called ligaments. **Joints** are places where bones fit together, and examples are shown in **Figure 20.32**. Joints differ in how much movement they allow. For example, *sutures* are joints between the bones of the skull allow no movement, while other kinds of joints allow movements in different planes. Some important kinds of joints include *ball-and-socket joints* such as those in your shoulders and hips that allow 360 degrees of movement. *Hinge joints* like those in fingers, elbows, and knees permit a bone to swing in just one plane. *Pivot joints* like those where the forearm and upper arm meet allows the palm to be turned up or down. **Ligaments** are strings of protein material including collagen and other elastic proteins that bind bones together. Muscles contract

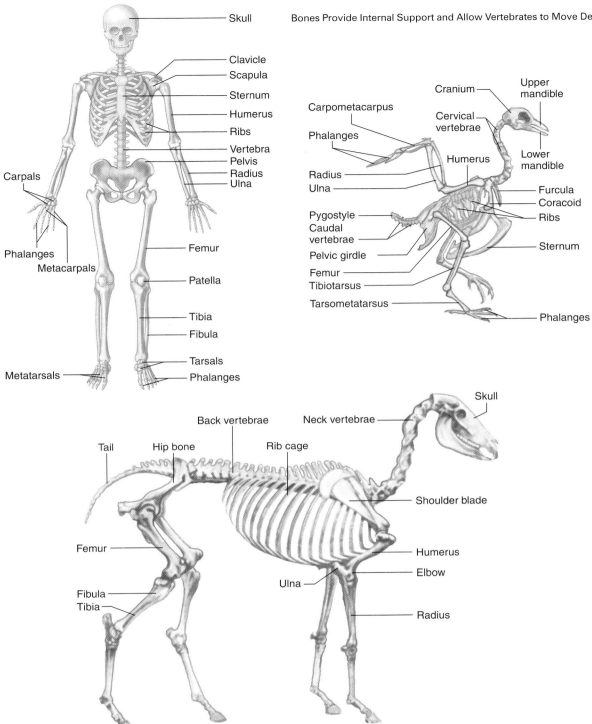

Figure 20.30
Tetrapod Skeletal Plan. Even though various vertebrate species have different skeletal specializations, there is a common plan to the skeletons of tetrapods.

to move your body, and when muscle contracts it actually shortens. Muscles are attached to bones by connective tissues called **tendons.** When muscles shorten, they pull on bones and cause them to move. Of course, bones can move only at the joints, so what muscle contraction really does is to flex, twist, or straighten joints.

Bone marrow is another important component of bones. Bone marrow is the soft, spongy mass inside many bones, but it is especially evident within the hollow cavities of long bones and large, flat bones. Bone marrow is rich in protein and fat and is responsible for making all of the cells that circulate in the bloodstream. Chapters 22 and 25 explore more about bone marrow.

QUICK CHECK

1. In what way is bone different from other kinds of tissues?

tendon connective tissue that attaches muscle to bone

Figure 20.31 **Bone Cells.** Bone cells secrete the hard matrix of bone that surrounds them. Blood vessels and nerves penetrate into bone.

■ *Are bone cells living cells or dead cells? Explain your answer.*

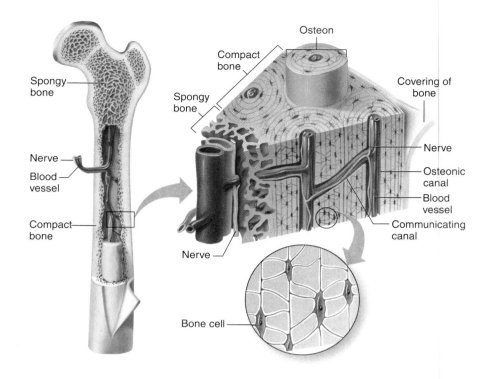

Figure 20.32 **Joints.** Bones fit together at joints.

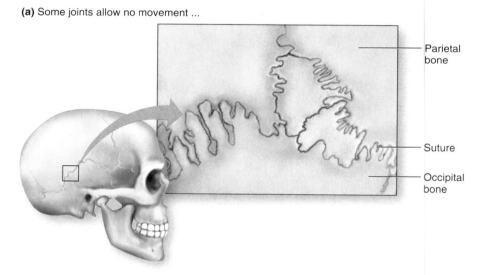

(a) Some joints allow no movement ...

(b) ... while others allow a range of movement.

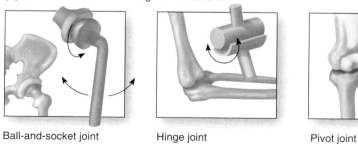

Ball-and-socket joint   Hinge joint   Pivot joint

20.8 An Added Dimension Explained
Understanding Jody's Addiction to Crack Cocaine

In the beginning of this journey, Jody was not very different from you. She had a normal brain that would respond to whatever the environment presented. She was 20 and came from a working-class family that always had managed to have enough. As a child, Jody never was overindulged, but she was never hungry either. After high school she planned to go to a large, four-year college; however, she was insecure about leaving her neighborhood. Her high school boyfriend seemed more comforting than the strangers at a big, impersonal campus, and soon they had a son. Realizing how meager her future might be, Jody enrolled at a nearby community college. Although her boyfriend seemed to love their child, the love between them rapidly died, and Jody found herself more and more on her own. Life was not much fun for Jody. One night she felt especially down and unhappy beneath the weight of her responsibilities and loss of freedom.

So when a friend offered Jody a hit of cocaine, she hesitated, but in a moment of abandon agreed to try it. When the cocaine hit her brain, it was like nothing she had ever experienced. The rush was nearly overwhelming and suddenly everything was brighter and more exciting. Life seemed so good again. At least for as long as the cocaine dominated her experience. What was happening in Jody's brain?

Inhaling the hot smoke gave the cocaine fast access to Jody's bloodstream. A high dose peaks rapidly, blasting the axon terminals in a tiny brain region called the *nucleus accumbens.* There the cocaine binds to and blocks nearly all of the reuptake transporters for a neurotransmitter called *dopamine.* Normally, the action of dopamine in the nucleus accumbens causes feelings of pleasure, but in the presence of cocaine Jody's dopamine repeatedly acted on the postsynaptic cells, and so Jody felt excessive pleasure, more than she had ever experienced naturally. Without a transporter to stop it, each dopamine molecule is like a crazy pinball. Each dopamine molecule binds to a receptor and bounces off, binding again and again. The intense action of so much dopamine triggered Jody's rush.

But Jody's rush receded as the cocaine gradually diffused out of synapses, diffused into capillaries, and was swept away in the bloodstream. Eventually, all the cocaine floating in Jody's brain was gone and none remained to have any impact on dopamine levels. The pipe was still full though, so Jody had the experience again and again.

Jody was not prepared for the strength of cocaine's effect. She also was not prepared for the bad feelings that developed soon afterward. The rush began to lose its intensity. Eventually, the negative feelings began to loom. Why did this happen? Bombarded with so much dopamine, the receptors on the surface of postsynaptic terminals in the nucleus accumbens gradually lost their sensitivity to dopamine. Tolerance to drugs is a common mechanism that acts to keep the brain and other body systems operating in a normal range. Some receptors are even pulled out of the membrane, reducing the unnatural bombardment from too much dopamine. By the time Jody had gone through several doses of cocaine, her dopamine responses were largely shut down. Up to a point she could take higher doses of cocaine to get the high, but eventually no amount of cocaine would produce any positive effect. Jody's binge was over. So were her good feelings.

What about those ugly feelings of anxiety and unhappiness? These feelings are more than just a comparison with the high produced by cocaine. After even one dose of cocaine most people feel down, depressed. Eventually, cocaine itself causes feelings of anxiety and unhappiness. The brain mechanisms that cause these feelings are not understood, but they are an inevitable part of cocaine use. They always accompany a cocaine binge, and as many cocaine users do, Jody turned to the sedating effects of alcohol to escape from them. Normally, alcohol would bring a rush and high as well, as it indirectly acted on the dopamine system. But Jody's dopamine system now was depleted. Alcohol affected other neurotransmitter receptors to produce an increase in inhibition in Jody's brain, and this allowed her to sleep.

Now think about the state that Jody was in when she finally awakened. The tolerance mechanisms that protected her from the effects of cocaine were still there. Her dopamine receptors were slow to recover, so Jody not only woke up remembering the high from her encounter with cocaine, but also she was more depressed and unhappy than usual. With her dopamine system turned down, Jody got less pleasure out of normal things. The antics of her son could not make her laugh and shopping for trinkets with her mother was no longer fun. Jody could not concentrate because her brain no longer was normal, so studying was just frustrating. *Jody had been changed by her exposure to a few high doses of cocaine.* Would she have been better off taking another drug? Amphetamine? Heroin? No. All drugs of abuse have these same basic effects. Some are more similar to cocaine than others. Methamphetamine acts directly at the dopamine synapse, in ways similar to the actions of cocaine. Heroin, alcohol, and barbiturates act directly on other neurotransmitter systems but have indirect effects on the dopamine system as well. Through the dopamine system all of these drugs lead to long-term changes in the brain that will dramatically alter who you are.

How long will it take Jody to recover? No one is quite sure. If she can resist the lure of cocaine, she eventually will return to a normal state, where things are fun and the world holds promise again. But each time she returns to cocaine, the changes in her dopamine and other brain systems will be deeper and harder to overcome. Continued cocaine use can even alter how

—Continued next page

Continued—

genes are expressed and can change such basic features as the shape and structure of the neurons affected by it. By the time Jody has invested a year in cocaine bingeing, her brain will have changed so much that recovery will take a long time. Jody truly is no longer herself.

Scientists hope that new knowledge about drug effects will lead to better treatments. For Jody, the best hope is treatment to help her to recover her normal dopamine functioning without causing further changes that will make it even harder to get back to normal. Presently, there is only one commonly used drug therapy: methadone for heroin addiction. Methadone is an opiate receptor agonist, but it does not produce the extreme euphoric rush generated by heroin. Addicts who take methadone say that it reduces their craving for heroin and eliminates withdrawal symptoms. It does leave the addict with an addiction to methadone, but most people can regain a degree of normal func-

tion with methadone addiction. While the replacement drug approach is controversial, most scientists and medical doctors feel that it has a place in a broad modern treatment approach.

Drug treatment options for Jody are limited because there are no good replacement drugs for stimulants like cocaine. Some drugs, such as antidepressants, seem to be able to stabilize neurotransmitter functions without having abuse potential or causing other long-term problems. They may be of some use to drug abusers like Jody, and some doctors might prescribe them. It still isn't clear, though, what the best drugs might be, or how they might work. So for now Jody will have to turn to treatment programs based more on "common sense" than science.

QUICK CHECK

1. How does cocaine change the levels and actions of dopamine in the brain?

Now You Can **Understand**

The Use of Brain Electrodes in Treatments of Disorders

In this chapter you have seen brief references to brain electrodes. These are small wires or delicate glass tubes inserted into the brain. The tips of the tubes or wires are smaller than the width of a cell. While brain electrodes might seem like science fiction, brain electrodes play an important role in scientific investigations and medicine. In scientific laboratories microelectrodes—the most delicate electrodes—are used to record the action potentials and synaptic potentials produced by neurons. These tiny electrodes can be gently lowered into an animal's brain and the responses of a single neuron can be measured—for example, to auditory or visual images that the scientist presents to the animal. Using electrodes scientists have learned a great deal about the functions of individual neurons in different brain regions. Electrodes also can be used to pass tiny electrical currents into the brain. These currents are on the same scale as action potentials, and neurons in the region of the electrode respond as if they are action potentials. Such stimulating electrodes can be used to determine what behaviors various brain regions control and influence. Stimulating electrodes were part of the evidence for the role of the temporal and parietal lobes in visual and auditory memories. Today, stimulation from implanted electrodes is used to control the tremor of Parkinson's disease or the extreme discomfort of chronic pain. One of the most intriguing recent stories about stimulating electrodes involved rats. After the 9-11 attacks many researchers began look-

ing for ways to explore dangerous places in emergency situations. One group of researchers used stimulating electrodes in the brains of rats to cause the rats to move in one direction or another. The thought was that such rats might be able to go into small areas— and if equipped with small microphones or cameras then would be able to send back information to their human counterparts.

What Do **You** Think?

"You Use Only 10% of Your Brain"

This is such a common belief that you can hear people say it all the time. Sometimes it's an excuse for poor performance, or sometimes it's an argument for the possibility of phenomenal brain abilities such as extrasensory perception or mental telepathy. What is the basis for this statement? Is it true? You have read in this chapter about various brain regions and what they do. Does it seem like only 10% of your brain is involved in your daily activities?

The origin of this statement is obscure, but it probably arose from early studies that stimulated the brain on the surface of the outer cortex. Early researchers were looking for gross motor behaviors, but electrical brain stimulation of most areas on the surface of the cortex did not produce any behavior. Yet further research, as you have seen in this chapter, has revealed that most brain areas have a concrete function. Nevertheless, this belief persists. Given what you have read in this chapter, what is the potential of the human brain for hidden, untapped abilities? What do *you* think about the claim that you use only 10% of your brain?

Chapter Review

CHAPTER SUMMARY

20.1 Sense Organs, Nervous System, Muscles, and Bones Coordinate Interactions with the Environment

Sense organs, the nervous system, the muscular system, and skeletal system allow organisms to respond to the environment.

20.2 Senses and Brains Have Prokaryotic Origins

Complexly integrated nervous systems and sensory systems have their roots in membrane ion channels of prokaryotes, such as the protein mechanoreceptor of the potassium channel. Accessory structures extend the range of a mechanoreceptor. Eyes also can be traced to prokaryotic beginnings. Bacteriorhodopsin in prokaryotes is a light-sensitive pigment that is similar to rhodopsin found in eyes of eukaryotes, from jellyfish larvae to image-forming eyes of vertebrates. Arrangement of nerves in nervous systems reflects adaptations to body organization. A nerve net is typical of radially symmetrical jellyfish, while bilaterally symmetrical animals have ganglia located in a head region. Actin and myosin also can be traced back to single-celled prokaryotes.

ganglion 478   nerve net 478

20.3 Sensory Receptors Detect Environmental Conditions

Sensory receptors are cells that detect environmental information and pass it on to neurons. Eyes are adapted to receive, focus, and send a nerve impulse to visual centers in the brain in response to light. Rods and cones in the retina are light-sensitive cells that contain rhodopsin. Photons cause rhodopsin to change shape. In turn, this causes the rod or cone cell to send information to a neuron that conveys information about light to the brain. Hair cells in the cochlea respond to vibrations transmitted to it by structures of the outer and middle ear. Hair cells send information about sounds to the brain. Touch, temperature, and pain receptors in skin are specialized sensory receptors. Taste and smell use receptors that respond to molecules in foods or in the air. Other animals have additional specialized receptors and senses that humans lack.

cochlea 480   rods 480
cones 480   sensory receptor 479
hair cell 480   taste bud 481
rhodopsin 480

20.4 Neurons Are Specialized Cells of the Nervous System

A neuron has dendrites, a cell body, and an axon. Sensory information is passed to the central nervous system via the peripheral nervous system, composed of axons of motor neurons and sensory neurons. Nuclei in the brain stem govern basic body functions; the thalamus is a processing station for sensory information; the hypothalamus helps regulate body temperature, intake of foods and fluids, and secretion of glands; the basal ganglia function in motor behavior and the perception of reward. The cerebellum governs complex, learned motor behaviors, while the cerebral cortex governs higher brain functions. Information from visual receptor cells travels to the thalamus, the primary visual cortex, and temporal lobes—each has a role in processing this information.

axon 482   dendrite 482
basal ganglia 486   hypothalamus 485
brain 484   neuron 482
brain stem 485   peripheral nervous system 484
cell body 482   spinal cord 484
central nervous system, 484   synapse 483
cerebellum 486   temporal lobe 486
cerebral cortex 486   thalamus 486

20.5 Neurons Communicate Using Chemical and Electrical Signals

Neurons communicate in synapses where a neurotransmitter chemical released by a presynaptic neuron carries the electrochemical signal of an action potential to a postsynaptic neuron. To respond to an action potential, a neuron sums excitatory and inhibitory inputs. Just a few neurotransmitters are used in the nervous system, and there are inhibitory and excitatory receptors for each neurotransmitter. Agonist drugs work similarly to neurotransmitters; antagonist drugs bind to neurotransmitter receptors and block them. Reuptake inhibitors are drugs that block the reuptake and inactivation of a neurotransmitter.

action potential 486   neurotransmitter receptor 489
agonist 490   reuptake inhibitor 490
antagonist 490   sodium-potassium pump 488
myelin 488   synaptic vesicle 489
neurotransmitter 489   voltage-dependent ion channel 487

20.6 Muscles Move Under Commands from the Nervous System

Muscle fibers are multinucleate cells. Sliding of myosin filaments past actin filaments allows a muscle fiber to shorten. ATP and calcium ions are needed for muscle contraction. Nerves communicate with muscles in neuromuscular junctions. The sarcoplasmic reticulum stores and releases calcium ions. Information from the motor cortex travels down motor neurons in the spinal cord to activate muscles.

acetylcholine 495   sarcoplasmic reticulum 495
neuromuscular junction 495

20.7 Bones Provide Internal Support and Allow Vertebrates to Move Despite the Pull of Gravity

Bone cells are surrounded by a matrix of calcium phosphate crystals and collagen fibers. Joints and ligaments bind bones together and tendons join muscles to bones. Different kinds of joints allow for different kinds of movements.

bone 496   joint 496
collagen 496   ligament 496
endoskeleton 496   tendon 497

20.8 An Added Dimension Explained: Understanding Jody's Addiction to Crack Cocaine

Cocaine blocks the dopamine transporter and increases the dopamine in brain synapses, causing euphoria. Adaptations to cocaine desensitize the brain to dopamine, creating a tolerance to the drug. This also causes depression that follows a cocaine binge.

TRUE or FALSE. If a statement is false, rewrite to make it true.

1. A neuron's cell body contains dendrites, nucleus, ribosomes, mitochondria, and axons.

2. Communication between neurons happens in synapses.

3. An action potential always results in a response.

4. Sensory receptors are nerve endings or modified neurons, but they are always unicellular.

5. Rods in the retina detect color, while cones detect different levels of light.

6. Taste receptors are multicellular.

7. The hypothalamus controls heart rate, breathing, pain perception, attention, and arousal.

8. The cerebellum is the largest component of the human brain.

9. The thalamus is a processing station for vision and hearing.

10. Acetylcholine is the neurotransmitter in the central nervous system and in axons that synapse with skeletal muscles.

11. The plan of multiple bones results in greater flexibility.

12. Tendons bind bones together in joints.

13. When a muscle contracts, myosin filaments are pulled more closely together.

14. ATP provides the energy for muscle contraction.

15. The sarcoplasmic reticulum stores and releases phosphate ions that are necessary for joint movement.

16. Many drugs mimic neurotransmitter action, or otherwise interact with neurotransmitter receptors.

MULTIPLE CHOICE. Choose the best answer of those provided.

17. Information reaches a neuron via its
    a. cell body.   d. dendrite.
    b. axon.   e. osteoid.
    c. myofibril.

18. Two choices will not be used. Sound waves vibrate the ___. This creates mechanical disturbances in fluid in the ____, which causes ____ to release ____ that sends information to the ____.
    a. brain   e. neurotransmitter
    b. retina   f. synapse
    c. hair cells   g. cochlea
    d. eardrum

19. If you were measuring your brain activity while riding a bike, most electrical impulses would be detected in the
    a. brain stem.   d. cerebellum.
    b. thalamus.   e. cerebral cortex
    c. hypothalamus.

20. Which neuron features end in presynaptic terminals?
    a. dendrites   d. peripheral nerves
    b. axons   e. glands
    c. cell bodies

MATCHING

21–25. Match the columns. (One choice will not match.)

21. neuron
22. action potential
23. neurotransmitter
24. rhodopsin
25. retina

a. chemical messenger molecule used in synapses
b. light-sensitive pigment
c. eye tissue that contains photoreceptors
d. a chemical/electrical signal in the nervous system
e. cellular unit of nervous system
f. structure that focuses an image

1. Putting together all that you know about the nervous system, how do you "see" the numbers on your alarm clock in the morning?

2. How do nerves, muscles, and bones work together to cause your body to move?

Information about objects that touch your skin is carried to the spinal cord by several kinds of neurons, and each kind has axons with distinctive characteristics. For example, information about how much pressure the object is putting on your skin is carried by large axons with thick myelin coatings. Information about pain from an object touching your skin is carried by thin axons with no myelin. So touch information is carried to the spinal cord faster than pain information. A typical touch axon will conduct action potentials at 80 meters per second, while a typical pain axon will conduct action potentials at 1.5 meters per second.

Assume that an object pushes on your skin, and also breaks the skin and causes pain. Assume also that the distance between the point of contact on your skin and the spinal cord is 1.2 meters. How long will it take information about touch and information about pain to reach the spinal cord?

1. Action potentials rely on sodium channels along the axon. Without these sodium channels, the axon will not conduct any action potentials. Many drugs used in surgery and dentistry block the sodium channels in axons and so block action potentials. To prepare for surgery, such drugs often are injected around the spinal cord, thereby affecting nerves that go to and from the spinal cord. Dentists inject such drugs into the nerves around the teeth. What kinds of effects would you expect from such action potential blocking drugs?

2. The actions of drugs often are used to help reveal the functions of neurotransmitters and brain regions. How could drugs be used in this way? What are potential problems with interpreting neurotransmitter function based on drug effects?

For additional study tools, visit www.aris.mhhe.com.

Nutrition and Digestion

100,000 Cases of Pellagra

Surgeon General Blue's Estimate of the Possible Number

October 10, 1915

The New Yorker Who Changed the Diet of the South

August 12, 2003

Pellagra was epidemic in parts of the United States from 1900 to 1940. A major focus of research at the time was determining what caused pellagra and how it could be avoided and cured.

If you were an impoverished southerner in the 1920s, grits, fatback, and corn bread were on the table for breakfast, lunch, and dinner. Fresh fruits, vegetables, and meats were rare, especially in the winter. Pellagra (pronounced "pell-ahg-rah") also was a fact of life. Each spring from about 1900 to 1940, in isolated shacks and in towns across the southern United States, this chronic disease would appear. Pellagra was epidemic in 15 southern states. Although exact numbers are difficult to pin down, several hundred thousand cases occurred in the United States, and tens of thousands of people died. Pellagra claimed victims in other countries too. For nearly 200 years (1730–1930) pellagra was widespread among the poor in Mediterranean countries, Latin America, Africa, and Asia. No one knew what caused it, but the symptoms were all too familiar. This distant history may seem irrelevant to your life, but it is only because of studies of the pain and sacrifice of pellagra victims of the past that today you are free of the disease.

People with pellagra (Italian for "rough skin") have a mosaic of dead-feeling, crusty scales and raw patches on the skin. Networks of deep, painful cracks open between the thickened scales, and victims say that their skin feels like it is on fire. Even before the skin changes, pellagra victims are tired, don't feel like eating, lose weight, and suffer bouts of diarrhea. Later, the skin of the hands, arms, neck, and legs becomes blistered and inflamed and the typical thickened, scaly patches develop (**see chapter-opening photo**).

If a painful and disfiguring skin inflammation were the only symptom of pellagra, the disease might be no more than an ugly annoyance—like bad sunburn or a severe case of poison ivy. But unlike sunburn or poison ivy, pellagra doesn't go away. If pellagra is left untreated, internal organs, especially the liver, are damaged. Worse yet are the mental effects. Victims of pellagra may suffer from suicidal depression, be aggressive, have hallucinations or delusions of persecution. "Pellagrins," as they once were called, often were confined to insane asylums where they had to be carefully watched to keep them from committing suicide. One newspaper description of a pellagrin's suffering was exceptionally vivid: "Victims die in great agony, the pain being like pouring boiling water on wounds already scalded."

Today pellagra is rare, but it has not disappeared. It still afflicts poor populations throughout the world. If you watch for pellagra's trademark—thickened, inflamed, and scaly skin that peels away to reveal glossy red skin beneath—you're apt to see it on the hands, wrists, legs, and ankles of street people and chronic alcoholics. What could cause these peculiar symptoms? What do street people and alcoholics have in common with southern sharecroppers and Italian villagers?

Later in this chapter you will unravel the mystery of pellagra and follow the adventures of the remarkable scientist who discovered its cause. For now, let pellagra be an example of how profoundly nutrition can affect the human body and mind. Before returning to this topic, this chapter will help you to understand the basics of nutrition. ■

21.1 Nutritional Needs Are Based on Biological Chemistry

Living organisms cannot manufacture all of the chemicals they need to sustain life. Critical chemicals that cannot be manufactured are called **nutrients,** and these must be obtained from the environment. Nutritional needs stem directly from the chemistry of life. Generally, organisms have two kinds of nutritional needs. First, organisms need a source of chemical energy to drive the chemical reactions that maintain life. Second, organisms need specific chemical nutrients to build biological structures and carry out biological processes. These will be the basic themes of this chapter.

Nutrients Provide Energy and Raw Materials for Life

One common nutritional theme is the need for a chemical source of energy. Organisms use the energy in the final phosphate bond of ATP to drive most cellular processes (see Section 6.3). Photosynthetic organisms capture the energy in sunlight and store it in the carbon-to-carbon bonds of glucose molecules. The process of cellular respiration breaks the bonds of glucose and captures its chemical energy in molecules of ATP. While photosynthetic organisms can manufacture glucose for ATP production, many organisms obtain chemical energy from other organisms (**Figure 21.1**).

A second nutritional theme is that, just as a factory must have raw materials to manufacture a product, cells require supplies of chemical elements to make chemical structures and carry out cellular processes. Carbon, oxygen, hydrogen, nitrogen, phosphorus, sulfur, calcium, and other elements are needed to make biological molecules such as lipids, carbohydrates, proteins, and nucleic acids. Organisms do not build these biological molecules from scratch, atom by atom. Rather, simpler biological molecules such as sugars, amino acids, and nucleotides are the raw material for larger biological molecules.

A cell can get these simpler biological molecules directly from the environment, or a cell can

nutrients chemicals that are necessary to maintain life, provide energy, or promote growth

convert an existing molecule into one it needs. For example, you can get the amino acid asparagine from your diet, or your metabolism can convert another amino acid such as aspartic acid into asparagine. These conversions require specific enzymes, and each species—even each individual— has enzymes to convert some molecules but not others. For example, humans can make the five nucleotides from other basic biological molecules but cannot make all of the 20 amino acids needed to make proteins.

In addition, every organism or cell needs some elements in their atomic form. For example, most cells need calcium, sodium, potassium, and other elements. Of course, these basic elements cannot be made from other elements. The diet must supply any atom or molecule required for normal development and health that the body's own metabolism cannot manufacture. Nutrients that must be supplied by the diet are called **essential nutrients.**

Essential amino acids provide a good example. If you search the Internet, or read the advertisements from health food stores, you will come across a list of eight to ten essential amino acids necessary to maintain good health. If you eat protein, your cells can make certain amino acids by converting one dietary amino acid into another. Essential amino acids cannot be manufactured at all and must be obtained from food. Most diets of modern cultures supply ample protein, and so usually there is no reason for concern about shortages of essential amino acids. Vegetarians can get good-quality protein from dairy products or soy products. But many people in the world have inadequate nutrition and fail to get a range of foods that will provide them with essential amino acids.

Several committees of the U.S. National Research Council try to sort through all of this information about nutrition and set nutritional guidelines. The U.S. government currently recognizes more than 40 nutrients as necessary for maintaining health, some of which are listed in **Table 21.1**. Nutrients required in gram amounts are called **macronutrients,** while nutrients required in milligrams or less are **micronutrients.** Macronutrients include lipids, carbohydrates, and proteins. Micronutrients include vitamins and minerals.

1. Compare and contrast essential nutrients, macronutrients, and micronutrients.

(a) Your peanut butter and jelly sandwich has nutrients that were made by plants.

(b) The mosquito that sucks your blood obtains nutrients and macromolecules from molecules in your red blood cells.

(c) In flight a swallow is a mosquito-catching trap. It digests mosquitos and other insects and obtains nutrients and macromolecules that are stored in their bodies.

(d) The black rat snake that eats young barn swallows obtains nutrients and macromolecules stored in the bodies of the young birds.

Figure 21.1 **Some Organisms Must Eat.** Although photosynthesizers and chemosynthetic prokaryotes can make their own food, many organisms must depend upon others for food.

essential nutrients nutrients that are necessary for life but cannot be manufactured by an organism

macronutrients large biological molecules such as lipids, carbohydrates, and proteins needed by the body in relatively large amounts, in the range of grams to kilograms

micronutrients small molecules and minerals needed by the body in small amounts, in the range of milligrams (mg = 1/1,000 gram) to micrograms (μg = 1/100,000 gram)

Table 21.1   Nutrients needed in the human diet

Nutrient	Typical sources	Daily amounts that promote good health in adults
Macronutrients		
Lipids	Fatty meats, fish, poultry, dairy products, avocados, coconuts	50–75 g
		very-low-fat diets have been recommended by some researchers (e.g. 20–30 g)
Carbohydrates	Fruit juices, grains, pasta, baked goods, fruits, vegetables	300–400 g
Fiber	Fruits and vegetables, grains	30 g
Proteins	Meats, fish, seafood, eggs, milk and dairy products	0.8 grams per kilogram of body weight
Micronutrients		
Water-soluble vitamins		
Thiamin	Pork, legumes, peanuts, whole grains	1.1 mg
Riboflavin	Dairy products, meats, enriched grains, vegetables	1.1 mg
Niacin	Nuts, meats, grains	14 mg
B_6	Meats, vegetables, whole grains	2 mg
Folic acid	Green vegetables, oranges, nuts, legumes, whole grains	400 µg
B_{12}	Meats, eggs, dairy products	2.4 µg
Biotin	Legumes, other vegetables, meat	30 µg
Pantothenic acid	Meats, dairy products, whole grains	5 mg
C	Citrus fruits, broccoli, cabbage, tomatoes, green peppers	75 mg
Fat-soluble vitamins		
A	Dark green and orange vegetables and fruits, dairy products	700 µg
D	Dairy products, egg yolk, also synthesized in human skin	5 µg
E	Vegetable oils, nuts, seeds	15 mg
K	Green vegetables, tea	90 µg
Minerals and other elements		
Calcium	Dairy products, dark green vegetables, legumes, sardines	1,000–1,300 mg
Phosphorus	Dairy products, meats, grains	700 mg
Magnesium	Whole grains, green leafy vegetables	310 mg
Zinc	Meats, seafood, grains	10 mg
Iodine	Seafood, dairy products, iodized salt	150 µg
Iron	Meats, eggs, legumes, whole grains, green leafy vegetables	18 mg
Fluorine	Fluoridated drinking water, tea, seafood	3 mg
Copper	Seafood, nuts, legumes, organ meats	900 µg
Chromium	Brewer's yeast, liver, seafood, meats, some vegetables	25 µg
Molybdenum	Legumes, grains, some vegetables	35–50 mg

*These estimates are for healthy females in their late twenties or early thirties. Requirements can vary depending on age, gender, and health of individual.

Source: National Academy of Science Food and Nutrition Board, 2004.

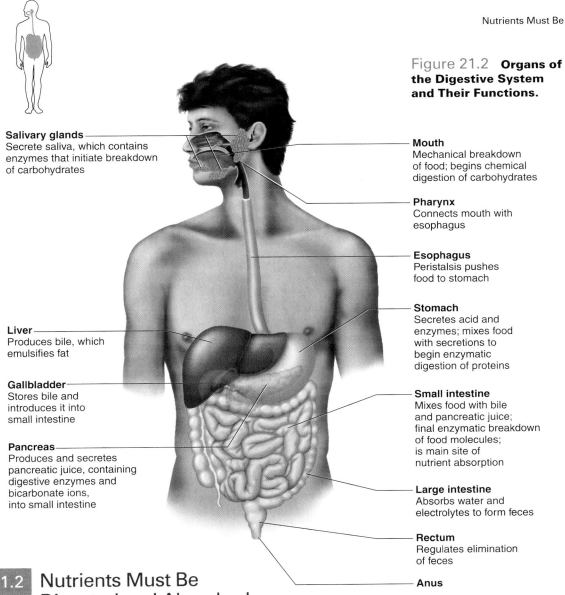

Figure 21.2 **Organs of the Digestive System and Their Functions.**

Salivary glands
Secrete saliva, which contains enzymes that initiate breakdown of carbohydrates

Mouth
Mechanical breakdown of food; begins chemical digestion of carbohydrates

Pharynx
Connects mouth with esophagus

Esophagus
Peristalsis pushes food to stomach

Stomach
Secretes acid and enzymes; mixes food with secretions to begin enzymatic digestion of proteins

Liver
Produces bile, which emulsifies fat

Gallbladder
Stores bile and introduces it into small intestine

Pancreas
Produces and secretes pancreatic juice, containing digestive enzymes and bicarbonate ions, into small intestine

Small intestine
Mixes food with bile and pancreatic juice; final enzymatic breakdown of food molecules; is main site of nutrient absorption

Large intestine
Absorbs water and electrolytes to form feces

Rectum
Regulates elimination of feces

Anus

21.2 Nutrients Must Be Digested and Absorbed

Of course, you obtain essential nutrients by eating or drinking them, but your cells cannot directly use the carbohydrates, proteins, or fats that are in your foods. This is because the specific kinds of macronutrients made by some other organism—the plant, animal, or fungus that you eat—will not be exactly the kind of biological molecules that the human body needs. Consider the avocado you may have had with your dinner last night. Certainly, the avocado contains some of the same organic molecules that your body does such as glucose and certain amino acids. Lipids, polysaccharides, proteins, and nucleic acids are not exactly the same as those in your body. If they were exactly the same, you would look much more like an avocado and less like a person. Most of the macronutrient molecules in the avocado must be digested down to their building blocks such as fatty acids, simple sugars, amino acids, and other simple components. Then

the body can take them in and chemically process them into human forms of biological molecules to use in cellular structures and processes.

The process of obtaining nutrients from food has two phases: **digestion** is the process of breaking down food particles, and **absorption** is the process of getting them from the gut into the blood that will carry them to body cells for use and further processing. Cells that need nutrients can take them up from the bloodstream. Let's look at the process of digestion in greater detail.

The Digestive System Digests Foods and Absorbs Nutrients

Figure 21.2 shows the organs of the human digestive system. You can think of the digestive tract as a long tube that runs from the mouth at one end to the anus at the other. Along its length the digestive

digestion the process that breaks down nutrients into small molecules

absorption the movement of digested nutrients across cell membranes and into the bloodstream

tract is differentiated into distinct organs, including the *esophagus, tongue, teeth, stomach, small intestine, large intestine,* and the *rectum* (Figure 21.2). At various points along the digestive tract, enzymes and other digestive substances are introduced from digestive glands and organs, including the *salivary glands,* glands located in the lining of the stomach, the *liver,* the *pancreas,* and the *gallbladder.* Food comes in through the mouth, is digested and absorbed, and any leftovers that cannot be used are excreted through the anus as waste. It usually takes six to eight hours for the small intestine to process a meal. The meal usually is eliminated from the large intestine after about 24 hours, but some of the meal may linger in the large intestine for several days.

The process of breaking food down into smaller components begins in the mouth. The tongue manipulates food, and teeth bite and grind it into smaller fragments and shreds. Swallowing pushes food into the esophagus, and the automatic wave of muscular contraction called *peristalsis* moves the food down the esophagus and into the stomach. In the stomach acids create a low pH environment that kills many bacteria and other foreign organisms. The proteins in food are denatured by the low pH, making them easier to break into smaller molecules and digest.

The stomach is made of three layers of muscles arranged in different orientations (**Figure 21.3**). The three layers of muscle churn food in the stomach, thoroughly mixing chewed food with stomach acids. If you suffer from heartburn, it results from acidic stomach contents splashing up onto the far end of the esophagus. The lining of the stomach has a deeply folded surface covered with a thick layer of protective mucus. In addition to glands that secrete stomach acids, mucous glands in the lining of the stomach provide a thick protective layer that prevents the stomach's acids from digesting the muscle of the stomach. People who suffer from stomach ulcers have a disorder in which bacteria are disrupting the normal protective layer of mucus, allowing stomach acids to damage the lining of the stomach. Ulcers now are treated with antibiotics. The entrance to and exit from the stomach are both controlled by circular muscles called *sphincters,* which regulate the flow of food into and out of the stomach.

The breakdown of food into smaller molecules occurs primarily in the small intestine. Digestive enzymes from the pancreas, bile made by the liver and stored and released from the gallbladder, and digestive enzymes made by glands in the lining of the small intestine all play roles in digesting different foods. The small intestine also has remarkable adaptations that cause the further breakdown of nutrient molecules and enable digested foods to be absorbed. The inner wall of the small intestine has many folds covered with even smaller finger-shaped projections called *villi* (**Figure 21.4**). Cells of villi have ciliated extensions on their surfaces called *microvilli* that provide a large surface area for the absorption of nutrients. The combined areas of folds, villi, and microvilli increase the surface area of the small intestine to that of a singles tennis court! Microvilli have enzymes that extend out of their cell membranes. These help with the final breakdown of foods into individual molecules that are small enough to cross cell membranes and be absorbed into the cells of microvilli. All these

Figure 21.3
Stomach Muscles and Stomach Lining.
(a) The stomach has layers of muscles that run in different directions, allowing it to contract in any direction. **(b)** The stomach lining has a ridged surface. Sphincter muscles control movement of food into and out of the stomach.

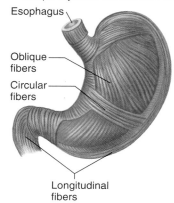

(a) Contractions of the circular, longitudinal, and oblique layers of the stomach churn food.

Esophagus

Oblique fibers

Circular fibers

Longitudinal fibers

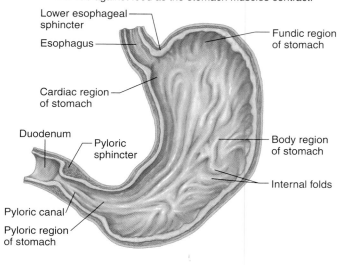

(b) The internal stomach wall has strong, muscular folds that rub against food as the stomach muscles contract.

Lower esophageal sphincter

Esophagus

Cardiac region of stomach

Duodenum

Pyloric sphincter

Pyloric canal

Pyloric region of stomach

Fundic region of stomach

Body region of stomach

Internal folds

one to three amino acids. Other enzymes from the small intestine then break these molecules into single amino acids.

21.3 Macronutrients Provide Energy and Have Other Diverse Cellular Functions

One obvious role of macronutrients—lipids, carbohydrates, and proteins—is to supply energy. The amount of energy in foods is measured in calories. Just as a meter is a standard unit of length, a **calorie** is a standard unit that tells how much chemical energy the body is likely to extract from a quantity of food. The best way to understand calories is compare the number of calories in different foods (**Table 21.2**). Each day women require between 1,300 and 2,200 calories, while men require between 2,300 and 3,000.

When they have been digested and absorbed nutrients from the diet enter the process of cellular respiration and provide cells with energy to break

Table **21.2** The number of calories in a sampling of common foods		
Food	**Amount**	**Calories**
Celery, diced, raw	1 cup (8 oz)	19
Mushrooms, sliced, raw	1 cup (8 oz)	18
Cauliflower, raw	1 cup (8 oz)	24
Apples, sliced, raw	1 cup (8 oz)	63
Blueberries, raw	1 cup (8 oz)	80
Cheerios, no milk	1 cup (8 oz)	90
Shrimp, cooked	1/2 cup (4 oz)	145
Cottage cheese, 1% fat	1 cup (8 oz)	165
Burrito, bean and cheese	Typical size	190
Rice, brown, cooked	1 cup (8 oz)	216
Turkey breast, roasted	1/2 cup (4 oz)	223
Beans, kidney, canned	1 cup (8 oz)	230
Salmon, broiled	1/2 cup (4 oz)	245
Milky Way bar	1 bar (2.05 oz)	245
Chicken breast, no skin, fried	1 whole chicken breast	322

down and build other molcules (**Figure 21.8a**) (see Section 5.2). Collectively, all the enzyme pathways that break down molecules and build up new ones are called **metabolism.** Digestion breaks each kind of macronutrient down into its basic chemical building blocks such as fatty acids, glucose, or amino acids. Glucose is processed by the four phases of glucose metabolism (Figure 21.8b). Fatty acids and amino acids are chemically converted to other molecules that enter various phases of glycolysis, the citric acid cycle, or oxidative phosphorylation. So all of the basic chemical building block molecules can be used as sources of energy in cellular respiration. Having many ways to process the array of biological molecules in food ensures that a multicellular eukaryote will have a reliable supply of ATP energy (Figure 21.8c).

One of the most important principles of nutrition is that any nutrient will play more than one role within an organism. In addition to their role in supplying energy, the chemical elements in nutrients also are used to build the large biological molecules that make up organelles and carry out processes within cells. Let's take a closer look at some of the ways that each type of nutrient is used.

Lipids Form Cell Membranes, Insulate Axons, and Move in the Bloodstream as Cholesterol

As you have seen, lipids are important components of the membranes that surround and compartmentalize cells. In addition, new membranes must be formed when new cells are made, when individual cells grow, or when damaged membranes are repaired.

Cell membranes also are important in nerve cell function. Like insulated electric wires, most nerve fibers are wrapped in thick, fatty *myelin sheaths* that confine the nerve signal to an individual nerve fiber. Myelin sheaths are made of several tightly wound layers of cell membrane, and their formation requires a supply of dietary lipids. Myelin sheaths form during childhood and adolescence, and their proper formation is one reason that young children need a higher-fat diet than adults do. Because mother's milk is high in lipids, it supplies ample raw materials to build cell membranes and other tissues that are high in lipids, including myelin sheaths. Phospholipids are most common in membranes (see Section 4.2), but several other kinds of lipids are also important, such as cholesterol. To make these various kinds of lipids, cells begin with a simple lipid and then use enzymes to add or subtract components, until the desired lipid is achieved.

calorie a standard unit of energy used to express the amount of energy available in food

metabolism all the enzymatic pathways that build up and break down molecules

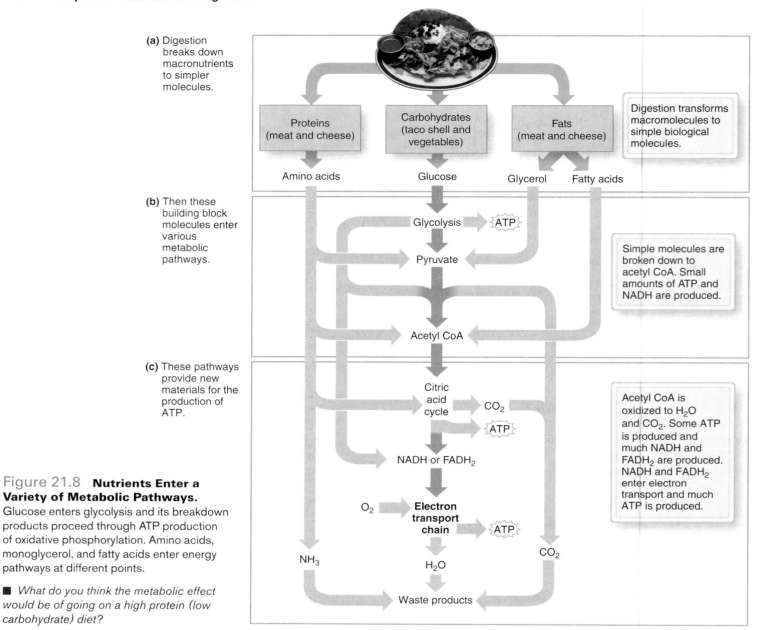

Figure 21.8 **Nutrients Enter a Variety of Metabolic Pathways.** Glucose enters glycolysis and its breakdown products proceed through ATP production of oxidative phosphorylation. Amino acids, monoglycerol, and fatty acids enter energy pathways at different points.

■ *What do you think the metabolic effect would be of going on a high protein (low carbohydrate) diet?*

Cholesterol is a small lipid that plays many crucial roles throughout the body. In addition to its role in cell membranes, cholesterol is the starting point for the production of steroid hormones such as cortisone, testosterone, and estrogen. When you hear the word *cholesterol,* you probably think of heart disease. Abnormally high levels of cholesterol in the blood cause blood vessels to become clogged, which can restrict the supply of blood to the heart. This increases the risk of heart disease. How are diet and blood cholesterol related? The answer probably will surprise you. Many people assume that eating a diet high in cholesterol will lead to high blood cholesterol. Not so. Actually, there is little relationship between dietary choles-terol and blood cholesterol levels. Your body synthesizes cholesterol, and when the amount of cholesterol in your diet increases, your body will make *less* cholesterol. On the other hand, levels of cholesterol in the blood *are* related to dietary fat: as fat intake increases, so does blood cholesterol. One reason is that dietary fat is the raw material for making cholesterol.

When you have your blood cholesterol levels measured, they will be reported as *HDL,* or high-density lipoprotein, and *LDL,* or low-density lipo-protein. HDL often is called the "good cholesterol" and LDL is called the "bad cholesterol." These two kinds of lipoproteins are involved in moving cho-lesterol around the body from one location to another. Cholesterol and other lipids are packed inside small spheres called **lipoproteins** (**Figure 21.9a**). The surface of a lipoprotein is made of a single layer of phospholipids embedded with proteins. The surface proteins allow the sphere to interact with cells throughout the body.

Lipoproteins vary in size, density, and function (Figure 21.9b). Both types of lipoprotein are made in the liver. High-density lipoproteins escort cholesterol to the liver, and from there cholesterol may be sent to the intestines and eliminated from the body. In contrast, low-density lipoproteins move cholesterol from the liver into the bloodstream and to various tissues such as blood vessels. Accordingly, high levels of HDLs mean that more cholesterol is being taken out of the bloodstream, while high levels of LDLs mean that there is more cholesterol in the bloodstream. A greater concentration of cholesterol in the bloodstream increases the likelihood that more of it will form deposits on the inner walls of arteries, making them narrower and decreasing their ability to carry blood. This is why LDL is called "bad cholesterol" and HDL is called "good cholesterol."

Now you can understand why your cholesterol numbers are so important. Your doctor will evaluate your total cholesterol and the relative amounts of HDLs and LDLs. If your HDL cholesterol is high, your doctor is likely to tell you not to worry. What can you do if your LDL cholesterol is too high? Aside from any medications your doctor may prescribe, the answer will sound familiar. Diet and exercise can influence cholesterol levels. High-fat diets, especially diets high in animal and dairy fats, lead to high levels of LDLs, as does a sedentary lifestyle. Low-fat diets and diets high in certain vegetable oils such as olive oil reduce LDLs and raise HDLs, as does regular exercise.

Carbohydrates, Proteins, and Nucleic Acids Provide Important Nutrition

Carbohydrates are an important dietary source of energy for the body. You also may recall that carbohydrates play roles in support structures, such as in the cell walls of plants and the exoskeletons of insects and other invertebrates. In addition, many proteins that extend out of cell membranes have sugar molecules attached to them (see Figure 4.7). These sugar molecules help to bind the cells of a tissue together, especially during early development, before more specialized cell contacts mature.

Proteins are more important in various cell functions than they are as energy sources. Using the 20 amino acids characteristic of life as building blocks, cells make their own proteins. Individual

lipoproteins spheres of lipids combined with proteins that carry cholesterol molecules in the bloodstream

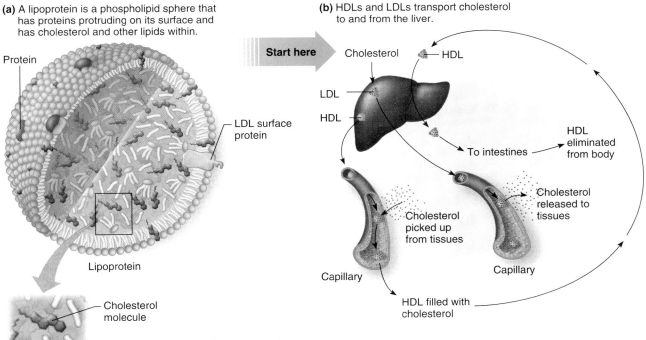

(a) A lipoprotein is a phospholipid sphere that has proteins protruding on its surface and has cholesterol and other lipids within.

Protein

LDL surface protein

Lipoprotein

Cholesterol molecule

Other lipids

(b) HDLs and LDLs transport cholesterol to and from the liver.

Start here

Cholesterol

HDL

LDL

HDL

HDL eliminated from body

To intestines

Cholesterol released to tissues

Cholesterol picked up from tissues

Capillary

Capillary

HDL filled with cholesterol

Figure 21.9 **Lipoproteins Control the Transport of Cholesterol from the Liver to the Rest of the Body.** **(a)** Structure of a lipoprotein. **(b)** Movement of HDLs and LDLs in the bloodstream.

proteins do not last very long in a cell, though, sometimes just a few days. So your cells continually use dietary proteins as one source of the amino acids needed to make new protein. Amino acids also can be chemically modified for special purposes. For example, many of the chemical signal molecules used within the brain are modified amino acids.

You never see nucleic acids listed on the food labels in the grocery store, but nevertheless, you need the building blocks of nucleic acids—nucleotides—to make new DNA. All foods contain nucleotides. Nucleic acids are so important, though, that the body can manufacture the basic nucleotide building blocks from other molecules in the diet.

Too Few Macronutrients Can Result in Malnutrition

When nutrition is so poor that there is a serious risk of death or disease, **malnutrition** has occurred. The most severe forms of malnutrition occur when the amount of macronutrients is extremely low, a condition that is called *undernutrition.* In undernutrition there is not enough energy to carry out life's processes, and there is not enough dietary protein to build the proteins that cells need. Severe shortages of energy and protein have serious consequences at any time of life, but they are more harmful early in life than in adulthood. Special nutritional needs for a new life actually begin before conception and continue all

malnutrition nutrition that is so far below optimal that there is serious risk of death

through pregnancy, childhood, and adolescence. During the early years of life body systems are forming, the number of cells in the body is increasing, the body is growing, bones and teeth are acquiring strength, and important higher brain regions are under construction. **Figure 21.10** shows some of the long-term problems that can result if a person experiences severe undernutrition as an infant or child. Even in adulthood undernutrition has serious consequences (Figure 21.10). Undernourished adults are more susceptible to disease, injuries, and accidents.

The World Health Organization estimates that about 167 million children under 5 years old get too little nutrition to grow properly. More than two-thirds of these hungry children are in Asia, about 20% are in Africa, and about 3% are in South America. Hunger is a genuine problem in the United States too. While only about 1% of American families experience severe hunger, about 12% of families, around 10 million people, in the United States do not get completely adequate nutrition. In the United States hunger has the greatest damaging impact on children. Undernutrition interferes with growth and development, as well as with the learning required if children are to become healthy, productive adults who can take care of their own families.

Eating Disorders Can Lead to Severe Malnutrition While severe hunger is not common in most developed societies, some individuals who have access to plenty of food are still at risk for malnutrition. In Western cultures, especially the United States, people with eating disorders limit their own nutritional intake. The most common eating disorders are anorexia nervosa, and bulimia nervosa. In *anorexia nervosa* individuals have a distorted body image that prevents them from seeing that they are underweight. In attempting to lose even more weight, anorexics starve themselves. Many anorexics also exercise excessively, in a further attempt to limit body fat. Anorexia puts a severe nutritional toll on body systems. Just a few of the possible consequences of anorexia are fatigue, dry skin, osteoporosis, risk of bone fractures, low blood pressure, low body temperature, heart arrhythmias, hair loss, headaches, dizziness, and abnormal menstrual periods. Health problems can persist in anorexics long after their eating has returned to normal. Anorexics are at much greater risk of death than people who eat normally, and suicide is a greater risk for anorexics. *Bulimia nervosa* is characterized by alternating periods of self-starvation and uncontrolled eating called bingeing.

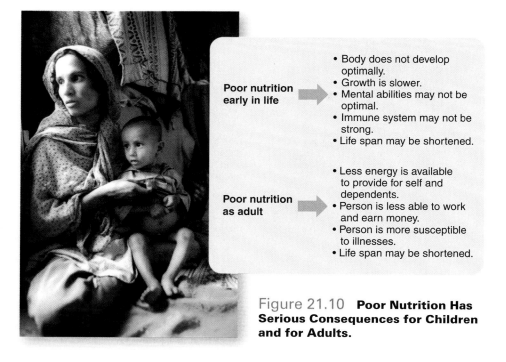

Poor nutrition early in life
- Body does not develop optimally.
- Growth is slower.
- Mental abilities may not be optimal.
- Immune system may not be strong.
- Life span may be shortened.

Poor nutrition as adult
- Less energy is available to provide for self and dependents.
- Person is less able to work and earn money.
- Person is more susceptible to illnesses.
- Life span may be shortened.

Figure 21.10 **Poor Nutrition Has Serious Consequences for Children and for Adults.**

Bulimics eat huge amounts of food and then take laxatives or make themselves vomit to purge their systems of food. Chronic episodes of bingeing and purging take an enormous toll on the body. They cause mouth sores and tooth decay from frequent exposure to stomach acids and disturbances of ion balances in body fluids.

It is difficult to get an accurate picture of how many people have eating disorders, but at any given time up to 5% of adolescent and young girls suffer from anorexia, while another 5% suffer from bulimia. Even though most young girls do not develop eating disorders, as many as half diet in unhealthy ways in response to dissatisfaction with their bodies, and such dieting is one factor along the road to an eating disorder. What causes such a severe aberration in normal eating patterns? The answers are complex.

From the time eating disorders were first recognized, people have blamed the culture of thinness that movies, TV, and advertisers promote. Young women who are movie stars and models are thinner than what would be a healthy weight, yet they are admired as ideals of beauty and attractiveness.

Beyond cultural expectations, there is increasing evidence that certain types of psychiatric disorders may be associated with eating disorders. Not only are girls with eating disorders likely to be depressed or have compulsive behaviors, families with a history of such disorders are more likely to have a girl with an eating disorder. The girl with an eating disorder also is likely to be a perfectionist in the extreme, is likely to strive to please others more than most people do, and is likely to view her control of eating as a triumph over her own impulses. As you can see, eating disorders are complex, and people with eating disorders can have complex motivations for maintaining them. The severe nutritional consequences to eating disorders extend far beyond simply not eating enough food.

Overnutrition Carries Serious Health Risks

While less prevalent from a world perspective, some countries are increasingly facing the costs associated with overnutrition of macronutrients. This simply means that many people in developed countries, especially the United States, consume more food energy than they need. In American culture it often is difficult to maintain a healthy diet. Americans are inundated by food choices. The size of servings in restaurants has dramatically increased until many restaurant portions are equivalent to more than a whole day's worth of calories (**Figure 21.11**). Food is synonymous with fun and pleasure, and when there

is such a variety and abundance of foods, it can be hard to restrict yourself to eating a reasonable amount of food. These circumstances lead many Americans to eat too much and become overweight.

The excess energy is stored as body fat, causing an increase in body weight. **Table 21.3** gives health risks as a function of a measure called BMI, or body mass index. BMI summarizes your weight relative to

Figure 21.11   Big Portions. Portion sizes at restaurants, stores, and at home have increased dramatically over the past 20 years.

■ *How many ounces is a typical steak? How many ounces is a typical hamburger?*

Table 21.3	Health consequences of obesity*—Health risks as a function of body mass index (BMI[†])	
BMI range	**Relative risk of disease**	**Specific examples**
19–24	Minimal	Lowest health risk for all major diseases.
25–26	Low	Small health risk across the board for a variety of conditions.
		Recommendation: do not gain weight but probably not necessary to lose weight.
27–29	Moderate	Increased risk for diabetes, heart disease, high blood pressure, and other conditions.
30–34	High	Marked increase in risk for diabetes, heart disease, and high blood pressure. Both quality and length of life are reduced.
35+	Very high	High risk of disorders associated with body weight. Mobility, quality of life, and life span dramatically impaired.

*From "Shape up America" website 19-24.
[†]BMI measures body weight relative to your height. To calculate your BMI use the following formula: BMI = body weight in kilograms/(height in meters), squared. Several websites have automatic BMI calculators. Here's how to calculate your own BMI: (1) multiply your weight in pounds by 0.45 to get your weight in kilograms; (2) Multiply your height in inches times 0.254 to get your height in meters; (3) Multiply your height in meters by itself; (4) Divide this number into your weight in kilograms. For example–if you weigh 130 lb and are 5'4" tall: 130 × 0.45 = 58.5 = your weight in kilograms; 64 inches × 0.0254 = 1.6 = your height in meters; 1.6 × 1.6 = 2.6425; 58.59/2.64 = 22.17; BMI = 22.17.

your height. As you can see, as BMI increases, the risk of many health problems also increases.

Three major risks are associated with obesity: high blood pressure, cardiovascular disease, and diabetes. High blood pressure means that there is too much pressure from blood on the walls of arteries, and this increased pressure raises the risk that a blood vessel will rupture, causing damage to surrounding tissue. If a blood vessel ruptures in the brain it is called a *stroke,* and strokes are one of the leading causes of death in the United States. Cardiovascular disease involves the heart and blood vessels and can take many forms. For example, arteries can have deposits of cholesterol that impede the flow of blood. If the arteries that supply the heart are clogged, then the heart muscle may become deprived of oxygen, which can cause damage to the heart, or a *heart attack.*

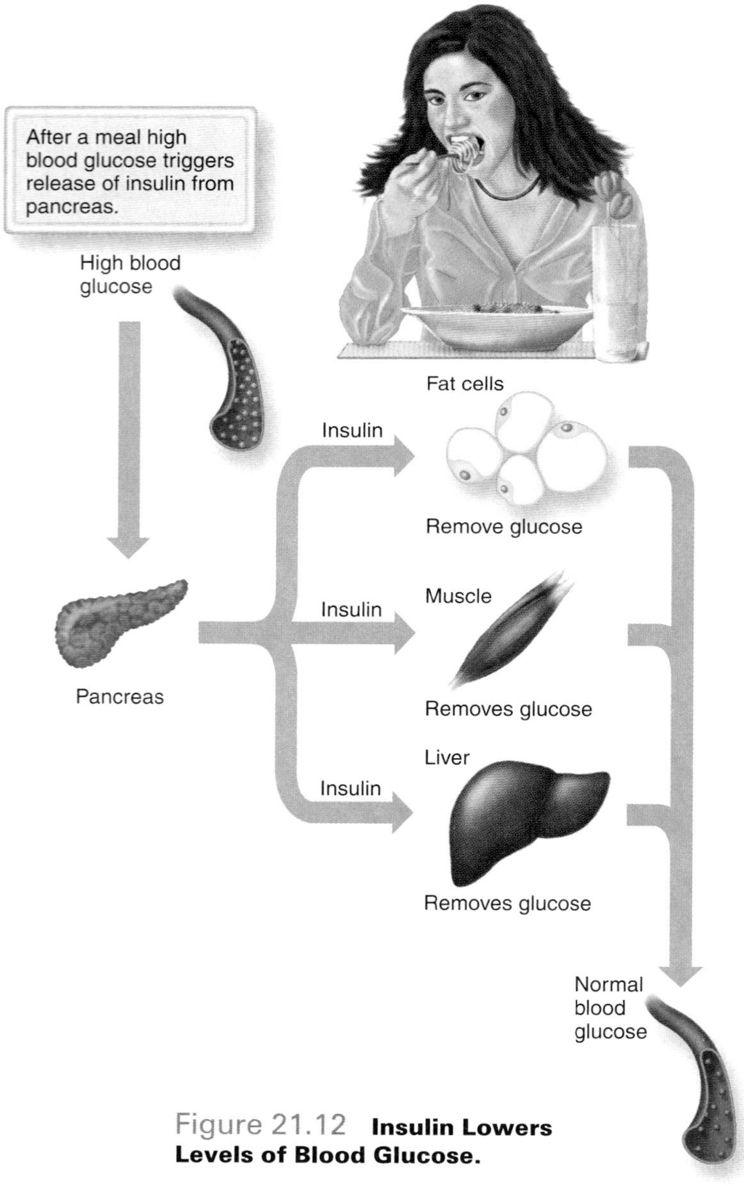

Diabetes Is a Disorder in How Glucose Is Used *Diabetes mellitus* is perhaps the most widespread effect of obesity. There are in fact two kinds of diabetes: Type 1 and Type 2. Type 1 usually occurs early in life and is not related to obesity, while Type 2 occurs later in life and is directly related to obesity. Over 15 million U.S. adults suffer from Type 2 diabetes, and it is increasingly a problem in children as well. Diabetes is a problem with the regulation of blood glucose. After you have eaten a meal, any carbohydrates in your food will be carried in the bloodstream as glucose. Humans evolved with a pattern of intermittent large meals, and one large meal supplies more than enough glucose to satisfy the body's immediate energy needs. Rather than have the extra glucose wasted by being excreted through the kidneys, mechanisms evolved to store those glucose molecules for a later time when food is not available. *Insulin* plays a major role in these mechanisms (**Figure 21.12**). Insulin is a hormone produced by cells of the pancreas, and it is released after a large meal. Insulin stimulates cells, especially those in the liver and muscle, to take up glucose and use it to build a larger storage molecule, such as glycogen.

In Type 1 diabetes the pancreas does not produce enough insulin, and so people with this disorder must take supplemental insulin to function normally. In contrast, people with Type 2 diabetes produce enough insulin, but their response to insulin is inadequate. Insulin acts through a membrane receptor. As a result of taking in excess calories for a long period of time, the insulin receptors change, and they no longer respond as strongly to the insulin signal. Both Type 1 and Type 2 diabetes lead to high blood glucose levels and these lead to a variety of problems (**Figure 21.13**). These health problems are uncomfortable and some are life threatening. One important lesson about Type 2 diabetes is that it can be prevented by eating a healthful, varied diet, by maintaining a healthy weight, and by exercising regularly. For example, regular exercise can increase the response of tissues to insulin and so help prevent Type 2 diabetes.

After a meal high blood glucose triggers release of insulin from pancreas.

High blood glucose

Insulin

Pancreas

Fat cells

Remove glucose

Insulin

Muscle

Removes glucose

Insulin

Liver

Removes glucose

Normal blood glucose

Figure 21.12   **Insulin Lowers Levels of Blood Glucose.**

QUICK CHECK

1. Why isn't there a minimum daily requirement for nucleic acids?

2. What are the risks of habitual overeating? Of habitually eating poorly?

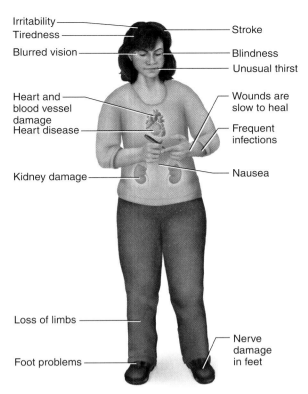

Irritability
Tiredness
Blurred vision
Heart and blood vessel damage
Heart disease
Kidney damage
Loss of limbs
Foot problems

Stroke
Blindness
Unusual thirst
Wounds are slow to heal
Frequent infections
Nausea
Nerve damage in feet

Figure 21.13 **Health Problems Associated with Diabetes.**

21.4 Micronutrients Are Vitamins and Minerals That Are Needed in Small Amounts

Life's complex and varied processes require a variety of chemicals beyond macronutrients. Currently, about 40 micronutrients are recommended for minimal health, and others probably will be discovered as more is learned about human nutrition. The known micronutrients can be divided into two general categories: vitamins and minerals. Let's consider the nutritional roles of vitamins in general and then focus on vitamins C and D before turning to the nutritional roles of minerals, highlighting how calcium contributes to health.

Vitamins Are Organic Micronutrients That the Body Cannot Manufacture

To qualify as a **vitamin** a molecule must be an organic compound that the body cannot make and that is essential in the diet in small amounts for minimal normal health. If a vitamin is missing from the diet, a typical individual will eventually experience severe health problems. Following this definition, any molecule that is useful for health but is not

essential for health is *not* a vitamin. For instance, if you look at the label for a vitamin B-complex supplement, you often will see choline, inositol, and PABA (short for para-aminobenzoic acid) listed as ingredients. But these substances are not vitamins. They are small organic compounds used in cellular processes, but the body can make them from amino acids. Because there have been no reports of health problems when these compounds are missing in the diet, at this time they are not defined as vitamins.

Vitamins come in two varieties: water-soluble vitamins and fat-soluble vitamins (**Table 21.4**). As the name implies, water-soluble vitamins are polar and so they dissolve readily in the bloodstream,

vitamin a small organic molecule that is an essential micronutrient

Table 21.4	Roles of vitamins in body processes	
Vitamin	**One chemical role**	**Example of role in overall health**
Water soluble		
C	Required for an enzyme that synthesizes collagen	Collagen is a protein excreted to the outside of cells. Collagen helps keep tissues firmly together, and keeps blood vessels from leaking.
Thiamin	Cofactor for enzymes in the citric acid cycle	Important for the production of ATP energy from glucose.
Riboflavin	Cofactor for enzymes in oxidative phosphorylation	Important for the release of energy from glucose.
Niacin	Cofactor for a variety of enzymes	Important for release of energy from fats and proteins, and for manufacture of certain hormones.
B_6	Cofactor for enzymes used in metabolism of amino acids	Important for any functions requiring interconversion of amino acids from one to another.
Fat soluble		
K	Cofactor for enzyme that modifies the amino acid glutamate in existing proteins; allows a protein to bind calcium ions	The proteins that bind calcium ions as a result of vitamin K activity are involved in blood clotting.
A	Acts during embryonic development to stimulate maturation of cells and tissues	Needed for normal development of fetus; involved in formation of visual pigments, maintenance of normal epithelial structures.
D	Acts as a hormone to stimulate the synthesis of a protein that increases calcium absorption from intestine.	Needed at moderate concentrations to promote integration of calcium into bone and promote bone density.
E	Uncertain	May be required for normal reproductive functions; powerful antioxidant, promotes wound healing, cofactor in electron transport, protects red blood cells from hemolysis.

(a) When the coenzyme is not present, X and Y do not react with the enzyme.

Enzyme

(b) When the coenzyme is bound to the enzyme, X and Y react with it.

Coenzyme

X
Y

(c) When the product, XY, is released, the reaction is complete.

XY

Figure 21.14 Many Enzymes Require Coenzymes. (a) In the absence of a coenzyme no chemical reaction occurs. **(b)** When a coenzyme is bound to an enzyme, a chemical reaction proceeds. **(c)** Enzyme and coenzyme are not incorporated into the product of the reaction.

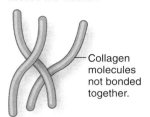
Without vitamin C collagen molecules are not bonded together and tissues are weaker.

Collagen molecules not bonded together.

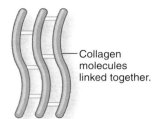
With vitamin C collagen molecules are tightly bonded into resilient tissues.

Collagen molecules linked together.

Figure 21.15 Vitamin C Is a Coenzyme in the Formation of Collagen.

rashes. In contrast, water-soluble vitamins such as vitamin C are polar molecules that are easily flushed out of the body in urine and sweat, and so these vitamins usually cannot accumulate to toxic levels.

What role do vitamins play in the normal processes of life? Unfortunately, there is no universal answer because each vitamin plays a different chemical role and therefore has a unique significance for overall health. The chemical roles of some of the currently identified vitamins are listed in Table 21.4. We will discuss a few examples.

Water-Soluble Vitamins Often Are Coenzymes Many water-soluble vitamins are required for certain enzymes to function properly. A chemical required for a specific enzyme to do its job is called a **coenzyme** (**Figure 21.14**). An enzyme can catalyze the reaction between two molecules when the coenzyme is present, but when the coenzyme is absent the enzyme does not function as well as a catalyst. Note that, like an enzyme, a coenzyme is not changed or used up in the reaction. This is one reason that vitamins are needed in such small amounts. They can be used again and again, in repetitions of the same reaction. Vitamin C and the various B vitamins are examples of coenzymes. When vitamin C is present, the molecules that form collagen are linked together, forming a strong, resilient, tissue (**Figure 21.15**). When vitamin C is absent, the molecules that form collagen are not strongly joined together and tissues are weaker.

Some Fat-Soluble Vitamins Act Like Hormones The fat-soluble vitamins often function differently from the water-soluble vitamins. For example, vitamin D acts more like a hormone than like a water-soluble vitamin. Hormones are substances that are made and released in one part of the body for the specific purpose of having an

body fluids, and cytosol, but they can't cross cell membranes without some special mechanism. Excess water-soluble vitamins are excreted by the kidneys in urine. By contrast, fat-soluble vitamins are nonpolar organic molecules that easily cross cell membranes and enter cells. Unlike water-soluble vitamins, fat-soluble vitamins are not carried readily in the bloodstream, and they do not dissolve in body fluids or in cytosol. Special lipid structures carry fat-soluble vitamins in the bloodstream, ferrying them from one place to another. When these lipid structures reach their destination, the fat-soluble vitamins are released to do their jobs.

The ease with which fat-soluble vitamins enter cells leads to a potentially dangerous characteristic: if you consume more than the recommended amounts of fat-soluble vitamins, they can accumulate to toxic levels and cause damage. For example, a single high dose of vitamin A can cause nausea, vomiting, headache, blurred vision, coma, and death. Most people would not get too much vitamin A from their diet, but you could get too much vitamin A by taking vitamins in pill form. The problems that develop with prolonged high doses of vitamin A include weight loss, muscle and joint pain, liver damage, eye damage, dry scaling lips, and skin

coenzyme a small chemical that binds to an enzyme and is required for the enzyme to catalyze a specific reaction

(a) Skin cells exposed to sunlight synthesize vitamin D.

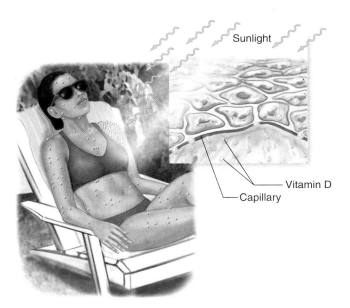

Sunlight

Vitamin D

Capillary

(b) Vitamin D is transported to the intestines.

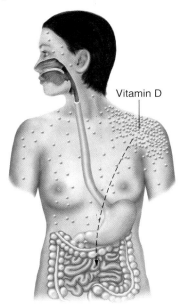

Vitamin D

(c) Vitamin D increases the absorption of calcium from foods.

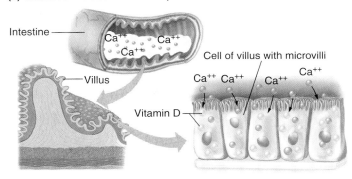

Intestine

Ca^{++} Ca^{++} Ca^{++}

Villus

Cell of villus with microvilli

Ca^{++} Ca^{++} Ca^{++} Ca^{++}

Vitamin D

Figure 21.16   **Vitamin D Promotes the Absorption of Calcium.**

effect in another part of the body. Vitamin D's major action is to promote the absorption of calcium, moving it first from food into cells of the small intestine and finally into the bloodstream (**Figure 21.16**). From there the absorbed calcium can be incorporated into bone. Vitamin D also is an unusual vitamin because, unlike most others, the human body can manufacture it. Sunlight stimulates certain enzymes in the skin to produce a molecule used to make vitamin D. If sunlight is all that is needed for the skin to produce the inactive form of vitamin D, why is it a vitamin? The answer is that many people do not get enough sunlight to make the vitamin D that they need. Because there are only a few dietary sources of vitamin D—salmon and tuna, as well as milk, egg yolks, and butter—in many countries vitamin D is added to milk. This practice has been fairly effective in making sure that people get sufficient amounts of vitamin D. Vitamin or calcium supplements often include vitamin D. Never-

theless, recent studies suggest that many adolescents in Western cultures have a vitamin D deficiency, perhaps because they are drinking lots of soda and not much milk.

An interesting twist to this story is that like other nutrients, one can have too much vitamin D. If vitamin D levels get too high, calcium is actually pulled out of bone, leading to bone weakness. This could be one reason why people whose ancestors evolved in tropical regions have darker skins. Darker skin filters sunlight, which could have many benefits in a sunny tropical environment, such as protection against skin cancer, but darker skin also would protect against the production of too much vitamin D under tropical sun.

Minerals Are Inorganic Substances Required for Normal Cell Function

In the context of nutrition a **mineral** is an inorganic substance needed in the diet in small quantities (**Table 21.5**). Many minerals such as sodium, chloride, potassium, calcium, and magnesium are dissolved as ions in the cytosol or extracellular fluid. If these ions are not kept at just the right levels, then cells and the entire organism will not function properly. Ions are lost through a number of normal

mineral an essential micronutrient that is a simple element

Table 21.5 Minerals and their roles in body processes

Mineral	One chemical role	Example of deficiency or excess
Calcium	Muscle contraction	Low levels cause loss of bone density.
Chloride	Nerve communication	None known.
Chromium	Response to ingested glucose	Low levels cause inability to respond to glucose.
Copper	Collagen synthesis	Low levels can stunt growth.
Fluorine	Building tooth enamel	High levels cause kidney damage, mottled teeth. Low levels increase risk of cavities.
Iodine	Synthesis of thyroid hormone	Low levels can cause abnormal development.
Iron	Binds oxygen in hemoglobin	Low levels cause weakness, lethargy.
Magnesium	Nerve communication	Low levels cause nausea.
Phosphorus	Bone structure	Low levels cause loss of bone density.
Potassium	Nerve communication	High levels lower blood pressure. Low levels cause muscle cramps, fatigue.
Selenium	Antioxidant	High levels cause nausea, fatigue. Low levels cause muscle weakness, pain.
Sodium	Nerve conduction	High levels increase blood pressure. Low levels cause muscle cramps.
Sulfur	Amino acid structure	None known as long as protein levels in diet are adequate.
Zinc	Regulation of protein synthesis, wound healing	High levels decrease absorption of other minerals. Low levels cause poor healing of wounds.

processes including sweating, urinating, bowel movements, and vomiting. These lost ions must be replaced to maintain normal body functions.

Many individual minerals serve more than one function in the body, and calcium is a good example. Calcium helps to make bones and teeth hard. It also is critical for muscle contraction. When calcium levels in the blood are high, extra calcium is incorporated into bones and teeth. In contrast, if there is not enough calcium in the bloodstream to provide all of the calcium that the body needs, bones and teeth are in trouble because calcium is pulled out of them to provide adequate levels to the rest of the body.

Minerals Play Other Important Roles Like some vitamins, minerals also contribute special functions to enzymes and other proteins. One familiar example is the iron that is a part of the protein hemoglobin in red blood cells. Hemoglobin is a large protein composed of four protein subunits. Each protein subunit is folded around an iron atom, and it is the iron that actually binds to and carries oxygen in the blood. Many cellular enzymes that catalyze chemical reactions require mineral cofactors that are in the form of ions. Cofactors include familiar minerals like iron, copper, zinc, and nickel as well as less familiar minerals like manganese, cobalt, molybdenum, vanadium, and selenium. All these minerals are essential because they give enzymes capabilities they would not otherwise have.

Too Much or Too Little of a Micronutrient Can Cause Health Problems

Every few years the National Academies of Science—an organization that advises the government on scientific issues—appoints a panel of scientists who examine and summarize all of the relevant scientific information about nutrition. One of the many guidelines produced by this panel is the **recommended dietary allowance (RDA)** for vitamins and minerals. The RDA is the amount of a vitamin or mineral needed by a typical person to avoid the symptoms of a vitamin or mineral deficiency. For instance, the RDA for vitamin C is 60 milligrams per day for a normal adult. This is a generous midpoint between the 10 milligrams necessary to prevent the symptoms of vitamin C deficiency (skin spots, bleeding gums, and loose teeth)

recommended dietary allowance (RDA) the amount of a nutrient that will allow a person to avoid diseases from deficiencies or excesses of that nutrient

and the 100 milligrams that will flood body fluids with vitamin C. In recent years, however, many additional functions for nutrients have been suggested, higher doses of some vitamins have been recommended by some people, and additional nutrients have been claimed to be essential for minimal health. The examples listed in **Table 21.6** demonstrate that people have become interested in whether nutrients can enhance health, even though these substances might not be necessary to maintain a minimal level of health and absence of disease. The sources of claims for additional functions are diverse and include cultural traditions, individual experiences, and spin-offs from new understandings of basic biology.

Even if your diet provides adequate calories and protein, you can still suffer health problems if you don't get enough of each micronutrient. For example, vitamin C deficiency causes a syndrome called *scurvy*. The symptoms are weakness, aching joints, bleeding gums, loose teeth, loss of appetite, and other severe problems that eventually result in death. In the 1790s James Lind, a British sea captain, observed that if ample citrus fruit were included in the seafaring diet, the symptoms of scurvy among sailors could be treated and prevented. We now know that a lack of vitamin C causes all of the symptoms of scurvy, and that citrus fruits contain a lot of vitamin C.

Scurvy's symptoms can be understood in light of the role of vitamin C in collagen formation. Collagen is a protein that is an important component of the extracellular matrix that holds tissues together, and vitamin C allows the formation of resilient cross-links that form stronger collagen fibers (see Figure 21.15). In scurvy blood seeps out of the skin and gums because the collagen around the cells that compose blood vessel walls does not adhere well and this allows blood to leak out of the blood vessels. Teeth become loose because without properly linked collagen, cells at the roots of teeth are not firmly attached to the underlying bone cells.

If vitamin D is lacking during early development, bones do not form properly, a condition called *rickets* (**Figure 21.17**). Children with rickets have abnormally shaped bones that will not support their weight. Another example of a vitamin deficiency, a syndrome called *beriberi*, results from too little thiamin, a B vitamin involved in the synthesis of neurotransmitter molecules. Beriberi involves nerve tingling, paralysis, and heart failure. Vitamin A deficiency is particularly harmful during embryonic development. Lack of vitamin A impairs eye development, so mothers whose diets are deficient in vitamin A often give birth to blind babies. Vitamin K deficiency can lead to problems with

Table 21.6	Additional roles proposed for essential nutrients and other chemicals in promoting health and longevity	
Nutrient	**Proposed usefulness**	**Status of scientific evidence**
Vitamin C	Reduces severity and duration of common cold	Small effect, if any
Echinacea	Reduces severity and duration of common cold	Mixed results, effect unclear
St. John's wort	Relieves depression	Small effect, if any; some evidence for adverse interactions, and interference with other medications
Selenium	Helps prevent cataracts	Complete lack and higher doses cause cataracts
Oat bran and other fiber	Reduces blood cholesterol, heart disease, and colon cancer	Conflicting evidence
Ginger	Reduces motion sickness and other nausea symptoms	Studies show mixed results
	Increases time needed to empty stomach	Some positive evidence
	Lowers cholesterol	Not much positive evidence
	Kills flu virus	Research not available
	Reduces inflammation	Mixed results

This child has the vitamin D deficiency disease, rickets.

 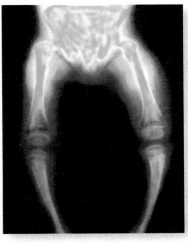

Figure 21.17 **Rickets Is Caused by a Lack of Vitamin D.**

blood clotting. If blood does not clot effectively, a minor injury may lead to excessive bleeding and even to death. It is unlikely that you will suffer from vitamin K deficiency, though, because every day some of the billions of bacteria that live in human intestines make about half of what you need.

In modern cultures a varied diet is relatively easy to come by, vitamins are added to many foods, and micronutrients are available in vitamin pills. Still, despite all of this abundance, vitamin deficiencies do occur. As previously noted, many U.S. teenagers show a vitamin D deficiency. Also right here in the U.S., older people, vegetarians, and people too poor to buy a variety of healthful foods often have a vitamin B^{12} deficiency. Many women do not get enough calcium to keep their bones strong throughout an otherwise long and healthy life. In developing countries around the world, vitamin deficiencies continue to be a problem.

At the other end of the spectrum, it's possible to take too many vitamins. Many alternative health sources promote taking megavitamins with hundreds to thousands of times greater than the RDA of various micronutrients. Few if any of the positive claims for the effects of megavitamins such as cancer prevention and treatment have been supported by research studies. More troubling are findings that large doses of vitamins may do harm. For example, vitamin E supplements were thought to reduce heart disease and cancer, but well-designed studies failed to show any such benefit. Instead, regular intake of high doses of vitamin E, 400 I.U. or more per day for several years, may increase a person's overall chance of dying from any cause.

QUICK CHECK

1. Why is it important to get enough calcium in your diet?

21.5 Claims Concerning Other Functions of Nutrients Must Be Evaluated Carefully

Because the scientific investigation of any nutritional claim requires years of intensive, careful work, as well as a source of financial support for that work, reliable studies are not available for every dietary supplement. In general, claims that are "too good to be true" probably are. Use your knowledge of biology and your critical-thinking skills to evaluate whether claims make sense. For example, if a product claims to keep you awake and alert but to be "all natural" and caffeine free, you should be suspicious that the product contains something similar to caffeine or some other strong stimulant and treat it with appropriate caution. Let's explore the biological and scientific status of some selected claims for alternate roles for basic nutrients and other chemicals.

fiber indigestible complex carbohydrate found in fresh fruits and vegetables

Is Vitamin C a Cure for the Common Cold?

As serious infectious diseases have come under control, people seem to be even more annoyed by the common cold. It is not surprising, therefore, that many products claim to prevent colds. The use of vitamin C to prevent and cure a cold is one of the most well-known examples. Several studies have found that for a normal, healthy person who eats a balanced diet, even large doses of vitamin C—1,000 mg or more—do not appear to do much to help get rid of cold symptoms, or to prevent colds from occurring. This makes sense because vitamin C is water soluble, and most of the excess is carried out of the body with urine.

Does Dietary Fiber Prevent Colon Cancer?

Nutritionists only recently have realized that polysaccharides, also called complex carbohydrates, do much more than provide energy. One important role of these molecules is to provide **fiber,** a complex carbohydrate that is not readily digested. Cellulose, lignin, and pectin are examples of indigestible dietary fiber. Diets rich in vegetables, fruits, and whole grains have more fiber than diets heavy in meat, starch, and refined sugar. You might wonder what good a nutrient can do if it is not digested and absorbed. For one example, as fiber passes through the intestines, it stimulates them to be active and thus shortens the amount of time that food stays in the intestines. This seems to be a good thing because diets high in fiber are associated with both lowered risk of colon and other cancers and lowered risk of heart disease and high blood pressure. People who live in cultures that eat a high-fiber diet have lower rates of all of these diseases. When such people change their diets—for example, when they adopt a more Western-style diet after moving to the United States or Europe—their risks of heart disease, colon cancer, and high blood pressure increase.

One question about dietary fiber is how you should change your diet to include it. In most studies the people who eat a lot of fiber get it through fruits, vegetables, and whole grains. It is common, though, for commercial food products to try to capitalize on possible health benefits of their products. For example, breakfast cereals and other foods are promoted by touting their value in preventing colon cancer, and you can buy fiber products in pill or powder form. But don't be fooled. Simply adding a high-fiber cereal to your existing diet is not likely to decrease your risk of colon cancer later in life. Only systematic changes in your overall diet are likely to lead to more health and less disease.

21.6 What Should You Eat?

The greatest nutritional concern in the United States is that the typical American diet is out of balance. The imbalance sometimes is driven by poverty, but more often it is simply the result of poor choices by people who could choose to eat a healthful, nutritious diet.

In America, diet is big business and trying to find the right approach can be costly. Advertisements for books, programs, herbal medications, exercise machines, and many other products claim that you *can* lose weight if only you buy their products. Some diets admonish you to eat ultra-low-fat foods; others emphasize ultra-low-carbohydrate foods. In reality, the answer to good nutrition, a healthy diet, and a reasonable weight is not that hard—or expensive—to come by.

Eat a Balanced, Varied Diet

Good nutrition involves just a few simple guidelines and a lot of common sense. These guidelines—for a typical American—can be summarized and easily remembered (**Figure 21.18**). Every day:

- Eat plenty of whole grains, fresh whole fruits, and a variety of vegetables.
- Eat low-fat or fat-free calcium-rich dairy products.
- Eat lean meats that are baked, broiled, or grilled. Eat a variety of proteins, including beans, fish, peas, nuts, and seeds.
- Keep fat, sugar, salt, and alcohol relatively low.

Add to this a vigorous exercise program that matches your abilities and overall conditioning, and you can be healthy and happy, at least where your weight is concerned.

How do these recommendations line up with the diet fads that sweep the country from time to time? The answer, again, is balance. Most diet fads encourage people to be extreme in one feature of their diet or another. Only by being extreme can they stand out and . . . sell books. It is not so exciting to try to sell a book that says eat a balanced diet, yet this is the real answer. Let's look at an example.

A recent popular diet encourages people to avoid carbohydrates and eat lots of protein and fat. The proponents argue that people have been told for years to eat a low-fat diet, and it has not worked because Americans are still obese. They argue that carbohydrates, not fat, cause weight gain. These arguments sound convincing and scientific, and people *have* lost weight on this kind of diet. But in the long run everything known about nutrition yells that this diet is not healthy. Why?

One thing to consider in this high-protein diet is the role of dietary carbohydrates. You have read that there are many kinds of carbohydrates, from sugar to chitin. The carbohydrates that most Americans eat have been modified to produce products like white flour and white sugar. These refined carbohydrates are high in calories and low in any other nutritional value. If your mother told you to eat whole-grain bread rather than white bread, she was right. An excess of refined carbohydrates certainly contributes to obesity. On the other hand, many healthy foods—fruits, vegetables, and whole grains—also have carbohydrates. The carbohydrates in these foods are diverse, provide a variety of nutrients, and contain many micronutrients that can contribute to overall health. At the same time there is abundant evidence that a diet high in animal fat is *not* healthy. So if you eat the bacon and eggs and skip the whole-grain toast and strawberries, you may lose weight in the short term, but you will not be enhancing your long-term health.

As simple as they are, these commonsense recommendations are not easy for many Americans to follow. Even without fad diets most Americans consume relatively large amounts of fat and protein—far more than are necessary to promote health. The U.S. government recommends that no more than 30% of your daily calories should come from fat, and some studies show dramatic health benefits if fat is cut to 10 or 20%. It is not uncommon for people in the United States to consume 50% or more of their daily calories as fat and much of the rest in refined carbohydrates. And, of course, the kind of fat that you eat makes a difference. From the perspective of maintaining good health, Americans tend to eat too much animal fat and too little of the fat that is in vegetables and nuts.

Your need for protein is relatively low. A typical adult female requires only 40 to 50 grams of protein per day and a male between 50 and 60 grams. People who exercise vigorously need somewhat more, but the amounts are far lower than most Americans eat every day. A chicken breast that weighs about 200 grams contains about 50 grams of protein. A typical lean hamburger patty contains about 20 grams of protein. So if you eat sausage for breakfast, a burger for lunch, and have

Figure 21.18 **What You Should Eat.**

MyPyramid
STEPS TO A HEALTHIER YOU
MyPyramid.gov

GRAINS	VEGETABLES	FRUITS	MILK	MEAT & BEANS
Make half your grains whole	Vary your veggies	Focus on fruits	Get your calcium-rich foods	Go lean with protein
Eat at least 3 oz. of whole-grain cereals, breads, crackers, rice, or pasta every day 1 oz. is about 1 slice of bread, about 1 cup of breakfast cereal, or 1/2 cup of cooked rice, cereal, or pasta	Eat more dark-green veggies like broccoli, spinach, and other dark leafy greens Eat more orange vegetables like carrots and sweetpotatoes Eat more dry beans and peas like pinto beans, kidney beans, and lentils	Eat a variety of fruit Choose fresh, frozen, canned, or dried fruit Go easy on fruit juices	Go low-fat or fat-free when you choose milk, yogurt, and other milk products If you don't or can't consume milk, choose lactose-free products or other calcium sources such as fortified foods and beverages	Choose low-fat or lean meats and poultry Bake it, broil it, or grill it Vary your protein routine — choose more fish, beans, peas, nuts, and seeds

For a 2,000-calorie diet, you need the amounts below from each food group. To find the amounts that are right for you, go to MyPyramid.gov.

| Eat 6 oz. every day | Eat 2 1/2 cups every day | Eat 2 cups every day | Get 3 cups every day; for kids aged 2 to 8, it's 2 | Eat 5 1/2 oz. every day |

Find your balance between food and physical activity
- Be sure to stay within your daily calorie needs.
- Be physically active for at least 30 minutes most days of the week.
- About 60 minutes a day of physical activity may be needed to prevent weight gain.
- For sustaining weight loss, at least 60 to 90 minutes a day of physical activity may be required.
- Children and teenagers should be physically active for 60 minutes every day, or most days.

Know the limits on fats, sugars, and salt (sodium)
- Make most of your fat sources from fish, nuts, and vegetable oils.
- Limit solid fats like butter, stick margarine, shortening, and lard, as well as foods that contain these.
- Check the Nutrition Facts label to keep saturated fats, *trans* fats, and sodium low.
- Choose food and beverages low in added sugars. Added sugars contribute calories with few, if any, nutrients.

Source. U. S. Dept. of Agriculture, April 2005.

chicken for dinner, you have taken in quite a lot of protein. There are however, many other healthful sources of protein. Beans, rice, whole-grain breads and cereals, nuts, pasta, and even fruits and vegetables all contain protein. If you add low-fat dairy products, it is easy to get the protein you need without the fat content of most meats.

Besides eating too much fat and protein, Americans eat far too few fresh fruits and vegetables. Diets high in these foods provide many of the micronutrients that maintain good health and combat disease. When you include whole grains, beans, and occasionally some nuts, you add the benefits of fiber, protein, many additional vitamins, and beneficial kinds of oils. Each different kind of food has a different combination of useful nutrients. Certainly, vitamin supplements have their place, especially for people with special conditions such as pregnancy, very young children, and the elderly, but eating a variety of foods is the best and most natural approach to health and longevity.

QUICK CHECK

1. How could you improve your own diet, based on what you have learned in this chapter?

An Added Dimension Explained
Niacin Deficiency Causes Pellagra

The story of pellagra in the United States is just one example of how out of balance the human diet can become, and how scientists can contribute to the solution to dietary problems. In the early years of the epidemic, explanations for pellagra were based on the geographic distribution of the disease. In the United States pellagra was most common in the southern states and in communities of poor and destitute people; pellagra rarely affected affluent people. These observations led to two early hypotheses that had numerous adherents. Many people suspected that a toxic substance in corn caused pellagra; others guessed that pellagra was an infectious disease caused by a microscopic organism. Those who thought that corn was the culprit encouraged people to eat only unspoiled, locally grown corn products or to give up corn altogether. Those who thought that a pathogen caused pellagra pushed hard for improved sanitation. These recommendations were sometimes helpful to overall health, but unfortunately they did nothing to stop pellagra. In the absence of a clear explanation and no treatment, pellagrins were often treated as outcasts. Many were consigned to state mental hospitals.

The real hero in the pellagra story was Joseph Goldberger, a scientist in charge of the pellagra investigation for the U.S. Public Health Service in 1914 (**Figure 21.19**). An early death in 1929 from cancer ended Goldberger's fight against pellagra, but his contributions still stand as both excellent science and important advances in public health. Goldberger's first step was an extensive field study of the affected areas in the South. Keeping an open mind, Goldberger observed pellagrins and their lives and the following observations formed the basis for his investigations:

- Pellagra was found among profoundly poor people. Whether these people were rural tenant farmers on cotton farms, workers in cotton mills, children in orphanages, or patients in mental institutions, their living conditions were abysmal.

- Working poor southerners seldom grew any of their own food. They didn't have gardens; the cotton fields that they picked for a living grew right up to their houses (**Figure 21.20**). In cotton mill towns, poor workers tended to buy all of their food from company stores that typically provided little variety.

- People at risk for contracting pellagra ate very limited diets, mostly meat, meal, and molasses. Unfortunately, the meat was salt pork, which is a slab of lard with only a thin strip of leaner meat in it. The meal was corn meal; often moldy or infested corn meal. And the molasses was, of course, mostly sugar. So the three *M*'s provided a meal that was high in simple carbohydrates and fats but low in proteins and essential micronutrients.

Figure 21.19 **Joseph Goldberger.** Shown here in his uniform as an officer of the United States Public Health Service, Goldberger discovered the cause of pellagra.

Figure 21.20 **Sharecroppers.** This 1938 photo shows how sharecroppers houses often were surrounded by crops, with little or no space for a kitchen garden.

Goldberger developed the hypothesis that diet was the key to pellagra and that poverty was the key to diet. Goldberger's idea was not a good one just because it turned out to be right. It was a good idea because he expressed it in ways that could be clearly tested. And, unlike others with pet explanations, Goldberger was unrelenting in his efforts to find out if he was right. In other words, he tried hard to prove himself wrong.

First, Goldberger carried out carefully designed experiments to find out if the apparent pellagra/diet connection was real. Goldberger chose two orphanages with high rates of pellagra and put the children there on a more varied and protein-rich diet. The results were dramatic: pellagra virtually disappeared from both orphanages. Affected children got better and new cases ceased to

—Continued next page

Continued—

appear. Goldberger had achieved what so many others had not: in a small setting he had eradicated pellagra.

Goldberger knew that the world would not be convinced unless he found exactly what was missing in the diet that led to pellagra. In parallel studies in people and animals, he systematically added foods—one at a time—to a basic "three M's" diet to see if pellagra's symptoms were relieved. A colleague painstakingly carried out these studies on humans in one mental institution. In his own labs Goldberger studied the effects of foods on "black tongue" disease in dogs, which he had discovered was a canine version of pellagra. To further combat the pathogen hypothesis, Goldberger even injected himself and other volunteers with all manner of "filth" from pellagrins. Preparations of blood, mucus, feces, and urine from pellagrins were administered to the volunteers and no one contracted pellagra. Using this systematic and determined approach, Goldberger was eventually successful in finding a way to combat pellagra. He made the startling discovery that simple brewer's yeast was a potent antipellagrin. The simple addition of brewer's yeast to a nutrient-poor diet had quick and dramatic results. A massive public health campaign to distribute brewer's yeast began in the late 1920s, and although the beginning of the end was in sight, the pellagra epidemic was not entirely over. Goldberger knew that if people were to be healthy in all respects, a varied, healthful diet was essential. But then as now, Americans preferred the quick fix. Getting people to change their unhealthful eating habits was something Goldberger never lived to see. In 1929 when Goldberger died, pellagra was still raging.

So where did pellagra end? Public health campaigns were helpful. People did begin to accept that the three M's did not form a healthful diet, but other factors were probably more important. For one, cotton declined as the South's main crop, and this led to greater variety in agriculture and thus to greater variety in diet. In the 1930s the New Deal helped alleviate the Depression, and then the increased prosperity following World War II combined to improve the standard of living in the South.

By the 1950s people shopped at grocery stores and learned to expect wider food choices, and more people had enough money to feed their families. Most importantly, through voluntary and mandatory guidelines, manufacturers added nutrients to many of their grain and cereal products. This allowed people to continue to eat foods they liked, such as white bread and corn meal, and still get the nutrients needed to prevent pellagra and other nutritional disorders. Finally, the pellagra epidemic was over.

Goldberger and others after him did isolate the antipellagra ingredient in brewer's yeast. It was niacin, one of the B vitamins. Later work solved some other interesting loose ends. For example, Goldberger had good evidence that both meat and milk were good foods for preventing and curing pellagra. It turns out that meat is a good source of niacin, but milk is not. Goldberger had isolated the amino acid tryptophan as the essential antipellagra ingredient in milk. Building on Goldberger's work, later researchers showed that humans can convert tryptophan to niacin. This reinforces the idea that the human body has astounding resources if you give it some basic materials to work with. In another interesting twist, later workers found that, after all, corn indeed did contribute to pellagra. For reasons that are still unclear, diets rich in corn actually *increase* the body's requirement for niacin. So, indirectly, those who originally pointed the finger at corn were right.

The story of pellagra reinforces the need for a varied diet. It demonstrates that when diet becomes too restrictive and narrow, even though calories are sufficient, the body loses it resilience and illness results. But consider the typical current American diet: it has too much fat, too much sugar, certainly more than enough protein, but not enough fresh fruits and vegetables. Americans have yet to fully understand that just as brewer's yeast could not eradicate pellagra in the absence of a healthful diet, vitamin pills and supplements cannot ensure good long-term health if the diet is unbalanced.

QUICK CHECK

1. What causes pellagra?

Now You Can **Understand**

Free Radicals, Reactive Oxygen Species, and Antioxidants

Free radicals are highly reactive chemicals that can combine with and damage biological molecules. Reactive oxygen species are free radicals that are especially reactive because they contain oxygen atoms. Finally, antioxidants are molecules that inactivate reactive oxygen species (**Table 21.7**). Reactive oxygen species play a significant role in health and disease. Evidence suggests that they may cause cancer, aging, Alzheimer's disease, and perhaps even hearing loss. How can one small class of reactive molecules wreak so much havoc?

The problem is that reactive oxygen species can and do interact with the important molecules of life. Damaged DNA can cause the production of abnormal proteins. Oxygen damage to DNA can cause aging as well as cancer and other disease states.

Reactive oxygen species have more positive roles to play, too. For instance, the immune system uses reactive oxygen species to destroy bacteria and viruses. Furthermore, no matter how healthful your diet and lifestyle, your body will produce reactive oxygen species. This appears to be the price that multicellular organisms pay for the ability to use oxygen to extract energy from glucose. And, of course, you can make matters worse by your own choices. For instance, you probably know that cigarette smoke causes

Table 21.7 Some antioxidants

Chemical	Description	Sources
Ascorbic acid	Vitamin C	Fresh fruits and vegetables
Carotenoids	Vitamin A, beta-carotene	Fresh fruits and vegetables
Tocopherol	Vitamin E	Vegetable oils, nuts, green vegetables
Chlorophylls	Plant photosynthetic compounds	Green plants
Bioflavonoids	Plant compounds	Olive oil
Selenium, zinc, copper	Minerals	Varied diet
Glutathione	Three-amino-acid string	Made in body
Lycopene	Similar to red and yellow plant photosynthetic compounds	Tomatoes

cancer, and the likely reason for this is the high concentration of reactive oxygen species and other free radicals in cigarette smoke. Other environmental agents such as ultraviolet light, air pollution, alcohol, and asbestos contain or trigger the production of reactive oxygen species, too.

Is there any solution to this problem? Maybe. Animals produce or obtain from their diet some molecules that react with reactive oxygen species in a variety of different ways, transforming them into molecules that are more stable and less harmful before they have a chance to do damage. Antioxidants interact with and inactivate reactive oxygen species, so antioxidants are important for long-term health and longevity.

The important question is: How should you get a good supply of antioxidants? Manufacturers of supplements will encourage you to buy their expensive products. They base their claims on research identifying vitamins C and E, beta-carotene, lycopene, and other molecules as antioxidants. Evidence is increasing that these substances, as they occur in foods, can help maintain health.

As you might expect from reading this chapter, however, there is little evidence that taking antioxidants in pill form will help. As with other nutritional roads to health, the path to antioxidants is a varied diet with lots of fresh fruits and vegetables.

What Do **You** Think?

Should the Government Regulate Dietary Supplements?

The U.S. government plays a big role in regulating prescription and over-the-counter drugs, and most Americans get upset if they find unexpected risks and side effects of drugs and medications. The Food and Drug Administration (FDA) is supposed to ensure that drugs are both safe and effective. The FDA helps to ensure not only that you know the risks of the drugs you take but also that you are not wasting your money by taking them. But what about all those vitamin and herbal supplements for sale in drug stores, health food stores, and on the Internet? The FDA does not have any formal review or regulation of these supplements. But should they? Proponents of government regulation say that these supplements are indeed drugs. For example, in the early 2000s supplements containing ephedra and other caffeinelike compounds were found to cause seizures and a number of deaths. More recently, many of the claims made by the sellers of supplements—for example, that the supplements prevent cancer or prolong life—have been contradicted by scientific research. Others argue that regulation of supplements is just one more way that the government wants to limit your freedom to choose alternative medicines. Some people argue any regulation of the supplement industry is really just a way to help big drug companies. The debate over regulation of supplements will certainly be influenced by the many scientific studies now under way and planned, but in the end you will have to decide your opinion on this issue. It may influence what candidates you vote for, or what products you buy. What do *you* think about government regulation of the manufacture and sale of dietary supplements?

Chapter Review

CHAPTER SUMMARY

21.1 Nutritional Needs Are Based on Biological Chemistry

Organisms need nutrients to supply energy and as a source of the basic building block molecules that life's structures and processes require. While humans can make many of the molecules required for life, they cannot manufacture essential nutrients and must obtain them from the diet. Macronutrients and micronutrients are needed in different amounts and play different roles, but all are important for proper nutrition.

21.2 Nutrients Must Be Digested and Absorbed

The human digestive system chemically prepares foods for absorption. The process of absorption moves the digested foods into the cells of villi. From there they move into the bloodstream, which distributes them to cells that require them. Organs of the digestive system and digestive glands are specialized for different phases of digestion. Peristalsis moves food along the digestive system. The mucous layer lining the stomach prevents strong stomach acids from damaging the stomach itself. Enzymatic digestion of carbohydrates begins in the mouth, with amylase in saliva, and continues in the small intestine. Secretions of glands in the lining of the small intestine, as well as the liver and pancreas, help break down foods to nutrient molecules that are small enough to be absorbed. The highly folded lining of the small intestine increases its surface area for absorption of nutrients. The large intestine absorbs water from what remains of digested food and compacts it into feces that are expelled from the rectum through the anus. The large intestine contains a culture of *E. coli* that break down nutrients and synthesize vitamin K. Digestion of proteins begins in the stomach and is completed in the small intestine, and lipid digestion takes place within the small intestine. Bile breaks fats into smaller droplets, and pancreatic enzymes break them down to fatty acids.

absorption   507   digestion   507

21.3   Macronutrients Provide Energy and Have Other Diverse Cellular Functions

Macronutrients contain energy in the form of calories. All of the chemical building block molecules can enter glucose metabolism and be processed to yield ATP. Each kind of macronutrient has specific roles within the body. Lipids form cell membranes, insulate axons, and move in the bloodstream as lipoproteins that incorporate cholesterol. HDL is the "good" cholesterol that is on its way to the liver for disposal, and LDL is the "bad" cholesterol that keeps circulating in the bloodstream where it is more likely to form arterial plaques. High-fat diets and a sedentary lifestyle are linked to higher levels of LDLs, while low-fat diets and regular exercise are linked to higher levels of HDLs. Malnutrition and undernutrition can be especially harmful during neonatal development and childhood, but adults that are malnourished are less able to care for themselves and their families. So undernutrition and malnutrition have ripple effects within societies. Hunger continues to be a significant problem in the world, especially in Asia and Africa. Western cultures have undernutrition problems associated with eating disorders such as anorexia and bulimia. Overnutrition is another significant problem in the United States, and obesity increases the likelihood that an individual will develop diabetes, cardiovascular disease, and high blood pressure. In Type 1 diabetes the pancreas produces insufficient insulin. In Type 2 diabetes insulin receptors are insensitive to insulin.

calorie   511   malnutrition   514
lipoproteins   513   metabolism   501

21.4   Micronutrients Are Vitamins and Minerals That Are Needed in Small Amounts

Vitamins are fat or water soluble. Water-soluble vitamins often function as coenzymes. Some fat-soluble vitamins act like hormones. Humans can make their own vitamin D. Minerals are inorganic chemical elements and ions that are required for normal body functions. Some are incorporated into enzymes and other proteins, while others are coenzymes required for enzymes to catalyze chemical reactions. RDAs are guidelines that tell how much of each micronutrient is needed for optimal health. Lack of vitamin C causes scurvy, a deficiency disease characterized by breakdown of collagen in skin, gums, and blood vessels. Lack of vitamin D early in development can cause the skeletal deformations of rickets. Too little vitamin B can cause beriberi. Too little vitamin A impairs eye development. Taking megadoses of micronutrients can be just as dangerous as not having enough of them.

coenzyme   518   recommended dietary allowance (RDA)   520
mineral   519   vitamin   517

21.5   Claims Concerning Other Functions of Nutrients Must Be Evaluated Carefully

Taking megadoses of vitamin C may help avoid colds, but does not cure symptoms more quickly. The role of fiber as a cancer preventative has not been supported by experimental trials, but eating a diet high in fiber is probably a healthful habit.

fiber   522

21.6   What Should You Eat?

You should eat a balanced, varied diet that has low amounts of fats, moderate amounts of proteins, and plenty of fresh fruits, vegetables, and whole grains. Eat fat from vegetable sources and avoid fat from animal sources. The new food pyramid recommends that regular exercise is part of a healthy lifestyle.

21.7   An Added Dimension Explained: Niacin Deficiency Causes Pellagra

Pellagra is a nutritional deficiency disease caused by lack of niacin, one of the B vitamins. Joseph Goldberger devoted much of his life to solving the problem of pellagra, and by adding brewer's yeast to poor diets, Goldberger dramatically reversed pellagra. Social changes and increased prosperity have changed the way that Americans eat, and as nutrition has become more varied, the number of cases of pellagra has decreased.

REVIEW QUESTIONS

TRUE or FALSE. If a statement is false, rewrite it to make it true.

1. The best diet is varied, low in protein and processed sugars, high in complex carbohydrates, with lots of fresh fruits and vegetables.

2. HDLs are the "good" cholesterol that take cholesterol to the liver for disposal; LDLs are the "bad" cholesterol that keeps it circulating in the bloodstream.

3. The body processes atoms, proteins, and fats into forms that can enter into glucose metabolism, the citric acid cycle, and electron transport.

4. If a child is malnourished in infancy, it won't matter because it has plenty of nutrients stored from prenatal life.

5. Many of the effects of malnutrition in adulthood impair normal functioning of families and society.

6. Villi and microvilli allow places for cultures of *E. coli* to lodge within the large intestine.

7. Within the body most minerals are in the form of ions dissolved in body fluids.

8. Vitamin C can prevent colds.

9. If you don't have the proper amounts of macronutrients, you will develop deficiency diseases like scurvy, rickets, and pellagra.

10. Joseph Goldberger solved the mystery of pellagra by administering brewer's yeast and prescribing a varied diet. Later researchers determined that pellagra is caused by a diet that is deficient in niacin, one of the B vitamins.

MULTIPLE CHOICE. Choose the best answer of those provided.

11. Once a substance has moved across the membrane of a villus, it has been
    a. absorbed.
    b. digested.
    c. excreted.
    d. ingested.

12. Once a substance has been chemically reduced to molecules that are small enough to cross a cell membrane, it has been
    a. absorbed.
    b. digested.
    c. excreted.
    d. ingested.

13. If secretions from your gallbladder cannot reach your small intestine, you will have difficulty digesting
    a. complex carbohydrates.
    b. simple carbohydrates.
    c. fats.
    d. proteins.
    e. amino acids.

14. The human body cannot directly use complex carbohydrates and proteins because
    a. they are not essential nutrients.
    b. they are micronutrients.
    c. these molecules are too large to cross cell membranes.
    d. they have no fiber.
    e. they lack calories.

15. Enzymatic digestion of which of the following large dietary molecules begins in the mouth and is completed in the small intestine?
    a. carbohydrates
    b. fats
    c. lipids
    d. amino acids
    e. proteins

16. Each day your body must obtain these from the foods you eat because it cannot manufacture them. They are
    a. metabolic pathways.
    b. vitamins.
    c. selenium.
    d. essential amino acids.
    e. calcium.

17. Even though you don't need these substances in large quantities, you will develop severe health problems if they are lacking from your diet.
    a. eggs, meat, dairy products
    b. pasta, rice, sugars
    c. butter, cheese, olive oil
    d. minerals and vitamins
    e. insulin and proteins

18. Which of the following statements is *true?*
    a. Taking large doses of vitamin C will prevent you from getting a cold and will help you get over a cold more quickly.
    b. Foods like butter, heavy cream, strips of fat on steaks, and high-fat hamburger contain large amounts of fat that are used to produce artery-clogging cholesterol.
    c. You should avoid cigarette smoke because it contains antioxidants.
    d. Pellagra reappears whenever people don't get enough complex carbohydrates.

19. Which of the following statements is *true?*
    a. One chicken breast a day supplies much more protein than an adult human needs.
    b. Americans eat a diet that is too rich in fats derived from plants.
    c. If you eat a variety of foods and eat lots of fresh produce and grains, it won't matter if you are overweight.
    d. Vitamins are important because they are free radicals.
    e. As body mass index decreases, the risk of health problems increases.

20. Many water-soluble vitamins are required so that certain enzymes will function properly. In this role these vitamins are called
    a. enzyme cofactors.
    b. essential nutrients.
    c. signaling molecules.
    d. antioxidants.
    e. protein supplements.

CONNECTING KEY CONCEPTS

1. How does food get turned into nutrients?
2. What is the difference between macronutrients and micronutrients, and what are some of the roles of each?

QUANTITATIVE QUERY

To maintain your weight you have to take in the same number of calories that you use. Calories are used in normal body processes and when you are physically active. The number of calories you need per day is called the basal metabolic rate. If you are a woman and do not exercise, your BMR can be estimated by the following equation:

$$\text{BMR} = 655 + (4.35 \times \text{your weight in pounds}) + (4.7 \times \text{your height in inches}) - (4.7 \times \text{your age in years}).$$

For men this equation is:

$$\text{BMR} = 66 + (6.23 \times \text{weight in pounds}) + (12.7 \times \text{height in inches}) - (6.8 \times \text{age in years}).$$

If you exercise moderately most days of the week your BMR is multiplied by 1.5. Calculate an estimate of your own BMR, assuming you are not physically active. Recalculate assuming you exercise nearly every day. Calculate it again assuming you are 10 years older, and still exercise.

THINKING CRITICALLY

1. With help from your library, use the large scientific database known as Medline to search for evidence that selenium supplements can prevent or treat cataracts. Look for articles that review the scientific studies to date, but also look at original research articles. Read some abstracts and access some full articles if you can. Then summarize what you were able to learn from these articles. Has your opinion of the value of the scientific literature changed as a result of this exercise? If so, how? Has your view of your own ability to understand the scientific literature changed as a result of this exercise? If so, explain how.

For additional study tools, visit www.aris.mhhe.com.

The Len Bias Tragedy
How Cocaine Stole a Life and Center Stage
November 13, 1989

The Death of Len Bias
A Chain of Tragedy
June 5, 1987

Len Bias. Len Bias was a basketball star at the University of Maryland in the mid 1980s.

What Killed Len Bias?

Even today, 20 years after his death, sports writers are drawn back to the day that Len Bias died. They recount his tragedy, and because Len Bias had such a promising future, they wonder how things might have turned out if he had lived. Len Bias was more than just a run-of-the-mill college basketball player. He was one of those remarkable star athletes that attract fans like an ice cream truck draws children on a hot summer afternoon. Len Bias often was compared to Michael Jordan and Larry Bird. He led the University of Maryland basketball team to great victories, even over North Carolina, the team that was ranked number one at that time. Len Bias's story now is a quarter of a century old, but it also is timeless. The drama of his success, ambitions, and early death have been repeated in other tragic life stories. Lessons still can be learned from the death of Len Bias.

Lenny's friends describe him as hard working and clean living. And those good habits paid off: June 19th, 1986, was the peak day of Lenny Bias' life. He was a senior at the University of Maryland, and the Boston Celtics had just offered him a contract. He looked forward to a spectacular career playing the game that he had always loved—and, of course, there would be money—lots of money and lots of fame in a fantastically exciting life to come. The photo of an elated Len Bias wearing his green new Celtics cap still peppers Internet sports sites, mute evidence of the importance of Len Bias to the sport of basketball (**see chapter-opening photo**).

On the night of June 19th Lenny celebrated the start of his career as a professional basketball player. At the University of Maryland Lenny still lived in the dorms, and he spent the early part of the evening there with friends, feasting on crab. Then it was out to parties until 2:30 AM or so. No one noticed anything unusual about Lenny that night, but early in the morning his dorm-mates found him lying unconscious on the floor. Len was not breathing, and his pulse was erratic. Using every available medical technique, the emergency room doctors tried to save Len Bias, but he never regained consciousness. So even before it had started, his spectacular future was snuffed out like a skyrocket abruptly swerving down instead of climbing into the night sky.

Len Bias' death is just one dramatic example of how a young life can be quickly ended. Even though he died years ago, his was not the first or last tragic death of a young person from a cause that could have been prevented. Young people today die from the same thing that killed Len Bias. What caused the death of someone in such peak physical condition? You will revisit this question at the end of this chapter, and perhaps Len Bias' sad death will be clarified once you understand how the heart works with the lungs and kidneys to maintain the normal internal environment of the human body. ■

22.1 Multicellular Organisms Control Their Internal Environments

Over billions of years of evolution organisms have become increasingly complex. Prokaryotes evolved into more complex single-celled eukaryotes, which in turn evolved into more complex multicellular fungi, animals, and plants. Being multicellular allows an organism to develop specialized parts and to grow much larger than any single cell can. These developments open the way for new functions to evolve and for complexity to increase even more.

Because the cells of a multicellular organism have the same physiological requirements as any independent cell, being multicellular has its own set of problems. Specifically, each eukaryotic cell must have a source of chemical energy, a source of oxygen to produce ATP, and a way to get rid of metabolic wastes. If an organism has just a single cell, all of these functions are carried out directly across its cell membrane. By contrast, a cell deep within a multicellular organism, such as a liver cell, cannot interact directly with the outside environment. In a multicellular animal the immediate environment for most cells is *within the organism*.

Three systems—the respiratory system, the circulatory system, and the excretory system—maintain the internal environment in mammals, including humans. These three systems are centered on blood, the fluid that circulates through arteries and veins and provides the immediate environment for every cell (**Figure 22.1**). In the **respiratory system** blood exchanges the gases involved in cellular respiration. Blood takes in oxygen and disposes of carbon dioxide for the whole organism. The **circulatory system** distributes blood to all tissues, bringing most of the substances that cells require. It also removes wastes and other surplus or unnecessary substances. Blood is filtered by the **excretory system,** which eliminates excess water as well as excess nitrogen, sodium, and other molecules, ions, and elements. These three systems work with other body systems to ensure that the chemical components of the internal environment remain in proper balance. Like a circus performer who teeters left and right, backward and forward to keep his balance on a tightrope, these three systems help keep the body in a stable state, even in fluctuating conditions. **Homeostasis** refers to this maintenance of a steady internal

respiratory system the lungs and associated breathing tubes and tissues responsible for the process of gas exchange in which oxygen enters the bloodstream and carbon dioxide exits the bloodstream

circulatory system the heart, blood vessels, and blood that collectively transport substances to cells and remove wastes from cells

excretory system the kidneys and associated organs and tubes responsible for removing wastes from the blood and eliminating them from the body in urine

homeostasis maintenance of constant internal conditions despite fluctuations in the external environment

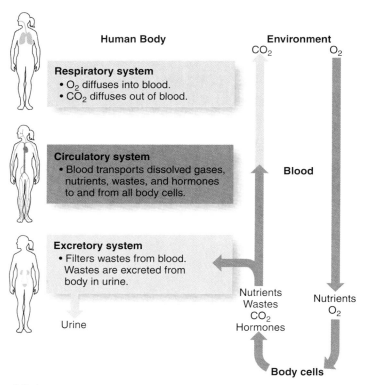

Figure 22.1 **Circulatory, Respiratory, and Excretory Systems Maintain the Internal Environment of Multicellular Organisms.**

■ *List several more substances that blood carries from the digestive system and respiratory systems and vice versa.*

(a) Just as a circus performer teeters back and forth to keep balanced on a tight rope . . .

(b) . . . the body's homeostatic controls keep physiological variables in balance in spite of fluctuating conditions.

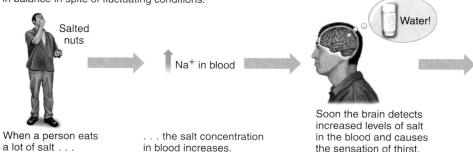

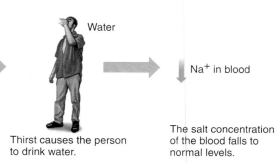

When a person eats a lot of salt . . .

. . . the salt concentration in blood increases.

Soon the brain detects increased levels of salt in the blood and causes the sensation of thirst.

Thirst causes the person to drink water.

The salt concentration of the blood falls to normal levels.

state (**Figure 22.2**). Through interactions of many body systems, including those described here, the internal conditions of a multicellular individual are homeostatically controlled.

22.2 Control of the Internal Environment Depends on Basic Cellular Processes

Multicellular organisms depend on diffusion, facilitated diffusion, active transport, and osmosis to get substances into and out of cells. Although these topics were covered in Section 4.2, they are so important that it is worth taking the time to quickly review them here. In *diffusion* a substance moves from a place where the substance is more concentrated to a place where the substance is less concentrated (**Figure 22.3a**). In other words, a substance diffuses down a concentration gradient. The lipid bilayer of the cell membrane does not allow polar substances, such as sodium or potassium ions, to enter or leave a cell by simple diffusion, but membrane proteins do allow these substances to cross (see Section 4.2). In *facilitated diffusion* a protein opens a channel in the membrane that allows diffusion of certain ions (Figure 22.3b). In *active transport* protein pumps use ATP to move a substance against the direction of diffusion, from its region of low concentration to its region of high

Figure 22.2 **Homeostasis.** Homeostatic mechanisms keep internal conditions stable despite external or internal fluctuations.

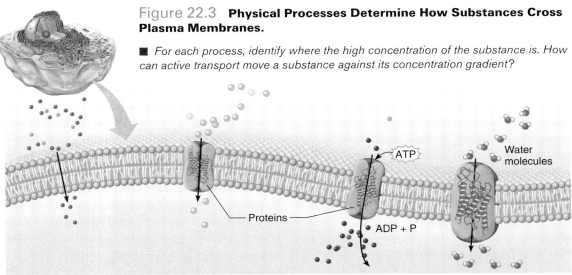

Figure 22.3 Physical Processes Determine How Substances Cross Plasma Membranes.

■ *For each process, identify where the high concentration of the substance is. How can active transport move a substance against its concentration gradient?*

(a) In **simple diffusion** a substance moves across a membrane. The direction of diffusion is from the area of greater concentration to the area of lesser concentration.

(b) In **facilitated diffusion** a substance moves across a membrane through a carrier protein.

(c) In **active transport** a substance is moved against its concentration gradient. Active transport requires a carrier protein and energy from ATP.

(d) **Osmosis** is the diffusion of water across a cell membrane. Osmosis is aided by protein channels that allow water to cross the membrane.

intracellular fluid fluid within plasma membranes, also known as cytosol

extracellular fluid fluid around cells and blood plasma collectively

concentration (Figure 22.3c). Finally, *osmosis* is the diffusion of water across a cell membrane (Figure 22.3d). Water also diffuses down its concentration gradient, although scientists usually talk about osmosis as the flow of water from a region that contains a low concentration of substances dissolved in water to a region with a high concentration of substances dissolved in water.

Extracellular Fluid Is Normally Much Different from Intracellular Fluid

In a multicellular animal the chemical environments inside and outside of a cell are usually different. The solution inside of cells is referred to as **intracellular fluid,** a synonym for cytosol (see Section 4.4). The solution immediately outside of a cell is called **extracellular fluid** (**Figure 22.4**). Cells can survive and thrive only when the concentration of ions and other chemicals inside and outside the cell are within certain ranges. At the same time, many cellular functions depend on changes in the intracellular and extracellular fluids. For example, during a nerve impulse the concentrations of sodium ions on the inside and outside of a neuron change rapidly and dramatically (see Section 20.5).

The intracellular and extracellular fluids are determined by several factors and are actively controlled by cellular processes. For example, ATP-driven pumps in cell membranes push sodium out

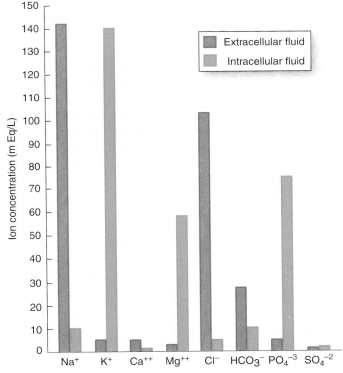

Figure 22.4 **Chemical Components of Extracellular and Intracellular Fluids.** Extracellular and intracellular fluids have different concentrations of ions.

of a cell and bring potassium into a cell, both against their concentration gradients. These pumps are the reason that animal cells have more sodium outside the cell and more potassium inside the cell. Under some conditions a cell can allow these ions to move along their concentration gradients. This is what happens when a neuron fires an action potential (see Section 20.5).

Figure 22.4 compares the chemical compositions of intracellular fluid and extracellular fluid for a typical animal cell. Note that intracellular fluid

and extracellular fluid almost always have opposite concentrations of substances. For instance, intracellular fluid has lower concentrations of sodium, chloride, and bicarbonate ions, while extracellular fluid has higher concentrations of them. Intracellular fluid has higher concentrations of potassium, phosphate, protein, calcium ions, magnesium, and sulfates, while extracellular fluid has lower concentrations of them. The chemistry of blood plasma is usually closer to, but not identical with, that of extracellular fluid.

Solutions Can Be Hypertonic, Hypotonic, or Isotonic When Compared to Intracellular Fluid

What would happen if this normal balance in the chemical compositions of extra- and intracellular fluids were upset? For example, what would happen if you ate a meal of nothing but extremely salty foods? Soon after digestion and absorption, extracellular fluid would have many more sodium and chloride ions, and a lower concentration of water, than intracellular fluid. A solution that contains more ions or other dissolved substances than are normally found inside a cell is **hypertonic** to the cytosol (**Figure 22.5a**). If a cell is exposed to a hypertonic solution, water will flow out of the cell and as it does the cell will shrivel and shrink. Or consider the reverse: what would happen if you drank a large amount of water? In this case the water would dilute the chemicals in your blood, and so the concentration of dissolved substances in extracellular fluid would be less than normal, and the concentration of water would be greater than normal. A solution with fewer ions or other dissolved substances than are normally found in a cell is **hypotonic** to the cytosol (Figure 22.5b). Following the laws of diffusion and osmosis, water will flow into a cell and as a result the cell will swell. Finally, when the concentrations of dissolved ions and other substances are in balance in the fluids on either side of a cell membrane, molecules will move in and out of the cell, but there will be no net accumulation of water on the inside or outside of the cell. This is an **isotonic** condition (Figure 22.5c).

The terms hypertonic, hypotonic, and isotonic may seem abstract, but they have practical applications. If body fluids become excessively hypotonic, the condition is called *hyponatremia*. Literally this term means "too little sodium." Under these conditions body cells and tissues will absorb water and swell. Death from hyponatremia usually is due to swelling of the brain stem, which controls many basic body functions such as breathing and heart rate. In the general population hyponatremia is rare,

(a) A **hypertonic solution** has a higher concentration of dissolved substances than is normally found inside of cells. In this solution more water moves out of a cell than moves in. The cell loses water and shrinks.

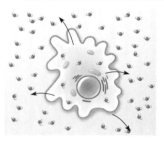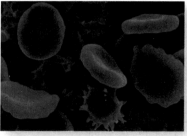

X1,515

(b) A **hypotonic solution** has a lower concentration of dissolved substances than normally found inside of cells. In this solution more water moves into a cell than moves out. Under these conditions an animal cell expands and bursts. Because the rigid cell wall keeps it from bursting, a plant cell becomes turgid.

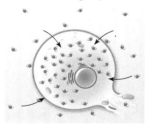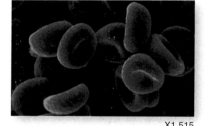

X1,515

(c) An **isotonic solution** has the same concentration of dissolved substances as normally found inside a cell. The movements of water into and out of the cell are equal and the cell does not shrink or swell.

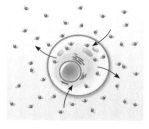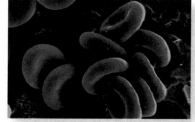

X1,515

Key:
= water molecules
= nucleus
⟶ = direction of movement of water

Figure 22.5 Hypertonic, Hypotonic, and Isotonic Solutions Have Different Effects on Cells.

but anyone who loses too much body sodium or consumes too much liquid may be in danger. Long periods of running or other physical exercise cause a person to lose water through sweating, but salts also are lost, especially sodium. Extreme exercise by itself can cause hyponatremia, and the situation can be made worse by drinking large amounts of water. For example, if a long-distance runner drinks only water during her run, this will intensify an already dangerously hypotonic state (**Figure 22.6**). Runners have died from this condition. This is why some experts recommend that long-distance runners *not* drink pure water, but drink something with a bit of sugar and ions. The sports drinks that are so popular with athletes aim to provide you with such a product. Another way that people can die from water intoxication is to drink a large amount of beer. Of course, beer contains alcohol, and an excess of alcohol itself can cause death, but in addition beer contains lots of water and is a hypotonic solution. So drinking excess beer can cause hyponatremia. Some of the saddest cases are high school and college age people who compete in beer-drinking contests and die because their body fluids become hypotonic.

Figure 22.6 **Hyponatremia.** Runners who drink large amounts of water may experience the potentially lethal condition of hyponatremia.

QUICK CHECK

1. Explain how passive and active transport are different. How are they similar?
2. Explain what happens to the water balance of a cell when it is placed in each of the following solutions: hypotonic, hypertonic, isotonic.

22.3 The Circulatory System Conveys Blood to and from All Tissues of the Body

The circulatory system is a set of organs and tissues that convey nutrients and other chemicals to cells throughout the body. The major components of the circulatory system are the blood, heart, and blood vessels.

Blood Is a Specialized, Complex Transport Tissue

Blood is a fluid tissue that contains many diverse cells, molecules, ions, and atoms. Blood transports needed chemicals to cells and removes unwanted chemicals. So blood defines the extracellular environment of each cell. Some people refer to blood as the "internal ocean," but it is a highly controlled ocean that contains cells that do some important, highly specific jobs. So blood is like an ocean that has fleets of many small ships—the blood cells and blood proteins that perform specific jobs.

Figure 22.7 shows some of the major components of blood. As you can see, blood is largely water. **Blood plasma** is the fluid component of blood that contains many dissolved proteins and other free molecules, ions, and atoms. The cellular component of blood contains a variety of cell types, including red blood cells (or RBCs, for short), various kinds of white blood cells (WBCs) that are part of the immune system, and platelets that help blood to clot. Blood interacts with other body tissues in the **capillaries,** the smallest blood

hypotonic solution when there are fewer ions and other dissolved substances than normal outside of the cell relative to the inside

isotonic solution when concentrations of dissolved ions and other dissolved substances are in balance in the fluids on either side of a plasma membrane

blood a fluid tissue that transports needed substances to each body cell and removes wastes

blood plasma the liquid component of blood that contains suspended blood cells and dissolved substances

capillaries the smallest blood vessels

Noncellular components of blood
55% of blood volume

Blood plasma	
Water	91.5%
Proteins	7%
Other	1.5%

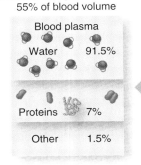

Whole blood

Cellular components of blood
45% of blood volume   (number / mm^3)

RBCs — 4.8 to 5.4 million

Platelets — 250,000 to 400,000

WBCs — 5,000 to 10,000

Figure 22.7
Composition of Blood. Fifty-five percent of whole blood is blood plasma; 45% of whole blood is various blood cells.

535

vessels that travel through tissues. Capillary walls are just one cell thick, allowing the diffusion of substances between blood and the extracellular fluid around cells in tissues (**Figure 22.8**). Oxygen, carbon dioxide, glucose, and other substances diffuse to and from the blood and body tissues. White blood cells can squeeze between cells of capillary walls and gain access to a body tissue that is infected.

Red blood cells, also called *erythrocytes,* carry oxygen to tissues and carry carbon dioxide away from tissues. Red blood cells are made in bone marrow, a dense, soft tissue located within bones. Bone marrow is one of those few body tissues that contain adult stem cells. These cells retain the ability to divide, and their daughter cells differentiate into the many kinds of blood cells including red blood cells. As they mature within bone marrow, the cells that will develop into red blood cells produce large amounts of *hemoglobin,* the protein that carries oxygen to and from tissues. Once the proper amount of hemoglobin has been produced, the red blood cell loses its nucleus. It is the only known body cell to do this. Without a nucleus a red blood cell cannot repair damage, and so red blood cells live only a few months. After it dies, its chemical components are recycled within the spleen, liver, or bone marrow. Cells in bone marrow must continually replace erythrocytes.

If a person produces too few red blood cells, or if red blood cells do not have enough hemoglobin, the condition is called **anemia.** As a result of anemia body tissues receive too little oxygen. Many different disorders can lead to anemia. For example, if a person's bone marrow does not supply enough red blood cells, then he or she may become anemic and require a blood transfusion. *Pernicious anemia* is a build up of immature erythrocytes that do not function normally in oxygen transport. It is caused by an inability to absorb vitamin B_{12}. *Sickle cell anemia* and *thalassemia* are forms of anemia caused by gene mutations.

Hemoglobin (**Figure 22.9a**) is a complex of four amino acid chains, with an iron group in the center of each one. Each iron group can bind one oxygen molecule (O_2), so each hemoglobin molecule can carry eight oxygen atoms (see Section 3.7). Hemoglobin also binds to carbon dioxide and carries 20 to 30% of the CO_2 away from tissues (Figure 22.9b). A small fraction (7 to 10% of the

red blood cells
specialized cells that contain large amounts of hemoglobin and so transport oxygen to body tissues

anemia a condition caused by too few red blood cells, or too little hemoglobin

(a) Hemoglobin transports oxygen and carbon dioxide.

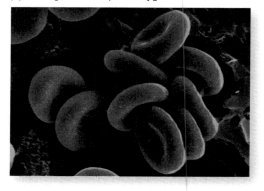

(b) Most carbon dioxide is transformed to bicarbonate ions within red blood cells.

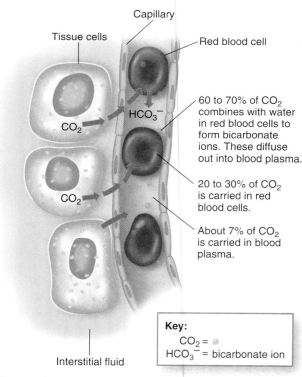

60 to 70% of CO_2 combines with water in red blood cells to form bicarbonate ions. These diffuse out into blood plasma.

20 to 30% of CO_2 is carried in red blood cells.

About 7% of CO_2 is carried in blood plasma.

Key:
CO_2 = ●
HCO_3^- = bicarbonate ion

Figure 22.8
Capillaries Are the Smallest Blood Vessels.

Capillary walls are thin and permeable to gases, nutrients, and cellular wastes.

Red blood cells must pass through capillaries in a single file.

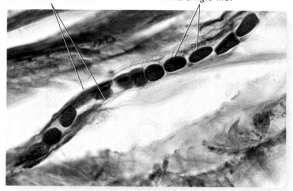

X400

Figure 22.9 **Hemoglobin.** (**a**) Red blood cells contain hemoglobin, a complex protein that can hold a maximum of four O_2 molecules. (**b**) Although blood plasma carries most carbon dioxide as bicarbonate ions, some carbon dioxide is carried in red blood cells.

carbon dioxide) is dissolved directly in blood plasma, but most carbon dioxide (60 to 70%) combines with water in blood plasma to form bicarbonate ions.

The structure and function of hemoglobin is also the key to understanding carbon monoxide poisoning. Each year thousands of people in the United States are made sick by this colorless, odorless gas, and several hundred people die. Carbon monoxide kills by binding to hemoglobin even more effectively than does oxygen. In an environment with high levels of carbon monoxide, less oxygen is carried to body tissues and a person may die of *hypoxia*, the condition in which too little oxygen is available to tissues. Short of death, carbon monoxide poisoning comes with symptoms such as dizziness, nausea, headache, and loss of consciousness. Short-term exposure to carbon monoxide can be treated by exposing the patient to pure oxygen. In an oxygen-saturated environment carbon monoxide does not bind to hemoglobin, and the symptoms of carbon monoxide poisoning usually reverse themselves. Often, though, there are other medical problems. Carbon monoxide also binds to one of the enzymes in mitochondria that is involved in energy metabolism. This causes damage to mitochondria that is responsible for the longer-lasting symptoms of carbon monoxide exposure.

White blood cells or *leukocytes* are a diverse population of cells involved in the immune response, the body's defense against disease-causing organisms. White blood cells attack and destroy invading viruses and bacteria. **Figure 22.10** shows one of the five kinds of white blood cells. Like red blood cells, most white blood cells come from the stem cells in bone marrow. So now the analogy of blood with the internal ocean has to be modified to add various kinds of warships to the ocean liners and transport vessels. White blood cells are like the warships, defending against attack. Chapter 25 will explore more about the cells of the immune system and how they function.

Blood also forms clots that protect against excessive blood loss. Think about the last time you got a cut, and recall what happened. Depending on the severity of the cut, blood fairly quickly forms an insoluble clot that blocks further bleeding. **Blood clotting** is a complex cascade of chemical reactions, and blood platelets play a major part in clotting (**Figure 22.11**). Blood platelets plug cuts and get caught in a sticky web of proteins, forming a dense, insoluble barrier that prevents blood cells and plasma from escaping. Blood clotting also can be a problem, though. A blood clot in one of the arteries that supply the heart muscle can cut off the flow of

Figure 22.10

A Macrophage. This kind of white blood cell attacks and kills bacteria and other invaders.

■ *How do you think white blood cells can distinguish between your own cells and the cells of bacteria?*

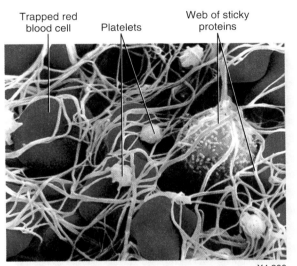

Trapped red blood cell · Platelets · Web of sticky proteins

X4,200

Figure 22.11 **Blood Clotting.** Blood clots through a complex chemical cascade that turns soluble blood proteins into a web of insoluble sticky proteins. Red blood cells and platelets are trapped within the clot.

blood and so cause a heart attack, while a blood clot in a blood vessel in the brain can cause a *stroke*. A stroke is the death of neurons in the brain, due to lack of oxygen. A blood clot in the lungs is called *pulmonary embolism* and is almost always fatal. A blood clot in the legs, called *deep vein thrombosis*, can lead to tissue damage, but the clot also can break out of the legs and move to the lungs.

A patient with a high risk of blood clot disorders may be prescribed one of several blood-thinning drugs, also called *anticoagulants*. Coumarin is one example. Common aspirin is also a powerful anticoagulant that interferes with the aggregation of blood platelets. Someone who has had a heart attack or a stroke, or has been diagnosed with heart disease, may be advised to take a very low dose of aspirin a day, about the equivalent of a baby aspirin.

white blood cells blood cells that function in defense of the body against disease-causing organisms

blood clotting the process of blood thickening that prevents blood from escaping when blood vessels are injured

Blood Transports Many Essential Substances

Blood plasma carries many important chemicals. Dissolved molecules and chemicals in blood plasma influence the composition of extracellular fluid. To understand the scope of the function of the circulatory system, let's briefly discuss the major substances carried by blood plasma.

Nutrients like glucose, amino acids, fatty acids, vitamins, and minerals come primarily from the digestive tract. The bloodstream absorbs them from the nutrient molecules derived from digestion and absorption. Nutrients also can come from other cells in the body. For example, when you need extra energy, liver cells will release stores of glucose that will diffuse from the extracellular fluid around liver cells into capillaries in the liver. Blood plasma transports glucose and delivers it to body cells. Blood transports hormones from the glands that release them to the target cells that they affect. Hormones allow the trillions of cells in the body to cooperate to make one organism. Chapter 23 presents much more about hormones.

A third important function of blood is to carry ions that are crucially important for normal cellular functions. If there are too many or too few specific ions cells will not function normally. For instance, too much potassium can cause muscle and nerve cells to lose the ability to produce action potentials. The ions carried in blood are the only source of external ions for cells, and so the correct concentrations of ions in blood are critical.

Blood also carries wastes away from cells. The complex metabolic pathways that break down proteins and other nutrients produce waste products. These wastes are carried away from cells and are processed into urine by the kidneys. Another important function of blood is to carry antibodies, large, complex proteins that defend the body against invading pathogens. Antibodies combine with proteins located on cell membranes of invading organisms and stimulate the immune system to destroy the invading cells. Finally, it is important to remember that blood carries water, the medium that all life requires. The concentration of water in the blood is critical, and all animals have a complex organ like a kidney that controls the amount of water in blood.

Contractions of the Heart Propel Blood Through Blood Vessels

Many animals have large bodies and must have a way to move blood from one part of the body to the other. Insects have *open circulatory systems* in which blood moves around within internal body spaces. Blood is collected within a major blood vessel and pumped out into tissue spaces again (**Figure 22.12a**). Many other animals have *closed circulatory systems* in which blood is contained within blood vessels (Figure 22.12b). In humans and other vertebrates the system that transports blood is made up of a heart connected to a network of blood vessels. The heart accumulates a mass of blood and then forcefully contracts, pushing blood out into the blood

Figure 22.12
A Comparison of Open and Closed Circulatory Systems. (**a**) In an open circulatory system blood directly bathes organs and tissues. (**b**) In a closed circulatory system blood is confined to blood vessels and the heart.

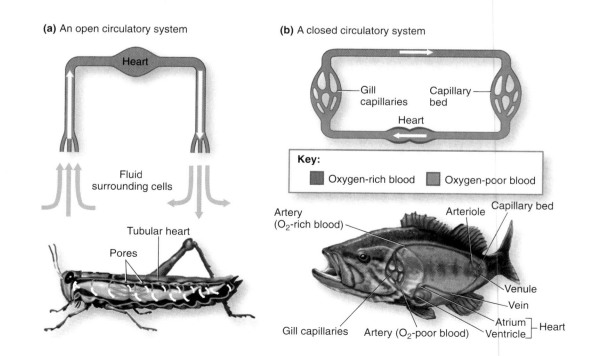

(a) An open circulatory system

Heart

Fluid surrounding cells

Tubular heart

Pores

(b) A closed circulatory system

Gill capillaries

Capillary bed

Heart

Key:
■ Oxygen-rich blood ■ Oxygen-poor blood

Artery (O₂-rich blood)

Arteriole

Capillary bed

Venule

Vein

Atrium
Ventricle — Heart

Gill capillaries

Artery (O₂-poor blood)

vessels that convey it to tissues in an organized way. Other blood vessels return blood to the heart. Let's take a closer look at both blood vessels and the heart.

Blood vessels are long tubes made of several layers of muscle tissues (**Figure 22.13**). Unlike the muscles that move the skeleton, the muscles of blood vessels are *smooth muscle* tissue. While the microscopic structure of smooth muscle is a bit different from that of skeletal muscle, smooth muscle tissue also contracts. When the walls of a blood vessel contract, the internal diameter of the tube gets smaller. Conversely, when the walls of a blood vessel relax, the internal diameter of the tube gets larger. As you can imagine, this is important for controlling the path of blood in the body. Blood vessels with relaxed walls will have more blood flow than will blood vessels with constricted walls. For example, when the body gets too warm, blood is routed to the surface of the skin where it can radiate body heat out into the environment. This routing is accomplished by relaxing the walls of blood vessels in the skin so that blood flows there more easily. The reverse happens when the body is cold: the walls of blood vessels in skin are constricted so that blood does not readily go to the skin—and as a result, body heat is conserved. Active muscles require an increased blood supply, so during exercise blood vessel walls relax, and this allows more blood to flow to them. Similarly, after a meal blood flow to the stomach and intestine is increased by relaxing the walls of arteries that supply them with blood.

Blood vessels come in three general kinds: those that carry blood away from the heart, those that carry blood to the heart, and those that connect the two. **Arteries** carry blood away from the heart (Figure 22.13a), while **veins** carry blood to the heart (Figure 22.13b). Capillaries connect veins and arteries (**Figure 22.14**). Almost all arteries carry oxygen-rich blood while almost all veins carry blood that is poor in oxygen and rich in carbon dioxide. These differences are reflected in the color of the blood in most arteries versus the color of blood in most veins. Most arterial blood is bright red because of the oxygen it carries; most venous blood is darker, almost purple, because it lacks oxygen. The exceptions are the pulmonary arteries and pulmonary veins. Pulmonary arteries carry blood that is low in oxygen, called *deoxygenated blood,* from the right side of the heart to the lungs. Pulmonary veins carry oxygen-rich blood, or *oxygenated blood,* from the lungs to the left side of the heart.

Blood vessels get progressively smaller as they approach their target tissues. The blood vessels that lead out of and into the heart are the largest,

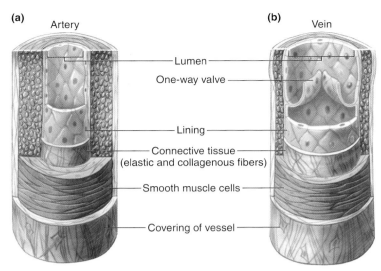

Figure 22.13   **Artery and Vein.** An artery (**a**) and a vein (**b**) have similar tissue layers, but different specializations. Notice that the walls of the artery are much thicker than those of the vein, and that the vein has one-way valves that the artery lacks.

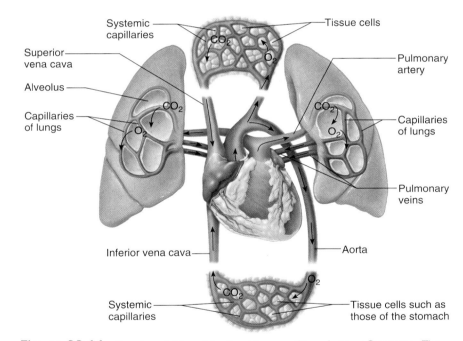

Figure 22.14   **Route of Blood in the Human Circulatory System.** The basic route of blood in the human circulatory system is aorta, arteries, systemic capillaries, veins, vena cava, heart, pulmonary artery, capillaries of lungs, pulmonary veins, and heart.

and those that directly connect to the heart are so large they are given special names—the *aorta* (the largest artery) and the *venae cavae* (the largest veins) (Figure 22.14). As the aorta moves away from the heart, it branches repeatedly into arteries that supply different organs and tissues, including the muscular walls of the stomach. *Capillaries* connect *arterioles* and *venules,* the smallest arteries

artery a blood vessel that carries blood away from the heart

vein a blood vessel that carries blood toward the heart

and veins. The work of circulation is accomplished across the walls of capillaries. Capillary walls are just one cell thick. Chemicals can cross them, and white blood cells can pass between them. White blood cells slip between cells of the walls of capillaries to interact with pathogens in the fluids in the extracellular spaces around body cells. In capillaries diffusion, facilitated diffusion, osmosis, or active transport shift substances into or out of the blood and cells. The venous circulation begins on the far side of the capillaries, and small veins drain blood from capillaries. Small veins merge into larger veins that receive blood from tissues and organs. The vena cavae carry blood from the body back to the heart.

The Left and Right Sides of the Heart Beat Together, But Have Separate Jobs

Blood must travel a long distance through blood vessels, and the heart provides most of the pumping force for it to do so. The heart is made of millions of individual *cardiac muscle* cells. In mammals, birds, and crocodiles and their close relatives the heart has four chambers, while other vertebrates may have hearts with two or three chambers. In a human the four-chambered heart sits just slightly to the left of center in the upper part of the chest (**Figure 22.15**). The heart's two upper chambers are called *atria* (singular: atrium), and the other two lower chambers are called *ventricles*. The left and right sides of the heart work together, but

they are separate from one another, and so the chambers of the heart are designated as right atrium, left atrium, right ventricle, and left ventricle (**Figure 22.16a**). Notice that the two atria are smaller chambers on the top of the heart, while the two ventricles are larger chambers and are on the bottom of the heart. Like one-way gates, valves lie between various chambers of the heart and at the entrances to the large blood vessels that lead out of the heart. The ventricles are the workhorses of the heart, and they have a great deal of pumping power. Let's follow the flow of blood as it travels through the heart's chambers.

Let's begin with deoxygenated blood that returns from the body, shown in blue in Figure 22.16b. This blood has been collected from the body and enters the heart through two enormous veins, the *inferior vena cava* and the *superior vena cava*. It flows into the *right atrium*. Stop for a moment to visualize where this would be in your body. This venous blood is pumped out of the right atrium. It passes through an *atrioventricular valve* and into the *right ventricle*. The right ventricle has the job of pumping blood to the lungs by way of the *pulmonary artery*. As the right ventricle contracts, the *semilunar valves* that allow entrance to the pulmonary artery open, and blood spurts into the pulmonary artery. The pulmonary artery divides, with one branch going to each lung. In the tissues of the lungs blood releases carbon dioxide and picks up oxygen. Then *pulmonary veins* carry oxygenated blood to the left atrium that pumps blood into the left ventricle. An *atrioventricular valve* opens as the *left atrium* contracts, allowing blood into the *left ventricle*. The left ventricle is the largest, strongest chamber of the human heart, and its contraction provides the force to pump blood out into the blood vessels. As it contracts, a *semilunar valve* opens and blood spurts into the aorta. This enormous blood vessel branches to supply blood to the head, arms, and the lower body.

When the heart beats, the heart muscle contracts and decreases the internal space where the blood is held and pushes it to the next destination. One-way atrioventricular valves connect the atrium and ventricle on the left and right sides of the heart, and one-way semilunar valves lead out of the heart to the aorta or pulmonary artery (Figure 22.16b). Only one valve is open when each chamber contracts. This prevents blood from flowing backward, and this is why blood flows in only one direction on each side of the heart. Of course, the chambers do not beat randomly, but in an order so that blood moves smoothly from one place to another. The two atria beat first, nearly in unison, pushing their blood to the ventricles. The ventricles beat a

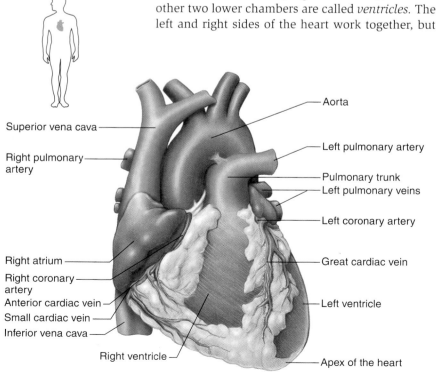

Superior vena cava

Right pulmonary artery

Right atrium

Right coronary artery

Anterior cardiac vein

Small cardiac vein

Inferior vena cava

Right ventricle

Aorta

Left pulmonary artery

Pulmonary trunk

Left pulmonary veins

Left coronary artery

Great cardiac vein

Left ventricle

Apex of the heart

Figure 22.15 **External View of the Heart and Major Blood Vessels.**

fraction of a second later, pushing their blood to the lungs and out to the body. You can hear the *lubb-dupp* sounds of a heartbeat through a stethoscope (**Figure 22.17**). These sounds are caused by vibrations set up in the valves and heart muscles when the valves close.

The pumping action of the heart is strong enough to move blood out to the body, but it is not strong enough to get blood back from the tissues, in the veins. Two mechanisms get this job done. First, skeletal muscles surrounding veins contract and push the blood back toward the heart. Why don't these contractions push blood away from the heart? Veins have one-way valves that allow blood to move toward the heart but prevent its moving away from the heart. These two mechanisms add to the force from the heartbeat to get deoxygenated blood back to the heart.

Action Potentials Coordinate the Heart's Contractions

How are the contractions of the heart's four chambers coordinated? The answers to these questions involve both heart muscle cells and nerves. Every individual cardiac muscle cell will beat all by itself, and you can see them beat if you have the chance to observe individual cardiac muscle cells in a culture dish. In a culture dish, though, individual muscle cells beat at random, whereas when the heart beats, many of its muscle cells beat in unison and in a particular pattern. Random beating would not produce an organized, forceful heartbeat that propels blood out of the heart and into arteries under pressure. The beating of the individual cardiac muscle fibers in the heart is coordinated by a special knot of cells located high up in the wall of the right atrium (**Figure 22.18**). The *sinoatrial node*, sometimes called the heart's pacemaker or SA node, is a group of cells that initiate the coordinated beat of heart muscle fibers. The sinoatrial node controls the heartbeat by generating electrical signals called action potentials (Section 21.5). When a person is at rest, the SA node generates about 70 to 80 action potentials each minute. Each action potential spreads a wave of electrical activity across the atria and causes them to contract nearly in unison. The wave of electrical activity encounters a barrier at the ventricles. There is a second node located between the atria and ventricles called the *atrioventricular node*, or AV node, that generates a second action potential, but after a short delay. This action potential then spreads along a bundle of associated nerve fibers located between the ventricles. These fibers penetrate down into the walls of the ventricles and cause

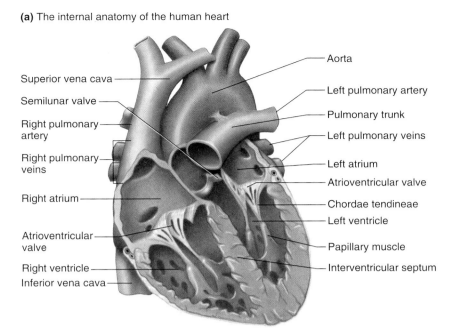

(a) The internal anatomy of the human heart

Superior vena cava
Semilunar valve
Right pulmonary artery
Right pulmonary veins
Right atrium
Atrioventricular valve
Right ventricle
Inferior vena cava

Aorta
Left pulmonary artery
Pulmonary trunk
Left pulmonary veins
Left atrium
Atrioventricular valve
Chordae tendineae
Left ventricle
Papillary muscle
Interventricular septum

(b) An overview of how blood flows through the heart

1 Blood fills the atria passively and is pumped into the ventricles.

2 Deoxygenated blood is pumped to the lungs.

3 Oxygenated blood is pumped to the body.

Figure 22.16 **Internal View of the Heart and Flow of Blood Through the Heart.** (**a**) A longitudinal section of the human heart shows the muscular walls that separate the heart into four chambers, as well as the valves that control flow of blood into the chambers, and the large blood vessels connected to the heart. (**b**) Although the left and right sides of the heart are separated and receive blood from different parts of the body and send it to different parts of the body, the two sides beat in unison.

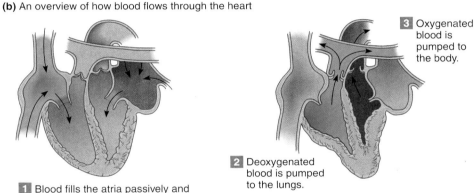

Atria contract	Atria relax		Atria contract	Atria relax	
Ventricles relax	Ventricles contract	Ventricles relax		Ventricles contract	Ventricles relax

Heart sounds

Lubb: AV valves close.   **Dupp:** Semilunar valves close.

Figure 22.17 **Heart Sounds Are Associated with Closing of Heart Valves.**

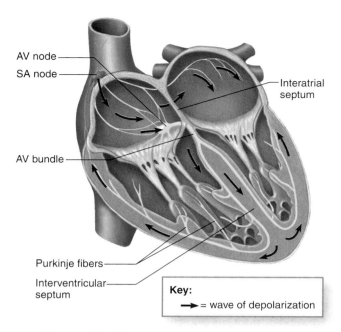

Figure 22.18 **Conduction System of the Heart.** The left and right sides of the heart beat in unison because of the nodal tissue and nerve fibers that conduct waves of depolarization from the SA node to the AV node, down the septum to the ventricles. Arrows show the path of conduction.

them to contract. As long as the extracellular fluid of heart muscle has the right ionic composition, these nodes and associated nerve fibers will keep the heart beating regularly. The heartbeat is an intrinsic feature of the heart. The only role for the nervous system is to speed up or slow down the heart rate by controlling the pace of action potentials in the SA and AV nodes.

cardiovascular disease
any one of many diseases affecting the heart and/or blood vessels

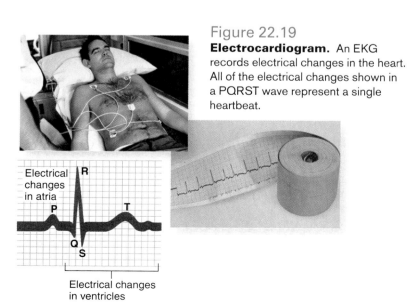

Figure 22.19

Electrocardiogram. An EKG records electrical changes in the heart. All of the electrical changes shown in a PQRST wave represent a single heartbeat.

These electrical changes within the heart can be measured when electrical wires—or electrodes—are placed on the skin and connected to a computer or other measuring device. The electrical recording is called an *electrocardiogram*, often abbreviated as ECG or EKG (**Figure 22.19**). Notice that a normal EKG has three peaks. The P wave reflects the electrical activity in the atria. The large QRS complex reflects the electrical activity of the ventricles, while the smaller T wave reflects electrical events that bring the heart muscles back to their resting electrical state. The heart then is ready to beat again. All of this happens in milliseconds. The EKG is one of the most valuable measurements in medicine because it is sensitive to a variety of heart problems. Even subtle changes in this electrical pattern can be indicators of heart disease or other health issues.

Cardiovascular Disease Is a Leading Cause of Death in the United States

Cardiovascular disease can involve either the heart or the blood vessels, but often involves both. Over 71 million Americans have some form of cardiovascular disease, and nearly a million die from their condition each year. One example will show how interrelated various aspects of cardiovascular disease can be. Coronary artery disease is one common form of cardiovascular disease. The coronary arteries supply the heart itself, and if these arteries do not function properly the heart may be damaged. One version of coronary artery disease involves the buildup of cholesterol in the coronary arteries, often called hardening of the arteries, or *atherosclerosis*. Atherosclerosis can affect any blood vessels in the body and often begins when chronic high blood pressure, another form of cardiovascular disease, damages the artery walls. Smoking cigarettes also can damage blood vessel walls and lead to atherosclerosis. The buildup of cholesterol in coronary arteries reduces blood flow to the heart, which can lead to heart damage or a *heart attack*. A heart attack happens when part of the heart muscle does not contract normally, and so blood does not flow normally through the body.

QUICK CHECK

1. What roles does the heart play in getting oxygen to cells in the body?
2. Use Figures 22.15 and 22.16 to trace the path of blood through the heart, around the routes of pulmonary and systemic circulations.

22.4 Blood Picks Up Oxygen and Releases Carbon Dioxide During Gas Exchange in the Lungs

Just as you know that your heart pumps blood through your body, you know that your lungs inhale air that has a high level of oxygen and exhale air that has a high level of carbon dioxide. In the lungs oxygen is taken up by hemoglobin in red blood cells while carbon dioxide is released from red blood cells and blood plasma. Both gases diffuse down their concentration gradients. When you inhale, oxygen enters the lungs because its concentration is lower inside the lungs. When you exhale, carbon dioxide leaves the lungs because its concentration is lower outside the lungs. This exchange of gases is part of the process commonly called **respiration.** Respiration in the lungs is not the same as the cellular respiration you read about in Section 6.3, but the two processes are related (**Figure 22.20**). You must breathe in oxygen because your cells must have oxygen to produce ATP. The production of ATP produces carbon dioxide, a waste product that must be eliminated. So it is not a coincidence that both processes are called respiration.

Respiration is a collaborative effort between heart, blood, lungs, and chest muscles. Blood comes to the lungs from the right ventricle via the *pulmonary arteries.* Within a bed of capillaries blood comes in close contact with cells in the lungs that, in turn, are in close contact with the environmental air (**Figure 22.21**). The lungs contain masses of tiny, thin-walled sacs called *alveoli* as well as the various tubules and larger tubes that lead to the bronchi, trachea, and the mouth and nose. Each *alveolus* (singular) is like a bubble filled with air. The alveoli are separated from their surrounding capillaries by just a thin layer of flattened cells, and the diffusion of oxygen and carbon dioxide gases occurs across the moist walls of the alveoli. The distance between the air in the alveoli and the blood in capillaries can be as little as 1 micrometer. In lung tissue oxygen diffuses into red blood cells, and carbon dioxide diffuses out of them. Both oxygen and carbon dioxide molecules are nonpolar, so they readily cross the cell membrane of red blood cells. The oxygen-rich blood circulates back to the left side of the heart and is pumped out to the body. Blood from the body returns to the right side of the heart and once again is pumped to the tissues of the lungs, eventually reaching capillaries that surround alveoli. There carbon dioxide and oxygen are once again exchanged, and the circuit continues.

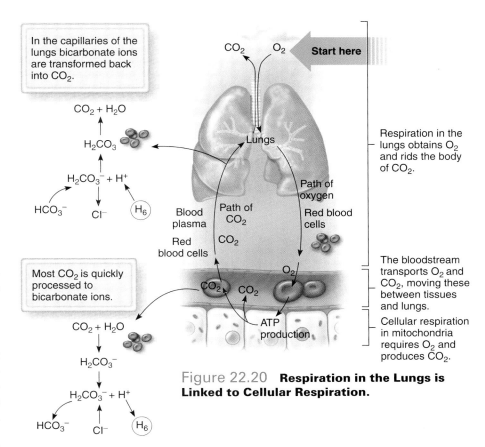

Figure 22.20 **Respiration in the Lungs is Linked to Cellular Respiration.**

Muscles, Nerves, and Brain Centers Control Breathing

Breathing is controlled indirectly, through nerves that innervate muscles around the chest. The muscles on the front and sides of the ribs are called intercostals (**Figure 22.22a**). There are two sets of intercostals, forming the *external* and *internal intercostal muscles.* Also important is the *diaphragm,* a muscle that forms the floor of the chest cavity. These muscles receive signals from nerve fibers that have their cell bodies in the brain stem, and these neurons control the rate of breathing. When the external intercostals and the diaphragm contract, the ribs are pulled outward, and the diaphragm moves down. Accordingly, the volume of the entire chest cavity increases. This creates a vacuum in the lungs that causes them to increase in volume and inflate. Air rushes into the vacuum through the mouth and/or nose, and air is drawn into the lungs (Figure 22.22b). When these same muscles relax, the chest cavity decreases in volume, and increased pressure in the chest cavity pushes air out of the lungs. Most of the time breathing is automatic, but when you consciously inhale or exhale, you intentionally relax or contract the intercostal muscles.

respiration gas exchange in the lungs during which oxygen is taken up from air in the environment and carbon dioxide is released from the body into the environment

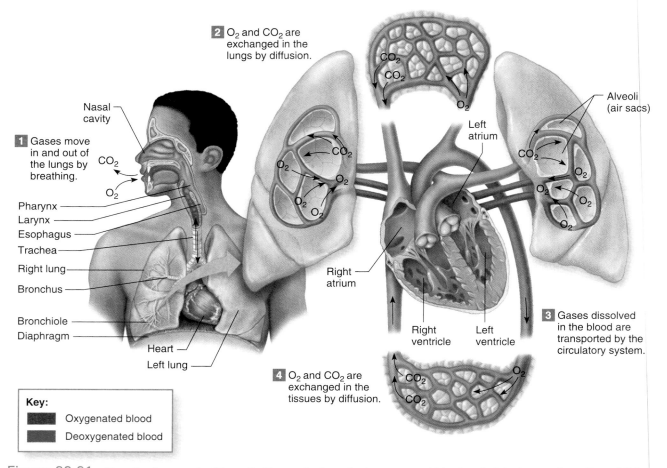

Figure 22.21 **Gas Exchange in Alveoli.** The work of respiration occurs within alveoli of the lungs where O_2 and CO_2 are exchanged.

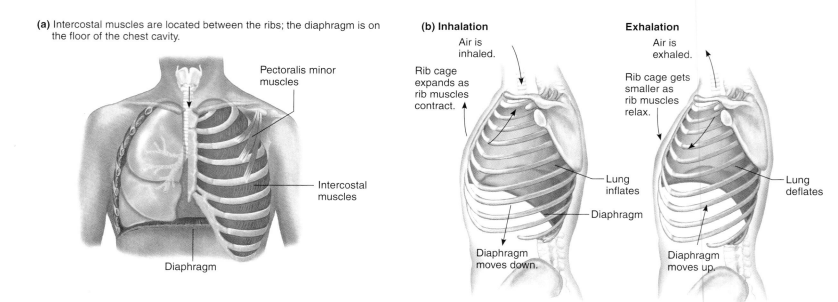

Figure 22.22 **Movements of the Chest During Breathing.** (**a**) Muscles that are active in breathing. (**b**) Chest movements during breathing.

Although breathing is maintained automatically, other factors can influence breathing. The most important factors are blood levels of carbon dioxide, and blood pH. Neurons in the brain stem monitors blood CO_2 and pH. When CO_2 levels or hydrogen ion concentrations rise, these neurons stimulate additional muscles to increase the rate and depth of respiration. Emotional excitement, cold temperatures, or just being surprised also can increase respiration. Surprisingly, blood oxygen levels do not influence breathing unless oxygen levels drop very low. The way that respiration is controlled can help you to understand what happens if you hyperventilate—that is, breathe hard and deep for several seconds. Hyperventilation increases the amount of oxygen in the blood, but also decreases the amount of carbon dioxide. It also causes the blood to be less acidic. Both of these factors decrease the stimulation of brain neurons that contribute to breathing, and after you hyperventilate you can voluntarily hold your breath longer. It takes longer for blood CO_2 to rise high enough to overcome your desire to hold your breath if you first hyperventilate. Unfortunately, hyperventilation also causes light-headedness, and even unconsciousness. This is why it is not a good idea to hyperventilate before swimming underwater, in the hopes of being able to stay underwater longer. The hyperventilation could lead to dizziness or unconsciousness underwater.

Certain drugs have strong effects on the brain stem neurons that control breathing. For instance, alcohol and other sedating drugs inhibit the activity of these neurons and so decrease breathing rate. A high enough dose of alcohol or similar drugs can inhibit these neurons so much that breathing stops, and a person may die.

QUICK CHECK

1. What role do chest muscles and the diaphragm play in gas exchange?

22.5 The Kidneys Maintain Homeostasis by Removing Water, Ions, and Nitrogen from the Blood

In vertebrates the chemical composition of blood is regulated in large part by the *kidneys*. The kidneys help control and maintain the amount of water in blood, and the concentration of other important chemicals, such as sodium and glucose.

The kidneys also get rid of toxic compounds. When proteins are converted to other biological molecules, nitrogen atoms are clipped off. The nitrogen-containing group that is removed from an amino acid is NH_3, or ammonia, a molecule that is toxic to cells. While many aquatic or marine organisms can just discard ammonia into the surrounding waters that will flush it away, cells of land animals cannot do this. In mammals ammonia is converted to urea, a less toxic molecule carried in the bloodstream. When blood circulates through kidneys, urea is removed and excreted as a component of urine. For these and other reasons kidneys are vital organs. Now let's explore how kidneys work.

Kidneys Are Composed of Millions of Nephrons That Filter Blood

The kidneys are part of the excretory system that filters wastes from blood and excretes them in urine. The basic anatomy of the human excretory system is relatively simple (**Figure 22.23**). It

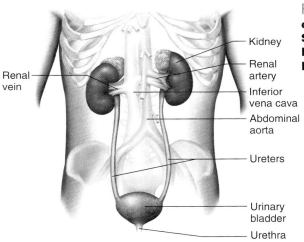

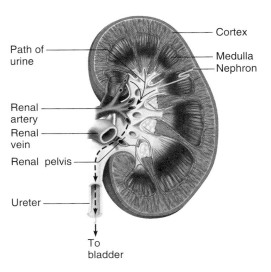

Figure 22.23 **Organs of the Excretory System Include Kidneys, Ureters, Bladder, and Urethra.**

includes paired kidneys and the *ureters,* the pair of tubes that connect kidneys to the *urinary bladder.* Kidneys have a plentiful blood supply, and normal blood pressure is essential to operate the excretory system. If blood pressure falls, kidney function becomes less efficient and may cease. Each kidney is a composite organ, made of about a million microscopic tubules that perform the function of filtering blood. Each kidney tubule is a **nephron,** and before you can understand how the kidneys maintain the proper mix of water, minerals, and ions in blood plasma, you must understand how a single nephron works.

To better grasp how a nephron works, think of a boat that catches shrimp in a large net dragged behind the boat. When the net is drawn on board, it contains many sea creatures other than shrimp. The crew must sort through this catch to find the shrimp. They toss back crabs, fishes, and sea stars, and separate out and save the shrimp. Nephrons work in a similar way. Just as the shrimpers strain shrimp out of ocean waters, nephrons filter the components of urine from blood. Shrimpers discard the unwanted catch and save the shrimp, while nephrons discard cellular wastes in urine and save substances that the body needs.

> **nephron** a microscopic kidney tubule that filters blood

The nephron is a microscopic-sized tubule with five segments; each segment has a specialized function (**Figure 22.24a**). The rounded end of the nephron has a double-walled structure called a *Bow-*

man's capsule, which encloses a ball of convoluted capillaries that lie on one another like a ball of snarled string. This ball of capillaries is called a *glomerulus.* The glomerulus and Bowman's capsule carry out the first step in the formation of urine: filtration. Notice that the diameter of the arteriole that leads blood into the glomerulus is greater than the diameter of the arteriole that carries blood away from the glomerulus (Figure 22.24). The difference in the sizes of these vessels creates the high pressure that allows the nephron to filter blood. Within the restricted passages of the contorted capillary knot of the glomerulus, the blood pressure increases. The increase in blood pressure squeezes about 20% of the blood plasma that enters the glomerulus *through* the wall of the capillary. This *filtrate* enters the space within the Bowman's capsule that encloses the glomerulus. Note that it is not diffusion, osmosis, or active transport that moves the blood plasma out of the capillary and into the Bowman's capsule. High-pressure filtration does this job.

The kidneys are blood filters and the total volume of blood is filtered about 60 times a day, which works out to be two and a half complete filtrations every hour! If all the filtrate liquid were collected each day, the kidneys would produce about 45 gallons (160 liters) of filtrate. Nevertheless, a person voids only between 1 and 2 liters of urine each day, so one of the more intriguing aspects of how kidneys work is that most of the filtrate is returned to the

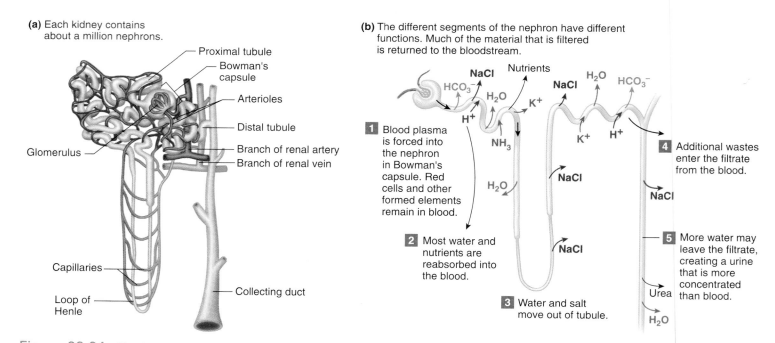

(a) Each kidney contains about a million nephrons.

Proximal tubule
Bowman's capsule
Arterioles
Distal tubule
Glomerulus
Branch of renal artery
Branch of renal vein
Capillaries
Loop of Henle
Collecting duct

(b) The different segments of the nephron have different functions. Much of the material that is filtered is returned to the bloodstream.

NaCl   Nutrients
HCO_3^- H_2O NaCl H_2O HCO_3^-
H_2O K^+
H^+

1 Blood plasma is forced into the nephron in Bowman's capsule. Red cells and other formed elements remain in blood.

NH_3
K^+ H^+
H_2O NaCl

4 Additional wastes enter the filtrate from the blood.

NaCl

2 Most water and nutrients are reabsorbed into the blood.

NaCl
NaCl

5 More water may leave the filtrate, creating a urine that is more concentrated than blood.

Urea

3 Water and salt move out of tubule.

H_2O

Figure 22.24 **Nephron Anatomy and Physiology.** (**a**) One of the approximately 2 million human nephrons that filters blood to make urine. (**b**) Different segments of a nephron have different functions.

bloodstream. This is the work of the other segments of the nephron. After the filtrate leaves Bowman's capsule, much of what was filtered out of the blood goes out of the tubule, back into the blood. A network of blood vessels surrounds the nephron. These blood vessels lie so close to the nephron that substances can be transferred from the enveloping blood vessels directly into the tubule or vice versa.

The different segments of the nephron have specific control over what stays in the filtrate and what goes back into the blood (Figure 22.24b). The nephron is not a homogeneous structure, and the cells in different lengths of the nephron take up and release substances in different ways. The first long stretch of tubule beyond Bowman's capsule lets water flow out but retains other substances. Surrounding capillaries pick up the water that flows out of this first segment of the tubule and return it to the blood. The next two portions of the tubule work primarily on salt. Sodium chloride passes out of the tubules here, along with more water. In the last segment of the nephron, a complex transfer of ions occurs. Water, sodium chloride, and bicarbonate ions leave the filtrate, while potassium and hydrogen ions enter it. This collecting duct is under the influence of hormones, and only after all of this processing does urine leave the nephron and travel through the ureter to the urinary bladder. Urine collects in the bladder, and eventually it is expelled from the body.

One final important issue is how the kidney can shift between producing lots of dilute urine when you have been drinking lots of fluids and producing sparse, concentrated urine when you have not. The walls of the collecting ducts are sensitive to *ADH* (short for antidiuretic hormone), which is made by the hypothalamus of the brain and stored in the posterior pituitary gland (**Figure 22.25**). The blood carries ADH to the kidneys, and when ADH is present, the walls of the collecting ducts become permeable to water. Intercellular spaces in the collecting ducts open, and by osmosis water moves out of the collecting ducts and into kidney tissues. It soon will return to nearby blood vessels via osmosis. So when ADH is present, water is removed from the filtrate, and a concentrated urine remains in the collecting ducts. In contrast, when ADH is not present, body water is lost, and a dilute urine is produced.

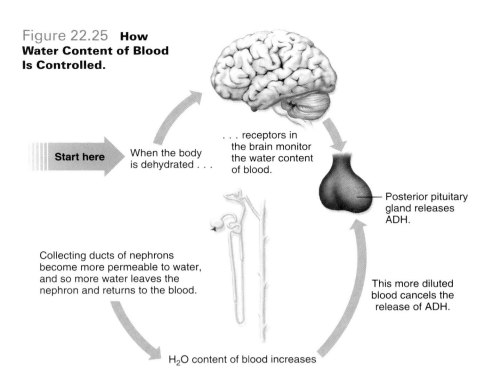

Figure 22.25 **How Water Content of Blood Is Controlled.**

Start here — When the body is dehydrated . . .

. . . receptors in the brain monitor the water content of blood.

Posterior pituitary gland releases ADH.

This more diluted blood cancels the release of ADH.

Collecting ducts of nephrons become more permeable to water, and so more water leaves the nephron and returns to the blood.

H_2O content of blood increases

22.6 Several Systems Help Control Blood Pressure

Blood pressure has become one of the most informative and important of all measures of human health. Every time you go to the doctor's office, a cuff is put around your arm and your blood pressure is measured. The nurse says, "130 over 80." What do those two numbers mean? What is blood pressure, how is it controlled, and what are the consequences if it gets out of balance?

Blood pressure is a measure of how strongly blood pushes against the walls of arteries. This pressure varies as the heart beats. When the left ventricle contracts and sends blood into the aorta, pressure in all blood vessels is high. Then, as the left ventricle relaxes, blood pressure in all the arteries reaches a low point. So the two numbers reported for blood pressure refer to these high and low values. They are called **systolic** and **diastolic blood pressure**, respectively (**Figure 22.26**).

Several other factors are important in controlling blood pressure. One factor that determines overall blood pressure is the volume of blood. Think about filling a balloon with more and more water. As the amount of water increases, the pressure on the balloon wall increases. So if the total volume of blood increases, then the pressure of blood on the blood vessel walls will increase accordingly. Many factors influence blood volume, and perhaps the most obvious of these is the amount of water in the blood. Another factor can be the amount of salt or other minerals in the diet. An excess of salt in the blood will cause

systolic blood pressure
the pressure exerted against artery walls during the active contraction of the left ventricle, when the pushing force is greatest

diastolic blood pressure
the pressure exerted against artery walls when the left ventricle is not actively contracting, and the pushing force is least

QUICK CHECK

1. What is the major functional unit of a kidney?
2. Use Figure 22.24b to describe how the nephron works to produce urine.

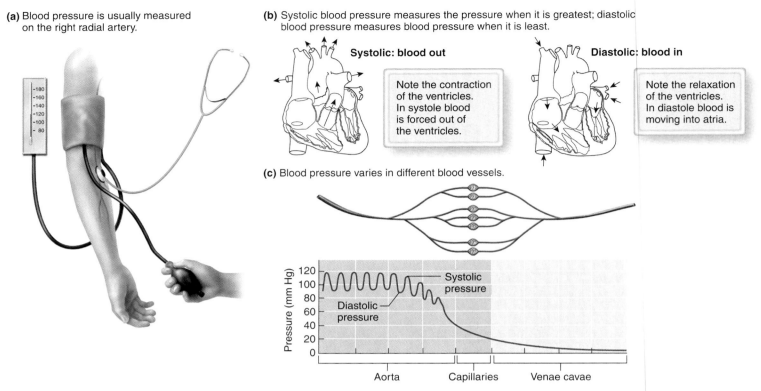

(a) Blood pressure is usually measured on the right radial artery.

(b) Systolic blood pressure measures the pressure when it is greatest; diastolic blood pressure measures blood pressure when it is least.

Systolic: blood out

Note the contraction of the ventricles. In systole blood is forced out of the ventricles.

Diastolic: blood in

Note the relaxation of the ventricles. In diastole blood is moving into atria.

(c) Blood pressure varies in different blood vessels.

Figure 22.26 **Systolic and Diastolic Blood Pressure.** **(a)** How blood pressure is measured. **(b)** Systolic blood pressure represents the pressure of blood on blood vessels when the ventricles contract. Diastolic blood pressure represents the pressure of blood on blood vessels when the ventricles relax. **(c)** Blood pressure is greatest in blood vessels nearest the heart and is lowest in veins.

water to move by osmosis into the blood from the extracellular fluid. So high-salt diets can lead to increased blood pressure. If the kidneys cannot get rid of the excess salt, or water, blood pressure may rise. Accordingly, the kidneys play a critical role in controlling blood pressure.

ADH stimulates the kidney to reduce excretion of water and sodium. Specialized cells in the hypothalamus of the brain release ADH, and it is stored and released from the posterior pituitary. Cells in the kidneys also are involved in the control of blood pressure. This is a complex system that involves a cascade of chemical reactions. When blood pressure drops, cells in the kidneys release a chemical called *renin*. In turn, renin causes the activation of a signal molecule called *angiotensin II* that, in turn, acts on membrane receptors in the adrenal glands that sit on top of the kidneys to produce a hormone called *aldosterone*. Then the bloodstream carries aldosterone to the kidneys, where it also stimulates the retention of water and sodium. As you can see, retention of water has

many control mechanisms, and all of them are involved in the control of blood pressure.

The Diameter of a Blood Vessel Affects Blood Pressure

A second important factor in determining blood pressure is the diameter of blood vessels. If blood volume stays constant, but blood vessels get smaller, then blood pressure will go up. How can this happen? Recall that blood vessels are tubes that have muscular walls. The muscle cells in the walls of blood vessels contract or relax when they are stimulated to do so by the nervous system or by hormones. In fact, the nervous system plays a crucial role in controlling blood pressure. For example, when you have been lying down for a long while, as when you are asleep, blood pressure stays at a resting level. If you suddenly stand up, the blood will fall toward your feet, drawn by gravity, and the blood pressure in your brain suddenly will drop. Luckily for you, centers in the brain detect this sudden change in posture and alter blood pressure so that your brain is not deprived of a steady supply of blood. Sometimes this does not happen fast enough, and you may feel dizzy or light-headed when you suddenly sit up after lying down for a while.

Why is the level of blood pressure so important? A normal blood pressure ensures that all parts of the body receive proper blood supply. The left ventricle is a strong and powerful muscle, but by itself it cannot completely overcome the effects of gravity. If there were no controls on blood pressure, blood would pool in the legs and drop from the brain. This and other negative consequences can happen if blood pressure drops too low, a condition called **hypotension.** This is one of the many problems with becoming dehydrated. Dehydration results in a drop in blood pressure, which means that body tissues do not get enough oxygen and nutrients. As you probably know, high blood pressure is a serious medical problem. Also called **hypertension,** this condition can cause devastating health problems if it continues untreated for many years. Hypertension causes damage to blood vessels and the heart and so can make a variety of *cardiovascular diseases* even worse. People with high blood pressure are more likely to suffer from heart attacks than are people with normal blood pressure. Hypertension also contributes greatly to **stroke,** a broken blood vessel in the brain that leads to death of brain cells. When blood pressure is too high, the tiny capillaries in the brain are more likely to break. Stroke is a debilitating disease. If you survive a stroke, depending on where in the brain the stroke happens, you may be paralyzed, unable to speak, or even unable to breathe on your own. Hypertension accounts for 20,000 deaths each year, even though many of these deaths are due to effects that are secondary to the high blood pressure itself.

Like many other modern ailments, high blood pressure can be prevented or reversed by changes in lifestyle. Certainly, there is a genetic component to hypertension, but even for those who have inherited a tendency to have high blood pressure, changes in lifestyle can make a big difference. Not surprisingly, the recommendations for avoiding or reversing high blood pressure are parallel to the recommendations for reducing heart disease, cancer, diabetes, and other serious health problems. You can probably recite them yourself:

■ Eat a healthful, balanced, low-fat, low-salt diet.

■ Maintain a healthy weight.

■ Exercise regularly and vigorously.

■ Don't use tobacco products and avoid secondhand smoke.

One additional recommendation that may not be as obvious is to reduce stress. In fact, stress reduction is recognized as a factor in combating all major diseases. This makes sense with respect to blood pressure. The regions of the nervous system that are activated in times of high stress are those that constrict blood vessels; the regions of the nervous system that are activated when you are relaxed contribute to the dilation of blood vessels. Exercise is one stress reducer, but other relaxation techniques also work to reduce stress. It is well documented that deep breathing, meditation, yoga, and visualization techniques act to reduce the effects of stress and reduce blood pressure.

Lifestyle choices can go a long way toward maintaining healthy blood pressure. Reducing high blood pressure is so important, though, that in many cases other treatments must be used as well. There are a variety of drugs that reduce hypertension. One standard treatment is called a **diuretic.** A diuretic drug increases the general action of the kidneys, pulling water and salts from tissues and from the blood and so reducing blood pressure. Other drugs act on the kidneys in other ways. At least two kinds of drugs inhibit the action of angiotensin II. One such class of drugs, called *ACE inhibitors,* prevent the activation of angiotensin II by renin. Another set of drugs blocks angiotensin II receptors in the adrenal glands and so blocks the production of aldosterone. *Beta-blockers* bind to cell membrane receptors that normally interact with the neurotransmitter, *norepinephrine.* Norepinephrine has many roles in the body, but one of its functions is to increase blood pressure. A beta-blocker drug binds to the norepinephrine receptor without activating it and so prevents the norepinephrine signal from increasing blood pressure. *Vasodilators* are in another class of antihypertension medications. A vasodilator is a drug that relaxes the muscles of blood vessel walls. When the blood vessel walls are relaxed, blood pressure drops. Along with careful lifestyle choices these and other kinds of drugs can control high blood pressure and prolong life.

QUICK CHECK

1. Explain what the two numbers of a blood pressure measurement mean.

hypotension abnormally low blood pressure

hypertension abnormally high blood pressure

stroke the rupture of a small blood vessel in the brain that causes damage to brain cells

diuretic a substance that increases the excretion of water by the kidneys

22.7 **An Added Dimension Explained**
What Killed Len Bias?

By all accounts Len Bias had lived a healthy life. He didn't drink alcohol and didn't take drugs. College sports teams are concerned about their players suffering sudden heart attacks and accordingly, they monitor them quite carefully. Len Bias had never shown any sign of heart disease, or any other disease. Yet he died of cardiac arrest. What caused his death? It is always hard to piece together an explanation for such a mysterious death, but one thing is clear: Len Bias had taken cocaine the night he died. No one claims to have seen him do this, and many people refused to believe it, but the autopsy reported cocaine in his system. Furthermore, the medical examiner speculated that cocaine probably had killed him. As you must know, cocaine is a highly addictive drug, but there is no evidence that Len Bias had an ongoing cocaine habit. How could taking cocaine a single time have killed Len Bias?

To understand how Len Bias died, you must revisit the control of heart rate. Recall that cells located in the sinoatrial (SA) and atrioventricular (AV) nodes keep the millions of cardiac muscle cells beating in unison, in the right order, to produce the complex and powerful contraction known as a heartbeat. The heart rate also is controlled by both nervous and hormonal inputs.

Two separate groups of cells in the brain stem control heart rate through bundles of nerve fibers that send signals to the SA and AV nodes (**Figure 22.27**). The *vagus nerve* releases the neurotransmitter acetylcholine from nerve terminals onto both the SA and AV nodes, and this slows the rate at which these nodes produce action potentials. The net result is that heart rate decreases. The *cardiac nerve* releases norepinephrine from nerve terminals onto the SA and AV nodes. Norepinephrine increases the rate of action potentials in the nodes and so increases heart rate. The vagus nerve is more active when a person is relaxed, meditating, or sleeping, while the cardiac nerve is more active when a person is excited, afraid, exercising, or alert. There are hormonal influences on the heart as well. For example, adrenaline is released from the adrenal glands in times of stress or danger, and it increases heart rate.

A normal heart is always in balance, whether the heart rate is increasing or decreasing. The different chambers of the heart normally beat in the right order, with the correct strength. The importance of this balance is seen when an imbalance occurs. If the heart beats abnormally or chaotically it will not pump blood to the body. A rapid heartbeat also can be a problem. If the heart beats too fast, there is not enough time between beats for the heart to effectively pump blood and body tissues will be deprived of oxygen and nutrients. Both irregular heart rate and rapid heart rate also can deprive the heart itself of oxygen and nutrients. In the extreme case the heart muscle stops beating altogether. This is cardiac arrest. When the heart stops the person will die unless

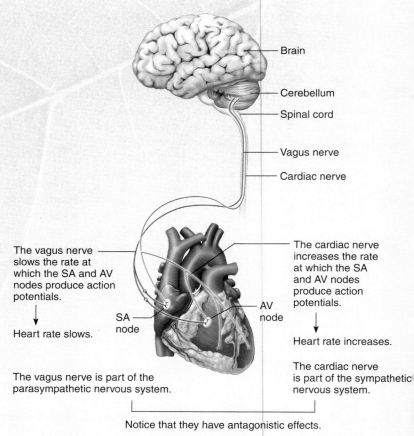

The vagus nerve slows the rate at which the SA and AV nodes produce action potentials.

↓
Heart rate slows.

The vagus nerve is part of the parasympathetic nervous system.

The cardiac nerve increases the rate at which the SA and AV nodes produce action potentials.

↓
Heart rate increases.

The cardiac nerve is part of the sympathetic nervous system.

Notice that they have antagonistic effects.

Figure 22.27   Nerve Supply of the Heart. Parasympathetic and sympathetic nervous systems control heart rate.

medical treatment is available. Heart arrhythmias are such a serious problem that many people who suffer from them have electrical devices, called heart *pacemakers,* implanted to maintain a coordinated heartbeat. These small devices deliver a regular electrical stimulus to the heart that coordinates the beating of heart muscle cells, just as the normal SA node does.

As you may have surmised, Len Bias died because of cocaine's dramatic effect on heartbeat. Cocaine blocks proteins called *neurotransmitter transporters* that take the neurotransmitter back into the nerve terminal (**Figure 22.28**). These transporters limit how long a neurotransmitter acts on its target cell. Cocaine enhances neurotransmitter action by blocking the action of the transporter (see Section 21.2). Cocaine blocks the transporter for at least two important neurotransmitters: dopamine and norepinephrine. In the brain these actions of cocaine lead to euphoria and motor activation, while in the other parts of the body cocaine can increase blood pressure, increase breathing, release glucose for use in the muscles, and increase muscle contraction. The deadly properties of cocaine in part are due to its actions on the heart. Cocaine increases the actions of norepinephrine on the

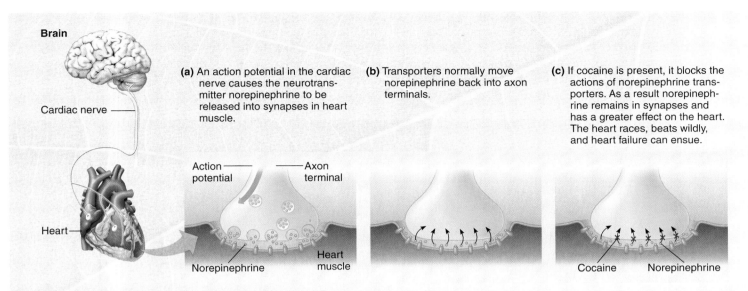

Brain

Cardiac nerve

Heart

(a) An action potential in the cardiac nerve causes the neurotransmitter norepinephrine to be released into synapses in heart muscle.

(b) Transporters normally move norepinephrine back into axon terminals.

(c) If cocaine is present, it blocks the actions of norepinephrine transporters. As a result norepinephrine remains in synapses and has a greater effect on the heart. The heart races, beats wildly, and heart failure can ensue.

Action potential — — Axon terminal

Norepinephrine   Heart muscle

Cocaine   Norepinephrine

Figure 22.28   Effect of Cocaine on the Heart. Because cocaine blocks norepinephrine transporters it can cause the heart to beat wildly and may cause cardiac arrest.

heart, and so throws the heartbeat out of balance. If there is too much cocaine in the bloodstream, the heart will beat abnormally. It will beat too fast, or beat out of rhythm, or will be abnormal in other ways. If someone dies from a cocaine overdose, it is likely that they die from the lack of coordinated contractions of a cardiac arrest. And of course, it is hard to predict what is "too much" cocaine for any individual. No matter how you look at it, cocaine

can be deadly—even if you take it only a single time. By the time Len Bias learned this lesson, it was too late.

QUICK CHECK

1. Explain how heart rate is normally controlled and tell what happens in cardiac arrest.

Now You Can **Understand**

The Effects of Diuretics on the Body

A diuretic is a substance that increases the amount of urine that the body produces. Many prescription drugs such as furosemide (Lasix) increase the body's output of urine, other common substances act as diuretics as well. For instance, alcohol works as a diuretic because it inhibits the release of ADH; because of this inhibition, water is lost in a dilute urine. The caffeine in coffee, tea, or cola is a diuretic too, but it works in a different way. Caffeine inhibits the uptake of sodium ions from filtrate, and so the kidney filtrate contains the water that passively follows sodium ions. How about beer? It has an interesting complex action on nephrons. The alcohol in beer inhibits the release of ADH, but several beers will cause a person to visit the restroom frequently. This is because in addition to alcohol, beer contains a lot of water. So drinking large quantities of beer raises blood pressure, and *more* than usual amounts of water are forced into nephrons. The combination of increased flow of filtrate and inhibition of ADH leads to the production of excess urine. With the excess water, sodium also is lost. This

can lead to *hyponatremia*, or low body sodium, which can be fatal. For many reasons it is wise to drink beer only moderately.

What Do **You** Think?

Do Magnets Worn on the Skin Increase Blood Flow and So Relieve Pain and Heal Injuries?

The Internet has been a boon for people to get their ideas out to the public and to try to sell products. Alternative medical products are all over the Internet, and one of the most common products advertised is magnetic therapy. The sellers claim that wearing a magnet on your skin will increase blood flow to that area and so reduce pain and promote healing. How can you know if such claims are valid, or if they are just another example of someone trying to part you from your money? One way to approach this topic is to look at what you know about blood flow, and see if that matches the common experience gained from wearing a magnet. For example, if blood flow increases under a skin magnet, what would you expect to see in the skin? Some researchers tested the simple hypothesis that if magnets increase blood flow, then the skin beneath magnets should become warmer. This study found no such effect. Other studies on the effects of magnets can be found by searching a *scientific* database, or search engine, with the term: *therapeutic magnets.* If you search using a regular search engine, you will likely find many testimonials and claims of those selling therapeutic magnets. Try different kinds of searches and find out how the scientific evidence compares with the sellers' claims. What do *you* think about the claims for magnetic therapy?

Chapter Review

CHAPTER SUMMARY

22.1 Multicellular Organisms Control Their Internal Environments

Just as a single-celled organism relies on the environment for nutrients, oxygen, and disposal of wastes, cells of a multicellular organism rely on the internal environment. Circulatory, respiratory, and excretory systems maintain homeostasis.

22.2 Control of the Internal Environment Depends on Basic Cellular Processes

The processes of passive and active transport that move substances in multicellular organisms are the same as those that move substances into or out of a single cell. Intracellular and extracellular solutions can be hypertonic, hypotonic, or isotonic. In hyponatremia cells become depleted of sodium. Hyponatremia is a potentially lethal condition brought about by loss of sodium and overconsumption of water, beer, or other fluids that mostly contain water.

22.3 The Circulatory System Conveys Blood to and from All Tissues of the Body

Blood and its components play active roles in delivering substances and removing wastes. Blood also has the protective functions of blood clotting and immune defenses. The heartbeat provides the force that circulates blood, eventually pushing it into capillaries where blood performs an array of circulatory system functions; some of these rely on diffusion, facilitated diffusion, osmosis, or active transport. Venous blood returns to the heart, is pumped to the lungs, and returns as oxygenated blood that is circulated to the body's tissues. The left and right sides of the heart are separate units, but both sides beat in unison, integrated by signals from nodal tissue and neurons within the heart muscle. The electrical fluctuations of heart action potentials are registered in an EKG. Cardiovascular disease is a leading cause of death in the United States that can involve blood vessels and/or the heart. In atherosclerosis the arteries are damaged, perhaps because of chronic high blood pressure, and may have deposits of cholesterol. This reduces blood flow to the heart that can lead to heart damage or a heart attack.

22.4 Blood Picks Up Oxygen and Releases Carbon Dioxide During Gas Exchange in the Lungs

Carbon dioxide is the metabolic waste released from blood in the capillaries of the lungs. Gas exchange occurs via diffusion within the capillaries that surround alveoli. There is a simultaneous diffusion of oxygen into the blood and carbon dioxide out of it. Oxygen-rich blood returns to the left side of the heart and is pumped to systemic capillaries. Oxygen-poor blood from the right side of the heart is oxygenated in the alveoli. Movements of rib cage and diaphragm passively inflate and deflate the lungs. Brain stem neurons control the rate of breathing, but there also is conscious control over respiratory movements of chest muscles. Various drugs sedate these neurons and stop breathing movements of the chest.

22.5 The Kidneys Maintain Homeostasis by Removing Water, Ions, and Nitrogen from the Blood

Nephrons of the kidneys filter the blood and maintain the proper balance of water, ions, and molecules in it. In a nephron blood plasma is filtered out of blood and chemically processed to maintain the correct balance of water, ions, and molecules. Nitrogenous wastes are eliminated in urine. Under the influence of ADH kidneys can produce a concentrated urine. When no ADH is present, a dilute urine is produced.

22.6 Several Systems Help Control Blood Pressure

Blood pressure is influenced by the amount of water and salt in the blood. ADH and aldosterone also affect blood pressure. In hypertension blood pressure is higher than normal, and it may lead to heart attacks or strokes. Lifestyle choices affect blood pressure. These include a healthful, balanced diet; a healthy body weight; regular vigorous exercise; avoidance of tobacco products, secondhand smoke, and stress.

22.7 An Added Dimension Explained: What Killed Len Bias?

Cocaine overdose causes cardiac arrest. This is what killed Len Bias.

TRUE or FALSE. If a statement is false, rewrite it to make it true.

1. Diffusion results from the intrinsic properties of atoms.
2. If a substance has a higher concentration of ions than a cell, water flows out of the cell. This describes a cell in a hypotonic solution.
3. Osmosis is the diffusion of salt across a plasma membrane.
4. Blood platelets begin the process of blood clotting.
5. When blood vessels dilate, blood pressure rises.
6. The walls of capillaries are one cell thick.
7. Gas exchange in lungs occurs in veins that surround alveoli.
8. Osmosis forces plasma into the Bowman's capsule.
9. Most kidney filtrate is returned to the bloodstream.
10. When ADH is present, a concentrated urine is produced.
11. When aldosterone is present, a concentrated urine is produced.

MULTIPLE CHOICE. Choose the best answer of those provided.

12. If the concentration of bicarbonate ions is greater within a cell than in extracellular fluid, how will bicarbonate ions diffuse out of the cell when membrane channels open?
    a. They will move into the cell.
    b. They will move out of the cell.
    c. There will be no net movement of bicarbonate ions.
    d. They will move by osmosis.

13. In this energy-requiring process a pump moves a substance in or out of a cell. It is called
    a. diffusion.
    b. active transport.
    c. osmosis.
    d. homeostasis.

14. The maintenance of constant internal conditions despite fluctuations in the external environment is characteristic of organisms. This dynamic balance is called
    a. hypotonic.
    b. hypertonic.
    c. isotonicity.
    d. homeostasis.
    e. osmosis.

15. The protein within erythrocytes is
    a. plasma.
    b. enucleate.
    c. hemoglobin.
    d. atria.
    e. nephron.

16. Blood transports oxygen bound to _____, while most carbon dioxide is carried in blood plasma as _____.
    a. hemoglobin; bicarbonate ions
    b. erythrocytes; sodium ions
    c. bicarbonate ions; hemoglobin
    d. sodium ions; cytosol
    e. calcium ions; plasma

MATCHING

17–22. Match the columns. (Each choice is used only once.)

17. pumps blood to the aorta
18. receives blood from the venae cavae
19. pumps blood to lungs
20. receives blood from the pulmonary vein
21. heartbeat originates here
22. measures electrical charges in the heart

   a. sinoatrial node
   b. EKG
   c. left ventricle
   d. right ventricle
   e. left atrium
   f. right atrium

CONNECTING KEY CONCEPTS

1. What factors control the composition of intracellular and extracellular fluid? How do these factors contribute to the functioning of the kidneys?
2. How does the heat pump blood through the body? What are the limitations of the heart's pumping actions, and how are these limitations overcome?

QUANTITATIVE QUERY

Research shows that about 10% of people over 70 will have a stroke by the time they turn 75. Of those who have a stroke in this age range, 40% had high blood pressure before the stroke. If your town has 26,750 people who are 70, how many of these will have a stroke in the next five years? How many of these had high blood pressure before the stroke?

THINKING CRITICALLY

1. Sometime in your life you may have to be given fluids in a doctor's office or a hospital. If your blood pressure is low, or you are dehydrated, the medical staff will insert a small needle into one of your veins and allow fluid to drip in. What kind of fluid should this be? What would be the consequence if, by mistake, you were given plain water instead of the appropriate fluid?

2. The principle of homeostasis explains much about how your body maintains a normal internal environment. If your external environment changes, then your internal environment also changes to compensate. Some runners and other athletes take advantage of homeostasis in the way they train. High-altitude training is one example. At high altitudes there is less oxygen than at sea level. In general, what do you think might happen during high-altitude training that helps athletic performance?

For additional study tools, visit www.aris.mhhe.com.

Biological science
is shedding light on
human sexuality.

Massachusetts Court Sets Limit
On Same-Sex Marriage Law

March 31, 2006

GAY ➡
LESBIAN ➡
⬅ TRANSGENDER
BISEXUAL ➡
⬅ HETEROSEXUAL
⬅ TRANSVESTITE

No Accord by Episcopal Clerics
After Long Session On Gay Issues

April 19, 2005

Most people grow up knowing which sex they belong to. Sometime during puberty most of us develop a strong, albeit awkward, sexual attraction to people of the opposite sex. But not everyone has confidence that his or her sexuality conforms to this conventional expectation. Throughout human history people from all walks of life have found that they or someone in their family is not strictly heterosexual. A recent "human interest" news story makes the point explicitly. A couple who met on the Internet took a cross-country trip together and fell in love. They now are married and by all accounts are a happy couple. "She" was born a male but was not quite happy that way and had a sex change operation and lived as a female. "He" was born a female and bore one child but always was a tomboy. His marriages failed because he was always "one of the boys" with his husbands, rather than a sexually intimate wife. Eventually, he had "sex change" surgery to become physically male. Amazingly, these two found each other and now are trying to put together a "normal" marriage.

What do you make of stories like this? How does a person "know" whether he or she is male or female? What determines whether a person is romantically and sexually attracted to men or to women or to both? This issue is the source of much debate, and some people have strong opinions about it. At the heart of this debate is the question just posed: what determines a person's sense of being male or female, their sexual orientation? Scientific studies have addressed these questions. After you have explored how hormones act to coordinate body functions, including reproduction, you will return to consider the research evidence about the specific question of *gender identity*. ■

23.1 Multicellular Organisms Depend on Cellular Communication

"Everybody needs somebody, sometimes" is a refrain from an old song, but it seems to be a basic biological principle, even when the "somebody" is a single cell. Even single-celled organisms communicate, exchange information, influence one another, and cooperate toward a common goal.

An illustration of the principles of cellular communication is shown in the life cycle of the cellular slime mold, *Dictyostelium discoideum* (**Figure 23.1**). This organism spends most of its life as an amoeboid unicellular eukaryote. Like a speck of living hair gel, each amoeba moves around, secreting a slimy external environment for itself and eating, growing, and dividing. Under some conditions amoebas send out the signal molecule, cyclic AMP, which binds to receptors on the surfaces of other amoebas, causing them to stream together into a large aggregation called a *slug*. Then other signals stimulate the slug to differentiate, forming a tall stalk that expands into a ball on top. The ball contains spores that are dispersed when it bursts; each spore can grow a new amoeba. As you can see, the basic principles that govern how multicellular organisms communicate, differentiate, and cooperate have their origins in much simpler organisms.

Unlike the amoebas of *Dictyostelium*, most cells in a multicellular organism do not normally survive on their own. They must have support from, and interactions with, the other cells in the individual. Millions of years of animal evolution have produced the **endocrine system** for body-wide communication. In vertebrates the endocrine system is made up of a set of **endocrine glands** that are located around the body (**Figure 23.2**). Endocrine glands are different from other glands because they release their secretions into the bloodstream, not into ducts that lead to internal or external body surfaces. Each endocrine gland releases one or several signal molecules called hormones. A **hormone** is a chemical released into the bloodstream by an endocrine gland and acts on cells in another tissue. Most hormones affect more than one target tissue, and this helps to coordinate the body's responses to a hormone. For example, adrenaline acts on many body tissues to coordinate a reaction to a potential threat or dangerous situation. The major endocrine glands, the hormones they release, and their target tissues are summarized in **Table 23.1**.

Hormones can be small molecules such as modified amino acids, or they can be large molecules such as strings of amino acids or *peptides*. Other hormones are lipids. Peptide hormones bind to a membrane protein receptor and change the internal functions of a cell (**Figure 23.3a**), while steroid hormones cross the plasma membrane to interact with receptors inside of the cell (Figure 23.3b). Regardless of whether it crosses the cell membrane or not, a hormone can have a variety of effects in a cell, and the same hormone can have different effects in different kinds of cells. For instance, a hormone might activate an enzyme that breaks a large molecule like glycogen into smaller molecules like glucose. Hormones can cause a cell to take up or release certain substances, even other hormones. For instance, hormones from one

endocrine system all the glands that produce hormones

endocrine gland a gland that secretes hormones directly into the bloodstream

hormone a chemical released into the bloodstream that acts on a tissue that is some distance away and initiates a response in cells of that tissue

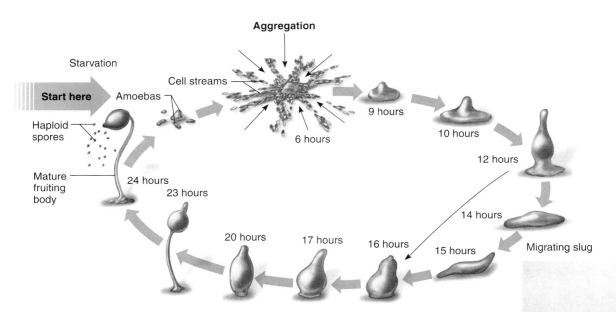

Aggregation

Starvation

Start here → Amoebas

Haploid spores

Cell streams

6 hours

9 hours

10 hours

12 hours

14 hours

15 hours

Migrating slug

16 hours

17 hours

20 hours

23 hours

24 hours

Mature fruiting body

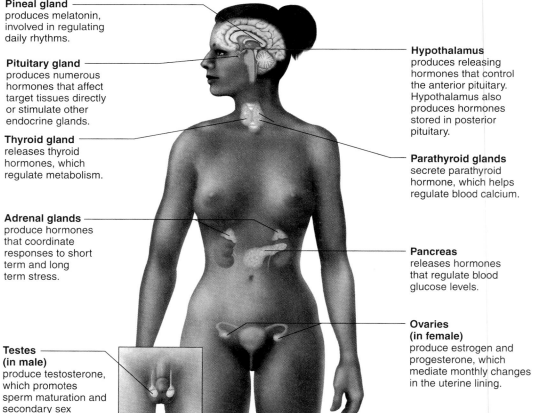

Figure 23.1 **Life Cycle of *Dictyostelium*.** *Dictyostelium* is an example of the principle that complex properties of multicellular organisms have their evolutionary roots in simpler organisms. *Dictyostelium* begins life as a single-celled eukaryotic organism. Once the amoebas run short of food they come together to form a multicellular reproductive structure. The fruiting body releases haploid spores. These develop into new amoebas.

Figure 23.2
Endocrine Glands. The endocrine system consists of glands where cells that produce hormones are concentrated. Other hormone-producing endocrine cells are widely scattered in body tissues.

Pineal gland produces melatonin, involved in regulating daily rhythms.

Pituitary gland produces numerous hormones that affect target tissues directly or stimulate other endocrine glands.

Thyroid gland releases thyroid hormones, which regulate metabolism.

Adrenal glands produce hormones that coordinate responses to short term and long term stress.

Testes (in male) produce testosterone, which promotes sperm maturation and secondary sex characteristics.

Hypothalamus produces releasing hormones that control the anterior pituitary. Hypothalamus also produces hormones stored in posterior pituitary.

Parathyroid glands secrete parathyroid hormone, which helps regulate blood calcium.

Pancreas releases hormones that regulate blood glucose levels.

Ovaries (in female) produce estrogen and progesterone, which mediate monthly changes in the uterine lining.

Table 23.1 Major Glands of the Endocrine System and the Actions of Their Hormones on Target Tissues

Endocrine gland	Hormone	Action of hormone
Hypothalamus (and posterior pituitary)	Antidiuretic hormone (ADH)	Promotes reabsorption of water by kidneys
	Oxytocin	Promotes reabsorption of water by kidneys; causes muscles of uterus to contract; causes ejection of sperm
Hypothalamus to anterior pituitary	Releasing hormones	Stimulate release of hormones from anterior pituitary
Anterior pituitary	Follicle-stimulating hormone (FSH)	Stimulates growth of follicle and secretion of estrogen; stimulates development of sperm
	Luteinizing hormone (LH)	Stimulates ovulation; stimulates testes to produce testosterone
	Thyroid-stimulating hormone (TSH)	Stimulates thyroid to release thyroxine
	Adrenocorticotropic hormone (ACTH)	Stimulates adrenal cortex to release glucocorticoids
	Growth hormone	Stimulates growth, protein synthesis, and fat metabolism
	Prolactin	Stimulates production and secretion of milk from mammary glands
	Endorphins	Reduce perception of pain
Thyroid	Thyroxine	Increases BMR, raises body temperature
	Calcitonin	Inhibits release of calcium from bones
Pancreas	Insulin	Decreases blood glucose levels; regulates fat metabolism
	Glucagon	Converts glycogen to glucose
Ovaries	Estrogen	Development of female secondary sexual characteristics and maturation of eggs; promotes growth of endometrium
	Progesterone	Stimulates development of endometrium and promotes formation of placenta
Testes	Testosterone	Stimulates development of male secondary sexual characteristics and production of sperm
Adrenal medulla	Epinephrine (adrenaline) and norepinephrine (noradrenaline)	Fight-or-flight response; prepare body for emergencies
Adrenal cortex	Glucocorticoids	Increase blood sugar; antiinflammatory
	Aldosterone	Increases reabsorption of salt in kidney
	Testosterone	Masculinizes body

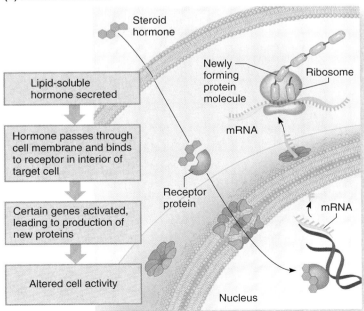

(a) Peptide Hormones

Peptide hormone
Activated protein complex

Water-soluble hormone secreted

Hormone binds to receptor on surface of target cell

Receptor protein

ATP

cAMP

Cascade of biochemical reactions ends by activating an enzyme

Activates enzymes

Effects on cell

Altered cell activity

(b) Steroid Hormones

Steroid hormone

Newly forming protein molecule

Ribosome

Lipid-soluble hormone secreted

mRNA

Hormone passes through cell membrane and binds to receptor in interior of target cell

Receptor protein

mRNA

Certain genes activated, leading to production of new proteins

Altered cell activity

Nucleus

Figure 23.3 **How Hormones Affect Cells.** (a) Peptide and (b) steroid hormones interact with cells in different ways.

hypothalamus a region of the brain that controls the secretions of endocrine glands

pituitary gland a major control gland for the endocrine system that releases hormones into the blood that act on other endocrine glands

endocrine gland can stimulate another endocrine gland to release a hormone. Alternatively, hormones can affect gene expression and so change the proteins produced by a cell. In all of these ways hormones coordinate the activities of diverse tissues and organs.

> **QUICK CHECK**
>
> **1.** How are endocrine glands different from other glands?

23.2 Hormonal Secretions Are Strictly Controlled

In vertebrates two structures control the endocrine system: the *hypothalamus,* which is a group of nerve cells in the brain, and the *pituitary gland,* which sits just outside the brain, beneath the hypothalamus (**Figure 23.4**).

The Hypothalamus and Pituitary Work Together to Regulate the Endocrine System

In many ways the **hypothalamus** is the master control center for the endocrine system. The hypothalamus is a complex of groups of cells, each group responsible for some aspect of basic body functions. Cells in the hypothalamus receive a rich flow of blood. By evaluating the blood's chemical content, cells of the hypothalamus monitor the body's state. For example the hypothalamus monitors the level of blood glucose, and gauges how much cellular energy is available, and in turn it can influence how hungry you feel. Cells of the hypothalamus also sense the temperature of blood, as well as its water and ion concentrations. Hormones released by endocrine glands also affect cells of the hypothalamus. By continually evaluating blood chemistry, the hypothalamus always is informed of the body's status and the actions of the various endocrine glands. Like the CEO of a huge corporation who has presidents and vice presidents to carry out specific jobs, the hypothalamus does not *directly* influence the endocrine glands. Instead, the hypothalamus interacts with the **pituitary gland,** the other major control point of the endocrine system. The pituitary gland has two lobes: the *posterior pituitary* and the *anterior pituitary* (**Figure 23.5**).

In a sense the posterior lobe of the pituitary gland is just an extension of the hypothalamus. Nerve cells that originate in the hypothalamus have extensions called nerve fibers that penetrate into the posterior pituitary. These nerve fibers carry and release two important hormones: oxytocin and ADH

Figure 23.4 Control Centers in the Brain. The hypothalamus and pituitary are important hormonal control centers in the brain.

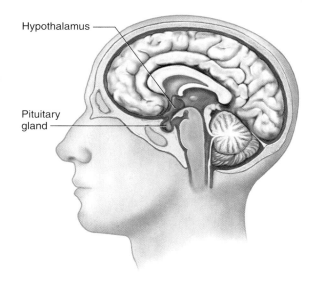

Hypothalamus

Pituitary gland

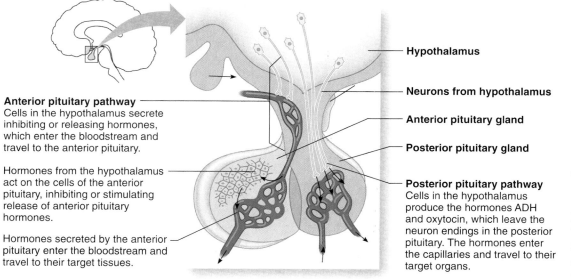

Anterior pituitary pathway
Cells in the hypothalamus secrete inhibiting or releasing hormones, which enter the bloodstream and travel to the anterior pituitary.

Hormones from the hypothalamus act on the cells of the anterior pituitary, inhibiting or stimulating release of anterior pituitary hormones.

Hormones secreted by the anterior pituitary enter the bloodstream and travel to their target tissues.

Hypothalamus

Neurons from hypothalamus

Anterior pituitary gland

Posterior pituitary gland

Posterior pituitary pathway
Cells in the hypothalamus produce the hormones ADH and oxytocin, which leave the neuron endings in the posterior pituitary. The hormones enter the capillaries and travel to their target organs.

Figure 23.5 Anterior and Posterior Pituitary Glands and the Hypothalamus. Although they are united in a single structure, the anterior and posterior pituitary glands operate separately. Both are under control of the hypothalamus, but this control differs.

(Figure 23.5). *Oxytocin* is a strong stimulator of contractions of the muscles of the uterus. During labor oxytocin initiates the forceful contractions that push a fetus out of the uterus. *ADH* (or antidiuretic hormone) acts on the kidneys to help control water and salt levels in the blood. Both hormones are made in cells within the hypothalamus and are released from the ends of nerve fibers of those same cells located in the posterior pituitary. Oxytocin and ADH enter blood vessels within the posterior pituitary, and the bloodstream carries them to their target tissues.

The anterior pituitary is an altogether different structure. It is separate from the hypothalamus and contains cells that make and release some important hormones. Anterior pituitary hormones are released only under the direction of the hypothalamus. How? The hypothalamus produces a variety of substances called *releasing hormones,* and each of these causes cells in the anterior pituitary to release a specific hormone. The bloodstream carries releasing hormones from the hypothalamus to their target cells in the anterior pituitary (Figure 23.5). **Table 23.2** lists the hypothalamic-releasing hormones, the pituitary hormones that each controls, and the endocrine gland that each anterior pituitary hormone acts on.

Negative Feedback Acts at Every Level in the Endocrine System

In addition to the master control system imposed by the hypothalamus, the endocrine system also is controlled at another, more local level. The release of a hormone by an endocrine cell usually is *inhibited* by the hormone that it produces. This form of control is called **negative feedback,** and during negative feedback the production of a hormone goes down as the level of the hormone in the body increases. **Figure 23.6** shows this relationship in an abstract way, using a room thermostat as an analogy. The thermostat detects changes in room temperature and turns the heater or air conditioner on and off to keep the preset room temperature stable. In living systems negative feedback is one example of a homeostatic mechanism. You will return to the effects of negative feedback when you consider how reproductive hormones control reproductive cycles in vertebrates.

QUICK CHECK

1. Explain the double role of the hypothalamus.

Table 23.2 Some of the Anterior Pituitary Hormones Released in Response to Signals from the Hypothalamus

Anterior pituitary hormone	Effect on body	Hypothalamic-releasing factor
Growth hormone (GH)	Mobilizes energy sources and promotes body growth	Growth hormone-releasing hormone (GHRH)
Thyroid-stimulating hormone (TSH)	Sets basal metabolic rate by stimulating production of ATP	Thyrotropin-releasing hormone (TRH)
Adrenocorticotropic hormone (ACTH)	Stimulates production of steroids by the adrenal cortex	Corticotropin-releasing hormone (CRH)
Follicle-stimulating hormone (FSH)	Promotes growth of mature egg in ovary	Gonadotropin-releasing hormone (GnRH)
Luteinizing hormone (LH)	Stimulates testes to produce testosterone; stimulates ovaries to produce estrogen, stimulates ovulation and formation of corpus luteum	Luteinizing hormone-releasing hormone (LHRH)

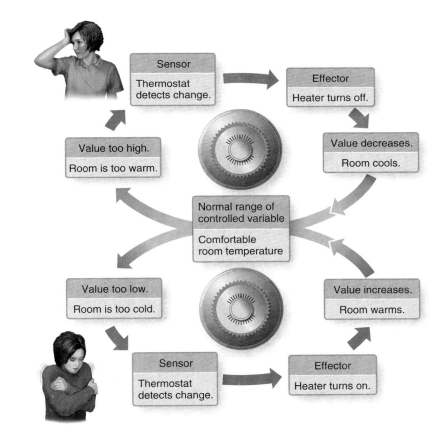

Figure 23.6 **Negative Feedback.** Physiological negative feedback systems in animals work in ways that are similar to how a household thermostat works.

■ *Give three examples of substances in the blood that the body maintains within certain limits.*

negative feedback
a hormone or other substance inhibits its own production

23.3 Hormones Coordinate Functions of Diverse Animal Systems

Under the influence of endocrine glands the diverse body systems in an animal can work together toward one common function. For example, the hormones secreted by the gonads (ovaries or testes) help to coordinate the physiological and behavioral aspects of reproduction. The adrenal glands coordinate the response to an external threat and stress. An exploration of the roles of some specific hor-

mones will help you to understand how these coordinated functions are achieved.

Insulin and Glucagon Are Active in Glucose Metabolism

Obtaining energy is an essential activity for any organism, but it is especially complex in animals. Much of the food that a human eats is converted to glucose, which is transported in the blood. Cells take up glucose from the bloodstream as they need it. The amount of glucose in the bloodstream determines how much energy is available to cells. If you think for a moment, though, you will realize that the eating patterns of many animals including humans aren't compatible with the body's continual need for glucose. Many animals eat one or a few large meals a day. Some animals eat only every few days. This provides a huge influx of glucose into the blood right after a meal, but little new glucose is available between meals. How do animals like humans deal with this inconsistent glucose supply? The answer is that animals have an intricate hormonal system that keeps glucose levels relatively stable.

The 3-pound human liver is the body's bulkiest organ and has important roles in detoxifying poisons in the blood, storing fat-soluble vitamins, synthesis of blood proteins, production of bile, and storing glycogen. The pancreas and liver work together to control blood glucose levels. Excess glucose available after a meal is not wasted but is taken up by liver cells, where it is strung together to form long molecules of *glycogen,* or "animal starch." In contrast, if a person has not eaten enough to provide for the immediate energy needs of cells, glycogen in the liver is broken down, and glucose is released into the bloodstream and carried to tissues that need it. Liver cells do not respond directly to blood glucose levels. Instead, special cells in the pancreas monitor blood glucose and release hormones—insulin and glucagon—that allow the liver and other cells to respond to blood glucose levels (**Figure 23.7**). When blood glucose levels are high, the pancreas makes and releases *insulin.* When blood glucose levels are low, the pancreas makes and releases *glucagon.* Insulin binds to receptors on the surface of liver cells and stimulates them to take up glucose and store it as glycogen molecules. Glucagon also binds to liver cells and has the opposite effect: glucagon stimulates liver cells to break down glycogen and release glucose. Both insulin and glucagon act on other tissues as well, such as muscle, but liver cells are the primary storage depot for glycogen when blood glucose levels are high. The glycogen molecules within liver cells also are the primary source of glucose when blood glucose is low.

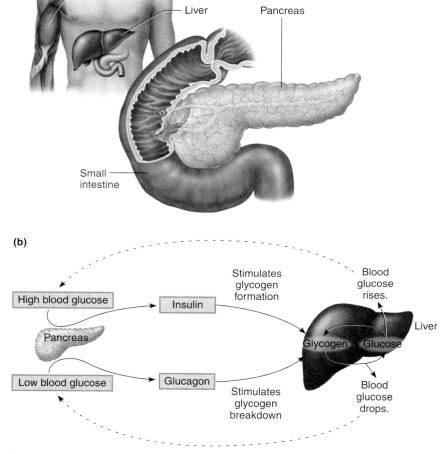

Figure 23.7 Insulin and Glucagon.
(**a**) Muscle cells and cells of the liver are especially active in the negative feedback loops that control blood glucose. The pancreas, located in a curve of the intestine, is the source of insulin and glucagon. (**b**) Insulin causes body cells to take up glucose and so reduces blood glucose levels. Glucagon causes muscle and liver cells to release glucose and so increases blood glucose levels. Both hormones affect blood glucose in negative feedback loops.

■ *What happens in type 2 diabetes to change this sequence of events?*

You may have heard of the practice of "carbo loading" that many athletes, but especially long-distance runners, use to prepare themselves for a big event. During a long-distance run, such as the 26 miles of a marathon, muscles must have a lot of glucose, and glucagon stimulates the liver and the muscles to break down their stored glycogen. It appears that loading up the liver and muscles with newly formed glycogen molecules in the days just prior to a long run provides a *more accessible* source of glycogen during the run. This is why runners and other endurance athletes "carbo-load" with pasta and other starches.

Thyroid Hormones Affect the Rate of Cellular Metabolism

The thyroid gland also plays a role in energy usage. This small endocrine gland sits astride the trachea and produces two hormones, *thyroxin* and *triiodothyronine*, that together are often called thyroid hormone. Thyroid hormone controls the amount of energy that cells burn at rest—when an animal is sleeping or quietly inactive. This resting level of energy use is called the *basal metabolic rate*, or *BMR*, and it differs from person to person. Thyroid hormone increases the resting levels of oxygen consumption and ATP production.

Thyroid hormone acts quite differently than insulin does (**Figure 23.8**). The two thyroid hormones are polar molecules, yet they cross the plasma membrane, transported across by a membrane protein. Once inside the cell thyroid hormone binds to a receptor molecule. This receptor is a transcription factor, a protein that binds to DNA and affects the transcription of a particular gene, turning it on or off (see Section 9.6). The thyroid receptor will not work as a transcription factor, however, unless thyroid hormone is bound to it. When thyroid hormone is bound to its internal receptor, the complex becomes a transcription factor that turns on key genes including some involved in energy metabolism.

Some Adrenal Hormones Prepare the Body for Stressful Emergencies

The adrenal glands sit like pointed caps on top of the much larger, bean-shaped kidneys, one on each side of the body (**Figure 23.9a**). Adrenal glands also influence energy metabolism, but they are active in times of stress and help mobilize your mind and body if you are in danger. Each adrenal is actually two endocrine glands in one: the *adrenal cortex* and the *adrenal medulla*, and they produce different hormones. The adrenal cortex is the outer tissue; it covers the inner adrenal medulla like chocolate coats a cream-filled candy.

(a) Location of thyroid gland

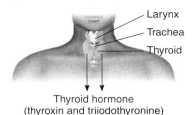

Larynx
Trachea
Thyroid

Thyroid hormone
(thyroxin and triiodothyronine)

Figure 23.8 **How Thyroid Hormone Affects a Cell.** (**a**) The thyroid gland is located on either side of the larynx, or voice box. (**b**) Thyroid hormone crosses the cell membrane and acts through receptors inside a cell to change gene expression in the cell.

(b) Thyroid hormone is taken into cells and re-directs protein synthesis.

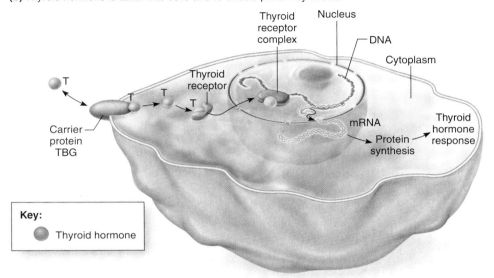

Thyroid receptor complex
Nucleus
DNA
Cytoplasm
Thyroid receptor
T
T
T
T
T
Carrier protein TBG
mRNA
Protein synthesis
Thyroid hormone response

Key:
Thyroid hormone

The adrenal medulla releases two related hormones: *adrenaline* (also called epinephrine) and *noradrenaline* (also called norepinephrine). These hormones orchestrate a coordinated response to a threatening situation and prepare the body either to flee to safety or to defend itself from the threat. Imagine you are walking home late at night, alone, and you sense that someone is following you (Figure 23.9b). Your heartbeat increases and you breathe more quickly. Tension builds in your stomach, and you prepare to either run or turn and confront your pursuer. You probably do not feel it, but the flow of blood to your internal organs, such as stomach, liver, and pancreas is reduced, while the flow of blood to your muscles and brain is increased. Your liver cells and muscles break down glycogen and release glucose. Because of these actions by your adrenal medulla and other endocrine glands, you feel alert and you may be able to do things you normally could not do—such as run faster, leap higher, or shove a heavy obstacle out of the way. Adrenaline and noradrenaline coordinate this complex set of reactions by binding to membrane receptors on cells throughout the body.

Too much adrenaline and noradrenaline can be dangerous. For example, too much adrenaline

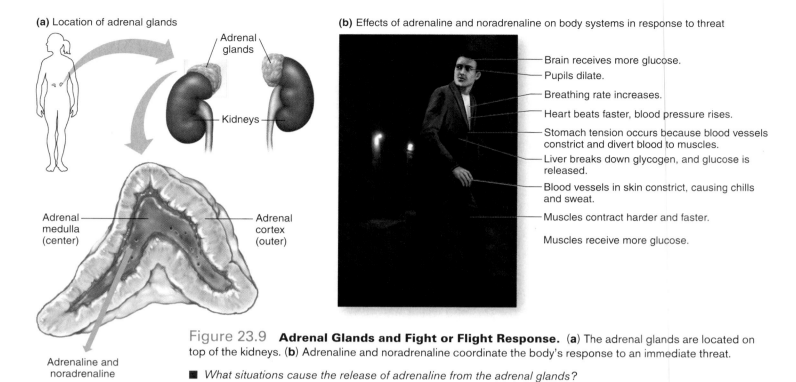

(a) Location of adrenal glands

Adrenal glands

Kidneys

Adrenal medulla (center)

Adrenal cortex (outer)

Adrenaline and noradrenaline

(b) Effects of adrenaline and noradrenaline on body systems in response to threat

Brain receives more glucose.

Pupils dilate.

Breathing rate increases.

Heart beats faster, blood pressure rises.

Stomach tension occurs because blood vessels constrict and divert blood to muscles.

Liver breaks down glycogen, and glucose is released.

Blood vessels in skin constrict, causing chills and sweat.

Muscles contract harder and faster.

Muscles receive more glucose.

Figure 23.9 **Adrenal Glands and Fight or Flight Response.** (**a**) The adrenal glands are located on top of the kidneys. (**b**) Adrenaline and noradrenaline coordinate the body's response to an immediate threat.

■ *What situations cause the release of adrenaline from the adrenal glands?*

can cause the heart to stop beating. But these hormones also are used as treatments for some life-threatening conditions. For example, you or someone you know may carry an *EpiPen* to use in the case of an extreme allergic reaction. The *EpiPen* contains a single injection of epinephrine (adrenaline is called epinephrine when it is injected). During an extreme allergic reaction many body systems react so adversely to the substance producing the reaction that the person may die. Air passages in the chest and lungs swell and cut off the oxygen supply. The injected epinephrine opens these airways and can save the life of a person who is having a severe allergic reaction.

The adrenal cortex releases a variety of hormones that fall into three major classes: hormones that affect water and mineral balance, hormones that affect sexual characteristics, and hormones that primarily affect how the body uses energy. Hormones from the adrenal cortex are made from cholesterol and are members of a group of compounds called steroid hormones. Other endocrine glands such as the testes and ovaries also produce steroid hormones. *Aldosterone* (see Section 22.6) acts on the kidneys and helps regulate the production of urine by the kidneys. It also exerts important control over blood levels of minerals such as potassium and sodium. The major sex hormones produced by the adrenal cortex are forms of testosterone. Because the testes of adult males produce large amounts of

testosterone, the testosterone made by the adrenal cortex probably has little impact on them. The effects of testosterone from the adrenal cortex are stronger in young males and in females. In young males adrenal testosterone can hasten and amplify the onset of puberty, while in females high levels of adrenal testosterone can produce mature masculine traits such as facial hair and increased muscle mass.

Cortisol is the dominant steroid hormone produced by the adrenal cortex, and it has many complex effects throughout the body (**Figure 23.10**). Cortisol primarily increases supplies of energy and other metabolic resources. It stimulates the transformation of a variety of organic molecules into glucose, a process called *gluconeogenesis*. This effect provides more energy to all cells in the body. Cortisol also stimulates the metabolism of fats for energy and the breakdown of proteins to provide amino acid building blocks for other body processes. An important function of cortisol is the response to stress. Stress can be injury, disease, starvation, or other conditions that will not kill a person immediately but will cause harm in the long run. While adrenaline is a response to immediate stress, cortisol kicks into action when stresses last longer. For example, more cortisol is released when a person fasts for several days or is exposed to a prolonged period of uncertain threat. The extra cortisol provides the body with increased resources. Cortisol also inhibits body processes that might use energy such as the immune response, the

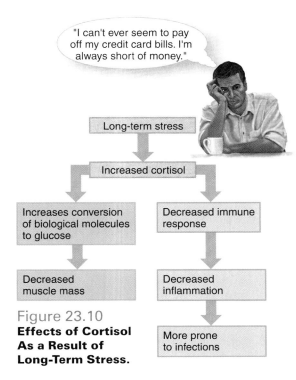

"I can't ever seem to pay off my credit card bills. I'm always short of money."

Long-term stress

↓

Increased cortisol

↓   ↓

Increases conversion of biological molecules to glucose

Decreased immune response

↓   ↓

Decreased muscle mass

Decreased inflammation

↓

More prone to infections

Figure 23.10
Effects of Cortisol As a Result of Long-Term Stress.

inflammatory response, and other growth processes such as bone formation.

Cushing's syndrome is a disorder caused by too much cortisol. People with Cushing's syndrome have symptoms such as high blood pressure, excessive hair growth, obesity, kidney stones, problems with glucose uptake, and menstrual irregularity. Cushing's syndrome is rare and usually is caused by a tumor in the adrenal glands or in the pituitary. In contrast, *Addison's disease* is caused by too little cortisol production, and many of the symptoms are opposite to those of Cushing's syndrome. People with Addison's disease experience weight loss, muscle loss, low blood pressure, and fatigue. Addison's disease can develop when a person's immune system abnormally attacks and destroys cells in the adrenal cortex. It can be treated with synthetic cortisol-like drugs such as prednisone.

Cortisol is a lipid hormone that crosses cell membranes. Receptors for cortisol are inside cells and, like the receptors for thyroid hormone, the cortisol receptor is a transcription factor. When cortisol binds to this protein receptor, the resulting complex molecule binds to specific sites along DNA and controls the transcription of specific genes. The list of genes that cortisol can affect is long. In some cases the cortisol-receptor complex inhibits transcription, while in other cases it activates transcription. These diverse effects on DNA are the reason why cortisol has such complex effects on the whole body.

The actions of cortisol can be useful in a medical setting. For example, many disorders are related to tissue inflammation. Allergic reactions, certain

kinds of arthritis, lupus, and even the effects of some cancers involve inflammation. Versions of cortisol such as prednisone often are prescribed to combat excess inflammation. When a person receives an organ or tissue from another person such as a liver or kidney transplant, cortisol-like drugs are given that inhibit the body's immune response. Prolonged use of cortisol-based drugs does carry risks. One of these is the risk of infection because the immune system is suppressed. Patients with chronic disorders that require long-term cortisol use must balance the benefits against these risks.

Some People Take Hormones in the Hope of Improving Normal Health

It is tempting to wonder whether you would be younger, stronger, or healthier if you had more of certain hormones. Some people go beyond wondering and take hormone supplements in the absence of any condition that a medical doctor would recognize and treat. Human growth hormone (HGH) is an increasingly popular product among people who believe it will help keep them young and fit into middle and old age. Is there any evidence for the claims that hormones will slow the aging process? In both sexes production of a variety of hormones, including growth hormone, does decline with age. Growth hormone is critical for normal development during childhood and adolescence. In response to growth hormone the muscle mass of young people increases, body fat decreases, and bones grow. Older adults who take growth hormone may be convinced it will have similar effects for them. The scientific literature shows mixed results, however, and points out risks to the use of growth hormone in otherwise healthy adults. Some studies have found changes in blood chemistry and joint problems, in older adults who take growth hormone for several months. There may be an increased risk of certain kinds of cancer. None of these conclusions can yet be made with confidence. As people seek ways to minimize the effects of aging, the study of, and debate about, hormone use by adults is likely to continue.

The most public and controversial use of hormones outside of a medical setting is the use of **anabolic steroids,** especially by athletes. Anabolic steroids are derivatives of male hormones such as testosterone. The term *anabolic* refers to metabolic pathways that build biological molecules, while the term *catabolic* refers to metabolic pathways that break down biological molecules. Testosterone stimulates cells to build molecules, especially the protein molecules of muscles, and so it is an anabolic steroid. Many athletes believe that anabolic steroids will increase their muscle mass and strength and so

anabolic steroid
a hormone that is a derivative of testosterone; anabolic steroids often are taken by athletes to increase muscle mass and improve athletic performance

Figure 23.11 **Use of Anabolic Steroids by High School Students.** Use of anabolic steroids by high school students has increased, while their perception that anabolic steroids are harmful has decreased.

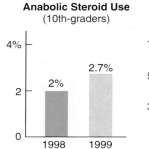

Anabolic Steroid Use
(10th-graders)

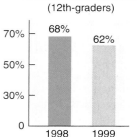

Perceived Risk of Harm
(12th-graders)

ple are concerned about the growing use of anabolic steroids to increase physical performance.

23.4 Hormonal Disorders Cause Major Disease Syndromes

Like any other body system, hormonal systems can go awry. In some cases a person is born with an inherited hormonal disorder that is apparent early in life. In other cases hormonal disorders can develop later in life, either as a result of a gradual disease process or as a result of experiences, infections, or lifestyle. Let's consider a few examples.

Problems in Insulin Function Result in Diabetes Mellitus

Insulin helps to regulate blood glucose levels by stimulating cells to take up glucose from the bloodstream (Figure 23.7). For many people, though, insulin does not function normally, and they have an excess of glucose in the bloodstream, a condition called **hyperglycemia.** There are many causes of hyperglycemia, but when this disorder is caused by problems with insulin, it is called **diabetes mellitus.** People with diabetes mellitus excrete glucose through their kidneys instead of storing it for leaner times. As a result, diabetics often are hungry and can lose weight even though they eat normally. The excess glucose also causes excessive water to be lost, so diabetics often are thirsty and urinate frequently. If the hyperglycemia of diabetes is not corrected, it can damage tissues and organs. For instance, people with chronic hyperglycemia suffer from poor circulation and as a result, sores on their legs and feet are slow to heal. Diabetics also can experience retinal damage; pains especially in the legs; and dry, itchy skin. In the long term diabetes damages the heart and also contributes to high blood pressure. Diabetes is the third leading cause of death in the United States. About 18 million people in the U.S. (about 6%) have diabetes, but only about 13 million are diagnosed and receiving treatment.

will enhance their athletic performance. Accordingly, they take injections of anabolic steroids. While athletes might be confident that high doses of these drugs are safe, most of the medical and scientific community has concluded that long-term use of anabolic steroids can produce a variety of health problems. For example, men on anabolic steroids may be subject to emotional mood swings and are more likely to be violent toward their domestic partners than are men who are not on steroids. Other studies document that growth can be stunted if anabolic steroids are taken before a person has achieved full growth, which for young men can be in the late teens or early twenties. Some younger athletes are drawn to steroid use (**Figure 23.11**). There also are effects on the heart and cardiovascular system, and high doses of anabolic steroids may be addictive. In the long run, anabolic steroid users may face an increased risk of cancer and a greater chance of an early death. These are just some of the reasons that many peo-

hyperglycemia excess glucose in the bloodstream

diabetes mellitus a condition in which the body either does not produce enough insulin (Type 1) or does not respond to insulin (Type 2)

Diabetes comes in two forms: Type 1 and Type 2. These two diseases have similar outcomes but different causes (**Table 23.3**). Type 1 usually develops early in life and is caused by a deficiency in the

Table 23.3 A Comparison of Type 1 and Type 2 Diabetes

	Type 1 diabetes	Type 2 diabetes
Cause	Insulin-producing cells in pancreas are destroyed by immune system	Pancreas produces normal amounts of insulin, but insulin receptors on body cells do not bind to insulin
Effect on blood sugar	Glucose is not taken up by body cells and stays in bloodstream	Glucose is not taken up by body cells and stays in bloodstream
Occurrence	Early in life; 10% of cases of diabetes in U.S.	Late in life; 90% of cases of diabetes in U.S.
Symptoms	Extreme thirst; frequent urination; drowsiness or lethargy; sugar in urine; sudden vision changes; increased appetite; sudden weight loss; fruity, sweet, or winelike odor on breath; heavy, labored breathing; stupor; and unconsciousness.	Fatigue, thirst, frequent urination, poor circulation in extremities, blurred vision, sexual dysfunction, brain dysfunction, kidney disease, coronary artery disease, diabetic retinopathy, diabetic neuropathy

amount of insulin available. Type 1 diabetes probably is caused by a genetic disorder of the immune system. The immune systems of people who have Type 1 diabetes target and destroy the cells in the pancreas that produce insulin. Because they do not produce insulin, people with Type 1 diabetes have uncontrolled levels of sugar in their blood. Today, people with Type 1 diabetes can inject themselves with insulin and so can stabilize their blood sugar. Type 1 diabetics must take supplemental insulin all through their lives.

More than 90% of diabetics have Type 2 diabetes, and this form of the disorder usually develops late in life. It often is associated with being overweight and not exercising. People with Type 2 diabetes have a much higher risk of death than the average person. Between 7 and 8 million people in the United States have this disorder. Many of these diabetics are 40 years old or older, but recently a growing number of children are being diagnosed with Type 2 diabetes.

Gestational diabetes is a form of Type 2 diabetes that can occur when a woman is pregnant. Most often gestational diabetes is not a permanent disease but goes away after the baby is born. Untreated gestational diabetes can harm both mother and baby, so it must be monitored carefully.

Type 2 diabetes is not as well understood as Type 1, but in at least some cases the problem is with the receptors for insulin located on the plasma membranes of liver and muscle cells. For some reason these receptors stop responding to insulin, and so even though people with Type 2 diabetes make enough insulin, it does them no good. A variety of medications can help control blood sugar in Type 2 diabetes, but a healthy diet, weight control, and exercise are critical parts of any treatment. How might these activities contribute to reducing the risk of Type 2 diabetes? Recent studies show the remarkable finding that even mild exercise—such as walking several times a week—can increase the number of insulin receptors on cells throughout the body.

Thyroid Hormone Abnormalities Affect Development and Metabolism

Thyroid hormones have strong effects on basal metabolic rates, but they also affect many other body processes. Proper levels of thyroid hormone are especially important during early development. Low levels of thyroid hormone during late pregnancy can have a variety of effects. Neurons may not differentiate, synapses may not form, and if there is no medical intervention, the child may be born mentally impaired. A deficiency of thyroid hormones during pregnancy also results in low birth weight and stunted development. The importance of thyroid hormones to normal growth and development is why thyroid hormone levels are routinely tested in newborns. Administering thyroid supplements right after birth can make up for the deficiency and can result in near-normal brain growth and development of near-normal intellect.

Later in life many people experience thyroid deficiency, a condition called *hypothyroidism*. Thyroid hormone production can fall at any time of life, but it is especially common when a person is 50 years old or older. Symptoms of thyroid deficiency include fatigue, depression, feeling cold, constipation, and low blood pressure. Paradoxically, low levels of thyroid hormones also can lead to anxiety and shortness of breath. The good news is that treatment with thyroid hormone supplements can reverse these symptoms.

Hyperthyroidism, the condition of secreting too much thyroid hormone, can lead to increases in basic metabolic rate, weight loss, irritability, changes in personality, heart palpitations, heart damage, muscle atrophy, diarrhea, and reduced sexual functions. Hyperthyroidism can be treated with drugs that inhibit the production of thyroid hormones, but this is not a long-term solution. Parts of the thyroid gland can be surgically removed, reducing hormone production. Another interesting treatment is radioactive iodine. The thyroid is the only tissue in the body that will take up iodine, and so radioactive iodine only affects the thyroid gland. At the proper doses the radioactive iodine kills thyroid gland cells, and so it reduces the amount of thyroid hormones produced.

Growth Hormone Abnormalities Can Produce Dwarfs or Giants

Growth hormone is released from the pituitary when the hypothalamus releases *growth hormone-releasing hormone* (*GHRH* for short). Growth hormone is not critical for embryonic and fetal growth, but instead begins to have an effect early in life and its influence continues throughout adolescence. One of the major effects of growth hormone is to increase cell proliferation. Too little growth hormone during childhood leads to short stature, or dwarfism (**Figure 23.12a**). Conversely, too much growth hormone during childhood will lead to tall stature, or giantism (Figure 23.12a). Even in adulthood growth hormone can continue to have effects on body structure. Figure 23.12b shows a woman with a condition called *acromegaly*, a disorder that involves excess secretion of growth hormone during adulthood. You can see that this woman's facial features and hands have been distorted. Deficiencies of growth hormone are

Figure 23.12 **Effects of Too Much or Too Little Growth Hormone.** (**a**) pituitary giant and pituitary dwarf; (**b**) In acromegaly growth hormone is overproduced during adulthood when the limb bones no longer respond to it, but bones of the face and hands do still respond. This woman was 16, 33, and 52 years old when these photos were taken.

sexual dimorphism the existence of traits that distinguish males from females

primary sexual characteristics the different reproductive organs of males and females

secondary sexual characteristics gender-related features that do not directly participate in the process of reproduction

treated with growth hormone supplements. Treatment of excessive growth hormone has lagged behind, although some new drugs show promise.

QUICK CHECK

1. How are Type 1 and Type 2 diabetes similar? How are they different?
2. What is the common hormonal problem involved in dwarfism, giantism, and acromegaly?

23.5 Hormones Coordinate Human Reproduction

Now that you have an overview of how hormones regulate important body systems, we can deepen your understanding by considering a specific example in greater detail. Let's look at the important role of hormones in the regulation of sexual reproduction.

Males and Females Are Distinguished by Primary and Secondary Sexual Characteristics

Sexual reproduction has evolutionary advantages because it leads to greater genetic diversity in a population than does asexual reproduction. Nearly all eukaryotic species can reproduce sexually. Animals have elaborate courting and mating behaviors, and many animals display **sexual dimorphism,** which is another way of saying that males and females look different and have different behavioral characteristics. The distinctive reproductive organs of male and female animals are called **primary sexual characteristics.** The many other characteristics associated with the sex or gender of an individual, that are not directly involved in reproduction, are called **secondary sexual characteristics.**

Figure 23.13 shows the primary sexual characteristics for male and female humans, and these are common to most mammals. A female has *ovaries, oviducts, uterus, cervix, clitoris, vagina,* and other important reproductive structures, while a male has *testes, seminiferous tubules, prostate, seminal vesicles, vas deferens, penis,* and other important reproductive structures. Noticeable external secondary sexual characteristics make it easy to tell human sexes apart. Females have enlarged breasts, higher voice, smaller size, and a rounded body shape; males lack enlarged breasts, and have lower voice, larger size, a more muscular body, and coarse facial hair. Certain behavioral traits also are correlated with gender, and so they also can be considered as secondary sexual characteristics. Let's put those aside for the moment, though, and examine how mammalian sexual reproduction in humans actually works.

The Production of Sperm and Eggs Are Hormonally Controlled

Sexual reproduction involves a complex set of mechanisms and behaviors, but central to its success is the production of the gametes: sperm and eggs. In both cases gamete production actually

(a)

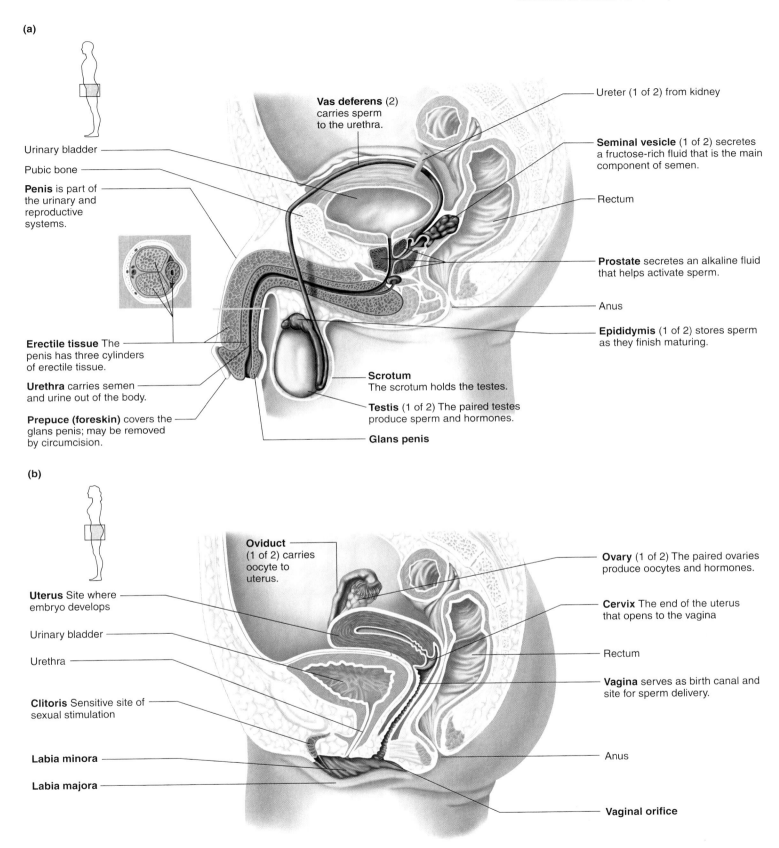

Vas deferens (2) carries sperm to the urethra.

Ureter (1 of 2) from kidney

Urinary bladder

Pubic bone

Penis is part of the urinary and reproductive systems.

Seminal vesicle (1 of 2) secretes a fructose-rich fluid that is the main component of semen.

Rectum

Prostate secretes an alkaline fluid that helps activate sperm.

Anus

Epididymis (1 of 2) stores sperm as they finish maturing.

Erectile tissue The penis has three cylinders of erectile tissue.

Urethra carries semen and urine out of the body.

Prepuce (foreskin) covers the glans penis; may be removed by circumcision.

Scrotum
The scrotum holds the testes.

Testis (1 of 2) The paired testes produce sperm and hormones.

Glans penis

(b)

Oviduct
(1 of 2) carries oocyte to uterus.

Ovary (1 of 2) The paired ovaries produce oocytes and hormones.

Uterus Site where embryo develops

Cervix The end of the uterus that opens to the vagina

Urinary bladder

Urethra

Rectum

Vagina serves as birth canal and site for sperm delivery.

Clitoris Sensitive site of sexual stimulation

Labia minora

Anus

Labia majora

Vaginal orifice

Figure 23.13   **Human Reproductive Systems.** (**a**) Male (**b**) Female.

begins in the embryo, stops during infancy and childhood, and resumes after puberty. Even though the separate processes have parallel phases, the details of sperm and egg production are different. Let's consider sperm production first.

Spermatogenesis is a lavish biological process that manufactures between 100 and 200 million sperm *a day*—for 365 days a year for as long as 75 years. The first sperm are produced in puberty, and sperm production continues throughout adult life. The whole process starts before the young male is born. Once the tissues of the embryonic testes are mature, the first steps in the production of sperm cells begin, but after that sperm production is suspended until puberty. **Figure 23.14a** shows the location of some of the important cells in the testes that contribute to the production of sperm. Specialized cells in the testes release testosterone, the primary male hormone, which controls sperm production. Sperm form within a set of tubules in the testes called the *seminiferous tubules.* Sperm mature from a population of stem cells called **spermatogonia.**

Figure 23.14b shows some of the important steps in sperm production. In the male embryo testosterone stimulates spermatogonia to go through mitotic cell division, but these cells do not develop any further until puberty, the age at which sexual reproduction becomes possible, usually around 14 years old for boys. The *descent of the testes* is another important event that occurs before birth. The testes first develop inside the abdominal cavity, much as the ovaries do. Sometime before birth the testes drop out of the abdominal cavity and rest in the *scrotum,* the external pouch that holds them. This developmental process is essential to production of viable sperm. Sperm can mature only at a temperature that is lower than normal body temperature; if the testes are too warm, sperm will not mature properly. The descent of the testes to a location outside the abdominal cavity lowers the temperature of the testes and so allows normal sperm development.

The next milestone events in sperm production come during puberty. At puberty two hormones are released from the anterior pituitary: luteinizing hormone (LH) and follicle-stimulating hormone (FSH), which have effects on the testes. In males LH stimulates cells in the testes to produce testosterone and FSH and testosterone act on other cells in the testes that support sperm development. Too much testosterone in the blood can inhibit sperm production. This is an example of negative feedback. If a man receives extra testosterone through injections, testosterone production in the testes is reduced. As a result, the testes actually are exposed to less testosterone, and sperm production

spermatogonia cells that produce sperm

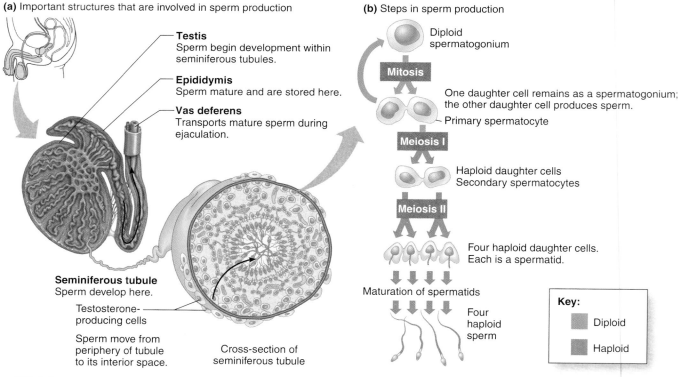

(a) Important structures that are involved in sperm production

Testis
Sperm begin development within seminiferous tubules.

Epididymis
Sperm mature and are stored here.

Vas deferens
Transports mature sperm during ejaculation.

Seminiferous tubule
Sperm develop here.

Testosterone-producing cells

Sperm move from periphery of tubule to its interior space.

Cross-section of seminiferous tubule

(b) Steps in sperm production

Diploid spermatogonium

Mitosis

One daughter cell remains as a spermatogonium; the other daughter cell produces sperm.
Primary spermatocyte

Meiosis I

Haploid daughter cells
Secondary spermatocytes

Meiosis II

Four haploid daughter cells.
Each is a spermatid.

Maturation of spermatids

Four haploid sperm

Key:
Diploid
Haploid

Figure 23.14 **Production of Sperm.**

drops. Some researchers are using this finding as a basis to try to develop a male contraceptive.

To make sperm cells, spermatogonia go through mitotic cell division and then through meiotic cell division. Spermatogonia have 46 chromosomes. The meiotic cell division produces cells with 23 chromosomes, half the normal number of human chromosomes. One spermatogonium eventually will give rise to four sperm. As each sperm matures, it becomes smaller, more specialized, and more condensed. Spermatids lose cytoplasm and develop the typical "tadpole shape": a compact head with a long tail (**Figure 23.15**). The nucleus is in the sperm's head, and the head has a cap that contains enzymes that allow it to digest a path into the egg's protective envelope. The sperm's body is packed with mitochondria that will provide the ATP necessary for the sperm's short independent life. Finally, the tail is made of pairs of microtubules arranged radially. It takes between 64 and 72 days for a sperm to develop from a spermatogonium.

Of course, to complete their reproductive function, the mature sperm must get out of the testes and into a female's reproductive tract. The seminiferous tubules where sperm are produced connect to the *epididymis*, a larger-diameter tube where sperm are stored. In the seminiferous tubules sperm begin to swim by lashing their flagellar tails. During *ejaculation*, the process that forcefully expels sperm from the male reproductive tract, sperm are transferred to the vas deferens and emitted through the penis. Several accessory glands also are components of the male reproductive tract. The prostate is a walnut-sized gland that encircles the urethra, just below the bladder. Secretions of the prostate help to activate sperm. Secretions of other glands provide a nutrient solution for sperm.

The Production of Eggs is Cyclical and Finite

Like the production of sperm, the production of eggs begins during embryonic development, then halts, and then resumes when a girl becomes sexually mature, at about 12 years old. Unlike sperm production, egg production largely stops when a woman is between 40 and 50 years old. Let's return to an early embryo and see how the female internal sexual organs develop, and follow egg production through puberty and adulthood.

Ovaries are the female reproductive organs and they complete development about the same time that a male embryo's testes do—around nine weeks after fertilization. In the ovaries the stem cells that will become eggs, called *oogonia* (**Figure 23.16**), begin the process of meiosis that eventually

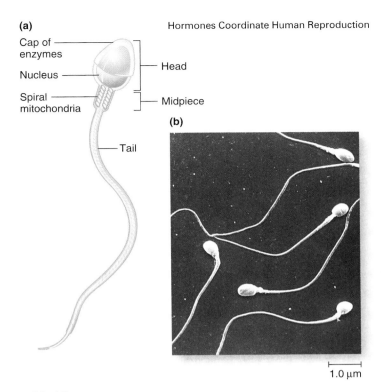

(a)

(b)

Figure 23.15 **Human Sperm.** The head with its cap of enzymes, midpiece with spiral mitochondria, and tail are the parts of a sperm.

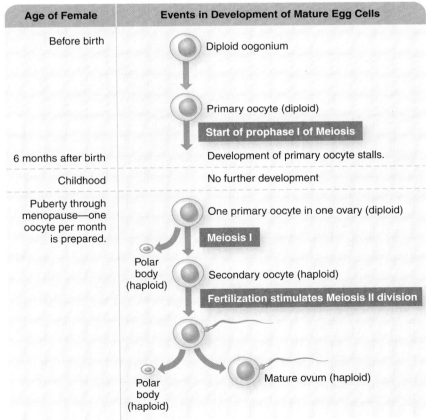

Age of Female	Events in Development of Mature Egg Cells
Before birth	Diploid oogonium
	Primary oocyte (diploid)
	Start of prophase I of Meiosis
6 months after birth	Development of primary oocyte stalls.
Childhood	No further development
Puberty through menopause—one oocyte per month is prepared.	One primary oocyte in one ovary (diploid)
	Meiosis I
Polar body (haploid)	Secondary oocyte (haploid)
	Fertilization stimulates Meiosis II division
Polar body (haploid)	Mature ovum (haploid)

Figure 23.16 **Cell Divisions That Lead to a Mature Human Egg Cell.** Notice that a mature ovum is not produced until after fertilization.

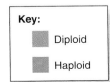

Key:

Diploid

Haploid

will produce a mature egg. Oogonia begin the process of meiosis when the female is an embryo, but meiosis stops shortly after it gets started. These stalled oogonia are called **primary oocytes,** and they do not resume meiosis until the girl has reached puberty. At birth a baby girl has about 400,000 primary oocytes, and she may not make any more for the rest of her life. At puberty the adult female human reproductive pattern begins, and each month one primary oocyte resumes meiosis. It matures into an ovum, or egg cell, but again the process is not complete even when the mature egg is released from the ovary. The primary oocyte completes meiosis I, producing one small cell called a *polar body* and one large cell that is released from the ovary as an egg. Even though the egg is no longer in the ovary, and is moving down an oviduct, toward the uterus, this egg has not yet completed meiosis! A mature egg has 23 *duplicated* chromosomes, meaning it has two copies of each of its 23 chromosomes. How can this be? Again, the development of the egg stalls, just before the second phase of meiosis. The final phase of meiosis that reduces the number of chromosomes from 46 to 23 does not happen until after the egg is fertilized by a sperm. This second phase of meiosis produces the *zygote* and another small polar body cell. This process of releasing about one egg once a month continues until *menopause,* the time in a woman's life when her ovaries stop producing eggs.

In females the onset of puberty marks the beginning of two, linked monthly cycles that govern reproduction: the ovarian cycle and the menstrual cycle. The **ovarian cycle** governs the monthly production of mature eggs from the oocytes that were formed in the embryo. The **menstrual cycle** coordinates with the ovarian cycle to prepare the uterus—each month—for the possibility of a pregnancy. Both of these female reproductive cycles are controlled by hormones. Let's look at the details of the ovarian cycle first.

Like the production of sperm, the production of mature eggs from oocytes after puberty is controlled by the pituitary hormones, LH and FSH (**Figure 23.17a**). The process also involves the female hormones, **estrogen** and **progesterone.** Figure 23.17b shows how these hormones interact to produce about one egg per month. The release of the egg from a mature follicle of the ovary is called **ovulation.** As shown in Figure 23.17c, the cycle begins about two weeks after the last ovulation. At this time the hypothalamus is in control of events because it releases *gonadotropin-releasing factor,* a hormone that stimulates the anterior pituitary to release LH and FSH. In different ways

these two hormones stimulate the maturation of a primary oocyte into a mature egg, and the stalled primary oocyte resumes meiosis. At the same time cells that surround the activated primary oocyte begin to produce estrogen, and so the level of estrogen in the bloodstream increases. Although estrogen has a feedback effect back on the anterior pituitary, that effect changes over time. At first the estrogen inhibits the release of LH and FSH, but these hormones continue to build up within the bloodstream and are monitored by cells of the anterior pituitary. When estrogen levels in the blood reach some critical level, the anterior pituitary is stimulated to release a big surge of LH and FSH. This surge causes the mature egg to be released from the ovary, and ovulation has occurred. Now another transition begins. After the egg is released, progesterone is released from the tissues around the site that once held the egg. As you will see shortly, progesterone links the ovarian cycle to the menstrual cycle and helps prepare the uterus for a possible pregnancy. Levels of all of the hormones—LH, FSH, estrogen, and progesterone—decline after ovulation. But at the same time the level of gonadotropin-releasing hormone from the hypothalamus is increasing, and so the ovarian cycle starts again, this time with a new primary oocyte in a new follicle in the female's other ovary.

The menstrual cycle operates in parallel with the ovarian cycle, and each month it prepares the uterus for a possible pregnancy (Figure 23.17b). If a newly ovulated egg is fertilized, the uterus must be prepared to receive the early embryo. The uterus must have a thick lining that has a rich blood supply for the embryo to attach to. The uterine lining will sustain the embryo until the *placenta* that provides the embryo's blood supply has developed. This is another effect of the estrogen released by tissues that surround the developing egg. Estrogen stimulates the uterine wall to thicken and store blood and nutrients. This blood-rich lining of the uterus is called the **endometrium.** Estrogen has similar effects on breast tissue, causing breasts to become enlarged and swollen. Both of these effects are caused, in part, by estrogen's stimulation of cell proliferation. They are also part of the reason why women are physically uncomfortable, with bloated abdomens and sore breasts, in the week before their menstrual periods. The effect of estrogen on cell proliferation is the primary reason why it plays a role in the development of cancer.

Progesterone is released after ovulation, and it continues the process of preparing the uterus for pregnancy. If fertilization does not occur, both

primary oocytes ovarian cells that may mature into egg cells

ovarian cycle the monthly cycle in ovaries that governs the production of mature eggs from oocytes

menstrual cycle the monthly cycle in the uterus that prepares it for a possible pregnancy

estrogen female hormone

progesterone the female hormone that helps ready and maintain the endometrium

ovulation the release of an egg from an ovary

endometrium the lining of the uterus that builds up each month and may be shed as menstrual fluid

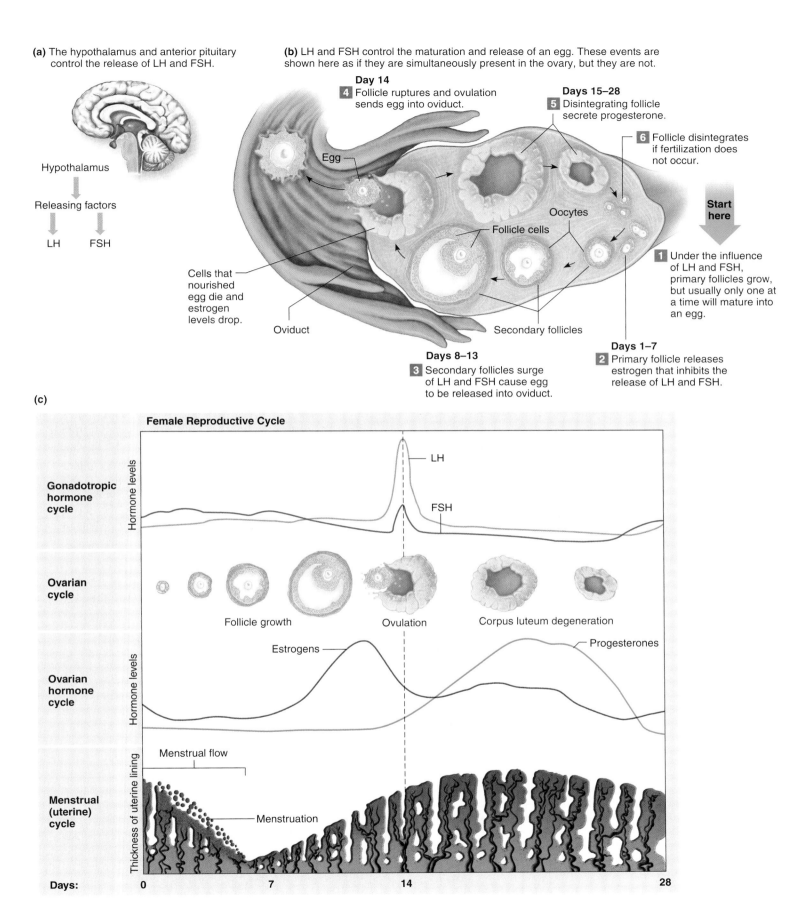

(a) The hypothalamus and anterior pituitary control the release of LH and FSH.

Hypothalamus

↓

Releasing factors

↓   ↓

LH   FSH

(b) LH and FSH control the maturation and release of an egg. These events are shown here as if they are simultaneously present in the ovary, but they are not.

Day 14
4 Follicle ruptures and ovulation sends egg into oviduct.

Days 15–28
5 Disintegrating follicle secrete progesterone.

6 Follicle disintegrates if fertilization does not occur.

Start here

1 Under the influence of LH and FSH, primary follicles grow, but usually only one at a time will mature into an egg.

Egg

Oocytes

Follicle cells

Cells that nourished egg die and estrogen levels drop.

Oviduct

Secondary follicles

Days 1–7
2 Primary follicle releases estrogen that inhibits the release of LH and FSH.

Days 8–13
3 Secondary follicles surge of LH and FSH cause egg to be released into oviduct.

(c)

Female Reproductive Cycle

Gonadotropic hormone cycle

Hormone levels

LH

FSH

Ovarian cycle

Follicle growth

Ovulation

Corpus luteum degeneration

Ovarian hormone cycle

Hormone levels

Estrogens

Progesterones

Menstrual (uterine) cycle

Thickness of uterine lining

Menstrual flow

Menstruation

Days: 0   7   14   28

Figure 23.17   **Hormonal Control of Egg Production (Ovarian Cycle) and Uterine Lining (Menstrual Cycle).**

menstruation the monthly loss of blood from the uterus as the endometrium sloughs off

cleavage the first few cell divisions of an embryo that have no growth between cell divisions

estrogen and progesterone levels fall. As a result, the uterine lining deteriorates and is lost during the monthly bleeding, or **menstruation.** If fertilization does occur, the developing embryo produces hormones that maintain the uterine environment.

Knowledge of how the female reproductive cycle is hormonally controlled has proven to be useful in a variety of medical settings. One of the most useful applications was the development of birth control pills for women. Birth control pills contain synthetic forms of the female hormones, estrogen and progesterone. These extra hormones even out the level of estrogen and progesterone in the body, and so they inhibit the release of LH and FSH. As a result, eggs do not mature and ovulation does not happen. In most versions of the pill no hormones are taken one week out of every four. This allows women to have a normal menstrual cycle without ovulation.

Fertilization in Humans Is a Beautifully Complex Process

Today, scientists have a fairly detailed picture of how fertilization actually occurs, but the route to that understanding is littered with some fairly amusing notions. Van Leeuwenhoek discovered sperm in 1678, as he trained his simple but effective magnifiers on everything around him, including his own semen. At first he considered sperm in semen to be parasites, and that is why he named them spermatozoa, or "little seed animals." Leeuwenhoek later came to believe that each sperm contained a perfectly formed human, and that the female's role was to provide a nurturing environment where the little one could develop (**Figure 23.18**). Then there were those who believed that the egg contained all the necessary material to make a baby human, leaving the sperm to the role of stimulating the start of development. It was not until 1824 that scientists showed that the nuclei of egg and sperm actually unite to produce a new individual. In humans the act of ejaculation gets about 280 million sperm cells on their way, but only about 200 of these ever get close to the egg as it meanders down the oviduct. For a sperm to get from the vagina to the egg is not just a matter of swimming with that flagellar tail. Muscular contractions of the uterus probably are essential for moving the sperm to the egg in a timely manner. Once in the general vicinity, the egg probably secretes a chemical that sperm home in on. Once they are close together, a sperm binds to the egg and releases enzymes that allow it to penetrate the egg cell membrane (**Figure 23.19**). Then the mem-

Figure 23.18
Homunculus. Once a sperm was thought to contain a preformed human.

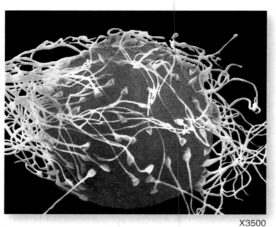

X3500

Figure 23.19 **Fertilization.**

■ *What do you think might happen to the embryo if more than one sperm were to fertilize the egg?*

branes of the two gametes fuse, and this opens a channel between the cells, and the sperm's DNA enters the egg. The two sets of DNA coalesce to form one nucleus, and fertilization is complete. The fertilized egg is now a zygote, ready to embark on the developmental paths presented next.

QUICK CHECK

1. Make a chart that compares formation of sperm with formation of eggs. How are these processes similar? How are they different?

23.6 Normal Development Involves Cell Division and Differentiation

Once the egg and sperm combine, processes are set in motion that transform the zygote into a multicellular organism. Soon after fertilization mitotic cell division produces an embryo in which each cell carries DNA with the same nucleotide sequence. Differential gene expression begins the process of cell differentiation. As development proceeds, gene expression becomes more specific in different cells, and embryonic cells differentiate into mature cell types that form tissues and organs.

Figure 23.20 diagrams the steps in early human development. These steps are generally the same in all mammals. The first few cell divisions are called **cleavage.** Like cutting a pie into smaller and smaller wedges, cleavage is a series of mitotic cell divsions with little or no cellular growth phase in between divisions. Cleavage divides the cytoplasm of the

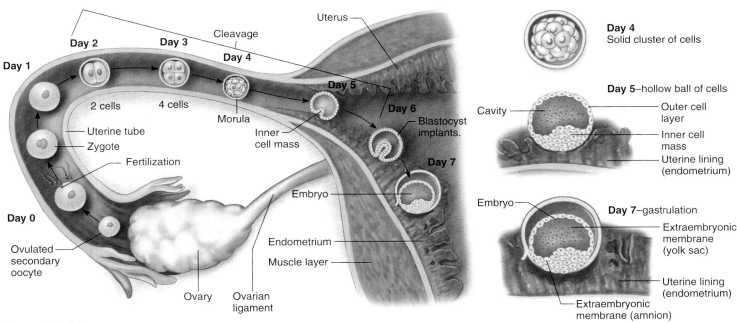

Figure 23.20   **Early Events in Human Development.**

zygote into smaller and smaller cells. Not only does cleavage provide the embryo with cells of the right size, it also divides the cytoplasm of the zygote into different cells. The cytoplasm of the egg contains mRNA and protein molecules that set up the initial phases of differential gene expression (see Section 9.5). These mRNA and proteins lay out the three body dimensions: anterior—posterior; back—front; right—left, as well as other features of the embryo.

After the rounds of cleavage the embryo forms a hollow ball of cells, commonly called a *blastocyst,* that has a cluster of cells at one end. This cluster of cells will become the embryo, while other cells of the blastocyst will become the *placenta* and various embryonic membranes. In humans the blastocyst forms around day 7 or 8 after fertilization, and the embryo implants into the lining of the uterus. After implantation, at about day 14, a series of cell movements begins that sets up the three embryonic layers: ecto-derm, endoderm, and meso-derm (see Section 16.4). The process is called **gastrulation,** and during gastrulation cells in the cluster at one end of the embryo move inside of the blasto-cyst (**Figure 23.21**) This is a compli-cated process, but you can better under-stand it if you think of a partially deflated balloon.

If you push your hand into the balloon, you have formed a structure that has an outer layer, an inner layer, a space in between the two, and a space where you pushed inward. The outer layer is the ectoderm, while the inner layer around your hand is the endoderm. Cell division continues through-out the process of gastrulation, and a third layer of cells, the mesoderm, develops between the endo-derm and ectoderm. Cells of endoderm, ectoderm, and mesoderm give rise to all of the tissues and organs of the body.

The rest of embryonic development is taken up with the formation of the various tissues and organs. **Figure 23.22** summarizes some of the major stages of human development as seen from the outside of the embryo. External structures are not really discernible until about the fourth week of development, when the embryo is only about

gastrulation the movement of cells in an early embryo that sets up the three embryonic cell layers of ectoderm, mesoderm, and endoderm

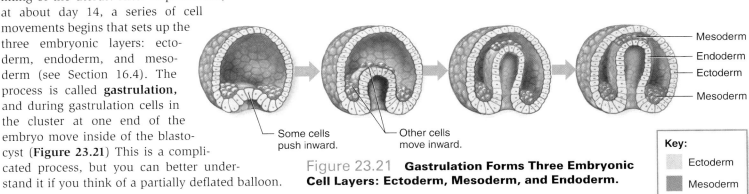

Figure 23.21   **Gastrulation Forms Three Embryonic Cell Layers: Ectoderm, Mesoderm, and Endoderm.**

Some cells push inward.

Other cells move inward.

Mesoderm
Endoderm
Ectoderm
Mesoderm

Key:
Ectoderm
Mesoderm
Endoderm

Figure 23.22 **Some of the Stages of Development of a Human Embryo.**

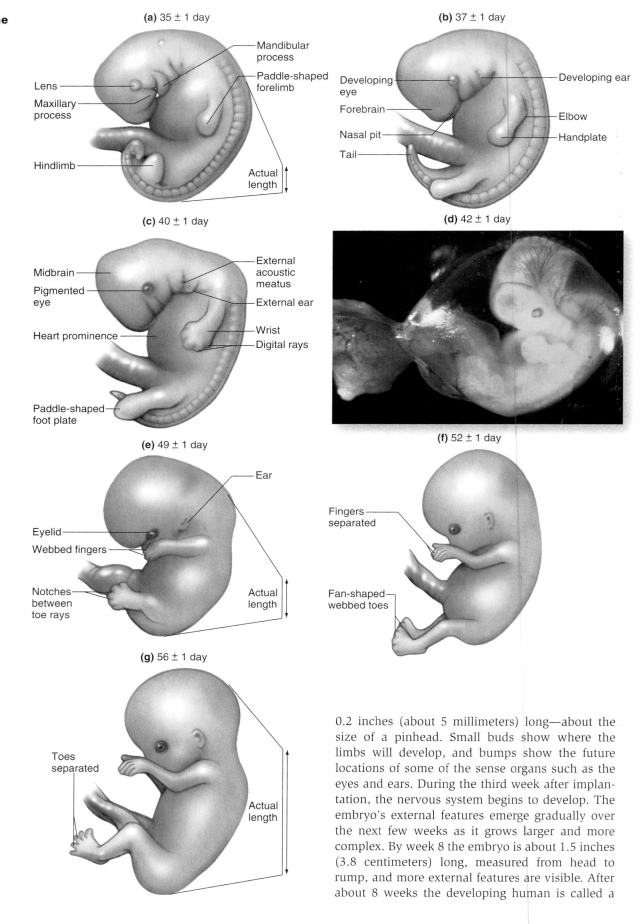

(a) 35 ± 1 day

Lens

Maxillary process

Hindlimb

Mandibular process

Paddle-shaped forelimb

Actual length

(b) 37 ± 1 day

Developing eye

Forebrain

Nasal pit

Tail

Developing ear

Elbow

Handplate

(c) 40 ± 1 day

Midbrain

Pigmented eye

Heart prominence

Paddle-shaped foot plate

External acoustic meatus

External ear

Wrist

Digital rays

(d) 42 ± 1 day

(e) 49 ± 1 day

Eyelid

Webbed fingers

Notches between toe rays

Ear

Actual length

(f) 52 ± 1 day

Fingers separated

Fan-shaped webbed toes

(g) 56 ± 1 day

Toes separated

Actual length

0.2 inches (about 5 millimeters) long—about the size of a pinhead. Small buds show where the limbs will develop, and bumps show the future locations of some of the sense organs such as the eyes and ears. During the third week after implantation, the nervous system begins to develop. The embryo's external features emerge gradually over the next few weeks as it grows larger and more complex. By week 8 the embryo is about 1.5 inches (3.8 centimeters) long, measured from head to rump, and more external features are visible. After about 8 weeks the developing human is called a

Table 23.4	Some Important Milestones During Human Embryonic and Fetal Development
Weeks after fertilization	**Functions in human embryo or fetus**
3	Cells of heart muscle begin to contract; tubular heart begins to beat; blood cells begin flow, but circulatory blood vessels not complete
6	Kidneys first produce urine
7–8	First electrical activity in brain; first involuntary muscle movement
12	Heartbeat can be detected; whole fetus moves
18	Phases of sleep and wakefulness
20	Sucks on thumb
24	Responses to light and sound
28	Rhythmic breathing movements
32	Eyes close during sleep; eyes open during wakefulness
36	Fetus turns toward light source
40	Baby is full term and ready to be born

fetus. During its development the fetus grows rapidly to reach a length of about 1 foot (about 35 centimeters) or longer. The newborn will be ready to survive outside the mother's body at about 9.5 months.

This description of normal human development is interesting and useful, but you may want to know what the embryo or fetus can *do* at various stages of development. When does the heart start beating, when do nerves first form synapses, and how well developed is the nervous system at various stages? **Table 23.4** summarizes some of these important milestones of development.

QUICK CHECK

1. Identify two important outcomes of the process of cleavage.

23.7 The Development of Gender Identity Reflects Biological Processes

One of the important things that happens during embryonic and fetal development is the determination of the child's sex. In mammals females carry XX chromosomes and have ovaries, a uterus, oviducts, and a vagina, while males are XY and have testes, associated sexual organs, and a penis—but there is a great deal more to gender than just physical traits. **Gender identity** refers to the entire repertoire of reproductive traits and behaviors that an animal displays. Gender identity includes whether you *think* you are a male or female.

An individual's external characteristics or their sex chromosomes aren't 100% reliable predictors of gender identity. For example, some people with XX chromosomes have strong male gender identity, while some with XY chromosomes have strong female identity. This puzzling state of affairs reflects the fact that the X and Y chromosomes are only part of the story of the development of gender identity.

The Biological Development of Gender Identity Involves the Gonads, the Adrenal Glands, and the Brain

Studies on rats in the 1970s produced some surprises about the development of gender identity. The early studies aimed to find out the effects of testosterone on the development of sexual behavior. Newborn male rats had their testes removed, and they were observed as they grew up. From the earliest ages, even before puberty, these animals did not behave like males. Instead, they behaved more like females. They groomed other rats as females do; they did not mount other rats as males do. Just like females they raised their tails and lowered their heads when males mounted them, and just like female rats they were not aggressive. Despite their male chromosomes and male external sexual organs, the castrated young rats behaved like females throughout their lives. Even when injected with testosterone, they did not show male behaviors.

Complementary results were found if female rat pups were exposed to testosterone shortly after birth. All through their early development and maturity these females behaved as males even though they had been exposed to testosterone only once, shortly after birth. They did little grooming of other rats, responded aggressively to mounting attempts by males, and attempted to mount other females. Subsequent studies have shed light on what is going on here, and these insights apply, to some extent, to all mammals including humans.

As you know, in mammals the X and Y chromosomes determine the genetic sex of the fetus, but X and Y chromosomes only get the process started. The critical gene is one that starts the process of determining maleness, and it sits on the

gender identity the expression of gender-related behaviors and thoughts, regardless of the genetic sex of male or female

tip of the Y chromosome. **Figure 23.23** summarizes the way that this gene sets the process of male sexual development into motion. The direct effect of this master gene is to cause the development of testes instead of ovaries (Figure 23.23a). If the master gene on the tip of the Y chromosome is absent, the immature gonads will become ovaries. From that point on the direct action of the Y chromosome on sexual development is largely finished, and the developing gonads assume control. If testes are present, male hormones are released—such as testosterone—that prevent formation of female organs and stimulate development of the penis, the scrotum, and vas deferens. If the master gene on the Y chromosome is absent, ovaries develop instead of testes. If no male hormones are released, female sexual organs develop.

Another surprising factor in this story of sexual development is that the early influence of testosterone does not stop with the development of sexual organs. In all mammals studied so far, including humans, the testes produce a surge of testosterone sometime after the sex organs have developed (Figure 23.23b). This testosterone surge masculinizes both the brain and the behavior of the individual. It affects developing muscles, bones, and other tissues, and stimulates their growth. The testosterone surge also stimulates the growth and rearrangement of certain brain and spinal cord areas that are crucial in the control of sexual behavior. Individuals who experienced a testosterone surge during early development play like males as children and act like males after puberty. In the absence of the testosterone surge, the basic brain and behavior pattern is female both in childhood and adulthood.

An important point is that the gender of a mammalian brain can be set only during this critical period early in development (Figure 23.23c). For example, if the embryonic testosterone surge is prevented or if drugs block its effects, the male will grow up behaving like a female, in spite of his Y chromosome and male genitals. This is what happened to the castrated newborn rats you just read about. Conversely, if a female is given an artificial surge of testosterone at the crucial developmental time, she will grow up behaving like a male.

After these male patterns of behavior are wired into the brain, they are still under hormonal control. Testosterone stimulates these brain regions to produce male behavior. Levels of testosterone are fairly low during childhood, so the male behaviors that emerge in childhood are weak and usually displayed in a context of play. During puberty the testes produce lots of testosterone (Figure 23.23d). At this stage of life the hormone causes all of the familiar human male secondary sexual characteristics—increased muscle mass, deepening of the voice, facial and body hair—that turn little boys into masculine young men. Having been sensitized by the early surge of testosterone, the male brain now responds to adult testosterone with strong, adult male behavior patterns. Similarly, when a female hits puberty, estrogen is the primary hormone that begins to circulate at higher levels.

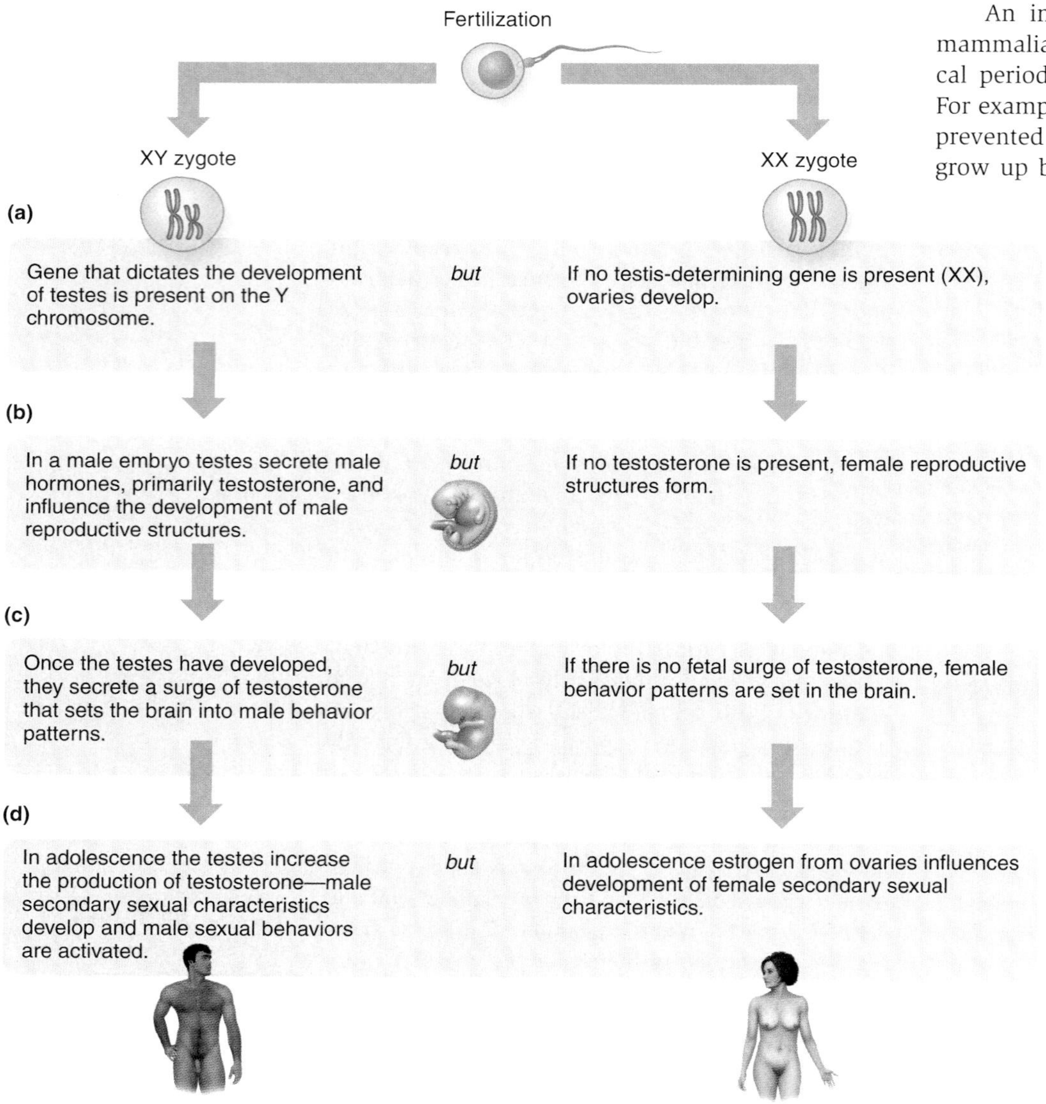

Fertilization

XY zygote   XX zygote

(a)

Gene that dictates the development of testes is present on the Y chromosome.   *but*   If no testis-determining gene is present (XX), ovaries develop.

(b)

In a male embryo testes secrete male hormones, primarily testosterone, and influence the development of male reproductive structures.   *but*   If no testosterone is present, female reproductive structures form.

(c)

Once the testes have developed, they secrete a surge of testosterone that sets the brain into male behavior patterns.   *but*   If there is no fetal surge of testosterone, female behavior patterns are set in the brain.

(d)

In adolescence the testes increase the production of testosterone—male secondary sexual characteristics develop and male sexual behaviors are activated.   *but*   In adolescence estrogen from ovaries influences development of female secondary sexual characteristics.

Figure 23.23   **Effects of Y Chromosome and Testosterone on Sexual Development.**

Setting the issue of estrogen cycling aside for now, the female brain responds to estrogen with typically female behaviors, whatever those behaviors are for her species. The strong differences in morphology and behavior that develop during puberty are referred to as *sexual dimorphism.* The degree of sexual dimorphism varies greatly across species, and humans actually show relatively modest differences between the sexes (**Figure 23.24**). Chimps, gorillas, and orangutans show considerable sexual dimorphism, while gibbons, a distantly related ape, show little sexual dimorphism.

The role of hormones in the behavior of males and females is not just a topic for textbook discussions. For example, through much of modern history people have recognized that the testes, and more specifically testosterone, are related to aggressive male behaviors. Removal of the testes, called *castration,* usually reduces aggressive and sexual behavior in a male. Castration has been used throughout human history to control the behavior of humans and other animals.

The Understanding of Gender Development Explains Some Puzzling Natural Behaviors

Understanding the development of gender identity sheds light on some puzzling situations. The spotted hyena provides a fascinating example. Female spotted hyenas have genitals that look like those of males, even though their ovaries produce estrogen just as they should (**Figure 23.25**). The female hyena's clitoris is enlarged to the size of a penis, and her urinary tract travels through this enlarged *pseudopenis,* just as the male's urinary tract travels through his penis. The labia of spotted hyenas are fused and seal the vagina. Copulation and birth actually happen through the urogenital tract that runs through the pseudopenis. Yet female spotted hyenas are XX and have fully functional ovaries. The transformation occurs when female spotted hyenas are still embryos. A pregnant female hyena produces large amounts of a hormone called *androstenedione,* and this crosses the placenta where it is converted to testosterone. High levels of testosterone during critical times of pregnancy masculinize both male and female hyenas. In essence, spotted hyenas demonstrate that nature figured out how to manipulate early testosterone surges long before biologists knew about them.

The adrenals are another source of testosterone, in both males and females of all mammal species. The amount of testosterone produced by the adrenals varies dramatically between species and between

Less sexual dimorphism **(a)**

(b)

(c)

(d)

More sexual dimorphism

Figure 23.24 **Degrees of Differences in Sexual Dimorphism Among Primates.** (**a**) gibbons, (**b**) humans, (**c**) orangutans, (**d**) gorillas.

■ *Explain how orangutans show more sexual dimorphism than gibbons?*

Figure 23.25 **Spotted Hyena.** The clitoris of a spotted hyena looks like a penis.

individuals of the same species. And it turns out that the few males who continue to exhibit male behaviors after castration have high levels of adrenal testosterone. As you soon will see, adrenal testosterone can have a big impact on a female.

23.8 An Added Dimension Explained
What Determines Human Gender Identity?

Human gender identity may be one of the most controversial and emotionally charged issues of our day. Is gender identity a choice, and can it be changed? Did the couple that you read about in the opening section of this chapter choose their sexual identities, or is gender identity beyond conscious control? What can biology tell us about these difficult questions? Human cultural expectations, parental input, and other early experiences clearly influence the development of human gender identity. Nevertheless, humans are mammals, and their basic biology is not fundamentally different from that of other mammals. Research is beginning to demonstrate the role of biology in human gender identity and to reveal its limits.

One issue that comes up as soon as you talk about gender identity is sexual orientation. Gender identity is related to sexual orientation, but the two are not exactly the same. Your sexual attraction to other people is called your **sexual orientation.** Most humans are **heterosexual** and are sexually attracted to members of the opposite sex. But some people are open to sexual experiences with either males or females and are so **bisexual.** Some people are **homosexual** and are exclusively sexually attracted to the same sex. Homosexuals usually have a strong sense that their gender identity matches their physical sex, even though they are attracted to their own gender. For example, many if not most homosexual women feel themselves to be women even though they are sexually attracted to women. Yet another pattern is when a person strongly believes they *belong to* the opposite sex, and these are **transsexual** people. A transsexual has a gender identity different from his or her genetic or physical sex. For example, a transsexual is physically a man, is sexually attracted to other men, and feels that he is a woman. From this description you can see just how complex human sexuality and gender identity can be.

You have read about animal studies that show a strong role for hormones in the emergence of gender identity. Is this also true for humans? A great deal more needs to be learned,

sexual orientation the nature of a person's sexual attraction

heterosexual sexually attracted to members of the opposite sex

bisexual sexually attracted to both sexes

homosexual sexually attracted to members of the same sex

transsexual an individual who believes she or he is of the opposite sex

but some interesting findings are emerging from studies of gender identity in humans. One way to measure gender identity is to examine childhood behaviors that are typical for little girls and little boys. Clearly, even when parents try to impose gender-neutral behaviors, girls and boys typically engage in distinctly different kinds of play. Most little girls really do prefer to play with dolls, play dress up, and have tea parties, while little boys generally prefer to play with trucks, play with construction sets, and wrestle. Of course there is a wide spectrum of behavior in both little girls and little boys, but gender differences in play are a reliable pattern. Even among nonhuman primates—monkeys, for example—young males will choose human toys that are typically male such as trucks, and young females will choose human toys that are typically female such as dolls (**Figure 23.26**).

Experiments that alter the amount of testosterone that a fetus is exposed to are not possible in humans, but there are some "natural experiments" that can illuminate this phenomenon. The amount of testosterone that a developing embryo is exposed to depends on many factors. The Y chromosome determines the formation of testes, which, in turn, produce testosterone. Of course, the amount of testosterone produced by the testes varies, and there are other sources of variation too. As in the case of the spotted hyenas, the adrenal glands are also a source of testosterone, and during pregnancy testosterone produced by the mother reaches the developing embryo or fetus. From both their mothers and their own adrenals, some girls are exposed to more testosterone than are others.

Figure 23.26 **Typical Play Behaviors.** Female (left) and male (right) monkeys like to play with different sorts of toys.

Several studies have shown that girl babies exposed to more testosterone, whether from their mothers or from themselves, display more male behaviors during childhood. One study measured prenatal testosterone levels in pregnant women and then rated childhood play behaviors in the children until the age of 3.5 years. Girls whose mothers had more testosterone played more with male toys than did girls whose mothers had less testosterone. Other studies have corroborated this finding. Interestingly, these kinds of results are found regardless of the views of the parents about how the children should play.

You probably have many questions at this point, some of which can be answered and many that, at this point in scientific study, cannot. You may wonder whether the girls with male play patterns grow up to "feel" like boys? In the end, is their sexual orientation toward physical boys, or physical girls? The answer is mixed. Many girls with high testosterone levels grow up to feel like girls and are attracted to boys. On the other hand, a girl with high testosterone levels is *more likely* to be attracted to other girls than is a girl with normal testosterone levels. This finding suggests that testosterone can play a role in a woman's gender identity.

But what about boys? A syndrome called **androgen insensitivity** provides some clues. In this condition the receptors for testosterone and other androgens do not bind efficiently, and so testosterone and other similar male hormones do not have as strong an effect on target tissues. Complete androgen insensitivity is similar to the early castration experiments in animals, and the results are similar as well. Many studies have shown that genetic males with androgen insensitivity feel that they are females and often live fulfilling lives as women.

androgen insensitivity
a condition in which testosterone receptors do not bind well to androgens; thus the hormones do not have normal effects on target tissues

These studies on the role of hormones in the development of gender identity are far from complete. Many questions remain unanswered, such as how gender identity relates to sexual orientation and how sexual orientation develops in people without obvious disorders such as over-expression of testosterone or androgen insensitivity. Such studies are time consuming, and it will take many years, but there is little doubt that studies of the biology of sexual behavior will provide insights into how people become who they are.

QUICK CHECK

1. What happens to a female embryo exposed to high levels of testosterone during development?

Now You Can **Understand**

Preventing Pregnancy

The two words "I'm pregnant" probably elicit as much extreme emotional intensity as any phrase in the language. If a woman or a couple wants a child, these words bring some of the greatest joys of a lifetime, but if a woman or a couple do not want a child—especially if the woman is young, has been raped, or if her life is endangered by a pregnancy—these words can bring despair and desperation. To a greater extent than ever before, though, women can have a choice over whether to have a child. In the early 1960s birth control pills became widely available. How do they work? The idea is fairly simple. Estrogen and progesterone are produced by the ovaries and inhibit the release of certain hormones from the pituitary, namely LH and FSH. These two hormones are responsible for stimulating the development and release of an egg from the ovary—a necessary condition for pregnancy. Normally, estrogen and progesterone are produced at a high level just for a few short days in the middle of a woman's menstrual cycle. Birth control pills provide a stable level of these hormones throughout the first three weeks of the cycle. As a result, LH and FSH are constantly inhibited and ovulation never happens. With no egg, there can be no pregnancy. A week off of the pills allows menstruation to occur and resets the cycle.

What Do **You** Think?

Is Chemical Castration an Appropriate Way to Control Sex Offenders?

You have read about the effects of testosterone on behavior. There is little doubt that testosterone is at least one cause of male aggressive behavior. In many situations this is a good thing. Male aggression sometimes can protect a female and her offspring from harm, although the threat is often from the aggression of other males. Sometimes, though, male aggression is inappropriate and even criminal. Rape and child molestation are crimes that most often are carried out by men. Neither punishment nor treatment programs are especially effective at preventing convicted sex offenders from committing additional sex crimes.

Recently, a chemical has been developed that blocks the production of testosterone, and many people believe it should be given to men who commit repeat sexual crimes. The drug must be given once a month in an injection. Studies have shown that only 5% of male sex offenders on the drug commit another sex-related crime. There are side effects of the drug, though, such as loss of bone density. These side effects can be prevented by other treatments, but for this and other reasons some people object to *chemical castration* as a way of controlling convicted sex offenders. You can find arguments in favor of and against chemical castration on the Internet. After looking into the issue, do you agree or disagree with this means of controlling sex offenders? What do *you* think about the use of chemical castration in our criminal justice system?

Chapter Review

CHAPTER SUMMARY

23.1 Multicellular Organisms Depend on Cellular Communication

Communication between cells is typical of life, and the endocrine glands produce hormones that act as long-distance chemical messages carried by the bloodstream. All vertebrates share a common set of endocrine glands. Hormones are signal molecules. Peptide hormones interact with receptors on plasma membranes, while steroid hormones cross the plasma membrane and interact with receptors inside the cell. All hormones redirect cellular activities. One hormone may affect different tissues in different ways.

endocrine gland 555   hormone 555
endocrine system 555

23.2 Hormonal Secretions Are Strictly Controlled

The pituitary gland and the hypothalamus regulate endocrine secretion and release through a system of monitoring hormone levels in the bloodstream and releasing hormones specific for each hormone. Inhibition by negative feedback also helps to regulate the secretion and release of hormones. The anterior and posterior pituitary are functionally separate endocrine glands. Both are controlled in different ways by the hypothalamus.

hypothalamus 558   pituitary gland 558
negative feedback 559

23.3 Hormones Coordinate Functions of Diverse Animal Systems

Insulin and glucagon are active in glucose metabolism, while thyroid hormones affect the rate of cellular metabolism, and adrenal hormones prepare the body for stressful emergencies. Anabolic steroids break down biological molecules, while catabolic steroids build them. Use of anabolic steroids is a dangerous practice.

anabolic steroid 563

23.4 Hormonal Disorders Cause Major Disease Syndromes

In hyperglycemia there is too much glucose in the bloodstream, a condition that may be caused by diabetes mellitus. Type 1 diabetes is an autoimmune disease in which pancreatic cells that produce insulin are destroyed. In Type 2 diabetes insulin receptors have decreased sensitivity to insulin. Abnormalities in thyroid hormones can cause abnormal brain development if they occur early in development and metabolic abnormalities if they occur later in life. Abnormalities in growth hormone secretion can produce giants, dwarfs, or if later in life, acromegaly.

diabetes mellitus 564   hyperglycemia 564

23.5 Hormones Coordinate Human Reproduction

Male and female primary and secondary sexual characteristics are governed by X and Y chromosomes acting on ovaries and testes and by hormones produced by the ovaries or testes. After puberty the interplay of pituitary and hormones produces sperm and mature ova. Sperm production continues throughout life, and egg production ceases at menopause. The monthly menstrual cycle prepares the uterus for the reception of a fertilized ovum.

endometrium 570   primary sexual characteristics 566
estrogen 570   progesterone 570
menstrual cycle 570   secondary sexual characteristics
menstruation 572   566
ovarian cycle 570   sexual dimorphism 566
ovulation 570   spermatogonia 568
primary oocytes 570

23.6 Normal Development Involves Cell Division and Differentiation

As a zygote goes through cleavage, the number of its cells increases and the mature cell size is reached. The blastocyst will give rise to both placenta and embryo. At about day 8 the embryo implants into the endometrium. Gastrulation begins at about 14 days, forming ectoderm, endoderm, and mesoderm. After about 8 weeks the early embryo is called a fetus. It will be born after about 40 weeks of development

cleavage 572   gastrulation 573

23.7 The Development of Gender Identity Reflects Biological Processes

Development of gender is a biological process, and all the varieties of gender identity also are largely biological phenomena. X and Y chromosomes are active early in development. After that, production of testosterone or absence of testosterone determines sexual orientation. Body and brain are both masculinized by a surge of testosterone. Feminine body and brain are formed when there is no testosterone surge.

gender identity 575

23.8 An Added Dimension Explained: What Determines Human Gender Identity?

Human gender identity and sexual orientation are probably influenced by hormone levels during development.

androgen insensitivity 579   homosexual 578
bisexual 578   sexual orientation 578
heterosexual 578   transsexual 578

TRUE or FALSE. If a statement is false, rewrite it to make it true.

1. All hormones exert their actions at the cell membrane and do not enter the cell.

2. The hypothalamus directly controls output of hormones.

3. ACTH is a pituitary hormone that controls secretion of hormones by the adrenal cortex.

4. Secondary sexual characteristics include penis and testes of males and vagina and uterus of females.

5. Too little of growth hormone in childhood results in acromegaly.

6. Adrenaline is released during periods of long-term stress, while cortisol is released during short emergencies.

7. In Type 1 diabetes mellitus the adrenal cortex makes enough insulin, but it is malformed.

8. Testosterone is an anabolic steroid.

9. FSH activates spermatogonia and oogonia and helps mature sperm in males and ova in females.

10. The scrotum keeps testes cool enough to produce mature sperm.

11. Males make sperm every day for about 75 years.

12. When an XX human embryo is 9 weeks old, its oogonia begin to develop into sperm.

13. Type 2 diabetes is associated with overweight and lack of exercise.

14. An ovum actually completes meiosis after it has been fertilized. This occurs in the uterus.

15. Mature follicles move toward the outer surface of an ovary; mature sperm move toward the hollow space of the epididymis.

16. In the menstrual cycle each ovary releases one egg per month.

17. More than one sperm fertilizes the ovum.

18. The adrenal cortex of females makes testosterone.

19. In the absence of testosterone a human embryo will develop as a female.

20. A surge of testosterone from testes in the early life of the embryo masculinizes the brain.

MULTIPLE CHOICE. Choose the best answer of those provided.

21. Liver and muscle cells store extra nutrients as
    a. glucose.
    b. insulin.
    c. glycogen.
    d. glucagon
    e. c and d.

22. When long-distance runners load up on carbs before a race they are
    a. reducing body fat.
    b. giving themselves hormones for the race.
    c. increasing the efficiency of muscle contraction.
    d. stocking the liver with readily accessible glucose.
    e. stocking the liver with readily accessible glycogen.

23. When there is plenty of glucose in your bloodstream, it will stimulate your pancreas to release _____ and your cells will remove sugar from your bloodstream.
    a. insulin
    b. glucagon
    c. cortisol
    d. thyroxine
    e. noradrenaline

24. Which hormones are produced by the hypothalamus and released from the pituitary gland?
    a. glucagon and insulin
    b. oxytocin and ADH
    c. thyroxine and noradrenaline
    d. epinephrine and cortisol

MATCHING

Match the columns.

25. insulin
26. glucagon
27. thyroxine
28. adrenaline
29. cortisol
30. releasing hormones

a. controls basal metabolic rate
b. removes glucose from bloodstream
c. triggers conversion of glycogen to glucose
d. prepares body for an emergency
e. helps mammals deal with long-term stress
f. travel from hypothalamus to anterior pituitary

1. How is the level of glucose in the blood regulated? How can regulation of blood glucose become abnormal?

2. How are the ovarian and menstrual cycles coordinated?

A diagnosis of diabetes is suggested by a fasting blood glucose level of 126 mg/dL of blood (deciliter), or higher. Prediabetes is a concern for fasting blood glucose levels between 100 and 124 mg/dL. Candace has a fasting blood glucose level that is 75% of the diabetic level. What is her fasting blood glucose level? Is she considered prediabetic?

1. Some promising new drugs are being investigated to control growth hormone–related gigantism. What might some of these drugs do?

2. Some of the early drugs to treat Type 2 diabetes actually stimulated the pancreas to release more insulin. Under what situation might this approach lose its effectiveness, after a time of treatment?

For additional study tools, visit www.aris.mhhe.com.

Moral Questions Amid
Medical Answers

July 28, 1997

Parents of severely disabled
or impaired children seek out
every possible option for a
cure or treatment.

Screening for Abnormal Embryos Offers
Couples Hope After Heartbreak

November 22, 2005

Sally and Jim Doe were expecting a baby. Of course, these are not their real names. And not much is known about where the couple lived or other details of their lives. It is known that before their baby was born, Sally and Jim were young and happily anticipating a life together as a perfect young family.

Sally gave birth to a baby girl, but the hidden fear of nearly all prospective parents was realized as soon as baby Jane Doe was born. She was not normal and healthy. For one thing she had no thumbs, her hips were not formed normally, and within a few months her parents realized that she was deaf in one ear. Her development revealed even more problems, and by the time little Jane was 3 years old she was deficient in all types of blood cells. It was

clear that Jane had an incurable inherited disorder called Fanconia anemia. Without help Jane probably would die before she was old enough to have a family of her own (**chapter-opening photo**).

What did Sally and Jim do? Certainly, they must have grieved and cried, but they loved their daughter and looked everywhere for a treatment. They searched for anything that would help, but treatments were difficult and they were not open to Jane. So when she was only 3 years old, little Jane's future was grim.

What happened to baby Jane Doe? The short answer is that, at least as of this writing, a treatment was found, and she has responded to it so well that now it seems her life has been saved. But this was an extraordinary medical intervention, and before you can fully appreciate what happened to baby Jane Doe, you must learn about the many ways that modern reproductive technologies have changed the ways that people have babies. ∎

24.1 Modern Reproductive Technologies Are Built on Knowledge About Normal Development

The development of a new individual is one of the most inspiring and mysterious of all biological events. You have learned enough about biology to appreciate that development must be understood in context of the chemical, molecular, and cellular processes that support all life. The basic laws of chemistry and physics described in Chapter 2 govern the processes of life such as protein synthesis and metabolism. Enzymes build complex chemical structures, using a source of energy and relying on the basic rules of chemical interactions. Cells influence one another through chemical signals and membrane protein receptors. In multicellular organisms new life is started by meiosis. Gametes unite to form a zygote and mitotic cell division and cell differentiation follow, producing an adult individual.

Despite our knowledge, conception, embryonic development, birth, and maturation are beyond

human control. We cannot replicate this delicate and intricate feat. At least not completely. This chapter explores how human knowledge of basic biology is changing the way that people make and have babies.

Louise Brown Was the World's First "Test-Tube Baby"

Louise Brown was the first child of this brave new world of technologically controlled reproduction, and she became a celebrity just by being born (**Figure 24.1**). Even before her birth in 1978, Louise's conception raised a flurry of ethical concerns and debates. Some headlines proclaimed that it represented the end of the human race as we know it. People were afraid that Louise would be born with deformities, or that other monstrous babies would be born, as a result of the technology associated with her birth. Many worried that Louise Brown's conception was unethical as well as immoral. Today, Louise Brown is a postal worker in England and has led a normal, productive life. Why did her birth raise such a fuss? Louise was the first

Figure 24.1 **Instant Celebrity.** (**a**) Louise Brown, the world's first test-tube baby, was born in 1978. (**b**) Louise Brown as an adult.

(a)
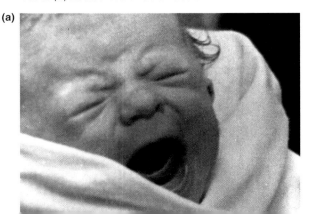

(b)

baby to be conceived and born using the techniques of *in vitro* fertilization. To the newspapers, Louise was the world's first "test-tube baby."

In vitro is a Latin term that means "in glass." In a scientific context, *in vitro* refers to any bodily or cellular process that happens in laboratory glassware, rather than in the body or in a natural environment. *In vitro* studies allow scientists to control the environment for biological experiments, and such studies have contributed greatly to our understanding of cellular processes.

As you might expect, *in vitro* **fertilization,** or **IVF,** is the fertilization of an egg by a sperm within laboratory glassware. IVF is now the cornerstone of treatments available to women who wish to bear children but are unable to conceive normally, and IVF has been remarkably successful. Twenty-five years after Louise Brown was born, nearly 15% of U.S. women of childbearing age seek medical help to conceive. Each year about 100,000 U.S. women undergo IVF procedures, resulting in thousands of IVF babies born yearly. It is hard to imagine that IVF will ever replace the more "normal" way of having a child, but as you will read in the following pages, IVF already is having a dramatic effect on the business of making babies.

The successes and limits of IVF are grounded in the modern understanding of normal human reproduction. To summarize what was presented in Chapter 23, a human baby is conceived when an egg is penetrated by a sperm in the process of fertilization. In males sperm production begins after puberty at about age 14. Sperm production continues throughout life, and each day a healthy adult male makes about 400 million sperm cells. When a female fetus is still developing within her mother's uterus, her ovaries begin to produce eggs. But instead of proceeding through the entire process, meiosis stalls. The eggs remain in this suspended state for 12 to 14 years, until the girl reaches puberty. At this time the development of ova resumes. Production of ova usually continues until menopause at about age 50. Each month hormonal signals cause one ovarian cell (or rarely two or three cells) to proceed through the first round of meiotic cell division and produce a mature ovum (or mature ova). Once fertilization has been successful, the zygote undergoes the final phase of meiosis and then begins the rounds of cell division and differentiation that produce an embryo, a fetus, and a newborn. One important feature of development is that the first divisions of the zygote occur in the oviducts, outside the mother's uterus. The embryo emerges from the oviduct as a hollow ball of cells that implants into the lining of the uterus. Researchers have taken advantage of the fact that human embryos begin their development independent of the uterus and have set up conditions that support fertilization and the first few rounds of cell division *in vitro.* These techniques are the basis for IVF.

> **QUICK CHECK**
>
> **1.** What is IVF?
> **2.** Normal human development occurs within a female's reproductive tract. How can IVF successfully copy those conditions?

24.2 IVF Is the Foundation for Modern Reproductive Technologies

IVF builds on the normal processes of egg and sperm production, but in practice it is complicated (**Figure 24.2**). The sequence of procedures in IVF includes:

- inhibition of ova production,
- stimulation of production of multiple ova,
- harvesting ova and sperm,
- mixing ova and sperm in a laboratory dish,
- incubation of zygotes for several days so they develop into embryos,
- visual examination of embryos under a microscope,
- choice of one or more embryos for implantation, and
- transfer of embryos into the prospective mother's uterus.

Let's take a closer look at each of these phases of IVF.

Even in normal human reproduction the odds that a woman will become pregnant during any particular ovulation cycle are not high. Only a tiny fraction of a woman's eggs ever become fertilized. And only a fraction of fertilized eggs go on to produce a healthy, viable embryo. The odds of fertilization and normal embryonic development in IVF can be even lower. To increase the odds that IVF will be successful, multiple ova must be available for fertilization. All of this must happen at just the right time in the woman's monthly menstrual cycle so that she is ready to carry a pregnancy when the embryos are ready to be implanted. To accomplish all of this, prospective mother, prospective father, medical doctors, and laboratory technicians

in vitro **fertilization (IVF)**
the fertilization of an ovum by a sperm within laboratory glassware

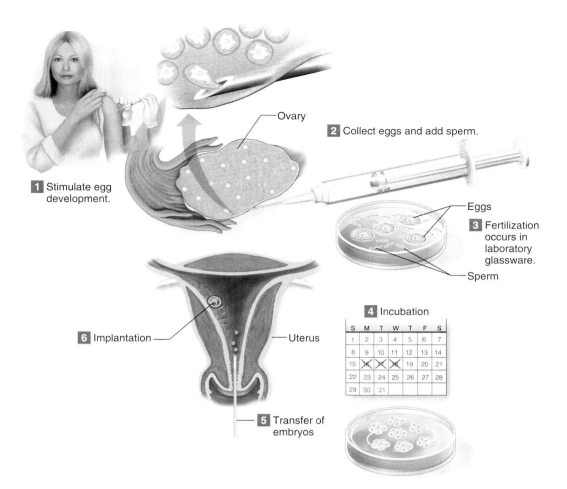

Figure 24.2 *In Vitro* **Fertilization.** (1) A woman's ovaries are stimulated to produce an abundant supply of eggs that are collected (2) and fertilized by sperm. (3) Embryos are incubated (4) for about three days before embryos are transferred to the mother's uterus (5) where they implant (6) and continue development.

1 Stimulate egg development.

Ovary

2 Collect eggs and add sperm.

Eggs

3 Fertilization occurs in laboratory glassware.

Sperm

4 Incubation

6 Implantation

Uterus

5 Transfer of embryos

embark on the careful timetable of coordinated IVF procedures.

First, the timing of ovum development is controlled with a series of hormone injections. Usually, these injections target the two major hormones involved in ovulation: luteinizing hormone (LH) and follicle-stimulating hormone (FSH). FSH stimulates the maturation of ova, and LH stimulates the release of ova (see Section 23.5). A surge of both hormones causes ovulation in about the middle of the ovarian cycle. In a typical preparation for IVF, a woman receives a long series of injections that inhibit LH production. She also is given injections of FSH to stimulate the maturation of several ova. The injections usually start after ovulation in one menstrual cycle and continue until ova are removed from her ovaries in about the middle of the next cycle. At that time another series of hormone injections is given to be sure that the uterus is ready to receive any viable IVF embryos.

To harvest the ova, the woman is lightly anesthetized, a long needle is inserted into her abdomen, and ultrasound is used to guide the needle to an ovary (**Figure 24.3**). Using ultrasound imaging, the doctor can actually see the regions on the ovary where the ova are maturing. Each mature ovum is gently sucked into the needle and

Figure 24.3 **Locating Eggs With Ultrasound.** An ultrasound image guides a physician's needle as eggs are harvested from an ovary.

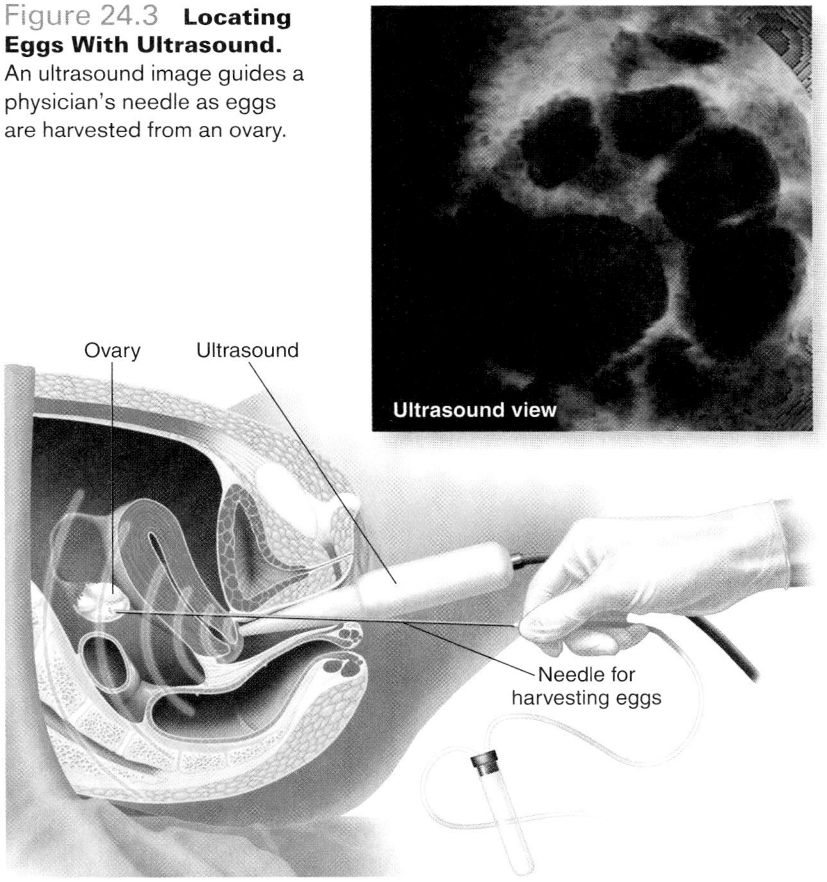

Ovary   Ultrasound

Ultrasound view

Needle for harvesting eggs

removed. When all of the mature ova have been harvested, the needle is withdrawn, and the ova are gently released into a dish of a special nutrient solution. Most IVF procedures require fresh sperm to fertilize the ova, so while the woman's ova are being extracted, the prospective father is taken to a private room to ejaculate a sperm sample. This also is placed in a special medium, and sperm may be "washed" with a solution that eliminates inactive, unhealthy sperm. Finally, several ova and several hundred sperm are mixed together, and fertilization happens.

In some cases a technique called ICSI (intracytoplasmic sperm injection) may be used (**Figure 24.4**). Using the mild suction of a thin glass tube, the ovum is stabilized. Then a single sperm is sucked into a fine glass needle with a tip that is less than 10 microns in diameter. Under a microscope this ultrafine needle is inserted into a healthy ovum, and the sperm is carefully injected. It may surprise you to learn that this sperm injection technique successfully produces a zygote that can develop into a healthy child.

In an IVF procedure the fertilized ova are incubated *in vitro* for one to several days, allowing the zygotes to go through several rounds of mitotic cell division and develop to an early embryonic stage (**Figure 24.5a&b**). This allows any embryos that look abnormal to be rejected and leaves only healthy-looking embryos to be implanted. When the embryos have between 8 and 64 cells, they are examined under the microscope, and one or more embryos are selected for implantation into the prospective mother. Figure 24.5c&d shows accepted and rejected embryos.

Implantation of the embryos is one of the simplest IVF steps. One or more embryos are sucked up into a thin tube and released into the woman's uterus. If she is at the correct phase of her menstrual cycle, the embryos will spontaneously implant. But nothing is certain, and at this point there is little that the doctor or technicians can do to ensure a pregnancy. Like any other attempt to get pregnant, the woman simply must wait and see if the procedure has worked. If she does become pregnant, a pregnancy test, like those used to verify a normal pregnancy, will reveal the presence of certain hormones characteristic of pregnancy.

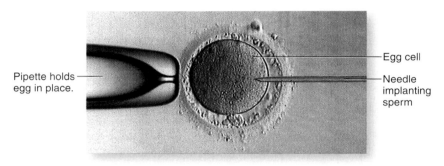

Figure 24.4 **Intracytoplasmic Sperm Injection.** ICSI uses the direct injection of a sperm into an egg to produce fertilization.

Labels on figure: Pipette holds egg in place. Egg cell. Needle implanting sperm.

The Techniques of Human IVF Are Derived from Hundreds of Research Studies on Other Mammals

For decades scientists have been studying the processes of normal and abnormal development in nonhuman animals. IVF would not be possible without the foundation of knowledge gained from these studies. In the 1950s, for example, researchers discovered that fertilization could occur in lab glassware, and that *in vitro* embryos could be implanted into a female mouse where they would develop normally. Step by step other researchers were discovering that an *in vitro* ovum or embryo could survive being manipulated with injections. Sometimes it may seem that basic research will lead nowhere, but basic research often yields important practical applications. It would have been hard to predict that these early studies on animal embryos would lead to the modern industry of IVF, but they did.

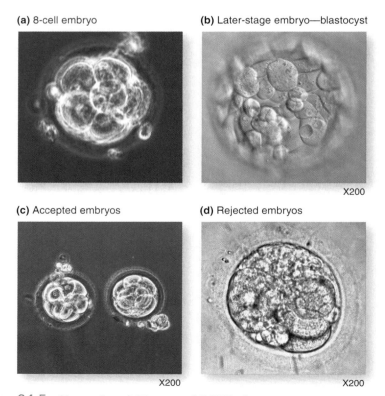

(a) 8-cell embryo

(b) Later-stage embryo—blastocyst

X200

(c) Accepted embryos

(d) Rejected embryos

X200

X200

Figure 24.5 **Normal and Abnormal IVF Embryos.**

IVF Is Not Always Successful and Does Carry Risks

IVF is a taxing experience. The normal emotional stresses of wanting a child, and the uncertainty of the IVF outcome, are compounded by the rigors of the IVF procedure. The hormone injections are far in excess of normal hormonal levels, and to maximize the number of ova produced, hormones are switched abruptly, resulting in mood swings and physical discomfort. So IVF treatment can put a woman on a physical and emotional roller coaster. She may gain weight, become depressed, become irritable, or have all of these symptoms at once. The daily injections are uncomfortable, and their emotional side effects often are disruptive. In addition, the procedure is expensive. Every cycle of IVF treatment costs as much as $10,000 and is rarely covered by insurance policies. It is a testament to the strong desire to have children that each year nearly 100,000 women in the United States alone undergo this ordeal.

What are the chances of success and what risks does IVF carry? Unfortunately, because of the complexity of IVF, it is difficult to know if a particular attempt will be successful. Things can go wrong at many points in the treatment. For instance, the harvested ova may not be healthy, or the sperm may be defective. The *in vitro* conditions may not be just right, or the implantation procedure may run into difficulties.

A woman's age is one major factor that affects her success as an IVF patient. IVF is much more successful for women in their twenties than for women in their forties, yet it is the 40-year-old who is more likely to feel IVF is her only option. Also, the chances of carrying a pregnancy successfully to term are less for an older woman. For these and other reasons the success of IVF decreases as the woman's age increases (**Figure 24.6**). Even for younger women the chances of a successful IVF procedure are less than 50%.

Anyone considering IVF also must consider the risks involved. The procedure itself is fairly safe, but sometimes more than one embryo is implanted into a woman during a given IVF attempt. A report on IVF by the Centers for Disease Control and Prevention concludes that implanting two embryos rather than one does increase the chances of achieving a pregnancy, but implanting more than two does *not* increase the chances any further. In fact, if four or more embryos are implanted, the chances of getting a healthy live birth decline.

Implanting multiple embryos increases the chance of multiple births. If you have known anyone who has given birth to twins or triplets, you know how difficult a multiple birth can be. The

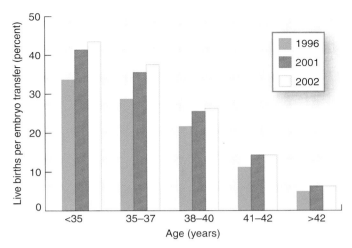

Figure 24.6 **The Success of IVF Decreases as the Age of the Woman Increases.**

■ *Do you think that the number of live births per transfer will continue to increase over the next few years? Explain your answer.*

mother usually is put on complete bed rest at some time during the pregnancy, and there is a high risk of a premature delivery. If the babies are born before they are fully developed, they may experience serious health complications (**Table 24.1**). The

Table 24.1	Health Complications of Babies Born Before Full Term
Health problem	**Explanation**
Respiratory distress syndrome (RDS)	Because their lungs lack surfactant that keeps alveoli from collapsing, premature babies require supplemental oxygen and drugs and may require other respiratory assistance.
Intraventricular hemorrhage (IVH)	Premature babies are at risk of bleeding into the brain. This can cause brain damage, learning problems, behavioral problems, and cerebral palsy.
Necrotizing enterocolitis (NEC)	Premature babies are at risk of inflammation of the intestine. This can lead to feeding problems and can slow growth.
Retinopathy of prematurity (ROP)	Premature babies may have abnormal growth of blood vessels in the eye. Vision loss and blindness can result.
Chronic lung disease	Premature babies are prone to developing a chronic lung problem that resembles asthma and can last for years
Infections	Because the immune system is immature, premature babies can develop sepsis, pneumonia, and meningitis.
Anemia	Because premature babies do not have enough red blood cells, they can become anemic. This can cause feeding problems, can retard growth, can make heart problems worse, and may require transfusions.
Patent ductus arteriosis (PDA)	The ductus arteriosus between the left and right sides of the heart may not close, and drugs or surgery will be needed to close it.
Apnea	Premature babies tend to stop breathing and must be monitored until their lungs have grown enough to operate independently.

Figure 24.7 **The Use of *In Vitro* Fertilization Is Increasing.**

■ *What impact might such procedures have on population growth and natural selection for humans?*

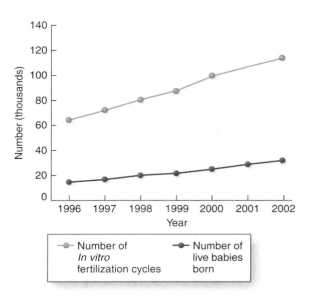

IVF Allows People Who Previously Could Not Have Been Parents to Have Children

IVF technology has opened a new world of reproductive opportunities for people to become parents. Today, reproductive disorders or diseases no longer limit having children. With IVF women who have blocked oviducts, men with low sperm counts, women past prime childbearing age, and even people with more severe limits on reproduction may all become parents.

IVF Allows the Use of Egg or Sperm Donors

Ova and sperm can be kept frozen indefinitely in liquid nitrogen and still produce healthy embryos. **Figure 24.8** shows some of the options for choosing egg donor, sperm donor, gestation mother, domestic mother, and domestic father. As you can see, there are many combinations that will produce a child, and as many as five different people can be involved. In an extreme case the couple that wants to have a baby are the domestic mother and domestic father. They choose an egg donor and a sperm donor. After IVF the developing embryos are implanted within the uterus of a gestational mother (sometimes called a surrogate mother) who carries the baby to term. So in one sense this child has five parents.

The use of egg and sperm banks opens some interesting possibilities for prospective parents. The keepers of ova and sperm maintain records on each donor's characteristics and a prospective parent can sort through hundreds of dossiers to find a donor with the characteristics most desired in a child. Prospective parents can select physical characteristics, behavioral traits, even religious or ethnic background. People who use sperm or egg donation for IVF go through a much more thorough review and selection process than do most couples who are contemplating marriage. Of course, the basic mechanisms of genetics (see Chapters 8 and 10) are still at work here. There is no guarantee that an IVF child will have the traits chosen in the parental ovum or sperm, but IVF has resulted in thousands of children born from carefully selected biological parents (**Figure 24.9**).

emotional cost of premature birth can be severe for the parents, and the economic cost of caring for premature babies is staggering. Many panels and reports have urged IVF practitioners not to resort to multiple implantations in an effort to achieve a successful pregnancy. Despite all of these difficulties, IVF is growing in popularity. **Figure 24.7** shows the growth in IVF treatments over the past 10 years and the associated increase in the number of IVF babies. The drive to have children is part of the reason that IVF is increasingly popular, but IVF offers a much greater prize to parents—the increased certainty of having a child with, or without, a particular gene. The next section will explore this completely new aspect of having children in the age of IVF.

QUICK CHECK

1. What are the steps involved in an IVF procedure?

24.3 IVF Provides Many Options for Having Children

IVF has opened up many new options not only for childless couples, but also for couples who otherwise would be unlikely to have a healthy baby. Although prenatal testing can tell parents if their child may by affected by certain diseases or disorders, IVF goes much further. Not only can IVF allow people who otherwise would not have children to become parents, variations on the IVF procedure can greatly increase the odds of having a child without a particular disorder.

IVF Allows Embryos to Be Genetically Screened and Selected

When IVF is combined with genetic testing and screening, the implications are profound. In these instances parents can actively select for or against a trait in their child. This is much more than just choosing the traits of the biological parents. Genetic screening allows parents to choose whether their children do or do not carry some specific genes. Technically, there are different ways to carry out genetic screening. Let's examine one possible scenario in some detail.

Genetic screening and selection begins like any other IVF procedure. The woman goes through a series of hormone treatments to produce many mature ova. These are removed and are placed in a special nutrient solution in a Petri dish and combined with sperm from the chosen biological father. In cases where the embryos will be genetically screened, many clinics use controlled injection of a single sperm into an ovum. ISCI ensures that DNA from sperm that did not fertilize the egg will not contaminate the DNA analysis. The fertilized eggs are allowed to develop to the 8-cell stage, and then one cell is removed from each of several embryos.

But wouldn't the removal of one embryonic cell cause problems for the developing embryo? The answer is no. Although cells of an early embryo begin to differentiate into specific adult cells (see Section 9.6), early cells retain the ability to become any cell type in the body. So removing one cell does not interfere with development: the remaining seven cells can produce the entire child.

Each cell that is removed is screened for whatever genes the parents are interested in. Because the amount of DNA in a single cell is so small, many screening protocols take advantage of selective PCR (polymerase chain reaction), summarized in **Figure 24.10**. PCR makes many copies of a sample of DNA (see Section 11.3). One big payoff of PCR is the ability to copy or amplify selected parts of the genome. PCR starts by splitting double-stranded DNA into single-stranded DNA. A primer, a short sequence of DNA that binds to a stretch of single-stranded DNA in the sample, produces a short region of double-stranded DNA in the sample. This is necessary because the enzymes used to copy DNA require a short stretch of double strand to get started. The trick for PCR is to use a primer that will bind close to a stretch of DNA that carries the gene you are interested in and want to copy. In a batch full of single-stranded DNA, only the DNA adjacent to this double-stranded region, carrying the gene of interest, will be copied. The end result is many copies of one single gene, if that gene is indeed in the sample DNA. Once PCR is complete, the amplified sample of DNA can be analyzed in a variety of ways to see if a particular gene was copied.

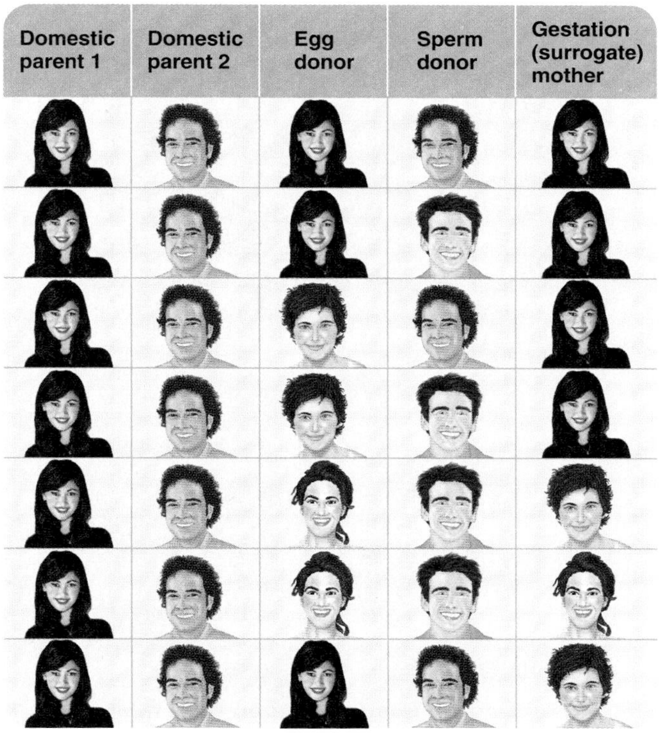

Domestic parent 1	Domestic parent 2	Egg donor	Sperm donor	Gestation (surrogate) mother

Figure 24.8 **Parental Arrangements to Produce a Baby Using IVF.** This table shows just some of the possible combinations of different people who might be involved in the birth and rearing of an IVF child. The table assumes two "domestic" parents in the home, but of course, other options are possible. The only apparent limits in this picture are that the egg donor and the gestation mother must be female; the sperm donor must be male.

Figure 24.9 **Same Sperm Donor.** Coincidentally, the mothers of these children discovered they all had children with sperm from the same donor.

(a) A mutated gene and the adjacent DNA are isolated, and the sequence of nucleotide bases is analyzed.

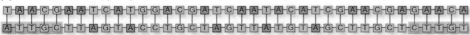

A-T-T-G-C-T-T-A-G-T-A-C-C-T-G-C-T-A-G-T-T-A-T-G-T-A-G-C-T-T-G-C-T-C-T-T-G-T

The gene is the nucleotide sequence between the shaded bases.

(b) A primer is attached to corresponding bases at one end of the sequence.

Primer—T-A-A-C-G

A-T-T-G-C-T-T-A-G-T-A-C-C-T-G-C-T-A-G-T-T-A-T-G-T-A-G-C-T-T-G-C-T-C-T-T-G-T

(c) PCR is used to copy only the sections of DNA in the sample that have this sequence.

T-A-A-C-G-A-A-T-C-A-T-G-G-A-C-G-A-T-C-A-A-T-A-C-A-T-C-G-A-A-C-G-A-G-A-A-C-A

A-T-T-G-C-T-T-A-G-T-A-C-C-T-G-C-T-A-G-T-T-A-T-G-T-A-G-C-T-T-G-C-T-C-T-T-G-T

(d) The process is repeated many times to make many copies of the gene.

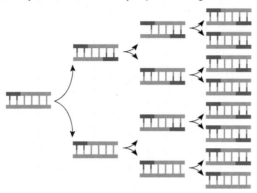

Figure 24.10 **PCR Can Be Used to Screen IVF Embryos for Mutations.** Because it multiplies the amount of DNA in a sample, PCR can multiply a small amount of DNA from an IVF embryo and test it for mutations before the embryo is selected for implantation.

Cystic Fibrosis Was One of the First Diseases Targeted by IVF Genetic Screening

Preimplantation genetic diagnosis, as the whole series of linked processes is sometimes called, has been used to eliminate embryos that carry alleles associated with specific diseases. Other techniques allow parents to choose the gender of their offspring. Many clinics will perform gender selection only to bring the ratio of male to female children in a family into balance, or to rule out certain disorders expressed only in males. A review of how IVF genetic screening has been used to prevent the inheritance of the disease cystic fibrosis will help you to appreciate the potential power of this new way to influence the traits of children.

The gene for cystic fibrosis (CF, for short) shows a simple recessive pattern, and a child who inherits two cystic fibrosis alleles will have the disease. Mucus will build up in the lungs, making breathing difficult. Treatments are expensive and even with them a CF child is not likely to live past age 20. Years of research have identified the protein and the gene that cause cystic fibrosis. The normal version of this gene codes for a protein that is a membrane channel for chloride ions in normal lung tissue (**Figure 24.11**). This membrane channel allows chloride to pass in and out of cells that line the air sacs in the lungs. Sev-

eral different abnormal alleles can cause CF, and the most common abnormal allele is missing three base pairs that code for one phenylalanine amino acid molecule in the protein. When this amino acid is missing, the protein does not function properly, chloride does not pass freely in and out of the cell, and mucus accumulates in the lungs. The accumulation of chloride in the fluid outside of cells gives the sweat of CF patients an extremely high salt content.

Cystic fibrosis is found in all populations in the United States, where it is the most common fatal inherited disorder, especially in Americans of European descent. At any given time CF affects about 30,000 people. Twelve million people are carriers, and 2,500 CF babies are born each year. Given the cost in human lives and expenses associated with caring for CF patients, it should not surprise you that millions of dollars have been invested to discover the CF gene and understand the mutations that cause CF. Funds for this research came mostly from the federal government and from private foundations. Since the discovery of the normal version of the CF gene in 1989, investigators have examined how the CF protein works, what causes it to be abnormal, and how to apply this information to the possible treatment of CF.

This massive research effort gave parents who carry the cystic fibrosis gene the opportunity to have healthy children. Consider your plight if you carry one cystic fibrosis allele and you fall in love with and marry someone who also carries a single cystic fibrosis allele. If you have children without technological intervention, each of your children has a 25% chance of inheriting two CF alleles and so developing CF. In contrast, IVF with preimplantation genetic diagnosis gives you the near certain assurance of *not* having a CF child. The first child chosen not to have CF was born in 1990. Since then many parents who might have had children with CF have had healthy children instead.

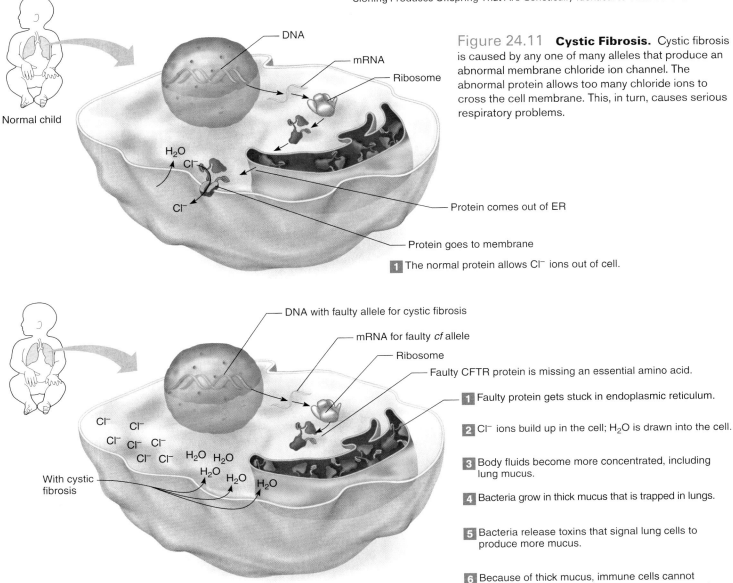

Figure 24.11 **Cystic Fibrosis.** Cystic fibrosis is caused by any one of many alleles that produce an abnormal membrane chloride ion channel. The abnormal protein allows too many chloride ions to cross the cell membrane. This, in turn, causes serious respiratory problems.

DNA

mRNA

Ribosome

Normal child

H_2O

Cl^-

Cl^-

Protein comes out of ER

Protein goes to membrane

1 The normal protein allows Cl^- ions out of cell.

DNA with faulty allele for cystic fibrosis

mRNA for faulty *cf* allele

Ribosome

Faulty CFTR protein is missing an essential amino acid.

1 Faulty protein gets stuck in endoplasmic reticulum.

2 Cl^- ions build up in the cell; H_2O is drawn into the cell.

3 Body fluids become more concentrated, including lung mucus.

4 Bacteria grow in thick mucus that is trapped in lungs.

5 Bacteria release toxins that signal lung cells to produce more mucus.

6 Because of thick mucus, immune cells cannot reach bacteria, so they attack lung tissue instead.

With cystic fibrosis

Cl^- Cl^- Cl^- Cl^- Cl^- Cl^- Cl^- H_2O H_2O H_2O H_2O H_2O

24.4 Cloning Produces Offspring That Are Genetically Identical to Their Parents

Reproductive cloning is probably the most controversial of the emerging techniques for producing a new human being. The aim of reproductive cloning is to produce a baby that is a genetic copy of a particular person. Of course, cloning does occur in nature, during asexual reproduction and in the development of identical twins, but researchers did not originally develop the laboratory techniques of cloning in order to produce a cloned baby. Although the popular discussion focuses on reproductive cloning, researchers developed these techniques to study biological processes in animal sub-jects that have a known genetic history. Because a population of clones is genetically uniform, clones are desirable for certain kinds of biological research. It was not long, however, before practical applications of cloning technology became obvious. Today, reproductive cloning is used to produce farm animals and even to produce pets. The big question is whether reproductive cloning will be a serious option for people seeking to have a child. Cloning techniques were presented in Section 11.6, but it is worth reviewing some of the important issues here, in the context of how humans make babies.

Researchers have developed a variety of cloning techniques, but *nuclear transfer cloning* is the most common and receives the most attention. In principle, the idea of this technique is simple: the nucleus from one cell is removed and placed into another cell whose DNA has been destroyed or removed. The donor cell is usually from adult epithelial tissues, and the host cell is almost always an unfertilized egg. Under the right conditions the cloned cell develops into an embryo and grows into a mature individual. Although the idea is simple, nuclear transfer cloning is a difficult procedure and often is not successful.

One of the first successful cloning experiments involved frogs and was done in the early 1970s. Frog eggs are easy to work with because they are large and hardy. Also, frogs normally undergo external fertilization and development, so the researchers did not have to pioneer any special techniques for getting the cloned cell to develop into an embryo and then into a tadpole. In one early experiment the host egg came from a normally colored green frog, while the donor nucleus came from the intestines of an albino frog tadpole—one that had no pigmentation in its cells (see Figure 11.16). The experiment was repeated many times, using nuclei from the same donor. The injected eggs developed normally, yielding clones of the original albino frog, even though the host egg was a normally pigmented green frog.

Cloning Mammals Is Much More Difficult Than Cloning Frogs

The success of cloning frogs encouraged scientists to try to clone other animals, including mammals. The procedures used are basically the same as those used with frogs, but cloning mammals has not been nearly as successful. Dozens of research teams around the world worked on the problem of cloning mammals for many years, but it was not until 1997 that the first report of a cloned mammal

made the front pages of newspapers. You may remember the birth of Dolly, the sheep who was cloned from a cell donated by her "mother." Since the cloning of Dolly there have been other successes in cloning mammals (**Figure 24.12**). Pigs, mice, cows, sheep, goats, cats, monkeys, dogs, and other mammals have been cloned. An important question is—how soon will society have to deal with a cloned human?

You can get an initial answer to this question by considering the actual success rates of cloning in research settings. Cloning is highly unreliable, given the current state of knowledge and technology. The success rate of cloning depends on a great many factors, but a success rate of 3% is not unusual. This means that only 3% of embryos produced by nuclear transfer cloning using an adult donor nucleus result in a healthy, live birth. The success rate may be higher if the donor nucleus comes from another embryo, but even then success rates are well below 50%. It is interesting that the success rate is much *lower* when primates are involved. As of this writing, out of many hundreds of attempts, only two cloned monkeys have been born.

Mammalian cloning fails for several reasons. In mammals development occurs inside the mother's uterus, which is much different from the external fertilization and development of frogs. Like IVF, cloning is started *in vitro,* and so you would not expect the success rate for cloning to be any higher than for IVF. Additional difficulties associated with cloning make the success rate in mammals even lower than for IVF. Many embryos that begin to develop may spontaneously abort, and abnormalities can be seen in the aborted embryos. Even if an embryo makes it through pregnancy and is born, many clones have postnatal abnormalities that lead to an early death. Even Dolly, the famous cloned sheep, did not live the

Figure 24.12 **Just a Few of the Species of Mammals that Have Been Successfully Cloned.**

Dolly (left) with her mother

Prometea, cloned horse (younger horse)

full life span of a normal sheep. She died when she was only 5 years old, far short of the 19- to 25-year life span usual for a sheep.

Knowledge of normal development is the only way to understand what can go wrong in cloning. In normal development mitotic cell division and cell differentiation combine to produce a mature individual that may have trillions of cells (see Section 9.1), each of which is dedicated to its own special job. One important aspect of development is the state of DNA in embryonic cells. During differentiation much of the DNA in a cell is shut down and is not expressed as functional proteins. DNA methylation and chromosome condensation are two ways that a cell can inactivate regions of DNA (see Section 9.3). The patterns of DNA inactivation are so important for the functioning of a mature individual that when a particular cell divides these patterns are transmitted even to daughter cells. Even the DNA of egg and sperm, which combine to form the new zygote, are not in their native state. The DNA of the zygote must be reset before normal development can occur. Fertilization does much more than just add male DNA to the ovum. Penetration of an ovum by a sperm sets in motion a sequence of events that reprograms the zygote's DNA, so that it is ready to direct development. These processes have evolved to work on egg and sperm. Methyl groups are removed and the DNA is uncoiled so that the RNA polymerases can get access to it and begin to transcribe mRNA. Fertilization resets the zygote's DNA so that it is able to direct the development of the whole embryo.

This knowledge of normal development gives a big clue as to why adult nuclear transfer cloning is so rarely successful. The nucleus from the donor cell contains all the DNA needed to produce a full organism, but much of it is turned off. Often the process of genetic reprogramming does not happen correctly in nuclear transfer cloning. But why does cloning ever work? For reasons that are not at all understood, sometimes the host egg cytoplasm *is* able to reprogram the adult donor nucleus. Then the zygote has a chance of developing into a normal embryo. Armed with this knowledge, researchers have already begun to study how genetic reprogramming happens normally, in the hopes of applying new understanding to the attempts to clone mammals.

You might be surprised to learn that most scientists are strongly opposed to any attempts to clone humans for reproductive purposes. Some scientists have personal ethical objections, but nearly all believe that the difficulties involved pose too great a risk that a cloned embryo would have severe birth defects. Whether these scientific reservations are respected remains to be seen.

Therapeutic Cloning Holds Out Real Hope for New Medical Treatments

The prospects for cloning a complete human may be remote, but other applications of cloning could be developed sooner. Modern medicine has many ways to treat the diseases that plague us, but as yet there is no way to fully replace the aging and damaged cells at the core of disease problems. Heart disease, Type 1 diabetes, stroke, Alzheimer's disease, and innumerable other diseases result from cells that do not function properly. As researchers have learned more about how normal cells function, many scientists and doctors have held out the hope that new, healthy cells could replace the old, worn-out ones. Section 9.6 looked at the potential for embryonic stem cells to differentiate and replace lost or diseased cells. A major problem with embryonic stem cells, however, is tissue rejection. The immune system of any mammal will attack and so reject tissue from another individual. This is a major problem with organ transplants, for example. While there are drugs that can suppress the immune response, this still may limit the use of embryonic stem cells derived from fertilized eggs. Therapeutic cloning offers an alternative.

In therapeutic cloning a nucleus from one of your own cells would be inserted into an ovum provided by an egg donor. This is the same process of nuclear transfer cloning that you just read about. In this case, though, there is no attempt to produce a whole organism. Instead, the idea is to take advantage of the genetic reprogramming that occurs during cloning and to produce a clone of healthy embryonic cells with your own genome. These embryonic cells could then be placed directly into your damaged organ. In principle your body will not reject the cells, because they carry your DNA. In addition, the normal signal molecules found in your own tissues will cause them to develop normally into the cells that you need. There is some evidence that therapeutic cloning can work, at least in nonhuman animals and in humans. Adult stem cells—cells that maintain the ability to divide but come from adult tissues instead of embryonic ones—have been used to treat heart disease in a few patients. A cloned human embryo could provide a source of stem cells.

Of course, no medical advance is without its risks. Can you think of one risk that might come with therapeutic cloning? Stem cells share many traits with cancer cells, particularly the ability to divide. Some researchers are concerned that treatments based on therapeutic cloning may increase the chances that the patient will develop cancer.

The ethical dilemma also remains with therapeutic cloning. Some people object to the idea of starting the growth of a human embryo only to use its cells to grow "spare parts." It should be noted, however, that some parents have already had a second child by IVF and preimplantation screening for the purpose of providing an organ or blood donor for their first child. So far these situations have not posed any health risk for the "chosen" child, while therapeutic cloning involves starting a human life and then stopping it, but it shows that people are willing to cross certain ethical lines to protect their children's health, if not their own.

QUICK CHECK

1. Why is it so difficult to clone mammals?

24.5 An Added Dimension Explained
Jane Doe's Life Was Saved by Modern Reproductive Technology

The introduction to this chapter posed the difficult but all too real story of parents who find that their eagerly awaited newborn has a serious, life-threatening disease. Fanconi anemia is an inherited, genetic disease caused by mutations in the genes that control blood cell differentiation. A person with Fanconi anemia has abnormal blood stem cells in his or her bone marrow that give rise to abnormal blood cells. The result is an anemia or lack of all blood cell types: red blood cells that carry oxygen, platelets that promote blood clotting, and white blood cells that fight infection. The symptoms of this form of anemia are widespread and include bleeding, weakness, and infections. Other body systems also are affected. For reasons not well understood, some limbs do not form properly, and a child might have sensory problems, mental retardation, short stature, or abnormal skin pigmentation. Fanconi anemia affects people from all backgrounds, and about five people out of every million people have the disorder. At this point the only effective treatment is to have a transplant of stem cells from a normal person's bone marrow. The problem is that these foreign blood stem cells often are attacked by the patient's own immune system. Only a transplant from someone whose blood cells closely match those of the patient will be successful, and close matches are difficult to find. So you can see why baby Jane Doe's parents were so distraught when they learned their child's diagnosis. There seemed to be little that they could do to help her.

In the early 1990s the new reproductive technologies that you read about in this chapter offered new avenues of hope. Jane Doe's parents and her doctors realized that preimplantation genetic screening also could be used to produce a tissue donor for their sick child. They decided to try to have a second child made to order—a child who did not express the deadly Fanconi anemia and who would be a compatible tissue donor for his or her sister.

For this family this high-tech approach to helping a sick child seems to have been successful, at least as of the time of this writing. A healthy sister was born who could and did provide a blood stem cell transplant for Jane Doe, and her anemia was largely corrected. The path to this success was far from easy, though.

You already know that IVF, with or without genetic screening, is difficult and not always successful. In this case the embryos had to be screened to find ones that did not carry the Fanconi anemia mutations, and they had to be screened to find a genetic match for Jane Doe's blood cells. Jane's parents had to go through five IVF cycles to produce one healthy child. Forty-one embryos were formed to produce seven that were implanted. This means that 34 embryos were created, but not used. Only one embryo was carried full term and resulted in a healthy live birth. The total cost of saving Jane Doe's life was about $100,000. Not surprisingly, other families also have tried this way of saving their children, and many have been successful.

Once Jane Doe's baby sibling was born, where did the transplant cells come from? The answer to this question is one of the easiest and most reassuring in this process. The placenta contains stem cells that can be used for bone marrow transplants (**Figure 24.13**). Blood from the sibling's placenta was extracted, frozen, and saved for the transfusion into little Jane. The sibling actually did not have to provide any of her own tissue. The process of transplantation was complicated, and there were many places where the attempt could have failed. For example, it took three years from the time the parents decided to try preimplantation genetic screening to the time they had cells to transplant into their sick child. During that time Jane Doe's Fanconia anemia progressed, and she might have become too sick to benefit from the procedure. Many of Jane Doe's symptoms have been reversed, and as of this writing she is a healthy young girl.

And then there are the ethical questions. The families and doctors involved, and many other people around the world, are grappling with whether this sort of procedure is ethical. Is it ethical to create a life in order to save another life? Most doctors and ethicists have concluded that such procedures are ethical if (1) the parents would have had another child anyway, (2) the procedure does not in any way endanger the life or health of the second child, and (3) there is no other medical option. These ethical guidelines are not binding, however, and like any other human endeavor, they could be abused. At this point most doctors are reluctant to tell parents that they cannot use preimplantation genetic screening to save the life of one of their children. You can be sure, though, that we will all face further ethical dilemmas as reproductive technology races ahead.

QUICK CHECK

1. List the different reproductive technologies that saved Jane Doe's life.

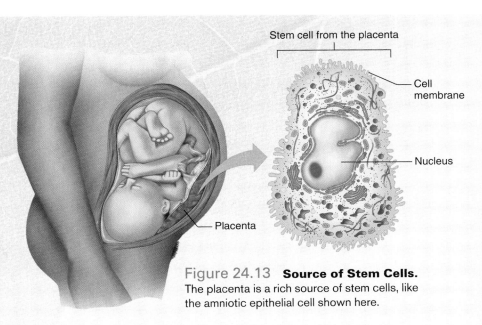

Figure 24.13 Source of Stem Cells.
The placenta is a rich source of stem cells, like the amniotic epithelial cell shown here.

Now You Can **Understand**

Concerns About Unequal Sex Ratios

One of the concerns often raised about IVF coupled with preimplantation genetic screening is the potential for selecting a child's sex. Why do people worry about this? In many cultures male children are favored over female children, and in many regions of the world there already is an imbalance in favor of males. China and India are often cited as cultures where female babies are less valuable than male babies and where female babies may be killed or neglected. Modern technology, such as ultrasound and amniocentesis, has allowed families to abort female fetuses before they are born. Today in China there are about 120 males born for every 100 females born, and in India there are about 115 males born for every 100 females. These findings show that given the choice, many couples will choose to have males and eliminate females. In addition to the ethical objection to favoring one sex over another, there is some evidence that excess males can lead to some cultural instabilities. Most Western doctors reject practices that choose males over females, but there is a concern that IVF with preimplantation screening will provide yet another way for couples to choose to have male children rather than female.

What Do **You** Think?

Human cloning

If you pay attention to headlines, you will see cloning crop up fairly often. With each new scientific breakthrough in cloning, people wonder about the possibility of human cloning. An Internet search will show you many organizations that either support human cloning or object to it. Not long after Dolly the sheep was cloned in 1997, at least one physician proclaimed that he intended to clone a human. And while many doubted the possibility of human cloning, another organization claimed to already have cloned humans. Known as the Raelians, this group has been on board with cloning for over 30 years. Founded in 1973 their whole premise and purpose for being is the belief that humans are the result of cloning by extraterrestrials. The Raelians sponsor UFOland, their Earth-based embassy, where they are ready to greet extraterrestrial travelers, and they also claim to have cloned a human baby. What do *you* think? Have the Raelians really cloned a human baby?

Chapter Review

CHAPTER SUMMARY

24.1   Modern Reproductive Technologies Are Built on Knowledge About Normal Development

IVF is the cornerstone of treatments that help women conceive who normally cannot conceive. The first test-tube baby was born in 1978, and each year in the United States about 100,000 women undergo IVF, and thousands of IVF babies are born. IVF takes advantage of the fact that the human embryo does not implant into the mother's uterus until it is a hollow ball of cells. Thus, its early development can occur within laboratory glassware.

in vitro fertilization   584

24.2   IVF Is the Foundation for Modern Reproductive Technologies

The IVF procedure is expensive and arduous for the woman, who must undergo repeated hormone injections. When an abundant supply of her ova has been produced by her ovaries, which have been artificially stimulated by the hormone injections, the ova are surgically removed. Her male partner contributes a sample of semen, and the ova and sperm are mixed and incubated. Careful laboratory work selects healthy zygotes that then are inserted into her uterus, where, if the procedure is successful, they implant, producing a viable pregnancy. IVF techniques were developed after years of basic research into various aspects of embryology. Although IVF does work, it is not a risk-free, fail-safe method. It works better with younger prospective mothers than older prospective mothers, who have a success rate of about 1%.

24.3   IVF Provides Many Options for Having Children

One interesting aspect of IVF is that when combined with preimplantation screening that uses PCR to amplify DNA, these processes can allow parents to pick and choose traits carried by various of their embryos. IVF allows parents who would not otherwise be able to be parents to become parents and can involve up to five people in the conception, gestation and birth, and rearing of a child. Egg and sperm donors and surrogate mothers are commonly employed in IVF techniques. IVF also gives prospective parents the option of avoiding a genetic disease if one of their embryos carries it. The gene for cystic fibrosis can be detected in an embryo, and parents can choose not to implant it, saving themselves and their child this heartbreaking and life-shortening genetic disease.

24.4   Cloning Produces Offspring That Are Genetically Identical to Their Parents

Cloned animals allow researchers to study the effects of human diseases in a controlled setting, while genetically engineered crops can help farmers all over the world reap greater harvests. To date no one has yet cloned a human embryo, so IVF combined with preimplantation genetic screening is currently the best way to decrease the chances of bearing a child who has a specific genetic disorder. Cloning creates genetically identical organisms, and although mammals have been cloned, success rates are extremely low. Therapeutic cloning techniques are aimed at replacing flawed cells with healthy ones. Embryonic stem cells that can develop into any kind of cells are necessary for therapeutic cloning. Growing a clone of cells from an embryo that grows from one of your own cells circumvents the problem of tissue rejection. Great ethical dilemmas surround the use of embryonic stem cells, harvesting cloned embryos for treatment of disease, rejection of embryos from preimplantation screening, and the whole IVF process.

24.5   An Added Dimension Explained: Jane Doe's Life Was Saved by Modern Reproductive Technology

After several failed IVF attempts, IVF combined with preimplantation screening of embryos achieved the successful birth of a sister to Jane Doe. The placenta of this new baby was used to prepare compatible stem cells that were transplanted into Jane's bone marrow cells. This combination treatment has allowed her body to make normal red blood cells and has reversed the effects of Fanconia anemia.

REVIEW QUESTIONS

TRUE or FALSE: If a question is false, rewrite it to make it true.

1. Production of human sperm begins before birth and continues until death.

2. Production of human ova begins at puberty and continues until death.

3. IVF is an inexpensive, easy, risk-free process that has a high success rate.

4. A woman in her mid-forties has about a 1% chance of success with IVF.

5. Implanting more than two embryos increases the chances of a healthy live birth.

6. IVF allows people who would not ordinarily be able to have children to become parents.

7. Fertilization removes methyl groups from a zygote's DNA.

8. Gene therapy is a safe and successful method.

9. Fertilization occurs within the uterus.

10. Preimplantation screening can determine the traits of an embryo.

MULTIPLE CHOICE. Choose the best answer of those provided.

11. In *in vitro fertilization* ova are fertilized in
    a. laboratory glassware.
    b. a female's oviducts.
    c. a female's uterus.
    d. virtual reality.

12. A human ovum completes meiosis
    a. at 28 days.
    b. every nine months.
    c. after fertilization.
    d. after puberty.

13. Production of sperm within testes
    a. begins with the first menstrual cycle.
    b. begins in embryonic life.
    c. is a daily process that continues throughout life.
    d. ceases at menopause.

14. A newly ovulated ovum has
    a. 23 single chromosomes.
    b. 23 replicated chromosomes.
    c. 15 chromosomes.
    d. 15 replicated chromosomes.
    e. 48 chromosomes.

15. An IVF patient is injected with hormones to allow her to
    a. ovulate an abundance of eggs.
    b. increase the speed of ovulation.
    c. increase the efficiency of fertilization.
    d. ensure fertilization.

MATCHING

16–20. Match the columns.

16. the most common, fatal inherited disorder
17. transgenic
18. clone
19. embryonic stem cell
20. human insulin

a. can differentiate into any cell type
b. cystic fibrosis
c. grown in genetically engineered *E. coli*
d. has a gene that originated in a different species
e. an organism that is genetically identical to another

CONNECTING KEY CONCEPTS

1. How is preimplantation genetic screening done, and what impact could it have on human populations?

2. Why do most scientists believe that the reproductive cloning of a human is unlikely to be accomplished any time in the near future?

THINKING CRITICALLY

1. Many people are concerned about the ethical and societal implications of preimplantation genetic diagnosis and screening. Some see these procedures as opening a new era of eugenics. Others support any options that increase the chances of birthing a healthy child. Based on your own discussions, list some of the potential concerns that surround preimplantation genetic screening. How would you balance these concerns? Should there be legislation or regulations that control how this process is used?

For additional study tools, visit www.aris.mhhe.com.

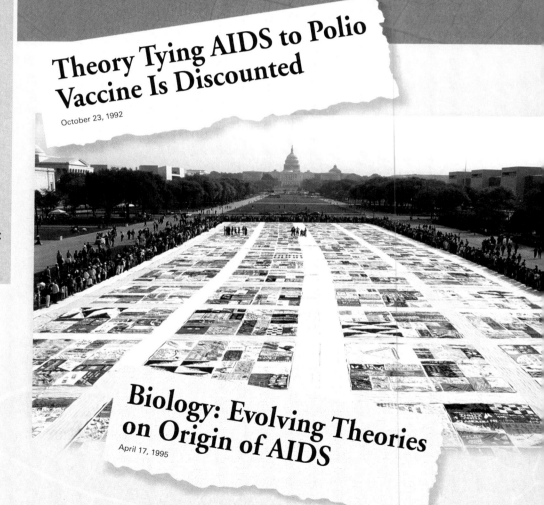

Theory Tying AIDS to Polio Vaccine Is Discounted

October 23, 1992

AIDS Quilt. The AIDS quilt displayed in Washington D.C. commemorated those who have died from AIDS.

Biology: Evolving Theories on Origin of AIDS

April 17, 1995

An Added Dimension
Where Did HIV Originate?

AIDS, Ebola, West Nile. Viruses cause all of these deadly diseases. At this writing people and governments are concerned that the virus that causes bird flu may "jump" to humans, causing an influenza pandemic that could kill millions worldwide. When your parents were young, the most well known viral diseases were childhood diseases—mumps, measles, and chickenpox. But today, new viral diseases with exotic-sounding names affect people of all ages. Experts predict that other new, virulent viruses will emerge in the future.

AIDS—acquired immune deficiency syndrome—is one of the most serious of the emergent viral diseases and is caused by a virus called HIV. A person infected with HIV who has developed AIDS suffers from many other infections, and if the person is not treated the outcome is nearly always a slow and miserable death. AIDS has killed tens of thousands of people in the United States and 20 million people worldwide (**see chapter-opening photo**). AIDS is so common in some parts of the world that in those regions life expectancy has dropped to under 40 years. AIDS even has reduced the rate at which the world population is expected to increase. Although tremendous progress has been made in understanding and treating AIDS, it remains an incurable disease and a serious threat to human health.

Despite how much is known about AIDS and HIV, misinformation still is common. For example, some people doubt that AIDS is caused by a virus, despite overwhelming scientific evidence that it is. People also wonder how and why AIDS suddenly appeared in the United States in the 1980s. Some claim that AIDS originated as a CIA experiment that went wrong. A different idea is that AIDS originated in a contaminated polio vaccine. The British journalist Edward Hooper makes this argument in his 1999 book, *The River*. Hooper claims that the polio vaccine used in Central Africa in the 1950s was contaminated with the virus that causes AIDS. Because it is easy to believe that a human error could cause a deadly disease, this idea sounds true to some people. In contrast, the scientific community proposes that on several different occasions the virus that causes AIDS was transmitted to humans from monkeys and chimpanzees in Africa.

Determining the origin of HIV and AIDS is important not only because such knowledge will help to understand the disease, but also because it is a part of a larger effort to explain how viral diseases emerge. How can you know which hypothesis for the origin of AIDS is correct? At the end of this chapter you will read some of the evidence for and against these hypotheses. Before you can evaluate this evidence, though, you will need to understand what viruses are, how they cause disease, and how a specific virus causes AIDS. ■

25.1 Infectious Diseases Have Affected Human Populations for All of Human History

Infectious diseases have plagued humans for all of recorded history (**Figure 25.1**). As human populations became more concentrated, the effects of an infectious disease could be devastating. One striking example is the black death, named because of the discolored skin characteristic of this disease. The earliest reports of black death were in the Mongolian region of China, and in about AD 46 two-thirds of the population of Mongolia died from black death. The disease slowly spread west and ravaged Europe around AD 1350. About 25 million Europeans died, and again this was a loss of over two-thirds of the population. Massive human deaths due to infectious diseases have occurred throughout history. Although

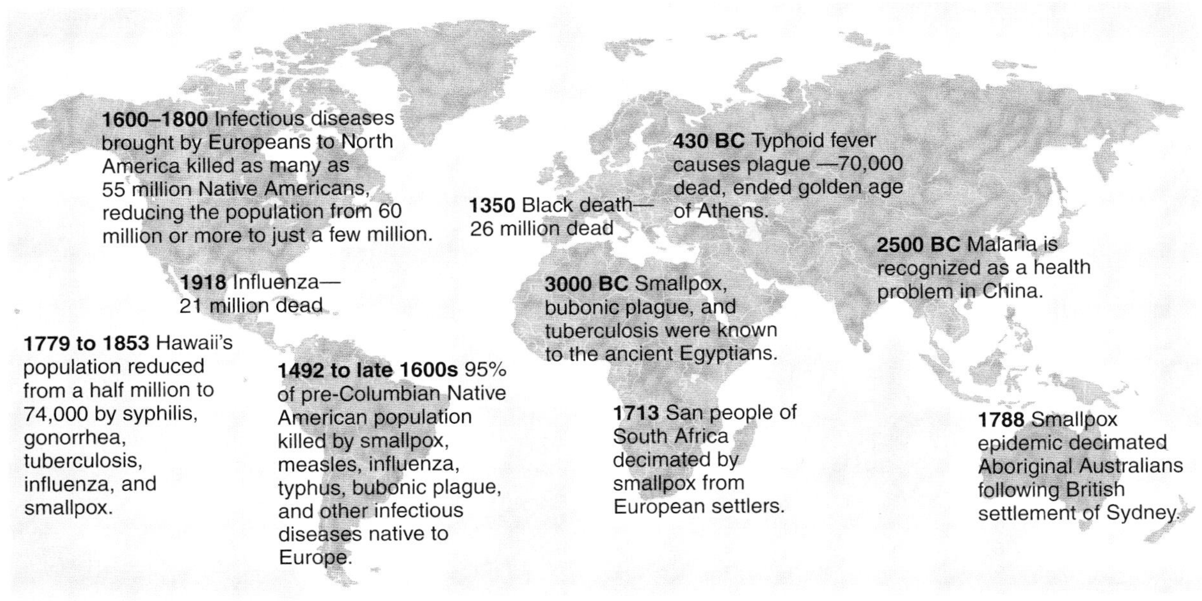

Figure 25.1
Infectious Diseases Have Killed Millions of People. Infectious diseases have affected human populations throughout human history.

1600–1800 Infectious diseases brought by Europeans to North America killed as many as 55 million Native Americans, reducing the population from 60 million or more to just a few million.

1350 Black death— 26 million dead

430 BC Typhoid fever causes plague —70,000 dead, ended golden age of Athens.

2500 BC Malaria is recognized as a health problem in China.

1918 Influenza— 21 million dead

3000 BC Smallpox, bubonic plague, and tuberculosis were known to the ancient Egyptians.

1779 to 1853 Hawaii's population reduced from a half million to 74,000 by syphilis, gonorrhea, tuberculosis, influenza, and smallpox.

1492 to late 1600s 95% of pre-Columbian Native American population killed by smallpox, measles, influenza, typhus, bubonic plague, and other infectious diseases native to Europe.

1713 San people of South Africa decimated by smallpox from European settlers.

1788 Smallpox epidemic decimated Aboriginal Australians following British settlement of Sydney.

no one is certain of how many people died in the influenza pandemic of 1918–1919, estimates range from 20 million to 100 million people worldwide. Smallpox was even more deadly than black death and probably affected human settlements as early as 10,000 BC, sometimes flaring up into an epidemic. Although the death toll from smallpox is impossible to reckon, smallpox epidemics brought down empires, made it possible for Europeans to establish colonies in the New World, and killed massive numbers of people worldwide. Today, modern vaccines have finally eradicated smallpox.

Later in this chapter you will read about medical success stories that have reduced the death toll from infectious diseases. In spite of these successes, some infectious diseases are still a threat. In addition, new infectious diseases spread as people push into previously uninhabited areas and as they travel the globe easily and frequently. For instance, HIV was first identified in the early 1980s, yet it already has killed 20 million people worldwide. Ebola hemorrhagic fever and SARS (short for severe acute respiratory syndrome) are infectious diseases that have killed far fewer people over the past decade, but only because rapid and effective public health campaigns have stopped their spread. Ebola now is a threat to wild gorilla populations and may push them to extinction. Other long-standing infectious diseases remain serious threats to health. For instance, each year malaria kills nearly 3 million people and debilitates hundreds of millions more. Even though the polio vaccine has been available for nearly half a century, and polio has been eradicated in most places in the world, new cases still occur in parts of Africa and Asia.

Because widespread infectious diseases kill people who are most susceptible and spare those who have some resistance, infectious diseases also have affected the genetic makeup of populations. So throughout human history infectious diseases have been part of the pressures of natural selection on human populations. One of the best-studied examples is the evolution of resistance to malaria in tropical countries. Over many generations alleles that confer some resistance to malaria became more prevalent in certain African populations. People who carry these protective alleles are less likely to become sick with malaria, and they survive longer if they do get malaria. Unfortunately, they also have an increased chance of having various kinds of anemia, such as sickle cell anemia.

pasteurization a method developed by Louis Pasteur in which a liquid is heated to a temperature high enough to kill the bacteria in it and held at that temperature for a specific period of time

QUICK CHECK

1. What are three of the most devastating infectious diseases in terms of the toll on human life?

25.2 Scientific Studies Have Discovered the Causes of Infectious Diseases

Have you ever walked through an old graveyard and read the inscriptions on headstones from the 1700s and 1800s? You might see families that had 10 children and lost eight to childhood infections (**Figure 25.2**). Babies often died in their first year of life, mothers often died of infections associated with childbirth, and soldiers died from infected wounds. Those who survived youth and middle age might live to be quite elderly, but relatively few reached those milestones because early death from infection was so common. Before the early twentieth century treatments for infections were ineffective, and people often were forced to rely on superstitions and folklore for remedies (**Figure 25.3**).

It was not until the late 1800s that it became clear that organisms cause many infectious illnesses. The great improvements in health and survival over the past century are due to discoveries about the causes of these diseases and the knowledge of how the body fights foreign invaders. Let's consider some of the scientific advances that provided the foundation for treating diseases, allowing people to live longer, healthier lives.

Early Studies Provided Important Knowledge About Infectious Diseases

One of the most critical scientific developments for the study of life was the invention of the compound microscope. This sophisticated research tool is the result of people learning to grind pieces of glass to make lenses, a skill perfected in Europe in the

Figure 25.2 **A Gravestone Remembering Children Who Died of Infectious Diseases.** Before the discovery and use of antibiotics many children—and adults—died of infectious diseases.

1600s (see Section 4.1). Early investigators such as Robert Hooke and Anton van Leeuwenhoek used simple microscopes to discover a universe of previously unseen organisms. Louis Pasteur (1822–1895) found that tiny living organisms abound in foods, drinks, and other common substances.

Pasteur also discovered ways to kill microorganisms. He learned that many of the microscopic organisms in a vial of laboratory nutrient solution were killed when the solution was heated to a certain temperature for a certain length of time. This was an important advance because killing microorganisms in liquids such as wine or milk made them safer to drink. Today, when you buy milk, fruit juice, or other liquids, you will see that they are **pasteurized** or ultrapasteurized (**Figure 25.4**). Because pasteurization heats liquids to a temperature that kills most microorganisms, pasteurized foods can be stored longer before they will spoil.

While basic research into microorganisms was progressing, important medical research focused on how to prevent the spread of infection in hospitals. For example, the Hungarian physician Ignaz Semmelweiss (1818–1865) concluded that doctors were spreading infections from the autopsy room to the delivery room. In his hospital he instituted a strict policy that all physicians must wash their hands in a strong solution before seeing any patient. This reduced the infections associated with childbirth. Another physician, Joseph Lister (1827–1912) hypothesized that microorganisms caused infections after surgery. Lister developed techniques for keeping hospitals and operating rooms clean and free of microbes, and his antiseptic techniques allowed the modern era of surgery to flourish. Neither Semmelweiss nor Lister initially was hailed as an innovator. Most of their colleagues did not believe their results and rejected their germ-killing procedures. It took years before their discoveries genuinely transformed medicine.

One of the most important researchers in early efforts to understand infectious diseases was the German medical doctor, Robert Koch (1843–1910), who is credited with being the first to show convincing proof that a particular microorganism causes a particular disease. In 1876 Koch identified the bacterium that causes anthrax, a devastating infection that once ravaged the livestock industry. This is the same organism that caused panic in 2002 when letters containing anthrax were sent through the U.S. mail. In addition to discovering the anthrax bacterium, Koch identified and isolated the bacterium that caused tuberculosis. Koch made these important contributions because he developed a set of postulates that must

Great Grandmother's Sure Fire Remedies for Colds and Sore Throats

To treat a sore throat. Rub the throat with warmed oil. Wrap the throat with a strip of flannel that has been soaked in hot water and wrung dry. Cover the hot flannel with a long, black, cotton stocking. Or just wrap the throat with a dirty sock.

For a head cold. Rub sage leaves in the nostrils in the morning.

To treat a cold. Peel and slice an onion and brown it in butter. Add 1 cup of strong broth and simmer until the onion is soft. Serve unstrained.

Figure 25.3 **Folk Cures for Common Infections.**

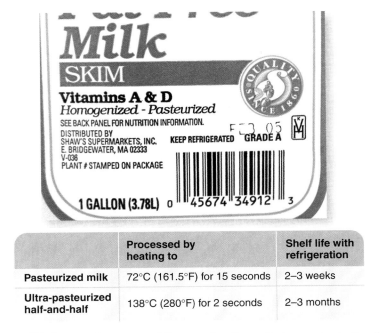

	Processed by heating to	Shelf life with refrigeration
Pasteurized milk	72°C (161.5°F) for 15 seconds	2–3 weeks
Ultra-pasteurized half-and-half	138°C (280°F) for 2 seconds	2–3 months

Figure 25.4 **The Process of Pasteurization.** During pasteurization liquids are heated to a high temperature and enough infectious bacteria are killed to prevent rapid spoilage. This allows food products to be kept and used longer before they spoil.

be verified to demonstrate that a particular organism causes a particular disease. Koch's postulates are as follows:

■ The microorganism must be detectable in the infected host at every stage of the disease.

■ The microorganism must be isolated from the diseased host and grown in culture.

■ When susceptible, healthy animals are infected with pathogens from the culture, the specific symptoms of the disease must occur.

■ The microorganism must be re-isolated from the infected animal and must correspond to the original microorganism in pure culture.

Each of Koch's postulates is a hypothesis that is tested by experimentation. His postulates mark the beginning of the realization that microorganisms can cause diseases, and that microorganisms can be transmitted from one person to another.

Some Bacteria and Viruses Are Pathogens

A *pathogen* is a parasite that causes disease (see Section 15.2). A pathogen lives in a *host* organism, obtaining nutrients from the host's living processes. Some common pathogens are shown in **Figure 25.5**, and others are listed in **Table 25.1**. As you can see, pathogens can be viruses, bacteria, protists, fungi, or invertebrate animals. Still, only a few of the millions of kinds of living organisms are pathogens. Bacteria and viruses are probably the most common pathogens, and the sections ahead will focus on them.

Viruses and bacteria are different in fundamental ways. Bacteria are living cells, while viruses are not alive (**Figure 25.6**). As long as bacteria have a useable supply of energy and other critical nutrients, they can thrive and reproduce independently, but a virus must take over the metabolic chemistry of a living cell to reproduce. Bacteria are prokaryotes whose DNA is not isolated in a membranous

(a)

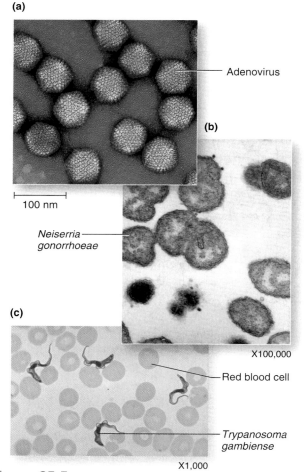

Adenovirus

100 nm

(b)

Neiserria
gonorrhoeae

X100,000

(c)

Red blood cell

Trypanosoma
gambiense

X1,000

Figure 25.5   **Some Common Pathogens.** (**a**) An adenovirus causes colds. (**b**) The bacterium that causes gonorrhea. (**c**) The single-celled eukaryote that causes sleeping sickness.

Table 25.1 Pathogens and Associated Diseases	
Pathogen	**Disease/symptoms**
Protists	
Cryptosporidium sp.	Inflammation of stomach lining
Toxoplasma gondii	Encephalitis
Fungi	
Candida albicans	Mouth thrush, vaginitis
Coccidioides immitis	Meningitis
Pneumocystis carinii	Pneumonia
Bacteria	
Mycobacterium tuberculosis	Tuberculosis
Proprionibacterium acnes	Acne
Streptococcus pyogenes	Strep throat
Streptococcus pneumoniae	Pneumonia
Haemophilus influenzae	Pneumonia
Chlamydia influenzae	Respiratory disease, cardiovascular disease
Neisseria meningitidis	Meningitis
Escherichia coli, strain K-1	Neonatal meningitis
Viruses	
Cytomegalovirus	Fever, hepatitis, retinitis, colitis
Varicella-zoster	Shingles (skin lesions), encephalitis
Respiratory syncitial virus	Pneumonia
Influenza virus A and B	Flu
Herpes virus and other viruses	Herpes, meningitis

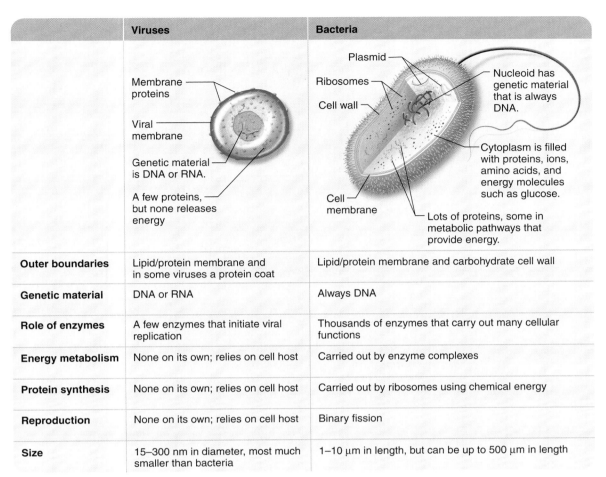

	Viruses	Bacteria
Outer boundaries	Lipid/protein membrane and in some viruses a protein coat	Lipid/protein membrane and carbohydrate cell wall
Genetic material	DNA or RNA	Always DNA
Role of enzymes	A few enzymes that initiate viral replication	Thousands of enzymes that carry out many cellular functions
Energy metabolism	None on its own; relies on cell host	Carried out by enzyme complexes
Protein synthesis	None on its own; relies on cell host	Carried out by ribosomes using chemical energy
Reproduction	None on its own; relies on cell host	Binary fission
Size	15–300 nm in diameter, most much smaller than bacteria	1–10 µm in length, but can be up to 500 µm in length

Figure 25.6 **Viruses and Bacteria Compared.** While viruses can replicate within living cells, they lack the organelles and enzymes necessary for independent life.

■ *Why do you think it is harder to produce anti-viral drugs compared to anti-bacterial drugs (antibiotics).*

nucleus (see Section 15.3). The single bacterial chromosome contains genes that are essential for survival, such as those that control metabolism or protein synthesis. In addition, a bacterium often will have additional DNA in the form of small, circular loops called *plasmids.* Think of plasmids as small, optional chromosomes that carry genes that help bacteria to survive in different environments. For example, genes found in plasmids may code for *toxins,* poisons that affect other organisms, or for genes that provide resistance to antibiotics.

Another important feature of a bacterium is the cell wall (Figure 25.6). Located outside of the cell membrane, the bacterial cell wall is made of sugar molecules and amino acids. The cell wall encloses the bacterium and protects it from the environment. As bacteria swim freely in a watery environment, their cell walls prevent the entry of unwanted chemicals. When a bacterium is in an extremely sugary or salty environment or in a dilute, watery environment, its thick cell wall keeps the bacterium from bursting or collapsing. While the cell wall is made of common biological

molecules, it has unique properties, and its composition is much different from the composition of an animal cell membrane.

The basic chemical processes of bacterial cells are similar to those of eukaryotic cells. Even though bacteria don't have internal membranous organelles, their cytosol contains enzymes and other cellular machinery similar to those of eukaryotic cells (see Section 15.3). For instance, bacterial DNA directs protein synthesis when a stretch of DNA is transcribed to mRNA. Protein synthesis of bacteria happens on ribosomes in the process of translation. Glucose metabolism is similar in bacteria and prokaryotic cells. In these basic ways bacterial cells are not awfully different from human cells.

Viruses Are Pathogens, But They Are Not Alive

Although viruses share many features with living cells, they are not alive (see Section 15.6). All viruses have a genetic molecule that is either DNA

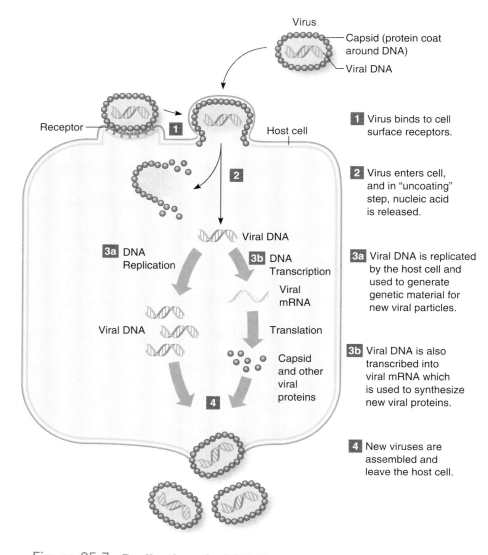

Figure 25.7 **Replication of a DNA Virus.**

1 Virus binds to cell surface receptors.

2 Virus enters cell, and in "uncoating" step, nucleic acid is released.

3a Viral DNA is replicated by the host cell and used to generate genetic material for new viral particles.

3b Viral DNA is also transcribed into viral mRNA which is used to synthesize new viral proteins.

4 New viruses are assembled and leave the host cell.

Even though viruses cannot use energy on their own, they can take over the organelles and energy of cells to make new viral particles. This process is directed by the virus' own genetic material, and this is what happens in a viral infection. Viruses are genetically quite diverse, and each kind of virus infects only certain kinds of living cells. For example, some viruses infect only bacteria, some infect only certain cells of plants, and others infect only certain cells of particular animals. The cell or organism that a virus infects is called the **host.** Most animal viruses normally infect only one or two major groups of organisms, but mutations in the viral genome can allow viruses to "jump species" and begin to infect a group of organisms that they did not infect in the past.

Figure 25.7 shows the process of viral infection for an animal virus that has DNA as its genetic material. First, the proteins on the surface of the virus interact with receptor proteins on the surface of a cell. These proteins determine specificity of a virus for a particular host cell, and if the proteins are incompatible, the virus cannot interact with the cell. The interaction of surface proteins brings the two membranes into close contact, and they fuse. During fusion the viral membrane opens, and the viral DNA enters the host cell. Enzymes normally present in the host cell dissolve the protective protein coat around the viral DNA. Once viral DNA is free, the cell's enzymes and organelles treat it just like host DNA. Host DNA polymerase and other enzymes copy viral DNA. Host RNA polymerase and other enzymes produce viral mRNA and, in turn, viral mRNA directs host ribosomes to produce viral proteins. Once all of these molecules are present in the host cell, they spontaneously assemble into new viruses. Some viruses accumulate within the cell until it bursts, while other viruses emerge out of the cell like buds. New viral particles go on to infect other cells, and some infect cells of new hosts.

Many viruses use RNA as genetic material. You might think that because its genetic material is RNA, an RNA virus might skip the DNA to mRNA transcription step and go straight to protein synthesis on ribosomes. In some viruses this is so, but this is not the case for **retroviruses,** a major group important in human disease. Some retroviruses can cause leukemia, or other cancers. The influenza virus is a retrovirus and so is HIV, the virus that causes AIDS. When a retrovirus replicates, information flows from viral RNA to viral DNA to viral mRNA. Let's follow this complex infection process in more detail, using HIV as an example.

Like other viruses, HIV begins its infectious process by binding to proteins on the membrane of a

or RNA (Figure 25.6). The genetic material of a virus is surrounded by a barrier, although unlike the nuclear membrane in eukaryotic cells, the viral barrier is made almost exclusively of proteins. Some viruses, including most animal viruses, have an additional outer membrane similar to a cell membrane. It is a lipid bilayer studded with proteins. Some viruses also have a few key proteins that will get the process of viral replication started once the virus has infected a cell. Despite this complexity, viruses lack the cellular machinery for both energy metabolism and protein synthesis. So viruses have no metabolic enzymes, mitochondria, ribosomes, chloroplasts, Golgi apparatus, and/or endoplasmic reticulum. They carry genes for only a few of the enzymes necessary for the processes of life.

host the cell or organism that a virus infects

retrovirus a virus that uses RNA as its genetic material and must transcribe its RNA into DNA before it can reproduce using the host cell's DNA

host cell (**Figure 25.8**). This causes the virus and cell to fuse and the viral genetic material and some key proteins enter the cell. The proteins on the surface of HIV are highly specific and interact only with certain kinds of cells in humans and some other primates. HIV replication begins with an HIV enzyme called **reverse transcriptase.** As its name suggests, reverse transcriptase works in the opposite direction from normal transcription. Following the base-pairing rules, this enzyme uses the information in the RNA strand to build a DNA strand. Because transcription requires energy, the virus can reproduce only within a cellular host, using energy that comes from the host's ATP. Once transcribed, the DNA of a retrovirus usually is integrated into the host cell's DNA. Later the viral DNA is transcribed to viral mRNA by the host cell's RNA polymerase. Translation of viral RNA, viral proteins, the assembly of viral particles, and the release of viral particles from the host cell all follow.

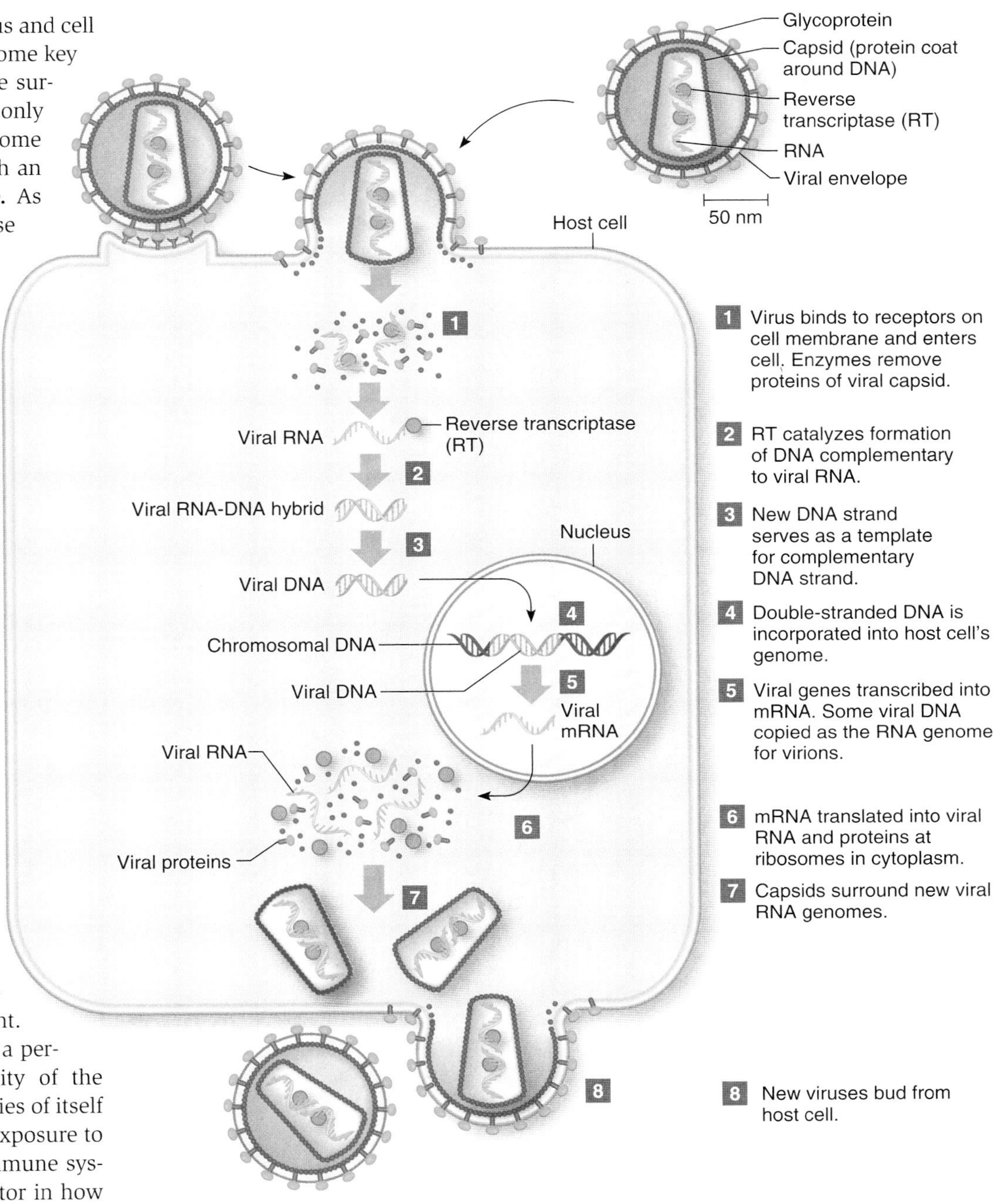

Pathogens Can Make a Person Feel Sick in Several Different Ways

Even though you are surrounded by bacteria and viruses, most of the time you do not get sick from them. Whether a particular pathogen makes a person sick at any given time depends on several factors. First, the toxic characteristics of the pathogen are important. Some pathogens are more likely to make a person sick than others. Second, the ability of the pathogen to multiply and make many copies of itself will affect how ill a person will get from exposure to the pathogen. Third, the ability of the immune system to kill the pathogen will be a big factor in how sick a person gets. And finally, strength of certain aspects of the immune response can determine how ill a person feels. Let's look at each of these briefly.

Some bacteria cause disease by secreting toxins. These damage surrounding tissues, and some bacteria secrete toxins into the environment as a way to protect themselves. Bacterial toxins often are proteins that interfere with other cells. An example is botulinum toxin produced by *Clostridium botulinum*. It causes a fatal form of food poisoning. Botulinum toxin interacts with other proteins on nerve cell membranes and blocks the release of the neurotransmitter acetylcholine that

1 Virus binds to receptors on cell membrane and enters cell. Enzymes remove proteins of viral capsid.

2 RT catalyzes formation of DNA complementary to viral RNA.

3 New DNA strand serves as a template for complementary DNA strand.

4 Double-stranded DNA is incorporated into host cell's genome.

5 Viral genes transcribed into mRNA. Some viral DNA copied as the RNA genome for virions.

6 mRNA translated into viral RNA and proteins at ribosomes in cytoplasm.

7 Capsids surround new viral RNA genomes.

8 New viruses bud from host cell.

Figure 25.8 **Replication of a Retrovirus.**

signals muscles to contract. When acetylcholine is not present in neuromuscular junctions, muscles do not contract; paralysis is one effect of botulism. If the toxic condition is not treated, death occurs soon after the toxin reaches the neuromuscular junctions of the muscles involved in breathing. Another example is the bacterial toxin that causes diphtheria, an infectious respiratory disease. Diptheria bacteria are breathed into the passages of the lungs and have their greatest effect there. The

reverse transcriptase
an enzyme carried in a retrovirus and activated inside the host cell that transcribes the viral RNA into viral DNA

diptheria toxin is a powerful inhibitor of protein synthesis that causes cells in the lining of the lungs to die. Accumulated dead lung cells can block respiratory passages, causing death.

Many infections have a similar set of symptoms. Fever, runny nose, and watery eyes, fluid and mucus in the lungs, diarrhea, rash, achy muscles, and fatigue can be associated with many different pathogens. Many of these symptoms are part of the immune system's general response to a pathogen. For example, fever likely is an adaptive mechanism that helps the body to fight infection. Fever may kill or inactivate pathogens that are sensitive to higher temperatures. Fever also can activate parts of the immune system, and so can speed recovery from an infection. If these general immune responses are too extreme, they can be harmful. For instance, too much lung congestion can impair breathing, and an extremely high fever can cause brain damage. Even if symptoms are not this extreme, they can be uncomfortable. For these reasons doctors usually treat the general symptoms of an infection.

Of course, pathogenic bacteria will have a bigger impact on a person's health if there are many of them. Once an infectious bacterium enters the body, it can cause sickness only if it multiplies. Normally, the immune system keeps harmful bacteria from multiplying. If a person is stressed, is taking certain medications, or has an impaired immune system, however, pathogenic bacteria that normally might not cause illness may do so. For example, the impaired immune system of a person infected with HIV leads to serious infections that would be fought off by a healthy immune system. The immune system is continually vigilant. On the job 24 hours a day, seven days a week, it protects a person from pathogens that can cause serious illness and death.

QUICK CHECK

1. How do viruses cause illness that is different from how bacteria cause illness?
2. Why were Koch's postulates important?
3. How is a retrovirus different from other viruses?

25.3 The Immune System Provides a Natural Defense Against Pathogens

All organisms have defenses against pathogens, and the most sophisticated examples of these defenses are found in mammals. Like other mammals, the human body has an army of defenders to fight disease. This military analogy is appropriate because the only way to prevent pathogens from making you sick is to kill them. It is not possible or even desirable to kill all of the pathogens in and around the body, but the body can keep the number of pathogens down to an acceptable level that will not cause illness. You can see the effectiveness of these natural defenses in how *infrequently* people actually do get sick, despite being surrounded by all kinds of pathogens in the environment.

The Immune System Has Two Major Components

The **immune system** is composed of cells, tissues, and organs that block, identify, and destroy pathogens. The human immune system has two components. First, *innate immune defenses* work against all pathogens. Second, *acquired immune defenses* react to the specific pathogens that are infecting the body at any particular time. Let's briefly consider both of these lines of immune defenses.

The Innate Immune Response Defends Against All Pathogens

As the name suggests, innate immunity involves inborn defenses against pathogens that operate regardless of a person's experience with any specific pathogen. One line of innate defenses involves *physical barriers* that prevent pathogens from entering the body and eliminates many pathogens that do gain entrance (**Figure 25.9**). Skin is the most important physical barrier to pathogens. Skin is formed of a continuous sheet of tightly connected cells that are bound together by proteins. Unless there is a break in the skin, it blocks the entry of bacteria, viruses, and other pathogens. Parts of the body that have more exposure to the outside world have additional defenses to block the entry of pathogens. For example, cells that line the lungs, the gut, and some other internal body surfaces secrete a thick mucus that forms a barrier that pathogens cannot penetrate easily. Other natural processes also get rid of pathogens. Stomach acids kill pathogens that are swallowed. Pathogens breathed into the nasal passages get trapped in mucus that is coughed up, sneezed out, or swallowed. These and other physical barriers prevent most pathogens from ever penetrating into the body's deeper tissues. Sometimes, though, pathogens do manage to get past the physical barriers of the innate immune system. A cut in the skin or a virus that has adapted to penetrate a mucous barrier are just two ways that pathogens can evade these barriers. Once inside tissues patho-

immune system cells, tissues, and organs devoted to identifying and destroying pathogens that can cause disease

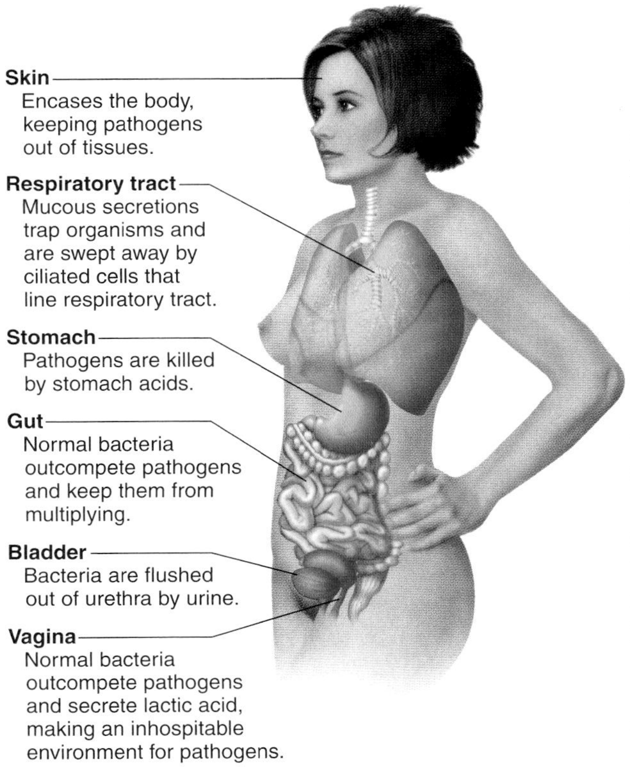

Skin
Encases the body, keeping pathogens out of tissues.

Respiratory tract
Mucous secretions trap organisms and are swept away by ciliated cells that line respiratory tract.

Stomach
Pathogens are killed by stomach acids.

Gut
Normal bacteria outcompete pathogens and keep them from multiplying.

Bladder
Bacteria are flushed out of urethra by urine.

Vagina
Normal bacteria outcompete pathogens and secrete lactic acid, making an inhospitable environment for pathogens.

Figure 25.9 **Physical Defenses Against Pathogens.**

gens can multiply quickly, using the host's energy and resources to live and reproduce. Usually, however, pathogens that reach tissues are stopped by other aspects of the innate immune response and by the acquired immune response. Let's consider each of these in more detail.

In addition to physical barriers, innate immunity also includes cells that react to and destroy pathogens. This more active part of innate immunity involves a variety of white blood cells, or **leukocytes,** that are made in the bone marrow and circulate in the blood. Leukocytes slip out of capillaries to places in tissues where infections are occurring. Leukocytes recognize pathogens by certain common proteins or carbohydrates that major classes of pathogens carry on their surfaces. While there are a variety of leukocytes, you can understand their function by learning about just three of them: mast cells, neutrophils, and macrophages.

Mast cells begin a series of actions that lead to an **inflammation** (**Figure 25.10**), a condition in which fluid and blood cells are drawn to an infected area. This leads to swelling, redness, and an infected part that feels hot to the touch. Inflammation is one of the responses of the innate immune system that causes many different pathogens to produce the same symptoms of illness. One way inflammation is triggered when a mast cell (**Figure 25.11**) encounters a pathogen and secretes *histamines.* These chemicals dilate blood vessels and also increase the permeability of capillary walls. As a result, blood flow to the infection increases, and white blood cells can more easily invade the infected area. With the white blood cells comes excess fluid. While histamines are essential to an effective innate immune response, they also cause discomfort. Histamines are responsible for the runny nose and watery eyes of a cold. They also cause itching when certain insect or plant poisons penetrate the skin. Over-the-counter *antihistamine* drugs block receptors for histamine and so reduce or eliminate these symptoms.

leukocyte a white blood cell involved in the innate immune response

inflammation a local response to infection in which fluid and blood cells are drawn into an infected area, causing swelling, redness, and fever in the infected part

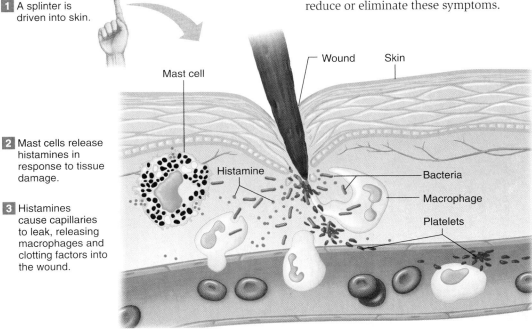

1 A splinter is driven into skin.

Mast cell

Wound   Skin

2 Mast cells release histamines in response to tissue damage.

Histamine

3 Histamines cause capillaries to leak, releasing macrophages and clotting factors into the wound.

Bacteria

Macrophage

Platelets

4 Macrophages engulf bacteria, dead cells, and cellular debris.

5 Platelets move out of the capillary to seal the wounded area.

Figure 25.10 **The Inflammatory Response to Pathogens.** The inflammatory response causes fluid and immune cells to enter the infected area.

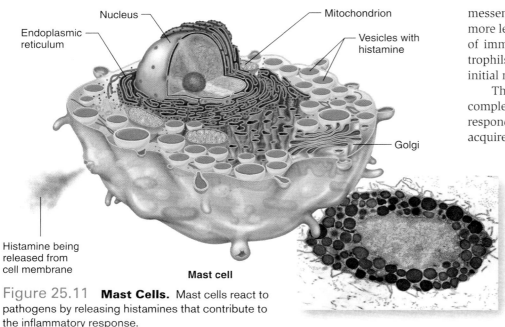

Figure 25.11 **Mast Cells.** Mast cells react to pathogens by releasing histamines that contribute to the inflammatory response.

lymphocyte a white blood cell involved in the acquired immune response

cellular immunity the aspect of the acquired immune system in which infected cells are attacked and killed by lymphocytes dedicated to that purpose

Neutrophils and macrophages are examples of *phagocytes,* cells that directly kill or destroy pathogens by engulfing them, much as an amoeba engulfs its food (**Figure 25.12**). Neutrophils often are more active early in an infection, while macrophages are active later in an infection, but they perform similar functions. If you feel sick and your doctor suspects that you have an infection, a neutrophil count may be done on a sample of your blood. If neutrophil numbers are unusually high, it suggests you have an active infection. Macrophages release chemicals called *cytokines.* These chemical

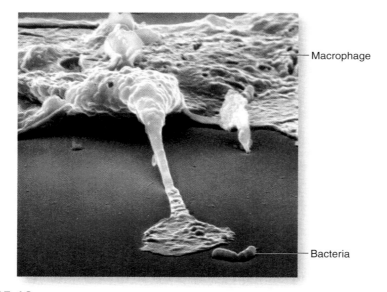

Figure 25.12 **A Macrophage Engulfs Pathogens.**

messengers stimulate the bone marrow to produce more leukocytes, and they also attract various kinds of immune cells to the site of the infection. Neutrophils and macrophages are critical players in the initial response to the presence of pathogens.

The acquired immune response is the most complex of the body's defense systems. Instead of responding in the same way to all pathogens, the acquired immune system is a highly specific response: it attacks only certain pathogens. This response to a certain pathogen develops only after the body has been exposed to that pathogen. The acquired immune response works side by side with the innate immune response.

The Acquired Immune Response Is Directed Toward Specific Pathogens

Figure 25.13 shows an overview of the tissues and cells involved in the acquired immune response. **Lymphocytes** are the cells involved in the acquired immune response. Lymphocytes are another kind of white blood cells that are made in the bone marrow. Some lymphocytes leave the bone marrow as mature cells, but others mature in the thymus, a gland located in the neck. Lymphocytes use two major ways to destroy or inactivate pathogens: *humoral immunity* and *cellular immunity.*

Cellular immunity is one major part of the acquired immune response. Cellular immunity involves lymphocytes such as *cytotoxic T cells.* Instead of destroying the pathogen, lymphocytes of the cellular immune system attack and kill *body cells* that are infected by a pathogen. When a body cell is infected, the pathogen usually inserts its own proteins or other molecules into the cell's membrane. Cytotoxic T cells recognize and bind to these antigens and are activated to kill the infected cell. One way they can kill a cell is to rip holes in its cell membrane. Each kind of cytotoxic T cell recognizes just one kind of antigen and so kills cells infected with just one kind of pathogen.

Humoral immunity involves free proteins, called **antibodies,** that are suspended in blood plasma. Antibodies are molecules that recognize and bind to parts of molecules located on the surfaces of pathogens. A molecule or part of a molecule that antibodies bind to is called an **antigen.** Most antigens are proteins, but a carbohydrate or specific kind of lipid also can be an antigen. Each kind of antibody will bind to only one specific antigen. An antibody molecule is shaped like a Y: it has a stem and two arms (**Figure 25.14**). The arms

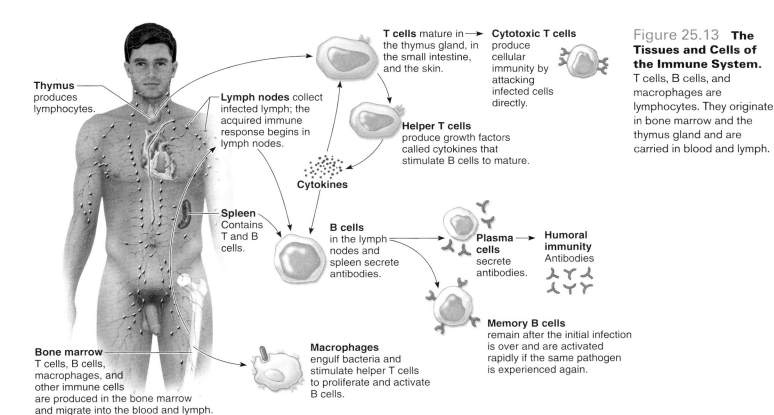

Figure 25.13 **The Tissues and Cells of the Immune System.** T cells, B cells, and macrophages are lymphocytes. They originate in bone marrow and the thymus gland and are carried in blood and lymph.

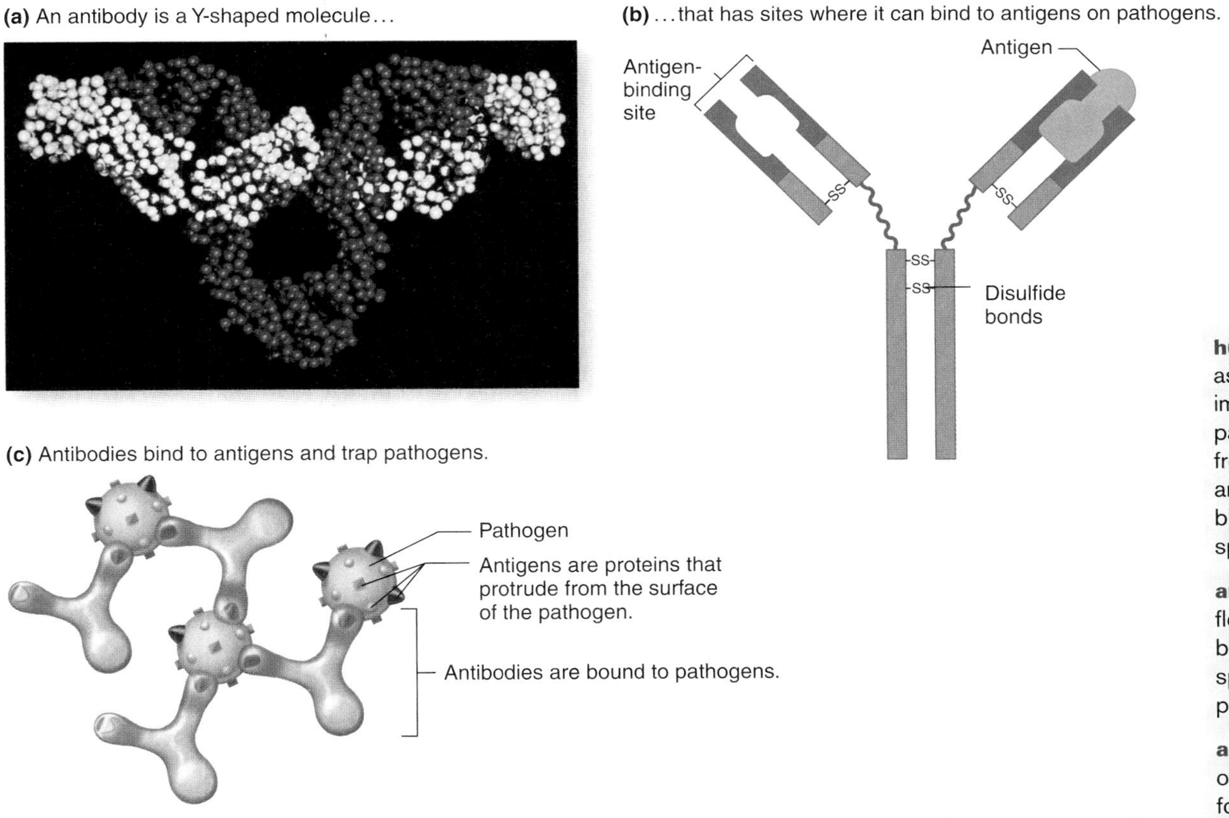

Thymus produces lymphocytes.

Lymph nodes collect infected lymph; the acquired immune response begins in lymph nodes.

Spleen Contains T and B cells.

Bone marrow T cells, B cells, macrophages, and other immune cells are produced in the bone marrow and migrate into the blood and lymph.

T cells mature in the thymus gland, in the small intestine, and the skin.

Cytotoxic T cells produce cellular immunity by attacking infected cells directly.

Helper T cells produce growth factors called cytokines that stimulate B cells to mature.

Cytokines

B cells in the lymph nodes and spleen secrete antibodies.

Plasma cells secrete antibodies.

Humoral immunity Antibodies

Memory B cells remain after the initial infection is over and are activated rapidly if the same pathogen is experienced again.

Macrophages engulf bacteria and stimulate helper T cells to proliferate and activate B cells.

(a) An antibody is a Y-shaped molecule...

(b) ...that has sites where it can bind to antigens on pathogens.

Antigen-binding site

Antigen

-SS-

Disulfide bonds

(c) Antibodies bind to antigens and trap pathogens.

Pathogen

Antigens are proteins that protrude from the surface of the pathogen.

Antibodies are bound to pathogens.

Figure 25.14 **Structure and Function of Antibodies.** Antibodies are complex Y-shaped proteins. The two arms of the Y bind to antigens. The stem of the Y sticks out from the antigens and stimulates other immune cells to attach the pathogen.

humoral immunity the aspect of the acquired immune system in which pathogens are attacked by free proteins, called antibodies, that float in the bloodstream and in the spaces between cells

antibody a protein that floats freely in tissue fluids, binds to an antigen, and speeds the destruction of pathogens

antigen a molecule or part of a molecule from a foreign organism that antibodies or cells of the immune system can bind to carry the active part of the antibody that binds to the antigen. When this has happened, the stem of the antibody sticks out from the surface of the pathogen like a pin sticks out of a pincushion. Because any particular pathogen such as a bacterium or a virus will have many antigen molecules on its surface, it can have hundreds of antibody molecules bound to them. Antibodies help to inactivate pathogens in at least three different ways.

First, when the surface of a viral pathogen is coated with antibodies, it is less likely to be able to infect a target cell. Second, the stems of the antibody molecules interact with leukocytes and lymphocytes and make it easier for them to destroy pathogens. Third, antibodies interact with **complement,** another system of blood proteins that interact with and kill pathogens that have antibodies bound to them. Complement builds a large protein structure on the pathogen, sometimes called a *membrane attack complex.* The complex sends a tube into the pathogen, across its lipid membrane. Water enters the tube by osmosis and the pathogen bursts. These three actions of antibodies make them powerful players in the acquired immune response. The story of how antibodies are made by special classes of lymphocytes is another phase of the acquired immune response.

Antibodies Are Made Against Specific Pathogens

The best way to understand how antibodies are made is to consider how the humoral immune response as a whole reacts to a particular pathogen. In this example you will see how the innate and acquired immune systems can work together to combat an infection. Consider the body's response to a cold or flu virus (**Figure 25.15**). When they are breathed into the lungs, the airborne cold or flu viruses encounter the protective mucous barrier that covers the cells that line respiratory passages. Many flu viruses, however, carry surface enzymes that can digest a path through the mucus. Once a virus enters lung tissue, mast cells and other leukocytes react to its presence. As part of the inflammatory response mast cells release histamines and other chemicals that draw more leukocytes to the infected area. Capillary walls become leaky, and leukocytes and extracellular fluid slip through the walls of capillaries. Inflammation is one reason that during the course of this infection a person coughs, sneezes, and may have a runny nose.

Macrophages are among the leukocytes attracted to the infection. They are highly mobile and can roam throughout the body, ready to fight an infection. While macrophages will engulf and kill any pathogen, they play another important role as well. When a macrophage ingests a pathogen, the antigens and other molecules in the pathogen are broken into small pieces and are displayed on the membrane of the macrophage. Special proteins embedded in the cell membrane hold the antigens away from the surface of the macrophage, displaying them like flags. Macrophages and other cells that do this are called **antigen-presenting cells,** meaning that they display antigens to other immune system cells (**Figure 25.16**).

What good does this do? Once it has digested a pathogen, the antigen-presenting cell leaves the infection site, carried in the bloodstream and or in lymph. The **lymphatic system** is a network of fluid-filled vessels that are much like blood

complement a set of proteins in blood that work together to kill cells that have antibodies attached to them

antigen-presenting cell any immune system cell that acts like a macrophage and displays foreign proteins from pathogens on its surface

lymphatic system a network of vessels that drain lymphatic fluid throughout the body

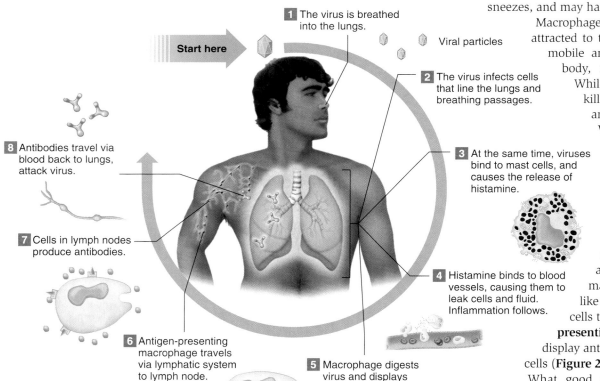

1 The virus is breathed into the lungs.

Start here

Viral particles

2 The virus infects cells that line the lungs and breathing passages.

3 At the same time, viruses bind to mast cells, and causes the release of histamine.

4 Histamine binds to blood vessels, causing them to leak cells and fluid. Inflammation follows.

5 Macrophage digests virus and displays viral antigens.

6 Antigen-presenting macrophage travels via lymphatic system to lymph node.

7 Cells in lymph nodes produce antibodies.

8 Antibodies travel via blood back to lungs, attack virus.

Figure 25.15 **The Response of the Immune System to a Cold Virus.**

(a) A macrophage with nonspecific antigen receptors

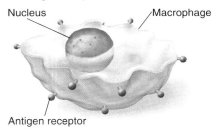

Nucleus

Macrophage

Antigen receptor

(b) A virus with surface proteins binds to macrophage receptors.

Virus

Figure 25.16
A Macrophage Is An Antigen Presenting Cell. Antigen presenting cells, such as macrophages, display antigens on their surface and alert the immune system to the presence of specific antigens.

(c) The macrophage engulfs the virus.

(d) The macrophage digests or breaks up the virus and displays viral proteins on its surface. These can stimulate the production of antibodies.

Remains of virus

Viral protein

vessels (**Figure 25.17**). The lymphatic system drains excess fluid from the entire body, and this fluid, called *lymph,* contains antigen-presenting macrophages from any sites of infections. The lymphatic system has no heart to pump the clear lymphatic fluid through the body; instead muscle contractions squeeze lymph vessels, and this keeps lymph moving. Lymphatic fluid from infected areas filters through **lymph nodes,** small glands scattered throughout the body. Concentrations of lymph nodes are found in the neck, armpits, and groin (Figure 25.17). If an infection is severe, the lymph nodes will swell as lots of lymph drains into them. When you have an infection, if you press on your neck, armpits, or groin, you can feel "swollen glands"—enlarged lymph nodes that are close to the infection. Once antigen-presenting cells have drained into lymph nodes, they activate the portion of the acquired immune response that takes place there. Antigen-presenting cells that are in the bloodstream accumulate in the **spleen.** The spleen is a lymph organ similar to lymph nodes, except that it filters blood rather than lymph. Antigen-presenting cells in the spleen also can activate an acquired immune response. Lymph drains into the large veins that bring blood back to the right side of the heart.

The activation of an acquired immune response in the lymph nodes or spleen involves several different cell types, including antigen-presenting cells, helper T cells, and B cells. **Figure 25.18** diagrams the process, using the production of antibodies as an example. In the lymph nodes and spleen, antigen-presenting cells make contact with lymphocytes called **helper T cells.** These cells get the *T* in their name because they mature in the thymus gland. Helper T cells bind to the antigens

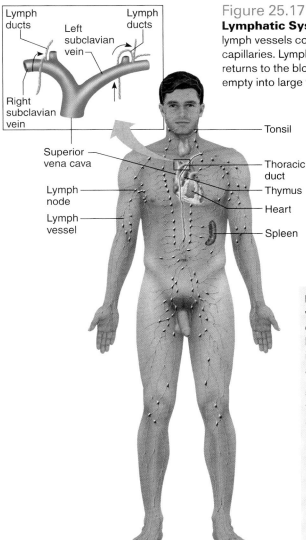

Lymph ducts

Left subclavian vein

Lymph ducts

Right subclavian vein

Superior vena cava

Lymph node

Lymph vessel

Tonsil

Thoracic duct

Thymus

Heart

Spleen

Figure 25.17 **The Human Lymphatic System.** A network of lymph vessels collects fluids that leak from capillaries. Lymph nodes filter lymph. Lymph returns to the bloodstream in ducts that empty into large veins that lead to the heart.

lymph node a small gland where lymph drainage is concentrated and one location where an acquired immune response is activated

spleen a lymphatic organ that drains blood and activates an acquired immune response

helper T cell a cell of the acquired immune system that is stimulated by a macrophage to divide and, in turn, stimulates B cells to divide, mature, and produce antibodies

(a) Helper T cells are activated by antigen-presenting macrophages.

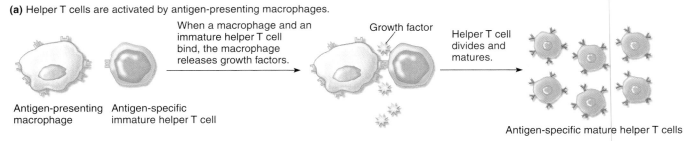

Antigen-presenting macrophage

Antigen-specific immature helper T cell

When a macrophage and an immature helper T cell bind, the macrophage releases growth factors.

Growth factor

Helper T cell divides and matures.

Antigen-specific mature helper T cells

(b) Helper T cells are activated by antitgen-presenting macrophages.

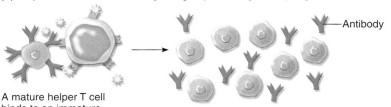

A mature helper T cell binds to an immature antigen-specific B cell. The T cell produces growth factors that bind to the B cell.

Antibody

This stimulates the antigen-specific B cell to divide, mature, and produce antibodies.

Figure 25.18 **The Activation of the Acquired Immune Response.**

already know the answer. Antigen-presenting macrophages carry bits of antigen to the lymph nodes and spleen. There they interact and activate only with helper T cells that carry membrane receptors for that particular antigen. Then the helper T cells activate *specific* B cells and *specific* cytotoxic T cells. A single activated helper T cell can multiply and create many specific activated helper T cells that, in turn, can activate hundreds of specific lymphocytes. At the end of this process there are millions of activated lymphocytes, especially B cells that make specific antibodies.

The process that results in acquired immunity has advantages and disadvantages. The biggest disadvantage is that it can take 7 to 10 days from an initial infection for an effective acquired immune response to be mounted. A person can die from an infection before the immune system has a chance to rally defenses that will fight it off. The biggest advantage is that the body can make lymphocytes to mount a vigorous attack on a pathogen without wasting energy on lymphocytes it does not need. Another advantage is that once B cells to a particular pathogen have been activated, some are held in reserve in case the body encounters that pathogen again (**Figure 25.19**). When these **memory B cells** encounter the pathogen a second time, they quickly begin to make antibodies. This is why a person usually has an infectious disease only once. Childhood chickenpox is a good example. Although the chickenpox virus frequently is encountered, most people get chickenpox just once in their lives. As you will see later, the activation of memory B cells also is the basis for vaccines.

The Immune System Identifies Self and Nonself

The immune system avoids killing the body's own cells by screening out any lymphocytes that might react to them. As lymphocytes mature in the thymus gland, immature T and B cells that react to the body's own molecules are killed or inactivated.

on the surface of the antigen-presenting cells. This stimulates helper T cells to divide many times. In turn, the mature helper T cells interact with other kinds of lymphocytes. Some helper T cells bind to cytotoxic T cells that are part of the innate immune response. These helper T cells stimulate the cytotoxic T cells to divide and mature so that they are ready to destroy the infecting pathogen.

Helper T cells also interact with B cells, another kind of lymphocyte. The name **B cell** stands for *bursa cell.* The bursa is a special organ in birds that make these cells. Even though mammals do not have a bursa, and B cells of mammals are made in the bone marrow, researchers have agreed to use the same name for these cells in mammals. B cells make antibodies. In both the lymph nodes and the spleen, helper T cells stimulate B cells to divide and mature. Once B cells are mature, they produce antibodies that are released into the bloodstream and lymphatic fluid. Antibodies attach to pathogens and speed their destruction by leukocytes and complement.

But how does this process produce antibodies against *specific* pathogens? During development and all through life, helper T cells arise that can bind to *every possible antigen a person ever could encounter.* This is astounding, but this conclusion is well documented by many research studies. If you think a moment, though, you will realize that it would take a lot of energy to maintain large numbers of each of these different kinds of helper T cells alive within the body. Instead, large numbers of any particular kind of helper T cell are made *only* when the pathogen that they can kill is present. How do helper T cells know if a pathogen is present? If you look back a few paragraphs, you will see that you

B cell a cell of the acquired immune system that produces antibodies

memory B cell a B cell that is partially activated by a pathogen and remains in reserve to respond more rapidly to that pathogen if it is encountered again

Sometimes, though, this screening process does not work perfectly, and some self-detecting lymphocytes survive. These can cause **autoimmune diseases,** such as Type 1 diabetes mellitus, multiple sclerosis, and lupus. Other diseases may turn out to have an autoimmune component.

The immune reaction also can pose problems in some medical situations. For example, someone who receives an organ transplant from another person, even a close relative, risks that the organ will be rejected as the immune system detects foreign proteins and attacks the cells of the transplanted organ that carry them. Transplant recipients must take cortisol and other drugs that reduce the effectiveness of the immune system.

QUICK CHECK

1. List the chronological events of a typical immune response and explain why the immune response is not an instantaneous reaction.

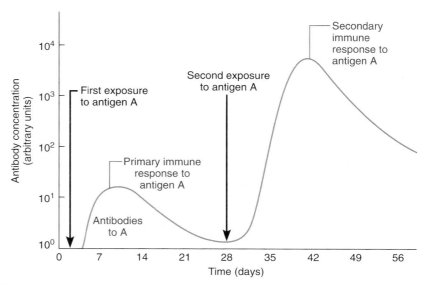

Figure 25.19 **Acquired Immunity.** The second time the immune system is exposed to an antigen the immune response is faster and more effective.

25.4 Humans Have Discovered Treatments That Control Infectious Diseases

Although the immune system is a marvel of biological evolution, pathogens still infect and kill millions of people. In Western cultures prior to the twentieth century, life expectancy was only 30 or 40 years in part because so many people died of infectious diseases early in life. This section explores three major approaches to treat or prevent infectious diseases: antibiotic drugs, vaccines, and antiviral drugs.

Serious Bacterial Infections Are Treated with Antibiotics

The last 50 years have produced a revolution in the treatment of bacterial infections. *Antibiotics* are the miracle drugs that have turned bacterial infections from deadly scourges to bothersome annoyances for most people. An antibiotic is a substance that kills bacterial cells. Antibiotics can kill other pathogens as well, but they cannot kill viruses. The most common use of antibiotics is to treat bacterial infections. Antibiotics are small molecules that interfere with the chemical reactions involved in key cellular processes. Each class of antibiotics works in a slightly different way. Penicillin blocks a critical enzyme that binds strings of sugars and amino acids together and causes bacteria to build faulty cell walls. As a result, the weak cell wall ruptures, and

the bacterium bursts and dies. In contrast, antibiotics called tetracyclines bind to ribosomes and keep tRNA molecules from attaching to them. The affected bacterium will die because it cannot synthesize proteins. Still other antibiotics interfere with the bacterial cell membrane, nucleic acid synthesis, energy metabolism, or different aspects of bacterial metabolism such as DNA synthesis. All antibiotics are effective with only minimal side effects because while bacterial biochemistry is similar to that of humans, it is not identical. So antibiotics do much more damage to bacterial cells than they do to human cells.

Despite the success of antibiotics, the development of antibiotic resistance is a major concern. **Antibiotic resistance** is the ability of a strain of bacteria to adapt so that it is unaffected by an antibiotic's killing effect. For example, penicillin is a powerful antibiotic that once was the preferred treatment for many bacterial infections. By the end of the last century, however, 30 to 50% of samples of bacteria that cause upper respiratory infections had become resistant to penicillin. Antibiotic resistance can happen because bacteria acquire a gene that allows them to destroy the penicillin molecule. Or resistant bacteria can have a slightly different chemical structure in their cell walls, or some other aspect of their biochemistry allows them to be unaffected by the antibiotic. Such genes are carried on plasmids.

Antibiotic resistance is a particular problem in hospitals. The Centers for Disease Control and Prevention estimates that over 2 million people acquire an infection in a hospital each year, and

autoimmune disease a disease in which the immune system attacks the body's own cells

antibiotic resistance the ability of some bacterial strains to avoid being killed by a particular antibiotic

90,000 people die of that infection. Most of these deaths are caused by bacterial pathogens that survive exposure to antibiotics. Over 70% of the bacterial species that cause hospital infections are resistant to one or more antibiotics. As you can see, antibiotic resistance is a serious public health problem.

Antibiotic resistance is an example of natural selection, and both frequent use of antibiotics as well as underuse of antibiotics can lead to antibiotic resistance. Routine or frequent antibiotic use by a person or by a population will kill any pathogens susceptible to the drug, leaving alive the less susceptible pathogens. If the surviving pathogens multiply, the resulting illness could become serious because available antibiotics will not kill the pathogens. Such a pattern was documented in eastern European counties in the 1970s. In some countries antibiotics were sold over the counter and people used them for even minor illnesses. As a result, widespread antibiotic resistance developed. More stringent controls later were placed on the sale of antibiotics, and antibiotic resistance declined. This is the reason that your doctor will be careful about prescribing antibiotics. If a physician believes that you will get well in a few days, or has reason to believe that your illness is caused by a virus and not by a bacterium, antibiotics probably will not be prescribed.

Underuse of antibiotics can happen when a person does not take all of a prescription. Once antibiotic treatment has been started, a week or two of a sustained dose is needed to kill most of the pathogens. If the treatment is cut short, the most susceptible bacteria will be killed, and those with some resistance will remain. The underuse of antibiotics is a particular problem in treating tuberculosis.

Vaccines Activate the Body's Immune Response

Treatments that kill infectious bacteria have been available for over 50 years and have revolutionized the health of modern humans. However, there are only a few effective antiviral treatments. Despite this lack of antiviral drugs, some viral diseases (notably smallpox and polio) have been eradicated.

How were these viral diseases eradicated without antiviral drugs? The answer is **vaccines,** an extremely clever medical treatment strategy that uses the immune system to prevent a viral infection before it occurs. **Figure 25.20** shows an example of how a measles vaccine works. Measles is a serious childhood disease that can cause severe illness and death. A vaccine administers a small amount of infectious material. The vaccine can contain a complete virus particle, a weakened virus that has been damaged in some way, a partial virus, or even just a protein from a virus. The viral agent in the vaccine is not enough to cause serious illness, but it will stimulate the immune system to mount a defense, just as it would against the actual pathogen. Antigen-presenting cells activate helper T cells, which, in turn, stimulate B cells to multiply and produce antibodies. At the same time memory B cells are produced and preserved, ready for the next encounter with the par-

vaccine a weakened pathogen that elicits a mild immune response and thereby confers immunity to that pathogen when it is encountered again

(a) The virus that causes measles

(b) The weakened virus in a measles vaccine stimulates mature B cells that produce antibodies and memory B cells. Most of the antibodies eventually are cleared from the body after the weakened virus is destroyed, but the memory B cells remain. Repeated vaccinations are given to maximize the number of memory B cells.

(c) If the vaccinated person is exposed to the real virus, the memory B cells produce a rapid antibody response, and the person does not become ill.

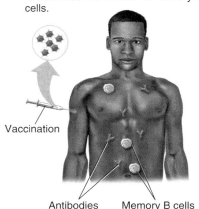

X150,000

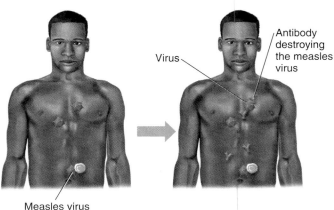

Vaccination

Antibodies   Memory B cells

Measles virus

Virus

Antibody destroying the measles virus

Figure 25.20   **How Vaccines Produce an Immune Response.**

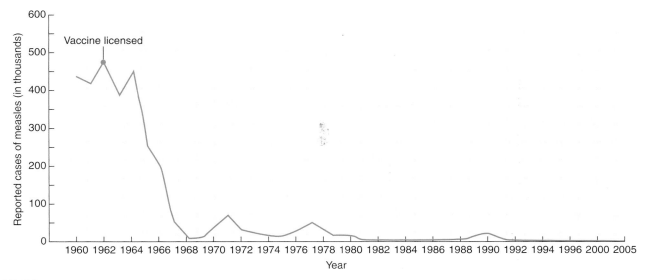

Figure 25.21 **Measles Cases After the Introduction of the Measles Vaccine.** After the introduction of the measles vaccine in 1962 the number of measles cases dropped dramatically.

ticular virus that the vaccine came from. After the measles vaccine was introduced in the United States in 1962, the number of cases of measles declined dramatically (**Figure 25.21**). Vaccines also have been developed against many bacterial diseases such as diptheria and typhoid fever.

In most cases, if a vaccine causes any symptoms, they are mild such as a low fever and body aches. In a few individuals, however, vaccines have caused full-blown disease. This worries some people and encourages others to avoid vaccines. Because you have a greater chance of becoming seriously ill from a virus if you are *not* vaccinated, it is not smart to avoid vaccinations. Worse, if many people avoid a vaccine, a viral disease can spread and become a major public health concern. There is little reason to avoid the vaccines that are widely recommended today. These vaccines provide protection against measles, mumps, diphtheria, whooping cough, hepatitis B, and other viral infections.

Antiviral Drugs Are New Weapons Against Infectious Diseases

Aside from vaccines, drug treatments for viral infections have lagged behind drug treatments for bacterial infections. Because viruses are not alive and have no internal metabolic processes of their own, the antibiotics that interfere with cell metabolism will not work on viruses. The recent advances in antiviral drugs have been achieved *not* by the trial-and-error testing of chemicals but by the direct application of knowledge of how viruses work. Each antiviral drug is targeted against a specific molecule associated with a specific kind of virus. Most viral infections still have no treatments,

but as knowledge of viruses increases, more antiviral drugs will become available.

How do antiviral drugs work? Like antibiotics, antiviral drugs are chemicals that interfere with a virus's basic chemical processes. One class of drugs acts on influenza viruses that cause many cases of flu each year. When these viruses infect a body cell, they reproduce using the cell's organelle's, energy, and metabolism (see Section 25.1). Once a new viral particle is assembled, it moves through the cell membrane and is released to the outside environment, ready to infect another cell. Before it is released, though, the virus is attached to the cell membrane by a chemical bond between a protein on the cell surface and a protein on the viral surface. The virus controls its own release from the cell by breaking this bond, using an enzyme called *neuraminidase* (**Figure 25.22**). From this description you can probably predict how this sort of antiviral flu drug works. These drugs bind to and inhibit the enzyme neuraminidase. Once the drug has been bound to the enzyme, neuraminidase no longer can cut the bond that holds the virus to the cell surface. So the virus is stuck there and cannot infect more body cells. These kinds of drugs prevent the spread of the virus but do not prevent initial infection. While these drugs cannot prevent people from getting sick, they can reduce symptoms and, more importantly, they can reduce the likelihood that the virus will spread. Other kinds of antiviral drugs target other aspects of viral reproduction.

QUICK CHECK

1. What is antibiotic resistance and how does it develop?

(a) The influenza virus has many proteins on its outside membrane, including neuraminidase.

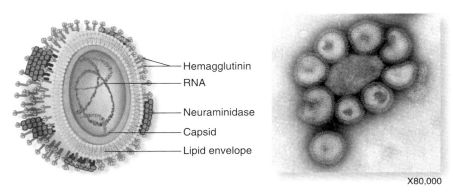

- Hemagglutinin
- RNA
- Neuraminidase
- Capsid
- Lipid envelope

X80,000

(b) Neuraminidase cuts the virus away from the host cell.

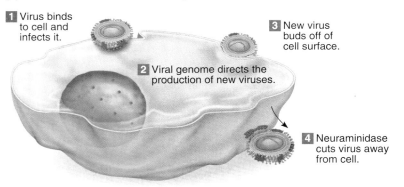

1 Virus binds to cell and infects it.

3 New virus buds off of cell surface.

2 Viral genome directs the production of new viruses.

4 Neuraminidase cuts virus away from cell.

(c) An antiviral drug blocks neuraminidase and prevents viruses from leaving infected cell.

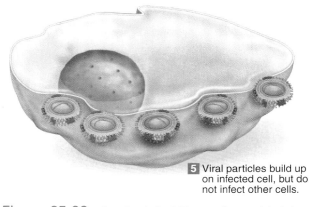

5 Viral particles build up on infected cell, but do not infect other cells.

Figure 25.22 **An Antiviral Drug.** One antiviral drug works by blocking the action of the viral enzyme neuraminidase.

■ *What happens to a virus-infected cell?*

acquired immune deficiency syndrome (AIDS) a disease defined by a collection of rare infections caused by a collapse of the immune system; AIDS is not an inherited disease

25.5 AIDS Is a Syndrome of Infections Caused by HIV

AIDS exploded onto American society in the early 1980s, when the Centers for Disease Control and Prevention (CDC) in Atlanta began to see odd patterns in reports of infectious diseases. These cases were unusual because they were rare infections that usually developed in people with impaired immune systems. As the number of these odd cases rose, the syndrome was given the name **acquired immune deficiency syndrome,** or **AIDS.** The name says three important things. First, the disease is a syndrome, or a mix of different symptoms. Second, the syndrome involves a problem in the immune response to infection. Third, the immune deficiency is acquired through some experience, not inherited. As the AIDS epidemic grew, it amply demonstrated what life without an immune system would be: short and filled with a relentless string of miserable infections. In the United States the AIDS epidemic spurred a national research effort like no other before it. Within a few years it was determined that AIDS is caused by a retrovirus—**HIV,** or **human immunodeficiency virus**—and the complete genome and protein components of the virus were described. New treatments recently have been developed, and several AIDS vaccines are under development.

New drug treatments have allowed people with AIDS to minimize effects of the virus, and many who have access to these drugs can live for years with AIDS, if they can bear the intense side effects of these drugs. Advances in drug treatments may make you think that AIDS no longer is a major health problem. Nothing could be further from the truth. In many parts of the world, particularly southern (sub-Saharan) Africa and parts of Southeast Asia, the AIDS epidemic kills tens of thousands of people each year, and in these countries AIDS has decimated the populations of young adults. In other parts of the world such as India, China, and Russia, AIDS infections are still growing. In the United States about 32,000 new cases of AIDS are reported each year, and many of these are teenagers. AIDS continues to present an important public health problem, so it is important to know how it spreads, how it is treated, and how to avoid getting AIDS. In addition, current knowledge of AIDS far surpasses knowledge of any other viral infectious disease, so you can learn a great deal about viruses by a deeper study of AIDS.

Several Lines of Evidence Support the Conclusion That HIV Causes AIDS

Because of the multiple infections that AIDS patients typically suffer, early researchers suspected that AIDS was caused by an immune deficiency, and counts of immune cells in blood samples of people with AIDS confirmed this suspicion. AIDS patients typically have an extremely low number of *CD4 cells,* a specific kind of helper T cell. Recall that helper T cells stimulate B cells to multiply, mature, and produce antibodies. Without these helper T cells the immune system cannot fight common pathogens. The pathogens eventually win the battle, and the patient dies. The CD4 cells that are lost in AIDS get their name from a protein that they have on their surfaces.

In the early years of the U.S. AIDS epidemic, biologists noticed that most cases of AIDS were concentrated in large population centers, which strongly suggested that AIDS was caused by an infectious pathogen. In 1983, just a few years after the first reports of AIDS, two scientists independently identified a retrovirus as the cause of illness and named it HIV. Despite the isolation of HIV in AIDS patients, some people in and out of the scientific community did not accept an infectious cause for AIDS. You might think that such disregard for scientific results would have been met with hostility from scientists, and perhaps there was some, but the more typical response was to test the ideas all the more rigorously and let the evidence eventually decide the case.

The most convincing evidence that HIV causes AIDS in humans is the persistent correlation between the presence of the virus, the loss of CD4 cells, and the symptoms of AIDS. Indeed, the virus has been isolated from *every* person tested who displays the AIDS infections, who has reduced numbers of CD4 T cells, and who does not have any other known reason for the symptoms of AIDS, such as an inherited disorder. **Figure 25.23** shows this relationship. HIV infection begins with a rapid rise in HIV particles in the blood, and then this number declines quickly. Then levels of HIV gradually rise, and levels of CD4 cells show a parallel decline. The symptoms of AIDS closely track this second rise in HIV and decline in CD4 cells. This pattern has been documented in millions of people infected with HIV. While there are a few exceptions, well over 99% of those infected with HIV eventually develop AIDS. Furthermore, people who reliably test negative for the virus do not develop the symptoms of AIDS. Indeed, the current diagnosis of AIDS relies on the presence of HIV antibodies coupled with loss of CD4 cells.

The argument that HIV causes AIDS also is supported by the way that AIDS is spread. Specifically, HIV is transmitted in genital fluids such as semen or vaginal secretions, in blood, and in breast milk. There is little evidence for HIV transmission in saliva, tears, or sweat. It turns out that genital secretions and blood have high concentrations of the virus, while other body fluids have negligible levels of HIV.

Worldwide, the patterns of infection reflect these underlying mechanisms. In the United States the AIDS epidemic initially was recognized in the male homosexual community, where it first spread rapidly. Anal intercourse is a particularly effective way of transmitting HIV because the high concentration of virus in semen has access to the recipient's bloodstream through small rips in the lining of the rectum. The ready transmission during anal intercourse, however, should not overshadow the fact that the virus can be transmitted in other ways too. Most HIV transmission in the world is between heterosexuals, not homosexuals. The important factor is not sexual orientation but the number of sexual partners. The more sexual partners an infected person has, the more other people are infected. Also, the more sexual partners an

human immunodeficiency virus (HIV) the retrovirus that causes the collapse of the human immune system and thus causes AIDS

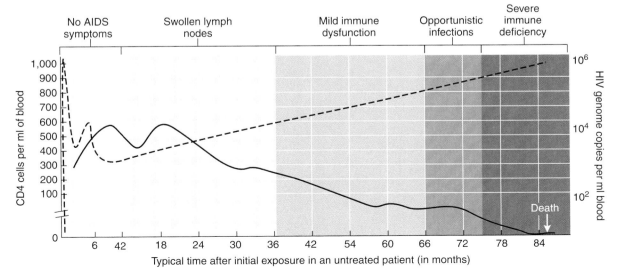

Figure 25.23 **Events in the Course of an Untreated HIV Infection.** The symptoms of AIDS correlate with the effect of HIV on CD4 cells.

uninfected person has, the greater are the chances that she or he will become infected with HIV.

Transmission by contaminated injection needles also is a major way that HIV is spread. Drug addicts who share needles readily transmit or acquire HIV. In countries that lack modern medical facilities, transmission of HIV even can occur in health care settings, where unsterilized injection needles are used repeatedly on different patients. Although the U.S. blood supply now is routinely tested for HIV, blood transfusions also have been sources of HIV transmission. The other major routes of HIV transmission involve infected mothers and their children. HIV can pass from an infected mother to her unborn child across the placenta. It also can pass from infected mother to infant in breast milk. Both methods of transmission lead to the tragic deaths of infants and children from AIDS.

Knowledge of how HIV is transmitted allows health officials to calculate the odds of contracting the virus (**Table 25.2**). First, as previously noted, HIV infection can be transmitted in both homosexual and heterosexual sexual encounters. Second, risk of transmission from infected males to their partners is greater than transmission from infected

females to their partners. This is because HIV is more concentrated in seminal fluid than in vaginal secretions and because the lining of the rectum and vagina are likely to suffer small rips during intercourse, giving the virus direct access to the bloodstream. Third, condoms dramatically reduce the chances of HIV transmission, but they do not eliminate HIV transmission completely. Finally, the best way to avoid HIV transmission is through sexual abstinence. The next best approach is to have just one exclusive sexual partner who is HIV-negative.

HIV Causes AIDS by Infecting and Killing Helper T Cells

The intense research effort directed toward the AIDS epidemic since the 1980s has yielded a deep understanding of how HIV causes the immune system to fail. Much of what has been learned is captured in the following example, in which HIV infection begins with a sexual encounter between an infected man and an uninfected woman. During intercourse small rips in the vaginal lining occur, allowing pathogens, including HIV, from the male to enter the woman. Once the virus enters cells of the vaginal lining, it activates macrophages and other similar cells (**Figure 25.24**). As in any infection, these antigen-presenting cells engulf the virus, display the HIV proteins on their surfaces, and begin activation of the immune response to fight that particular pathogen. In the case of HIV, however, something else happens. *HIV also infects the macrophages* because the macrophages carry CD4 receptors on their surface.

The antigen-presenting macrophages infected with HIV travel to the lymph nodes. This starts an immune response to HIV. Helper T cells are stimulated to proliferate and, in turn, they stimulate B cells to produce antibodies against HIV. Indeed, the presence of such antibodies is the basis for the test that determines whether a person is HIV-positive. If you are HIV-positive, you have antibodies to HIV in your bloodstream.

Unfortunately, an infected macrophage acts like a Trojan horse because it will harbor thousands of new HIV particles. These emerge within a lymph node, where the real battle begins. Each HIV particle released from an incoming macrophage finds a galaxy of CD4 helper T cells to infect and kill. Each infected helper T cell releases thousands more HIV particles, and these infect other helper T cells. The process can take years, but gradually the population of infection-free helper T cells will be reduced to lower and lower levels. As HIV infection kills more and more helper T cells, the infected person loses the ability to fight infections.

Table **25.2** The Odds of Contracting HIV Under Different Conditions		
HIV status of your partner	**Approximate risk in 1 sexual encounter**	**Approximate risk in 500 sexual encounters**
HIV status of partner is not known		
Partner not in high risk group*		
Using a condom	1 in 50 million	1 in 110,000
Not using a condom	1 in 5 million	1 in 16,000
Partner in high risk group		
Using a condom	1 in 50,000	1 in 100
Not using a condom	1 in 5,000	1 in 25
Partner is HIV-negative		
No history of high-risk behavior		
Using a condom	1 in 5 billion	1 in 11 million
Not using a condom	1 in 500 million	1 in 1.6 million
History of high risk behavior		
Using a condom	1 in 500,000	1 in 1100
Not using a condom	1 in 50,000	1 in 160
Partner is HIV-positive		
Using a condom	1 in 5,000	1 in 11
Not using a condom	1 in 500	2 in 3

*High-risk group or behavior is defined as your partner being in groups such as homosexual or bisexual males, injection drug users, hemophiliacs, or female prostitutes.

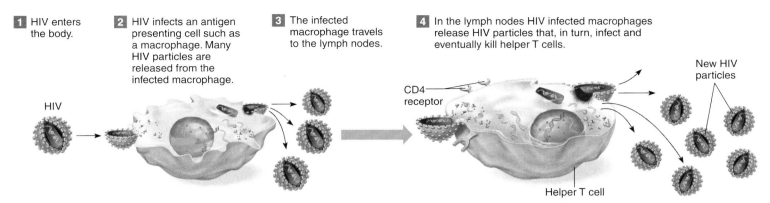

1 HIV enters the body.

2 HIV infects an antigen presenting cell such as a macrophage. Many HIV particles are released from the infected macrophage.

3 The infected macrophage travels to the lymph nodes.

4 In the lymph nodes HIV infected macrophages release HIV particles that, in turn, infect and eventually kill helper T cells.

HIV

CD4 receptor

New HIV particles

Helper T cell

Figure 25.24   **HIV Infection.** HIV is a retrovirus that first infects antigen-presenting cells and then infects helper T cells.

Knowledge About HIV Explains Some Puzzling Aspects of AIDS

One intriguing difference between people is how long it takes for HIV infection to develop into AIDS. Typically, it takes a year or two for the symptoms of AIDS to develop, but in some people it takes much longer. Indeed, some people live for 15 years or longer with minimal symptoms while carrying the virus, a condition known as a *latent infection.* Several aspects of HIV infection contribute to long latent periods that some people with HIV experience.

First, the body has millions of helper T cells, and bone marrow makes more on a regular basis. In the early stages of the infection, the bone marrow can keep up, and lost helper T cells are continually replenished. Only when the balance tips and the number of helper T cells drops dramatically, do *opportunistic infections* become overwhelming. Opportunistic infections are those that would not affect a person with a normal immune system, but do affect a person with a compromised immune system. Second, viral DNA from HIV is not always transcribed immediately. It usually is spliced into the host cell's genome. When this happens, the spliced HIV DNA may not be transcribed until that region of the host DNA is transcribed. For example, helper T cells do not normally transcribe and translate much of their DNA until they are activated by a macrophage. If HIV infects an unactivated helper T cell, the viral DNA may not be transcribed until the specific antigen-presenting macrophage activates it. Thus, the viral DNA can lie dormant within a cell for many years.

Finally, the initial HIV infection does stimulate an immune response, especially the production of antibodies. The effectiveness of this immune system response will partially determine how long an infected person will stay free of symptoms. Despite the immune system's attack on HIV, the virus usually ultimately wins because it is killing the cells needed to fight off other kinds of infections. Even though antibodies are produced, eventually helper T cells are killed at an ever-increasing rate, and the immune system's ability to fight other infections is lost.

Another mystery is that some people are infected easily, while others are not. For instance, some people are infected by a single exposure to HIV, while others can be in a sexual relationship with an HIV-positive individual for many years without becoming infected. Recent findings on how the virus infects cells give some hints about the mechanism behind their resistance to the virus. During HIV infection a protein called *gp120,* located on the surface of the virus, binds to CD4 on the host cell. Another protein, called *CCR5,* also is required for HIV to bind to the host cell. The specific function of the CCR5 protein is not known, but its relationship to HIV infection is interesting. Most people have lots of CCR5, while others have less of it, and a small percentage of people completely lack CCR5 protein. The important thing is that people without CCR5 are *not susceptible* to HIV infection. Such people can be exposed to HIV, but it never infects any of their immune cells. It turns out that people of European descent are more likely to lack CCR5 than other populations, but of course this does not mean that you are immune to AIDS if your ancestors came from Europe. Most people have the CCR5 receptor and will be infected if exposed to HIV.

Treatments for HIV Infection Have Cut U.S. Death Rates from AIDS in Half

At the time of this writing there are three kinds of treatments that can control HIV infection, and all are antiviral drugs: reverse transcriptase inhibitors,

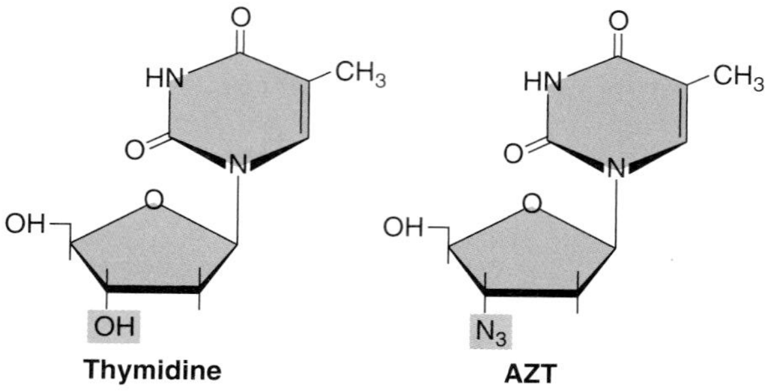

Figure 25.25 AZT. The HIV drug AZT is similar to the nucleotide thymidine. It is put in place of thymidine in replicating viral DNA molecule and stops further viral DNA replication.

Thymidine

AZT

protease inhibitors, and viral fusion inhibitors. All three are offshoots of knowledge about the mechanisms of HIV infection, and the best treatment now available combines all three approaches. While not a perfect solution, the use of this combination therapy has cut death rates from AIDS in half, at least in the United States.

The first drugs that were used to treat HIV infection were the **reverse transcriptase inhibitors.** As a retrovirus HIV uses a reverse transcriptase enzyme to transcribe viral RNA to viral DNA. Reverse transcriptase inhibitors block the actions of this enzyme. The shape and size of the molecules of these drugs are similar to a normal nucleotide base but are not identical (**Figure 25.25**). The reverse transcriptase will place azidothymidine (AZT) and similar drugs in the growing viral DNA chain. But, because the drugs are not normal nucleotides, the reverse transcriptase mole-

cule gets stuck after the drug molecule has been administered and transcription stops. This effectively prevents the virus from replicating.

The reverse transcriptase inhibitors all reduce the amount of virus in the blood and increase the number of CD4 helper T cells in AIDS patients. They are not, however, an ideal treatment because each has serious side effects. For instance, AZT can cause severe nausea and vomiting, headaches, rashes, insomnia, and general loss of energy. Also, reverse transcriptase inhibitors do not eliminate HIV completely, so the infection is slowed but not stopped. Finally, and most importantly, within a year or so HIV develops resistance to the drugs, and then the infection proceeds rapidly again.

This stalemate in HIV treatment was broken in 1996 with the approval of a completely new kind of HIV drug. Again, the approach was based on knowledge of how HIV infection works. After the viral mRNA is translated into protein by the host ribosomes, viral proteins must be modified before they can be made part of a new virus. One important modification is to cut the original proteins into smaller ones, and this process is accomplished by an enzyme called a *protease* (**Figure 25.26**). The second major advance in HIV treatment was the development of a specific *HIV protease inhibitor.* This drug binds to and blocks the action of HIV protease, which, in turn, blocks HIV replication because the proteins that have been made are too large to be incorporated into the virus's protein coat. Neither drug regimen (reverse transcriptase blocker or protease inhibitor) is effective at significantly lowering HIV infection levels over the long term, but when they are taken together, the two

Figure 25.26 How Proteases Work. Proteases cut proteins in a defined place along the amino acid sequence.

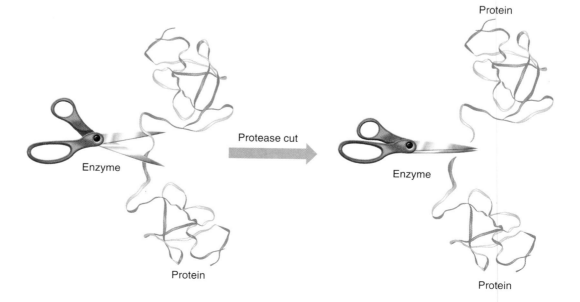

Protein

Enzyme

Protein

Protease cut

Protein

Enzyme

Protein

drugs produce a dramatic decline in viral levels in many patients. Even more promising is that resistance to the combined drug regimen does not appear to develop as rapidly as it does for each drug alone. The success of this combination approach is credited in part with the decline in U.S. deaths from AIDS starting in 1997 (**Figure 25.27**).

Recently, a new drug—a *fusion inhibitor*—has been approved for use in treating HIV patients. This class of drugs blocks the CD4 receptor and so prevents fusion of HIV with the CD4 helper T cell. Only one fusion inhibitor is presently approved for use, but certainly more will follow. Having three classes of drugs to block HIV infection will lead to another big improvement in the duration and quality of life for people infected with HIV.

Unfortunately, the news about drug treatments for HIV is not all rosy. The drug regimen required to affect this level of HIV eradication is severe and requires that the drugs be taken on a precise schedule. In addition, there are side effects that some patients find intolerable. Thus, it is not clear how long a patient can stay on such a regimen. The need for long-term maintenance on the drug combination is reinforced by recent findings that even though HIV levels are extremely low in patients who are taking these drugs, HIV is still present, and it can resume rapid, active replication if the drugs are stopped. Thus, while the use of this combination drug therapy offers hope, it is not the end point in HIV treatment. Research into the development of a HIV vaccine continues. A vaccine has been a research goal since the

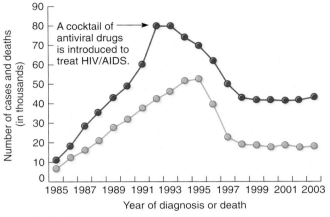

Estimated Number of AIDS Cases and Deaths Among Adults and Adolescents with AIDS, 1985–2003 United States

Key:
- AIDS
- Deaths

Figure 25.27 **AIDS Deaths Have Declined Since the Introduction of Antiviral Drugs.**

identification of HIV, but the development of a safe vaccine has been elusive. Thus far biologists have had little success in generating a vaccine that does not itself cause HIV infection.

QUICK CHECK

1. Describe how reverse transcriptase inhibitors and protease inhibitors stop the action of HIV.

25.6 An Added Dimension Explained
Where Did HIV Originate?

In the opening paragraphs of this chapter, you read several different hypotheses about the emergence of HIV. Some of the hypotheses can be rejected fairly easily. For example, the idea that no virus is involved at all was considered and rejected earlier in this chapter—but to reiterate, every person who has died from AIDS has had HIV infection. HIV infection clearly causes AIDS. Another hypothesis is that the AIDS virus was produced as part of a CIA plot to control population growth in Africa and other countries. According to this hypothesis, the virus was genetically engineered and intentionally released. The first actual documented case of HIV/AIDS is of a man who died in 1959, and there may have been cases of infection before that. Human knowledge of viruses, genetic engineering, and the immune system was primitive at that time, and the production of a genetically engineered virus would not have been possible.

Two remaining hypotheses appear to be the most likely at this point, and both assert that HIV "jumped" from chimps to people in Central Africa sometime before 1960. But there the agreement between these competitive hypotheses ends. Let's call the first hypothesis the *contaminated vaccine hypothesis*. It suggests that a contaminated polio vaccine introduced HIV into Central Africa. According to this hypothesis, the massive polio vaccination campaign introduced HIV to humans. Let's call the alternative hypothesis the *chimpanzee blood hypothesis*. It is favored by several groups of scientists. The chimpanzee blood hypothesis maintains that while chimpanzees are one source of HIV infections, the virus was transferred to people who hunted, butchered, ate, and otherwise had contact with the blood of infected chimpanzees. Let's examine the history of HIV and apply that knowledge to evaluate these two hypotheses.

The latest possible date for introduction of HIV into humans seems to be the late 1950s. This timing is perfect support for the

—Continued next page

Continued—

contaminated vaccine hypothesis, but there is another way to find out when the virus was transferred from chimps to humans. This approach rests on knowledge of viral RNA and will take a little explaining. HIV carries RNA as its genetic material. The viral genome is small enough so that the complete sequence of the nucleotides that compose HIV RNA can be determined. HIV from different infected people or animals can be compared and rates of mutation can be used to estimate a date of first introduction for the HIV virus.

The first finding of these sorts of studies is that human HIV is not a single homogeneous virus. The RNA of viruses from different people is different enough so that several strains of HIV can be identified. Two major groups are called HIV-1 and HIV-2. Interestingly, while HIV-1 causes AIDS, HIV-2 causes no damage or only mild damage to the immune system. These two major strains of HIV can be divided into smaller groups. Most human infections are caused by HIV-1 group M. Two other HIV-1 groups, N and O, cause fewer known cases of the disease. Showing further complexity, the M group of HIV-1 can be separated into 26 smaller subgroups.

Why is HIV such a diverse virus? The enzyme that copies viral RNA to viral DNA—reverse transcriptase—is sloppy. It makes errors so that every new generation of viruses has slight differences in RNA sequence. Many variations are not viable—but nevertheless, the virus will survive and continue to mutate through each generation.

The second discovery to come out of these RNA sequencing studies is that HIV-1 and HIV-2 are a subset of a larger family of SIV viruses that infect monkeys and chimpanzees in Africa. At this time 18 distinct SIVs have been identified that infect 20 different primate species in Central Africa. The best guess is that SIVs have been present in African simians for a long time, perhaps 100,000 years. The name SIV is a bit of a misnomer, though, because the virus does not actually make monkeys, chimps, or other simians ill. The animals harbor the virus, and it reproduces within their bodies, but it does not deplete their immune cells. No one is sure why SIV is benign in monkeys and apes, while HIV is a deadly pathogen. But the discovery of multiple SIVs is the basis for the idea that HIV came from chimpanzees. Among all the SIVs, chimpanzee SIV is most similar to HIV.

But we still must consider how chimp SIV got transferred to people. Was it in a contaminated vaccine or by contact with chimpanzee blood? The answer could lie in the timing of the transference. For instance, if SIV were transferred to humans around the late 1950s, then the vaccine hypothesis has support. On the other hand, if SIV found its way to humans much earlier, the vaccine could not be the culprit. So what do RNA sequences and the array of viral families reveal about the time of the transfer? A careful examination of the degrees of similarity and differences between known strains of viruses provides an estimate of how fast the virus mutates. Several different computer models have been developed to determine when SIV and HIV-1 (M) diverged. The assumption here is that when SIV jumped to humans, the human environment allowed SIV to diverge further, forming new strains and subgroups. The estimates from all of these computer calculations project a date of about 1930—nearly 30 years before the polio vaccine. If just one computer calculation had set the date of transference that far back, you might not be convinced. But when *all* the calculations agree, the suggestion that the transfer occurred around 1930 is strongly supported.

Of course, all of this is indirect evidence. It does provide lots of information about SIVs and HIVs, but a direct test of the polio vaccine would be more satisfying. Unfortunately, most of the records of the vaccine production are no longer available, and only a few people who were involved in its development are still alive. Those few flatly deny that any chimp tissue was used in the manufacture of the vaccine. They state that chimps were used to test the vaccine but not to produce it. But human assertions are not considered as scientific proof. A better approach would be to test samples of the 1950 polio vaccine for HIV. Only a few samples of the vaccine are known to exist. One was tested several years ago and found to contain no HIV or SIV. More recently, new techniques that are more sensitive have been used to test other samples of the original vaccine, and no evidence for SIV or for chimp DNA can be found in the samples. While the proponents of the contaminated vaccine hypothesis may not be convinced, this seems to rule out the idea that chimp SIV contaminated the polio vaccines used in the 1950s. The evidence more closely supports the idea that HIV was transferred to humans when they had contact with SIV in chimpanzee meat, or chimpanzee blood, or other body fluids.

QUICK CHECK

1. What is the scientifically accepted hypothesis of how HIV was transmitted to humans?

Now You Can **Understand**

Anthrax Infections

Fall of 2001 was a tragic and painful year for the United States. The attack on the World Trade Centers and the Pentagon shocked and traumatized people the world over and for some on the East Coast that attack was followed by the threat of death through the mail. Letters laced with anthrax were found in Florida, Washington DC, and New York City. After having watched two of the nation's landmark buildings destroyed by airplanes used as weapons, people had to worry each time they opened their mail. The mystery of the anthrax case still has not had any public resolution, but the attacks stimulated the public to learn more about anthrax, a pathogen that the scientific community has known about since the time of Koch. What is anthrax and how does it pose such a threat?

Anthrax is a pathogenic species of bacteria that can cause a variety of symptoms, from annoying to lethal. Domestic farm animals are usually the target of anthrax infections and historically, infections in humans have been rare, and usually have been confined to people who work with infected animal products. For example, a collector of exotic African drums made from animal hides recently returned from one of his travels abroad with an anthrax infection. When humans are infected, anthrax can be lethal and difficult to treat. Anthrax has a complex life history that includes the formation of spores when conditions for reproducing more anthrax are not optimal. The spores are bacterial cells with a protective shell that disintegrates when the spore finds an optimal environment. They are resistant to heating or drying or other means of destroying them, and so they can persist for a long time. People can be infected by bacterial cells or bacterial spores. Infection can be through the skin, the lungs, or the intestines. The lungs are the infectious route when anthrax spores are inhaled and this form of anthrax infection is usually fatal.

Regardless of how a person becomes infected, the illness is caused largely by a toxin that is released by the bacterial cells. Since the 2001 attacks researchers have learned more about how the anthrax toxin works. One piece of the toxin binds to a cellular protein receptor and as a result the toxin is shuttled inside the cell. Once inside, another piece of the toxin binds to an important cellular protein. The anthrax toxin clips a string of amino acids off of this protein and without it cells die. This important information about how anthrax works should lead to new treatments for anthrax infections.

What Do **You** Think?

Should HIV Testing Be Mandatory?

Over the past decade there has been a great deal of debate about mandatory testing for HIV. Mothers who are infected with HIV can pass the virus to their children, both before and after the baby is born. Many states urge women to have an HIV test before the baby is born and if a mother refuses to be tested, some states require testing of her newborn as well as the initiation of treatment if the baby is HIV positive. As you can imagine, some people agree with these policies while others don't.

One criticism of such mandatory testing is that it might discourage women who are likely to have HIV from getting prenatal care. Some studies have shown that mandatory *reporting* of the names of HIV positive people to a government agency does not decrease the percentage of the population who are tested for HIV. Whether these findings apply to pregnant women is not clear.

Mandatory HIV testing of mothers or infants is an important social issue that will be influenced by public opinion. Do some reading about this issue and decide which side you agree with. Should expectant mothers be required to have HIV testing? What do *you* think?

Chapter Review

CHAPTER SUMMARY

25.1 Infectious Diseases Have Affected Human Populations for All of Human History

History is marked by dramatic episodes of epidemic infectious diseases. The black death killed about two-thirds of the population of Mongolia and spread west to kill over two-thirds of the European population. The influenza epidemic of 1918–1919 killed 20 million to 100 million people worldwide. Although smallpox now has been eradicated because of an effective vaccine, smallpox epidemics also stalked through human history. Today there is a mix of old infectious diseases like malaria and polio and new infectious diseases like AIDS, Ebola, SARS, and West Nile virus. New diseases emerge as people who previously had been isolated come into contact with new pathogens. Ease of travel and dense human populations increase the likelihood of spread of an infectious disease. Infectious diseases act as the agents of natural selection that change the genetic makeup of populations.

25.2 Scientific Studies Have Discovered the Causes of Infectious Diseases

Profound advances in biological knowledge that contributed to new understandings of infectious diseases included the development of the compound microscope, the discovery of cells and of bacteria, pasteurization, sterile surgical techniques, and Koch's standard methods for identifying the causes of infectious diseases. Pathogens include any disease causing agents. Viruses are not alive and the flow of genetic information in retroviruses is from viral RNA to viral DNA that is incorporated into host cell DNA. Then new viral mRNA and new viral proteins and nucleic acids are synthesized. Reverse transcriptase copies viral RNA to make viral DNA. Some bacteria cause illness by secretion of toxins, while others cause illness by invoking responses of the immune system, like fever, runny eyes and nose, mucus in lungs, diarrhea, achy muscles and fatigue.

host   604
pasteurization   600

retrovirus   604
reverse transcriptase   605

25.3 The Immune System Provides a Natural Defense Against Pathogens

Immune defenses include innate and acquired immune defenses. White blood cells are key players in both. Innate defenses include physical barriers and the inflammatory response. Mast cells secrete histamines that allow phagocytes access to infected tissues. Phagocytes such as neutrophils and macrophages engulf and destroy invaders. Some are antigen-presenting cells that travel through the lymphatic system to nodes and initiate an acquired immune response. Antibodies are part of the humoral immune response and circulate in the blood and attach to specific antigens. Phagocytes are part of the cellular immune response and attack specific invaders. B cells produce antibodies against specific invaders. Helper T cells join to antigen-presenting macrophages and initiate events that lead to exponential multiplication of both B cells and helper T cells and production of more antibodies. Cytotoxic cells bind to helper T cells and begin to mature and proliferate. Then they move to an infected area, where they directly attack and kill host cells that are infected by pathogens. Memory B cells are partially activated and will quickly begin an immune response when the particular pathogen invades once more.

antibody   609
antigen   609
antigen-presenting cell   610
autoimmune disease   613
B cell   612
cellular immunity   608
complement   610
helper T cell   611
humoral immunity   609

immune system   606
inflammation   607
leukocyte   607
lymphatic system   610
lymph node   611
lymphocyte   608
memory B cell   612
spleen   611

25.4 Humans Have Invented Treatments That Control Infectious Diseases

Antibiotics interfere with cellular processes within bacteria and cause them to die. Today, bacteria have evolved resistance to antibiotics, an excellent example of natural selection in action that stems from incorrect use or abuse of antibiotics. Vaccines invoke a protective immune response and create a shortcut to a future immune response. Antiviral drugs short circuit viral chemical processes and interrupt viral reproduction.

antibiotic resistance   613   vaccine   614

25.5 AIDS Is a Syndrome of Infections Caused by HIV

AIDS is caused by HIV and has characteristic low levels of helper T cells, high levels of HIV virus, and waves of opportunistic infections. Male transmission of HIV is greater than female transmission of HIV. Condoms reduce HIV transmission, but they are not 100% effective. There is no cure for AIDS, but there are treatments that may be too expensive or intolerable for patients. The sloppiness of reverse transcriptase creates many errors or mutations in HIV and thus makes the virus an elusive target for a vaccine or drug therapy. Combinations of reverse transcriptase inhibitors and protease inhibitors have cut the U.S. AIDS death rate in half, but AIDS still kills and infects millions of people worldwide.

acquired immune deficiency   human immunodeficiency virus (HIV)   617
  syndrome (AIDS)   616   reverse transcriptase inhibitor   620

25.6 An Added Dimension Explained: Where Did HIV Originate?

HIV probably originated in Africa in about 1930 as humans were exposed to SIVs in the blood of infected chimpanzees.

REVIEW QUESTIONS

MULTIPLE CHOICE. Choose the best answer of those provided.

1. Bacteria exchange genetic material using
   a. pathogens.   d. reverse transcriptase.
   b. plasmids.   e. antibiotics.
   c. viruses.

2. Retroviruses, notably HIV, have this difference from other viruses.
   a. They have a protein coat around their DNA core.
   b. They are living organisms that live independent lives.
   c. Using cellular machinery of the host, they use RNA to produce viral DNA.
   d. They increase genetic diversity by exchanging DNA with bacteria using pilli.
   e. They are killed by antibiotics.

3. By taking antibiotics indiscriminately and too often, you run the risk of developing
   a. AIDS.   d. resistant bacteria.
   b. HIV.   e. an inflammatory response.
   c. an immune response.

4. The portion of the immune system called humoral immunity involves
   a. macrophages.   d. antibodies.
   b. helper T cells.   e. B cells.
   c. antibiotics.

5. Making a clone of antibodies or helper T cells is like making
   a. a pair of shoes using no pattern and hand tools.
   b. a pair of shoes using a pattern and machinery.
   c. 100,000 pairs of shoes using a pattern and machinery.

6. Memory B cells allow your cells to
   a. quickly destroy an invading pathogen the first time that your cells encounter it.
   b. be more efficient at phagocytosis.
   c. make antibodies more quickly in the future.
   d. activate your helper T cells.
   e. produce reliable vaccines.

7. The point of vaccinations is to trick the immune system into making
   a. antibodies.   d. memory B cells.
   b. helper T cells.   e. B cells.
   c. phagocytes.

8. HIV causes AIDS by causing
   a. infected helper T cells to commit cellular suicide.
   b. cytotoxic cells to attack infected helper T cells.
   c. enormous losses in helper T cells.
   d. bone marrow to stop making helper T cells.
   e. all of the above except for d.

9. What is the function of cytotoxic T cells?
   a. Cytotoxic T cells cause the formation of a clone of helper T cells.
   b. Cytotoxic T cells kill invading organisms in the bloodstream.
   c. Cytotoxic T cells kill invading organisms that are in the spleen.
   d. Cytotoxic T cells kill infected body cells.
   e. All of the above except for d.

10. Which statement best explains the origin of HIV?
    a. HIV originated from contaminated polio vaccine.
    b. HIV originated from SIV in the milk of African cows.
    c. HIV originated from SIVs in blood of chimpanzees that jumped species.
    d. SIVs originated from HIV that jumped species.
    e. HIV originated from the blood of wild birds.

MATCHIING

11–15. Match the columns. (One choice will not match.)

11. Anton van Leeuwenhoek
12. Robert Hooke
13. Robert Koch
14. Louis Pasteur
15. Semmelweiss and Lister

a. discovered antibiotics
b. established protocols to prove that a pathogen causes a disease
c. instituted policies that prevented infections in hospitals
d. observed cells and coined the term "cells"
e. established a method for killing microbes in foods
f. observed many microscopic creatures

Match the columns.

16. antigen
17. pathogen
18. histamine
19. macrophage
20. antibiotic

a. causes blood vessels to leak fluid into spaces around cells
b. a cell that engulfs and eats invaders
c. a drug that kills bacteria
d. a foreign protein that will evoke an immune response
e. a parasite that causes disease in its host

CONNECTING KEY CONCEPTS

1. How do bacteria and viruses make you sick? In what ways are bacterial and viral infections similar and in what ways are they different?

2. What do the innate and acquired immune responses contribute to your reaction to a pathogen?

3. What is the evidence that HIV causes AIDS?

QUANTITATIVE QUERY

One way to study how pathogens might grow in the human body is to study how they grow in a dish. Here is a typical graph of the growth of bacterial cells in a culture dish. Describe four general phases of growth that you see in this growth pattern and offer a hypothesis for each phase.

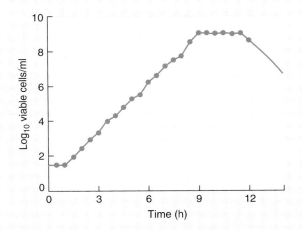

THINKING CRITICALLY

1. Many people are hopeful that an HIV vaccine will dramatically reduce the number of AIDS victims around the world. Other scientists are not sure a vaccine is the answer. Based on your knowledge of the nature of HIV, describe at least one major problem that could limit the effectiveness of any HIV vaccine.

For additional study tools, visit www.aris.mhhe.com.

Ecology

POPULATIONS AND COMMUNITIES

Songbirds Go Silent
June 16, 1997

7 State Birds Threatened
March 5, 2002

A yellow warbler feeds a
cowbird chick.

Like a nervous shadow she waits behind a screen of leaves. She has been watching this place for days, making sure it is just right. She must be ready to take advantage of any opportunity. Instinct shapes her actions. She will not raise her young one herself, but her commitment to its survival is intense. The right moment comes, and she rushes in, leaves her offspring, and darts away, silent and unnoticed.

Many questions are raised in these few seconds. What has happened? How will the foster parents react to the abandoned young?

This mother is a cowbird, a brown, robin-sized bird. Instead of building her own nest and raising her own young, a female cowbird lays her eggs in the nests of other birds. In most cases instinctive behaviors force the foster parents to feed and rear the cowbird chick, even though it may grow to be twice the size of their own young (**see chapter-opening photo**). The larger, stronger young cowbird outeats the smaller, more delicate, songbird nestlings. It may even push the songbird babies out of their own nest. When the cowbird grows up, if it is a female, instinct will guide it to lay eggs in the nests of songbirds. The net result is fewer songbirds and more cowbirds.

For most of human history few people have known or cared much about cowbirds and their strange reproductive habits, but now the toll taken by interactions with cowbird populations is combining with other events to seriously reduce the populations of many species of songbirds. Bird watchers have noticed that songbird populations seem to be declining, and scientists have confirmed their observations. Some species seem to be scarce in places where they used to be plentiful. There are many complex reasons for the disappearance of songbirds. While cowbirds are an important part of the story of the decline of songbird populations, there is much more to this story, involving destruction of habitats in the United States and abroad. At the end of this chapter you will return to this topic, but first you will read about populations and other kinds of interactions between species. ■

26.1 Ecology Is the Study of Patterns of Interactions Between Organisms

Ecology is the branch of biology that focuses on interrelationships between organisms and their environments and how these relationships affect the distribution and abundance of organisms. Ecology is one of the newer branches of biology. Many people do not realize the distinctions between ecology, environmental science, and environmental activism. While many ecologists support environmental causes, the scientific work of ecologists focuses on the interactions between organisms and their environments. It may be helpful if you think of the relationship between ecology, environmental science, and environmental activism as similar to the relationship between the laws of physics, the science of engineering, and the marketing and sales of engineered products. Ecologists are interested in questions such as: How do predators affect the populations of their prey? What is the minimum area required for a healthy forest? How do materials move through the biosphere? In contrast, environmental science applies ecological principles to practical human concerns such as handling solid wastes, movements of toxins through environments, acid rain and greenhouse gases. Environmental activists work to prevent environmental destruction and raise public consciousness of current environmental problems.

From its earliest beginnings ecology has been concerned with patterns of relationships among organisms. In the 1920s along with botanist V. S. Summerhayes, Charles Elton studied the feeding relationships among the organisms on Bear Island, near Spitsbergen, in the North Atlantic Ocean. They compiled a summary of "who ate whom" on Bear Island (**Figure 26.1**). While you may not recognize the names of all of the species mentioned in this figure, notice that species rely on one another for food.

A **population** consists of all of the organisms of a species that occupy a particular place at a particular time. The populations in the Bear Island study formed at least two **communities,** ecological units that involve interactions *between* the populations of different species that inhabit a location. **Ecosystems** are composed of the living communities within a defined area and the nonliving components of their environments. **Biomes** are regional ecosystems that are recognized by their dominant vegetation. Biomes cover wide geographic areas and are controlled by broad climatic zones. The broadest level of ecological organization is the **biosphere,** which includes all of Earth's ecosystems and biomes. **Figure 26.2** shows the relationships of these ecological levels.

ecology the study of the interrelationships of organisms and their environments that affect the distribution and abundance of organisms

population all of the members of a species that live in an area; populations are defined by interbreeding within the group

community all of the populations that interact within an area

ecosystem a spatially defined interacting ecological community plus the nonliving parts of its environment

biome a regional ecosystem characterized by dominant vegetation and controlled by a similar climatic regime

biosphere the global ecosystem composed of all of Earth's ecosystems and biomes

QUICK CHECK

1. What is the difference between environmental activism and ecology?

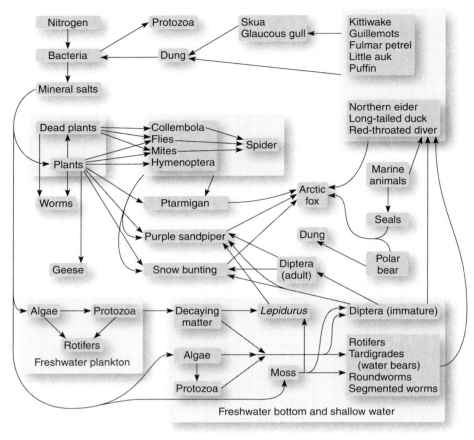

Figure 26.1 **Who Eats Whom on Bear Island.** This diagram summarizes feeding relationships of populations of organisms on Bear Island, near Spitzbergen Island in 1927. (→ = movement of substances; arrowhead points to the organism that consumes a substance at the other end of the arrow)

■ *Why does it matter if a single species goes extinct?*

26.2 Populations Have Characteristics Related to Their Rate of Increase

Populations are not uniformly distributed within their geographic ranges but are limited to **habitats,** specific places where they usually live that are characterized by a physical setting or a kind of plant. For example, although eastern red salamanders are distributed from New York State to the Gulf Coast, you will find them only in moist glades along streams. In contrast, eastern mud salamanders are most often found beneath rocks in swiftly flowing streams.

The distribution of individuals in a population can vary in other ways too. Some populations are sparse, while others are dense. Some have a *uniform distribution* of individuals across the geo-

habitat the specific place where an organism usually lives that is characterized by a physical setting or a kind of plant

The biosphere contains all of earth's ecosystems.

An ecosystem is the functional unit of ecological studies. It includes biotic and abiotic factors.

A community is an association of populations of different species.

A population includes all of the individuals of one species in an area.

An organism is the whole body of a living thing.

Start here

Figure 26.2 **Population to Biosphere.**

Figure 26.3 **Population Distributions.**
Populations are distributed in characteristic
patterns. Populations can be spaced randomly
like these species of lichens (**a**), uniformly like
these penguins (**b**), or in clumps like the human
population of the United States (**c**).

(a) Random spacing

Increasing uniformity

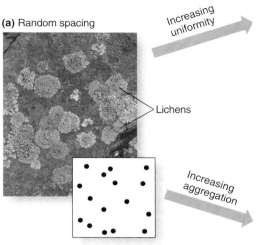

Lichens

Increasing aggregation

(b) Uniform spacing

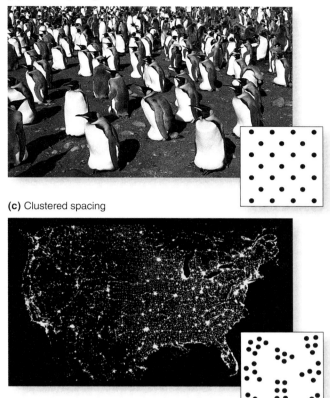

(c) Clustered spacing

graphic range of the population, others have
a *clumped distribution,* and others are distributed at *random* (**Figure 26.3**). One of the
first questions an ecologist may have about a
population is: How many individuals are in
this population? A variety of field methods
allow census of a population. For organisms that
don't move around much, a direct count is the easiest way to determine population size. A census of
a population of mobile animals can involve capture, mark, release, and recapture studies, as well
as indirect census methods such as noting animal
signs like droppings or tracks, entrances to underground burrows, or calls or songs. Once a scientist
has the basic information about how many individuals belong to a population, other population characteristics and trends can be investigated.

Populations Grow at Different Rates

One focus of ecological research is to determine the
rates at which populations grow and then to project
how they will grow in the future. Before these calculations can be made, you must know the distribution of sexes within the population and determine
how many individuals are in their reproductive
years. The distribution of ages in a population is
one of the prime characteristics of a population,
and it is summarized in a **life table** that tracks the
numbers of individuals that live and die during various age intervals. **Figure 26.4b** shows a life table
compiled for Dall's mountain sheep on Mount

McKinley (now renamed Denali) National Park. In
this study the oldest sheep died at between 13 and
14 years old. The results of these studies were used
to set up the life tables for this population of Dall's
mountain sheep.

The data in a life table can be used to calculate
how many individuals of each age class typically
survive to the next age interval. Plotting these calculations on graph paper allows you to see a second
characteristic of a population: its **survivorship
curve,** a graphic representation of when death
occurs in a population. **Figure 26.5** shows the survivorship curve for this population of Dall's mountain sheep. The flat portion of the curve indicates
that few deaths occurred when sheep were "middle
aged," and the steep slope of the curve indicates
that most deaths occurred later in life. Survivorship
curves for populations differ, and Figure 26.5 contrasts this survivorship curve with those for populations of other animals. While most elephants live to
old ages, the gray squirrel population shows steady
losses throughout the life span, and most oysters die
early in life. Study of mortality may sound morbid,
but the patterns of death within a population are an
important characteristic that greatly influences
future population growth. You may be wondering

life table a summary of
age data for a population
that allows predictions of
when death occurs in the
population

survivorship curve
a graph that shows
when death occurs in a
population

(a)

Figure 26.4 **Dall's Mountain Sheep and Their Life Table.** Researchers used a collection of the continuously growing horns of male Dall's Mountain Sheep (a) to construct this life table of a population of sheep in Mount McKinley National Park (b).

(b)

Age interval (years)	Number dying during age interval	Number surviving at beginning of age interval	Number surviving as a fraction of newborn (l_s)
0–1	121	608	1.000
1–2	7	487	0.801
2–3	8	480	0.789
3–4	7	472	0.776
4–5	18	465	0.764
5–6	28	447	0.734
6–7	29	419	0.688
7–8	42	390	0.640
8–9	80	348	0.571
9–10	114	268	0.439
10–11	95	154	0.252
11–12	55	59	0.096
12–13	2	4	0.006
13–14	2	2	0.003
14–15	0	0	0.000

Life table for the Dall's mountain sheep (*Ovus dalli*) constructed from the age at death of 608 sheep in Mount McKinley (now Denali) National Park

Source: Based on data in Murie, 1944, quoted by Deevey 1947.

Figure 26.5
Survivorship Curves.
(a) The survivorship curve for Dall's Mountain sheep. (b) Populations tend to have three general patterns to their survivorship curves. Most members of some populations live to old ages, other populations have a regular mortality rate throughout the lifespan, still other populations experience heavy mortality early in life, with relatively few individuals surviving to old age.

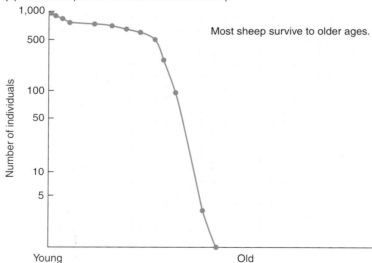

(a) Survivorship curve for Dall's mountain sheep:

Most sheep survive to older ages.

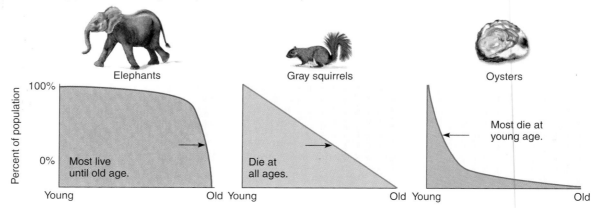

(b) There are three general patterns to survivorship curves:

what survivorship curves for human populations look like. Human survivorship curves vary by region and time (**Figure 26.6**), but many resemble the survivorship curve for elephants, with many individuals living relatively long lives.

The Age Structure of a Population Allows Predictions of How Fast It Will Grow

Once an ecologist has learned how many individuals of each sex are in each age group, a bar graph of the population's age structure can be constructed. **Age structure** is a useful population characteristic used to project whether the population size will increase, decrease, or remain stable in the future

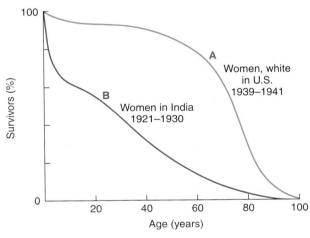

Figure 26.6 **Variation in Survivorship Curves of Human Populations.** A comparison of the survivorship curve of two populations of women shows that survivorship curves are characteristic of populations, not of species.

(**Figure 26.7**). A population that has the greatest percentage of its members in the prereproductive years, will increase as these individuals enter their reproductive years. In contrast, a population with fewer members in the prereproductive years will decline in the future. Finally, a stable population has equal or nearly equal percentages in each of the reproductive phases of life. You would expect populations of different species to have different life tables, survivorship curves, and age structures, but it is important to emphasize that, depending on situations, different populations of the *same species* also can have different life tables, different survivorship curves, and different age structures.

Some Populations Grow Exponentially, While Others Grow Logistically

Anyone who has ever kept gerbils or mice or rabbits knows how quickly a pair of small mammals can multiply if conditions are perfect: if there is plenty of food, water, and space, and if all individuals reproduce at the maximum rate. Ecologists call how fast a population *can* reproduce its *biotic potential,* and northern bobwhites can help to illustrate this concept.

Northern bobwhites are chunky birds that live in 38 U.S. states (**Figure 26.8**). Bobwhites have a high rate of reproduction, and if these quails were able to reproduce at their biotic potential, in 10 years a pair would have grown into a population of over 2 billion! Of course, nearly all of the quail offspring die, and the biotic potential of northern bobwhites never is expressed in nature. Ecologists have analyzed the factors that prevent organisms from achieving their biotic potential and have identified patterns of population growth.

age structure a graph that shows the distribution of ages of individuals in a population

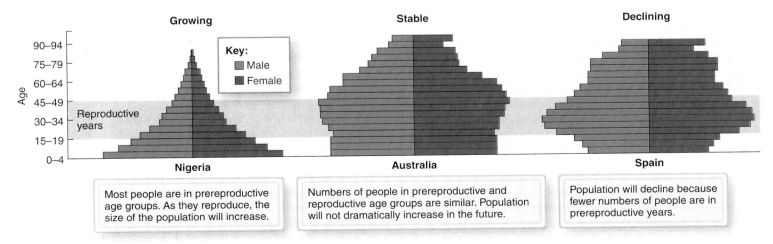

Figure 26.7   **Population Growth Is Predicted by Population Age Structures.**

Figure 26.8
Huddling to Stay Warm. Like most organisms, bobwhites never achieve their biotic potential. Accidents of weather kill many of them.

Environments are not equal in their ability to support populations of different organisms. Some environments have appropriate food, nutrients, water, light, climatic factors, and space, while others do not. The unequal distribution of these basic requirements for life helps determine the geographic range of a species. An environment's ability to support a species is summarized in the ecological concept of **carrying capacity** (abbreviated as **K**). Just as a storm shelter can house and support only a finite number of people, an environment only has resources for populations of finite sizes. Just as conditions in an overcrowded storm shelter deteriorate and endanger the lives of refugees, once a population increases beyond the environment's carrying capacity, individuals are negatively affected. For instance, animals may starve, die of thirst or exposure, be picked off by predators, or be weakened by parasites. Plants may be shaded out by taller competitors or may not have sufficient water or nutrients. Carrying capacity is a constant environmental factor that does not change unless the environment changes. To return to the bobwhite example, a population of bobwhites will increase until it overshoots the environment's carrying capacity. After this point the quails

Table 26.1	A Comparison of *K*- and *r*-selected species	
	K-selected species	**r-selected species**
Limits population size	Carrying capacity (*K*)	Reproductive rate (*r*)
Factors that control population size	Density-dependent factors	Density-independent factors
Organism size and longevity	Larger, with longer lives	Smaller, with shorter lives
Age at first reproduction	Late	Early
Number of reproductions per life span	Often several	Usually one
Number of young per reproductive event	Fewer offspring produced	Many offspring produced
Care of young	Care of young provided	No care for young
Examples	Elephants	Insects
	Sequoias	Small rodents
	Chimpanzees	Frogs
	Humans	Many "weeds"
Curves of population growth	Populations numbers are stable at or around *K*	Population rapidly rises, overshoots *K*, and then collapses

carrying capacity (K) the ability of an environment to provide the conditions necessary to sustain and support a population of a certain size

will not be able to reproduce maximally, and many will not have enough food and water to survive. Then starvation, illness, exposure, predators, and parasites will reduce the population until it once again is at or below K for bobwhites.

Species whose numbers are limited by the environment's carrying capacity have relatively stable populations that fluctuate around K (**Table 26.1**). These organisms tend to be large and live long lives. They have few offspring and provide parental care until their offspring can live independently. Elephants, humans, chimpanzees, large plants, and beavers are a few examples of **K-selected species.** The word *selected* in this phrase means that these species are selected for adaptations that allow them to successfully compete when their population levels are near the environment's carrying capacity.

Not all species follow this pattern. The lives and habits of other species have different adaptations that allow their populations to grow rapidly for brief times (Table 26.1). Small size, short life span, rapid generation time, production of many offspring in one reproductive effort, and little parental care are some adaptations for quick reproduction. Species that have these adaptations are called **r-selected species** (r is short for reproduction), and insects and small mammals are examples of r-selected species. As you might imagine, r-selected species cannot maintain their biotic potential indefinitely. Populations of r-selected species sometimes do grow rapidly, but once they overshoot the environment's carrying capacity, their populations collapse. So their numbers fluctuate dramatically with periodic highs and lows.

This discussion has made it seem as if all species are either r- or K-selected, but this is not the case. Instead, r- and K-selected traits are the end points of a continuum. Some species exhibit a blend of r- and K-selected traits. For instance, a maple tree produces abundant seeds—far more than an oak, avocado, or palm tree. So even though a maple tree has many attributes of a K-selected species such as large size, and a 150-year life span, it is r-selected when compared with oak, avocado, or palm trees.

When the growth of a population that is increasing at its biotic potential is recorded on a graph, a J-shaped curve results (**Figure 26.9**). This J-shaped curve is typical of **exponential population growth.** How is an exponential rate of increase different from a simple mathematical rate of increase? One way to think about exponential rate of increase is to compare it to a savings account that compounds interest yearly on the money invested in the account. If you start with a $900 deposit and an

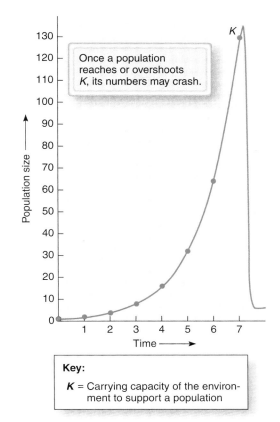

> Once a population reaches or overshoots K, its numbers may crash.

Key:

K = Carrying capacity of the environment to support a population

Figure 26.9 **Exponential Population Growth.** A J-shaped curve is typical of a population that is growing at its biotic potential.

interest rate of 7.5% per year, at the end of a year your investment will have increased to $967.50. Because interest is compounded on the *entire sum* in the account, at the end of the second year the account will have earned an additional 7.5% of $967.50 or $72.56. At the end of 50 years the savings account would have grown to $33,470.77! In contrast, if the deposit were growing arithmetically, you would earn only $67.50 each year, and in 50 years the initial investment of $900 would have increased to $3,375. Populations are like interest-compounded bank accounts because the individuals that are added by reproduction can reproduce and add to the population size.

Once a population that is growing at an exponential rate approaches the carrying capacity of the environment to support it, environmental factors that are part of carrying capacity slow its rate of increase. Lack of food, nutrients, mates, space, light, or other resources kill some individuals or otherwise prevent them from reproducing. When a population that is growing exponentially encounters environmental carrying capacity, its J-shaped curve

K-selected species a species selected for traits that favor competitive ability at population densities near carrying capacity

r-selected species species that have adaptations for rapid reproduction and high population densities that allow them briefly to overshoot the environment's carrying capacity

exponential population growth a geometric rate of increase of a population, shown in a J-shaped curve of population growth

Figure 26.10
Exponential and Logistic Growth.

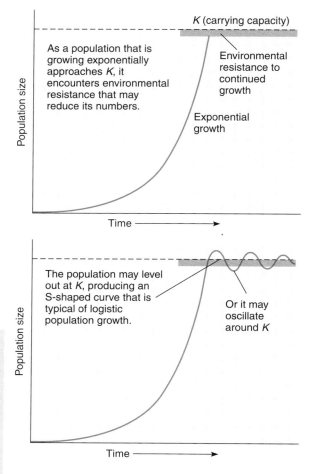

As a population that is growing exponentially approaches *K*, it encounters environmental resistance that may reduce its numbers.

K (carrying capacity)

Environmental resistance to continued growth

Exponential growth

Time →

The population may level out at *K*, producing an S-shaped curve that is typical of logistic population growth.

Or it may oscillate around *K*

Time →

logistic population growth a leveling off of population increase as the environment's carrying capacity is reached, shown as an S-shaped curve

density-dependent population control a population control factor that increases in intensity as population size increases

density-independent population control a population control factor that acts without regard to the size of a population

of population growth begins to level out, forming an S-shaped curve (**Figure 26.10**). Alternately, the population size may bounce around the environment's carrying capacity for the population, sometimes overshooting it, sometimes being a bit under it. When a population has leveled off or is oscillating around the environment's carrying capacity, **logistic** population growth has been achieved. Logistic growth is typical of *K*-selected species.

Some Population Controls Intensify as Population Size Increases

The effects of **density-dependent population controls** intensify as the size of a population increases. For instance, infectious diseases can sweep more effectively through a crowded population, and parasites can spread more easily in dense populations. Stresses from crowding may produce hormonal changes that alter reproductive patterns and reproductive behaviors, resulting in fewer young produced and/or surviving. Competition for food, water, and space also intensify as populations increase. Lack of nutrients may cause some individuals to cease breeding. In animal species that maintain territories, when the population density increases, some males may be unable to hold and defend territories. A male without a territory usually will not be able to attract mates and may not breed (**Figure 26.11a**). Predation is a density-dependent population control factor because predators are attracted to high-density prey populations. As the numbers of prey increase, the numbers of predators also increase. For plants competition for light increases as population densities increase. Some plants that are shaded out will not survive or may not reproduce. Population sizes of *K*-selected species are controlled by density-dependent factors.

Other population control factors are independent of population size. Accidental deaths from extreme weather or other natural disasters eliminate individuals in sparse as well as in dense populations. Mudslides, volcanic eruptions, flooding, fires, drought, earthquakes, and unusual extremely cold or hot weather are some of the **density-independent population control** factors

Figure 26.11 **Density-Dependent and Density-Independent Population Controls.** (**a**) A male red-winged blackbird that does not have a territory will be less able to attract a female and may not be able to mate. This is a density-dependent population control. (**b**) Drought kills individuals irrespective of the density of their populations. Drought and other accidents of weather are density-independent population controls.

(a)

(b)

that reduce populations irrespective of their numbers (Figure 26.11b). Humans also cause density-independent deaths from environmental disasters such as oil spills, excessive ultraviolet radiation caused by the thinning of the ozone layer, and acid precipitation. Populations of *r*- and *K*-selected species are influenced by density-independent population controls.

QUICK CHECK

1. Specify the characteristics of a population and explain in general how the characteristics of a population are different from those of a species.

26.3 Analysis of Human Populations Helps Predict Their Future Growth

Now that you understand some of the important characteristics of biological populations, let's turn to the human population, learn about its patterns of growth, and see how scientists predict it will grow in the future.

The J-shaped curve in **Figure 26.12** shows that after thousands of years of stable, small human populations, in the last 200 years the human population has experienced exponential growth. For the last 20 years population growth experts have been unsure about future human population growth and now the consensus view is that logistic growth seems to be slowing exponential growth. Nevertheless, it is uncertain when Earth's human population will stabilize. For most Americans the growth of Earth's human population is an abstract concept. To better appreciate how fast the human population is growing, log in to **www.census.gov/main/www/popclock.html.** Click on the U.S. POPClock and World POPClock links and print the pages that come up or jot down the information there so that you will have a baseline reference for the rates at which the world and the U.S. human populations are increasing. For comparison, I am typing these words at 12 noon on August 28, 2005, and the U.S. population is increasing at the rate of one birth every 8 seconds and decreasing by one death every 13 seconds. In addition, every 25 seconds one international migrant enters the United States. So there is a net gain of one person to the U.S. population every 11 seconds. The world's human population also is increasing (**Figure 26.13**). In the last 10 minutes 2,127 people have been added to the world population. This is an amazing rate of

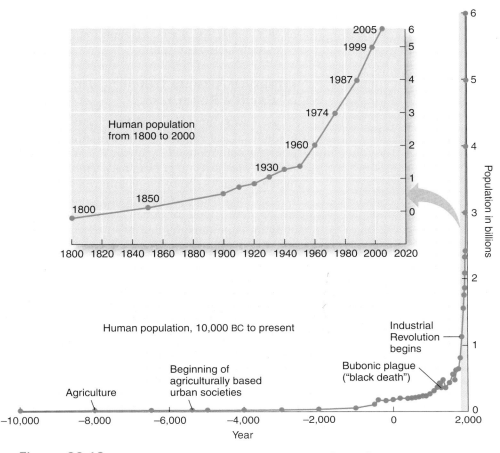

Figure 26.12   **The History of Human Population Growth.**

■ *Do you think the large population growth in the last 200 years was the result of the Industrial Revolution or improved health care? Explain your answer. Source:* Data from United States Census Bureau.

Monthly world population figures	
07/01/05	6,451,058,790
08/01/05	6,457,380,056
09/01/05	6,463,701,322
10/01/05	6,469,818,677
11/01/05	6,476,139,943
12/01/05	6,482,257,297
01/01/06	6,488,578,564
02/01/06	6,494,899,830
03/01/06	6,500,609,361
04/01/06	6,506,930,627
05/01/06	6,513,047,982
06/01/06	6,519,369,248
07/01/06	6,525,486,603

World POPClock Projection	
Total population of the world	Date
6,456,707,996	7/28/05

In _____ the world population has
(time in months)
grown by_____ people.

Figure 26.13   **The Growth of the World Human Population.**

growth, indeed: about 3.5 people are added to the world population every minute. Now consult the POPClock information again and see how much the U.S. and world populations have grown in the time it has taken you to read these few lines. How

(a)

does the rate of increase compare to that in late July of 2005? Enter today's information from the POP-Clock website in Figure 26.13.

Lifestyle Affects Consumption of Earth's Resources

Today, Earth's human population stands at about 6.5 billion, and experts speculate that it is in the last stages of exponential growth. The human population is projected to grow to between 8.8 and 9.3 billion by 2050—probably within your lifetime. This will add perhaps half as many people as now are on Earth. From a global perspective the big question is how the human population will grow after that. Will it continue to increase, decrease, or stabilize? This question can be restated more precisely as: What is Earth's carrying capacity for humans?

Unfortunately, it is difficult to measure Earth's carrying capacity for humans. For one thing, people constantly change the environment's carrying capacity by introducing new ways of getting food (see Section 21.6). Improvements in sanitation, health care, medicine, and technology also help humans to evade density-dependent population controls. Some experts believe that genetically engineered crops and new food sources will increase the carrying capacity of needy regions, saving the lives of millions or even billions of hungry people. Nevertheless, scientists are fairly certain that if the human population does not stabilize, we will face starvation and disease, just like any other species that exceeds the carrying capacity of its environment.

A newer concept related to carrying capacity is the *ecological footprint,* defined as the amount of resources that an individual or group consumes. Ecological footprints are expressed in the area required to support the demand for resources. Individuals who consume more resources have larger ecological footprints than do those whose consumption of resources is more modest. Accessing **www.ecofoot.org** allows you to assess your own ecological footprint. This calculation can give you a rough idea of whether you are living within or beyond Earth's ability to support your lifestyle and can help you to become more ecologically aware and environmentally responsible. **Figure 26.14** compares the household goods of a typical

(b)

(c)

(d)

Figure 26.14 **Comparison of Household Goods.** (**a**) Getu family, Ethiopia; (**b**) Skeen family, U.S.A.; (**c**) Yadev family, India; (**d**) Wu family, China.

American family with those of families in other cultures, and the differences are obvious. Ecologists are concerned that in 50 years the heavy consumption characteristic of the American lifestyle will have been adopted by millions of people all around the globe. It seems certain that this level of consumption cannot be sustained. The only reason that Americans now can live this kind of lifestyle is that resources are relatively cheap. This situation is beginning to change as millions in China and India become more affluent and adopt the American heavy-consumptive lifestyle.

Regional Human Populations Have Different Rates of Growth

The prospects for future population growth are dramatically different in different parts of the world (**Figure 26.15**), because of unique local conditions. Many countries, particularly those in western Europe as well as Japan, are actually experiencing population declines. For these countries the explosive period of population growth is over, and a true logistical pattern of population growth has been established. In these countries young people are marrying later, and having fewer children when they do marry. For some regions exponential growth has slowed because of recent effective access to birth control and education programs. In India, for example, the average number of children per woman has declined from around four children in 1990 to around three children in 2005. A decline in fertility due to the devastating effects of AIDS is the tragic situaton in some African countries. In southern Africa AIDS has killed many of the people of childbearing age. Partly as a result of this, in southern Africa the average number of children per woman has decreased from over six in the 1950s to fewer than four in 2005. It is expected to decline even more by 2050. In other developing countries human population growth still is rapid and approaches exponential growth. Countries in Middle, East, and West Africa as well as the Middle East have exponential population growth.

Despite local declines the overall human population is expected to increase by approximately 3 billion people over the next 50 years (**Figure 26.16**). These population increases will put great stress on resources that already are stretched thin. Water, food, and housing may become even scarcer in some regions, and the hope for a normal, comfortable, modern life will be beyond the reach of many.

These local patterns of human population growth hold some interesting lessons. For example,

(a)

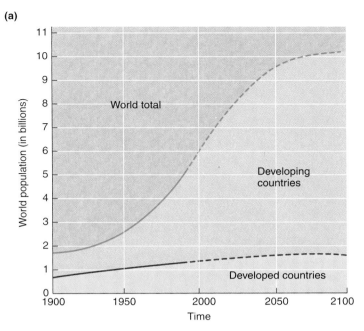

(b)

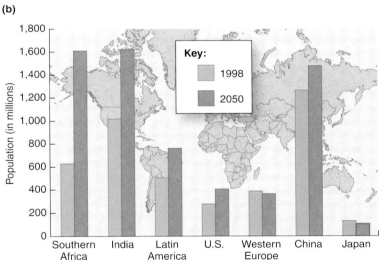

Figure 26.15 **Predictions of Future Growth of the Human Population.** (**a**) It is likely that most future human population growth will occur in developing countries. (**b**) Estimated populations in various regions, 1998 compared with projections of growth for 2050.

■ *Why do you think that the populations of western Europe and Japan are predicted to decline in the next 45 years?*

a high birthrate that leads to rapid population growth also is associated with a high death rate, high levels of poverty, and an overall lower quality of life. Different governments have taken different approaches to solve the problem. In China over 30 years ago the government saw a population that was approaching 1 billion with no end in sight. A stringent birth control policy was instituted that forced parents to restrict themselves to having just one

World Population Growth
UN Projection, 2003–2100

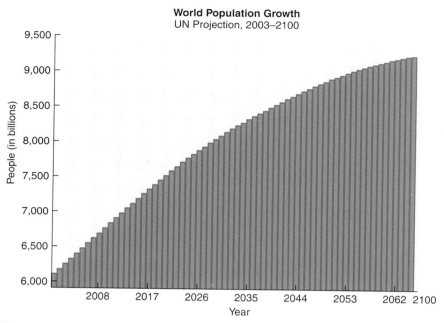

Figure 26.16   **Increase in World Population.** This is one projection for the growth of the world population through 2100.

Figure 26.17   **China's One-Family One Child Policy.**

■ *What problems could be caused by population growth? In France the government is encouraging people to have* more *children. What advantages are there in having a growing population?*

---

predator an animal that eats other organisms and removes them from the population

prey an organism that is consumed by a predator

carnivore a predator that eats other animals

herbivore a predator that eats plant tissues

omnivore a predator that eats both animals and plants

detritivore an organism that eats dead tissues of plants or animals

producer an organism that produces its own food supply

consumer an organism that does not produce its own food supply but obtains it by eating other organisms

food chain the transfer of food energy from producer through various levels of consumers

---

child (**Figure 26.17**). It is hard to say if China's population might have slowed its growth without government intervention, but China's rate of population growth has slowed. Most other countries have chosen more persuasive rather than enforced methods to accomplish the same goal, and several approaches seem to work when they are combined. For instance, providing better access to birth control methods helps to decrease population numbers. Improving the overall economic well-being and general health of people also seems to reduce population growth, because when people have economic security they become more confident that each child will grow up healthy and cared for. Finally, improvements of women's economic and political conditions are critically important. When women cannot protect themselves, own property, have access to education, or build their own wealth, their only security is to have many children in the hopes that one or two of their children will provide those things for them. It is documented that improvements in conditions for women lower both birthrates and population growth rates. All of these are uniquely human interventions, but they may help to bring human populations to sustainable, stable levels.

So How Will the Global Human Population Grow?

It seems likely that by 2050 Earth will hold close to 9 billion humans. Is 9 billion too many humans? What will happen after that? These answers depend on what we do now and in how rapidly the American lifestyle is adopted as the world ideal. And, as you will read in Chapter 28, the human population already has had profound effects on the biosphere; a larger human population will undoubtedly continue this trend.

QUICK CHECK

1. What difference will it make to you if Earth's human population reaches 9 billion people in 2050?

26.4 Populations Interact Within Ecological Communities

An *ecological community* is defined as a group of populations that interact as each individual pursues the resources it needs to survive and reproduce. A community might consist of just a few species or of many different species. Through the interactions of its members, an ecological community may extend over a large geographic area, or it may be confined to a few inches of soil. Regardless of the details, every individual is a part of one or more living communities. The boundaries of a community are not always distinct, and neighboring communities can have overlapping boundaries.

Community interactions affect every aspect of survival and successful breeding. For example, finding food, acquiring a mate, defending a territory, building a shelter, and protecting offspring all involve aspects of the biological community. Now comes the interesting part: even though organisms differ greatly, there is a limited array of general ways that organisms interact. The themes of consuming, cooperating, and competing describe most community interactions.

Predators feed on other organisms, collectively called their **prey**. Arctic foxes, seals, polar bears, skuas, and glaucous gulls are just a few of the many predatory animals that Elton and Summerhayes documented in the communities on Bear Island. Today, many ecologists extend the definition of predator to include animals that consume seeds and whole plants. Another way to think about food relationships reflects what kind of food is being eaten. **Carnivores** eat other animals, and usually carnivores are synonymous with predators. Seals and polar bears are just two examples of carnivores on Bear Island. **Herbivores** are animals that eat plant tissues, while **omnivores** are animals that eat both plants and meat. On Bear Island geese were herbivores, while ptarmigans, purple sandpipers, and snow buntings were omnivores. **Detritivores** obtain energy from dead plants or dead animals. Worms, flies, mites, and springtails were Bear Island detritivores.

In a community organisms are linked in relationships related to how they obtain energy. Photosynthesizers are **producers** that make food through the process of photosynthesis. The different roles that animals play in their environments can be thought of as ways to consume nutrients, and all of these roles are grouped as **consumers**. Herbivores are called *primary consumers,* while carnivores that eat herbivores are called *secondary consumers.* Carnivores that feed on secondary consumers are called *tertiary consumers,* or *top consumers.* Detritivores and decomposers reduce the wastes and dead tissues of community members to simpler chemical compounds and make them available to plants. Within a community organisms interact in feeding relationships called **food chains.** All of the food chains in a community are represented in more complex, hypothetical feeding relationships called **food webs. Figure 26.18** shows Antarctic food chains and a food web. Note that an organism can be a member of more than one food chain.

A species' habitat plus all of the resources that it uses in its life, including all of the other species

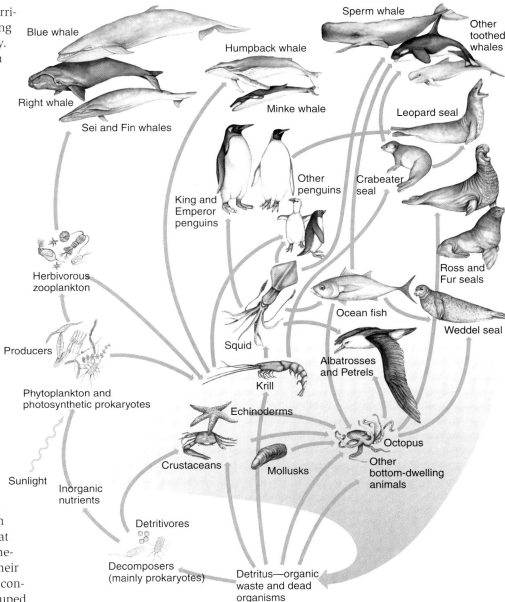

Figure 26.18 Food Web in Antarctica. The arrows trace the complex patterns of food relationships of producers, consumers, scavengers, and decomposers of the Antarctic. Because there are relatively few species in the Antarctic, this is a relatively simple food web. For each pair of species connected by an arrow, the base of the arrow indicates the species that is eaten, and the arrowhead indicates the species doing the eating.

that it interacts with, are parts of its **ecological niche.** This term can be confusing because it sounds as though an ecological niche is a place in the environment. A species' ecological niche includes the biotic and abiotic features of its habitat, but habitat is just one of the components of an ecological niche. Another way to think about ecological niche is to imagine it as an organism's ecological role or job in the environment. For instance, white-tailed

food web all of the food relationships within a community

ecological niche a species habitat, plus all of the abiotic and biotic resources it requires for its life and survival

Figure 26.19 **Ecological Niche.** The ecological niche of a white-tail deer includes the plants it eats at different seasons, its predators, parasites, and its habitat.

(a) In mutualistic relationships both partners benefit from the relationship.

Corals benefit from extra food; algae have a safe place to live.

Photosynthetic algae

(b) In commensalistic relationships one partner benefits, but the other is unaffected.

Follicle mites get nutrients and a safe place to live; human gets no benefit and is not harmed.

Follicle mite

X278

(c) In parasitic relationships one partner benefits and the other is harmed.

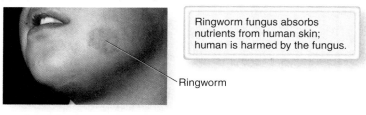

Ringworm fungus absorbs nutrients from human skin; human is harmed by the fungus.

Ringworm

Figure 26.20 **Mutualism (a), Commensalism (b), and Parasitism (c) Are Basic Kinds of Symbiotic Relationships.**

■ *Explain why a coral that has been bleached will die.*

deer are large herbivores that mostly eat leaves and twigs of shrubs and trees (**Figure 26.19**). Whitetail deer also eat quantities of acorns, mushrooms, and fruits. Deer used to be preyed on by wolves and cougars, but as these have largely been eliminated, human hunters now are the major predators of white-tailed deer. The typical external and internal parasites of white-tailed deer also are parts of their ecological niche.

Food relationships are just one way that organisms are closely tied to one another. **Symbiosis** is an umbrella term that describes close and sometimes necessary associations between two species. When symbiotic partners live together and both partners benefit, this is the pattern of symbiosis called **mutualism.** For instance, the single-celled algae and the coral animals whose tissues they inhabit have a mutualistic relationship (**Figure 26.20a**). The algae receive a secure place in sunny waters where they can photosynthesize, while the coral animals supplement their diet with nutrients produced by these live-in photosynthetic partners. Not all symbiotic relationships result in mutual benefits for both symbiotic partners. **Commensalism** is a one-sided form of symbiosis in which one partner benefits, while the other partner receives neither benefit nor harm. For example, the mites that live within the follicles of your eyebrows and eyelashes are one example of this rarer form of symbiosis (Figure 26.20b). Follicle mites do you no harm, but they benefit greatly from the protection of your oily hair follicles where they live, mate, and breed. **Parasitism** is a final form of symbiosis in which one partner benefits at the expense of the other. A parasite can cause its host to lose nutrients, become stunted, ill, unable to resist infectious diseases; parasites even can cause death. Parasites attack their hosts while they are still alive. Parasites are most often thought of as animals, but members of other kingdoms also can be parasites. For instance, dodder is a parasitic plant, and ringworm is a parasitic fungus (Figure 26.20c).

Adaptations Play Important Roles in the Interactions of Predators and Their Prey

Adaptations help prey animals to evade and escape from predators. Many adaptations allow prey to blend in with the background. For example, *camouflaged* shapes, colors, and patterns make some moths look like lichens on tree bark, transform caterpillars into twigs, and allow grasshoppers to disappear against a leafy background (**Figure 26.21**).

Figure 26.21 **Hiding in Plain Sight.** This marbled tree frog has perfect camouflage with lichen-covered bark of a tree in the Amazon rain forest of Peru. The frog's color and pattern blend in so well it is nearly invisible.

Most predators also have camouflaged coloration. This is true of sit-and-wait predators like spiders, rattlesnakes, and herons and stalking predators like cougars. So relatively few animals are brightly colored and readily visible in their environments. Male birds in bright breeding plumage are an exception to this pattern, and many molt their gaudy feathers and replace them with drab-colored feathers at the end of the breeding season.

Not all animals are camouflaged, though, and many animals can risk attracting the attention of predators because they have toxic chemicals in their tissues, stings, or venom. These animals often have bright colors or eye-catching patterns that serve as *warning coloration* that the animal is dangerous if eaten or disturbed. Some of Earth's most toxic animals are tropical rain forest frogs that have extremely brightly colored skins (**Figure 26.22**). Other animals adopt a protective strategy that alerts other animals that they are dangerous. Usually, this involves black-and-white, or black-and-yellow, or black-red-and-yellow warning coloration. For instance, the black-and-yellow bands of bees and wasps advertise that they will inflict painful stings, and the black-and-white striped fur of skunks advertises the nasty-smelling chemical that they spray into a predator's face. Warning coloration is an excellent defense, and whole groups of harmless animals copy the appearance and sometimes even the behaviors of animals that are dangerous. For example, hoverflies have no stings or venom, but they are difficult to tell from bees because they copy bees' body shape, colors, and loud, buzzing flight. When disturbed, many nonvenomous snakes will rattle their tails against dry leaves or soil, making a sound that may fool a predator into thinking that the snake is a venomous rattler. Just a few seconds of hesitation may allow the snake to escape.

Species Have Varying Levels of Importance to Their Communities

If you visit an old-growth forest on the Oregon coast, you will see many individuals of a few dominant species and will be exceptionally lucky to see any tertiary or top consumers like owls, eagles, cougars, or wolves. The same is true in other communities. Dominant species in a community are those that are most noticeable and numerous. Even more ecologically important are **keystone species,** those that modify the community and enrich it for other community members but may not be present in large numbers. For instance, by eating large sea urchins that graze on kelps, sea otters are a keystone species in the underwater kelp forests off the Pacific coast, from Alaska to Baja California (**Figure 26.23**). When sea otters return to a kelp forest, the populations of sea urchins and abalone decrease, while populations

symbiosis the association of two organisms that live together in a close and often necessary relationship

mutualism a symbiotic interaction that involves two species that both benefit from the interaction

commensalism a one-sided symbiotic interaction between two species in which only one benefits from the interaction, while the other is not harmed

parasitism a symbiotic relationship in which one organism receives a benefit at the expense of another

keystone species a species that modifies the environment and enriches it for other community members

Figure 26.22 **Advertising Danger.** The bright colors of this poison arrow frog advertise its highly toxic skin.

Figure 26.23 **A Keystone Species.** By eating sea urchins that graze on kelp and destroy kelp forests, sea otters foster the community of many invertebrates and vertebrates of kelp forests.

of kelps and animals associated with kelp forests increase. Just as the keystone of an arch supports the weight of the structure, a keystone species is critically important to the health of a community. When a keystone species is removed, the community is drastically altered. Alligators are another American keystone species because they dig "gator holes" in southern swamps. In effect "gator holes" serve as nurseries and deep-water refuges where other aquatic species can survive during droughts. When alligators are removed from a swamp, it more rapidly fills in and may become a wholly different kind of plant and animal community.

Competition Occurs When Two Individuals Seek the Same Resource

In an everyday sense, **competition** occurs when more than one individual wants the same thing and only one can get it. In the natural world competition is commonplace, both within a species and between species. Individuals may compete over anything that is valuable to them including food, shelter, nutrients, water, mates, and territory. Sometimes competition ends with clear winners and losers, but in other instances the disputed resources end up being shared between the top winners. In either case competition often means that one individual will get less of a valuable resource than it would have if there had been no competition. So because competition increases the survival and breeding of some individuals at the expense of others, it becomes a part of natural selection.

The fiercest competition occurs between individuals that require exactly the same resources from the environment; generally, these are individuals that belong to the same species. It might seem odd that individuals of the same genetic group would compete rather than cooperate, but for each individual the basic strategy is to survive and reproduce. For example, competition for mates often ends with one male being able to mate and the other males not mating. In some species competition among males for mates is so severe that one party is injured or killed. This is the case in many animal species but is especially well documented for elephant seals (**Figure 26.24**). Many species have evolved more ritualized forms of competition for mates, in which fewer males die, and so more males live to have the opportunity to mate another day. In competitive rituals the stronger male symbolically establishes his preeminence, and the weaker male gives up before a serious injury occurs. In some species the larger, more colorful, or louder male automatically wins the right to mate.

competition a interaction that occurs when two or more individuals need the same resource and each works to acquire it

Figure 26.24   **Male Elephant Seals Compete for Mates.**

Nevertheless, being a male in a competitive species carries considerable risk.

Individuals of different species may compete for space, food, water, or other nutrients. This common form of ecological interaction is all around you. When weeds invade your lawn, it is a case of competition for resources between two or more species. Invasions by alien species such as the Africanized bees, imported South American fire ants, or zebra mussels are examples of competition between different species for local resources. When these alien species find themselves on American soil, they compete with native bees, ants, or mussels for food, space, shelter, and other resources. Freed of their normal environment that includes their usual competitors, predators, and parasites, some alien species flourish and become serious nuisance or pest species (**Figure 26.25**).

In many instances humans have fostered the dominance of one species over all others. This can happen when humans introduce a new species to an area. The incredibly fast-growing kudzu vine is a notorious example (**Figure 26.26**). Kudzu was introduced to the United States at the 1876 Centennial Exposition in Philadelphia, Pennsylvania. The Japanese exhibit featured the kudzu vine, and Americans liked it so much they began to plant it in gardens all through the South. Kudzu grows so quickly that southerners sometimes call it "mile-a-night vine." While that is an overstatement, the growth rate of kudzu is up to 1 foot per day and

Figure 26.25 Pest Species. The brown tree snake has been inadvertently introduced on the island of Guam and has proliferated to become a widespread pest species. Brown tree snakes have depleted bird populations on Guam. They also eat small mammals, lizards, bats, frogs, and birds' eggs.

over 60 feet per year. In the 1930s, people thought that kudzu would rescue overgrazed, overculti-vated, or clear-cut lands where soil was drying up and blowing away. Able to thrive in a wide variety of environmental conditions, kudzu did stabilize the soil, but it wasn't long before kudzu showed its aggressive side. The vine scrambles to the tops of mature trees, shuts out sunlight, stifles their growth, and kills whole forests. Kudzu's huge roots go deep, and it takes an enormous amount of effort to dig even one vine from the ground. Kudzu is emblematic of quick ecological fixes that have developed into ecological nightmares. Experiments with kudzu's natural predators are on-going, and it is hoped that soon they may be imported to control or eradicate kudzu. After kudzu gets a foothold, it exerts near complete dominance over the other plants in a community. Ecologists call complete dominance **competitive exclusion.**

QUICK CHECK

1. Why do alien species so often exhibit competitive exclusion?

Figure 26.26 Competitive Exclusion. Because kudzu grows so fast and has no predators, it easily can swallow up whole habitats, killing forest trees in the process.

26.5 Ecological Succession Is the Pattern of Community Change

Imagine this all too common scenario: it is spring. The weather is balmy, all threat of frost is past, the sun is shining, and you are eager to break ground for your new vegetable garden. You till the soil for the garden beds and plant the tomatoes, onions, broc-coli, cucumbers, zucchinis, beans, peas, melons, and herbs. You decorate the edges of the beds with insect-discouraging chrysanthemums and nastur-tiums. You add a fence that will keep out squirrels, rabbits, deer, and the neighbor's dog. Once all this work is done, you are so proud of the garden that you take its picture and imagine the baskets of beau-tiful produce that will grow in your garden (**Figure 26.27**). For about a week all you have time to do is look to see if the automatic watering system is work-ing properly. The next time you visit your garden, to your dismay something seems to be going awry: little seedlings have sprouted all over. Which ones are the vegetables and which are weeds? You pull out some but are unsure that you have guessed correctly.

For the next month your life gets too hectic for gardening. You decide to wait until the plants get big-ger so that you can tell with more certainty which ones are weeds. A week goes by, and now your once-pristine garden soil is a dense patch of weeds. Some of them are already putting out flowers. You are dis-couraged and decide to give up on gardening and buy your vegetables from the supermarket.

competitive exclusion
the complete dominance of one species over another, to the point where the population of the nondominant species approaches zero

A garden just after it is first cleared, fenced, and planted.

Two weeks later little seedlings are sprouting all over.

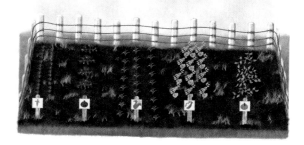
About a month later weeds are beginning to show up.

Several months later weeds have taken over the garden.

Figure 26.27 **Secondary Succession in a Garden.**

succession a predictable process of community change over time

secondary succession ecological succession in places where plants have grown

primary succession ecological succession in places where no plants have ever grown

What has happened here? Your garden is a victim of a natural, predictable process of change in communities over time that is called **succession.** In succession some species move in and grow large populations that flourish for a while and then are replaced by other species that do the same. Ecological succession is driven by the impact of species on their own environments. Species change their habitats in subtle or dramatic ways as plants compete for light, moisture, space, and nutrients. For instance, the weeds that monopolized your garden and choked out the vegetables require lots of sunlight. Other plants such as the seedlings of trees will grow up in the shade that the weeds provide. Eventually, small trees may grow above and shade out the weeds. Figure 26.27 shows what can happen to your failed garden plot in time. Notice how the assemblage of species and the dominant species change as succession proceeds. In your garden you have experienced a small-scale version of what farmers must deal with as they grow crops. Just as you would have had to expend effort or money to have kept your garden weed free, the cost of keeping crop fields free of weeds includes the farmer's time, the fuel to run machines that eliminate weeds, and the expenses associated with spraying pesticides and herbicides. So even though you may not ever have a vegetable garden, every day your life is directly affected by the force of succession.

There are two general patterns of succession that differ in the ecological conditions before succession begins. **Secondary succession** takes place in disturbed areas where plants have previously grown. This includes abandoned farm fields, road edges, landfills, gardens, and places where hurricanes or tsunamis have destroyed communities. **Primary succession** is the same process of community change, except it occurs in places where no plants have ever grown and no animals have ever lived. A prime example of primary succession is on lava fields where volcanoes recently have erupted. The first organisms to enter an area in either form of succession are called *pioneer species.* Both kinds of succession eventually lead to a plant community where change slows or stops. When this happens, a *climax community* has developed. Climax communities have dominant plant species and reflect the climate of an area.

QUICK CHECK

1. Explain how primary succession and secondary succession are similar and how they are different.

26.6 An Added Dimension Explained
Why Are There Fewer Songbirds?

More than 525 species of songbirds breed in the United States. Now it seems that the numbers of many of these species are declining dramatically. The exact numbers are hard to come by. Songbirds are small, often hard to observe, and individuals are difficult to distinguish. Nevertheless, people are so captivated by songbirds that each year thousands of birders fan out to participate in annual bird counts. The results are fed into a national database. While this kind of counting may not be as reliable as a well-controlled scientific study, it has been carried out for decades, and so annual bird counts can give a rough idea of changes in the sizes of bird populations. **Figure 26.28** shows the distribution of selected songbirds in the United States in the mid-1900s and in the early 2000s. As you can see, the populations of these specific songbirds have declined over this time, and many other songbird populations show similar declines.

To more completely understand songbird decline, let's consider the life history of a songbird. Most songbirds are migratory, wintering in southern areas, flying north in spring, and breeding in the northern areas in spring and summer. Sometimes songbirds raise two successive nests of chicks before they return to their winter homes. This complex lifestyle has many stress points, especially in a world filled with humans. Let's pick up the story at the end of the summer breeding cycle.

At the end of summer songbirds are tired from the rigors of raising their young. It has cost a great deal of energy to establish and defend a territory, find a mate and go through the appropriate courtship behaviors, build a nest, lay eggs, and feed and protect the young until they are old enough to go it alone. Songbirds are insect eaters that glean bugs from leaves, flowers, bark of trees, soil. After the breeding season, songbirds face another problem: northern winters are too cold for insects to survive. This forces songbirds to fly back south, to winter in Mexico, Caribbean islands, and Central and South America. In these tropical climates they find abundant food and can rest and recover from the breeding season. When the season turns again, once more songbirds migrate north.

Each species of songbird has evolved in a certain kind of environment, and each species has ideal summer breeding grounds and winter homes for recuperation. Most songbirds prefer a spacious, densely wooded habitat. If their winter homes are disrupted, songbirds may not be able to feed well enough to gather the strength necessary to fly back to their nesting grounds. As a consequence they may die during migration. If their summer breeding grounds are disrupted, fewer birds will hatch and survive to make the return trip to southern climates. Disruptions and environmental destruction in their northern and southern habitats or on their flyways, as migration pathways are called, are taking a toll on migrating songbirds.

You probably are aware that the world's forests are being cut down at an alarming rate. Most of the world's forests are found in the equatorial regions, and while *deforestation,* or loss of forests, is most intense in Central and South America and in Southeast Asia, deforestation is happening in forested biomes everywhere. The shrinking forests force populations of songbirds into smaller spaces where they compete for a smaller pool of insects. Not surprisingly, many songbirds are not strong enough to survive the trip back north, and the loss of habitat in tropical forests is one major reason for their decline.

But you should not think that deforestation is a problem restricted to developing countries. A great deal of deforestation has occurred in the United States too. **Figure 26.29** shows the amount of forest cover in the United States when the Europeans first came here in the 1600s, compared with forest cover at times since then. As you can see, our era of cutting down forests began long ago. You might think that there is not much more damage to be done, but you would be wrong. When forests first were cut in the United States, much of that land became agricultural land interspersed with patches of woodlands. Eventually, some farmlands returned to woodlands. In the last few decades people have fled from farms to cities, and the space occupied by cities and suburbs has expanded amazingly quickly. **Figure 26.30** shows this effect for the region around Atlanta, Georgia, one of the fastest-growing metropolitan areas in the country. What you see here is growth of urban and suburban areas, but what this means is that forests, farms, parks, and undeveloped areas have been paved and developed for human uses. The trend is that forests have become fragmented or have disappeared altogether beneath asphalt. Not surprisingly, if the songbirds come back to their breeding or wintering grounds to find streets and houses, they will not stay to breed. Worse yet, there is a good chance that displaced birds may never breed at all.

Even when developers try to save trees, the result often is a fragmentation of the forest or field environments. *Habitat fragmentation,* as this is called, is just as much a problem as complete loss of a habitat. Many songbirds need a minimal amount of space to breed, and if the developers leave just a token block of trees in a central square, this may not be enough to support breeding birds. Habitat fragmentation has other consequences, and this brings us back to how cowbirds affect songbird populations.

Cowbirds have taken their toll on some songbird populations for millennia. The normal cowbird habitat, though, is not the forests favored by most songbirds. Songbirds thrive in large, dense woods, while cowbirds are an open-country species that prefers the edges of woods where there is plenty of open space for feeding and displaying on the ground. This allows a balance between cowbirds and songbird species that can be maintained. This balance is upset when people intrude into these natural

—Continued next page

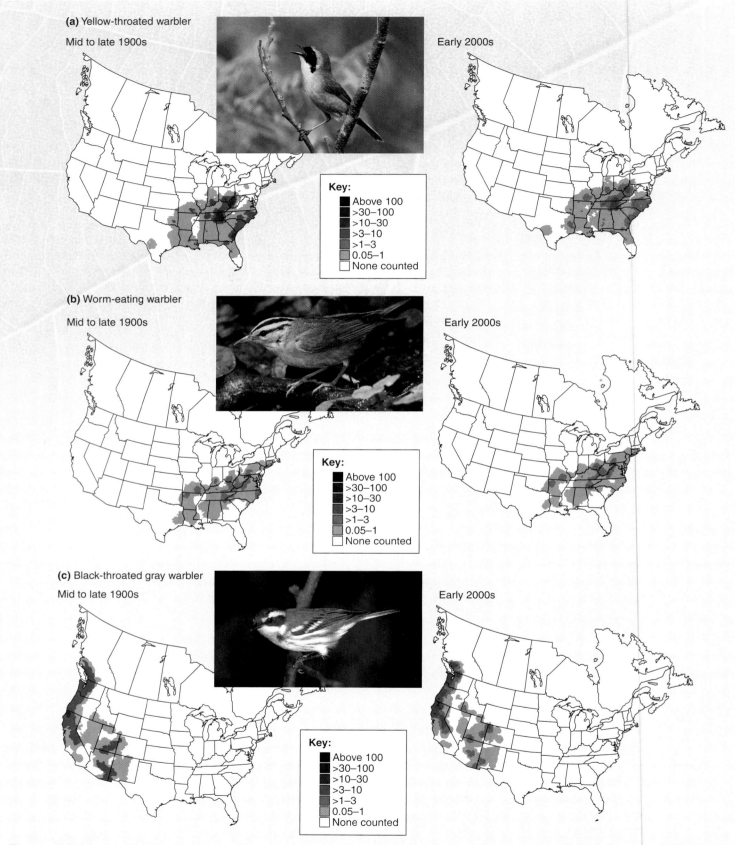

Figure 26.28 **Songbird Declines.** (a) Decline in numbers of yellow-throated warbler, (b) worm-eating warbler, (c) black-throated gray warbler. *Source:* Breeding Bird Survey.

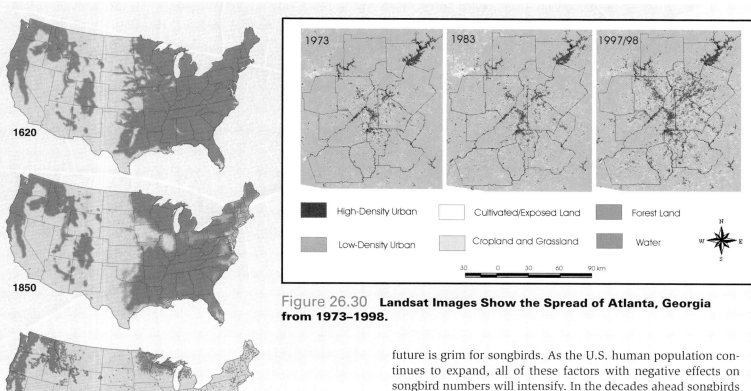

Figure 26.30 **Landsat Images Show the Spread of Atlanta, Georgia from 1973–1998.**

Figure 26.29 **Deforestation in the United States.** By the 1920s forests in the continental United States had mostly been cut down.

environments. As large forests are cut up, even only by roads, edge species such as cowbirds can move in. Habitat fragmentation produces lots of forest edges. Because no dense woodland separates songbirds and cowbirds, cowbird populations can increase at the expense of songbirds. This interference is not insignificant. Just a single female cowbird can prevent 20 to 30 warblers, finches, and vireos from raising their young.

Habitat fragmentation, deforestation in the tropics and in songbird breeding grounds, and cowbird parasitism are three reasons that songbird populations are declining or disappearing altogether. Songbirds that nest in small fragments of forest are vulnerable to increased predation too. Depending on whether they are tree- or ground-nesting songbirds, their eggs and young can become the prey of blue jays, American crows, and common grackles as well as feral and domestic cats, raccoons, rats, gray squirrels, and dogs. If this weren't bad enough, environmental toxins from insecticides also can kill songbirds. In short, the future is grim for songbirds. As the U.S. human population continues to expand, all of these factors with negative effects on songbird numbers will intensify. In the decades ahead songbirds may become even more rare or even may disappear altogether.

People are fond of songbirds and often are willing to exert efforts to save them. An upbeat story will make the point. The black-capped vireo once was an abundant songbird across much of the United States, but now it is found only in regions of Mexico, Texas, and Oklahoma (**Figure 26.31**). In the heart of

—Continued next page

Figure 26.31 **Black-capped Vireo.** Efforts of the U.S. Army at Fort Hood to foster the black-capped vireo by habitat restoration and removal of cowbirds have had dramatic results in increasing the population of this endangered songbird.

Continued—

black-capped vireo breeding grounds is the U.S. Army base at Fort Hood, Texas. You might be thinking, "Uh oh, the birds are in trouble," but read on. In between training exercises army personnel at Fort Hood help the black-capped vireo population to recover. Perhaps they cannot do much about habitat fragmentation, but they can enhance breeding locations by planting stands of the kinds of grasses and trees that the vireos choose for breeding. This is not just an effort of individual committed soldiers; it is U.S. Army policy and so is carried out at all levels. The Fort Hood soldiers also try to minimize the impact of cowbirds on black-capped vireos. Soldiers remove cowbird eggs from songbird nests, and they set up traps to catch and kill cowbirds. All of this has paid off remarkably well. With a concerted and officially supported effort, the population of black-capped vireos has gone from 85 pairs in 1987 to 3,000 pairs in 2003. This kind of effort may not save the songbirds across America, but it shows that local efforts can make a big local difference.

QUICK CHECK

1. How are habitat fragmentation and deforestation similar? How are they different? How are the two connected to cowbird parasitism and loss of songbirds?

Now You Can **Understand**

Some Countries Are Close to Negative Population Growth

A look at the details of population growth in countries around the world shows some surprising trends. While the focus has been on increases in the world human population, individual countries show much different population growth rates. Many countries are showing growth rates of about 1 to 2% per year, and these rates are declining. Some regions of Africa and Asia, however, still have growth rates above 2% per year. Still other regions, though, such as Japan and northern Europe, Italy, Germany, and Russia are close to or at zero-population growth, and their populations are expected to decrease over the next few decades. The societal impacts of such negative population growth are unclear.

What accounts for these varying trends? One big factor is economics, in particular the economic status of women. Population growth rates fall when women have economic security that is independent of men, and this translates to access to education and job opportunities. On top of these factors, access to birth control for women also drives population growth rates down. These factors related to the status of women account for much of the difference in growth rates between different countries. The United States is interesting in this regard because it is among those countries where women have economic power, yet the U.S. population is growing at a modest rate. How can this be? The answer is largely immigration. Not only does immigration add people to the U.S. population, but also immigrants bring their own cultural values. In general, birthrates are much higher among populations of immigrant women than among women born in the U.S.

What Do **You** Think?

How Many Children Do You Plan to Have?

People who are concerned that the Earth already has too many people are adamant that everyone should limit family size to one or two children. Others see this as an issue of personal rights and argue that people should be able to have as many children as they can support. The controversial, but apparently successful, efforts by the Chinese to limit family size add an additional dimension. If the United States were anticipating a population of 1 billion in 30 years—rather than the expected 420 million—would efforts to limit family size be more justified? Given its immigration pressures, how far should the United States go in contributing to the control of birth and growth rates in other countries? These are complex questions and ones that you will certainly have to face in your lifetime. You will find more information about trends, opinions, and policies worldwide online. After consulting a few websites, formulate your own views on if and how human population growth should be managed. What do *you* think about how many children you should have?

Chapter Review

CHAPTER SUMMARY

26.1 Ecology Is the Study of Patterns of Interactions Between Organisms

Ecology is the study of patterns of interactions of organisms that affect their distribution and abundance. Ecology is different from environmental science or environmental activism. Environmental science uses ecological principles to solve practical human problems such as the production of greenhouse gases; environmental activism aims to prevent environmental destruction.

biome 627
biosphere 627
community 627
ecology 627
ecosystem 627
population 627

26.2 Populations Have Characteristics Related to Their Rate of Increase

Populations live in typical habitats and are characterized by demographic summary data such as life tables, survivorship curves, age structures, and population growth rates. The carrying capacity (K) of the environment is its ability to support a population. K-selected species have adaptations that keep their population numbers at or near environmental carrying capacity, while r-selected species have adaptations for rapid reproduction. Populations also have characteristic population growth curves. J-shaped curves are typical of populations with exponential growth, while S-shaped curves are characteristic of populations with logistic growth. The effects of density-dependent population controls increase as population numbers increase, while density-independent population controls are unrelated to population numbers.

age structure 631
carrying capacity (K) 632
density-dependent population
   control 634
density-independent population
   control 634
exponential population growth 633
habitat 628
K-selected species 633
life table 629
logistic population growth 634
r-selected species 633
survivorship curve 629

26.3 Analysis of Human Populations Helps to Predict Their Future Growth

By 2050 the global human population is projected to increase to around 9 billion people. Ecological footprints allow assessment of the share of resources that individuals consume. The heavy-consuming typical American lifestyle is unsustainable. Regional human populations grow at different rates. AIDS is having a negative effect on human population growth, especially in southern Africa.

26.4 Populations Interact Within Ecological Communities

Populations interact in various ways in ecological communities. Populations are associated in food chains and food webs. A species' ecological niche includes its habitat and all of the abiotic and biotic resources that sustain it. Mutualism, commensalism, and parasitism are forms of symbiosis. Camouflage and warning coloration encompass adaptations

that enhance survival. Keystone species influence their habitats, altering them in ways that promote the survival of other species. Competition is a common interaction between individuals and between species. Introduced species have fewer natural population controls and may exert competitive exclusion.

carnivore 638
commensalism 641
competition 642
competitive exclusion 643
consumer 638
detritivore 638
ecological niche 639
food chain 638
food web 639
herbivore 638
keystone species 641
mutualism 641
omnivore 638
parasitism 641
predator 638
prey 638
producer 638
symbiosis 641

26.5 Ecological Succession Is the Pattern of Community Change

Primary and secondary succession are two patterns of changes in plant communities.

primary succession 644
secondary succession 644
succession 644

26.6 An Added Dimension Explained: Why Are There Fewer Songbirds?

Songbirds are declining due to habitat fragmentation, deforestation in tropical regions as well as in the United States, and increased predation. Nest parasitism by cowbirds has intensified declines in populations of many songbirds. When they are protected from cowbirds, and their habitats are enriched, songbird numbers can rebound.

REVIEW QUESTIONS

TRUE or FALSE. If a statement is false, rewrite it to make it true.

1. In the year 2050 it is predicted that Earth will hold close to 9 billion humans.

2. A goldfinch is an herbivorous predator that eats seeds.

3. Plants function as consumers in ecological communities.

4. Humans and the gut bacteria that synthesize vitamin K are in a parasitic relationship.

5. Plants growing in landfills is an example of primary succession.

6. Songbirds are declining only because of intense parasitism by cowbirds.

MULTIPLE CHOICE. Choose the best answer of those provided.

7. Collecting data on the size and kinds of fishes caught by breeding roseate terns, an endangered species of seabird, is a typical of
   a. environmental activism.
   b. ecology.
   c. environmental science.
   d. environmental medicine.

8. The pioneering studies of Elton and Summerhayes identified
   a. relationships of cowbirds and songbirds.
   b. the limits of the biosphere.
   c. feeding relationships between populations.
   d. exponential population growth.

9. Ecological units in which populations interact are
   a. superpopulations.   d. the biosphere.
   b. biomes.   e. communities.
   c. ecosystems.

10. The habitat of a species is
    a. the same as its ecological niche.
    b. a facet of its distribution and can be clumped, uniform, or random.
    c. the place in its geographic range where it usually is found.
    d. its job in the community.

11. Life tables do not show
    a. when death occurs in a population.
    b. survivorship of each age class in a population.
    c. age structure in a population.
    d. relationships to other members of a community.

12. Predict the future growth of a population that has an age structure shaped like an inverted pyramid. In the future the population will
    a. increase in size.
    b. decrease in size.
    c. remain the same.
    d. Age structure does not allow such predictions about a population.

13. What is the best description of a population that has reached biotic potential?
    a. It is growing as fast as it can.
    b. It is growing at a low rate.
    c. It is growing moderately.
    d. It is in danger of going extinct.

14. What does K stand for?
    a. biotic potential
    b. the ability of the environment to support a population of a certain size
    c. a population experiencing exponential growth
    d. survivorship

15. What happens when a population of a K-selected species reaches K?
    a. The population continues to grow indefinitely.
    b. The population switches to become an r-selected species.
    c. The population begins to grow at its biotic potential.
    d. The population decreases.

16. How is population growth similar to growth of a savings account that has compound interest?
    a. A standard number of organisms is regularly added to the population.
    b. The individuals added reproduce and, in turn, add to the population.
    c. A standard number of organisms is regularly subtracted from the population.
    d. The population decreases.

MATCHING.

17–20. Match the columns. (One choice will not be used.)

17. J-shaped curve.
18. density-dependent control
19. ecological footprint
20. women's lives improve

a. population size decreases
b. how many resources an individual uses
c. exponential population growth
d. infectious disease
e. logistic population growth

CONNECTING KEY CONCEPTS

1. What role does carrying capacity play in the population growth of K-selected and r-selected species?

2. How does what an organism eats contribute to its ecological niche?

QUANTITATIVE QUERY

The mathematics used to predict population sizes can be complex, but you can get a feel for population growth by doing simpler calculations of a hypothetical population. Imagine that you have just bought a pair of gerbils, and you would like to anticipate how many gerbils you might have in a year. To do the calculations, you must make some simplifying assumptions. These assumptions are not completely accurate, but you still will get an idea of how a gerbil population might grow. You could change the assumptions and see how your final population would change. Assume the gerbils breed every 3 months, and that the young are able to breed when they are 3 months old. So every 3 months every female in your population has an average of four offspring. Every offspring survives to breed, and there are no deaths in the year. Assume also that in each generation the number of female offspring equals the number of male offspring. Finally, assume that each female gerbil has four offspring every 3 months. How many gerbils will you have after one year?

THINKING CRITICALLY

1. Human population size has been increasing rapidly in the past 100 years and now is in an exponential growth phase. Some people have argued that human population size will overshoot the Earth's carrying capacity for our species, and that at some point in the future the human population will experience an extreme crash. According to this hypothesis, the human species is doomed to experience a period when many people die, and the human population is reduced to just a fraction of its peak size. Ignoring what you read from the experts, and based on what you have read in this chapter about K-selected and r-selected species, what would you predict about the future of human population growth? Why do you reach this conclusion?

For additional study tools, visit www.aris.mhhe.com.

Ecosystems and Biomes

5 States to Confer With Federal
Aides on Lake Erie Pollution
August 1, 1965

WALTER E. WATSON

Lake Erie Called a Health Danger
U.S. Pollution Report Calls for Remedial Action

July 31, 1965

Cuyahoga River, 1969. This oil-choked river
carried industrial pollution into Lake Erie. It is
the river that was so polluted that it caught on
fire in 1969.

Lake Erie in the 1960s and 1970s was the most badly polluted of the five Great Lakes. Parents did not allow their children to run along Lake Erie's beach because it stank of dead fish and overgrown algae. And of course, you would never *eat* anything that came out of Lake Erie because everyone knew that the fish were loaded with toxins. As if all of this were not bad enough, in Ohio in 1969 the Cuyahoga River, which empties into Lake Erie, actually caught fire, burned, and set fire to a bridge (**see chapter-opening photo**). The Cuyahoga was so choked with oil and other pollutants that a tossed match set it aflame. Johnny Carson, then the renowned host of the *Tonight Show*, quipped, "Lake Erie is where fish go to die."

At the time all the Great Lakes were polluted ghosts of their former glory, but Lake Erie was dead. It was ugly. Plain and simple. Kids no longer splashed in Lake Erie's waves, tried to catch fishes, built sand castles, or raced along the shore. Idyllic summer days at Lake Erie's beaches were just a memory.

But conditions weren't always so bleak. The enormous Great Lakes, which contain 20% of the world's freshwater, were once clean and beautiful. When European settlers first arrived at the shores of Lake Michigan, they thought they had reached the Pacific Ocean, but then realized that the water wasn't salty. Native Americans called the Great Lakes "sweetwater seas." For generations the lakes were unpolluted and filled with life. Fishing fleets netted big fishes like pike, sturgeon, salmon, and herring. The Great Lakes were a jewel of North America's natural heritage and should have been cherished and protected for future generations to enjoy. Yet by the 1960s or 1970s all of the Great Lakes were a mess, and Lake Erie was the worst of them.

What has happened to Lake Erie since the 1970s? At the end of this chapter you will refocus your attention on Lake Erie and read about its current state. First you will explore the subject of this chapter—larger ecological units: ecosystems, biomes, and the biosphere. ■

27.1 An Ecosystem Is the Functional Ecological Unit

Living organisms interact with one another to form ecological communities, and they also interact with the nonliving environment. An **ecosystem** is an ecological community plus the **abiotic** or nonliving features of its environment—such as water, soil and soil nutrients, rocks, and atmospheric gases.

The concept of an ecosystem may be easier to understand if you compare it to a town. The people, animals, plants, and other organisms that live in this town are its biotic components; the buildings, streets, roadways, soil, rocky landscape, and atmosphere are some of its abiotic components. So are its electrical supply, water and sewage facilities, and landfill. Just as the town cannot function without people, the town cannot function without the buildings and utilities. *Ecology* concentrates on the interactions between organisms and their environments. The interdependent biotic communities of an ecosystem rely on, and are affected by, the abiotic environment. In ecosystems populations affect one another in immediate or distant ways as they obtain energy and use and/or recycle the chemical elements essential to life.

Each geographic region has many ecosystems, and you can see a variety of ecosystems even within a relatively short distance. As examples, consider just two different ecosystems that you would see on a 50-mile "eco-trip" from Jasper County, South Carolina, across the Georgia state line and east to the Atlantic coast (**Figure 27.1a**).

In Jasper County you would travel through long-leaf pine flatwoods: a distinctive forest ecosystem where tall, long-leaf pine trees grow in soil with a thick layer of clay beneath it (Figure 27.1b). You might see a fox squirrel scramble up the trunk of a tree, or hear the call of a red-bellied woodpecker. The thin, sandy soil of the forest floor is littered with fallen pine needles and pine cones. You notice deer and turkey tracks and watch your step because diamondback rattlesnakes also live in flatwoods. Here and there are openings in the pines where sunny, wet meadows are dotted with carnivorous plants that live only in the acidic bogs associated with the pine flatwoods ecosystem.

Next you drive about 50 miles (80 kilometers) east to the Atlantic coast to an entirely different ecosystem—the coastal dunes. The grass-tufted dunes that front the beach are the dominant feature of this ecosystem—that of the sand dunes near the ocean. Here the wind and salt spray blown off the ocean shape the vegetation (Figure 27.1c). Behind the dunes ground-hugging shrubs and trees grow in low, wind-twisted groves. These plants have small, tough leaves and can withstand windblown salt spray, scouring by sand, and extremely dry conditions. You see the tracks of mice, rabbits, and foxes and hear the cries of gulls.

ecosystem all of the biotic and abiotic factors in a defined area

abiotic nonliving components of the environment

QUICK CHECK

1. What are the biotic and abiotic components of the ecosystem that surrounds you right now?

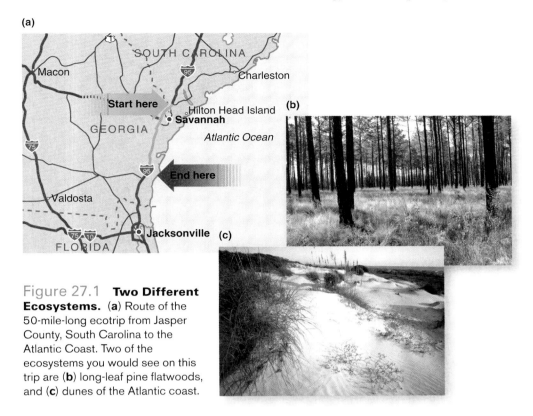

Figure 27.1 **Two Different Ecosystems.** (**a**) Route of the 50-mile-long ecotrip from Jasper County, South Carolina to the Atlantic Coast. Two of the ecosystems you would see on this trip are (**b**) long-leaf pine flatwoods, and (**c**) dunes of the Atlantic coast.

27.2 Energy Moves Through Ecosystems, and Chemicals Cycle Within Ecosystems

Ecosystems are characterized by the movement of energy and chemicals. Natural processes and ecological relationships move energy and chemicals between and among abiotic and biotic components of an ecosystem. The movement of chemicals through and between ecosystems is different from the movement of energy. Any given chemical—either an atom or a molecule—can be indefinitely cycled through ecosystems. Although chemicals are changed as they move from one part of an ecosystem to another, chemicals are re-used. The Sun's energy also is passed from one organism to another, but unlike chemicals, energy is not recycled indefinitely. Let's consider how energy moves through ecosystems, and then see how chemicals are cycled and re-cycled.

Energy Is Lost at Each Link in a Food Chain

The Sun's energy flows in a one-way direction through an ecosystem: from the Sun to producers to various levels of consumers (**Figure 27.2a**). Producers capture the energy in sunlight and convert it to energy stored in the bonds of biological molecules. Some of the energy in the molecules of producers is used by the consumers that eat them. When the producers and consumers die, some of their chemical energy is taken up and used by detritivores such as fungi and bacteria. It may sound like energy is recycled in this food chain, but at each link energy is lost to the environment as heat. Eventually none of the solar energy, captured by a given producer is available for other organisms to use. This is why life requires a constant supply of new energy.

Photosynthesis carried out by producers captures and uses *less than 2%* of the solar energy that reaches Earth's surface. Nevertheless, once it has been captured, transformed by photosynthesis, and fixed in the carbon-to-carbon bonds of glucose molecules, this tiny fraction of solar energy supports nearly all life. As this energy is transferred from one link in a food chain to another, energy is lost. Because it costs energy to run an organism's metabolism and life activities, about 90% of the energy that an average organism takes in is used up. Only about 10% of the energy is stored in the chemical bonds in an organism's tissues. So a snail that eats a plant's leaves does not have access to all of the solar energy that the plant absorbed (Figure 27.2b). This *rule of 10%* continues up the food chain. For instance, the shrew that eats a snail gets only about 10% of the energy that the snail got from the leaves

(a) Flow of energy in an ecosystem

(b) A pyramid of energy

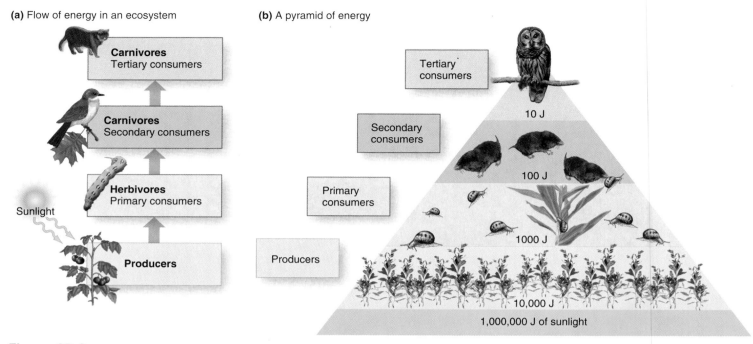

Figure 27.2 **Energy Flow in an Ecosystem.** (**a**) Energy flows in a one way direction from producers to various levels of consumers. (**b**) Because on average only 10% of energy available to an organism is stored in its tissues, pyramids of energy are typical of organisms that are linked in food chains. J stands for Joule, a unit of energy.

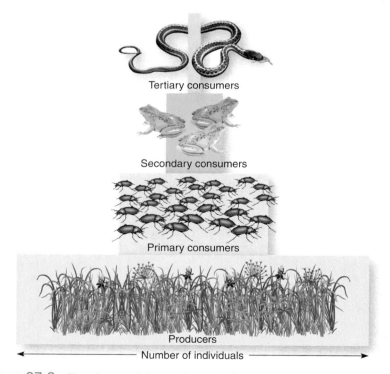

Figure 27.3 **Numbers of Organisms Reflect Available Energy.** Because more energy is available to lower links of a food chain, in any community there are more individual producers than primary consumers, more primary consumers than secondary consumers, and more secondary consumers than tertiary consumers.

it has eaten. The metabolic processes and activities of the shrew also have a high energy cost. Again, about 90% of the energy that the shrew takes in is used up, and only about 10% is stored in its tissues. This drastic loss of energy at each link in a food chain is the reason that there are few terrestrial food chains with more than three or four links. *Pyramids of energy* show the energy available to each link in a food chain. They demonstrate the loss of energy in metabolic costs as energy flows through a food chain.

As you move up the links of a food chain, the number of individuals in each population decreases. A *pyramid of numbers* counts the individuals at each level or link in a food chain (**Figure 27.3**). In any ecosystem there are many more individual producer organisms than individual primary consumers. Similarly, there are more primary consumers than secondary consumers and more secondary consumers than top consumers. As you move up the links of a food chain, the number of individuals in each population decreases. The loss of energy at each link of a food chain restricts the amount of energy available to higher-level consumers—which is why you'll see mostly producers when you look at a landscape. There are fewer primary consumers and even fewer top consumers.

Natural Processes Recycle Chemical Elements

Although the energy for Earth's life comes primarily from the Sun, everything else that life must have to survive must be obtained here on Earth. The chemical elements necessary for life, notably C, H, N, O, P, S, Ca, and K plus many more (see Table 2.1) were available to the first cells that evolved nearly 4 billion years ago; they still are required by living cells. But because chemicals are recycled, many living cells, including some of the cells in your body, contain the *exact same* atoms and molecules that were present billions of years ago! For instance, although some water molecules are broken down and re-formed during metabolism, many water molecules pass unchanged from one cell to another or from one organism to another. Furthermore, water in rivers and streams as well as the water that comes out of the kitchen faucet contains some of the same water molecules that dinosaurs and other prehistoric animals drank and excreted millions of years ago.

Biogeochemical cycles move chemicals through the biosphere, passing them through organisms and abiotic components of the biosphere such as atmosphere, oceans and freshwaters, soils, and rocks. Like a wheel with no beginning and no end, a cycle is a continual process of transformations. The basic components of a cycle may be used over and over in slightly different forms, but they always return to the original form to begin the cycle again.

Each living organism is a part of many different biogeochemical cycles. Although the details of these cycles differ, they follow similar patterns. First, biogeochemical cycles are driven by energy from the Sun that powers biological and physical processes such as photosynthesis, evaporation, condensation, and precipitation. Second, each biogeochemical cycle involves *reservoirs* where chemicals are stored or concentrated for long or short periods of time. Reservoirs of various chemicals can be within the bodies of organisms, or within abiotic components of the environment. Third, biogeochemical cycles function on both local and global levels, linking distant ecosystems. Earth has many biogeochemical cycles, and they demonstrate that Earth is a closed system. Elements are recycled, not replenished from some outside source. This recycling of materials is one reason that environmentalists often call our planet Spaceship Earth. Biogeochemical cycles demonstrate how apt this name actually is.

Physical and Chemical Processes Cycle Water

Life must have water. All of the biochemical interactions required for life take place in a watery environment (see Section 4.2). You might think that because water covers three-quarters of Earth's surface, getting water would not be a problem for organisms, but you would be wrong. Ocean water is called salt water because it has high concentrations of dissolved sodium, potassium, chloride, and other chemicals. In contrast, freshwater has low levels of these chemicals. While ocean water is fine for the species adapted to live in it, terrestrial or freshwater species lack adaptations necessary to tolerate salt water. About 97% of the water on Earth is salty; only 3% is fresh. In the water cycle freshwater is continually purified by evaporation and recycled by other natural processes.

Let's begin the water cycle with all that salty ocean water, the largest reservoir in the water cycle (**Figure 27.4**). Even though the oceans are salty, they are the primary source of freshwater on Earth. The heat of the Sun causes ocean water to evaporate into the atmosphere. The process of **evaporation** transforms liquid water molecules into *water vapor*, the gaseous form of water. As ocean water evaporates, the dissolved salts are left behind. This means that water evaporated out of the ocean is freshwater. Evaporation from the oceans contributes to atmospheric freshwater that can come down on land as rain, sleet, hail, or snow. Water molecules in water vapor are spread out because the hydrogen bonds that once held them together have been broken. If these diffuse water molecules are to come down as rain, snow, sleet, or hail, they must get closer together; in other words, they must *condense*. In the process of condensation molecules in water vapor get close enough together that hydrogen bonds form, resulting in liquid water or solid ice. **Condensation** can occur when a mass of warm air holding water vapor hits a mass of colder air. At the interface between warm and cold air, water vapor condenses, forming drops of rain. If the temperature is above freezing, rain will fall. If it is freezing or close to it, hail, sleet, or snow will fall. These forms of water that fall from the sky are grouped under the technical term, **precipitation.** All solid forms of precipitation can melt into liquid water that can run off into lakes, rivers, and oceans, and seep into the ground, accumulating in **aquifers.** These are porous layers of underground rock or sand that hold large quantities of water. Run off also is taken up by plants and animals. The process of *transpiration* in plants also contributes

biogeochemical cycle the circulation of a chemical as it passes through biotic and abiotic components of the biosphere

evaporation the process in which molecules of liquid water are transformed into water vapor

condensation the process in which water molecules come close together, form hydrogen bonds, and turn into liquid water or ice

precipitation water that falls from the atmosphere to Earth in the form of rain, snow, sleet, or ice

aquifer a porous layer of underground rock or sand that holds large quantities of water

Figure 27.4 **The Water Cycle.** Water that evaporates from the ocean falls as precipitation over land, and flows through lakes, rivers, and aquifers back to the ocean.

■ *What two processes of the water cycle purify water?*

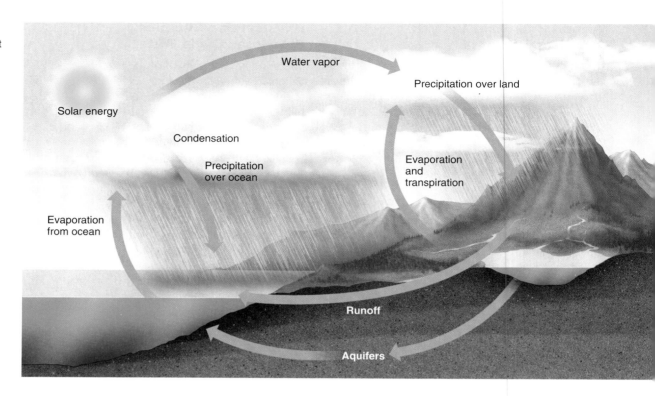

water vapor to the water cycle as water passes through the bodies of plants and evaporates from leaves (see Section 18.2). Because Earth is a closed system, precipitation eventually returns to its starting point in the oceans.

Most of the water vapor that evaporates from the ocean falls right back into it as precipitation, but wind drives some clouds over land, and so precipitation reaches Earth's continents. Some precipitation is absorbed into the soil. There it may stay or collect in aquifers. Other rainwater flows over the land, perhaps finding its way to creeks, lakes, or rivers.

Photosynthesis and Respiration Cycle Carbon

Carbon is the central element in all biological molecules (see Section 3.2). Biological carbon is found in molecules such as sugars, lipids, amino acids, and nucleotides. Other carbon molecules associated with organisms include carbon dioxide (CO_2) in the atmosphere and calcium carbonate ($CaCO_3$) in *limestone* rocks. Carbon cycles among five major compartments (**Figure 27.5**): dissolved carbon dioxide in water reservoirs such as oceans, lakes, and rivers; carbon dioxide in the atmosphere; carbon compounds in tissues of living organisms; carbon compounds in limestone and other rocks; and carbon compounds in deposits of fossil fuels produced from the remains of living organisms. Notice that only one of these compartments is biotic. Exchanges between

the biotic and abiotic compartments of the carbon cycle result from several processes.

The largest reservoirs of carbon are the carbon dioxide in the ocean and in the atmosphere. Other processes add or subtract carbon dioxide from these reservoirs. For instance, occasional volcanic eruptions contribute 3% of the carbon dioxide in the atmosphere. Biotic processes move a large amount of carbon dioxide into and out of the atmosphere and oceans. Aerobic respiration returns carbon dioxide to the environment, while photosynthesis removes carbon dioxide from the environment and uses it to form carbon-based biological molecules. The metabolic processes of photosynthesis and respiration keep carbon rapidly cycling to and from reservoirs in waters and the atmosphere.

Other metabolic processes store carbon compounds in tissues of living organisms for longer periods. For example, the glycogen and fat deposits in animals and the starch deposits in plants remain for long-term use and are not rapidly metabolized to carbon dioxide. When energy is needed, stored carbon compounds can be broken down, and carbon dioxide is released in the process. Carbon-based compounds are incorporated into chitin and calcium carbonate in exoskeletons, bones, shells, and coral skeletons. When marine and freshwater organisms die, their calcium carbonate shells settle to the bottom of a body of water. Over eons this sediment accumulates, forming limestone rocks that are extremely long-term storage for carbon dioxide. It

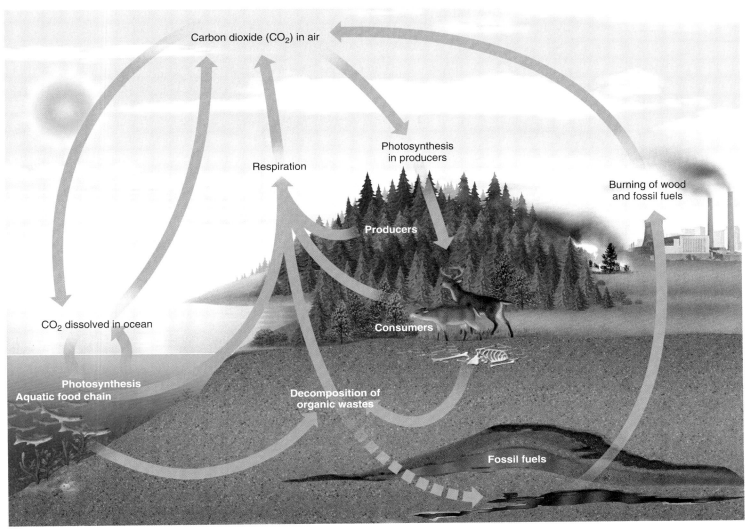

Figure 27.5   **The Carbon Cycle.** Carbon dioxide in the atmosphere and in bodies of water is taken up by photosynthesizers and used to make carbon-based biological molecules. Consumers eat photosynthetic organisms, and eat each other, and make their own carbon-based biological molecules. Cellular respiration burns glucose and other biological molecules and so returns carbon dioxide to the atmosphere and to oceans, rivers, and lakes. Human burning of fossil fuels also returns carbon dioxide to the atmosphere.

may be millions or even billions of years before chemical reactions move some of the stored carbonates back to the atmosphere as carbon dioxide.

Decomposers are crucial parts of biogeochemical cycles. In the carbon cycle decomposers extract energy from the carbon molecules of dead organisms and release carbon dioxide. Biological molecules not immediately degraded may enter other long-term storage compartments of the carbon cycle, where physical and chemical processes may transform them into fossil fuels such as peat, coal, oil, or natural gas. Peat is compressed organic matter that forms over centuries. Coal, oil, and natural gas are incompletely decomposed deposits of organic material that have been compressed to form substances composed largely of hydrocarbons. It takes millions of years to form deposits of coal, oil, and natural gas.

When fossil fuels are burned, their carbon-to-carbon bonds are broken. This releases energy and returns carbon to the atmosphere as carbon dioxide. Burning fossil fuels adds other greenhouse gases to the atmosphere and contributes to global warming. It also adds acidic compounds to the atmosphere and can contribute to acid precipitation and smog. Levels of carbon dioxide in the atmosphere have fluctuated over geologic history, reflecting geological processes and biological events (see Figure 6.24). In the last 200 years or so, however, levels of atmospheric carbon dioxide have increased dramatically (see Section 6.5). The overwhelming scientific consensus is that increased levels of carbon dioxide and other greenhouse gases are heating up Earth's atmosphere. Most scientists are convinced that global climate change is real. Although no one is certain what

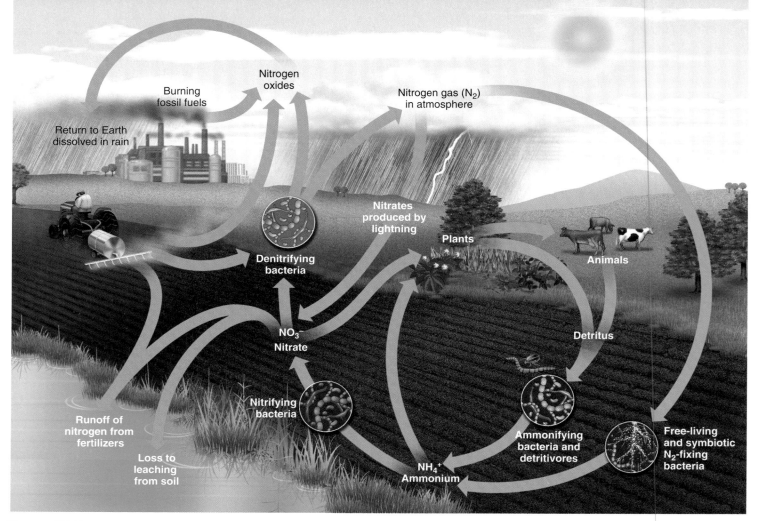

Figure 27.6 **The Nitrogen Cycle.** Atmospheric nitrogen cannot be directly used by living organisms, but some soil bacteria can convert nitrogen to useable compounds. Nitrogen fixing bacteria convert atmospheric nitrogen to ammonia and other compounds. Nitrifying bacteria convert ammonia to nitrate compounds that can be used by plants, animals, fungi, and other organisms. Denitrifying bacteria return nitrogen to the atmosphere.

all of the effects of global climate change will be, high-altitude glaciers, ice caps, and ice sheets are melting at record rates, global temperatures have risen by an average of 0.3 to 0.6°C (0.5 to 1.0°F), and biological seasons are shifting.

Prokaryotes Are Key Players in the Nitrogen Cycle

Nitrogen is a critical component of nucleic acids and proteins, and each day organisms require fresh supplies of these biomolecules. Because nitrogen is the most abundant gas in Earth's atmosphere, obtaining nitrogen could be as simple as breathing. But most organisms cannot use nitrogen gas directly from the atmosphere. Only certain bacteria can split molecules of atmospheric nitrogen (N_2) into single atoms of nitrogen for use in their own metabolic processes. Directly or indirectly most other organisms depend on these bacteria to incorporate atmospheric nitrogen into compounds like

nitrates (NO_3^-) and ammonia (NH_3) that contain single nitrogen atoms. Bacteria that can do this include cyanobacteria and certain soil bacteria that are the focus of the nitrogen cycle.

The technical term for incorporating an atom into a chemical form suitable for cycling through living organisms is *fixing*. For example, some bacteria called **nitrogen-fixing bacteria** take in molecules of atmospheric nitrogen and fix the nitrogen into ammonia molecules that are released into the soil (**Figure 27.6**). While many nitrogen-fixing bacteria live in soil and water, swellings on the roots of some plants contain colonies of nitrogen-fixing bacteria. This is a mutualistic relationship: the plant receives a supply of ammonia, while the bacteria have a sheltered environment and direct access to small carbon-based molecules to use as sources of energy and building materials. Plants in the legume family such as peas and clover commonly have swellings on their roots that contain nitrogen-fixing bacteria, and other plants also have them.

nitrogen-fixing bacteria soil bacteria that incorporate atmospheric nitrogen into nitrogen-containing compounds that other organisms can use

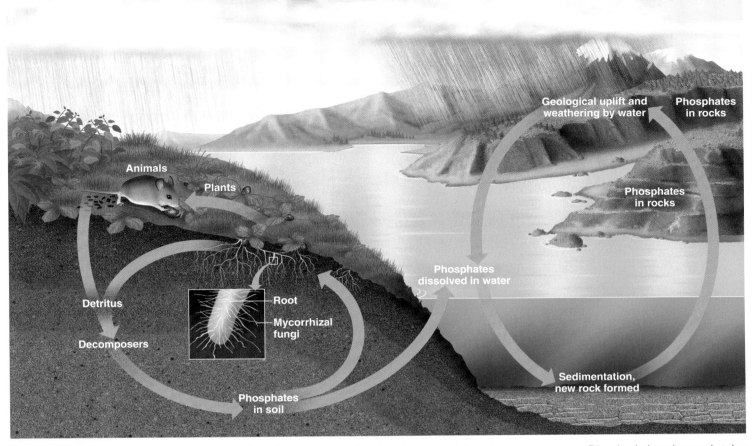

Figure 27.7 **The Phosphorus Cycle.** Phosphorus in rocks dissolves in oceans, lakes, rivers, and streams. Dissolved phosphorous is taken up by photosynthetic organisms. Animals and fungi gain phosphorous by digesting photosynthetic organisms.

Many eukaryotes cannot use ammonia as a nitrogen source and need another form of nitrogen as the basis for their biological molecules. They get these other forms of nitrogen from metabolic processes in bacteria. Some bacteria convert ammonia (NH_3) to nitrites (NO_2^-) that are released into the soil. Still other bacteria take up nitrites and convert them to nitrates (NO_3^-). Plant roots take up nitrates and can use this form of nitrogen in the synthesis of amino acids, which are then used to make plant proteins. Animals get most of their nitrogen in the form of amino acids from plants. Some animals eat plants and use plant amino acids to build their own proteins. These animals are food for yet other animals.

Through these biogeochemical transformations of nitrogen from one molecule to another, atmospheric nitrogen becomes available to different life-forms. In this half of the nitrogen cycle, different kinds of bacteria remove nitrogen from the atmosphere and transform it into nitrogen-containing compounds that organisms use. The remaining half of the nitrogen cycle is completed by other organisms *returning* nitrogen gas to the atmosphere. Although there is an enormous volume of nitrogen in Earth's atmosphere, if nitrogen were continually drained from it, drastic chemical changes eventually would occur. Chemical processing within a dif-ferent sort of bacteria returns nitrogen to the atmosphere. These bacteria live in mud or other anaerobic environments. They take in nitrates, nitrites, and other compounds that contain nitrogen and oxygen, use the oxygen in their own cellular processes, and release nitrogen back to the atmosphere as a waste product. Because these bacteria *remove* nitrates and nitrites from soil, they are called **denitrifying bacteria.** Notice that bacteria do most of the recycling of nitrogen. Also notice that if one component of the nitrogen cycle is missing, the nitrogen cycle cannot operate.

The Phosphorus Cycle Has No Atmospheric Component

Phosphorus is an essential component of nucleic acids, ATP, and the phospholipids that form cell membranes. The phosphorus cycle has no atmospheric component and takes place within food chains on land and in water. Rocks and soils are important reservoirs of phosphorus (**Figure 27.7**), but until rocks erode, phosphate ions are locked up within them and are unavailable to organisms. When rocks and sediments erode, phosphates are released. Plants must have phosphates to grow well. Mycorrhizae absorb phosphates and pass

denitrifying bacteria bacteria that transform nitrites and nitrates and release nitrogen to the atmosphere

them on to plants. Because phosphates are not very soluble in water and are not found in all rocks, lack of phosphates limits plant growth. Once phosphates are incorporated into plant tissues, they enter food chains and are passed from organism to organism. When organisms die, decomposers return phosphorus to soil and water. From there geological processes incorporate phosphorus into soils, rocks, and water.

QUICK CHECK

1. Make a chart that compares the long- and short-term processes of the carbon cycle, nitrogen cycle, water cycle, and the phosphorus cycle.

27.3 Ocean Life Is Influenced by Food, Light, Currents, and Pressure

To see the most varied forms of life, you've got to go to sea. Some animal phyla have no representatives that live in freshwater or on land. For instance, echinoderms such as sea stars, sand dollars, and sea cucumbers live only in the oceans. Other animal phyla are mostly marine but have a few freshwater representatives. For example, the flamboyant diversity of marine sponges and cnidarians makes the few freshwater species of each phylum seem pale by comparison. Still other animal phyla such as mollusks live in oceans, in freshwater, and on land—but are wildly diverse in the oceans. While there are snails and slugs on land and clams and snails in freshwater, mollusks like chitons, nudibranchs, octopi, and various species of nautilus live only in the oceans. And the comparatively few species of land and freshwater mollusks cannot rival the spectacular diversity of oceanic clams and snails.

Insects are one major terrestrial group that is not found in the oceans. Instead, crustaceans like crabs, lobsters, copepods, and barnacles are common and varied in oceans. Ocean vertebrates include sharks, rays, skates, bony fishes, snakes, crocodiles, turtles, penguins, shore birds, seals, walruses, and whales. Why should more animal phyla live in the oceans than in other environments? One reason is that life first evolved in the oceans. The transition to life in freshwater came later, and the move to dry land happened even later. So the oceans are life's earliest home, and some forms of life never have left. Although the oceans are home to the greatest *variety* of animals, the oceans do not have the greatest *number* of animal species. Insects and seed plants evolved on land and proliferated into many species there; later

they made the transition to life in freshwater. Only a few species of seed plants or insects are adapted to life in the oceans.

Phytoplankton Are the Basis of Ocean Food Chains

Another way that life in water is different from life on land concerns the organization of food chains. Plants are the producers in land-based food chains (see Section 26.4). In the oceans the major producers are free-floating microscopic algae called *phytoplankton* (**Figure 27.8**). Since the 1950s scientists have documented that phytoplankton are not uniformly distributed in the oceans but are infinitely more numerous along coastlines of continents and in high latitudes, such as waters off of the Newfoundland coast and in Arctic and Antarctic oceans.

Food chains in oceans begin with tiny organisms called *nanoplankton* and *ultraplankton* that are thought to be responsible for 50 to 70% of photosynthesis in the oceans (**Figure 27.9**). Other types of plankton eat these tiny organisms. Small fishes eat plankton and are eaten by larger fishes, which are eaten by larger animals such as marine mammals.

Bacteriovores are another level of ocean food chains that are especially important. While bacteriovores are found in terrestrial ecosystems, they often are ignored because they are microscopic or small in size. Bacteriovores include marine single-celled eukaryotes such as ciliates, foraminiferans, radiolarians, and flagellates that eat bacteria. In turn, these microscopic single-celled eukaryotes are eaten by *zooplankton,* organisms large enough to be seen with the naked eye. Zooplankton include larvae of many kinds of marine invertebrates, as

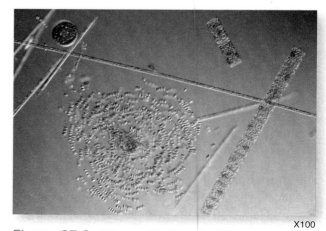

X100

Figure 27.8 **Phytoplankton.** Free-floating microscopic algae such as these diatoms are at the base of ocean food chains.

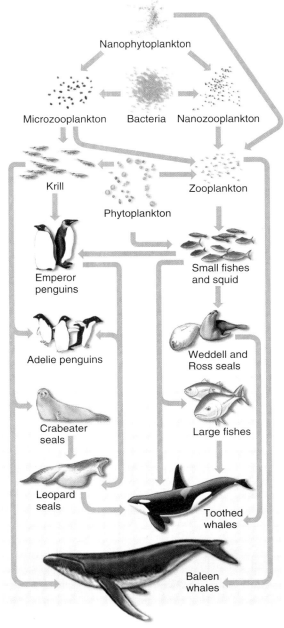

Figure 27.9 **A Marine Food Web is Based on Microbes and Various Sizes of Phytoplankton.**

■ *What do leopard seals eat? What do baleen whales eat?*

well as small crustaceans. Most zooplankton have independent movement and are large enough to swim against the current. The shrimplike *krill* that swarm in huge clouds in cold ocean waters are important zooplankton.

More recently, satellites have monitored concentrations of chlorophyll in the oceans, allowing us to view global distributions of phytoplankton (**Figure 27.10**). These satellite images have substantiated the pattern of huge populations of phytoplankton around continents and in high latitudes, north and south, and sparse populations of phytoplankton in the open oceans. Satellite imaging also enables scientists to see seasonal changes in phytoplankton populations.

Why are phytoplankton thickly distributed along coastlines and in high latitudes? Two factors are involved. First, runoff from land carries nutrients that allow dense populations of phytoplankton to bloom in the waters near coastlines. Second, at the edges of continents ocean currents collide with the solid continental shelf. Water is forced upward, and this *upwelling* brings nutrients up into surface waters. Ocean waters lack some of the metal ions such as iron that plants require. When open ocean waters that lacked phytoplankton were experimentally seeded with iron, blooms of phytoplankton resulted (**Figure 27.11**).

Oceans Contain a Variety of Habitats and Ecosystems

Marine environments are the largest portion of the biosphere and contain about 97% of its habitats. Because many ocean environments are inaccessible to humans without special gear, life in many marine ecosystems is much less well known than is life on land. Nevertheless, biologists have identified different ocean zones differentiated by depth, presence of light, and distance from land. The amount of nutrients, temperature, and currents—rather than

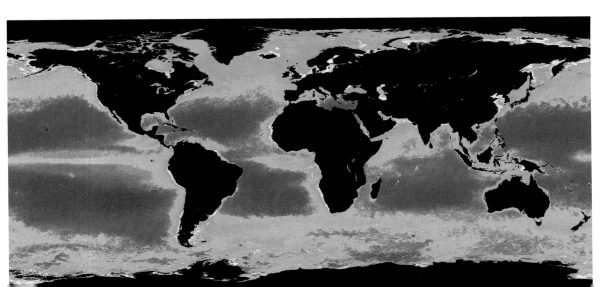

Figure 27.10
Distribution of Phytoplankton.
Phytoplankton populations shown in green are dense around continents and sparse in the open ocean.

Figure 27.11 **Effect of Adding Iron on the Growth of Phytoplankton.** The growth of phytoplankton is controlled by the availability of nutrients. In this experiment researchers showed that adding iron to an area of ocean water increased the growth of phytoplankton in that area. The iron was added in the shape of a C, and the area of high density phytoplankton matches that shape.

moisture and precipitation—influence the distribution and abundance of ocean life. Most photosynthetic activity happens near the ocean's surface where sunlight is most intense. Sunlight fades as water gets deeper. More than 100 feet (about 30 meters) below the surface it is dark and difficult to see. Below 1,000 feet (about 300 meters), it is so dark that you cannot see your hand in front of your eyes. Sporadic lights come from fascinating, often nightmarish, *bioluminescent* creatures.

Oceans can be subdivided into horizontal and vertical zones (**Figure 27.12**). The *littoral zone* is located between the high tide and low tide lines.

The *neritic zone* is the portion of the ocean above the continental shelf. The neritic zone receives abundant sunlight as well as nutrients from the coasts. So the neritic zone has vast populations of photosynthesizers and supports abundant marine life. Beyond the continental shelf is the *pelagic zone* or oceanic zone that includes deeper ocean waters. The upper, sunlit portion of the pelagic zone is the *photic zone*. The *oceanic zone* is the portion of the ocean over deep water. While its upper portions have abundant light, the pelagic zone has less life because it is farther from shoreline nutrient sources. Pelagic zones can be like deserts, with sparse populations of photosynthesizers and other organisms. Larger organisms like jellyfish and larvae of shrimps that can swim against the current make several daily *vertical migrations* that carry them downward and upward in the water. As these organisms move up and down in the water, they bring food and nutrients to other species that live at different levels. Dead organisms and wastes from swarms of krill drift down to the seafloor, and bring nutrients to the lower zones. As ocean water gets deeper, light and temperature decrease, leading to vertically arranged life zones that have characteristic organisms. The deepest parts of the ocean are the *hadal zone;* above this is the *abyssal zone.* Both are dark, cold, high-pressure waters with few nutrients.

As you might expect, ocean temperatures in regions near the equator are higher than are temperatures in regions closer to the poles, and surface waters are warmer than deeper waters. There also are seasonal changes in ocean water temperatures. This distribution of temperatures is complicated by the flow of major ocean currents (**Figure 27.13**). For example, warm water that originates offshore of

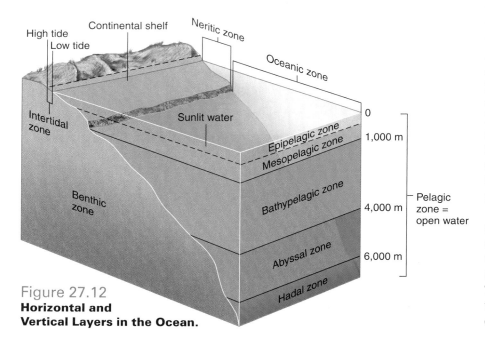

Figure 27.12
Horizontal and Vertical Layers in the Ocean.

Japan crosses the Pacific Ocean and reaches the American Northwest coast, bringing warm water and warmer air with it. This provides the moisture for the foggy Northwest summers that saturate coastal forests with dripping mist. Other currents take colder water to more equatorial regions. The icy Humboldt Current wells up along the coast of South America, bringing cold, nutrient-rich water from Antarctica. Ocean currents offset the effects of local temperatures and make some northern regions warmer, and more biologically productive, than they otherwise would be.

Just like their counterparts on dry land, the ocean's photosynthesizers require phosphorus, nitrogen, potassium, and other nutrients. So life in oceans also is limited by the nutrients available to photosynthesizers. Places where natural nutrient levels are high are the ocean's most productive ecosystems, while places where nutrients are lacking have only sparse life. Life is plentiful along ocean coasts where nutrients enter from land, and waves and currents stir up sediments that contain nutrients. Here water tends to be shallow, and penetrated by light. Coastal areas make up about only about 10% of the ocean, but they are extremely productive and are home to about 90% of ocean life. In the open ocean life is most abundant in the upper sunlit portions where photosynthesis takes place and tends to diminish with depth. In **estuaries,** rivers enter oceans, carrying nutrients from land into oceans. Estuaries and river deltas are generally areas of great biodiversity. Not only do phytoplankton of all sizes bloom in these waters, but also large populations of zooplankton, crustaceans, marine invertebrates, and small fishes provide ample food for birds, fishes, reptiles, and mammals.

Let's look first at some coastal marine ecosystems and then consider an unusual marine ecosystem located in deeper waters.

Life in the Littoral Zone Is Adapted to Rough Water and Periods of Dryness and Heat
Consider for a moment how difficult it must be for a barnacle or rock crab to live in the littoral zone (**Figure 27.14**). High tides occur twice a day, about $12\frac{1}{2}$ hours apart, and can involve gentle flooding or vigorous pounding and scouring. Organisms that live here must have ways to prevent being damaged and swept away. When the tide is out,

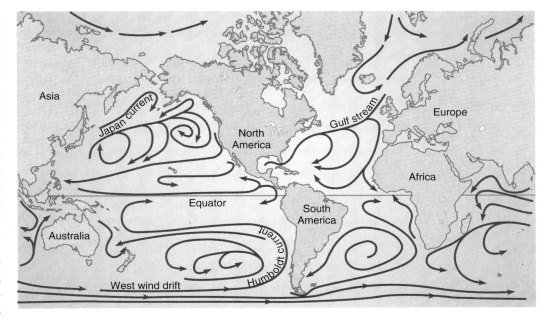

Figure 27.13 **Ocean Currents.** The Japan current carries cooler ocean water to the west coast of North America. The Humboldt current carries cooler ocean water to the west coast of South America. The Gulf Stream carries warmer ocean water toward the British Isles and west Africa.

■ *Use one sentence to correlate the information in Figure 27.11 with the information here.*

Figure 27.14 **Life in the Littoral Zone.** If these seaweeds weren't firmly attached to rocks, the pounding and scouring of waves would wash them away.

estuary a place where fresh and salt waters mix; generally where rivers enter the oceans

they must be able to cope with dry conditions and intensely hot sunlight. Some mollusks have rootlike structures of tough threads that anchor them to rocks. Other mullosks burrow into mud or sand or contract into shells that are clamped to rocks. Sea anemones draw in their tentacles and pull their bodies down to a tough nub. Many types of seaweed grow in the littoral zone, including kelps with holdfasts that cling to rocks. Kelp beds are home to a wide variety of marine invertebrates including crabs, worms, sea anemones, and starfish.

Estuaries and Mangrove Swamps Are Nurseries for Ocean Life

You may know estuaries as tidal marshes or salt marshes. These are extremely productive environments that occur on tide-covered flat land where freshwater from a river empties into an ocean (**Figure 27.15**). Tidal marshes are washed daily by high tides. The presence of freshwater dilutes ocean water, forming less salty, **brackish water** that allows salt marsh grasses and other plants to gain a foothold and thrive. Cyanobacteria cover the surface of the mud of a salt marsh. Because these prokaryotes can fix atmospheric nitrogen into nitrogen-containing compounds, the nitrogen cycle is especially active in salt marshes. Water continuously filters through the matted plant roots, and organic materials that have been carried into the marsh settle out and decompose. All of this provides abundant nutrients for food chains of detritivores as well as for photosynthetic algae and protists. These form the base of food chains that include shrimp, lobsters, horseshoe crabs, and the young of many other kinds of crustaceans and ocean fishes. In essence, estuaries are muddy, nutrient-rich traps for organic matter that become nurseries for marine invertebrates and vertebrates. Salt marshes also are

brackish water a mixture of salt water and freshwater that occurs in an estuary

critical environments for migratory birds and a huge variety of shorebirds.

Mangroves are tropical trees that look like they are walking on stilts. *Mangrove* is an umbrella term for more than 40 species of trees and shrubs that can live in shallow, tidal, salty water and can form huge, often impenetrable swamps around coastlines (**Figure 27.16**). Their "stilts" actually are aerial roots. Like the roots of grasses in a salt marsh, the roots of mangrove plants trap nutrients. Submerged roots provide shelter for diverse populations of marine invertebrates, algae, and small fishes, while aerial roots provide habitats for other kinds of animals. Many birds, lizards, snails, insects, and snakes live in the understory and canopy of the mangrove swamps where alligators and crocodiles also find shelter. Mangrove trees are sensitive to disturbance. Once they have been cut down, mangroves seldom regrow. This is because other plants move into the area, outcompete the mangroves, and replace them. Mangroves are keystone species, and when the mangroves are gone, many of the populations that they shelter also disappear. Both salt marshes and mangrove swamps occupy prime coastal real estate, so these wetlands are perennially in danger of disappearing as human seaside communities are developed.

Coral Reefs Shelter Organisms and Coastlines

Coral reefs are spectacularly diverse, wildly colorful ecosystems found in shallow, clear, tropical waters (**Figure 27.17**). The reef itself is made of the intricate, calcium carbonate exoskeletons grown by coral polyps. In addition to the coral polyps, however, many other species contribute to the complex ecosystem of a coral reef. For example, the individual corals often are cemented together

Figure 27.15 **Salt Marsh.** Glossy ibis, white ibis, and great egret are just a few species of birds that find food and shelter in this salt marsh.

Figure 27.16 **Mangrove Swamp.** The interlaced roots of these mangroves makes a mangrove swamp impenetrable to humans, but provides shelter for many organisms that make their homes in mangrove swamps.

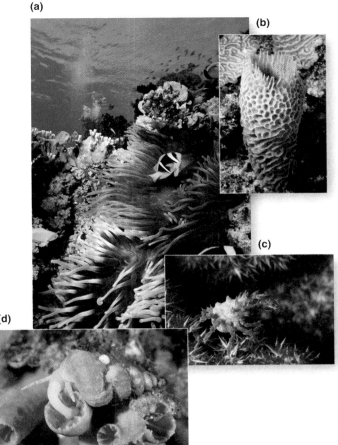

Figure 27.17 **Coral Reef.** (**a**) A sample of the diversity of life in coral reefs: (**b**) a purple sponge, (**c**) a candy crab camouflaged on a soft coral, (**d**) a wentletrap snail feeds on a cup coral.

when tropical forests are cut, silt washes into rivers and creeks. When these empty into the oceans, silt eventually accumulates on coral reefs. The mutualistic algae within coral animals must have clear water for efficient photosynthesis. If a deposit of silt blocks light, it can reduce photosynthesis and may kill the corals. Collectors hack out corals from reefs, and some divers hunt fishes and invertebrates in coral reefs. Corals are damaged by boats and anchors. Corals are subject to *coral bleaching*, a worrisome phenomenon in which corals lose their beautiful colors and turn completely white. When this happens, the symbiotic algae have abandoned the tissues of the coral polyps. Coral bleaching is a response to environmental stress, and if the corals are not recolonized by symbiotic algae, they may die. While the exact causes of coral bleaching are being investigated, coral bleaching has been linked to global climate change or increased ultraviolet radiation caused by the thinning of Earth's ozone layer. Coral reefs also are threatened by the increase in the amount of carbon dioxide in the atmosphere. Research has shown that corals do not grow well when atmospheric carbon dioxide levels are high, as they are now due to the burning of fossil fuels.

by *corraline algae.* Coral polyps are carnivores that feed on smaller organisms and organic particles suspended in the water. Mutualistic single-celled algae live within their tissues and provide the polyps with an additional food source. The resident algae photosynthesize and produce sugars making corals the primary producers in coral reef ecosystems, even though they are animals. Sponges, squid, snails, and crustaceans are just a few of the organisms that live within the crevices of a coral reef. Fishes of all sizes and colors dart about the corals, and schools of other fishes wheel as shifting walls of living silver. Sharks, sea turtles, rays, and skates glide about, all looking for food. Coral reefs are one of Earth's most productive and diverse ecosystems. Coral reefs also shelter shorelines from pounding waves. This is especially important during hurricanes, typhoons, and other violent storms.

Coral reefs are important in the ecology of tropical oceans. Yet they are threatened. For example,

The Ocean Floor Is Rich in Diversity   You may think of the ocean as an expanse of water, but the geography of the ocean floor is as varied as that of land. Valleys, plains, and mountains provide a landscape for life, and some undersea mountains are higher than any found on Earth's dry surface. Volcanoes erupt under the oceans. Some regions of the ocean floor are muddy, some are sandy, and some are rocky. Life on the ocean floor varies with depth. Although life is most plentiful in shallow ocean floor environments, life also flourishes on the floor of the deepest part of the ocean.

Many ocean species live suspended or swimming in the water like the birds and butterflies of life on land. On the ocean floor organisms settle, walk, slither, and crawl much as they do on land. Lobsters, sea anemones, oysters, and starfish inhabit shallow waters, while animals of deeper ocean floors have adaptations to cold, dark waters where physical pressures are enormous. Many are found nowhere else, and if they are brought to the surface their delicate bodies will become physically distorted. Because humans can visit the deep ocean floor only in specially designed submarines and cannot stay for long periods of time, the ocean depths are the least known of any of Earth's ecosystems.

Until just a few years ago it was thought that only glass sponges, brittle stars, and odd kinds of bioluminescent shrimp, squid, and fishes survived in the ocean's depths, as scavengers on the remains of fallen organisms. Thanks to deep-sea submarines it is now clear that the alien realm of the ocean floor is marked by pockets of remarkable life-forms.

Exploring the abyss of one of the ocean's deepest trenches, in the deep-sea submarine *Alvin,* investigators discovered a place where the water was filled with what looked like a black plume of smoke. External sensors showed that the temperature within the plume was 270° to 380°C (518° to 716°F), while typical ocean floor waters are close to freezing. Gradually, the explorers realized that the plume of the "black smoker" was seawater rich with minerals superheated by the hot magma from below Earth's crust (**Figure 27.18**). The heat and nutrients dramatically altered conditions on the barren seafloor, creating a dense ecosystem of unexpected life. The explorers saw enormous red-lipped worms that turned out to belong to a completely new phylum. These worms have no mouths or digestive systems, and cells of the lipstick-red "lips" that protrude from tough chitinous tubes contain hemoglobin. Later studies revealed that these worms have mutualistic colonies of chemosynthetic prokaryotes living within their bodies. Their chemosynthetic bacteria harvest minerals from the black smoker and provide the tube worms with food that allows them to grow quickly. The 9-foot-tall tube worms are among Earth's fastest-growing invertebrates and can grow nearly 6 feet (2 meters) a year. The tube worms were not the only life at the black smoker. An entirely new ecosystem was found in deep ocean trenches. White crabs scuttled away from *Alvin's* lights, and there were many filter feeders, including mussels,

clams, and feather duster worms. Scavengers like octopi and fishes are part of the *hydrothermal vent communities,* as are about 500 other species.

How does life survive without light for photosynthesis? The unusual hydrothermal vent organisms have evolved to take advantage of the ability of chemosynthetic bacteria to use hydrogen sulfide (H_2S) to produce glucose. These bacteria produce organic compounds that pass directly to the worms' tissues. Hydrothermal vent communities often are limited in life span. In one or two years the vents can become exhausted, or can close, cutting off the source of energy. Then the organisms die, leaving a ghost community for scientists to study. Some offspring of organisms that live at hydrothermal vents may by chance drift off. A few will find another hydrothermal vent where new populations will form.

QUICK CHECK

1. Explain how life in ocean ecosystems is influenced by the presence of light and nutrients.

27.4 Freshwater Ecosystems Link Marine and Terrestrial Ecosystems

Freshwater ecosystems have components found at sea and on land, and they bridge the two. Aquatic invertebrates include diverse forms of insects and insect larvae, worms that belong to several phyla, and a wide variety of mollusks. All groups of vertebrates—including amphibians that are absent from the oceans—live in aquatic ecosystems. In addition to ferns, mosses, and liverworts, many aquatic seed plants are adapted to life in freshwater ecosystems. As in ocean waters, many of the major producers in freshwater systems include phytoplankton of various sizes. Detritivores and food chains based on bacteria are common and important in freshwater ecosystems. Beyond these biotic factors, freshwater ecosystems are influenced by differences in water temperature, water movement, and levels of oxygen. Freshwater ecosystems are broadly divided into *still water,* or lakes and ponds, and *flowing water,* or brooks, creeks, streams, and rivers.

Lakes and Ponds Can Become Oxygen Depleted

Because the temperature of water influences its density, deep lakes and ponds tend to form layers that have different temperatures. In summer a lake is heated by sunlight, and its upper layers grow

Figure 27.18 **Black Smoker.**

■ *What hydrothermal vent organisms can you see in this photo? What hydrothermal vent organisms are* not *visible?*

warm and become less dense than deeper, colder, lower waters. At the bottom of the lake is the coldest, most dense layer of water (**Figure 27.19**). The upper layers have abundant light, so photosynthesizers are active. Thus the warmer layers have lots of oxygen. As the phytoplankton in the lake die, they settle to the bottom of the lake. There aerobic bacteria begin to decompose them and use much oxygen in the process. There is little or no mixing of the layers of a lake, so in summer the lower layers can become oxygen depleted, while the upper layers can become depleted of nutrients by phytoplankton. As seasons turn colder, water temperatures fall. Soon the distinct layers of temperatures disappear, and from top to bottom the lake water reaches about 4°C (39.2°F), the temperature at which water is most dense. Winds blow across the surface and cause the water to *overturn*. The overturn moves nutrients up from the bottom of the lake and moves oxygenated waters down to the bottom. As temperatures continue to fall, water at

the top of the lake freezes. This traps slightly warmer and denser water below it. Aquatic organisms live through the winter beneath the ice. Some burrow down into the mud and continue life at a low ebb, while others remain active through the winter but generally do not feed. Ice melts in spring. When water warms to 4°C, it is at its densest. Once again, pushed by winds, the lake water overturns. Again, nutrients are brought up from the bottom, setting the stage for blooms of phytoplankton. Population increases of zooplankton and other members of aquatic food chains typically follow the spring overturn.

In summer the upper, warmer layer of water can become depleted of oxygen, producing die-offs of fishes. Summer stagnation and "summer kills" are typical of many ponds and lakes. This natural **eutrophication,** the process of a body of water developing an overabundance of nutrients, can be worsened by nutrients that originate from human activities. In winter a different sort of die-off can

eutrophication the process in which a body of water becomes enriched with nutrients

Figure 27.19
Seasonal Cycle of a Freshwater Pond.
The small black arrows show movements of water at different seasons. These movements are related to the temperature and density of water and are influenced by winds.

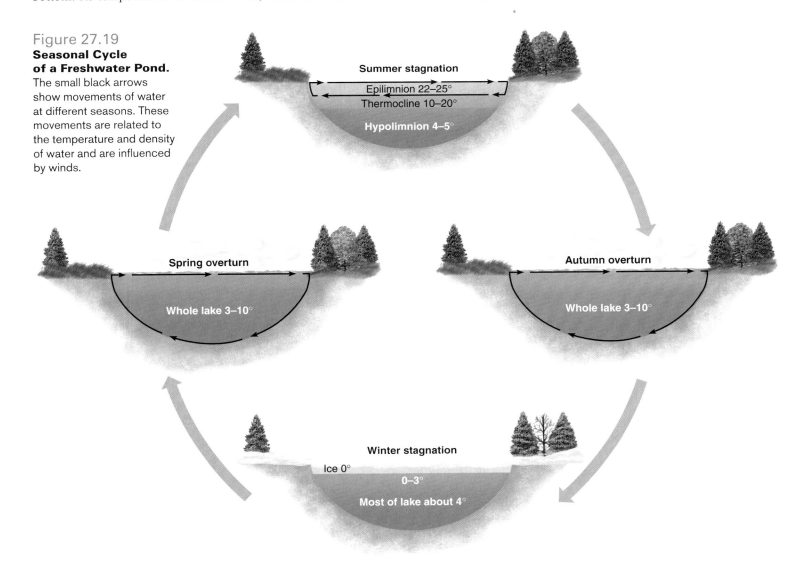

Summer stagnation
Epilimnion 22–25°
Thermocline 10–20°
Hypolimnion 4–5°

Spring overturn
Whole lake 3–10°

Autumn overturn
Whole lake 3–10°

Winter stagnation
Ice 0°
0–3°
Most of lake about 4°

occur. A heavy cover of snow over winter ice can block light so that photosynthesis slows or even stops. Under these conditions winter lakes can become oxygen depleted, and a "winter kill" of fishes and other aquatic organisms can occur.

Just as light is a critical factor that influences the distribution of life in ocean waters, light also influences the distribution of life in still-water ecosystems. Lakes and ponds have littoral zones in which plants grow (**Figure 27.20**). Cattails and rushes are part of the *emergent vegetation* that protrudes above the water. So are leaves and flowers of water lilies and other plants that float on the water's surface. A feature of still freshwater that is absent or minimally exploited in still salt water is the water's *surface film.* Water striders, mosquito larvae, whirligigs, water boatmen, and other aquatic arthropods use the water's surface film as a habitat. Farther out from the shoreline is the limnentic zone, which is penetrated by light. The *profundal zone* is at the bottom of the lake where like deep oceans, it is darker and colder, and there is little or no photosynthesis. Each of these zones has typical life-forms.

Creeks, Streams, and Rivers Contain Life Adapted to Moving Water

Creeks, brooks, and streams are small bodies of flowing water, while rivers are larger, generally 3 meters or more wide. Creeks, brooks, and streams can flow all year or can be intermittent, flowing only seasonally or after heavy rains, while rivers are more permanent ecosystems. Streams have two basic habitats: *pools* of deeper water and *riffles* where water is shallow and flows rapidly over pebbles or rocks. Water in riffles has a high oxygen content and relatively few planktonic photosynthesizers.

In moving water life must be able to resist the currents or be swept downstream. Some animals that live in these turbulent waters hide beneath pebbles and stay out of the current. Mayfly larvae and salamander larvae use this strategy. Others like trout or dace are strong swimmers that can make headway against the current. Still other riffle organisms have holdfasts that allow them to hang on as the water moves past them. Caddisfly larvae cement grains of sand, twigs, and pebbles into protective cases. Life in pools includes crayfish, fishes, various kinds of burrowing worms, and insect larvae. Frogs, salamanders, and turtles also may be present in freshwater pools.

Flowing-water ecosystems merge into one another as creeks grow wider, become rivers, and empty into the sea. Along this continuum there is a corresponding shift in the physical characteristics and organisms that live in flowing waters. Oxygen levels, water clarity, amount of nutrients, water temperature, and speed of flow all shift as a creek widens into a river that flows to the ocean. Oxygen levels are highest in colder, rapidly moving water upstream and lower in warmer, slowly moving river water. Streams usually are clear, while rivers are murky. Stream fishes usually strike their prey after they have seen it, while scent and taste are much more important to river fishes. Streams have relatively few nutrients when compared to rivers that have abundant populations of photosynthesizers and receive many nutrients from runoff and from all of the streams that enter them.

Flowing water is greatly influenced by organic matter that enters the water from the surrounding environments. Part of this ecological phenomenon in rivers is fishes like salmon and alewives that spend much of their lives in the oceans and "run upstream" to breed. Although some of the adults make the return trip downstream back to the ocean and live for other years, many of these fishes die upstream. As they move upstream, their wastes and dead bodies enrich the stream (**Figure 27.21**).

Figure 27.20   **Freshwater Ecosystem.** Cattails and water lilies are some of the emergent vegetation in this northern pond.

Figure 27.21   Fishing Grizzly. Salmon swimming upstream bring nutrients gathered in the oceans into freshwater ecosystems and surrounding biomes.

They also draw foraging land-based carnivores like bears and scavengers like bald eagles. Eventually, these animals enrich the surrounding terrestrial biomes with wastes that originated as the flesh of oceangoing fishes.

QUICK CHECK

1. Compare the factors that shape life in still water as opposed to flowing water.

27.5 Biomes Are Characterized by Typical Vegetation

In broad terms the kinds of life found in a region on land are related to the climate, and similar climates have similar kinds of life. For instance, the Australian outback, the plains and prairies of North America, the pampas of Argentina, and the African plains are primarily grasslands. They have similar climates and plants and animals with similar adaptations, even though the plant and animal species differ. Terrestrial ecosystems like grasslands that cover large geographical areas and share similar climates are called **biomes.** You will read about six major biomes here. **Figure 27.22** shows the distri-

bution of arctic tundra, taiga, temperate deciduous forest, grassland, desert, and tropical rain forest. Notice that each biome is found in more than one geographic region. Three features largely define a terrestrial biome: availability of freshwater, temperature, and dominant plant species. Because these features greatly influence natural selection in a region, they influence the kinds of plant and animal adaptations found in each biome. **Figure 27.23** shows the distribution of temperature and rainfall in relationship to biomes. Let's start in high northern latitudes with arctic tundra and work toward the equator and tropical rain forests.

Arctic Tundra Has Short Summers and Permanently Frozen Subsoil

Arctic tundra (**Figure 27.24**) is a northern biome that can be quite dry but terribly cold. Arctic tundra covers regions close to the Arctic Circle, where temperatures are almost always below freezing, and six months of sunshine are followed by six months of near to total darkness. Because the tundra is so cold, even in summer the subsoil about 3 feet (about 1 meter) from the surface is permanently frozen and is called **permafrost.** The long summer days and the presence of permafrost have a profound impact on the patterns of life in the tundra. For one thing, there are no tall trees. Instead,

biome a major kind of ecosystem that covers a large geographical region and has a similar climate

arctic tundra a northern biome characterized by extremely cold temperatures, restricted precipitation, permafrost, short summers with extremely long days, and low, ground-hugging vegetation primarily composed of mosses, lichens, and grasses

permafrost tundra subsoil that is permanently frozen

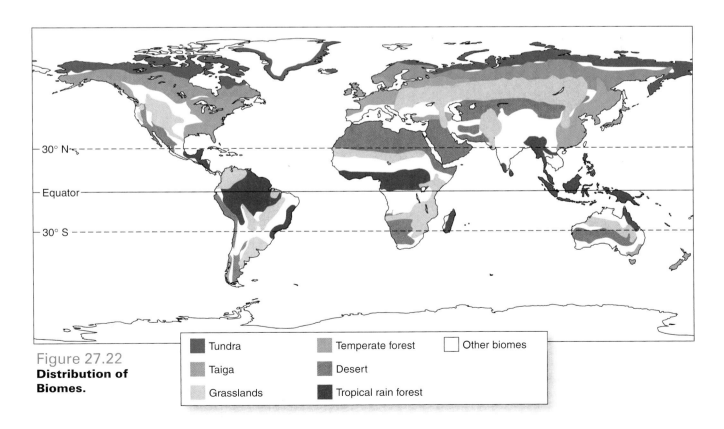

Figure 27.22
Distribution of Biomes.

Tundra • Taiga • Grasslands • Temperate forest • Desert • Tropical rain forest • Other biomes

Figure 27.23 **Rainfall and Temperature Influence the Distribution of Biomes.**

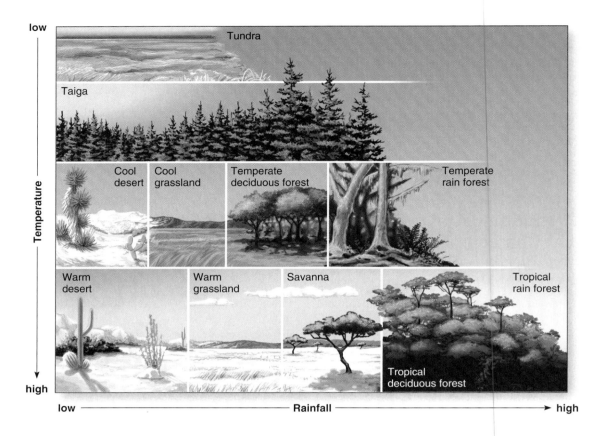

Figure 27.24 **Tundra.** Permafrost, cold, snowy winters, short summers that can have 24 hours of daylight, low plant growth, and populations of animals that seasonally migrate are characteristic of tundra.

flowers, small bushes, shrubs, lichens, and mosses carpet the landscape. Why are the plants so short? A combination of permafrost and strong winds is responsible. Because roots cannot penetrate permafrost, root systems are forced to be shallow. Tall trees must have deep roots to stabilize them, and so the combination of permafrost and fierce winds eliminates tall trees from the tundra.

While the tundra is generally dry, if you visit the tundra during its two-month-long summer, it looks extremely wet. In the warm summer months the snow and ice on the surface melts, but water cannot seep down through the permafrost. As a result pools and bogs form, and a few species of insects such as mosquitoes, deer flies, and black flies multiply rapidly and swarm in huge numbers.

Rates of photosynthesis are high during the long summer days, and reproduction is fast. Flocks of birds that have spent the winter in southern latitudes migrate north to mate and raise their young on the tundra. Parents feed their young on an accelerated schedule, and the young grow rapidly. After the brief summer months, juvenile and adult birds gather into flocks and fly south before winter sets in. When the winter darkness returns, the tundra's permanent residents must have enough stored energy to last the long winter. Tundra mammals typically gorge during the summer and then either migrate or stay active beneath the snow during the winter. Few tundra animals hibernate because the permafrost limits deep underground refuges from the cold. Small mammals dig burrows beneath the snow where they shelter from the bitter temperatures and fierce winds. Mammals like brown bears, caribou, and musk oxen grow thick coats of fur. The coats of some mammals like lemmings, hares, and wolves change color to match the snow.

The future of the arctic tundra biome is uncertain. As global temperature change warms the atmosphere, the permafrost across the Arctic is thawing, and warmer temperatures are creating a more humid and moderate climate. Already whole villages that were built on solid permafrost have had to be relocated because the permafrost is melting beneath them. Landscapes that once were covered by only short, thin vegetation are now growing larger shrubs. As permafrost melts, organic matter contained within it also thaws and becomes accessible to bacterial decomposers. Most of these decomposers produce methane as a by-product of their metabolism. For example, since 1950 in parts of Sweden methane release from permafrost regions has increased anywhere from 20 to 60%. Methane is an even *more* powerful greenhouse gas than is carbon dioxide, and it is likely that melting permafrost will intensify the global warming.

Another concern is human exploitation of resources in the tundra. Beneath the Alaskan tundra is a large, economically valuable oilfield, and its development is a concern to ecologists. Arctic haze is a weather phenomenon that carries air pollutants northward to the Arctic, depositing them thousands of miles from their U.S. origins. The tundra also experiences *bioaccumulation,* the phenomenon in which environmental toxins accumulate in the tissues of organisms and move upward in food chains. Carnivores are most vulnerable to bioaccumulation of toxins in their fat deposits. Humans are directly affected when they eat animal fat that contains toxins. Fat is a food source that traditionally is prized by many northern people.

Taiga Is Northern Coniferous Forest

South of the tundra are large regions of coniferous forests, which are among Earth's most common forests (**Figure 27.25**). Recall that conifers are trees such as pines and firs that produce their seeds within cones. **Taiga** is the Russian word for this biome. It also is called *boreal forest.* Winter in the taiga is long, snowy, and extremely cold, and the cool summer growing season is short—typically less

Figure 27.25 **Taiga.** Dense forests of a few species of conifers are typical of taigas that have cold, snowy winters, but no permafrost.

30° N
Equator
30° S

than 130 days. There is a bit more water in the taiga than in the arctic tundra, and this combined with the absence of permafrost means that trees have the resources to grow tall. In the taiga one or two species of conifers like spruces and firs cover acre after acre, giving the taiga a uniform appearance. Because these trees are so tall, they block out much of the sunlight, and beneath the trees there is little underbrush and few shrubs. A walk through mature taiga can be a meditative experience: there is plenty of space between trees and the forest floor is quiet and dim. There also are few signs of birds or small mammals, and footsteps are muffled by a thick layer of fallen needles. Caribou, minks, beavers, and arctic hares are typical taiga mammals.

Taiga is the largest biome, and although it is not heavily inhabited, humans are beginning to have a negative impact on it. Large tracts of taiga have been opened to logging, especially in northern Russia. Scientists are concerned about loss of this unique biome and about the contribution to global environmental problems if this vast northern forest is lost.

taiga a northern biome characterized by cold, snowy winters, cool summers, moderate precipitation, and conifer forests

Trees of Temperate Deciduous Forests Drop Their Leaves in Fall

Temperate deciduous forests (Figure 27.26) are found in the eastern United States, across much of Europe, and in China, Japan, the eastern coast of Australia, and New Zealand. Although some conifers are present, most of the trees in this biome have broad, flat leaves that die and drop from the tree when the cool temperatures and short days of fall and winter arrive. Before they die, the leaves of deciduous trees reveal the colors of their accessory photosynthetic pigments, giving a glorious show of color in autumn. Having deciduous leaves is an adaptation to a strongly seasonal climate, allowing a tree to conserve energy and prevent heat loss by dropping its leaves. In winter the groundwater of a deciduous forest is frozen. To avoid damage from the formation of ice crystals within its water transport system, a deciduous tree withdraws liquids and useful chemicals from its shoots and stores them in its roots. After a leaf drops, a tough scale forms over the place where the leaf was attached. In spring, as temperatures grow warm, liquids rise within the water transport systems of the trees. Photosynthesis and growth begin again. Buds at the tips of branches that have been protected from freezing by tough, waxy *bud scales* begin to expand. Soon the tree blossoms, and fresh green leaves unfurl on its branches.

Temperate deciduous forests have stratified layers. The **canopy** is the covering formed by the crowns of trees. Below this is an **understory** formed by younger trees growing up into the canopy as well as smaller species of trees that do not grow as high as the canopy trees. Closer to the ground is the *shrub layer*, and farther down are layers of *herbs, grasses,* and *mosses.* Sunlight can reach the forest floor in the spring and fall before the canopy has leafed out and after leaves have fallen; at these seasons wildflowers bloom on the forest floor. There usually is a deep layer of rotting leaves. As these decay, they enrich the forest soil.

Many animals of temperate deciduous forests have adaptations for life in trees. For instance, tree frogs have adhesive discs on their toes, flying squirrels have flaps of skin for gliding from perches, opossums and mice have grasping tails. Because dense leaves of canopy and understory block vision, animals that live in trees tend to have keen hearing, and their voices and calls are important adaptations. In temperate deciduous forests many animals like chipmunks hibernate during the winter to avoid food shortages. Migratory birds are typical of temperate deciduous forests. Nectar-sipping hummingbirds, insect-eating warblers, and caterpillar-eating tanagers migrate to warmer climates when their young are independent. They return to raise their young in temperate deciduous forests in the spring.

From a human perspective the temperate forests have provided a wealth of resources. Northern Europe and North America once were largely covered by temperate forests of one kind or another, but much of this forest was cut down and used for building materials as well as a source of energy (see Figure 27.30). You certainly are aware of worldwide efforts to save rain forests in other parts of the world, but American forests were largely cut down in the first 300 years after the Europeans arrived. A similar loss of forests was seen in Europe as far back as the 1700s and 1800s. Today, efforts at conservation and restoration of forests are being increased, but forests are still being cleared for housing developments and shopping malls. As forest acreage is lost or fragmented, habitats are disrupted and species become locally extinct.

temperate deciduous forest a midlatitude biome where moisture falls about equally all year; distinct warm and cold seasons favor trees that drop their leaves and go dormant in winter

canopy the tops or crown layer of forest trees

understory a layer of forest growth that is higher than the tallest shrubs but does not reach into the heights of the canopy

30° N
Equator
30° S

Figure 27.26 **Temperate Deciduous Forest.** Autumn displays of colored leaves are typical of temperate deciduous forests.

Figure 27.27 **Grasslands.** Bison graze on a North American prairie.

Savannas, Pampas, Prairies, Veldt, and the Outback Are Grasslands

Grasslands are biomes that have only a bit more rainfall than deserts—enough to grow a ground cover of different species of grasses and wildflowers (**Figure 27.27**). Because there is relatively modest water, grasslands usually have no trees except along river or creek banks. The 10 to 30 inches (25 to 75 centimeters) of rain that does fall on grasslands is seasonal, and much of the year grasslands are dry. Grasses are adapted to these arid conditions. Their long, thin leaves minimize evaporation, and grasses have dense, matted root systems that quickly absorb any rainfall. Grasses provide food and moisture for large populations of insects as well as for herds of grazing mammals. Grasses have adaptations that allow them to be gnawed right down to the ground and still survive. As long as the roots and the base of the plant are intact, a grass can grow again. Periodic fires are an important feature of life in grasslands. Fires that usually are started by lightning burn grasses down to the soil. After rains, scattered seeds and roots will sprout a new cover of fresh grasses. Many young deciduous shrubs and trees are not able to withstand regular fires, and fire quickly eliminates most windblown or animal-transported seeds of trees that sprout.

Grasslands usually are found in the interior of a continent. The Australian outback, the South American pampas, the African veldt or savanna, the Russian steppe, and the North American plains and prairies are all grasslands. Early settlers who crossed the United States in large covered wagons called "prairie schooners" described vast seas of grasses, with mile after mile of 3- to 5-foot-high grasses swaying in the breeze. The enormous herds of bison that once roamed North America lived on these prairies. Animals that live on grasslands have adaptations that include traveling in herds or large flocks, teeth specialized for grazing, use of burrows, and life cycles that feature adaptations to help them escape from cold or dry seasons. Grassland animals hibernate, migrate, and *estivate*, the term for becoming inactive or dormant in hot and dry conditions.

Grasslands have deep, rich soils, and they are extremely valuable for agriculture. Today, farms have replaced prairies and plains across most of the United States. One consequence of the loss of the original North American prairie is the loss of biodiversity. The diversity of grass species in the prairies maintains the richness of the soil as diverse grasses die back in winter and are decomposed by bacteria and fungi. Modern ecologists and farmers have a lot to learn from the prairies, and conservation focuses on replanting native prairie grasses and reestablishing small pockets of the biome. Little native prairie remains in the United States and Canada.

Deserts Can Be Hot or Cold, But They Are Always Dry

Deserts are some of the most extreme environments on Earth. While some deserts are hot and others are cold, all deserts are dry. To better define this environmental quality, ecologists classify a **desert** as any area that receives less than about 10 inches (25 centimeters) of rainfall a year (**Figure 27.28**). Major deserts are found on all continents, usually about midway between the equator and the poles (30° N and S). Deserts are not dry all year long. In most deserts rain may fall abundantly in some seasons, but in other seasons there is no rainfall. When you think of a desert, you probably imagine camels and sand dunes or cacti and burning hot sunlight, but many deserts don't follow this pattern. For instance, the interior of Antarctica is a

grassland a biome characterized by wet and dry seasons, hot summers and cold winters, and year-round cover of grasses

desert a biome characterized by sparse moisture, great daily fluctuations in temperature, poor soils, and sparse vegetation with adaptations to conserve water

Figure 27.28 **Desert.** Scarce rainfall in a desert influences the water-conserving adaptations of desert plants like these saguaro cacti.

desert where the temperature can drop to nearly –100°C (–148°F). There is lots of water, but because it is frozen, it is unavailable to life.

Extremes of temperature variations are characteristic of deserts. Sunlight raises daytime temperatures, but at night temperatures fall dramatically, sometimes dropping as much as 10°C (50°F). Temperatures fluctuate widely because at night there is little water vapor in the air to retain the Sun's heat. The heat of the day is quickly lost to space.

These conditions of scarce water and fluctuations of temperature limit the kinds of life that can survive in a desert. Instead of broad leaves, most desert plants have thorns or spines that conserve water. Desert plants that do have leaves tend to lose them during dry seasons. Another adaptation of desert plants is that their stems usually are green. The chloroplasts that normally would be found in leaves of other plants are located in the outer tissue layer of modified stems of desert plants. In addition, desert plants have a thick, tough, waxy cuticle and stomata sunken into pits. Both of these adaptations limit water loss. Many desert *succulents* have thickened leaves, stems, or roots where water is stored. These and other adaptations help desert plants conserve water and survive in places where only a little rain falls.

Desert animals also have adaptations that allow them to take advantage of water in wet times and to survive on sparse water in dry times. One example is the kangaroo rat that lives in the Southwest desert of the United States (**Figure 27.29**). These small rodents have silky, sand-colored fur that matches the soils where they live. Like other desert animals, kangaroo rats are nocturnal. By day they sleep in cool, moist underground burrows and avoid high temperatures and hot sunlight. They are active at night, when it is cooler. Like most noctur-

nal mammals, kangaroo rats have large, dark eyes. Like pet hamsters, kangaroo rats have large, fur-lined check pouches. The kangaroo rats stuff these pouches full of seeds that they carry down into food caches within their underground burrows. Unlike many other desert rodents, kangaroo rats do not hibernate.

Although it seems unbelievable, kangaroo rats *never* drink water or eat green plant tissue that contains water. Instead they get water as a by-product of the metabolism of their food. Kangaroo rats have several adaptations that preserve their metabolic water, so they do not have to drink. For example, kangaroo rats have highly specialized kidneys that resorb most of the water in their urine. They produce almost no sweat. During the day kangaroo rats block the many entrances to their burrows with soil. This helps retain the cooler, more humid atmosphere in their burrows. Kangaroo rats cannot avoid losing water vapor as they breathe, but while they are within their burrows their exhalations are trapped underground, adding moisture to the air they breathe.

Figure 27.29 **Kangaroo Rat.** At night this desert rodent emerges from its burrow to gather the seeds that supply it with food and water.

tropical rain forest the most productive terrestrial biome, characterized by warm to hot year-round temperatures, extreme humidity, large amounts of rainfall, and highly varied vegetation

No matter how highly adapted desert animals and plants are to their arid homes, they cannot survive long when humans move in and change the ecological equation. Many U.S. desert regions are under pressure because they have become desirable places to live; towns and cities are expanding into desert regions. Along with an increased human population comes an increased demand for water. Expanding communities divert water from rivers. This leaves less water for human populations that are downriver and creates bitter disputes about who "owns" rivers and has the right to their waters. Most of the migrants to U.S. desert areas come from other places in the United States, Canada, and Mexico, where they are accustomed to having plentiful water. So many of these new residents are not as sensitive to the need for water conservation in their new desert homes. As humans move into desert areas, habitats are fragmented, and wild species are displaced. Increased human presence in deserts has other associated problems including herds of livestock that are released to freely graze on desert vegetation, use of off-road vehicles that damage deserts, and the introduction of alien and invasive species of plants that replace native perennial grasses. Cheatgrass brome is an invasive alien grass. It outcompetes native grasses, reduces the natural biodiversity of desert environments, and creates fire hazards.

Tropical Rain Forests Have the Greatest Biodiversity of Any Biome

The most lush and productive terrestrial biome is the **tropical rain forest** (**Figure 27.30**). Here water and warmth combine to produce an explosion of plant growth and an enormous number of species. Tropical rain forests are distributed in a belt that encircles Earth at the equator. Parts of Central America, South America, western Africa, Indonesia, and Australia all have tropical rain forests. Here the climate is warm to hot, a great deal of rain falls—80 inches (200 cm) or more each year—and it is extremely humid. Because of their equatorial position, tropical rain forests have remarkable uniformity of climate and day length. The temperature is usually about 27°C (80°F) and days and nights are generally 12 hours long, all year long.

Tropical rain forests are extremely active biologically. The vegetation removes a huge amount of carbon dioxide from the air and releases an equally large amount of oxygen. The number of plant species is astounding. In the tropical rain forest neighboring trees are usually different species, and you might have to search for *a mile or two*

Figure 27.30
Tropical Rain Forest. Tropical rain forests have abundant rainfall and plants grow luxuriously. Most of the animals live in the tree canopy. The tropical rain forests have the highest diversity of organisms of any terrestrial biome.

30° N
Equator
30° S

before you would find another tree of the *same* species. This is markedly different from taiga, where acre after acre will have only one or a few species of conifers. It is also different from temperate deciduous forests, where there will be a few dominant species, with perhaps a dozen to 20 other species present. Water is plentiful in tropical rain forests, so trees can grow very tall, and many have expanded bases that stabilize their tall trunks.

With so much growth, though, there is great competition for sunlight. Tall trees monopolize sunlight, and shorter trees are adapted to use the reduced light that filters through the canopy. In tropical rain forests the canopy is nearly continuous. It blocks light, and the forest floor is dim. Understory plants often have adaptations that maximize the use of whatever sunlight they get. These include large, broad, dark green leaves that have lots of chloroplasts. *Vines* are common in the tropical rain forests but are less common in other biomes. Tropical rainforest vines snake up tree trunks, growing toward the sunlit canopy. *Epiphytes*, plants that grow on other plants, are another unusual feature of the canopy, where tall trees, vines, and epiphytes compete for a place in the Sun. In tropical rain forests entire aerial gardens grow on tree limbs. Researchers have learned that epiphytes may provide a source of nourishment for their host trees. It long was thought that epiphytes sent their roots into the transport tissues of their host plants. Now it seems as though the reverse is true: the supporting plants send accessory roots out of their limbs and into tissues of epiphytes. Because they serve as homes for whole communities of animals within the treetops, and because they trap leaves and other debris, epiphytes are an aerial source of nutrients for trees.

Although the luxuriant vegetation might make you think that soil beneath a tropical rain forest is Earth's most fertile soil, this is a misconception. Tropical rain forest soils are some of the poorest and *least* fertile soils. These red-orange soils are little more than oxidized clay, and they contain few of the mineral nutrients that plants need. Tropical rain forest soils contribute little to a plant's growth other than stabilizing it and providing water. Most of the nutrients in a tropical rain forest are within the shoots and roots of plants and in the living bodies of other organisms. The soil beneath a tropical rain forest has a scanty covering of dead leaves because dead organic matter decays quickly in the warm, humid climate, and the carbon and nitrogen cycles turn quickly. The forest floor has fewer species of understory trees than does the temperate deciduous forest. It is nothing like the "tropical jungle" that you might imagine. A jungle actually is a place where the canopy has been broken and light streams down to the forest floor. Plants respond with thick, often impenetrable growth. Jungle usually develops along riverbanks, in windbreaks where old trees have fallen, or in areas that humans have cleared.

Most of the animals in the rain forest live in the canopy and seldom descend to the forest floor. Many animals have grasping tails, adhesive toes, or other adaptations that make life in the trees easier. There are gliders, leapers, hangers, acrobatic swingers, and of course fliers in tropical rain forests. Although there are many animal species, like the plants the density of each species is low. So there aren't enormous flocks of birds or huge herds of mammals typical of the tundra, taiga, or grasslands. Being hidden among the leaves, animals often have to rely on sound to communicate. This makes the rain forest a noisy place, especially at dawn and dusk. Like animals in deciduous forests, rain forest animals often have keen hearing and sophisticated ways to produce sounds.

Tropical rain forests are rapidly being cut down. It is estimated that an acre of forest is cleared each second, worldwide, and many biologists predict that at this rate of destruction, all tropical rain forests may be gone in 30 to 50 years. This is worrisome for a number of reasons. Tropical rain forests contain more biological diversity than any other region of Earth. Furthermore, unlike deciduous forests or taiga, there is little evidence that intact, complex tropical rain forests with all of their thousands of species can regenerate once a forest has been cut down. Biologists estimate that *over half* of living species are found only in tropical rain forests. Many of these species are plants and insects, but vertebrates also are diverse in tropical rain forests. And most of these species are still unknown, unstudied, and unidentified. As tropical rain forests disappear, the fascinating tropical rain forest animals will dwindle, disappear, and may even go extinct. It is not an exaggeration to state that a major extinction event is now taking place. Unlike other major extinctions in the history of life, this one is caused by human actions.

QUICK CHECK

1. Describe important aspects of the tropical rain forest biome. Why is it so extremely productive when other biomes are less productive?

27.6 The Biosphere Encompasses All of Earth's Biomes

Experience has taught you that life is prolific. For instance, if you set out a bucket of table scraps on a hot summer afternoon, in several days it will be seething with maggots, the larvae of flies. That zucchini forgotten at the back of the refrigerator soon rots into a squishy blob spotted with fuzzy, black mold. After a soaking rain mushrooms pop out of the lawn and, if you aren't vigilant, in only a few hot summer days the swimming pool can turn a lovely shade of algal green.

You also know that life is resilient—it flourishes in nearly every region and habitat on Earth. Organisms live under rocks, within rocks, and deep in mine shafts. Pale, blind fishes, blind salamanders, and blind crayfishes lurk in pools deep within caves; insects have been netted as high as 1.86 miles (3,000 meters) into the atmosphere. Extremes of cold and heat are not barriers to life either. In the Antarctic Ocean, where temperatures hover close to the freezing mark for much of the year, algae thrive beneath the ice. They provide a banquet for krill that, in turn, are gobbled by hordes of other animals, including penguins, squids, fishes, seals, and whales. In hot, dry deserts if spring rains are plentiful, delicate flowers will carpet sand dunes (**Figure 27.31**). Hosts of tiny organisms spend their entire lives in the spaces between grains of beach sand. This list could go on and on, but the point is clear—Earth is full of prolific, resilient life.

Even though life is prolific and resilient, it is found only on a thin shell at the Earth's surface. Earth has a diameter of 7,926 miles (12,756 kilometers), but all life is confined to a thin surface coating that is only about 12.4 to 15.5 miles (20 to 25 kilometers) thick. This includes prokaryotes that live more than 1 mile (more than 1.5 kilometers) under the Earth's surface. Just for comparison, Mt. Everest, Earth's highest mountain peak, is 5.5 miles (8.8 kilometers) high. Let's put it another way: if Earth were a peach, its fuzzy covering would hold all known life *as well as* all the environments that support life. This veneer of life permeates Earth's waters, soils, rocks, and atmosphere; it extends to the deepest ocean trenches and the lower portion of the atmosphere (**Figure 27.32**). The portion of Earth that includes all life is called the *biosphere*, and it contains all ecological units: populations, communities, ecosystems, and biomes.

Figure 27.31 **Desert in Bloom.**

Figure 27.32 **Biosphere.** From space the ecological details of different biomes are barely visible and their atmospheric, geological, and living components merge into the biosphere, Earth's largest ecological level.

Why does life thrive in the biosphere and nowhere else on Earth? Life can flourish only under a limited set of physical conditions. Energy from the Sun must be available for life to use, but harmful ultraviolet solar radiation must be minimized. For at least part of the year temperatures must allow chemical reactions to proceed at a reasonable pace. Earth's atmosphere helps shield the surface of the planet from ultraviolet radiation, and the atmosphere also traps heat. Nearly all forms of life must have oxygen to release chemical energy. In addition, other atoms and molecules must be available at the right levels. Although scientists are actively searching for life on Mars as this book is being written, the biosphere is the only known place where all of these physical conditions are met and where life exists. The final chapter of this text will focus on human impact on species, populations, ecosystems, biomes, and the biosphere.

QUICK CHECK

1. What is the biosphere? Can you see it from space? Explain your answer.

27.7 An Added Dimension Explained
The Death and Life of Lake Erie

The horror stories of polluted Lake Erie and the other Great Lakes that opened this chapter are not exaggerations. Across the globe humans have degraded environments to the point where they are nearly uninhabitable. In Chattanooga, Tennessee during the 1970s air pollution was so bad that the sky was dark by midafternoon and was filled with soot. Alien grasses are choking the life out of the Mediterranean. The marine environment off the mouth of the Mississippi in the Gulf of Mexico is devoid of all life. It is dead because of pollution added to the river by human activities. From the American Midwest to the plains of China to Africa, grasslands are turning into deserts. And, of course, in the early 1970s the popular press declared that Lake Erie was dead. Can anything bring life back to a lifeless, polluted environment? The answer is a qualified "Yes." It takes a great deal of effort, and it is not easy to combat the natural human tendency to foul our surroundings. But the present state of Lake Erie shows that it can be done.

In the 1960s and 1970s many people in five U.S. states and Canada enjoyed Lake Erie and depended on it including fishermen, vacationers, business and industry, and people from all walks of life. So as Lake Erie became more and more polluted, a rare convergence of public and government opinion declared that enough was enough. No one knew if Lake Erie could be saved, but all agreed that it was worth the effort to try. The result shows that, when given a chance, ecosystems are resilient and will recover. But only if given a chance.

What Polluted Lake Erie?

Many of the problems faced by Lake Erie now are familiar to you. Algae were growing rampantly, and schools of fishes were dying. It took years for researchers to figure out the problem. Pollution from many sources entered Lake Erie. In the end excess nutrients were identified as the major culprits, specifically phosphorus. It may be hard for you to grasp that too much phosphorus could cause such devastating effects. This is an example of how one small ecological factor can have profound, global effects. Phosphorus is a nutrient required by photosynthetic organisms, and too much phosphorus in lakes, streams, or oceans causes algae to overgrow. If you have ever been around an algal bloom, you know it can *stink*. Lake Erie did smell bad, but that was the least of its problems. Algae themselves can harm fishes and other large animals, but excess algae can cause even more damage. Dead algae sank to the bottom where bacteria fed on them. These decomposers are aerobic bacteria that use oxygen in their metabolic pathways. The overgrowth of bacteria depleted the oxygen in Lake Erie. Without oxygen other organisms could not survive. Soon larger fish suffocated, and lots of dead fish washed up on the beach. This effect is called *anthropogenic eutrophication,* and it has been repeated in many waters throughout the world. Lake Erie was just one early example. Eutrophication is a normal event in the life of still bodies of freshwater, but this was something different.

Where did the excess nutrients come from? Detergents were one source. Phosphates make for more effective detergents, so phosphate detergents were popular when they first were marketed. But, of course, phosphates in laundry detergents ended up in wastewater, and this eventually fed into lakes and streams. Phosphorus also entered the lakes from wastes of manufacturing plants and in human sewage—treated or not—that was dumped into natural waterways. So many sources added excess phosphorus to Lake Erie. Each helped to kill the lake.

Phosphorus was not the only problem. Mercury, a major toxic pollutant, entered all the Great Lakes from many manufacturing sources. Mercury accumulates in tissues of fishes and can be deadly for humans who eat mercury-tainted fishes. In 1969 mercury levels in Lake Erie's fishes were so high that fishing was banned. Even if the "dead" lake had any life in it, it was deadly life.

As if all this were not enough, pollution of the Great Lakes, especially Lake Erie, also came from imported species. Exotic or alien species introduced from other regions often cause havoc. Of course, it is usually humans that inadvertently do the

Figure 27.33 **Lampreys Attached to a Lake Trout.**

Figure 27.34 **Lake Erie Today.** Although Lake Erie is still in danger, now that most sources of pollutants have been eliminated, the lake is repairing itself and becoming beautiful once again.

importing. In 1921 sea lampreys invaded Lake Erie. Lampreys are parasitic eel-like fishes that attach to other fishes to rasp their bodies and suck nutrients from them. Sea lampreys are aggressive parasites that are not native to the Great Lakes (**Figure 27.33**). They probably swam into Lake Erie after the Erie Canal was completed in 1819, connecting the Atlantic Ocean to Lake Ontario via the Hudson River. By 1946 sea lampreys had moved all the way to Lake Superior, and they flourished in the Great Lakes. Lampreys decimated the populations of many of the species of larger fishes and contributed to the collapse of the Great Lakes fishing industry.

Lake Erie Is Rehabilitated, If Not Recovered

Beginning in the early 1970s many different forms of legislation addressed these problems. The Clean Water Act empowered the federal government to force companies to clean up wastewater before dumping it into waterways that eventually fed into Lake Erie. Canada agreed with this approach, and a treaty was signed that committed both nations to keep inputs to the Great Lakes clean. Under pressure from legislation, ships began the practice of emptying their ballast containers far from sensitive waterways. Detergent manufacturers eventually found substitutes for phosphates that satisfied consumers' preference for "whiter than white" laundry. Sewage treatment plants upgraded their facilities to reduce nutrients that flowed into waterways. Manufacturing plants removed mercury from their processes or cleaned mercury from water before it was released into the environment. Certainly not all parties made these changes willingly or enthusiastically, and threats of fines and penalties were needed to prod many to comply with new regulations. In the end, though, Lake Erie has come back to life (**Figure 27.34**). The offensive algae are gone. People can fish again. A milestone was the return of mayflies, small insects that had disappeared from the lake when it was so degraded. People now enjoy Lake

Erie, and the bordering states can boast of its attractions to lure tourists and industry.

The result is not complete recovery, though. Many of the large fish species that once made their home in Lake Erie have not come back, or their populations are struggling to return to former levels. And the kinds of fishes that live in Lake Erie have changed too. The lake is much cleaner but not as pristine as it was even 100 years ago. In the last five years fish catches have declined again. Pollutants still enter the lake, and some parties still push to have controls and regulations relaxed. For those who do not remember the dead lake, the stringent controls seem burdensome. Perhaps we will all have to learn the same lesson all over again if Lake Erie returns to its previous state.

The lake also faces new challenges. Despite efforts to control the release of ballast water from ships, exotic species are still a problem. Zebra mussels are the newest pest. These thumbnail-size shellfish have a voracious appetite and high biotic potential. Zebra mussels eat single-celled eukaryotes, phytoplankton, and zooplankton and threaten to deplete the lake of food sources for fishes. Ecologists believe that zebra mussels could cause the collapse of fishing in Lake Erie, wiping out all the efforts people have made for the last 30 years. So the fight for Lake Erie is not over. As with all efforts to take back what we have lost, the fight is never won. Only with a sustained effort will the positive results be maintained.

QUICK CHECK

1. What three important concepts have you learned from the story of Lake Erie?

Now You Can **Understand**

Ecologically, It Is Impossible to Do Just One Thing

Because of the interconnectedness of all of Earth's systems, even if you want to, it is impossible to do just one thing. Two examples will make the point, starting with a simple one. Let's imagine that you want to eat an apple. And, to make the scenario extremely simple, let's assume that you are fortunate enough to have an apple tree in your own backyard, *and* that it is covered with ripe fruit. You walk outside and pluck an apple. In the kitchen you rinse it off and bite into it. Soon all that is left is the core that you throw into the garbage. Sounds like you have done just one thing, but consider the following:

- There has been an effect on the tree of having an apple plucked. The tree will have to heal over the raw spot where its living tissues are exposed.

- Rinsing the apple will remove a lot of bacteria and yeasts that live on the apple's skin, but it won't remove all of them. By eating them you introduce them into the culture of normal bacteria that live within your mouth and digestive tract. As they mix with your stomach acids, most will die, but some will multiply and add to the bacterial populations whose home world is inside your body.

- Rinsing the apple has linked you to the water cycle. Breathing, cellular respiration, and consuming the apple link you to the carbon cycle. Eating the flesh of the apple introduces new chemicals into your mouth that affect the organisms in the biofilm that coats your teeth and tongue. The sugars will feed populations of rapidly multiplying bacteria there. Eating the apple introduces new nutrients into your body and begins the process of digestion that has ripple effects on your own health and on the communities of microorganisms living within your intestines.

- An apple core contains five shiny seeds, and each has the potential to grow into a whole, new apple tree. When you throw the apple core into the garbage or compost, you probably are snuffing out the life in these seeds and are reducing the reproductive potential of the parent apple tree.

If the simple act of eating an apple has so many ecological consequences, imagine how they multiply in a complicated action that introduces a new product or a new idea into the marketplace where ecology and economy intersect. Henry Ford had such an idea in 1896 when he hand built his first motor car, the Quadricy-cle. In 1908 the first Ford Model T was introduced, and in 1913 Ford innovated the first continuously moving assembly line. Ford Motor Company became the largest automobile manufacturer in the world, and the innovations of Henry Ford's company began a series of ecological ripple effects that continue to this day. Global warming is just one of them. Revolutionizing manufacturing and increasing demand for manufactured instead of handmade goods was another. Increased demand for fossil fuels is another. And, of course, because all of Earth's systems are linked in cycles that transform and move chemical elements and compounds, environmental repercussions have continued to follow and have intensified as the human population has increased.

What Do **You** Think?

Wetlands Conservation

The ecological cycles you have learned about here are critical for the health of the biosphere. Nearly all scientists agree on that. Most of the rest of us agree too, in the abstract, but getting support for projects to restore damaged environments is much more difficult when it conflicts with the self-interest of the people involved. Wetlands are a good example. Wetlands are ecosystems such as swamps, marshes, and salt marshes that are adjacent to major oceans, bays, rivers, or lakes. Wetlands filter and purify water as it goes into the larger body of water, and they provide a habitat for a wide variety of species. Wetlands also provide buffer zones that protect areas inland from floods and storms. If the wetlands around New Orleans and along the Mississippi Gulf Coast had been intact, the damage from Hurricane Katrina in 2005 would have been much reduced. Wetlands have many additional beneficial aspects; not only do they clean runoff, they also are nursery grounds for fisheries, and they provide resting places for many migrating birds. Yet it is hard to get localities to approve the establishment of *new* wetlands in areas where they are needed. In Southern California, for example, a project to establish extensive wetlands met strong objections from people who did not want to lose their beach. In areas across the country developers drain wetlands and refuse to abide by orders to stop, citing their right to develop their land as they see fit. These arguments pit one definition of common good against another, and often the wetlands are the losers. Have you encountered any similar debates in your own community? How would you respond? What do *you* think about the need to protect environmental regions such as wetlands versus the rights of people to develop and use land as they choose?

Chapter Review

27.1 An Ecosystem Is the Functional Ecological Unit

In ecosystems energy moves in the biotic components of the system, and chemicals cycle between the biotic and abiotic components of the ecosystem. Abiotic ecosystem components include air, water, soil, and rocks. Biotic components are the communities of organisms in the area.

abiotic 652   ecosystem 652

27.2 Energy Moves Through Ecosystems, and Chemicals Cycle Within Ecosystems

Because of the metabolic processes of organisms, energy is lost at each link in a food chain, leading to pyramids of numbers of organisms in a community and pyramids of energy. Natural processes cycle materials between biotic and abiotic components of ecosystems, and these cycles have long- and short-term components. Evaporation, condensation, precipitation, and transpiration are active in the water cycle. Precipitation not absorbed by plants or soil runs off the land or may be stored within aquifers. Photosynthesis, respiration, and metabolism are active short-term processes of the water cycle, while fossilization and formation of deposits of limestone and fossil fuels are long-term storage of carbon. When fossil fuels are burned, carbon reenters the atmosphere as the greenhouse gas, carbon dioxide. The overwhelming evidence is that carbon dioxide has increased in the atmosphere, and that global climate change is under way. Prokaryotes like nitrogen-fixing bacteria are responsible for the transformation of atmospheric nitrogen into nitrogen-containing compounds that can enter food chains. Other bacteria process ammonia into nitrates that plants can absorb. Denitrifying bacteria transform nitrates and release nitrogen back to the atmosphere, completing the nitrogen cycle. The phosphorus cycle takes place within rocks, soil, water, and organisms. Mycorrhizae help plants absorb phosphorus from soil.

aquifer 655   evaporation 655
biogeochemical cycle 655   nitrogen-fixing bacteria 658
condensation 655   percipitation 655
dentrifying bacteria 659   water vapor 655

27.3 Ocean Life Is Influenced by Food, Light, Currents, and Pressure

Distribution of life in oceans is regulated by temperature, availability of nutrients, and availability of light. Most ocean life is in warm waters of the littoral and neritic zones. Ocean currents can bring warm water that is rich in nutrients to waters that are offshore of continents. Organisms that live in the littoral zone must have adaptations that allow them to survive rough waters and daily exposure to extremes of temperature, dryness, and sunlight. Estuaries like salt marshes and mangrove swamps are nurseries for many forms of ocean life that protect coastlines and slowly release trapped nutrients to offshore waters. Coral reefs are biologically active ecosystems created by the mutualistic association of coral polyps and green algae. Chemosynthetic prokaryotes are the keystone species of deep-sea hydrothermal vents, and the communities that develop there are among Earth's most unusual.

brackish water 664   estuary 663

27.4 Aquatic Ecosystems Link Marine and Terrestrial Ecosystems

Aquatic ecosystems have some animal groups, notably amphibians and insects, that are missing from oceans. Plants are important photosynthesizers in aquatic ecosystems. Aquatic ecosystems bridge terrestrial and oceanic ecosystems. Water movement, or the lack of it, contributes important characteristics to aquatic ecosystems. Ponds and lakes have seasonal cycles in which their water overturns, enriching upper waters and carrying dissolved oxygen down to the depths. Ponds and lakes are especially sensitive to oxygen depletion that can occur as a natural process of eutrophication. Stream and river organisms must have means that allow them to maintain their position in rapidly moving currents. Flowing-water ecosystems vary in their oxygen levels, temperatures, turbidity, amount of nutrients, and animal and plant adaptations.

eutrophication 667

27.5 Biomes Are Characterized by Typical Vegetation

Biomes are defined by dominant plant communities and include tundra, taiga, temperate deciduous forest, grassland, tropical rain forest, and desert. Each biome has characteristic organisms, and each biome is threatened by human activities. Habitat fragmentation and species loss are the most prevalent threats to Earth's biomes.

arctic tundra 669   permafrost 669
biome 669   taiga 671
canopy 672   temperate deciduous forest 672
desert 673   tropical rain forest 674
grassland 673   understory 672

27.6 The Biosphere Encompasses All of Earth's Biomes

The biosphere contains all of Earth's life and the abiotic components of ecosystems.

27.7 An Added Dimension Explained: The Death and Life of Lake Erie

Lake Erie is rehabilitated from the depths of pollution in the 1970s. Introduced alien species continue to be a problem, as does the tendency of some people to weaken or circumvent the legislation that has rehabilitated Lake Erie. Only constant vigilance will keep Lake Erie clean.

TRUE or FALSE. If a statement is false, rewrite it to make it true.

1. Rocks, soil, atmosphere, nutrient cycling, water, and organisms are characteristic components of ecological communities.

2. A pyramid of energy explains why there are more producers than there are top consumers in an ecological community.

3. The most varied forms of life are found in freshwater ecosystems.

4. Upwelling, coastal run off and abundant light make the pelagic zone highly productive.

5. Tropical rain forests are analogous to the neritic zone of the ocean.

6. Hydrothermal vent communities depend on mycorrhizae.

7. As streams merge into rivers, levels of oxygen increase, while turbidity decreases.

8. If something doesn't drastically change, biologists predict that tropical rain forests may be gone in 30 to 50 years.

9. Most tropical rain forest animals are familiar and well studied.

10. The biosphere is made of all of Earth's biotic features.

MULTIPLE CHOICE. Choose the best answer of those provided.

11. On average, how much of the energy that a swallow gets from the insects that it eats are stored in its tissues?
   a. none
   b. 1%
   c. 2%
   d. 10%
   e. 50%

12. If a green plant receives 257,000 watts of solar energy in an afternoon, how much will it be able to capture and use in photosynthesis?
   a. all of it
   b. 128,500 watts
   c. half of it
   d. about 2%
   e. about 10%

13. Life in oceans depends on these photosynthesizers:
   a. plants
   b. bacteriovores
   c. phytoplankton and zooplankton
   d. nanophytoplankton, ultraphytoplankton, and phytoplankton
   e. Portuguese men-o'-war

14. In what oceanic zone would bioluminescent organisms be plentiful?
   a. neritic
   b. abyssal
   c. photic
   d. pelagic
   e. littoral

15. Estuaries, mangroves, salt marshes, marshes, and lake bottoms are all rich in
   a. photosynthesizers.
   b. trapped and decomposing nutrients.
   c. light.
   d. top carnivores.
   e. dissolved oxygen.

MATCHING.

16–20. Match the columns. (One choice will be used twice.)

16. evaporation, condensation, precipitation
17. soil bacteria
18. erosion, mycorrhizae
19. limestone, coral reefs
20. photosynthesis, respiration, consumption

   a. nitrogen cycle
   b. carbon cycle
   c. phosphorus cycle
   d. water cycle

21–26. Match the columns. (One choice will be used more than once.)

21. arctic tundra
22. taiga
23. temperate deciduous forest
24. grassland
25. desert
26. tropical rain forest

   a. seasonally colorful leaves, spring and fall wildflowers; migratory birds
   b. threatened by human activities
   c. permafrost, mosses, lichens, migratory birds, mosquitoes, bioaccumulation
   d. pines and firs, boreal forest, long snowy winter, caribou
   e. uniform climate, epiphytes, infertile soils; calls are important
   f. periodic fires, grazing mammals, hibernators, migrators, estivators
   g. succulents, spiny leaves, nocturnal mammals

CONNECTING KEY CONCEPTS

1. Using specific examples, contrast the characteristics of ocean habitats with terrestrial biomes.

2. Living organisms play a key role in the cycling of elements in the biosphere. Choose two examples of cycles and explain the roles of living organisms in each.

QUANTITATIVE QUERY

Assume that a square meter of your backyard has received 1,370 watts of energy in the form of sunlight per day. This square meter is covered with vegetation that captures and uses 2% of this energy. Assume the grass has been growing for 100 days. Metabolic processes of grass plants consume 90% of the solar energy that reaches the grass, and only 10% of the available energy is stored in the organic molecules of the plant. A deer comes along and eats all of the vegetation in the square meter. (a) How much energy was available to the vegetation over the 100 days? (b) How much energy did the vegetation capture and use? (c) How much usable energy does the deer get from the vegetation?

THINKING CRITICALLY

1. How would you redesign the way that plants use photosynthesis to allow plants to capture more than about 2% of the Sun's energy? Suggest a few adaptations that would allow plants to capture more light and then refer to Figure 5.0 for an additional answer.

2. Fossil fuels such as gas and oil are hydrocarbons derived from organisms that lived long ago. When fossil fuel is burned, carbon dioxide is returned to the atmosphere. The burning of fossil fuels as a source of energy is increasing the levels of carbon dioxide in the atmosphere and is a major cause of global climate change. Many people suggest that massive tree-planting efforts could reduce atmospheric carbon dioxide and lessen global climate change. What is the scientific basis for this suggestion? Others suggest that planting lots of trees is only a short term solution to the atmospheric carbon dioxide problem. What is the basis for this argument? Which side do you agree with?

For additional study tools, visit www.aris.mhhe.com.

Impact of Humans on the Earth

CHAPTER

28

Declaration Signed on Great Apes

September 12, 2005

Chimpanzees are in danger of going extinct.

Chimps Nearing Extinction, Study Warns
Orphan Data Suggests One Species Could Be Gone in 20 Years

June 8, 2004

What Does It Mean to Be Human?

One of the biggest lessons learned from modern biology is that humans evolved from other living things, and so humans share traits with other living things. According to this view no human traits are unique to humans. This includes characteristics that thinkers have claimed set us apart from other animals. This lesson comes into focus when you examine the traits of our closest living relatives, the chimpanzees (**see chapter-opening photo**).

What comes to your mind when you try to list traits that define what it means to be human? Your list might include: Humans have large brains and are intelligent. Humans use tools and language. Humans provide lavish care for their children and derive a great deal of pleasure and satisfaction from doing so. Humans live in socially supportive groups, including families and communities. Humans have a sense of self and recognize different individuals in their group. Humans have a sense of right and wrong.

Unfortunately, there are dark sides to being human as well. For instance, humans fight wars, rape women, kill babies, and murder one another. Humans steal from each other and try to dominate one another. Some humans are abusive parents and are cruel to their children. Some humans ignore the difference between right and wrong and simply pursue their own selfish interests.

For most of human history many if not all of these traits were thought to be distinctly human, but in the past half century almost all have fallen as defining features of our species. Much of this insight has come from studies of chimpanzees in the wild, and Jane Goodall was an early pioneer in this field. Today we know that chimps are indeed very human, and humans are in many ways much like chimps. Both scientists and nonscientists now realize that only by understanding chimps more deeply can we better define what it means to be human.

Unfortunately, this opportunity for self-knowledge may be short-lived. For the last 50 years, chimpanzee numbers have been declining. When Dr. Goodall began to study chimpanzees in 1960, nearly 2 million chimps lived in African forests. Today, there are only around 150,000. Threatened from all sides, some scientists predict that if something drastic is not done, chimpanzees will be extinct in the wild in 50 years or less.

What sorts of self-knowledge will we lose if the world's wild chimpanzee populations are lost? What are the threats to chimps, and what is being done to protect them? You will return to these topics later in the chapter. But first you will consider many of the other ways that humans have changed the biosphere, and how some people are working to protect the living world that remains. ■

28.1 The Impact of Humans on the Biosphere Is Widespread and Profound

Humans have had profound effects on the biosphere that stretch from local losses of biodiversity to the broad effects of global climate change (**Table 28.1**). The impact of humans on Earth's environments is intensified by the growing number of people in the human population and the disproportionately large amount of resources that some humans use. Any population of animals that overshoots the carrying capacity, or K, of the environment can exhaust resources and degrade the environment (see Section 26.2). As the environment's carrying capacity is approached or overshot, more individuals in the population die (see Section 26.2). Soon deaths offset births, and the population stops growing. These natural controls limit the impacts that nonhuman populations have on their environments.

For most of human prehistory people were not as numerous as they are now, and human impact probably was restricted by the environment's carrying capacity for humans. This pattern ended around 10,000 years ago when humans first began to domesticate plants and animals, providing more reliable food sources that allowed human populations to increase (**Figure 28.1**). Today, humans are expert farmers, and advances in science, hygiene, medicine, and technology dramatically have changed the way that humans use and dispose of resources.

Currently, there are about 6.5 billion humans on Earth, and by 2050 Earth may hold 9 billion humans (see Section 26.3). The impact of 6.5 billion humans on Earth's ecological systems is undeniable; in the view of many ecologists, the human population *already* may be beyond Earth's carrying capacity. These scientists conclude that right now humans have an **overpopulation** problem. Even if all humans lived in tents and ate simple foods, a population of 6.5 billion humans would stress natural environments. With the spread of the American heavy-consumption lifestyle, the impact of each person, especially in more developed cultures, is even greater than it otherwise would be.

The size of the human population combined with technological advances mean that humans have a greater impact on the biosphere than do other species. Although all species produce bodily wastes, technology adds chemical wastes and toxic pollutants that can degrade and pollute the environment. Negative effects of human populations

overpopulation when a population grows beyond the environment's carrying capacity for that species

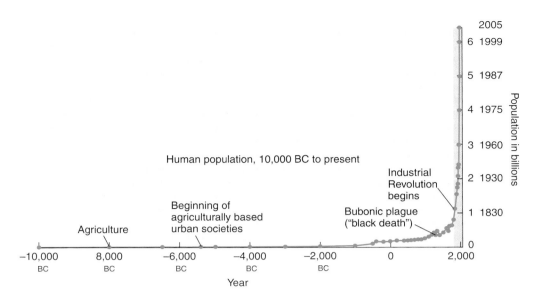

Human population, 10,000 BC to present

Agriculture

Beginning of agriculturally based urban societies

Industrial Revolution begins

Bubonic plague ("black death")

Figure 28.1
Human Population Growth Related to Civilization's Advance. Human population growth plotted on a log/log scale shows three increases. Each is associated with a revolution in culture, agriculture, or science and technology.

include reduction of green space, chemical pollution of streams and rivers, smog and air pollution, and soil contamination near factories. In addition, the activities of humans in one region can have consequences in distant places. For example, in 2002 up to 75% of the air pollution in Seattle, Washington, had originated in China. About 30% of the mercury in U.S. air originates in Chinese power plants that burn coal. But, of course, that does not mean that U.S. cities and factories also do not produce pollutants. Overall, the U.S. produces more of its own air pollution than it gets from other countries, and without doubt some U.S. air pollution is exported to other countries. Eastern Canada has been particularly negatively affected by acid precipitation that originates in the United States. Pollution now reaches some of the most pristine and previously untouched landscapes. **Figure 28.2** contrasts the air over the Grand Canyon on good and bad air days. This pollution comes from many sources, including cars, factories, power plants, and lawn mowers.

Table 28.1	Some Examples of Human Impact on the Biosphere

Global warming is caused by the heat-trapping action of greenhouse gases that accumulate in the upper atmosphere. Greenhouse gases include carbon dioxide and methane.

Depletion of the ozone layer is caused by the release of chlorofluorocarbons and related ozone-destroying gases.

Loss of topsoil is related to farming methods that disturb soil and deplete soils of water-retaining organic matter.

Pollution of waterways with silt and fertilizer runoff can be traced to habitat destruction, farming methods that lead to wind and water erosion, and overapplication of chemical fertilizers.

Acid precipitation is caused by industrial processes that introduce sulfur and nitrogen oxides into the atmosphere, where they react to form acidic rain, snow, sleet, and fog.

Introduction of bioamplified chemicals and toxic wastes may damage or kill organisms outright or may be passed up food chains and affect the top carnivores.

Disruption of habitats caused by clear-cutting forests, road-building, farming, settlements, and draining of wetlands leads to loss of habitats, species loss, and loss of biodiversity.

Introduction of alien species that have no natural predators disrupts communities and may result in loss of native species.

Overuse of pesticides and fungicides has produced pesticide- and fungicide-resistant organisms.

Loss of species and loss of biodiversity has occurred mainly through loss of habitats.

(a)

(b)

Figure 28.2 Air Pollution in Grand Canyon. (a) On a good air day the layers of rock in the walls across the Grand Canyon are plainly visible. (b) On a bad air day pollution obscures the view across the Grand Canyon.

The effects of human overpopulation even extend to remote, nearly uninhabited biomes such as the icy deserts of Antarctica and the high Arctic, places where there are no industries to introduce pollutants. The Inuit people of Northern Quebec provide a dramatic example. Inuit do little to degrade their own environment. Yet some of their traditional foods, seal and beluga whale blubber, contain high levels of the pesticide DDT (short for diethyl-diphenyl trichloroethane) and PCBs (short for polychlorinated biphenyls) that originate in U.S. manufacturing plants. DDT and PCBs from the United States are carried north on winds and in water vapor within clouds (**Figure 28.3**). Once the pollutants are deposited in the north, they enter the food chain. Because DDT and PCBs are nonpolar molecules, they accumulate in the fatty tissues of carnivores such as marine mammals, and then in the fatty tissues of humans who eat them. The levels of these pollutants in the breast milk of Inuit women are four to seven times higher than the breast milk of Quebec women who do not eat blubber. PCB-contaminated breast milk may impair infant brain development, and high rates of infectious diseases in Inuit infants may be caused by damage to their immune systems related to their consumption of PCBs. In fact, there is *no place* on Earth that is pristine and free of pollutants.

"Okay," you may be thinking, "the human population is growing, and we are polluting the environment. But does this really matter?"

Before you can begin to evaluate this question, you must learn much more about the nature of the impact that humans have had on Earth. By the end of this chapter you will know whether human impact on Earth's natural systems is something to be concerned about. You also will have read what can be done to help alleviate some pressing environmental problems. Let's start by examining a few cases of human impact on other animal species.

(a) Pollutants from the U.S. enter upper level airstreams, and are carried north and are deposited in Nunavut.

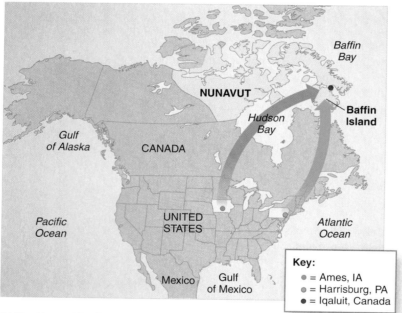

(b) The Nunavut landscape.

Figure 28.3 **United States Air Pollution Is Deposited in the Canadian Arctic.** (a) Weather systems carry pollutants north and deposit them in the Canadian arctic. Even though it seems a pristine Arctic landscape (b) the breast milk of Inuit women who live in Nunavut contains up to six times the levels of PCBs of breast milk of women who live in southern Canada.

> **QUICK CHECK**
>
> 1. How can overpopulation be defined?
> 2. How have some human populations reset *K*?

28.2 Humans Can Hunt with Deadly Efficiency

One reason for human success is the human ability to use tools to hunt and kill other animals for food and other resources. Prehistoric hominins probably began using tools to hunt at least a half million years ago in certain areas. For many reasons large animals are more susceptible to extinction than small animals. Combined with pressures from climate change, the effective hunting of large mammals by even relatively small populations of humans probably contributed to their extinction. The story of the moas of New Zealand is an example.

Maori Hunters Eliminated New Zealand's Moas

Forget Tweety Bird, Daffy Duck, Woody Woodpecker, and the Road Runner. Moas will challenge your mental image of a bird. Picture an ostrich, but remove its wings and even its wing bones, and make it *twice* as large. The giant moa was a wingless bird that stood 13 feet (4 meters) tall and weighed about 500 pounds (275 kg). Now give this giant a long neck, a beak like a broad spear point, and cover its body with soft feathers, and you're getting closer to a moa (**Figure 28.4**). Once you've created this creature, you'd better stand back, because giant moas had enormous legs and feet that were powerful weapons. Their drumsticks were as big as king-size pillows and their feet had claws the size of bananas. Before the Maoris arrived from Polynesia in about AD 1000, New Zealand was home to about 150,000 to 200,000 moas in about 10 species. These were doomed to extinction at the hands of humans 100 to 200 years later.

The record of the role of humans in moa extinctions is found in caves. Scrap heaps left by early Maoris give a record of when they arrived and what they ate. The earliest moa remains in these caves are in the northern part of New Zealand where the Maoris first landed. As the Maoris migrated south, they left behind piles of moa bones that mark their path. Analysis of these scrap heaps shows that moas and their eggs sustained the Maoris. Surprisingly, the early Maori population was perhaps about 600 individuals by the time the moas were gone. Between eating the moas for food and destroying their habitats for living space, the Maori rapidly eliminated the large flightless birds of New Zealand.

Civilized humans who have ample supplies of other food sources cause extinctions too. In the past, limited human hunting technologies meant that human actions could add to other factors such as climate change that were already threatening species, but it was unlikely that hunting would drive a large, thriving population to extinction. The development of more effective hunting tools such as bows and arrows, guns, harpoons, and massive nets has allowed humans to kill off even well-established species numbering in the millions. Next, let's consider the case of the passenger pigeon.

Americans Eliminated the Passenger Pigeon

The story of the extinction of passenger pigeons includes the lonely image of Martha, the last surviving passenger pigeon. She died in the Cincinnati Zoo on September 1, 1914, at 1 P.M. (**Figure 28.5**). Martha had been hatched in captivity and was 29 years old when she died. Passenger pigeons had been Earth's *most numerous bird species*. Read that again to make sure that you've taken it in. It is incredible but true that a relatively few Americans exterminated all of the passenger pigeons. Reliable accounts of passenger pigeon populations stagger

Dinornis maximus
Extinct *c.* 1850

Pachyornis septentrionalis
Extinct *c.* 1500

Emeus huttoni
Extinct *c.* 1500

Megalapteryx didinus
Extinct *c.* 1850?

Figure 28.4   **Different Kinds of Moas.** All of these species of moas are extinct. Some of them were extremely large birds. A 6 foot-tall (2-meter-tall) person is included for scale.

Figure 28.5   **Martha, the Last Passenger Pigeon, Died in Captivity in 1914.**

the imagination, but American settlers wiped them out. How could this have happened?

Put simply, passenger pigeons were delicious and cheap, and Americans liked to eat them. For several reasons passenger pigeons were easy to kill, and Americans became experts at killing them. The huge flocks of passenger pigeons were easy targets. They flocked together in such huge numbers that the noise of their beating wings sounded like threshing machines, steam locomotives, and sleigh bells all rolled into one thunderous roar. Nothing alive today is a suitable comparison to their enormous flocks. In the fall of 1813 the ornithologist and famous painter of birds, John James Audubon, saw a column of passenger pigeons flying overhead. It filled the sky and darkened a bright noontime sun. When the flock of passenger pigeons first appeared, Audubon was traveling to Louisville, Kentucky, in an open wagon. By sunset, when he had reached his destination *55 miles* (88.5 kilometers) away, the birds were *still* flying overhead in a solid mass. And for *three more days* other huge flocks appeared and passed overhead. Audubon estimated that the first flock had more than a billion individuals. (His actual estimate was 1,015,035,000 birds, but who wants to quibble?) Passenger pigeons earned their name because they traveled from place to place, seeking food as well as roosting and nesting sites (**Figure 28.6**).

It is worth noting that Seneca tribes who had been living in North America for thousands of years had harvested passenger pigeons every year.

Senecas would eat only fully grown nestlings, and hunters would use poles to knock them out of their nests. Younger nestlings, eggs, and adults were not harmed. The Seneca hunting strategy preserved the species, but allowed people to enjoy the tasty birds and the yearly festival that accompanied their arrival at the nesting grounds. By comparison, after just 50 years of intensive hunting, European Americans had wiped out passenger pigeons. The newly opened railroad lines allowed market hunters to ship the dead birds to cities back East. Unfortunately, this was before refrigerated cars were invented and, even though the birds were packed in ice, a huge portion spoiled before they got to market. This meant that even more birds were killed to keep pace with the demand. When Martha died on her perch in the Cincinnati Zoo, her species disappeared with her. The patterns that you can observe from these historical anecdotes are clear: humans tend to overuse resources, and unrestricted human hunters can drive species to extinction.

A modern version of the extinction of the moas and the passenger pigeons is happening in Africa, home of a remaining *megafauna* that lives on the plains and many more species that live in forests. In some parts of Africa people eat *bushmeat*—wild-caught animals that are shot or snared by hunters—rather than domestically grown meat sold in supermarkets (**Figure 28.7**). The volume of trade in bushmeat is astounding. For instance, in Ghana it is estimated that 75% of the population depends at least partly on wild-caught food. Over 1 million

Figure 28.6   **Passenger Pigeons.**

SHOOTING WILD PIGEONS IN IOWA.

Shooting Passenger Pigeons.

Figure 28.7 **Bushmeat in an African Market.**

metric tons of bushmeat is taken from West African forests each year. The situation is similar in other African nations, and bushmeat markets also are active in Southeast Asia and in South and Central America. Bushmeat markets are found not only in Africa but also in European and U.S. cities with West African populations. New York, Paris, and San Francisco are just a few of the cities outside Africa with thriving bushmeat markets. Because of the demand for bushmeat, many forest regions of Africa are now nearly empty of mammals including some primate species. So the historical human behavior pattern seems to hold: unless humans are restricted, they will overuse resources and will drive animal species to extinction. A similar situation is under way in the oceans, where overfishing has depleted the populations of many kinds of fishes.

Overfishing Has Depleted Ocean Life

Fishing provides high-quality protein for human populations around the world. Fishes have always have been part of the human diet, especially for people who live around water. Fishing also involves other species that have value for humans. Crustaceans, such as crab, shrimp, and lobster, are popular food items in many human cultures as are mollusks like clams, mussels, and oysters, and even marine mammals. All of these marine species have been threatened with extinction because of human activities.

The regions where specific populations of fish are caught are called *fisheries,* and each fishery is defined by a particular species of fish. In the past people could only fish in local waters near settlements and so human impact on fisheries was minimal. Prehistoric and early historic fishing technology limited how many fish could be caught, and how many people could be supported by a local fishery. The technology used in modern fishing has changed the balance between people and fisheries. Today fleets of factory ships travel world wide in search of specific species of fishes. Modern fishing technology is so effective that now about one bil-

lion humans get most of their protein from seafood. Worldwide, fishes and other marine species provide about 19% of the protein that people eat, while in Asia nearly 30% of dietary protein comes from fishes. Over the past 100 years fishing harvests have increased dramatically. For example, in 1900 people around the world probably harvested about 3 million tons of fishes and other sea creatures; by comparison, in 1989 *86 million tons* were harvested. The wild fish harvest peaked at 96 million tons in 2000 and declined to 90 million tons in 2003. As the term implies, **overfishing** depletes stocks of fishes and other sea creatures. The world's demand for fish is increasingly being met by fish farms, where fish are bred, incubated, and raised. This solution is not without its own problems, however. Fish farms pollute local waters and damage ecosystems. Despite the rise in consumption of farm-raised fish, the effective management of wild fisheries is critical for the long-term health of people and ocean ecosystems.

The Grand Banks off of Canada's East Coast (**Figure 28.8**) provide a dramatic example of the effects of overfishing and the challenges of dealing with the collapse of a fishery. Five hundred years ago cod were so plentiful in these waters that sailors could pull them in just by dipping baskets into the sea. Cod and many other fishes were caught on this enormously productive fishing ground and sold all over the world. Over the last decades, however, the

overfishing harvesting that exceeds the replacement rate of a species

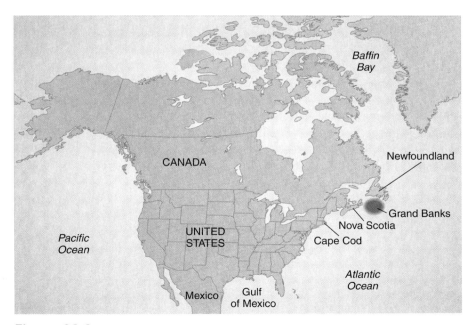

Figure 28.8 **Canada's Grand Banks.** These fishing grounds are south of Newfoundland. Cod and flounder once were extremely common in these waters. Now they are rarer and the fisheries have been closed to fishing for cod and flounder.

cod and flounder catches have dwindled. To protect the diminished populations of cod and flounder, the Canadian government closed the Grand Banks fishing ground for these species in 1992 and severely restricted fishing for other species. As of this writing, fishing still is restricted in this once-productive area. This drastic step has caused much economic hardship, and many people have been thrown out of work. On the other hand, had the Canadian government not banned fishing for cod and flounder in the Grand Banks, these species would have been fished to extinction.

Banning fishing of selected species for some length of time now is a common way to restore fishing stocks, and it is often successful. For example, swordfish populations in the North Atlantic have rebounded as a result of fishing restrictions. Canadian cod have not recovered, though. Scientists are not sure why, but some ideas have emerged. For one thing, the World Wildlife Fund recently reported that many fishing vessels are getting around the ban by declaring that cod or other restricted species were caught by accident and so

(a) Status of the world's marine fisheries

Marine Fisheries

Key:
- ☐ Fully exploited
- ■ Underexploited
- ☐ Overfished
- ■ Recovering
- ■ Depleted
- ☐ Moderately exploited

50% 20% 15% 6% 2% 6%

(b) Status of commercially fished species

Figure 28.9 **Status of the World's Fishers and the Status of Individual Commercial Fish Species.** (**a**) Status of marine fisheries. 50% of the world's marine fisheries are overfished. (**b**) Status of species that are commercially fished.

■ *Is it ecologically responsible to eat broiled shrimp? Orange roughy? Caviar? Atlantic cod?*

Species is relatively abundant. Fishing and farming methods cause little damage to habitat and other wildlife.

- Farmed clams, mussels, oysters, and bay scallops
- Alaska salmon
- Striped bass, wild and farmed
- Mahimahi, pole- and troll-caught
- Albacore, bigeye, yellowfin, and skipjack tuna, pole- and troll-caught
- American (Maine") lobster, Maine and Canada

- Squid
- Pacific soles
- Dungeness, king, and stone crabs
- Catfish, U.S.-farmed
- Shrimp, U.S.-farmed
- Tilapia, U.S.-farmed
- Pacific cod
- Pacific halibut

Some problems occur with fishing or farming methods, or insufficient information available for evaluation.

- Albacore, bigeye, yellowfin, and skipjack tuna, canned or longline-caught
- Monkfish
- Sea scallops

- Atlantic flounders and soles
- Rainbow trout
- Swordfish
- Blue, snow, and tanner crabs

Overfished: fishing methods damage other species and habitats, or farming methods have serious environmental impacts.

- Atlantic cod
- Sharks
- Shrimp, imported
- Farmed (Atlantic) salmon
- Caviar, from wild-caught sturgeons

- Groupers
- Orange roughy
- Chilean seabass
- Atlantic bluefin tuna
- Rockfish, U.S. West Coast
- Atlantic halibut

Key:

 Fishery is certified as sustainable to the Marine Stewardship Council's environmental standard. Learn more at www.msc.org.

▷ One or more consumption advisories exist from state agencies, the U.S. Food and Drug Administration, the Environmental Protection Agency, or scientific studies. For more information, see *www.blueocean.org/seafood.*

can be kept and sold under the current rules. This sort of problem is one of many reasons that scientists are suggesting a new approach to fisheries management. In most cases bans or restrictions on fishing focus on a particular species. Some scientists now are arguing that restrictions should be applied to a whole ecosystem. This approach takes into account the complex ecological relationships between species and—scientists argue—is more likely to preserve and nurture the specific species that people catch, sell, and eat. This ecological approach leads to the closure of a whole geographical region to fishing. Recent experiences show that when a marine reserve is established to protect one or a few species, and no fishing is allowed in the region, the numbers of many species increase in waters around the reserve. If a complete ecosystem in a region is protected, then all species in the region flourish and their numbers spill over into nearby unprotected areas.

The pattern of depletion of fish stocks is seen across the world. Twenty-one percent of the world's fishing grounds are now *overfished* or depleted, meaning that population losses caused by fishing exceed the reproductive rate of fishes (**Figure 28.9a**). Another 50% of the world's fishing grounds are *fully fished,* meaning that reproduction can just keep pace with demand. Only 26% of fishing grounds are being fished at a relatively low level, and 2% are recovering from overfishing or depletion. One study has suggested that all of the world's wild fishery species could collapse to near extinction by 2050. Figure 28.9b shows some of the species most in peril. You may recognize some of these as fishes that you like to eat.

You may be left wondering whether you should ever eat fish. Wild-caught fish are being depleted, and there are reports that farm-raised fish contain dangerous levels of toxins. What should you do if you wish to contribute to the ecological stability of the world's oceans but not give up eating fish forever? The major point of this presentation is to make you aware of the issues and to let you make your own decisions. Some suggestions might help you reason your way through this ethical dilemma. You might want to become aware of which fish species are endangered and avoid buying them. Your political voice also can play a role, especially if you live in a region where fishing is part of the local economy. Becoming involved in political decisions can mean more than just supporting or opposing local legislation. The right blend of sensitivity to the state of marine fisheries and the state of your local economy could provide a balanced point of view.

Bison Nearly Were Exterminated for Political Reasons

Humans kill animal species for reasons besides obtaining food or clothing. The wild bison herds that used to roam the midwestern plains are a prime example of animals that have nearly been eliminated for political reasons. In the 1700s millions of bison grazed on the midwestern American prairies, moving along migratory routes used by bison for tens of thousands of years (**Figure 28.10**). Before about 1800 settlers hunted bison for their meat, skins, and tongues, while Native Americans, especially tribes living in the Great Plains, used bison meat, fat, skins, sinews, bones, and even stomachs, hooves, and horns. In those days the

(a)

(b)

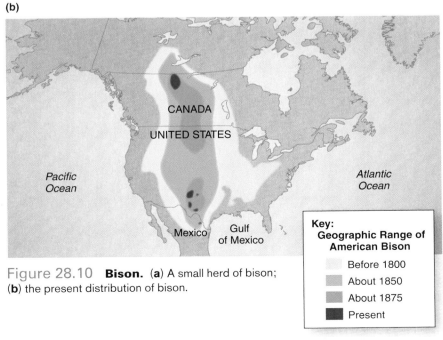

Figure 28.10 **Bison.** (**a**) A small herd of bison; (**b**) the present distribution of bison.

number of bison killed was balanced by the ability of bison to reproduce, and the herds remained large and stable. This balance changed after the 1803 Louisiana Purchase added lands west of the Mississippi to the United States. Much of this land was occupied by native tribal people. As long as the bison were plentiful, native tribes would stay, and European-American settlers would not want to move in. So the U.S. government encouraged the elimination of the bison herds. In just two years, 1871 and 1872, about 8,500,000 bison were shot; most of them were left to rot. One observer called the plains of Kansas a "putrid desert." The campaign to eliminate bison was remarkably successful: in 1860 there were about 60 million bison; just 29 years later only 150 individuals remained.

Once Native American tribes had been removed to reservations, the American government and conservation organizations launched legislation and breeding programs that eventually have saved the

bison. Today, there are a few thousand bison in America, but this is a far cry from the millions that thrived on the American plains only 200 years ago.

QUICK CHECK

1. What similarities are there in the impact of human hunters on moas, passenger pigeons, bison, and codfishes and other ocean species?

28.3 Combating Deforestation Is a Continuing Challenge

The forests of the world emerged as dominant biomes after the last ice age, around 10,000 years ago. Before the development of agriculture many continents were covered by forests. Of course, prehistoric humans used fire to hunt prey and clear forests and as a result converted some forests to grasslands. Until several hundred years ago, however, the impact of humans was still minimal. With time humans learned more effective ways to clear forests and convert them to farmland. Europeans were especially good at this. For example, before Europeans came to the Americas about half of the United States was forested. By the early 1900s almost all of those forests were gone.

The United States is not the only region to have lost most of its forests. At one time the Amazon rain forest covered the northern half of South America. India, China, and Europe also were covered by forests (**Figure 28.11a**). The biodiversity of these forests was astounding. Humans have cut down nearly all of these forests. It may look as though Earth still has extensive forest cover (Figure 28.11b), but today's forests are only a fraction of what they were when human civilizations first began. About 17 million square miles of forest has been lost in the past 4,000 years. This is an area about the size of Asia. In Europe and the United States deforestation occurred hundreds of years ago. Other regions have seen their forests almost completely disappear in the last century. By the middle of the 1900s large forests remained in the tropics, notably the huge Brazilian rain forest and rain forests in Indonesia. Deforestation now is acute in these regions. Between 2000 and 2005 a forest area larger than twice the size of Alaska (410 million hectares or over 1.5 million square miles) was lost. Most of this loss was in the South American rain forests, but parts of Africa and Asia also showed big losses. In 2005 Russia and Brazil had the largest shares of the world's forests; the Brazilian rain forest has one of the highest rates of deforestation.

(a) Frontier forests 8,000 years ago

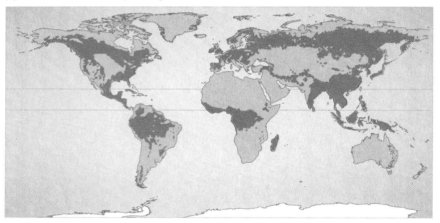

(b) Frontier forests in 1997

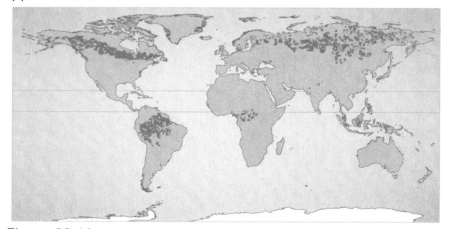

Figure 28.11 **Deforestation Worldwide.** Notice the difference between amount of frontier forest cover 8,000 years ago (**a**) and in 1997 (**b**). Frontier forests are largely undisturbed by human activity and are large enough to maintain biodiversity.

The causes of deforestation are complex and are related to human activities that increase as human populations grow. Poor farmers look to forests for farmland and fuel wood. The demand for wood products leads logging companies to cut down forests for lumber and pulpwood. Corporate farms cut down forests to expand their operations. Forests give way to development of roads, housing, shopping malls, schools, factories, and airports.

Cutting down any forest has profound ecological consequences. Farmers who have cut down forests for farmland soon find that the land is not nearly as productive as it once was. Without the forest to absorb rainfall, water runs off, swelling local rivers and causing flooding. The mudslides and floods in the hills of new California towns demonstrate that an altered landscape can cost human lives as well as money. As another example, in regions of Central Africa clearing rain forests has led to the expansion of the Sahara Desert. In China loss of forests has led to pollution and flooding of major rivers, and many lives have been lost as a result. We cannot know all of the consequences of cutting down forests, but some major consequences of rain forest deforestation are listed in **Table 28.2**.

Of course, loss of forests and other green spaces means loss of habitat for the species that once lived there. With each acre of forest that is cut down, scores of species edge closer to extinction. Many of these species are not even known to sci-

ence. One of the irreparable consequences of clearing forest is the loss of **endemic species,** those that are found nowhere else (**Figure 28.12**). Thousands of endemic species are thought to go extinct each year, and this trend will continue as long as forests continue to be destroyed.

Deforestation alarms many people including environmental groups, governments, and industries, and efforts to slow and reverse deforestation are finally showing results. Two major approaches are protection of existing forests and restoration of damaged forests including planting of whole new forests. As a result of all of these efforts, in 2005 the overall

endemic species species that have a restricted geographic range and are found only in a small area

(a)

Table **28.2**	Some Consequences of Clearing Rain Forest
Effect	**Consequence**
Loss of endemic species	Biodiversity is diminished
Loss of forest cover	Soil is unprotected and forest may not regenerate
Loss of soil protection	When soil dries out, it may harden into rocklike surface; unprotected soil is prone to erosion
Soil erosion	Silt is deposited in waterways and on coral reefs; corals may die
Disruption of weather patterns	Lack of rainfall
Loss of forest cover	Rainfall floods the area because of reduced plant cover to absorb it
Loss of wildlife	Disruption of lives of local tribal people

(b)

Figure 28.12 **Endemic Species.** (**a**) Galapagos tortoises are endemic to the Galapagos Islands; (**b**) Pandas are endemic to the bamboo forests of China.

■ *Go online and find out what endemic species live in your local area.*

loss of forests finally slowed. Forests still are being lost, but not at the same rate as they were in the late 1900s. Here are some examples of these efforts.

In some cases the return of forests has been an unintended consequence of changes in land use. In the northeastern United States and in Europe the amount of land devoted to farming has decreased over the past 50 years. While human developments occupy some of this land, other land is returning to forest. Of course, secondary succession may not produce the same kind of forest that existed several hundred or 1,000 years ago, but both the United States and Europe are experiencing a net increase in forests in part because of these trends.

Protection of forest land is the best way to preserve the forests in their original, natural state. Many U.S. national parks and preserves do just this, although there continues to be controversy over the extent to which protected areas should be open to logging. Environmental organizations also contribute to the protection of native forests. For example, The Nature Conservancy uses member contributions to buy ecologically important lands and shield them from development. Other countries also have increased their protection of natural forests. Brazil and China, two countries with high rates of deforestation, have launched major initiatives to protect some of their native forests. The government of Madagascar has announced an aggressive forest protection plan that should save up to 6 million hectares (23,166 square miles) of Madagascar's remaining forests. Related to efforts at protection are *sustainable forestry* programs. This approach recognizes that humans can use forest resources and still preserve the forests. The nature of sustainable forestry depends on what forest it is practiced in. In the Pacific Northwest advocacy groups are pushing to require that logging be carried out selectively, rather than cutting down every tree on a hill and then replanting the whole area with young trees. Selective logging is more expensive in the short run, but it preserves valuable resources and is kinder on the environment in the long run. In the rain forest sustainable forestry might mean encouraging local communities to harvest resources from the forest—such as nuts, or rubber, or medicinal plants—without cutting the whole forest down. In the big scheme of things sustainable forestry projects may seem small and insignificant, but they are changing the way people relate to forests in many parts of the world.

Once forests are gone, the only way to increase forest areas is to plant trees. In many instances such efforts are essentially forest farming. For example, it is standard practice for logging companies to replant trees after they have cleared an area. Unfortunately, these trees are the equivalent of a tree plantation. Because they are essentially a monoculture of a single, fast-growing tree species, they will not restore the native forest. It also is likely that once they are mature they will be cut down. Nevertheless, such efforts help to maintain a balance between forest growth and forest loss. Many governments have launched massive reforestation campaigns. China has planted millions of acres of tree plantations in an effort to reverse the disastrous environmental damage of widespread deforestation. The Chinese government recently has announced a new plan to plant more complex, healthier forests by introducing a diverse set of native trees.

Efforts also are ongoing to save or restore the biodiversity lost due to deforestation and the giant pandas in China are one example. Pandas live exclusively in bamboo forests and eat only bamboo. In the last part of the 1900s bamboo forests were being lost at an astounding rate, and people worried that pandas also might disappear. The Chinese government began to establish bamboo forest reserves where the pandas and their environment are protected. The effort was partly financed by Western environmental organizations and Western zoos with pandas on loan from the Chinese government. By 2004 there were thought to be about 1,600 wild pandas. Given all the safeguards that protect them, panda populations now should remain stable and perhaps increase over the next decades.

Sometimes extraordinary efforts are required to save a species whose home is lost to deforestation. The golden lion tamarins of Brazil's Atlantic rain forests of Brazil are an example. Golden lion tamarins are small bright golden monkeys that are among the most endangered of all primates (**Figure 28.13a**). They have become emblems of hope that the trends toward loss of rain forests and loss of irreplaceable species can be reversed. Golden lion tamarins tend to live in monogamous pairs and raise just one or two babies at a time. A family usually lives in and defends an 8.1-hectare (about 20 acres) forest territory. When Europeans first came to Brazil, the spectacular rain forest covered 330 million acres. Because of logging, it now is only about 7% of its original size, and much of the remaining rain forest is fragmented into smaller regions. So golden lion tamarins are reduced to only about 2% of their original geographic range (Figure 28.13b). Biologists have been tracking and studying the golden lion tamarins for years. They can identify just about every individual alive today. Efforts have been made to reduce habitat destruction and to rear tamarins in zoos and reintroduce them into their

(a)

(a)

Figure 28.13 **Golden Lion Tamarin.** (a) The Golden Lion Tamarin (b) now only occupies a tiny fragment (green) of its previous geographic range (red) in the Atlantic rain forest in Brazil.

(b)

South America

Brazil

Key:
- ▢ Original Atlantic rain forest
- ▪ Remaining Atlantic rain forest
- ● Golden lion tamarin range

wild habitat. Ranchers and farmers have agreed to plant forest corridors that link fragmented habitats, allowing the tamarins to roam more widely and have a greater choice of mates. While progress is slow, it does appear that these efforts are working. In 1980 there were fewer than 100 golden lion tamarins in the wild; today there are more than 1,200. It is not clear if golden lion tamarins have been saved, but the example shows that conservation efforts can have a positive impact. And with every acre of rain forest set aside as a refuge for golden lion tamarins, uncounted numbers of other species are protected.

Some effects of the loss of green space are more subtle. For example, because farmland traditionally is not 100% used for croplands or pastures, farmland can be a haven for wildlife. Usually there are *hedgerows,* edges of fields where bram-bles, shrubs, weeds, and wildflowers grow, and uncultivated areas where native species can survive alongside of domesticated plants and animals (**Figure 28.14**). When farms and fields are small and family owned, there tend to be lots of areas left as wild. In hedgerows native trees, bushes, grasses, and other plants form a small habitat. As farms get bigger, however, hedgerows get smaller and fewer or entirely disappear. While this farming practice allows the huge machinery of factory farms to operate with increased efficiency, the loss of "wild corners" and hedgerows has an enormous

Figure 28.14 **Hedgerows Are Disappearing.** Even though they shelter many species of wildlife, hedgerows (**a**) are being cleared from farmers' fields. Their outlines persist in the new, larger fields (**b**).

(a)

(b)

negative effect on local wildlife. In one study in Iowa, the wild borders between farms were home to an average of 36 species of birds per every 6 miles (about 10 kilometers) of border. When these wild areas were removed or reduced, the areas could sustain no more than 9 bird species per 6 miles (10 kilometers).

QUICK CHECK

1. Make a chart that summarizes the "ripple effects" of logging in tropical rain forests.

28.4 Humans Introduce Pollutants into Environments

So far you have read ways that humans deliberately alter the environment by removing components through hunting, overuse, or habitat destruction. While these human activities have a profound effect on the biosphere, they are not the only ecological impact that humans have. Pollutants that originate from human activities also have profound negative impacts on the biosphere. Let's take a closer look at some of these pollutants that are found in water, soil, and air.

Human Biological Wastes Can Threaten Waterways and Human Health

Each of Earth's 6.5 billion humans need an average of nearly four liters (one gallon) of water a day to drink and much more water is needed for cooking, washing, irrigation, and many other jobs. But because only 3% of Earth's water is fresh water, (**Figure 28.15**), it is always in short supply. In many regions available fresh water is wasted. For example, inefficient use of water for irrigation may lead to water loss as run-off. In more affluent countries water is used for many non-essential purposes, such as filling swimming pools. But human biological waste is probably the biggest threat to a supply of fresh water.

Human waste is a prime habitat for many pathogenic viruses, bacteria, and protists, so any environment that is contaminated may have an excess of pathogenic microbes. Because human waste that enters the water supply can cause deadly or serious diseases like cholera, amoeboid dysentery, viral hepatitis A, typhoid fever, and chronic diarrhea. Wastewater treatment is one

(a) Safe drinking water is a precious resource.

(b) Much of the developing world does not have access to safe drinking water.

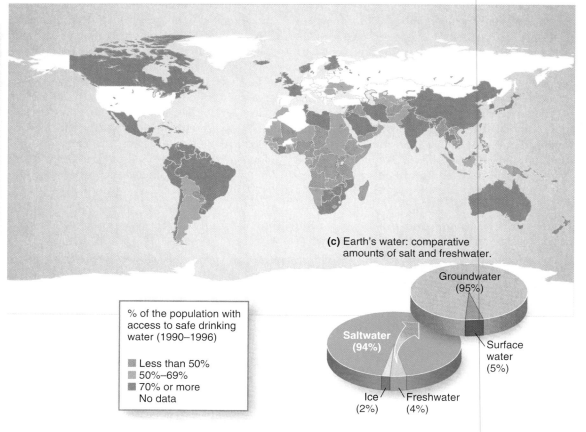

(c) Earth's water: comparative amounts of salt and freshwater.

% of the population with access to safe drinking water (1990–1996)

- Less than 50%
- 50%–69%
- 70% or more
- No data

Groundwater (95%)

Saltwater (94%)

Surface water (5%)

Ice (2%) Freshwater (4%)

Figure 28.15 **Earth's Supplies of Freshwater.** (**a**) Safe drinking water is a necessity of life. (**b**) While safe drinking water is available in much of the northern hemisphere, its availability varies in most of the southern hemisphere. (**c**) Only 4% of Earth's water is fresh water.

Figure 28.16 **A Simple Idea Saves Life and Safeguards Health.** Filtering water through eight layers of sari fabric purifies it and makes it safe to drink. After filtration the cloth is spread in the sunlight to dry.

■ *What kills pathogens trapped in sari material by this filtration method of water purification?*

developing countries and is worth learning about in greater detail.

Researchers have discovered that filtering drinking water using eight layers of finely woven cotton cloth removes up to 99% of pathogens from the water. Sari material is used in this study, but almost any kind of finely woven cloth also will successfully purify water (**Figure 28.16**). After filtration the cloth is rinsed and spread to dry in the sun and be sterilized by strong sunlight. Then it can be used again. While purifying drinking water will not provide new sources of freshwater, this practical method easily can be incorporated into poor people's lives. It is just this kind of highly practical thinking that is needed to solve many of Earth's environmental problems. One thing is certain: freshwater is in short supply now, and the struggle for freshwater only will grow in importance in years to come.

technological answer to this problem, and in developed countries governments spend billions of dollars a year restoring wastewater to a drinkable state. Not all countries can afford this necessity, though. In some parts of the world, water treatment is not readily available and human and industrial wastes contaminate water supplies.

Recently an easy, cost-free water purification method has been developed in Bangladesh and is currently being tested for its efficacy. It is an example of the sort of low-tech solutions that must be applied to solve many environmental problems in

Acid Precipitation Falls When Chemical Pollutants Are Washed Out of the Air

You may have heard about the problems of acid precipitation, a combination of air and water pollution that washes chemical pollutants out of the air and deposits them on the landscape below. It is called acid precipitation because the rain, snow, sleet, or fog has a lower than normal pH. Normal rainwater is slightly acidic, with a pH between 5 and 6. Acid rain can have a pH of 4.4, 4.2, or even lower. Fossil fuels are not pure hydrocarbons and also contain sulfur, nitrogen, and other impurities that are released when fossil fuels are burned in power plants, cars, and by industries (**Figure 28.17**). These

Figure 28.17 **How Acid Rain Forms.** Industrial emissions of sulfur dioxide (SO_2) and nitrous oxide (NO) are converted to acids and are deposited in forms of precipitation. Runoff of acid precipitation acidifies streams, ponds, and lakes.

combine with oxygen and form chemical compounds such as nitric oxide and sulfur dioxide. In the atmosphere these contaminants combine with water, forming nitric and sulfuric acids. Acidified precipitation lowers the pH of bodies of water, and this kills acid-sensitive organisms like fishes, frogs, and salamanders. Nearly lifeless acidified lakes can be found across North America, from Maine to Washington State, although they are more common in the Northeast (**Figure 28.18**). Forests also suffer damage when they are bathed in acid precipitation. Acid precipitation kills mycorrhizae (see Section 16.2), diminishing the ability of a tree to take up nutrients. Eventually, whole hillsides of acidified forests are covered with dead trees (Figure 28.18b). This effect was first observed in western Europe, but it also has been identified in the Appalachian and Blue Ridge Mountains and in Canada.

Excess Nutrients in Water Pose Additional Problems

Excess nutrients originate mostly from fertilizers and animal wastes. Fertilizers are spread on farms, fields, lawns, orchards, golf courses, and other places where people want plants to grow. Of course, plants must have certain nutrients to thrive and in particular they need nitrogen, phosphorus,

and potassium. You will see these nutrients listed on bags of plant fertilizers. If the fertilizers would stay put, nothing would be amiss, but unfortunately rains wash fertilizers out of the soil and into local streams, rivers, and lakes. Depending on geography, nutrient-rich runoff eventually may reach estuaries and oceans.

Animal wastes also contain nutrients, and large factory farms and feedlots that supply chicken, beef, and pork produce enormous quantities of animal wastes. Much of this waste is allowed to run off into local waterways, and eventually the water cycle returns it to lakes, rivers, and oceans. If animal wastes enter the human water supply, and people inadvertently consume pathogenic bacteria, people may become desperately ill and may die. Human wastes also add excess nutrients to waterways and can threaten human health. In spite of large waste treatment plants, human wastes do end up in streams, rivers, and eventually the ocean.

Think about how extra nutrients affect aquatic habitats. As you would expect, increased nutrients promote excess growth of plants and algae. Is that bad? Sometimes. Excess nutrients can allow particular plants to grow wildly and crowd out other species. This happens dramatically in aquatic systems when green algae bloom or when water hyacinths form mats that cover the water's surface. One of the most devastating effects of excessive growth of plants is a drop in the amount of oxygen dissolved in the local waters (**Figure 28.19**). Oxygen levels drop because as algae die they fall to the bottom where bacteria decompose them. These bacteria use oxygen in their metabolic processes. Populations of bacteria increase as bacteria feed on dead plant tissues and with so much bacterial growth, local oxygen levels fall dramatically. Every summer over half of the U.S. coastal waterways experience declines in oxygen levels. Many coastal water environments experience a *complete* loss of oxygen during this time. There is little doubt that this is a direct result of fertilizer runoff and excess animal wastes entering coastal waters.

What happens when water is depleted of oxygen? Organisms will die if they cannot escape to a healthier environment. An especially profound example is the dead zone that forms each year in the Gulf of Mexico where the Mississippi River enters it (**Figure 28.20**). Runoff from agricultural fields into the Mississippi is deposited in the Gulf of Mexico and it is intense during the spring and summer months. As extreme as this sounds, the dead zone has little or no animal life. Scientists describe swimming with scuba gear for miles underwater without seeing a single live fish, or

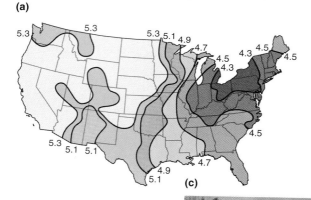

(a)

(b)

Figure 28.18 **Acid Precipitation.** (**a**) In the United States acid precipitation is concentrated in the eastern half of the country. (**b**) An acid-killed lake being chemically treated to raise its pit; (**c**) an acid killed forest.

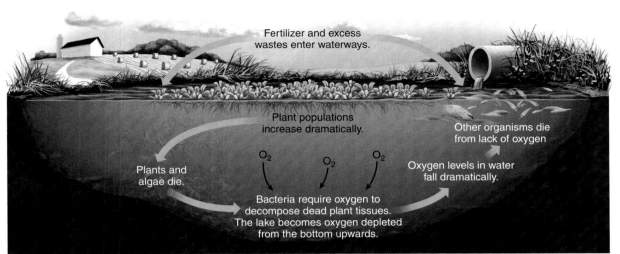

Fertilizer and excess wastes enter waterways.

Plant populations increase dramatically.

Other organisms die from lack of oxygen

Plants and algae die.

O_2 O_2 O_2

Oxygen levels in water fall dramatically.

Bacteria require oxygen to decompose dead plant tissues. The lake becomes oxygen depleted from the bottom upwards.

Figure 28.19

Oxygen Depletion of a Pond or Lake. Oxygen depletion of ponds or lakes is linked to an excess supply of plant nutrients that cause plants to grow luxuriantly. When plants die, bacteria require oxygen to decompose them. This creates oxygen depletion that moves upwards from the bottom of the water. Eventually other organisms die from lack of oxygen.

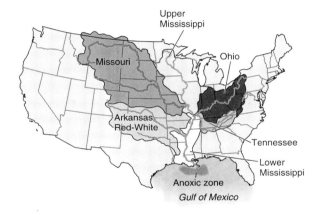

Upper Mississippi

Missouri

Ohio

Arkansas Red-White

Tennessee

Lower Mississippi

Anoxic zone

Gulf of Mexico

Figure 28.20   The Dead Zone. The dead zone in the Gulf of Mexico is an anoxic area of the water that results from too many nutrients draining into the Mississippi River. The colored areas on the map are the drainage areas of the major river systems that enter the Mississippi River.

crab, or shrimp. Rather, the bottom of the Gulf of Mexico is littered with the bodies of dead animals. In 2005 the dead zone was the size of Connecticut; in other years it has been nearly twice that size. The size of the dead zone is directly related to how much river water flows into the Gulf from the Mississippi. In years when heavy rains produce a large river flow, the dead zone is at its largest. It shrinks dramatically during drought years when there is little flow from the Mississippi. Measurements show that the water that flows from the Mississippi is polluted with excess nutrients.

Fishing in the Gulf of Mexico is big business. Nearly 2 billion pounds of seafood are caught in the Gulf each year. The dead zone in the Gulf not only destroys a delicate ecosystem, it threatens the

jobs of thousands. It also eliminates a major source of seafood for people of the United States. As you will see shortly, there are things that can be done to avoid creating dead zones in the Gulf of Mexico and other estuaries.

Many Toxic Contaminants Become Concentrated as They Move Up Food Chains

Many toxic chemicals including pesticides, heavy metals, and PCBs (a group of synthetic chemicals used in electrical insulators, plastics, and hydraulic fluids). Instead of being diluted as they move away from a source of pollution such as a toxic waste dump, town dump, or a point where they are discharged from a manufacturing plant, some toxic chemicals become *more concentrated* as they move through the biosphere. These toxic pollutants are **carcinogens** and/or **mutagens.** Carcinogens are chemicals implicated in the formation of cancers; mutagens are chemicals that cause chemical transformations in DNA. Mutagens also can cause neurological problems, damage the kidneys, and affect hormone functions. But how can the concentrations of toxic substances increase once they have been released into the biosphere from a manufacturing plant, or have been sprayed onto crops?

This phenomenon is called **bioamplification** and it is a direct result of the way that animals are linked in food chains and the way that these chemicals are processed by organisms. Let's explore bioamplification using the example of DDT, a notorious pesticide.

When the pesticide DDT is sprayed onto any environment, it is not **biodegraded,** broken down by bacteria, fungi, or other organisms. Instead,

carcinogen a chemical implicated in the formation of cancers

mutagen a chemical that causes chemical transformations in DNA

bio-amplification the process by which a toxic compound becomes increasingly concentrated as it moves up a food chain

biodegrade to break down by biological means such as the effects of bacteria, fungi, or other organisms

when organisms consume DDT, tiny amounts of it are absorbed and *stored* in their reserves of body fat. Following the mathematical proportions of food pyramids, each organism in a food chain consumes *many* individuals at a lower level in the chain. This means that the DDT can become concentrated within the tissues of higher-level consumers. This pattern of bioamplification can magnify the amount of DDT or other toxic compound 10 to 100 times at each link in a food chain (**Figure 28.21**). So a top carnivore receives a whop-

ping dose of DDT that can have been bioamplified a million times. Although widespread use of DDT has resulted in many insects that are resistant to it, top carnivores have not yet developed this resistance. Often they are killed or made sick when they eat prey laced with DDT.

Largely because of the warnings of Rachel Carson in her 1962 book *Silent Spring,* Americans realized the dangers of continued use of DDT. In 1972 DDT was banned for use in the United States. Most Western countries also banned the use of DDT

Figure 28.21
Bioamplification of DDT. The concentration of DDT in organisms' bodies increases up food chains. The blue pyramid in the middle of the diagram represents the numbers of organisms on each of the levels (primary producers, primary consumers, etc.). The number next to each organism shows the concentration of DDT per unit of its tissues, measured in parts per million. Notice that DDT is most concentrated in tertiary consumers.

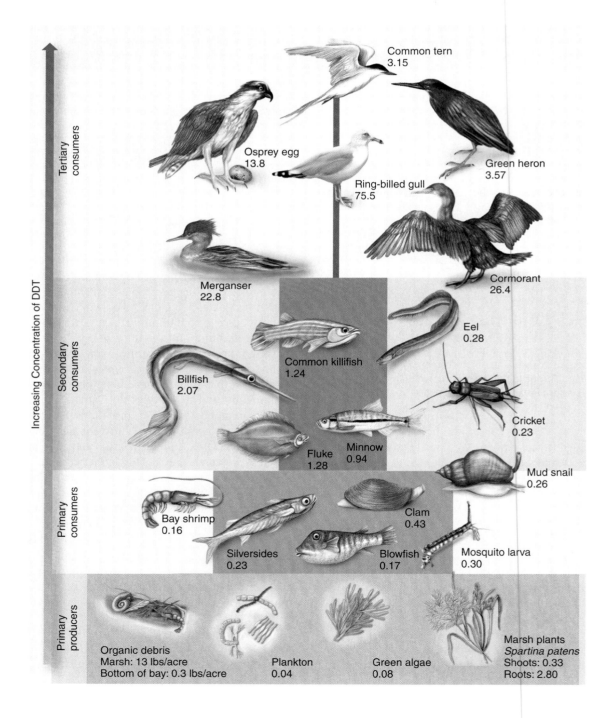

about that time too. This ban has had a positive impact. For instance, the U.S. national symbol—the bald eagle—and other predatory birds like ospreys were endangered and near the brink of extinction in the 1970s, mainly because bioamplified DDT had caused the shells of their eggs to become thinner than normal. Eggshells of birds affected by DDT were too weak to support the weight of incubating parent birds. Researchers found nest after nest of crushed eggs (**Figure 28.22**). This had disastrous results for populations of bald eagles and ospreys. Thirty years later these populations have largely recovered, thanks in large part to the ban on DDT as well as other efforts to increase their populations. Nevertheless, DDT has not disappeared because U.S. manufacturers *still* make and ship DDT to developing countries, where governments have no such regulations to protect people and wildlife from DDT. Tests of tissue samples of agricultural workers in Central America have revealed that they contain 11 times as much DDT as the tissues of average U.S. citizens.

A classic example of how DDT affects food chains occurred in the 1950s when the World Health Organization (WHO) heavily sprayed DDT on a part of the island of Borneo in an effort to eradicate mosquitoes that carried the parasites that cause malaria. This is an excellent lesson in the errors of applying simple solutions to complex ecological problems. After spraying with DDT, the mosquito populations fell dramatically, but then other, odd things began to happen. DDT also killed small wasps that preyed on caterpillars that infested the thatch roofs of houses. With their predators gone, the caterpillar populations exploded, and soon, all over Borneo, thatch roofs were collapsing because of the ripple effects of spraying DDT. But that wasn't too bad—thatch roofs are fairly easy to repair, and so the spraying program continued, and here bioamplification began to show itself.

Large lizards called geckos eagerly lapped up dead or dying insects killed by DDT. In turn, as the geckos became sick from all the DDT-laced insects they had eaten, the poisoned lizards became easy prey for housecats. In this way bioamplified DDT moved farther up the food chain, passing from mosquito to wasp to gecko to housecat. Soon housecats were dying. You probably can guess what happened next: relieved of their predators, the populations of rats skyrocketed. Now rats raided fields and ate crops, and they entered houses. As rats came into close contact with people, their fleas began to bite humans, causing the incidence of both bubonic plague and typhus to rise. Finally, in an effort to reduce the rat population, the government of Borneo requested that housecats be dropped by para-

(a)

(b) Eggshell Thinning in Merlin

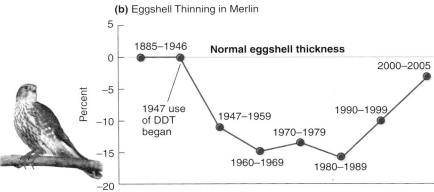

Figure 28.22 **Lethal Eggshell Thinning.** (**a**) DDT makes eggshells too weak to support the weight of an incubating parent bird. (**b**) Eggshell thinning in the merlin, a small species of falcon. The normal value was determined by measuring thickness of eggs collected between 1885 and 1946.

chute into the areas where bubonic plague had broken out. The image of 14,000 live housecats dangling from parachutes as they floated down to do commando work in remote villages in Borneo is comical, but Operation Cat Drop has a serious message. It shows how connected things are in the biosphere and of how impossible it is to do just one thing. This is especially true when toxic chemicals are introduced. As a postscript to the story of Operation Cat Drop, you should know that by introducing thousands of housecats into remote ecosystems in Borneo, the government was wagering the survival of its human populations against the survival of its populations of small mammals, birds, lizards, and snakes. They were hoping that the cats would concentrate on the abundant rat populations; however, cats are instinctive predators and opportunistic feeders. It is not yet clear how much housecats, dogs, and feral cats and dogs damage natural populations, but they do have a large negative impact on them. Island populations of small animals are especially vulnerable to predation by introduced cats, dogs, rats, goats, and pigs.

The ban on DDT is not without controversy. In regions of the world where malaria is a serious human health problem, governments use DDT-soaked netting and limited DDT spraying to protect people from malaria-carrying mosquitoes. Malaria kills over 1 million people a year, and around 500 million cases develop annually. It is not surprising that people in these countries are less enthusiastic about the ban on DDT than are people in the United States and Europe.

Table 28.3 summarizes the sources of, and problems caused by many of the toxic chemicals that you will read about almost daily in the newspapers. Consider that, like DDT, they are bioamplified and are dangerous additions to the biosphere.

Many people are concerned about how human actions affect the atmosphere. You have read about the realities of global warming (see Section 6.4). The final topic of this chapter is how human activities have altered the composition of the upper atmosphere, depleting Earth's protective layer of ozone. This is an important story because it contains the seeds of hope that humans can eliminate or repair the environmental damage they have done.

Ozone Depletion Reduces Earth's Protective Layer

In 1984 British climatologists announced the stunning news that there was a huge hole in the layer of ozone that encircles the globe (**Figure 28.23a**). This was worrisome because the ozone layer pro-tects Earth and all living creatures from sunlight damaging ultraviolet (UV) energy. Ozone absorbs ultraviolet radiation. At first the ozone hole appeared to be limited to the region around the South Pole, but later a second ozone hole was spotted above the North Pole. Subsequently, ozone holes were found over other areas too. The sizes of the ozone holes change from year to year, but in 1987 the Antarctic hole was as deep as Mount Everest and as large as the entire United States. After the ozone hole was first reported, American climatologists revealed that satellites had been sending back similar information since 1981, but because it seemed so unlikely, their computers had been programmed to disregard the information as "noise."

Thinning and deterioration of Earth's ozone layer is caused by the release of chemical compounds that combine with ozone (O_3) and transform the three-oxygen molecule to pairs of oxygen atoms (O_2). Many of the human-made molecules that take part in this reaction are organic molecules with either chlorine or fluorine atoms—or both—attached. An example is the Freon gas that previously was used as coolant in refrigerators or air conditioners. The general name for these compounds is *halocarbons*. The thinning of the ozone layer is a prime demonstration of how far-reaching the effects of pollutants can be, of how interconnected all the parts of the biosphere are, and of how quick action can repair some environmental catastrophes.

Table 28.3 Toxic Chemicals That You Should Know About

Toxic chemical	Source	Problems it causes	Biomagnifies in food chain	Bioaccumulates within an organism
DDT and other chlorinated hydrocarbons	Pesticides; all are now banned for widespread use and may only be used with EPA approval on an emergency basis	Thins eggshells of birds; kills helpful as well as harmful insects; kills amphibians	Yes	Yes
PCB	Insulation in transformers, used as a fire retardant chemical, plastics manufacture; humans ingest foods that contain PCBs, especially contaminated fishes	Impairs reproduction; suspected carcinogen	Yes	Yes
Mercury	Papermaking, building dams, burning fossil fuels	Damages brain and nerves, toxic to organisms in higher doses	Yes	Yes
Cadmium	Cigarette smoke	Permanently damages kidneys; a carcinogen	No	Yes
Dioxin	Waste burning, paper mills, production of PVC plastics; found in animal foods like meats, dairy products, fishes	Carcinogen; extremely toxic	Yes	Yes

(a) Photos of the Antarctic ozone hole 1979 to 1984

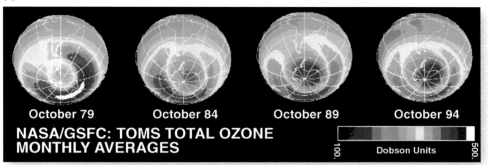

October 79   October 84   October 89   October 94

NASA/GSFC: TOMS TOTAL OZONE MONTHLY AVERAGES

Dobson Units

100.   500.

Figure 28.23 **Ozone Depletion.** **(a)** A satellite photograph for 1984 shows the hole in the ozone layer; **(b)** The level of ozone in the atmosphere influenced by the level of halocarbons in the atmosphere. When halocarbon levels are high (lower on the graph), ozone levels fall.

(b) Antarctic ozone and halocarbon levels

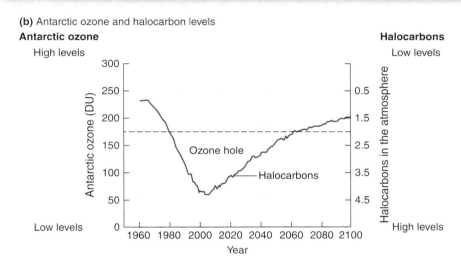

The ozone hole is one environmental problem that humans just *might* be able to solve (Figure 28.23b). In 1987 countries around the world agreed to ban the manufacture and use of halocarbons. For example, Freon is no longer manufactured and can no longer be legally used as a refrigerant. Other ozone-damaging chemicals also have been phased out. The good news is that the levels of harmful ozone-depleting chemicals in the atmosphere have declined dramatically, and in 2005 scientists finally reported that the thinning of the ozone layer had slowed. The bad news is that the levels of halocarbons in the atmosphere have stopped falling because older appliances and equipment—made before the 1987 ban—continue to leak ozone-depleting chemicals into the atmosphere. Scientists now think the ozone layer will recover fully by 2065.

QUICK CHECK

1. You are asked to explain the problem of acid rain to a class of third-graders. Write the outline of your lesson plan for them.
2. Why aren't predatory birds at the top of land-based food chains affected by bioaccumulation of DDT?

28.5 How Can Human Impact on the Biosphere Be Lessened?

Human actions have had many damaging effects on Earth's natural systems. What can you—and society—do to lessen this impact and preserve a healthier habitat for all living creatures, including humans? Several key strategies apply in a variety of situations, and some of these have had some success.

Legislation Can Help Solve Environmental Problems

Legislation is probably the most visible way that governments try to control people's impact on the environment, and you can influence the kind of legislation passed by voting for or against candidates and ballot measures. You also can get involved in organizations that work toward legislative goals that you share. From local zoning laws to national legislation on clean air and water, people often try to pass laws that will stop others from harming the environment. If you look at this effort from a case-by-case perspective, it may seem like one endless battle, with no winners. But if you step back and survey the overall effort, laws do seem to have improved environmental quality.

Laws to limit water pollution actually date back to 1948 when the U.S. Congress looked to the surgeon general for ways to clean up waterways. In 1972 and 1987 laws affecting waterways underwent major revisions. At least in some regions the Clean Water Act, as it first was called, has had a positive impact. Part of the goal of the Clean Water Act was to restore "fishable and swimmable waters" throughout the country, and as a result many regions have been motivated to clean up local waters. In parts of the Hudson River, for example, between the 1970s and the later 1990s, levels of metal pollutants declined from 20 to 90%. In addition, fishes that had disappeared from the Hudson have returned, and their populations are thriving. Fishes from the Hudson still contain dangerous levels of PCBs, though. Until PCBs are removed from the river's sediments, they will continue to bioamplify, and the Hudson River will not have completely recovered. This success story promises to get better if pressure to clean up the Hudson River continues. The Environmental Protection Agency claims that 700 billion pounds of pollutants have been kept out of our waterways as a result of the Clean Water Act.

One impact of environmental laws is the protection of **wetlands,** regions of marshes and swamps that help to keep lakes and rivers clean because they act as buffers for major waterways. The protection of wetlands is controversial. When developers are not allowed to build in wetlands, they complain that their rights are being restricted, while environmentalists worry that wetlands are disappearing. Wetlands are more than sanctuaries for animals, though. They provide essential protection against floods. Natural wetlands are usually located in floodplains—the regions around a body of water that normally flood during storms or heavy rains. If people build in floodplains, they risk flooded homes. In addition, wetlands absorb flood waters and slowly release them to the surrounding environment, something that the asphalt, concrete, and steel infrastructure of cities cannot do. So wetlands can help prevent floods and actually can protect human lives and property. For example, after Hurricane Katrina ravaged the Gulf Coast in 2005, many scientists pointed out that if the region had preserved its extensive wetlands instead of allowing them to be developed, there would have been much less flooding. In many cases it takes the force of law to ensure that people do not build in wetland regions.

Laws to govern clean air came a bit later, after the internal combustion engine brought air pollution to major cities in the middle of the last century. By the 1990s, though, every car owner in the country was affected by the Clean Air Act. Lead in gasoline was banned, new cars were designed to get better gas mileage, and cars were equipped with catalytic converters that destroy some of the impurities produced when gasoline is burned. Electric and hybrid cars are also part of this effort, and tax credits for them are one way that the U.S. government encourages environmentally friendly cars. Unfortunately, another trend began in the 1990s: the popularity of larger cars, such as vans, trucks, and SUVs. These trends have offset the benefits derived from improved technology, and today many U.S. cities have air badly polluted with automobile exhaust. And now, of course, technology may ride to the rescue again, this time in the form of cars that run on hydrogen rather than on fossil fuels. As a consumer you are aware of all of these trends. The important question is whether you will make environmental concerns a priority and let them guide your choices as a consumer.

Fishing is another major area where laws are enacted. For example, in the United States and Canada laws and regulations have restricted fishing. Earlier you read about legislation that severely curtailed fishing in Canada's Grand Banks. In the United States every state has a complex set of regulations that govern what fish species can be caught, and the number, size, sex, and age of legal catches. For instance, Maryland regulates the commercial fishing of bass, herring, eel, flounder, croakers, as well as oysters, blue crabs, and many other species.

Laws and regulations on fishing, air pollution, and water quality require enforcement to be effective. Is all this hassle worth it? Have these laws had a positive impact on the quality of air and water and the diversity of species in bays and oceans? You know that it is hard to draw a cause and effect conclusion just by looking at a correlation. In the case of these laws, though, all you can do is see if there is a correlation between the passage of the laws and the quality of the environment. For example, the levels of major pollutants in the atmosphere and water have dropped since the Clean Air Act and Clean Water Act were passed and implemented. This correlation suggests that legislation can cause improvements in the environment. It also suggests that directly and indirectly, these laws have caused people, local governments, and industries to change their behavior. People now must buy unleaded gas. Most local governments monitor air quality and require inspections of car emissions. Industries must alter their manufacturing operations to minimize pollution. These are

wetlands marshes, swamps, or estuaries that form transitional environments between larger bodies of water and drier uplands; wetlands absorb flood waters and slowly release them

large-scale examples of how laws and regulations can lessen human impact on the environment. Laws that act locally concern things like recycling, hunting, or where you may build a new apartment complex, housing development, or industrial plant.

Smaller-Scale Projects Can Help to Recover Species and Habitats

Smaller-scale projects also can change human impact on the environment. Many nongovernmental groups devote time and money to pursue a variety of goals to protect the natural world and to restore damaged areas, species, and ecosystems. One encouraging example is the effort to restore tall grass prairies to North America. Tall grass prairies covered the North American Great Plains before they were settled and roadways, farms, cities, and towns sprawled across the country. Prairies were diverse grassland ecosystems that harbored many species of 3- to 5-foot-tall grass species as well as birds, insects, and mammals. Now, all across North America local people are working with scientists to replant prairies (**Figure 28.24**). It is not likely that

Figure 28.24 **Prairie Restoration.** Prairie before **(a)** and after **(b)** restoration of native plant species.

(a)

(b)

more than a fraction of North America's agricultural fields will be replaced by native grasses, but it is encouraging that some acres of tall grass prairie have been successfully restored.

The dead zone in the Gulf of Mexico also could benefit from small-scale restoration projects. Many groups are trying to restore the natural habitat of the wetlands and floodplains along the Mississippi so that excess nutrients and pollutants will naturally be removed before reaching its waters. After Hurricane Katrina no one is sure what the state of the coastline and the dead zone will be. Many people hope that the reconstruction efforts along the Gulf Coast will include habitat reconstruction, which not only will improve the ecology of the Gulf of Mexico but also will help protect the coastline from powerful storms in the future.

An interesting approach taken by some conservation groups, such as The Nature Conservancy, is to bypass laws and directly purchase valuable natural lands. In California, for example, The Nature Conservancy maintains over 100 preserves that it owns outright. The group takes care of these lands and offers access to them for nature tours. This approach not only protects wild areas from development, it gives ecologists a chance to experiment with various management strategies to see which effectively nurture endangered species and repair habitats. To minimize habitat fragmentation and provide green corridors for wildlife, The Nature Conservancy convinces private landowners to cooperate in efforts to protect natural landscapes. They have had success by working toward cooperation rather than confrontation in resolving disputes about how land should be used.

Personal Choices Make a Difference

Environmental problems might seem so enormous that you may wonder what you can do to contribute to a cleaner and healthier biosphere. Nevertheless, individual choices do make a difference and from your shopping habits, to your driving habits, to choices in your home furnishings, you can make a difference—even if it is a small one. One common phrase used by environmental groups is "reduce, reuse, recycle," and these three *R*'s can help you to make better choices. One area is product packaging. If you have a choice between buying a product in a heavy plastic container versus a lighter cardboard container, you would reduce consumption, and thus pollution, if you chose the lighter one. If you find ways to reuse your discarded things, instead of throwing them out, you reduce your consumption of goods and so reduce pollution. Most municipalities have recycling efforts, but a little searching probably will show that you could recycle much more efficiently if you take items to a recycling center. And, of course, it helps to support the whole recycling effort if you buy products made from recycled materials. Choices in your home might influence your overall energy budget and so the amount of pollution you create. Consider adjusting the thermostat and being a bit cooler in the winter and a bit warmer in the summer. Consider an investment in solar energy devices or added insulation. If you think about it, you will be surprised about how many choices you have that can positively influence the health of the biosphere.

28.6 An Added Dimension Explained
The Effort to Save Chimpanzees

The introduction to this chapter asked you to consider the question, What does it mean to be human? Definitions of our uniqueness have changed over the centuries that people have been asking this question, and it may be that there is nothing unique about humans. Over the past 100 years revelations about chimpanzees, our closest relatives, have forced each new definition of humanity to be revised. For example, the definition of humans in a 1957 book entitled *Man the Tool-Maker* had to be abandoned with the recognition that chimps make and use a variety of tools. A definition of humans that involves our ability to use language, produce art, or use mathematical concepts also must be rejected because chimps do all of these too. The differences in abilities between chimps and humans on these tasks are a matter of degree.

If you really look into the eyes of a chimp, you sense something almost human. As you watch chimps, you get the feeling that you are watching people, yet not quite people. Human and chimp DNA is so similar that it is hard to draw the line between the two species. From every perspective chimps are our closest living relatives. If we want to know ourselves, we have to know our closest cousins.

The lesson from studies of chimps is that distinctive human behavior has its roots in evolutionary history as much as it does in human culture. For instance, like humans chimpanzees can be peaceful, playful, or aggressive. They kiss and embrace one another in situations where humans would do the same. Members of a troop call to each other in the night, apparently to give assurance that others of their kind are close by. Chimps like creature comforts, especially when they sleep. They make comfortable beds from leaves and branches in the treetops each night. If you were camping in the forests and eager to get up off the ground and away from predators, you might learn to make just such a bed. Chimpanzees communicate with calls, gestures, postures, and facial expressions. They glare when they are angry and have a variety of facial expressions that are uncannily human. Chimpanzees defend and protect their young, elderly or sick relatives, and their friends. It is obvious that chimps have friends. Each individual develops close relationships with some other chimps but ignores or is hostile toward others. Most chimps do not have pair bonding—or marriage—like humans do, but some chimp couples do form long-lasting mating relationships. They make and use tools that help them to drink, eat, and clean themselves. They also make and use weapons. Chimpanzee "warfare" occurs in which raiding parties attack and kill nearby competitors. Chimpanzees can commit murder, and they can brutally attack chimps inside and outside of their own troupe.

One important thing to appreciate is that there are two species of chimpanzees—and that they differ in some interesting ways. Some researchers believe that humans have a combination of the traits seen in these two different kinds of chimps. The common chimpanzee, *Pan troglodytes,* and the bonobo chimpanzee, *Pan paniscus,* have only slight physical differences. Bonobos are smaller and more lightly built than are common chimpanzees (**Figure 28.25**). Overall, common chimps are more aggressive and have male-dominant societies, while bonobo chimps are more social, peaceful, and have more female-dominant social groups. The male-dominated common chimpanzees sometimes form bands that defend their territories and *invade* the territories of other chimpanzee groups. If they find a lone chimp, they may attack it. These territorial raids of chimpanzees are similar to human wars, gang rivalries, and other forms of aggression between human groups.

In contrast, the female-dominated bonobo societies apparently do not engage in this kind of intergroup aggression and spend more time in social activities like grooming one another. Both species are sexually promiscuous. Both common and bonobo chimp males mate with multiple females. In addition to heterosexual matings, though, bonobo females and males engage in homosexual activities. While both kinds of promiscuity might seem at odds with human monogamous societies, a close look shows that both promiscuity and homosexuality are common in human societies. Both chimpanzee species show cultural advances that are passed down over generations. When one individual in a population learns to use a new tool to crack nuts, develops a better way to build a sleeping platform, or even learns sign language taught by humans, the chimp that

(a)

(b)

Figure 28.25 **Two Species of Chimpanzees.** (**a**) Bonobo; (**b**) Common chimpanzee.

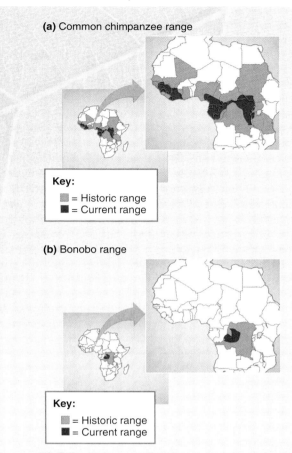

(a) Common chimpanzee range

Key:
= Historic range
= Current range

(b) Bonobo range

Key:
= Historic range
= Current range

Figure 28.26 **Historic and Current Geographic Ranges of Chimpanzees.**

learned the new ability teaches it to others. Many studies have documented that chimpanzees develop and transmit culture.

Recently, the chimpanzee genome has been sequenced, and scientists now know that DNA of humans and chimps is almost identical. If scientists can understand how these subtle gene differences relate to differences in behavior, intelligence, culture, and other traits, then *you* will know a lot more about what it means to be human and will have a better understanding of the basis for your own behaviors. There is only one obstacle to realizing these insights. In 50 years or less there may

be no more chimpanzees left to observe and study. The 150,000 chimps left in the wild are likely to go extinct.

Chimpanzee extinction is being driven by two familiar causes: destruction of habitats and hunting by humans. Chimpanzees never were widely distributed in Africa. They remain largely in their ancestral homes, the same forests that gave rise to the common ancestors that chimps and humans share. The forests where they live are being cut down for human living space. Today, chimpanzees occupy just a small fraction of their original geographic ranges (**Figure 28.26**). Even more disturbing is the hunting of chimpanzees for bushmeat. Poor communities in Central Africa have few resources to draw on. Chimpanzee meat, along with the meat of other primates and other mammals, is one of the few sources of protein in Central Africa. Day by day African forests are being stripped of chimpanzees and other wild animals.

The conservation efforts to save the chimps involve laws that protect the remaining chimpanzee populations and set land aside in national forests especially for refuges for chimpanzees.

—*Continued next page*

Continued—

Equally important are public awareness campaigns that teach people what is happening to wildlife. Jane Goodall, the pioneer chimpanzee researcher (**Figure 28.27**) now devotes her time to conservation efforts. These range from programs that increase the awareness of schoolchildren and adults to the plight of vanishing wildlife to lobbying African governments to protect chimpanzees. The problem of bushmeat is difficult to address, but there are ways to reduce it. Many organizations are pushing governments to enforce existing hunting laws, to educate their populations about the importance of maintaining all animal populations, to extend sanctuaries and reserves, and, most importantly, to provide local people with alternative food sources. Only time will tell if these efforts will be successful.

Even in the United States there has been a shift in attitude about chimpanzees and their role in our society. Chimpanzees have long been used as experimental subjects in medical research. Chimpanzees have been used to study cancers, tuberculosis, HIV, and many other diseases. Because chimpanzees easily can learn human sign language, they have been used to investigate how people learn language. Even the first "person" sent into space in the United States was a chimpanzee, not a human. Chimpanzees have given much to human society and have received little in return, but that is changing. Today, scientists are much more careful about using chimps in scientific studies, and they do so much less often than in the last century. Many scientists now believe it is unethical to use chimps for any type of invasive research. In the late 1990s Congress passed legislation to provide research chimpanzees with comfortable communities where they can live out their old age.

These various initiatives and projects show that humans can make the efforts necessary to lessen their impact on the

Figure 28.27 **Jane Goodall and Chimpanzees.**

biosphere and on their fellow species. We all can contribute to these efforts with time, money, or public support. If we will do this, then it is possible that we may just save the chimpanzees and other natural populations, communities, and ecosystems, and find a way preserve the natural world that we all depend on.

QUICK CHECK

1. What human activities are driving chimpanzees closer to extinction?
2. What behavioral and social differences can be seen between common chimpanzees and bonobos?

Now You Can **Understand**

Killer Fog and the Fight Against Air Pollution

London fog. The phrase may bring up images of soft mists and the fuzzy glow of streetlights, but for hundreds of years London fog has meant illness and death for Londoners. You probably think of air pollution as a modern problem, but the effect of human population growth on air quality has a long history in Britain. Until about AD 1200 most people in Britain burned wood to heat their homes, but the increasing population eventually burned all the available wood, eradicating most of the forest. Then people turned to soft coal for heat, but without special technology coal burns "dirty." In a typical fireplace or coal stove the temperature does not get hot enough to break most of the carbon bonds, and so large particles are released into the air as dense smoke. The smoke carries a variety of pollutants, but the humid air, the natural fog of London and other areas of Britain, makes matters worse by capturing and accumulating the coal-driven pollution. London has lived

with this problem for centuries. As early as the mid-1800s periods of "pea soup fog" killed hundreds of people each year. In 1952 the situation became extreme when a dense polluted fog hung around the city for about a week and disrupted normal life. It was nearly pitch dark in the middle of the day, and even with headlights on drivers could see only about a foot ahead of their cars. People abandoned their cars and stayed home. The worst effect of this nasty London fog had yet to be seen, though. Within a week of the time when the fog had lifted, blown away by drier winds, over 4,000 Londoners had died as a direct result of the fog. Many more suffered illnesses, and the long-term death toll from the fog of '52 was certainly much higher. Although this emergency situation had occurred, it took a decade or more to get stringent legislation enforced and to reduce the black London fog. Today, London's air is much clearer, and people can joke about actually having sunny days in London. But, of course, the pollution from automobile exhaust now affects many urban areas, and so the battle for clean air in London and many other places is certainly not over.

What Do **You** Think?

Human Impact on the Environment

Throughout this book, from the early chapters on chemistry to these later chapters on ecology, you have grappled with the impact that humans have on the environment. Local, national, and international issues related to this complex topic will face you for the rest of your life. For example, in the 1980s there was the fight over the spotted owl and its endangered status. Some saw this small bird as a focus for preserving a whole ecosystem of old-growth forest in the Pacific Northwest, while others found it absurd that one small bird would slow human progress and prosperity. There are continual challenges that try to weaken successful environmental programs. Even efforts that have been lauded as successful like the National Park System, the Clean Air Act and Clean Water Act, and the Superfund for toxic waste cleanup are continually challenged. Some groups want to weaken them or roll them back, while others wish to strengthen them. New initiatives are guaranteed to face fierce battles. How have you expressed your concerns about issues that face your local area, and how have you contributed to solutions? What do *you* think about the hope that humans can live and prosper and maintain a clean, healthy, diverse natural environment?

Chapter Review

CHAPTER SUMMARY

28.1 The Impact of Humans on the Biosphere Is Widespread and Profound

As human populations grow larger, they exert more impact on the biosphere. The current exponential growth of Earth's human population is a recent phenomenon, and for most of human history human populations probably remained within the environment's carrying capacity. The domestication of plants and animals, the development of farming villages, and the industrial and scientific revolutions have allowed K to be reset for many human populations. In the view of many ecologists, there is a definite human overpopulation problem. Humans have a much greater negative impact on the biosphere than other animals do because they produce pollutants that spread in unforeseen ways and degrade and destroy environments.

overpopulation   684

28.2 Humans Can Hunt with Deadly Efficiency

Humans can be deadly and efficient hunters. Humans certainly have caused the extinctions of other animals including New Zealand's moas, North America's passenger pigeons, and they nearly wiped out bison in North America. The consumption of bushmeat is a force to be reckoned with in Africa, and it is depleting African lands of mammals of all kinds. Overfishing has dramatically reduced the numbers and sizes of ocean fishes, and some fishing grounds are restricted to fishing fleets until their fish populations recover from overfishing.

overfishing   689

28.3 Combating Deforestation Is a Continuing Challenge

Deforestation reduces biodiversity and damages ecosystems in many ways. The general trend has been that humans cut down forest, using the land for agriculture and many other purposes. Over the last 4,000 years deforestation in all regions has led to a loss of forest that would be the size of Asia. The largest remaining forests are in Russia and Brazil. The Brazilian rain forest has an especially high rate of deforestation. Habitat destruction is the major way that humans eliminate biodiversity from an area. Endemic species are particularly endangered by habitat destruction. Cutting down rain forests is a poor practice because the soils are nearly useless for long-term agriculture. The aftermath of cutting down a rain forest is floods, mudslides, damage to offshore coral reefs, and of course, loss of species that may be unknown to science. Efforts are being made to protect and replant forests in many regions. Protection of specific animal species leads to protection of forests. Once forests are restored, species like golden lion tamarins can be reintroduced, and so far these reintroductions seem to be successful. When hedgerows are eliminated, many species of plants and animals are displaced, and species diversity is decreased.

endemic species   693

28.4 Humans Introduce Pollutants into Environments

Human biological wastes can enter waterways and pollute them and can carry many virulent diseases. Small-scale projects like teaching people to purify water with eight layers of sari material are especially helpful in combating water pollution in developing countries. Acid rain stems from burning of fossil fuels, and many fishes and amphibians cannot survive in acidified waters. Acidified lakes can be found all across North America. Acid rain also kills forest trees by damaging mycorrhizae. Acid-killed forests are found in western Europe and in eastern North America. Excess nutrients enter waterways and cause algal blooms that can result in lack of oxygen and many dead animals. This commonly happens as rivers run through agricultural lands and empty into oceans. The dead zone at the mouth of the Mississippi River has been produced by lack of oxygen caused by excessive runoff of fertilizers and animal wastes. Toxic contaminants tend to bioaccumulate and poison or otherwise interfere with the health of populations of top carnivores. Eggshell thinning in birds is caused by the bioaccumulation of DDT in their tissues that affects the way that calcium is deposited in eggshells. DDT is illegal in the United States but still finds its way into the biosphere because it is still sold in developing countries. The hole in Earth's ozone layer is created by human-generated chemicals that erode ozone. Now that Freon has been made illegal as a refrigerant, the hole in the ozone layer is shrinking—but nevertheless, it still is a problem.

bio-amplification   699   carcinogen   699
biodegrade   699   mutagen   699

28.5 How Can Human Impact on the Biosphere Be Lessened?

Environmentally protective legislation has helped control the negative impacts that humans have on the biosphere. In the U.S. the Clean Water Act and Clean Air Act have led to dramatic improvements in water and air quality, have protected many species, and are allowing endangered environments such as Canada's Grand Banks to recover from overuse. Small-scale environmental projects can be extraordinarily successful, as can private efforts to preserve natural habitats and endangered species. The economic choices made by individuals also can lessen human impact on the biosphere. Reducing consumption, reusing materials, and recycling are ways that individuals can help lessen their own negative impacts on the biosphere.

wetlands   000

28.6 An Added Dimension Explained: Lessons in Humanity from Chimpanzees

Even though chimpanzees are the animals most closely related to humans, they may become extinct in the wild before we have the chance to further define what it means to be human. Both species of chimpanzees are under pressure from deforestation and the bushmeat trade that threaten to push them into extinction in the wild within 50 years or less. Chimpanzees show many similarities to humans, including facial expressions, social organizations, and behaviors. The female-dominant societies of bonobos are peaceful and social, while the male-dominant societies of common chimpanzees are aggressive, and they may attack neighboring troupes and may kill competitors. Setting aside forest refuges to protect the dwindling populations of chimpanzees, writing laws that protect chimpanzees from hunters and enforcing those laws, and finding new sources of protein for Africa's poor are some of the measures that can help preserve Earth's chimpanzees.

REVIEW QUESTIONS

TRUE or FALSE. If a statement is false, rewrite it to make it true.

1. Humans tend to overhunt animals, and without restrictions they will hunt species to extinction.

2. Moas were Australian dinosaurs.

3. Passenger pigeons were Earth's largest bird; Americans hunted them to extinction.

4. The bushmeat trade is a serious threat to Africa's remaining wildlife.

5. Overfishing is defined as taking more than the replacement level of a species.

6. In the 1880s bison had been hunted until there were only about 150 individuals left alive.

7. Acid precipitation comes from burning of fossil fuels.

8. Acid precipitation can kill invertebrates, frogs, salamanders, fishes, but trees are unaffected.

9. One problem created by excess nutrients entering streams is that bodies of water become oxygen depleted.

10. DDT and many other nonbiodegradable substances grow stronger as they pass up a food chain, eventually killing or damaging the top carnivores.

11. DDT is no longer in the environment.

12. The hole in the ozone layer allows toxic chemicals to enter Earth's atmosphere.

13. Legislation is ineffective in curbing humanity's negative impact on the environment.

14. When something is biodegraded, it is broken down by living organisms.

15. International cooperation has led to a treaty that is having positive results in stopping depletion of Earth's protective ozone layer.

16. The development of tools, agriculture, and science and modern technology have made no change in K for humans.

17. Although cutting down rain forests has led to loss of biodiversity, it has been compensated for by the gain in world food production.

18. Chemicals that are bioamplified include DDT, mercury, and dioxin.

19. Chimpanzees are the animals that are closest to humans. We share many physical traits and social behaviors.

20. Chimpanzee populations are dwindling because of habitat destruction and the demand for meat of the African bushmeat trade.

CONNECTING KEY CONCEPTS

1. Using three examples, discuss how the development of specific technologies or advances have led humans to have a greater impact on biodiversity and the environment.

2. How have the regulation of DDT and halocarbons been successful in reversing environmental damage? What problems have been encountered that have made it difficult to completely eliminate these compounds from the biosphere?

THINKING CRITICALLY

1. Green lawns are the pride of a suburban neighborhood. How are green lawns usually achieved? What impact do you think a neighborhood of green lawns might have on the stream that runs nearby? Do you think that the striving for green lawns might pose a health problem for humans that wade in the stream? Explain your answers.

2. As the threat of worldwide deforestation has gotten more attention, U.S. activists have urged developing countries such as Brazil to conserve their remaining forests. Such countries often accuse U.S. activists of hypocrisy. What do you think is the basis for this charge? What might the United States and other developed nations do to counteract this charge and make conservation efforts in the Earth's remaining forests more effective?

For additional study tools, visit www.aris.mhhe.com.

Appendix A

Answer Key

Chapter 1

Quick Check Questions

Section 1.1
1. The four major themes emphasized in this text include emergent properties, unity and diversity of life, evolution by means of natural selection, and the processes of science.

Section 1.2
1. The cell membrane serves to separate cell processes from the environment, while it allows the cell to communicate with the environment. 2. Homeostasis refers to all of the stable internal conditions of an organism that support life's cellular and chemical processes. When a human body is suddenly exposed to a cold environment, mechanisms like shivering are activated that keep the body temperature stable.

Section 1.3
1. The levels of biological organization are atoms, biochemical molecules, organelles, cells, tissues, organs, organ systems, organisms, population of a particular species, communities, ecosystems, biomes, and biosphere. 2. A car has properties, such as motion and carrying passengers, that its individual parts do not possess.

Section 1.4
1. Unity and diversity are the two biological themes that characterize living organisms. 2. Natural selection occurs when individuals with different traits survive and have offspring at different rates. These traits are passed on to their offspring. The environment determines which traits lead to successful reproduction and which do not. This can lead to the evolution of new kinds of organisms.

Section 1.5
1. Probability is the chance that something will happen, expressed as a decimal between 0.0 and 1.0, or as a fraction or a percentage. 2. A hypothesis is a testable explanation or answer to a scientific question; a scientific theory is an explanation for a broad set of events that is strongly supported by a great deal of evidence.

Section 1.6
1. Hantavirus causes Four Corners Disease. 2. Researchers suspected that Four Corners Disease might be spread by rats and mice because they knew of hantavirus infections in other countries and recognized the connection of hantaviruses and rodent infestations.

Review Questions

True/False
1. **F** The common features of living organisms lead to an understanding of the common principles that govern life. **2. T; 3. F** The major chemical elements needed by all living things include carbon, hydrogen, oxygen, and nitrogen. **4. T; 5. T**

Multiple Choice
6. a, **7.** d, **8.** b, **9.** b, **10.** e, **11.** d, **12.** c, **13.** a, **14.** e, **15.** e, **16.** b, **17.** a, **18.** e, **19.** d, **20.** e

Chapter 2

Quick Check Questions

Section 2.1
1. Studying biology begins with chemistry because living organisms are made of chemicals and chemical interactions are basic to biological processes. 2. Emergent properties are new things that occur when two or more chemicals, components, or subunits are combined. The result is not a mix of the original two chemicals but is something else entirely. A simple example of emergent properties that occur when two or more chemicals combine is the combination of hydrogen atoms and an oxygen atom to produce a molecule of water.

Section 2.2
1. An element is composed of only one kind of atom. It is a substance that cannot readily be broken down into other substances by ordinary processes such as burning, evaporation, or filtration. 2. An atom or group of atoms with a positive or negative electrical charge is called an ion.

Section 2.3
1. A covalent bond involves one or more electrons from each of two atoms that encircle both atomic nuclei. Thus, a covalent bond involves the sharing of electrons between two atoms. A covalent bond is relatively strong and stable. A hydrogen bond is a weak temporary bond that forms between a partially positively charged hydrogen atom on one group or molecule and a partially negatively charged atom on another group or molecule. 2. A polar covalent bond is a variant of a covalent bond in which the shared electrons are pulled more toward one atom involved in the bond than the other. The result is a polar molecule with partial charges on different regions.

Section 2.4
1. First, water is the only common substance that exists as a solid, a liquid, and a gas, at temperatures frequently found on Earth. Second, it can hold a great deal of heat relative to its volume. This makes coastal areas of continents and the edges of large lakes a bit cooler in the summer and warmer in the winter than is land far from water. Third, the hydrogen bonds that form between water molecules cause water to have surface tension. Fourth, because the covalent bonds within a water molecule are a special type of polar covalent molecule, water is able to dissolve a huge number of substances. 2. A solution with a pH of 2 is acidic, while a solution with a pH of 10 is basic. A solution with a pH of 2 has many more hydrogen ions than does a solution with a pH of 10.

Section 2.5
1. The water molecules in ice are held together by a maximum number of hydrogen bonds. In liquid water the water molecules are held together by hydrogen bonds that are made and broken over time. In water vapor the water molecules have no hydrogen bonds between them. 2. At this time there are three major problems with cryopreservation of humans. First, as water molecules form ice crystals, the regular arrangement of the water molecules disrupts the relationships of water molecules with the many other molecules inside each cell. Second, as the water molecules distance themselves from one another to form ice, they take up more space and may break cells open. Finally, ice crystals have many sharp, pointed edges that can damage cells and structures within cells.

Review Questions

True/False

1. F An ionic bond is formed between two atoms (ions) as one has gained one or more electrons and is negatively charged, while the other has lost one or more electrons and is positively charged. **2. F** Water is called the universal solvent because it dissolves so many things. **3. F** A covalent bond involves two atoms sharing a pair of electrons; each of the atoms donated one electron to the single covalent bond. **4. F** A molecule of water is the smallest amount of this chemical compound. **5. T**

Multiple Choice

6. d, **7.** d, **8.** d, **9.** b, **10.** c, **11.** c, **12.** b, **13.** a, **14.** a, **15.** d, **16.** e, **17.** d, **18.** b, **19.** d, **20.** c

Chapter 3

Quick Check Questions

Section 3.1

1. C, H, O, N, P **2.** Lipids: a major component of the membranes of every cell; Carbohydrates: the main energy storage molecules for short-term storage; Proteins: a major component of many body parts, such as muscles; Nucleic acids: the information storage molecules found in every cell

Section 3.2

1. A single carbon atom is able to form up to four covalent bonds with other atoms because it has four electrons in its outer electron shell that could hold eight electrons. **2.** A hydroxyl group: $-OH$; A carboxyl group: $-COOH$; An amino group: $-NH_2$; A phosphate group: $-PO_4$

Section 3.3

1. A lipid is an organic molecule that does not readily dissolve in water. **2.** A phospholipid molecule has a large phosphate group attached to two long strands of hydrocarbons. Phospholipids are important components of cell membranes.

Section 3.4

1. A carbohydrate is a molecule composed only of C, H, and O, usually in a ratio of 1:2:1, and having at least three carbon atoms. **2.** Whole-grain carbohydrates are a more varied source of nutrition and contain relatively less easily digestible carbohydrates than do refined carbohydrates.

Section 3.5

1. Guanine: a double ring with an amino group and a double-bonded oxygen attached (a purine); Adenine: a double ring with an amino group attached (a purine); Cytosine: a single ring with an amino group and a double-bonded oxygen attached (a pyrimidine); Uracil: a single ring with two double-bonded oxygens attached (a pyrimidine); Thymine: a single ring with two double-bonded oxygens and a hydrocarbon attached (a pyrimidine) **2.** DNA is double-stranded; RNA is single-stranded. DNA contains the sugar deoxyribose; RNA contains the sugar ribose. DNA contains the nucleotide thymine; RNA contains the nucleotide uracil. DNA is a very long molecule; RNA is much shorter. DNA carries instructions for the cell; RNA executes the instructions. DNA structure is the same in all cells of an organism; RNA molecules are varied, depending on their tasks.

Section 3.6

1. Primary structure: the sequence of amino acids in a protein is its primary structure. Secondary structure: the string of amino acids folds into either helices or pleated sheets, as nearby amino acids form hydrogen bonds between their amino groups and carboxyl groups that pull them into these two shapes. Tertiary structure: interactions between charges on R-groups help pull the entire protein together into a shape that is unique to that type of protein. Quaternary structure: the structure of proteins composed of more than one amino acid chain. **2.** Extreme heat or extreme pH changes can cause a protein to denature.

Section 3.7

1. Hemoglobin is the protein molecule found in red blood cells. It is made of four subunits, chains of amino acids, each containing one atom of iron in a special, folded pocket in the structure. The iron combines with oxygen molecules in the lungs, allowing red blood cells to carry oxygen to all the cells of the body. **2.** A molecule of hemoglobin must have an exact shape to be able to pick up and transport oxygen molecules. In an individual with sickle cell anemia, the hemoglobin has one important amino acid difference that keeps the molecule from assuming its proper shape. This deforms the entire red blood cell. Hemoglobin not only does not pick up oxygen properly, the jagged shape of the red blood cell can cause it to become stuck in tiny blood vessels, blocking off the flow of blood, nutrients, and oxygen to some tissues.

Section 3.8

1. A prion is a protein molecule that has folded improperly and is able to transform other protein molecules into nonfunctional forms. **2.** The uninfected organism eats the meat or the brain tissue of an infected organism. **3.** Sheep that were infected with scrapie died and were made into cattle feed. Cattle ate infected feed and contracted BSE.

Review Questions

True/False

1. F Lipid molecules do not readily dissolve in water. **2. F** Proteins are the most complex of the biological molecules. **3. F** RNA is a nucleic acid; it is single-stranded, is made of ribose sugar, and contains uracil. **4. F** In sickle cell anemia the hemoglobin molecule has one incorrect amino acid. **5. T**

Multiple Choice

6. b, **7.** c, **8.** d, **9.** c, **10.** e, **11.** e, **12.** e

Matching

13. d, **14.** c, **15.** b, **16.** a, **17.** c, **18.** a, **19.** d, **20.** b

Chapter 4

Quick Check Questions

Section 4.1

1. Leeuwenhoek and Hooke developed and used the earliest microscopes and were able to see single cells. **2.** All living things are made of cells. All cells arise from preexisting cells.

Section 4.2

1. Your drawing should show that a cell membrane is made of a lipid bilayer with embedded proteins. **2.** Membrane proteins let a cell communicate with the environment, allow substances to enter or leave a cell, and operate facilitated diffusion and active transport. **3.** Diffusion is the movement of a substance from an area of high concentration to an area of low concentration. Osmosis is the movement of water across a semipermeable membrane, also from the side of high concentration to the side of low concentration.

Section 4.3

1. A signal molecule is a molecule outside the cell that brings information to the cell. A membrane receptor binds to the signal molecule and changes the inside of the cell. **2.** Facilitated diffusion is diffusion across the cell membrane that happens when a signal molecule binds to a receptor and opens a membrane channel for ions to flow through. **3.** Signal transduction is when signal molecules cause a change in the shape of the membrane protein, resulting in a change in some chemical function inside the cell, although the signal molecule itself remains on the outside of the membrane.

Section 4.4

1. Eukaryotes have highly structured, membrane-bound organelles, while prokaryotes do not. Eukaryotes have their DNA surrounded by a nuclear membrane similar to the bilayer cellular membrane. Prokaryotic cells have their DNA loosely collected into a nucleoid, but there is no surrounding nuclear membrane. In most cases prokaryotic cells are smaller than eukaryotic cells. Most prokaryotes are single-celled organisms. Many eukaryotic organisms are multicellular. **2.** The cytoskeleton gives a eukaryotic cell its characteristic shape. Ribosomes carry out protein synthesis. Mitochondria produce ATP. **3.** Plant cells have a cell wall, which gives structural support, while animal cells rely on a cytoskeleton. Plant cells have chloroplasts, which provide nutrients through the process of photosynthesis; animal cells must obtain their nutrients from external sources.

Section 4.5

1. Spinal cord axons do not regrow. **2.** Sensory neurons obtain information from within and outside the body and carry that information to the spinal cord. Motor neurons are cells that carry messages from the spinal cord out to the muscles to tell them to contract. Interneurons are located between the sensory neurons and the motor neurons. They receive messages from the sensory neurons and decide whether to activate one or more muscles and, if so, which ones.

Review Questions

True/False

1. F Leeuwenhoek and Hooke were two pioneering scientists who studied *microscopic cells*. **2. F** *Lipids* in the cell membrane prevent ions and polar molecules from crossing the cell membrane. **3. T; 4. F** Smooth ER has the primary job of manufacturing and modifying *lipids*. **5. F** Photosynthesis happens within the *chloroplasts* of a eukaryotic cell.

Multiple Choice

6. c, **7.** b, **8.** d, **9.** a, **10.** d, **11.** d, **12.** b, **13.** a, **14.** e, **15.** b

Matching

16. c, **17.** d, **18.** b, **19.** a, **20.** e

Chapter 5

Quick Check Questions

Section 5.1

1. Energy is the potential to do work and often is carried in the motion of an object. **2.** Potential energy is the energy that is stored in some object that is at rest. Kinetic energy is the energy in an object that is in motion. **3.** First law: Energy can change its form—for instance, from electrical energy to sound energy or heat energy, but you cannot make more energy and you can't destroy energy. Second law: Any orderly system will become disordered over time. Things fall apart, dust settles, even the finest house, car, or tree, over enough time, becomes old and finally crumbles. Only an input of energy will keep the house clean, car running, or tree growing and prevent disorder from taking over.

Section 5.2

1. a. electrical energy—the energy carried in the motion of electrons; **b.** light energy—the energy carried by photons; **c.** heat energy—the energy of random particle motion; **d.** chemical energy—the energy in chemical bonds **2.** Biological organisms use the energy in chemical bonds; most use the energy in one of the phosphate bonds of ATP. **3.** Activation energy is the

amount of energy required to begin a chemical reaction, whether making or breaking bonds in a molecule.

Section 5.3

1. First, the energy from the photons of sunlight is used to make molecules of ATP. Then the energy from several ATP molecules is used to link carbon atoms together to form molecules of glucose. **2.** Photosynthetic organisms can make ATP only on sunny days; without a storage molecule they would not be able to carry on many activities at night or on rainy or cloudy days because it is difficult for cells to store many ATP molecules. It's also hard to move ATP molecules from one place to another, such as from leaf cells to roots or flowers. It is much more efficient to pack the energy of many ATP molecules into a single glucose molecule that is much easier to store or to transport to other cells. **3.** A diagram that shows how photosynthesis and glucose metabolism are related should look like Figure 5.12.

Section 5.4

1. A catalyst is a chemical that speeds up a reaction. but does not take part in the reaction. **2.** Enzymes act as catalysts to speed up the chemical reactions inside a living organism. **3.** ATP provides the energy for enzymes to perform chemical reactions in a cell. Nearly all known enzymes of all organisms use ATP for their energy source.

Section 5.5

1. The earliest fossils resemble present-day bacteria, and some resemble present-day photosynthetic bacteria. **2.** Four billion years ago Earth had an atmosphere of ammonia, methane, carbon dioxide, and water vapor. Lightning strikes were frequent, and ultraviolet radiation was intense. Temperatures were warm in the early oceans that were forming. **3.** The earliest biological molecules on Earth could have formed using the energy in lightning, UV light from the Sun, the energy from the core, or the heat that comes up in deep sea vents.

Review Questions

True/False

1. T; 2. T; 3. F Organisms use the energy in ATP to directly drive the making and breaking of chemical bonds. **4. T; 5. T**

Multiple Choice

6. c, **7.** a, **8.** b, **9.** d, **10.** e, **11.** a, **12.** b, **13.** a, **14.** b, **15.** d, **16.** e, **17.** b, **18.** e, **19.** e, **20.** d

Chapter 6

Quick Check Questions

Section 6.1

1. The process of photosynthesis captures and transforms the energy of sunlight into glucose molecules. In turn, the energy in the carbon-to-carbon bonds of glucose is chemically removed and stored in the chemical energy of ATP. Enzymes use the energy in ATP.

Section 6.2

1. Your sketch should look like the chloroplast in Figure 6.2. The major parts of a chloroplast are: (a) thylakoid stacks, sets of membranous disks each of which contains the molecules responsible for photosynthesis; (b) the thylakoid membrane, the lipid bilayer membrane of the thylakoid disks that contains the antenna complex, the reaction center, and the electron transport chain; and (c) the stroma, the space outside the thylakoid disks that contains water and hydrogen ions used in photosynthesis. **2.** $6CO_2 + 6H_2O \rightarrow C_6H_{12}O_6 + 6O_2$; This equation means that six molecules of carbon dioxide and six molecules of water are broken apart and used to make one molecule of glucose and six molecules of oxygen. This overall reaction is driven by the energy in light. The reaction does not show that the first step is to make ATP using light energy. The energy in ATP is used to make glucose out of carbon dioxide. **3.** An electron transport chain is a set of proteins in the membrane of a chloroplast that gradually released energy in a controlled way. The energy is used to push hydrogen ions across the thylakoid membrane, which produces a concentration of hydrogen ions inside the thylakoid disk. The concentrated hydrogen ions then flow through the ATP-synthase molecule by diffusion, providing the energy needed to make ATP. **4.** Carbon fixation is the phase of photosynthesis during which the carbon atoms in carbon dioxide are incorporated into a molecule of carbohydrate.

Section 6.3
1. a. The first step is glycolysis; breaking a glucose molecule in half. **b.** Next a molecule of acetyl coenzyme A is generated from each half. **c.** Each of the acetyl CoA molecules enters the Krebs cycle (also called the citric acid cycle). **d.** The remaining molecular products then enter oxidative phosphorylation through the electron transport chains. The final result is molecules of ATP. **2.** Oxidative phosphorylation occurs in the mitochondria. Two molecules from the citric acid cycle, FADH and $NADH_2$, are used. Each transfers electrons to a chain of proteins which are on the inner membrane within a mitochondrion. This is another electron transport chain similar to the ones we have seen elsewhere. The electrons move down the chain, with enzymes causing them to release energy in a series of reactions. The energy then is used to move hydrogen ions, one at each step, across the internal membrane. The concentration of hydrogen ions acts as potential energy. As the hydrogen ions are allowed to leave the membrane, they pass through an enzyme that uses the kinetic energy to make ATP from ADP. In the final step of the electron transport chain, the electron acceptor is oxygen. The oxygen then combines with the hydrogen ions and produces molecules of water. **3.** Fats in your diet are broken down through a different pathway and converted into acetyl CoA molecules that can then be used in the citric acid cycle. If there is not sufficient fat, then dietary proteins are used. They are broken down to their amino acids, which are changed to acetyl CoA. If your diet is sufficiently limited that there is still insufficient materials for glucose metabolism, then your body begins to convert first stored fats, and second the proteins in your muscles, in order to use them.

Section 6.4
1. Aerobic metabolism uses oxygen as the final electron acceptor in the full metabolism of glucose to form up to 38 molecules of ATP per glucose molecule. Aerobic metabolism in eukaryotes uses the electron transport chain and ATP synthase molecule in mitochondria to accomplish this. Anaerobic metabolism uses only fermentation reactions like glycolysis to metabolize glucose and does not use oxygen. Anaerobic metabolism allows only the formation of up to four molecules of ATP from a single glucose molecule. **2.** Yeasts produce alcohol and ATP as a result of fermentation metabolism. **3.** Early photosynthetic organisms removed carbon dioxide from the atmosphere and added oxygen. Millions of years of this process eventually altered the atmosphere of Earth and produced the modern atmosphere with about 20% oxygen and less than 1% carbon dioxide.

Section 6.5
1. A greenhouse gas is an atmospheric gas that traps solar energy and so causes an increase in the surface temperature of a planet. Examples are carbon dioxide and methane. **2.** If the level of greenhouse gases in the atmosphere increases, then global temperatures will increase. This appears to be happening on Earth right now. **3.** A fossil fuel is a hydrocarbon formed by physical and chemical processes from the bodies of dead organisms. When a fossil fuel is burned, the process is much like glucose metabolism but much faster. The bonds of the fossil fuel are broken and energy is released, along with carbon dioxide and water. During the past 100 years fossil fuels have been burned in human societies at a rapid pace. The

result is an increase in the level of carbon dioxide in the atmosphere, causing an increase in global surface temperatures.

Review Questions
True/False
1. T; 2. T; 3. F Oxidative phosphorylation occurs in mitochondria. **4. F** Carbon dioxide is a major greenhouse gas. **5. T**

Multiple Choice
6. d, **7.** d, **8.** c, **9.** d, **10.** e, **11.** a, **12.** e, **13.** c, **14.** d, **15.** b, **16.** b, **17.** b, **18.** a, **19.** c, **20.** a

Chapter 7
Quick Check Questions
Section 7.1
1. (a) Fred Griffith found that nucleic acids could alter traits when transferred from one organism to another. (b) Oswald Avery determined that DNA was the genetic material. (c) Alfred Hershey and Margaret Chase found that viral DNA could infect bacteria (not viral proteins).

Section 7.2
1. (a) Linus Pauling and his team built molecular models and found that proteins had a helical structure. Watson and Crick adopted his model-building techniques. (b) Maurice Wilkins and Rosalind Franklin worked on X-ray crystallography, and Franklin took the excellent images instrumental in deciphering the final structure of DNA. Watson and Crick used Franklin's images to help build their model. (c) Francis Crick and James Watson did not do original DNA research. They looked at the research of others and built models based on what they had learned from the work of Franklin, Chargoff, and others, with Rosalind Franklin giving them feedback on their models.

Section 7.3
1. DNA is made of a double string of nucleotides wound into a helix, rather like a twisted ladder. Each nucleotide is made up of a sugar, a phosphate group, and a nitrogenous base—guanine, thymine, adenine or cytosine. The sugar and phosphate are linked together, attaching to the sugar and phosphate of neighboring molecules, and form the sides of the ladder. The bases stretch into the center of the ladder, with two bases attached in the center with a hydrogen bond, forming one rung of the ladder. **2.** A chromosome is formed of DNA and a set of associated proteins. DNA coils around the proteins into structures called nucleosomes. DNA keeps coiling tighter and tighter to form chromosomes.

Section 7.4
1. DNA polymerase makes two complete copies of DNA by taking advantage of the base-pairing rules. Once DNA helicase separates the two complementary strands of the original DNA molecule, DNA polymerase uses each strand as a template to build a new strand. The original plus the new strand combine to form a new DNA double helix. At each position on the original template strand, DNA polymerase puts in place the complementary nucleotide. For example, if the original template strand has a sequence CGTA, then DNA polymerase puts in place a new complementary strand with the sequence GCAT. In this way each template strand from the parent molecule can be used to generate a complete, new DNA double helix.

Section 7.5
1. The genetic code is the translation of DNA to RNA to protein. The sequence of DNA is transcribed to mRNA. When transcription is carried out on the ribosome, every three nucleotides along the strand of mRNA codes for one amino acid. **2.** tRNAs bind to and carry amino acids to the ribosome

to be used for protein synthesis. mRNA is copied from mRNA and is used as the sequence to dictate the synthesis of a sequence of amino acids. rRNA is the RNA found in ribosomes involved in the actual synthesis of amino acid chains. **3.** In transcription a stretch of nucleotide sequence on one side of the DNA is copied to mRNA. Several enzymes such as DNA helicase, DNA polymerase, and DNA ligase are involved. **4.** The mRNA, small ribosome subunit, and large ribosome subunit come together as one structure. Within this structure are several sites, or bays, where tRNA molecules that have amino acids attached to them can sit. The tRNA molecules bind to mRNA via the anticodons on their bases. When two tRNA molecules sit in these bays, the ribosome binds the two amino acids together. Then the ribosome moves down the mRNA strand to allow another tRNA to come to one of the bays, and binds the next amino acid to the strand.

Section 7.6

1. Two important points summarize how DNA function provides evidence for the unity of life. First, the genetic code is nearly universal. Almost every known kind of organism uses the same genetic code. Only a handful of species of prokaryotes use a slightly different code. This means that all living things are made by the same rules, and this produces organisms based on the same basic principles. Second, all life is genetically related because DNA has been copied and handed down from cell to cell and from parent to offspring over all the generations of life. **2.** DNA can change over generations by mutation—a change in nucleotide sequence, by the switching of DNA between one chromosome and another, and by the duplication of whole stretches of DNA.

Section 7.7

1. Tuberculosis is an infectious disease that primarily affects the respiratory system, but it can spread throughout the body to affect other organs and can be fatal. **2.** The causal bacterium, *Mycobacterium tuberculosis,* quickly becomes resistant to a single antibiotic. The current treatment is to take as many as four different antibiotic medicines at once for six months or longer. This still does not cure some people who are infected with extremely resistant strains.

Review Questions

True or False

1. F In the DNA molecule sugar and phosphate molecules form the *uprights* of the ladder. The nitrogenous bases form the rungs of the DNA ladder. **2. T; 3. F** Watson and Crick *did* realize the implications of the structure of DNA as a means for copying genes. **4. T; 5. F** RNA polymerase helps transcribe *DNA into RNA.*

Multiple Choice

6. d, **7.** e, **8.** e, **9.** a, **10.** b, **11.** a, **12.** b, **13.** b, **14.** c, **15.** e, **16.** c, **17.** a, **18.** c, **19.** d, **20.** b

Chapter 8
Quick Check Questions

Section 8.1

1. In asexual reproduction a single individual parent produces offspring by giving a complete copy of its DNA to the offspring. In sexual reproduction two parents each contribute half of their DNA to an individual offspring. Asexual reproduction produces offspring genetically identical to the parent, while sexual reproduction produces offspring with genetic variation.

Section 8.2

1. Homologous chromosomes are the pairs of chromosomes, found in eukaryotes. Each complete set of chromosomes is a complete copy of the genome of the organism. For example, humans for example have 23 homologous pairs of chromosomes, making a total of 46 chromosomes, while fruit flies have 4 homologous pairs for a total of 8 chromosomes. **2.** Alleles are different versions of a gene that control a particular trait. There may be two or more alleles for a trait. **3.** The haploid number is one set of homologous chromosomes. The diploid number is two complete sets of homologous chromosomes.

Section 8.3

1. Mitotic cell division is the process of cell division in eukaryotic cells that includes copying DNA, moving the copies away from each other, and dividing the parent cell into two. It results in two identical daughter cells. In contrast, meiotic cell division is the process of cell division that copies DNA and then allocates half of the DNA to each daughter cell. In the end of the process the final daughter cell has half of the DNA of the parent cell and each daughter cell can become a gamete.

Section 8.4

1. The cell cycle is the orderly sequence of events in the life of a eukaryotic cell that includes the processes of cellular function and cell division. The phases of the cell cycle are (a) G_1, or gap-1, during which the cell grows and carries out a variety of normal functions; (b) S-phase, or synthesis, during which the entire DNA, in the form of chromosomes, is copied; (c) G_2, or gap-2, during which the cell prepares to divide and checks the DNA for errors; (d) M, or mitosis, during which the duplicated chromosomes condense and the sister chromatids are pulled apart to opposite sides of the cell; and (e) C, or cytokinesis, during which the cell physically divides into two cells, each with a complete set of DNA. **2.** The sister chromatids are pulled apart by the spindle apparatus that arises from the centrioles. **3.** Growth factors are signal molecules that bind to membrane receptors during G_1 and communicate to the cell to start the cell division process.

Section 8.5

1. Gametes are cells that contain half of the DNA of other body cells. They are the haploid cells, produced through meiosis, that fuse during sexual reproduction to produce a new individual. **2.** Sexual reproduction produces the genetic variation across generations that help individuals and species to survive in a variety of environments. **3.** Offspring usually must have the exact number of genes and chromosomes required to form that kind of organism. If gametes carry the same amount of DNA as all other body cells, each new generation produced by the union of egg and sperm will have *twice* the DNA of its parents. Such a drastic change in the amount of genetic material occasionally can produce a whole new species, and this has happened with many plant species. In most other organisms, however, and especially in most animals, having twice the amount of DNA is not likely to produce a viable offspring. Sometimes organisms get too many copies of some chromosomes or no copies at all of a chromosome. Both kinds of errors can cause the organism to die or to have serious problems.

Section 8.6

1. In meiosis the duplicated chromosomes pair up with their homologous counterparts and line up, side by side, along the center of the cell, with spindle fibers attached to one side of the sister chromatid. The homologous pair is then pulled apart, one set of sister chromatids staying attached and going to one side of the cell, the other set remaining attached and moving to the other side of the cell. Cytokinesis now occurs, resulting in two cells, each with half of the normal set of chromosomes. Now a second cell division occurs, with the sister chromatids lining up single file along the center of each cell, just as in mitosis. The difference is that this is a haploid cell, with only half of the original chromosomes. The sister chromatids separate and two new cells form—a total of four new haploid gametes. **2.** Early during the first round of meiotic division, the duplicated homologous chromosomes line up. The two homologous chromosomes actually touch, and at the places where they touch alleles can break out and trade places. Crossing-over results in eggs or sperm that have chromosomes with different sequences of alleles than those in the rest of the parent's cells. This is the process in which a series of alleles on homologous chromosomes are physically switched, producing a new combination of alleles on chromosomes. **3.** Mitosis can occur in all cells of a multicellular organism, and it ends with two new, identical diploid cells that have exactly the same information as the original cell. Its purpose is for growth and/or replacement of cells. Meiosis only occurs in certain cells in the ova of a female or the testes of a male, to create haploid eggs and sperm. Meiosis results in gametes that have half of the DNA of the parent cell. Gametes from different parents can then combine to form a new organism.

Section 8.8

1. As individual cells age, their functions deteriorate. The accumulation of cellular problems eventually gets so great that a cell dies. In a multicellular organism the accumulation of cellular aging eventually causes the whole organism to die. Aging is the accumulation of cellular abnormalities that leads to a decline in functions and to the eventual death of the individual cell or organism. **2.** Fibroblast cells produce collagen and elastic fibers in skin, muscle and tendons. They help give tissues structure and stiffness. They divide throughout a person's life, but as a person ages the fibroblasts divide less frequently and finally stop dividing.

Review Questions

True or False

1. T; **2.** F Crossing-over is one type of *random mutation*, which occurs during meiosis; it increases variability in offspring. **3.** F Alleles are *different versions* of a single gene. **4.** T; **5.** F *Sexual reproduction* involves the fusion of two haploid gametes into a single diploid zygote. Asexual reproduction involves the fission of a diploid cell into two new diploid cells.

Multiple Choice

6. e, **7.** b, **8.** e, **9.** a, **10.** d, **11.** b, **12.** c, **13.** b, **14.** d, **15.** c

Matching

16. d, **17.** c, **18.** b, **19.** a, **20.** e

Chapter 9
Quick Check Questions

Section 9.1

1. Cell division is the process that turns one parent cell into two daughter cells. In embryonic development repeated rounds of mitotic cell division turns the single-celled zygote into a mature organism with thousands, millions, or trillions of cells. In contrast, cell differentiation is the process that produces different kinds of cells in the developing embryo.

Section 9.2

1. The gradual steps during development that produce the limbs of a vertebrate include the following: a small set of chemical changes occurs in cells along the embryo's trunk; small swellings occurs on the embryo's trunk; internal limb structures begin to appear; and then bones and other structures develop, first close to the body and then farther away from the body. **2.** The genome is the total DNA nucleotide sequence of an organism; the proteome is the total kinds of proteins expressed by a particular cell or organism.

Section 9.3

1. DNA methylation and chromosome condensation are similar in that both are ways to turn off large stretches of DNA and prevent them from being transcribed. DNA methylation involves putting methyl groups along the

stretch of DNA; DNA condensation involves coiling the DNA. In both processes the RNA polymerase enzymes cannot get access to the DNA and so transcription cannot occur. **2.** The 25,000 to 30,000 human genes can make about 100,000 human proteins because of mRNA splicing and posttranslational modifications.

Section 9.4

1. The promoter is the stretch of DNA ahead of a gene where RNA polymerase binds. RNA polymerase is the molecule that transcribes DNA to RNA. A transcription factor is a protein that binds to DNA ahead of the promoter and enhances the function of RNA polymerase for a particular gene, and so speeds up transcription of that gene.

Section 9.5

1. The *Pax6* gene is common among different animal species and plays a role in the control of eye development in all animal species. If the *Pax6* gene from one species is injected into another species, it can stimulate the development of an eye. For example, if *Pax6* from a mouse is injected into the tail region of a developing fly embryo, a fly eye will develop at that location. This shows that *Pax6* operates in a similar way in diverse animal species.

Section 9.6

1. A stem cell has the ability to divide and produce daughter cells that can differentiate into any one of several mature cell types. **2.** Embryonic stem cells have the potential to develop into many, perhaps all, body cell types. Adult stem cells are found in specific tissues of mature individuals—in contrast to embryonic stem cells, each adult stem cell has the potential to develop into just a few different body cell types.

Section 9.7

1. Dioxin acts as a lipid-soluble signal molecule that is similar to estrogen. It crosses the cell membrane and binds to proteins inside the cell, much as estrogen does. The dioxin-protein complexes are transcription factors that bind to DNA and either stimulate or inhibit transcription.

Review Questions

True/False

12. F When chromatin is condensed, it is not readily transcribed. **13.** T; **14.** T; **15.** T

Multiple Choice

1. d, **2.** e, **3.** d, **4.** a, **5.** a, **6.** e, **7.** b, **8.** a, **9.** d, **10.** c, **11.** a

Matching

16. b, **17.** e, **18.** d, **19.** c, **20.** a

Chapter 10
Quick Check Questions

Section 10.1

1. Genetic crosses are carried out to test hypotheses about inheritance. Once a hypothesis is stated clearly, individuals with specific traits are bred together to test the hypothesis. The offspring from the mating are carefully observed and their traits recorded. The hypothesis is evaluated by comparison with the test-cross results and modified. Further crosses are done to test the modified hypothesis.

Section 10.2

1. To cross a dwarf pea plant with a tall pea plant Mendel would first have developed two populations of true-breeding plants, one population that was true breeding for each trait. Let's assume Mendel used pollen from a dwarf plant to fertilize a tall plant. First, Mendel would cut the pollen-producing organs from the tall plant. Then he would carefully gather some pollen from a dwarf plant and brush it onto the female organs of the tall plant. To prevent the tall plant from fertilizing itself he would place a bag around the flower of the fertilized tall plant. Then he would gather the seeds after they developed, let them grow into adult plants, and determine whether the F_1 plants were dwarf or tall.

Section 10.3

1. Mendel studied crosses of pea plants having either yellow or green seeds to determine how these traits are inherited. The parent generation was composed of two populations of true-breeding plants with either yellow or green seeds. When individuals from these two populations were crossed, all of the offspring, called the F_1 generation, had yellow seeds. This told Mendel that the yellow trait was dominant over the green trait. Further crosses revealed that the green trait had not disappeared. When individuals of the F_1 generation were crossed, the F_2 generation had a mix of seed colors. Three-quarters of the F_2 generation were yellow, but one-quarter was green. This told Mendel that the green trait could be expressed under some conditions, and so the green trait was recessive. Expressed in modern genetic language, each trait is controlled by two alleles. In the case of seed color the yellow allele is dominant over the green allele. This means that if a plant carries two yellow alleles, or if it carries one yellow and one green allele, then in both cases the plant will have yellow seeds. Only if the plant carries two green alleles will it have green seeds. The two alleles that a plant carries for a given trait can be called homozygous if they both represent the same value for a trait, or heterozygous if they are different. This is the plant's genotype. Mendel's yellow seeded plants could be either homozygous dominant or heterozygous. The green-seeded plants are all homozygous recessive.

Section 10.4

1. An allele that is fully dominant is expressed regardless of the nature of the second allele. In full dominance a heterozygous individual will have the same phenotype as a homozygous dominant individual. In incomplete dominance the heterozygous individual has a phenotype that is intermediate between the homozygous dominant and heterozygous phenotypes. In codominance the traits for both alleles are seen in the phenotype, and so the phenotype is a mix of both allele traits. 2. Cat fur color can show the effects of the environment when the cat carries a temperature-sensitive allele involved in melanin production. The protein from this allele works at low temperatures but not at high temperatures. In Siamese cats this produces the pattern of light body color and dark pigmentation in the tips of the ears, tail, and paws. Schizophrenia is a complex disease controlled by many factors. Studies of identical twins show that if one twin is schizophrenic, the other twin has a 50% chance of also being schizophrenic. This is much higher than the average chances a person will be schizophrenic. This shows that genetics is an important factor in schizophrenia but the environment plays a big role as well.

Section 10.5

1. Sickle cell disease is a disorder in which blood is not carried adequately to peripheral tissues, especially under low oxygen conditions. It causes fatigue, pain, and even death. The disease is caused by a recessive allele for the hemoglobin gene. The abnormal allele causes production of abnormal hemoglobin molecules, and a person has the full-blown disease if he or she is homozygous recessive for the allele. A heterozygous individual is largely normal but can show some disease symptoms under extreme low-oxygen conditions. So while it looks like sickle cell disease is inherited in a dominant/recessive fashion, it really looks more like incomplete dominance. At the cellular level the disease is an example of codominance because in the heterozygous condition cells will produce both normal and abnormal hemoglobin. Because abnormal red blood cells can affect so many different functions, the sickle cell allele is pleiotropic.

Section 10.6

1. The eugenics movement made the assumption that complex human traits such as intelligence or laziness were inherited in a simple dominant/recessive fashion and so could be readily bred into or out of human populations. This assumption was wrong because such complex traits are determined by pleiotropy and polygenic inheritance and have a strong environmental component.

Review Questions

True/False

14. **F** An individual who has the blood type AB is *heterozygous* for this trait because these alleles do not match. **15. T; 16. T**

Multiple Choice

1. b, 2. d, 3. a, 4. c, 5. d, 6. e, 7. d, 8. c, 9. a, 10. b, 11. a, 12. a, 13. c

Matching

17. d, 18. c, 19. e, 20. a

Chapter 11
Quick Check Questions

Section 11.1

1. Biotechnology is the group of techniques used to isolate, characterize, manipulate, and control biological molecules such as DNA or proteins. Biotechnology has many practical applications in medicine, research, and agriculture. It not only is used in the general society, but also it is used to provide further knowledge about how DNA works.

Section 11.2

1. Restriction enzymes cut DNA at a particular point in the DNA sequence. They are used to make small lengths of DNA from a larger sample. The small lengths can be separated on a gel, and the pattern is a DNA fingerprint.
2. The polymerase chain reaction is a technique that makes many copies of one stretch of DNA. It uses a polymerase enzyme from a prokaryote that lives at very high temperatures. It is useful when there is only a tiny sample of DNA, too little to use in the standard DNA fingerprinting or other tests. Small samples of DNA can be amplified using PCR, allowing other tests to be carried out.

Section 11.3

1. Nuclear DNA is found in the nucleus and is the genome of a eukaryotic cell. It carries the instructions for making the proteins involved in nearly all functions of a cell. Nuclear DNA is copied during S-phase of the cell cycle. Mitochondria carry their own DNA that contains many of the genes used in energy metabolism. Mitochondrial DNA is copied on its own independent cell division cycle. It is inherited only via eggs through the mother.
2. Mitochondrial DNA is inherited through the mother and also changes little from one generation to the next. For this reason, people with the same mother or grandmother have nearly identical mitochondrial DNA. This allows mitochondrial DNA testing to be used to identify the genetic relationships between individuals. In addition, the rate of mitochondrial mutations is known, so if the mitochondrial DNA of two individuals is compared, you can calculate how long ago they shared a common ancestor.

Section 11.4

1. A DNA microarray is a device used to detect the presence of any one of many genes in a sample. A microarray contains thousands of spots of single-stranded DNA, each an allele for a gene of interest. Using a DNA microarray, you can take a sample of mRNA from a person, label it with some marker, and determine if any of the labeled strands bind to the test spots of single-stranded DNA. If they do bind, then that allele is present and is expressed in

the individual being tested. DNA microarrays are useful in genetic testing for the presence of specific disease alleles.

Section 11.5

1. A transgenic organism is one that has DNA from another species inserted into its genome. The DNA from any organism can be inserted into any other organism, and the proteins from the foreign DNA will be expressed. Transgenic organisms are used in agriculture, medicine, and research. In agriculture, genetically modified crops provide resistance to pests and so increase crop yields. In medicine transgenic bacteria produce medicines such as insulin. In research transgenic organisms provide animal models for the study of human diseases.

Section 11.6

1. Nuclear transfer cloning produces an embryo or whole organism that is genetically identical to another organism. The usual procedure involves inserting the nucleus from an adult cell of one organism into an unfertilized egg whose DNA has been removed or destroyed. The embryo is allowed to develop through several rounds of cell division and then is implanted into a surrogate mother. **2.** In reproductive cloning the cloned embryo is allowed to develop into a mature individual, although this happens in only about 3% of the attempts in animals. In therapeutic cloning the cloned embryo is only allowed to develop in a culture dish, and cells from embryo are used to treat diseases of the individual who donated the nucleus. Neither reproductive nor therapeutic cloning has been successful in humans.

Section 11.7

1. Genes can be gotten into cells for gene therapy via viruses, liposomes, injection of DNA, or by direct uptake of naked DNA. **2.** It is still difficult to get DNA into cells. It is also difficult to ensure that the DNA will be under the control of the right promoters and transcription factors and be expressed when it is supposed to. Also, there is a concern about an immune response to the procedure that can cause illness or death.

Review Questions

True/False

1. T; **2. F** Restriction enzymes come from *bacteria* and are used by *bacteria* to help fight infection by *viruses*. **3.** T; **4. F** The biotechnology revolution began with the discovery of the structure of *DNA* in *1953*. **5. F** The Human Genome Project has demonstrated that there are between *25,000 and 30,000* genes.

Multiple Choice

6. e, **7.** e, **8.** b, **9.** e, **10.** b, **11.** a, **12.** d, **13.** a, **14.** e, **15.** b, **16.** a, **17.** d, **18.** c, **19.** a, **20.** e

Chapter 12

Quick Check Questions

Section 12.1

1. Death rates from cancer have just started to decline in the last years of the twentieth century and early twenty-first century.

Section 12.2

1. Cancer is a set of diseases that affect different body systems, but it has some features in common. In particular, cancer is a disease of excess proliferation of cells that also have the ability to leave their neighbors and spread the cancer to other body locations. **2.** A benign tumor is a mass of cells that results from abnormal proliferation, but the mass is confined to the location where it originated. The cells in a benign tumor have not developed the ability to leave their neighbors and spread cancer to other

locations. A malignant tumor is a mass of cells that results from abnormal proliferation in which some cells have left the tumor and started a tumor somewhere else. **3.** A stage IV cancer is one that has spread through the lymph nodes to some distant location.

Section 12.3

1. Cancer is genetic because it is caused by mutations in DNA. Cancer is not always inherited because those mutations come from many sources apart from parents. **2.** Four abnormalities in cancer cells are abnormally high rates of cell proliferation, reduced adhesion, reduced apoptosis in response to abnormalities, and stimulation of angiogenesis. Abnormal proliferation leads to the accumulation of a mass of abnormal cells. Reduced adhesion allows cancer cells to leave their neighbors and take up residence elsewhere in the body, where they proliferate and establish a new tumor. Reduced apoptosis means that abnormal cancer cells do not kill themselves. They proliferate and produce more abnormal cells. Stimulation of angiogenesis means that cancer cells secrete growth factors that cause local blood vessels to grow into the tumor. This provides the tumor with increased nutrients for further growth.

Section 12.4

1. A tumor suppressor protein binds to a cell division related transcription factor and prevents it from binding to DNA and stimulating cell division. Abnormal tumor suppressor proteins fail to carry out this function, allowing a cell to divide when it should not. Abnormal tumor suppressor proteins are found in many types of cancers. **2.** p53 is a protein that is involved in detecting cellular abnormalities and preventing those abnormalities from being passed on to future generations of cells. In an abnormal cell p53 can activate a chemical pathway that stops cell division, and it can activate a chemical pathway that causes apoptosis. Mutations in the gene that codes for p53 lead to abnormal cells being able to survive and divide. **3.** A Pap smear is a test for cervical cancer. A swab is taken from the cervix, and the cells are stained and examined under the microscope. The test is based on the shape of the cells. Cells with abnormal shapes are more likely to indicate cancer. Further tests are required to confirm or reject evidence from Pap smears.

Section 12.5

1. Many risk factors increase cancer risk. Five examples are inheriting a gene that contributes to cancer, smoking tobacco, exposure to UV radiation from the Sun, eating a high-fat diet, and lack of exercise. **2.** A carcinogen is an environmental chemical that causes DNA mutations that lead to cancer. **3.** Many cancers are caused by lifestyle choices that could be avoided. Smoking cigarettes is one major lifestyle choice that increases the chances of developing cancer. Eating a high-fat diet and not getting exercise are also factors. Risk can be reduced by eating a variety of fresh fruits and vegetables.

Section 12.6

1. Both chemotherapy and radiation damage cells so severely that they undergo apoptosis. **2.** Her2 is a growth factor receptor on breast cells that responds to signal molecules that stimulate cell division. Many breast cancer cells produce too much Her2 receptor and show abnormally high proliferation. A drug called Herceptin binds to and blocks the Her2 receptor and so prevents the excess cell proliferation.

Section 12.7

1. Laetrile therapy is based on the hypothesis that cancer cells produce cyanide when metabolizing the key ingredient in laetrile. The logic is that the cyanide will differentially kill cancer cells and be otherwise safe. The first line of evidence against this hypothesis is that studies have shown that while liver cells produce cyanide when metabolizing laetrile, cancer cells do not. Second, examination of records of cancer patients does not show any evidence that laetrile affects cancer. Finally, many studies show that the cyanide produced by the liver during laetrile metabolism is toxic.

Review Questions

True/False
1. T; **2.** F A benign tumor is *localized to the tissue where it originated.*
3. F Cancer is *a genetic disease, but it is not always inherited.* **4.** T; **5.** T

Multiple Choice
6. a, **7.** d, **8.** b, **9.** c, **10.** b, **11.** a, **12.** b, **13.** e, **14.** c, **15.** e, **16.** e, **17.** e, **18.** d, **19.** d, **20.** e

Chapter 13
Quick Check Questions
Section 13.1
1. Human memory is an example of an adaptation. Other examples are the sharp teeth and special chemicals in the saliva of vampire bats and the ability of some plants to live in harsh deserts. Adaptations come from modifications of the traits of a population from one generation to the next. The allele frequencies for a particular gene change, and so the traits of the population change. This is evolution, and evolution produces the diverse traits of organisms.

Section 13.2
1. The nineteenth century was a time when many people believed that the universe was constant and static, and so was Earth's life. It was thought that species never change and that species are distinct and different kinds. Also, though, this time saw the emergence of thoughts about evolution and about how individuals survive. Some writers started to explore the idea that life changes over generations and came up with explanations. Lamarck proposed that when organisms change as a result of experience, these changes are passed on to offspring. Other writers focused on the human condition. Malthus saw competition for resources and was concerned that the Earth's resources would not be able to support a growing human population.

Section 13.3
1. The *Beagle* voyage showed Darwin that although the new areas he visited had animals and plants new to him, areas that were close geographically had species that tended to resemble one another. So he could characterize South American, African, or Australian animals. Darwin was able to collect fossils and to appreciate how different they were from their present day relatives who still lived in the same geographic areas. He was able to study geology and see firsthand the effects of tidal waves and earthquakes. He paid attention to the geological formations in which the fossils were found. Darwin saw many different kinds of organisms and noted how their appearance was related to how close to one another they lived. Writings by prominent geologists, which Darwin read while on the voyage, also greatly influenced Darwin's thinking. **2.** Natural selection involves differential breeding success. In any population the individuals will vary along any given trait, and these traits are at least in part inherited. These individuals live in a particular environment, and some versions of any trait will be more successful in that environment than other versions of the trait. The individuals with the successful traits are more likely to survive and will have more offspring. In this way the environment determines which individuals survive and breed—and so which individuals pass on their traits to the next generation. If the environment in which a population lives changes, then the traits that are successful will be different. This means that the traits passed on to future generations will change, and so the population traits will change.

Section 13.4
1. An example of natural selection seen in nature is the shapes of beaks of species of Galápagos finches. During and after dry periods the beaks of the finches get larger and stronger because the seeds that the birds eat get harder. This leads to better breeding success by individuals with larger and stronger beaks and less breeding success by birds with smaller beaks. An example of natural selection in an experiment is the study of alcohol dehydrogenase in flies. One group of flies was given plain food, while another was given food with alcohol added. The flies were allowed to breed normally, but a few flies were removed from each group and used to start a new population. Again, these flies had either plain or alcohol food. Again, a few flies were removed to start a new generation. After several generations of this procedure, the flies that had lived on food containing alcohol had much higher levels of the effective enzyme that breaks down alcohol into nontoxic compounds. The genetic traits of the population of flies were changed over generations by the environment they lived in.

Section 13.5
1. The evidence from cytochrome *c* supports Darwin's principle of common descent. Species with similar cytochrome *c* proteins are those that are more closely related by other measures. Species with less similar cytochrome *c* are less closely related. All cytochrome *c* molecules share the same important features, and the cytochrome *c* from one species can work in another species.

Section 13.6
1. Intelligent Design has two primary assertions. The first is that the many aspects of life are too complex to be explained by natural processes. The second is that there must be a designer involved in the evolution of life. Neither of these assertions is testable scientifically. You can test a hypothesis about a particular explanation of a feature of life, but you cannot test the idea that some aspect of life is too complex to explain. Similarly, the notion of an Intelligent Designer is beyond the realm of science.

Section 13.7
1. Birds are most closely related to therapod dinosaurs. Birds and therapod dinosaurs share many features, such as neck joint that swivels, long, flexible neck, hop or walk on 2 hind legs, same digits on forelimbs, 5 toes in adult, hollow bones, large pelvis, first toe turned backward, warm bloodedness, and feathers.

Review Questions

True/False
1. F *Many fossils* show precisely what prehistoric plants or animals looked like. **2.** T; **3.** F *Charles Darwin did not originate the idea of evolution; his contribution was natural selection as the mechanism that directs evolution.* **4.** T; **5.** F The purpose of the *Beagle* voyage was *to map coastlines of southern continents.* **6.** T; **7.** F Darwin's finches live in the *Galápagos Islands.* **8.** T; **9.** F The fossil record shows that *life changes through time.* **10.** T; **11.** T; **12.** F Mammalian traits evolved through modifications of a *general primitive reptilian design.* **13.** F Chihuahuas, broccoli, seedless watermelons, and super-sweet ears of corn are just a few examples of the results of *artificial selection.* **14.** F House sparrows *vary in color and size* across the United States and down into Central America. **15.** F Populations of *Drosophila melanogaster* that are fed food with alcohol in it develop *higher than normal frequencies* of the enzyme that detoxifies alcohol. **16.** T; **17.** F A change in the nucleotide sequence of DNA is defined as a *mutation.* **18.** F Evolution *can proceed rapidly under some circumstances.* **19.** F Natural selection *is not always a slow, gradual process that takes hundreds of millions of years.* **20.** T

Chapter 14
Quick Check Questions
Section 14.1
1. The three definitions of a species are: morphological, biological, and phylogenetic. The morphological species concept focuses on the similarities and differences in the anatomical structures of organisms. The biological species concept says that a species is a population of individuals that has the potential to breed with one another and produce viable offspring. The phylogenetic species concept defines a species as the smallest group of organisms that share many similar traits and all descend from one common ancestor. **2.** Species arise from reproductive isolation. Reproductive isolation can develop from geographic isolation, exploitation of a new aspect of the environment, selection, or polyploidy. All of these prevent gene flow between groups of individuals. Geographic isolation happens when some feature of the geography physically isolates one group of a population from another. Sometimes an aspect of the environment of a population can change, and a subset of the population takes advantage of that change, establishing a breeding population that eventually evolves into a new species. Sexual selection usually involves females selecting mates, and so determining the traits of the next generation. If some females start choosing a different kind of mate, then a new species can arise. Polyploidy is an error of meiosis that results in more than the diploid number of chromosomes. Many plant species came about because of polyploidy.

Section 14.2
1. The three domain scheme divides living organisms into Archaea, Bacteria, and Eukarya. It is based on analyses of the DNA of various organisms that codes for ribosomal RNA. Older classification schemes were based more closely on morphological traits. The older schemes described five kingdoms. While the DNA analyses support three kingdoms within the larger domain of eukaryotes, they also show that two of the traditional five kingdoms are not phylogenetic groups.

Section 14.3
1. Bacteria are prokaryotes that don't have organelles with membranes around them. Bacteria are small cells shaped like rods, spheres, or corkscrews—they live in and around us and some live in extreme environments. The Archaea are prokaryotes that have many features in common with Bacteria but also are related to Eukarya. Archaeans are found only in extreme environments such as hot springs or salt flats. All other living organisms are in domain Eukarya. They have organelles with membranes around them, including a nucleus. Eukaryotes can be unicellular or multicellular.

Section 14.4
1. The first major change in Earth's environment was due to photosynthetic organisms putting out oxygen as a product of photosynthesis. This resulted in an increase in atmospheric oxygen that led to the evolution and expansion of eukaryotes. Later changes involved extinctions of major groups of organisms. These extinctions allowed other groups of organisms to survive and flourish. For instance, a mass extinction of marine invertebrates allowed the rise of vertebrates, the extinction of many groups of ferns allowed the rise of flowering plants, and the extinction of the dinosaurs allowed mammals and birds to flourish.

Section 14.5
1. Overall, the scientific data do not support the division of people into distinct races. Genetic studies show that there are more differences between people of one race than there are between people of different races. When specific genes are examined, there are genetic differences between certain populations of humans, but these genetic differences do not always correlate with the typical races identified in American society. Depending on what gene one is looking at, there can be no races, or scores of races. And, these genetic differences are not absolute. Even if a particular gene is more common in one group of people than another, it does not mean the gene is not found at all in other groups.

Review Questions
True/False
1. F The three domains are Archaea, Bacteria, and Eukarya. **2. T; 3. T; 4. F** Invertebrates have no backbone. **5. F** Hawthorne flies and apple flies are two species. **6. T; 7. F** Classification is something that everyone does. **8. F** Life first evolved in an environment that had little oxygen. **9. F** Because human races are genetically overlapping and indistinct, human races are not valid biological categories.

Multiple Choice
10. e, **11.** e, **12.** c, **13.** b, **14.** e, **15.** a, **16.** b, **17.** c, **18.** a, **19.** e, **20.** d

Chapter 15
Quick Check Questions
Section 15.1
1. It is difficult to define species of prokaryotes because details of the structures of prokaryotes are difficult to see without electron microscopy; they freely swap genes, changing their genetic makeup; and they reproduce asexually and do not conform to the definition of a biological species.

Section 15.2
1. An organism living on an iceberg in Antarctica would face these environmental constraints: extreme cold, high winds, little or no light for half of the year, few organisms to eat other than what is in the ocean, no shelter, no fresh water.

Section 15.3
1. Archaeans are not bacteria because they have distinctly different lipids, complex carbohydrates that are incorporated into their cell walls, different DNA that codes for ribosomal RNA, and other biochemical differences. Archaeans tend to be extremophiles, while bacteria are found in more general environments. Archaeans and bacteria have small size and prokaryotic cellular organization in common.

Section 15.4
1. If all of Earth's algae were suddenly wiped out, there would be less oxygen in Earth's waters as well as in Earth's atmosphere. There also would be disruptions in the lives of organisms that feed on algae. These organisms would have to find other sources of food. If they could not, their numbers would decrease, and some probably would go extinct. Examples that come to mind include blue whales that filter feed on crustaceans that, in turn, have fed on single-celled eukaryotes and algae. Other species that might be disrupted would be some species of ocean fishes that depend on algae—either when the fishes are newly hatched, or fishes that eat smaller organisms that feed on algae. **2.** Your sketch of alternation of generations should look like Figure 15.23. The important understanding that you should illustrate is that a haploid individual alternates with a diploid individual. The multicellular haploid life cycle stages produce eggs and/or sperm that fuse to form a diploid individual. The multicellular diploid individual undergoes meiosis and produces haploid spores. Each of these grows into a haploid life cycle stage that produces eggs and/or sperm.

Section 15.5
1. It seems fairly sure that the first life was prokaryotic, but there is not enough evidence to determine if it was similar to bacteria or archaeans or some now-unknown group.

Section 15.6

1. Just as garden plants grow more luxuriously when there are plenty of nutrients, populations of single-celled algae experience algal blooms when lots of nutrients are available. If less nutrients are available, there will be fewer harmful algal blooms.

Review Questions

True/False

1. F *Plasmodium* causes malaria. *Trypanosoma* causes sleeping sickness. Both are protists. **2. F** Most algae are not plants. They use different forms of chlorophyll. **3. T; 4. T; 5. F** Red algae are unusual eukaryotes. **6. T; 7. F** The division of prokaryotes into two domains is based on comparisons of the DNA for ribosomal RNA that they contain. **8. T; 9. T; 10. F** Bacteria have no nuclei; *Paramecium* has macronuclei and micronuclei. **11. T; 12. F** Gram stains allow us to identify different kinds of bacteria. **13. F** Most bacteria are decomposers. **14. F** Protists are eukaryotes. **15. T**

Multiple Choice

16. d, **17.** c, **18.** d, **19.** c, **20.** c

Chapter 16

Quick Check Questions

Section 16.1

1. The evidence that supports the idea that fungi are more closely related to animals than to plants includes the presence of chitin in cell walls of fungi and animals, more similar DNA that codes for ribosomal RNA, glycogen as the storage molecule for excess carbohydrates, similar cytoskeleton proteins, and similar amino acid sequences in other proteins. Fungi and animals are nonphotosynthetic and use other organisms for food. Both have external digestion of food.

Section 16.2

1. A mycelium is the fine mass of feeding tubules that extends into a fungal food source. It functions as a means to obtain nutrients.

Section 16.3

1. A fungus such as *Penicillium* would benefit from producing an antibiotic compound because it could use the antibiotic to kill off bacteria that compete for its food sources.

Section 16.4

1. Some examples of animals that use asexual reproduction include budding by hydras and parthenogenesis by some arthropods and vertebrates. In budding a small individual develops from the body of the parent. In parthenogenesis eggs that are unfertilized develop into new individuals.

Section 16.5

1. Kinds of animals that infect or infest humans include various sorts of parasitic worms, such as roundworms, tapeworms, pinworms, *Trichina* worms, and hookworms. These are internal parasites. External parasites include arthropods such as head lice. **2.** Disease vectors such as ticks and mosquitoes transmit diseases such as malaria, Lyme disease, West Nile virus, and bubonic plague.

Section 16.6

1. Humans aren't descended from chimpanzees, gorillas, or orangutans, because those primates are alive today. Humans and chimps share a common ancestor, and both groups are closely related—but neither descended from the other.

Section 16.7

1. In Scotland there were many instances of similar outbreaks of witchcraft while in Ireland there were few. People in Scotland and in Salem ate bread

made from rye that can be infected with ergot, while in Ireland bread was made from other grains.

Review Questions

True/False

1. F Sperm and ova are haploid. **2. T; 3. T; 4. T; 5. T; 6. T; 7. F** Skin is made of ectoderm; gut lining is made of endoderm. **8. F** Your belly button is located on your ventral surface. **9. F** Humans have bilateral symmetry. **10. T; 11. T; 12. F** Animals and fungi are both heterotrophs, but only animals ingest and then digest and absorb food. Fungi digest their food and then absorb it with hyphae. **13. T; 14. F** Race is not a biological reality. All human groups are very closely related.

Multiple Choice

15. e

Matching

16. d, **17.** a, **18.** c, **19.** e, **20.** f

Chapter 17

Quick Check Questions

Section 17.1

1. Most animals depend directly or indirectly on plants, so it is nearly impossible to think of one that does not. An animal that exclusively fed on algae, single-celled eukaryotes, or fungi would be one whose life did not depend on plants.

Section 17.2

1. DNA studies show that land plants and green algae are closely related. Also land plants and green algae share the same photosynthetic pigments.

Section 17.3

1. Land plants need cuticles for restricting water loss from their cells and stomata to allow gases to move in and out of the plants. Green algae do not need either cuticle or stomata because they live in watery places.

Section 17.4

1. If the green mass has plant organs like roots, stems, leaves, and vascular tissues, it is probably a land plant.

Section 17.5

1. Mosses are distinguished from ferns by their smaller size and densely packed "leaves" that hold water. Ferns have vascular tissues that include rhizomes and roots. Ferns have fronds that are not tightly bunched together like the densely packed "leaves" of mosses.

Section 17.6

1. The sporophyte of a magnolia tree is large and showy, while the gametophyte generation is tiny and inconspicuous and lives within the tissues of the sporophyte. **2.** Adaptations of a dandelion to life on dry land include cuticle and stomata, pollen, flowers, vascular tissue, and roots.

Section 17.7

1. The combination of factors in the peat bog that preserved Lindow Man included the acidic waters that inhibit bacterial growth in the bog, phenols excreted by *Sphagnum* mosses, and the layering of dead plant tissues that created the peat above the body of Lindow Man.

Review Questions

True/False

1. F If all of Earth's plants died off, animals would quickly die off too. **2. F** A plant is an autotrophic eukaryote that has chlorophylls *a* and *b* and carotenoids as accessory pigments. **3. F** There are many unicellular green algae. **4. T; 5. T; 6. F** Liverworts, hornworts, and mosses lack vascular tissue. **7. T; 8. F** Eggs and sperm of liverworts, hornworts, mosses, and ferns unite in female tissues. **9. T; 10. F** Cycads and ginkgos have seeds. **11. F** Ninety percent of all plant species are flowering plants.

Multiple Choice

12. b, **13.** d, **14.** c, **15.** b, **16.** a, **17.** b, **18.** a, **19.** e, **20.** d

Chapter 18

Quick Check Questions

Section 18.1

1. Ground tissue is made of parenchyma, sclerenchyma, and collenchyma cells. Parenchyma cells are the most abundant cells. They have thin cell walls and carry out a variety of plant functions, such as photosynthesis and food storage. Parenchyma cells can dedifferentiate, divide, and mature into mature parenchyma cell types. Collenchyma cells have thicker cell walls but are still flexible. They provide support and flexibility to the plant. Sclerenchyma cells are dead as mature cells. They have a secondary cell wall fortified with lignin and provide the plant with strength and support. **2.** Xylem cells come in two types: tracheids and vessel elements. Both stack one on top of another to form long tubes. The ends of tracheid cells have pits, which are regions where the secondary cell wall is absent. The ends of vessel element cells have perforations, where both the primary and secondary cell walls are absent. Both of these adaptations allow water to flow through xylem. **3.** Vascular cambium is a lateral meristem tissue that produces secondary growth, resulting in an increase in the width of the stem or trunk of the plant. The cells in vascular cambium divide to produce xylem and phloem. The cells that divide toward the outside of vascular cambium differentiate into phloem. The cells that divide toward the inside of the vascular cambium differentiate into xylem. The xylem accumulates over the life of the plant to produce wood.

Section 18.2

1. Phloem tubes are different from xylem tubes because phloem cells and companion cells are alive at maturity, while xylem cells are dead. Phloem transports dissolved sugars from photosynthetic sites to other plant parts; xylem transports water and dissolved nutrients up from roots. Liquids move in xylem drawn by transpiration. In phloem liquids are moved by active transport, osmotic water pressure, and bulk flow.

Section 18.3

1. When a plant is pinched back, the apical bud and its auxin is removed. This allows the auxin produced by lateral buds to influence the growth of lateral stems, and so a bushier plant develops instead of a taller plant. **2.** Auxins cause cells of stems to lengthen and affect other aspects of plant growth and development; gibberellin causes internodes to lengthen; ethylene causes fruits to ripen; abscisic acid slows down growth of plants, keeps seeds dormant, and closes stomata when conditions are dry.

Section 18.4

1. Plants obtain nitrogen from soils that contain it. Although nitrogen is plentiful in Earth's atmosphere, only some organisms can break the triple covalent bond in nitrogen molecules. Nitrogen-fixing bacteria found in soils can do this, and plants obtain surplus nitrogen in the form of nitrates that remain in the soil from the actions of these bacteria.

Section 18.5

1. Primary metabolites are chemicals produced by a plant that are used in growth and development. Secondary metabolites are varied chemicals used to defend the plant and communicate with other organisms. **2.** Secondary metabolites can make organisms that eat the plant sick, or can even make

them die. Secondary metabolites also can attract other organisms that eventually will kill herbivores. They also can inhibit the growth of other plants.

Section 18.6

1. Thujone affects the nerve cells that inhibit brain activity, so the brain becomes more active. **2.** The high alcohol content of absinthe acts as a sedative.

Review Questions

True/False

1. T; 2. T; 3. F Stomata open to let in carbon dioxide, and in doing so they release water vapor that has traveled upward in xylem. **4. F** Summer wood is typically darker and denser than lighter spring wood. **5. F** The companion cells of phloem are alive. Xylem cells are dead, and phloem cells are alive when mature. **6. F** Growth in length of a plant is called primary growth; growth in thickness of a plant is called secondary growth. **7. F** Plants do not use secondary metabolites for growth. Secondary metabolites function in defense and communication between plants. **8. T**

Multiple Choice

9. b, **10.** b, **11.** a, **12.** b, **13.** d, **14.** a

Matching

15. d, **16.** e, **17.** a, **18.** b, **19.** c, **20.** g

Chapter 19

Quick Check Questions

Section 19.1

1. Your diagram that demonstrates the concept of alternation of generations in the life cycle of a seed plant should look like Figure 19.1. **2.** Conifers and angiosperms share these reproductive structures: microspores, microgametophytes, pollen grains, pollen tubes, sperm cells, megaspores, megagametophytes, ovules, egg cells, polar bodies.

Section 19.2

1. Pollination is the transfer of pollen from anthers to stigma of angiosperms, or the arrival of pollen in the sticky sap around the ovules of an immature female cone. Fertilization occurs later, after the pollen tube has grown into the megagametophyte and sperm nucleus has united with the egg cell. **2.** Ways that pollination and fertilization of angiosperms and conifers differ include (a) pollination and fertilization of conifers take longer than pollination and fertilization of angiosperms; (b) in conifers pollen is transported by wind, while in angiosperms pollen is transported by wind, water, and animals; (c) different structures are involved: flowers in angiosperm and pollen and seed cones in conifers; and (d) angiosperms have double fertilization, while conifers do not.

Section 19.3

1. A flower that has a fragrance is advertising its presence in the environment to its pollinators. Some pollinators also may eat the fragrance and use it in their metabolic or reproductive processes. A pollinator that is especially drawn to a fragrance will be drawn to other flowers that have the same fragrance, and with each visit the insect will distribute pollen, cross-pollinating and increasing the genetic diversity of these flowers. Giving sweet rewards of nectar also ensures cross-pollination of individuals that offer it to their pollinators.

Section 19.4

1. "Dry sperm" refers to the fact that pollen is a means to transporting sperm to eggs without the use of water. The tough pollen coat protects sperm cells and prevents their drying out. This is an adaptation to life on dry land.

Section 19.5

1. Plants reproduce asexually by sending out new roots from parts that are removed from the main plant. This is what happens with plant cuttings. Rhizomes of mints and runners of strawberry plants are asexual reproductive methods, as are suckers and plantlets that develop on rims of leaves and fragmentation of plants like sagebrush. **2.** Selfing in plants means that their flowers pollinate themselves with their own pollen.

Section 19.6

1. By having its flowers located within the walls of its fruits where most pollinators cannot reach them, a fig is a secure location where young fig wasps can pass their early life cycle stages.

Review Questions

True/False

1. F Anthers form pollen grains that are immature microgametophytes. **2. F** Pollen and seeds have allowed plants to become fully terrestrial. **3. F** Pollination occurs before fertilization. They are not simultaneous events. **4. T; 5. F** Most seeds die before germinating and growing into new plants. **6. F** Pollen cones produce pollen. **7. F** Flowers of bird-pollinated plants typically have red flowers that open during the day. **8. F** At fertilization an angiosperm ovule is completely enclosed in tissue. **9. F** Carpels contain megagametophytes that eventually will give rise to eggs that if fertilized will develop into seeds that contain plant embryos. **10. T**

Multiple Choice

11. e, **12.** a, **13.** b, **14.** b, **15.** c

Matching

16. b, **17.** d, **18.** e, **19.** a, **20.** f

Chapter 20
Quick Check Questions

Section 20.1

1. The four body systems that coordinate your response to the environment are the nervous system, sensory system, muscular system, and skeletal system. The sense organs respond to signals from the external or internal environment and relay information to the nervous system; the nervous system analyzes the information and responds, often by sending signals to the muscles that move bones.

Section 20.2

1. In mechanoreceptors proteins are deformed, and this causes an ion channel to open, resulting in a response from a cell. Bacteriorhodopsin is a pigment that captures light. It is the ancestor of rhodopsin in animal eyes. **2.** The nervous system of a bilateral animal has a head end with a cluster of ganglia and chains of ganglia that go down the body. The design of a nerve net has a net of neurons connected to one another so that nerve tissue is distributed in 360 degrees. Animals with radial symmetry have no head ends.

Section 20.3

1. A sensory receptor cell is an entire cell adapted for responding to a stimulus, such as a taste bud that responds to a particular chemical stimulation. In contrast, a membrane receptor is a protein molecule embedded in the membrane of a cell. **2.** The retina is a sheet of light-sensitive cells at the back of the eye that contains rhodopsin and other light-sensitive pigments. Rods respond to intensity of light; cones respond to color wavelengths of light.

Section 20.4

1. Dendrites, cell body, and axon(s) are the three basic parts of a neuron. **2.** A synapse is a gap between two neurons where they communicate. **3.** Four regions of the central nervous system are (a) brain nuclei—such as the hypothalamus, which monitors eating, drinking, temperature control, endocrine glands—and the thalamus, which processes sensory information; (b) cerebral cortex—controls sophisticated information processing; (c) brain stem—the cerebellum, which controls motor functions; and (d) spinal cord—ganglia and axons of motor neurons and axons of peripheral nerves.

Section 20.5

1. An action potential is a brief change in the voltage across the cell membrane. Initiated by an electrical signal in the cell body, sodium channels open allowing sodium to rush into the cell from the outside. This causes the inside of the cell to change from about –70 mV relative to the outside to about +40 mV relative to the inside. Potassium channels then open, potassium moves out of the cell, and the –70 mV resting state is restored. **2.** At a synapse the electrical signal of the action potential is converted to a chemical signal that crosses the gap between neurons. The chemical signal interacts with membrane receptor proteins on the postsynaptic terminal and either increases or decreases the chances that the postsynaptic cell will fire an action potential. **3.** A neuron "decides" to fire an action potential based on the strength of the excitatory and inhibitory signals from various synapses. The neuron sums up these signals, and if the sum is positive enough to reach some threshold, then the neuron will fire an action potential.

Section 20.6

1. The two major proteins involved in muscle contraction are actin and myosin. The process of muscle contraction works by repeated cycles of making, moving, and breaking connections between these two muscle proteins. Heads of the myosin molecules use ATP energy to form, move, and break connections to myosin and are pulled toward the center of each myofibril, shortening the muscle. In response to a signal from a nerve, sodium flows into the muscle fiber, which in turn causes calcium to be released from the internal sarcoplasmic reticulum. The calcium causes the myosin head to contact the actin-containing filament, and the two sets of proteins slide past one another, causing the muscle to contract.

Section 20.7

1. Bone is different from other tissues in that bone cells secrete the hard calcium-containing matrix around themselves. The bony tissue envelops the collagen molecules that also are part of bone, and the nerve and blood supply of bone.

Section 20.8

1. Cocaine binds to and blocks the dopamine transporter. The transporter takes up dopamine from the synapse back into the presynaptic terminal, and so limits the actions of dopamine released into the synapse to a short time period. When the transporter is blocked, there is more dopamine in the synapse, and so dopamine has a larger and longer effect on the postsynaptic cell.

Review Questions

True/False

1. F A neuron's cell body contains nucleus, ribosomes, and mitochondria, but not dendrites or axons. **2. T; 3. F** An action potential can send an excitatory or an inhibitory signal. **4. F** Sensory receptors are nerve endings or modified neurons, and they can be unicellular or multicellular. **5. F** Cones in the retina detect color, while rods detect different levels of light. **6. T; 7. F** The brain stem controls heart rate, breathing, pain perception, attention, and arousal. The hypothalamus regulates body functions like eating, drinking, temperature, and endocrine glands. **8. F** The cerebral cortex is the largest component of the human brain. **9. T; 10. T; 11. T; 12. F** Ligaments bind bones together in joints. **13. F** When a muscle contracts, actin filaments are pulled more closely together. **14. T; 15. F** The sarcoplasmic reticulum stores and releases calcium ions that are necessary for muscle contraction. **16. T**

Multiple Choice
17. d; **18.** d, g, c, e, a; **19.** d; **20.** b

Matching
21. e, **22.** d, **23.** a, **24.** b, **25.** c

Chapter 21
Quick Check Questions

Section 21.1
1. Essential nutrients are nutrients that the body must have but cannot manufacture on its own. These include the essential amino acids. Macronutrients are those nutrients that humans require in relatively large amounts ranging from grams to kilograms daily. Carbohydrates, lipids, and proteins are macronutrients. Micronutrients are those nutrients that humans require in tiny amounts such as milligrams and less per day. Vitamins and minerals are micronutrients.

Section 21.2
1. Digestion, absorption, and metabolism are similar in that they all contribute to internal processing of macronutrients. The kind of processing differs. For instance, digestion chemically simplifies and reduces foods to nutrient molecules that are small enough to cross cell membranes. Absorption is the physical movement of nutrient molecules across cell membranes and into the bloodstream. Metabolism is the sum of all chemical processing by enzymatic pathways that harvests ATP from nutrient molecules. The three processes also differ in where they take place. Digestion occurs within the digestive tube, which is technically outside of the body; absorption occurs across the wall of cells of the small intestine; metabolism is the sum total of enzymatic pathways in all cells of the body.

Section 21.3
1. There isn't a minimum daily requirement for nucleic acids because every time a person eats a cell of another organism, the person consumes nucleic acids. If ingestion of macronutrients is normal, then the person will consume sufficient nucleic acids, and there is little danger of not getting enough of them. In addition, the body can make nucleotides that are not found in the diet. **2.** Habitual overating can lead to obesity, which, in turn, can lead to other diseases, especially diabetes. Habitually eating poorly can lead to undernutrition or lack of specific nutrients. Undernutrition can impair growth, and lead to loss of immune system functions so a person may be more susceptible to infectious diseases.

Section 21.4
1. Calcium is an ion that is important in many cellular processes. It is essential in muscle contraction and also is a crucial component of bones, teeth, and other tissues in the skeleton.

Section 21.5
1. First, you should consider whether the supplement has any side effects or any cross-reactions with any medicines you take. Second, you should consider whether there is actually a good reason to use the supplement. What evidence is there that it will achieve the result that you desire? You should look at the dosage of the supplement, compare it with the RDA for the supplement, and find out what evidence there is that the dosage is safe.

Section 21.6
1. In general, your diet can be improved by eating more fresh fruits and vegetables, less saturated fats, and more complex carbohydrates and limiting sweets, sodas, and fast foods. It's also good to eat more regular meals and include a wider variety of foods.

Section 21.7
1. Pellagra is caused by a niacin deficiency that can be traced to a diet that lacks fresh fruits and vegetables and fresh meats. Social conditions also are linked to pellagra because extremely poor people have a more limited access to a diet rich in nutrients than do people in better financial conditions.

Review Questions
True/False
1. T; **2.** T; **3.** T; **4. F** Malnourishment in infancy will have significant and potentially fatal results. **5.** T; **6. F** Villi and microvilli are found in the small intestine; they help absorb nutrient molecules into the bloodstream. **7.** T; **8. F** There is no evidence that vitamin C can prevent colds. **9. F** If you don't have proper amounts of micronutrients you will develop deficiency diseases. **10.** T

Matching
11. a, **12.** b, **13.** c, **14.** c, **15.** a, **16.** d, **17.** d, **18.** b, **19.** a, **20.** a

Chapter 22
Quick Check Questions

Section 22.1
1. Common homeostatic mechanisms that help the human body to rid itself of excess heat and maintain a stable body temperature include panting, sweating, flushing of face, and dilation of surface blood vessels.

Section 22.2
1. Active transport uses the energy in the final bond of the phosphate group of ATP to move a substance against its concentration gradient, while passive methods of transport move substances from high to lower concentrations, following their concentration gradients. Passive transport does not require ATP energy. **2.** A cell in a hypotonic environment will gain water and swell. A cell in a hypertonic environment will lose water and shrivel. A cell in an isotonic environment will be unchanged.

Section 22.3
1. The pumping action of the heart forcefully propels blood into the blood vessels, creating a forward momentum that helps push blood into the smallest capillaries. **2.** The path of blood through the heart is right atrium, atrioventricular valve, right ventricle, semilunar valve, pulmonary artery, capillaries of the lungs, pulmonary veins, left atrium, atrioventricular valve, left ventricle, semilunar valve, aorta, arteries, systemic capillaries, vena cava, right atrium.

Section 22.4
1. Chest muscles and diaphragm enlarge and constrict the volume of the chest cavity, creating negative or positive pressure on the lungs. The lungs inflate or deflate in response.

Section 22.5
1. The nephron is the major functional unit of a kidney. **2.** Blood plasma is filtered out in Bowman's capsule and enters the nephron. In the next segments of the nephron, water returns to blood, salt and water pass out of the nephron, water, salt, bicarbonate ions leave the nephron, while potassium and hydrogen ions enter the nephron. If ADH is present, water

leaves the collecting ducts of the nephron, producing a concentrated urine. If no ADH is present, water is retained in urine, producing a dilute urine.

Section 22.6

1. The first blood pressure number is the blood pressure in an artery when the heart contracts. The second number is the blood pressure in the same artery when the heart relaxes. These are the highest and lowest values for blood pressure for an individual under normal, resting conditions.

Section 22.7

1. Heart rate normally is controlled by the nodal tissue that sets the pace of the heartbeat. Nerve centers in the brain stem can slow or speed up the rate at which the nodal tissue fires. Hormones like adrenaline can speed up the heart. In cardiac arrest the heart beats chaotically, and blood is not pumped normally out of the heart. This deprives all tissues including brain and heart of blood supply. Or the heart muscle may stop altogether.

Review Questions

True/False

1. T; 2. F This describes a cell in a hypertonic solution. 3. F Osmosis is the diffusion of water across a plasma membrane. 4. T; 5. F When blood vessels dilate, blood pressure falls. 6. T; 7. F Gas exchange in lungs occurs in capillaries that surround alveoli. 8. F High pressure in the glomerulus forces plasma into the Bowman's capsule. 9. T; 10. T; 11. F When aldosterone is present, a dilute urine is produced.

Multiple Choice

12. b, 13. b, 14. d, 15. c, 16. a

Matching

17. c, 18. f, 19. d, 20. e, 21. a, 22. b

Chapter 23

Quick Check Questions

Section 23.1

1. Endocrine glands are different from other glands because they deposit secretions directly into the bloodstream, not into ducts that lead to some organ or other body surface.

Section 23.2

1. The hypothalamus monitors blood chemistry and through the anterior pituitary is the master control center for anterior pituitary hormones. The hypothalamus also is an endocrine gland, secreting posterior pituitary hormones and signaling their release.

Section 23.3

1. An anabolic steroid uses biochemical pathways that build biological molecules, while a catabolic steroid breaks biological molecules down. 2. Insulin and glucagon are metabolic opposites because insulin stimulates the removal of glucose from the bloodstream, while glucagon stimulates the conversion of glycogen into molecules of glucose.

Section 23.4

1. Type 1 and Type 2 diabetes are similar in that they are both disorders that affect the level of glucose in the bloodstream and are diseases associated with insulin supply. Type 1 diabetes is an early-onset immunological disease, while Type 2 diabetes is a late-onset disease associated with insulin resistance, sedentary lifestyle, and overweight. 2. The common problem in dwarfism, giantism, and acromegaly is with growth hormone. In dwarfism

there is too little GH in childhood, in giantism there is too much GH in childhood, and in acromegaly, there is too much growth hormone during adulthood.

Section 23.5

1. Your chart comparing formation of sperm and eggs should look like this:

	Formation of sperm	**Formation of eggs**
Length of time the process continues	Puberty to the end of life	Puberty to menopause
Number of mature reproductive cells produced each month	Millions	One or two
Pituitary hormones that control the process	LH, FSH	LH, FSH
Steroid hormones that influence the process	Testosterone	Estrogen, progesterone

Section 23.6

1. Cleavage changes the zygote into a blastocyst. It creates a multicellular organism and produces the correct cell size.

Section 23.7

1. An XX embryo normally doesn't develop into a male because she lacks the Y chromosome that influences the gonads to develop into a testis and secrete testosterone.

Section 23.8

1. A female embryo that is exposed to high levels of testosterone during development may display more male behaviors during childhood and may be more attracted to other girls than is a girl with normal testosterone levels.

Review Questions

True/False

1. F Not all hormones do this, some interact with DNA. 2. F The hypothalamus indirectly controls output of most hormones. 3. T; 4. F Secondary sexual characteristics include facial hair, deep voice, larger muscles, and larger size of males and enlarged breasts, fat deposits, higher voice, and rounded body shape of females. 5. F Too little growth hormone in childhood produces dwarfism. 6. F Adrenalin is released during periods of short emergencies, while cortisol is released during periods of long-term stress. 7. F In diabetes mellitus the pancreas does not make enough insulin. 8. T; 9. T; 10. T; 11. T; 12. F When an XX human embryo is 9 weeks old, its oogonia begin to develop into ova. 13. T 14. F An ovum actually completes meiosis after it has been fertilized. This occurs in the oviduct. 15. T; 16. F In the menstrual cycle only one ovary releases one egg per month. 17. F Only one sperm fertilizes the ovum. 18. T; 19. T; 20. T

Multiple Choice

21. c, 22. e, 23. a, 24. b

Matching

25. b, 26. c, 27. a, 28. d, 29. e, 30. f

Chapter 24
Quick Check Questions
Section 24.1
1. IVF is *in vitro* fertilization, the fertilization of an ovum by a sperm within laboratory glassware. **2.** IVF can successfully copy the beginnings of normal human development because for the first 2 to 4 weeks after fertilization, the embryo is free within the oviduct and only implants later. So if fertilization occurs in glassware, and the embryo is later introduced into the female's uterus, implantation and development will proceed normally after that.

Section 24.2
The steps of IVF are: inhibiting production of ova, stimulating production of multiple ova, harvesting ova and sperm, mixing them in a laboratory dish, incubating zygotes, examining zygotes, choosing embryos for implantation, inserting embryos into mother's uterus.

Section 24.3
1. The purpose of preimplantation screening of embryos is to look for genetic diseases and to remove embryos that have genes for diseases from the pool of embryos that may be implanted. **2.** The part of the DNA of an embryo that carries a particular gene is screened. **3.** A transgenic organism is one that has had one or more genes from another organism inserted into its genome, and the transgenic organism expresses that gene.

Section 24.4
1. It is difficult to clone mammals because, unlike frogs, their development occurs within a mother's uterus. There are other difficulties that are incompletely understood.

Section 24.5
1. The reproductive technologies involved in saving Jane Doe's life include five IVF attempts, preimplantation screening of embryos to find one with a matching blood type, and preimplantation screening of embryos who had no Fanconia anemia.

Review Questions
True/False
1. F Production of sperm begins at puberty and continues until death.
2. F Production of human eggs begins before birth, is suspended until puberty, and continues until menopause. **3. F** IVF is an expensive, tricky, somewhat risky process that is not 100% sure. **4. T; 5. F** Implanting more than two embryos does not increase the chance of a healthy live birth. **6. T; 7. T; 8. F** Gene therapy has yet to be proven safe. **9. F** Fertilization occurs within an oviduct. **10. T**

Multiple Choice
11. a, **12.** c, **13.** c, **14.** b, **15.** a

Matching
16. b, **17.** d, **18.** e, **19.** a, **20.** c

Chapter 25
Quick Check Questions
Section 25.1
1. Black death, influenza, and smallpox are the three infectious diseases that killed the most people worldwide.

Section 25.2
1. Viruses multiply within cells and take over cells to do this. Bacteria also multiply within multicellular organisms often secreting toxins that also damage cells. **2.** Koch's postulates were important because they set up a rigorous scientific procedure to identify the agent causing a disease. **3.** A retrovirus has RNA, not DNA. Viral DNA is incorporated into the host's genome and when it is transcribed, it makes viruses.

Section 25.3
1. The chronological events of a typical immune response are as follows: injury; mast cells release histamine; capillary walls become more permeable; lymph and white blood cells leak out to tissues; local temperature increases; and movements of immune cells speed up; cytokines attract more white blood cells; white blood cells engulf invaders; some white blood cells are antigen-presenting cells, and they go to the lymph nodes, where they cause helper T cells to bind and form a clone; helper T cells cause cytotoxic T cells to divide; helper T cells cause B cells to divide, mature and form antibodies; some of the B cells become memory cells. This process takes between 7 and 10 days.

Section 25.4
1. In antibiotic resistance pathogens have become immune to the effects of an antibiotic that used to kill them. It develops because people overuse or underuse antibiotics. It is an example of natural selection because the bacterial cells that carry genes for resistance to antibiotics pass those on to their daughter cells. In contrast, bacterial cells that do not have resistance to antibiotics die and do not reproduce.

Section 25.5
1. Reverse transcriptase inhibitors are similar in size and shape to normal nucleotides. They are put into the growing DNA strand by the reverse transcriptase molecule, but they stop the enzyme from further copying the genetic RNA molecule of the virus. In this way reverse transcriptase inhibitors stop production of HIV by preventing it from generating viral DNA. Protease inhibitors block HIV replication because proteins produced are not cut into small enough portions to fit within a virus. Viral particles cannot assemble in the host cell because the required proteins are not present.

Section 25.6
1. The scientifically accepted hypothesis about the origin of HIV is that it passed from the blood of slaughtered chimps to humans in Africa sometime in the 1930s.

Review Questions
Multiple Choice
1. a, **2.** c, **3.** d, **4.** d, **5.** c, **6.** a, **7.** e, **8.** e, **9.** d, **10.** c

Matching
11. f, **12.** d, **13.** b, **14.** e, **15.** c, **16.** c, **17.** e, **18.** a, **19.** b, **20.** c

Chapter 26
Quick Check Questions
Section 26.1
1. Environmental activism focuses on combating all forms of environmental destruction and increasing public awareness of environmental destruction; ecology focuses on the science of the effects that populations have on one another.

Section 26.2
1. Characteristics of populations include typical life tables, survivorship curves, age structures, and rates of reproduction. The characteristics of a

species that differ from those of a population include biotic potential (populations may not be able to achieve it in their individual environments), whether it is *r*- or *K*-selected (this refers to a whole species); typical distribution and typical habitat also are more characteristic of species and usually don't vary from population to population.

Section 26.3
1. You may see no difference, or you may be concerned about greater demands on Earth's natural resources and destruction of resources and natural environments.

Section 26.4
1. Alien species often are able to exclude all competitors and take over areas because their natural population controls such as predators, parasites, and competitors are lacking in the environments where they have been introduced.

Section 26.5
1. Primary and secondary succession are similar in that both are patterns of community change. Both are natural processes. They are different in that primary succession takes place where no plants have previously grown, while secondary succession takes place where plants previously have grown.

Section 26.6
1. Habitat fragmentation and deforestation are similar in that both involve the loss of habitat. It gets chopped into smaller parts in habitat fragmentation and may be completely destroyed in deforestation. Habitat fragmentation allows cowbirds paths to invade because it increases the number of edge habitats in forests. As a result, cowbird numbers increase. Deforestation depletes songbird populations by decreasing their feeding and nesting resources. This makes individual songbirds less able to survive migration and the rigors of raising young.

Review Questions

True/False
1. T; **2.** T; **3. F** Plants function as producers in ecological communities. **4. F** Humans and the gut bacteria that synthesize vitamin K are in a mutualistic relationship. **5. F** Plants growing in landfills is an example of secondary succession. **6. F** Songbirds are declining because of deforestation, habitat fragmentation, cowbird parasitism, and increased predation.

Multiple Choice
7. b, **8.** c, **9.** e, **10.** c, **11.** d, **12.** b, **13.** a, **14.** b, **15.** d, **16.** b

Matching
17. c, **18.** d, **19.** b, **20.** a

Chapter 27
Quick Check Questions

Section 27.1
1. Your answer should include various organisms (animals, plants, bacteria, fungi, and perhaps even archaea) as the biotic components of the ecosystem. Abiotic components will include atmosphere, soil, rocks, and may include waters.

Section 27.2
1. This chart compares long- and short-term processes of four biogeochemical cycles.

Biogeochemical cycle	Short-term processes	Long-term processes
Carbon cycle	Respiration; photosynthesis; decay; incorporation of carbon compounds into corals, bones, shells, exoskeletons	Formation of fossil fuels, formation of limestone
Nitrogen cycle	Nitrogen-fixing by bacteria, processing to nitrates by various bacteria, return of nitrogen to atmosphere by bacteria, assimilation of nitrogen into biological molecules	
Water cycle	Evaporation, condensation, precipitation, transpiration	Deposition of water into underground aquifers, runoff of waters to oceans, circulation of water in oceans
Phosphorus cycle	Incorporation of phosphorus-containing compounds into plant and animal tissues, decay of both and return of phosphorus to soil	Erosion, sedimentation, and rock formation

Section 27.3
1. Life in ocean ecosystems depends on populations of phytoplankton whose numbers are controlled by the amounts of nutrients available and the amount of light. Places in the oceans where nutrients and light are plentiful tend to have large populations of phytoplankton. These spread a banquet of food for zooplankton and a host of other consumers.

Section 27.4
1. Life in still freshwater is shaped by amounts of dissolved oxygen and water temperatures that seasonally produce stratified layers that can become oxygen-depleted. Life in flowing water has plenty of oxygen but must have behavioral or structural means to avoid being swept away in the current.

Section 27.5
1. Important aspects of the tropical rain forest biome include its high diversity of species, many of which are not found anywhere else. Tropical rain forests are layered forests with a nearly continuous canopy over a dim forest floor. The tropical rain forest is so extremely productive when other biomes are less productive because it has uniform conditions of temperature, high rainfall, high humidity, and little seasonality.

Section 27.6
1. The biosphere is all of Earth's life and the physical environments that support it. You can see hints of the biosphere from space in the green of continents and in seasonal changes that you can observe in amount of chlorophyll in the oceans.

Section 27.7
1. Three lessons that can be learned from the story of Lake Erie are (a) habitats that are ecologically damaged can be rehabilitated, (b) a combination of governmental legislative approaches has been effective in cleaning up badly polluted areas, and (c) many sources were involved in the pollution of Lake Erie.

Review Questions
True/False
1. F Rocks, soil, atmosphere, nutrient cycling, water, and organisms are characteristic components of *ecosystems.* **2. T; 3. F** The most varied forms of life are found in oceanic ecosystems. **4. F** Upwelling, coastal runoff, and abundant light make the *neritic* zone highly productive. **5. T; 6. F** Hydrothermal vent communities depend on *chemosynthetic bacteria.* **7. F** As streams merge into rivers, levels of oxygen *decrease,* while turbidity *increases.* **8. T; 9. F** Most tropical rain forest animals are *unidentified and unknown.* **10. F** The biosphere is made of all of Earth's *biotic features and all the abiotic features that support life.*

Multiple Choice
11. d, **12.** d, **13.** d, **14.** b, **15.** b, **16.** d, **17.** a, **18.** c, **19.** b, **20.** b

Matching
21. b and c, **22.** b and d, **23.** b and a, **24.** b and f, **25.** b and g, **26.** b and e

Chapter 28
Quick Check Questions
Section 28.1
1. In overpopulation a population has grown beyond the carrying capacity (*K*) of the environment to support it. **2.** Some human populations have reset *K* through advances in agriculture, hygiene, medicine, and technology.

Section 28.2
1. The similarity in the impact of human hunting of moas, passenger pigeons, bison, cod and other ocean species is that when hunting is unrestricted by cultural practices or legislation, humans will hunt animals to extinction.

Section 28.3
1. Here is a list of some of the "ripple effects" of logging tropical rain forests: soils are exposed → soils erode → silt accumulates in waterways → may smother coral reefs; soils become hardened and dry → plants may not recolonize → animals desert the area → may not find new homes → species diversity and species number are diminished; native people's lives are disrupted.

Section 28.4
1. An outline of lesson plan for third-graders on acid rain could include the following: **a.** Animals and plants need water with the right chemicals in it.

b. Industries like power plants can add chemicals to air. **c.** Some of these chemicals combine with water in clouds to change the nature of rain, snow, sleet, or hail. They make it more *acidic.* **d.** Explain what an acid is on a simple level. Collect rainwater and test pH with test strips. **e.** You can see acid rain in damage to some stone statues and buildings. **f.** Acid precipitation also kills animals, acidifies lakes. **g.** Acid precipitation also can kill whole forests. **h.** Students discuss what might happen to the animals in the forest killed by acid rain. **2.** Predatory birds at the top of land-based food chains aren't as affected by bioaccumulation of DDT because although they do receive doses of it, they are parts of shorter food chains, so they don't receive as much as do predatory birds at the top of water-based food chains that involve more animal links and thus more DDT bioaccumulation.

Section 28.5
1. Environmental legislation protects species by providing protected refuges, regulating hunting, and preventing pollution. There is a good correlation between passing and implementation of the Clean Water Act and the Clean Air Act and reductions in levels of major pollutants.

Section 28.6
1. Habitat destruction and hunting for the bushmeat trade are driving chimpanzees to extinction. **2.** Bonobos have female-dominant societies that rarely engage in aggressive interactions while common chimpanzees have male-dominant societies which show aggression between individuals, and sometimes attack other chimp groups.

Review Questions
True/False
1. T; 2. F Moas were large, flightless *birds* that once lived in *New Zealand.* **3. F** Passenger pigeons were *Earth's most numerous bird;* Americans hunted them to extinction. **4. T; 5. F** Species of fishes whose numbers are declining are called *overfished.* **6. T; 7. T; 8. F** Acid precipitation can kill invertebrates, frogs, salamanders, fishes, *trees, and whole forests.* **9. T; 10. T; 11. F** DDT *is still* in the environment and continues to be an environmental problem. **12. F** The hole in the ozone layer allows *harmful energy* to enter Earth's atmosphere. **13. F** Legislation *is effective* in curbing humanity's negative impact on the environment. **14. T; 15. T; 16. F** The development of tools, agriculture, and science and modern technology have helped to *reset* K *for humans.* **17. F** Rain forest soils are *unsuited* for agriculture. Cutting down rain forests has led to loss of biodiversity *that nothing can compensate for.* **18. T; 19. T; 20. T**

Appendix B

Table of Amino Acid Structures

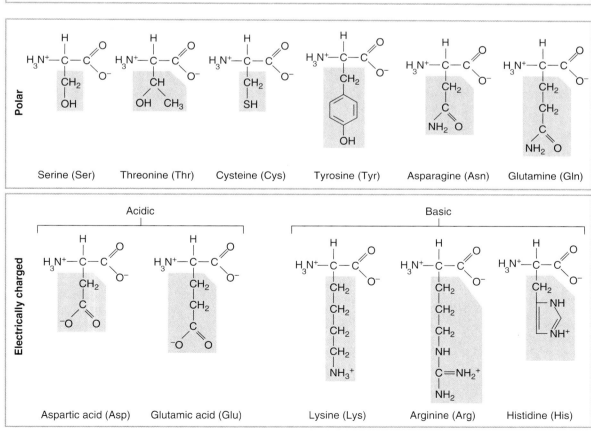

Amino Acids Can Be Grouped According to the Change of the R Group.

Nonpolar: Glycine (Gly), Alanine (Ala), Valine (Val), Leucine (Leu), Isoleucine (Ile), Methionine (Met), Phenylalanine (Phe), Tryptophan (Trp), Proline (Pro)

Polar: Serine (Ser), Threonine (Thr), Cysteine (Cys), Tyrosine (Tyr), Asparagine (Asn), Glutamine (Gln)

Electrically charged

Acidic: Aspartic acid (Asp), Glutamic acid (Glu)

Basic: Lysine (Lys), Arginine (Arg), Histidine (His)

Glossary

A

abiotic nonliving components of the environment

abscisic acid a plant hormone that slows growth of a plant, causes seeds to become dormant, and closes stomata during dry conditions

absorption the movement of digested nutrients across cell membranes and into the bloodstream

acetylcholine a neurotransmitter used in motor synapses

acid a solution with a higher concentration of H^+ ions than plain water

acquired immune deficiency syndrome (AIDS) a disease defined by a collection of rare infections caused by a collapse of the immune system; AIDS is not an inherited disease

action potential an electrical signal that travels along a neuron

activation energy the amount of energy needed to start the chemical reactions of making or breaking bonds

active transport the process in which a protein uses energy to move ions into or out of a cell toward the side of higher concentration

adaptation an inheritable trait that allows an individual with the trait to survive and reproduce in a certain environment

adhesion the tendency for cells to stick together

ADP (adenosine diphosphate) a nucleotide molecule containing two phosphate groups that is generated when ATP is used by an enzyme to power a chemical reaction

adult stem cells adult cell that divides repeatedly to produce a long line of daughter cells, each of which produces a restricted set of one or a few adult cell types

aerobic metabolic pathways and conditions that use oxygen

age structure a graph that shows the distribution of ages of individuals in a population

agonist a drug that binds to a membrane receptor and has the same effect as the normal signal or neurotransmitter

algal bloom an explosion in the population of one or more species of photosynthetic, single-celled eukaryotes

alleles different versions of genes that control the same trait but are not identical

alternation of generations a life cycle that alternates between a multicellular haploid generation and a multicellular diploid generation

amino acid a molecule with a central carbon to which an amino group, a hydrogen atom, a carboxyl group, and an R-group are attached

amino group a functional group derived from ammonia that has a nitrogen and two hydrogen atoms

anabolic steroid a hormone that is a derivative of testosterone; anabolic steroids often are taken by athletes to increase muscle mass and improve athletic performance

anaerobic metabolic pathways and conditions that do not use oxygen

anal pore a membrane pore through which wastes are expelled from *Paramecium*

androgen insensitivity a condition in which testosterone receptors do not bind well to androgens; thus the hormones do not have normal effects on target tissues

anemia a condition caused by too few red blood cells, or too little hemoglobin

angiogenesis the growth of blood vessels

animal a multicellular, aerobic eukaryote that obtains energy by ingesting food

anion a negatively charged ion

antagonist a drug that binds to a receptor and has no effect, blocking the normal signal or neurotransmitter function

anterior end the head end of an animal

antibiotic an antibacterial substance that kills bacteria produced by other bacteria or molds

antibiotic resistance the ability of some bacterial strains to avoid being killed by a particular antibiotic

antibody a protein that floats freely in tissue fluids, binds to an antigen and speeds the destruction of pathogens

anticodon a three-base sequence that is the complement of the codon for the specific amino acid attached to a tRNA molecule

antigen a molecule or part of a molecule from a foreign organism that antibodies or cells of the immune system can bind to

antigen-presenting cell any immune system cell that acts like a macrophage and displays foreign proteins from pathogens on its surface

apical meristem meristem tissue at the tips of shoots or roots that allows for growth in length of shoots and roots

apoptosis the automatic process that causes the death of abnormal cells

aquifer a porous layer of underground rock or sand that holds large quantities of water

Archaea a domain of prokaryotes characterized by the nature of the DNA that codes for ribosomal RNA and other traits

arctic tundra a northern biome characterized by extremely cold temperatures, restricted precipitation, permafrost, short summers with extremely long days, and low, ground-hugging vegetation primarily composed of mosses, lichens, and grasses

artery a blood vessel that carries blood away from the heart

artificial selection the production of a new strain from an existing species by controlling which individuals mate and produce offspring

asexual reproduction a form of biological reproduction in which a single parent cell or parent organism produces an offspring whose genetic makeup is identical to the parent's

atavism   a feature of an individual that reflects an ancestral structure or form

atom   the smallest unit of a chemical element that has all of the properties of that element

ATP (adenosine triphosphate)   a nucleotide molecule containing three phosphate groups that is a common energy source for living organisms

autoimmune disease   a disease in which the immune system attacks the body's own cells

auxin   a plant hormone that causes cells of stems to lengthen and affects other aspects of plant growth and development

axon   a thin extension that leads away from the cell body of a neuron and carries information to other neurons

B

B cell   a cell of the acquired immune system that produces antibodies

Bacteria   a domain of prokaryotes defined by the nature of the DNA that codes for ribosomal RNA and other traits

bacterial conjugation   the joining of two prokaryotic cells by a short cytoplasmic bridge through which DNA is exchanged

bacterial spore   a resting stage of some bacteria that allows the organism to survive poor environmental conditions

bacterial transformation   process in which living prokaryotes take up DNA from the environment and incorporate it into their own genomes

basal ganglia   small group of neurons above the brain stem that is involved in motor behavior and perception of reward

base   a solution with a lower concentration of H^+ ions than plain water

benign tumor   a mass of cells that is confined to a single area and can often be entirely removed by surgery

bilateral symmetry   a kind of symmetry in which identical halves result only from a cut along the midline of an object or an organism

binary fission   the process of a prokaryotic cell dividing into two; each daughter cell receives a copy of the parent's chromosome

bioamplification   the process by which a toxic compound becomes increasingly concentrated as it moves up a food chain

biodegrade   to break down by biological means such as the effects of bacteria, fungi, or other organisms

biodiversity   the total diversity of Earth's life, including genetic, species, and ecosystem diversity

biogeochemical cycle   the circulation of a chemical as it passes through biotic and abiotic components of the biosphere

biological molecules   complex chemicals unique to living organisms but not commonly found in nonliving natural systems

biological species concept   defines a species as a group of individuals that naturally interbreed and produce fertile offspring

biology   the study of life

biome   a regional ecosystem characterized by dominant vegetation and controlled by a similar climatic regime

biophilia   the natural affinity that humans have for nature

bioremediation   the general term for the use of biological organisms to clean up waste sites produced by human activities

biosphere   the global ecosystem composed of all of Earth's ecosystems and biomes

biotechnology   techniques used to isolate, characterize, manipulate, and control biological molecules such as DNA or proteins

bisexual   sexually attracted to both sexes

blood   a fluid tissue that transports needed substances to each body cell and removes wastes

blood clotting   the process of blood thickening that prevents blood from escaping when blood vessels are injured

blood plasma   the liquid component of blood that contains suspended blood cells and dissolved substances

bone   a tissue made of living cells, extracellular collagen, and calcium deposits that forms the hard, supportive skeleton of vertebrate animals

brackish water   a mixture of salt water and freshwater that occurs in an estuary

brain   the large mass of nervous system tissue that sits atop the spinal cord, inside the skull

brain stem   the structure at the base of the brain, just above the spinal cord, that governs basic body functions

C

calorie   a standard unit of energy used to express the amount of energy available in food

cancer   a disease involving a mass of cells produced by abnormally high levels of cell division; cancerous cells can detach from their neighbors and invade other tissues

canopy   the tops or crown layer of forest trees

capillaries   the smallest blood vessels

carbohydrate   a sugar or starch molecule that contains carbon, hydrogen, and oxygen—often in a ratio of 1:2:1

carboxyl   a chemical group with one carbon atom, two oxygen atoms, and sometimes a hydrogen atom

carcinogen   a chemical implicated in the formation of cancers

cardiovascular disease   any one of many diseases affecting the heart and/or blood vessels

carnivore   a predator that eats other animals

carpel   the innermost ring of modified leaves of a flower; carpels are the ovule-producing female structures of a flower, including the ovary and its contents, as well as style and stigma; can be a simple or a compound structure

carrying capacity (K)   the ability of an environment to provide the conditions necessary to sustain and support a population of a certain size

cation   a positively charged ion

cell   the smallest structural unit of life

cell body   the central portion of a neuron where mitochondria, ribosomes, endoplasmic reticulum, nucleus, and other organelles are located

cell culture   a population of cells growing in a laboratory dish or tube

cell cycle   the orderly sequence of events in the life of a cell that includes the processes of cellular function and cell division

cell differentiation   the development of specific cell traits that occurs during the growth of an embryo

cell membrane   the boundary that separates a cell from other cells and from the environment, while it allows the cell to communicate with its environment

cell proliferation   repeated rounds of mitotic cell division that increase the number of cells in an organism, or in a part of an organism

cell theory   the cell is the universal, basic structural unit of organisms; cells arise only from preexisting cells

cell wall   a rigid supporting structure that is outside of the cell membrane

cellular immunity   the aspect of the acquired immune system in which infected cells are attacked and killed by lymphocytes dedicated to that purpose

cellular respiration   the combined processes of the citric acid cycle and oxidative phosphorylation

central nervous system   the parts of the nervous system that are enclosed in the spinal column and the skull

centriole   a pair of ridged, cylindrical organelles found in animal cells that give rise to the spindle apparatus during cell division

centromere   the point on a duplicated chromosome where the two sister chromatids join

cerebellum   a large, convoluted structure sitting astride the brainstem that is in involved in complex, learned motor behaviors

cerebral cortex   a large convoluted structure that covers most of the brain in mammals; higher functions are governed by the cerebral cortex

chemical bond   the electrical interactions that hold two atoms or ions together

chemical catalyst   a chemical that speeds up a chemical reaction but does not take part in that chemical process

chemical element   a basic kind of matter

chemical reaction   an interaction between atoms or molecules that transforms substances from one kind into another through the making and breaking of chemical bonds

chemotherapy   the use of a chemical, or drug, treatment to kill cancer cells

chlorophyll   the pigment that helps capture the energy in sunlight to provide the power for photosynthesis

chloroplast   an organelle within cells of green plants and certain other

organisms where the energy in sunlight is captured and stored in molecules that provide the energy to produce glucose

chromosome   DNA and associated proteins found in the nucleus of a eukaryotic cell and in the cytoplasm of a prokaryotic cell; chromosomes carry genetic information; each organism has a characteristic number of chromosomes

circulatory system   the heart, blood vessels, and blood that collectively transport substances to cells and remove wastes from cells

citric acid cycle   the chemical process that extracts energy from the two-carbon acetyl CoA molecule and provides energy-rich molecules for the process that makes ATP

classification   the process of sorting things based on their similarities and differences

cleavage   the first few cell divisions of an embryo that have no growth between divisions

clone   genetic replica of an individual organism or cell

cochlea   the complex structure in the ear that detects and analyzes sounds

codominance   the type of inheritance in which the traits of both true-breeding parents are seen in each offspring

codon   a sequence of three nucleotides that codes for an individual amino acid

coenzyme   a small chemical that binds to an enzyme and is required for the enzyme to catalyze a specific reaction

coevolution   the mutual evolutionary influences that pairs of species have on one another

collagen   a large protein that forms long, elastic strands used in bone and many other body structures

colony   a group of cells that are held together by some external matrix but that retain the identity of individual cells

commensalism   a one-sided symbiotic interaction between two species in which only one benefits from the interaction, while the other is not harmed

community   interacting populations of different species that live in a local area

competition   a interaction that occurs when two or more individuals need the

same resource and each works to acquire it

competitive exclusion   the complete dominance of one species over another, to the point where the population of the non-dominant species approaches zero

complement   a set of proteins in blood that work together to kill cells that have antibodies attached to them

compound   a substance made of the atoms of two or more elements

compound microscope, or light microscope   a scientific instrument that uses lenses to bend light rays in a way that makes objects appear closer to the eye and therefore larger

condensation   the process in which water molecules come close together, form hydrogen bonds, and turn into liquid water or ice

cone   a light-sensitive cell in the retina that detects different colors of light; in plants a specialized reproductive structure of a conifer that produces pollen or seeds

consumer   an organism that does not produce its own food supply but obtains it by eating other organisms

contractile vacuole   an organelle in *Paramecium* that regulates cellular water balance

control group   any group or condition included in a study to rule out other interpretations

controlled experiment   a scientific study in which some subjects (experimental subjects) are assigned to experience a specific experimental condition, while similar subjects (the controls) do not experience the experimental condition

correlation   when two events reliably happen together in space and time

covalent bond   a bond between two neutral atoms that involves the sharing of two or more electrons

crosses   the mating of selected individuals (usually of different kinds) to reveal the rules underlying inheritance

crossing over   the process in which the alleles on homologous chromosomes switch places, producing new combinations of alleles on each chromosome involved

curare   any compound of biological origin that paralyzes vertebrates

cuticle   a protective layer on the outer surface of plants that prevents water loss

cyanobacteria photosynthetic bacteria able to use energy from sunlight to convert carbon dioxide and water to sugar

cynodont an extinct early mammal-like species from about 250 million years ago that had skeletal characteristics that closely resemble those of modern mammals

cytokinesis the cell cycle phase immediately after mitosis, when the cell with separated chromosomes is split into two daughter cells

cytoplasm the organelles and cytosol taken together

cytoskeleton a network of protein fibers that lends support and shape to a cell

cytosol a jellylike substance within a cell that contains various molecules and ions

D

daughter cells the two cells produced when a parent cell divides

denaturation when the hydrogen bonds that hold a protein together are broken, and the protein's three-dimensional shape is lost

dendrite an extension of the neuron cell body that usually receives input from other neurons

denitrifying bacteria bacteria that transform nitrites and nitrates and release nitrogen to the atmosphere

density-dependent population control a population control factor that increases in intensity as population size increases

density-independent population control a population control factor that acts without regard to the size of a population

desert a biome characterized by sparse moisture, great daily fluctuations in temperature, poor soils, and sparse vegetation with adaptations to conserve water

detritivore an organism that eats dead tissues of plants or animals

diabetes mellitus a condition in which the body either does not produce enough insulin (Type 1) or does not respond to insulin (Type 2)

diastolic blood pressure the pressure exerted against artery walls when the left ventricle is not actively contracting, and the pushing force is least

diffusion the movement of a substance from a place where it is more concentrated to a place where it is less concentrated

digestion the process that breaks down nutrients into small molecules

diploid a cell or organism that carries two complete sets of each homologous

directional selection individuals at one extreme or another on a trait do not survive and breed, so the population average moves in one direction or another

disaccharide a carbohydrate with two subunits

disruptive selection individuals with values of a trait around the mean do not survive and breed, so the population is pushed into two groups, each with extreme values of the trait

diuretic a substance that increases the excretion of water by the kidneys

DNA (deoxyribonucleic acid) the molecule that carries the instructions for life

DNA fingerprint an analysis of a person's DNA that can be used to uniquely identify that person

DNA helicase the enzyme that catalyzes the unzipping of the DNA molecule by breaking the hydrogen bonds between the paired bases

DNA ligase the enzyme that binds short segments of DNA together during DNA replication of the 5′ to 3′ strand

DNA microarray a glass slide that carries thousands of dots of single-stranded DNA, used to detect complementary DNA or mRNA in a tissue sample

DNA polymerase the enzyme that brings new nucleotides into place to form a new strand of DNA on a single unzipped strand

domain a phylogenetic group of organisms identified by DNA analysis

dominant allele an allele expressed regardless of the other allele being carried for that trait

dorsal surface the back or upper surface of an animal

duplicated chromosome a joined set of chromosomes; each has two copies of a strand of DNA, or a pair of sister chromatids

E

ecological niche a species' habitat, plus all of the abiotic and biotic resources it requires for its life and survival

ecology the study of the interrelationships of organisms and their environments that affect the distribution and abundance of organisms

ecosystem communities and the nonliving aspects of their environments

ectoderm outer layer of cells of an embryo

electrical force the force that causes particles with opposite charges to be mutually attracted, and the force that causes particles with a similar charge to repel one another

electromagnetic force the force that causes particles with opposite charges to be attracted to one another, while particles with the same charges repel one another

electron microscope (EM) a scientific instrument that uses a beam of electrons to give greater magnification than can be obtained with a light microscope

electron transport chain a series of proteins that handles energy-laden electrons by passing them from one protein to another and releasing some of their energy with each step

electron a negatively charged subatomic particle that moves around an atom's nucleus

element a substance that cannot readily be broken down into other substances by ordinary processes such as burning, evaporation, or filtration

embryonic stem cells embryonic cell that divides repeatedly to produce a long line of daughter cells, each of which has the ability to differentiate into many adult cell types

emergent properties the principle that simple components interact to produce more complex systems with more complex traits and behaviors

endemic species species that have a restricted geographic range and are found only in a small area

endocrine gland a gland that secretes hormones directly into the bloodstream

endocrine system all the glands that produce hormones

endoderm inner layer of cells of an embryo

endometrium the lining of the uterus that builds up each month and may be shed as menstrual fluid

endoplasmic reticulum a network of membranous tunnels in the cytosol that comes in two varieties, smooth endoplasmic reticulum and rough endoplasmic reticulum

endoskeleton a hard structure inside an animal that provides body support and allows movement

endosperm a nutrient-rich tissue within an angiosperm seed that nourishes the plant embryo

energy the ability to do work

enzyme a protein or nucleic acid that acts as a catalyst to speed the chemical reactions inside a living organism by lowering the activation energy of a specific chemical reaction

epidermal tissue the outer layer that covers a plant's surface and protects it from injury, but is lost in woody plants

essential nutrients nutrients that are necessary for life but cannot be manufactured by an organism

estrogen female hormone

estuary a place where fresh and salt waters mix; generally where rivers enter the oceans

ethylene a plant hormone that causes fruits to ripen

eugenics the philosophical and political belief system that asserts that the human race can and should be improved by selective breeding

Eukarya the domain of eukaryotic life defined by the nature of the DNA that codes for ribosomal RNA and other traits

eukaryotic cell a cell that has organelles surrounded by membranes

eutrophication the process in which a body of water becomes enriched with nutrients

evaporation the process in which molecules of liquid water are transformed into water vapor

evolution the change in the traits of a population of organisms over generations, resulting from changes in the frequencies of alleles found in that population

excretory system the kidneys and associated organs and tubes

responsible for removing wastes from the blood and eliminating them from the body in urine

exon the portion of a gene that actually codes for a protein

exponential population growth a geometric rate of increase of a population, shown in a J-shaped curve of population growth

extracellular fluid fluid around cells and blood plasma collectively

F

facilitated diffusion the diffusion of ions across the cell membrane through a protein channel

fermentation reaction an anaerobic metabolic pathway, such as glycolysis, that does not use electron transport

fertilization union of sperm and egg

fiber indigestible complex carbohydrate found in fresh fruits and vegetables

filter feeding a process that moves large amounts of water through an organism and filters out food particles

filter-feeder an animal that strains its food out of the environment, using a special feeding organ that water flows over

first filial generation (F_1): the offspring that result from the mating of the parent generation

first law of thermodynamics energy cannot be created or destroyed, only changed in form

flagellum (plural : flagella) a tail-like cytoplasmic extension used to propel a cell

flower a specialized reproductive structure of an angiosperm that produces pollen and/or ovules that develop into seeds

food chain the transfer of food energy from producer through various levels of consumers

food web all of the food relationships within a community

forensics the application of scientific knowledge and techniques to legal issues

fossil a naturally preserved trace of an organism

fossil fuel a hydrocarbon fuel such as peat, coal, oil, and natural gas

founder effect when a small number of individuals with limited genetic

diversity are isolated from the larger population of the species and become the ancestors of a new species

fruiting body a showy structure involved in the reproduction of a fungus

functional group a group of atoms that is bound to an organic molecule and influences the properties of that molecule

fungi mostly multicellular, nonphotosynthetic eukaryotes that digest food outside of their bodies and absorb digested nutrients

G

G_0 (G-zero) the phase of the cell cycle immediately after cell division; some cells are permanently mature and never return to cell division

G_1 (gap-1) the phase of the cell cycle between mitosis and DNA synthesis when a cell carries out diverse, mature cellular functions

G_2 (gap-2) the cell cycle phase immediately after S-phase, when the cell biochemically prepares itself for division

gamete a reproductive cell with a haploid number of chromosomes

gametophyte a life cycle phase in a plant that has alternation of generations, a gametophyte is a haploid plant that produces gametes by mitosis

ganglion (plural: ganglia) a cluster of neurons that have connections with one another

gastrulation the movement of cells in an early embryo that sets up the three embryonic cell layers of ectoderm, mesoderm, and endoderm

gel electrophoresis a process used to separate different-sized pieces of DNA

gender identity the expression of gender-related behaviors and thoughts, regardless of the genetic sex of male or female

gene expression the process of expressing specific protein from DNA

gene mutation a change in the nucleotide sequences in the DNA of a cell

gene therapy a general term for the replacement of a defective gene with a normal one in the affected tissues

gene a sequence of DNA that codes for a single protein

genetic code the complete list of translations between each triplet codon on mRNA and the amino acid each one represents

genetics the scientific study of inheritance

genome the entire nucleotide sequence of all the genetic material in an organism

genotype the alleles that an individual carries

geographic isolation an isolating mechanism that involves a geographic barrier that separates individuals of a single species into two or more different environments

gibberellin a plant hormone that causes internodes to lengthen

global climate change relatively rapid, significant, and dramatic changes in the Earth's climate that are happening as a result of human activities

glucose metabolism the process of building up and breaking down glucose

glycolysis the chemical process that breaks glucose into two three-carbon molecules and releases two to four ATP molecules

Golgi complex a mass of membranous spaces where proteins are altered before being exported from the cell

Gram stain a staining technique that distinguishes between two kinds of cell walls of bacteria

grassland a biome characterized by wet and dry seasons, hot summers and cold winters, and year-round cover of grasses

gravity a force that causes two masses to be attracted to one another

greenhouse gas an atmospheric gas that causes an increase in a planet's surface temperature

ground tissue tissue that carries out metabolic and other specific functions, and aside from wood, makes up most of the body of a plant: parenchyma, collenchyma, and sclerenchyma are the various kinds of ground tissues

growth factor a small molecule that is released by one cell and that stimulates both itself and surrounding cells to divide

H

habitat the specific place where an organism usually lives that is characterized by a physical setting or a kind of plant

hair cell a specific sensory cell inside the cochlea that detects sounds

haploid a cell or organism that carries one complete set of each homologous chromosome

helper T cell a cell of the acquired immune system that is stimulated by a macrophage to divide and, in turn, stimulates B cells to divide, mature, and produce antibodies

hemoglobin the protein in red blood cells that binds to oxygen and so transports it to body cells

herbivore a predator that eats plant tissues

hermaphrodite an individual that can produce both eggs and sperm

heterosexual sexually attracted to members of the opposite sex

heterozygous when the pair of alleles that govern a particular trait are not the same; one allele is located on each homologous chromosome

homeostasis maintenance of constant internal conditions despite fluctuations in the external environment

homologous chromosome one of a pair of chromosomes; each carries one allele for a particular gene

homosexual sexually attracted to members of the same sex

homozygous when the pair of alleles that govern a particular trait are the same; one allele is located on each homologous chromosome

hormone chemical released into the bloodstream that acts on a tissue that is some distance away and initiates a response in cells of that tissue

host the cell or organism that a virus infects

Human Genome Project the research effort to identify all genes in the human genome and to determine the nucleotide sequences of the entire human genome

human immunodeficiency virus (HIV) the retrovirus that causes the collapse of the human immune system and thus causes AIDS

humoral immunity the aspect of the acquired immune system in which pathogens are attacked by free proteins, called antibodies, that float freely in the bloodstream and in the spaces between cells

hybrid the result of a mating between two species

hydrocarbon a molecule that contains only carbon and hydrogen atoms

hydrogen bond a weak and temporary chemical bond that forms between a partially positively charged hydrogen atom in one polar molecule or chemical group and a partially negatively charged atom in another nearby molecule or chemical group

hydroxyl a functional group with a hydrogen atom and an oxygen atom

hyperglycemia excess glucose in the bloodstream

hypertension abnormally high blood pressure

hypertonic solution when there are more ions and other dissolved substances than normal outside of the cell relative to the inside

hyphae individual threads or tubes of a mycelium or fruiting body

hypotension abnormally low blood pressure

hypothalamus a cluster of structures on the underside of the brain, helps to regulate functions such as body temperature, eating, drinking, and glandular secretions

hypothesis a possible explanation of an event or an answer to a scientific question that can be tested by formal studies or experiments

hypotonic solution when there are fewer ions and other dissolved substances than normal outside of the cell relative to the inside

I

immune system cells, tissues, and organs devoted to identifying and destroying pathogens that can cause disease

incomplete dominance the type of inheritance when two true-breeding parents have offspring with traits that are intermediate between those of their parents

inferential statistics mathematical tools that can determine the probability that the results of a study are real or just chance differences

inflammation a local response to infection in which fluid and blood cells are drawn into an infected area, causing swelling, redness, and fever in the infected part

ingest   the process of bringing food inside of an organism for digestion

intermediate form   an individual, fossil or living, that has traits usually associated with more than one kind of organism

intracellular fluid   fluid within cell membranes, also known as cytosol

intron   a noncoding sequence of DNA located within the nucleotide sequence of a gene

invertebrate   an animal with no backbone or cartilage to protect the nerve cord

in vitro **fertilization (IVF)**   the fertilization of an ovum by a sperm within laboratory glassware

ion   an atom (or group of atoms) that carries a net electrical charge

ionic bond   a bond that forms when a pair of oppositely charged ions are attracted to each other and held together by electrical forces that result from a transfer or gain or loss of an electron

isotonic solution   when concentrations of dissolved ions and other dissolved substances are in balance in the fluids on either side of a plasma membrane

isotope   a form of an element defined by the number of neutrons in the nucleus

J

joint   a structure where two bones fit together

K

karyotype   a photograph of chromosomes taken when each chromosome is duplicated; a karyotype shows sets of two completely condensed sister chromatids

keystone species   a species that modifies the environment and enriches it for other community members

kinetic energy   the energy carried by a moving object

*K***-selected species**   a species selected for traits that favor competitive ability at population densities near carrying capacity

L

lateral meristem   actively dividing layer of cells that produce the increase in the thickness of a tree trunk or branch

leaf   a photosynthetic plant organ that also functions in gas exchange

leukocyte   a white blood cell involved in the innate immune response

life table   a summary of age data for a population that allows predictions of when death occurs in the population

ligament   a structure that connects one bone to another; made of collagen and other elastic proteins

lignin   a complex organic alcohol molecule that is part of the cell wall of woody plants

lipid   a molecule that is not highly soluble in water

lipid bilayer   a double layer of lipid molecules oriented so that the molecules' polar heads are surrounded by water molecules on both the outside and the inside of the bilayer, while their nonpolar tails are intertwined and away from water molecules

lipoproteins   spheres of lipids combined with proteins that carry cholesterol molecules in the bloodstream

logistic population growth   a leveling off of population increase as the environment's carrying capacity is reached, shown as an S-shaped curve

lymph node   a small gland where lymph drainage is concentrated and one location where an acquired immune response is activated

lymphatic system   a network of vessels that drain lymphatic fluid throughout the body

lymphocyte   a white blood cell involved in the acquired immune response

lysosome   an organelle that contains enzymes that degrade wastes

M

macronucleus   contains the copy of the *Paramecium* genome used for day-to-day transcription

macronutrients   large biological molecules such as lipids, carbohydrates, and proteins needed by the body in relatively large amounts in the range of grams (g) to kilograms (kg)

malignant tumor   tumor that contains cells that have the ability to leave their neighbors, travel through the body, and start a tumor in a new location

malnutrition   nutrition that is so far below optimal that there is serious risk of death

marsupial mammal   a mammal that nurtures its young within a pouch

mass extinction   a widespread extinction event that wipes out large numbers of species

matter   any material substance that takes up space and has mass

mean   the average of a group of numbers; to calculate the mean you divide the sum by the total number of values

megaspore   a spore in an ovule that will develop into a megagametophyte that will give rise to an egg cell

meiosis   the process of separating condensed coiled chromosomes that results in gamete formation

meiotic cell division   the form of cell division that allocates half of the DNA in the parent cell to each daughter cell and forms gametes

membrane channel   a membrane protein that has a hole down the center through which ions and water will flow

membrane protein   a protein located among the lipids in cell membranes

memory B cell   a B cell that is partially activated by a pathogen and remains in reserve to respond more rapidly to that pathogen if it is encountered again

menstrual cycle   the cycle in the uterus that prepares it for a possible pregnancy

menstruation   the monthly loss of blood from the uterus as the endometrium sloughs off

meristem   plant tissue containing cells that divide and produce daughter cells that differentiate into mature plant cells, and so are responsible for plant growth; meristem tissues are similar to stem cells of animals

mesoderm   a layer of cells of an embryo that is sandwiched between ectoderm and endoderm

messenger RNA (mRNA)   ribonucleic acid that carries DNA's instructions for making proteins

metabolic pathways   chains of related chemical reactions

metabolism   all the enzymatic pathways that build up and break down

metastasis   spread of cancerous cells from the site of the original tumor to other areas of the body

micronucleus contains the copy of the *Paramecium* genome used for sexual reproduction

micronutrients small molecules and minerals needed by the body in small amounts in the range of milligrams (mg = 1/1000 gram) or micrograms (µg = 1/100,000 gram)

microspore a spore in a reproductive structure that will develop into a pollen grain

mineral an essential micronutrient that is a simple element

mitochondria organelles where energy stored in glucose and various other molecules is used to produce ATP molecules

mitochondrial genome the total genome of a single mitochondrion

mitosis (M) the cell cycle phase immediately after G_2 when the duplicated chromosomes are separated so that each daughter cell can receive an exact copy of the DNA from the parent cell

mitotic cell division the process of cell division in eukaryotic cells that includes copying DNA, moving the copies away from each other, and dividing the parent cell into two

molecular biology techniques tools and procedures used to isolate and identify the DNA in a sample of cells

molecule a chemical structure that results when two or more atoms combine through covalent chemical bonds

monosaccharide a carbohydrate with just one subunit or building block

monotreme a mammal that lays and incubates shelled eggs

morphological species concept defines a species as a group of individuals that share distinctive anatomical characteristics

mRNA splicing the process in which introns are cut out of a mRNA transcript and the cut pieces are spliced back together to make a functional mRNA molecule

multicellular organism an organism made of two or more cells

mutagen a chemical that causes chemical transformations in DNA

mutualism a symbiotic interaction that involves two species that both benefit from the interaction

mycelium a mass of fine, threadlike fungal tissues that invade a food source and provide the fungus with nutrition

mycorrhizae symbiotic associations of plant roots and fungal hyphae that provide plants with an increased supply of nutrients

myelin a wrapping of several layers of cell membrane around an axon; prevents loss of ions and increases the speed of the action potential traveling down the axon

N

natural selection the effects of environmental factors that allow some individuals in a population survive to breed, while others do not

negative feedback a hormone or other substance inhibits its own production

nephron a microscopic kidney tubule that filters blood

nerve net a diffuse nervous system in which neurons connect to one another, but there is no concentration of nerve cells in any one location

neuromuscular junction the synapse between a neuron and the muscle fiber that it controls

neuron the cellular unit of the nervous system

neurotransmitter a messenger molecule that will carry an action potential across a synapse

neurotransmitter receptor a specialized membrane protein receptor in a postsynaptic neuron that binds to a specific neurotransmitter

neutron an uncharged subatomic particle in an atom's nucleus

nitrogen-fixing bacteria soil bacteria that incorporate atmospheric nitrogen into nitrogen-containing compounds that other organisms can use

nitrogenous base an organic ring structure containing of nitrogen, carbon, hydrogen, and oxygen

nonpolar the property of having an even charge distribution across the atoms of a molecule

nuclear genome the genome of an organism carried within the nucleus of each cell

nuclear membrane the barrier that separates DNA from the rest of a eukaryotic cell

nucleic acid a long chain of nucleotides

nucleoid the region of a prokaryotic cell where DNA is found

nucleolus a dense area within the nucleus that contains DNA responsible for making ribosomal RNA

nucleosome a little ball of protein with DNA wound around it, a nucleosome is active at the first stage of coiling and condensing of DNA molecules

nucleotide an organic molecule that contains a phosphate group and a nitrogenous base attached to a sugar ring

nucleus the area surrounded by the nuclear membrane that contains the genetic material of a eukaryotic cell

nutrients chemicals that are necessary to maintain life, provide energy, or promote growth

O

observational study a thoughtful, planned examination of organisms in their natural environment

omnivore an animal that eats both animals and plants

oncogene a cancer gene that codes for an abnormal version of a protein that promotes cell division, or that overexpresses a protein that promotes cell division

oogonia ovarian cells that will mature into primary oocytes

oral groove a depression in the surface of *Paramecium* that funnels food particles down to a place in the membrane where food enters the cell

orbital the space around the nucleus where a particular electron is likely to be found

organ a grouping of different types of tissues that work together to perform a specific task

organ system a group of organs that work together to perform a specific body function

organelle a molecular assemblage within a cell that carries out a specific function

organic chemistry the study of the interactions of carbon atoms and carbon-containing molecules

organism the entire body of a living thing

osmosis   the diffusion of water across a semipermeable membrane

ovarian cycle   the monthly cycle in ovaries that governs the production of mature eggs from oocytes

ovary   the portion of a carpel where ovules that contain eggs develop

overfishing   harvesting that exceeds the replacement rate of a species

overpopulation   when a population grows beyond the environment's carrying capacity for that species

ovulation   the release of an egg from an ovary

ovule   a reproductive structure that contains the megagametophyte generation, including the egg cell

oxidative phosphorylation   the process that occurs in mitochondria that produces ATP, using the energy intermediates from the citric acid cycle

P

parasitism   a symbiotic relationship in which one organism receives a benefit at the expense of another

parental generation (P)   the first parents crossed in a given genetic experiment

pasteurization   a method developed by Louis Pasteur in which a liquid is heated to a temperature high enough to kill the bacteria in it and held at that temperature for a specific period of time

pathogen   an organism that causes a disease

pelycosaur   an extinct reptile from 300 million years ago that had the beginnings of mammalian jaw and skull characteristics

penicillin   the first antibiotic to be isolated and used in clinical practice, effective against fatal diseases such as pneumonia

peptide bond   the bond in a protein between the carboxyl carbon atom of one amino acid and the amino nitrogen of another amino acid

peripheral nervous system   the parts of the nervous system that are outside of the spinal column and skull

permafrost   tundra subsoil that is permanently frozen

petal   ring of modified leaves of a flower that gives a visual or olfactory signal to pollinators

pH   how acidic or basic a substance is

phenotype   the traits that an individual expresses that result from expression and interaction of alleles

phloem   a vascular tissue that transports sugars and other biological molecules within a plant

phosphate   a functional group with a central phosphorus atom bound to four oxygen atoms

phospholipid   a kind of lipid that incorporates a phosphate group and includes a polar head region and a nonpolar tail region

photosynthesis   the process of using light energy to produce ATP molecules that provide the energy required to make glucose from carbon dioxide and water

phylogenetic classification
a classification scheme that reflects genetic and evolutionary relationships

phylogenetic species concept
a species is defined as the smallest group of organisms with similar features that comes from one common ancestor

phytoremediation   the use of plants to remove toxic chemicals from soil

pistil   the portion of a flower that contains female reproductive parts; a simple pistil is made up of an ovary and its contents, and a style, and a stigma

pituitary gland   a major control gland for the endocrine system that releases hormones into the blood that act on other endocrine glands

placental mammal   a mammal that has a placenta to nourish its young during pregnancy

plankton   small organisms that float or drift suspended in oceans, lakes, rivers, and streams

plant   a photosynthetic eukaryotic organism that uses both chlorophyll *a* and chlorophyll *b*, has carotenoids as accessory pigments, has cellulose in its cell walls, uses starch as a storage carbohydrate, and has paired anterior flagella on its flagellated cells

plant hormone   a substance produced by certain plant cells that causes other cells in the plant to respond biochemically

plasmid   a small loop of bacterial DNA that can be transferred to another bacterial cell

pleiotropy   the name for when a single gene has effects on more than one phenotype

polar covalent bond   a covalent bond in which the shared electrons are pulled more toward one atom involved in the bond than the other

polar molecule   a molecule with regions of negative and positive charge, but no net charge across the whole molecule

pollen cone   a reproductive structure of a conifer that produces pollen grains

pollen grain   an immature male microgametophyte encased within a tough coating

pollen tube   a tube that grows out of a pollen grain; sperm travel through this tube to reach the egg

pollination   transfer of pollen from the structures that produced it to structures where it will germinate

polygenic   the type of inheritance in which a trait is controlled by more than one gene; polygenic traits are expressed along a continuum

polymerase chain reaction (PCR)
a technique used to make thousands of copies of a small sample of DNA

polyploid   a cell or organism that carries more than two complete sets of each homologous chromosome

polysaccharide   a carbohydrate with many subunits

population   all of the members of a species that live in an area; populations are defined by interbreeding within the group

posterior end   the hind end of an animal

potential energy   energy carried by a stationary object that has the potential to cause motion

precipitation   water that falls from the atmosphere to Earth in the form of rain, snow, sleet, or ice

predator   an animal that hunts and eats other organisms

prey   an organism that is hunted and eaten by a predator

primary metabolite   a basic metabolic compound produced by a plant that supports its growth and development

primary oocytes   ovarian cells that may mature into egg cells

primary sexual characteristics   the different reproductive organs of males and females

primary structure   the amino acid sequence of a protein

primary succession ecological succession in places where no plants have ever grown

prion an infectious protein that can transform other proteins into defective, nonfunctional molecules

probability a number that represents the chances that something will happen; expressed as a decimal between 0.0 and 1.0 or as a fraction or a percentage

producer an organism that produces its own food supply

products the substances that are produced in a chemical reaction

progeria an umbrella term for several rare genetic disorders in which the symptoms of aging occur prematurely

progesterone the female hormone that helps ready and maintain the endometrium

prokaryotic cell a cell that lacks cellular organelles surrounded by membranes, including a nucleus

promoter a noncoding DNA sequence that is located before the DNA sequence of a gene; to initiate the transcription of a gene, RNA polymerase first binds to the promoter and then transcribes the gene

prostaglandin a chemical produced in tissues in response to damage; stimulates blood vessels and nerves to respond to the damage in ways that promote healing

protein a large biological molecule made of one or more chains of amino acids

proteome an umbrella term for all of the protein expressed by a cell

proton a positively charged subatomic particle found in an atom's nucleus

proto-oncogene protein a protein that actively promotes cell division

pseudoscience a statement that sounds scientific but is based on faulty or incomplete evidence

Q

quaternary structure a composite molecule made of two or more protein molecules

R

radial symmetry a kind of symmetry in which identical halves result from any cut that bisects the middle of an object or organism

radiation therapy the use of high-energy rays to kill cancer cells

reactants the substances that interact to produce a different substance during a chemical reaction

receptor a membrane protein specialized to interact with a signal molecule and transmit the signal's message to the inside of the cell

recessive allele an allele expressed only when the other allele for that trait is identical

recommended dietary allowance (RDA) the amount of a nutrient that will allow a person to avoid diseases from deficiencies or excesses of that nutrient

red blood cells specialized cells that contain large amounts of hemoglobin and so transport oxygen to body tissues

reproductive cloning the production of a fully developed, cloned individual

reproductive isolating mechanism a process or trait that results in barriers to reproduction and thus prevents interbreeding between species

respiration gas exchange in the lungs during which oxygen is taken up from air in the environment and carbon dioxide is released from the body into the environment

respiratory system the lungs and associated breathing tubes and tissues responsible for the process of gas exchange in which oxygen enters the bloodstream and carbon dioxide exits the bloodstream

restriction enzyme a bacterial enzyme that cuts DNA at a particular nucleotide sequence

restriction fragment length polymorphism (RFLP) the pattern of different-sized DNA pieces produced by restriction enzymes that is unique to each individual

retina the complex tissue at the back of the eye composed of photoreceptors and neural cells

retrovirus a virus that uses RNA as its genetic material and must transcribe the RNA to DNA before it can reproduce in the host cell

reuptake inhibitor a drug that binds to a reuptake transporter, blocking the normal reuptake and inactivation of a neurotransmitter

reverse transcriptase an enzyme carried in a retrovirus and activated inside the host cell that transcribes viral RNA into viral DNA

reverse transcriptase inhibitor a drug that blocks the action of reverse transcriptase and prevents a virus from reproducing

rhizoid nonvascular threadlike structure that anchors liverworts, mosses, and hornworts into soil

rhizome a horizontal belowground stem; shoots grow upward from it and roots grow downward from it

rhodopsin the pigment molecule found in retinal photoreceptors that detects light

ribosomal RNA (rRNA) RNA found in the structure of ribosomes

ribosome an organelle that bonds amino acids together to form a protein

ribozymes RNA molecules that function as biological catalysts

RNA polymerase an enzyme that catalyzes the formation of mRNA from DNA

rod light-sensitive cell in the retina that detects different levels of light

root a plant organ specialized to absorb fluids and minerals from the soil, anchor the plant, and sometimes serve as a storage depot for excess carbohydrates

root hair a delicate extension of the epidermis of a root that increases its surface area and thus allows the root to absorb more water

rough endoplasmic reticulum (RER) a network of membranous tubules that are closely associated with ribosomes and that modify proteins by adding important carbohydrate groups

***r*-selected species** species that have adaptations for rapid reproduction and high population densities that allow them briefly to overshoot the environment's carrying capacity

S

sarcoplasmic reticulum a labyrinth of membranous tubes that surrounds a muscle fiber and stores and releases calcium ions

scala naturae a ranking of all organisms on a scale of closeness to God

scientific law   a reliable and precise mathematical description of an event or set of events

second filial generation (F_2):   the offspring that result when members of the F_1 generation are crossed

second law of thermodynamics   without an input of energy, the order in any system declines over time

secondary metabolite   a compound made by a plant that functions in plant defense and communication

secondary sexual characteristics   gender-related features that do not directly participate in the process of reproduction

secondary structure   the folding of an amino acid chain into a helix or pleated sheet that results from hydrogen bonds between the carboxyl group of one amino acid and the amino group of another

secondary succession   ecological succession in places where plants have grown

seed   a dormant sporophyte embryo and its food supply stored within a protective coat

seed cone   a reproductive structure of a conifer that produces ovules

sensory receptor   a cell or group of cells specialized to detect a particular kind of environmental information

sepal   the outermost ring of modified leaves of a flower; sepals enclose and protect a flower bud

sex-linked trait   a trait that is carried on a sex chromosome

sexual dimorphism   the existence of traits that distinguish males from females

sexual orientation   the nature of a person's sexual attraction

sexual reproduction   the method of reproduction used by multicellular organisms in which each of two parents contributes DNA to the offspring, resulting in a new, completely unique combination of DNA

sickle cell disease   an inherited form of anemia associated with red blood cells that assume an abnormal, sickle shape under conditions when the oxygen content of the blood is low

signal molecule   a small molecule in a cell's environment that carries information about some aspect of the environment

signal transduction   binding of a signal molecule to a membrane receptor produces changes in the internal chemical reactions of a cell

sister chromatids   the name given to the identical chromosome copies that are joined together after DNA replication in S-phase

smooth endoplasmic reticulum (SER)   a network of membranous tubules that manufacture and modify lipids

sodium-potassium pump   a membrane protein that uses the energy in ATP to pump sodium out of the cell and potassium into the cell

sperm cells   the pair of cells within a pollen grain; one will fertilize the egg to form the diploid zygote, while the other fertilizes the polar cells to form polyploid endosperm

spermatogonia   cells that produce sperm

S-phase (synthesis phase)   the cell cycle phase immediately after G_1, when DNA is copied

spinal cord   the long rope of nervous system tissue that runs down the back of a vertebrate, inside the spinal column

spindle apparatus   an array of fibers that spread out from each pole of a cell and overlap at the equator of the cell, where they attach to sister chromatids

spleen   a lymphatic organ that drains blood and activates an acquired immune response

spore   a haploid cell produced by meiosis that divides mitotically to produce the sporophyte generation of an alga or a plant

sporophyte   a life cycle phase in a plant that has alternation of generations, a sporophyte is a diploid plant that produces spores by meiosis

stabilizing selection   individuals with extreme values of a trait do not successfully breed, and so the population values tend to cluster around the average value

stamen   a pollen-producing organ of a flower; it has a supporting filament and anther

stem   a plant organ that supports leaves and contains vascular tissues that transport fluids from roots to aboveground portions of the plant

stoma (plural: stomata)   pores in the cuticle that are opened and closed by guard cells to admit carbon dioxide and release water vapor and oxygen

stroke   the rupture of a small blood vessel in the brain that causes damage to brain cells

stroma   the space between the stacks of thylakoids and the inner membrane of the chloroplast

stromatolite   large columnar masses of cyanobacteria and other prokaryotes

succession   a predictable process of community change over time

survivorship curve   a graph that shows when death occurs in a population

symbiosis   the association of two organisms that live together in a close and often necessary relationship

synapse   the small gap between two neurons, usually between an axon terminal and a dendrite, where information is sent from one neuron to another

synaptic vesicle   a tiny, membranous sac in a presynaptic terminal that stores and releases neurotransmitters

systolic blood pressure   the pressure exerted against artery walls during the active contraction of the left ventricle, when the pushing force is greatest

T

taiga   a northern biome characterized by cold, snowy winters, cool summers, moderate precipitation, and conifer forests

taste bud   a multicellular sensory organ that detects molecules in food and produces the sensation of taste

taxonomist   a biologist who specializes in classification of organisms

telomere   a stretch of noncoding DNA at either end of a chromosome

temperate deciduous forest   a midlatitude biome where moisture falls about equally all year; distinct warm and cold seasons favor trees that drop their leaves and go dormant in winter

temporal lobe   a region of the cerebral cortex devoted to complex aspects of vision and to memory

tendon   connective tissue that attaches muscle to bone

tertiary structure the unique three-dimensional shape of a protein determined by the folding of the protein's secondary structure

testosterone male hormone

thalamus a structure roughly in the middle of the brain that is a processing station for sensory information

theory a scientific explanation that is strongly supported by a large body of scientific research and is a highly likely explanation of a broad set of related events

therapeutic cloning developing cells or tissues that can be used to treat disease

therapsid an extinct mammal-like reptile from about 300 to 275 million years ago that was probably ancestral to modern mammals

theropod a predatory dinosaur that walked on two powerful hind legs, had small forelegs with sharp claws, and a neck joint that allowed the head to swivel from side to side

thylakoid a membranous sac within chloroplasts where light energy is captured and stored as chemical energy

tissue a group of similar cells gathered together into a unit that performs a specific function

transcription transfer of information from DNA to messenger RNA

transcription factor a protein that binds to DNA and through interactions with RNA polymerase increases transcription of a particular gene

transfer RNA (tRNA) an RNA molecule that transfers amino acids from the cytosol to a ribosome, where it ensures that the correct amino acid is used to make a specific protein

transgenic organism bioengineered organism that combines the genes of two or more organisms

translation transformation of messenger RNA's instructions into a chain of amino acids

transpiration loss of water vapor from leaf tissues

transsexual an individual who believes she or he is of the opposite sex

tropical rain forest the most productive terrestrial biome, characterized by warm to hot year-round temperatures, extreme humidity, large amounts of rainfall, and highly varied vegetation

true-breeding traits identical parental characteristics that appear in offspring when both of their parents have that same trait

tumor a mass of cells that results from abnormal cell proliferation

tumor suppressor protein a protein that inhibits cell division by binding to and inhibiting transcription factors needed for cell division

U

understory a layer of forest growth that is higher than the tallest shrubs but does not reach into the heights of the canopy

unicellular organism an organism made of just one cell

V

vaccine a weakened pathogen that elicits a mild immune response and thereby confers some immunity to that pathogen when it is encountered again

vacuole a storage organelle

vascular bundle the water-conducting and sugar-conducting tissues of plants

vascular plant plant that has internal systems of tubules that transport fluids and dissolved nutrients

vascular tissue strands of xylem and phloem that transport water, minerals, sugars, and other important molecules throughout the plant

vegetative propagation a form of asexual reproduction that allows the growth of a whole new plant from a portion of a parent plant

vein a blood vessel that carries blood toward the heart

ventral surface the belly or lower surface of an animal

vertebrate an animal with a backbone around the spinal cord

virus a small nonliving structure that incorporates biological molecules

vitamin a small organic molecule that is an essential micronutrient

voltage-dependent ion channel membrane ion channel that opens when the voltage across the membrane changes

W

wetlands marshes, swamps, or estuaries that form transitional environments between larger bodies of water and drier uplands; wetlands absorb flood waters and slowly release them

white blood cells blood cells that function in defense of the body against disease-causing organisms

X

xylem a plant tissue that conducts water and minerals from the roots throughout the plant

Z

zygote the cell produced when two gametes fuse to form a new individual

Credits

Text and Line Art

Chapter 12

Table 12.1: Data from the National Cancer Institute: http://seer.cancer.gov/csr/1975_2000/sections.html; **Table 12.2:** Data from the American Cancer Society; **Table 12.4:** Data from the National Cancer Institute: http://seer.cancer.gov/csr/1975_2000/sections.html. **Table 12.5:** Data from the National Cancer Institute: http://seer.cancer.gov/csr/1975_2000/sections.html

Chapter 13

Fig. 13.29: Cladogram data based on "When Is a Bird Not a Bird?" *Nature,* June 25, 1998, p. 729.

Chapter 21

Table 21.1: Data from the National Academy of Science Food and Nutrition Board, 2004; **Fig. 21.8:** U.S. Department of Agriculture, 2005.

Chapter 26

Fig. 26.4b: Based on data in Murie, 1944, quoted in Deevey, 1947. **Fig. 26.12:** United States Census Bureau.

Photographs

Author Photo © Sears Portrait Studios

Chapter 1

Opener: © V.C.L./Taxi/Getty Images; **1.1a,c:** © Dr. Dennis Kunkel/Visuals Unlimited; **1.1b:** © M. I. Walker/Photo Researchers, Inc.; **1.1d:** © Biophoto Associates/Photo Researchers, Inc.; **1.2 (top):** © Susumu Nishinaga/SPL/Photo Researchers, Inc.; **1.2 (bottom):** Dr. Joseph Hoffman, Dept. of Cellular & Molecular Physiology, Yale University; **1.5:** © Michael Abbey/Visuals Unlimited; **1.6:** © Digital Vision/PunchStock; **1.11:** © Michael & Patricia Fogden/Corbis; **1.14:** © Thierry Thomas/Peter Arnold, Inc.; **1.15:** © Michael Newman/PhotoEdit; **1.17:** © David Young-Wolff/Photographer's Choice/Getty Images; **1.18a:** © Ed Young/Corbis; **1.18b:** © Michael Abbey/Photo Researchers, Inc.; **1.18c:** © Steve Allen/Getty Images; **1.18d:** Davidson & Steller: Blocking apoptosis prevents blindness in Drosophila retinal degeneration mutants. *Nature*, Vol. 391, 5 February 1998. © 1998 Nature Publishing Group.; **1.18e:** © Inga Spence/Visuals Unlimited; **1.18f:** © Andrew Syred/SPL/Photo Researchers, Inc.; **1.19 (right):** © Parke H. John, Jr./Visuals Unlimited; **1.19 (left):** Centers for Disease Control/Cynthia Goldsmith, Luanne Elliott.

Chapter 2

Opener: © Stockbyte/PunchStock; **2.1 (left):** © Darwin Dale/Photo Researchers, Inc.; **2.1 (right):** © Fletcher & Baylis/Photo Researchers, Inc.; **2.2 (Helium):** © David Buffington/Getty Images; **2.2 (Neon):** © Royalty-Free/Corbis; **2.2 (Gold):** © Janis Christie/Getty Images; **2.2 (Iron):** © TexPhoto/iSTOCKphoto.com; **2.9 (Sodium):** © B. RUNK/S. SCHOENBERGER/Grant Heilman Photography; **2.9 (Chlorine):** © Charles D. Winters/Photo Researchers, Inc.; **2.9 (Salt):** © GSO Images/Photographer's Choice/Getty Images; **2.10 (left):** © Bananastock; **2.10 (right):** © Royalty-Free/Corbis; **2.12:** © Digital Vision; **2.13:** © PhotoLink/Photodisc/Getty Images; **2.14:** © Bilderberg/Photonica/Getty Images; **2.15:** © Herman Eisenbeiss/Photo Researchers, Inc.

Chapter 3

Opener: Dr. Art Davis/U.S. Dept. of Agriculture—Animal and Plant Health Inspection Service, APHIS; **Table 3.1 (top):** © Charles D. Winters/Photo Researchers, Inc.; **Table 3.1 (middle):** © Maximilian Stock Ltd/SPL/Photo Researchers, Inc.; **Table 3.1 (bottom):** © PhotoEdit; **Table 3.2 (Gasoline):** © The McGraw-Hill Companies, Inc./Gary He, photographer; **Table 3.2 (Butane):** © Jeff Daly/Grant Heilman Photography; **Table 3.2 (Lemon):** © Mauritius/SuperStock; **Table 3.2 (Rubber):** © David R. Frazier/Photo Researchers, Inc.; **Table 3.3 (all), 3.5, Table 3.5 (all):** © The McGraw-Hill Companies, Inc./John Thoeming, photographer; **3.14b:** © Dr. Dennis Kunkel/Visuals Unlimited; **3.14c:** © Walter Reinhart/Phototake NYC; **3.15:** © Holger Leue/Lonely Planet Images; **3.17a,b:** Collinge et al, Depleting Neuronal PrP in Prion Infection Prevents Disease and Reverses Spongiosis, *Science* (31 October 2003: Vol. 302. no. 5646, pp. 871–874). © 2003 American Association for the Advancement of Science.

Chapter 4

Opener (right): © Cannon/DC Comics/The Kobal Collection; **Opener (left):** © AP/Wide World Photos; **4.1 (Tree):** © Frans Lanting/Corbis; **4.1 (Woman):** © Kevin Peterson/Getty Images; **4.1 (Eye):** © Phil Jude/Photo Researchers, Inc.; **4.1 (Photoreceptors):** © Dr. Fred Hossler/Visuals Unlimited; **4.2:** © John Reader/Photo Researchers, Inc.; **4.12:** © Photodisc/Getty Images; **4.13b:** © Biophoto Associates/Photo Researchers, Inc.; **4.16:** © K.G. Murti/Visuals Unlimited; **4.18:** © Dr Elena Kiseleva/SPL/Photo Researchers, Inc.; **4.19:** © R. Bolender & D. Fawcett/Visuals Unlimited; **4.20:** © Prof. P. Motta & T. Naguro/SPL/Photo Researchers, Inc.; **4.21a:** © Biophoto Assoc./Photo Researchers, Inc.; **4.21b:** © Bill Longcore/Photo Researchers, Inc.; **4.22:** © Dr. Don W. Fawcett/Visuals Unlimited; **4.23:** Centers for Disease Control/Cynthia Goldsmith; Dr. Erskine. L. Palmer; Dr. M. L. Martin; **4.27:** © Dr. Torsten Wittmann/Photo Researchers, Inc.

Chapter 5

Opener: © Universal/The Kobal Collection; **5.2a:** © Royalty-Free/Corbis; **5.2b:** © Jonathan Daniel/Getty Images; **5.3:** © James Balog/Getty Images; **5.4 (left):** © Royalty-Free/Corbis; **5.4 (center):** © Kent Knudson/PhotoLink/Vol. 19/Getty Images; **5.4 (right):** © GDT/The Image Bank.Getty Images; **5.5:** © Peter M. Wilson/Corbis; **5.7 (top):** © Charles Stirling/Alamy; **5.7 (bottom):** © ImagePix/Alamy; **5.11 (top):** © Nature Scenes/V36/Getty Images; **5.11 (middle):** © Herbert Kehrer/zefa/Corbis; **5.11 (bottom):** © Hope Ryden/National Geographic/Getty Images; **5.15a:** © J. William Schopf; **5.15b:** © Sinclair Stammers/Photo Researchers, Inc. **5.16:** © Chris Butler/Photo Researchers, Inc.

Chapter 6

Opener: © Johnny Johnson/Photographer's Choice/Getty Images; **6.2:** © Bob Stefko/the Image Bank/Getty Images; **6.4 (left):** © Dr. Jeremy Burgess/Photo Researchers, Inc.; **6.4 (right):** © ISM/Phototake NYC; **6.17a:** © Alfred Pasieka/SPL/Photo Researchers, Inc.; **6.17b:** © Dr. T.J. Beveridge/Visuals Unlimited; **6.17c:** © Michael Abbey/Photo Researchers, Inc.; **6.17d:** © Biophoto Associates/Photo Researchers, Inc.; **6.18a:** © Ed Reschke; **6.20:** © Reg Morrison/Auscape; **6.21:** © Theo Allofs/zefa/Corbis.

Chapter 7

Opener (top): © Meredith Birmingham; **Opener (bottom):** © Jon Jones/Sygma/Corbis; **7.3a:** Reproduced from The Journal of Experimental Medicine, 1944, 79: 137–158. Copyright 1944 The Rockefeller University Press; **7.4b:** © Lee D. Simon/Photo Researchers, Inc.; **7.5:** © A. Barrington Brown / Photo Researchers, Inc.; **7.6 (top):** © CSHL Archives/Peter Arnold, Inc.; **7.6 (bottom):** © National Portrait Gallery, London; **7.8:** Photo © and courtesy of Dr. Victoria Foe, from *Molecular Biology of the Cell* by Bruce Alberts et al, Garland Publishing, 1994.; **7.18 (left):** © CESAR RANGEL/AFP/Getty Images; **7.18 (right):** © John Sahn/Stone/Getty Images; **7.19:** © CNRI/Photo Researchers, Inc.; **7.20:** © The Center for Molecular Biology of RNA, University of California, Santa Cruz.

Chapter 8

Opener: © McClatchy-Tribune Information Services. All Rights Reserved. Reprinted with Permission; **8.1:** © Carolina Biological Supply Company/Phototake.com; **8.2a:** © CNRI/Photo Researchers, Inc.; **8.2b:** © Science Pictures Limited/Photo Researchers, Inc.; **8.3 (top left & top right):** © David Young-Wolff/Photographer's Choice/Getty Images; **8.3 (middle left):** © Biophoto Associates/Photo Researchers, Inc.; **8.3 (middle right):** © Yorgos Nikas/Stone/Getty Images; **8.3 (bottom):** © David Young-Wolff/Photographer's Choice/Getty Images; **8.4:** Jannan Jenner; **8.8 (top):** © Pascal Goetgheluck/Photo Researchers, Inc.; **8.8 (both in middle):** © Rawlins/Custom Medical Stock Photo; **8.8 (bottom):** © Family and Lifestyles/V15/Getty Images; **8.13:** © Ed Reschke; **8.13 (Telophase):** © Ed Reschke/Peter Arnold, Inc.; **8.14a&b:** University of Washington, Department of Pathology (http://www.pathology.washington.edu); **8.14c:** © Adrian T. Sumner/Photo Researchers, Inc.; **8.15a:** © Dr. Conly L. Rieder and Dr. Alexey Khodjakov/Visuals Unlimited; **8.16a:** Dr. Mark W. Kirschner, Harvard University Medical School; **8.16b:** © R. Valentine/Visuals Unlimited; **8.23:** © Wessex Regional Genetics Centre/The Wellcome Trust Photographic Library.

Chapter 9

Opener: © Tek Image/Photo Researchers, Inc.; **9.1a–f:** Modified from: Kimmel et al., 1955. *Developmental Dynamics* 203:253–310. Copyright © 1995 Wiley-Liss, Inc. Reprinted only by permission of Wiley-Liss, a subsidiary of John Wiley & Sons, Inc.; **9.1g:** © Jean Claude Revy—ISM/Phototake NYC; **9.3:** Images courtesy of Dr. S. J. DiMarzo & Prof. Kohei Shiota and the Kyoto Collection of Human Embryos. © 2006 Dr. Mark Hill. Cell Biology Lab, University of New South Wales. http://embryology.med.unsw.edu.au/; **9.4a,b:** © & Courtesy of Dr. Judith Cebra-Thomas; **9.5 (top):** Courtesy of Kohei Shiota, M.D. © Kyoto Collection of Human Embryos; **9.5 (bottom):** © Dr. Don W. Fawcett/Visuals Unlimited; **9.6a:** © Dr. Gopal Murti/Visuals Unlimited; **9.6b:** © Dr. Don W. Fawcett/Visuals Unlimited; **9.7c:** From: Barr, ML 1963 The Sex Chromatin. pp. 48–71, In: Intersexuality (C. Overzier, ed.). Academic Press, NY; **9.7d:** © Juniors Bildarchiv/Alamy; **9.9:** © Peter Lansdorp/Visuals Unlimited; **9.14a:** © Stephen Small, New York University; **9.14b:** © David Kosman, John Reinitz. Mount Sinai School of Medicine; **9.16b:** © Thomas Eisner; **9.16c:** © Fred Bavendam/Peter Arnold, Inc.; **9.18:** © AFP/Getty Images; **9.19a:** © UPI; **9.19b:** Reproduced with permission and copyright © of the British Editorial Society of Bone and Joint Surgery. Bar-Maor JA, Kesner KM, Kaftori JK. Human tails. J Bone Joint Surg [Br] 1980;62–B:508–510, Figure 3.

Chapter 10

Opener (top): Copyright 1999–2004: Cold Spring Harbor Laboratory; American Philosophical Society; Truman State University; Rockefeller Archive Center/Rockefeller University; University of Albany, State University of New York; National Park Service, Statue of Liberty National Monument; University College, London; International Center of Photography; Archiv zur Geschichte der Max-Planck-Gesellschaft, Berlin-Dahlem; and Special Collections, University of Tennessee, Knoxville; **Opener (bottom):** © AP/Wide World Photos; **10.2 (left):** © Business and Industry/V01/Getty Images; **10.2 (right):** © Sylvia Cordaiy Photo Library Ltd/Alamy; **10.3 (left):** © Hulton Archive/Getty Images; **10.3 (right):** © American Philosophical Society; **10.5:** © Fred Habegger/Grant Heilman Photography, Inc.; **10.18a:** © Brad Wilson/Ionica/Getty Images; **10.18b:** © Bill Varie/Corbis; **10.18c:** © David Young-Wolff/PhotoEdit, Inc.; **10.22 (both):** © & Courtesy of Gillian Heighley; **10.24:** © Skjold Phototographs/PhotoEdit, Inc.; **10.25c:** © & Courtesy of T. E. Wellems and R. Josephs, University of Chicago; **10.27:** Copyright 1999–2004: Cold Spring Harbor Laboratory; American Philosophical Society; Truman State University; Rockefeller Archive Center/Rockefeller University; University of Albany, State University of New York; National Park Service, Statue of Liberty National Monument; University College, London; International Center of Photography; Archiv zur Geschichte der Max-Planck-Gesellschaft, Berlin-Dahlem; and Special Collections, University of Tennessee, Knoxville; **10.28:** © Michael S. Yamashita/Corbis.

Chapter 11

Opener: © & Courtesy of Paul Gelsinger; **11.1:** © AP/Wide World Photos; **11.4a:** © Mauro Fermariello/Photo Researchers, Inc.; **11.5:** © & Courtesy of Orchid Cellmark; **11.6:** © Dr. Don W. Fawcett/Visuals Unlimited; **11.7 (left):** © Dr. David Phillips/Visuals Unlimited; **11.7 (right):** © Prof. P. Motta/Dept Of Anatomy/University "La Sapienza", Rome/Photo Researchers, Inc.; **11.8:** © Bettmann/Corbis; **11.9a:** © Rykoff Collection/Corbis; **11.9b:** © Hulton Archive/Getty Images; **11.10:** Courtesy of Dr. Nicole Hauser, Fraunhofer IGB, Germany; **11.13:** © Volker Steger/Photo Researchers, Inc.; **11.14:** © Eye of Science/Photo Researchers, Inc.; **11.15:** © Najlah Feanny/Corbis SABA; **11.16:** Provided by J B Gurdon, Cambridge, UK.

Chapter 12

Opener: © MIRISCH/UNITED ARTISTS/THE KOBAL COLLECTION; **12.5:** © Dr. A. Liepins/SPL/Photo Researchers, Inc.; **12.9 (left):** © E. Walker/Photo Researchers, Inc.; **12.9 (right):** © Dr. Frederick Skvara/Visuals Unlimited; **12.10b:** Courtesy of Joanne R. Less, Mitchell C. Posner, Thomas C. Skalak, Norman Wolmark, and Rakesh Jain; **12.12a:** © Dr. Jeremy Burgess/Photo Researchers, Inc.; **12.12b:** © Dr. F. C. Skarva/Visuals Unlimited; **12.13:** © Elizabeth Kreutz/NewSport/Corbis; **12.16b:** © Dr. P. Marazzi/Photo Researchers, Inc.; **12.16c:** © Dr. P. Marazzi/Photo Researchers, Inc.; **12.18a:** © Health and Medicine 2/V40/Getty Images; **12.18b:** © Kings College School of Medicine/Photo Researchers, Inc.

Chapter 13

Opener (top): © Renee Morris/Alamy; **Opener (bottom):** © Joe Tucciarone/Photo Researchers, Inc.; **13.1:** © Jim Clare/naturepl.com; **13.2a:** © Science Source/Photo Researchers, Inc.; **13.2b:** © Natural History Museum, London; **13.2c:** © National Maritime Museum, London; **13.3 (left):** © TUI DE ROY/Minden Pictures; **13.3 (right):** © Zigmund Leszczynski/Animals Animals—Earth Scenes; **13.4a:** © Natural

History Museum, London; **13.4b:** © Simon King/naturepl.com; **13.5:** © Steven J. Kazlowski/Alamy; **13.6a:** © Martin Creasser; **13.8 top:** © Janet Horton; **13.8 bottom:** © Monte M. Taylor; **13.12:** © Medioimages/PunchStock; **13.17:** © FRANS LANTING/Minden Pictures; **13.18a:** © Scientifica/Visuals Unlimited; **13.18b:** © Ed Degginger/Animals Animals—Earth Scenes; **13.26:** © Museum für Naturkunde Berlin; **13.28a:** © Todd Marshall; **13.28b left:** Photo courtesy of Terry D. Jones & John A. Ruben; **13.28b (middle):** © O. Louis Mazzantena/NGS Image Collection; **13.28b right:** © Corbis Royalty Free; **13.29:** Stenonychosaurus inequalis and a Hypothetical Dinosauroid reproduced with the Permission of The National Museum of Natural Science, Ottawa, Canada.

Chapter 14

Opener: © Everyday People/SS16 & SS33/Getty Images & Faces of the World/Corbis; **14.1b:** © Robin Rudd/Unicorn Stock Photos; **14.4b:** © Raymond Blythe/OSF/Animals Animals—Earth Scenes; **14.6:** © Dr. Kenneth Y. Kaneshiro; **14.8a:** © Tom Murray; **14.8b:** © Andrew Forbes; **14.9:** © MARK MOFFETT/Minden Pictures; **14.10:** © ONE WORLD IMAGES/Alamy; **14.11 (top & bottom):** © Kevin Bauman; **14.12:** North American Tortoise-Shells and Angel-Wings, a rendering by William H. Howe; **14.19:** © Dr. Tony Brain & David Parker / Photo Researchers, Inc.; **14.20a:** © Raymond B. Otero/Visuals Unlimited; **14.20b:** © Wim van Egmond/Visuals Unlimited; **14.20c:** © Michael Abbey/Visuals Unlimited; **14.21a:** © Herbert Zettl/zefa/Corbis; **14.21b:** © Papilio/Alamy; **14.21c:** © David R. Frazier Photolibrary, Inc./Alamy; **14.23b:** Milwaukee Public Museum, Silurian Reef Diorama; **14.24 (left):** Narbonne, G.; Modular Construction of Early Ediacaran Complex Life Forms, *Science* (20 August 2004: Vol. 305. no. 5687, pp. 1141–1144). © 2004 American Association for the Advancement of Science; **14.24 (middle & right):** © Ken Lucas/Visuals Unlimited.

Chapter 15

Opener: © Mark Deeble & Victoria Stone/Oxford Scientific Films/photolibrary.com; **15.3a:** © David M. Phillips/Visuals Unlimited; **15.3b:** © George J. Wilder/Visuals Unlimited; **15.3c:** © Thomas Tottleben/Tottleben Scientific Company; **15.5a** © American Society for Microbiology; **15.5b:** © George Chapman/Visuals Unlimited; **15.7:** © Jim Wark/AirPhotoNA.com; **15.8:** © Dennis Kunkel/Phototake NYC; **15.9:** © Dr. T.J. Beveridge/Visuals Unlimited; **15.10:** © David Scharf/Photo Researchers, Inc.; **15.11a:** © Dr. Kari Lounatmaa/Photo Researchers, Inc.; **15.11b:** © Science VU/S. W. Watson/Visuals Unlimited; **15.13:** Photograph by Prof. Dr. Hans Reichenbach, Helmholtz Centre for Infection Research, Braunschweig; **15.14:** © Raymond B. Otero/Visuals Unlimited; **15.15:** © Michael Abbey/Visuals Unlimited; **15.16a:** © Wim van Egmond/Visuals Unlimited; **15.16b:** © eBioMedia; **15.16c:** Tokophrya lemnarum, trophont ingesting Tetrahymena thermophila, x 400, Japan, 1997 by Y. Tsukii **15.17:** © Michael Abbey/Visuals Unlimited; **15.18a:** © Ed Degginger/Color-Pic; **15.18b:** © Alfred Pasieka/Photo Researchers, Inc.; **15.19:** © Norman Nicoll/Natural Visions; **15.21:** © Royalty-Free/CORBIS; **15.22:** © Ed Degginger/Animals Animals—Earth Scenes; **15.25a:** © Bill Bachman/Photo Researchers, Inc.; **15.25b:** © Wim van Egmond/Visuals Unlimited; **15.26a:** Florida Fish and Wildlife Conservation Commission; **15.26b:** © Larry Jon Friesen.

Chapter 16

Opener: © Peabody Essex Museum, Salem, Massachusetts/The Bridgeman Art Library; **16.5:** Courtesy of N. Allin & G.L. Barron, University of Guelph; **16.6a (left):** John Dennis/Canadian Forest Service **16.6a (right):** © Dr. Gerald Van Dyke/Visuals Unlimited; **16.6b:** © B. RUNK/S. SCHOENBERGER/Grant Heilman Photography Inc.; **16.7:** © Viard/Jacana/Photo Researchers, Inc.; **16.9:** © Harvest Seasons CD042/Pixtal/age fotostock; **16.10a (left):** © Food Icons /X023/Getty Images;**16.10a (right):** © John Durham/Photo Researchers, Inc.; **16.15:** © Doug Perrine/Seapics.com; **16.21a–c:** © Dwight Kuhn; **16.24:** © & Courtesy of Dr. Hans Kerp, Forschungsstelle für Paläobotanik Münster **16.25 (Platypus):** © Jean-Phillipe Varin/Photo Researchers, Inc.; **16.25 (Kangaroo):** © Timothy Laman/National Geographic/Getty Images; **16.25 (Elephant):** © Vol. 6/PhotoDisc/Getty Images; **16.25 (Anteater):** © The McGraw-Hill Companies, Inc./Jill Braaten, photographer; **16.25 (Bat):** © RF/Corbis; **16.25 (Rabbit):** © Image Source/PunchStock; **16.28a:** © Peter Brown; **16.29:** © David Cavagnaro/Peter Arnold, Inc.

Chapter 17

Opener: © The British Museum/HIP/The Image Works; **17.1:** © Erich Lessing/Art Resource, NY; **17.4:** © Wim van Egmond/Visuals Unlimited; **17.5:** © George P. Chamuris, Ph.D. Bloomsburg University of PA; **17.6a:** © Ed Reschke/Peter Arnold, Inc.; **17.6b (1&2):** © Ray Simon/Photo Researchers, Inc.; **17.8a,b:** © Wim van Egmond/Visuals Unlimited; **17.8c:** © Dr. James W. Richardson/Visuals Unlimited; **17.9:** © Andrew J. Martinez/Photo Researchers, Inc.; **17.12, 17.13, 17.14:** © B. RUNK/S. SCHOENBERGER/Grant Heilman Photography Inc.; **17.15:** © Rod Planck/Tom Stack & Assocs.; **17.17:** © Ed Reschke/Peter Arnold, Inc.; **17.20a:** © Kathy Merrifield/Photo Researchers, Inc.; **17.20b:** © Walter H. Hodge/Peter Arnold, Inc.; **17.20c:** © Royalty-Free/CORBIS; **17.21:** © C. Munoz-Yague/Eurolios/Phototake NYC; **17.22b:** © Chris Gornersall/naturepl.com.

Chapter 18

Opener: © Samuel Courtauld Trust, Courtauld Institute of Art Gallery/The Bridgeman Art Library; **18.5b:** © George Wilder/Visuals Unlimited; **18.6a (left):** © Cabisco/Visuals Unlimited; **18.6a (right):** © Dwight Kuhn; **18.6b (right):** © Science VU/Visuals Unlimited; **18.7:** © M.I. Walker/Photo Researchers, Inc.; **18.8:** © Jack M. Bostrack/Visuals Unlimited; **18.9:** © John D. Cunningham/Visuals Unlimited; **18.11:** © Sheila Terry/Photo Researchers, Inc.; **18.12:** © Dr. Jeremy Burgess/Photo Researchers, Inc.; **18.19a:** © Wally Eberhart/Visuals Unlimited; **18.20 (left) :** © Christie Carter/Grant Heilman Photography Inc.; **18.20 (middle):** © Dr. Phillip Barak, Univeristy of Wisconsin–Madison; **18.20 (right):** © Nigel Cattlin/Photo Researchers, Inc.; **18.21:** Photo is copyright 2003 by John Neystadt, http://www.neystadt.org/john/album/; **18.22:** © Esther Beaton Wild Pictures; **18.23:** © Keith Douglas/Alamy; **18.24:** © Musee d'Orsay, Paris/SuperStock; **18.25:** © Mary Ellen (Mel) Harte, www.forestryimages.org; **18.26:** © James Strawser/Grant Heilman Photography Inc.

Chapter 19

Opener: © Perennou Nuridsany/Photo Researchers, Inc.; **19.2a:** © Wally Eberhart/Visuals Unlimited; **19.2b:** © Herve Conge/ISM/Phototake, NYC; **19.3a:** © Chris Evans, The University of Georgia, www.forestryimages.org; **19.3b:** © B. RUNK/S. SCHOENBERGER/Grant Heilman Photography Inc.; **19.3c:** © Andrew Syred/SPL/Photo Researchers, Inc.; **19.5:** © Biology Media/Photo Researchers, Inc.; **19.9a** © Dwight Kuhn; **19.9b:** © Dr. Jeremy Burgess/SPL/Photo Researchers, Inc.; **19.9c:** Dr Jeremy Burgess/Photo Researchers, Inc.; **19.11a:** © Ed Reschke/Peter Arnold, Inc.; **19.11b** © Alan & Linda Detrich/Grant Heilman Photography Inc.; **19.12a:** © David Cappaert, www.insect images.org; **19.12b:** © Bill Beatty/Painet.com; **19.12c:** Dartmouth Electron Microscope Facility; **19.13a:** © Dwight Kuhn; **19.13b:** © Ed Reschke; **19.14:** © Dwight Kuhn; **Table 19.2 (Peach):** © Inga Spence/Visuals Unlimited; **Table 19.2 (Tomato):** © Alan & Linda Detrick/Photo Researchers, Inc.; **Table 19.2 (Apple):** © Zig Leszczynski/Animals Animals Earth Scenes; **Table 19.2 (Blackberry):** © Dwight Kuhn; **Table 19.2 (Pineapple):** © Phil Degginger; **Table 19.2 (Beans):** © Dwight Kuhn; **Table 19.2 (Acorn):** © Ed Degginger; **Table 19.2 (Maple):** © R. J. Erwin/Photo Researchers, Inc.; **19.15b:** © Neil Emmerson/Robert Harding World Imagery/Getty Images; **19.16a:** © Adam Hart-Davis/SPL/Photo Researchers, Inc.; **19.16b:** © William Radcliffe/Science faction/Getty Images; **19.17a:** © Wally Eberhart/Visuals Unlimited; **19.17b:** © Jerome Wexler/Photo Researchers, Inc.; **19.18:** © Rod Planck/Photo Researchers, Inc.; **19.19:** © Wayne P. Armstrong.

Chapter 20

Opener: © Robin Nelson/PhotoEdit, Inc.; **20.2 (left):** © Bryan Reynolds/Phototake NYC; **20.2 right:** Courtesy of S. Gorb, J. Berger, Max Planck Society, Germany; **20.3a:** © David Wrobel/SeaPics.com; **20.4a:** © Ed Reschke/Peter Arnold, Inc.; **20.6c:** © Omikron/Photo Researchers, Inc.; **20.9:** © CNRI/Photo Researchers, Inc.;**20.16b:** © Fawcett/Coggeshall/Photo Researchers, Inc.; **20.17b:** © Dr. Dennis Kunkel/Visuals Unlimited; **20.23:** © Ed Reschke/Peter Arnold, Inc.; **20.24:** © H.E. Huxley; **20.32:** Courtesy of John W. Hole, Jr.

Chapter 21

Opener: © Amedeo John Engel Terzi/The Wellcome Trust Photographic Library; **22.1a:** © Michael Newman/PhotoEdit, Inc.; **22.1b:** Centers for Disease Control/James Gathany; **21.1c:** © Dwight Kuhn; **21.1d:** © Larry Miller/Photo Researchers, Inc.; **21.10:** © Lynsey Addario/Corbis; **21.11:** © The McGraw-Hill Companies, Inc./John Thoeming, photographer; **21.17 (left):** © Jeffrey L. Rotman/Corbis; **21.17 (right):** © Dr. LR/Photo Researchers, Inc.; **21.19:** Centers for Disease Control; **21.20:** Library of Congress.

Chapter 22

Opener: © Richard Mackson/Sports Illustrated; **22.2a:** © Bruce Dale/National Geographic/Getty Images; **22.5a–c:** © Dr. Dennis Kunkel/Visuals Unlimited; **22.6:** © Rudi Von Briel/PhotoEdit, Inc.; **22.8:** © Ed Reschke/Peter Arnold, Inc.; **22.9a:** © Dr. Dennis Kunkel/Visuals Unlimited; **22.10:** © Dr. David M. Phillips/Visuals Unlimited; **22.11:** © CNRI/Photo Researchers, Inc.; **22.19a:** © Annabella Bluesky/Photo Researchers, Inc.; **22.19b:** © Modern Technologies/V29/Getty Images.

Chapter 23

Opener: © Stockbyte Silver/Getty Images; **23.1:** © Copyright, Mark Grimson and Larry Blanton, Electron Microscopy Laboratory, Department of Biological Sciences, Texas Tech University; **23.12a:** © Ewing Galloway/Index Stock Imagery; **23.12b:** Clinical Pathological Conference on Acromegaly, Diabetes, Hypemetabolism, Protein Use & Heart Failure, American Journal of Medicine 20: 133 (1956); **23.15b:** © David M. Phillips/Visuals Unlimited; **23.19:** © Dr. David M. Phillips/Visuals Unlimited; **23.22f:** © Camera M.D. Studios; **23.24a:** © GERARD LACZ/Animals Animals—Earth Scenes; **23.24b:** © Johner Images/Getty Images; **23.24c:** © Art Wolfe/The Image Bank/Getty Images; **23.24d:** © Nick Greaves/Alamy; **23.25:** © MITSUAKI IWAGO/Minden Pictures; **23.26:** Gerianne M. Alexander. From, Alexander & Hines, Evolution and Human Behavior 23 (2002) 467–479.

Chpter 24

Opener: © Corbis SYGMA; **24.1a:** © Keystone/Getty Images; **24.1b:** © Adrian Arbib/Corbis; **24.3:** © DR NAJEEB LAYYOUS/Photo Researchers, Inc; **24.4:** © Mauritius, GMBH/Phototake, NYC; **24.5a:** © Rawlins/Custom Medical Stock Photo; **24.5b:** © Pascal Goetgheluck/Photo Researchers, Inc. **24.5c:** © Rawlins/Custom Medical Stock Photo; **24.5d:** © The Wellcome Trust Photographic Library; **Table 24.1:** © Royalty-Free/Corbis; **24.9:** © Jim Wilson/The New York Times/Redux; **24.12 (left):** © Roslin Institute; **24.12 (right):** © VINCENZO PINTO/AFP/Getty Images.

Chapter 25

Opener: © Susan Steinkamp/Corbis; **25.2:** © Kim Karpeles; **25.4** © Susan Van Etten/PhotoEdit, Inc.; **25.5a:** © E.O.S./Gelderbloom/Photo Researchers, Inc.; **25.5b:** Centers for Disease Control/VD, SCSD; **25.5c:** Centers for Disease Control/Dr. Myron G. Schultz; **25.11:** © Don W. Fawcett/Photo Researchers, Inc.; **25.12:** © Lennart Nilsson/Albert Bonniers Forlag AB; **25.14a:** © Lennard Lessin/Peter Arnold, Inc.; **25.20a:** © Dr Linda Stannard, Uct/Photo Researchers, Inc.; **25.22a:** Centers for Disease Control/Dr. F. A. Murphy.

Chapter 26

Opener: © Ed Degginger/Color-Pic; **26.3a:** © Wayne P. Armstrong; **26.3b:** © John A. Novak/Animals Animals/Earth Scenes; **26.3c:** © NOAA/NGDC, DMSP Digital Archive/Photo Researchers, Inc.; **26.4a:** © Nature, Wildlife and the Environment/V06/Getty Images; **26.8:** © Tom Tietz/Stone/Getty Images; **26.11a:** © Richard Day/Daybreak Imagery; **26.11b:** © Chris Jackson/Getty Images; **26.14a:** © Shawn G. Henry; **26.14b:** © Peter Menzel, www.menzel photo.com, from the book project Material World, a Global Family Portrait, Sierra Club books; **26.14c:** © Peter Ginter; **26.14d:** © Leong Ka Tai; **26.17:** © Yann Layma/The Image Bank/Getty Images; **26.19:** © Alaska Stock; **26.20a (left):** © Peter Parks/imagequestmarine.com; **26.20a (right):** © Tom & Therisa Stack/Tom Stack ; **26.20b:** © Andrew Syred/Photo Researchers, Inc.; **26.20c:** © Wellcome/Custom Medical Stock Photo; **26.21:** © Gregory G. Dimijian, M.D./Photo Researchers, Inc.; **26.22:** © Dr. James L. Castner; **26.23:** © Bud Lehnhausen/Photo Researchers, Inc.; **26.24:** © Ingo Arndt/naturepl.com; **26.25:** © Miles Barton/naturepl.com; **26.26:** © AP/Wide World Photos; **26.28a:** © Greg Lasley; **26.28b:** © Deborah Allen; **26.28c:** © Tim Zurowski/Corbis; **26.30:** Reprinted with permission from The American Society for Photogrammetry & Remote Sensing. Lo, C.P. and Dale A. Quattrochi, 2003. "Land use and land cover change, urban heat island phenomenon, and health implications: a remote sensing approach." Photogrammetric Engineering and Remote Sensing, 60: 1053–1063. **26.31:** © R. & N. Bowers/VIREO.

Chapter 27

Opener: © John Launois/Stockphoto.com; **27.1b:** © Chris Evans, The University of Georgia, www.forestryimages.org; **27.1c:** © David Muench/Corbis; **27.8:** © Wim van Egmond/Visuals Unlimited; **27.10:** NASA Goddard Space Flight Center; **27.11 (left):** Tim Stanton and Jim Stocke, Oceanography Department, Naval Postgraduate School; **27.11 (right) :** Image courtesy of MODIS Rapid Response Project at NASA/GSFC; **27.14:** © FRANS LANTING/Minden Pictures; **27.15:** © Marc Epstein/DRKphoto.com; **27.16:** © Theo Allofs/Corbis; **27.17a:** © Altrendo/Getty Images; **27.17b:** © Sea Life/V112/Getty Images; **27.17c:** © Tim Laman/National Geographic/Getty Images; **27.17d:** © Jeff Jaskolski/SeaPics.com; **27.18:** © Dr. Verena Tunnicliffe; **27.20:** © JIM BRANDENBURG/Minden Pictures; **27.21:** © Animal Life/EP004/Getty Images; **27.24:** © Frank Krahmer/Bruce Coleman, Inc.; **27.25:** © Thomas Kitchin/Photo Researchers, Inc.; **27.26:** © Larry Lefever/Grant Heilman, Inc.; **27.27:** © Photodisc Inc./Getty Images; **27.28:** © Ed Reschke; **27.29:** © Michael & Patricia Fogden/Corbis; **27.30:** © Michael J. Doolittle/Peter Arnold, Inc.; **27.31:** © TIM FITZHARRIS/Minden Pictures; **27.32:** © Brand X Pictures/PunchStock; **27.33:** US Fish and Wildlife Service; **27.34:** © Jim Baron/Image Finders.

Chapter 28

Opener: © CYRIL RUOSO/JH EDITORIAL/Minden Pictures; **28.2a,b:** National Park Service; **28.3b:** © Geray Sweeney/Corbis; **28.5:** © Museum of Natural History, Smithsonian Institution; **28.6 left:** © Thomas Bennett; **28.6 right:** © Corbis; **28.7:** © J-F Lagrot/Galbe.Com; **28.11a:** © Royalty-Free/Corbis; **28.12a&b:** © Digital Vision/PunchStock; **28.13a:** © FRANS LANTING/Minden Pictures; **28.14a:** © Colin Raw/Stone/Getty Images; **28.14b:** © John William Banagan/Riser/Getty Images; **28.15a:** © JOERG BOETHLING/Peter Arnold, Inc.; **28.16:** © Johan Siebke/Alamy; **28.18a:** © MARK EDWARDS/Peter Arnold, Inc.; **28.18b:** © CLYDE H. SMITH/Peter Arnold, Inc.; **28.22a:** © Robert T. Smith/ardea.com; **28.23a:** NASA, Goddard Space Flight Center; **28.24a,b:** © Richard Hamilton Smith; **28.25a:** © Martin Harvey/Peter Arnold, Inc.; **28.26b:** © Creatas/PunchStock; **28.27:** © Michael Nichols/National Geographic Image Collection.

Index

Page numbers followed by a *t* designate tables; page numbers followed by an *f* refer to figures; page numbers set in *italics* refer to definitions of terms or introductory discussions.